STRUCTURE REPORTS

for 1985

Volume 52B

Part 1

Structure Reports is prepared under the guidance of a Commission of the International Union of Crystallography. The members of the Commission sometime concerned with the preparation of this volume are listed below.

STRUCTURE REPORTS

for 1985

Volume 52B (Part 1)

ORGANIC SECTION

General editor

G. Ferguson

Section editor

G. Ferguson

Published for the

INTERNATIONAL UNION OF CRYSTALLOGRAPHY

SPRINGER-SCIENCE+BUSINESS MEDIA, B.V.

First published in 1993

ISSN 0166–7033

Originally published by International Union of Crystallography in 1992
Softcover reprint of the hardcover 1st edition 1992

Printed on acid-free paper

ISBN 978-94-017-3159-1 ISBN 978-94-017-3157-7 (eBook)
DOI 10.1007/978-94-017-3157-7

TABLE OF CONTENTS

Introduction, VI

Organic Compounds, 1

Main-group Compounds, 695

Transition-metal Compounds, 909

INTRODUCTION

The present volume continues the aim of Structure Reports to present critical accounts of all crystallographic structure determinations. Details of the arrangement in the volumes, symbols used etc. are given in previous volumes (e.g. 41B or 42A, pages vi-viii).

University of Guelph,
Guelph, Ontario, Canada

G. FERGUSON

1 May, 1993

STRUCTURE REPORTS

SECTION III

ORGANIC COMPOUNDS

Edited by

G. Ferguson

with the assistance of

C. H. Morgan
D. F. Rendle
S. J. Rettig
S. N. Scrimgeour
T. J. R. Weakley
C. C. Wilson
D. W. Young

To find a particular organic or organometallic compound consult one of the indices (subject, permuted subject, formula or permuted formula) at the end of Part 2 of this volume. The general arrangement follows the classification used in the Cambridge Crystallographic Data Base and is: aliphatic or open chain compounds; open chains with N, S; benzene derivatives; cyclic hydrocarbons; condensed ring systems; heterocyclic compounds; carbohydrates; amino acids; natural products; molecular complexes; organometallic compounds - B, Si, P, As, Sb, groups IA, IIA, III, IV, VI; transition metal complexes - π-complexes, other ligands; inorganic ions which have organic counter ions.

DIDEUTEROOXALIC ACID DI(DEUTERIUM OXIDE) SOLVATE
$C_2D_2O_4,2(D_2O)$

M.-L. PUTKONEN, R. FELD, C. VETTIER and M.S. LEHMANN, 1985. *Acta Cryst.*, **B41**, 77-79.

Monoclinic, $P2_1/n$, a = 6.156, b = 3.616, c = 12.115 Å, β = 106.618°. Neutron radiation, R = 0.117 for powder data.

The structure was studied as a function of pressure using the Rietveld refinement procedure with neutron powder data. For previous X-ray work on α-oxalic acid see (1). R_1 values on the powder profile after constrained refinement are in the range 0.12 to 0.18. The only observable structural changes up to 0.5 GPa hydrostatic pressure are small changes in the hydrogen-bond arrangement. Of the three hydrogen bonds in the structure, the two longer bonds contract around 0.04 Å, whereas the shortest bond is unchanged. This is found to be in qualitative agreement with an estimate based on force constants from spectroscopic observations.

1. Structure Reports, **34B**, 3-5; see also the following Report.

α-OXALIC ACID DIHYDRATE
$C_2H_2O_4,2(H_2O)$

Y. WANG, C.J. TSAI, W.L. LIU and L.D. CALVERT, 1985. *Acta Cryst.*, **B41**, 131-135.

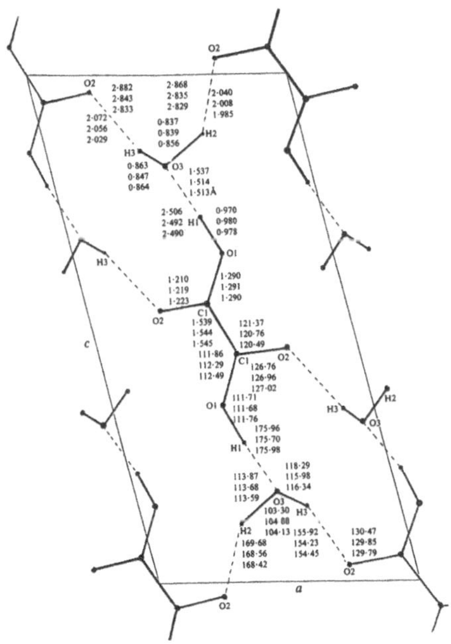

Fig. 1. Bond lengths and angles at 300 K (top), 170 K (middle) and 100 K (bottom) in α-oxalic acid dihydrate.

I. Monoclinic, $P2_1/n$, a = 6.102, b = 3.5011, c = 11.964 Å, β = 105.79°, Z = 2. Mo radiation, R = 0.035 for 1680 reflections (at 100 K).

II. Monoclinic, $P2_1/n$, a = 6.102, b = 3.5190, c = 11.973 Å, β = 105.84°, Z = 2. Mo radiation, R = 0.038 for 1502 reflections (at 130 K).

III. Monoclinic, $P2_1/n$, a = 6.102, b = 3.5298, c = 11.978 Å, β = 105.89°, Z = 2. Mo radiation, R = 0.042 for 1439 reflections (at 170 K).

IV. Monoclinic, $P2_1/n$, a = 6.109, b = 3.5595, c = 12.008 Å, β = 106.02°, Z = 2. Mo radiation, R = 0.040 for 1271 reflections (at 225 K).

V. Monoclinic, $P2_1/n$, a = 6.120, b = 3.6058, c = 12.057 Å, β = 106.29°, Z = 2. Mo radiation, R = 0.033 for 643 reflections (at 300 K).

Single-crystal X-ray analyses were made on oxalic acid dihydrate at five different temperatures- 300, 225, 170, 130, 100 K. A few selected peak profiles at various 2θ angles were monitored; the net intensities of the reflections increase on lowering the temperature as expected and this increase is more prominent for high-2θ-angle reflections than for low ones. The cell parameters contract anisotropically on lowering the temperature in the range studied. Molecular dimensions are shown in Fig. 1. Deformation density maps were investigated at 100 K: there are bonding electron densities at the midpoints of bonds C-C and C-O, as expected; the lone-pair densities around the O atoms are revealed; the hydrogen bond between the oxalic acid molecule and the water molecule is interestingly demonstrated.

DITHIODIGLYCOLIC ACID
$C_4H_6O_4S_2$ $(\text{-S-CH}_2\text{-COOH})_2$

C.P. NASH, M.M. OLMSTEAD, B. WEISS-LOPEZ, W.K. MUSKER, N. RAMASUBBU and R. PARTHASARATHY, 1985. *J. Am. Chem. Soc.*, **107**, 7194-7195.

I. Monoclinic, *I2/a*, a = 10.211, b = 7.794, c = 9.501 Å, β = 102.81°, Z = 4. R = 0.050.

II. Triclinic, $P\bar{1}$, a = 4.769, b = 4.837, c = 17.112 Å, α = 92.70, β = 91.47, γ = 114.79°, Z = 2. R = 0.060 (at 140 K).

In (I) the molecule lies about a twofold axis with torsion angles C-C-S-S -167.4(2), C-S-S-C -86.3(1)°. Form (II) has no crystallographic symmetry elements and the torsion angles are C-C-S-S -76.2 and -64.4(6)°, C-S-S-C 92.3(4)°. In (I) the carboxyl group planes are essentially coincident with the SCC planes; in (II) the S-C-C=O torsion angle is -79.1(7)° and the other carboxyl group is disordered over two orientations, one that is coplanar with SCC and one that is rotated out of planarity by 64(2)°.

(±)-MALIC ACID
$C_4H_6O_5$

P. VAN DER SLUIS and J. KROON, 1985. *Acta Cryst.*, **C41**, 956-959.

Monoclinic, $P2_1/c$, a = 4.889, b = 8.815, c = 13.036 Å, β = 102.93°, Z = 4. Mo radiation, R = 0.049 for 774 reflections.

The carbon chain of the molecule (Fig. 1) is in the extended form. The conformation is anti, ψ = -175.5(2)°, with the carboxyl group of the glycolic acid moiety approximately planar with C(2) and O(5). The carboxyl group conformation around the terminal C-C bond is anticlinal, χ = -61.5(4)°. In the crystal, the molecules form infinite chains in which they are coupled via cyclic hydrogen-bonded carboxyl dimers. The aliphatic hydroxyl groups link these chains together by hydrogen bond interactions, so that layers are formed parallel to (101) (O(2)...O(1') 2.707(3), O(3)...O(4'') 2.641(4), O(5)...O(4''') 2.859(3) Å).

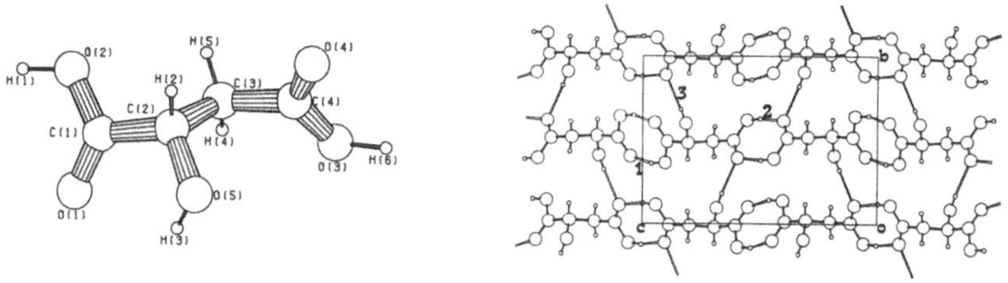

Fig. 1. $C_4H_6O_5$: the molecular structure and packing viewed down **a**.

1,1,1,3,3,3-HEXAFLUORO-2-PROPYLAMINO-1-THIOOXAMIDE
$C_5H_4F_6N_2OS$

H.W. ROESKY, H. HOFMANN, M. NOLTEMEYER and G.M. SHELDRICK, 1985. *Z. Naturforsch.*, **B40**, 124-126.

Monoclinic, *Cc*, a = 28.653, b = 9.244, c = 7.029 Å, β = 93.3°, Z = 4. Mo radiation, R = 0.066 for 614 reflections.

The molecule (Fig. 1) has atoms C(3),N(2),C(2),S,C(1),O,N(1) nearly in a plane and the C(2)-N(2)-C(3)-H torsion angle is -15.9°. Main bond lengths are: N(1)-C(1) 1.330(13), N(2)-C(2) 1.335(12), N(2)-C(3) 1.428(12), C(2)-S 1.627(10), C(1)-O 1.220(15), C-F 1.305-1.323(18) Å.

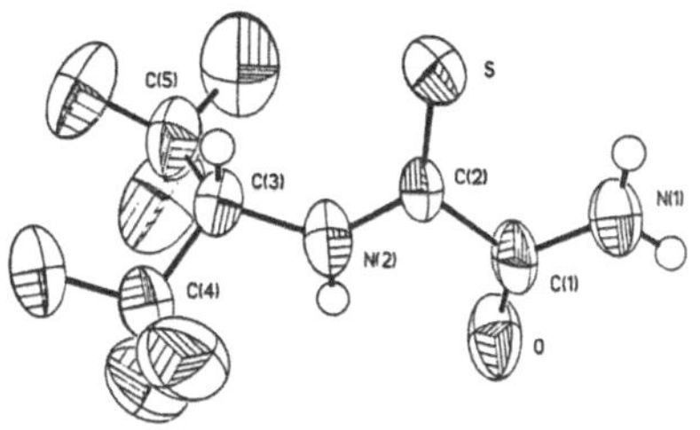

Fig. 1. A view of $C_5H_4F_6N_2OS$.

(Z)-2-BROMOMETHYL-2-BUTENOIC ACID
$C_5H_7BrO_2$

F. AMEER, S.E. DREWES, J.S. FIELD and P.T. KAYE, 1985. *S. Afr. J. Chem.*, **38**, 35-36.

Triclinic, $P\overline{1}$, a = 11.553, b = 7.143, c = 4.388 Å, α = 106.24, β = 104.00, γ = 79.41°, Z = 2. Mo radiation, R = 0.092 for 899 reflections.

The molecule is shown to have the Z configuration.

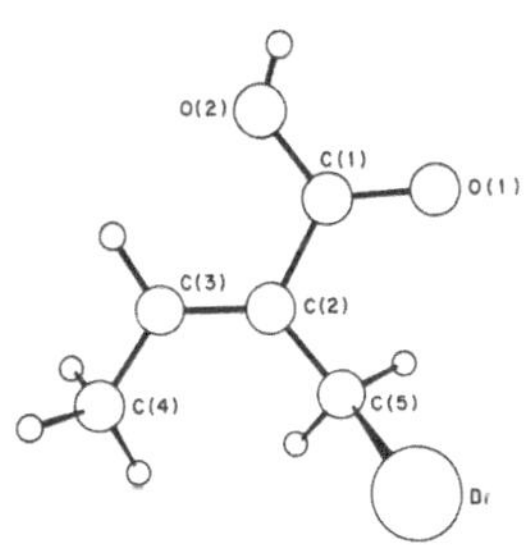

Fig. 1. A view of $C_5H_7BrO_2$.

γ-GUANIDINO-β-HYDROXYBUTYRIC ACID
$C_5H_{11}N_3O_3$

Y.B. KIM, T. FUJIWARA and K.-I. TOMITA, 1985. *Acta Cryst.*, **C41**, 68-69.

Triclinic, $P\overline{1}$, a = 8.519, b = 9.909, c = 4.879 Å, α = 92.69, β = 112.00, γ = 69.90°, Z = 2. Mo radiation, R = 0.091 for 1622 reflections.

The molecule (Fig. 1) occurs as a zwitterion with the H atom of the carboxyl group transferred to the N atom of the guanidyl group. The molecule has a fully extended trans-zigzag form with a torsion angle around the C(4)-N(1) bond of 186.4(4)°. The C-N bond lengths in the guanidyl group, with average value 1.335 Å, indicate a degree of double-bond character. All the non-H atoms in the molecule, with the exception of the hydroxyl O atom, are essentially coplanar. The molecules are held together by a three-dimensional network of N-H...O and O-H...N hydrogen bonds.

Fig. 1. $C_5H_{11}N_3O_3$: the molecular conformation and the structure projected down the c-axis with hydrogen bonds shown as dotted lines.

BIS(GUANIDINIUM) TARTRATE MONOHYDRATE
$C_6H_{16}N_6O_6,H_2O$ $C_4H_4O_6^{2-},2(CH_6N_3^+),H_2O$

W. KRUMBE, S. HAUSSUHL and R. FROHLICH, 1985. *Z. Kristallogr.*, **170**, 112-113.

Monoclinic, $P2_1$, a = 4.760, b = 14.151, c = 9.852 Å, β = 102.97°, Z = 2. Mo radiation, R = 0.043 for 8850 reflections.

The structure contains isolated ions linked by hydrogen bonding.

2-ISOPROPYLCROTONIC ACID
$C_7H_{12}O_2$

P.C. CHENCHAIAH, H.L. HOLLAND and M.F. RICHARDSON, 1985. *Acta Cryst.*, **C41**, 95-96.

Monoclinic, $P2_1/c$, a = 5.507, b = 10.206, c = 13.717 Å, β = 97.15°, Z = 4. Mo radiation, R = 0.045 for 884 reflections.

In the molecule (Fig. 1) the alkene part is planar with the carboxylate not quite coplanar. C(1), O(1) and O(2) deviate from this plane by 0.060(2), 0.147(2) and 0.019(2) Å respectively. Bond lengths and angles in the molecule are normal. The molecules form hydrogen-bonded centrosymmetric dimers with O(2)...O(1') distances of 2.663(3) Å.

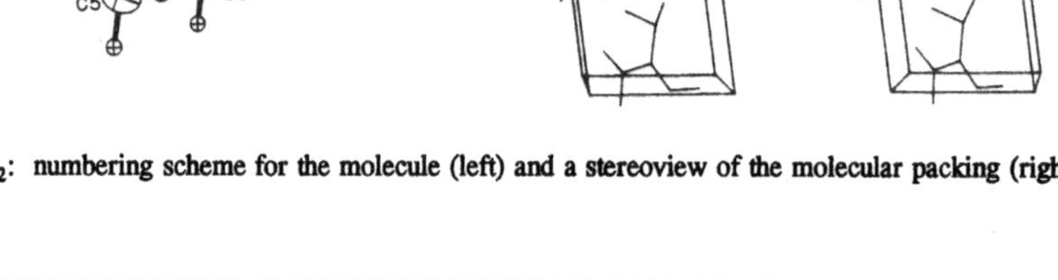

Fig. 1. $C_7H_{12}O_2$: numbering scheme for the molecule (left) and a stereoview of the molecular packing (right).

2,2-DICHLORO-N-(CHLORO(DIETHYLAMINO)METHYLENE)MALONAMOYL CHLORIDE
$C_8H_{10}Cl_4N_2O_2$

W. RIED, J. NENNINGER and J.W. BATS, 1985. *Chem. Ber.*, **118**, 4707-4715.

Monoclinic, $P2_1/n$, a = 11.413, b = 10.309, c = 11.578 Å, β = 102.75°, Z = 4. Mo radiation, R = 0.036 for 1728 reflections.

The analysis confirms the proposed constitution (Fig. 1). Bond lengths in the central portion of the molecule are C(3)-N(1) 1.346(2), N(1)-C(4) 1.297(2), C(4)-N(2) 1.314(2), N(2)-C(5) 1.471(2), N(2)-C(7) 1.477(2) Å. There is a short intramolecular O(2)...Cl(4) contact (2.996(1) Å) and an increase in the N(1)-C(3)-O(2) angle to 131.1(2)°.

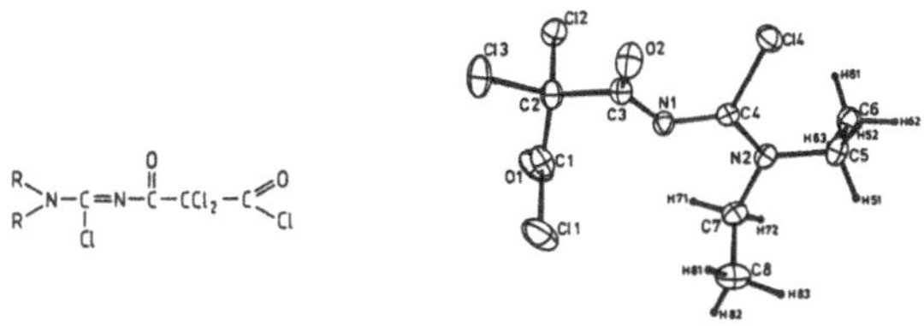

Fig. 1. The $C_8H_{10}Cl_4N_2O_2$ molecule and conformation.

NITRILOTRIPROPIONIC ACID
$C_9H_{15}NO_6$

A.V. GASPARYAN, L.M. SHKOL'NIKOVA, N.V. TSIRUL'NIKOVA, N.A. EGORUSHKINA, L.G. MAKAREVICH and N.M. DYATLOVA, 1985. *Zh. Strukt. Khim.*, **26**, 3-153-157 [J. Struct. Chem., **26**, 447-451].

Monoclinic, $P2_1/c$, a = 5.667, b = 23.229, c = 8.094 Å, β = 105.35°, Z = 4. Mo radiation, R = 0.044 for 1394 reflections.

The molecule (Fig. 1) exists as a zwitterion. The O(5) atom of the carboxylate group and the carbonyl O(4) of one of the carboxy groups are involved in the formation of a bifurcate intramolecular hydrogen bond with the betaine nitrogen atom (N...O 2.803(4), 3.003(4). H...O 2.11(4), 2.31(4) Å, N-H..O 123(3), 124(3)°). Intermolecular O-H...O hydrogen bonds (O...O 2.527(4) and 2.533(4) Å) link the molecules into ribbons.

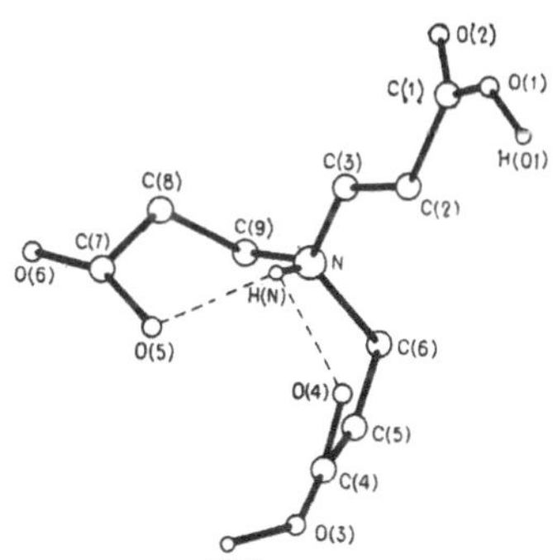

Fig. 1. The structure of the $C_9H_{15}NO_6$ molecule.

3,3-BIS(2-CARBOXYETHYL)-3-HYDROXYAMINOPROPIONIC ACID BETAIN
$C_9H_{15}NO_7$

G. BRACHTEL and M. JANSEN, 1985. *Z. Naturforsch.*, **B40**, 574-577.

Monoclinic, $P2_1/n$, a = 8.097, b = 9.343, c = 14.647 Å, β = 91.35°, Z = 4. Mo radiation, R = 0.041 for 1636 reflections.

The chains are maximally extended and O-H...O hydrogen bonds (H...O 1.56, 1.67, 1.86(3) Å) link the molecules in the crystal (Fig. 1). Main bond lengths are N-O 1.424(2), N-C 1.513-1.521(3), C-C 1.498-1.515(4) Å, O-N-C 103.6-110.6(2), C-N-C 108.9-112.6(2)°.

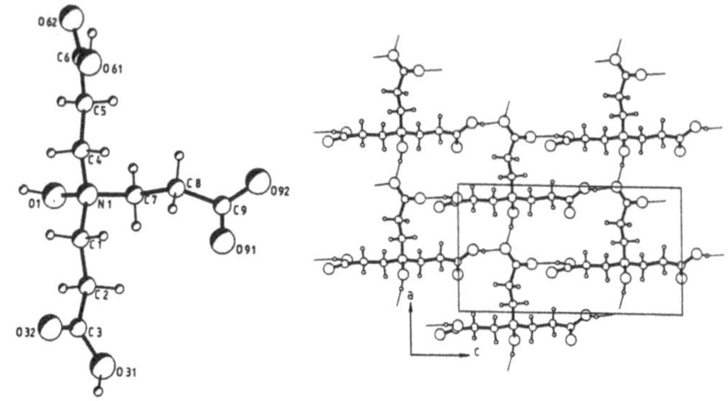

Fig. 1. The $C_9H_{15}NO_7$ molecule and a view of the hydrogen bonding.

p-FORMYLCINNAMIC ACID
$C_{10}H_8O_3$

H. NAKANISHI, M. HASEGAWA and T. MORI, 1985. *Acta Cryst.*, **C41**, 70-71.

Monoclinic, $P2_1/n$, a = 6.261, b = 4.825, c = 27.614 Å, β = 91.54°, Z = 4. Cu radiation, R = 0.058 for 1286 reflections.

The structure (Fig. 1) consists of nearly planar molecules forming a parallel plane-to-plane stack along the b-axis. Bond lengths ad angles in the molecule are in good agreement with generally accepted values. The molecules are associated as dimers and are hydrogen-bonded around inversion centres with O...O separations of 2.641(4) Å. The plane-to-plane perpendicular distance between the reactive molecules in the stack is 3.388 Å.

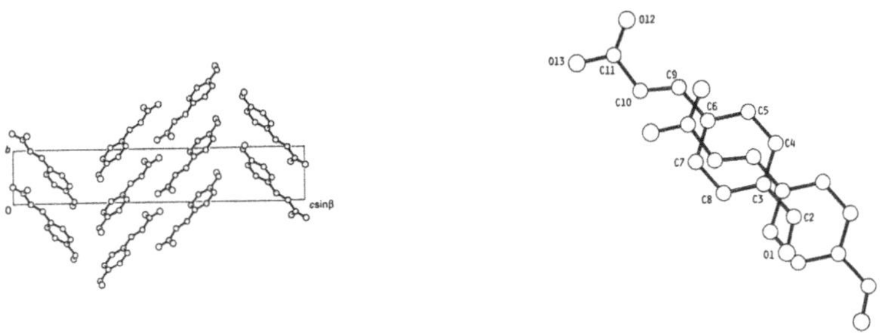

Fig. 1. $C_{10}H_8O_3$: crystal structure viewed along the a-axis (left) and the overlapped reactive molecules viewed along the normal to the mean molecular plane (right).

2-METHYL-5-OXO-5-PHENYLPENTANAMIDE
$C_{12}H_{15}NO_2$

G.D. ANDREETTI, G. BOCELLI, J. CYBULSKI, Z. DABROWSKI and J.T. WROBEL, 1985. *J. Mol. Struct.*, **128**, 259-267.

Monoclinic, $P2_1/c$, a = 11.781, b = 5.095, c = 19.699 Å, β = 103.7°, Z = 4. Cu radiation, R = 0.055 for 1328 reflections.

The molecule has the extended conformation shown in Fig. 1 with normal dimensions. The n-pentane side chain is practically planar and forms a 7.5(3)° dihedral angle with the phenyl ring. Molecules are linked by N(1)-H...O(2') hydrogen bonds (N...O 2.882(8) Å).

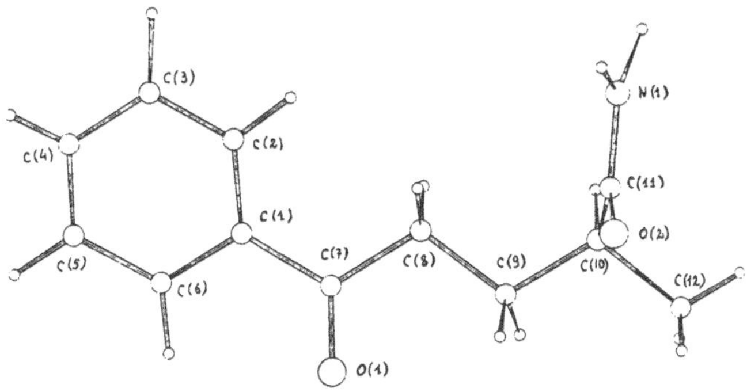

Fig. 1. The conformation of $C_{12}H_{15}NO_2$.

(Z)-3-ETHOXYCARBONYL-4-PHENYLPENT-3-ENOIC ACID
$C_{14}H_{16}O_4$

M. SORIANO-GARCIA, R.A. TOSCANO, G. GARCIA and F. GARCIA, 1985. *Acta Cryst.*, **C41**, 1064-1066.

Orthorhombic, *Pbcn*, a = 35.934, b = 7.502, c = 10.056 Å, Z = 8. Cu radiation, R = 0.050 for 1600 reflections.

The C(4)-C(7) bond in the molecule (Fig. 1), of length 1.522(3) Å, is significantly longer than the normal C(sp^2)-C(sp^2) length indicating that there is no conjugation of C(3)-C(4) with the phenyl ring. This ring is planar and has normal geometry. The penteneoic acid and ethoxycarbonyl groups are oriented in such a way that the transannular interactions between their O atoms is minimised. The O(2)...O(4) separation is 3.254(4) Å. The dihedral angle between the ethoxycarbonyl group and the plane C(1,2,3,4), O(1,2) is 110.4(4)°. The molecules form centrosymmetric hydrogen-bonded dimers in the crystal with O(1)-H...O(2) distances of 2.690(3) Å.

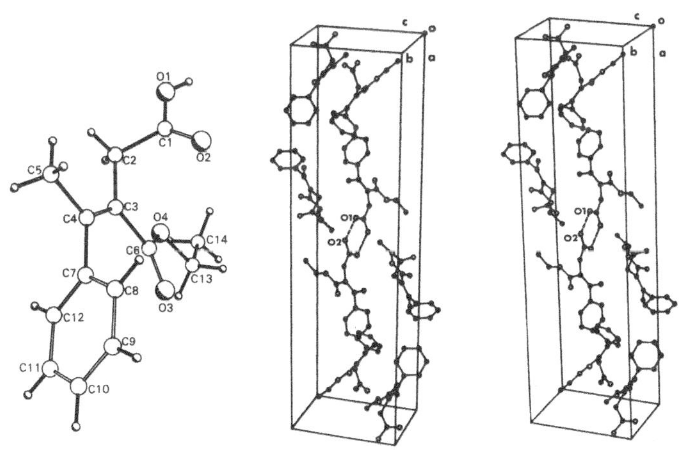

Fig. 1. $C_{14}H_{16}O_4$: a view of the molecule showing the atomic labelling and a stereoview of the unit cell showing the hydrogen-bonded dimers.

2-HYDROXYMETHYL-2-(4-ISOBUTYLPHENYL)PROPANOIC ACID
$C_{14}H_{20}O_3$

A. GINEBREDA, J.F. PINIELLA, J.L. BRIANSO and X. SOLANS, 1985. *Afinidad*, **42**, 177-180.

Monoclinic, $P2_1/a$, a = 12.38, b = 5.780, c = 19.32 Å, β = 103.5°, Z = 4. Mo radiation, R = 0.096 for 1492 reflections.

Atoms C(12), C(13), and C(14) are thermally disordered. The atom C(4) lies in the plane of the benzene ring, and the angles around C(2) range from 105.3(5) to 114.7(6)°. Distances of note are: C(1)-O(15) 1.196(7), C(1)-O(16) 1.318(8), C(1)-C(2) 1.562(9), C(3)-C(2) 1.542(9), C(4)-C(2) 1.537(9), C(5)-C(2) 1.532(8) and C(3)-O(17) 1.425(8) Å. The C-C bonds within the benzene ring range from 1.366(11) to 1.409(12) Å. Enantiomeric pairs are linked by O(17)-H...O(15) (2.66(10 Å) hydrogen bonds.

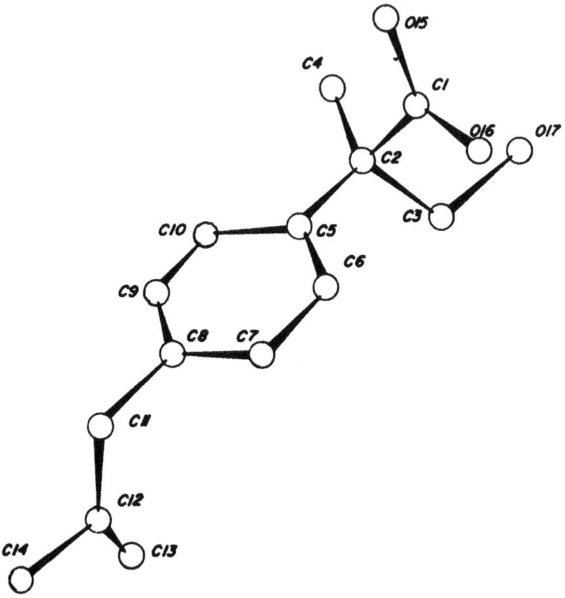

Fig. 1. Molecular structure of $C_{14}H_{20}O_3$.

(μ_2-OXO)BIS(TRIFLUOROACETATOPHENYLIODINE)
$C_{16}H_{10}F_6I_2O_5$

J. GALLOS, A. VARVOGLIS and N.W. ALCOCK, 1985. *J. Chem. Soc. Perkin Trans. 1*, 757-763.

Monoclinic, $P2_1/c$, a = 15.013, b = 5.809, c = 23.455 Å, β = 91.80°, Z = 4. Mo radiation, R = 0.043 for 1787 reflections.

The crystal structure analysis established the chemical identification of the molecule (Fig. 1). The I-O bond distances vary from 1.999 to 2.287 Å. The O-I-O angle is 172.7(3)°.

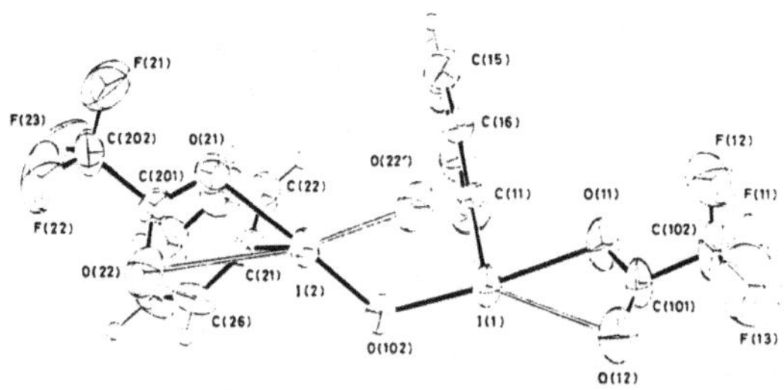

Fig. 1. The molecule of $C_{16}H_{10}F_6I_2O_5$; long I-O interactions are shown with outline bonds.

(E)-METHYL N-t-BUTYL-N-(2-CYANO-1-METHYLTHIO-2-PHENYLVINYL)CARBAMATE
$C_{16}H_{20}N_2O_2S$ (I)

(Z)-METHYL N-t-BUTYL-N-(2-CYANO-1-METHYLTHIO-2-PHENYLVINYL)CARBAMATE
$C_{16}H_{20}N_2O_2S$ (II)

L. TOUPET, G. MOREL, E. MARCHAND and A. FOUCAUD, 1985. *Acta Cryst.*, **C41**, 1121-1123.

I. Monoclinic, $P2_1/n$, a = 8.755, b = 20.802, c = 9.394 Å, β = 93.17°, Z = 4. Mo radiation, R = 0.060 for 1631 reflections.

II. Triclinic, $P\bar{1}$, a = 8.811, b = 9.432, c = 11.04 Å, α = 77.19, β = 75.21, γ = 71.16°, Z = 2. Mo radiation, R = 0.036 for 1795 reflections.

In the two isomers (Fig. 1) the nitrogen atom N(2) and the three C atoms bonded to it are coplanar. The plane of the double bond, C(2,3,11) and that of the aromatic ring have interplanar dihedral angles of 47.2(2) and 43.1(2)° for (I) and (II) respectively. The dihedral angle between the NC_3 plane and the C(2,3,11) plane are 91.9(2) and 84.3(2)° in (I) and (II). These twists are ascribed to steric interactions between, in (I), the S atom and the phenyl ring and, in (II), the t-butyl and phenyl groups. Both isomers have very similar bond lengths and angles and these are in good agreement with normal values.

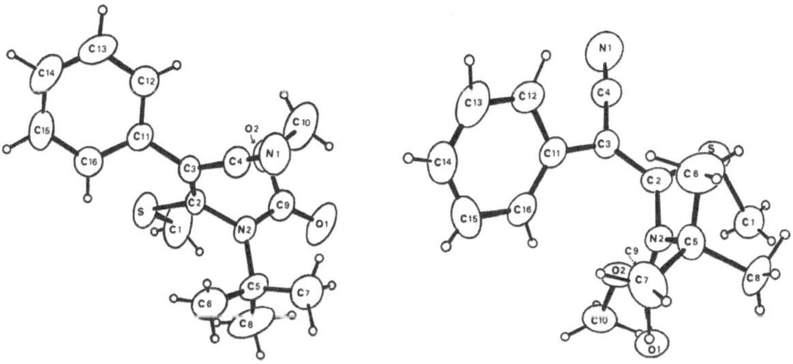

Fig. 1. $C_{16}H_{20}N_2O_2S$: views showing the atomic numbering in (left) isomer (I) and (right) isomer (II).

TETRAMETHYL 4,5-BIS(DIMETHYLAMINO)-2,3,6,7-TETRAAZAOCTA-1,3,5,7-TETRAENE-1,1,8,8-TETRA-CARBOXYLATE
$C_{16}H_{24}N_6O_8$

B. TINANT, J.-P. DECLERCQ, M. VAN MEERSSCHE, A. BOUVY, Z. JANOUSEK and H.G. VIEHE, 1985. *J. Chem. Soc. Perkin Trans.* 2, 1419-1421.

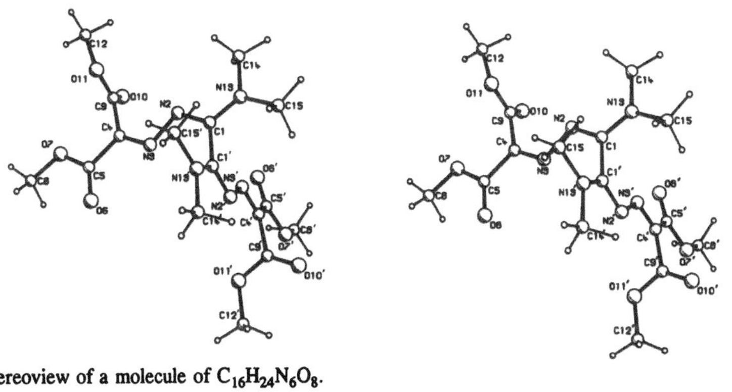

Fig. 1. Stereoview of a molecule of $C_{16}H_{24}N_6O_8$.

Orthorhombic, $P2_12_12_1$, a = 20.448, b = 12.657, c = 8.339 Å, Z = 4. Mo radiation, R = 0.067 for 844

reflections.

Both amide azine moieties are almost perpendicular (Fig. 1). All the atoms of the molecule, with the exception of hydrogen, lie in one of four planes: the first comprises C(1),N(2),N(3),C(4), the dimethylamino group on C(1),N(13),C(14),C(15) and one ester substituent (C(5),O(6),O(7),C(8)); the same atoms numbered with primes comprise the second plane which forms a dihedral angle of 93° with the first; the two other methyl ester substituents (C(9),O(10),O(11),C(12) and C(9'),O(10'),O(11')C(12')) are planar and perpendicular to planes 1 and 2 respectively (dihedral angles of 88 and 87°). The central C(1)-C(1') bond length is 1.56(2) Å.

(2S,1'S)-N-(1-(HYDROXYMETHYL)-2-METHYLPROPYL)-2-PHENYLMETHYL-4-PENTENAMIDE
$C_{17}H_{25}NO_2$

M.J. KURTH, O.H.W. DECKER, H. HOPE and M.D. YANUCK, 1985. *J. Am. Chem. Soc.*, **107**, 443-448.

Orthorhombic, $P2_12_12_1$, a = 5.054, b = 20.342, c = 16.170 Å, Z = 4. Mo radiation, R = 0.045 for 2466 reflections (at 140 K).

The absolute configuration of the title compound was established. The structure shows no exceptional features.

α-FLUORO-trans-CINNAMOYL PEROXIDE
$C_{18}H_{12}F_2O_4$

D. BETHELL, D.J. CHADWICK, M.M. HARDING, G.Q. MALING and M.D. WILEY, 1985. *Acta Cryst.*, **C41**, 470-472.

Monoclinic, $C2/c$, a = 18.78, b = 10.91, c = 7.50 Å, β = 94.4°, Z = 4. Cu radiation, R = 0.062 for 607 reflections.

The molecule adopts an extended conformation (Fig. 1) in which the two α-fluorocinnamoyl groups have the s-cis conformation and lie in nearly mutually orthogonal planes. They are joined by a peroxide bond of length 1.449(7) Å. The mid-point of this bond lies on a crystallographic diad axis and thus the asymmetric unit consists of half of the molecule. The carbonyl bond length, 1.185(11) Å, is short and appears to be little affected by its coplanarity with the conjugated styryl moiety.

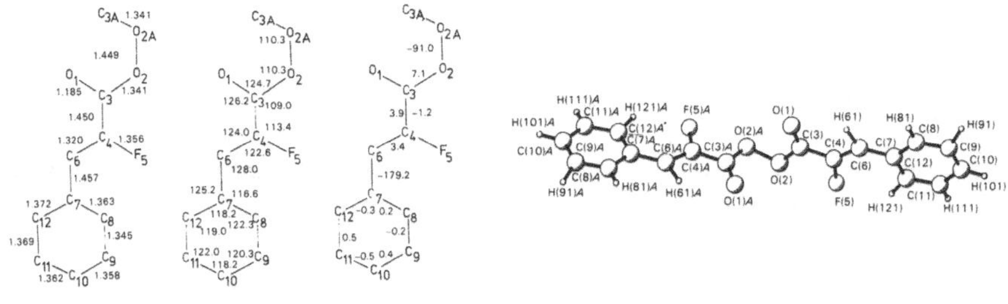

Fig. 1. Bond lengths, valence angles, selected torsion angles in the asymmetric unit and a view of the $C_{18}H_{12}F_2O_4$ molecule.

(2R,3S)-N-((t-BUTYLOXY)CARBONYL)-3-AMINO-2-ACETOXY-4-PHENYLBUTYL ACETATE
$C_{19}H_{27}NO_6$

G.J. HANSON and T. LINDBERG, 1985. *J. Org. Chem.*, **50**, 5399-5401.

Orthorhombic, $P2_12_12_1$, a = 9.828, b = 11.732, c = 17.457 Å, Z = 4. Mo radiation, R = 0.043 for 1873 reflections.

The crystal structure analysis established the stereochemistry of the molecule (Fig. 1) and hence those of related systems.

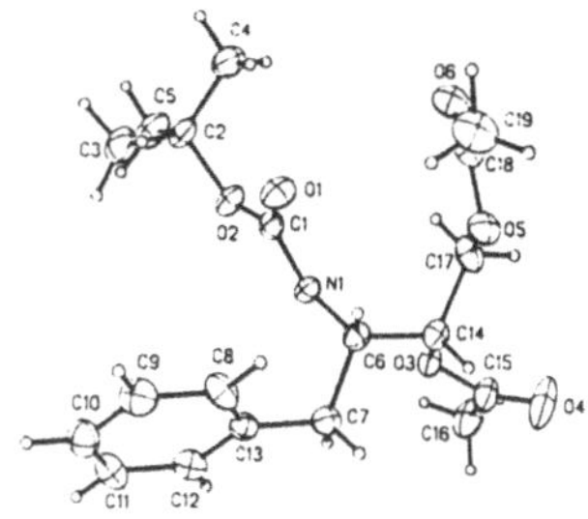

Fig. 1. ORTEP diagram of $C_{19}H_{27}NO_6$.

2,2-DIMESITYLVINYL DIMESITYLACETATE
$C_{40}H_{46}O_2$

D.A. NUGIEL and Z. RAPPOPORT, 1985. *J. Am. Chem. Soc.*, **107**, 3669-3676.

Monoclinic, $P2_1/c$, a = 15.137, b = 15.079, c = 15.570 Å, β = 110.02°, Z = 4. Mo radiation, R = 0.094.

The molecule has a propeller conformation (Fig. 1) with all rings twisted in the same direction. The twist angle of the double bond is 11° and the plane of the carbonyl group deviates from that of the double bond by 25°.

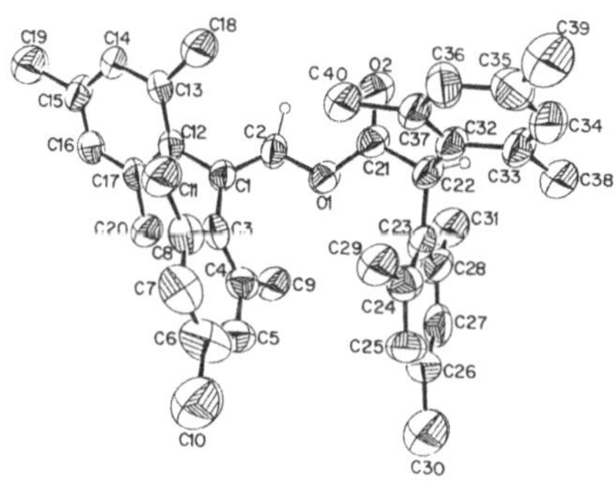

Fig. 1. A view of $C_{40}H_{46}O_2$.

POTASSIUM N-FORMYLDITHIOCARBAMATE
$C_2H_2KNOS_2$ $C_2H_2NOS_2^-,K^+$

R. GERNER, G. KIEL and G. GATTOW, 1985. *Z. Anorg. Allg. Chem.*, **523**, 76-88.

Monoclinic, Cc, a = 13.187, b = 12.928, c = 13.962 Å, β = 101.75°, Z = 16. Mo radiation, R = 0.034 for 1857 reflections.

The compound crystallizes by building a super-structure based on a 16-membered hydrogen-bridged ring system linking the four independent anions (Fig. 1) (N-H...S 3.472 and 3.522(9), N-H...O 3.004 and 3.006(8) Å). Two different binuclear K^+ coordination polyhedra are formed each with two O and two S atoms in common (Fig. 1). K-S distances are in the range 3.25-4.29 Å and K-O distances are 2.73-2.83 Å.

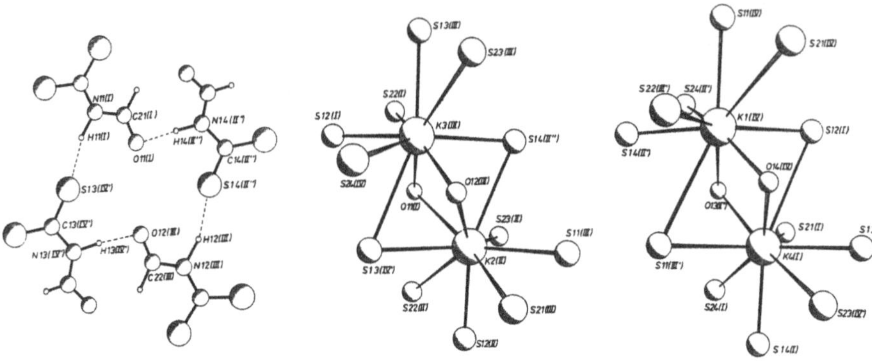

Fig. 1. The hydrogen-bonded anions and the K^+ coordination in $C_2H_2NOS_2^-,K^+$.

N^1-HYDROXYFORMAMIDINIUM HYDROGEN OXALATE (N'-HYDROXYMETHANIMIDAMIDE HYDROGEN OXALATE)

$C_3H_6N_2O_5$ $C_2HO_4^-,CH_5N_2O^+$

I.K. LARSEN, 1985. *Acta Cryst.*, C41, 749-752.

Monoclinic, *Pc*, a = 3.530, b = 9.642, c = 16.706 Å, β = 91.07°, Z = 4. Mo radiation, R = 0.028 for 2259 reflections (at 105 K).

The crystals are composed of hydrogen oxalate anions (A) and of N^1-hydroxyformamidinium cations (B), the protonated form of a tautomer of formamide oxime (Fig. 1). There are two crystallographically independent ions of each type in the asymmetric unit. Both nearly planar cations have synperiplanar conformations with N(1)-C(3)-N(2)-O(5) torsion angles of 8.5(2) and 2.2(2)° for B and B' respectively. This conformation is stabilized by intramolecular N(1)-H...O(5) hydrogen-bonding interactions. Geometrical parameters in the two cations are in good mutual agreement. The geometry of both anions is similar, with dihedral angles between the two COO planes of 1.7(1) and 2.3(1)° for A and A' respectively, and bond lengths and angles agree well with expected values. The crystal structure is stabilized by a network of hydrogen bonds in which all the H atoms in both pairs of ions participate. The anions are linked into infinite chains by strong asymmetric O-H...O bonds (2.471(1) and 2.514(2) Å). These chains are interconnected by hydrogen bonds to the cations.

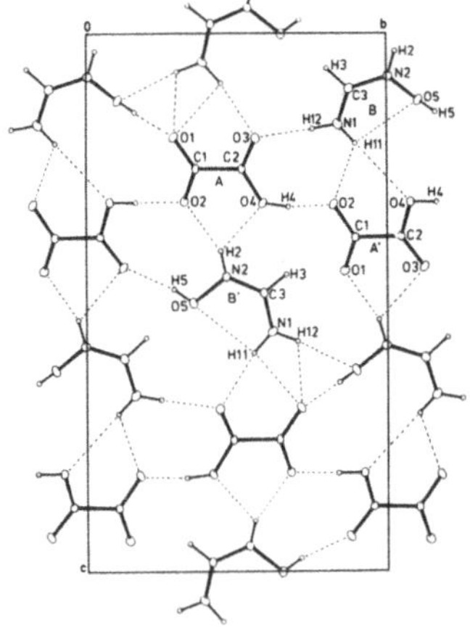

Fig. 1. A projection of the structure of $C_3H_6N_2O_5$ onto (100) with hydrogen bonds shown as broken lines.

1-METHYLHYDRAZINIUM TRIFLUOROACETATE

$C_3H_7F_3N_2O_2$ $C_2F_3O_2^-,CH_7N_2^+$

V.E. SHKLOVER and Yu.T. STRUCHKOV, 1985. *Acta Cryst.*, **C41**, 734-736.

Monoclinic, $P2_1/c$, a = 7.841, b = 13.191, c = 6.149 Å, β = 92.80°, Z = 4. Mo radiation, R = 0.044 for 1308 reflections (at 150 K).

The crystal is composed of $[F_3CCOO]^-$ anions and $[H_2NN(H_2)Me]^+$ cations linked together by N-H...O hydrogen bonds to form zigzag chains along the c-axis. The anion has pseudo C_s symmetry with F(2),C(1,2) atoms on a pseudo mirror plane. The F-C distances are 1.319(2), 1.347(2) and 1.326(2) Å and the mean F-C-F valence angle is 108.3°. The C-C and mean C-O bond lengths are 1.545(2) and 1.241(3) Å respectively. In the cation the N(1)-N(2) and N(2)-C(3) lengths are 1.444(2) and 1.475(3) Å respectively. In addition to the hydrogen bonds of lengths in the range 2.733(2) to 3.107(2) Å there are several F...N inter-ion contacts shorter than the sums of the van der Waals radii.

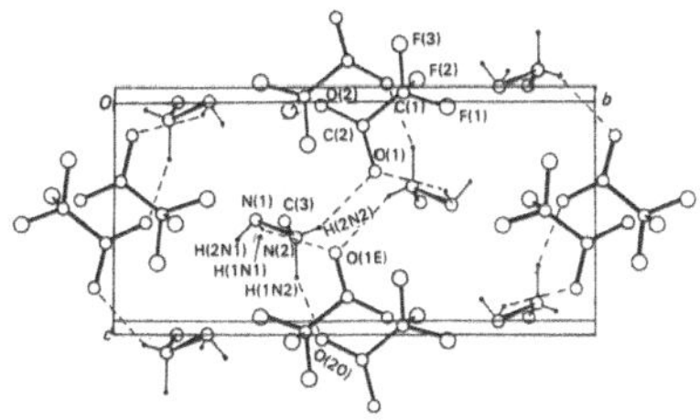

Fig. 1. $C_3H_7F_3N_2O_2$: a projection of the structure onto (100) in which N-H...O hydrogen bonds are shown as dashed lines.

BARIUM 1,1-DICYANOETHYLENE-2-OXO-2-THIOLATE TETRAHYDRATE
$C_4BaN_2OS,4(H_2O)$ $C_4N_2OS^{2-},Ba^{2+},4(H_2O)$

H.-U. HUMMEL, 1985. *Z. Naturforsch.*, **B40**, 722-725.

Orthorhombic, $Pn2_1a$, a = 6.575, b = 14.873, c = 9.808 Å, Z = 4. Ag radiation, R = 0.0285 for 938 reflections.

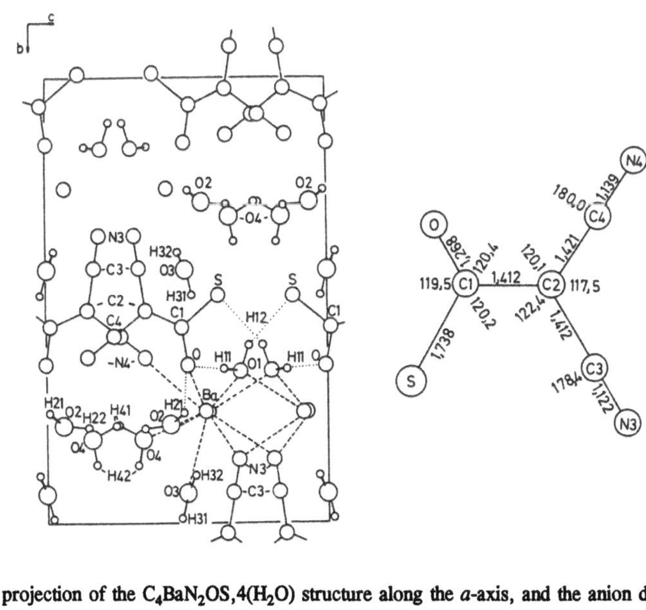

Fig. 1. The projection of the $C_4BaN_2OS,4(H_2O)$ structure along the *a*-axis, and the anion dimensions.

The Ba coordination polyhedron is a tricapped trigonal prism (Fig. 1) with Ba bound to three N (Ba-N 2.2855

3.043(6) Å) and six O atoms (Ba-O 2.719-2.888(5) Å). The $[O(S)CC(CN)_2]^{2-}$ anion (Fig. 1) is involved in O-H...O and O-H...S hydrogen bonding with the water molecules (O...O 2.807-2.858(8), O...S 3.163 and 3.281(7) Å).

POTASSIUM SUCCINATE TRIHYDRATE
$C_4H_4K_2O_4,3(H_2O)$ $C_4H_4O_4^{2-},2(K^+),3(H_2O)$

I. FONSECA, S. MARTINEZ-CARRERA and S. GARCIA-BLANCO, 1985. *Z. Kristallogr.*, **172**, 245-250.

Orthorhombic, *Pnma*, a = 15.695, b = 7.592, c = 8.508 Å, Z = 4. Mo radiation, R = 0.051 for 846 reflections.

The succinate ion and water molecules lie on a crystallographic mirror plane (Fig. 1). The K^+ ion has distorted octahedral coordination with K-O 2.716(3) to 3.070(4) Å.

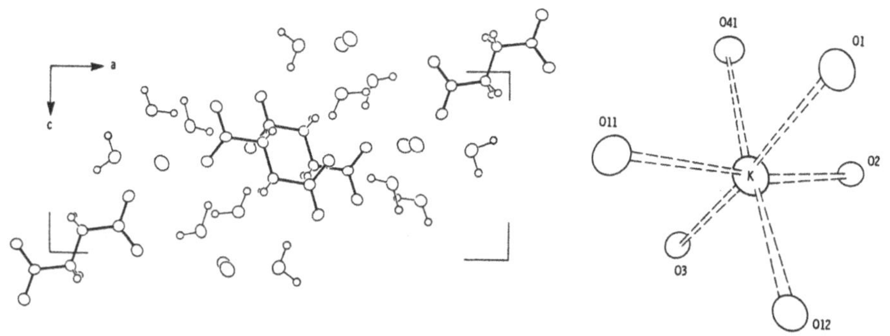

Fig. 1. A view of the $C_4H_4K_2O_4,3H_2O$ structure and the potassium coordination.

DISODIUM meso-TARTRATE
$C_4H_4Na_2O_6$ $C_4H_4O_6^{2-},2(Na^+)$

A.J.A.R. BLANKENSTEYN and J. KROON, 1985. *Acta Cryst.*, **C41**, 182-184.

Monoclinic, $P2_1/n$, a = 4.999, b = 6.348, c = 10.135 Å, β = 97.49°, Z = 2. Mo radiation, R = 0.029 for 2056 reflections.

The meso-tartrate anion is located at a crystallographic centre of symmetry (Fig. 1). This unfavourable conformation results in large deviations of the C and O atoms of the glycolic acid moieties from planarity (O(2)-C(1)-C(2)-O(3) -27.2(1)°). The anions are joined together via a two-dimensional network of hydrogen bonds (O(3)...O(1) 2.695(1) Å). The Na^+ ions are irregularly coordinated to six O atoms (Na-O 2.338(1)-2.526(1) Å).

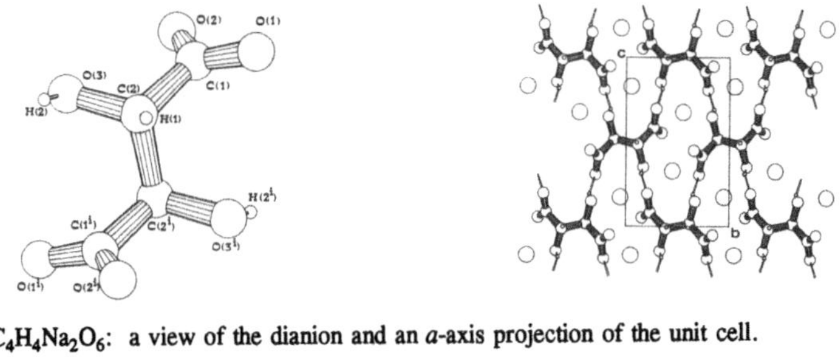

Fig. 1. $C_4H_4Na_2O_6$: a view of the dianion and an *a*-axis projection of the unit cell.

catena-(POTASSIUM HYDROGEN-L(+)-TARTRATE)
$C_4H_5KO_6$

J. BUSCHMANN and P. LUGER, 1985. *Acta Cryst.*, **C41**, 206-208.

Orthorhombic, $P2_12_12_1$, a = 7.749, b = 10.582, c = 7.604 Å, Z = 4. Mo radiation, R = 0.0196 for 1499 reflections (at 100 K).

This study confirms that the structure is isomorphous with that of ammonium hydrogen tartrate (**1**). The structure consists of layers of hydrogen-bonded tartaric anions separated by layers of K^+ ions. Each K^+ ion is surrounded by eight O atoms in a deformed cubic array, average K^+-O 2.867(1)Å. Each tartaric ion is engaged in six intermolecular O...O hydrogen bridges, of which three are crystallographically independent (O...O 2.532(2), 2.782(2), 2.735(2) Å). In the anion the angle between the two least squares planes of atoms O(1)O(2)C(1)C(2)O(3) and O(4)C(3)C(4)O(5)O(6) is 59.3(25)°. The two C-C-O angles of the deprotonated carboxy group are equal at 117.6(1)°, while for the other carboxy group they are 111.9(1) and 123.5(1)°.

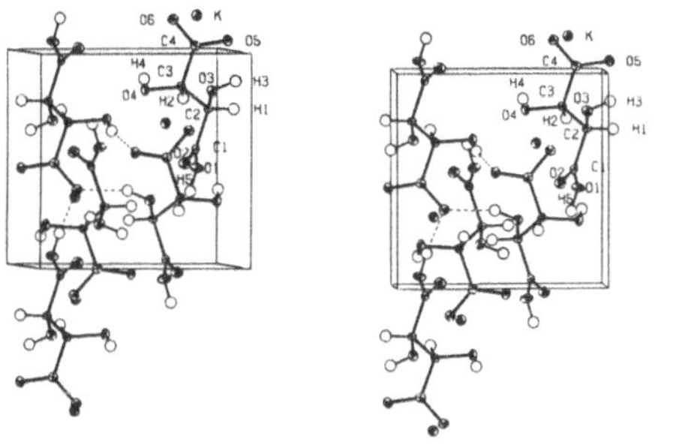

Fig. 1. $(C_4H_5KO_6)_n$: stereoview of the cell.

1. P. Luger, 1980. Modern X-ray Analysis on Single Crystals. Berlin, New York: Walter de Gruyter.

CALCIUM α-ETHYLMALONATE
$C_5H_6CaO_4$ $C_5H_6O_4^{2-},Ca^{2+}$

A. ZELL, H. EINSPAHR and C.E. BUGG, 1985. *Biochemistry*, **24**, 533-537.

Monoclinic, $P2_1/c$, a = 11.003, b = 6.595, c = 9.723 Å, β = 110.14°, Z = 4. Cu radiation, R = 0.060 for 1021 reflections.

Details of the dimensions of the asymmetric unit are in Fig. 1 along with a view of the calcium coordination. The dicarboxylate anion (Fig. 1) represents the functional side-chain of γ-carboxyglutamic acid (GLA) residues, which are implicated in Ca^{2+} ion binding in a variety of proteins. The anion chelates the Ca^{2+} in a bidentate fashion (Fig. 1) leading to an eight-membered coordination polyhedron formed by O atoms from seven carboxylate groups contributed by six symmetry related anions (Ca-O 2.2-2.75 Å). There are three modes of Ca^{2+}-carboxylate interaction; a unidentate mode, a bidentate mode with two O atoms from the same carboxyl group, and a bidentate mode with two O atoms from each of the two carboxylate groups of the anion.

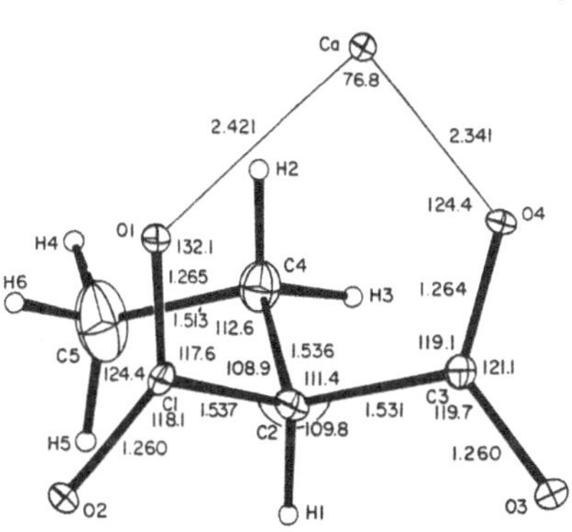

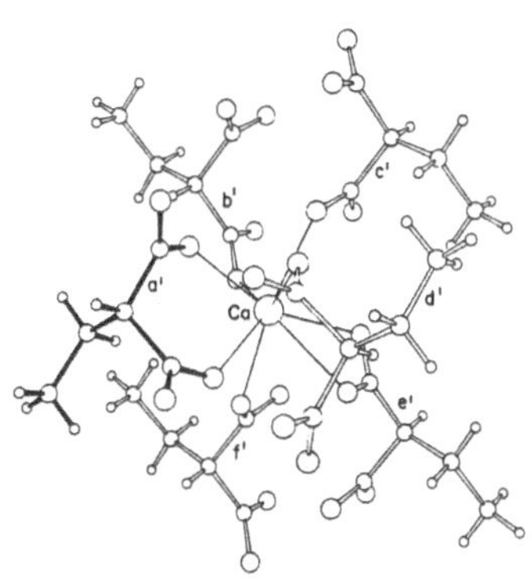

Fig. 1. The asymmetric unit of $C_5H_6O_4^{2-},Ca^{2+}$ and a view of the Ca^{2+} coordination.

GUANIDINIUM HYDROGEN MALEATE
$C_5H_9N_3O_4$ $C_4H_3O_4^-,CH_6N_3^+$

L. GOLIC, I. LEBAN, S. DETONI, B. OREL and D. HADZI, 1985. *J. Cryst. Spect. Res.*, **15**, 215-228.

I. Orthorhombic, *Pca*2$_1$, a = 19.012, b = 3.758, c = 11.119 Å, Z = 4. Mo radiation, R = 0.065 for 488 reflections.

II. Orthorhombic, *Fdd*2, a = 18.505, b = 20.007, c = 8.406 Å, Z = 16. Mo radiation, R = 0.043 for 605 reflections.

Both forms have planar guanidinium cations $C(NH_2)_3^+$ with C-N 1.31(2)-1.34(1) in (I) and 1.311(7)-1.329(9) Å in (II). In the anions there is an intramolecular O-H...O hydrogen bond of length 2.40(1) Å in (I) and 2.445(7) Å in (II). The crystal packing (Fig. 1) is completely different in the two forms; the cations and anions are connected by N-H...O hydrogen bonds (N...O 2.83-3.21(1) Å in (I), 2.853(7)-3.302(8) Å in (II)).

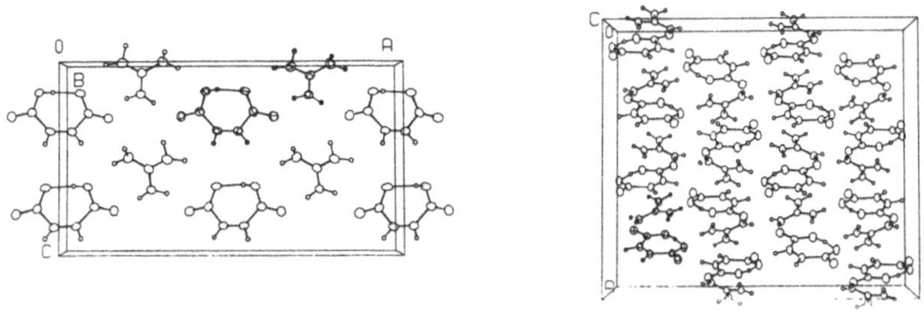

Fig. 1. The crystal packing in form (I) (left) and (II) (right) of $C_5H_9N_3O_4$.

BIS(GUANIDINIUM) FUMARATE DIHYDRATE
$C_6H_{14}N_6O_4,2(H_2O)$ $C_4H_2O_4^{2-},2(CH_6N_3^+),2(H_2O)$

R. FROHLICH, G. MATTERN and W. KRUMBE, 1985. *Z. Kristallogr.*, **172**, 315-317.

Monoclinic, *P*2$_1$/*n*, a = 7.867, b = 8.980, c = 9.646 Å, β = 93.19°, Z = 2. Mo radiation, R = 0.025 for 658 reflections.

The structure consists of isolated guanidinium and fumarate ions and water molecules, linked by hydrogen bonds as shown in Fig. 1. The fumarate ion lies on an inversion centre with mean C-O 1.254(9), C(2)-C(3) 1.491(4), C(3)-C(3') 1.293(6) Å. The guanidinium ion is planar with mean C-N 1.325(4) Å.

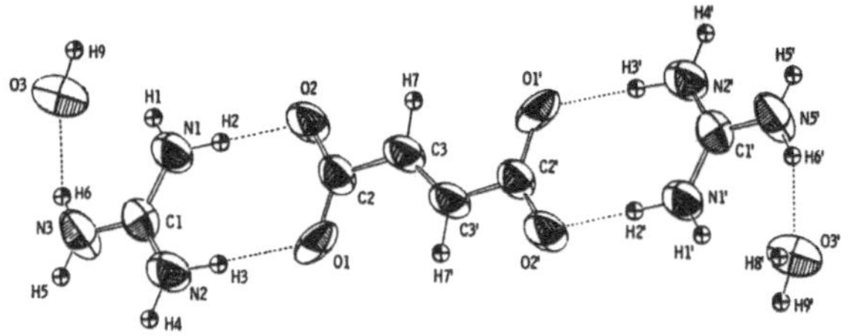

Fig. 1. A projection of the bis(guanidinium) fumarate dihydrate structure along *b*.

TETRAMETHYLAMMONIUM HYDROGEN MALEATE
$C_8H_{15}NO_4$ $C_4H_3O_4^-,C_4H_{12}N^+$

S. DROBEZ, L. GOLIC and I. LEBAN, 1985. *Acta Cryst.*, **C41**, 1503-1505.

Monoclinic, *P*2$_1$/*c*, a = 5.789, b = 9.718, c = 18.362 Å, β = 90.83°, Z = 4. Mo radiation, R = 0.047 for 1109

reflections.

The structure consists of discrete tetramethylammonium ions, and hydrogen maleate residues (Fig. 1). The skeletal chain of C atoms in the hydrogen maleate residue is planar with the end carboxyl groups rotated out of the C-atom plane by 6.0(4) and 5.4(4)°. There is a strong intramolecular hydrogen bond between O(2) and O(3) (2.403(4) Å). The cation is tetrahedral with N-C bond distances in the range 1.483(4) to 1.497(4) Å and C-N-C angles in the range 108.6(2) to 110.7(2)°. There are no unusually close contacts between the cations and anions.

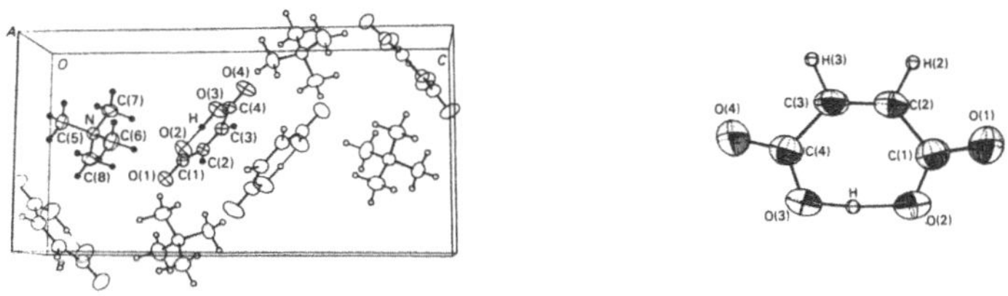

Fig. 1. $C_8H_{15}NO_4$: the unit cell and the hydrogen maleate ion.

N[1]-ISOPROPYL-2-METHYL-1,2-PROPANEDIAMMONIUM MALATE

$C_{11}H_{24}N_2O_5$ $(C_7H_{20}N_2)^{2+}(C_4H_4O_5)^{2-}$

J. KANSIKAS, 1985. *Acta Chem. Scand.*, B39, 563-567.

Triclinic, $P\bar{1}$, a = 7.685, b = 9.122, c = 11.432 Å, α = 69.13, β = 104.52, γ = 112.31°, Z = 2. Mo radiation, R = 0.090 for 1683 reflections.

The structure consists of $(C_7H_{20}N_2)^{2+}$ cations and $(C_4H_4O_5)^{2-}$ anions (Fig. 1) which form a hydrogen bonded network. The amine conformation is synclinal with a N...N distance of 3.139(8) Å and torsion angle N(1)-C(1)-C(2)-N(2) 73.7(5)°; the malate ion conformation is antiperiplanar with a torsion angle C(8)-C(9)-C(10)-C(11) 178.3(5)°.

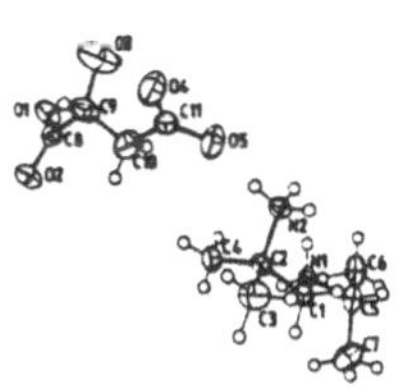

Fig. 1. ORTEP drawing of $(C_7H_{20}N_2)^{2+}(C_4H_4O_5)^{2-}$.

BIS(1,1,1,3,3,3-HEXAFLUORO-2-((HEXAFLUOROISOPROPYLIDENE)AMINO)-2-PROPANOLATO)DISODIUM 1,2-DIMETHOXYETHANE SOLVATE

$C_{12}F_{24}N_2Na_2O_2,2(C_4H_{10}O_2)$ (I) $2(C_6F_{12}NO^-)(Na^+),2(C_4H_{10}O_2)$

TETRAPHENYLPHOSPHONIUM 1,1,1,3,3,3-HEXAFLUOROPROPANEDIOL 1,1,1,3,3,3-HEXAFLUORO-2-HYDROXY-2-PROPANOLATE

$C_{30}H_{23}F_{12}O_4P$ (II) $C_{24}H_{20}P^+,C_3HF_6O_2^-,C_3H_2F_6O_2$

2,2,2',2',8,8,8',8',10,10'-DECAKIS(TRIFLUOROMETHYL)-3,7':7,3'-BIS(DITHIO)BIS(1-THIA-4,6,9-TRIAZA-SPIRO[4.5]DECA-3,6,9-TRIENE)

$C_{22}F_{30}N_6S_6$ (III)

POTASSIUM 2,5-DIHYDRO-2,2,5,5-TETRAKIS(TRIFLUOROMETHYL)-4-OXAZOLATE 1,2-DIMETHOXY-ETHANE SOLVATE

$C_7H_{12}KNO_2,0.5(C_4H_{10}O_2)$ (IV) $C_7F_{12}NO_2^-,K^+,0.5(C_4H_{10}O_2)$

H.W. ROESKY, J. LUCAS, K. KELLER, K.S. DHATHATHREYAN, M. NOLTEMEYER and G.M. SHELDRICK, 1985. *Chem. Ber.*, 118, 2659-2670.

I. Orthorhombic, *Fdd2*, a = 19.487, b = 39.690, c = 9.039 Å, Z = 8. Mo radiation, R = 0.055 for 1221 reflections.

II. Tetragonal, $I4_1/a$, a = 12.579, c = 38.582 Å, Z = 8. Mo radiation, R = 0.067 for 1433 reflections.

III. Monoclinic, $P2_1/c$, a = 19.212, b = 11.462, c = 17.633 Å, β = 109.21°, Z = 4. Mo radiation, R = 0.053 for 3837 reflections.

IV. Orthorhombic, *Pbcn*, a = 20.093, b = 10.572, c = 14.443 Å, Z = 8. Mo radiation, R = 0.074 for 1297 reflections.

Molecule (I) has twofold crystallographic symmetry with Na-O 2.243 and 2.226(5) Å, O-Na-O 89.7(1), Na-O-Na 90.0(1)°, C(1)-O 1.329(6) Å. Hydrolysis of (I) leads to (II) which has dimeric anions (Fig. 1) with H(1) on a fourfold axis; the anions are connected via four asymmetrical and two symmetrical O...H...O bonds (O(1)-H(1) 1.234(4), O(2)-H(2) 0.86(7) Å). The five- and six- membered rings in (III) (Fig. 2) are individually planar and the S-S distance in the disulphide bridges are 2.026 and 2.027(3) Å. The structure of the potassium salt (IV) is shown in Fig. 2; the K^+ ion is five-coordinate with K...O 2.671-2.781(8) Å.

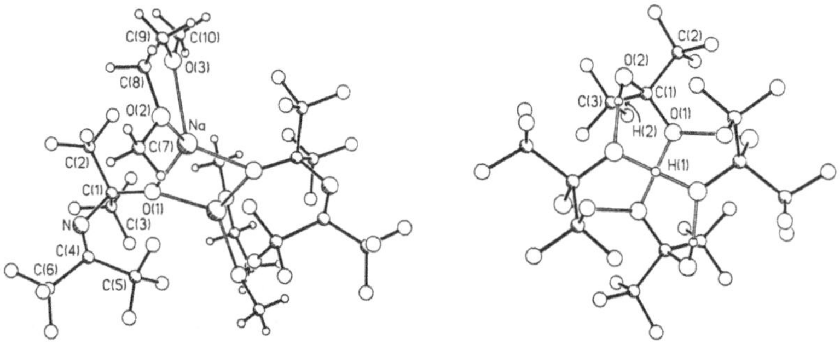

Fig. 1. Views of $C_{12}F_{24}N_2Na_2O_2,2(C_4H_{10}O_2)$ (I) (left) and the anion $C_3HF_6O_2^-,C_3H_2F_6O_2$ in (II) (right).

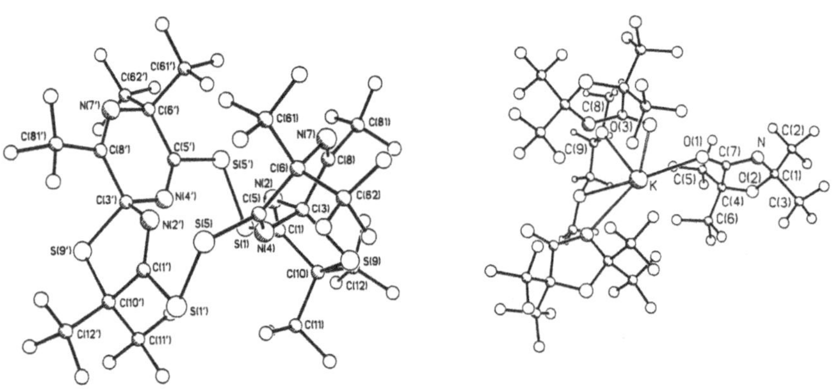

Fig. 2. Views of $C_{22}F_{30}N_6S_6$ (III) (left) and $C_7F_{12}NO_2^-,K^+,0.5(C_4H_{10}O_2)$ (IV) (right).

BIS(POTASSIUM HYDROGEN L-MALATE) MALIC ACID
$C_{12}H_{16}K_2O_5$ $2(C_4H_5O_5^-),C_4H_6O_5,2(K^+)$

W.K.L. VAN HAVERE, P.T. BEURSKENS and A.T.H. LENSTRA, 1985. *J. Cryst. Spect. Res.*, **15**, 45-52.

Orthorhombic, $P2_12_12_1$, a = 7.719, b = 7.916, c = 29.863 Å, Z = 4. Mo radiation, R = 0.059 for 962 reflections.

The central carbon skeleton of the three crystallographically independent malate moieties is anti (Fig. 1). The K^+ ions are eight-coordinate forming face-sharing distorted square antiprisms with K...O 2.749(11)-3.113(11) Å. The packing is stabilized by a three-dimensional network of hydrogen bonds (O...O 2.455-2.823(14) Å).

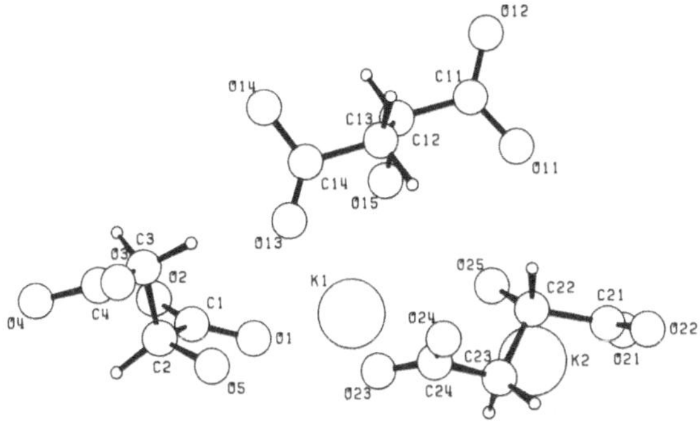

Fig. 1. The asymmetric unit of $2(C_4H_5O_5^-),C_4H_6O_5,2(K^+)$.

BIS(μ_3-α-METHYLMALONATO)(μ_2-α-METHYLMALONATO)TETRAAQUATRICALCIUM
$(C_{12}H_{20}Ca_3O_{16})_n$ (I)

TETRAAQUA(α-METHYLMALONATO-O,O')MAGNESIUM
$C_4H_{12}MgO_8$ (II)

M.E. CURRY, D.S. EGGLESTON and D.J. HODGSON, 1985. *J. Am. Chem. Soc.*, **107**, 8234-8238.

I. Monoclinic, *C2/c*, a = 16.689, b = 18.951, c = 6.635 Å, β = 90.80°, Z = 4. Mo radiation, R = 0.077 for 812 reflections.

II. Orthorhombic, *Pnma*, a = 7.455, b = 7.985, c = 14.279 Å, Z = 4. Mo radiation, R = 0.035 for 848 reflections.

In (I) one Ca atom (Ca(2)) lies on a twofold axis and the other is in a general position. The geometry at Ca(1) is seven-coordinate and the methylmalonate ions show binding modes which are unidentate, bidentate and a bidentate mode involving two O atoms from two different carboxylate groups (Fig. 1). Geometry at Ca(2) is eight-coordinate. The Ca(1)-O distances are in the range 2.319-2.568(7) and Ca(2)-O are in the range 2.355-2.562(6) Å. The Mg complex (II) has nearly regular octahedral geometry with Mg-O 2.050-2.069(1) Å.

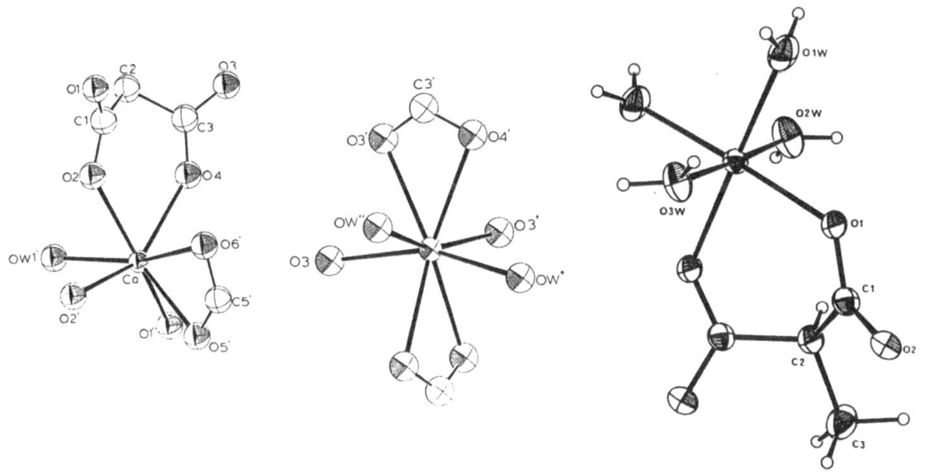

Fig. 1. The coordination at Ca(1) and Ca(2) in (I) (left, centre) and the Mg coordination in $C_4H_{12}MgO_8$ (II).

BIS((+)-(R)-α-METHYLBENZYLAMMONIUM) meso-TARTRATE MONOHYDRATE

$C_{20}H_{28}N_2O_6,H_2O$ $C_4H_4O_6^{2-},2(C_8H_{12}N^+),H_2O$

W. MOERMAN, M. OUWERKERK and J. KROON, 1985. *Acta Cryst.*, **C41**, 1205-1208.

Triclinic, *P*1, a = 9.758, b = 10.552, c = 11.726 Å, α = 82.37, β = 66.21, γ = 88.50°, Z = 2. Cu radiation, R = 0.095 for 3859 reflections.

The meso-tartaric acid molecules (Fig. 1) are nearly centrosymmetric, the two independent moieties being related by a pseudo-centre of symmetry at 0.39, 0.75, 0.57. The anions are sandwiched between α-methylbenzylammonium cations and water molecules. The structure is characterised by a strong hydrogen bonding network (O...O 2.621(6)-2.978(1), N...O 2.760(6)-3.016(5) Å). Bond lengths and angles are normal.

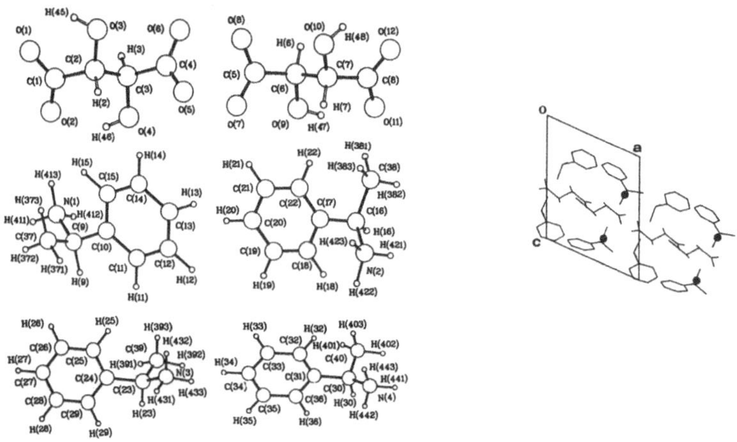

Fig. 1. $C_{20}H_{28}N_2O_6,H_2O$: the cations and anions (left) and projection of the unit-cell contents down the *b*-axis.

METHYLAMMONIUM NITRATE

$CH_6N_2O_3$ $CH_3NH_3^+,NO_3^-$

M. MYLRAJAN, T.K.K. SRINIVASAN and G. SREENIVASAMURTHY, 1985. *J. Cryst. Spect. Res.*, **15**, 493-500.

Orthorhombic, *Pcmn*, a = 6.414, b = 6.581, c = 10.412 Å, Z = 4. Mo radiation, R = 0.048 for 296 reflections.

Both ions lie across crystallographic mirror planes (Fig. 1) and are linked by N-H...O hydrogen bonds (N...O 2.885(3) and 2.915(5) Å). Main dimensions are: N-O 1.221(3) and 1.232(2), C-N 1.453(4) Å, O-N-O 120.6 and 118.6(3)°.

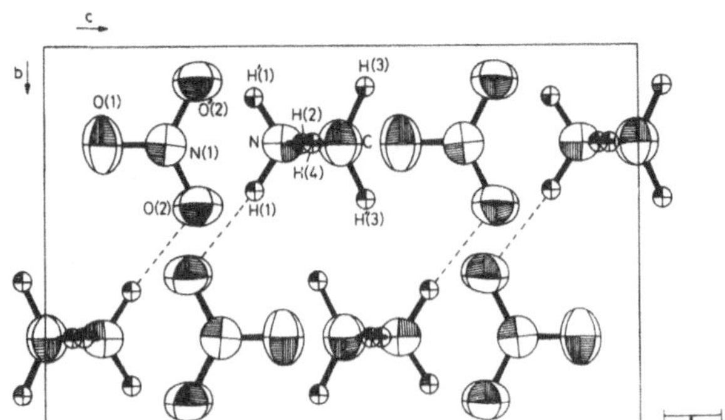

Fig. 1. The crystal structure of $CH_3NH_3^+,NO_3^-$ viewed along the a-direction.

2-AMINOPROPENENITRILE
$C_3H_4N_2$

P. SEILER and J.D. DUNITZ, 1985. *Helv. Chim. Acta*, **68**, 2093-2099.

Orthorhombic, $P2_12_12_1$, a = 5.889, b = 10.943, c = 11.805 Å, Z = 8. Mo radiation, R = 0.027 for 1860 reflections (at 97 K).

There are two independent molecules in the crystallographic asymmetric unit and these are shown with molecular dimensions in Fig. 1. The crystal is built from approximately centrosymmetric dimers stabilised by hydrogen bonding between the amino group of each molecule and the nitrile group of its partner. The dimers are linked into chains by further hydrogen bonds.

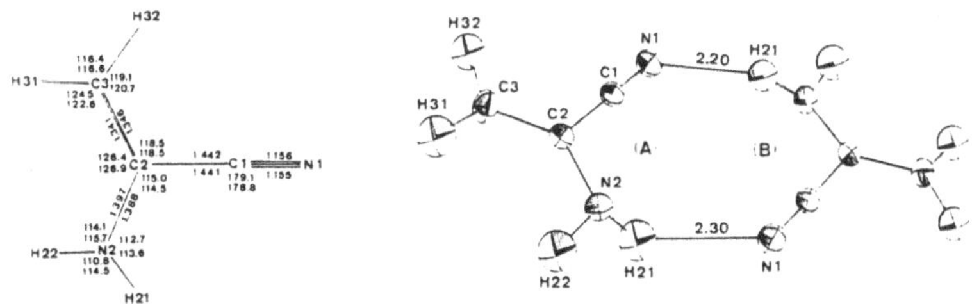

Fig. 1. $C_3H_4N_2$: bond lengths (Å) and angles (°) for the two independent molecules with the upper figures for molecule A and the lower for molecule B (left); perspective drawing of the dimer formed by molecules A and B linked by N-H...N bonds (distances in Å) (right).

N-METHYL-2-NITROETHENAMINE
$C_3H_6N_2O_2$

E.N. GATE, M.A. MEEK, C.H. SCHWALBE, M.F.G. STEVENS and M.D. THREADGILL, 1985. *J. Chem. Soc. Perkin Trans.* 2, 251-255.

Monoclinic, $P2_1/c$, a = 3.915, b = 12.009, c = 10.271 Å, β = 95.46°, Z = 4. Mo radiation, R = 0.0622 for 384 reflections.

As shown by torsion angles the molecule is virtually planar; the nitro and methylamino groups are arranged trans about the carbon-carbon double bond; the stereochemistry about the C(1)-N(3) bond is Z; the C(1)-N(3) distance is 1.303(7) Å and the C(2)-N(5) distance is 1.378(6) Å. Intermolecular hydrogen bonding exists between molecules in the crystal related by the screw axis (N(3)-H(3)...O(7) 2.937 Å with a bond angle of 164°).

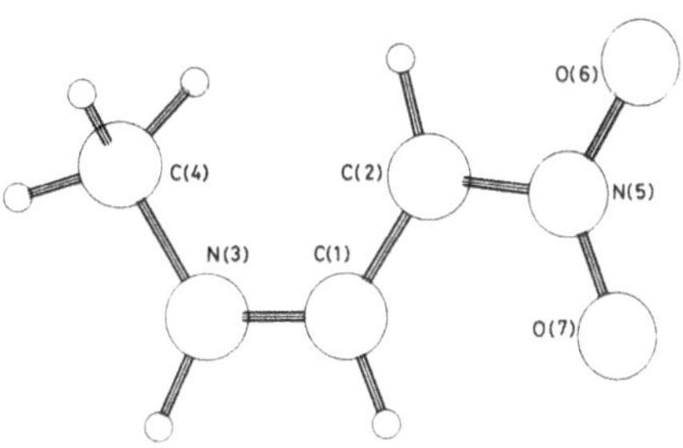

Fig. 1. Structure of $C_3H_6N_2O_2$ in the crystal.

trans-N,N-DIMETHYL-2-NITROVINYLAMINE
$C_4H_8N_2O_2$

M.J. DIANEZ, A. LOPEZ-CASTRO and R. MARQUEZ, 1985. *Acta Cryst.*, **C41**, 981-983.

Monoclinic, $P2_1/n$, a = 10.053, b = 7.547, c = 7.995 Å, β = 105.20°, Z = 4. Mo radiation, R = 0.053 for 1202 reflections.

The molecule as a whole (Fig. 1) is nearly planar and the twist angle around the C=C bond is -179.1(2)°. The N(1)-C(1) and N(2)-C(2) bond distances of 1.390(3) and 1.325(3) Å indicate partial double bond character. The C=C bond length is 1.350(3) Å and N(1)-O(11) and N(1)-O(12) of the nitro group 1.236(3) and 1.250(3) Å respectively. Packing is governed by van der Waals interactions.

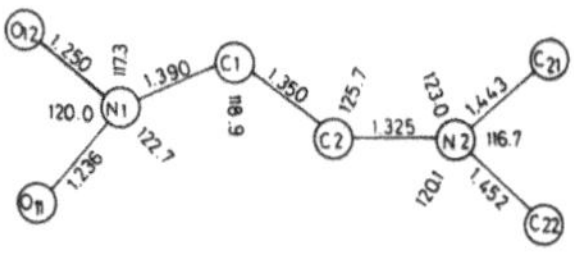

Fig. 1. $C_4H_8N_2O_2$: bond lengths and angles.

SODIUM N,N-DIETHYLDITHIOCARBAMATE TRIHYDRATE
$C_5H_{10}NNaS_2,3(H_2O)$ $Na[S_2CNEt_2],3(H_2O)$

K. MEREITER, A. PREISINGER, W. MIKENDA and H. STEIDL, 1985. *Inorg. Chim. Acta*, **98**, 71-78.

I. Orthorhombic, *Pbca*, a = 7.577 b = 28.356, c = 10.365 Å, Z = 8. Mo radiation, R = 0.024 for 2582 reflections (at 175 K).

A feature of the structure (Fig. 1) is the presence of pairs of $NaS(OH_2)_5$ octahedra sharing an edge (two H_2O), further linked into layers by sharing corners (H_2O). There is an O-H...O bond (2.830(1) Å; all bond lengths at 175 K) and five independent O-H...S bonds (3.244-3.330(1) Å). Bond lengths about Na: S 3.033(1), H_2O 2.333-2.469(1), average 2.409 Å.

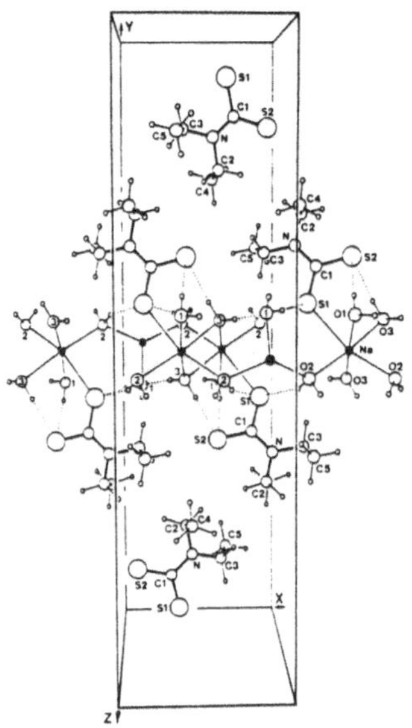

Fig. 1. $Na[S_2CNEt_2],3(H_2O)$: *b*-axis projection.

AMPHI-2-DIMETHYLAMINO-1-METHYLGLYOXIME
$C_5H_{11}N_3O_2$

A.F. MISHNEV, Ya.Ya. BLEIDELIS, V.G. ANDRIANOV and A.V. EREMEEV, 1985. *Zh. Strukt. Khim.*, **26**, 1-140-142 [J. Struct. Chem., **26**, 119-121].

Triclinic, $P\bar{1}$, a = 5.334, b = 8.128, c = 9.521 Å, α = 80.41, β = 85.33, γ = 74.25°, Z = 2. Cu radiation, R = 0.054 for 1026 reflections.

The molecular geometry is shown in Fig. 1. The glyoxime fragment is nonplanar; the N(2)-C(3)-C(4)-N(3) torsion angle is 110.5°. The amidoxime group has the E configuration.

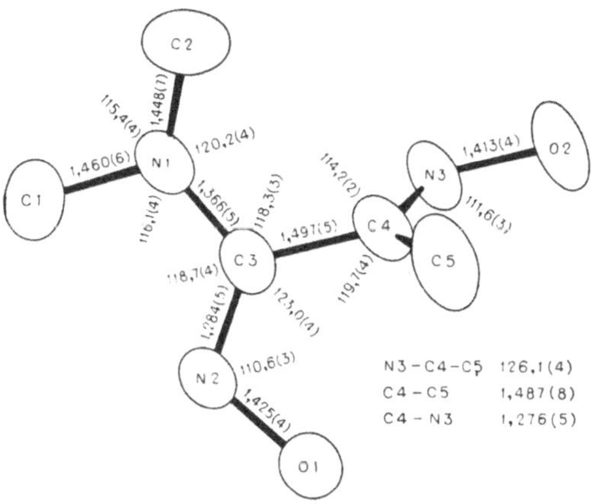

Fig. 1. The molecular geometry of $C_5H_{11}N_3O_2$.

TETRAKIS(AMMONIOMETHYL)METHANE TETRACHLORIDE

$C_5H_{20}Cl_4N_4$ (I) $C_5H_{20}N_4{}^{4+}$,4(Cl^-)

TETRAKIS(AMMONIOMETHYL)METHANE DISULPHATE

$C_5H_{20}N_4O_8S_2$ (II) $C_5H_{20}N_4{}^{4+}$,2(O_4S^{2-})

T.S. CAMERON, W.J. CHUTE and O. KNOP, 1985. *Canad. J. Chem.*, **63**, 586-592.

I. Tetragonal, $P4_2/n$, a = 9.503, c = 6.549 Å, Z = 2. Mo radiation, R = 0.0392 for 465 reflections.

II. Orthorhombic, $P2_12_12_1$, a = 8.382, b = 10.508, c = 14.400 Å, Z = 4. Mo radiation, R = 0.056 for 908 reflections.

The cation in (I) has $\bar{4}$ crystallographic symmetry and the cations and anions are arranged in layers parallel to (001) (Fig. 1). In (II) the cation/anion packing is of the very distorted fluorite type. The mean S-O distance is 1.465(9) Å. Both compounds have strong N-H...X hydrogen bonds linking cations and anions.

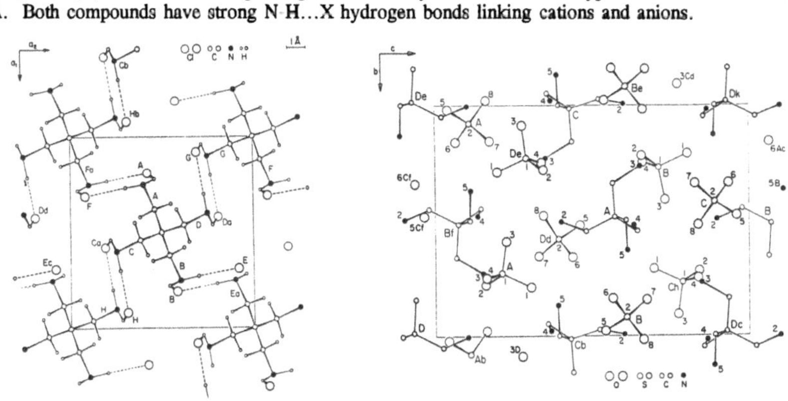

Fig. 1. Views of the crystal structures of $C_5H_{20}Cl_4N_4$ (I) (left) and $C_5H_{20}N_4O_8S_2$ (II) (right).

PENTAERYTHRITYLTETRAMMONIUM NITRATE
$C_5H_{20}N_8O_{12}$ $C_5H_{20}N_4^{4+},4(NO_3^-)$

Y. OYUMI, T.B. BRILL, A.L. RHEINGOLD and C. LOWE-MA, 1985. *J. Phys. Chem.*, **89**, 2309-2315.

Tetragonal, $P4_1$, a = 10.156, c = 14.629 Å, Z = 4. Mo radiation, R = 0.0548 for 1179 reflections.

The cation has close to S_4 symmetry (Fig. 1) with maximally extended chains. Strong N-H...O hydrogen bonds are responsible for the lattice cohesion. The C-C distances are in the range 1.535-1.549(7) Å, with C-N 1.482-1.503(6) Å. The nitrate ions all have one N-O bond shorter than the other two (1.206-1.217 and 1.233-1.262(6) Å).

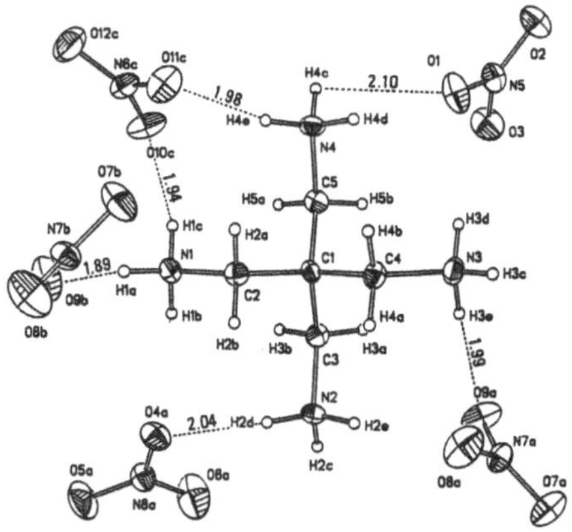

Fig. 1. An ORTEP plot of $C_5H_{20}N_4^{4+},4(NO_3^-)$.

TRIETHYLAMMONIUM CHLORIDE
$C_6H_{16}ClN$ (I) $C_6H_{16}N^+,Cl^-$

TRIETHYLAMMONIUM BROMIDE
$C_6H_{16}BrN$ (II) $C_6H_{16}N^+,Br^-$

TRIETHYLAMMONIUM IODIDE
$C_6H_{16}IN$ (III) $C_6H_{16}N^+,I^-$

M.A. JAMES, T.S. CAMERON, O. KNOP, M. NEUMAN and M. FALK, 1985. *Canad. J. Chem.*, **63**, 1750-1758.

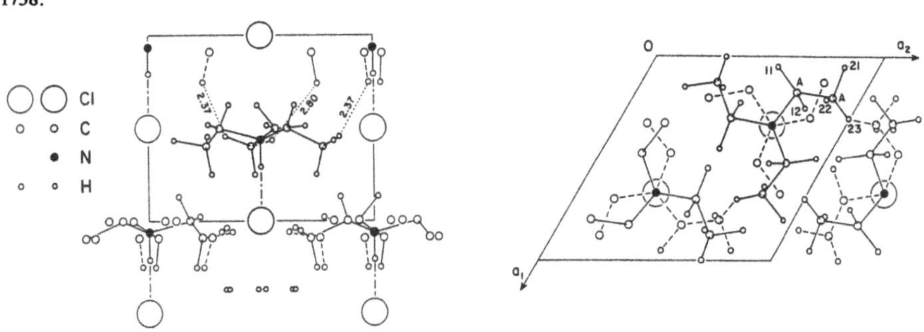

Fig. 1. Two projections of the $C_6H_{16}ClN$ (I).

I. Hexagonal, $P6_3mc$, a = 8.329, c = 7.058 Å, Z = 2. Mo radiation, R = 0.0433 for 179 reflections.

II. Hexagonal, $P6_3mc$, a = 8.449, c = 7.373 Å, Z = 2. Mo radiation, R = 0.030 for 125 reflections.

III. Hexagonal, $P6_3mc$, a = 8.700, c = 7.883 Å, Z = 2. Mo radiation, R = 0.0504 for 132 reflections.

The ethyl groups in these isomorphous structures are orientationally disordered. The N-H...X hydrogen bonds lie along threefold axes with N...Cl 3.107(6), N...Br 3.261(14), N...I 3.595(21) Å (Fig. 1).

N,N'-DI-t-BUTYLETHANEDIIMINE
$C_{10}H_{20}N_2$

C.J.M. HUIGE, A.L. SPEK and J.L. DE BOER, 1985. *Acta Cryst.*, **C41**, 113-116.

Monoclinic, $P2_1/n$, a = 9.662, b = 10.643, c = 11.472 Å, β = 110.15°, Z = 4. Mo radiation, R = 0.0441 for 2593 reflections (at 98 K).

The asymmetric unit of the structure consists of halves of two crystallographically independent molecules each located on a centre of inversion. In both molecules the C=N bonds are approximately coplanar with one of the C-CH_3 bonds. Thus in one molecule C(4), C(6), C(4') and C(6') and in the other molecule C(10), C(12), C(10') and C(12') are displaced from the planes defined by the other non-H atoms. The central C-C bond lengths are 1.467(2) and 1.470(2) Å respectively. The C=N bond lengths are 1.267(2) and 1.267(1) Å. Intermolecular distances are essentially normal van der Waals contacts.

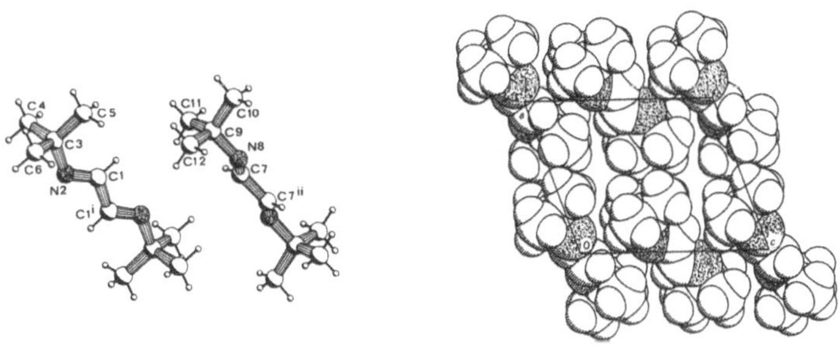

Fig. 1. $C_{10}H_{20}N_2$: views of and numbering in the two independent centrosymmetric molecules as viewed along the b-axis and molecular packing of the layer at y = 0.

(S,S)-N,N'-BIS(1-HYDROXY-2-BUTYL)ETHYLENEDIAMINE DIHYDROCHLORIDE ((S,S)-ETHAMBUTOL DIHYDROCHLORIDE)
$C_{10}H_{26}Cl_2N_2O_2$ $C_{10}H_{26}N_2O_2^{2+}$,2(Cl^-)

R. HAMALAINEN, M. LEHTINEN and M. AHLGREN, 1985. *Arch. Pharm.*, **318**, 26-30.

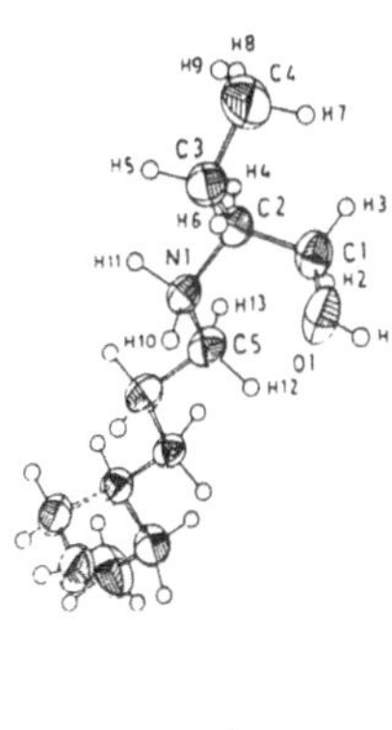

Fig. 1. Views of the $[C_{10}H_{26}N_2O_2]^{2+}$.

Orthorhombic, $P2_12_12$, a = 23.18, b = 6.555, c = 5.176 Å, Z = 2. Mo radiation, R = 0.037 for 762 reflections.

The ethambutol cation has crystallographic twofold symmetry (Fig. 1). The C-O distances of 1.412(5) Å and

N-C distances of 1.508(5)-1.507(5) Å are within the usual range of values within error limits. The C-C distances of 1.519(6)-1.540(7) Å are likewise normal. The angle between the planes calculated through C(2),N(1),C(5) and C(2*),N(1*),C(5*) is 33.6(3)°, where * refers to the position -x+1, -y+1, z. In the unit cell, diamine molecules are joined to the three dimensional network through the chloride ions, which are hydrogen-bonded to both N and O atoms. The relevant distances are: N(1)...Cl(1) 3.066(4), N(1)...Cl(1[i]) 3.174(4) and O(1)...Cl(1[ii]) 3.073(4) Å, where i refers to the position x, y, z+1 and ii to the position x, y+1, z+1.

E-N,N-DIMETHYL-3-PHENYL-2-PROPENYLIDENIMINIUM PERCHLORATE

$C_{11}H_{14}ClNO_4$ (I) $C_{11}H_{14}N^+,ClO_4^-$

E-N,N-DIMETHYL-3-(p-METHOXYPHENYL)-2-PROPENYLIDENIMINIUM PERCHLORATE

$C_{12}H_{16}ClNO_5$ (II) $C_{12}H_{16}NO^+,ClO_4^-$

R.F. CHILDS, B.D. DICKIE, R. FAGGIANI, C.A. FYFE, C.J.L. LOCK and R.E. WASYLISHEN, 1985. *J. Cryst. Spect. Res.*, **15**, 73-87.

I. Orthorhombic, *Pmcn*, a = 6.595, b = 18.288, c = 10.216 Å, Z = 4. Mo radiation, R = 0.0790 for 833 reflections.

II. Triclinic, $P\bar{1}$, a = 6.862, b = 9.830, c = 13.376 Å, α = 119.05, β = 114.99, γ = 90.79°, Z = 2. Mo radiation, R = 0.0736 for 1670 reflections.

The cation in (I) lies on a crystallographic mirror plane (although there is the possibility of some slight disorder about the plane); the cation in (II) shows some small deviations from planarity with N(1) (-0.233(6)) and C(11) (-0.274(8) Å) (Fig. 1) being furthest from the plane defined by the aromatic ring. Both cations have s-trans arrangements. The bond lengths are consistent with a highly localized structure with the positive charge situated principally in the iminium moiety.

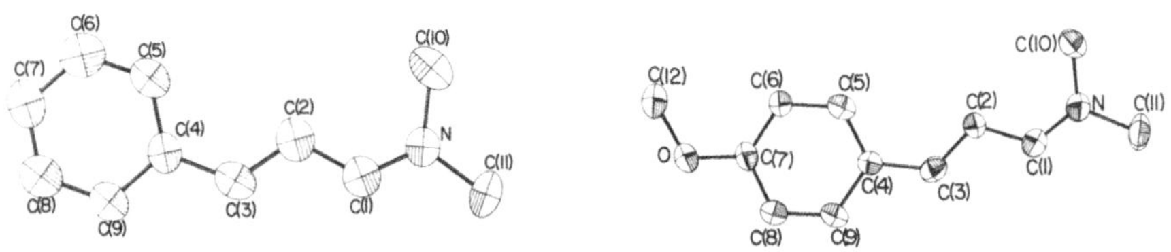

Fig. 1. The cations $C_{11}H_{14}N^+$ (left) and $C_{12}H_{16}NO^+$ (right).

TETRAETHYLAMMONIUM O-ETHYLXANTHATE HEMIHYDRATE

$C_{11}H_{25}NOS_2,0.5(H_2O)$ $C_8H_{20}N^+,C_3H_5OS_2^-,0.5(H_2O)$

B.F. HOSKINS, E.R.T. TIEKINK and G. WINTER, 1985. *Inorg. Chim. Acta*, **101**, L11-L13.

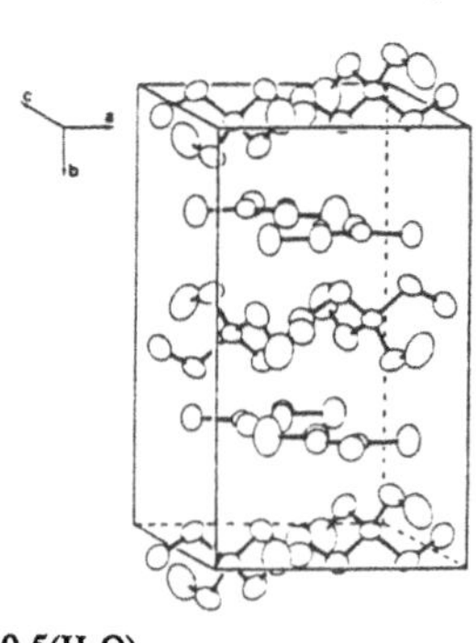

Fig. 1. Unit cell of $C_{11}H_{25}NOS_2,0.5(H_2O)$.

Monoclinic, $P2_1/m$, a = 7.343, b = 13.456, c = 15.674 Å, β = 98.92°, Z = 4. Cu radiation, R = 0.071 for 2229 reflections.

There are two independent anions, each lying in a crystallographic mirror-plane. The cation lies on a general position. Cations and anions form alternate layers parallel to the ac plane (Fig. 1). Bond lengths and angles include: cation, C-N 1.521-1.530(6), anion, S-C 1.657-1.688(7), C(S)-O 1.342, 1.397(8) Å, S-C-S 125.7, 127.4(4), S-C-O 111.8-121.4(4)°.

BIS(BIS(DIMETHYLAMINO)METHYL) ETHER BIS(TRIFLUOROMETHANESULPHONATE)

$C_{12}H_{24}F_6N_4O_7S_2$ $\{[(Me_2N)_2C]_2O\}^{2+}(CF_3SO_3^-)_2$

T. GRAMSTAD, S. HUSEBYE, K. MAARTMANN-MOE and J. SAEBØ, 1985. *Acta Chem. Scand.*, **B39**, 35-45.

Monoclinic, $P2_1/n$, a = 8.246, b = 23.830, c = 11.371 Å, β = 101.65°, Z = 4. Mo radiation, R = 0.052 for 2279 reflections (at -135°C).

The dication ether (Fig. 1) is bent with a central C(3)-O-C(4) angle of 122.4(4)°; it is very nonplanar. The anions have normal molecular geometries.

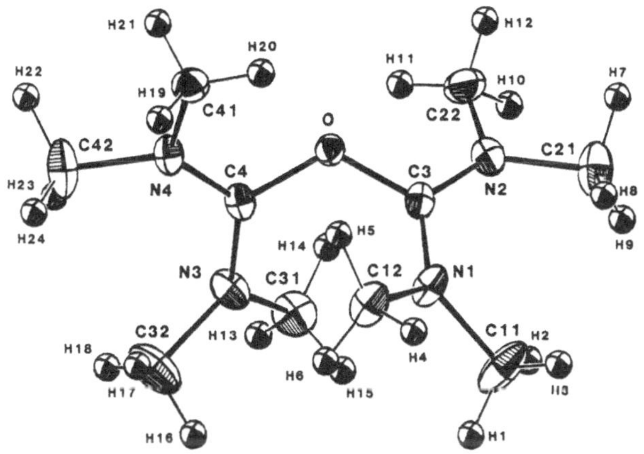

Fig. 1. The structure of the dication ether in $C_{12}H_{24}F_6N_4O_7S_2$ as viewed along the normal to the central C-O-C plane.

3-DIMETHYLAMINO-1,1-DIPHENYL-2-AZAALLENIUM HEXACHLOROANTIMONATE

$C_{16}H_{17}Cl_6N_2Sb$ $C_{16}H_{17}N_2^+,Cl_6Sb^-$

M. AL-TALIB, I. JIBRIL, G. HUTTNER and J.C. JOCHIMS, 1985. *Chem. Ber.*, **118**, 1876-1886.

Monoclinic, $P2_1/c$, a = 9.789, b = 22.37, c = 16.800 Å, β = 144.26°, Z = 4. Mo radiation, R = 0.035 for 2040 reflections (at 238 K).

Details of cation conformation and dimensions are in Fig. 1. The planes defined by the sp^2 hybridised C atoms of the C=N=C are almost perpendicular to one another and the C-N(1)-C angle is 128.2(9)°.

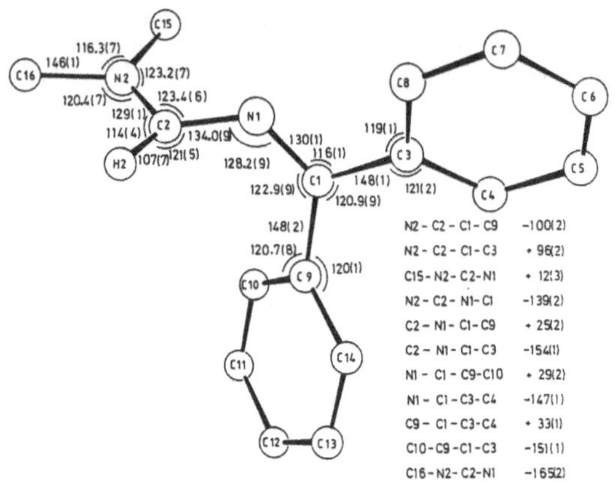

Fig. 1. Molecular plot for the $C_{16}H_{17}N_2^+$ ion with selected distances (pm) and angles (°).

2,N,N-TRIMETHYL-4,4-BIS(PHENYLSULPHONYL)BUTYLAMINE
$C_{19}H_{25}NO_4S_2$ (I)

2-(2-METHYL-3-PIPERIDINOPROPYL)-1,3-DITHIAN 1,1,3,3-TETRAOXIDE
$C_{13}H_{25}NO_4S_2$ (II)

R.L. HARLOW, CHUEN LI and M.P. SAMMES, 1985. *J. Chem. Res.*, S44-45, M701-748.

I. Triclinic, $P\bar{1}$, a = 10.625, b = 12.745, c = 8.295 Å, α = 106.13, β = 98.55, γ = 66.39°, Z = 2. Mo radiation, R = 0.039 for 2451 reflections.

II. Monoclinic, $P2_1/c$, a = 6.713, b = 24.677, c = 9.895 Å, β = 107.34°, Z = 4. Mo radiation, R = 0.040 for 2576 reflections.

The molecules of (I) (Fig. 1) have a compact structure with an N...H distance (intramolecular) of 2.34(3) Å. The molecules of (II) have a more open structure without a short N...H distance but having an intermolecular C-H...O interaction to an axial sulphone oxygen atom (H(1)...O(1) 2.26(2) Å).

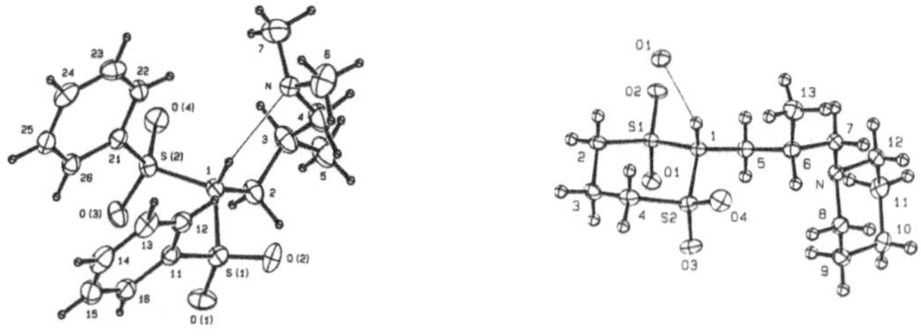

Fig. 1. ORTEP drawings of molecules of $C_{19}H_{25}NO_4S_2$ (left) and of $C_{13}H_{25}NO_4S_2$ (right).

N,N'-(CHLOROSULPHONIUMYLIDENE)BIS(N^1,N^1-DIMETHYLCHLOROFORMAMIDINE) TRICHLORIDE
$C_6H_{12}Cl_6N_4S$ (I) [{$Me_2NC(Cl)N$}$_2$SCl](Cl_3)

N,N'-(CHLOROSULPHONIUMYLIDENE)BIS(N^1,N^1-DIMETHYLCHLOROFORMAMIDINE) CHLORIDE DICHLOROETHANE
$C_6H_{12}Cl_4N_4S,C_2H_4Cl_2$ (II) [{$Me_2NC(Cl)N$}$_2$SCl]Cl,$C_2H_4Cl_2$

N,N-DIMETHYLCHLOROFORMAMIDINIUM CHLORIDE HYDRATE
$C_3H_8Cl_2N_2,H_2O$ (III) [$Me_2NC(Cl)NH_2$]Cl,H_2O

T. CHIVERS, J.F. RICHARDSON and N.R.M. SMITH, 1985. *Inorg. Chem.*, **24**, 2453-2458.

I. Triclinic, $P\bar{1}$, a = 10.275, b = 11.170, c = 6.767 Å, α = 104.98, β = 95.41, γ = 89.82°, Z = 2. Mo radiation, R = 0.043 for 1716 reflections.

II. Triclinic, $P\bar{1}$, a = 10.224, b = 12.158, c = 7.188 Å, α = 95.74, β = 106.64, γ = 93.97°, Z = 2. Mo radiation, R = 0.038 for 3439 reflections.

III. Monoclinic, $P2_1/c$, a = 9.186, b = 5.599, c = 14.745 Å, β = 95.09°, Z = 4. Mo radiation, R = 0.038 for 1017 reflections.

The same cation (Fig. 1) is present in (I) and (II), together with, respectively, Cl_3^- ions and Cl^- ions. Bond lengths and angles include (I, II): S-Cl 2.218(4); 2.233(1); S-N 1.590, 1.610(8); 1.607, 1.607(3); (S)N-N 1.327, 1.356(11); 1.321, 1.324(4) Å; C-N-S 117.2, 118.4(7), 117.0, 117.3(2); N-S-N 96.9(4); 97.3(1); N-S-Cl 100.6, 104.2(4); 99.1, 100.4(1)°. The Cl-Cl bond lengths in (I) are 2.249, 2.340(4) Å and the Cl-Cl-Cl angle is 177.5(2)°. The C_3N_2Cl moieties in (I) and (II) are effectively planar. The dimensions of the planar cation in (III) include: Cl-C 1.703(4), C-NH_2 1.325(5), C-NMe_2 1.282(6) Å, Cl-C-N 115.2, 119.7(3), N-C-N 125.0(4)°. The anion participates in Cl...H-O and Cl...H-N bonds (respectively 3.30, 3.35 and 3.16, 3.22(1) Å).

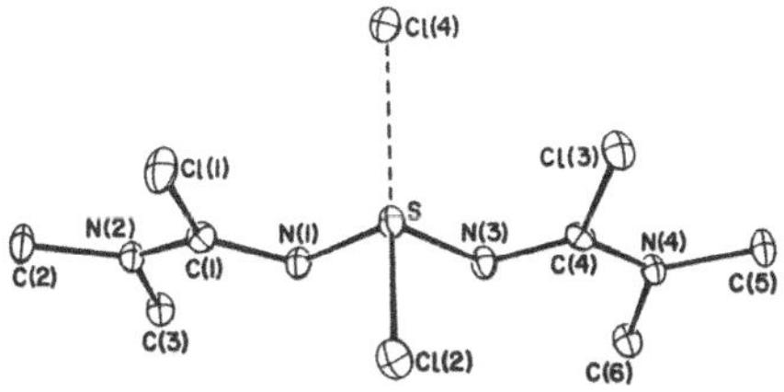

Fig. 1. The cation in (I) $C_6H_{12}Cl_6N_4S$.

TRIS(DIMETHYLAMINO)SULPHONIUM TRIFLUOROMETHOXIDE
$C_7H_{18}F_3N_3OS$ $C_6H_{18}N_3S^+,CF_3O^-$

W.B. FARNHAM, B.E. SMART, W.J. MIDDLETON, J.C. CALABRESE and D.A. DIXON, 1985. *J. Am. Chem. Soc.*, **107**, 4565-4567.

Orthorhombic, *Pnma*, a = 14.547, b = 11.125, c = 7.310 Å, Z = 4. Mo radiation, R = 0.036 for 895 reflections (at -100°C).

Both cation and anion lie on crystallographic mirror planes (Fig. 1). The C-F bond lengths are long and the C-O bond length is shorter than expected.

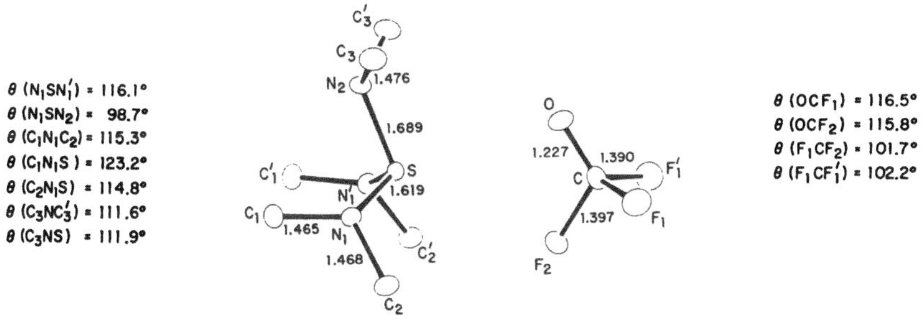

Fig. 1. The $C_6H_{18}N_3S^+$ and CF_3O^- ions with details of geometry.

BIS(t-BUTYLTHIODIIMIDO)SULPHIDE
$C_8H_{18}N_4S_3$

W. ISENBURG, R. MEWS and G.M. SHELDRICK, 1985. *Z. Anorg. Allg. Chem.*, **525**, 54-66.

Monoclinic, $P2_1/c$, a = 8.73, b = 15.48, c = 10.70 Å, β = 90.6°, Z = 4. Mo radiation, R = 0.171 for 882 reflections.

The conformation of the molecule is shown in Fig. 1. Main dimensions are S=N 1.50(2)-1.56(2), S-N 1.65(3) and 1.70(3) Å, N(2)-S(2)-N(3) 97.4(1)°.

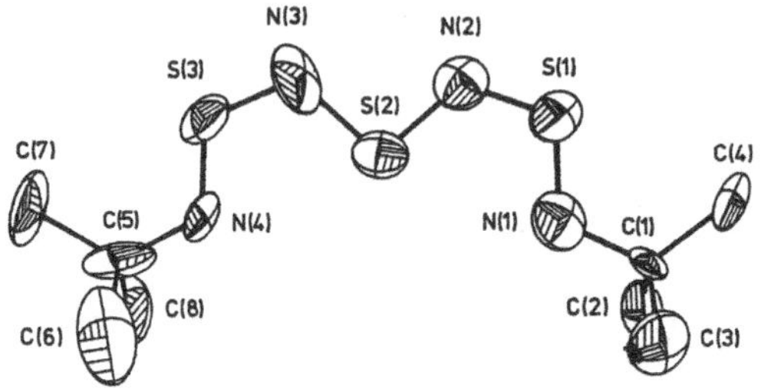

Fig. 1. A view of $C_8H_{18}N_4S_3$.

DIPHENYLMETHYLENEAMINO(THIONYLIMINO)SULPHIDE
$C_{13}H_{10}N_2OS_2$

T. CHIVERS, R.T. OAKLEY, R. PIETERS and J.F. RICHARDSON, 1985. *Canad. J. Chem.*, **63**, 1063-1067.

Triclinic, $P\bar{1}$, a = 9.9078, b = 10.0967, c = 15.1682 Å, α = 78.646, β = 71.065, γ = 63.449°, Z = 4. Mo radiation, R = 0.033 for 2932 reflections.

In the two independent molecules, the CNSNSO unit is an essentially planar cis-trans chain with mean dimensions S(1)-N(1) 1.680(3), S(1)-N(2) 1.644(2), S(2)-N(2) 1.529(2), S(2)-O 1.453(3), C(1)-N(1) 1.489(4) Å.

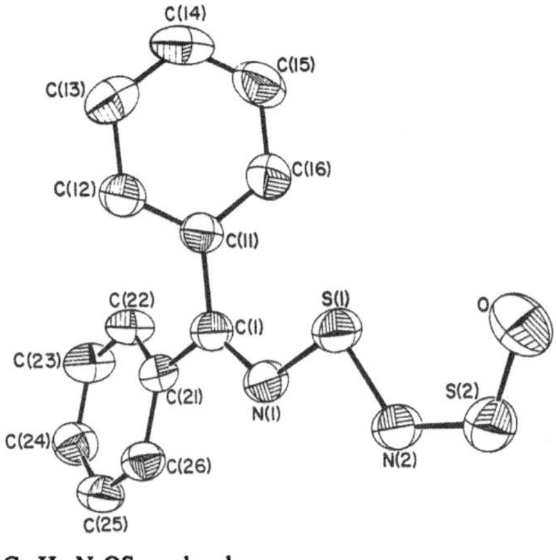

Fig. 1. A drawing of a $C_{13}H_{10}N_2OS_2$ molecule.

TETRA-n-BUTYLAMMONIUM N-THIOFORMYLDITHIOCARBAMATE
$C_{18}H_{38}N_2S_3$ $C_2H_2NS_3^-,C_{16}H_{36}N^+$

R. GERNER, G. KIEL and G. GATTOW, 1985. *Z. Anorg. Allg. Chem.*, **525**, 101-111.

Triclinic, $P\bar{1}$, a = 9.694, b = 11.478, c = 12.551 Å, α = 63.03, β = 66.42, γ = 85.60°, Z = 2. Mo radiation, R = 0.061 for 2249 reflections.

The two anions are linked to form centrosymmetric dimers by N-H...S bridges (Fig. 1) (S...H 2.47(7) Å). The cation has three of the n-butyl chains maximally extended and the fourth is in a gauche conformation. Main anion dimensions are C(1)-S 1.671 and 1.683(5), C(2)-S 1.624(8), N(1)-C(1) 1.406(9), N(1)-C(2) 1.340(8) Å, S(1)-C(1)-S(2) 125.9(4), N(1)-C(2)-S(3) 127.3(6)°.

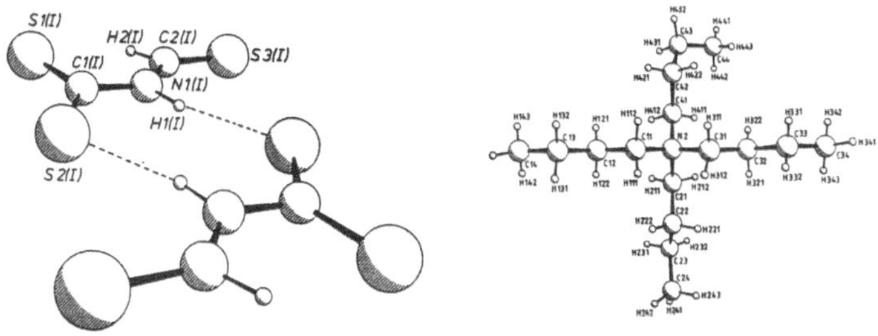

Fig. 1. The hydrogen-bonded anion dimer and the cation in $C_{18}H_{38}N_2S_3$.

CHLOROTRIFLUOROMETHANE
$CClF_3$

G.S. PAWLEY and A.W. HEWAT, 1985. *Acta Cryst.*, **B41**, 136-139.

I. Orthorhombic, $Cmc2_1$, a = 7.1725, b = 9.5689, c = 4.6791 Å. Neutron radiation, R = 0.053 for powder data (at 5 K).

II. Orthorhombic, $Cmc2_1$, a = 7.2295, b = 9.5987, c = 4.7101 Å. Neutron radiation, R = 0.061 for powder data (at 57 K).

III. Orthorhombic, $Cmc2_1$, a = 7.2789, b = 9.6156, c = 4.7355 Å. Neutron radiation, R = 0.136 for powder data (at 81 K).

The structure (Fig. 1) was solved by a simple packing procedure. Constrained refinement involved only four structural parameters (two bond lengths, a position along the y-axis and a molecular rotation in the symmetry plane). The molecules have crystallographic m symmetry.

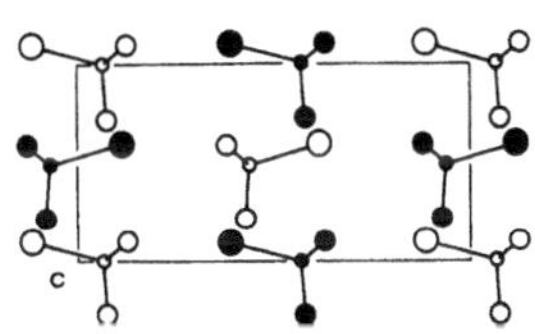

Fig. 1. Projection of the chlorotrifluoromethane structure at 5 K. Bold molecules are centred at a/2.

3,4,5-TRIIODOTOLUENE
$C_7H_5I_3$

YOU-XI LIU, C.B. KNOBLER, K.N. TRUEBLOOD and R. HELGESON, 1985. *Acta Cryst.*, **C41**, 959-961.

Tetragonal, $P\bar{4}2_1m$, a = 15.407, c = 4.348 Å, Z = 4. Mo radiation, R = 0.033 for 572 reflections.

A crystallographic mirror plane, normal to the molecular plane (Fig. 1) bisects the molecule. With the exception of the H atoms of the CH_3 group, the molecule is planar. Within the molecule the shortest I...I contact is 3.613(1) while between molecules the shortest I...I contact is 3.746(1) Å. This latter contact involves (I) atoms in the 3 and 5 positions only and links molecules in sheets perpendicular to the c-axis.

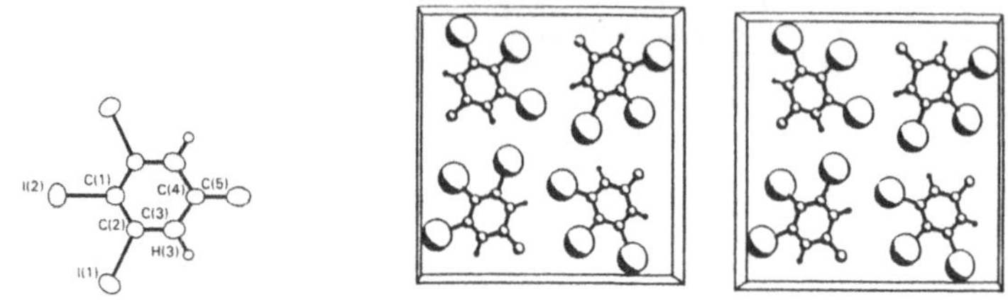

Fig. 1. $C_7H_5I_3$: the molecule and a stereoview of the structure viewed down the **c**-axis.

1,3-DIAMINOUREA (CARBOHYDRAZIDE)
CH_6N_4O

G.A. JEFFREY, J.R. RUBLE, R.G. NANNI, A.M. TURANO and J.H. YATES, 1985. *Acta Cryst.*, **B41**, 354-361.

Monoclinic, $P2_1/c$, a = 3.618, b = 8.789, c = 12.487 Å, β = 106.43°, Z = 4. Neutron radiation, R = 0.031 for 1701 reflections (at 15 K).

The molecule (Fig. 1) has the cis/trans/cis/trans conformation, with significant distortions of the carbon, oxygen and nitrogen atoms from coplanarity. The structures, energies and dipole moments of the eight lowest-energy conformers with approximate C_s symmetry were calculated for the isolated molecules, at rest, by molecular-orbital theory at the HF/3-21G level, using GAUSSIAN80. The lowest-energy conformer is that observed in the crystal structure. However, the molecular distortions in the crystal correspond to a calculated energy that is 22.8 kJ mol^{-1} above the minimum energy. These distortions arise from a complex system of intra- and intermolecular hydrogen bonds in the crystal.

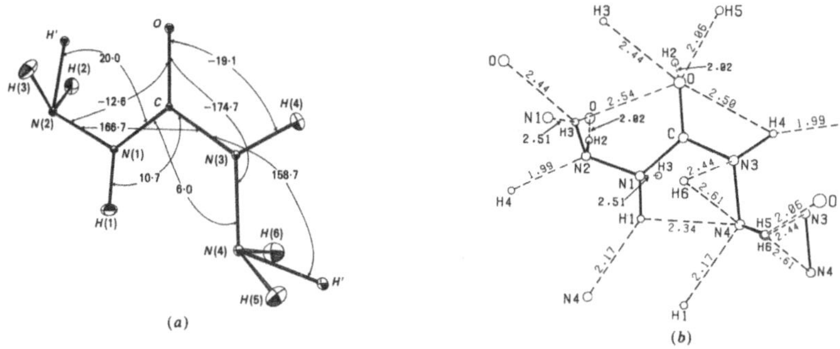

Fig. 1. Carbohydrazide showing (a) torsion angles and (b) the hydrogen bonding environment.

1-CARBAMOYLGUANIDINE ETHANOL
$C_2H_6N_4O,C_2H_6O$ (I)

1-CARBAMOYLGUANIDINIUM PERCHLORATE
$C_2H_7ClN_4O_5$ (II) $C_2H_7N_4O^+,ClO_4^-$

M.J. BEGLEY, P. HUBBERSTEY and C.H.M. MOORE, 1985. *J. Chem. Res.*, S380-381, M4035-4088.

I. Orthorhombic, $P2_12_12_1$, a = 750.9, b = 889.6, c = 1156.5 Å, Z = 4. Mo radiation, R = 0.046 for 715 reflections.

II. Monoclinic, $P2_1/c$, a = 802.0, b = 999.9, c = 961.1 Å, β = 105.54°, Z = 4. Cu radiation, R = 0.041 for 1108 reflections.

The two guanidine systems are effectively planar. Dimensions are in Fig. 1.

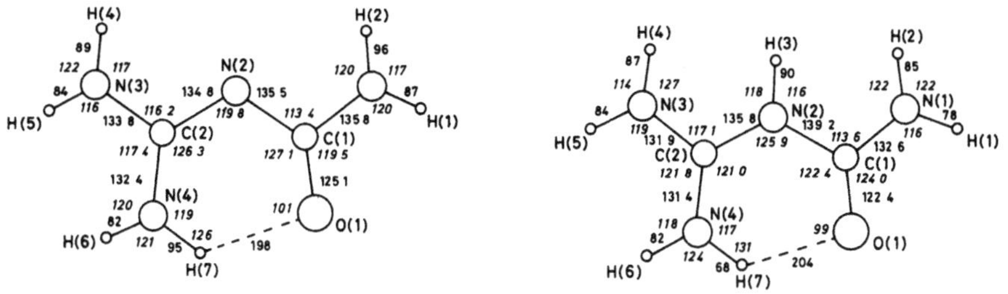

Fig. 1. Molecular geometries of $C_2H_6N_4O,C_2H_6O$ (I) (left) and $C_2H_7ClN_4O_5$ (II) (right); bond distances in pm and angles (italics) in degrees.

BIS(GUANIDINIUM) TETRABORATE DIHYDRATE
$C_2H_{16}B_4N_6O_9,2(H_2O)$ $2(CH_6N_3{}^+),H_4B_4O_9{}^{2-},2(H_2O)$

T.J.R. WEAKLEY, 1985. *Acta Cryst.*, C**41**, 377-379.

Triclinic, $P\bar{1}$, a = 8.443, b = 12.560, c = 7.290 Å, α = 106.75, β = 103.87, γ = 90.38°, Z = 2. Cu radiation, R = 0.052 for 1712 reflections.

The structure (Fig. 1) consists of guanidinium cations, tetraborate anions and water of solvation. The anion has pseudo-C_{2v} symmetry, if hydrogen atoms are ignored, with atoms B(1,2) tetrahedrally bonded and B(3,4) trigonally bonded. The respective average B-O bond lengths for the two types of boron atom are 1.475 and 1.369 Å. The BO_3 groups are virtually planar with an interplanar dihedral angle of 68°. The valence angle B(1)-O(2)-B(2) is 110.2(3)° compared with values of from 118.0 to 118.6(3)° at other bridging O atoms. There is a three-dimensional network of O-H...O and N-H...O hydrogen bonds involving all the hydrogen atoms in the anion and all except two of the hydrogen atoms of the cation.

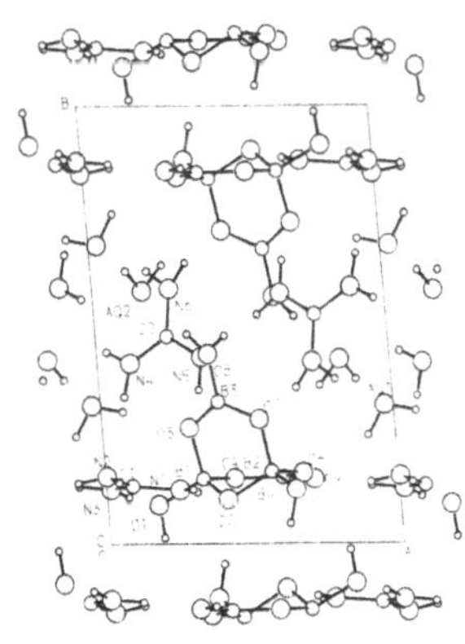

Fig. 1. $C_2H_{16}B_4N_6O_9,2(H_2O)$: a projection of the crystal structure onto (001).

TRIS(FORMAMIDINIUM) HEXACHLOROFERRATE(III)
$C_3H_{15}Cl_6FeN_6$ $3(CH_5N_2{}^+),Cl_6Fe^{3-}$

U. DEMANT, E. CONRADI, U. MULLER and K. DEHNICKE, 1985. *Z. Naturforsch.*, **B40**, 443-445.

Tetragonal, $P4_2/m$, a = 9.616, c = 8.764 Å, Z = 2. Mo radiation, R = 0.047 for 473 reflections.

The crystal structure is depicted in Fig. 1. One of the two formamidinium ions has twofold symmetry and is disordered; the C-N distance in the non-disordered cation (with m symmetry) is 1.281(9) Å. The $FeCl_6{}^{3-}$ anion has 2/m crystallographic symmetry and Fe-Cl 2.390 and 2.375(2) Å.

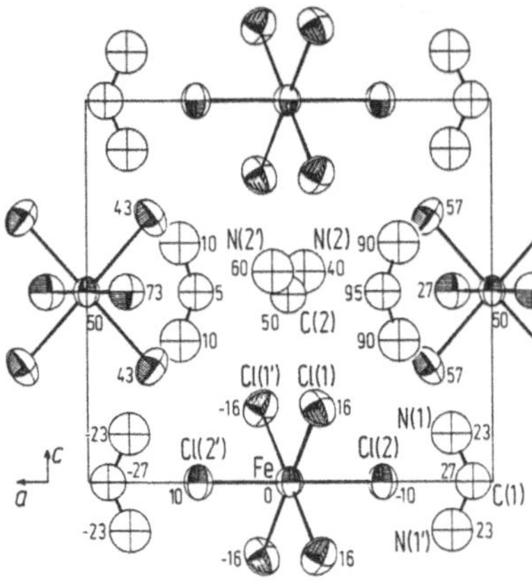

Fig. 1. The structure of $3(CH_5N_2^+),Cl_6Fe^{3-}$ in its *b*-axis projection.

ARCAINE SULPHATE

$C_6H_{14}N_6O_4S$ $C_6H_{18}N_6^{2+},O_4S^{2-}$

V.G. THAILAMBAL, V. PATTABHI, F.L. LEE, Y. LE PAGE and E.J. GABE, 1985. *Acta Cryst.*, **C41**, 105-107.

Monoclinic, $P2_1/n$, a = 7.263, b = 35.140, c = 15.424 Å, β = 101.70°, Z = 12. Cu radiation, R = 0.062 for 2649 reflections.

There are three crystallographically independent sulphate ions and three arcaine ions in the asymmetric unit (Fig. 1). The sulphate ions all have tetrahedral coordinations. The bulk of the amine ions have all-trans conformations: from N(16) to N(24) in molecule (I), from N(29) to N(36) in molecule (II) and from N(41) to N(48) in molecule (III). The torsion angles around bonds C(23)-N(24) of molecule (I), C(35)-N(36) of molecule (II) and C(47)-N(48) of molecule (III) are 93, -85 and 80° respectively. The guanidyl groups are trigonal planar. The amine ions form layers nearly parallel to (100) with the sulphate ions interleaved. There is a complex network of N-H...O hydrogen bonds stabilizing the layered structure.

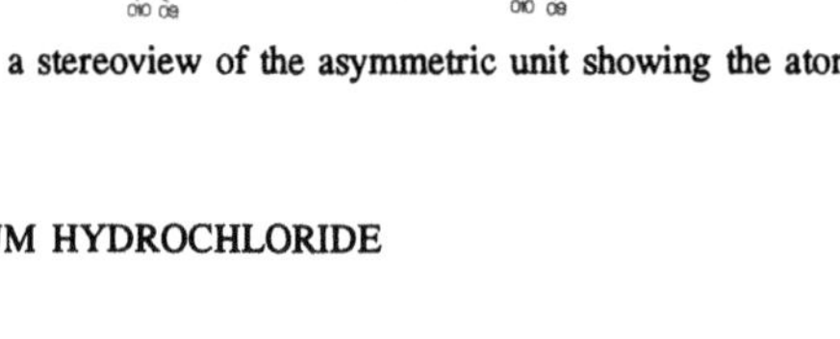

Fig. 1. $C_6H_{14}N_6O_4S$: a stereoview of the asymmetric unit showing the atomic numbering.

PHENYLBIGUANIDINIUM HYDROCHLORIDE

$C_8H_{12}ClN_5$ $C_8H_{12}N_5^+,Cl^-$

J.M. AMIGO, J. MARTINEZ-CALATAYUD and T. DEBAERDEMAEKER, 1985. *Bull. Soc. Chim. Belg.*, **94**, 119-121.

Monoclinic, $P2_1/n$, a = 17.079, b = 6.665, c = 9.306 Å, β = 104.705°, Z = 4. Mo radiation, R = 0.115 for 1137 reflections.

The relatively high R-factor is accounted for by the poor quality of the crystals. The biguanide group (Fig. 1) is far from planar; the C(8)-N(8)-N(9) and N(9)-C(10)-N(11) planes are inclined at 50.3° to each other. The N...Cl⁻ contacts have a mean value 3.33(8) Å. The average C-N distance in the biguanide group is 1.34(2) Å.

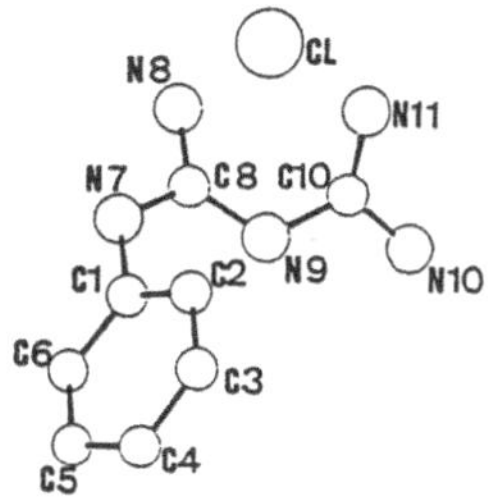

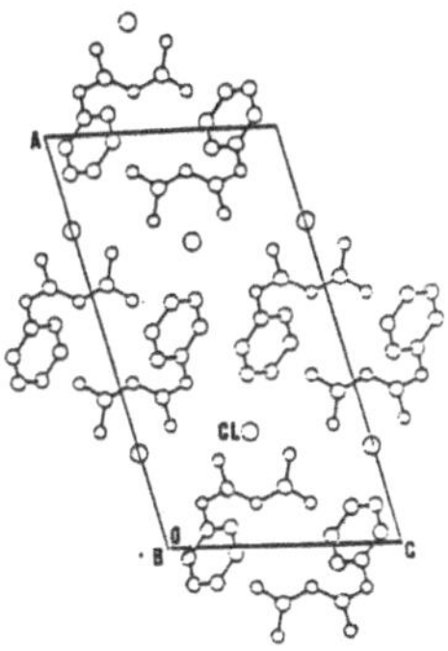

Fig. 1. The $C_8H_{12}N_5^+,Cl^-$ system and the molecular structure projected on the (010) plane.

(METHYLENEBIS(N,N'-DIMETHYLUREA))HYDROGEN(I) TRIFLUOROMETHANESULPHONATE
$C_8H_{17}F_3N_4O_5S$ $C_7H_{17}N_4O_2^+,CF_3O_3S^-$

G. MAAS, 1985. *Acta Cryst.*, **C41**, 1133-1136.

Triclinic, $P\bar{1}$, a = 7.055, b = 9.582, c = 11.955 Å, α = 83.92, β = 82.14, γ = 70.36°, Z = 2. Mo radiation, R = 0.0451 for 2160 reflections.

The cation (Fig. 1) is symmetric with respect to bond lengths and angles involving corresponding atoms. There are, however, small but significant differences in the torsion angles around the N(1)-C(4) and N(3)-C(4) bonds. The pseudo symmetric halves of the cation are connected by a very strong, nearly symmetric intramolecular hydrogen bond O(4)...O(5) of length 2.420(3) Å connecting the carbonyl groups and forming an eight-membered ring. The anion has normal bond lengths and angles and an anion is connected to two cations through hydrogen bonds N(2)-H...O(3) and N(4)-H...O(1) of lengths 2.908(3) and 2.855(3) Å respectively. These bonds in the crystal build up infinite chains (..cation-anion-cation..) along the *c*-axis.

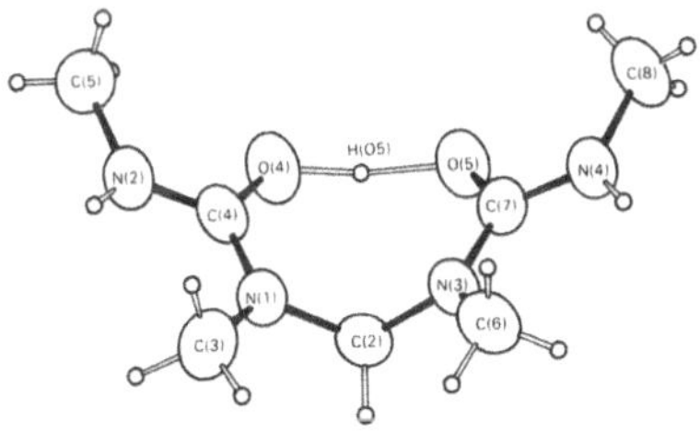

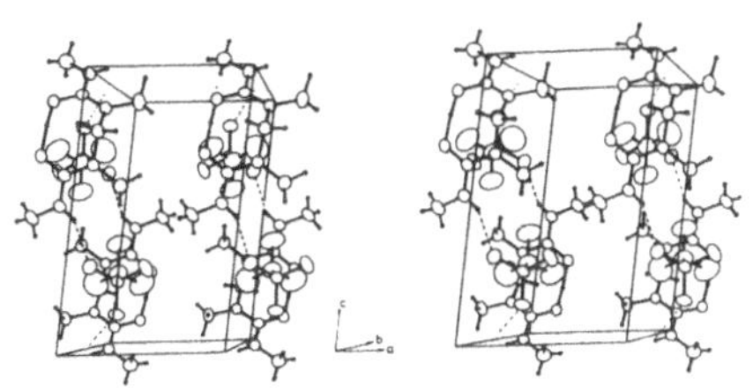

Fig. 1. $C_8H_{17}F_3N_4O_5S$: a view of the cation and a stereoview of the unit cell contents.

trans-AZOBENZENE trans-STILBENE
$0.74(C_{12}H_{10}N_2),0.26(C_{14}H_{12})$ (I)

trans-AZOBENZENE trans-STILBENE
$0.44(C_{12}H_{10}N_2),0.56(C_{14}H_{12})$ (II)

trans-AZOBENZENE trans-STILBENE
$0.54(C_{12}H_{10}N_2),0.46(C_{14}H_{12})$ (III)

J.A. BOUWSTRA, A. SCHOUTEN, J. KROON and R.B. HELMHOLDT, 1985. *Acta Cryst.*, **C41**, 420-426.

I. Monoclinic, $P2_1/c$, a = 15.306, b = 5.7716, c = 12.230 Å, β = 112.04°, Z = 4. Cu radiation, R = 0.056 for 1706 reflections.

II. Monoclinic, $P2_1/c$, a = 15.481, b = 5.749, c = 12.300 Å, β = 111.98°, Z = 4. Mo radiation, R = 0.051

for 1586 reflections.

III. Monoclinic, $P2_1/c$, a = 15.409, b = 5.759, c = 12.284 Å, β = 111.97°, Z = 4. Neutron radiation (λ = 1.304 Å), R = 0.082 for 910 reflections.

In these mixed crystal structures (Fig. 1) not only is there substitutional disorder but both molecules also show configurational disorder at one or both of the two independent centrosymmetric sites, A and B. The least-squares structure refinements, in which both kinds of disorder were taken into account, were performed with constraints and restraints. This appears to indicate that the trans-stilbene molecules show a slight preference for substitution at the disordered sites.

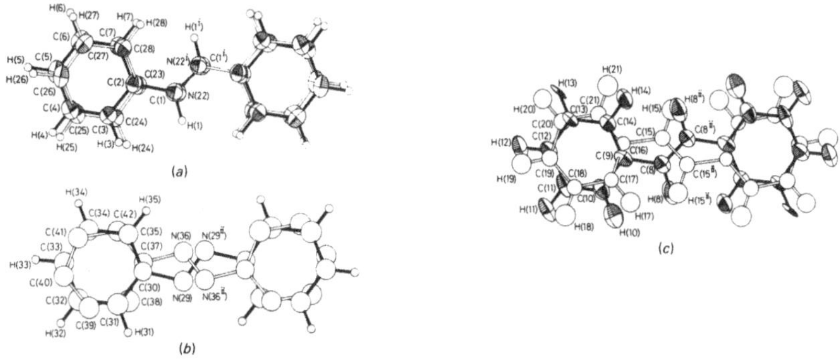

Fig. 1. Views of the disordered composite trans-stilbene:trans-azobenzene molecules (a) at sites A1/A2 in (II), (b) at sites B3/B4 in (II) and (c) a composite view of the disordered stilbene molecules at sites B in (III).

POTASSIUM 1,5-DIPHENYL-3-MERCAPTOFORMAZANATE

$C_{13}H_{11}KN_4S$ $K^+,C_{13}H_{11}N_4S^-$

H. NODA, L.M. ENGELHARDT, J.M. HARROWFIELD, C. PAKAWATCHAI, J.M. PATRICK and A.H. WHITE, 1985. *Bull. Chem. Soc. Jpn.*, **58**, 2385-2388.

Orthorhombic, *Pbca*, a = 28.942, b = 13.247, c = 7.163 Å, Z = 8. Mo radiation, R = 0.045 for 1101 reflections.

In the crystal structure (Fig. 1) there are pseudo-planar anion stacks parallel to the *ab* face of the cell with considerable delocalisation as shown by the anion dimensions. The K^+ environment is sparse with K...N 2.793-3.305(5) and K...S 3.150(2) and 3.426(3) Å.

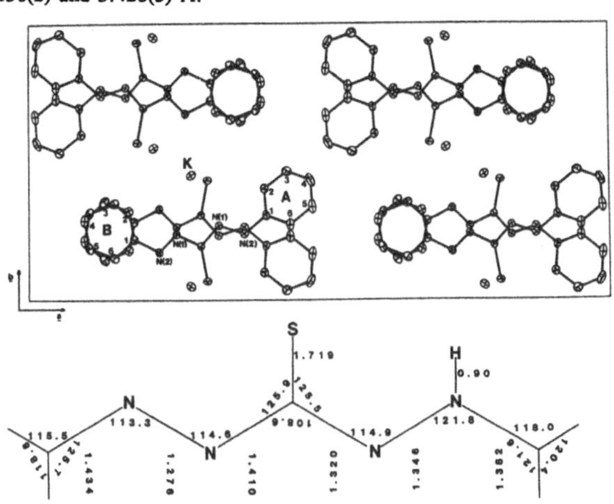

Fig. 1. The crystal structure of $C_{13}H_{11}KN_4S$ projected down *c* and non-phenyl anion dimensions.

TRIDEUTERONITROMETHANE
CD_3NO_2

G.A. JEFFREY, J.R. RUBLE, L.M. WINGERT, J.H. YATES and R.K. McMULLAN, 1985. *J. Am. Chem. Soc.*, **107**, 6227-6230.

Orthorhombic, $P2_12_12_1$, a = 5.185, b = 6.237, c = 8.507 Å, Z = 4. Neutron radiation, R = 0.036 for 636 reflections (at 15 K).

The conformation of the molecule (Fig. 1) is such that the N-C-D(2) plane makes an angle of 89.7(1)° with the mean CNO_2 plane. The N atom is displaced 0.0112(7) Å out of the carbon and oxygen plane. The N-O bond lengths are 1.2270(9) and 1.2225(9) Å. There are short C-D...O contacts in the range 2.39-2.58 Å for one O atom and 2.69-3.09 Å for the other. For previous work see (**1**).

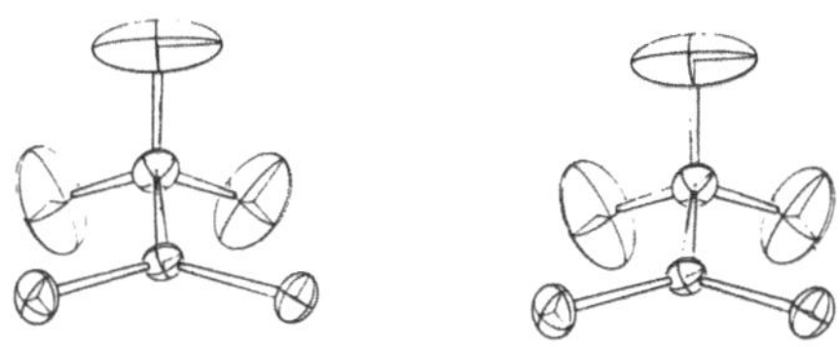

Fig. 1. A stereoview of CD_3NO_2.

1. Structure Reports, **46**B, 39.

NITROMETHANE
CH_3NO_2

D.T. CROMER, R.R. RYAN and D. SCHIFERL, 1985. *J. Phys. Chem.*, **89**, 2315-2318.

I. Orthorhombic, $P2_12_12_1$, a = 5.182, b = 6.259, c = 8.645 Å, Z = 4. Mo radiation, R = 0.026 for 92 reflections (at 0.3 GPa).

II. Orthorhombic, $P2_12_12_1$, a = 5.192, b = 6.278, c = 8.675 Å, Z = 4. Mo radiation, R = 0.026 for 86 reflections (at 0.33 GPa).

III. Orthorhombic, $P2_12_12_1$, a = 5.140, b = 6.191, c = 8.553 Å, Z = 4. Mo radiation, R = 0.035 for 114 reflections (at 0.6 GPa).

IV. Orthorhombic, $P2_12_12_1$, a = 5.090, b = 6.115, c = 8.464 Å, Z = 4. Mo radiation, R = 0.021 for 103 reflections (at 1.0 GPa).

V. Orthorhombic, $P2_12_12_1$, a = 4.994, b = 5.994, c = 8.272 Å, Z = 4. Mo radiation, R = 0.030 for 97 reflections (at 2.0 GPa).

VI. Orthorhombic, $P2_12_12_1$, a = 4.890, b = 5.885, c = 8.099 Å, Z = 4. Mo radiation, R = 0.040 for 91 reflections (at 3.5 GPa).

VII. Orthorhombic, $P2_12_12_1$, a = 4.884, b = 5.878, c = 8.116 Å, Z = 4. Mo radiation, R = 0.043 for 99 reflections (at 3.5 GPa).

The crystal structure at high pressures is similar to that observed in a low-temperature ambient-pressure study (**1**). At 3.5 GPa the methyl group is rotated about 45° from the low-temperature form.

1. Structure Reports, **46**B, 39.

1,1,1,3,6,8,8,8-OCTANITRO-3,6-DIAZAOCTANE
$C_6H_8N_{10}O_{16}$

Y. OYUMI, T.B. BRILL and A.L. RHEINGOLD, 1985. *J. Phys. Chem.*, **89**, 4824-4828.

Monoclinic, $P2_1/n$, a = 5.9716, b = 12.0170, c = 11.8407 Å, β = 97.114°, Z = 2. Mo radiation, R = 0.032 for 1109 reflections.

The molecule lies about a crystallographic inversion centre (Fig. 1). The C(3)-N(4)-N(5)-O(8) torsion angle is 13.1(2)°. The average N-O bond length of the CNO_2 units (1.203(3) Å) is slightly shorter than that of the NNO_2 unit (1.216(2) Å). The N-N bond length is 1.378(2) Å. The conformation of the $C(NO_2)_3$ group does not define a propeller form; two of the NO_2 groups lie approximately in a plane defined by the respective C_2N linkage while the third is nearly perpendicular to it.

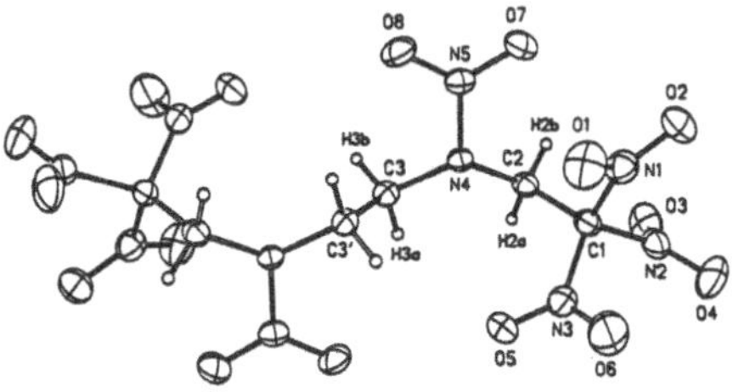

Fig. 1. The $C_6H_8N_{10}O_{16}$ molecule.

TRIFLUOROMETHYLDICHLOROSULPHONIUM HEXAFLUOROARSENATE(V)

$CAsCl_2F_9S$ $CCl_2F_3S^+,AsF_6^-$

R. MINKWITZ, U. NASS, A. RADUNZ and H. PREUT, 1985. *Z. Naturforsch.*, **B40**, 1123-1129.

Monoclinic, $P2_1/c$, a = 7.553, b = 9.381, c = 12.091 Å, β = 100.97°, Z = 4. Ag radiation, R = 0.022 for 1376 reflections (at 133 K).

The AsF_6^- anion has octahedral geometry and the S atom of the cation also achieves distorted octahedral coordination through contacts with F atoms of adjacent AsI_6^- ions as shown in Fig. 1. Main cation dimensions are C-F 1.288(5), C-S 1.978(3), S-Cl 1.959(1) Å, F-C-F 111.8(3), Cl-S-Cl 104.22(5), C-S-Cl 99.8(1)°. As-F distances are in the range 1.710-1.731(2) Å. The inter-ion S...F contacts are in the range 2.720-2.972(2) Å.

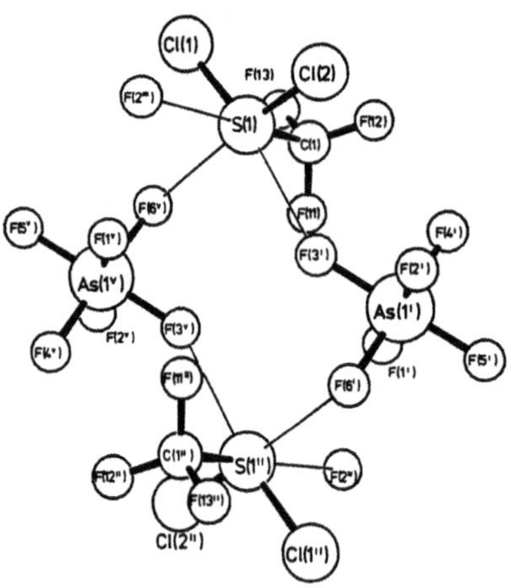

Fig. 1. A view showing the $CF_3SCl_2^+$ and AsF_6^- ions and the contacts between them.

DIPOTASSIUM HYDRAZINE-1,2-BIS(DITHIOFORMATE)

$C_2H_2K_2N_2S_4$ $C_2H_2N_2S_4^{2-},2(K^+)$

G. KIEL, G. GATTOW and S. LOTZ, 1985. *Z. Anorg. Allg. Chem.*, **531**, 89-96.

Orthorhombic, *Pbna*, a = 6.635, b = 10.825, c = 12.866 Å, Z = 4. Mo radiation, R = 0.034 for 969 reflections (at -85°C).

The $[S_2C\text{-}NH\text{-}NH\text{-}CS_2]^{2-}$ ions are linked by N-H...S hydrogen bonds as shown in Fig. 1. The C-N-N-C torsion angle is 85.5(6)°. The K^+ ion is surrounded by seven S atoms in irregular coordination with K-S 3.176-3.453(1) Å.

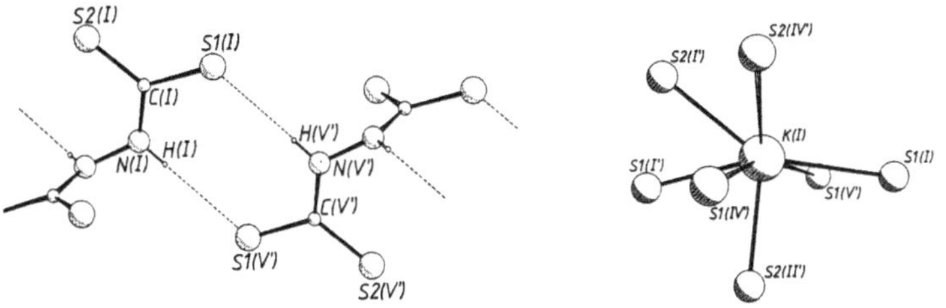

Fig. 1. The hydrogen bonded $[S_2C\text{-}NH\text{-}NH\text{-}CS_2]^{2-}$ ions and the K^+ coordination in $C_2H_2K_2N_2S_4$.

BIS(TRICHLOROMETHYL)PENTATHIODIPERCARBONATE
$C_3Cl_6S_5$ (I)

CHLOROTRIS(TRICHLOROMETHYLDITHIO)METHANE
$C_4Cl_{10}S_6$ (II)

CHLOROTRIS(TRICHLOROMETHYLTRITHIO)METHANE
$C_4Cl_{10}S_9$ (III)

H.C. HANSEN, A. SENNING and R.G. HAZELL, 1985. *Tetrahedron*, **41**, 5145-5158.

I. Monoclinic, $P2_1/a$, a = 23.054, b = 10.468, c = 5.895 Å, β = 105.29°, Z = 4. Mo radiation, R = 0.075 for 1266 reflections.

II. Monoclinic, $P2_1/n$, a = 19.522, b = 5.963, c = 17.149 Å, β = 103.43°, Z = 4. Mo radiation, R = 0.035 for 2640 reflections.

III. Rhombohedral, $R3c$, a = 18.697, c = 11.460 Å, Z = 6. Mo radiation, R = 0.025 for 971 reflections.

Drawings of the three complexes are shown in Fig. 1. Molecule (I) has a localized C=S bond (1.581(12) Å) and all other bond lengths are typical for single bonds. Molecule (III) has threefold crystallographic symmetry with C-Cl 1.763(6), C-S(1) 1.820(2), S(1)-S(2) 2.046(1), S(2)-S(3) 2.030(1) Å. Molecule (II) has one of its branches twisted differently and has no symmetry axis.

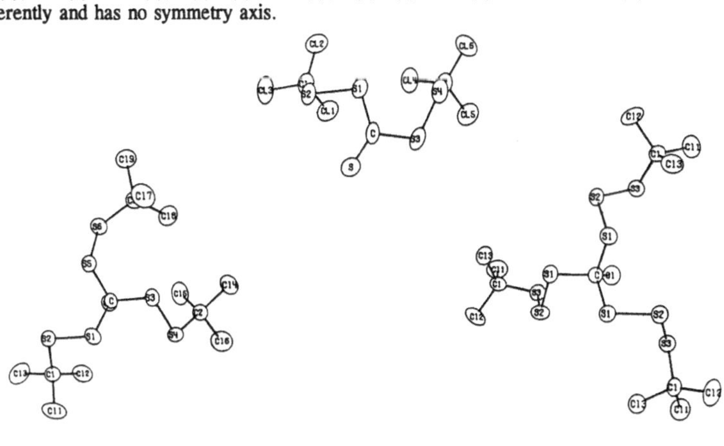

Fig. 1. Views of $C_3Cl_6S_5$ (I) (top), $C_4Cl_{10}S_6$ (II) (lower left) and $C_4Cl_{10}S_9$ (III) (lower right).

BARIUM 1,1-DICYANOETHYLENE-2,2-DITHIOLATE TRIHYDRATE

$C_4BaN_2S_2,3(H_2O)$ $C_4N_2S_2^{2-},Ba^{2+},3(H_2O)$

H.-U. HUMMEL and C. WOLF, 1985. *Z. Naturforsch.*, **B40**, 1638-1643.

Monoclinic, *C2/c*, a = 13.937, b = 8.842, c = 16.221 Å, β = 103.13°, Z = 8. Ag radiation, R = 0.0341 for 6836 reflections.

The crystal contains Ba^{2+} ions, $[S_2C\text{-}C(CN)_2]^{2-}$ anions and lattice water (Fig. 1). The Ba coordination is distorted tricapped trigonal prismatic to four O atoms (2.81-2.87 Å), three N atoms (2.84-3.04 Å) and two S atoms (3.243 and 3.263(3) Å). The atoms of the anion deviate from planarity (-0.28 Å (S(1)) to +0.19 Å (C(1))). The anions and water molecules take part in an extensive H-bonded network with S...O 3.162(8)-3.36(1), O...O 2.75 and 3.04(1) Å.

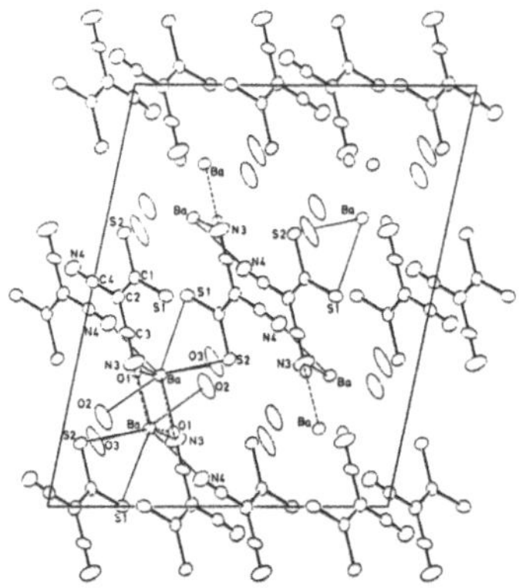

Fig. 1. The *b*-axis projection of $C_4N_2S_2^{2-},Ba^{2+},3(H_2O)$.

DIPOTASSIUM 1,1-DICYANOETHYLENE-2,2-DITHIOLATE MONOHYDRATE

$C_4K_2N_2S_2,H_2O$ $C_4N_2S_2^{2-},2(K^+),H_2O$

H.-U. HUMMEL, 1985. *Acta Cryst.*, **C41**, 1591-1592.

Orthorhombic, *Pnma*, a = 7.215, b = 13.162, c = 9.449 Å, Z = 4. Ag radiation, R = 0.036 for 587 reflections.

The unique K^+ ion in the asymmetric unit of the structure (Fig. 1) is coordinated by three S, three N and one O atom forming a monocapped trigonal antiprism with average K-S distance of 3.26(3) Å. Each S and N atom coordinates to two other K^+ ions while each O atom is linked with one other K^+ ion. The roughly planar 1,1-dicyanoethylene-2,2-dithiolate ion has approximate C_{2v} symmetry with the central atoms C(1) and C(2) located on a mirror plane of symmetry. The C(1)-S, C(1)-C(2), C(2)-C(3) and C(3)-N bond lengths in this symmetric ion are 1.713(3), 1.405(7), 1.420(4) and 1.143(4) Å respectively. The water molecules form two sets of two equidistant O-H...S hydrogen bonds with O...S separations of 3.395(3) and 3.568(3) Å.

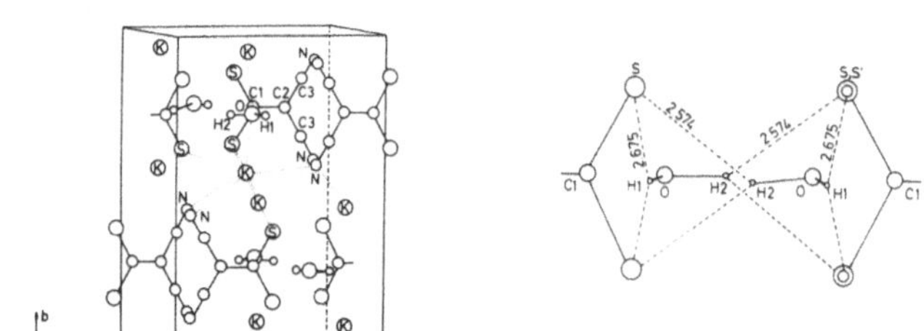

Fig. 1. $C_4K_2N_2S_2,H_2O$: a view of the structure in which the K^+ coordination is shown by faintly dotted lines and the coordination of the water molecule.

1,2-DIMESYL-1,2-DIMETHOXYHYDRAZINE
$C_4H_{12}N_2O_6S_2$

K. BRINK and R. MATTES, 1985. *Chem. Ber.*, **118**, 564-573.

Monoclinic, $P2_1/n$, a = 7.598, b = 12.795, c = 11.366 Å, β = 109.94°, Z = 4. Mo radiation, R = 0.032 for 2127 reflections (at -150°C).

The molecule has approximate C_2 symmetry and the N atoms are shown to have pyramidal coordination (Fig. 1). Principal mean dimensions include: N-N 1.381(2), N-S 1.718(1), N-O 1.437(2), S-O 1.428(1), C-S 1.759(2), C-O 1.449(2) Å, S-N-O 107.7(1), S-N-N 113.6(1), N-N-O 113.0(1)°.

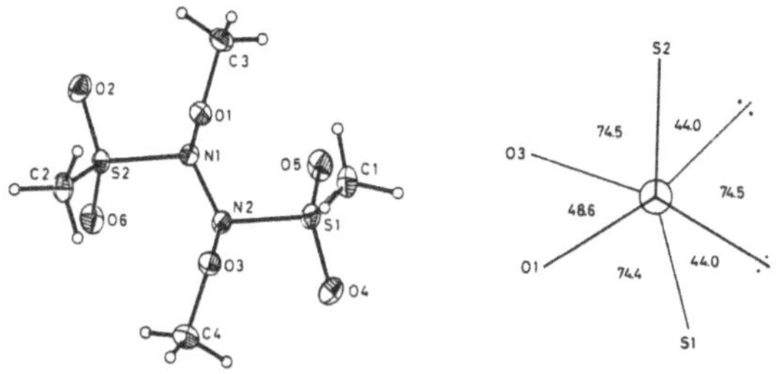

Fig. 1. The $C_4H_{12}N_2O_6S_2$ molecule and a Newman projection along the N-N bond.

1,2-BIS(METHYLTRITHIOCARBONATO)ETHANE
$C_6H_{10}S_6$

G. KIEL, G. GATTOW and U. SCHUBERT, 1985. *Z. Anorg. Allg. Chem.*, **530**, 109-116.

Triclinic, $P\bar{1}$, a = 6.769, b = 8.334, c = 11.222 Å, α = 80.93, β = 72.01, γ = 78.58°, Z = 2. Mo radiation, R = 0.035 for 2004 reflections (at 40°C).

The unit cell contains two independent similar molecules each lying about inversion centres (Fig. 1). The S-CS-S groups are planar, with C=S 1.623 and 1.637(3), Csp^2-S 1.738-1.753(2), Csp^3-S 1.805 and 1.807(2) Å. The S-C-C-S torsion angles are required to be 180° by symmetry.

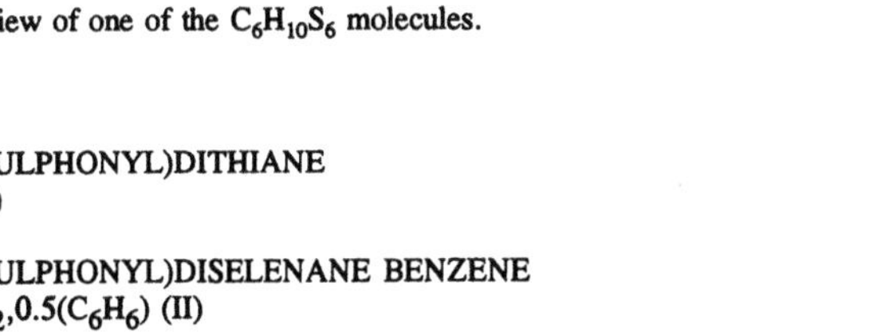

Fig. 1. A view of one of the $C_6H_{10}S_6$ molecules.

BIS(PHENYLSULPHONYL)DITHIANE
$C_{12}H_{10}O_4S_4$ (I) $S_2(SO_2Ph)_2$

BIS(PHENYLSULPHONYL)DISELENANE BENZENE
$C_{12}H_{10}O_4S_2Se_2,0.5(C_6H_6)$ (II) $Se_2(SO_2Ph)_2,0.5C_6H_6$

BIS(p-TOLYLSULPHONYL)DITHIANE
$C_{14}H_{14}O_4S_4$ (III) $S_2(SO_2C_6H_4Me)_2$

BIS(p-TOLYLSULPHONYL)DISELENANE
$C_{14}H_{14}O_4S_2Se_2$ (IV) $Se_2(SO_2C_6H_4Me)_2$

BIS(p-TOLYLSULPHONYL)THIANE
$C_{14}H_{14}O_4S_3$ (V) $S(SO_2C_6H_4Me)_2$

BIS(p-TOLYLSULPHONYL)SELENANE
$C_{14}H_{14}O_4S_2Se$ (VI) $Se(SO_2C_6H_4Me)_2$

BIS(p-TOLYLSULPHONYL)TRITHIANE
$C_{14}H_{14}O_4S_5$ (VII) $S_3(SO_2C_6H_4Me)_2$

BIS(p-TOLYLSULPHONYL)TRISELENANE
$C_{14}H_{14}O_4S_2Se_3$ (VIII) $Se_3(SO_2C_6H_4Me)_2$

O. FOSS, F. KVAMMEN and K. MAROY, 1985. *J. Chem. Soc. Dalton Trans.*, 231-237.

I. Monoclinic, $P2_1/c$, a = 12.339, b = 6.107, c = 20.804 Å, β = 109.43°, Z = 4. Mo radiation, R = 0.045 for 2071 reflections.

II. Monoclinic, $P2_1/c$, a = 5.531, b = 14.221, c = 23.164 Å, β = 108.41°, Z = 4. Mo radiation, R = 0.045 for 2071 reflections.

III. Monoclinic, $P2_1/c$, a = 9.947, b = 15.050, c = 11.215 Å, β = 94.94°, Z = 4. Mo radiation, R = 0.034 for 3058 reflections.

IV. Monoclinic, $P2_1/c$, a = 10.123, b = 15.387, c = 11.242 Å, β = 94.18°, Z = 4. Mo radiation, R = 0.043 for 2141 reflections.

V. Monoclinic, $C2/c$, a = 18.837, b = 5.844, c = 16.480 Å, β = 119.90°, Z = 4. Mo radiation, R = 0.035 for 1882 reflections.

VI. Monoclinic, $C2/c$, a = 19.014, b = 5.752, c = 16.786 Å, β = 118.59°, Z = 4. Mo radiation, R = 0.034 for 1384 reflections.

VII. Tetragonal, $P4_12_12$ = 7.700, c = 29.305 Å, Z = 4. Mo radiation, R = 0.044 for 1571 reflections.

VIII. Tetragonal, $P4_12_12$, a = 7.787, c = 29.673 Å, Z = 4. Mo radiation, R = 0.041 for 1822 reflections.

Molecules lie on general positions in (I)-(IV) and on crystal diad axes in (V-VIII) ((I), (II), (III), Fig. 1; (VI), (VIII), Fig. 2). The ranges of bond lengths are: O_2S-S 2.101-2.141(1); other S-S 1.987-2.037(1); O_2S-Se 2.242-2.268(1); Se-Se 2.246-2.304(1) Å. Most Y-X-S-C torsion angles (X, Y = S or Se) are in the range 70-89°. The S-SO_2 bond length varies inversely with the average of the S-S-C and S-S-O angles.

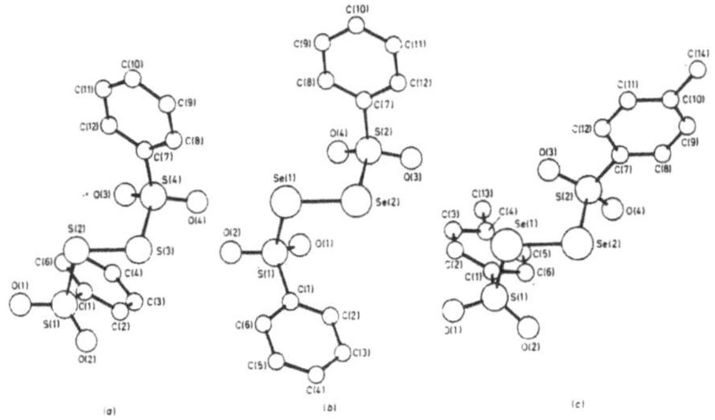

Fig. 1. (I), $C_{12}H_{10}O_4S_4$; (II), $C_{12}H_{10}O_4S_2Se_2$; and (III), $C_{14}H_{14}O_4S_4$.

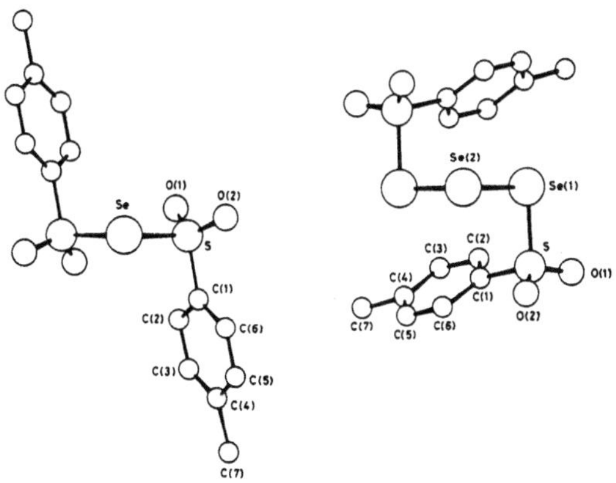

Fig. 2. (VI), $C_{14}H_{14}O_4S_2Se$, and (VIII), $C_{14}H_{14}O_4S_2Se_3$.

(Z)-1,2-BIS(BENZENESULPHONYL)ETHYLENE
$C_{14}H_{12}O_4S_2$ (I)

(E)-1,2-BIS(BENZENESULPHONYL)ETHYLENE
$C_{14}H_{12}O_4S_2$ (II)

1,1-BIS(BENZENESULPHONYL)ETHYLENE
$C_{14}H_{12}O_4S_2$ (III)

O. DE LUCCHI, L. PASQUATO, G. MODENA and G. VALLE, 1985. *Z. Kristallogr.*, **170**, 267-274.

I. Orthorhombic, *Pbca*, a = 10.726, b = 15.864, c = 33.801 Å, Z = 16. Mo radiation, R = 0.053 for 3038 reflections.

II. Orthorhombic, *Pbca*, a = 15.073, b = 11.044, c = 8.201 Å, Z = 4. Mo radiation, R = 0.035 for 1018 reflections.

III. Orthorhombic, $P2_12_12_1$, a = 14.229, b = 10.998, c = 8.950 Å, Z = 4. Mo radiation, R = 0.038 for 892 reflections.

Views of the molecules are in Fig. 1. The (Z)-isomer (I) has two distinct conformers in its cell, with the central double bond slightly puckered (3° and 6°). The (E) isomer (II) lies about an inversion centre (with a consequently planar double bond). There is no appreciable distortion in (III). The C=C bond lengths are (I), 1.302(8) and 1.327(8); (II), 1.309(5); (III), 1.325(8) Å.

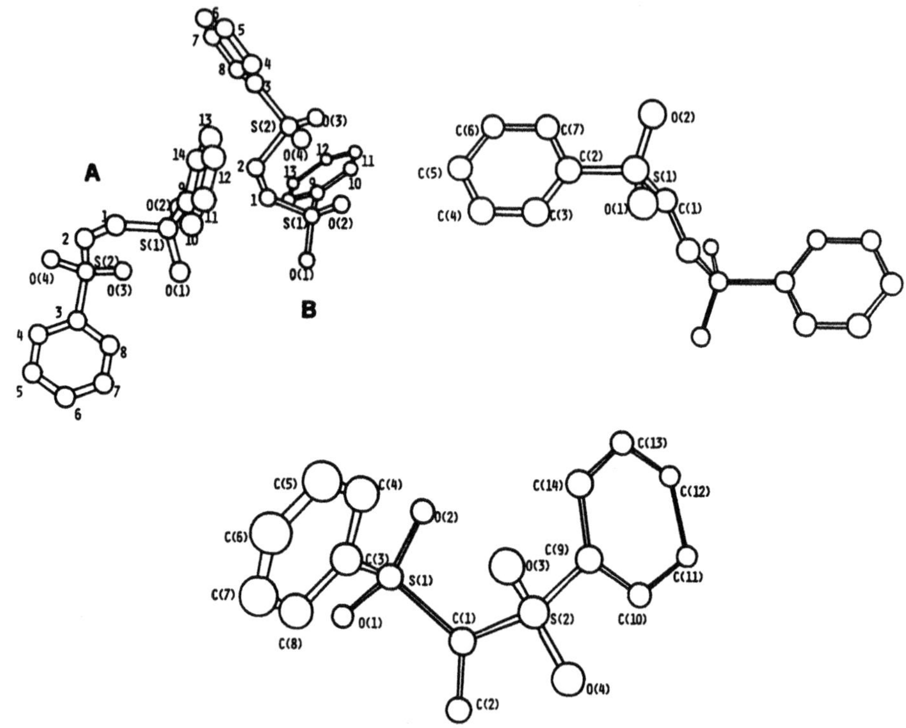

Fig. 1. The $C_{14}H_{12}O_4S_2$ molecules: (I) (top left), (II) (top right) and (III) (lower).

1-PHENYL-2,2,2-TRIFLUOROETHYLTOSYLATE
$C_{15}H_{13}F_3O_3S$ (I)

1-PHENYL-1-METHYL-2,2,2-TRIFLUOROETHYLTOSYLATE
$C_{16}H_{15}F_3O_3S$ (II)

1-CYANO-1-PHENYL-2,2,2-TRIFLUOROETHYLTOSYLATE
$C_{16}H_{12}F_3NO_3S$ (III)

1-PHENYL-1-(TRIFLUOROMETHYL)-2,2,2-TRIFLUOROETHYLTOSYLATE
$C_{16}H_{12}F_6O_3S$ (IV)

V.M. KANAGASABAPATHY, J.F. SAWYER and T.T. TIDWELL, 1985. *J. Org. Chem.*, **50**, 503-509.

I. Monoclinic, $P2_1/c$, a = 16.807, b = 5.712, c = 17.329 Å, β = 112.15°, Z = 4. Mo radiation, R = 0.0518 for 1393 reflections.

II. Monoclinic, $P2_1/n$, a = 12.298, b = 6.029, c = 21.452 Å, β = 100.92°, Z = 4. Mo radiation, R = 0.0543 for 1667 reflections.

III. Triclinic, $P\bar{1}$, a = 7.467, b = 8.084, c = 15.007 Å, α = 78.64, β = 83.78, γ = 66.79°, Z = 2. Mo radiation, R = 0.0458 for 1775 reflections.

IV. Monoclinic, $P2_1/c$, a = 15.126, b = 8.402, c = 13.172 Å, β = 101.84°, Z = 4. Mo radiation, R = 0.0809 for 1511 reflections.

In the paper the molecular geometries of these four molecules are analyzed and correlated with their reactivities.

1,4-DITOSYL-1,4-DIOXABUTANE
$C_{16}H_{18}O_6S_2$ (I)

1,7-DITOSYL-1,4,7-TRIOXAHEPTANE
$C_{18}H_{22}O_7S_2$ (II)

1,10-DITOSYL-1,4,7,10-TETRAOXADECANE
$C_{20}H_{26}O_8S_2$ (III)

P. GROTH, 1985. *Acta Chem. Scand.*, **A39**, 587-591.

I. Monoclinic, *C2/c*, a = 10.572, b = 10.213, c = 15.704 Å, β = 90.02°, Z = 4. Mo radiation, R = 0.031 for 1378 reflections (at -130°C).

II. Monoclinic, *C2/c*, a = 23.385, b = 5.442, c = 15.282 Å, β = 91.10°, Z = 4. Mo radiation, R = 0.045 for 1225 reflections (at -130°C).

III. Monoclinic, $P2_1/c$, a = 8.351, b = 11.435, c = 11.112 Å, β = 93.12°, Z = 2. Mo radiation, R = 0.087 for 1075 reflections (at -130°C).

The molecules are shown in Fig. 1. The conformation of the $TsOCH_2CH_2O$-unit is aga for (I) and (II), and agg for (III). Molecule (I) has a twofold axis through the mid-point of the C(8)-C(8'), (II) has a corresponding axis through O(4) while (III) has a centre of symmetry at the midpoint of the C(10)-C(10') bond.

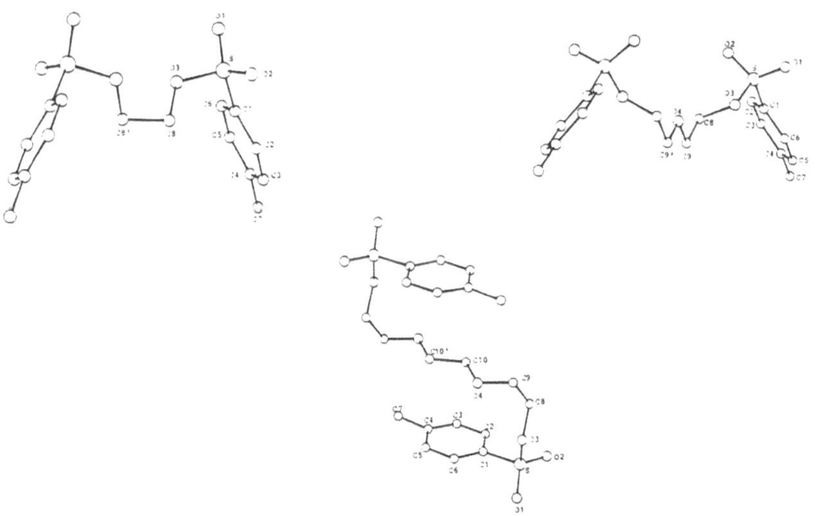

Fig. 1. Perspective drawings of molecules of $C_{16}H_{18}O_6S_2$ (top left), $C_{18}H_{22}O_7S_2$ (top right) and $C_{20}H_{26}O_8S_2$ (bottom).

2-(α-(METHYLTHIO)BENZYLIDENEAMINO)-2-PHENYL-1,1-ETHYLENEDICARBONITRILE
$C_{18}H_{13}N_3S$

J.L. BALCAZAR, F. FLORENCIO and S. GARCIA-BLANCO, 1985. *Acta Cryst.*, **C41**, 1795-1797.

Triclinic, $P\bar{1}$, a = 18.853, b = 8.843, c = 10.545 Å, α = 111.39, β = 83.77, γ = 102.05°, Z = 4. Cu radiation, R = 0.057 for 4093 reflections.

There are two independent molecules (A,B) in the asymmetric unit (Fig. 1). The phenyl groups in each molecule are planar with dihedral angles between the least-squares planes of 32.7(2) and 40.2(2)° respectively, for A and B. There is no conjugation in the molecular chain C(2)=N(3)=C(4)=C(5), torsion angles 68.8(6) (A) and 81.3(5)°. The C(2)-N(3) bond distance is the same in both molecules (1.275(5) Å). The conformation of the molecule is E with respect to the C(2)=N(3) bond. The methylthio methyl group is cis to N(3) in both molecules.

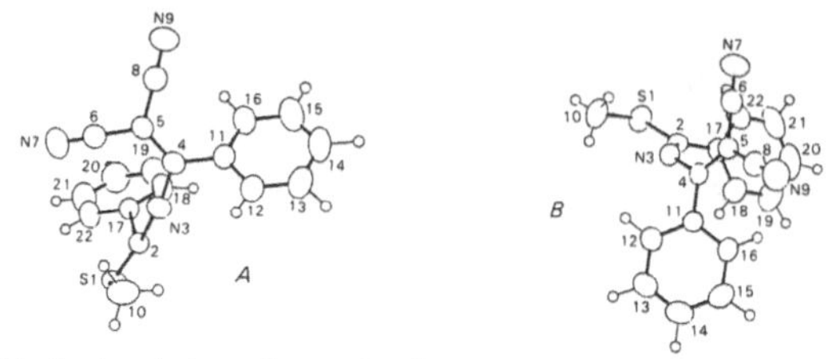

Fig. 1. $C_{18}H_{13}N_3S$: the two independent molecules.

2,4-HEXADIYNYLENE BIS(BENZENESULPHONATE)
$C_{18}H_{14}O_6S_2$

D.J. ANDO, D. BLOOR, M.B. HURSTHOUSE and M. MOTEVALLI, 1985. *Acta Cryst.*, **C41**, 224-226.

Monoclinic, $P2_1/c$, a = 13.079, b = 5.280, c = 13.977 Å, β = 111.19°, Z = 2. Cu radiation, R = 0.0434 for 1271 reflections.

The diacetylene molecules (Fig. 1) pack such that the potentially reacting C atoms, for trans-1,4-addition polymerisation of adjacent molecules, are 3.746(6) Å apart. The angle between the axis of the diacetylene moiety and b is 45.2(4)°. Molecular dimensions are normal.

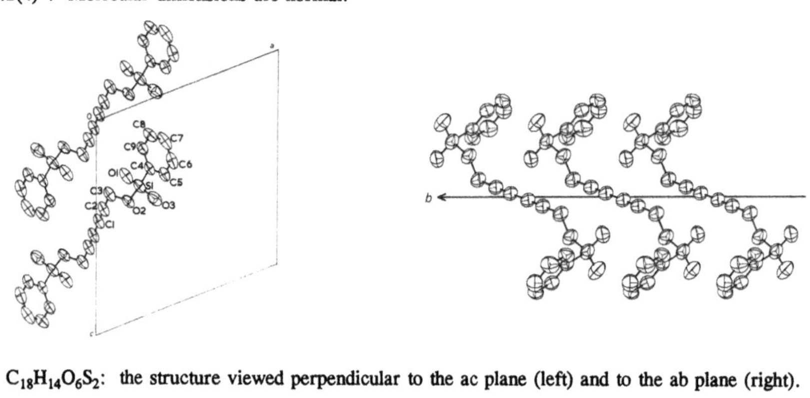

Fig. 1. $C_{18}H_{14}O_6S_2$: the structure viewed perpendicular to the ac plane (left) and to the ab plane (right).

(E)-α-PHENYL-β-(p-TOLYLTHIO)CINNAMALDEHYDE
$C_{22}H_{18}OS$ (I)

(E)-α-PHENYL-β-METHOXYCINNAMALDEHYDE
$C_{16}H_{14}O_2$ (II)

(E)-2,4-DIPHENYL-1,3-OXAZOLE
$C_{15}H_{11}NO$ (III)

(E)-5-BENZYL-1-PHENYLTETRAZOLE
$C_{14}H_{12}N_4$ (IV)

Z. RAPPOPORT and A. GAZIT, 1985. *J. Org. Chem.*, **50**, 3184-3194.

I. Monoclinic, $P2_1/c$, a = 12.789, b = 9.504, c = 15.729 Å, β = 110.81°, Z = 4. Mo radiation, R = 0.053 for 2234 reflections.

II. Triclinic, $P\bar{1}$, a = 8.707, b = 9.615, c = 7.736 Å, α = 95.91, β = 99.55, γ = 85.52°, Z = 2. Mo radiation, R = 0.049 for 2213 reflections.

III. Orthorhombic, $P2_12_12_1$, a = 11.147, b = 17.545, c = 5.692 Å, Z = 4. Mo radiation, R = 0.0972 for 1066 reflections (at -155°C).

IV. Monoclinic, $P2_1/n$, a = 13.190, b = 8.072, c = 11.410 Å, β = 96.64°, Z = 4. Mo radiation, R = 0.041 for 1598 reflections.

The crystal structure analyses established the stereochemistries of the molecules (Fig. 1).

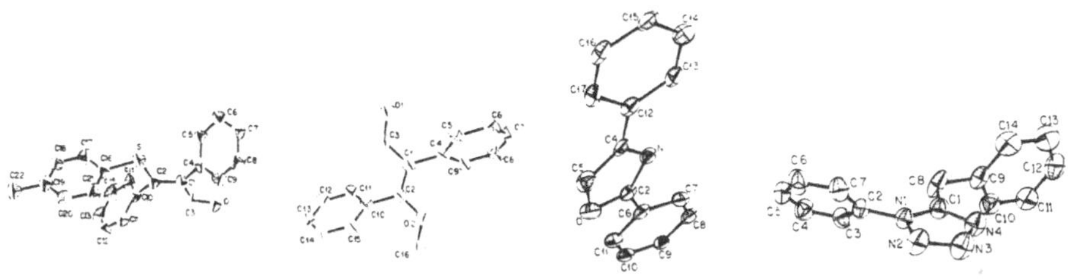

Fig. 1. Drawings (from left to right) of $C_{22}H_{18}OS$, $C_{16}H_{14}O_2$, $C_{15}H_{11}NO$ and $C_{14}H_{12}N_4$.

2,4-HEXADIYNYLENE BIS(2-MESITYLENESULPHONATE)
$C_{24}H_{26}O_6S_2$

A.R. WERNINCK, E. BLAIR, H.W. MILBURN, D.J. ANDO, D. BLOOR, M. MOTEVALLI and M.B. HURSTHOUSE, 1985. *Acta Cryst.*, **C41**, 227-229.

Monoclinic, $P2_1/a$, a = 5.774, b = 25.864, c = 7.982 Å, β = 95.05°, Z = 2. Mo radiation, R = 0.0433 for 1418 reflections.

The diacetylene molecules (Fig. 1) pack so that the potentially reacting C atoms, for trans-1,4-addition polymerisation of adjacent molecules are separated by a distance of 5.473(2) Å. The angle between the a axis and the diacetylene moieties is 66.3(4)°. Molecular dimensions have acceptable values.

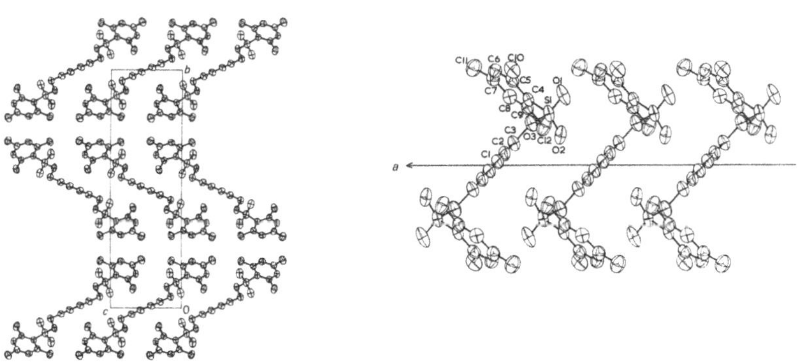

Fig. 1. $C_{24}H_{26}O_6S_2$: the structure viewed perpendicular to the bc plane (left) and to the ab plane (right).

4,6-DECADIYNYLENE BIS(1,2,4,5-TETRAMETHYLPHENYLSULPHONATE) (4,6-DECADIYNYLENE BIS(3-DURENESULPHONATE))
$C_{30}H_{38}O_6S_2$

R.J. DAY, D.J. ANDO, D. BLOOR, P.A. NORMAN, E. BLAIR, A.R. WERNINCK, H.W. MILBURN, M. MOTEVALLI and M.B. HURSTHOUSE, 1985. *Acta Cryst.*, **C41**, 1456-1458.

Monoclinic, $P2_1/n$, a = 6.311, b = 17.862, c = 12.921 Å, β = 96.94°, Z = 2. Cu radiation, R = 0.0406 for 2184 reflections.

The aromatic rings lie on either side of the diacetylene unit forming an S-shape (Fig. 1). The C≡C bond distance is 1.195(4) Å and the C≡C-C angle 179.2(2)°. The molecular conformation adopted by the monomer sterically inhibits a topochemical reaction of diacetylene units in nearest-neighbour molecules.

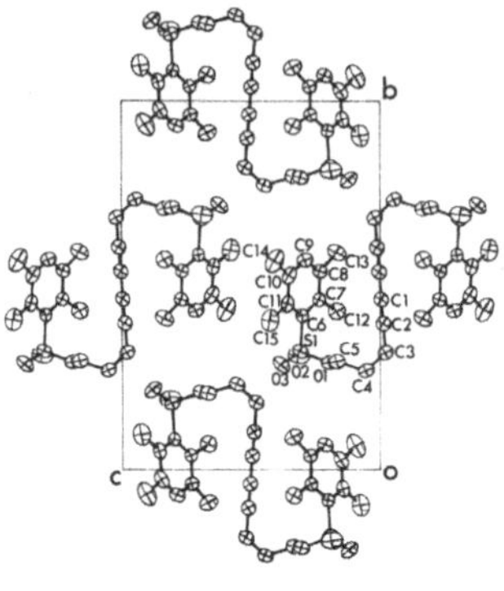

Fig. 1. $C_{30}H_{38}O_6S_2$: view of the structure perpendicular to the bc plane.

DIOCTADECYL TRISULPHIDE
$C_{36}H_{74}S_3$ (I)

DIOCTADECYL TETRASULPHIDE
$C_{36}H_{74}S_4$ (II)

R. GILARDI and J.L. FLIPPEN-ANDERSON, 1985. *Acta Cryst.*, C41, 72-76.

I. Triclinic, $P\bar{1}$, a = 4.810, b = 9.596, c = 43.270 Å, α = 93.5, β = 91.2, γ = 102.9°, Z = 2. Cu radiation, R = 0.092 for 3206 reflections.

II. Monoclinic, *C2/c*, a = 5.443, b = 9.427, c = 78.731 Å, β = 90.6°, Z = 4. Cu radiation, R = 0.082 for 1588 reflections.

Both structures are linear non-branched nonplanar alkyl polyphosphides (Fig. 1). In (II) the central S-S bond (2.060(4) Å) is longer than the terminal S-S bonds (2.027(3) Å). The C-S bond lengths in both (I) and (II) are slightly shorter than the normal single bond length. In (I) one of the central C-S-S-S torsion angles is 67.3° while in (II) the central S-S-S-S torsion angle is -65.3°. In (I) the molecule has an essentially extended trans configuration. Molecule (II) exhibits a trans-trans geometry and has point group symmetry 2 with the diad axis passing through the mid-point of the central S-S bond. The symmetry related octadecyl chains are oriented in such a way that the molecules have a V shape. In both compounds the molecules stack in columns along the b-axis direction. Contiguous stacks in (II) contain molecules going in opposite directions.

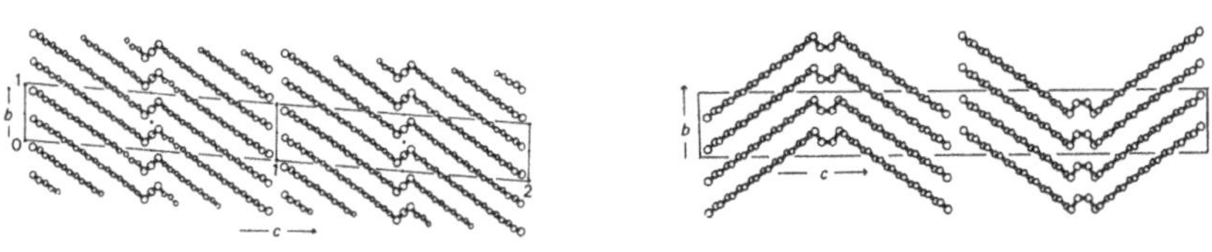

Fig. 1. Packing of $C_{36}H_{74}S_3$ viewed down the a-axis (left) and of $C_{36}H_{74}S_4$ viewed down the b-axis (right).

NONASODIUM OCTAKIS(p-TOLUENESULPHONATE) TRIIODIDE DIHYDRATE
$C_{56}H_{56}I_3Na_9O_{24}S_8,2(H_2O)$ $8(C_7H_7O_3S^-),9(Na^+),I_3^-,2(H_2O)$

F.H. HERBSTEIN, G.M. REISNER and W. SCHWOTZER, 1985. *Acta Cryst.*, C41, 510-513.

Triclinic, $P\bar{1}$, a = 19.214, b = 18.526, c = 10.946 Å, α = 88.69, β = 81.52, γ = 87.36°, Z = 2. Mo radiation, R = 0.089 for 7975 reflections.

The structure comprises alternating cationic and anionic (non-interacting triiodides) sheets parallel to (010) (Fig. 1). The methyl groups of the p-toluenesulphonate layers point away from the Na,O of the cationic layers forming hydrophilic and hydrophobic layers. The independent triiodide ions are both centrosymmetric with I-I distances of 2.942(2) and 2.920(2) Å.

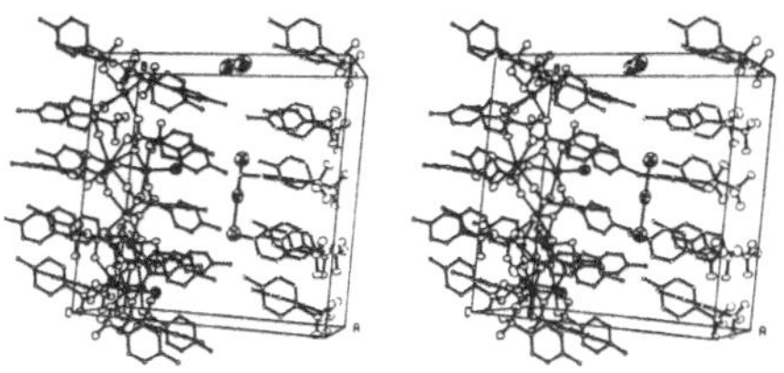

Fig. 1. A stereoview of the crystal structure $C_{56}H_{56}I_3Na_9O_{24}S_8,2(H_2O)$.

1-OXA-3-AZABUTATRIENIUM HEXACHLOROANTIMONATE
$C_4H_7Cl_6N_2OSb$ $C_4H_7N_2O^+,Cl_6Sb^-$

E. MULLER, O. ORAMA, G. HUTTNER and J.C. JOCHIMS, 1985. *Tetrahedron*, **41**, 5901-5912.

Monoclinic, $P2_1/c$, a = 7.128, b = 11.968, c = 15.765 Å, β = 95.83°, Z = 4. Mo radiation, R = 0.048 for 938 reflections.

The structure contains discrete cations and anions; a view of the cation with dimensions is given in Fig. 1. All non-hydrogen atoms of the cation lie in one plane.

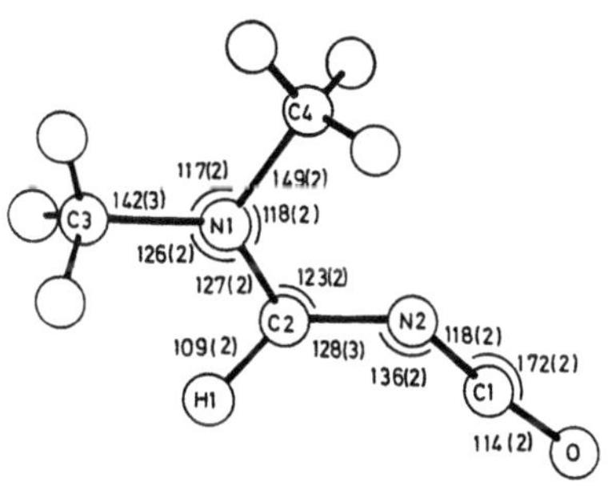

C3-N1-C2-N2	-178(2)
C3-N1-C2-H2	- 8(2)
C4-N1-C2-N2	+ 2(3)
C4-N1-C2-H1	+172(1)
N1-C2-N2-C1	-176(2)
H1-C2-N2-C1	+ 17(4)

Fig. 1. Molecular dimensions and conformation for $C_4H_7N_2O^+$.

1,3-DIETHOXY-1,3-DIPHENYL-2-AZAALLENIUM TETRAFLUOROBORATE
$C_{18}H_{20}BF_4NO_2$ $C_{18}H_{20}NO_2{}^+,BF_4{}^-$

E.-U. WURTHWEIN, R. KUPFER, R. ALLMANN and M. NAGEL, 1985. *Chem. Ber.*, **118**, 3632-3642.

Monoclinic, $A2/a$, a = 29.691, b = 17.777, c = 10.693 Å, β = 92.20°, Z = 12. Mo radiation, R = 0.063 for 1214 reflections.

The analysis shows that the alkoxy substituents are exo and in s-cis-conformations (Fig. 1). There are two independent molecules in the structure, one lying on a crystallographic twofold axis. Mean dimensions are C-N-C 132.3(13)°, N-C 1.294(14) Å. The mean angle between planes of the type N(21),C(21),C(22),O(21) and N(21),C(31),C(32),O(31)is 70.7(5)°. The phenyl rings are rotated between 9.1 and 29.9° out of their N-C-C-O planes.

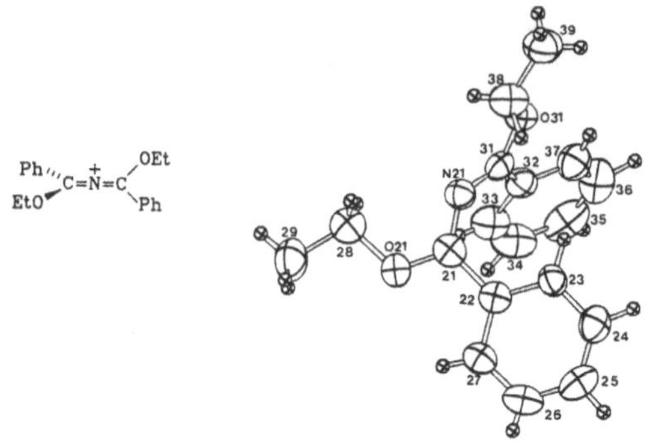

Fig. 1. The $C_{18}H_{20}NO_2^+$ ions.

METHYLTRIPHENYLPHOSPHONIUM TRINITROMETHANIDE
$C_{20}H_{18}N_3O_6P$ (I) $C_{19}H_{18}P^+,CN_3O_6^-$

IODOMETHYLTRIPHENYLPHOSPHONIUM TRINITROMETHANIDE
$C_{20}H_{17}IN_3O_6P$ (II) $C_{19}H_{17}IP^+,CN_3O_6^-$

K.D. SCHERFISE, F. WELLER and K. DEHNICKE, 1985. *Z. Naturforsch.*, **B40**, 906-912.

I. Monoclinic, $P2_1/c$, a = 8.250, b = 14.039, c = 17.706 Å, β = 97.39°, Z = 4. Mo radiation, R = 0.063 for 1277 reflections.

II. Monoclinic, $P2_1/c$, a = 8.476, b = 13.642, c = 18.688 Å, β = 99.91°, Z = 4. Mo radiation, R = 0.036 for 1968 reflections.

In the anions of these isomorphous compounds (Fig. 1) the C atoms are almost coplanar with the N atoms of the nitro groups with C-N 1.362-1.413(10) Å, mean 1.39 Å, and mean N-C-N 120°. In (I) two of the NO_2 groups are almost coplanar with the CN_3 moiety (dihedral angles 0.3 and 8.1°) and the third is disordered over two orientations (dihedral angles 15.8 and 74.7°); in (II) there is no disorder and corresponding dihedral angles are in the range 15 to 32°.

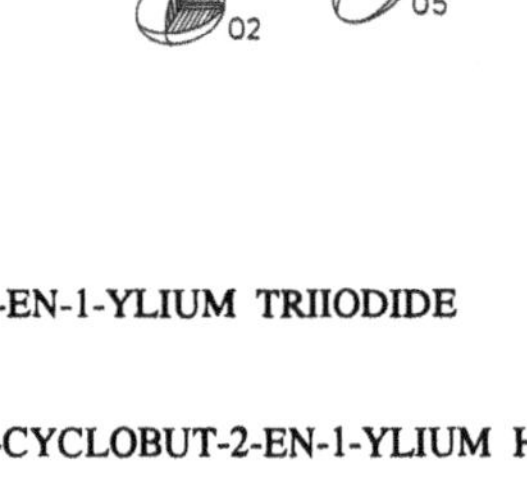

Fig. 1. The $[C(NO_2)_3]^-$ ion.

1,2,3,4-TETRA-t-BUTYLCYCLOBUT-2-EN-1-YLIUM TRIIODIDE
$C_{20}H_{37}I_3$ (I) $C_{20}H_{37}^+,I_3^-$

1,2,3,4-TETRA-t-BUTYL-4-HYDROXY-CYCLOBUT-2-EN-1-YLIUM HEXAFLUOROANTIMATE
$C_{20}H_{37}F_6OSb$ (II) $C_{20}H_{37}O^+,F_6Sb^-$

G. MAIER, R. EMRICH, K.-D. MALSCH, K.-A. SCHNEIDER, M. NIXDORF and H. IRNGARTINGER, 1985. *Chem. Ber.*, **118**, 2798-2810.

I. Monoclinic, $C2/m$, a = 11.629, b = 13.243, c = 16.317 Å, β = 100.39°, Z = 4. Mo radiation, R = 0.048

for 2413 reflections.

II. Monoclinic, $P2_1/c$, a = 8.990, b = 20.419, c = 14.294 Å, β = 133.33°, Z = 4. Mo radiation, R = 0.041 for 3920 reflections.

The cation in (I) lies on a crystallographic mirror plane while that in (II) lies in a general position (Fig. 1). Both cations are folded (37.3° in (I), 36.4° in (II)) because of 1,3-bonding interactions; the 1-3 distances across the rings are short (1.806(6) Å in (I), 1.833(4) Å in (II)).

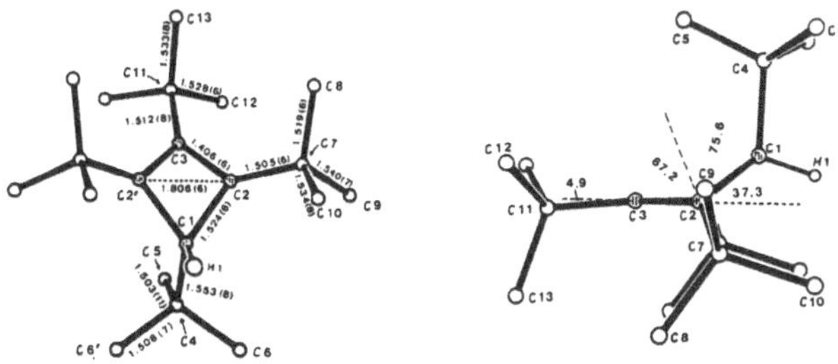

Fig. 1. Cation conformation and dimensions for $C_{20}H_{37}^+$.

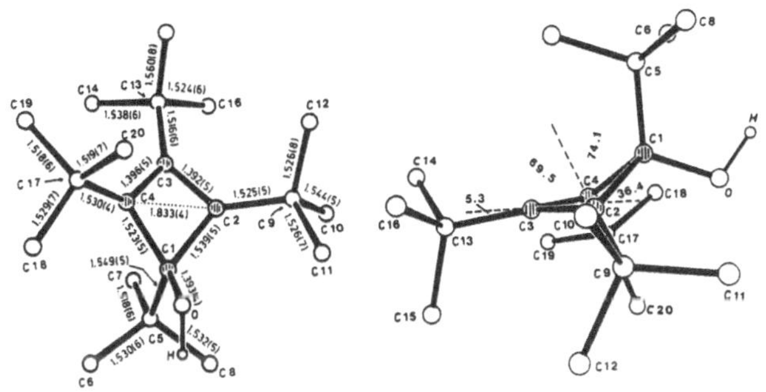

Fig. 2. Cation conformation and dimensions for $C_{20}H_{37}O^+$.

TETRAPHENYLARSONIUM TRIFLUOROACETYLIDE-DICYANOMETHANIDE
$C_{29}H_{20}AsF_3N_2O$ $C_{24}H_{20}As^+,C_5F_3N_2O^-$

H.W. ROESKY, M. NOLTEMEYER and G.M. SHELDRICK, 1985. *Z. Naturforsch.*, **B40**, 883-885.

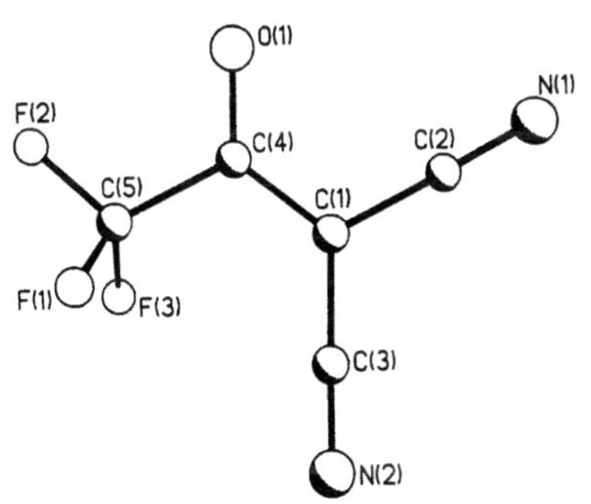

Fig. 1. The structure of the $C_5F_3N_2O^-$ anion.

Monoclinic, $P2/c$, a = 19.058, b = 7.786, c = 17.604 Å, β = 97.97°, Z = 4. Mo radiation, R = 0.058.

The anion is shown in Fig. 1 and has dimensions C(1)-C(2) 1.407(13), C(1)-C(3) 1.405(10), C(1)-C(4) 1.378(11), C(2)-N(1) 1.153(12), C(3)-N(2) 1.161(9), C(4)-C(5) 1.535(13), C(4)-O(1) 1.235(9), C(5)-F 1.27-1.33(1) Å, C(1)-C(4)-C(5) 117.9(7), C(4)-C(1)-C(2) 117.1(7), C(4)-C(1)-C(3) 125.4(8), C(2)-C(1)-C(3) 117.5(7)°.

2,4-DINITROBENZOIC ACID
$C_7H_4N_2O_6$

T. WIECKOWSKI and T.M. KRYGOWSKI, 1985. *Croat. Chem. Acta,* **58**, 5-14.

Monoclinic, $P2_1/c$, a = 6.2150, b = 14.393, c = 9.487 Å, β = 97.175°, Z = 4. Cu radiation, R = 0.050 for 1043 reflections.

The molecules occur as centrosymmetric dimers (Fig. 1) with O...O 2.656(3) Å. The vertical separation of the two parallel carboxyl group planes is 0.113 Å.

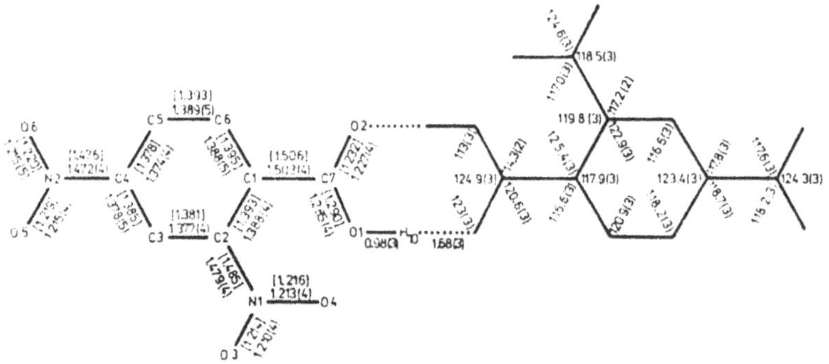

Fig. 1. Bond lengths (upper values corrected for libration) and bond angles in $C_7H_4N_2O_6$.

2,5-DINITROBENZOIC ACID
$C_7H_4N_2O_6$

S.J. GRABOWSKI and T.M. KRYGOWSKI, 1985. *Acta Cryst.*, **C41**, 1224-1226.

Monoclinic, $C2/c$, a = 12.551, b = 11.883, c = 11.588 Å, β = 108.88°, Z = 8. Cu radiation, R = 0.0359 for 962 reflections.

The deviations of ring geometry from D_{6h} symmetry are highly significant and affect bond distances (σ 0.003 Å) as well as angles (σ 0.2°) (Fig. 1). The dihedral angle between the mean planes of the benzene ring and the carboxyl group is 4.0°, while the ortho-NO_2 substituent is nearly perpendicular to the benzene ring (86.0°). The angle between the planes of the carboxy and nitro groups in the ortho position is 89.8°. The molecules are paired by hydrogen bonding between the carboxy groups (O...O 2.621(2) Å), the two halves of the dimer being related by a centre of symmetry.

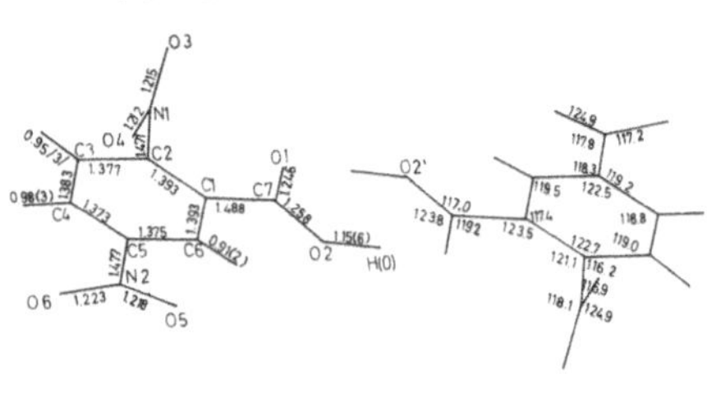

Fig. 1. $C_7H_4N_2O_6$: bond lengths and angles.

m-FLUOROBENZOIC ACID
$C_7H_5FO_2$

T. TAGA, N. YAMAMOTO and K. OSAKI, 1985. *Acta Cryst.*, **C41**, 153-154.

Monoclinic, $P2_1/c$, a = 6.809, b = 3.808, c = 25.904 Å, β = 109.20°, Z = 4. Cu radiation, R = 0.045 for 859 reflections.

The benzene ring in the molecule (Fig. 1) is significantly distorted from hexagonal regularity with C-C bonds varying from 1.364(2) to 1.388(3) Å. The internal valence angle of the ring at the F-substituted position is 122.9(2)° and the C-F bond length is 1.358(3) Å. The COOH group is rotated about the C(1)-C(7) bond by 6.0(1)° out of the benzene ring plane. The molecules form hydrogen-bonded dimers with an O...O separation of 2.631(2) Å and these form stacks along the b-axis. In the stacks the mean separation between the benzene rings is 3.512(7) Å.

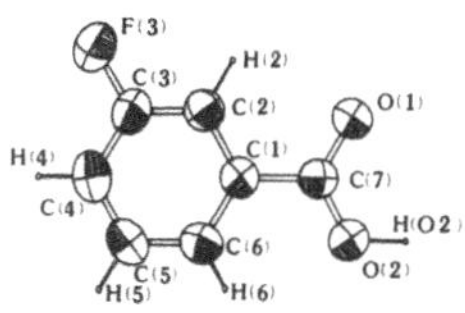

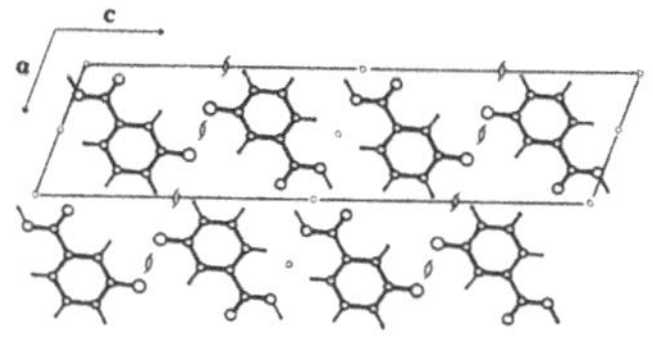

Fig. 1. $_7H_5FO_2$: atomic numbering (left) and (right) a projection of the structure onto (010).

m-IODOBENZOIC ACID
$C_7H_5IO_2$

A.A. PATIL, D.Y. CURTIN and I.C. PAUL, 1985. *Israel J. Chem.*, **25**, 320-326.

Monoclinic, $P2_1/c$, a = 6.220, b = 4.689, c = 26.67 Å, β = 101.80°, Z = 4. Mo radiation, R = 0.026 for 1288 reflections.

The molecule crystallizes as centrosymmetric essentially planar hydrogen-bonded dimers (Fig. 1) with significant intermolecular I...I contacts (3.9848(4) Å). The structure is compared with other o-, m- and p-halogenobenzoic acids.

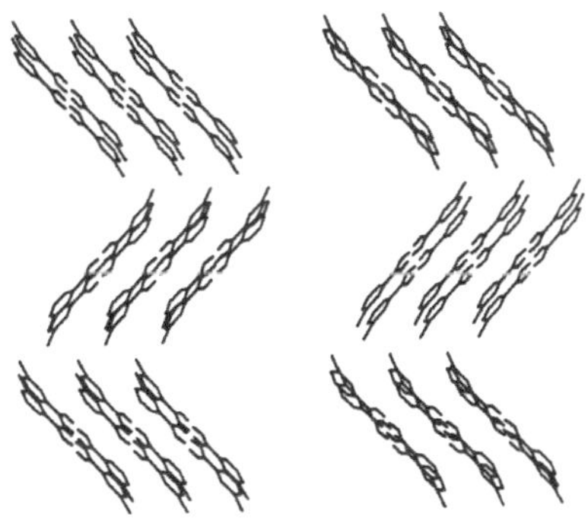

18CR figs.

Fig. 1. Dimensions for m-iodobenzoic acid and the packing viewed down a^*.

3,5-DIAMINO-2,4,6-TRINITROBENZOIC ACID
$C_7H_5N_5O_8$

H.L. AMMON and S.M. PRASAD, 1985. *Acta Cryst.*, **C41**, 921-924.

Monoclinic, $P2_1/c$, a = 7.246, b = 10.309, c = 13.962 Å, β = 101.02°, Z = 4. Cu radiation, R = 0.039 for 1574 reflections.

Four of the six substituents on the ring (Fig. 1) deviate substantially from the ring plane. The carboxyl group, the two o-nitro groups and the p-nitro group are twisted out of the ring plane by 68.7(3), 22.0(3), 27.3(3) and 14.6(3)° respectively. In the ring the average C-C distance for the four bonds associated with the NH_2 group is 1.431(3) Å, longer than expected. The average C-NH_2 distance is 1.324(5) Å. The carboxyl group participates in the formation of a hydrogen-bonded dimer across a centre of symmetry (O(2)...O(1) 2.605(2) Å). There are also N-H...O interactions between NH_2 and NO_2 groups (N(4)...O(4) 2.964(3), N(2)...O(6) 3.018(2) Å).

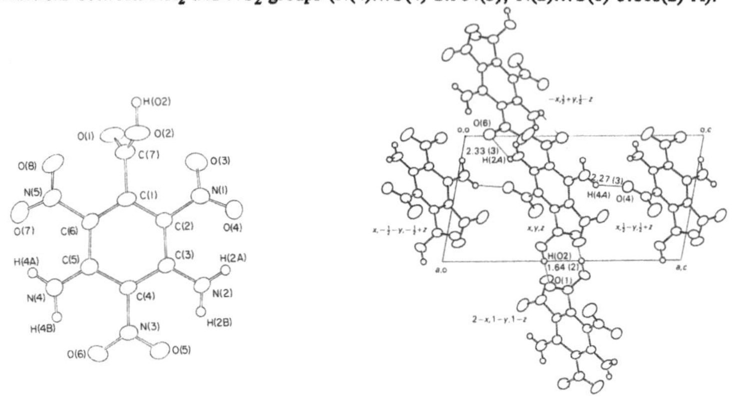

Fig. 1. The molecular structure of $C_7H_5N_5O_8$ and the packing viewed down b.

ANTHRANILIC ACID (2-AMINOBENZOIC ACID)
$C_7H_7NO_2$

C.J. BROWN and M. EHRENBERG, 1985. *Acta Cryst.*, **C41**, 441-443.

Orthorhombic, $P2_1cn$, a = 12.864, b = 10.790, c = 9.314 Å, Z = 8. Neutrons, λ = 1.168 Å, R = 0.048 for 1207 reflections.

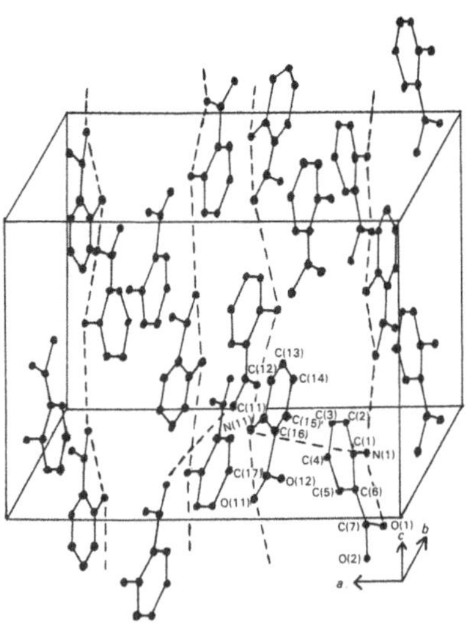

Fig. 1. $C_7H_7NO_2$: the unit cell contents showing the atomic numbering and with the hydrogen bonds indicated by dashed lines.

In the asymmetric unit there are two molecules of which one is neutral and the other is a zwitterion broadly confirming the structure reported in **1**. The phenyl rings of the molecules are essentially planar and the carboxyl groups are rotated by 5.38(5) and 1.71(5)° out of the planes of the rings to which they are attached. There are some noticeable differences between the bond lengths and angles from this determination and those reported previously. The molecules in the cell (Fig. 1) are linked by a three-dimensional network of hydrogen bonds of lengths in the range 2.497(7) to 2.894(6) Å.

1. C.J. BROWN, 1968. Proc. Roy. Soc., London, **302A**, 185-199.

5-SULPHOSALICYLIC ACID OXONIUM HYDRATE
$C_7H_{14}O_{10}S,H_2O$ $(H_9O_4^+)[C_6H_3(COOH)(OH)SO_3^-],H_2O$

D. MOOTZ and A. MERSCHENZ-QUACK, 1985. *Z. Kristallogr.*, **170**, 131.

Monoclinic, *Im*, a = 6.856, b = 28.294, c = 7.210 Å, β = 99.78°, Z = 4 (at 113°).

Very short strong O-H...O hydrogen bonds (O...O 2.465-2.571 Å) link independent $H_9O_4^-$ ions which lie on mirror planes. Hydrogen bonds to the water molecule are longer (2.770-2.919 Å).

THALLIUM PHTHALATE
$C_8H_5O_4Tl$ $C_6H_4(COOH)COOTl$

LIU SHIXIONG and ZHANG ZHENKE, 1985. *Jiegou Huaxue (J. Struct. Chem.)*, **4**, 93-95.

Orthorhombic, $P2_1ab$, a = 6.596, b = 10.012, c = 12.835 Å, Z = 4. Mo radiation, R = 0.063 for 966 reflections.

The crystal structure (Fig. 1) contains thallium cations and acid phthalate anions. The Tl^+ ion is surrounded by six O atoms (Tl...O 2.76-3.06 Å). The anions are linked to form dimers by O-H...O hydrogen bonds.

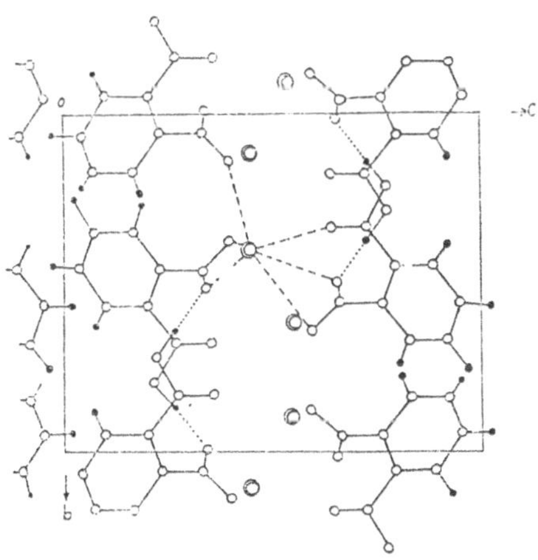

Fig. 1. A drawing of the $C_6H_4(COOH)COOTl$ crystal structure.

3-(4-CARBAMOYLPHENYL)-1-METHYLTRIAZENE 1-OXIDE
$C_8H_{10}N_4O_2$

R. KURODA and D.E.V. WILMAN, 1985. *Acta Cryst.*, **C41**, 1543-1545.

Monoclinic, $P2_1/a$, a = 9.708, b = 6.7806, c = 14.707 Å, β = 101.30°, Z = 4. Cu radiation, R = 0.037 for 1028 reflections.

The molecule (Fig. 1) adopts the N-oxide tautomeric form. The N(1)-N(2) and N(2)-N(3) bond distances are 1.340(3) and 1.271(3) Å respectively. The four atoms of the triazene N-oxide group are planar and make an angle of 11(1)° with the plane of the phenyl ring. The amide group is also planar and makes an angle of 27(1)° with the plane of the phenyl ring. Hydrogen bonds are formed between amide H atoms and O(2) of a centrosymmetrically related molecule (H(N41)...O(2) 1.83(4)) and O(2) of an a-glide-plane-related molecule (H(N42)...O(2) 2.11(5) Å).

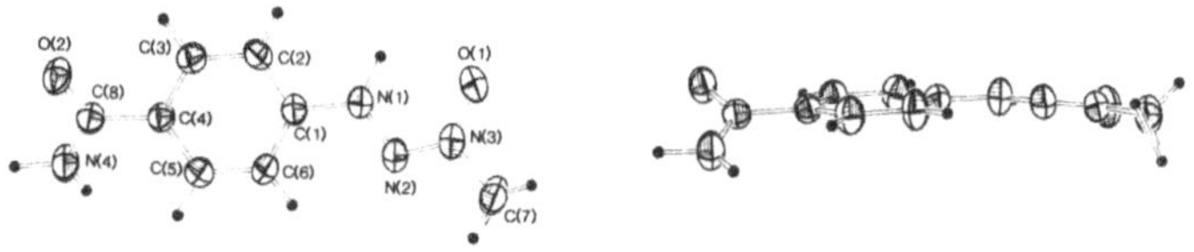

Fig. 1. $C_8H_{10}N_4O_2$: plan and side views of the molecule.

4-METHYL-5-SULPHOSALICYLIC ACID DIAQUAOXONIUM MONOHYDRATE

$C_8H_{14}O_9S,H_2O$ $C_8H_7O_6S^-,H_7O_3^+,(H_2O)$

T. GUSTAFSSON, 1985. *Acta Cryst.*, **C41**, 443-446.

Monoclinic, $P2_1/c$, a = 7.425, b = 25.353, c = 8.291 Å, β = 118.09°, Z = 4. Mo radiation, R = 0.073 for 2860 reflections.

The molecular structure (Fig. 1) is best formulated as the diaquaoxonium salt $H_7O_3^+.CH_3C_6H_2COOH(OH)SO_3^-.H_2O$. The 4-methyl-5-sulphosalicylic anions are linked by hydrogen bonds to the lattice water molecule, OW(2), and to the oxygen atoms OW(1,3,4) of the disordered $H_7O_3^+$ complex. Each O atom in an $H_7O_3^+$ is bonded to three H atoms and each disordered O is sited approximately 0.5 Å above and below the H_3 plane.

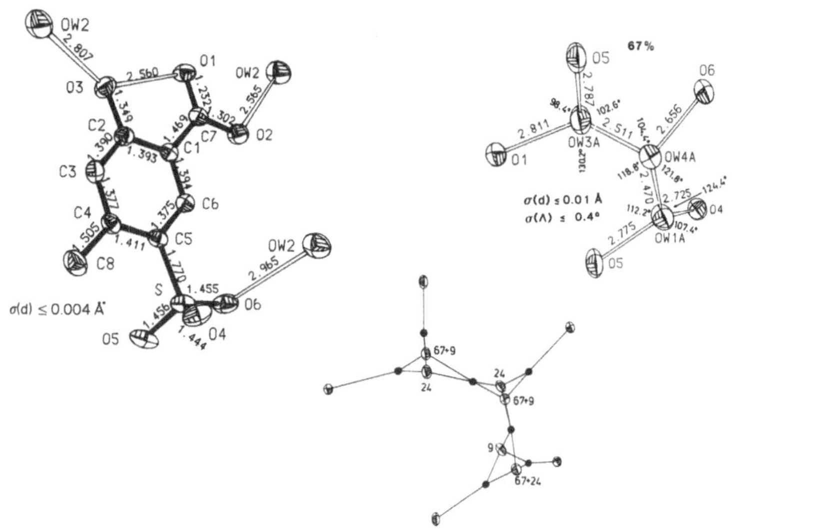

Fig. 1. $C_8H_{14}O_9S,H_2O$: (left) bond lengths with intra- and intermolecular hydrogen bonds shown as open lines and (right) geometry of the most probable $H_7O_3^+$ complex (upper) and a superposition of the probable complex conformations showing the site occupancies (lower).

METHYL 2-(METHYLSULPHINYL)BENZOATE

$C_9H_{10}O_3S$ (I)

S-(2-METHOXYCARBONYLPHENYL)-S-METHYL-N-(4-METHOXYPHENYLSULPHONYL)SULPHILIMINE

$C_{16}H_{17}NO_5S_2$ (II)

DIMETHYL 2,2'-SULPHINYLDIBENZOATE

$C_{16}H_{14}O_5S$ (III)

METHYL 2-(2-NITROPHENYLSULPHINYL)BENZOATE
$C_{14}H_{11}NO_5S$ (IV)

2,2'-DINITRODIPHENYLSULPHOXIDE
$C_{12}H_8N_2O_5S$ (V)

METHYL 2-(2-NITROPHENYLSULPHINYL)PHENYLACETATE
$C_{15}H_{13}NO_5S$ (VI)

A. KUCSMAN, I. KAPOVITS, I. KOVESDI, A. KALMAN and L. PARKANYI, 1985. *J. Mol. Struct.*, **127**, 135-148.

I. Monoclinic, $P2_1/n$, a = 8.418, b = 9.407, c = 12.457 Å, β = 100.60°, Z = 4. Mo radiation, R = 0.043 for 2107 reflections.

II. Monoclinic, $P2_1/c$, a = 13.895, b = 10.155, c = 13.416 Å, β = 115.66°, Z = 4. Mo radiation, R = 0.038 for 2541 reflections.

III. Monoclinic, $P2_1/c$, a = 8.211, b = 17.500, c = 14.079 Å, β = 132.63°, Z = 4. Mo radiation, R = 0.039 for 2535 reflections.

IV. Monoclinic, $P2_1/n$, a = 14.049, b = 8.348, c = 12.013 Å, β = 107.75°, Z = 4. Mo radiation, R = 0.047 for 2728 reflections.

V. Monoclinic, $C2/c$, a = 14.116, b = 7.915, c = 22.739 Å, β = 104.68°, Z = 8. Mo radiation, R = 0.053 for 2517 reflections.

VI. Triclinic, $P\bar{1}$, a = 7.589, b = 7.707, c = 12.664 Å, α = 98.55, β = 93.60, γ = 95.60°, Z = 2. Mo radiation, R = 0.043 for 3226 reflections.

Solid state conformations for (I)-(VI) are in Fig. 1. The observed axial ((I) and (II)) and axial-twist ((III)-(VI)) conformations are interpreted by taking into account steric effects, conjugation and S^{IV}...O interaction. Linear and non-linear X=S^{IV}...O contacts (2.664-2.778 and 2.847-3.618 Å respectively) are related to favourable bond direction and unfavourable lone-pair direction approaches. Molecular dimensions are not appreciably affected by S^{IV}...O interactions.

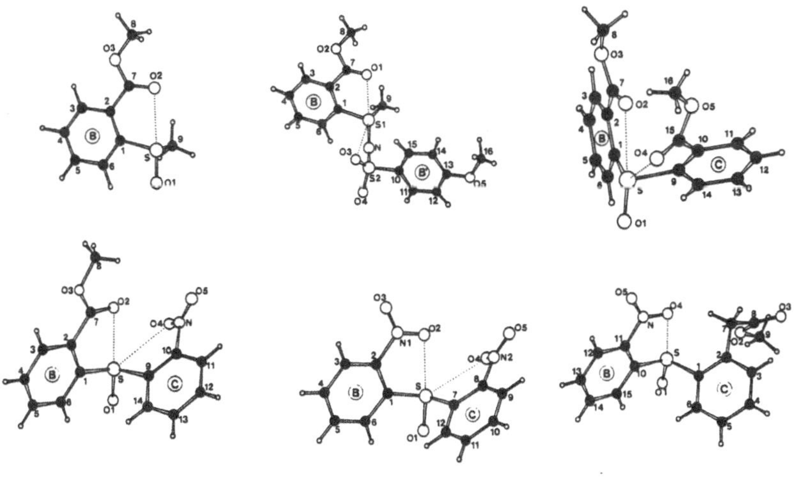

Fig. 1. The conformations of the S^{IV} compounds (I)-(VI) (top row, left to right, (I)-(III); bottom row, left to right, (IV)-(VI)). Rings marked with a **B** are close to perpendicular to the C_{ar}-S-C plane and those marked with a **C** have intermediate ring orientations.

N-ISOBUTYRYLBENZAMIDE
$C_{11}H_{13}NO_2$

R.B. BATES, K.D. JANDA and M.E. WRIGHT, 1985. *Acta Cryst.*, **C41**, 263-264.

Orthorhombic, $Pca2_1$, a = 8.878, b = 13.464, c = 9.009 Å, Z = 4. Mo radiation, R = 0.048 for 385 reflections.

The compound (Fig. 1) is an acyclic imide in which the O atoms adopt a twisted 'U' conformation. Electrostatic repulsions between O atoms are relieved by twisting the π system and by opening the O-C-N and C-N-C angles which are all greater than 120°. The molecules in the crystal are held together by an infinite series of bifurcated hydrogen bonds in the c direction from the HN of one molecule to the two carbonyl O atoms of the next (N...O(7) 2.92, N...O(8) 3.15, HN...O(7) 2.10 Å).

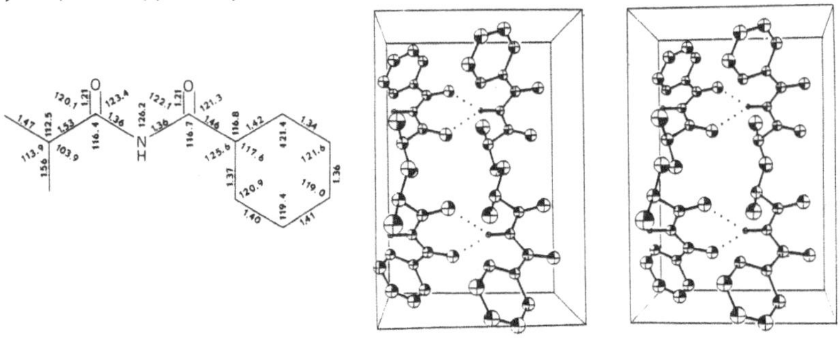

Fig. 1. $C_{11}H_{13}NO_2$: bond lengths and angles and a stereoview of the unit cell.

(-)-METHYL 2-(4-BROMOBENZYLSULPHINYL)BENZOATE
$C_{15}H_{13}BrO_3S$

A.E. KOZIOL, W. MAJEWSKI and M. JANCZEWSKI, 1985. *Acta Cryst.*, **C41**, 144-146.

Monoclinic, $P2_1$, a = 12.081, b = 5.740, c = 12.956 Å, β = 122.48°, Z = 2. Mo radiation, R = 0.044 for 3228 reflections.

In the molecule (Fig. 1) the chiral sulphur atom has the S configuration. Both benzene rings are planar and the dihedral angle between their least-squares planes is 36.3(9)°. The $COOCH_3$ group is planar and is twisted out of the plane of the benzene ring to which it is bonded by 6.3(9)°. The S-O bond is in the transoid conformation with respect to the methoxycarbonyl group. The S...O(1) intramolecular distance is 2.776(5) Å. The S-C(7) and S-C(8) bonds are of significantly different lengths, 1.854(6) and 1.809(6) Å respectively. Except for two short intermolecular contacts involving the methoxycarbonyl group all other intermolecular distances are consistent with van der Waals contacts.

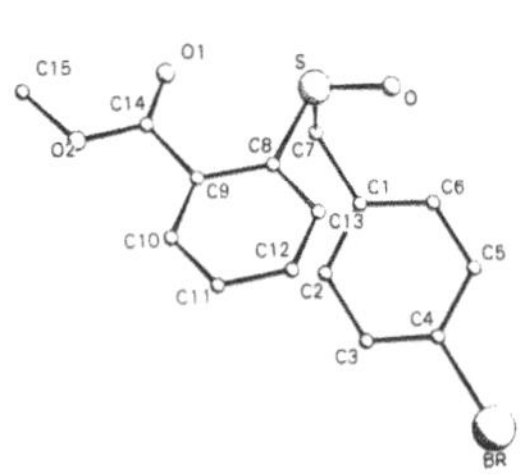

Fig. 1. $C_{15}H_{13}BrO_3S$: a view of the molecule.

N-BENZOYLBENZIMIDATE ETHYL ESTER
$C_{16}H_{15}NO_2$

R. KUPFER, M. NAGEL, E.-U. WURTHWEIN and R. ALLMANN, 1985. *Chem. Ber.*, **118**, 3089-3104.

Monoclinic, $P2_1/n$, a = 13.728, b = 9.449, c = 11.148 Å, β = 104.91°, Z = 4. Mo radiation, R = 0.056 for 1024 reflections.

The C=N- and C=O fragments of the molecule (Fig. 1) are twisted significantly (torsion angle O(1)-C(1)-N-C(2) -71.6°) with bond lengths C(2)-N 1.262(5), N-C(1) 1.380(4), C(1)-O(1) 1.212(3) Å. The aromatic ring orientations are defined by the torsion angles O(1)-C(1)-C(11)-C(12) 169.4° and O(2)-C(2)-C(21)-C(22) 157.8°.

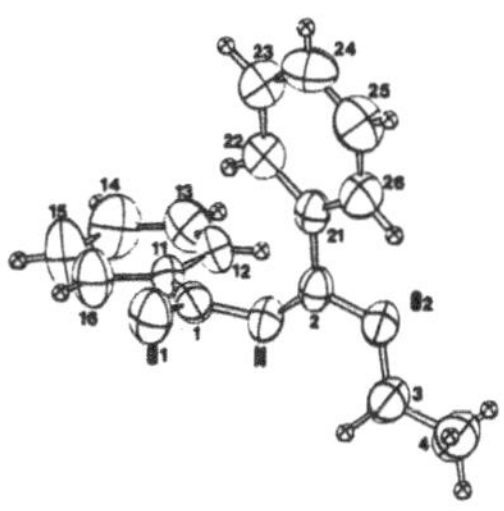

Fig. 1. The $C_{16}H_{15}NO_2$ molecule.

METHYL 3,6-DIHYDROXY-2-(2-HYDROXY-6-METHOXYBENZYL)-4-METHYLBENZOATE
$C_{17}H_{18}O_6$

G. FERGUSON, B.L. RUHL, J. GILMORE, T. IBRAHIM and W.B. WHALLEY, 1985. *J. Chem. Res.*, S284-285, M3101-3137.

Monoclinic, $P2_1/n$, a = 7.292, b = 23.949, c = 8.749 Å, β = 101.88°, Z = 4. Mo radiation, R = 0.0361 for 2057 reflections.

There is an intramolecular O-H...O hydrogen bond (O(2)-H...O(3), O...O 2.715(2), H...O 1.98(3) Å) linking the two individually planar aromatic rings (Fig. 1). The rings make angles of 77.3 and 100.5° (for rings C(1-6) and C(9-14) respectively) with the bridging C(1)-C(8)-C(9) plane. There is another intramolecular hydrogen bond (O(4) H...O(5), O...O 2.633(2), H...O 1.85(3) Å). The plane C(14), C(16), O(5), O(6) is rotated 32.5(2)° out of the C(9-14) plane. The methoxy group C(7)-O(1) is coplanar with its aromatic ring. In the crystal, molecules are linked into chains along the z-direction by hydrogen bonds.

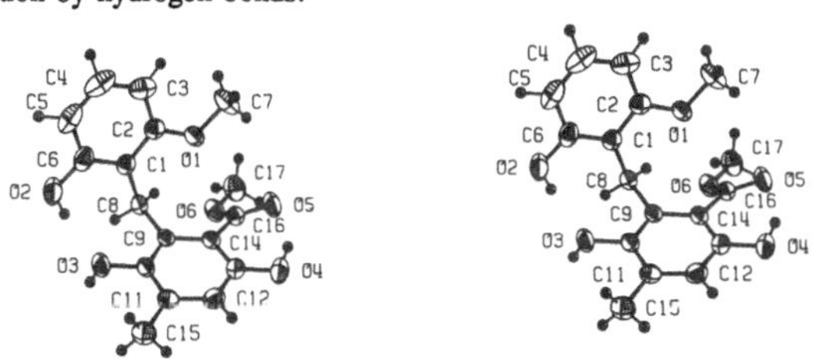

Fig. 1. A stereoview of $C_{17}H_{18}O_6$.

1,3-BUTANEDIYL BIS(p-CHLOROBENZOATE)
$C_{18}H_{16}Cl_2O_4$ (I)

2,4-PENTANEDIYL BIS(p-CHLOROBENZOATE)
$C_{19}H_{18}Cl_2O_4$ (II)

G. BOCELLI and M.F. GRENIER-LOUSTALOT, 1985. *Acta Cryst.*, C**41**, 249-252.

I. Monoclinic, $P2_1/n$, a = 16.935, b = 6.045, c = 16.887 Å, β = 95.01°, Z = 4. Cu radiation, R = 0.054 for 2049 reflections.

II. Monoclinic, $P2_1/n$, a = 15.274, b = 11.099, c = 11.559 Å, β = 106.24°, Z = 4. Cu radiation, R = 0.048 for 2573 reflections.

The conformation of the propanediyl segment of the chain is gauche-gauche in the two compounds (Fig. 1). In (II) a rotation of one aromatic ring with respect to the other results in an endo conformation for the molecule. The two carboxylic groups are tilted by 8.3(1), 13.1(1)° and 3.6(1), 6.7(1)° with respect to their phenyl rings in the two compounds. Molecular dimensions are normal and there are no unusually close intermolecular contacts.

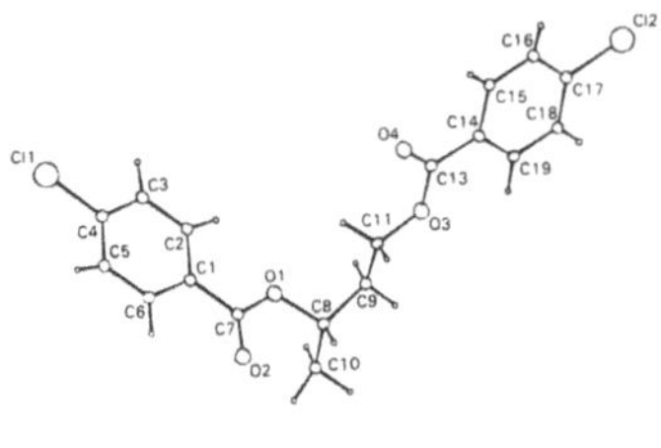

Fig. 1. Projections of the molecules: $C_{19}H_{18}Cl_2O_4$ (left) and $C_{18}H_{16}Cl_2O_4$ (right).

TETRAMETHYLENE GLYCOL DIBENZOATE
$C_{18}H_{18}O_4$ (I)

TETRAMETHYLENE GLYCOL DI-p-CHLOROBENZOATE
$C_{18}H_{16}Cl_2O_4$ (II)

TETRAMETHYLENE GLYCOL DI-p-NITROBENZOATE
$C_{18}H_{16}N_2O_8$ (III)

A. PALMER, S. POULIN-DANDURAND and F. BRISSE, 1985. *Canad. J. Chem.*, **63**, 3079-3088.

I. Triclinic, $P\bar{1}$, a = 7.870, b = 8.574, c = 12.993 Å, α = 83.76, β = 89.92, γ = 64.68°, Z = 2. Cu radiation, R = 0.042 for 2687 reflections.

II. Triclinic, $P\bar{1}$, a = 5.916, b = 7.599, c = 10.404 Å, α = 67.81, β = 77.47, γ = 81.63°, Z = 1. Cu radiation, R = 0.046 for 976 reflections.

III. Monoclinic, $P2_1/c$, a = 6.086, b = 11.475, c = 13.162 Å, β = 101.93°, Z = 2. Mo radiation, R = 0.045 for 781 reflections.

Molecules (II) and (III) lie about inversion centres; molecule (I) has no imposed crystallographic symmetry. The conformation of the methylenic sequence varies and depends on the substituent on the benzoate group (Fig. 1); the observed conformations are tg⁻ttt for (I), ttttt for (II) and tg⁻tgt for (III) (t = trans, g = gauche). The O-C(sp^3) distances range from 1.451(2)-1.457(2) Å, mean 1.454 Å.

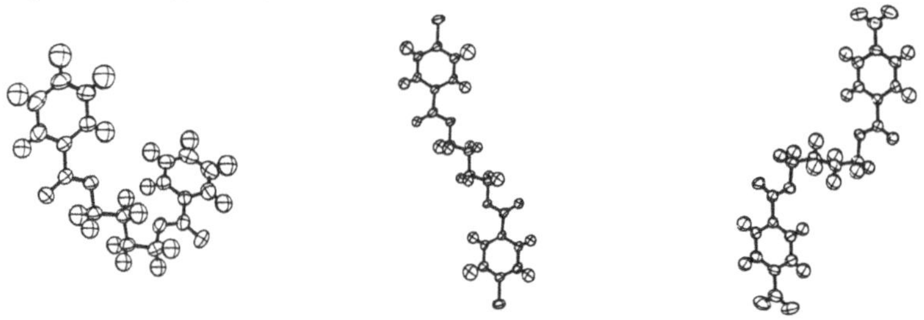

Fig. 1. Views (left to right) of $C_{18}H_{18}O_4$ (I), $C_{18}H_{16}Cl_2O_4$ (II) and $C_{18}H_{16}N_2O_8$ (III).

2-METHYL-1,4-BUTANEDIYL DIBENZOATE
$C_{19}H_{20}O_4$

G. BOCELLI, F. UGOZZOLI and M.F. GRENIER-LOUSTALOT, 1985. *Acta Cryst.*, **C41**, 485-487.

Triclinic, $P\bar{1}$, a = 13.211, b = 8.059, c = 8.704 Å, α = 114.57, β = 97.98, γ = 91.57°, Z = 2. Cu radiation, R = 0.062 for 2056 reflections.

In the molecule (Fig. 1) the carboxylic groups are planar and the average dihedral angle between these groups and the planes of their phenyl rings is 2.4(1)°. The methyl group is axial with respect to the C(8,9,10,12) sequence. The conformation of the aliphatic chain is trans-trans-gauche and the lengths of its C(sp^3)-C(sp^3) bonds are all less than 1.525 Å. The bond lengths and angles of the aromatic rings are normal and intermolecular distances are essentially van der Waals contacts.

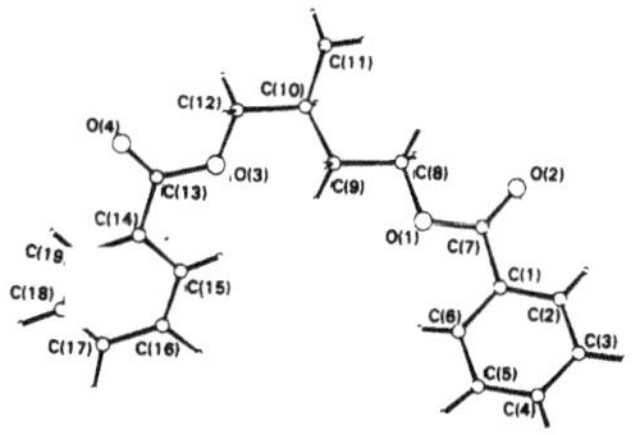

Fig. 1. $C_{19}H_{20}O_4$: a projection of the molecule along the b-axis showing the atomic numbering.

TRIS(AMMONIUM) PENTAKIS(DIOXONEPTUNIUM) BIS(BENZENE-1,2,4,5-TETRACARBOXYLATE) HEPTAHYDRATE

$C_{20}H_{16}N_3Np_5O_{26},7(H_2O)$ $2(C_{10}H_2O_8{}^{4-}),3(H_4N^+),5(NpO_2{}^+),7(H_2O)$

A. COUSSON, 1985. *Acta Cryst.*, **C41**, 1758-1761.

Monoclinic, $P2_1/c$, a = 19.315, b = 19.640, c = 10.295 Å, β = 95.27°, Z = 4. Mo radiation, R = 0.039 for 9458 reflections.

The structure consists of infinite chains of neptunyl cations (Fig. 1) separated by 1,2,4,5-benzenetetracarboxylate ligands. The mean Np-O(1-5) distances are 2.293, 2.266, 2.282, 2.253 and 2.287 Å respectively and Np...Np distances range from 4.0395(7) to 4.1409(5) Å. The organic moiety has acceptable bond lengths and angles. A new type of cation...cation interaction is observed (Fig. 1).

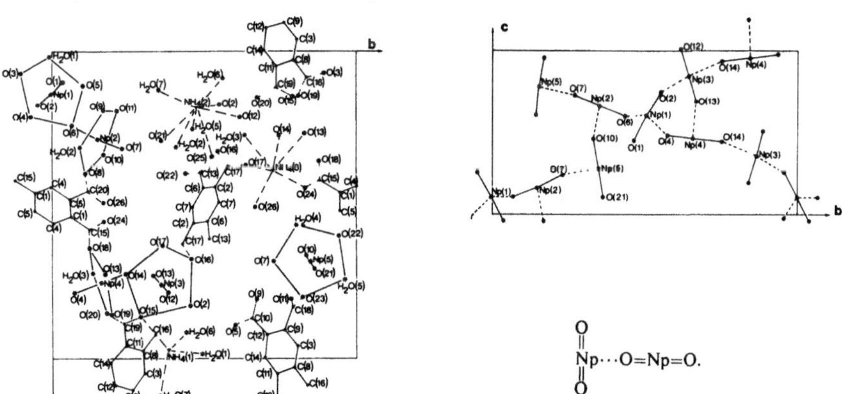

Fig. 1. $C_{20}H_{16}N_3Np_5O_{26},7(H_2O)$: projection of the structure parallel to *c* for cations O < Z < 0.25 (left), the neptunyl groups projected along *a* (top right) and the new cation-cation interaction (bottom right).

DIETHYLENEGLYCOL BIS(p-METHYLBENZOATE)

$C_{20}H_{22}O_5$

B. REMILLARD and F. BRISSE, 1985. *J. Cryst. Spect. Res.*, **15**, 473-481.

Monoclinic, $P2_1/c$, a = 11.408, b = 5.808, c = 28.542 Å, β = 109.50°, Z = 4. Cu radiation, R = 0.049 for 1400 reflections.

The geometry and conformation are shown in Fig. 1. The aromatic rings are individually planar and show the usual distortions attributable to para-substitution. The carboxylate groups are rotated 2.2(4) and 4.7(4)° from the benzene planes. The conformation of the di(ethyleneglycol) moiety is $tg^{-}gtg^{-}g^{-}$.

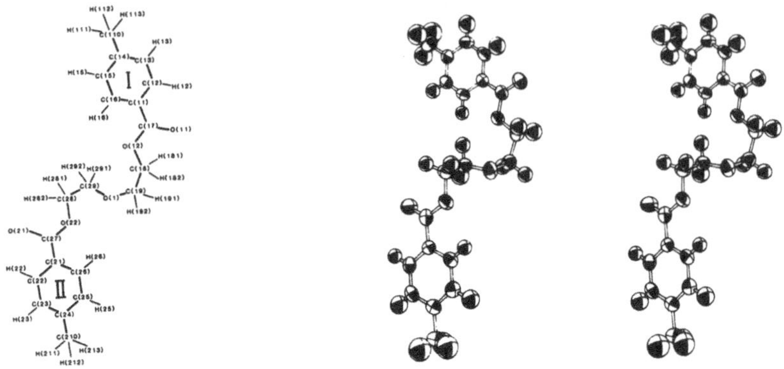

Fig. 1. The $C_{20}H_{22}O_5$ molecule and a stereoview.

N,N'-(p-PHENYLENE)DIBENZAMIDE TETRAKIS(SULPHURIC ACID)

$C_{20}H_{24}N_2O_{18}S_4$ $C_{20}H_{18}N_2O_2^{2+}$,2(HO_4S^-),2(H_2O_4S)

J.C. CALABRESE and K.H. GARDNER, 1985. *Acta Cryst.*, **C41**, 389-392.

Triclinic, $P1$, a = 9.747, b = 10.313, c = 7.879 Å, α = 108.67, β = 111.41, γ = 89.20°, Z = 1. Mo radiation, R = 0.026 for 2877 reflections (at 173 K).

Two of the four sulphuric acid groups nearest to the organic molecule (PPDB) have caused protonation of carbonyl oxygen atoms of the amide groups (Fig. 1). These C-O bonds are here elongated to 1.30 Å on average and the mean amide bond length is shortened to 1.30 Å compared with the values observed in the uncomplexed molecule (1). The phenyl rings are nearly coplanar with interplanar dihedral angles of 7.7(4) and 8.2(4)° between neighbouring pairs of rings and with the amide groups rotated out of this plane. This contrasts with the uncomplexed molecule where the corresponding dihedral angles are 64 and 79°. The S-O distances range from 1.404(3) to 1.552(2) Å and are consistent with this coordination. The sulphuric acid and bisulphate groups form hydrogen-bonded layers which alternate with sheets of overlapped PPDB molecules. These alternating layers are interconnected by strong amide-sulphate hydrogen bonds.

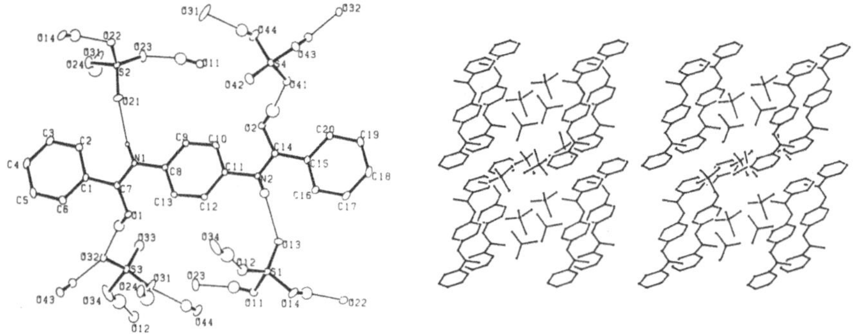

Fig. 1. $C_{20}H_{24}N_2O_{18}S_4$: a view of the 4:1 complex (left) in which the fine lines represent hydrogen bonds and (right) a stereoview along the c-axis of the molecular packing.

1. Structure Reports, **43B**, 124; **44B**, 82.

4-AMINO-5-CHLORO-2-METHOXY-N-(2-(N-(p-METHOXYBENZYL)METHYLAMINO)-1,1-DIMETHYLETHYL)BENZAMIDE
$C_{21}H_{28}ClN_3O_3$

T. FURUYA, S. IWANAMI, A. TAKENAKA and Y. SASADA, 1985. *Acta Cryst.*, C**41**, 1507-1509.

Monoclinic, $P2_1/c$, a = 6.144, b = 26.512, c = 13.741 Å, β = 94.76°, Z = 4. Mo radiation, R = 0.065 for 3926 reflections.

In the molecule, an analogue of a potent neuroleptic drug, ring A (Fig. 1) is essentially planar and coplanar with the amino methoxyl and carbamoyl groups attached to it. The distance between the centre of this ring and the tertiary amino N atom is 5.83 Å. There is an intramolecular hydrogen bond between the amide N(2) and methoxyl O(1) (N...O 2.662(3) Å) which forms a six-membered ring fused with the benzene ring A. Ring B is planar. Bond lengths and angles are normal. An intermolecular hydrogen bond is found between the amino N atom N(1) and carbonyl oxygen O(2) related by a *c* glide (N...O 3.019(3) Å).

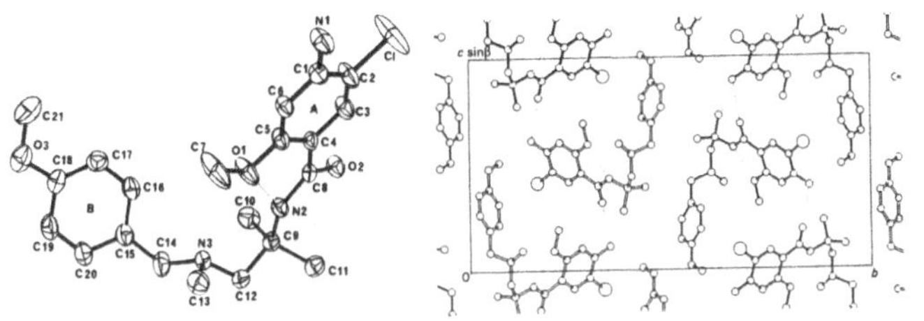

Fig. 1. $C_{21}H_{28}ClN_3O_3$: the molecular structure and crystal structure viewed down the *a* axis.

2,2'-(BENZYLIDENEMETHYLENEBIS(CARBONYLAMINO))DIBENZOIC ACID MONOHYDRATE
$C_{24}H_{18}N_2O_6,H_2O$ (I)

2,2'-(BENZYLIDENEMETHYLENEBIS(CARBONYLAMINO))DIBENZOIC ACID
$C_{24}H_{18}N_2O_6$ (II)

T. TAGA, N. YAMAMOTO, J. RIKU, J. YOSHINAGA and Y. KATO, 1985. *Acta Cryst.*, C**41**, 1235-1237.

I. Triclinic, $P\bar{1}$, a = 11.687, b = 10.088, c = 10.519 Å, α = 105.79, β = 110.25, γ = 97.83°, Z = 2. Cu radiation, R = 0.060 for 2873 reflections.

II. Monoclinic, $P2_1/a$, a = 16.469, b = 11.334, c = 11.312 Å, β = 99.49°, Z = 4. Cu radiation, R = 0.069 for 2726 reflections.

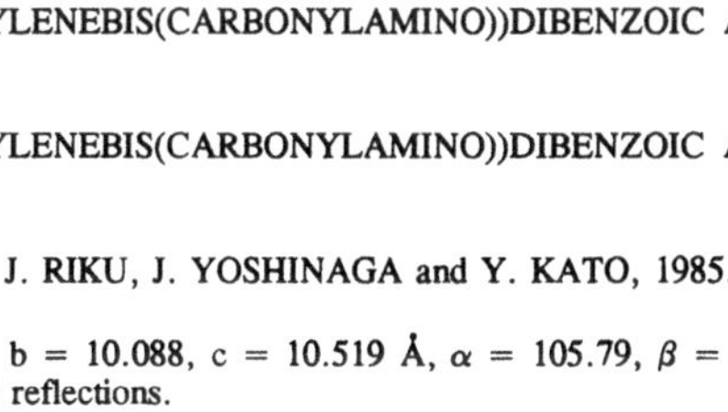

(I) (II)

Fig. 1. The molecular structures: $C_{24}H_{18}N_2O_6,H_2O$ (left) and $C_{24}H_{18}N_2O_6$ (right).

In both molecules all phenyl rings are planar. The dihedral angles between the C=C plane and the phenyl rings are 18.6(2) (C(1) ring), 36.3(1) (C(11)) and 26.5(1)° (C(27)) for (I) and for (II) are 5.6(2), 93.6(1) and 20.5(2)° respectively. N-H...O hydrogen bonds are formed between all NH and COOH groups (N...O 2.653(6)-2.755(12) Å). The main difference between (I) and (II) (Fig. 1) is in the rotation about the C(21)-C(25) bond, torsion angle

166.8(3)° for (I) and -29.9(5)° for (II). One NH group in (I) forms a bifurcated intramolecular hydrogen bond (H(10)...O(8) 1.86(5), H(10)...O(24) 2.23(4) Å). The water molecule in (I) forms three hydrogen bonds with O(19), O(22) and O(24) of neighbouring molecules.

(Z)-α-(2,2-DIPHENYLETHYLIDENE)-2,4,6-TRIMETHYLBENZENEMETHANOL BENZOATE
$C_{31}H_{28}O_2$

C.E. PFLUGER, A.G. PINKUS, A.-B. WU and P.W. HURD, 1985. *Tetrahedron,* **41**, 4417-4424.

Triclinic, $P\bar{1}$, a = 9.058, b = 11.028, c = 12.713 Å, α = 104.60, β = 90.20, γ = 93.55°, Z = 2. Cu radiation, R = 0.049 for 3801 reflections.

The molecule is shown to have the Z-configuration (Fig. 1); all molecular dimensions are normal. The conformation adopted, serves to minimise repulsive interactions between phenyl groups as well as the benzoyl phenyl group.

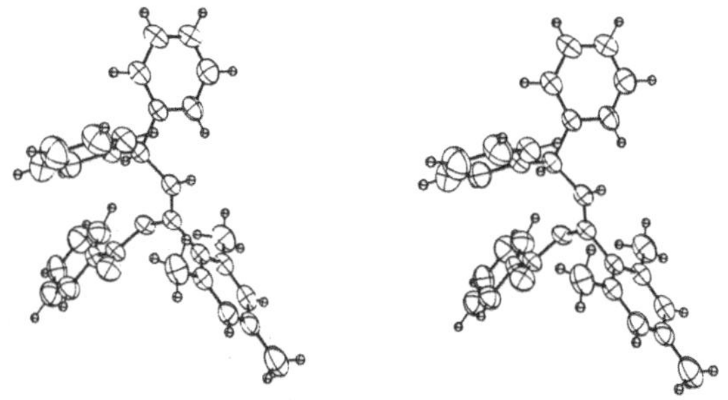

Fig. 1. A stereoview of $C_{31}H_{28}O_2$.

2,4-HEXADIYNE-1,6-BIS(p-n-HEXYLOXYBENZOATE)
$C_{32}H_{38}O_6$

V.E. SHKLOVER, Yu.T. STRUCHKOV, I.A. BAGIROV, I.I. KONSTANTINOV and Yu.B. AMERIK, 1985. *Z. Kristallogr.,* **173**, 25-31.

Monoclinic, $P2_1/c$, a = 19.753, b = 7.7531, c = 19.794 Å, β = 100.09°, Z = 4. Cu radiation, R = 0.060 for 3223 reflections.

The molecule (Fig. 1) has a skew conformation with the pseudotorsion angle relative to the linear diacetylene moiety equal to -110,0(9)°. The non-ability to polymerization in the solid state is explained on the basis of the crystal structure.

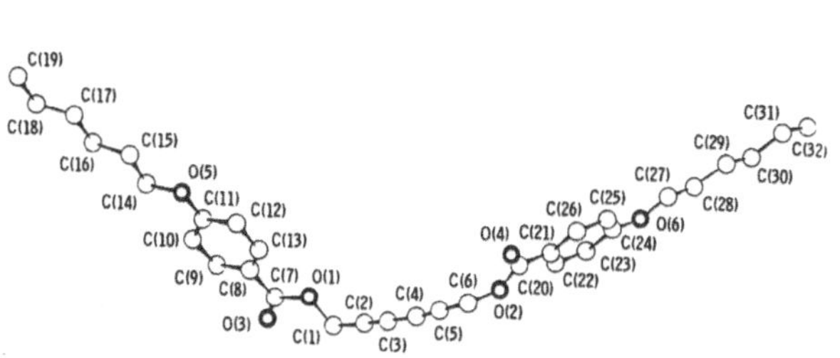

Fig. 1. A view of the $C_{32}H_{38}O_6$ molecule.

OCTAFLUOROBIPHENYL-4,4'-DIYL BIS(OXY-1,3-PHENYLENE)DIBENZOATE
$C_{38}H_{18}F_8O_6$ (I)

BIS((3-PENTAFLUOROPHENOXY)PHENYL)-4,4'-OXY DIBENZOATE
$C_{38}H_{16}F_{10}O_7$ (II)

S.V. LINDEMAN, Yu.T. STRUCHKOV, B.F. MALICHENKO, V.E. SHKLOVER and V.A. VASNEV, 1985.
Makromol. Chem., **186**, 2087-2102.

I. Monoclinic, $P2_1/n$, a = 14.994, b = 10.050, c = 20.343 Å, β = 90.162°, Z = 4. Cu radiation, R = 0.045 for 2423 reflections.

II. Monoclinic, $C2/c$, a = 62.819, b = 5.4580, c = 9.403 Å, β = 92.577°, Z = 4. Cu radiation, R = 0.063 for 1903 reflections.

Molecule (II) has twofold crystallographic symmetry. Both compounds exhibit almost completely stretched conformations (Fig. 1). In the crystal structures there are no short intermolecular contacts, and rather unexpectedly there is no parallel overlapping of benzene rings with stack formation. The fluorinated and non-fluorinated molecular segments are packed separately which is another important feature. The majority of chemically-equivalent bond lengths and angles are equal (within the error limits).

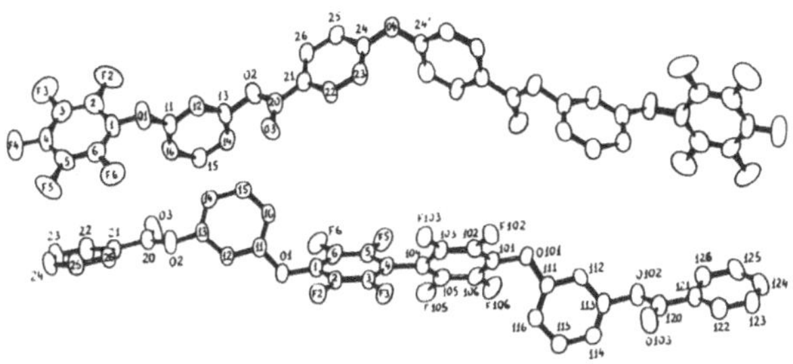

Fig. 1. Views of $C_{38}H_{18}F_8O_6$ (I) (lower) and $C_{38}H_{16}F_{10}O_7$ (II) (upper).

LITHIUM HYDROGEN PHTHALATE MONOHYDRATE
$C_8H_5LiO_4,H_2O$ (I) $C_8H_5O_4^-,Li^+,H_2O$

H. KUPPERS, F. TAKUSAGAWA and T.F. KOETZLE, 1985.
J. Chem. Phys., **82**, 5636-5647.

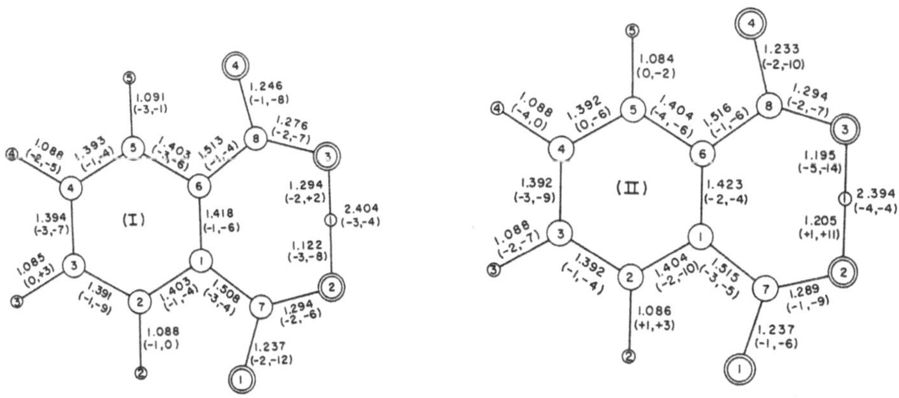

Fig. 1. Bond distances in hydrogen phthalate ions at 15 K. Also shown are the differences $\Delta D(T) = D(T) - D(15)$ in units of 0.001 Å for T = 100 and 298 K respectively in parentheses below the 15 K values. Esd's: 0.001-0.002 Å.

I. Triclinic, $P\bar{1}$, a = 11.601, b = 8.971, c = 7.886 Å, α = 96.83, β = 96.01, γ = 94.94°, Z = 4. Neutron radiation, R = 0.044 for 4114 reflections (at 15 K).

II. Triclinic, $P\bar{1}$, a = 11.667, b = 8.942, c = 7.895 Å, α = 96.81, β = 95.91, γ = 94.53°, Z = 4. Neutron radiation, R = 0.0358 for 4128 reflections (at 100 K).

III. Triclinic, $P\bar{1}$, a = 11.937, b = 8.881, c = 7.945 Å, α = 97.20, β = 95.94, γ = 93.08°, Z = 4. Neutron radiation, R = 0.0451 for 4083 reflections.

The two independent phthalate ions both have very short intramolecular O...H...O hydrogen bonds with O...O 2.394 and 2.404 Å and O...H...O ≈ 170°. One of these O...H...O linkages is markedly asymmetric (Figs. 1 and 2). The Li coordination is distorted tetrahedral with Li-O 1.908-2.056 (at 15 K). For details of the X-ray study, see (1).

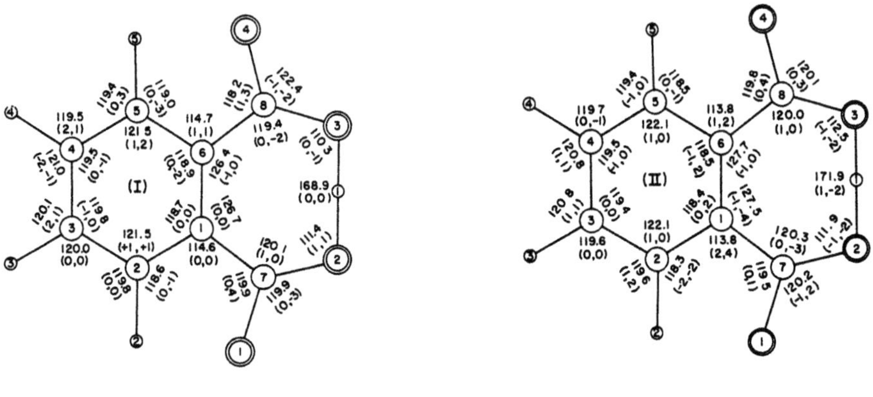

Fig. 2. Bond angles in hydrogen phthalate ions at 15 K. Also shown are the differences ΔA(T) = A(T) - A(15) in units of 0.1 Å for T = 100 and 298 K respectively in parentheses below the 15 K values. Esd's are less than 0.1°.

1. Structure Reports, **44B**, 86.

4-CHLORO-1,2-DINITROBENZENE (γ-form)
$C_6H_3ClN_2O_4$ (I)

4-CHLORO-1,2-DINITROBENZENE (β-form)
$C_6H_3ClN_2O_4$ (II)

A. WILKINS and R.W.H. SMALL, 1985. *Acta Cryst.*, **C41**, 1509-1512.

I. Orthorhombic, $P2_12_12_1$, a = 12.11, b = 5.57, c = 11.94 Å, Z = 4. Mo radiation, R = 0.033 for 1053 reflections.

II. Monoclinic, $P2_1/c$, a = 7.64, b = 7.82, c = 14.82 Å, β = 114.0°, Z = 4. Mo radiation, R = 0.065 for 762 reflections.

Molecules (I) and (II) (Fig. 1) differ only slightly in shape and dimensions. However packing in the γ-form consists of interleaved molecules, while that in the β-form consists of layers. Both molecules deviate significantly from planarity due to rotations of the nitro groups relative to the aromatic rings arising from repulsion between O atoms of adjacent groups. The intramolecular contacts O(2)...O(3) are similar in the two forms (γ, 2.793(3); β, 2.798(10) Å). Both structures involve close N...O (γ, 3.049(3); β, 2.95(1), 3.04(1) Å) and Cl...O intermolecular contacts (γ, 3.06(1); β, 3.073(3) Å).

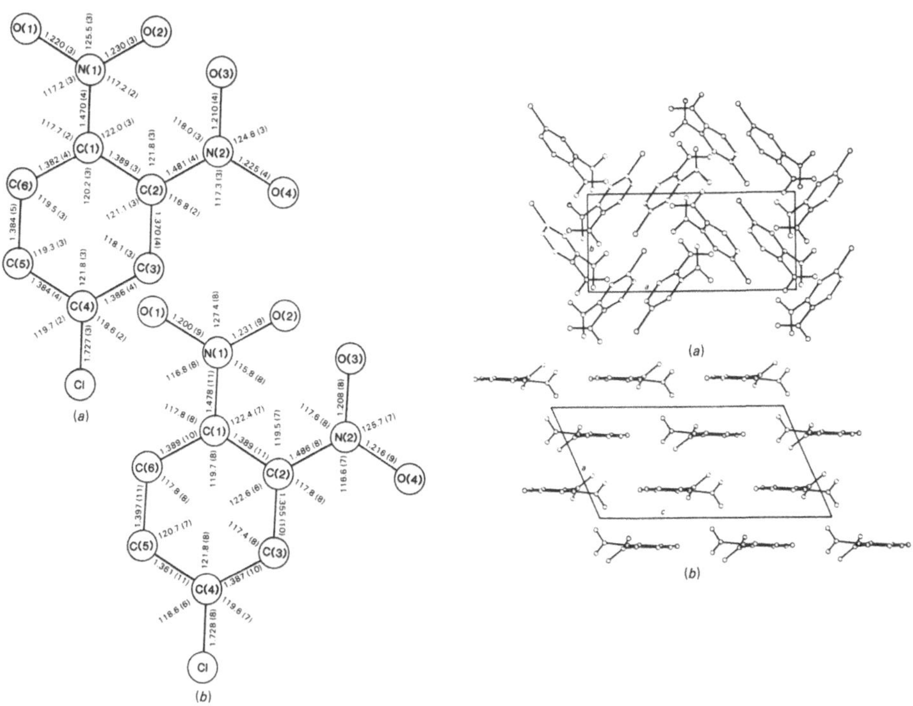

Fig. 1. $C_6H_3ClN_2O_4$: bond distances, angles and the packing, γ-form (top), β-form (bottom).

1-CHLORO-3-NITROBENZENE (m-CHLORONITROBENZENE)
$C_6H_4ClNO_2$

S. SHARMA, H. PAULUS, N. WEIDEN and A. WEISS, 1985. *Z. Kristallogr.*, **171**, 101-112.

Orthorhombic, *Pna*2_1, a = 21.1490, b = 5.9320, c = 5.2900 Å, Z = 4. Mo radiation, R = 0.031 for 1381 reflections.

For previous work see (**1**). The molecule is almost planar, with a small out-of-plane bending of the NO_2 group (2.9°). Ring bond lengths are in the range 1.376(3)-1.383(3), C-Cl 1.736(2), C-N 1.481(3), N-O 1.222(2) and 1.225(2) Å.

1. Structure Reports, **30**B, 63.

(E)-o-METHYL-p-NITROBENZOHYDROXIMOYL CHLORIDE
$C_8H_7ClN_2O_3$

J.E. JOHNSON, A. GHAFOURIPOUR, Y.K. HAUG, A.W. CORDES, W.T. PENNINGTON and O. EXNER, 1985. *J. Org. Chem.*, **50**, 993-997.

Orthorhombic, *Pna*2_1, a = 19.735, b = 3.827, c = 12.302 Å, Z = 4. Mo radiation, R = 0.051 for 636 reflections.

The dihedral angle between the benzene ring and the hydroximoyl group is 50° (Fig. 1). The crystal structure analysis confirmed the stereochemistry of the molecule. The N=C-Ar bond angle is 131°.

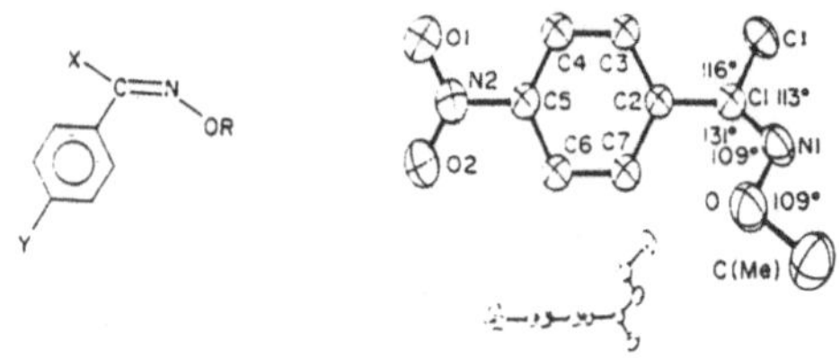

Fig. 1. ORTEP drawing of a molecule of $C_8H_7ClN_2O_3$ with a view parallel to the plane of the nitrobenzene ring.

2-NITRO-1,3-BIS(PHENYLTHIO)BENZENE
$C_{18}H_{13}NO_2S_2$

V.M. LYNCH, S.H. SIMONSEN, R.F. MILLER, J.C. TURLEY and G.E. MARTIN, 1985. *Acta Cryst.*, **C41**, 1240-1242.

Monoclinic, $P2_1/c$, a = 7.640, b = 13.499, c = 15.587 Å, β = 96.818°, Z = 4. Mo radiation, R = 0.041 for 2856 reflections (at 163 K).

The molecule (Fig. 1) has approximate twofold symmetry with the twofold axis lying along the C-N bond axis. The phenyl rings are all planar and approximately at right angles to each other. The nitro group is twisted out of the plane of the central phenyl ring by 40.69(7)°. The C(2)-N(1) distance is 1.460(2) Å and the C-S distances range from 1.776(2) to 1.786(2) Å. Close, non-bonded interactions of 2.733(2) and 2.748(2) Å are observed between the S atoms and the O atoms.

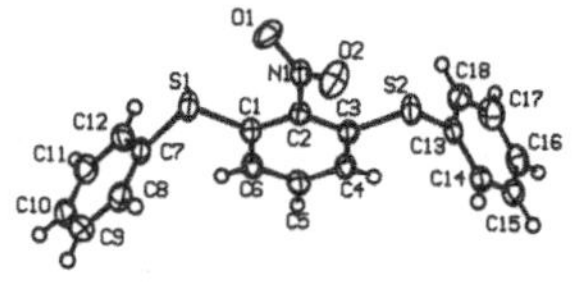

Fig. 1. The $C_{18}H_{13}NO_2S_2$ molecule.

p-CHLOROANILINE HYDROCHLORIDE
$C_6H_7Cl_2N$ (I) $C_6H_7ClN^+,Cl^-$

p-CHLOROANILINE
C_6H_6ClN (II)

G. PLOUG-SORENSEN and E.K. ANDERSEN, 1985. *Acta Cryst.*, **C41**, 613-615.

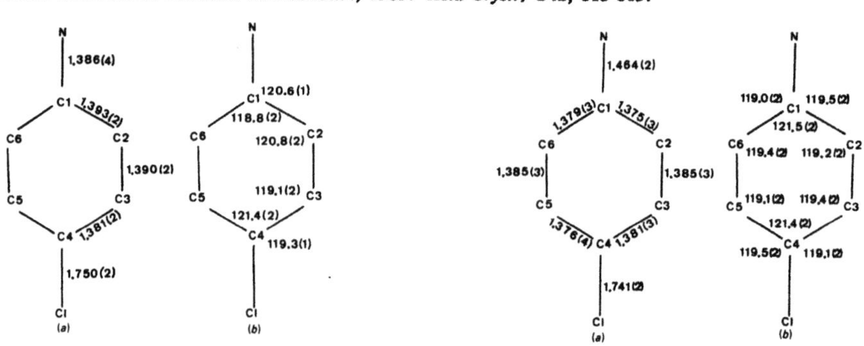

Fig. 1. Dimensions in p-chloroaniline (left) and in the p-chloroanilinium ion (right).

I. Monoclinic, $P2_1/c$, a = 8.728, b = 9.649, c = 9.823 Å, β = 108.6°, Z = 4. Mo radiation, R = 0.042 for 1959 reflections.

II. Orthorhombic, *Pnma*, a = 8.668, b = 7.409, c = 9.289 Å, Z = 4. Mo radiation, R = 0.030 for 671 reflections.

The structure of (I) (Fig. 1) is composed of puckered layers of p-chloroanilinium ions held together by ionic forces and hydrogen bonds to the Cl^- ions. The NH_3^+ group acts as a donor in hydrogen bonds to three Cl^- ions (average N...Cl^- 3.136 Å). The six-membered rings of both (I) and (II) are planar and the N and Cl atoms in both lie out of the ring planes on the same side by ca. 0.02 Å. For previous work on (II), see (1).

1. Structure Reports, 31B, 69.

ACETANILIDE
C_8H_9NO

H.J. WASSERMAN, R.R. RYAN and S.P. LAYNE, 1985. *Acta Cryst.*, C**41**, 783-785.

Orthorhombic, *Pbca*, a = 19.509, b = 9.364, c = 7.778 Å, Z = 8. Mo radiation, R = 0.033 for 1301 reflections (at 113 K).

In this low-temperature redetermination of the structure of the molecule (Fig. 1) the only significant change in the molecular geometry from that reported previously (1) is in the lengthening of the C(7)-O distance by 0.015 Å. The asymmetry of the angles at the nitrogen atom- C(1)-N-H(1) 115(1), C(7)-N-H(1) 117(1) and C(1)-N-C(7) 127.6(1)°- is due to steric hindrance between the phenyl and acetyl groups. The length of the linear intermolecular hydrogen bond N-H...O is 2.914(3) Å, some 0.029 Å less than the room-temperature value.

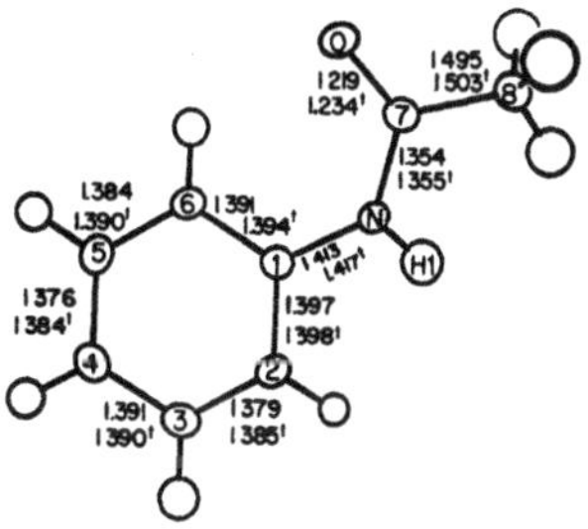

Fig. 1. C_8H_9NO: a view of the molecule in which the upper values of the bond lengths correspond to the room-temperature study.

1. Structure Reports, **31B**, 70; **18**, 705.

(E)-2,3-BUTANEDIONE-2'-NITROPHENYLHYDRAZONE
$C_{10}H_{11}N_3O_3$ (I)

(E,E)-2,3-BUTANEDIONEBIS(2'-NITROPHENYLHYDRAZONE)
$C_{16}H_{16}N_6O_4$ (II)

(E)-2,3-BUTANEDIONE-2',4'-DINITROPHENYLHYDRAZONE
$C_{10}H_{10}N_4O_5$ (III)

G.R. WILLEY and M.G.B. DREW, 1985. *Acta Cryst.*, C**41**, 589-594.

I. Orthorhombic, *Pmnb*, a = 6.730, b = 11.305, c = 14.160 Å, Z = 4. Mo radiation, R = 0.061 for 624 reflections.

II. Monoclinic, $P2_1/n$, a = 6.078, b = 5.122, c = 25.800 Å, β = 95.6°, Z = 2. Mo radiation, R = 0.091 for 488 reflections.

III. Monoclinic, $P2_1/c$, a = 10.013, b = 11.630, c = 10.529 Å, β = 108.0°, Z = 4. Mo radiation, R = 0.065 for 1377 reflections.

Molecules of (II) lie on a crystallographic inversion centre. All three structures (Fig. 1) exhibit the E form. Thus in (I) and (II) carbonyl O atom O(21) is not hydrogen bonded to H(7). All three structures contain a strong N(7)-H(7)...O(1) intramolecular hydrogen bond (N...O 2.62(1), 2.61(1) and 2.64(1) Å for (I), (II) and (III) respectively). Bond lengths and angles follow normal patterns in all three compounds. All the molecules are slightly distorted from planarity there being small twists of the o- and p-nitro groups with respect to the benzene ring planes. All intermolecular contacts are van der Waals in nature.

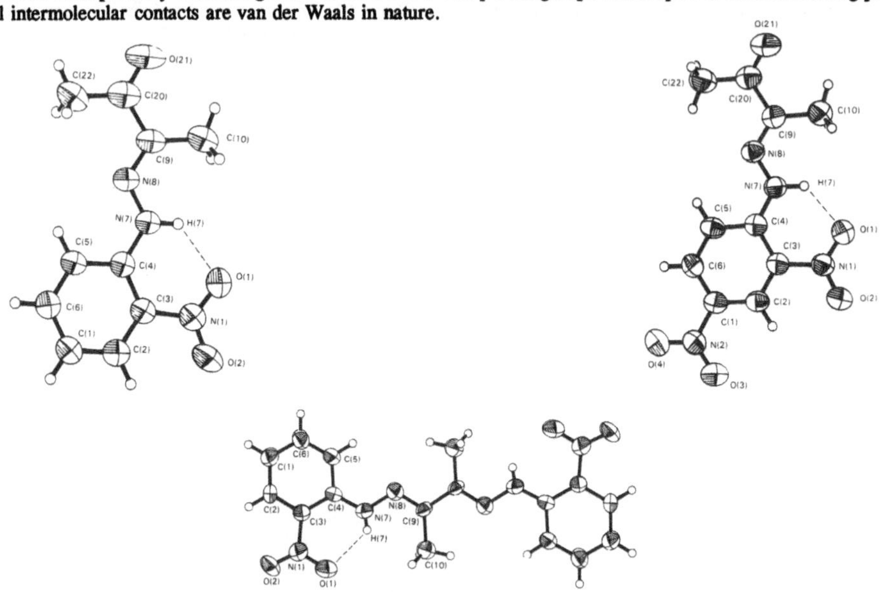

Fig. 1. The molecular structures: $C_{10}H_{11}N_3O_3$ (top left), $C_{10}H_{10}N_4O_5$ (top right) and $C_{16}H_{16}N_6O_4$ (bottom).

p-DIETHYLAMINOBENZENEDIAZONIUM HEXAFLUOROPHOSPHATE

$C_{10}H_{14}F_6N_3P$ $C_{10}H_{14}N_3{}^+,F_6P^-$

R.G. BALL and R.M. ELOFSON, 1985. *Canad. J. Chem.*, **63**, 332-335.

Triclinic, $P\bar{1}$, a = 12.105, b = 12.340, c = 10.439 Å, α = 96.53, β = 104.11, γ = 64.44°, Z = 4. Mo radiation, R = 0.054 for 1917 reflections (at -70°C).

The two independent cations (Fig. 1) have similar dimensions with mean N-N 1.096(6), mean C-N 1.357(7) Å. The cation dimensions are consistent with a major contribution from a quinoidal resonance form in which the N≡N bond is essentially intact.

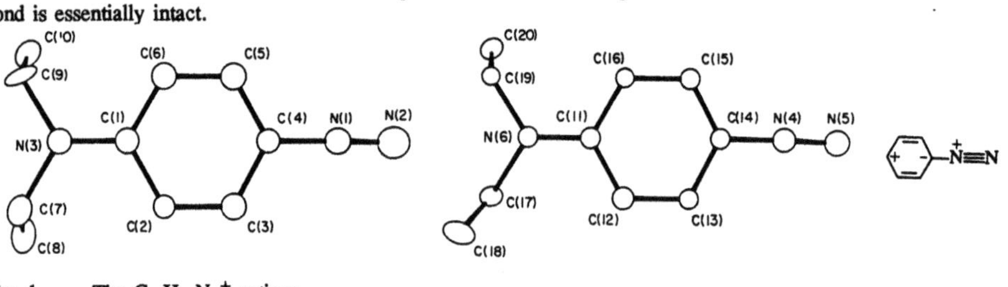

Fig. 1. The $C_{10}H_{14}N_3{}^+$ cations.

4-DIMETHYLAMINO-3-NITROCINNAMIC ACID

$C_{11}H_{12}N_2O_4$ (I)

1-(4-DIMETHYLAMINOCINNAMOYL)IMIDAZOLE

$C_{14}H_{15}N_3O$ (II)

C.P. HUBER, 1985. *Acta Cryst.*, **C41**, 1076-1079.

I. Monoclinic, $P2_1/c$, a = 18.213, b = 7.111, c = 8.736 Å, β = 93.88°, Z = 4. Cu radiation, R = 0.047 for 1697 reflections.

II. Monoclinic, $P2_1/c$, a = 8.375, b = 6.565, c = 23.783 Å, β = 101.81°, Z = 4. Cu radiation, R = 0.050 for 1861 reflections.

The molecule in (I) (Fig. 1) is significantly non-planar with the nitro and dimethylamino groups twisted by 45.3(2) and 20.3(2)° out of the mean plane of the benzene ring. This ring is itself warped due to steric hindrance between the adjacent p-nitro and dimethylamino substituents. The acid side chain is planar. The molecule in (II) is much more nearly planar. The imidazole ring is planar and makes an angle of 8.47(3)° with the least-squares plane through the remaining non-hydrogen atoms. Bond lengths in (II) show a more quininoid character than those in (I) with a significant lengthening and shortening of the ethylenic double and adjacent single bonds, respectively. Molecules in (I) form hydrogen-bonded dimers in which the carboxyl groups are disordered. In (II) there is a C(15)-H...N(16) interaction in which the H...N contact is of length 2.40(2) Å. All other intermolecular distances in both crystals are consistent with van der Waals contacts.

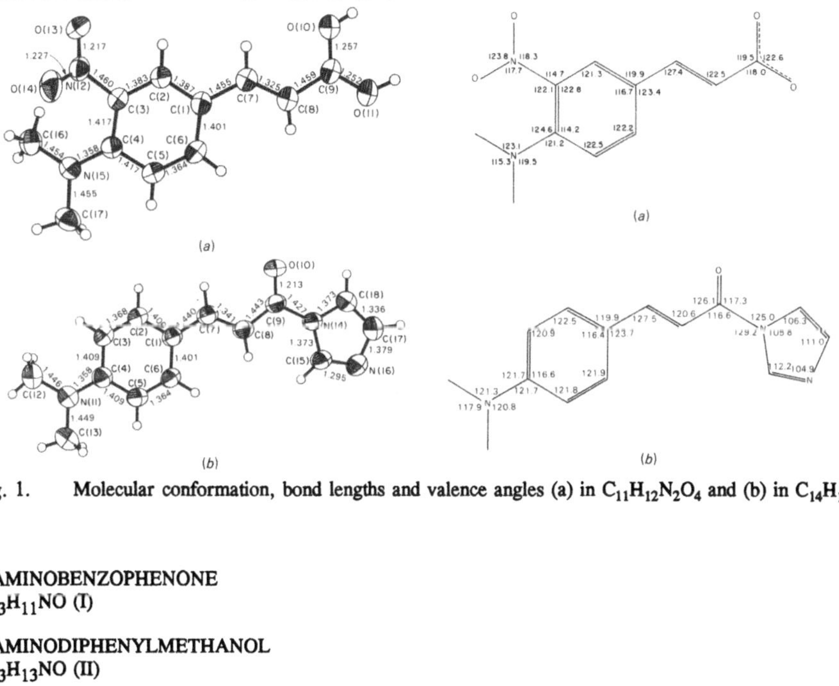

Fig. 1. Molecular conformation, bond lengths and valence angles (a) in $C_{11}H_{12}N_2O_4$ and (b) in $C_{14}H_{15}N_3O$.

2-AMINOBENZOPHENONE
$C_{13}H_{11}NO$ (I)

2-AMINODIPHENYLMETHANOL
$C_{13}H_{13}NO$ (II)

L. ANTOLINI, I.M. VEZZOSI, L.P. BATTAGLIA and A.B. CORRADI, 1985. *J. Chem. Soc. Perkin Trans. 2*, 237-239.

I. Monoclinic, $P2_1/c$, a = 10.247, b = 10.985, c = 9.113 Å, β = 99.03°, Z = 4. Mo radiation, R = 0.042 for 959 reflections.

II. Rhombohedral, $R\bar{3}$, a = 13.418 Å, σ = 113.23°, Z = 6. Mo radiation, R = 0.064 for 916 reflections.

In (I) the planes of the phenyl rings make dihedral angles of 19.1 and 56.1° with the planar carbonyl group and are inclined at 67.6° to each other. The molecular packing (Fig. 1) is due to intramolecular and intermolecular N-H...O hydrogen bonds which produce a linear chain structure. In (II) the molecules are linked into polymeric units by strong hydrogen bonds between the OH functions.

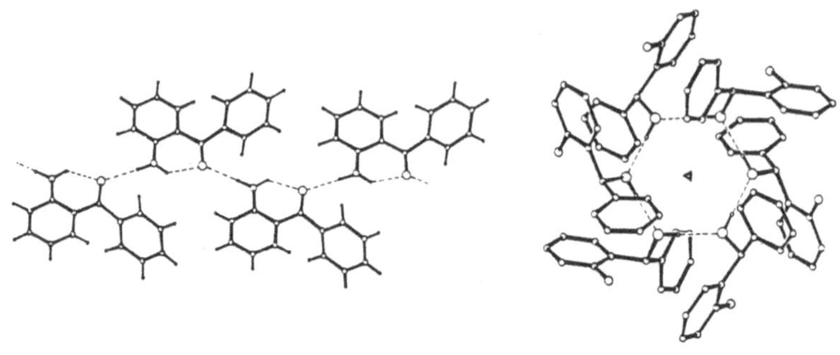

Fig. 1. Molecular packing in crystals of $C_{13}H_{11}NO$ viewed down the c-axis, and the arrangement of the $C_{13}H_{13}NO$ molecules around the $\bar{3}$. In each case hydrogen bonds are shown by broken lines.

(E)-BENZALDEHYDE PHENYLHYDRAZONE
$C_{13}H_{12}N_2$ (I)

(E)-BENZALDEHYDE 4'-NITROPHENYLHYDRAZONE
$C_{13}H_{11}N_3O_2$ (II)

B. VICKERY, G.R. WILLEY and M.G.B. DREW, 1985. *Acta Cryst.*, C41, 1072-1075.

I. Monoclinic, $P2_1/c$, a = 6.053, b = 17.643, c = 15.207 Å, β = 92.2°, Z = 6. Mo radiation, R = 0.077 for 1069 reflections.

II. Monoclinic, Cc, a = 6.049, b = 23.320, c = 8.506 Å, β = 96.80°, Z = 4. Mo radiation, R = 0.071 for 789 reflections.

The asymmetric unit of (I) consists of a molecule (Fig. 1) in a general position and a molecule disordered around a crystallographic inversion centre. Both molecules and that in (II) take up an E conformation. In the ordered molecule of (I) the dihedral angle between the two benzene ring planes is 8.1(1)° and the N(7)=N(8) and N(8)-C(9) bond lengths are 1.330(8) and 1.312(9) Å respectively. In (II) these bond lengths are 1.358(8) and 1.278(8) Å. In the crystal there is weak intermolecular hydrogen bonding between the p-nitro group and the NH group with N(7)-H...O(1) of length 3.10(1) Å. There appears also be a weak C(3)-H...O(2) interaction of length 3.40(1) Å. There is no hydrogen bonding in the crystal structure of (I) where molecular packing is governed by van der Waals forces.

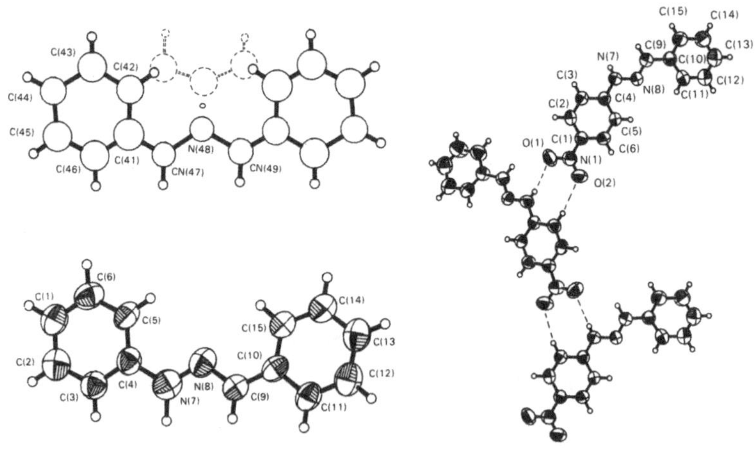

Fig. 1. $C_{13}H_{12}N_2$: views of the disordered molecule (upper) and of the ordered molecule (lower) and a view of the structure of $C_{13}H_{11}N_3O_2$ with weak intermolecular interactions shown as dotted lines.

p-NITROPHENACYLANILINE
$C_{14}H_{12}N_2O_3$

R.F. ABDULLA, D.B. BOYD, N.D. JONES and J.K. SWARTZENDRUBER, 1985. *J. Org. Chem.*, **50**, 3502-3505.

Monoclinic, $P2_1/c$, a = 6.130, b = 13.301, c = 14.992 Å, β = 93.45°, Z = 4. Cu radiation, R = 0.065 for 1886 reflections.

The molecule is almost planar (Fig. 1). The crystal packing shows stacks of coplanar molecules, spaced 3.54 Å apart. Adjacent p-nitrophenyl and aniline rings are interleaved in a continuous, alternating stack of paired molecules (Fig. 1).

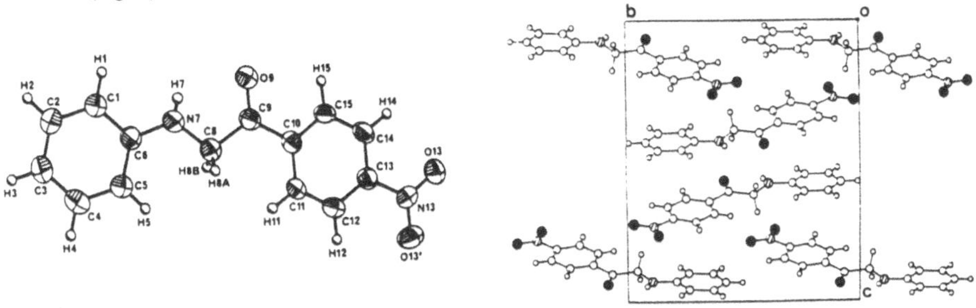

Fig. 1. ORTEP plot of a molecule of $C_{14}H_{12}N_2O_3$ with crystal packing, viewed down the *a* axis.

α-(2-NITROBENZENEHYDRAZONO)-α-ACETYLACETANILIDE
$C_{16}H_{14}N_4O_4$

A. WHITAKER, 1985. *Z. Kristallogr.*, **171**, 17-22.

Monoclinic, $P2_1/n$, a = 7.593, b = 20.029, c = 10.217 Å, β = 101.87°, Z = 4. Cu radiation, R = 0.066 for 2547 reflections.

The molecule exists as the hydrazone tautomer and intramolecular hydrogen bonds keep the molecule approximately planar (Fig. 1). Molecules are packed in layers approximately parallel to the ($\bar{1}01$) plane and in columns parallel to the *a*-axis. Alternate molecules within the columns are antiparallel.

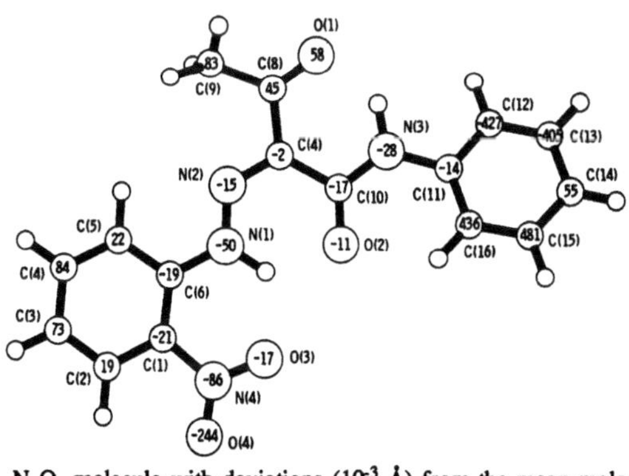

Fig. 1. The $C_{16}H_{14}N_4O_4$ molecule with deviations (10^{-3} Å) from the mean molecular plane.

α-(4-NITROBENZENEHYDRAZONO)-α-ACETYLACETANILIDE
$C_{16}H_{14}N_4O_4$

A. WHITAKER and N. WALKER, 1985. *Z. Kristallogr.*, **171**, 7-15.

Triclinic, $P\bar{1}$, a = 7.691, b = 10.247, c = 11.195 Å, α = 109.00, β = 73.80, γ = 112.21°, Z = 2. Cu radiation, R = 0.054 for 2716 reflections.

The molecule exists as the hydrazone tautomer (Fig. 1) and is approximately planar due partly to intramolecular hydrogen bonds and partly to steric hindrance.

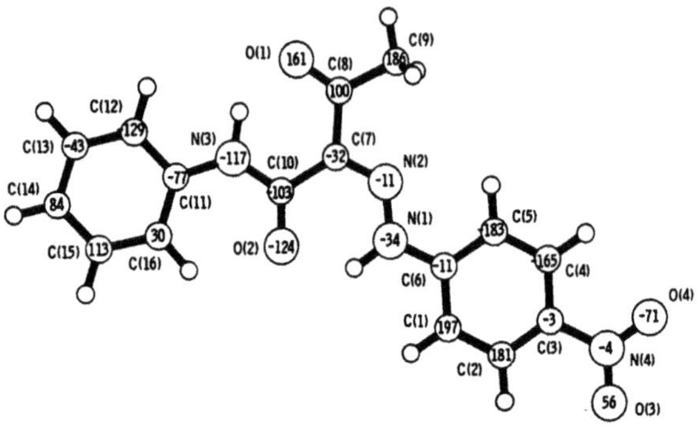

Fig. 1. A view of $C_{16}H_{14}N_4O_4$ with deviations (10^{-3} Å) from the molecular plane.

(1-N-(ACETYL(N-PHENYLCARBAMOYL)METHYLENE)AZO)-4-METHYL-2-NITROBENZENE $C_{17}H_{16}N_4O_4$

A. WHITAKER, 1985. *Z. Kristallogr.*, **170**, 213-223.

Monoclinic, $P2_1/n$, a = 7.567, b = 20.420, c = 10.345 Å, β = 99.29°, Z = 4. Cu radiation, R = 0.077 for 2291 reflections.

The molecule exists as the hydrazone tautomer and intramolecular hydrogen bonds keep it approximately planar (Fig. 1). The crystal structure is a mixture of two pigments (in one case a methyl group on C(3) and in the other a hydrogen atom).

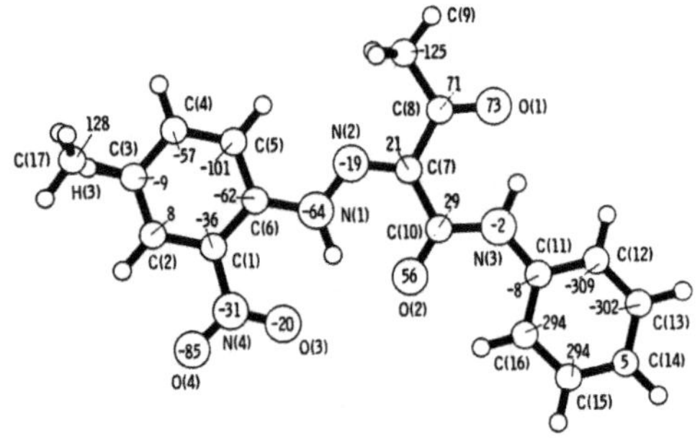

Fig. 1. A view of "$C_{17}H_{16}N_4O_4$" with deviations (10^{-3} Å) from the best plane.

N-PHENYL-2-(2-METHYLPHENYLHYDRAZONE)-3-OXO-BUTYRAMIDE $C_{17}H_{17}N_3O_2$

E.F. PAULUS and W. RIEPER, 1985. *Z. Kristallogr.*, **171**, 87-100.

Monoclinic, *I2/c*, a = 20.962, b = 11.623, c = 26.881 Å, β = 110.80°, Z = 16. Mo radiation, R = 0.036 for 3115 reflections.

There are two molecules in the asymmetric unit with different conformations (Fig. 1). In one molecule the methylphenyl group is twisted out of the hydrazono group by 1.7° whereas in the other molecule, the corresponding angle is 11.4°. The N-H groups take part in intramolecular N-H...O hydrogen bonds. The molecules are found in the hydrazone tautomers. In the crystal lattice the molecules lie in layers with distances between 3.27 and 3.44 Å.

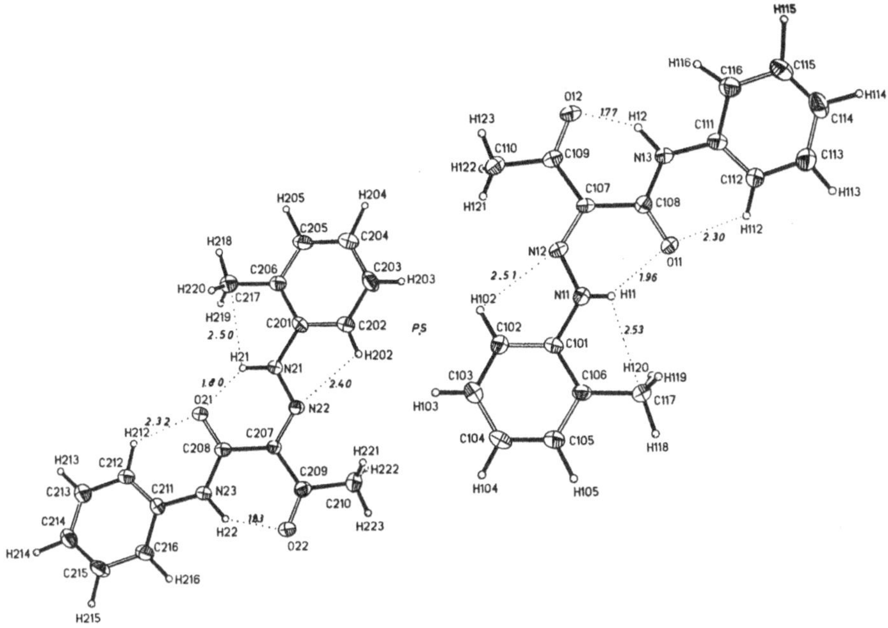

Fig. 1. The two independent $C_{17}H_{17}N_3O_2$ molecules.

TRIPHENYLAMINE
$C_{18}H_{15}N$

A.N. SOBOLEV, V.K. BELSKY, I.P. ROMM, N.Yu. CHERNIKOVA and E.N. GURYANOVA, 1985. *Acta Cryst.*, **C41**, 967-971.

Monoclinic, *Bb*, a = 15.655, b = 22.257, c = 15.807 Å, γ = 91.04°, Z = 16. Mo radiation, R = 0.036 for 2902 reflections.

Four independent molecules in the crystal (Fig. 1) are related pairwise by supersymmetry elements. The value of the CNC bond angle in the four molecules varies from 118.5 to 120.8° and the N-C bond lengths are in the range 1.408 to 1.427 Å. The N atom deviates from the plane defined by the three C atoms to which it is attached (mean 0.08(4) Å) and the phenyl rings are all planar. The dihedral angles between phenyl-ring planes and the plane of the N-bonded C atoms vary from 37.0 to 50.5°.

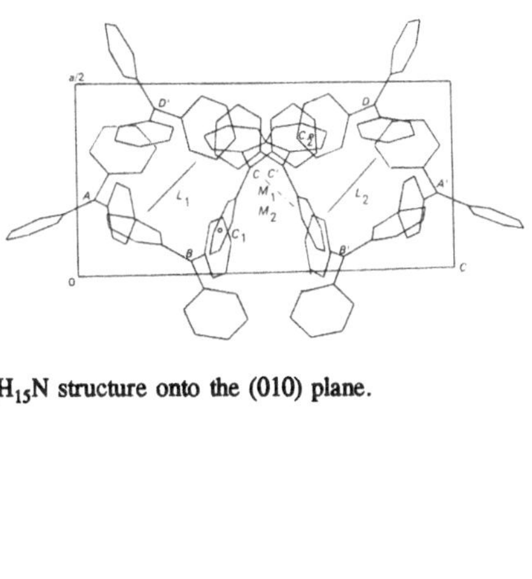

Fig. 1. Projection of the $C_{18}H_{15}N$ structure onto the (010) plane.

N-ACETYL-4,N-DIMETHYL-6-(N-ACETYL-p-TOLUIDINOMETHYL)ANILINE
$C_{20}H_{24}N_2O_2$

P. CHAKRABARTI, K. VENKATESAN, T.S. CAMERON, T. DEMIR and R.A. SHAW, 1985. *J. Cryst. Spect. Res.*, **15**, 229-245.

Triclinic, $P\bar{1}$, a = 12.999, b = 8.911, c = 9.199 Å, α = 105.56, β = 113.50, γ = 92.53°, Z = 2. Mo radiation, R = 0.066 for 2585 reflections.

The analysis provides data on the conformation of the amine (Fig. 1) when there is no intermolecular hydrogen-bonding. The amide moieties are disordered corresponding to rotation of 180° about the C(1')...(O1') vector, and atoms N(1') and ND(1') are -0.35 and +0.71 Å from their benzene ring plane. The molecule adopts an exo conformation and the angles between the amide planes and the benzene ring to which they are attached are 80° and 83° (for the ordered and disordered amide groups respectively) showing little amide-ring resonance. The C(8)-N(1) distance is 1.363(4) Å, longer than normally found in peptides.

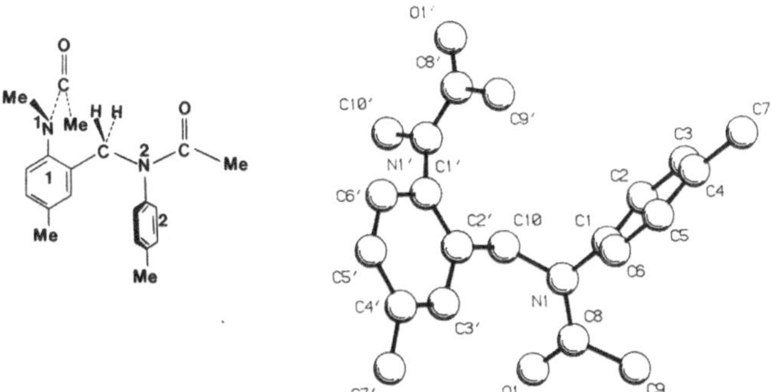

Fig. 1. A view of the $C_{20}H_{24}N_2O_2$ molecule.

2,3-DIPHENYL-1-(p-TOLYL)-1-AZA-1,3-PENTADIENE
$C_{23}H_{21}N$

C. RUIZ-VALERO, E. GUTIERREZ-PUEBLA and A. MONGE, 1985. *Acta Cryst.*, **C41**, 1791-1792.

Triclinic, $P\bar{1}$, a = 9.927, b = 10.143, c = 10.487 Å, α = 63.29, β = 96.26, γ = 108.97°, Z = 2. Mo radiation, R = 0.056 for 2095 reflections.

The two phenyl groups of the molecule (Fig. 1) are synclinal with respect to the C(1)-C(2) bond (1.486(4) Å). In the double-bond system, the two double bonds are oriented synclinal about C(1)-C(2). The C(1)-N and C(2)-C(3) bond lengths are 1.280(4) and 1.327(6) Å respectively. Packing is governed by van der Waals interactions.

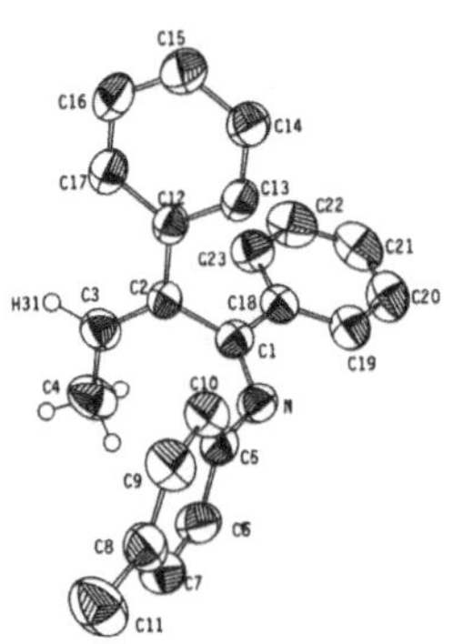

Fig. 1. The $C_{23}H_{21}N$ molecule.

2-ANILINO-N,N'-DI-t-BUTYL-2-(N-PHENYLSULPHENAMOYL)MALONAMIDE
$C_{23}H_{32}N_4O_2S$

A. KALMAN, G. ARGAY and E. FISCHER, 1985. *Acta Cryst.*, **C41**, 1648-1650.

Monoclinic, $P2_1/n$, a = 13.814, b = 10.354, c = 16.980 Å, β = 96.22°, Z = 4. Cu radiation, R = 0.051 for 4416 reflections.

In the molecule (Fig. 1) the prochiral atom C(2) sits at the centre of a distorted tetrahedron, the smallest angle of which C(1)-C(2)-S(1) is 104.5(3)°. The S(1)-C(2) bond is elongated to 1.904(3) Å, probably due to the high electrophilicity of C(2). The C(2)-N(3) bond, of length 1.430(3) Å, is significantly shorter than the normal single-bond length. There is a close intramolecular S(1)...O(1) contact of length 2.819(2) Å, considerably shorter than the normal van der Waals separation of 3.25 Å. There is an associated shortening of the length of the S(1)-N(4) bond in the N-phenylsulphenamoyl group from a normal value of about 1.735 to 1.656(2) Å. The observed molecular conformation is stabilized by N-H...O and N-H...N intramolecular hydrogen bonds. Intermolecular distances are consistent with essentially van der Waals interactions.

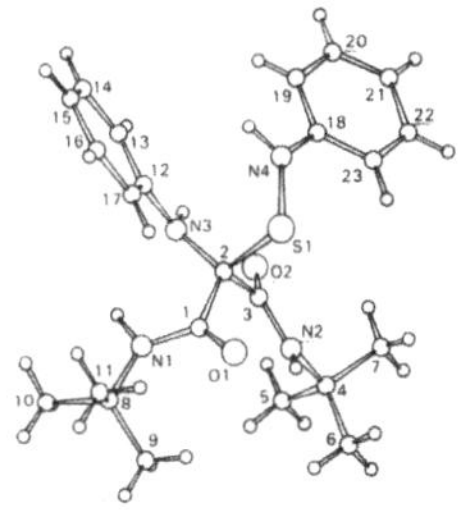

Fig. 1. A view of the $C_{23}H_{32}N_4O_2S$ molecule.

POLY-1,2-BIS(4-(p-BROMOPHENYLCARBAMOYLOXY)-n-BUTYL)BUT-1-EN-3-YNYLENE
$(C_{26}H_{26}Br_2N_2O_4)_n$

P. SPINAT, C. BROUTY, A. WHULER and M.-C. SICHERE, 1985. *Acta Cryst.*, **C41**, 1452-1455.

Triclinic, $P\bar{1}$, a = 5.670, b = 22.47, c = 4.864 Å, α = 97.40, β = 98.98, γ = 83.26°, Z = 1. Cu radiation, R = 0.080 for 1118 reflections (at 205 K).

The molecule (Fig. 1) is centrosymmetric and exhibits all the characteristics of an acetylenic-type structure with C(1)-C(1') 1.16(1) Å. With the exception of the phenyl group the polymer chain is planar and lies parallel to the $\bar{1}40$ plane. The phenyl group is rotated 41° about C(P1)...Br with respect to the polymer chain. The polymer has a repeat distance of 4.864 Å along *c* and is oriented by 14.5(3)° to the translation vector.

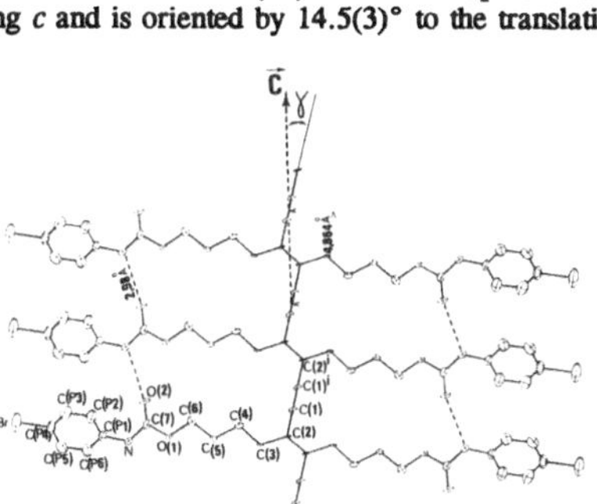

Fig. 1. $(C_{26}H_{26}Br_2N_2O_4)_n$: the $\bar{1}40$ plane of the molecule.

4-AMINO-2-HYDROXY-ω-DICHLOROACETOPHENONE
$C_8H_7Cl_2NO_2$

R.R. BANDYOPADHYAY, S.C. KUNDADAS and B.S. BASAK, 1985. *Z. Kristallogr.*, **171**, 195-200.

Monoclinic, $P2_1/a$, a = 7.458, b = 9.109, c = 13.400 Å, β = 99.86°, Z = 4. Mo radiation, R = 0.044 for 1181 reflections.

Molecular dimensions are shown in Fig. 1. There is an intramolecular O-H...O hydrogen bond.

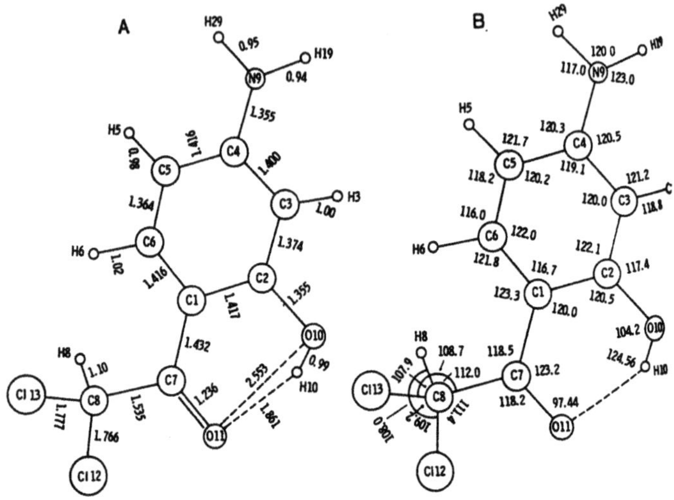

Fig. 1. Molecular dimensions for $C_8H_7Cl_2NO_2$.

(2-NITROPHENOXY)ETHANOIC ACID
$C_8H_7NO_5$ (I)

catena-(BIS(μ_2-2-NITROPHENOXYETHANOATO-O,O')DIAQUAMANGANESE(II))
$(C_{16}H_{16}MnN_2O_{12})_n$ (II)

BIS(2-NITROPHENOXYETHANOATO-O)TETRAAQUACOPPER(II) DIHYDRATE
$C_{16}H_{20}CuN_2O_{14},2(H_2O)$ (III)

BIS(2-AMINOPYRIMIDINE)AQUABIS(2-NITROPHENOXYETHANOATO-O,)COPPER(II)
$C_{24}H_{24}CuN_8O_{11}$ (IV)

C.H.L. KENNARD, S.W. STEWART, E.J. O'REILLY, G. SMITH and A.H. WHITE, 1985. *Polyhedron*, **4**, 697-705.

I. Monoclinic, $C2/c$, a = 11.496, b = 7.783, c = 19.577 Å, β = 98.35°, Z = 8. Mo radiation, R = 0.085 for 774 reflections.

II. Triclinic, $P\bar{1}$, a = 14.319, b = 7.912, c = 4.558 Å, α = 114.69, β = 90.47, γ = 90.36°, Z = 1. Mo radiation, R = 0.049 for 1430 reflections.

III. Monoclinic, $P2_1/c$, a = 6.349, b = 5.117, c = 35.820 Å, β = 99.04°, Z = 2. Mo radiation, R = 0.062 for 927 reflections.

IV. Monoclinic, $C2/c$, a = 25.418, b = 5.718, c = 18.785 Å, β = 99.73°, Z = 4. Mo radiation, R = 0.049 for 997 reflections.

Molecules of (I) are essentially planar, but with the NO_2 group rotated 35° from the C_6 ring plane. They are present as centrosymmetric, hydrogen-bonded pairs (O...O 2.606(8) Å). The Mn atom in (II) (Fig. 1) lies on an inversion centre. The MnO_6 octahedron is made up of two H_2O ligands together with one O atom from each of four carboxylate groups which thus bridge the Mn atoms. Bond lengths include: Mn-OH_2 2.213(3), Mn-O 2.156, 2.172(3) Å. The Cu atom in (III) lies on an inversion centre and has tetragonally-distorted octahedral coordination, with two equatorial carboxylate O atoms (Cu-O 1.943(5) Å) and two equatorial and two axial H_2O (Cu-O 1.954, 2.578(7) Å respectively). The axial ligand H_2O is hydrogen-bonded to the lattice H_2O (O...O 2.66 Å). The Cu atom in (IV), on a crystal diad axis, has square-pyramidal coordination geometry with an axial H_2O ligand (Cu-O 2.211(7) Å); two pyrimidine N atoms (Cu-N 2.078(5) Å) and two carboxylate O atoms (Cu-O 1.947(5) Å) lie in the basal plane.

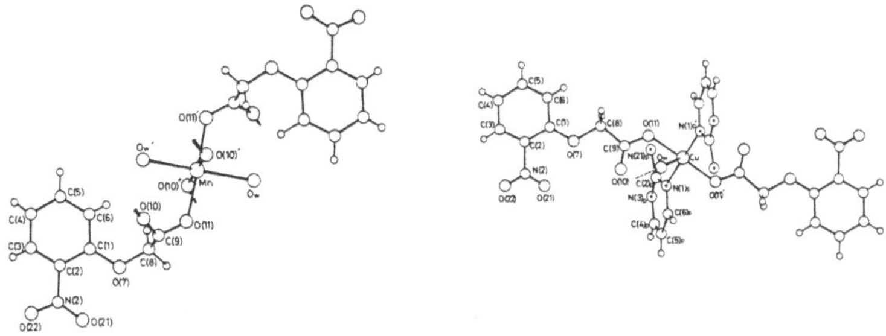

Fig. 1. The $(C_{16}H_{16}MnN_2O_{12})_n$ molecule (left) and the $C_{24}H_{24}CuN_8O_{11}$ molecule (right).

1-METHOXYMETHOXY-3,5-DINITROBENZENE
$C_8H_8N_2O_6$

P.G. JONES, G.M. SHELDRICK, A.J. KIRBY and A.J. BRIGGS, 1985. *Acta Cryst.*, C**41**, 1377-1379.

Orthorhombic, $P2_12_12_1$, a = 5.387, b = 11.053, c = 16.783 Å, Z = 4. Mo radiation, R = 0.040 for 1654 reflections.

The conformation about the acetal centre in the molecule (Fig. 1) is close to (-sc,-sc) with torsion angles C(12)-O(2)-C(11)-O(1) and O(2)-C(11)-O(1)-C(1) of magnitude -70.4(4) and -64.1(4)° respectively. The C(11)-O(1) bond is of length 1.431(3) Å. The aromatic ring is planar and its immediate substituents are essentially coplanar. The N atoms have trigonal planar geometries and the bond lengths and angles are in good agreement with generally acceptable values. No unusually short intermolecular contacts are reported.

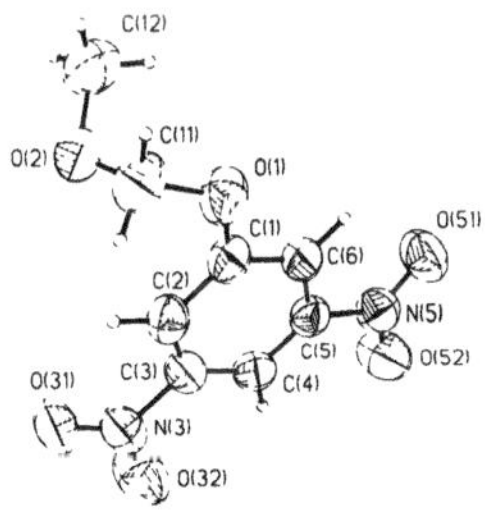

Fig. 1. $C_8H_8N_2O_6$: a view of the molecule showing the atom-numbering scheme.

4-(2-AMINOETHYL)-2-HYDROXYPHENYL HYDROGEN SULPHATE (DOPAMINE-4-O-SULPHATE)
$C_8H_{11}NO_5S$ (I)

5-(2-AMINOETHYL)-2-HYDROXYPHENYL HYDROGEN SULPHATE (DOPAMINE-3-O-SULPHATE)
$C_8H_{11}NO_5S$ (II)

D.S. EGGLESTON, D.F. CHODOSH, T. JAIN, C. KAISER and D.M. ACKERMAN, 1985. *Acta Cryst.*, C**41**, 76-82.

I. Monoclinic, $P2_1/n$, a = 9.866, b = 10.454, c = 19.799 Å, β = 95.78°, Z = 8. Mo radiation, R = 0.041 for 2799 reflections.

II. Monoclinic, $P2_1/c$, a = 8.706, b = 12.749, c = 9.214 Å, β = 102.97°, Z = 4. Mo radiation, R = 0.043 for 1530 reflections.

Both dopamine sulphate molecules (Fig. 1) crystallize as zwitterions with the ethylamine side chains protonated and the sulphate groups ionised. The two crystallographically independent molecules of (I) have essentially identical geometries and are conformational isomers. Bond lengths and angles in these isomers are in good agreement with generally accepted values. The aromatic rings are planar and in one of these isomers the ethylamine side chain atoms and C(4) are coplanar while in the other they are not. The torsion angles C(3)-C(4)-C(7)-C(8) for the molecules in (I) are 83.8 and -47.4°. The crystal structure of (I) is stabilized by an extensive three-dimensional network of hydrogen bonds. In addition to this there is an intramolecular hydrogen bond between the hydroxyl and sulphate groups in each molecule. Bond lengths and angles in (II) are very similar to those in the molecules of (I). The ethylamine side chain in (II) is not planar and the torsion angle C(3)-C(4)-C(7)-C(8) is 62.9°. A hydrogen-bonding network stabilizes the crystal structure of (II).

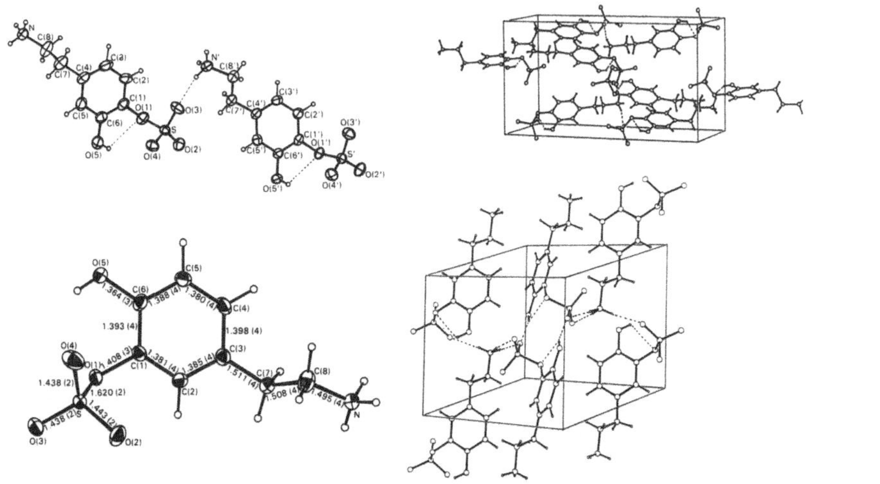

Fig. 1. $C_8H_{11}NO_5S$: (Upper) the numbering scheme in the crystallographically independent molecules of (I) (left) and (right) a view of the unit cell contents of (I), and (lower) the geometry of (II) (left) and (right) a view of the unit cell contents of (II).

2,2-DICHLORO-2'-HYDROXY-4'-METHOXYACETOPHENONE
$C_9H_8Cl_2O_3$

D. CHATTOPADHYAY, T. BANERJEE, S.K. MAZUMDAR and G. PODDER, 1985. *Acta Cryst.*, C41, 287-289.

Orthorhombic, $P2_12_12_1$, a = 6.981, b = 11.901, c = 12.061 Å, Z = 4. Mo radiation, R = 0.038 for 946 reflections.

The molecule (Fig. 1) is approximately planar and is in the endo conformation. An intramolecular O-H...O (keto) hydrogen bond (O(2)...O(1) 2.589(6) Å) stabilises the conformation. The π-acceptor and π-donor substituents take part in conjugation with the phenyl rings. The C-Cl distances in the dichloroacetyl group are 1.798(7) and 1.750(6) Å. Packing is governed by van der Waals forces.

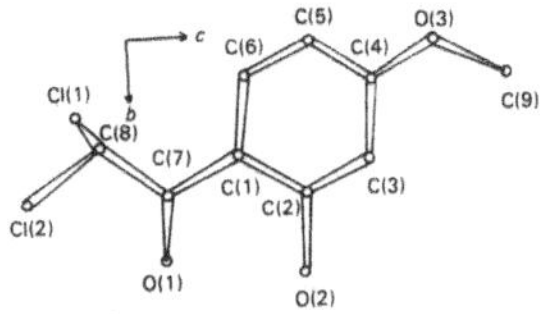

Fig. 1. $C_9H_8Cl_2O_3$: a view of the molecule.

2-FORMYLPHENOXYACETIC ACID
$C_9H_8O_4$ (I)

TETRAAQUABIS(2-FORMYLPHENOXYACETATO-O)COBALT(II)
$C_{18}H_{22}CoO_{12}$ (II)

TETRAAQUABIS(2-FORMYLPHENOXYACETATO-O)ZINC(II)
$C_{18}H_{22}O_{12}Zn$ (III)

C.H.L. KENNARD, E.J. O'REILLY, G. SMITH and T.C.W. MAK, 1985. *Aust. J. Chem.*, **38**, 1381-1389.

I. Monoclinic, $P2_1/n$, a = 8.066, b = 7.375, c = 14.206 Å, β = 98.12°, Z = 4. Mo radiation, R = 0.039 for 1197 reflections.

II. Monoclinic, $P2_1$, a = 5.036, b = 13.463, c = 15.033 Å, β = 96.88°, Z = 2. Mo radiation, R = 0.058 for 1178 reflections.

III. Monoclinic, $P2_1$, a = 5.044, b = 13.481, c = 15.055 Å, β = 96.95°, Z = 2. Mo radiation, R = 0.045 for 1230 reflections.

The planar molecules in (I) (Fig. 1) form unusual hydrogen-bonded chain polymers (O...O 2.694(3) Å) linked through carboxyl groups. Complexes (II) and (III) are isomorphous. The metal atoms have octahedral geometry, being coordinated to four water molecules (2.102-2.188(8) (II), 2.113-2.173(8) (III)) and two trans-related unidentate formylphenoxyacetate ligands (2.09(1), 2.088(8) (II) and 2.046(8), 2.113(7) Å (III)). The two independent acetato ligands are not symmetry-related and are significantly different conformationally. The torsion angles about the O(7)-C(8) vector in the two ligands (A,B) are 86, 178° (II) and 87, 177° (III). The formyl groups retain their coplanarity with the phenoxy groups. These crystal structures are stabilised by associations of the complex monomers with adjacent monomers through the water molecules (O...O 2.99, 2.93 Å).

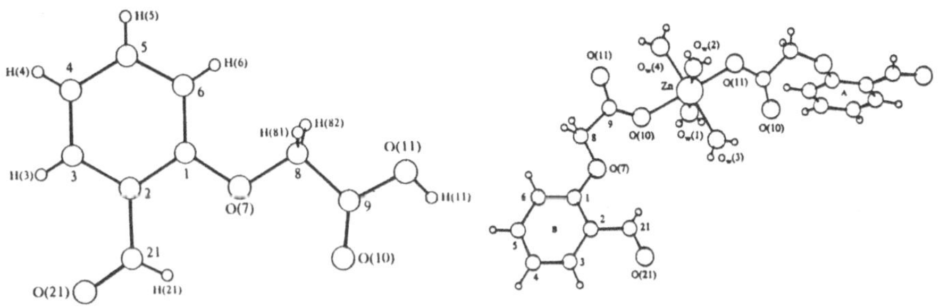

Fig. 1. The molecular structures: $C_9H_8O_4$ (left) and $C_{18}H_{22}CoO_{12}$ (right).

2-(N,N-DIMETHYLAMINOMETHYL)-4-NITROPHENOL
$C_9H_{12}N_2O_3$

A. KOLL and T. GLOWIAK, 1985. *J. Cryst. Spect. Res.*, **15**, 411-421.

Monoclinic, $C2/c$, a = 23.437, b = 5.334, c = 17.444 Å, β = 119.90°, Z = 8. Cu radiation, R = 0.032 for 1095 reflections.

The molecules exist as zwitterions which are linked about an inversion centre to form N-H...O hydrogen-bonded dimers (N...O 2.667(2) Å). The dimethylaminomethyl side chain is fully extended (torsion angle C(2)-C(7)-N(2)-C(9) -177.2(4)°). The nitro group makes an angle of 3.5° with the phenyl ring plane.

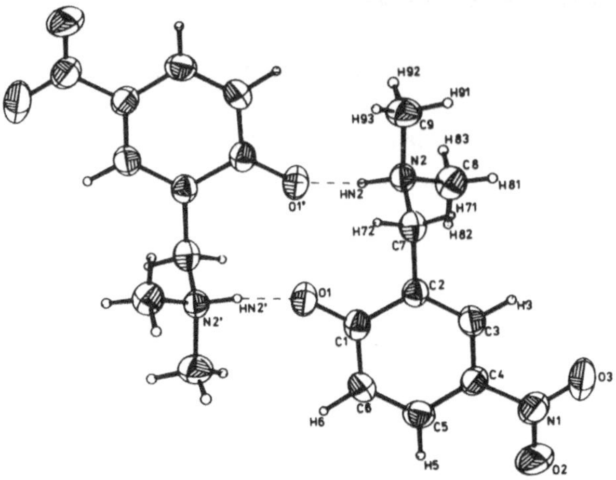

Fig. 1. The centrosymmetric H-bonded dimer formed by two $C_9H_{12}N_2O_3$ molecules.

2,6-BIS(HYDROXYMETHYL)-4-ISOPROPYLPHENOL
$C_{11}H_{16}O_3$

D. OEHLER, A. THOZET and M. PERRIN, 1985. *Acta Cryst.*, C**41**, 1766-1768.

Orthorhombic, *Pcab*, a = 12.897, b = 18.978, c = 8.497 Å, Z = 8. Cu radiation, R = 0.047 for 1207 reflections.

The two hydroxymethyl groups have different orientations with respect to the ring plane, one being nearly coplanar (-0.133(3) Å) and the other tilted (-1.253(3) Å) out of the plane (Fig. 1). The dihedral angle between the ring plane and the plane of the isopropyl group is 98.6(5)°. The C-C (aromatic) distances have a mean value of 1.393(4) Å and the benzene ring has the expected geometry. C-O bond distances have values of 1.380(3) for the phenolic substituent and average 1.430(3) Å in the hydroxymethyl groups. Molecules are held together by hydrogen bonds (O...O 2.698(4)-2.818(4) Å) forming sheets inter-related by van der Waals contacts.

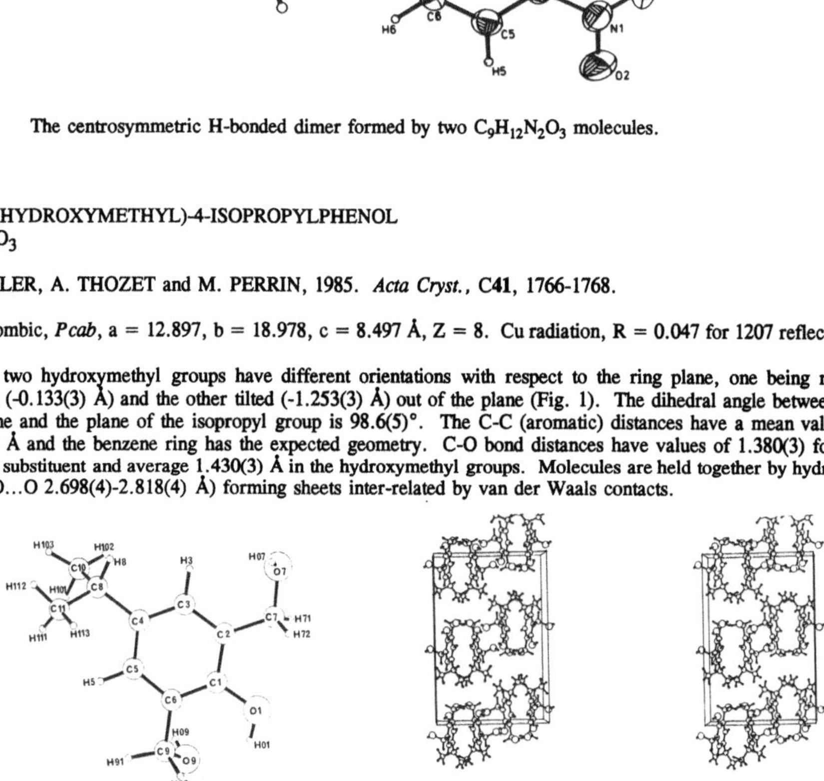

Fig. 1. $C_{11}H_{16}O_3$: the molecular structure viewed along [100] and the packing viewed along [001].

3-(o-HYDROXYPHENYL)-1-PHENYLTRIAZENE 1-OXIDE
$C_{12}H_{11}N_3O_2$

C. SAMANTA, P.K. DE, S.B. SARKAR, S.C. SAHA and S.K. TALAPATRA, 1985. *Acta Cryst.*, C**41**, 142-144.

Monoclinic, $P2_1/n$, a = 6.569, b = 11.870, c = 13.889 Å, β = 94.95°, Z = 4. Mo radiation, R = 0.040 for 1449 reflections.

The molecule is nearly planar having a dihedral angle of only 6.7(1)° between the phenyl ring planes (Fig. 1).

There is an intramolecular N(3)-H...O(1) hydrogen bond of length 2.476(2) Å closing a planar ring which makes angles of 12.8(1) and 8.3(1)° with the phenyl rings attached at N(1) and N(3) respectively. The C(9)-N(1) bond of length 1.443(3) Å is appreciably longer than the C(5)-N(3) bond (1.400(3) Å). The N(1)-N(2) distance, 1.280(2) Å, is significantly shorter than the N(2)-N(3) bond of length 1.328(3) Å, and shows appreciable double-bond character. There are also O(2)-H...O(1) intermolecular hydrogen bonds of length 2.738(2) Å stabilizing the crystal structure.

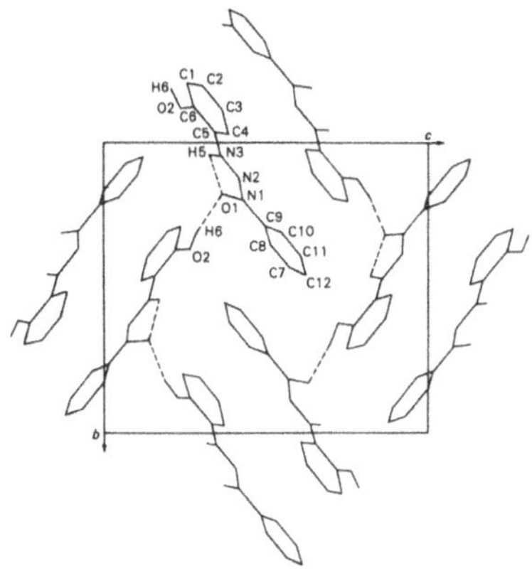

Fig. 1. $C_{12}H_{11}N_3O_2$: a projection of the structure onto (100).

(±)-2-(2-CHLORO-4,5-DIMETHOXYPHENYL)-2-HYDROXY-N-ISOPROPYLETHYLAMINE HYDROCHLORIDE

$C_{13}H_{21}Cl_2NO_3$ $C_{13}H_{21}ClNO_3{}^+,Cl^-$

J. KOORTS and M. CAIRA, 1985. *Acta Cryst.*, **C41**, 1372-1374.

Monoclinic, $P2_1/c$, a = 24.19, b = 8.782, c = 7.430 Å, β = 93.06°, Z = 4. Mo radiation, R = 0.046 for 1804 reflections.

The phenyl ring in the cation (Fig. 1) is planar and its methoxy substituents are essentially coplanar. The ethylamine side chain adopts a perpendicular-trans conformation with respect to the phenyl ring and the isopropyl group is gauche to the chiral centre of the cation. The intramolecular non-bonded N(9)...O(7) contact is of length 2.856(4) Å. The conformation about the C(8)-N(9) bond is (-)-synclinal. There are no unusual bond lengths or angles in the cation. In the crystal, each cation acts as a donor in one O-H...Cl^- bond of length 3.098(3) Å and in two N-H...Cl^- bonds with lengths 3.142(3) and 3.154(3) Å. This network of hydrogen bonds stabilizes the crystal structure.

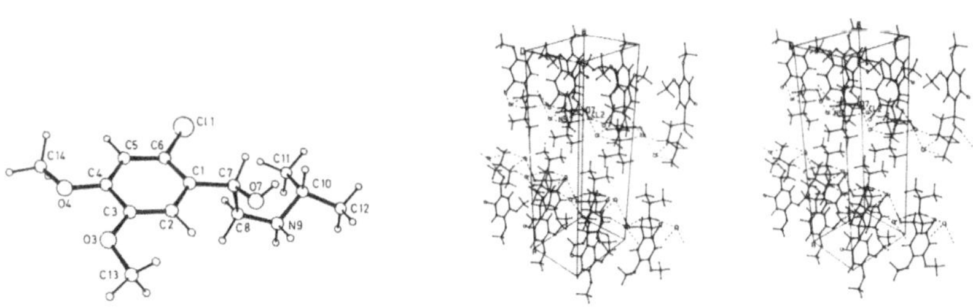

Fig. 1. $C_{13}H_{21}Cl_2NO_3$: conformation of and atomic numbering in the R-enantiomeric cation and a stereoview of the crystal packing in which hydrogen bonds are shown as broken lines.

(±)-2-(2-CHLORO-3,4-DIMETHOXYPHENYL)-N-ISOPROPYL-2-METHOXYETHYLAMINE HYDROCHLORIDE

$C_{14}H_{23}Cl_2NO_3$ $C_{14}H_{23}ClNO_3^+,Cl^-$

J. KOORTS and M. CAIRA, 1985. *Acta Cryst.*, **C41**, 552-554.

Triclinic, $P\bar{1}$, a = 7.537, b = 9.174, c = 12.625 Å, α = 103.81, β = 76.24, γ = 101.89°, Z = 2. Mo radiation, R = 0.040 for 1689 reflections.

The substituted ethylamine side chain is extended and is oriented approximately perpendicular to the phenyl ring (Fig. 1). The phenyl ring is planar with the OMe group nearest the Cl atom oriented out of the plane and the other approximately coplanar. The N(9)...O(7) distance is 2.831(3) Å. Atom C(10) of the isopropyl group is approximately gauche to the chiral centre of the cation. The N atom is 5.16 Å from the centre of the phenyl ring. Hydrogen bonds of the type N-H...Cl^- link pairs of inversion-related cations (3.209(3), 3.106(3) Å).

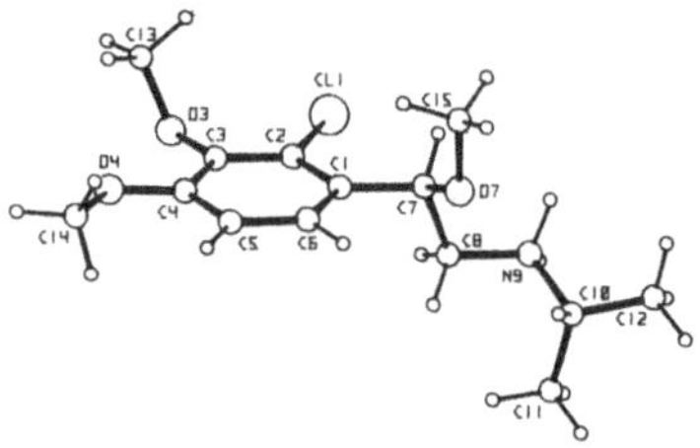

Fig. 1. $C_{14}H_{23}Cl_2NO_3$: conformation of the (R)-enantiomeric cation.

BIS(o-METHOXYPHENYL) KETONE (BIS(o-ANISYL) KETONE)
$C_{15}H_{14}O_3$

C. FAERMAN, S.C. NYBURG, G. PUNTE, B.E. RIVERO, A.A. VITALE and N.S. NUDELMAN, 1985. *Canad. J. Chem.*, **63**, 3374-3377.

Monoclinic, $P2_1/n$, a = 9.893, b = 10.719, c = 12.136 Å, β = 90.56°, Z = 4. Mo radiation, R = 0.046 for 1383 reflections.

The molecule is shown to have a twist conformation with molecular dimensions given in Fig. 1; the separation of oxygens O(1) and O(3) is 2.89 Å.

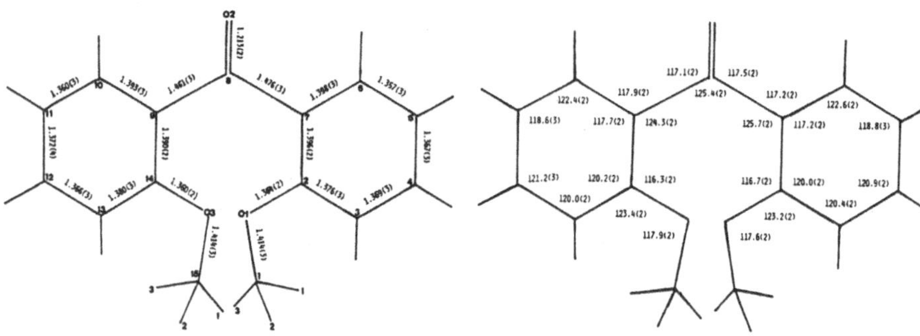

Fig. 1. Bond lengths and angles in $C_{15}H_{14}O_3$.

1,1-DICHLORO-2,2-BIS(p-METHOXYPHENYL)ETHYLENE
$C_{16}H_{14}Cl_2O_2$

A.K. PAL, R.R. BANDYOPADHYAY and B.S. BASAK, 1985. *Z. Kristallogr.*, **171**, 159-163.

Monoclinic, $C2/c$, a = 18.290, b = 7.963, c = 11.096 Å, β = 115.74°, Z = 4. Mo radiation, R = 0.035 for 991 reflections (see also (1)).

The molecule has twofold crystallographic symmetry (C(2)=C(3) on twofold axis, Fig. 1). The C=C distance is 1.333 Å.

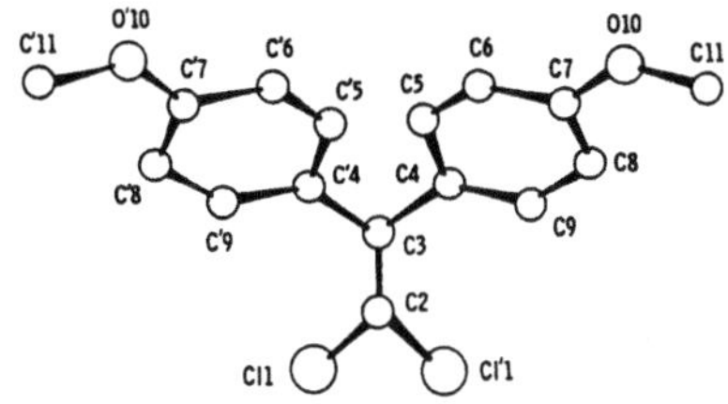

Fig. 1. A view of $C_{16}H_{14}Cl_2O_2$ [the molecule has twofold symmetry, so the bond tapers in the above diagram are not all correct].

1. Following Report.

p-ANISOIN
$C_{16}H_{16}O_4$

S. KASHINO, M. HAISA and I. SURYANARAYANA, 1985. *Acta Cryst.*, **C41**, 1066-1069.

Orthorhombic, $P2_12_12_1$, a = 12.073, b = 18.917, c = 5.977 Å, Z = 4. Cu radiation, R = 0.042 for 1242 reflections.

The molecular conformation (Fig. 1) is similar to that found for benzoin (1). There is an intramolecular hydrogen bond between the carbonyl and hydroxyl O atoms with an O(1)...O(2) separation of 2.611(3) Å. In benzoin these O atoms were involved in intermolecular bifurcated hydrogen bonds. Both phenyl rings are planar with an interplanar dihedral angle of 98.1(1)°. Torsion angles O(1)-C(13)-C(14)-O(2) and C(1)-C(13)-C(14)-C(7) are 15.1(4) and 72.4(3)° respectively. Molecules form stacked sheets in the crystal with intermolecular distances consistent with essentially van der Waals contacts.

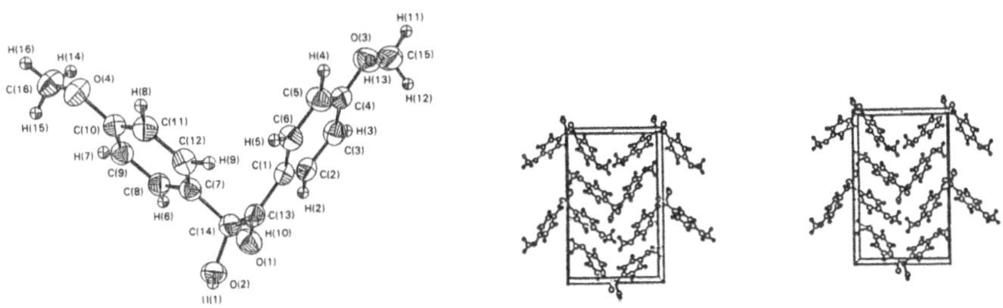

Fig. 1. $C_{16}H_{16}O_4$: a view of the molecule showing the atom numbering and a stereoview of the molecular packing.

1. Structure Reports, **46B**, 116.

N-(4-ETHOXYSALICYLIDENE)-4-METHYLANILINE (α-form)
$C_{16}H_{17}NO_2$ (I)

N-(4-ETHOXYSALICYLIDENE)-4-METHYLANILINE (β-form)
$C_{16}H_{17}NO_2$ (II)

A.E. OBODOVSKAYA, Z.A. STARIKOVA, B.M. BOLOTIN, T.N. SAFONOVA and N.B. ETINGEN, 1985. *Zh. Strukt. Khim.*, **26**, 1-111-119 [J. Struct. Chem., **26**, 92-99].

I. Monoclinic, $P2_1/n$, a = 14.416, b = 8.257, c = 12.901 Å, β = 113.90°, Z = 4. Mo radiation, R = 0.040 for 1057 reflections.

II. Monoclinic, $P2_1/n$, a = 10.093, b = 16.623, c = 16.658 Å, β = 94.14°, Z = 8. Mo radiation, R = 0.040 for 1671 reflections.

The α form of $C_{16}H_{17}NO_2$ is luminescent while the β form is not. The α molecule and the two independent β molecules have similar geometric parameters, but differ in the conformation of the plane of the ethoxy group at the (Ph)C-O bond: trans (α) and cis (β) relative to the OH group of the phenyl ring (Fig. 1).

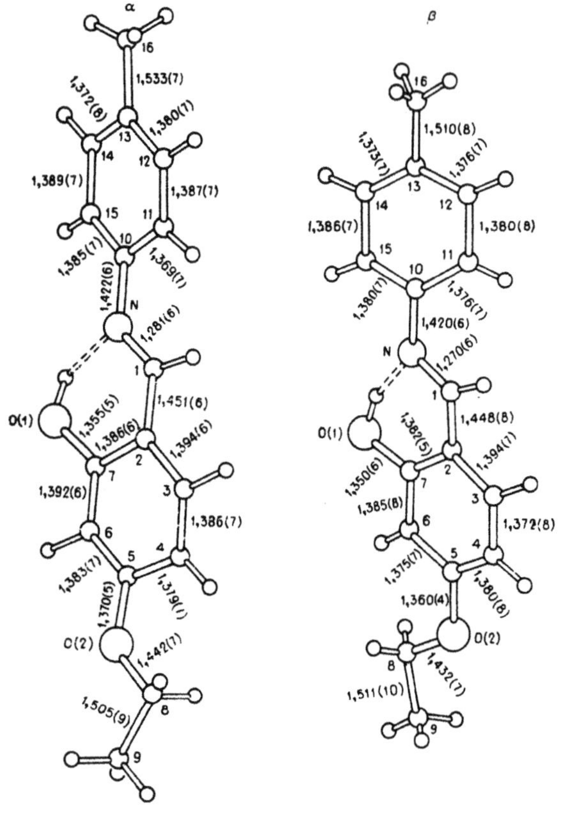

Fig. 1. The α (left) and β (right) forms of $C_{16}H_{17}NO_2$ (average values are shown for the β form).

1-(3'-CHLOROPHENOXYACETYL)-4-PROPYLBENZENE
$C_{17}H_{17}ClO_2$ (I)

1-(2'-CHLOROPHENOXYACETYL)-4-PROPYLBENZENE
$C_{17}H_{17}ClO_2$ (II)

T. ATABAEV, K.Sh. KARAEV, N.V. PODBEREZSKAYA, S.V. BORISOV, A. ASHIROV and A. KAZAKOV, 1985. *Zh. Strukt. Khim.*, **26**, 5-188-191 [J. Struct. Chem., **26**, 826-829].

I. Monoclinic, $P112_1/b$, a = 17.968, b = 9.434, c = 9.508 Å, γ = 102.33°, Z = 4. Mo radiation, R = 0.053 for 1190 reflections.

II. Monoclinic, $P112_1/b$, a = 14.201, b = 10.045, c = 10.931 Å, γ = 105.033°, Z = 4. Mo radiation, R = 0.058 for 1672 reflections.

Bond lengths and angles for (I) and (II) are shown in Figs. 1 and 2 respectively. Corresponding dimensions for the two isomers are similar.

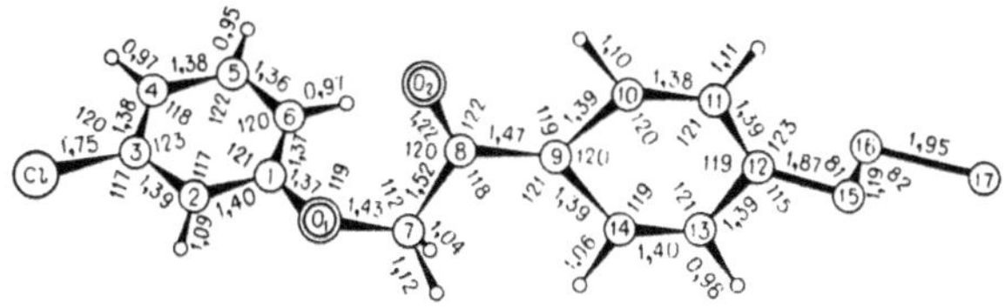

Fig. 1. Geometry of $C_{17}H_{17}ClO_2$ (I).

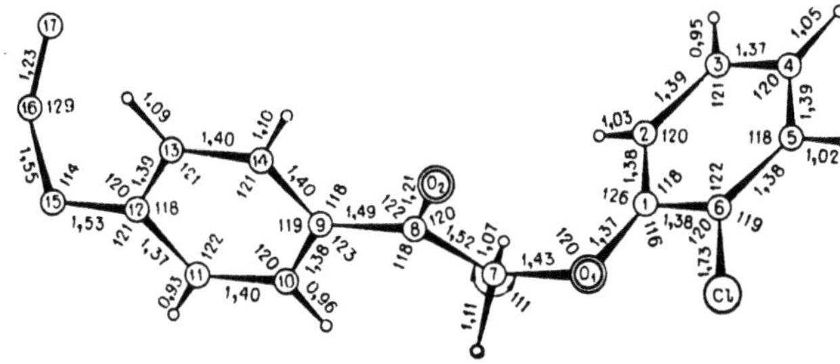

Fig. 2. Geometry of $C_{17}H_{17}ClO_2$ (II).

threo-2-(p-METHOXYPHENYL)-2-(p-NITROPHENYLTHIOLATO)-1-(METHOXYCARBONYL)ETHANOL
$C_{17}H_{17}NO_6S$

T. HASHIYAMA, H. INOUE, M. TAKEDA, K. AOE and K. KOTERA, 1985. *J. Chem. Soc. Perkin Trans. 1*, - 421-427.

Triclinic, $P\bar{1}$, a = 7.728, b = 16.574, c = 7.152 Å, α = 103.623, β = 79.640, γ = 102.544°, Z = 2. Cu radiation, R = 0.058 for 2351 reflections.

The crystal structure analysis established the stereochemistry of the molecule (Fig. 1).

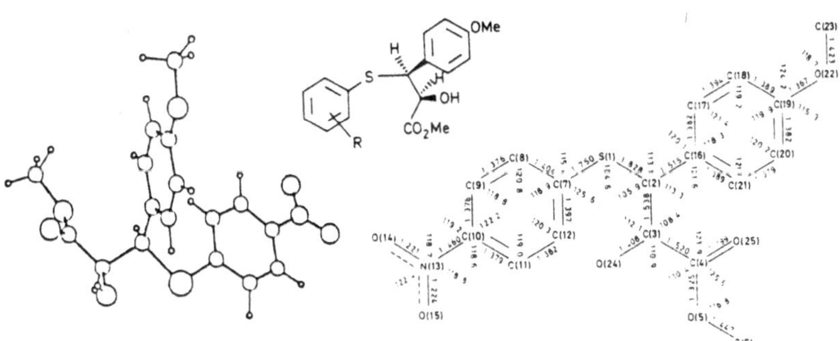

Fig. 1. Perspective drawing of a molecule of $C_{17}H_{17}NO_6S$ (left) with bond lengths (Å) and angles (°) (right).

2'-HYDROXY(1-PHENYL-1-(N-FORMYL)-n-PROPYLAMINO)ACETOPHENONE
$C_{18}H_{19}NO_3$

I. YOKOE, K. NAKAMURA, K. HIGUCHI, Y. SHIRATAKI, M. KOMATSU and H. MIYAMAE, 1985. *Heterocycles*, **23**, 291-293.

Monoclinic, $B112/b$, a = 26.884, b = 22.640, c = 9.465 Å, γ = 33.69°, Z = 8. Cu radiation, R = 0.072 for

2375 reflections.

The analysis established the structure and stereochemistry as shown in Fig. 1.

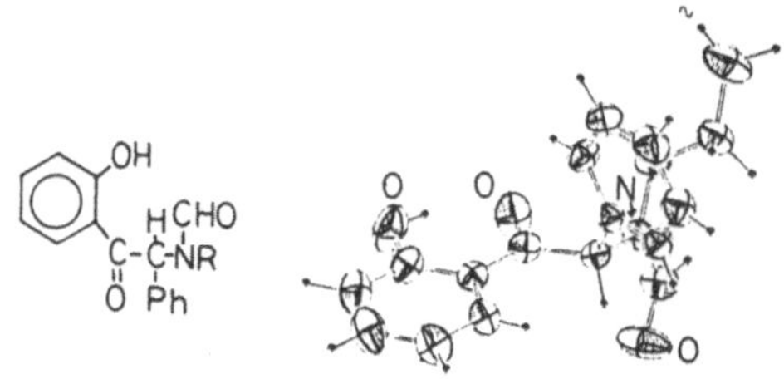

Fig. 1. The $C_{18}H_{19}NO_3$ molecule (R = n-propyl).

1,1-BIS(p-ETHOXYPHENYL)ETHYLENE
$C_{18}H_{20}O_2$ (I)

1,1-DICHLORO-2,2-BIS(p-METHOXYPHENYL)ETHYLENE
$C_{16}H_{14}Cl_2O_2$ (II)

T. BANERJEE, S. CHAUDHURI and P.T. MUTHIAH, 1985. *Acta Cryst.*, **C41**, 269-271.

I. Orthorhombic, *Pbcn*, a = 7.714, b = 6.656, c = 29.682 Å, Z = 4. Cu radiation, R = 0.044 for 867 reflections.

II. Monoclinic, *C2/c*, a = 18.290, b = 7.964, c = 11.103 Å, β = 115.76°, Z = 4. Mo radiation, R = 0.042 for 1290 reflections.

Both structures (Fig. 1) retain their diad axes in the lattice and are held together by van der Waals forces. The C-C ethylene distance is 1.342(3) and 1.326(4) Å in (I) and (II) respectively. In (II) the C-Cl distance is 1.728(2) Å. The twist about the ethylene bond is 0.9(7)° in (I) and 9.7(1)° in (II).

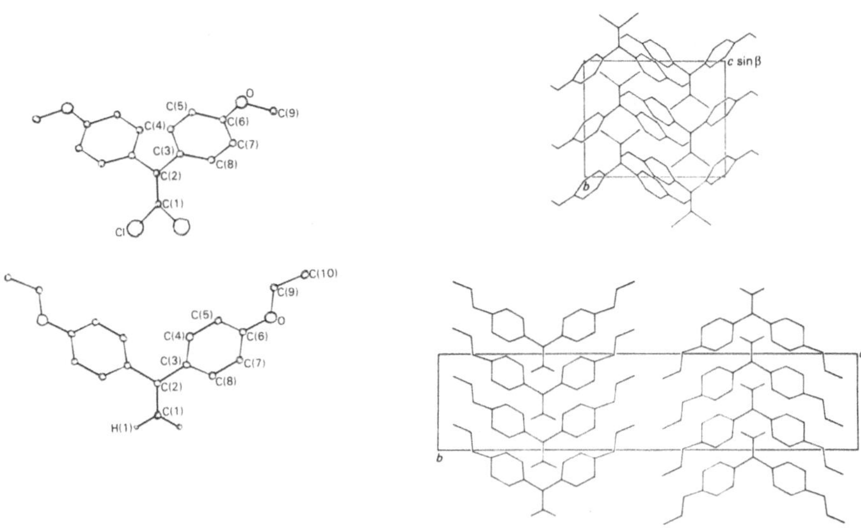

Fig. 1. A view of the molecule and the molecular packing: $C_{16}H_{14}Cl_2O_2$ (top) and $C_{18}H_{20}O_2$ (bottom).

BIS((o-HYDROXYPHENOXY)ETHOXY)ETHANE MONOHYDRATE

$C_{18}H_{22}O_6,H_2O$

IL-HWAN SUH, HAE NAMGUNG, YOUNGKU YOON, W. SAENGER and F. VOGTLE, 1985. *J. Inclusion Phenomena,* **3**, 21-26.

Orthorhombic, $P2_12_12_1$, a = 25.909, b = 8.542, c = 8.187 Å, Z = 4. Cu radiation, R = 0.0805 for 1538 reflections.

The polyether adopts an annular structure (Fig. 1) with the water molecule at the centre, and is slightly folded. All torsion angles about C-O are trans and about C-C are gauche. The complex has three three-centre (bifurcated) and one two-centre hydrogen bonds between phenolic hydroxyl and water oxygen.

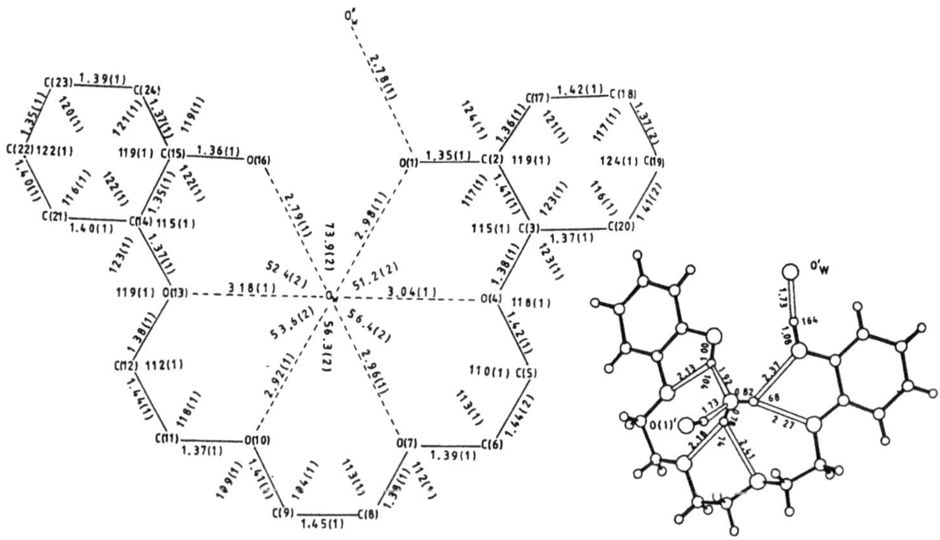

Fig. 1. Molecular dimensions and a view of the $C_{18}H_{22}O_6,H_2O$ hydrogen-bonded complex.

4-(α,α-DIMETHYLBENZYL)PHENYL METHACRYLATE
$C19H_{20}O_2$

F.J. VIERSEN, Y.Y. TAN, F. VAN BOLHUIS and R.J.M. ZWIERS, 1985. *Makromol. Chem.*, **186**, 1987-1993.

Monoclinic, $P2_1$, a = 6.622, b = 12.578, c = 9.892 Å, β – 99.86°, Z = 2. Mo radiation, R = 0.084 for 1132 reflections.

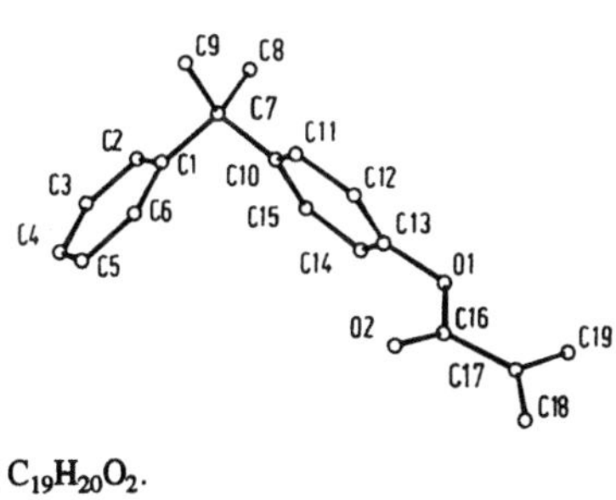

Fig. 1. Molecular structure of $C_{19}H_{20}O_2$.

The two phenyl rings are clearly not coplanar (Fig. 1), and the dihedral angle between them is 93.1°. Neither C(17)-C(18) nor C(17)-C(19) can be clearly assigned to a single or double bond since the respective bond lengths of 1.44(2) and 1.40(2) Å are between that of a single and that of a double bond. Other parameters of interest are: O(1)-C(13) 1.42(1), O(1)-C(16) 1.35(1), O(2)-C(16) 1.21(1) Å, C(16)-C(17)-C(18) 114(1), C(16)-C(17)-C(19)

119(1), C(18)-C(17)-C(19) 126(1)°. The distances in the benzene rings range from 1.37(2)-1.45(3) Å (C(1)-C(6)), and from 1.35(1)-1.42(2) Å (C(10)-C(15)). The angles around C(7) range from 106.6(9)-111.2(9)°. The C(16)-C(17) bond length is 1.54(2) Å.

HEXA-2,4-DIYNE-1,6-DIYL BIS(o-METHOXYPHENYLCARBAMATE)
$C_{22}H_{20}N_2O_6$

A. WHULER, M.-C. SICHERE, P. SPINAT and C. BROUTY, 1985. *Acta Cryst.*, **C41**, 736-739.

Orthorhombic, *Pbca*, a = 21.051, b = 8.281, c = 11.364 Å, Z = 4. Cu radiation, R = 0.050 for 1225 reflections (at 205 K).

The bond lengths and angles in this monomeric compound (Fig. 1) are in good agreement with generally accepted values. The centre of the C(1)-C(1') bond in the molecule, of length 1.377(3) Å, coincides with a crystallographic inversion centre. The dihedral angle between the plane of the phenyl ring and that of the urethane function is 13(1)°. There is an intramolecular N-H...O(3) hydrogen bond of length 2.573(3) Å. The angle between the urethane and diacetylene functions is 84(2)° giving the molecule a bent conformation. All intermolecular distances are essentially van der Waals contacts.

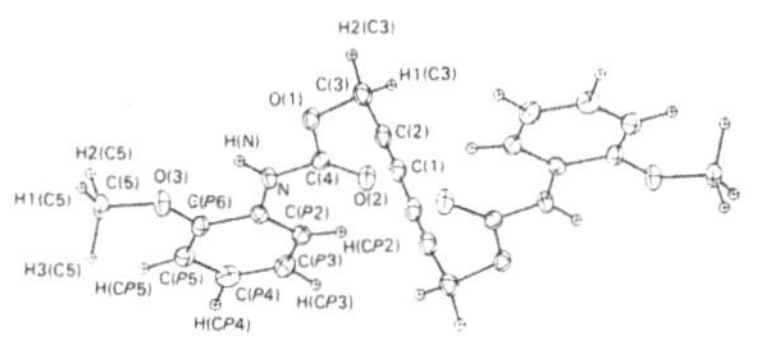

Fig. 1. $C_{22}H_{20}N_2O_6$: a projection of the molecule perpendicular to [001].

AZINOBIS(ETHYLIDYNE-p-PHENYLENE) DIPROPIONATE
$C_{22}H_{24}N_2O_4$

M.R. CIAJOLO, A. SIRIGU and A. TUZI, 1985. *Acta Cryst.*, **C41**, 483-485.

Monoclinic, $P2_1/c$, a = 19.836, b = 5.450, c = 24.068 Å, β = 128.37°, Z = 4. Cu radiation, R = 0.055 for 875 reflections.

The asymmetric unit of the structure consists of halves of two crystallographically independent centrosymmetric molecules, A and B. These molecules (Fig. 1) have very similar bond lengths and angles. In both molecules the-C(Me)=N-N=C(Me)- atomic grouping at the inversion centres is planar. The planar phenyl rings are rotated out of this plane with torsion angles N-C(10)-C(9)-C(8) having values of 171(1) and -168(2)°, for molecules A and B respectively. The N-C(10)-C(11) angle, of mean value 126(2)°, is slightly widened due to steric interference between atoms N' and C(11) which have intramolecular separations of 2.71(1) and 2.69(1) Å. Molecules pack in a herring-bone fashion forming layers of symmetry-related molecules. The intermolecular distances are consistent with essentially van der Waals contacts.

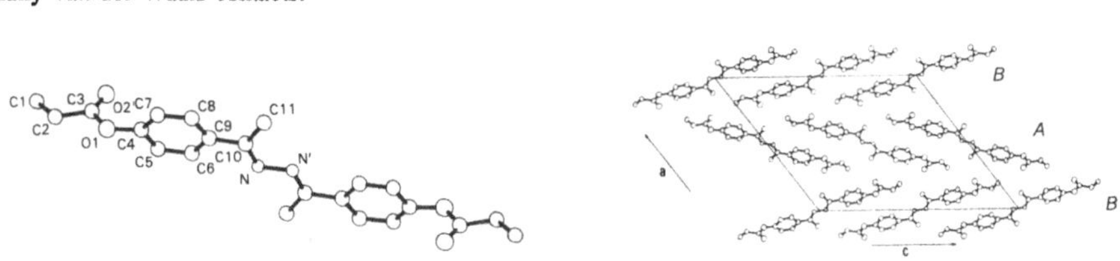

Fig. 1. A view of one of the centrosymmetric $C_{22}H_{24}N_2O_4$ molecules and a projection of the crystal structure onto (010).

4,4'-BIS(PENTYLOXY)AZOXYBENZENE
$C_{22}H_{30}N_2O_3$

N.C. SHIVAPRAKASH, M.M.M. ABDOH and J.S. PRASAD, 1985. *Z. Kristallogr.*, **172**, 79-87.

Triclinic, $P\bar{1}$, a = 7.428, b = 8.267, c = 17.868 Å, α = 97.73, β = 98.66, γ = 101.25°, Z = 2. Mo radiation, R = 0.0427 for 1960 reflections.

The molecules (Fig. 1) are in their most extended trans conformation. They are nearly planar and are packed in alternate sheets in which all the molecular long axes are either in head or tail arrangement.

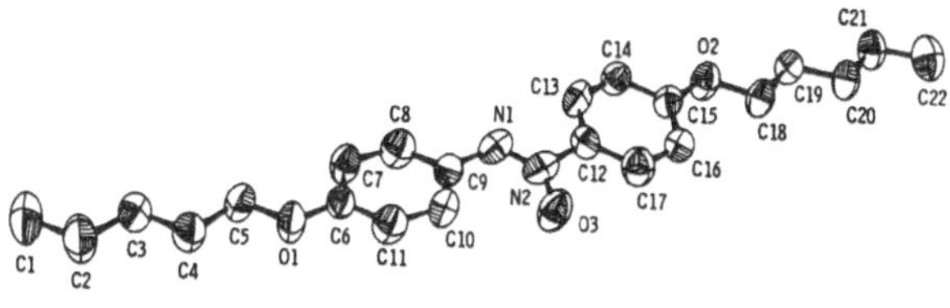

Fig. 1. A view of $C_{22}H_{30}N_2O_3$ (H atoms omitted).

TETRA-n-BUTYLAMMONIUM 1,2-DIHYDROXYBENZENE CHLORIDE (TETRA-n-BUTYLAMMONIUM CATECHOL CHLORIDE)
$C_{22}H_{42}ClNO_2$ (I) $C_6H_6O_2,C_{16}H_{36}N^+,Cl^-$

TETRA-n-BUTYLAMMONIUM 1,2-DIHYDROXYBENZENE BROMIDE (TETRA-n-BUTYLAMMONIUM CATECHOL BROMIDE)
$C_{22}H_{42}BrNO_2$ (II) $C_6H_6O_2,C_{16}H_{36}N^+,Br^-$

M.A. KHAN, A.W. McCULLOCH and A.G. McINNES, 1985. *Canad. J. Chem.*, **63**, 2119-2122.

I. Orthorhombic, *Pbca*, a = 15.323, b = 16.166, c = 19.713 Å, Z = 8. Cu radiation, R = 0.048 for 1361 reflections.

II. Orthorhombic, *Pbca*, a = 15.432, b = 16.244, c = 19.824 Å, Z = 8. Cu radiation, R = 0.041 for 1958 reflections.

These isomorphous structures have discrete cations and halide anions hydrogen bonded to catechol moieties (Fig. 1). Each halide ion accepts two essentially linear hydrogen bonds from the same catechol with O...Cl 3.014, 3.033(4), O...Br 3.182, 3.215(4) Å.

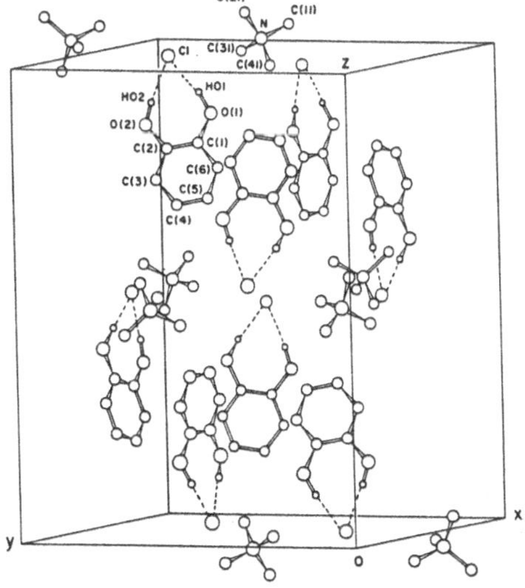

Fig. 1. A packing diagram for $C_6H_6O_2,C_{16}H_{36}N^+,Cl^-$.

E-1-(4-FLUOROPHENYL)-1-(4-METHOXYPHENYL)-2-PHENYL-1-BUTENE
$C_{23}H_{21}FO$

J. SHANI, A. GAZIT, T. LIVSHITZ and S. BIRAN, 1985. *J. Med. Chem.*, **28**, 1504-1511.

Triclinic, $P\bar{1}$, a = 10.107, b = 10.683, c = 9.682 Å, α = 114.00, β = 105.56, γ = 77.42°, Z = 2. Mo radiation, R = 0.051 for 2286 reflections.

The structure and stereochemistry are shown in Fig. 1. The molecule (a fluorotamoxifen precursor) has a propeller like arrangement.

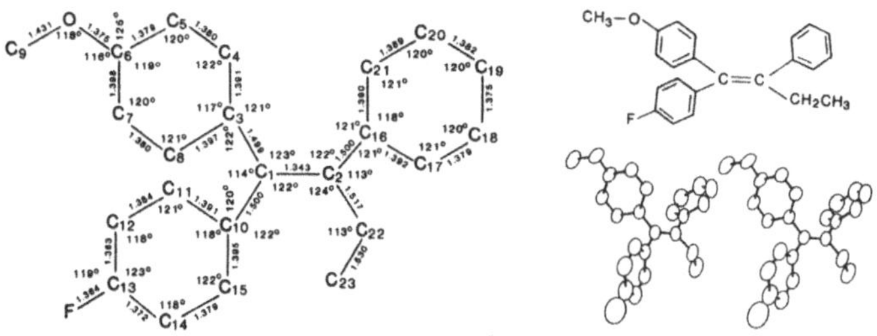

Fig. 1. The $C_{23}H_{21}FO$ molecule and details of dimensions and conformation.

2-(p-BROMOPHENOXY)-N-ETHYL-2-PHENYL-N-(p-TOLYL)ETHANESULPHONAMIDE
$C_{23}H_{24}BrNO_3S$

K. SEKIDO, Y. YOKOMORI, S. HIROKAWA, K. OKUMA and H. OHTA, 1985. *Acta Cryst.*, C**41**, 461-463.

Triclinic, $P\bar{1}$, a = 11.560, b = 10.356, c = 10.549 Å, α = 90.15, β = 111.94, γ = 107.92°, Z = 2. Mo radiation, R = 0.0526 for 2916 reflections.

In the molecule (Fig. 1) the benzene rings are planar and bond lengths and valence angles are in good agreement with generally accepted values. The molecule is a racemate and the two enantiomers are related by a crystallographic inversion centre. The shortest intermolecular contact is 3.29 Å between C(2) and C(3) and the molecules are held together in the crystal by van der Waals forces.

Fig. 1. $C_{23}H_{24}BrNO_3S$: a perspective view of the molecule and the unit cell contents projected onto (010) showing intermolecular contacts.

2,2'-(o-PHENYLENEBIS(METHYLENEOXY))DIACETOPHENONE
$C_{24}H_{22}O_4$

W.J. COLUCCI, S.J. JUNGK, J.W. NAIL, H. GHANE, K. KARAGIORGOS, S.F. WATKINS, F.R. FRONCZEK and R.D. GANDOUR, 1985. *Acta Cryst.*, C**41**, 554-556.

Monoclinic, $P2_1$, a = 10.842, b = 5.131, c = 17.390 Å, β = 92.12°, Z = 2. Mo radiation, R = 0.039 for 1383 reflections.

The molecule (Fig. 1) adopts an asymmetric conformation in the crystal, with the two acetophenone rings forming dihedral angles of 98.5(2) and 159.5(2)° with the o-phenylene ring. Bond lengths and angles are normal.

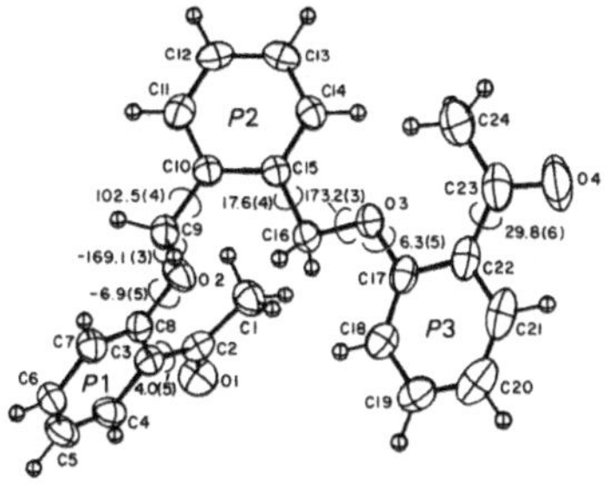

Fig. 1. The $C_{24}H_{22}O_4$ molecule showing selected torsion angles.

2-HYDROXYTAMOXIFEN
$C_{26}H_{29}NO_2$ (I)

3-HYDROXYTAMOXIFEN
$C_{26}H_{29}NO_2$ (II)

2-METHYL-4-HYDROXYTAMOXIFEN
$C_{27}H_{31}NO_2$ (III)

1-(p-METHOXYPHENYL)-2-PHENYL-1-(2-(TETRAHYDROPYRAN-2-YLOXY)PHENYL)-1-BUTANOL
$C_{28}H_{32}O_4$ (IV)

R. KURODA, S. CUTBUSH, S. NEIDLE and ON-TAI LEUNG, 1985. *J. Med. Chem.*, **28**, 1497-1503.

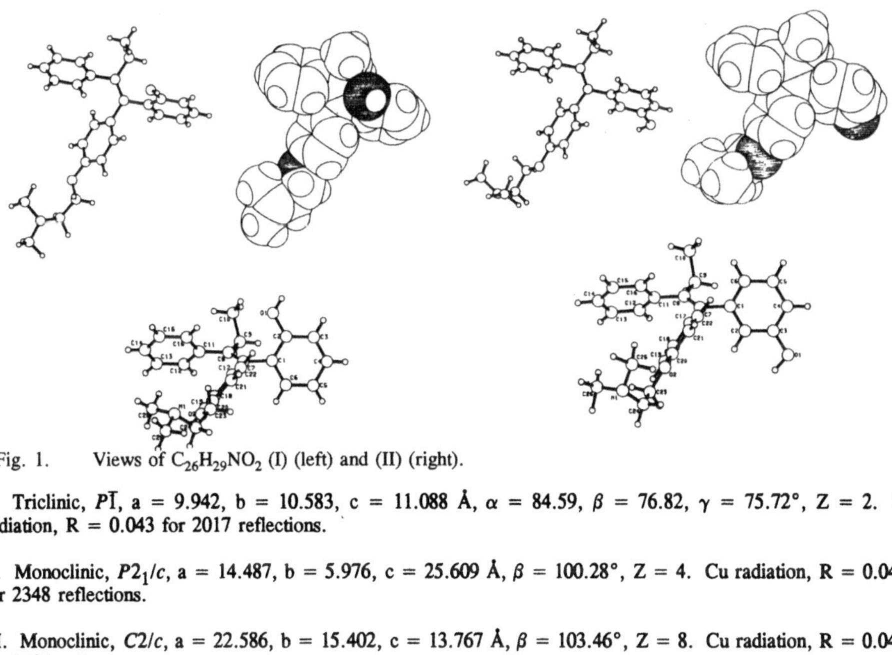

Fig. 1. Views of $C_{26}H_{29}NO_2$ (I) (left) and (II) (right).

I. Triclinic, $P\bar{1}$, a = 9.942, b = 10.583, c = 11.088 Å, α = 84.59, β = 76.82, γ = 75.72°, Z = 2. Cu radiation, R = 0.043 for 2017 reflections.

II. Monoclinic, $P2_1/c$, a = 14.487, b = 5.976, c = 25.609 Å, β = 100.28°, Z = 4. Cu radiation, R = 0.0464 for 2348 reflections.

III. Monoclinic, $C2/c$, a = 22.586, b = 15.402, c = 13.767 Å, β = 103.46°, Z = 8. Cu radiation, R = 0.0442 for 2585 reflections.

IV. Triclinic, $P\bar{1}$, a = 9.970, b = 10.657, c = 12.400 Å, α = 94.76, β = 78.22, γ = 109.14°, Z = 2. Cu radiation, R = 0.0471 for 3826 reflections.

Details of structure and conformation for (I)-(IV) are shown in Figs. 1 and 2. All four compounds have the ethyl group and the 4-substituted phenyl ring attached to the central ethylene moiety, in a trans orientation with respect to each other. The four molecules all adopt a propeller conformation with respect to the triphenylethylene system. The non-coplanarity of the phenyl rings with the ethylene bond is a consequence of non-bonded interactions between the rings and with the ethyl group. The conformation of the (dimethylamino)ethoxy side chain shows a high degree of flexibility; molecule (I) has this chain in a gauche$^-$ conformation, while (II) and (III) have gauche$^+$ arrangements.

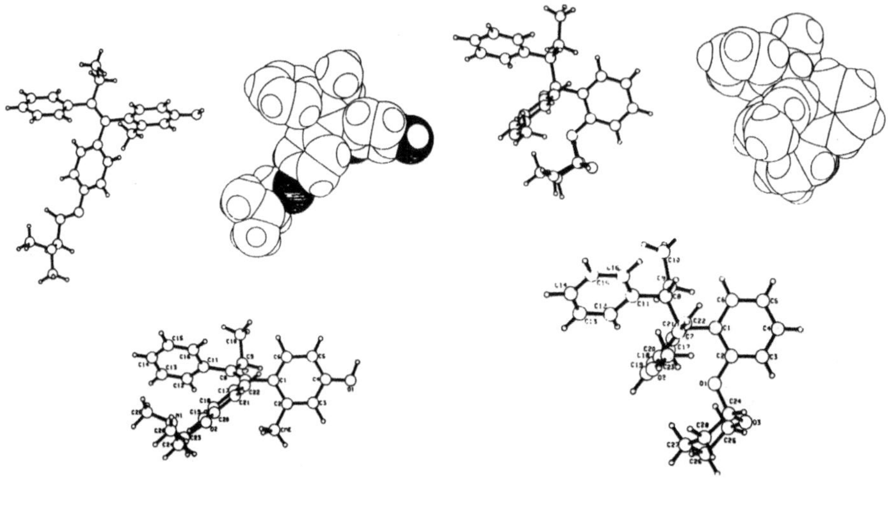

Fig. 2. Views of $C_{27}H_{31}NO_2$ (III) (left) and $C_{28}H_{32}O_4$ (IV) (right).

3-ACETAMIDO-1,1-BIS(4-ACETOXYPHENYL)-2-PHENYLPROPYLIDENE
$C_{27}H_{27}NO_5$

G. PRECIGOUX, M. HOSPITAL, J.-F. MIQUEL and J. GILBERT, 1985. *Acta Cryst.*, C41, 1244-1246.

Orthorhombic, *Pbca*, a = 9.458, b = 32.090, c = 16.359 Å, Z = 8. Cu radiation, R = 0.053 for 2212 reflections.

Bond lengths and angles in the molecule (Fig. 1) lie within normal ranges. The C(7)-C(19) (1.527(5)) and C(13)-C(33) (1.519(5) Å) distances indicate there is no conjugation between the aromatic rings. The angle between the mean planes of the adjacent phenyl rings is 77.6(6)° and that between the 'trans' phenyl rings is 51.5(5)°. There is an intermolecular hydrogen bond between the acetamido O atom and the N atom (N...O 2.817(7) Å).

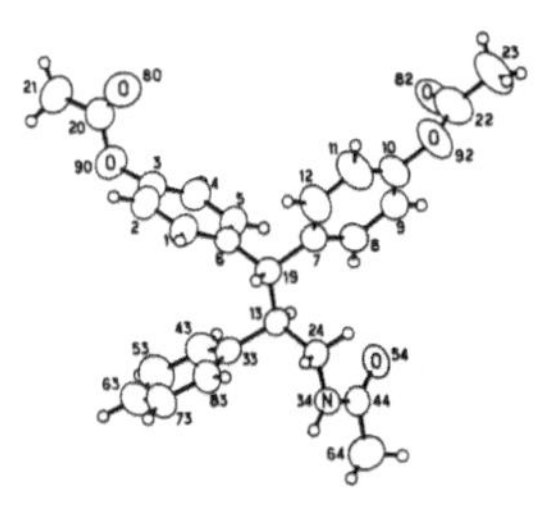

Fig. 1. $C_{27}H_{27}NO_5$: the molecular structure.

2,2',2''-TRIS(SALICYLIDENEIMINO)TRIETHYLAMINE (TRENSAL)

$C_{27}H_{30}N_4O_3$

N. GUNDUZ, T. GUNDUZ, M.B. HURSTHOUSE, H.G. PARKES, L.S. SHAW, R.A. SHAW and M. TUZUN, 1985. *J. Chem. Soc. Perkin Trans.* 2, 899-902.

Monoclinic, $P2_1/c$, a = 9.857, b = 11.208, c = 22.910 Å, β = 98.81°, Z = 4. Cu radiation, R = 0.073 for 3113 reflections.

The hydroxy hydrogen atoms are hydrogen-bonded to the azomethine nitrogen atoms (O...N 2.573 Å) giving an almost planar six-membered ring (Fig. 1), which is coplanar with the adjacent benzene ring. These fused bicyclic systems are perpendicular to the plane through C(11)-C(21)-C(31); looking down the unique nitrogen atom through this three carbon plane one sees an approximate threefold axis with the hydrogen bonds, O-H...N uppermost in the bicyclic structural units (Fig. 1).

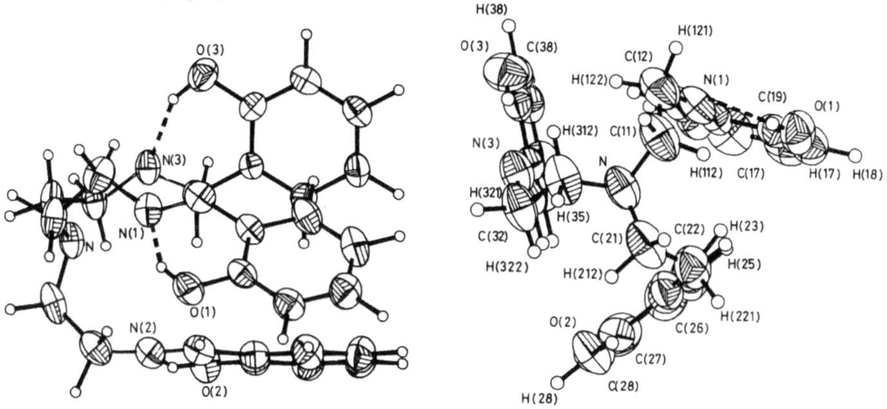

Fig. 1. Molecular diagrams of $C_{27}H_{30}N_4O_3$ showing hydrogen-bonding (left) and view down threefold axis (right).

α-ISOPROPYL-α-((N-METHYL-N-HOMOVERATRYL)-γ-AMINOPROPYL)-3,4-DIMETHOXYPHENYL-ACETONITRILE HYDROCHLORIDE (VERAPAMIL HYDROCHLORIDE)

$C_{27}H_{39}ClN_2O_4$ $C_{27}H_{39}N_2O_4{}^+,Cl^-$

A. CARPY, J.-M. LEGER and C. MELCHIORRE, 1985. *Acta Cryst.*, C41, 624-627.

Triclinic, $P\bar{1}$, a = 7.086, b = 10.591, c = 19.196 Å, α = 100.10, β = 93.73, γ = 101.55°, Z = 2. Cu radiation, R = 0.048 for 1182 reflections.

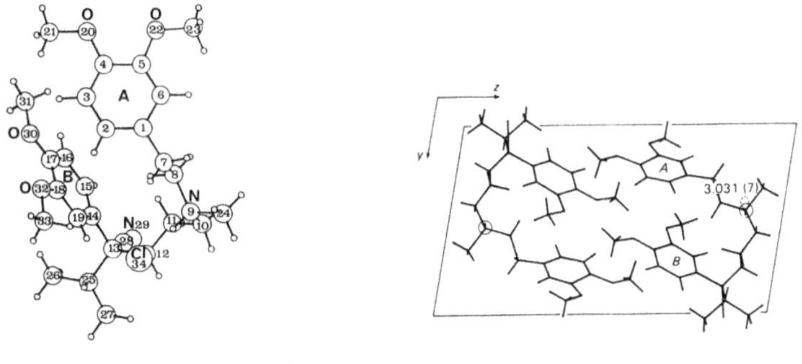

Fig. 1. $C_{27}H_{39}ClN_2O_4$: a view of the molecule and packing of the molecules projected on (100).

Bond lengths and angles have acceptable values. The bridge chain of the Ca^{2+} channel antagonist is folded at the N atom (Fig. 1). The angle between rings A and B of the molecule is 69(1)° and the distance between N(9) and the centre of ring A is 5.18(1) Å and between N(9) and the centre of ring B is 5.09(1) Å. Crystalline cohesion is achieved through hydrogen bonding (N(9)...Cl(34) 3.031(7) Å).

p,p'-BIS(OCTYLOXY)BENZOPHENONE

$C_{29}H_{42}O_3$

T. SUGAWARA, S. MURATA, K. KIMURA, H. IWAMURA, Y. SUGAWARA and H. IWASAKI, 1985. *J. Am. Chem. Soc.*, **107**, 5293-5294.

Orthorhombic, *Pnca*, a = 52.8, b = 7.32, c = 7.05 Å, Z = 4. Mo radiation, R = 0.093 for 1616 reflections.

The crystal structure contains a multilayered structure (Fig. 1). Each layer is separated by the all-trans octyl chains which are aligned parallel to one another. The molecules have crystallographic twofold symmetry.

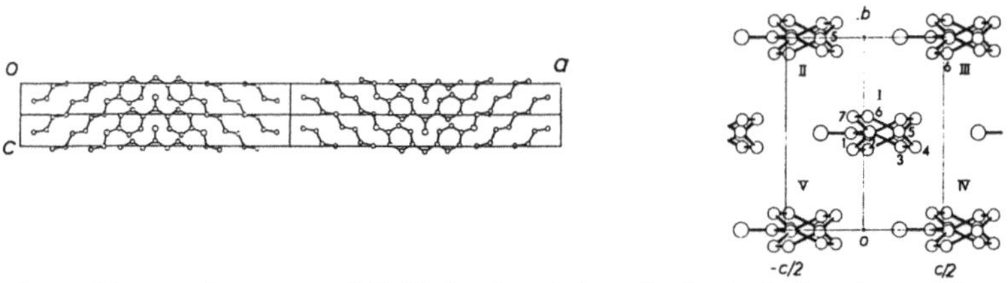

Fig. 1. The crystal structure of $C_{29}H_{42}O_3$ viewed along the *b*-axis (left) and a partial packing arrangement (chains and H atoms omitted) viewed along the *a*-axis (right).

TETRAKIS(GALLAMINE) UNDECABROMIDE IODIDE METHANOL ETHANOL SOLVATE UNDECAHYDRATE

$C_{120}H_{240}Br_{11}IN_{12}O_{12},CH_4O,C_2H_6O,11(H_2O)$ $4(C_{30}H_{60}N_3O_3^{3+}),11(Br^-),I^-,CH_4O,C_2H_6O,11(H_2O)$

J. HUSAIN, I.J. TICKLE and R.A. PALMER, 1985. *Acta Cryst.*, **C41**, 1491-1496.

Triclinic, *P*1, a = 14.216, b = 14.081, c = 20.895 Å, α = 104.90, β = 92.60, γ = 94.50°, Z = 1. Cu radiation, R = 0.083 for 2352 reflections.

There are four independent gallamine molecules in the asymmetric unit, pairs of which are related by non-crystallographic twofold axes or mirror planes. The crystal structure thus comprises both left- and right-handed pseudo-equivalent conformations. Of the three side chains A, B and C in each gallamine molecule (Fig, 1), A and C are almost mirror related, while B is disposed asymmetrically between them. The A and C conformations are such as to rotate their N^+ atoms as far as possible from the benzene-ring plane, while for the B side chains the N^+ atoms are roughly coplanar with the benzene rings. The complex solvent structure is distributed along channels running parallel to the crystallographic *c* direction. Of the twelve heavy counterions, eight refined as well-behaved Br^- anions. The others are disordered.

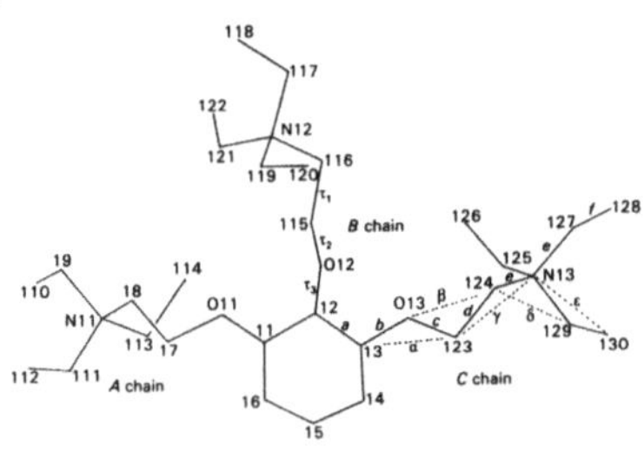

Fig. 1. The gallamine molecule.

2,6-DIPHENYL-1,4-BENZOQUINONE

$C_{18}H_{12}O_2$

I.A. KROL, V.M. AGRE and B.D. SVIRIDOV, 1985. *Zh. Strukt. Khim.*, **26**, 1-124-128 [J. Struct. Chem., **26**, 103-107]

Monoclinic, $P2_1/c$, a = 18.602, b = 3.855, c = 21.462 Å, β = 121.61°, Z = 4. Mo radiation, R = 0.046 for 999 reflections.

Both of the phenyl rings in the molecule (Fig. 1) are planar, but the quinone ring is nonplanar and has a sofa conformation.

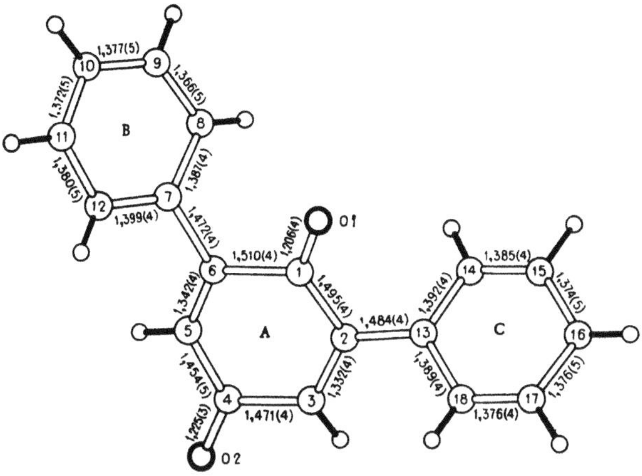

Fig. 1. Molecular geometry of 2,6-diphenyl-1,4-benzoquinone.

1,4-DIIODOBENZENE (p-DIIODOBENZENE)
$C_6H_4I_2$

A. HINCHLIFFE, R.W. MUNN, R.G. PRITCHARD and C.J. SPICER, 1985. *J. Mol. Struct.*, **130**, 93-96.

Orthorhombic, *Pbca*, a = 16.942, b = 7.299, c = 6.160 Å, Z = 4. Mo radiation, R = 0.079 for 408 reflections.

The C-I distance was found to be 2.05(1) Å with C-C 1.41-1.42(1) Å and all C-C-C and C-C-I angles 119.8(2)° implying a 2° deviation from planarity. The crystal used in the analysis was grown from solution. For details of the structure of a melt-grown crystal, see (1).

1. L. DUN-CHAI and Yu.T. STRUCHKOV, 1959. *Izv. Akad. Nauk SSR Otd. Khim. Nauk*, **12**, 2095.

2,2',6,6'-TETRACHLOROBIPHENYL
$C_{12}H_6Cl_4$ (I)

4'-BROMO-2,4,6-TRICHLOROBIPHENYL
$C_{12}H_6BrCl_3$ (II)

2,2',4,4',6,6'-HEXABROMOBIPHENYL
$C_{12}H_4Br_6$ (III)

2,4,4',6-TETRABROMOBIPHENYL
$C_{12}H_6Br_4$ (IV)

L.D. FIELD, B.W. SKELTON, S. STERNHELL and A.H. WHITE, 1985. *Aust. J. Chem.*, **38**, 391-399.

I. Orthorhombic, *Pnca*, a = 12.35, b = 10.521, c = 9.390 Å, Z = 4. Cu radiation, R = 0.040 for 528 reflections.

II. Monoclinic, *C2/c*, a = 19.27, b = 7.940, c = 33.315 Å, β = 90.93°, Z = 16. Cu radiation, R = 0.036 for

2621 reflections.

III. Orthorhombic, *Pnca*, a = 14.48, b = 12.71, c = 8.534 Å, Z = 4. Cu radiation, R = 0.044 for 693 reflections.

IV. Triclinic, $P\bar{1}$, a = 8.14, b = 10.74, c = 35.42 Å, α = 87.87, β = 79.39, γ = 63.55°, Z = 8. Cu radiation, R = 0.049 for 3962 reflections.

Molecules in (I) and (III) lie on a crystallographic twofold axis. There are two molecules in the asymmetric unit in (II) and four in the asymmetric unit in (IV). In all cases the structure is composed of discrete molecular units with the expected geometry (Fig. 1, Fig. 2). The main interest in the compounds was in the X...X, H...X (X = halogen) intramolecular contacts and the effect of these on the dihedral angle between the two phenyl rings. In (I), the lowest Cl...Cl contact is 3.746(3) Å and the dihedral angle is 81.7°. A similar situation is observed in (III) where the lowest Br...Br contact is 3.973(4) Å and the interplanar dihedral angle is 83.0°. In (II), the lowest H...Cl contacts across the central bond are 3.54 Å while a similar value, 3.58 Å, is observed for the shortest Br...H contact in (IV). Interplanar angles in (II) are 84.3 and 86.0°, while in (IV) these range from 80.7 to 89.8°.

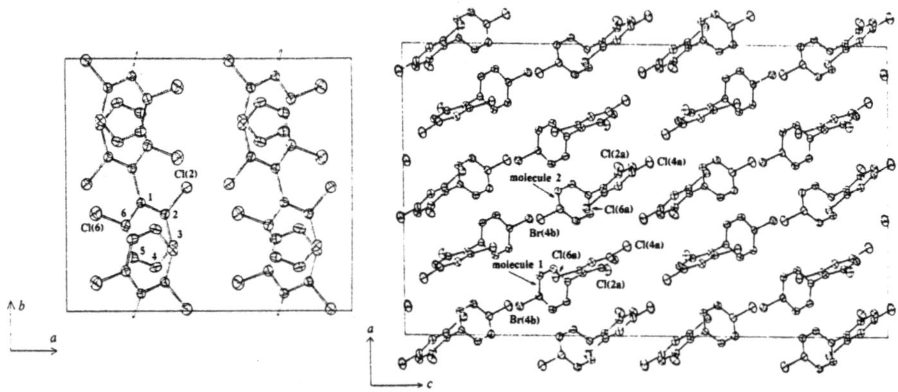

Fig. 1. The unit cell contents of $C_{12}H_6Cl_4$ (left) and $C_{12}H_6BrCl_3$ (right).

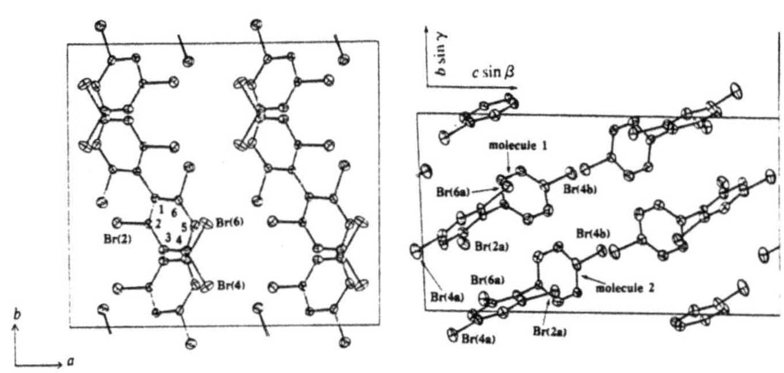

Fig. 2. The unit cell contents of $C_{12}H_4Br_6$ (left) and $C_{12}H_6Br_4$ (right).

4,4'-DICHLOROBENZOPHENONE
$C_{13}H_8Cl_2O$

M.M. GRANGER and M.F. COILLOT, 1985. *Acta Cryst.*, **C41**, 542-543.

Monoclinic, *C2/c*, a = 25.161, b = 6.104, c = 7.535 Å, β = 101.98°, Z = 4. Mo radiation, R = 0.079 for 635 reflections.

There is one half of a molecule per asymmetric unit. The Cl and C(7) atoms (Fig. 1) lie approximately in the plane of the benzene ring. The C-C distances in the ring range from 1.363 to 1.398(4) and C-O is 1.213(4) Å.

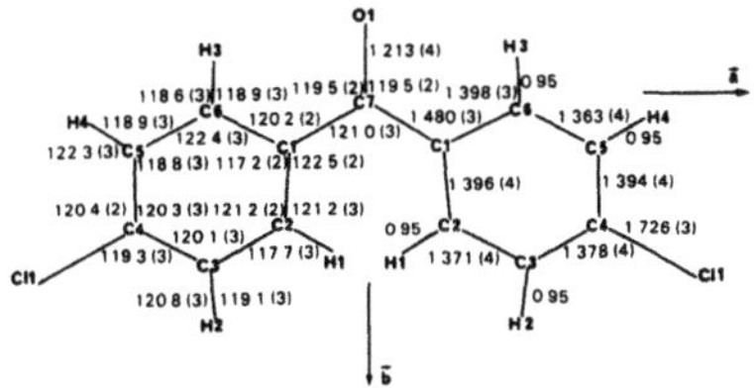

Fig. 1. $C_{13}H_8Cl_2O$: bond lengths and angles.

4-ACETYL-3'-CHLOROBIPHENYL
$C_{14}H_{11}ClO$

H.H. SUTHERLAND, A. RAWAS and M.J. MOTTRAM, 1985. *Acta Cryst.*, C41, 926-927.

Monoclinic, $P2_1/c$, a = 3.88, b = 9.03, c = 32.82 Å, β = 98.5°, Z = 4. Cu radiation, R = 0.068 for 1212 reflections.

The phenyl rings in the molecule (Fig. 1) are planar with an average C-C distance of 1.391 Å. The molecule is non-planar with an angle of 37.7(3)° between the phenyl rings and the acetyl group is rotated about its C-C bond by 4.1(2)°. The C-Cl distance is 1.745(6) Å and makes an angle of 1.3(2) with the phenyl plane. The acetyl groups form non-bonded dimers across centres of symmetry in the cell.

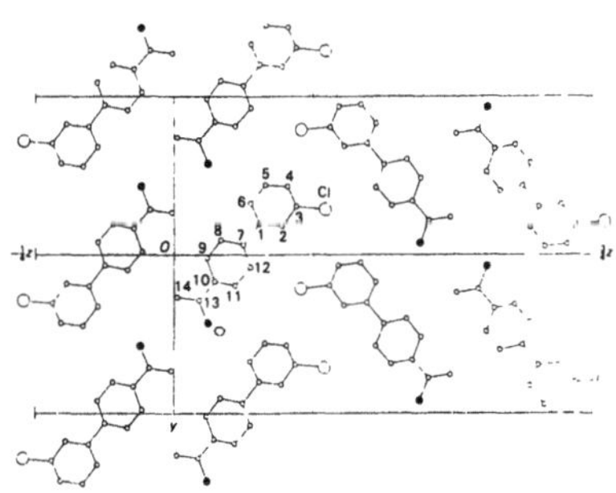

Fig. 1. The unit cell of $C_{14}H_{11}ClO$ viewed along a.

1,1-DIPHENYLETHANOL
$C_{14}H_{14}O$

B.Yu. SULTANOV, A.N. SHNULIN and Kh.S. MAMEDOV, 1985. *Zh. Strukt. Khim.*, 26, 6-163-166 [J. Struct. Chem., 26, 980-982].

Tetragonal, $P\bar{4}2_1c$, a = 17.27, c = 7.665 Å, Z = 8. Mo radiation, R = 0.065 for 636 reflections.

The structure is shown in Fig. 1. Bond lengths (σ 0.01-0.02 Å) are as expected. Hydrogen bonding (O...O 2.89 Å, O-H...O 178.4°) links the molecules into tetramers in the crystal.

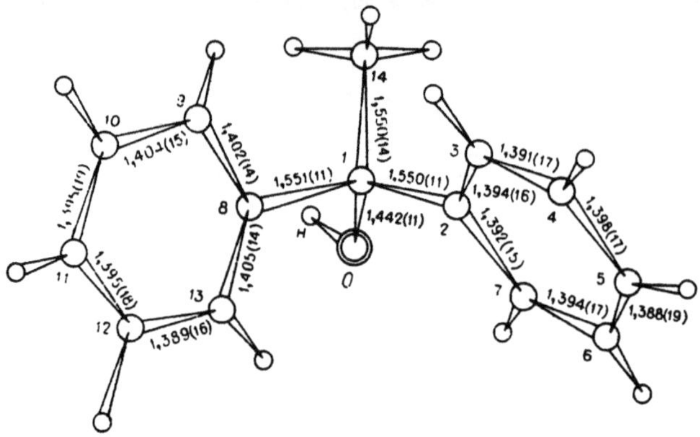

Fig. 1. The $CH_3C(Ph)_2(OH)$ molecule.

1,1-DIPHENYLPROPANOL
$C_{15}H_{16}O$

B.Yu. SULTANOV, A.N. SHAULIN and Kh.S. MAMEDOV, 1985. *Zh. Strukt. Khim.*, **26**, 2-187-190 [J. Struct. Chem., **26**, 313-315].

Monoclinic, $P2_1/c$, a = 8.811, b = 11.891, c = 12.007 Å, β = 104.57°, Z = 4. Mo radiation, R = 0.070 for 595 reflections.

Bond lengths are shown in Fig. 1 (σ(C-C) 0.007-0.009 Å in the phenyl rings and 0.018-0.025 Å for the others). The most important feature of the structure is an interaction between the hydroxyl hydrogen and the π-system of a phenyl ring (H...C 2.36 and 2.45 Å).

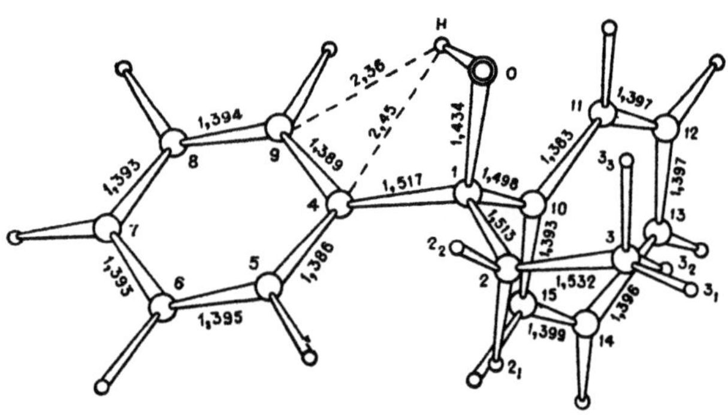

Fig. 1. The structure of 1,1-diphenylpropanol.

(E)-1,2-BIS(2-METHYLPHENYL)ETHENE ((E)-2,2'-DIMETHYLSTILBENE)
$C_{16}H_{16}$

J. TIRADO-RIVES, F.R. FRONCZEK and R.D. GANDOUR, 1985. *Acta Cryst.*, C**41**, 1327-1329.

Monoclinic, $P2_1/n$, a = 8.680, b = 6.687, c = 10.886 Å, β = 98.55°, Z = 2. Mo radiation, R = 0.044 for 658 reflections (at 299 K).

In the molecule (Fig. 1), which is located on a crystallographic inversion centre, the olefinic C(7)=C(7') double bond is unusually short and of length 1.284(4) Å. The planar phenyl ring is twisted out of the plane of the double bond by 10.7(3)°. The C(1)-C(7) bond length is 1.469(2) Å and the C(1)-C(7)-C(7') valence angle is 128.7(3)°.

Bond lengths in the aromatic ring which involve atoms C(4) and C(5) show an apparent systematic shortening from expected values. No unusually short intermolecular distances are reported.

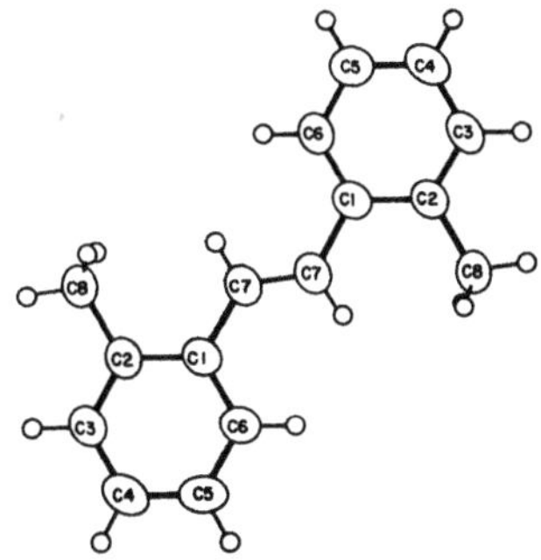

Fig. 1. A view of the $C_{16}H_{16}$ molecule.

o-t-BUTYLBENZOPHENONE
$C_{17}H_{18}O$

P.J. WAGNER, B.P. GIRI, J.C. SCAIANO, D.L. WARD, E. GABE and F.L. LEE, 1985. *J. Am. Chem. Soc.*, **107**, 5483-5490.

Orthorhombic, *Pbca*, a = 12.770, b = 16.082, c = 13.706 Å, Z = 8. Mo radiation, R = 0.051 for 1153 reflections.

The t-butylphenyl ring (Fig. 1) is twisted 69° from coplanarity with the carbonyl and two of the t-butyl H atoms are 2.2-2.7 Å from the O atom. The phenyl ring is only twisted 11° from coplanarity with the carbonyl group.

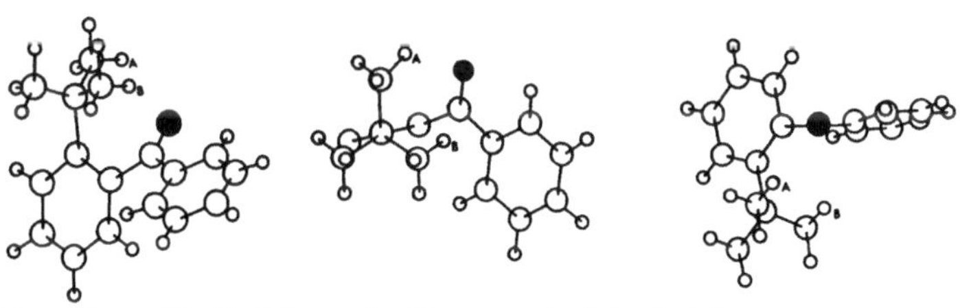

Fig. 1. Three views of $C_{17}H_{18}O$ (some ring H atoms omitted for clarity).

2,2',4,4',6,6'-HEXAMETHYLBIPHENYL (BIMESITYL)
$C_{18}H_{22}$ $(Me_3C_6H_2)_2$

R. FROHLICH and H. MUSSO, 1985. *Chem. Ber.*, **118**, 4649-4651.

I. Monoclinic, $P2_1/c$, a = 8.215, b = 8.548, c = 22.059 Å, β = 95.92°, Z = 4. Mo radiation, R = 0.084 for 1646 reflections.

II. Monoclinic, $P2_1/c$, a = 8.168, b = 8.515, c = 21.743 Å, β = 96.67°, Z = 4. Mo radiation, R = 0.079 for 1040 reflections (at 200K).

The interplanar angle between the two aromatic rings is 89.7° at 295 K and 89.6° at 200 K. Mean dimensions are: (295 K) C-C(arom) 1.391(10), C(arom)-C(sp^3) 1.525(6) Å; the corresponding values at 200 K are 1.392(9) and 1.517(9) Å.

2,2',4,4',6,6'-HEXAMETHYLBENZIL (MESITIL)
$C_{20}H_{22}O_2$

S.C. BISWAS and R.K. SEN, 1985. *Z. Kristallogr.*, **171**, 235-241.

Orthorhombic, *Pbca*, a = 23.982, b = 16.593, c = 8.683 Å, Z = 8. Mo radiation, R = 0.052 for 1901 reflections.

In the molecule (Fig. 1) the two carbonyl groups have a transoid conformation (dihedral angle 178.3° between "carbonyl" planes). The planes of the adjacent mesityl groups form dihedral angles of 75.4 and 78.3° with the carbonyl systems.

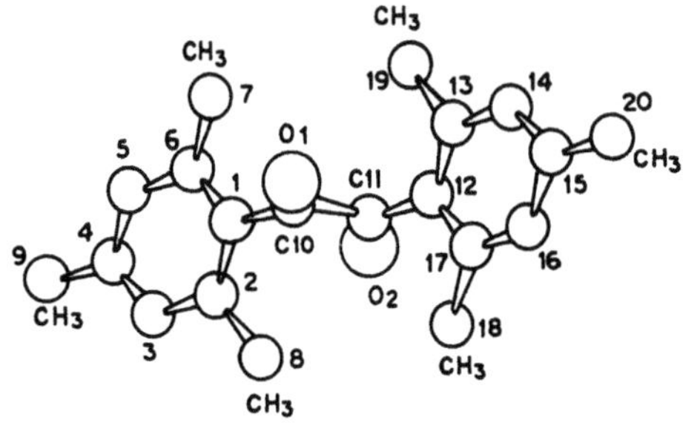

Fig. 1. A view of the $C_{20}H_{22}O_2$ molecule.

1-ETHOXY-3,3-DIPHENYL-1-p-TOLYL-2-AZAALLENIUM HEXACHLOROANTIMONATE
$C_{23}H_{22}Cl_6NOSb$ $C_{23}H_{22}NO^+,Cl_6Sb^-$

R. KUPFER, E.-U. WURTHWEIN, M. NAGEL and R. ALLMANN, 1985. *Chem. Ber.*, **118**, 643-652.

Orthorhombic, *Fdd2*, a = 31.613, b = 33.611, c = 10.456 Å, Z = 16. Mo radiation, R = 0.058 for 1601 reflections.

The cation (Fig. 1) is shown to have features of both the 2-azaallyl and 2-azaallenium types and is characterized by a central C-N-C system with C-N 1.28 and 1.32 Å, C-N-C 141°, and a torsion angle of 83° between the two methylene planes. The anion has Sb-Cl 2.35-2.38 Å.

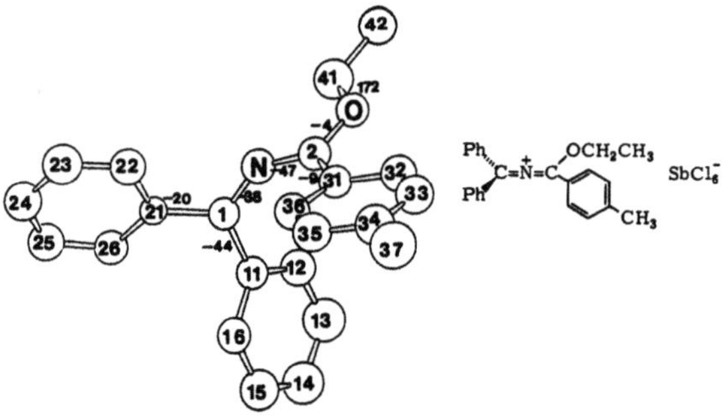

Fig. 1. The $C_{23}H_{22}NO^+$ ion with selected torsion angles.

TRIMESITYLETHENOL
$C_{29}H_{34}O$ (I)

2,2-DIMESITYL-1-(9-ANTHRYL)ETHENOL ETHANOL SOLVATE
$C_{34}H_{32}O,C_2H_6O$ (II)

1,2-DIMESITYL-2-PHENYLETHENOL
$C_{26}H_{28}O$ (III)

TRIMESITYLETHANONE
$C_{29}H_{34}O$ (IV)

1,2-DIMESITYL-2-PHENYLETHANONE
$C_{26}H_{28}O$ (V)

TRIMESITYLETHYLENE
$C_{29}H_{34}$ (VI)

M. KAFTORY, S.E. BIALI and Z. RAPPOPORT, 1985. *J. Am. Chem. Soc.*, **107**, 1701-1709.

I. Triclinic, $P\bar{1}$, a = 15.621, b = 15.774, c = 11.468 Å, α = 100.51, β = 82.70, γ = 119.68°, Z = 4. Mo radiation, R = 0.065 for 5675 reflections.

II. Monoclinic, $P2_1/c$, a = 8.919, b = 27.185, c = 12.153 Å, β = 93.77°, Z = 4. Mo radiation, R = 0.073 for 2627 reflections.

III. Triclinic, $P\bar{1}$, a = 11.960, b = 12.311, c = 8.014 Å, α = 109.09, β = 97.53, γ = 72.80°, Z = 2. Mo radiation, R = 0.076 for 2356 reflections.

IV. Orthorhombic, $Pca2_1$, a = 17.183, b = 8.413, c = 16.503 Å, Z = 4. Mo radiation, R = 0.065 for 1332 reflections.

V. Orthorhombic, *Pbca*, a = 22.927, b = 15.359, c = 11.664 Å, Z = 8. Mo radiation, R = 0.083 for 1948 reflections.

VI. Triclinic, $P\bar{1}$, a = 15.367, b = 10.323, c = 8.288 Å, α = 107.64, β = 84.45, γ = 109.09°, Z = 2. Mo radiation, R = 0.077 for 3644 reflections.

In the enols (I), (II), (III) and (VI) the three aryl rings are rotated in the same sense from the plane of the double bond, giving a helical propeller conformation. The relevant torsion angles agree well with expected values except in (III) where C(1)-C(2)-Ar(1) and C(1)-C(2)-Ar(2) are 38.3 and 74.4° respectively. Bond lengths and angles are normal. There is evidence of conformational dependence on hydrogen bonding, with the torsion angle C=C-O-H of the hydroxy group being 14.1° in (I), but 166° in (II) when the OH group is involved in a hydrogen bond. The ketones (IV) and (V) have normal bond lengths but show some deviations from tetrahedral bond angles at C(2). The C=O and C-H bonds are anti. The Ar_3 and H are nearly eclipsed and the C=O group is bisected by Ar_2CH. The ketones have a distorted propeller shape and are chiral. There are some differences in bond angles and non-bonded contacts due to steric overcrowding within the sequence of structures. The two independent molecules of (I) are similar and there is some disorder of the solvent in (II) and the ethylene moiety in (III).

1,2,2-TRIMESITYLETHANOL
$C_{29}H_{36}O$

S.E. BIALI and Z. RAPPOPORT, 1985. *J. Am. Chem. Soc.*, **107**, 1007-1015.

Monoclinic, $P2_1/c$, a = 12.131, b = 9.553, c = 21.318 Å, β = 105.59°, Z = 4. Mo radiation, R = 0.094 for 2107 reflections.

The mesityl groups (Fig. 1) exhibit an HCCH torsion angle of 74°, and an HCCO angle of 167°. The central C(1)-C(2) bond length is 1.572(7) Å. Two pairs of enantiomers form the unit cell.

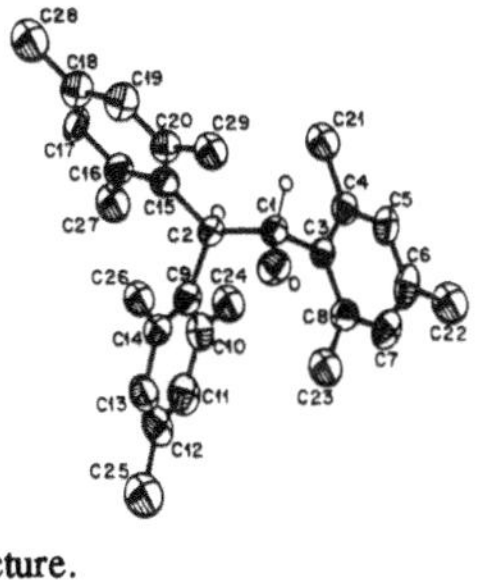

Fig. 1. $C_{29}H_{36}O$: the molecular structure.

TETRABENZYLETHYLENE
$C_{30}H_{28}$

L. ANDERSEN, U. BERG and I. PETTERSSON, 1985. *J. Org. Chem.*, **50**, 493-499.

Monoclinic, *C2/c*, a = 26.05, b = 13.450, c = 21.360 Å, β = 113.18°, Z = 12. Mo radiation, R = 0.058 for 1922 reflections.

There are two independent molecules in the crystallographic unit cell and these have slightly different conformations.

meso-2,5-DIMETHOXY-1,1,3,4,6,6-HEXAPHENYL-1,5-HEXADIENE
$C_{44}H_{38}O_2$

M.W. KLETT and R.P. JOHNSON, 1985. *J. Am. Chem. Soc.*, **107**, 6615-6620.

Triclinic, $P\bar{1}$, a = 12.955, b = 13.558, c = 10.623 Å, α = 105.03, β = 108.34, γ = 75.78°, Z = 2. Mo radiation, R = 0.072 for 3124 reflections.

The structure of this meso isomer was established as shown in Fig. 1.

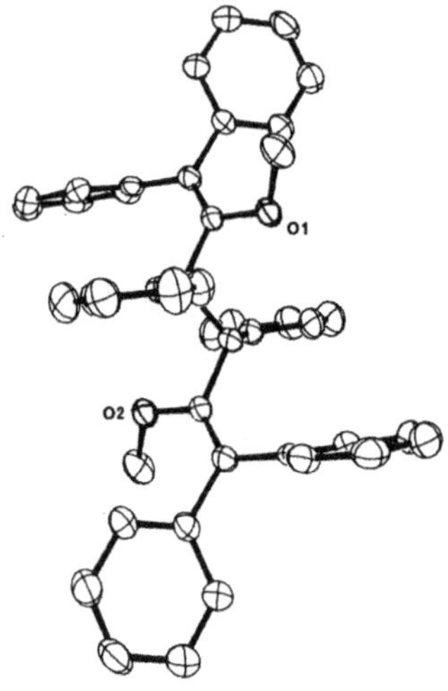

Fig. 1. A view of $C_{44}H_{38}O_2$.

3,3,4,4-TETRAFLUORO-N-METHYL-2-(cis,S-trans-METHYL-N,N,O-AZOXY)-S-cis-1-CYCLOBUTENE-1-AMINE
$C_6H_7F_4N_3O$

H. HOPE, C.B. SHOEMAKER, D.P. SHOEMAKER, H.M. MARSDEN and J.M. SHREEVE, 1985. *J. Org. Chem.*, **50**, 1136-1137.

Monoclinic, $P2_1/c$, a = 6.438, b = 13.185, c = 10.010 Å, β = 92.55°, Z = 4. Mo radiation, R = 0.036 for 941 reflections (at 140 K).

The crystal structure analysis unambiguously identified the molecule (Fig. 1). All the carbon, nitrogen and oxygen atoms are essentially coplanar; the molecules are associated through N-H...O links to weakly hydrogen bonded, centrosymmetric dimers.

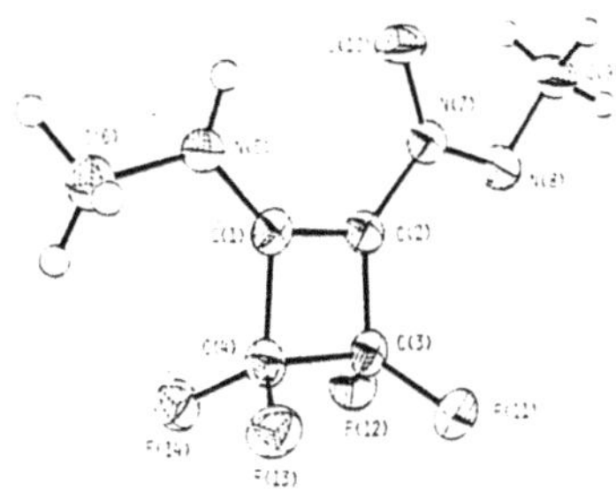

Fig. 1. A molecule of $C_6H_7F_4N_3O$.

2-IMINOCYCLOPENTANEDITHIOIC ACID
$C_6H_9NS_2$

H. MIYAMAE and T. OIKAWA, 1985. *Acta Cryst.*, C**41**, 1489-1490.

Monoclinic, *A2/a*, a = 13.265, b = 10.310, c = 12.198 Å, β = 114.03°, Z = 8. Cu radiation, R = 0.051 for 1135 reflections.

All the non-H atoms of the molecule lie on a plane (Fig. 1). The N-C(3) distance of 1.305(4) Å is consistent with a double bond and C(1)-C(2) and C(2)-C(3) (1.394(4), 1.400(4) Å) are short enough to indicate that the C(2) atom is sp^2 hybridized. The S(1)-C(1) and S(2)-C(1) distances are 1.690(3) and 1.762(3) Å. In the crystal, a couple of centrosymmetrically related molecules almost overlap each other.

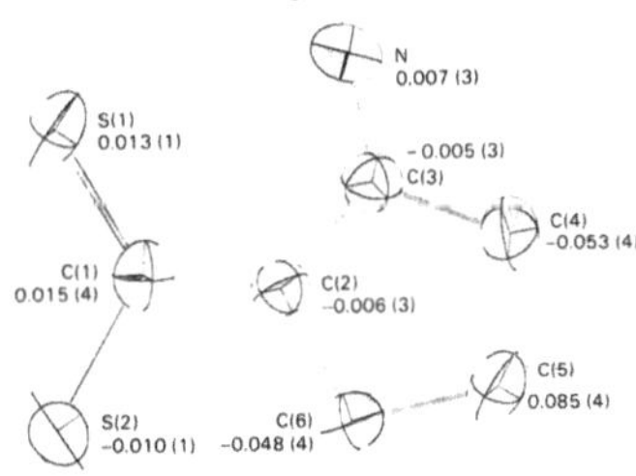

Fig. 1. $C_6H_9NS_2$: the molecular structure.

DISODIUM (3,4-BIS(CYANIMINO)-1-CYCLOBUTENE-1,2-DIYL)BIS(CYANAMIDATE) TETRAHYDRATE
$C_8N_8Na_2,4(H_2O)$ $C_8N_8^{2-},2(Na^+),4(H_2O)$

K. KOHLER, W. MASSA, G. OFFERMANN, G. SEITZ and R. SUTRISNO, 1985. *Chem. Ber.*, **118**, 1903-1916.

Monoclinic, $P2_1/c$, a = 3.603, b = 11.037, c = 17.301 Å, β = 90.37°, Z = 2. Mo radiation, R = 0.037 for 1410 reflections.

The anion lies about an inversion centre (Fig. 1) and consequently the four-membered ring is planar. The ring C-C distances are 1.447(2) and 1.452(2) Å with C(1)-N(1) 1.316(2), C(2)-N(3) 1.310(2) Å. The Na^+ ion is seven-coordinate by four O atoms 2.340-2.419(2) Å and three N atoms (2.507-2.582(2) Å). The water molecules also hydrogen bond to the anion nitrogens (Fig. 1).

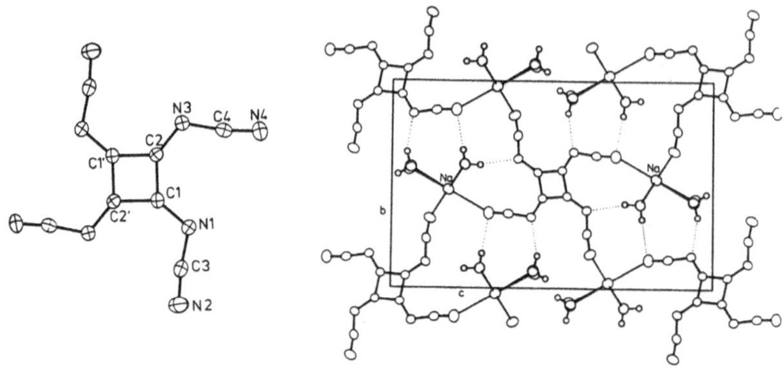

Fig. 1. The $C_8N_8^{2-}$ cation and the crystal packing viewed down the short *a*-axis.

(Z)-2-CHLORO-3,3-DIMETHYL-2-(TRICHLOROVINYL)CYCLOPROPANECARBOXYLIC ACID $C_8H_8Cl_4O_2$

W. WEBER and A. DE MEIJERE, 1985. *Chem. Ber.*, **118**, 2450-2471.

Monoclinic, $P2_1/n$, a = 9.701, b = 12.397, c = 9.840 Å, β = 104.94°, Z = 4. Mo radiation, R = 0.076.

Details of structure and dimensions are shown in Fig. 1. The trichlorovinyl group forms a dihedral angle of 88° with the C_3 ring plane.

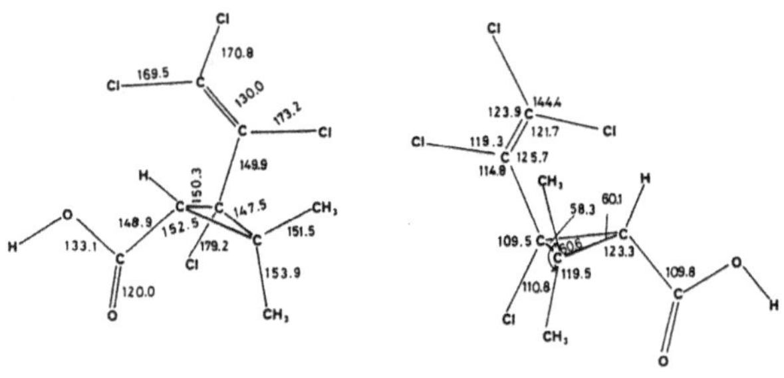

Fig. 1. The $C_8H_8Cl_4O_2$ molecule with bond lengths (pm) and angles (°).

METHYL 2-t-BUTYLTHIO-1-CHLORO-2-CYANO-1-CYCLOPROPANECARBOXYLATE $C_{10}H_{14}ClNO_2S$

B. TINANT, J.P. DECLERCQ and M. VAN MEERSSCHE, 1985. *Acta Cryst.*, **C41**, 597-599.

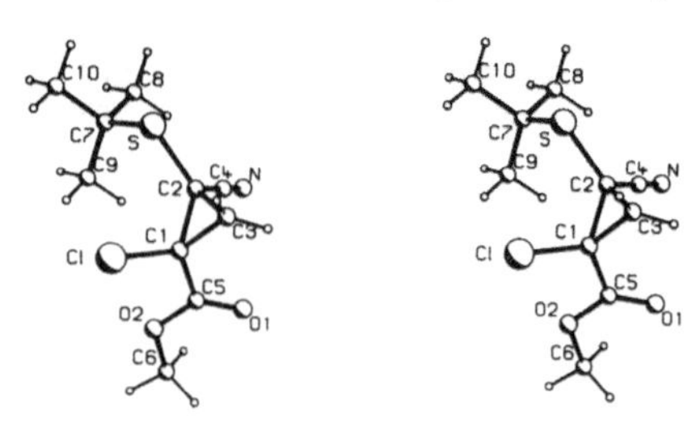

Fig. 1. $C_{10}H_{14}ClNO_2S$: stereoscopic view of the molecule.

Monoclinic, $P2_1/n$, a = 8.953, b = 23.338, c = 6.041 Å, β = 99.66°, Z = 4. Mo radiation, R = 0.044 for 1313 reflections.

The cyclopropyl ring bond lengths in the molecule (Fig. 1) are significantly different: C(1)-C(2) 1.545(5), C(1)-C(3) 1.477(6) and C(2)-C(3) 1.506(3) Å. The carbonyl group of the methoxycarbonyl moiety has an s-cis conformation (torsion angle O(1)-C(5)-C(1)-M(23) 26(1)°, M = midpoint of C(2)-C(3)).

2-t-BUTYLTHIO-1-r,2-t-DICYANO-1-METHOXYCARBONYLCYCLOPROPANE
$C_{11}H_{14}N_2O_2S$ (I)

2-t-BUTYLTHIO-1-r,2-t-DICYANO-1-PHENYLTHIOCYCLOPROPANE
$C_{15}H_{16}N_2S_2$ (II)

2-t-BUTYLTHIO-1-r,2-t-DICYANO-1-PHENYLCYCLOPROPANE
$C_{15}H_{16}N_2S$ (III)

B. TINANT, S. WU, J.-P. DECLERCQ, M. VAN MEERSSCHE, A. DE MESMAEKER, W. MASAMBA, R. MERENYI and H.G. VIEHE, 1985. *J. Chem. Soc. Perkin Trans.* 2, 535-540.

I. Monoclinic, $P2_1/n$, a = 8.551, b = 24.940, c = 6.124 Å, β = 101.77°, Z = 4. Mo radiation, R = 0.047 for 1572 reflections.

II. Monoclinic, $P2_1/c$, a = 11.821, b = 10.479, c = 25.605 Å, β = 99.80°, Z = 8. Cu radiation, R = 0.064 for 3729 reflections.

III. Triclinic, $P1$, a = 8.910, b = 13.703, c = 6.252 Å, α = 93.18, β = 80.29, γ = 102.92°, Z = 2. Mo radiation, R = 0.040 for 1768 reflections.

In crystals of (II) and (III) there are two molecules in each asymmetric unit but in each case the molecules have very similar geometries. The nitrile groups are in the trans configuration in all three compounds (Fig. 1). The C(1)-C(2) bond is long in all three molecules: 1.580(4) (I), 1.557(5) (II) and 1.555(5) Å (III). The average values of the C(2)-S(2) and S(2)-C(8) distances are 1.778(2) and 1.853(2) Å respectively and the C(2)-S(2)-C(8) angle is 106.9(1)°. In (II) the C(1)-S(1) and S(1)-C(12) distances are 1.798(4) and 1.778(4) Å respectively with a value of 101.5(1)° for the angle C(1)-S(1)-C(12).

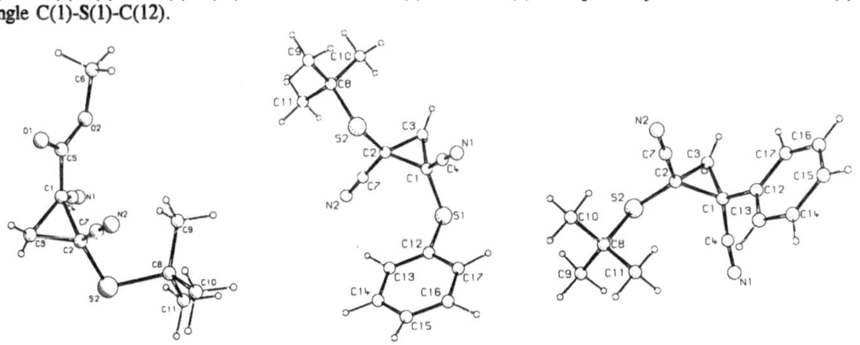

Fig. 1. Perspective drawing of molecules of $C_{11}H_{14}N_2O_2S$ (left), $C_{15}H_{16}N_2S_2$ (centre) and $C_{15}H_{16}N_2S$ (right).

2-ACETYLCYCLOPENTADIENONE-1-PHENYLHYDRAZONE
$C_{13}H_{12}N_2O$

S. HUSEBYE, K. MAARTMANN-MOE, R.E. BOZAK and K.L. RINEHART Jr., 1985. *Acta Chem. Scand.*, B39, 55-64.

Triclinic, $P\bar{1}$, a = 8.424, b = 8.719, c = 15.875 Å, α = 85.24, β = 86.22, γ = 68.81°, Z = 4. Mo radiation, R = 0.077 for 1660 reflections (at -135°C).

There are two crystallographically independent, but quite similar molecules in the asymmetric unit (Fig. 1). The molecules consist of a methyl group and three roughly planar rings: a phenyl ring, a cyclopentadienyl ring and a seven-membered-C-O...H-N-N-C-C- ring held together by a strong hydrogen bond between the acetyl oxygen and hydrazone hydrogen atom. Except for the methyl hydrogens, the molecules are nearly planar.

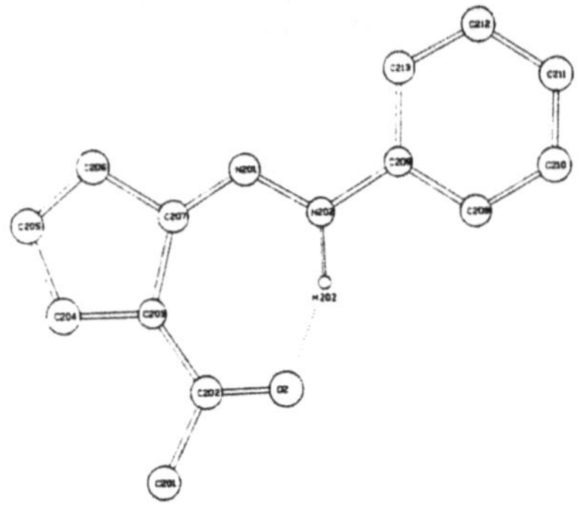

Fig. 1. One of the two crystallographically independent molecules of $C_{13}H_{12}N_2O$ in the crystal.

r-2,c-1,t-3-TRIMETHYL-1-PHENYLSULPHONYLCYCLOBUTANE
$C_{13}H_{18}O_2S$ (I)

r-2,c,c-TRIMETHYL-1-PHENYLSULPHONYLCYCLOBUTANE
$C_{13}H_{18}O_2S$ (II)

Y. GAONI, A. TOMAZIC and E. POTGIETER, 1985. *J. Org. Chem.*, **50**, 2943-2947.

I. Monoclinic, $P2_1/n$, a = 8.125, b = 12.913, c = 11.703 Å, β = 85.42°, Z = 4. Cu radiation, R = 0.110 for 1260 reflections.

II. Monoclinic, $P2_1/n$, a = 6.529, b = 13.933, c = 14.554 Å, β = 88.95°, Z = 4. Cu radiation, R = 0.072 for 2117 reflections.

The crystal structure analyses established the identities of the two molecules (Fig. 1).

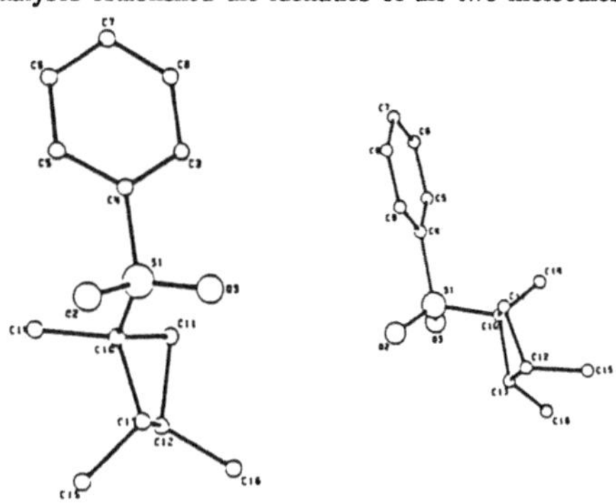

Fig. 1. Molecular structures of $C_{13}H_{18}O_2S$ (I) (left) and $C_{13}H_{18}O_2S$ (II) (right).

BIS(p-CHLOROPHENYL)CYCLOPROPENONE
$C_{15}H_8Cl_2O$

K. PETERS and H.G. VON SCHNERING, 1985. *Chem. Ber.*, **118**, 2147-2149.

Monoclinic, $C2/c$, a = 10.884, b = 12.828, c = 10.279 Å, β = 120.28°, Z = 4. Mo radiation, R = 0.034 for

1217 reflections.

A refinement with a new data set of a structure reported previously (1). The molecule (Fig. 1) has twofold crystallographic symmetry and approximate mm2 symmetry; the maximum displacement of a heavy atom from the molecular plane is 0.19 Å.

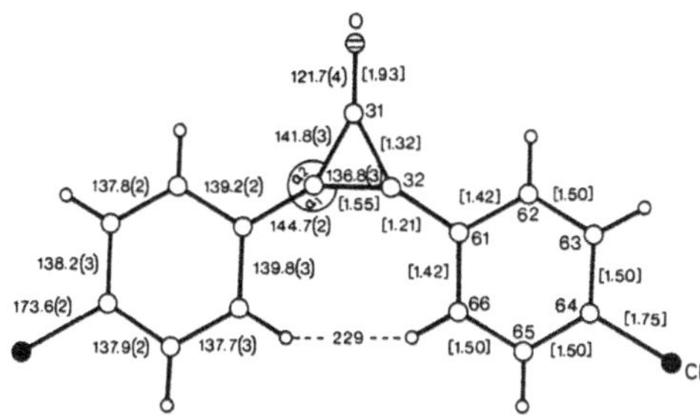

Fig. 1. A view of $C_{15}H_8Cl_2O$ with dimensions (pm). The angles α_1 and α_2 are 149.1 and 149.7° respectively. Numbers in square brackets denote bond-orders.

1. Structure Reports **39B**, 94.

cis-1-(2-METHYL-PROP-1-ENYL)-3-(p-TOLYLSULPHONYL)CYCLOBUTANE
$C_{15}H_{20}O_2S$ (I)

(±)-4-PHENYLSULPHONYLTRICYCLO[3.3.0.0^{1,4}]OCTANE
$C_{14}H_{16}O_2S$ (II)

Y. GAONI and A. TOMAZIC, 1985. *J. Org. Chem.*, **50**, 2948-2957.

I. Monoclinic, *P2/c*, a = 17.069, b = 5.831, c = 17.299 Å, β = 60.83°, Z = 4. Mo radiation, R = 0.092 for 1832 reflections.

II. Monoclinic, $P2_1/c$, a = 11.851, b = 9.276, c = 11.553 Å, β = 91.62°, Z = 4. Cu radiation, R = 0.042 for 2267 reflections.

The crystal structure analyses established the stereochemistries of the two molecules (Fig. 1).

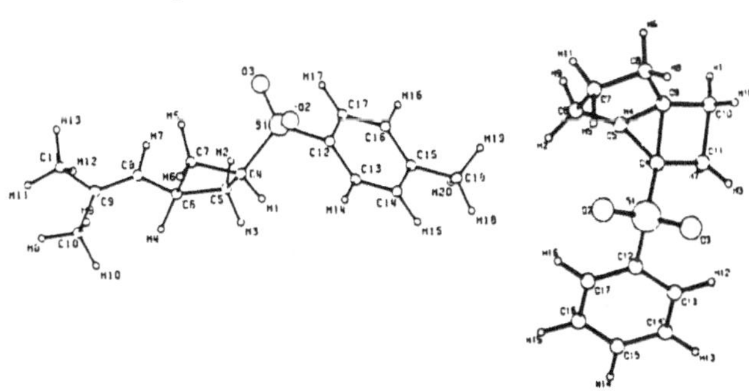

Fig. 1. Molecular structures of molecules of $C_{15}H_{20}O_2S$ (left) and of $C_{14}H_{16}O_2S$ (right).

(R)-(+)-3-FLUORO-2,2-DIPHENYLCYCLOPROPANECARBOXYLIC ACID
$C_{16}H_{13}FO_2$

W.H. RIES and I. BERNAL, 1985. *J. Mol. Struct.*, **131**, 371-381.

Orthorhombic, $P2_12_12_1$, a = 6.172, b = 7.890, c = 27.091 Å, Z = 4. Mo radiation, R = 0.0329 for 739 reflections.

The plane of the carboxyl group (C(1),O(1),O(2)) (Fig. 1) is nearly orthogonal to the plane of the cyclopropane ring (dihedral angle 99.05°) while the dihedral angles between the two phenyl rings and the cyclopropane ring are 117.77 and 63.18° respectively. Because the molecule is chiral, the normal centrosymmetric hydrogen-bonded carboxylic acid dimer is not possible; instead chains of O-H...O linked molecules are formed about twofold screw axes. The bond lengths in the cyclopropane ring are C(2)-C(3) 1.496(9), C(2)-C(4) 1.563(9), C(3)-C(4) 1.494(8) Å. The lengthening of the C(2)-C(4) bond is attributed to an electronic effect of the F atom on C(2).

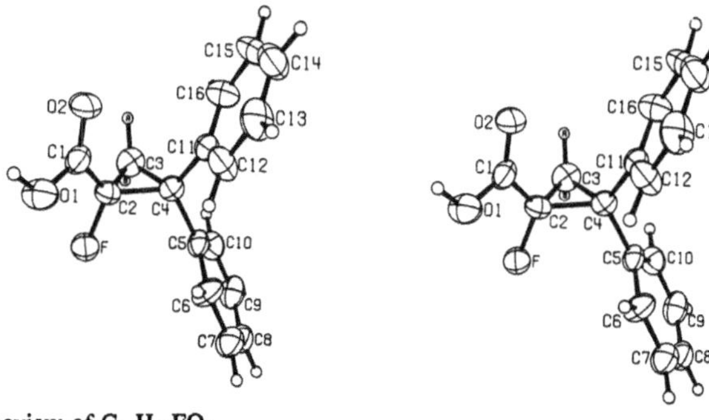

Fig. 1. A stereoview of $C_{16}H_{13}FO_2$.

2,2-DIMETHYL-5-(N-ISOPROPYL)AMINO-6-(2,2-DIMETHYL-1-HYDROXY-CYCLOPROPYL)HEX-4-EN-3-ONE
$C_{16}H_{29}NO_2$

J.P. DECLERCQ, N. DE KIMPE, M. PALAMAREVA and N. SCHAMP, 1985. *Bull. Soc. Chim. Belg.*, **94**, 301-302.

Monoclinic, $P2_1/c$, a = 5.826, b = 19.328, c = 15.293 Å, β = 95.85°, Z = 4. Mo radiation, R = 0.095 for 1240 reflections.

The analysis establishes the structure and conformation as shown in Fig. 1. There is an intramolecular N-H...O hydrogen bond (2.67 Å) between N(10) and O(9).

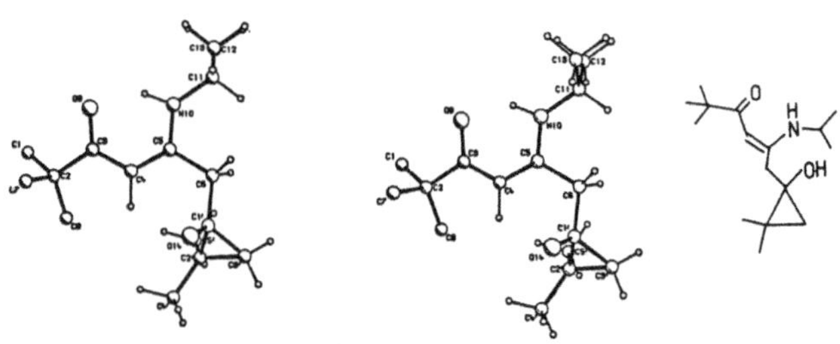

Fig. 1. A stereoview of $C_{16}H_{29}NO_2$.

cis-1,2-DICYANO-1,2-BIS(PHENYLTHIO)CYCLOPROPANE
$C_{17}H_{12}N_2S_2$

B. TINANT, J.-P. DECLERCQ and M. VAN MEERSSCHE, 1985. *J. Chem. Soc. Perkin Trans. 2*, 541-543.

Monoclinic, $P2_1/n$, a = 17.775, b = 6.965, c = 12.994 Å, β = 101.57°, Z = 4. Mo radiation, R = 0.046 for 1507 reflections.

There is a long C(1)-C(2) bond of length 1.561(6) Å as observed in similar cyclopropanes (1). The molecule is shown in Fig. 1.

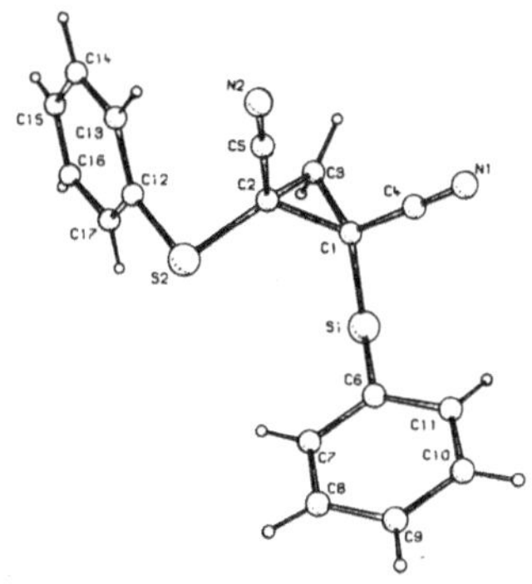

Fig. 1. Perspective drawing of a molecule of $C_{17}H_{12}N_2S_2$.

1. This volume, p 109.

2,2,2',2',5,5,5',5'-OCTAMETHYL-2,2',5,5'-TETRAHYDROFULVALENE
$C_{18}H_{28}$

T. PILATI and M. SIMONETTA, 1985. *Acta Cryst.*, C**41**, 147-148.

Monoclinic, $P2_1/n$, a = 9.735, b = 8.364, c = 9.985 Å, β = 108.61°, Z = 2. Mo radiation, R = 0.045 for 1062 reflections.

The asymmetric unit consists of half of the centrosymmetric molecule (Fig. 1) which also has approximate 2/m symmetry. The tetrasubstituted ethylene is strictly planar and in the cyclopentene ring, which has an envelope conformation, the C(3)-C(4) double bond is shortened to 1.289(3) Å. The ethylene double bond C(1)-C(1') is normal and of length 1.349(3) Å while steric hindrance extends the neighbouring single bonds to an average length of 1.583(2) Å. Steric hindrance also affects the valence angles involving methyl groups which range from 105.8(2) to 118.0(2)°.

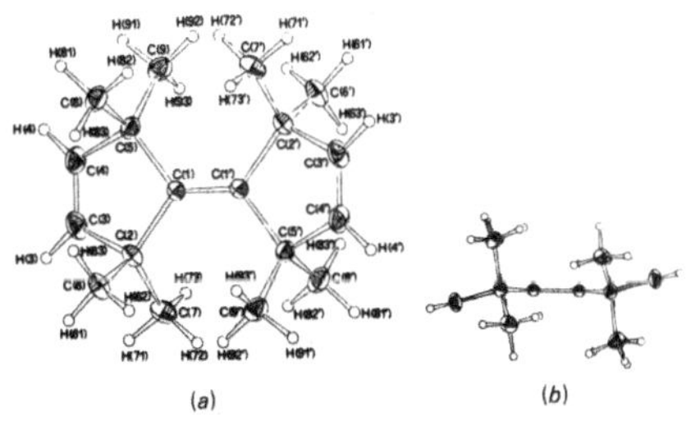

Fig. 1. $C_{18}H_{28}$: (a) a frontal view of the molecule showing the atomic numbering scheme and (b) a side view.

13,14-DIDEHYDRO-11-DEOXY-15-KETO-(11R,13E,15S)-11,15-DIHYDROXY-9-OXOPROST-13-ENOIC ACID (13,14-DIDEHYDRO-11-DEOXY-15-KETOPROSTAGLANDIN E1)
$C_{20}H_{30}O_4$

J.D. OLIVER and L.C. STRICKLAND, 1985. *Acta Cryst.*, C**41**, 1477-1480.

Monoclinic, $P2_1/n$, a = 5.544, b = 12.834, c = 27.587 Å, β = 92.22°, Z = 4. Mo radiation, R = 0.059 for 2463 reflections (at 173 K).

The molecule (Fig. 1) has an 'L'-shaped conformation, which is not the conformation usually observed for this type of molecule (hairpin). The alkyl chains have the fully extended all-trans conformation and the five-membered ring adopts a C(12)-envelope conformation. The molecule as a whole is relatively planar and has acceptable bond lengths and angles. Packing is dominated by strong H bonds between the carboxyl groups of adjacent molecules (O(1)...O'(2) 2.652(4) Å).

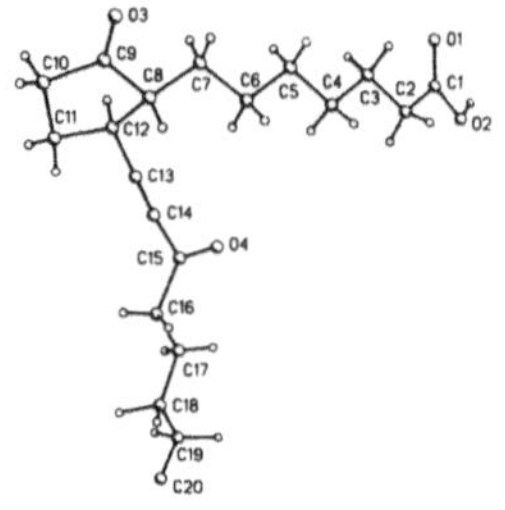

Fig. 1. A view of the $C_{20}H_{30}O_4$ molecule.

BIS(PENTAMETHYLCYCLOPENTADIENYL) SULPHIDE
$C_{20}H_{30}S$ $(Me_5C_5)_2S$

A.J. BARD, A.H. COWLEY, J.K. LELAND, G.J.N. THOMAS, N.C. NORMAN, P. JUTZI, C.P. MORLEY and E. SCHLUTER, 1985. *J. Chem. Soc. Dalton Trans.*, 1303-1307.

Monoclinic, $P2_1/c$, a = 10.375, b = 11.274, c = 16.132 Å, β = 91.92°, Z = 4. Mo radiation, R = 0.073 for 1863 reflections.

The molecule contains two cyclopentadiene rings σ-bonded to S via the tetrahedral ring atom (Fig. 1). Bond lengths and angles include: S-C 1.848, 1.864(11), ring C-C 1.315-1.329, 1.454, 1.479, 1.497-1.509(6), C-Me 1.478-1.509(6) Å, C-S-C 111.7(2), S-C-C (ring) 113.7-115.4(3), S-C-Me 102.5, 102.9(3), ring C-C-C (sp^3) 102.5, 102.8, (sp^2) 108.3-110.4(4)°. The rings are slightly staggered, and the dihedral angle between them is 12°.

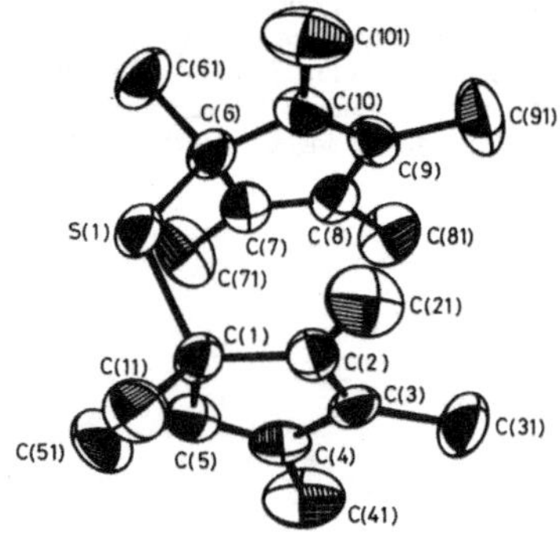

Fig. 1. $C_{20}H_{30}S$: the molecular structure.

1,2,3-TRIPHENYLCYCLOPROPENE
$C_{21}H_{16}$ (I)

3-ISOPROPYL-1,2,3-TRIPHENYLCYCLOPROPENE
$C_{24}H_{22}$ (II)

I.N. DOMNIN, J. KOPF, S. KEYANIYAN and A. DE MEIJERE, 1985. *Tetrahedron*, **41**, 5377-5382.

I. Monoclinic, $I2/c$, a = 15.127, b = 8.119, c = 24.586 Å, β = 93.61°, Z = 8. Mo radiation, R = 0.040 for 1952 reflections.

II. Monoclinic, $C2$, a = 15.799, b = 11.767, c = 9.708 Å, β = 92.06°, Z = 4. Mo radiation, R = 0.028 for 1688 reflections.

Both molecules (Fig. 1) adopt geometries with bisected conformations of the 3-phenyl groups (dihedral angles with the plane C(1),C(2),C(3) of 89.8 and 86.6° for (I) and (II) respectively) and a completely planar arrangement of the 1- and 2-phenyl groups with the three-membered rings. In the cyclopropene ring the C-C distances are 1.293(4), 1.513(3) and 1.518(3) Å in (I) and 1.298(5), 1.518(5) and 1.514(7) Å in (II).

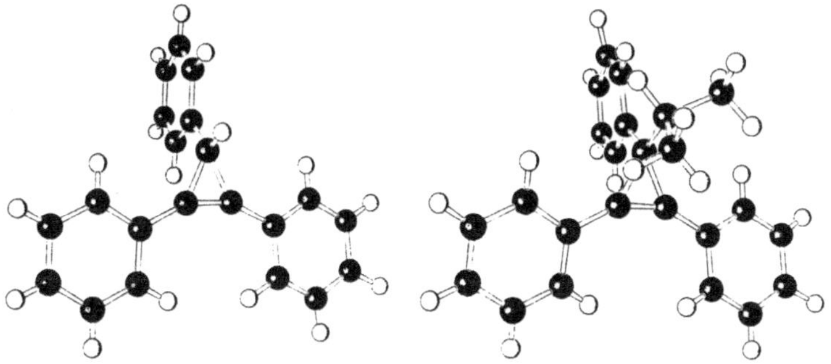

Fig. 1. Views of $C_{21}H_{16}$ (I) (left) and $C_{24}H_{22}$ (II) (right).

DIETHYL trans-2-PHENYL-3-PHENYLSULPHONYLCYCLOPROPANE-1,1-DICARBOXYLATE $C_{21}H_{22}O_6S$ (I)

ETHYL 2-METHYL-4-PHENYL-5-PHENYLSULPHONYL-4,5-DIHYDROFURAN-3-CARBOXYLATE $C_{20}H_{20}O_5S$ (II)

I. YAMAMOTO, T. SAKAI, K. OHTA, K. MATSUZAKI and K. FUKUYAMA, 1985. *J. Chem. Soc. Perkin Trans. 1*, 2785-2787.

I. Triclinic, $P\bar{1}$, a = 11.402, b = 9.870, c = 11.198 Å, α = 64.65, β = 76.95, γ = 65.22°, Z = 2. Cu radiation, R = 0.071 for 2586 reflections.

II. Triclinic, $P\bar{1}$, a = 11.783, b = 7.939, c = 11.441 Å, α = 111.65, β = 91.56, γ = 108.00°, Z = 2. Cu radiation, R = 0.074 for 2512 reflections.

The geometry between the phenylsulphonyl and aryl groups is trans in each molecule (Fig. 1). The crystal structure analyses established the identities of the molecules.

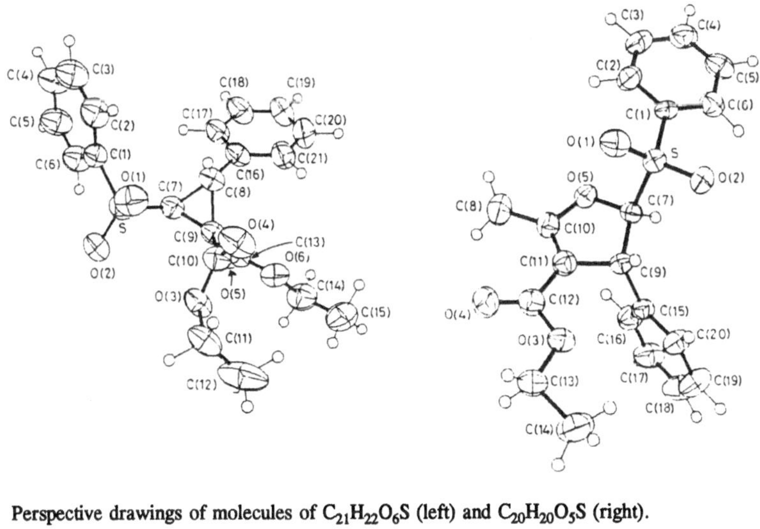

Fig. 1. Perspective drawings of molecules of $C_{21}H_{22}O_6S$ (left) and $C_{20}H_{20}O_5S$ (right).

BIS(sym-TRIPHENYLCYCLOPROPENYLIUM) HEXABROMOTELLURATE(IV)

$C_{42}H_{30}Br_6Te$ $2(C_{21}H_{15}^{+}),Br_6Te^{2-}$

B.A. BORGIAS, R.C. SCARROW, M.D. SEIDLER and W.P. WEINER, 1985. *Acta Cryst.*, **C41**, 476-479.

Monoclinic, $P2_1/c$, a = 13.663, b = 13.360, c = 21.718 Å, β = 93.612°, Z = 4. Mo radiation, R = 0.026 for 3500 reflections.

The structure (Fig. 1) contains nearly octahedral $TeBr_6^{2-}$ anions sandwiched between pairs of independent essentially identical monocations. The Te-Br bond lengths vary between 2.69 and 2.72 Å and the Br-Te-Br angles between 87.6 and 92.1°. The average C-C bond length in the cyclopropenylium ring is 1.377(12) Å and the planar phenyl rings are twisted by approximately 5° out of the plane of this C_3 ring. There are no unusually short interion contacts.

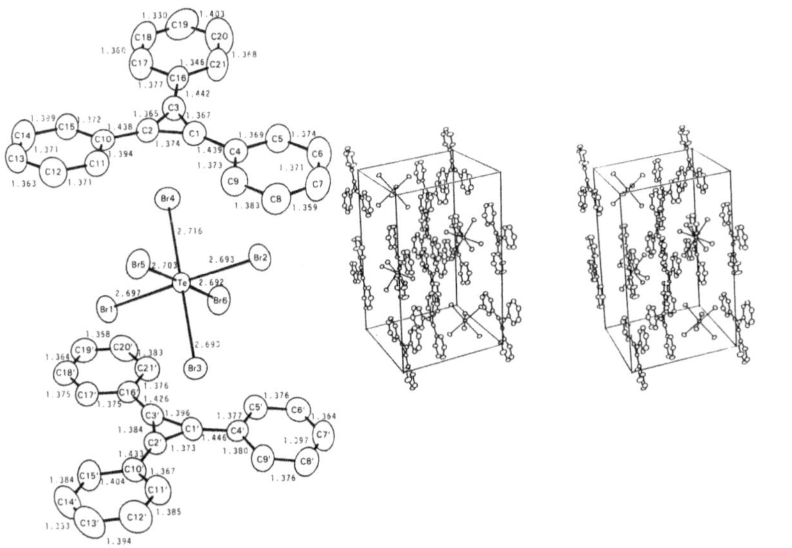

Fig. 1. $C_{42}H_{30}Br_6Te$: a view with bond lengths in the anion and independent cations and a stereoview of the unit cell contents.

6-HYDROXY-2,6-DIMETHYL-2,4,5-TRINITROCYCLOHEX-3-ENONE

$C_8H_9N_3O_8$ (I)

2,6-DIHYDROXY-2,6-DIMETHYL-4,5-DINITROCYCLOHEX-3-ENONE

$C_8H_{10}N_2O_7$ (II)

4-BROMO-6-HYDROXY-2,6-DIMETHYL-2,5-DINITROCYCLOHEX-3-ENONE

$C_8H_9BrN_2O_6$ (III)

M.P. HARTSHORN, J.M. READMAN, W.T. ROBINSON, J. VAUGHAN and A.R. WHYTE, 1985. *Aust. J. Chem.*, **38**, 1693-1704.

I. Orthorhombic, $Pna2_1$, a = 26.606, b = 6.1685, c = 13.815 Å, Z = 8. Cu radiation, R = 0.046 for 1255 reflections.

II. Monoclinic, $P2_1/c$, a = 5.804, b = 25.47, c = 14.05 Å, β = 93.84°, Z = 8. Mo radiation, R = 0.053 for 1965 reflections.

III. Triclinic, $P\bar{1}$, a = 6.332, b = 8.040, c = 11.188 Å, α = 97.58, β = 94.58, γ = 95.03°, Z = 2. Mo radiation, R = 0.0485 for 1724 reflections.

The crystals of (I) and (II) contain two crystallographically independent molecules in the asymmetric unit. In (I) (Fig. 1), each molecule exists with the alicyclic ring in a flattened skew-boat conformation. The two molecules differ in the orientation of the 2-nitro group, the C(7)-C(2)-N(2)-O(22) torsion angle being 31(1)° for one molecule and 49(1)° for the other. The 4-nitro group eclipses the C(4)-C(3) bond in both molecules. In (II), the alicyclic ring exists in a flattened skew-boat conformation in both molecules. The torsion angles C(3)-C(4)-C(5)-N(5), C(3)-C(4)-N(4)-O(42), H(5)-C(5)-N(5)-O(52) and O(1)-C(1)-C(2)-O(2) are 87.6(5), 11.5(6), -2.4(4) and 74.3(5)°, respectively, in one molecule and 90.0(5), 7.9(6), -4.7(4) and 73.0(5)°, respectively, in the other. In (III) (Fig. 2) the ring system adopts a flattened skew-boat conformation.

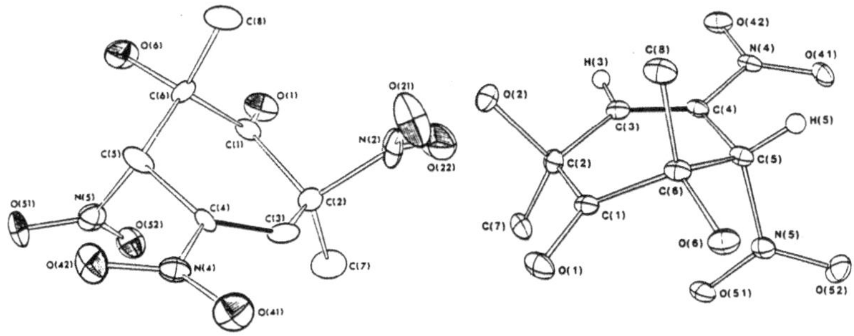

Fig. 1. Views of $C_8H_9N_3O_8$ (I) (left) and $C_8H_{10}N_2O_7$ (II) (right).

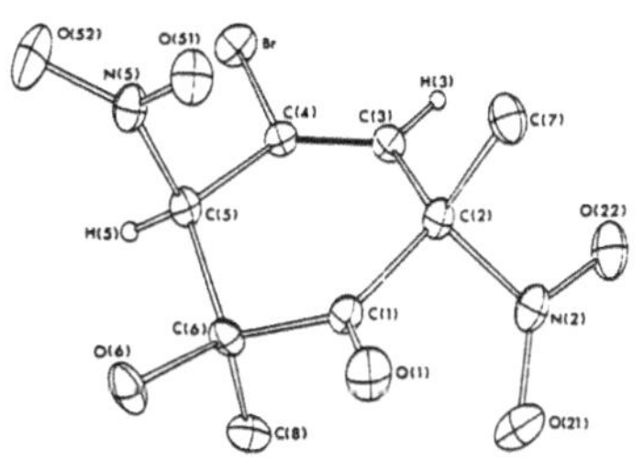

Fig. 2. A view of $C_8H_9BrN_2O_6$ (III).

(-)-cis-2-ETHYLCYCLOHEXANAMINE HYDROCHLORIDE ETHANOL SOLVATE

$C_8H_{18}ClN,0.39(C_2H_6O)$ $C_8H_{18}N^+,Cl^-,0.39(C2H_6O)$

G. KNUPP, A.W. FRAHM, A. KIRFEL, T. FROHLICH and G. WILL, 1985. *Acta Cryst.*, C**41**, 468-470.

Orthorhombic, *I*222, a = 10.804, b = 13.173, c = 15.275 Å, Z = 8. Mo radiation, R = 0.031 for 1866 reflections.

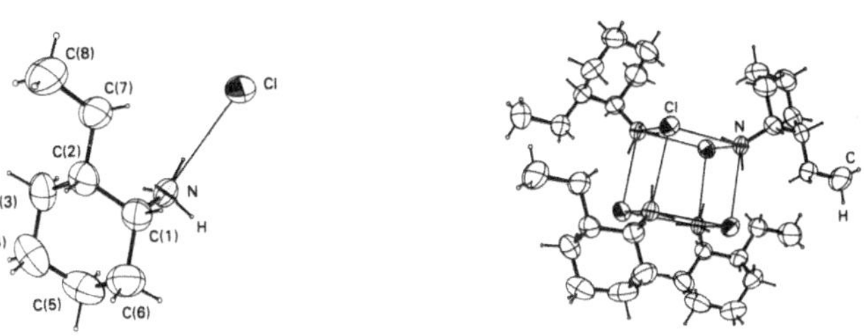

Fig. 1. $C_8H_{18}ClN,0.39(C_2H_6O)$: Views of the unsolvated asymmetric unit and of the salt cluster arrangement of four molecules.

In the crystals the molecule (Fig. 1) possesses the (1R,2S)-cis configuration with the amino group in an axial and the ethyl group in an equatorial position. The cyclohexane ring is in the chair conformation. Bond lengths and valence angles are in good agreement with generally accepted values. Four molecules pack together with four Cl atoms forming a salt cluster through $H_2N...H\text{-}Cl$ hydrogen bonds. The N and Cl atoms form a distorted cubic arrangement with N...Cl separations lying in the range 3.162(3) to 3.192(2) Å. The disordered ethanol molecules are located between the N_4Cl_4 clusters.

3,4-DIBROMO-2,5,6-TRIMETHYL-r-2,c-5,c-6-TRINITROCYCLOHEX-3-ENONE
$C_9H_9Br_2N_3O_7$ (I)

3,4-DIBROMO-2,5,6-TRIMETHYL-r-2,t-5,t-6-TRINITROCYCLOHEX-3-ENONE
$C_9H_9Br_2N_3O_7$ (II)

3,4-DIBROMO-c-6-HYDROXY-2,5,6-TRIMETHYL-r-2,c-5-DINITROCYCLOHEX-3-ENONE
$C_9H_{10}Br_2N_2O_6$ (III)

3,4-DIBROMO-t-6-HYDROXY-2,5,6-TRIMETHYL-r-2,t-5-DINITROCYCLOHEX-3-ENONE
$C_9H_{10}Br_2N_2O_6$ (IV)

M.V. CHAMBERS, M.P. HARTSHORN, W.T. ROBINSON and J. VAUGHAN, 1985. *Aust. J. Chem.*, 38, 133-144.

I. Monoclinic, $P2_1/c$, a = 8.525, b = 26.78, c = 6.229 Å, β = 108.62°, Z = 4. Mo radiation, R = 0.033 for 1268 reflections.

II. Monoclinic, $P2_1/n$, a = 11.915, b = 7.458, c = 15.291 Å, β = 95.37°, Z = 4. Mo radiation, R = 0.040 for 1583 reflections.

III. Monoclinic, $P2_1/c$, a = 10.521, b = 7.057, c = 17.846 Å, β = 100.31°, Z = 4. Mo radiation, R = 0.057 for 1646 reflections.

IV. Monoclinic, $P2_1/c$, a = 12.630, b = 6.228, c = 16.622 Å, β = 91.14°, Z = 4. Mo radiation, R = 0.078 for 948 reflections.

Compound (I) (Fig. 1) exists with the ring in a flattened half-chair conformation and the 5-nitro group in a pseudo-axial orientation. In (II) (Fig. 1), the ring has a skew-boat with the 5-nitro group in a 'flagpole' orientation. In both (I) and (II) the plane of the 5-nitro group is nearly aligned with the C(5)-C(8) bond. This is not true of the 2- and 6-nitro groups and the O(2)-C(7) and C(6)-C(9) bonds. The structures of (III) and (IV) (Fig. 2) are similar to those of (I) and (II), respectively. In (III) the C(1)-O(1) and C(6)-O(6) bonds are approximately coplanar and the O(1)...H(6) distance is 2.10 Å.

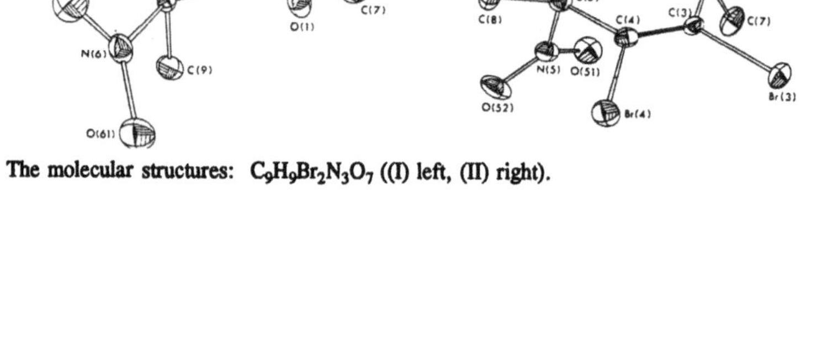

Fig. 1. The molecular structures: $C_9H_9Br_2N_3O_7$ ((I) left, (II) right).

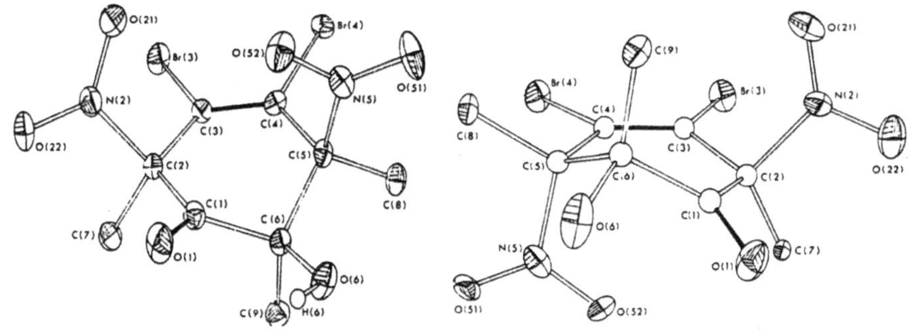

Fig. 2. The structure of $C_9H_{10}Br_2N_2O_6$ (III) (left), (IV) (right).

trans-1,5,5-TRIMETHYLCYCLOHEXANE-1,3-DIOL HEMIHYDRATE
$C_9H_{18}O_2$,0.5(H_2O) (I)

c-3-AMINO-1,5,5-TRIMETHYL-r-CYCLOHEXAN-1-OL
$C_9H_{19}NO$ (II)

F.W. NADER, W. HEINRICH, M. BAAR-SCHAFER and E. HANGEL, 1985. *Chem. Ber.*, **118**, 4314-4329.

I. Monoclinic, *C2/c*, a = 12.184, b = 16.245, c = 10.971 Å, β = 114.21°, Z = 8. Mo radiation, R = 0.060 for 1186 reflections.

II. Monoclinic, $P2_1/n$, a = 9.250, b = 10.880, c = 10.371 Å, β = 108.46°, Z = 4. Mo radiation, R = 0.031 for 1803 reflections.

Views of the two ring systems are shown in Fig. 1 with some intramolecular H...H contacts. Molecule (I) adopts the trans-sc-ap conformation and (II) has the cis-ap-ap form with no intramolecular hydrogen bonding. Molecules are linked by O-H...O (2.728-2.760 Å) hydrogen bonds in (I) and by O-H...N (2.837 Å) and N-H...O (3.030 Å) hydrogen bonds in (II).

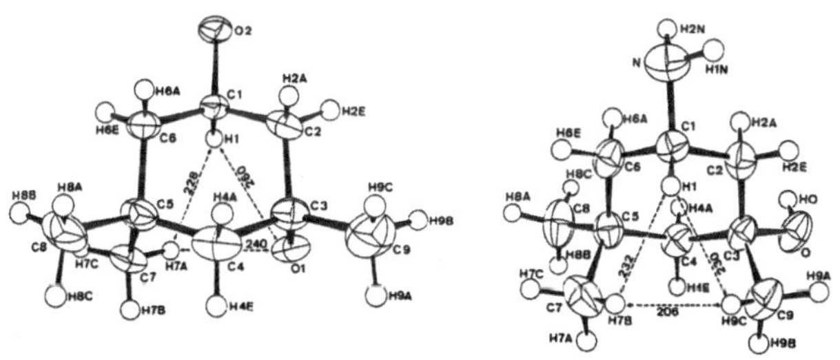

Fig. 1. Views of $C_9H_{18}O_2$ in (I) (left) and $C_9H_{19}NO$ (II) (right).

2,3,5,6-TETRAMETHYL-r-2,4,t-5,c-6-TETRANITROCYCLOHEX-3-ENONE
$C_{10}H_{12}N_4O_9$ (I)

2,3,5,6-TETRAMETHYL-r-2,4,c-5,t-6-TETRANITROCYCLOHEX-3-ENONE
$C_{10}H_{12}N_4O_9$ (II)

r-2-HYDROXY-2,3,5,6-TETRAMETHYL-4,t-5,t-6-TRINITROCYCLOHEX-3-ENONE
$C_{10}H_{13}N_3O_8$ (III)

c-6-HYDROXY-2,3,5,6-TETRAMETHYL-r-2,4,c-5-TRINITROCYCLOHEX-3-ENONE
$C_{10}H_{13}N_3O_8$ (IV)

t-6-HYDROXY-2,3,5,6-TETRAMETHYL-r-2,4,t-5-TRINITROCYCLOHEX-3-ENONE
$C_{10}H_{13}N_3O_8$ (V)

1-ACETYL-2,3,5-TRIMETHYL-4,c-5-DINITROCYCLOPENT-3-ENE-r-1,t-2-DIOL
$C_{10}H_{14}N_2O_7$ (VI)

M.P. HARTSHORN, W.T. ROBINSON, J. VAUGHAN and J.M. WHITE, 1985. *Aust. J. Chem.*, **38**, 145-160.

I. Orthorhombic, $Pca2_1$, a = 13.143, b = 8.537, c = 12.669 Å, Z = 4. Mo radiation, R = 0.051 for 1333 reflections.

II. Monoclinic, $P2_1/c$, a = 13.478, b = 6.341, c = 16.295 Å, β = 95.51°, Z = 4. Mo radiation, R = 0.053 for 1511 reflections.

III. Monoclinic, $P2_1/c$, a = 6.628, b = 8.402, c = 23.941 Å, β = 96.23°, Z = 4. Mo radiation, R = 0.040 for 1341 reflections.

IV. Orthorhombic, $Pna2_1$, a = 15.635, b = 6.132, c = 13.981 Å, Z = 4. Mo radiation, R = 0.046 for 984 reflections.

V. Monoclinic, $P2_1/c$, a = 10.648, b = 15.206, c = 8.363 Å, β = 95.95°, Z = 4. Mo radiation, R = 0.061 for 1112 reflections.

VI. Triclinic, $P\bar{1}$, a = 6.100, b = 8.031, c = 13.619 Å, α = 75.06, β = 83.52, γ = 70.79°, Z = 2. Mo radiation, R = 0.053 for 589 reflections.

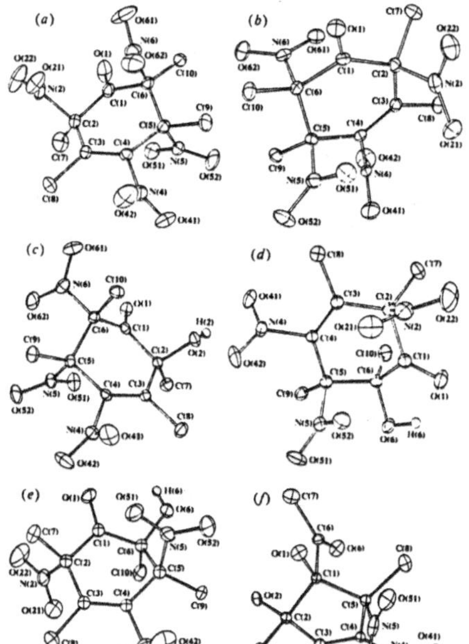

Fig. 1. The molecular structures: (a) $C_{10}H_{12}N_4O_9$ (I), (b) (II), (c) $C_{10}H_{13}N_3O_8$ (III), (d) (IV), (e) (V) and (f) $C_{10}H_{14}N_2O_7$.

Compounds (I) and (II) (Fig. 1) both exist with the cyclohexenone ring in a skew-boat conformation and the 5-nitro group in a 'flagpole' orientation. The orientations of the 4- and 5-nitro groups in (I) and (II) are also very similar. Differences do exist between the orientations of the 2- and 6-nitro groups in the two compounds ((I), O(21)-N-2-C(2)-C(3) 27.0(5)°, O(62)-N(6)-C(6)-C(5)-3.8(5)°; (II), O(21)-N(2)-C(2)-C(3)-11.0(4), O(62)-N(6)-C(6)-C(5) 97.0(5)°). (III) exists with the ring in the skew-boat conformation with the 5-nitro group in a 'flagpole orientation while (IV) has a half-chair ring conformation with the 5-nitro group pseudo-axial. There is a short (2.19 Å) O(1)...H(6) intermolecular contact in (IV) and the C(1)-O(1) and C(6)-O(6) bonds are almost coplanar. In (V) the ring has a skew-boat conformation, O(1)...H(6) is 2.29 Å and the C(1)-O(1) and C(6)-O(6) bonds are approximately coplanar. Compound (VI) is a cyclopentenediol. In this molecule there is no intramolecular hydroxyl-hydroxyl or hydroxyl-carbonyl hydrogen bonding. The 5-nitro group eclipses the C(5)-C(4) bond.

2,4,5,6-TETRAMETHYL-r-2,t-5,c-6-TRINITROCYCLOHEX-3-ENONE
$C_{10}H_{13}N_3O_7$ (I)

2,4,5,6-TETRAMETHYL-r-4,t-5,t-6-TRINITROCYCLOHEX-2-ENONE
$C_{10}H_{13}N_3O_7$ (II)

2,4,6,6-TETRAMETHYL-r-2,t-5-DINITROCYCLOHEX-3-ENONE
$C_{10}H_{14}N_2O_5$ (III)

2,4,6,6-TETRAMETHYL-r-4,t-5-DINITROCYCLOHEX-2-ENONE
$C_{10}H_{14}N_2O_5$ (IV)

c-6-HYDROXY-2,4,5,6-TETRAMETHYL-r-2,c-5-DINITROCYCLOHEX-3-ENONE
$C_{10}H_{14}N_2O_6$ (V)

r-2-HYDROXY-2,4,6,6-TETRAMETHYL-t-5-NITROCYCLOHEX-3-ENONE
$C_{10}H_{15}NO_4$ (VI)

M.P. HARTSHORN, J.M. READMAN, W.T. ROBINSON and J. VAUGHAN, 1985. *Aust. J. Chem.*, **38**, 587-603.

I. Monoclinic, $P2_1/c$, a = 14.405, b = 7.212, c = 12.271 Å, β = 94.56°, Z = 4. Mo radiation, R = 0.054 for 1300 reflections.

II. Monoclinic, $P2_1/n$, a = 11.924, b = 8.491, c = 13.016 Å, β = 101.67°, Z = 4. Mo radiation, R = 0.063 for 1445 reflections.

III. Triclinic, $P\bar{1}$, a = 6.440, b = 13.919, c = 14.413 Å, α = 68.91, β = 85.37, γ = 88.07°, Z = 4. Mo radiation, R = 0.052 for 2510 reflections.

IV. Monoclinic, Pc, a = 11.549, b = 6.705, c = 15.376 Å, β = 90.32°, Z = 4. Mo radiation, R = 0.071 for 1252 reflections.

V. Monoclinic, $P2_1/c$, a = 12.715, b = 8.620, c = 10.784 Å, β = 93.05°, Z = 4. Mo radiation, R = 0.059 for 1437 reflections.

VI. Triclinic, $P\bar{1}$, a = 6.580, b = 7.324, c = 12.169 Å, α = 82.70, β = 88.00, γ = 74.19°, Z = 2. Mo radiation, R = 0.064 for 962 reflections.

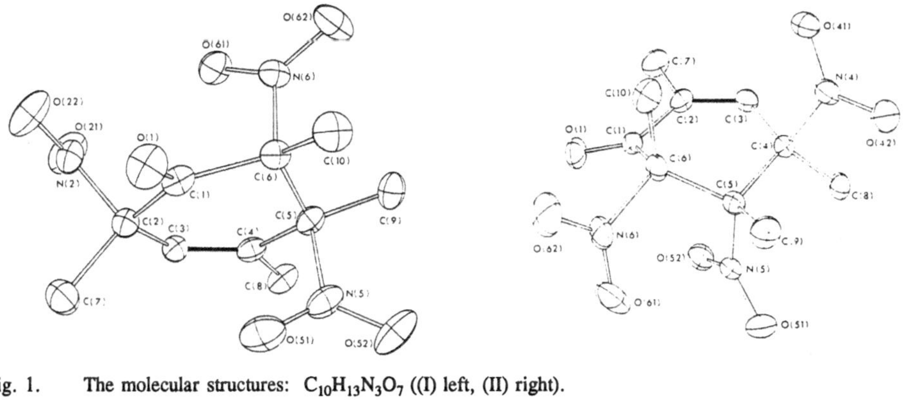

Fig. 1. The molecular structures: $C_{10}H_{13}N_3O_7$ ((I) left, (II) right).

The molecules are shown in Figs. 1-3. In (I), the cyclohex-3-enone ring exists in a skew-boat conformation with the 5-nitro group in a flagpole orientation. The 2-nitro group plane is perpendicular to the C(2)-C(7) bond, the 5-nitro group is aligned with the C(5)-C(9) bond and the torsion angle at the 6-nitro group is -28.6(3)° (C(1)-C(6)-N(6)

O(61)). The ring in compound (II) also exists in a skew-boat conformation with the 5- and 6-nitro groups aligned with the C(5)-C(9) and C(6)-C(10) bonds respectively and the 4-nitro group oriented such that the O(42)-N(4)-C(4)-C(8) torsion angle is -38.5(4)°. (III) and (IV) each contain two, well-separated, crystallographically independent molecules in the asymmetric unit. The differences between the two molecules are minor in both, affecting the 5-nitro group in (III) and the conformation of the ring in (IV). Molecules of (III) exist in a skew-boat conformation with the 5-nitro group in a flagpole orientation and the plane of this group aligned with the geminal C-H bond. This also occurs in (IV) and (V). Notable in (IV) is the fact that the plane of the 4-nitro group is aligned with the C(3)-C(4) bond in both molecules. There is no intramolecular hydroxyl-carbonyl hydrogen bonding in (V). In (VI), the ring system has a flattened half-chair conformation with the 5-nitro group pseudoaxial. The torsion angle O(1)-C(1)-C(6)-O(6) 13.8(4)°, and the O(1)...H(6) distance of 2.12 Å are consistent with intramolecular hydroxyl-carbonyl hydrogen-bonding.

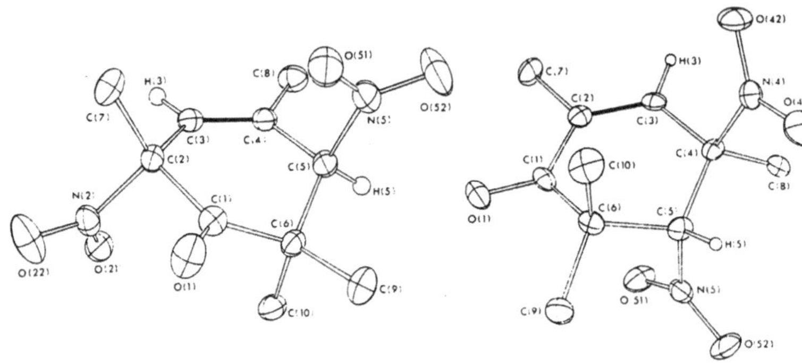

Fig. 2. The structures of $C_{10}H_{14}N_2O_5$ (III) (left) and (IV) (right).

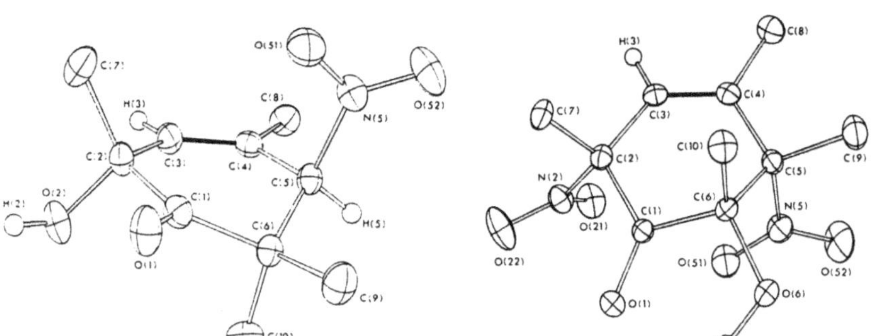

Fig. 3. The structures of $C_{10}H_{14}N_2O_6$ (V) (left) and $C_{10}H_{15}NO_4$ (VI) (right).

trans-2-t-BUTYLCYCLOHEXANE-1-CARBOXYLIC ACID
$C_{11}H_{20}O_2$

R. VIANI, M. COSSU, J. LAPASSET, J.-P. AYCARD and H. BODOT, 1985. *Acta Cryst.*, **C41**, 1267-1270.

Monoclinic, *C2/c*, a = 21.583, b = 7.821, c = 13.618 Å, β = 99.78°, Z = 8. Cu radiation, R = 0.064 for 1511 reflections.

The cyclohexane ring (Fig. 1) adopts a chair conformation, which is only slightly distorted by interaction between the t-butyl and carboxyl groups. Bond lengths and angles are normal. In the crystal, molecules are paired by association between carboxyl groups of neighbouring molecules.

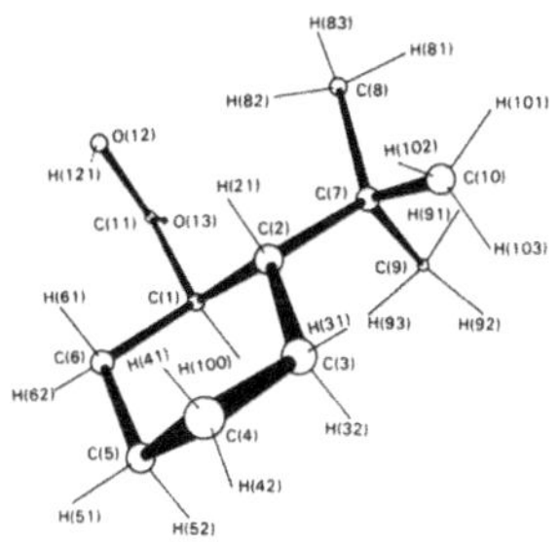

Fig. 1. The $C_{11}H_{20}O_2$ molecule.

2-t-BUTYL-trans-2-METHOXYCYCLOHEXANOL
$C_{11}H_{22}O_2$

G. BERTI, P. CROTTI, F. MACCHIA and P. DOMIANO, 1985. *Gazz. Chim. Ital.*, **115**, 315-318.

Monoclinic, $P2_1/n$, a = 6.712, b = 15.122, c = 11.199 Å, β = 100.30°, Z = 4. Mo radiation, R = 0.052 for 1697 reflections.

The molecule has an almost undistorted chair conformation with the t-butyl group equatorial and hydroxy and methoxy groups axial. Molecules are linked to form chains by O-H...O hydrogen bonds (Fig. 1) with O...O 2.884(2) Å between the hydroxyl group and methoxy O atom of an adjacent molecule. The orientation of the t-butyl methyl groups is shown in Fig. 1 via a Newman projection. For details of the very similar 2-trimethylsilyl derivative see (1).

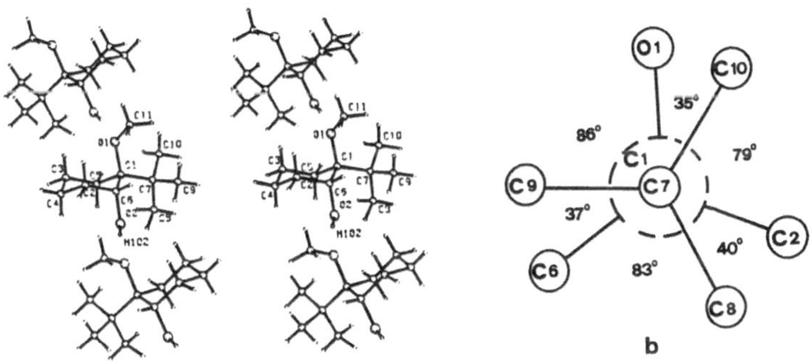

Fig. 1. A stereoview of the hydrogen-bonded system in $C_{11}H_{22}O_2$ and a Newman projection along C(1)-C(7) with rotation angles.

1. Structure Reports, 51B, 752.

2-t-BUTYL-4,6-DIMETHYL-r-4,c-5,t-6-TRINITROCYCLOHEX-2-ENONE
$C_{12}H_{17}N_3O_7$ (I)

2-t-BUTYL-4,6-DIMETHYL-r-4,t-5,c-6-TRINITROCYCLOHEX-2-ENONE
$C_{12}H_{17}N_3O_7$ (II)

2,4-DI-t-BUTYL-6-METHYL-r-2,c-5,t-6-TRINITROCYCLOHEX-3-ENONE
$C_{15}H_{23}N_3O_7$ (III)

2,4-DI-t-BUTYL-6-METHYL-r-2,t-5,t-6-TRINITROCYCLOHEX-3-ENONE
$C_{15}H_{23}N_3O_7$ (IV)

2,4-DI-t-BUTYL-6-METHYL-r-2,c-5,c-6-TRINITROCYCLOHEX-3-ENONE
$C_{15}H_{23}N_3O_7$ (V)

2,4-DI-t-BUTYL-r-2-HYDROXY-6-METHYL-t-5,c-6-DINITROCYCLOHEX-3-ENONE
$C_{15}H_{24}N_2O_6$ (VI)

2,4-DI-t-BUTYL-r-2-HYDROXY-6-METHYL-t-5,t-6-DINITROCYCLOHEX-3-ENONE
$C_{15}H_{24}N_2O_6$ (VII)

2,4-DI-t-BUTYL-C-6-HYDROXY-6-METHYL-r-2,c-5-DINITROCYCLOHEX-3-ENONE
$C_{15}H_{24}N_2O_6$ (VIII)

M.P. HARTSHORN, W.T. ROBINSON, K.H. SUTTON and J. VAUGHAN, 1985. *Aust. J. Chem.*, **38**, 161-177.

I. Orthorhombic, $Pna2_1$, a = 17.642, b = 13.806, c = 6.074 Å, Z = 4. Mo radiation, R = 0.044 for 934 reflections.

II. Monoclinic, $P2_1/n$, a = 13.716, b = 6.137, c = 18.731 Å, β = 110.59°, Z = 4. Mo radiation, R = 0.048 for 1264 reflections.

III. Orthorhombic, *Pbca*, a = 13.870, b = 25.012, c = 10.564 Å, Z = 8. Mo radiation, R = 0.068 for 1081 reflections.

IV. Monoclinic, $P2_1/c$, a = 17.259, b = 12.659, c = 18.269 Å, β = 115.84°, Z = 8. Mo radiation, R = 0.056 for 2499 reflections.

V. Orthorhombic, *Pbca*, a = 14.459, b = 12.461, c = 19.863 Å, Z = 8. Mo radiation, R = 0.046 for 1099 reflections.

VI. Monoclinic, $P2_1/c$, a = 9.528, b = 21.472, c = 9.517 Å, β = 116.79°, Z = 4. Mo radiation, R = 0.046 for 1095 reflections.

VII. Monoclinic, $P2_1/n$, a = 13.585, b = 7.534, c = 16.664 Å, β = 92.45°, Z = 4. Mo radiation, R = 0.048 for 1173 reflections.

VIII. Orthorhombic, *Pbca*, a = 13.462, b = 13.567, c = 18.125 Å, Z = 8. Mo radiation, R = 0.044 for 1315 reflections.

The molecules are shown in Figs. 1-4. Compound (IV) contains two similar independent molecules in the asymmetric unit. In (I), the six-membered ring adopts a flattened skew-boat conformation while the ring system of (II), its C(4)-epimer, has a modified half-chair conformation. In both structures the planes of the 5-nitro and 6-nitro groups are nearly aligned with the C(5)-C(6) and C(6)-C(9) bonds, respectively. In contrast, the 4-nitro group is aligned with the C(4)-C(5) bond in (I), but with the C(4)-C(8) bond in (II). In neither structure are the olefinic and carbonyl functions coplanar. Compounds (III) and (V) exist with the ring in the skew-boat conformation, while the two molecules of (IV) exist in a flattened half-chair conformation. In these four molecules the C(2)-C(7) bond distance shows a wide variation (1.570(6) in molecule 2 of (IV) to 1.605(8) Å in III). Compound (VI) exists with the ring in a flattened half-chair conformation, with the 2-hydroxy and 5-nitro groups pseudoaxial. (VII), in contrast, adopts a skew-boat conformation. Neither (VI) nor (VII) exhibit any hydroxyl-carbonyl intramolecular hydrogen bonding. The C(2)-C(7) bond is 1.554(5) in (VI) and 1.565(6) Å in (VII). Compound (VIII) exists in a skew-boat conformation with the 5-nitro group in the 'flagpole' orientation. The O(1)...H(6) distance is 2.12 Å and the C(1)-O(1) and C(6)-O(6) bonds are almost coplanar. The C(2)-C(7) bond length is 1.576(5) Å.

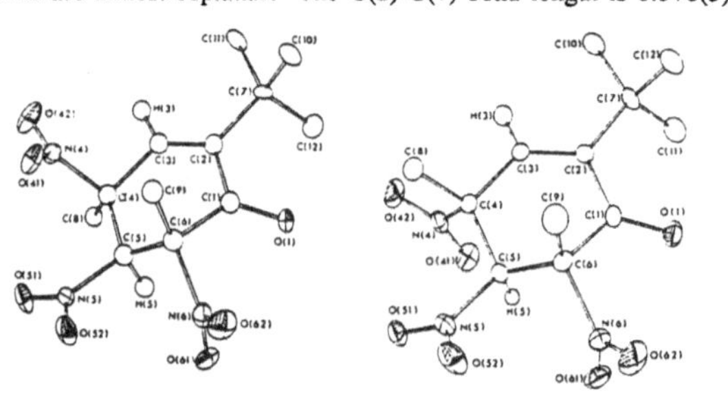

Fig. 1. Views of $C_{12}H_{17}N_3O_7$ (I) (left) and (II) (right).

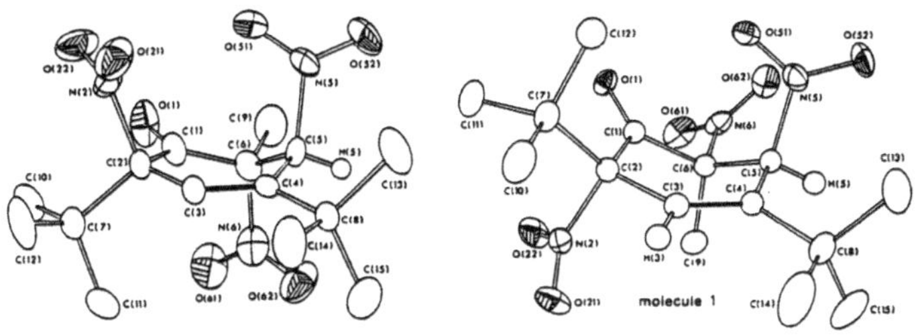

Fig. 2. Views of $C_{15}H_{23}N_3O_7$ (III) (left) and (IV) (right).

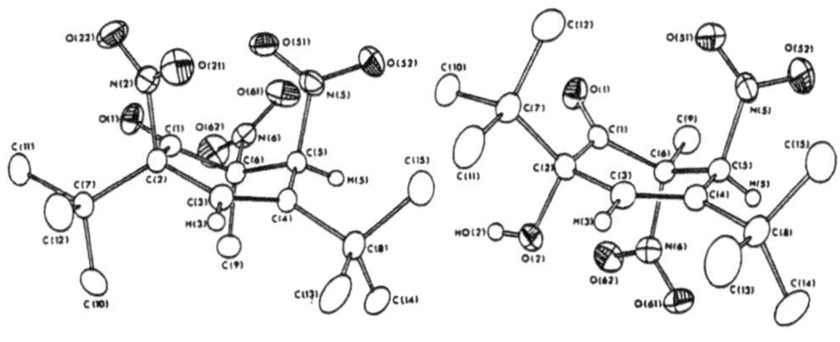

Fig. 3. Views of $C_{15}H_{23}N_3O_7$ (V) (left) and $C_{15}H_{24}N_2O_6$ (VI) (right).

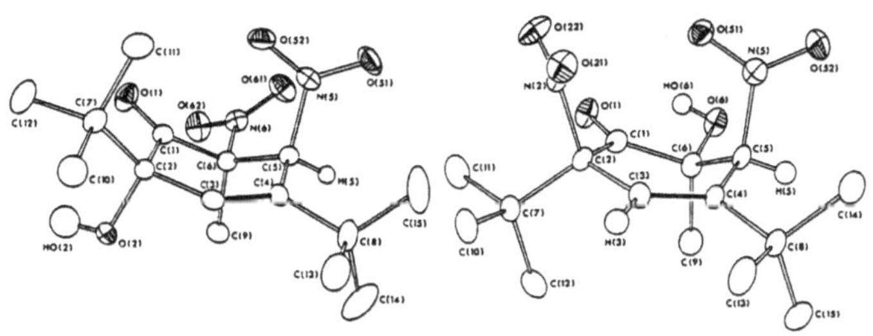

Fig. 4. Views of $C_{15}H_{24}N_2O_6$ (VII) (left) and (VIII) (right).

1,3,5-TRIMETHYLCYCLOHEXANE-1,3,5-TRICARBOXYLIC ACID
$C_{12}H_{18}O_6$ (I)

1,3,5-TRIMETHYLCYCLOHEXANE-1,3,5-TRIMETHYLCARBOXYLATE
$C_{15}H_{24}O_6$ (II)

3-(3'-(N,N-DIISOPROPYLAMINOCARBONYL)-6'-METHYLPHENYL)-7-CARBOXYLATO-1,5,7-TRIMETHYL 3-AZABICYCLO[3.3.1]NONANE-2,4-DIONE MONOHYDRATE CYCLOHEXANE SOLVATE
$C_{26}H_{36}N_2O_5,H_2O,C_6H_{12}$ (III)

2,2'-(μ_2-4,6-DIMETHYL-m-PHENYLENE)BIS(7-CARBOXYLATO-1,5,7-TRIMETHYL-3-AZABICYCLO[3.3.1]-NONANE-2,4-DIONE) MONOHYDRATE BUTANOIC ACID SOLVATE
$C_{32}H_{40}N_2O_8,H_2O,C_4H_8O_2$ (IV)

J. REBEK Jr., L. MARSHALL, R. WOLAK, K. PARRIS, M. KILLORAN, B. ASKEW, D. NEMETH and N. ISLAM, 1985. *J. Am. Chem. Soc.*, **107**, 7476-7481.

I. Monoclinic, $P2_1/n$, a = 8.471, b = 12.138, c = 12.902 Å, β = 101.80°, Z = 4. Mo radiation, R = 0.047 for 1571 reflections.

II. Monoclinic, $P2_1/n$, a = 9.020, b = 18.930, c = 10.106 Å, β = 110.47°, Z = 4. Mo radiation, R = 0.064 for 2038 reflections.

III. Triclinic, $P\bar{1}$, a = 8.153, b = 12.577, c = 15.019 Å, α = 88.145, β = 87.675, γ = 70.739°, Z = 2. Mo radiation, R = 0.064 for 3229 reflections.

IV. Monoclinic, $P2_1/n$, a = 14.313, b = 13.448, c = 19.03 Å, β = 92.18°, Z = 4. Mo radiation, R = 0.096 for 3030 reflections.

Molecules (I) and (II) are shown to have a U shape (Fig. 1). In (III) the methyl group is shown to be directed away from the carboxyl group. Molecule (IV) has the substituents directed as shown in the figure.

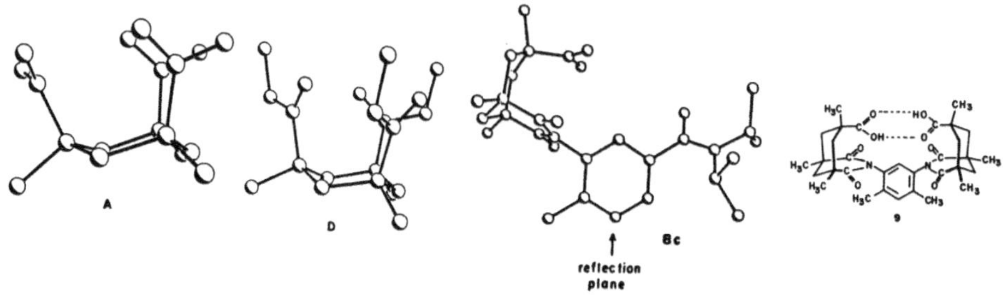

Fig. 1. Views of $C_{12}H_{18}O_6$ (I) (left), $C_{15}H_{24}O_6$ (II) (centre) and $C_{26}H_{36}N_2O_5$ (III) (right). Also shown is the skeleton for $C_{32}H_{40}N_2O_8$ (IV).

BIS(1-NITROSO-2-NITROCYCLOHEXANE)
$C_{12}H_{20}N_4O_6$ (I)

BIS(1-NITROSO-2-NITRO-1-PHENYLETHANE)
$C_{16}H_{16}N_4O_6$ (II)

K.W. CHIU, P.D. SAVAGE, G. WILKINSON and D.J. WILLIAMS, 1985. *Polyhedron*, **4**, 1941-1945.

I. Monoclinic, $C2/c$, a = 14.438, b = 12.382, c = 11.706 Å, β = 131.38°, Z = 4. R = 0.038 for 971 reflections.

II. Triclinic, $P\bar{1}$, a = 5.494, b = 8.742, c = 9.507 Å, α = 70.20, β = 78.28, γ = 89.60°, Z = 1. R = 0.053 for 1029 reflections.

The pseudo-nitrosite molecules have similar structures, except that (I) has an imposed diad axis and (II) an imposed inversion centre (Fig. 1). The NO groups are linked by a N=N bond. Bond lengths and angles include (I, II): N-N 1.302(4), 1.301(4), O-N(N) 1.274(3), 1.269(3), C-N(N) 1.487(2), 1.479(3), C-NO_2 1.507(3), 1.494(3) Å, C-N-N 117.4(3), 118.6(3), O-N-N 121.2(2), 120.5(3)°. The ONNO and CNNC torsion angles in (I) are 180.0(2), 178.9(2)° respectively; the values in (II) are both 180° by symmetry.

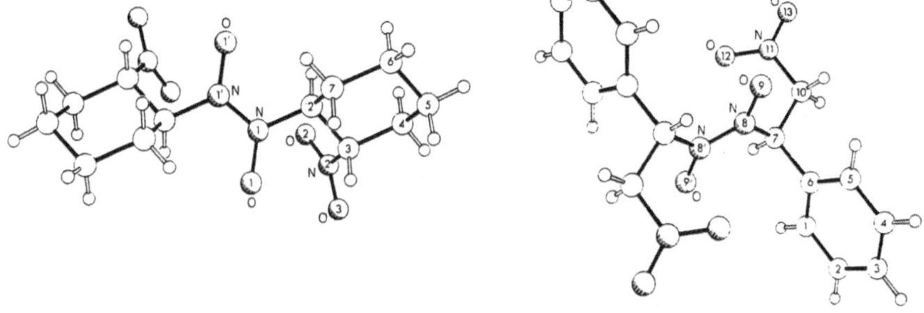

Fig. 1. $C_{12}H_{20}N_4O_6$ (left) and $C_{16}H_{16}N_4O_6$ (right): the molecular structures.

r-2,t-6-DIHYDROXY-6-METHYL-4,t-5-DINITRO-2-PHENYLCYCLOHEX-3-ENONE
$C_{13}H_{12}N_2O_7$ (I)

r-2-HYDROXY-6-METHYL-4,c-5,t-6-TRINITRO-2-PHENYLCYCLOHEX-3-ENONE
$C_{13}H_{11}N_3O_8$ (II)

M.P. HARTSHORN, W.T. ROBINSON, J. VAUGHAN and J.M. WHITE, 1985. *Aust. J. Chem.*, **38**, 575-585.

I. Monoclinic, $P2_1/c$, a = 10.749, b = 21.165, c = 6.075 Å, β = 105.05°, Z = 4. Mo radiation, R = 0.055 for 859 reflections.

II. Monoclinic, $P2_1/n$, a = 9.184, b = 6.917, c = 22.984 Å, β = 96.72°, Z = 4. Mo radiation, R = 0.042 for 1717 reflections.

In (I) (Fig. 1), the alicyclic ring system exists in a flattened half-chair conformation with the 5-nitro group pseudoaxial to the ring. The 4- and 5-nitro groups are aligned with the C(3)-C(4) and C(5)-H(5) bonds respectively. The O(1)...H(6) distance is 2.29 Å and the C(1)-O(1) and C(6)-O(6) bonds are approximately coplanar. The C(3)-C(2)-C(7)-C(75) torsion angle is -75.6(5)°. The 2-OH group is pseudoaxial to the alicyclic ring with the 6-methyl group 1:3 syn-axial to it. The alicyclic ring in (II) is similar to that in (I) with the 4- and 5-nitro groups in similar orientations in the two molecules. The plane of the 6-nitro group is aligned with the C(5)-C(6) bond. There is no intramolecular hydroxyl-carbonyl group hydrogen-bonding in (II). The orientation of the 2-phenyl group is pseudoaxial.

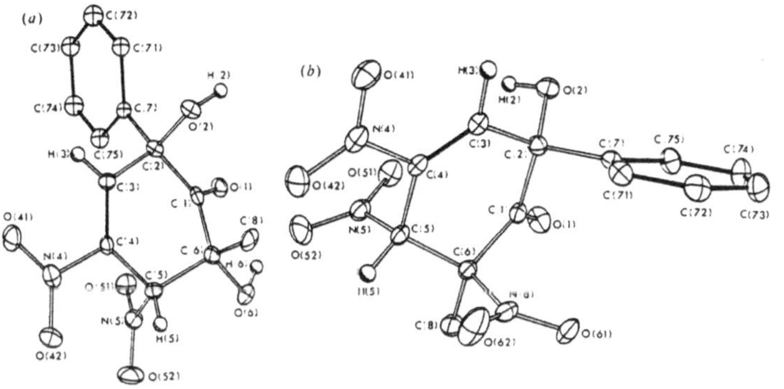

Fig. 1. The molecular structures: $C_{13}H_{12}N_2O_7$ (left) and $C_{13}H_{11}N_3O_8$ (right).

t-3,4-DIMETHYL-r-4-PHENYLCYCLOHEXANONE
$C_{14}H_{18}O$ (I)

t-1-METHYL-c-1-PHENYL-r-2-((p-NITROBENZOYL)OXY)CYCLOHEXANE
$C_{20}H_{21}NO_4$ (II)

c-4-METHYL-1-PHENYLCYCLOHEXAN-r-1-OL
$C_{13}H_{18}O$ (III)

D.J. HODGSON, U. RYCHLEWSKA, E.L. ELIEL, M. MANOHARAN, D.E. KNOX and E.M. OLEFIROWICZ, 1985. *J. Org. Chem.*, **50**, 4838-4843.

I. Monoclinic, $P2_1/c$, a = 14.810, b = 6.316, c = 13.187 Å, β = 110.17°, Z = 4. Mo radiation, R = 0.035 for 1399 reflections.

II. Monoclinic, $P2_1/a$, a = 14.950, b = 7.457, c = 16.759 Å, β = 111.41°, Z = 4. Mo radiation, R = 0.057 for 1264 reflections.

III. Triclinic, $P\bar{1}$, a = 11.069, b = 11.142, c = 9.630 Å, α = 107.57, β = 97.66, γ = 86.59°, Z = 4. Mo radiation, R = 0.064 for 1672 reflections.

In (III) there are two independent molecules in the crystallographic asymmetric unit, but they are very similar. In (I) (Fig. 1) the phenyl group is within 3.4° of the bisector plane of the cyclohexane ring; in (II) it is at 47° to the plane; and in (III) it is 15.5 and 25.3° for the two independent molecules.

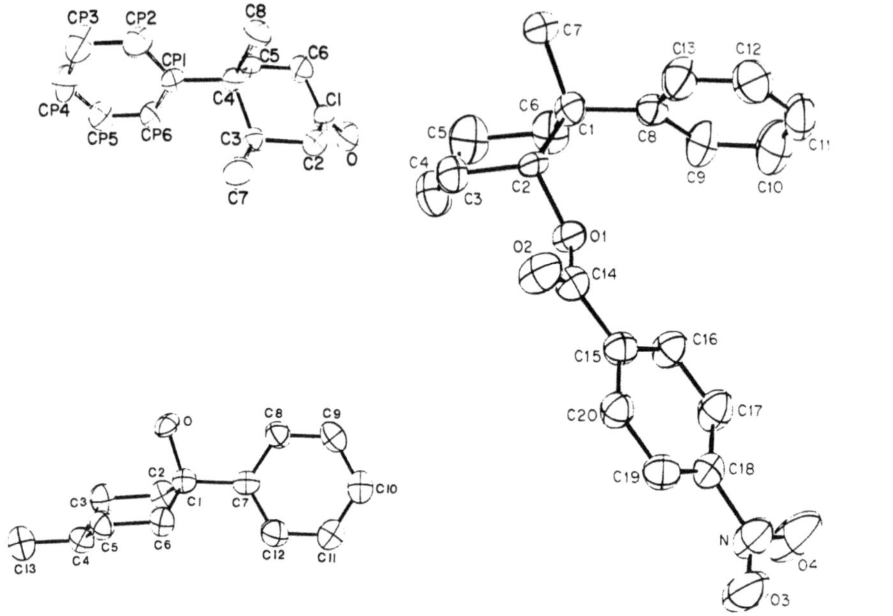

Fig. 1. Perspective drawings of molecules of $C_{14}H_{18}O$ (I) (top left), $C_{20}H_{21}NO_4$ (II) (right) and $C_{13}H_{18}O$ (III) (bottom left).

cis-METHYL 4-ACETOXY-2-t-BUTYLCYCLOHEX-3-ENE-1-CARBOXYLATE
$C_{14}H_{22}O_4$

R. VIANI, J. LAPASSET and J.-P. AYCARD, 1985. *Acta Cryst.*, C41, 791-793.

Monoclinic, $P2_1/c$, a = 8.576, b = 7.834, c = 22.291 Å, β = 110.34°, Z = 4. Cu radiation, R = 0.054 for 1777 reflections.

The ring conformation in the molecule (Fig. 1) is intermediate between half-chair and sofa. The t-butyl and methoxycarbonyl substituents are respectively in pseudo-equatorial and axial positions. The planar group of atoms C(1,2,4,6) makes dihedral angles of 84(1) and 88(1)° respectively, with the planes of the methoxycarbonyl and acetoxy moieties. Bond lengths and angles in the molecule are normal. There is one short intermolecular contact C(12)=O(13)...C(14) of length 3.06(1) Å which might constitute a weak oxygen-methyl hydrogen interaction. Other intermolecular distances are typical van der Waals contacts.

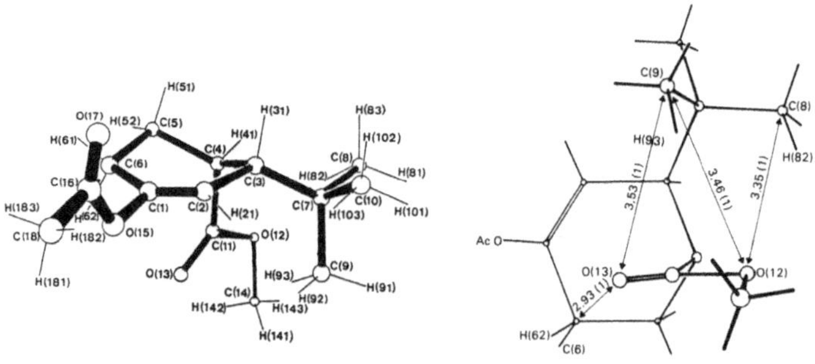

Fig. 1. $C_{14}H_{22}O_4$: a view of the molecule and the principal short non-bonding intramolecular contacts between the t-butyl and methoxycarbonyl groups.

DIHYDROGEN (±)-trans-1,2-CYCLOHEXANEDIAMINETETRAACETIC ACID TETRACHLOROPALLADATE(II) PENTAHYDRATE
$C_{14}H_{24}Cl_4N_2O_8Pd,5(H_2O)$ (I) $C_{14}H_{24}N_2O_8^{2+},Cl_4Pd^{2-},5(H_2O)$

DIHYDROGEN (±)-trans-1,2-CYCLOHEXANEDIAMINETETRAACETIC ACID TETRACHLOROPLATINATE(II) PENTAHYDRATE
$C_{14}H_{24}Cl_4N_2O_8Pt,5(H_2O)$ (II) $C_{14}H_{24}N_2O_8^{2+},Cl_4Pt^{2-},5(H_2O)$

E.N. DUESLER, R.E. TAPSCOTT, M. GARCIA-BASALLOTE and F. GONZALEZ-VILCHEZ, 1985. *Acta Cryst.*, **C41**, 678-681.

I. Monoclinic, $P2_1/n$, a = 12.456, b = 11.427, c = 18.833 Å, β = 106.11°, Z = 4. Mo radiation, R = 0.035 for 4672 reflections.

II. Monoclinic, $P2_1/n$, a = 12.499, b = 11.371, c = 18.885 Å, β = 106.09°, Z = 4. Mo radiation, R = 0.048 for 5223 reflections.

These isostructural crystals contain dications, protonated at all four carboxylic acid positions and at the two nitrogen atoms, and $PdCl_4^{2-}$ or $PtCl_4^{2-}$ anions. The crystals contain alternating layers of ions and water molecules parallel to the ab-plane. In the dianions, the metal-Cl bond lengths lie in the range 2.300(1) to 2.309(3) Å. Bond lengths and angles in the dications are in good agreement with each other and with generally accepted values. Due to the presence of chiral centres on the cyclohexane ring the two acetate groups on each nitrogen atom of the dications (Fig. 1) are non-equivalent. The N-C(acetate)-C(CO) valence angles differ on average by nearly 2° while the C(cyclohexane)-N-C(acetate) angles differ by some 6°, for the two types of non-equivalent acetate groups. The conformation of the acid cations is close to that required for six-or seven-coordinate ligands to be attached to metal atoms. With the exception of O(5), each O atom in the dication is associated with one water molecule and the structure is stabilized by an extensive hydrogen-bonding network.

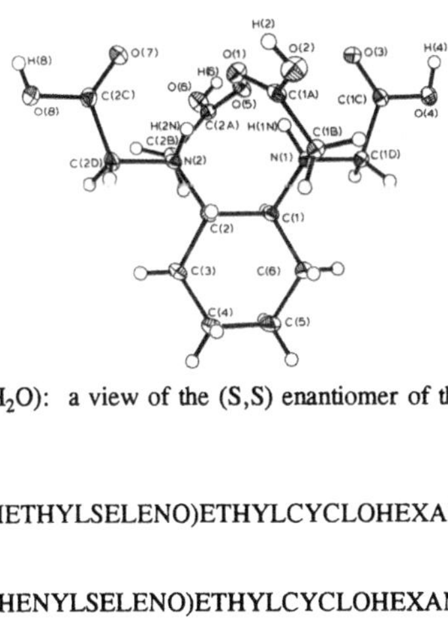

Fig. 1. $C_{14}H_{24}Cl_4N_2O_8Pt,5(H_2O)$: a view of the (S,S) enantiomer of the dication.

4-t-BUTYL-1-(1'-METHYL-1'-METHYLSELENO)ETHYLCYCLOHEXAN-1-OL
$C_{14}H_{28}OSe$ (I)

4-t-BUTYL-1-(1'-METHYL-1'-PHENYLSELENO)ETHYLCYCLOHEXAN-1-OL
$C_{19}H_{30}OSe$ (II)

D. LABAR, A. KRIEF, B. NORBERG, G. EVRARD and F. DURANT, 1985. *Bull. Soc. Chim. Belg.*, **94**, 1083-1100.

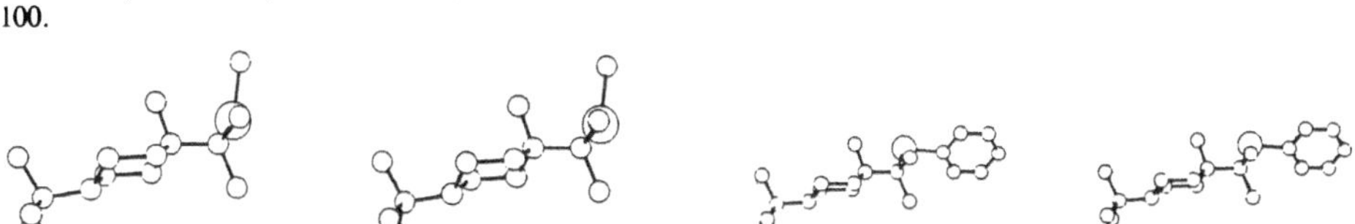

Fig. 1. Stereoviews of $C_{14}H_{28}OSe$ (I) (left) and $C_{19}H_{30}OSe$ (II) (right).

I. Monoclinic, $P2_1/c$, a = 11.453, b = 11.050, c = 12.145 Å, β = 105.15°, Z = 4. Mo radiation, R = 0.040 for 2079 reflections.

II. Triclinic, $P\bar{1}$, a = 11.132, b = 11.363, c = 7.966 Å, α = 70.46, β = 106.25, γ = 93.27°, Z = 2. Mo radiation, R = 0.020 for 2449 reflections.

The structures were determined as shown in Fig. 1.

2-CYCLOHEXYL-1-(p-TOLYL)ETHANONE
$C_{15}H_{20}O$ (I)

1-(4-CHLOROPHENYL)-2-CYCLOHEXYLETHANONE
$C_{14}H_{17}ClO$ (II)

2-CYCLOHEXYL-1-(4-METHOXYPHENYL)ETHANONE
$C_{15}H_{20}O_2$ (III)

1-(4-CARBOXYPHENYL)-2-CYCLOHEXYLETHANONE (4-CYCLOHEXYLMETHYLCARBONYLBENZOIC ACID)
$C_{15}H_{18}O_3$ (IV)

1-(4-CYANOPHENYL)-2-CYCLOHEXYLETHANONE (4-CYCLOHEXYLMETHYLCARBONYLBENZONITRILE)
$C_{15}H_{17}NO$ (V)

S. ARIEL and J. TROTTER, 1985. *Acta Cryst.*, **C41**, 446-450.

I. Monoclinic, $P2_1/c$, a = 8.065, b = 11.977, c = 13.281 Å, β = 92.84°, Z = 4. Mo radiation, R = 0.085 for 810 reflections.

II. Monoclinic, $P2_1/a$, a = 7.944, b = 10.789, c = 14.898 Å, β = 95.81°, Z = 4. Mo radiation, R = 0.053 for 1749 reflections.

III. Triclinic, $P\bar{1}$, a = 6.377, b = 10.613, c = 10.735 Å, α = 111.13, β = 100.62, γ = 97.02°, Z = 2. Mo radiation, R = 0.070 for 959 reflections.

IV. Monoclinic, $P2_1/n$, a = 13.450, b = 5.470, c = 17.800 Å, β = 93.42°, Z = 4. Mo radiation, R = 0.053 for 1351 reflections.

V. Triclinic, $P\bar{1}$, a = 7.161, b = 7.945, c = 11.927 Å, α = 89.21, β = 79.34, γ = 80.77°, Z = 2. Mo radiation, R = 0.059 for 1131 reflections.

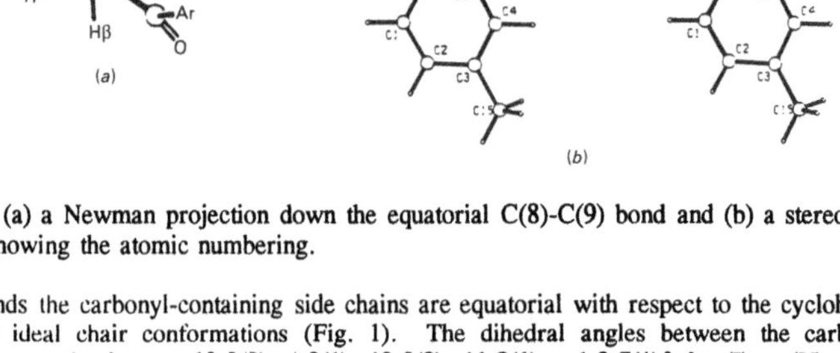

Fig. 1. $C_{15}H_{20}O$: (a) a Newman projection down the equatorial C(8)-C(9) bond and (b) a stereoview of the molecule showing the atomic numbering.

In all five compounds the carbonyl-containing side chains are equatorial with respect to the cyclohexane rings which all have nearly ideal chair conformations (Fig. 1). The dihedral angles between the carbonyl group C(6,7,8),O(1) and the aromatic ring are 13.3(2), 4.2(1), 12.9(2), 11.2(1) and 2.7(1)° for (I) to (V) respectively. Dimensions for the molecules agree well with generally accepted values. In all the compounds there is a close intramolecular contact of average length 2.61 Å between the carbonyl O atom and the equatorial hydrogen atom H_e

of C(10). In (IV), the crystal structure is composed of hydrogen-bonded centrosymmetric carboxylic acid dimers with an O(2)...O(3) separation of 2.634(4) Å. The interdimer distances in (IV) and the intermolecular in the other four compounds correspond to normal van der Waals interactions.

METHYL (1'R,2S)(1'S,2R)-2-BENZOYLAMINO-2-(2-OXOCYCLOHEXYL)ACETATE
$C_{16}H_{19}NO_4$ (I)

(-)-8-PHENYLMENTHYL (1'R,2S)-2-BENZOYLAMINO-2-(2-OXOCYCLOHEXYL)ACETATE
$C_{31}H_{39}NO_4$ (II)

R. KOBER, K. PAPADOPOULOS, W. MILTZ, D. ENDERS, W. STEGLICH, H. REUTER and H. PUFF, 1985. *Tetrahedron*, **41**, 1693-1701.

I. Triclinic, $P\bar{1}$, a = 8.229, b = 9.322, c = 10.582 Å, α = 107.89, β = 94.03, γ = 99.13°, Z = 2. Mo radiation, R = 0.077 for 3039 reflections.

II. Monoclinic, $P2_1$, a = 11.041, b = 9.717, c = 12.396 Å, β = 91.59°, Z = 2. Mo radiation, R = 0.058 for 2805 reflections.

These esters were shown to have structure and stereochemistry as detailed in Fig. 1. In (I) the unit cell has two enantiomers; in (II) the absolute stereochemistry was derived from the known configuration of the 8-phenylmenthyl moiety.

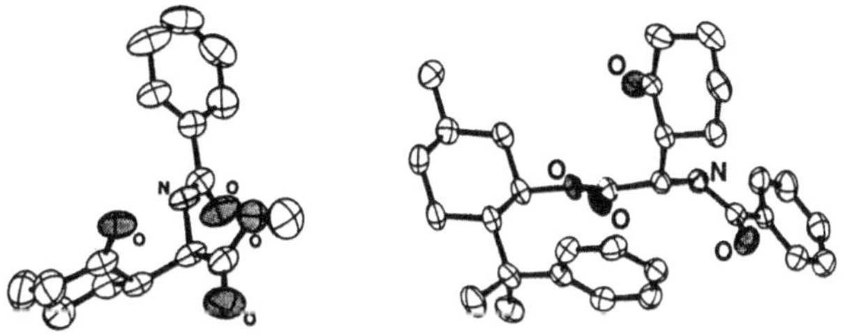

Fig. 1. Views of $C_{16}H_{19}NO_4$ (I) (left) and $C_{31}H_{39}NO_4$ (II) (right).

r-4-HYDROXY-4-METHYL-c-5,c-6-DINITRO-2,6-DIPHENYLCYCLOHEX-2-ENONE
$C_{19}H_{16}N_2O_6$ (I)

r-4-HYDROXY-4-METHYL-c-5,c-6-DINITRO-2-(2'-NITROPHENYL)-6-PHENYLCYCLOHEX-2-ENONE
$C_{19}H_{15}N_3O_8$ (II)

4-METHYL-r-4,c-5,c-6-TRINITRO-2,6-DIPHENYLCYCLOHEX-2-ENONE
$C_{19}H_{15}N_3O_7$ (III)

t-6-HYDROXY-4-METHYL-r-4,t-5-DINITRO-2,6-DIPHENYLCYCLOHEX-2-ENONE
$C_{19}H_{16}N_2O_6$ (IV)

4-METHYL-r-4,t-5,t-6-TRINITRO-2,6-DIPHENYLCYCLOHEX-2-ENONE
$C_{19}H_{15}N_3O_7$ (V)

r-2,c-6-DIHYDROXY-4,c-5-DINITRO-2,6-DIPHENYLCYCLOHEX-3-ENONE
$C_{18}H_{14}N_2O_7$ (VI)

M.P. HARTSHORN, R.J. MARTYN, W.T. ROBINSON, K.H. SUTTON, J. VAUGHAN and J.M. WHITE, 1985. *Aust. J. Chem.*, **38**, 1613-1630.

I. Monoclinic, $P2_1/c$, a = 7.354, b = 13.823, c = 17.779 Å, β = 99.24°, Z = 4. Mo radiation, R = 0.055 for 1026 reflections.

II. Orthorhombic, *Pbca*, a = 14.841, b = 13.951, c = 17.554 Å, Z = 8. Mo radiation, R = 0.100 for 1049 reflections (at 173 K).

III. Orthorhombic, $Pna2_1$, a = 9.202, b = 14.736, c = 13.688 Å, Z = 4. Mo radiation, R = 0.046 for 1330 reflections.

IV. Monoclinic, $P2_1/c$, a = 7.725, b = 27.317, c = 16.844 Å, β = 97.77°, Z = 8. Mo radiation, R = 0.058 for 2594 reflections.

V. Monoclinic, $P2_1/c$, a = 13.621, b = 13.498, c = 10.885 Å, β = 112.90°, Z = 4. Mo radiation, R = 0.039 for 1226 reflections.

VI. Orthorhombic, $P2_12_12_1$, a = 13.185, b = 13.951, c = 18.615 Å, Z = 8. Mo radiation, R = 0.062 for 1541 reflections (at 140 K).

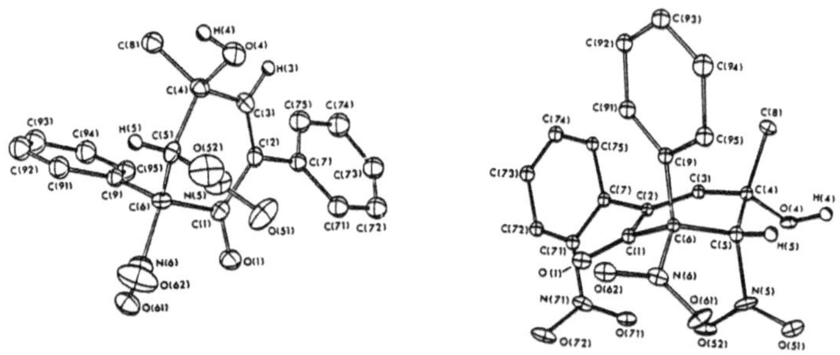

Fig. 1. Views of $C_{19}H_{16}N_2O_6$ (I) (left) and $C_{19}H_{15}N_3O_8$ (II) (right).

Structures (IV) and (VI) each contain two crystallographically independent molecules in the asymmetric unit. Compounds (I) and (II) (Fig. 1) have almost identical conformations with the alicyclic rings adopting half-chair conformations and the 6-phenyl group plane aligned with the C(5)-C(6) bond in each. The torsion angle C(3)-C(2)-C(7)-C(75) is -46.9(8) in (I) and -59(2)° in (II). In each structure the plane of the 5-nitro group is aligned with the C(5)-H(5) bond and the torsion angle C(8)-C(4)-O(4)-H(4) is 40°. In (III), (IV) and (V) (Fig. 2, Fig. 3) the cyclohex-2-enone ring system adopts a half-chair conformation with the 6-phenyl ring pseudo-axial. The exception is molecule 2 of (IV). In (IV), a distance of 2.24 Å between O(1) and H(6) and the coplanarity of the C(1)-O(1) and C(6)-O(6) bonds suggest an intramolecular hydrogen bond. The two molecules of (VI) (Fig. 3) have nearly identical half-chair alicyclic ring conformations. The main differences between the two molecules are in the orientations of the 2-phenyl and 4-nitro groups relative to the alicyclic ring. The C(3)-C(2)-C(7)-C(12) and C(3)-C(4)-N(4)-O(41) torsion angles are 4.6(8) and -6(1)°, respectively, in molecule 1 and 13.0(8) and -13(1)° in molecule 2. There is no evidence for intramolecular hydroxy-carbonyl group hydrogen bonding.

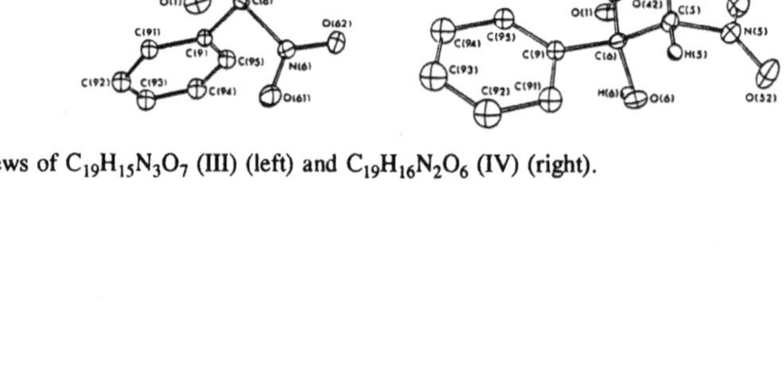

Fig. 2. Views of $C_{19}H_{15}N_3O_7$ (III) (left) and $C_{19}H_{16}N_2O_6$ (IV) (right).

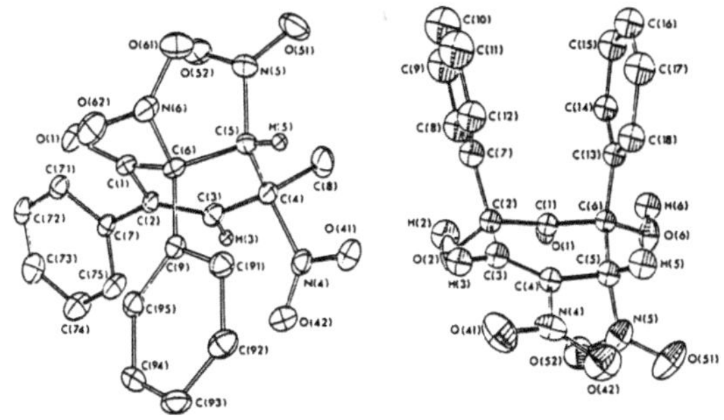

Fig. 3. Views of $C_{19}H_{15}N_3O_7$ (V) (left) and $C_{18}H_{14}N_2O_7$ (VI) (right).

(2,4,6-CYCLOHEPTATRIEN-1-YLIDENE)MALONALDEHYDE (8,8-DIFORMYLHEPTAFULVENE) $C_{10}H_8O_2$

C. REICHARDT, K.-Y. YUN, W. MASSA, R.E. SCHMIDT, O. EXNER and E.-U. WURTHWEIN, 1985. *Liebigs Ann. Chem.*, 1997-2011.

Monoclinic, $P2_1/c$, a = 3.792, b = 16.375, c = 12.150 Å, β = 90.95°, Z = 4. Mo radiation, R = 0.036 for 1154 reflections (at -80°C).

The molecular conformation and bond lengths are shown in Fig. 1. The malonaldehyde residue is rotated 18.1° out of the cycloheptatriene plane to relieve overcrowding.

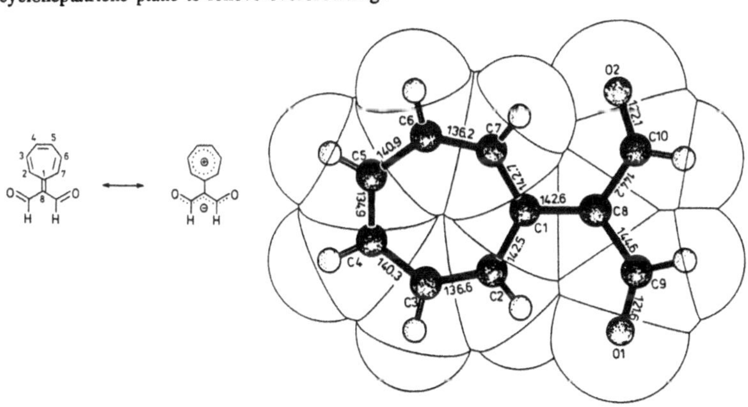

Fig. 1. Possible resonance forms for $C_{10}H_8O_2$ and a view of the molecule.

METHYL 3,3,7,7-TETRAMETHYL-2-OXOCYCLOHEPTANESULPHINATE $C_{12}H_{22}O_3S$

M.N. PONNUSWAMY and J. TROTTER, 1985. *Acta Cryst.*, **C41**, 915-917.

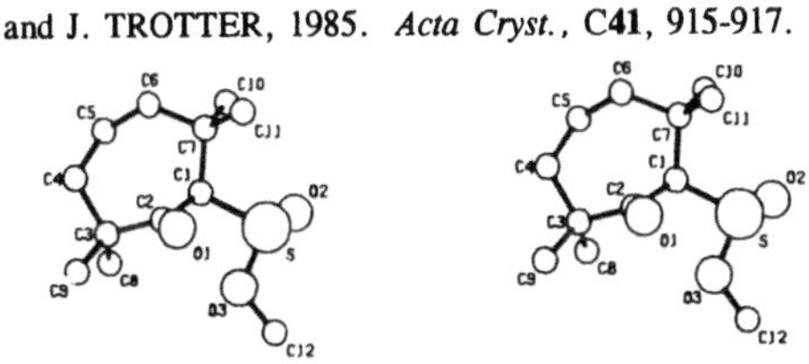

Fig. 1. A stereoview of the $C_{12}H_{22}O_3S$ molecule.

Triclinic, $P\bar{1}$, a = 8.680, b = 10.643, c = 8.306 Å, α = 110.45, β = 111.20, γ = 79.97°, Z = 2. Mo radiation, R = 0.036 for 2971 reflections.

The cycloheptane ring (Fig. 1) has a twist chair conformation, with mean C-C bond lengths of 1.538(6) Å, and mean C-C-C bond angles of 113.5(16)°. The C=O distance is 1.215(2) Å. The S atom has a pyramidal configuration with C-S 1.824(1), S=O 1.466(1) and S-OCH_3 1.631(1) Å. Intermolecular distances correspond to van der Waals interactions.

syn-PERCHLOROHEPTAFULVALENE
$C_{14}Cl_{12}$

H.M. HUGEL, E. HORN and M.R. SNOW, 1985. *Aust. J. Chem.*, **38**, 383-389.

Monoclinic, *C2/c*, a = 12.376, b = 10.980, c = 15.242 Å, β = 100.54°, Z = 4. Mo radiation, R = 0.042 for 1581 reflections.

Each asymmetric unit consists of two basic C_7Cl_6 rings disordered into opposing orientations. The two rings in the heptafulvalene structure are in a syn geometry and each ring exhibits the boat conformation (Fig. 1). The resulting U-shaped structure prevents cross conjugation between the two rings and hence lack of colour. The average C-Cl, C-C and C=C bond lengths are 1.737, 1.437 and 1.356 Å respectively.

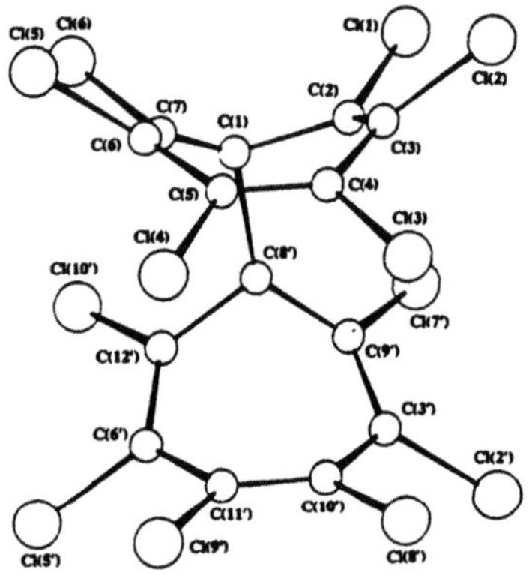

Fig. 1. $C_{14}Cl_{12}$: the molecular structure.

3-ANILINO-2-(2,4,6-CYCLOHEPTATRIEN-1-YL)-2-PROPENAL
$C_{16}H_{15}NO$ (I)

3-(2,4,6-CYCLOHEPTATRIEN-1-YL)-1,5-BIS(1,3,3-TRIMETHYL-2-INDOLINYL)PENTAMETHINIUM TETRAFLUOROBORATE CHLOROFORM SOLVATE
$C_{34}H_{37}BF_4N_2,2(CH_2Cl_2)$ (II) $C_{34}H_{37}N_2^+,BF_4^-,2(CHCl_3)$

C. REICHARDT, KYEONG-YEOL YUN, W. MASSA, M. BIRKHAHN and R.E. SCHMIDT, 1985. *Liebigs Ann. Chem.*, 1969-1986.

I. Orthorhombic, *Pbcn*, a = 17.411, b = 11.816, c = 13.236 Å, Z = 8. Mo radiation, R = 0.043 for 1148 reflections.

II. Monoclinic, *C2/c*, a = 29.823, b = 11.365, c = 24.361 Å, β = 102.38°, Z = 8. Mo radiation, R = 0.051 for 3834 reflections (at -50°C).

The structures were determined to be as shown in Figs. 1 and 2. In (I), the seven-membered ring is in tub-form; the N-C(3) distance (1.350 Å) is relatively short and the phenyl ring is rotated 22.5° out of the propenal plane. The pentamethinium cation in (II) (Fig. 2) has an all-trans configuration. The indolinyl groups are rotated 24.8 and 8.8° out of the pentamethine plane. The seven-membered ring has a similar conformation to that found for (I) with α =

53.6 and β = 25.7° (see Fig. 1).

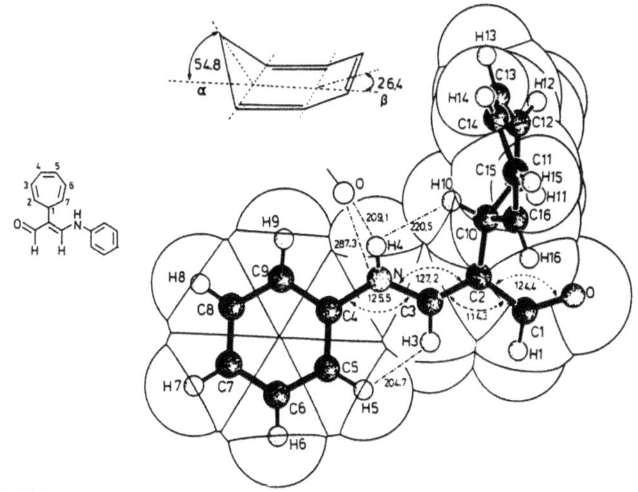

Fig. 1. The $C_{16}H_{15}NO$ structure.

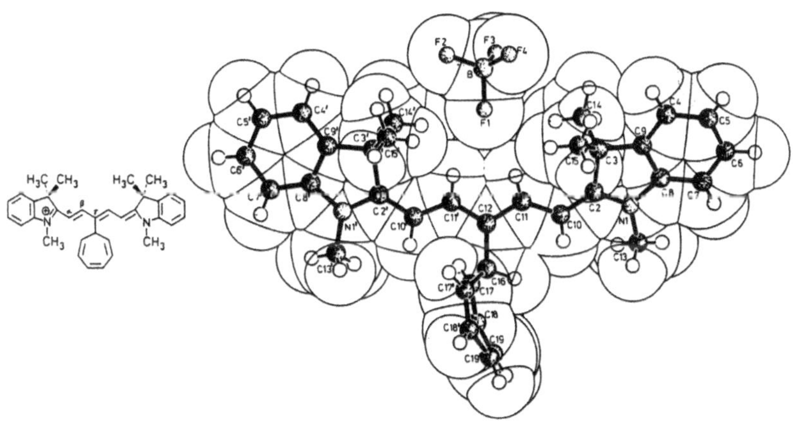

Fig. 2. The $C_{34}H_{37}N_2{}^+$ cation.

5-ACETYL-7-(N-METHYLANILINO)CYCLOHEPTA-2,4-DIEN-1-ONE
$C_{16}H_{17}NO_2$

C.B. KNOBLER, D.E. BUGNER, O.L. CHAPMAN and K.N. TRUEBLOOD, 1985. *Acta Cryst.*, **C41**, 1335-1337.

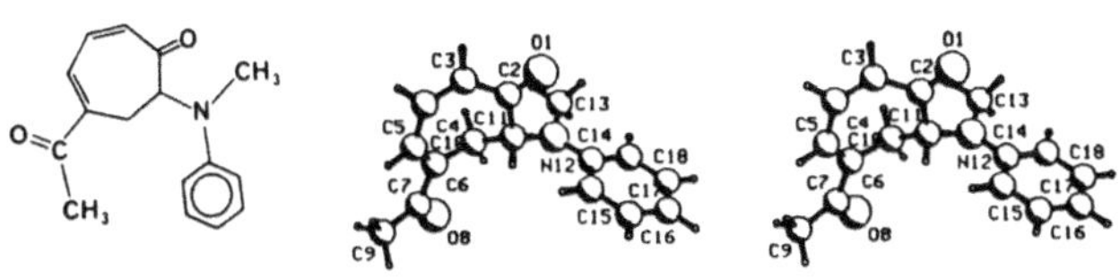

Fig. 1. $C_{16}H_{17}NO_2$: structural formula and a stereoview of the molecule.

Orthorhombic, $P2_12_12_1$, a = 6.219, b = 14.176, c = 15.432 Å, Z = 4. Mo radiation, R = 0.077 for 1133 reflections.

The methylanilino group in the molecule (Fig. 1) is essentially planar and there is a strong intramolecular non-bonding interaction between the carbonyl oxygen atom O(1) and the nitrogen atom N(12). The O...N separation is 2.69(1) Å. Bond lengths in the diene-dione system, which extends from O(1) to O(8), indicate that there is no significant electron delocalization. The methylanilino group is in a pseudo-equatorial position with its plane nearly perpendicular (87°) to the least-squares plane of the diene-dione system. There are no unusually short intermolecular distances.

meso-N,N,N',N',N'',N'',N''',N'''-OCTAMETHYLBIS(CYCLOHEPTA-2,4,6-TRIENYL)-3,3'-DICARBOXAMIDINIUM BIS(TRIIODIDE)
$C_{24}H_{36}I_6N_4$ $C_{24}H_{36}N_4^{2+},2(I_3^-)$

J. DAUB, A. HASENHUNDL, E. SALBECK, A. GIEREN and T. HUBNER, 1985. *Z. Naturforsch.*, **B40**, 1390-1397.

Triclinic, $P\bar{1}$, a = 8.394, b = 9.117, c = 13.848 Å, α = 67.93, β = 70.35, γ = 65.36°, Z = 1. Mo radiation, R = 0.029 for 3393 reflections.

The cation (Fig. 1) and both I_3^- ions lie about independent inversion centres. In the cation, the cycloheptatriene rings have boat conformation and the amidinium groups are twisted out of the C(2)-C(3) double bond plane (torsion angles C(2)-C(3)-C(8)-N(1) -67°, C(4)-C(3)-C(8)-N(2) -63°). The I-I distances in the I_3^- ions are 2.909(1) and 2.914(1) Å.

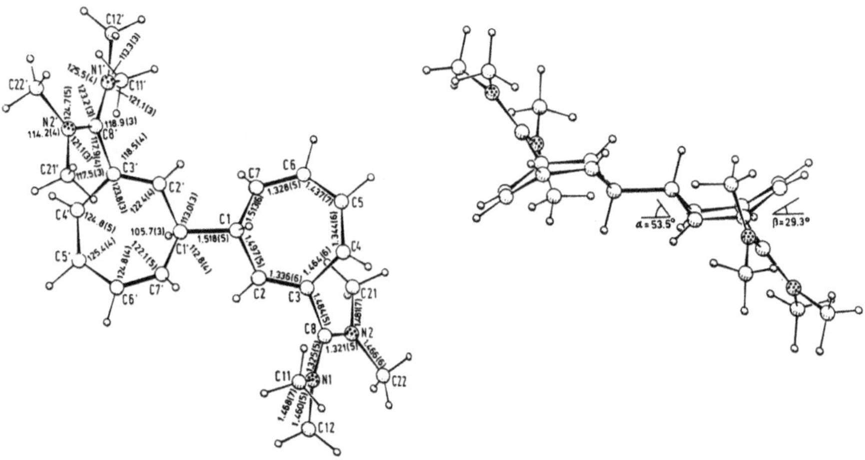

Fig. 1. The $C_{24}H_{36}N_4^{2+}$ cation with dimensions and a side-on view.

3,4,6-TRIS(BENZYLIDENE)CYCLOHEPTANONE
$C_{28}H_{24}O$

D.J. PASTO, NAIZHONG HUANG and C.W. EIGENBROT, 1985. *J. Am. Chem. Soc.*, **107**, 3160-3172.

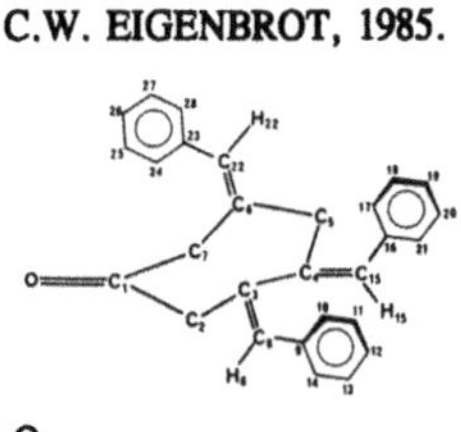

Fig. 1. The stereochemistry of $C_{28}H_{24}O$.

Monoclinic, $C2/c$, a = 23.785, b = 14.216, c = 12.659 Å, β = 98.64°, Z = 8. Cu radiation, R = 0.039 for 1671 reflections.

As the stereochemistry shows (Fig. 1) there is a highly congested environment about the benzylidene groups leading to structural distortions of which the following are the most severe: C(4)-C(3)-C(8) 126.6(2), C(3)-C(8)-C(9) 130.4(2), C(3)-C(8)-H(8) 112(1)° about the 3-benzylidene group, C(5)-C(4)-C(15) 124.3(2), C(4)-C(15)-C(16) 129.7(2)° about the 4-benzylidene group, and C(7)-C(6)-C(22) 123.3(2), C(6)-C(22)-C(23) 129.5(2)° about the 6-benzylidene group. In addition to these bond angle distortions, each double bond is 4.7-5.4° from planarity, the phenyl rings are 31.8-38.2° from coplanarity with the double bonds and there is a 53.1° skew about the C(8)-C(3)-C(4)-C(16) diene system (dihedral angle between the planes C(8)-C(3)-C(4) and C(3)-C(4)-C(16)).

DECAMETHYLCYCLOPENTADECA-1,4,7,10,13-PENTAYNE (DECAMETHYL[5]PERICYCLYNE)
$C_{25}H_{30}$ (I)

5,5,6,6,11,11,12,12-OCTAMETHYL-1,3,7,9-CYCLODODECATETRAYNE
$C_{20}H_{24}$ (II)

K.N. HOUK, L.T. SCOTT, N.G. RONDAN, D.C. SPELLMEYER, G. REINHARDT, J.L. HYUN, G.J. DECICCO, R. WEISS, M.H.M. CHEN, L.S. BASS, J. CLARDY, F.S. JORGENSEN, T.A. EATON, V. SARKOZI, C.M. PETIT, L. NG and K.D. JORDAN, 1985. *J. Am. Chem. Soc.*, **107**, 6556-6562.

I. Monoclinic, $P2_1/c$, a = 20.457, b = 9.512, c = 11.313 Å, β = 92.68°, Z = 4. Cu radiation, R = 0.055 for 2609 reflections.

II. Monoclinic, $P2_1/n$, a = 7.053, b = 10.436, c = 10.821 Å, β = 92.91°, Z = 2. Mo radiation, R = 0.064 for 1090 reflections (at -160°C).

The pentayne (I) has an envelope C_s conformation (Fig. 1) with C15 as the flap. Average bond lengths are: C≡C 1.190(5), Csp-Csp^3 1.480(7), Csp^3-Csp^3 1.533(8) Å. At the quaternary Csp^3 atoms the internal angles are 108.0(8)°. The tetrayne (II) (Fig. 1) adopts a non-planar conformation with C_{2h} symmetry. Repulsions between the adjacent pairs of gem-dimethyl groups produces torsion angles around the Csp^3-Csp^3 ring bonds (C4-C5-C6-C7 and C10-C11-C12-C1) of 40.7° and a lengthening of those bonds to 1.602(3) Å. Around the remainder of the ring, average bond lengths are C≡C 1.200(4), Csp-Csp 1.387(3), Csp-Csp^3 1.482(3) Å.

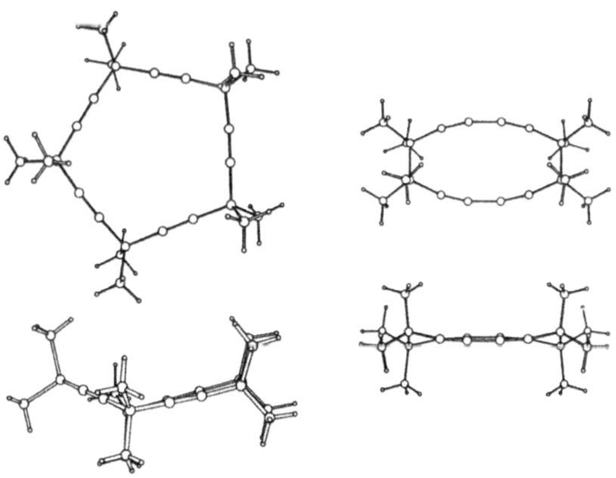

Fig. 1. Views of $C_{25}H_{30}$ (I) (left) and $C_{20}H_{24}$ (II) (right).

(±)-2-(6-METHOXY-2-NAPHTHYL)PROPIONIC ACID (NAPROXEN)
$C_{14}H_{14}O_3$

K. RAVIKUMAR, S.S. RAJAN, V. PATTABHI and E.J. GABE, 1985. *Acta Cryst.*, **C41**, 280-280.

Monoclinic, $P2_1$, a = 13.3150, b = 5.7765, c = 7.8732 Å, β = 93.88°, Z = 2. Mo radiation, R = 0.053 for 1037 reflections.

The methoxy group of the molecule (Fig. 1) is coplanar with the naphthalene moiety which in turn is somewhat distorted there being an angle of 5.2(2)° between the benzene ring planes. The carboxyl group is rotated with respect to the benzene ring (torsion angles C(9)-C(8)-C(13)-C(15) 70.5(8), C(8)-C(13)-C(15)-O(16) 90.3(8)°). The structure is stabilized by an intermolecular hydrogen bond between O(16) and O(17) of a neighbouring molecule (O(16)...O(17) 2.636(8) Å).

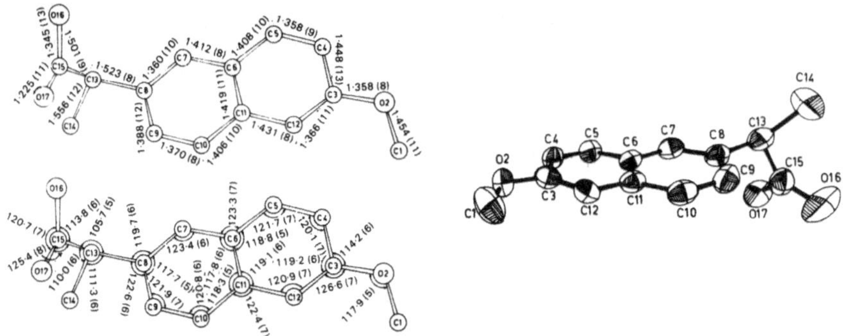

Fig. 1. $C_{14}H_{14}O_3$: bond lengths and angles and a view of the molecule.

1,4,5,8-TETRAMETHYLNAPHTHALENE
$C_{14}H_{16}$

F. IMASHIRO, K. TAKEGOSHI, A. SAIKA, Z. TAIRA and Y. ASAHI, 1985. *J. Am. Chem. Soc.*, **107**, 2341-2346.

Monoclinic, $P2_1/n$, a = 7.948, b = 5.190, c = 12.887 Å, β = 104.95°, Z = 2. Mo radiation, R = 0.0548 for 904 reflections.

The carbon atoms in the structure are planar to within r.m.s. deviation of 0.022 Å, and the molecules are stacked (Fig. 1) along the b-axis in a stair-like formation with 3.76 Å between stacked molecules. The peri-methyl groups adopt a clashed-gear conformation and stack at about 3.542 Å from the plane spanned by the adjacent naphthalene ring. There is a distinct deviation of the methyl carbons from the molecular mean-plane, by 0.07 Å (C(11) and 0.01 Å (C(12)), toward a 'boat-like' conformation. The C(9)-C(10) bond, at 1.447 Å, is somewhat longer than expected, and the C(2)-C(3) bond, 1.382 Å, somewhat shorter.

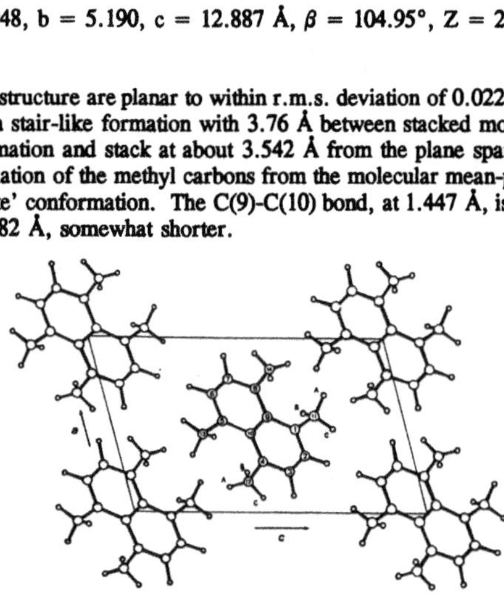

Fig. 1. $C_{14}H_{16}$: view of the molecular packing.

p-NITROBENZYL (+)-1,5-NITRONAPHTHYLSULPHINYLACETATE
$C_{19}H_{14}N_2O_7S$

A.E. KOZIOL, K. KURYS and M. JANCZEWSKI, 1985. *Z. Kristallogr.*, **171**, 165-171.

Monoclinic, $P2_1$, a = 12.429, b = 4.762, c = 15.689 Å, β = 97.27°, Z = 2. Cu radiation, R = 0.0414 for 1189 reflections.

The absolute configuration at the sulphur atom was shown to be R (Fig. 1). Some bond lengths are: S-C(1)

1.806(6), S-C(11) 1.824(11), S-O 1.490(4).

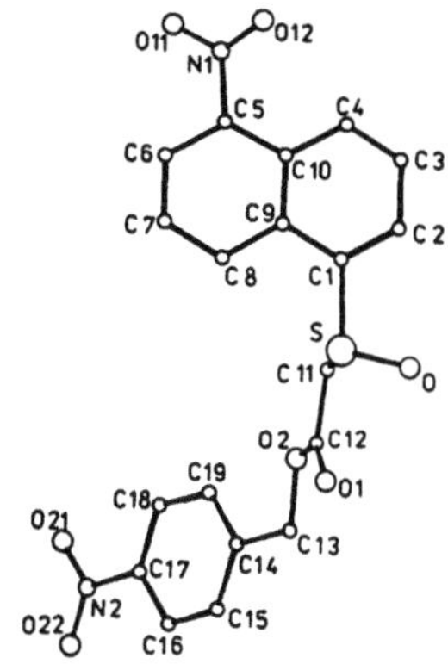

Fig. 1. The $C_{19}H_{14}N_2O_7S$ molecule.

2,3-DICHLORO-5,8-DIHYDROXY-1,4-NAPHTHOQUINONE (2,3-DICHLORONAPHTHAZARIN)
$C_{10}H_4Cl_2O_4$

P. RUBIO, F. FLORENCIO, S. GARCIA-BLANCO and J.G. RODRIGUEZ, 1985. *Acta Cryst.*, **C41**, 1797-1799.

Monoclinic, $P2_1/n$, a = 14.110, b = 7.070, c = 9.810 Å, β = 102.09°, Z = 4. Cu radiation, R = 0.058 for 1531 reflections.

The crystal is composed of molecules (Fig. 1) which possess 2m symmetry and are stacked along a diagonal axis. The molecules overlap to form a charge-transfer complex with an interplanar separation of 3.40(3) Å. The annular system consists of a benzenoid and quininoid ring with bond lengths and angles in normal ranges.

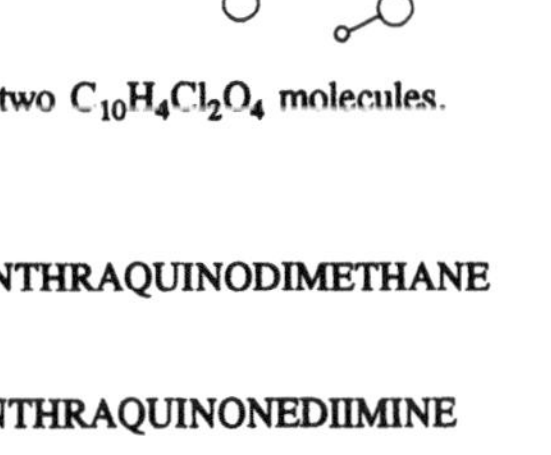

Fig. 1. A view of the overlap of two $C_{10}H_4Cl_2O_4$ molecules.

11,11,12,12-TETRACYANO-9,10-ANTHRAQUINODIMETHANE
$C_{20}H_8N_4$ (I)

N,N'-DICYANO-2,3-DIMETHYLANTHRAQUINONEDIIMINE
$C_{18}H_{12}N_4$ (II)

U. SCHUBERT, S. HUNIG and A. AUMULLER, 1985. *Liebigs Ann. Chem.*, 1216-1222.

I. Triclinic, $P\bar{1}$, a = 9.082, b = 10.013, c = 10.094 Å, α = 109.59, β = 116.60, γ = 95.50°, Z = 2. Mo radiation, R = 0.042 for 1935 reflections.

II. Monoclinic, $I2/a$, a = 7.969, b = 12.676, c = 13.806 Å, β = 96.89°, Z = 4. Mo radiation, R = 0.061 for 870 reflections.

Molecule (I) has a distinctly non-planar geometry (Fig. 1) with a 145° angle between the end aromatic rings leading to a boat conformation for the central ring. The dicyanomethylene units are folded in the opposite direction and their planes form angles of 30.7 and 30.1° with the C(3),C(5),C(10),C(12) plane. Molecule (II) has twofold crystallographic symmetry (Fig. 1) and is nearly planar; steric interactions between the NCN moieties and the anthracene H atoms are reduced by rotation of the N-CN substituent within the molecular plane (C(3)-C(4)=N(1) 127.0°, C(5)-C(4)=N(1) 114.3°, C(4)=N(1) 1.289(5) Å, N(1)-C(8) 1.340(5) Å).

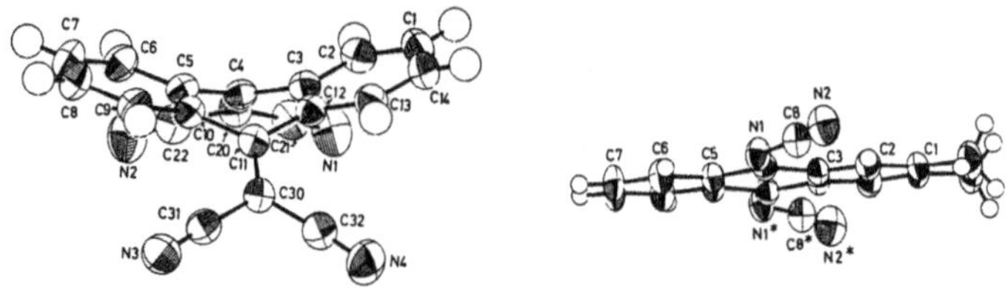

Fig. 1. Views of $C_{20}H_8N_4$ (I) (left) and $C_{18}H_{12}N_4$ (II) (right).

1-ANISYL-2,2-DIPHENYLVINYL BROMIDE
$C_{21}H_{17}BrO$ (I)

α-BROMO-9-(p-METHOXYBENZYLIDENE)ANTHRONE
$C_{22}H_{15}BrO_2$ (II)

α-TOSYLOXY-9-(p-METHOXYBENZYLIDENE)ANTHRONE
$C_{29}H_{22}SO_5$ (III)

p-ACETAMIDO-α-BROMOSTYRENE
$C_{10}H_{10}BrNO$ (IV)

M. KAFTORY, Y. APELOIG and Z. RAPPOPORT, 1985. *J. Chem. Soc. Perkin Trans. 2*, 29-37.

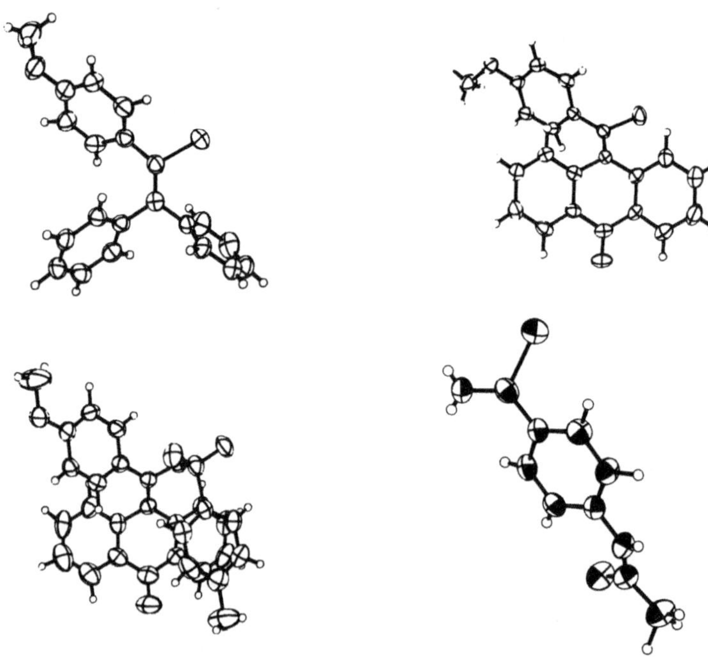

Fig. 1. Perspective diagrams of $C_{21}H_{17}BrO$ (top left), $C_{22}H_{15}BrO_2$ (top right), $C_{29}H_{22}SO_5$ (bottom left) and $C_{10}H_{10}BrNO$ (bottom right).

I. Monoclinic, $P2_1/c$, a = 5.644, b = 17.678, c = 17.431 Å, β = 97.80°, Z = 4. Mo radiation, R = 0.065 for 2371 reflections.

II. Monoclinic, $P2_1$, a = 12.550, b = 6.824, c = 10.715 Å, β = 109.93°, Z = 2. Mo radiation, R = 0.031 for 1634 reflections.

III. Monoclinic, $P2_1/c$, a = 15.118, b = 10.813, c = 14.850 Å, β = 96.14°, Z = 4. Mo radiation, R = 0.073 for 2568 reflections.

IV. Monoclinic, $P2_1/c$, a = 11.151, b = 9.237, c = 10.142 Å, β = 106.19°, Z = 4. Mo radiation, R = 0.057 for 1412 reflections.

The crystal structure analysis established the structure of the tosylate (III). The molecules are shown in Fig. 1. Most of the torsional angles of the aryl groups with the double bond are 44.3±3.3°, except that for the angle of the β-ring cis to the bromine atom in (I) which is 68.6° and that for the α-ring in (II) which is 75.6°. The central rings of the anthronylidene moiety in (II) and (III) are puckered. The C-Br bond lengths in (I) (1.927 Å) and (II) (1.923 Å) are longer than in (IV) (1.907 Å) and than the mean value of other =C-Br bonds. The XC=C angle is wider for (II) (X=Br) than for (III) (X=O).

9-ACETOXY-10-(2-ACETOXY-5-BROMOPHENYL)ANTHRACENE
$C_{24}H_{17}BrO_4$

A. ROSZAK and Z. SKRZAT, 1985. *Acta Cryst.*, **C41**, 1483-1486.

Triclinic, $P\bar{1}$, a = 11.321, b = 11.212, c = 16.454 Å, α = 94.32, β = 87.54, γ = 81.83°, Z = 4. Mo radiation, R = 0.050 for 4874 reflections.

The two crystallographically independent molecules in the asymmetric unit are stacked in two alternating columns along the *a* and *b* axes, respectively (Fig. 1). Bond lengths and angles in the two molecules are not significantly different. The anthracene ring system is slightly buckled in both molecules and is also twisted along its long axis. The phenyl ring is planar and this plane forms an angle of 69.9 (A) and 64.8° (B) with the mean plane of the anthracene moiety. The acetoxy groups are planar and that connected with the phenyl ring makes an angle of 52.2 (A) and 52.9° (B) with the ring. The acetoxy group bonded to C(9) is approximately perpendicular to the anthracene moiety. Br atoms lie on the puckered ribbons extending in the c direction. A short Br...Br intermolecular distance is observed (3.693 Å).

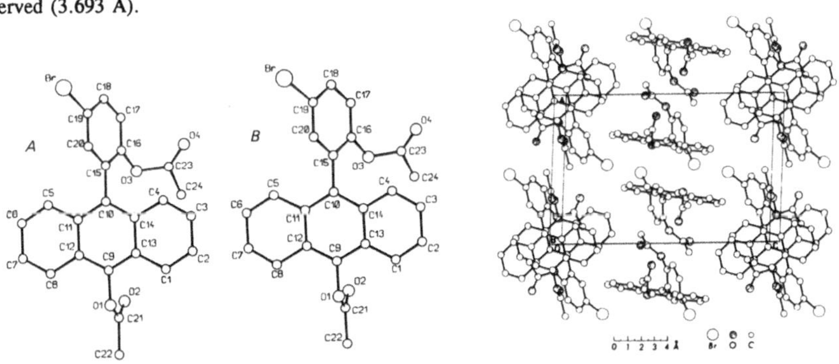

Fig. 1. $C_{24}H_{17}BrO_4$: the two independent molecules and the packing.

1,2-BIS(9-ANTHRYL)ACETYLENE
$C_{30}H_{18}$

H.-D. BECKER, B.W. SKELTON and A.H. WHITE, 1985. *Aust. J. Chem.*, **38**, 1567-1570.

Monoclinic, $P2_1/c$, a = 12.432(5), b = 5.112(1), c = 18.758(8) Å, β = 126.58°, Z = 2. Mo radiation, R = 0.042 for 566 reflections.

The molecule (Fig. 1) is centrosymmetric, with the two anthracene moieties coplanar. The distance between the H(1) and H(8) atoms and their centrosymmetrically related equivalents is 2.1 Å. The acetylenic C-C distance is 1.19(1) Å. In the crystal, intermolecular interactions are observed between anthryl and acetylenic moieties.

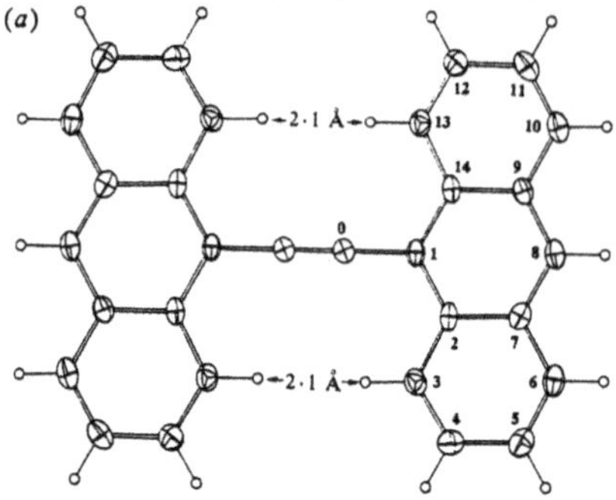

Fig. 1. $C_{30}H_{18}$: the molecular structure.

(E)-1-(9-ANTHRYL)-2-(10-METHYL-9-ANTHRYL)ETHENE
$C_{31}H_{22}$

H.-D. BECKER, L. HANSEN, B.W. SKELTON and A.H. WHITE, 1985. *Aust. J. Chem.*, **38**, 809-815.

Triclinic, $P\bar{1}$, a = 12.95, b = 9.316, c = 9.098 Å, α = 86.17, β = 72.26, γ = 74.61°, Z = 2. Mo radiation, R = 0.046 for 1059 reflections.

In the molecule (Fig. 1), the planes of the two anthracene moieties form an angle of 70.8°, the plane of the ethene bond being twisted out of the planes of the anthracenes by an angle of 55°. The interplanar distance between anthracene π-systems in parallel adjacent molecules is 3.5 Å. Short intramolecular contacts, 2.1 and 22 Å, involve H atoms attached to the ethene and the anthracene systems. The intermolecular packing pattern is characterised by head-to-tail pairwise-arranged anthracene moieties.

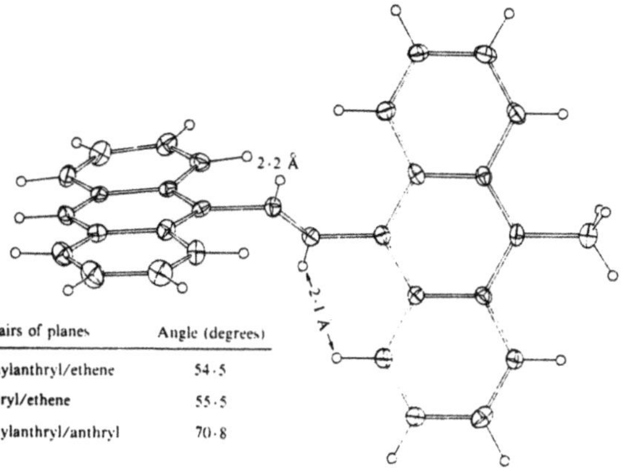

Pairs of planes	Angle (degrees)
Methylanthryl/ethene	54·5
Anthryl/ethene	55·5
Methylanthryl/anthryl	70·8

Fig. 1. $C_{31}H_{22}$: the molecular structure.

meso-BIS(α-(9-ANTHRYL)ETHYL) ETHER
$C_{32}H_{26}O$ (I)

rac-(R,R,S,S)BIS(α-(9-ANTHRYL)ETHYL) ETHER
$C_{32}H_{26}O$ (II)

H.-D. BECKER, V.A. PATRICK, B.W. SKELTON and A.H. WHITE, 1985. *Aust. J. Chem.*, **38**, 1417-1423.

I. Triclinic, $P\bar{1}$, a = 19.03, b = 14.21, c = 9.451 Å, α = 79.46, β = 89.68, γ = 68.97°, Z = 4. Mo radiation, R = 0.041 for 3568 reflections.

II. Orthorhombic, *Pcab*, a = 23.07, b = 19.85, c = 10.241 Å, Z = 8. Mo radiation, R = 0.053 for 1868 reflections.

The asymmetric unit of the meso isomer comprises two molecules with similar geometry. The angle between the short axes of the two anthracene moieties in each molecule (Fig. 1) is approximately 90°, the angles between the anthracene planes are 46° and 43° for the two molecules, and the dihedral angles between the methyl groups along the ether bonds are 90 and 93°. In the racemate, the dihedral angle between the methyl groups along the ether bonds is 12°, and the short axes of the anthracene moieties lie at an angle of about 120°. The molecular dimensions are normal in both (I) and (II).

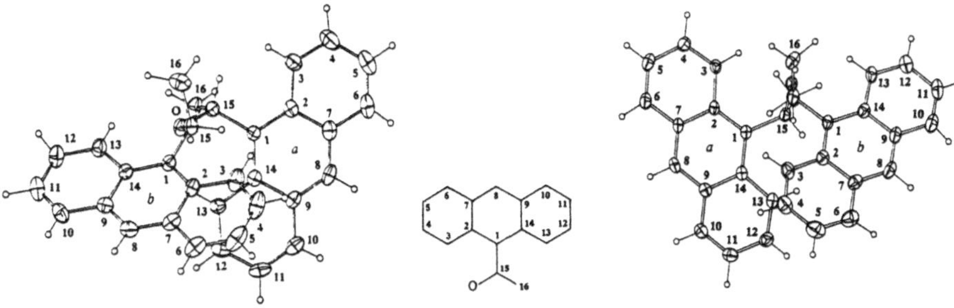

Fig. 1. $C_{32}H_{26}O$: a molecule of the meso-isomer (left) and of the rac-isomer (right).

(1R*,3R*,4R*,6S*)-BICYCLO[4.1.0]HEPTANE-3,4-DICARBOXYLIC ACID
$C_9II_{12}O_4$

N. EBBY, J. LAPASSET, L. PIZZALA, J.-P. AYCARD and H. BODOT, 1985. *Acta Cryst.*, **C41**, 567-570.

Triclinic, $P\bar{1}$, a = 14.986, b = 11.358, c = 8.480 Å, α = 98.81, β = 77.00, γ = 98.91°, Z = 6. Cu radiation, R = 0.072 for 3039 reflections.

The three independent molecules in the asymmetric unit (Fig. 1) are associated through conventional hydrogen bonds (O...O 2.67(1) Å, mean) to form a trimer, each molecule being bonded to the other two. Consequently there is a large cavity in the centre of the trimer. In each of A, B and C the cyclopropane ring has the same orientation with respect to the carbon skeleton. Bond lengths and angles show some variation between the three molecules but the mean values are acceptable.

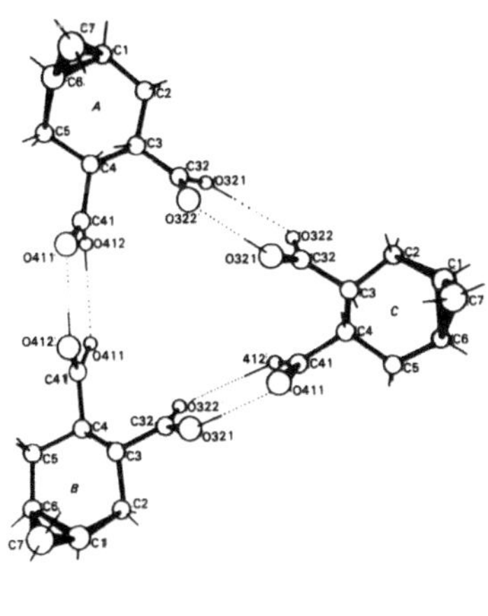

Fig. 1. $C_9H_{12}O_4$: the trimer.

ETHYL 2-AMINO-1-OXO-INDEN-3-CARBOXYLATE
$C_{12}H_{11}NO_3$

G.W. KIRBY, H. McGUIGAN, J.W.M. MacKINNON and P.R. MALLINSON, 1985. *J. Chem. Soc. Perkin Trans. 1*, 405-408.

Monoclinic, $P2_1/a$, a = 14.988, b = 5.133, c = 15.192 Å, β = 112.73°, Z = 4. Cu radiation, R = 0.089 for 976 reflections.

The four molecules in the unit cell are arranged in pairs, each pair consisting of centrosymmetrically related molecules (Fig. 1), so that an eight-membered ring is formed which has two sides consisting of hydrogen bonds. Atoms C(6), C(10), O(1) and O(3) deviate significantly from the plane of the other non-hydrogen atoms.

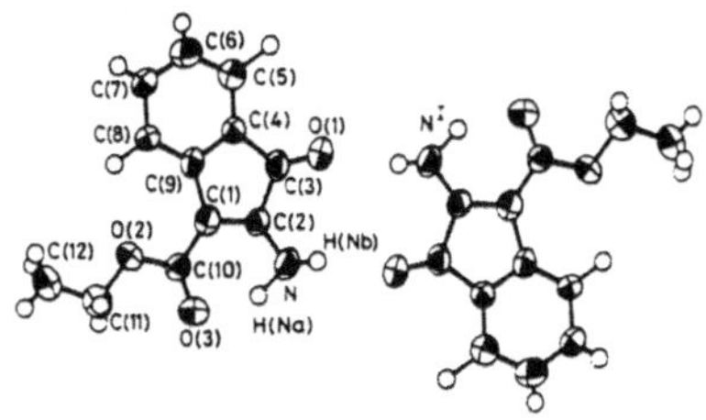

Fig. 1. The centrosymmetric hydrogen-bonded dimeric unit in the crystal of $C_{12}H_{11}NO_3$.

(6R*,7S*)-METHYL 4-(6-endo-HYDROXYBICYCLO[3.2.0]HEPT-2-EN-7-endo-YLAMINO)-4-OXOBUTYRATE
$C_{12}H_{17}NO_4$

B. SHELDRICK, D. AKRIGG, M.I. PAGE and D. AGATHOCLEOUS, 1985. *Acta Cryst.*, **C41**, 942-944.

Monoclinic, $P2_1/c$, a = 11.594, b = 6.138, c = 17.512 Å, β = 102.35°, Z = 4. Cu radiation, R = 0.077 for 1714 reflections.

The fused rings of the molecule (Fig. 1) make an angle of 116.5(7)° with each other. The amide group is planar and the carbonyl nearly so. The chain twists around C(8)-C(9) allowing the H atoms of the CH_2 to adopt a staggered conformation and the chain to curve round to form a compact shape. Bond lengths and angles are normal. Both the NH and OH groups form hydrogen bonds with lengths of 3.35(1) and 2.74(1) Å respectively, linking molecules in the b-axis direction.

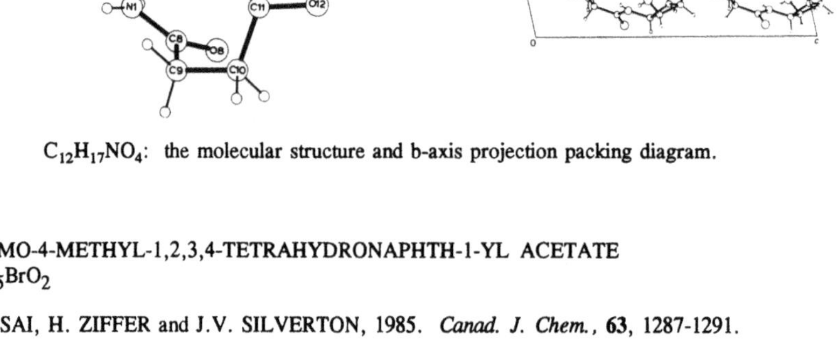

Fig. 1. $C_{12}H_{17}NO_4$: the molecular structure and b-axis projection packing diagram.

2-BROMO-4-METHYL-1,2,3,4-TETRAHYDRONAPHTH-1-YL ACETATE
$C_{13}H_{15}BrO_2$

M. KASAI, H. ZIFFER and J.V. SILVERTON, 1985. *Canad. J. Chem.*, **63**, 1287-1291.

Orthorhombic, $P2_12_12_1$, a = 8.810, b = 9.661, c = 14.852 Å, Z = 4. Cu radiation, R = 0.049 for 1275

reflections.

The relative stereochemistry of this key intermediate was established to be as shown in Fig. 1. Molecular dimensions are unexceptional. As five crystals were used for data collection, no attempt at absolute stereochemistry determination was made.

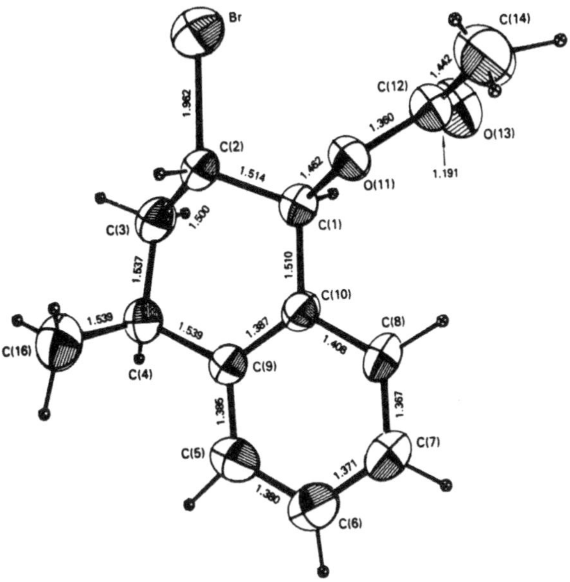

Fig. 1. The conformation and relative stereochemistry of $C_{13}H_{15}BrO_2$.

2-HYDROXY-1,6,10-TRIMETHYLBICYCLO[5.3.0]DECA-3,9-DIEN-5-ONE
$C_{13}H_{18}O_2$

P.T. LANSBURY, D.J. MAZUR and J.P. SPRINGER, 1985. *J. Org. Chem.*, **50**, 1632-1636.

Monoclinic, $P2_1/n$, a = 6.762, b = 24.323, c = 7.115 Å, β = 91.21°, Z = 4. Cu radiation, R = 0.039 for 1328 reflections.

The seven-membered ring (Fig. 1) has a boat conformation. The crystal structure analysis confirmed the stereochemistry of the molecule.

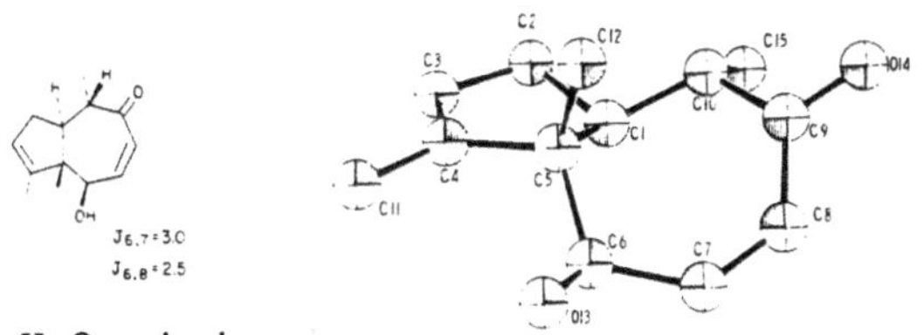

Fig. 1. The $C_{13}H_{18}O_2$ molecule.

cis-3,8a-DIHYDRO-3-METHOXY-1,2-AZULENEDICARBOXYLIC ACID DIMETHYL ESTER
$C_{15}H_{16}O_5$ (I)

trans-1,2,3,8a-TETRAHYDRO-3-METHOXY-1,2-DIAZA-1,2-AZULENEDICARBOXYLIC ACID DIMETHYL ESTER

$C_{13}H_{16}N_2O_5$ (II)

J. DAUB, G. HIRMER, L. JAKOB, G. MAAS, W. PICKL, E. PIRZER and K.M. RAPP, 1985. *Chem. Ber.*, **118**, 1836-1856.

I. Monoclinic, $P2_1/n$, a = 18.363, b = 12.712, c = 5.995 Å, β = 92.33°, Z = 4. Mo radiation, R = 0.0480 for 1432 reflections.

II. Triclinic, $P\bar{1}$, a = 7.7692, b = 13.287, c = 14.4406 Å, α = 103.080, β = 100.008, γ = 97.881°, Z = 4. Cu radiation, R = 0.0428 for 4134 reflections.

The structures of these cycloadducts were established as shown in Fig. 1. There are two independent molecules in (II) with similar dimensions. The seven-membered rings in (I) and (II) have boat conformations with fold angles α 52.8°, β 28° in (I) and α 54.6, 53.5°, β 27.7, 26.1° in (II). The five-membered ring in (I) is essentially planar, but in (II) the five-membered ring is an N(2) envelope.

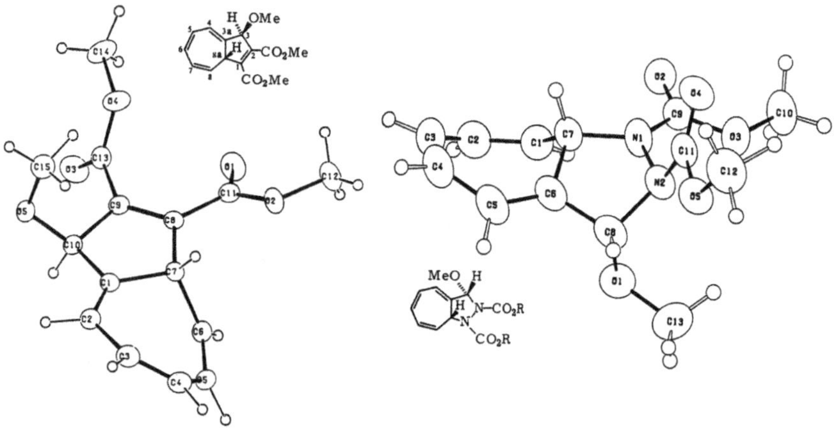

Fig. 1. Views of $C_{15}H_{16}O_5$ (I) (left) and $C_{13}H_{16}N_2O_5$ (II) (right).

CARYOPHYLLENE NITROSITE
$C_{15}H_{24}N_2O_3$ (I)

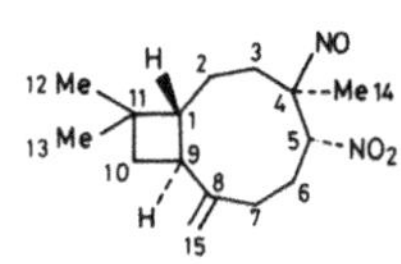

DINITROCARYOPHYLLENE
$C_{15}H_{24}N_2O_4$ (II)

A.A. FREER, D.K. MacALPINE, J.A. PEACOCK and A.L. PORTE, 1985. *J. Chem. Soc. Perkin Trans.* **2**, 971-982.

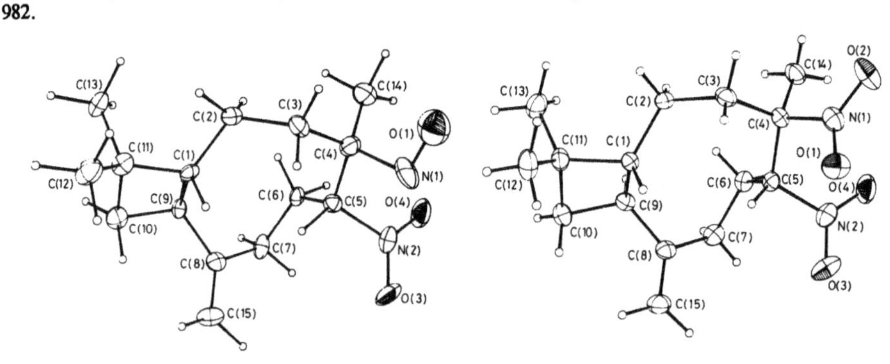

Fig. 1. Molecular conformations of $C_{15}H_{24}N_2O_3$ (left) and $C_{15}H_{24}N_2O_4$ (right).

I. Orthorhombic, $P2_12_12_1$, a = 6.476, b = 13.199, c = 18.784 Å, Z = 4. Mo radiation, R = 0.085 for 714

reflections.

II. Monoclinic, $P2_1$, a = 6.405, b = 12.991, c = 9.797 Å, β = 93.04°, Z = 2. Mo radiation, R = 0.036 for 1453 reflections.

The molecular conformations of (I) and (II) are very similar and are shown in Fig. 1.

(1RS,2RS,6RS,7SR,1'RS)-6-(1'-HYDROXYETHYL)-7-ISOPROPENYL-2,5,5-TRIMETHYLBICYCLO[4.1.0]-HEPTAN-2-OL
$C_{15}H_{26}O_2$

U. GOLDENER, M.E. SCHELLER, P. MATHIES, B. FREI and O. JEGER, 1985. *Helv. Chim. Acta*, **68**, 635-650.

Monoclinic, $P2_1/n$, a = 13.294, b = 13.029, c = 17.602 Å, β = 105.32°, Z = 8. Mo radiation, R = 0.038 for 1728 reflections.

The crystallographic asymmetric unit contains two independent molecules (Fig. 1). The crystal structure analysis unambiguously identified the molecule.

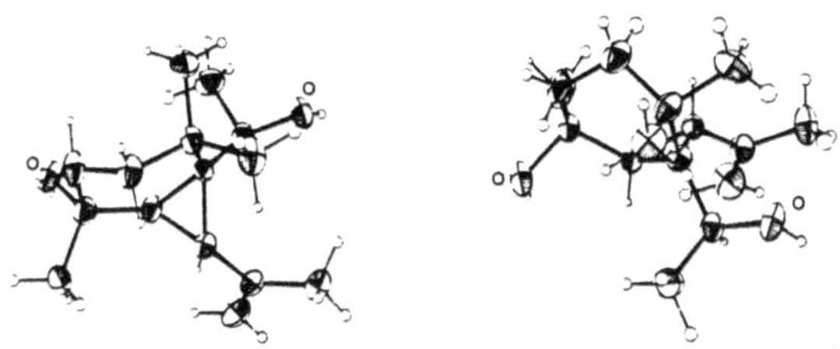

Fig. 1. Perspective drawings of the two independent molecules in the crystal of $C_{15}H_{26}O_2$.

(1R,2R,3R,6S,10R)-1-FLUORO-3-(2-HYDROXY-2-PROPYL)-6,10-DIMETHYLBICYCLO[4.4.0]DECAN-2-OL
$C_{15}H_{27}FO_2$

R.G. PRITCHARD, R. RAMAGE and I.A. SOUTHWELL, 1985. *Acta Cryst.*, **C41**, 771-773.

Orthorhombic, $P2_12_12_1$, a = 7.458, b = 13.436, c = 14.736 Å, Z = 4. Mo radiation, R = 0.036 for 1856 reflections.

The cis-bridged bicyclodecane skeleton of the molecule (Fig, 1) is in a slightly distorted double chair conformation. All the substituents except the bulky 2-propanol group are in axial positions. The hydroxyl O(1) interacts with hydrogen atoms on C(7) and C(9), on one side of the ring, tending to flatten the molecule in the bridge region. Further steric hindrance is noted on the opposite side of the ring. There is an intramolecular O(2)-H...O(1) hydrogen bond of length 2.698(2) Å bridging the two hydroxyl groups. Another hydrogen bond between O(1) and a symmetry related hydroxyl oxygen atom O(2) of length 2.687(2) Å links the molecules into infinite chains parallel to the a-axis.

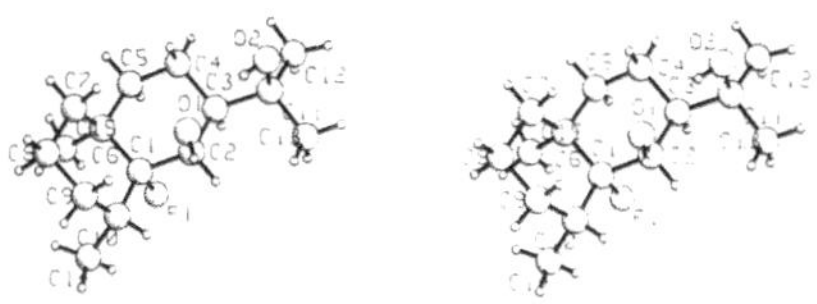

Fig. 1. A stereoview of the $C_{15}H_{27}FO_2$ molecule.

(±)-5-METHYL-1β-(PHENYLMETHOXY)-4-NITRO-2,3,3aβ,4α,5α,7aα-HEXAHYDRO-1H-INDENE
$C_{17}H_{21}NO_3$ (I)

(±)-5-METHYL-1α-((4-METHYLPHENYL)SULPHONYL)-4-NITRO-2,3,3aβ,4α,5α,7aα-HEXAHYDRO-1H-INDENE
$C_{17}H_{21}NO_4S$ (II)

M.J. KURTH, M.J. O'BRIEN, H. HOPE and M. YANUCK, 1985. *J. Org. Chem.*, **50**, 2626-2632.

I. Monoclinic, $P2_1/c$, a = 11.766, b = 5.101, c = 24.18 Å, β = 90.36°, Z = 4. Mo radiation, R = 0.092 for 1307 reflections (at 140 K).

II. Monoclinic, $P2_1/c$, a = 5.321, b = 14.057, c = 21.484 Å, β = 91.65°, Z = 4. Mo radiation, R = 0.066 for 1129 reflections (at 140 K).

The crystal structure analyses established the trans ring fusion of (I) and the stereochemistry of (II) (Fig. 1).

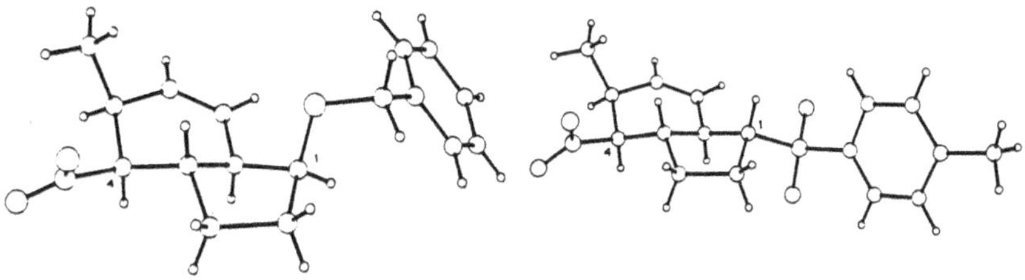

Fig. 1. Drawings of $C_{17}H_{21}NO_3$ (I) (left) and $C_{17}H_{21}NO_4S$ (II) (right).

2,3,4aβ,6,7,8aβ-HEXAMETHYL-4a,5,8,8a-TETRAHYDRO-1-NAPHTHOQUIN-4α-YL ACETATE
$C_{18}H_{26}O_3$ (I)

2,3,4aβ,6,7,8aβ-HEXAMETHYL-4a,5,8,8a-TETRAHYDRO-1-NAPHTHOQUIN-4β-YL ACETATE
$C_{18}H_{26}O_3$ (II)

S. ARIEL and J. TROTTER, 1985. *Acta Cryst.*, **C41**, 295-298.

I. Monoclinic, $P2_1/c$, a = 14.826, b = 6.900, c = 17.511 Å, β = 109.87°, Z = 4. Mo radiation, R = 0.052 for 2066 reflections.

II. Triclinic, $P\bar{1}$, a = 8.993, b = 9.228, c = 10.770 Å, α = 100.29, β = 94.26, γ = 106.71°, Z = 2. Mo radiation, R = 0.046 for 2639 reflections.

The molecular structures (Fig. 1) show that the two molecules have fused six-membered rings with half-chair conformations which are folded and twisted about the C(4a)-C(8a) bond. They have different ring-flipped conformations and the acetate groups occupy the more sterically favoured pseudo-equatorial positions in both molecules. The bridgehead torsion angles in (I) are -61.3(3) and 62.4(2)° while in (II) the corresponding angles are 59.6(2) and -58.9(2)°. In (I) there is a close intramolecular approach between H(51) and C(2) (2.74(2)) and C(3) (2.70(2)) while in (II) the close approach is between H(81) and C(2) (2.79(2)) and C(3) (2.84(2) Å).

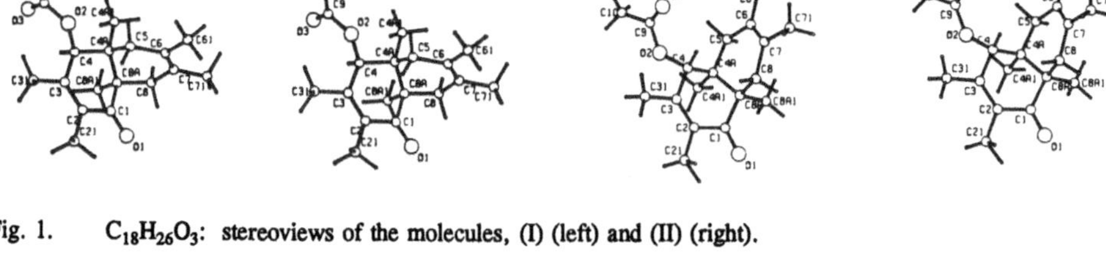

Fig. 1. $C_{18}H_{26}O_3$: stereoviews of the molecules, (I) (left) and (II) (right).

8-((2,2-DIMETHYL-3-OXOCYCLOHEXYL)HYDROXYMETHYL)-1-METHYLBICYCLO[4.2.0]OCTAN-7-ONE
$C_{18}H_{28}O_3$

C.K. FAIR, G.R. CLARK and M.M. NIKAIDO, 1985. *Acta Cryst.*, **C41**, 155-157.

Monoclinic, $P2_1/c$, a = 7.095, b = 11.584, c = 20.191 Å, β = 95.05°, Z = 4. Mo radiation, R = 0.041 for 1665 reflections (at 140 K).

In the molecule (Fig. 1) both cyclohexane rings are in the chair conformation. The cyclobutanone is in an envelope conformation having a dihedral angle of 26.1(2)° between the C(5,6,7) and C(5,8,7) planes. Bond distances and angles in the molecule are normal. Though the compound is an aldol there is no intramolecular hydrogen bond between the cyclobutanone carbonyl and the hydroxy group. Instead. the molecules form hydrogen-bonded dimers in which this hydroxy group is involved; the O(2)-H...O(3) separation is 2.771(4) Å.

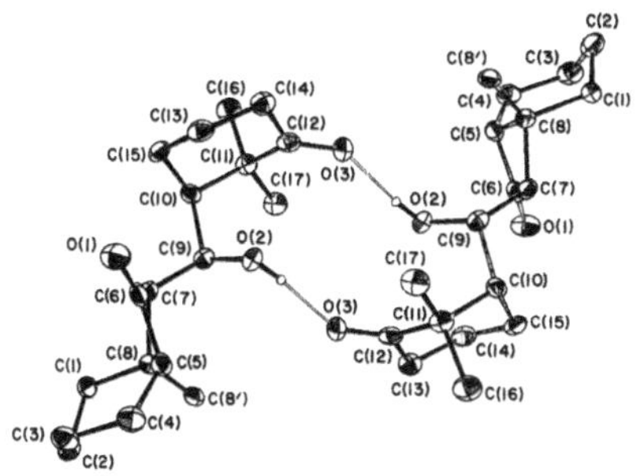

Fig. 1. $C_{18}H_{28}O_3$: a view of a hydrogen-bonded dimer.

(±)-1,1',3,3'-TETRAMETHYL-1,1'-BIINDENE
$C_{22}H_{22}$

Y. MAOUCHE, G. GERMAIN, M.C. BRIANSO, J.Y. SANCHEZ, P. NICOLET and M.J.M. ABADIE, 1985. *Acta Cryst.*, **C41**, 99-101.

Orthorhombic, $Pna2_1$, a = 12.204, b = 16.854, c = 7.985 Å, Z = 4. Mo radiation, R = 0.085 for 839 reflections.

In the molecule (Fig. 1) both indene groups are planar with an interplanar angle of 120°. The compound corresponds to the diastereomer (±) which exists in the two enantiomorphic forms S,S and R,R. The bond distances show the ethylenic nature of the C(4)-C(5) and C(16)-C(17) bonds of the benzene rings in each of the indene groups. The torsion angle C(19)-C(7)-C(10)-C(21) about the elongated bond joining the indene moieties is -50.2°. There are several short intermolecular H...H contacts indicative of the strong van der Waals forces between the molecules.

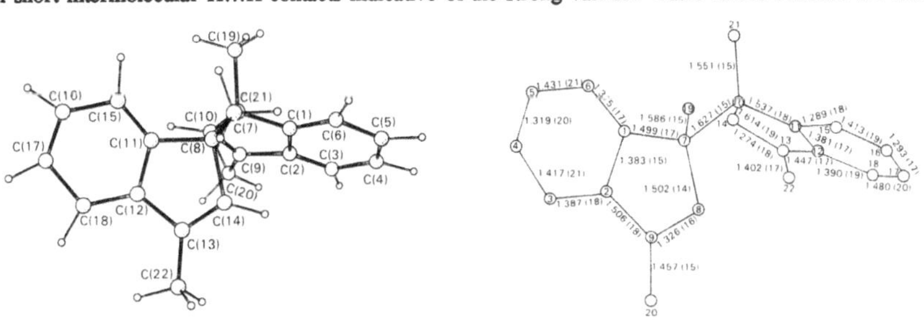

Fig. 1. $C_{22}H_{22}$: a view of the molecule (left) and bond lengths (right).

1,3-BIS(DICYANOMETHYLENE)-2-(4'-(N,N-DIETHYLAMINO)PHENYLIMINO)INDANE
$C_{25}H_{18}N_6$

L.A. CHETKINA, V.K. BEL'SKII, E.G. POPOVA and B.P. BESPALOV, 1985. *Zh. Strukt. Khim.*, **26**, 5-115-119 [J. Struct. Chem., **26**, 748-751].

Triclinic, $P\bar{1}$, a = 8.725, b = 8.238, c = 16.335 Å, α = 74.50, β = 89.61, γ = 69.50°, Z = 2. Mo radiation, R = 0.049 for 1706 reflections.

The molecule is shown in Fig. 1. Due to steric factors, the electron-withdrawing dicyanomethyleneindane fragment and the electron-donating diethylaminophenyl fragment are twisted 38.2° relative to one another. Bond length changes indicate intramolecular charge transfer. The five-membered ring of the indane system has an envelope conformation, while the phenylene ring is a compressed boat.

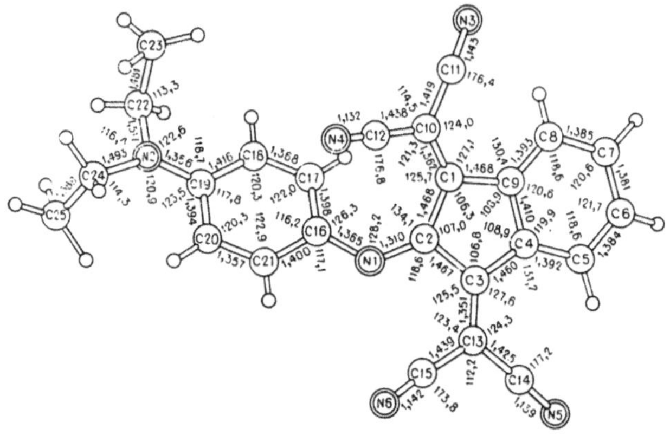

Fig. 1. The structure of $C_{25}H_{18}N_6$.

3-(DIPHENYLMETHYL)-3-PHENYL-1-INDANONE
$C_{28}H_{22}O$

W. ADAM, A. BERKESSEL, E.-M. PETERS, K. PETERS and H.G. VON SCHNERING, 1985. *J. Org. Chem.*, **50**, 2811-2814.

Monoclinic, $P2_1/n$, a = 15.253, b = 13.494, c = 10.067 Å, β = 99.00°, Z = 4. Mo radiation, R = 0.073 for 1291 reflections.

The crystal structure analysis unambiguously determined the stereochemistry of the molecule.

ACENAPHTHYLENE
$C_{12}H_8$

R.A. WOOD, T.R. WELBERRY and A.D. RAE, 1985. *J. Chem. Soc. Perkin Trans. 2*, 451-456.

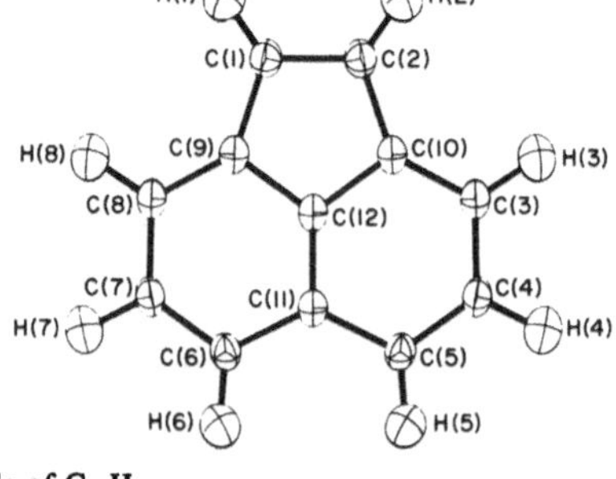

Fig. 1. Drawing of a molecule of $C_{12}H_8$.

Orthorhombic, $P2_1nm$, a = 7.584, b = 7.589, c = 27.851 Å, Z = 8. Neutron radiation (λ = 1.238 Å), R =

0.069 for 950 reflections (at 80K).

The structure is disordered. A full description and discussion of the sites of disorder and occupancy levels is given in the paper, together with the molecular geometry (Fig. 1).

2-NITROFLUORENONE
$C_{13}H_7NO_3$

N.G. FURMANOVA, V.I. ANDRIANOV, A.M. ANDRIEVSKII and A.N. POPLAVSKII, 1985. *Kristallografiya*, **30**, 82-85.

Monoclinic, $P2_1$, a = 16.937, b = 7.694, c = 3.790 Å, β = 92.42°, Z = 2. Mo radiation, R = 0.046 for 740 reflections.

The fluorene nucleus (Fig. 1) is practically planar and the nitro group is rotated 6.6° out of this plane. Molecular dimensions agree with those in fluorenone (1). In the crystal, stacks are formed in the direction of the *x*-axis and the stacks are packed in a parquet motif.

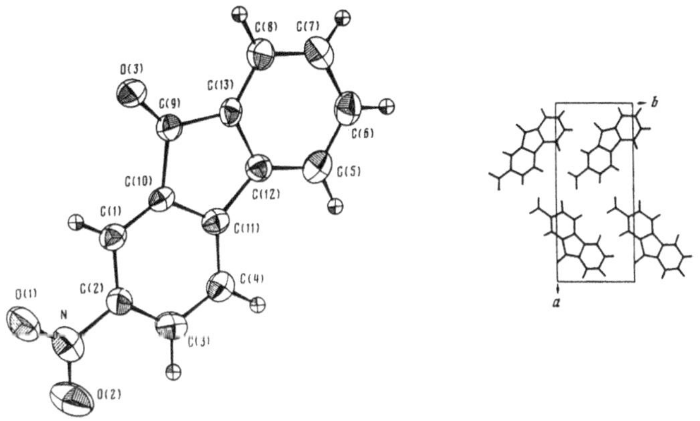

Fig. 1. The 2-nitrofluorenone molecule and the packing in the ab plane.

1. Structure Reports, 38B, 211.

4-CYANOTRICYCLO[6.4.0.0^{2,4}]DODECA-1(12),6,8,10-TETRAEN-5-ONE
$C_{13}H_9NO$

H. SUIGINOME, CHIFU LIU, M. TOKUDA and A. FURUSAKI, 1985. *J. Chem. Soc. Perkin Trans. 1*, 327-329.

Monoclinic, $P2_1/c$, a = 9.351, b = 12.242, c = 9.217 Å, β = 105.78°, Z = 4. Mo radiation, R = 0.052 for 1560 reflections.

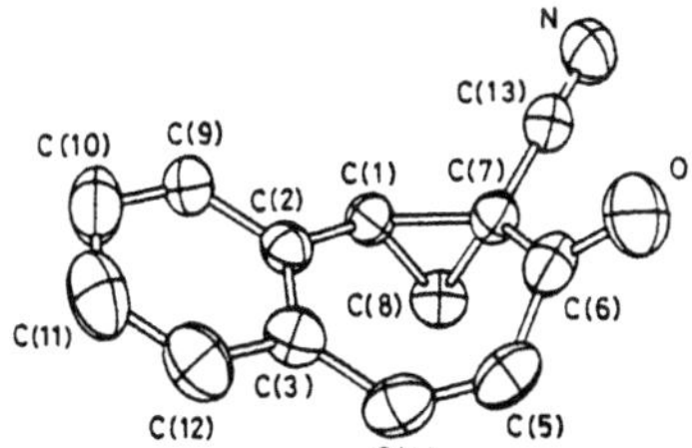

Fig. 1. Perspective view of a molecule of $C_{13}H_9NO$ with hydrogen atoms omitted.

The seven-membered ring (Fig. 1) has a flattened boat conformation with an approximate mirror plane through the carbonyl group. The cyclopropane ring is approximately bisected by each of the benzene-ring and carbonyl planes whilst the C(4)=C(5) bond deviates from these planes; as a result of this, the α-hydrogen atom at C(8) approaches the C(5) atom, with a separation of 2.67(3) Å.

cis-4b,6,7,8,9,9a-HEXAHYDRO-5H-BENZO[3,4]CYCLOBUTA[1,2]CYCLOHEPTEN-1-OL
$C_{13}H_{16}O$

A. COURTOIS, V. PESME-PICHON, M.C. CARRE and P. CAUBERE, 1985. *Acta Cryst.*, **C41**, 793-795.

Orthorhombic, $P2_12_12_1$, a = 18.126, b = 11.910, c = 4.887 Å, Z = 4. Mo radiation, R = 0.038 for 762 reflections.

This compound has cis geometry at the saturated ring-benzocyclobutenic ring junction (Fig. 1). The benzocyclobutenic moiety is planar to within 0.036 Å and the seven-membered ring has a chair conformation. The C(1)-C(8) bond fusing these moieties is of length 1.605(6) Å: all other bond lengths have normal values, though valence angles exhibit significant deviations from ideal values. Hydrogen bonds (O-H...O) of length 2.813(4) Å, between the alcohol functions link the molecules into infinite chains parallel to the c-axis.

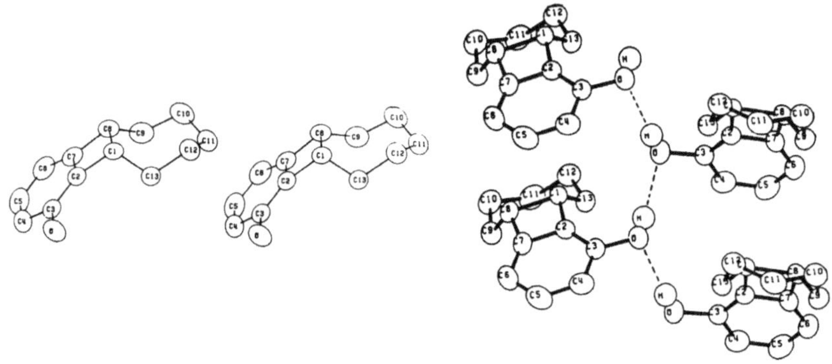

Fig. 1. $C_{13}H_{16}O$: a stereoview of the molecule and a projection of part of the crystal structure onto (100) showing the hydrogen-bonded chain of molecules.

(3aR*-(3aα,5aβ,9aβ,9bα))-DODECAHYDRO-9b-NITRO-5H-BENZ[e]INDEN-5-ONE
$C_{13}H_{19}NO_3$

A. MEZZETTI, P. NITTI, G. PITACCO and E. VALENTIN, 1985. *Tetrahedron*, **41**, 1415-1422.

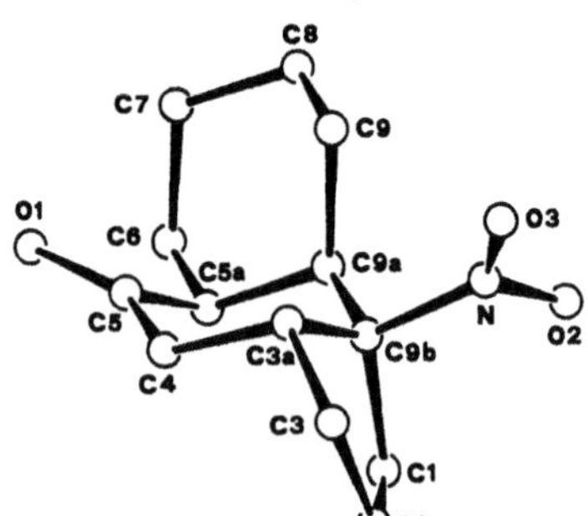

Fig. 1. The $C_{13}H_{19}NO_3$ molecule.

Monoclinic, $P2_1/c$, a = 11.056, b = 6.512, c = 17.670 Å, β = 101.99°, Z = 8. Mo radiation, R = 0.047 for 1785 reflections.

The analysis established the structure and stereochemistry as shown in Fig. 1.

9-BROMO-9-NITRO-10-TRINITROMETHYL-9,10-DIHYDROANTHRACENE
$C_{15}H_9BrN_4O_8$ (I)

9-NITRO-9-PHENYL-10-TRINITROMETHYL-9,10-DIHYDROANTHRACENE
$C_{21}H_{14}N_4O_8$ (II)

9-HYDROXY-9-PHENYL-10-TRINITROMETHYL-9,10-DIHYDROANTHRACENE DICHLOROMETHANE SOLVATE
$C_{21}H_{15}N_3O_7,0.5(CH_2Cl_2)$ (III)

J.M. MASNOVI and J.K. KOCHI, 1985. *J. Org. Chem.*, **50**, 5245-5255.

I. Monoclinic, $P2_1/a$, a = 17.486, b = 13.443, c = 7.838 Å, β = 118.02°, Z = 4. Mo radiation, R = 0.0562 for 2427 reflections (at -160°C).

II. Monoclinic, $P2_1/n$, a = 9.119, b = 13.211, c = 16.037 Å, β = 101.12°, Z = 4. Mo radiation, R = 0.0465 for 1983 reflections (at -162°C).

III. Triclinic, $P\bar{1}$, a = 14.530, b = 14.793, c = 10.486 Å, α = 110.45, β = 95.90, γ = 102.85°, Z = 4. Mo radiation, R = 0.0834 for 4631 reflections (at -160°C).

The crystal structure analyses established the stereochemistries of molecules (I), (II) and (III) (Fig. 1).

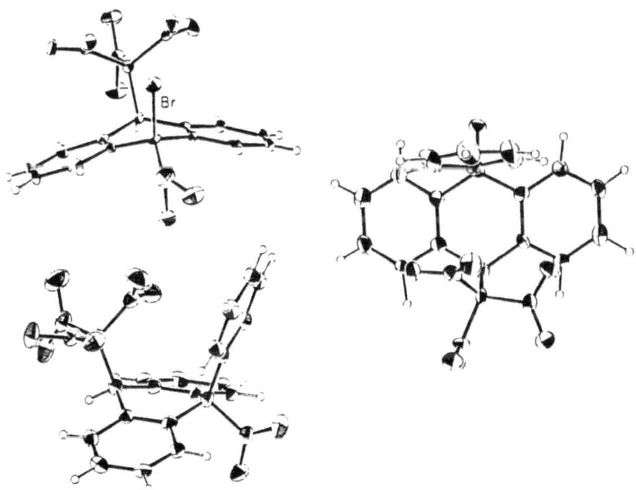

Fig. 1. ORTEP diagrams of $C_{15}H_9BrN_4O_8$ (I) (top left), $C_{21}H_{14}N_4O_8$ (II) (bottom left) and $C_{21}H_{15}N_3O_7,0.5$-(CH_2Cl_2) (III) (right).

14β-HYDROXY-5-METHOXY-DE-A-OESTRA-5,7,9,16-TETRAEN-15-ONE
$C_{15}H_{16}O_3$

R.A. HOWIE, A.B. TURNER and P.J. COX, 1985. *J. Chem. Soc. Perkin Trans. 2*, 127-130.

Monoclinic, $P2_1$, a = 8.591, b = 7.969, c = 10.089 Å, β = 113.90°, Z = 2. Mo radiation, R = 0.055 for 863 reflections.

The molecules (Fig. 1) are held together in the crystal by intermolecular hydrogen bonding (O(2)...O(3)' 2.910(9), HO(2)...O(3)' 2.01(5) Å, O(2)-HO(2)...O(3)' 156(5)°). The C/D ring junction is cis with the torsion angle C(17)-C(13)-C(14)-O(2) equal to 139.0(7)°. Ring C has a 12β-sofa conformation whilst ring D has a flattened envelope conformation with C(14) 0.315(6) Å below the plane of C(13,17,16,15).

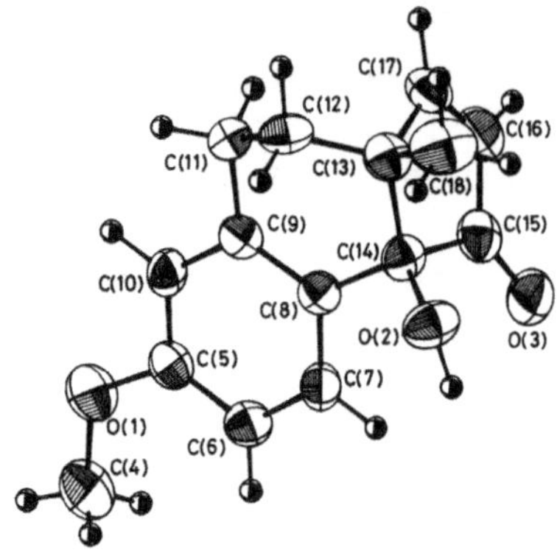

Fig. 1. Atomic arrangement in molecules of $C_{15}H_{16}O_3$.

(3aα,4β,5α,6aα,9aβ,10aβ)-DODECAHYDRO-5-IODO-4,9a-DIMETHYLDICYCLOPENTA[a,d]CYCLOOCTEN-1-(2H)-ONE
$C_{16}H_{25}IO$

R.M. COATES, J.W. MUSKOPF and P.A. SENTER, 1985. *J. Org. Chem.*, **50**, 3541-3557.

Monoclinic, $P2_1/c$, a = 8.659, b = 29.75, c = 6.551 Å, β = 107.37°, Z = 4. Mo radiation, R = 0.029 for 2942 reflections.

The crystal structure analysis confirmed the stereochemistry of the molecule (Fig. 1). The eight-membered ring has a crown-like conformation which is between chair-chair and crown conformations for cyclooctane.

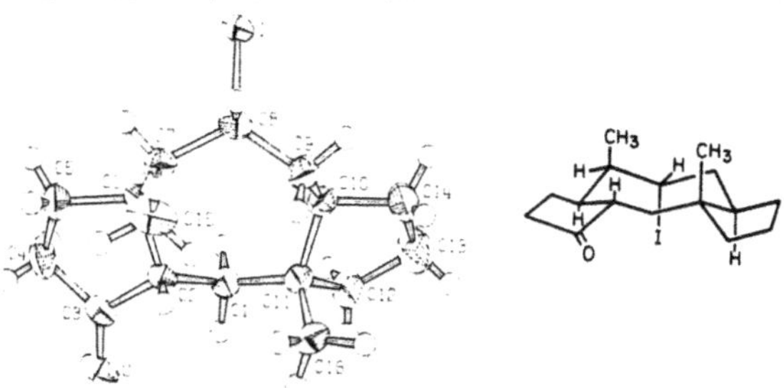

Fig. 1. ORTEP plot of a molecule of $C_{16}H_{25}IO$.

cis-cisoid-TRICYCLO[8.6.0.0^{2,9}]HEXADECANE
$C_{16}H_{28}$

A.L. SPEK, A.J.M. DUISENBERG, P.J.J.A. TIMMERMANS and A. MACKOR, 1985. *Acta Cryst.*, **C41**, 111-113.

Monoclinic, $C2/c$, a = 19.723, b = 5.269, c = 12.957 Å, β = 97.79°, Z = 4. Cu radiation, R = 0.055 for 1090 reflections.

The central cyclobutane ring in the molecule, which has a crystallographic twofold axis of symmetry, is significantly puckered (Fig. 1). This is characterised by the torsion angles C(1)-C(2)-C(2')-C(1') and C(4)-C(3)-C(3')-C(4') of value 24.3(2) and 23.9(2)° respectively. Atoms C(1) and C(4), due to this puckering, are axial and equatorial substituents respectively, of the four-membered ring. The cyclooctane rings have boat-type conformations. There are no unusually short intermolecular contacts.

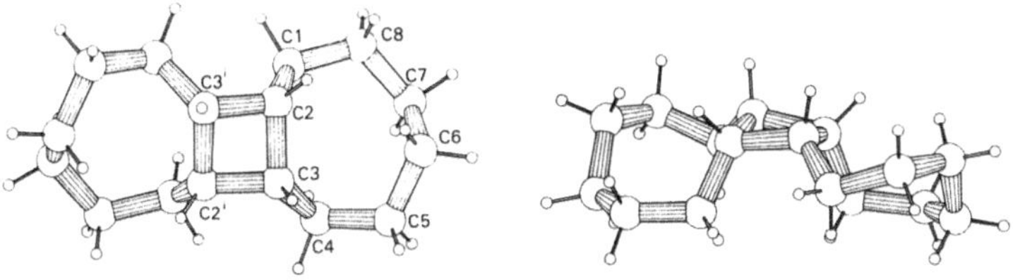

Fig. 1. $C_{16}H_{28}$: views of the molecule showing the atomic numbering.

cis-METHYL 3,4,4a,9,10,10a-HEXAHYDROPHENANTHREN-2(1H)-ONE-4a-ACETATE
$C_{17}H_{20}O_3$

A.K. PAL, S.C. KUNDADAS and B.S. BASAK, 1985. *Z. Kristallogr.*, **172**, 55-62.

Triclinic, $P\bar{1}$, a = 6.413, b = 10.501, c = 10.862 Å, α = 82.71, β = 87.73, γ = 85.29°, Z = 2. Mo radiation, R = 0.055 for 944 reflections.

In the molecule, the aromatic ring (A) is planar and the two cyclohexane rings have (B) half-chair and (C) chair conformations. Bond lengths are shown in Fig. 1.

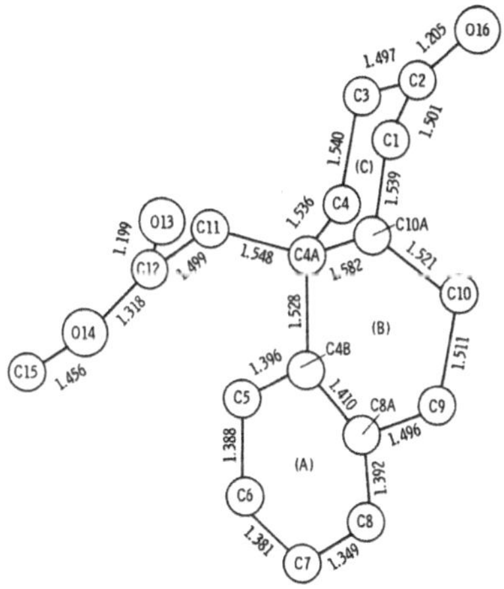

Fig. 1. A line drawing of $C_{17}H_{20}O_3$ with bond lengths.

(4aRS,4SR,10aSR,11RS)-4a-HYDROXY-4-CARBOXY-4-METHYL-1,2,3,4,4a,9,10,10a-OCTAHYDROPHENANTHREN-10a-YL(HYDROXY)ACETIC ACID 4a,10a-LACTONE
$C_{18}H_{20}O_5$ (I)

ETHYL 4-METHYL-7-METHOXY-1,2,3,9,10,10a-HEXAHYDROPHENANTHREN-10a-YL(HYDROXYMETHYLENE)ACETATE
$C_{21}H_{26}O_4$ (II)

B. SAHA, S.C. ROY, G.O.S.V. SATYANARAYANA, U.R. GHATAK, A. SEAL, S. RAY, R. BANDYOPADHYAY, M. GHOSH, B. DAS and B.S. BASAK, 1985. *J. Chem. Soc. Perkin Trans. 1*, 505-513.

I. Orthorhombic, $P2_12_12_1$, a = 10.210, b = 15.674, c = 9.407 Å, Z = 4. Cu radiation, R = 0.049 for 1622 reflections.

II. Orthorhombic, $Pn2_1a$, a = 22.818, b = 12.333, c = 6.577 Å, Z = 4. Mo radiation, R = 0.041 for 1131 reflections.

The X-ray crystal determinations of the two molecules served to establish their structures and stereochemistries (Fig. 1).

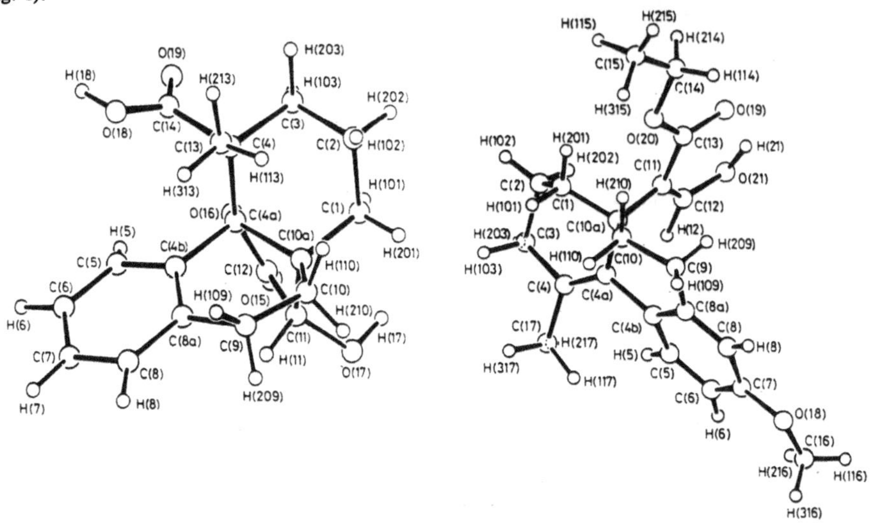

Fig. 1. Perspective drawings of the molecules of $C_{18}H_{20}O_5$ (left) and $C_{21}H_{26}O_4$ (right).

TRIPHENYLNITRONE
$C_{19}H_{15}NO$ (I)

N-FLUOREN-9-YLIDENEPHENYLAMINE-N-OXIDE
$C_{19}H_{13}NO$ (II)

C.P. FALSHAW, N.A. HASHI and G.A. TAYLOR, 1985. *J. Chem. Soc. Perkin Trans. 1*, 1837-1843.

I. Orthorhombic, *Pbca*, a = 17.831, b = 8.840, c = 18.177 Å, Z = 8. Cu radiation, R = 0.0484 for 944 reflections.

II. Monoclinic, $P2_1/c$, a = 5.205, b = 23.492, c = 11.232 Å, β = 100.38°, Z = 4. Cu radiation, R = 0.0522 for 1185 reflections.

The structure of (I) has been reported previously (1); the present results agree with those found before. The main difference between the two molecules occurs at the N-phenyl bond (Fig. 1). The torsion angles O(1)-N(1)-C(14)-C(15) are 67.1° for (I) and 90.2° for (II).

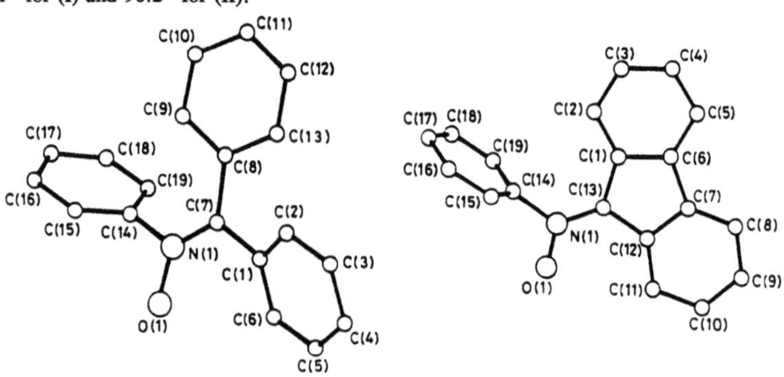

Fig. 1. The molecular structures of $C_{19}H_{15}NO$ (left) and $C_{19}H_{13}NO$ (right).

1. Structure Reports, 39B, 43.

cis-9,10-DIHYDRO-9,10-DIPROPYLANTHRACENE-9,10-DIOL
$C_{20}H_{24}O_2$ (I)

trans-9,10-DIHYDRO-9,10-DIPROPYLANTHRACENE-9,10-DIOL
$C_{20}H_{24}O_2$ (II)

cis-9,10-DIETHYL-9,10-DIHYDROANTHRACENE
$C_{18}H_{20}$ (III)

N. AHMAD, R.J. GODDARD, I.K. HATTON, J.A.K. HOWARD, N.J. LEWIS and J. MacMILLAN, 1985. *J. Chem. Soc. Perkin Trans. 1*, 1859-1863.

I. Monoclinic, $P2_1/c$, a = 7.854, b = 22.277, c = 9.904 Å, β = 101.86°, Z = 4. Mo radiation, R = 0.054 for 1188 reflections.

II. Monoclinic, $P2_1/c$, a = 8.774, b = 16.414, c = 12.396 Å, β = 104.59°, Z = 4. Mo radiation, R = 0.061 for 1779 reflections.

III. Monoclinic, $P2_1/c$, a = 12.405, b = 7.658, c = 15.046 Å, β = 106.73°, Z = 4. Mo radiation, R = 0.042 for 2332 reflections (at 200 K).

Both cis-isomers have the central ring in a boat conformation (Fig. 1) with the 9,10-dialkyl substituents in the pseudo-axial positions. The trans-isomer has a planar ring structure in which the n-propyl group is staggered with the hydrogens β to the 9- and 10- positions eclipsing the C-C bonds of the central ring. In (II) there are two molecules in the crystallographic asymmetric unit.

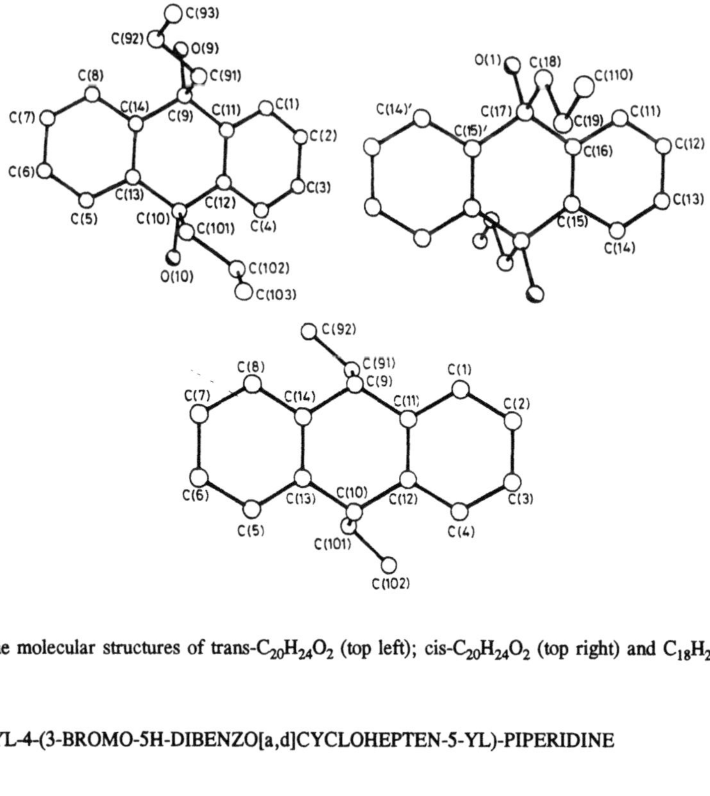

Fig. 1. The molecular structures of trans-$C_{20}H_{24}O_2$ (top left); cis-$C_{20}H_{24}O_2$ (top right) and $C_{18}H_{20}$ (bottom).

(+)-1-METHYL-4-(3-BROMO-5H-DIBENZO[a,d]CYCLOHEPTEN-5-YL)-PIPERIDINE
$C_{21}H_{22}BrN$

S.D. YOUNG, J.J. BALDWIN, D.W. COCHRAN, S.W. KING, D.C. REMY and J.P. SPRINGER, 1985. *J. Org. Chem.*, **50**, 339-342.

Orthorhombic, $P2_12_12_1$, a = 9.663, b = 16.034, c = 11.622 Å, Z = 4. Cu radiation, R = 0.036 for 1313 reflections.

The crystal structure analysis established the absolute configuration of the (+)-enantiomer and hence that of the (-)-enantiomer (Fig. 1).

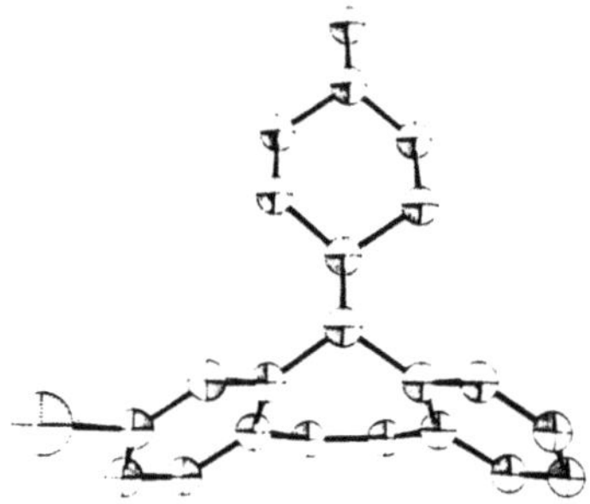

Fig. 1. ORTEP drawing of a molecule of the (-)-enantiomer of $C_{21}H_{22}BrN$.

1,2,r-3a,4,5,c-9b-HEXAHYDRO-7-METHOXY-t-5-(p-METHOXYPHENYL)-3H-BENZ[e]INDEN-3-ONE
$C_{21}H_{22}O_3$

K. HANTAWONG, W.S. MURPHY, G. FERGUSON and M. PARVEZ, 1985. *J. Chem. Soc. Perkin Trans. 2*, 1577-1582.

Orthorhombic, $Pna2_1$, a = 18.013, b = 5.546, c = 17.224 Å, Z = 4. Mo radiation, R = 0.040 for 1426 reflections.

The five-membered ring (Fig. 1) has a C(1)-envelope conformation. The six-membered ring, which is cis-fused to the five-membered ring, has a half-chair conformation. The p-methoxyphenyl ring is equatorially attached to C(5) with the hydrogen atom of C(5) cis to the hydrogen atoms on C(3a) and C(9b). Both aromatic rings are individually planar; the two methoxy groups (O(7),C(71) and O(4'),C(41')) adopt conformations such that they are close to being coplanar with the appropriate ring system.

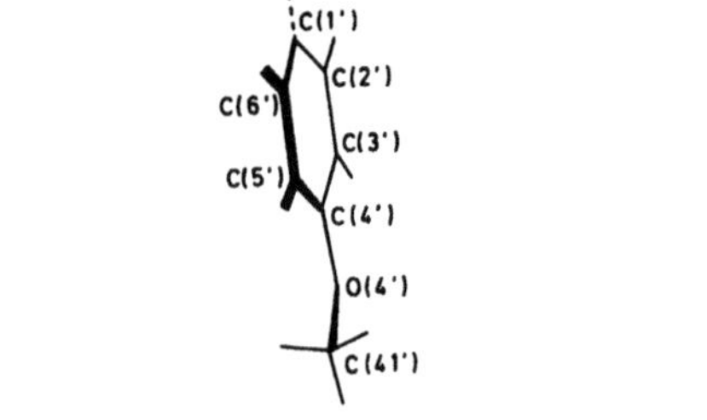

Fig. 1. A view of a molecule of $C_{21}H_{22}O_3$.

trans-9,10-DIHYDROXY-9,10-DIPHENYL-9,10-DIHYDROANTHRACENE

$C_{26}H_{20}O_2$ (I)

trans-9,10-DIHYDROXY-9,10-DIPHENYL-9,10-DIHYDROANTHRACENE BIS(METHANOL) CLATHRATE
$C_{26}H_{20}O_2,2(CH_4O)$ (II)

F. TODA, K. TANAKA, S. NAGAMATSU and T.C.W. MAK, 1985. *Israel J. Chem.*, **25**, 346-352.

I. Triclinic, $P\bar{1}$, a = 6.814, b = 8.119, c = 9.053 Å, α = 103.66, β = 96.30, γ = 106.55°, Z = 1. Mo radiation, R = 0.069 for 1139 reflections.

II. Monoclinic, *C2/c*, a = 12.165, b = 12.484, c = 16.349 Å, β = 110.33°, Z = 4. Mo radiation, R = 0.070 for 1412 reflections.

Molecule (I) lies about an inversion centre and its central six-membered ring is planar within experimental error. In the hydrogen bonded complex with methanol (II) (Fig. 1) the central six-membered ring is also planar and the corresponding dimensions in both (I) and (II) are in good agreement. The crystal structure of (I) has only van der Waals interactions between molecules. In (II) a centrosymmetric planar $(OH)_4$ ring system is formed by two methanol molecules and two hydroxy groups from adjacent DDDA molecules (O...O 2.769(5) and 2.765(5) Å, O...O...O 85.7(4)°); a chain and layer structure is thus developed.

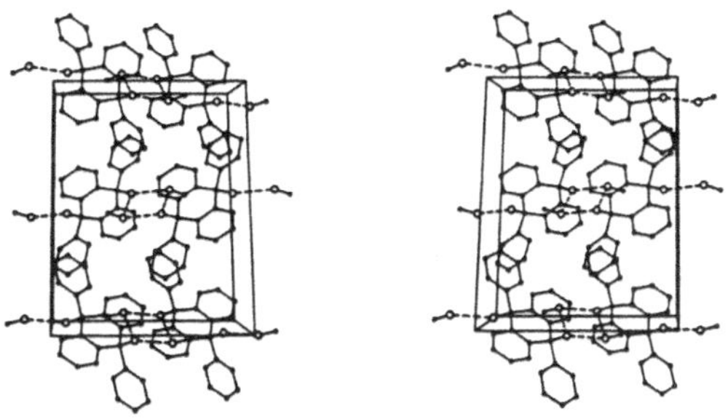

Fig. 1. A stereoview of the crystal structure of $C_{26}H_{20}O_2,2(CH_4O)$ (II).

2-BUTANOATO-7-OCTYLDECAHYDROPHENANTHRENE
$C_{26}H_{46}O_2$

W. SUCROW, H. MINAS, H. STEGEMEYER, P. GESCHWINDER, H.-R. MURAWSKI and C. KRUGER, 1985. *Chem. Ber.*, **118**, 3332-3349.

Triclinic, $P\bar{1}$, a = 5.538, b = 14.004, c = 16.749 Å, α = 108.16, β = 92.95, γ = 93.10°, Z = 2. Cu radiation, R = 0.066 for 2649 reflections.

The rings decahydrophenanthrene moiety are in chair conformation and the side chains are equatorial and maximally extended. Molecular dimensions are normal.

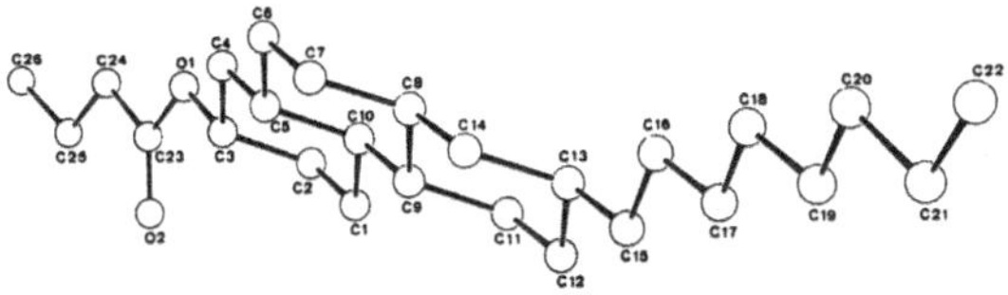

Fig. 1. A view of $C_{26}H_{46}O_2$.

1,2,3,4,7,8-HEXA-t-BUTYLTRICYCLO[4.2.0.0^{2,5}]OCTA-3,7-DIENE
$C_{32}H_{56}$

G. MAIER, K. EULER, H. IRNGARTINGER and M. NIXDORF, 1985. *Chem. Ber.*, **118**, 409-412.

Monoclinic, $P2_1/c$, a = 10.391, b = 31.025, c = 9.422 Å, β = 104.83°, Z = 4. Mo radiation, R = 0.150 for 1446 reflections.

The analysis was severely complicated by disorder but served to establish that dimerization of tri-t-butylcyclobutane gives rise to the *anti* isomer, as shown in Fig. 1.

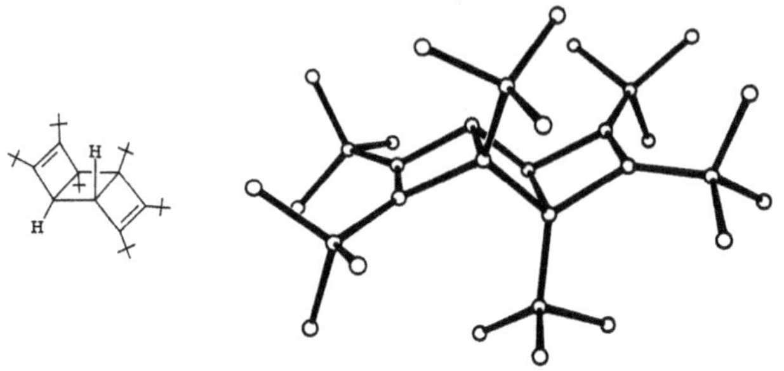

Fig. 1. The $C_{32}H_{56}$ molecule.

trans,trans-1,2,3,3a,4,5,9,10,10a,10b-DECAHYDROPYRENE
$C_{16}H_{20}$

T. SATO, K. NISHIYAMA, A. MORITA and Y. IITAKA, 1985. *Bull. Chem. Soc. Jpn.*, **58**, 2366-2369.

Monoclinic, $C2/c$, a = 19.139, b = 9.970, c = 15.311 Å, β = 123.34°, Z = 8. Cu radiation, R = 0.060 for 2467 reflections.

The geometry at the ring junctions in the molecule was established as shown in Fig. 1.

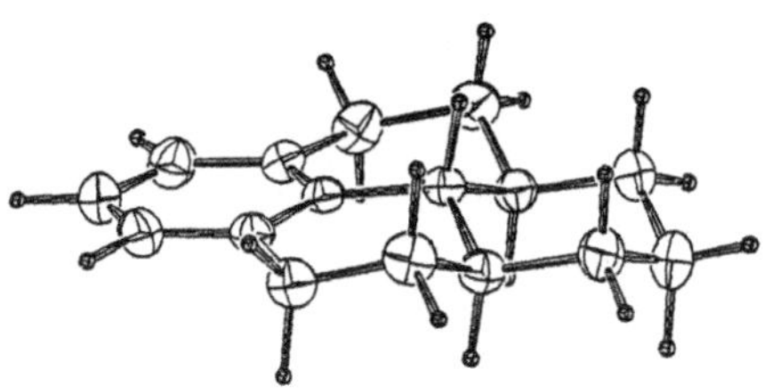

Fig. 1. A view of $C_{16}H_{20}$.

1-(trans-2-BROMOETHENYL)PYRENE ((E)-1-(2-BROMOVINYL)PYRENE)
$C_{18}H_{11}Br$

L.I. FOSS and E.D. STEVENS, 1985. *Acta Cryst.*, **C41**, 1114-1116.

Monoclinic, $P2_1/c$, a = 20.506, b = 3.877, c = 16.236 Å, β = 106.79°, Z = 4. Mo radiation, R = 0.0423 for 2265 reflections (at 115 K).

The rings in the pyrene part of the molecule (Fig. 1) are relatively planar with the exception of ring D which is slightly twisted. The bond lengths and angles in the pyrene moiety are in good agreement with previously determined values. The torsion angle between the bromoethenyl group and the least-squares plane of the pyrene ring system is 24.9(4)°. Layers of molecules in the crystal stack in a herring-bone fashion with a spacing of 3.487(4) Å between neighbouring layers.

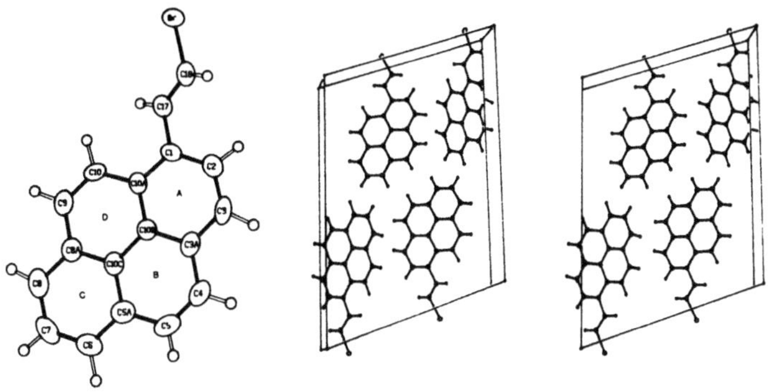

Fig. 1. $C_{18}H_{11}Br$: atomic numbering in the molecule and a stereoview of the molecular packing in the unit cell.

(+)-(1R,3S,3'S)-3,3'-DI-t-BUTYL-1,1'-SPIROBIINDAN
$C_{25}H_{32}$ (I)

(±)-rel-(1R,3R,3'R)-3,3'-DI-t-BUTYL-1,1'-SPIROBIINDAN
$C_{25}H_{32}$ (II)

(±)-rel-(1R,3R,3'S)-3,3'-DI-t-BUTYL-1,1'-SPIROBIINDAN
$C_{25}H_{32}$ (III)

H. NAKAI, M. SHIRO, S. IMAJO, H. KURITANI and K. SHINGU, 1985. *J. Chem. Soc. Perkin Trans. 2*, 147-152.

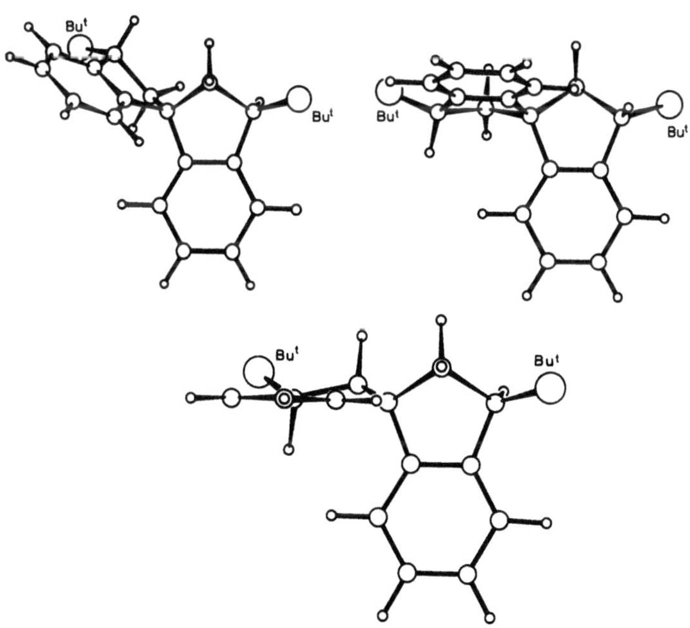

Fig. 1. Schematic views of the three isomers of $C_{25}H_{32}$: (I) (left), (II) (centre) and (III) (right).

I. Orthorhombic, $P2_12_12_1$, a = 12.412, b = 14.842, c = 11.287 Å, Z = 4. Cu radiation, R = 0.032 for 1455 reflections.

II. Monoclinic, $P2_1/c$, a = 16.230, b = 11.719, c = 11.400 Å, β = 110.23°, Z = 4. Cu radiation, R = 0.034 for 2146 reflections.

III. Monoclinic, $P2_1/c$, a = 13.220, b = 13.877, c = 11.897 Å, β = 109.81°, Z = 4. Cu radiation, R = 0.044 for 2437 reflections.

The benzene rings are nearly planar in all three molecules; the cyclopentene rings are all in envelope conformations; the tertiary butyl groups are in quasi-equatorial positions in (I) and (III) but in a quasi-axial position in (II). The dihedral angles between the two benzene ring planes are 69.5(1), 91.4(1) and 94.9(1)° for (I), (II) and (III) respectively. The relative stereochemistries of the three molecules are shown in Fig. 1.

7,12-DIPHENYL-4a,12-DIHYDROINDENO[7a,1-b]NAPHTHALENE
$C_{29}H_{22}$

MAZHAR-UL-HAQUE, J. AHMED and W. HORNE, 1985. *Acta Cryst.*, **C41**, 406-409.

Orthorhombic, *Pbca*, a = 9.968, b = 17.163, c = 23.583 Å, Z = 8. Cu radiation, R = 0.037 for 2304 reflections.

The core fragment of the molecule (Fig. 1) consists of a fused tetracyclic moiety in which rings A, B and C are roughly coplanar, with ring D cis-fused and disposed pseudo-axially to ring C. Ring A is planar and ring B has C(12) and C(13) displaced on opposite sides of the least-squares plane of the other four ring atoms by 0.31 and 0.42 Å respectively. The interplanar dihedral angle between the phenyl groups attached at C(7) and C(12) is 54.5°. Rings C and D have flattened envelope conformations with C(13) and C(14) as their respective flaps. The C(13)-C(14) and C(12)-C(13) elongated bonds, of average length 1.554 Å, and the reduction in the valence angles C(14)-C(13)-C(15) and C(13)-C(14)-C(5) to 103.1° are evidence of the strain in the fused ring system. There are no unusually short intermolecular distances.

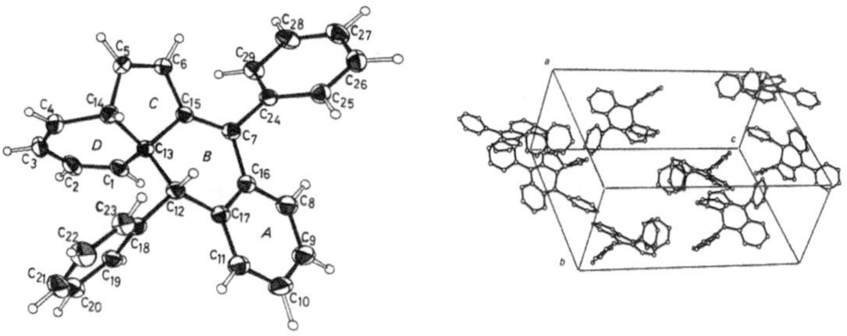

Fig. 1. $C_{29}H_{22}$: a perspective view of the molecule (left) and of the unit cell contents (right).

1,1,2,3,4,5,6-HEPTAPHENYL-1,4-DIHYDROBENZ[e]-as-INDACENE
$C_{58}H_{40}$

C.T. GRAINGER, 1985. *Acta Cryst.*, **C41**, 157-158.

I. Triclinic, $P\bar{1}$, a = 11.903, b = 18.433, c = 10.035 Å, α = 107.05, β = 95.42, γ = 98.76°, Z = 2. Cu radiation, R = 0.042 for 5993 reflections.

With a different set of reflection data and a different unit cell, the structure essentially agrees with that reported previously (**1**).

1. Structure Reports, **46B**, 162.

7H-DIBENZ[a,kl]ANTHRACEN-7-ONE
$C_{21}H_{12}O$

S. FUJISAWA, I. OONISHI, J. AOKI, Y. OHASHI and Y. SASADA, 1985. *Bull. Chem. Soc. Jpn.*, **58**, 3356-3359.

Orthorhombic, $Pn2_1a$, a = 23.325, b = 14.149, c = 3.986 Å, Z = 4. Mo radiation, R = 0.044 for 656 reflections.

The molecule has a non-planar conformation as shown in Fig. 1 and the angle C(1)-C(13)...C(11)-C(21) is 33.4°. In the crystal lattice, molecules are stacked face-to-face along the short *c*-axis (Fig. 1).

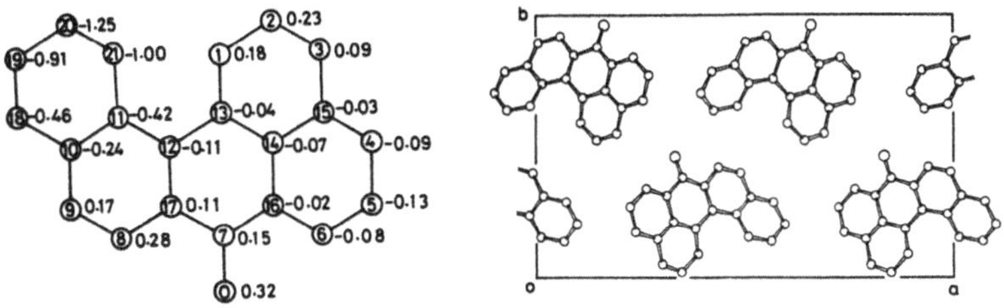

Fig. 1. Deviations (Å) of the atoms from the mean molecular plane of $C_{21}H_{12}O$ and a view of the crystal structure along the *c*-axis.

PICENE
$C_{22}H_{14}$

A. DE, R. GHOSH, S. ROYCHOWDHURY and P. ROYCHOWDHURY, 1985. *Acta Cryst.*, **C41**, 907-909.

Monoclinic, $P2_1$, a = 8.480, b = 6.154, c = 13.515 Å, β = 90.46°, Z = 2. Cu radiation, R = 0.037 for 759 reflections.

The molecule (Fig. 1) is slightly distorted from planarity due to intramolecular overcrowding. The C-C bond distances range from 1.358(9) to 1.479(8) Å, the bond distances in the two outer rings being less than those in the other three rings. The dihedral angle between rings (I) and (V) is 3.72°. Packing is governed by van der Waals interactions.

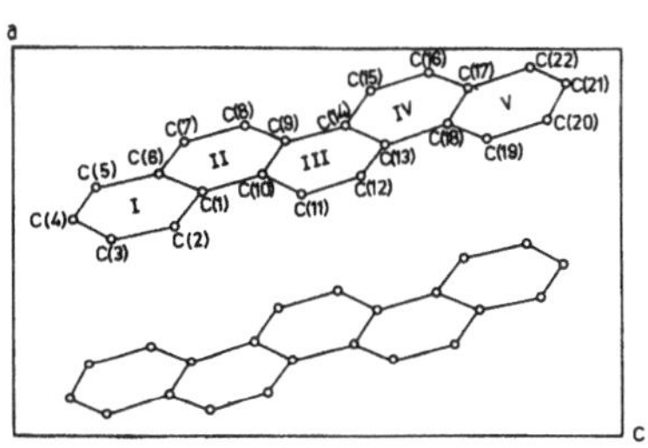

Fig. 1. $C_{22}H_{14}$: b-axis projection of the crystal structure.

2-METHOXY-6-METHYL-6-TRICHLOROMETHYLBICYCLO[3.1.0]HEXAN-3-ONE
$C_9H_{11}Cl_3O_2$

A.C. BRISIMITZAKIS, D.I. SCHUSTER and J.M. VAN DER VEEN, 1985. *Canad. J. Chem.*, **63**, 685-687.

Triclinic, $P\bar{1}$, a = 5.963, b = 9.958, c = 10.106 Å, α = 105.47, β = 105.52, γ = 81.91°, Z = 2. Mo radiation, R = 0.074 for 1905 reflections.

The analysis establishes the structure and stereochemistry (as shown in Fig. 1) which is related to that of a previously postulated intermediate.

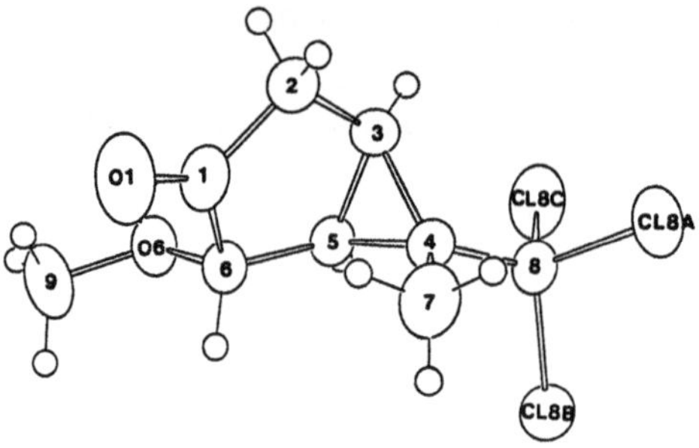

Fig. 1. The $C_9H_{11}Cl_3O_2$ molecule.

TRICYCLO[4.3.0.0^{3,7}]NONANE-6,7-DIOL (BREXAN-6,7-DIOL)
$C_9H_{14}O_2$

V.G. ANDRIANOV and Yu.T. STRUCHKOV, 1985. *Zh. Strukt. Khim.*, **26**, 6-143-144 [J. Struct. Chem., **26**, 959-960].

Trigonal, $P3_121$, a = 10.988, c = 5.6690 Å, Z = 3. Mo radiation, R = 0.145 for 432 reflections.

The molecule (Fig. 1) has crystallographic twofold symmetry. The ring skeleton contains three five-membered rings and two six-membered rings, the former having half-chair conformations and the latter distorted boat conformations. Bond lengths are normal.

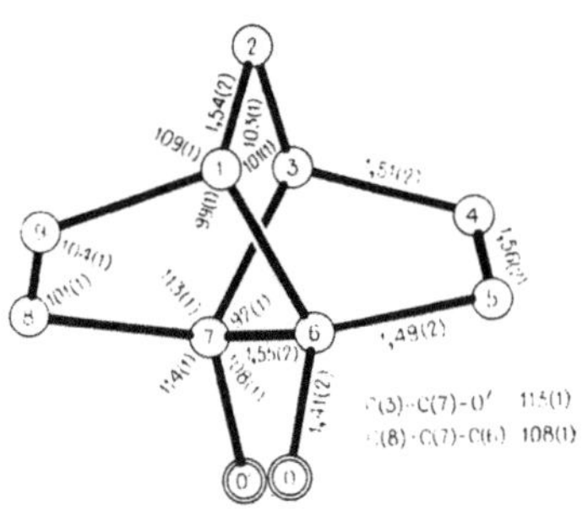

Fig. 1. The molecular structure of brexan.

1-exo,2-endo,3-exo,4,5,6,7,8,8-NONACHLORO-3a,4,7,7a-TETRAHYDRO-4,7-METHANOINDANE
$C_{10}H_5Cl_9$

C.H.L. KENNARD, G. SMITH and G. SCHILLER, 1985. *Z. Kristallogr.*, **171**, 303-311.

Orthorhombic, $P2_1cn$, a = 7.721, b = 11.379, c = 18.507 Å, Z = 4. Mo radiation, R = 0.030 for 999 reflections.

The molecule is shown (Fig. 1) to have the endo-ring configuration with the exo, endo, exo (aaa) conformation for the three chlorine atoms on the attached cyclopentane ring. The molecule is compared with related 'cyclodiene' insecticides.

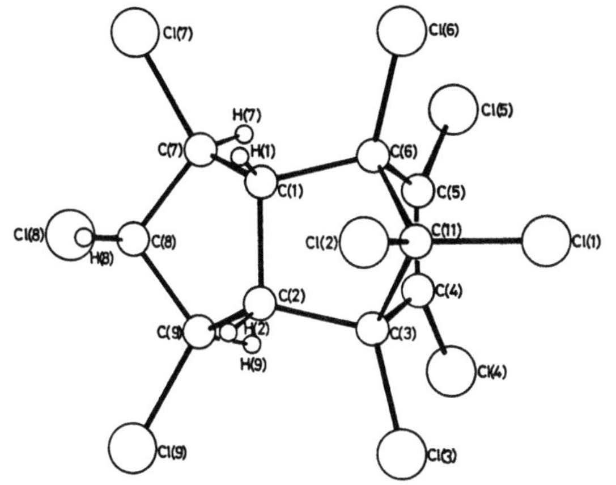

Fig. 1. A view of $C_{10}H_5Cl_9$.

2,3,3a,4,5,8-HEXACHLORO-3a,6,7,7a-TETRAHYDRO-1,6-METHANO-1H-INDENE (γ-CHLORDENE)
$C_{10}H_6Cl_6$

C.H.L. KENNARD and G. SMITH, 1985. *Acta Cryst.*, **C41**, 1079-1081.

Monoclinic, $P2_1/n$, a = 8.936, b = 14.372, c = 9.774 Å, β = 100.72°, Z = 4. Mo radiation, R = 0.039 for 1326 reflections.

The molecule (Fig. 1) is a bridged chlorinated cyclodiene and contains a six-membered and two five-membered rings fused together. It does not possess the hexachloronorbornene nucleus common to other cyclodiene insecticides. The ethene C(2)=C(3) bond is of length 1.330(5) Å and the C(sp^2)-Cl and C(sp^3)-Cl average bond lengths are 1.716(6) and 1.803(4) Å respectively. The bridging angle, C(1)-C(6A)-C(3A) is distorted to 102.9(3)°. There are several short Cl...Cl intermolecular contacts with lengths ranging from 3.335(2) to 3.552(2) Å. This molecule is structurally very similar to those reported in **1**, for the two crystallographically independent ozonized molecules.

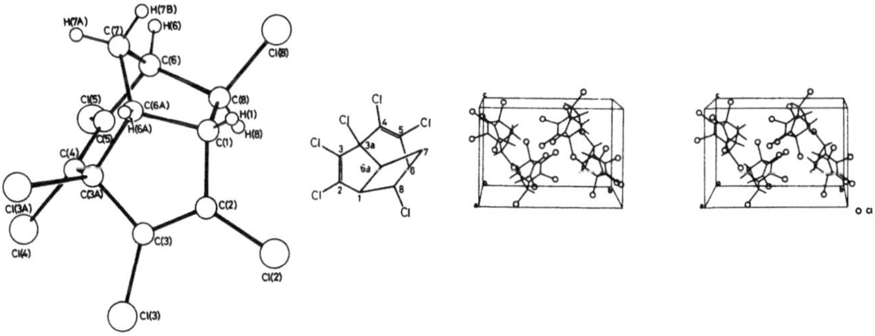

Fig. 1. $C_{10}H_6Cl_6$: molecular configuration, structural formula and a stereoview perpendicular to the bc plane.

1. Structure Reports, **44B**, 281.

6,6,10,10-TETRANITROPENTACYCLO[5.3.0.0^{2,5}.0^{3,9}.0^{4,8}]DECANE
$C_{10}H_8N_4O_8$

C. GEORGE, R. GILARDI, J.L. FLIPPEN-ANDERSON, C.S. CHOI, A.P. MARCHAND and D.S. REDDY, 1985. *Acta Cryst.*, **C41**, 788-791.

Monoclinic, *Pc*, a = 7.761, b = 11.500, c = 14.277 Å, β = 108.63°, Z = 4. Mo radiation, R = 0.041 for 2457 reflections.

The carbon skeleton of the molecule (Fig. 1) consists of four puckered five-membered rings and two four-membered rings which are bent about their diagonals by approximately 20°. There are two crystallographically independent molecules in the asymmetric unit having essentially identical bond lengths and angles. These agree well with the values observed in other related compounds. The primary difference between these molecules is in the orientations of the NO_2 groups. The molecules pack in the unit cell with the oxygen atoms of molecule 1 opposing those of molecule 2. The nearest O...O contacts lie in the range 2.84 to 3.26 Å and all intermolecular contacts are consistent with essentially van der Waals packing forces.

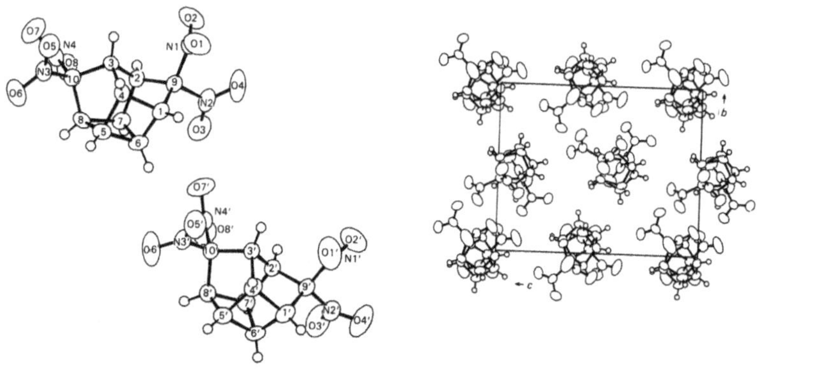

Fig. 1. $C_{10}H_8N_4O_8$: views of the independent molecules and the molecular packing projected onto (100).

6,6,8-TRINITROPENTACYCLO[5.3.0.0^{2,5}.0^{3,9}.0^{4,8}]DECANE
$C_{10}H_9N_3O_6$

H.L. AMMON, D. ZHANG, C.S. CHOI, O. SANDUS, A.P. MARCHAND and S.C. SURI, 1985. *Acta Cryst.*, C41, 404-406.

Monoclinic, *P*2$_1$/*c*, a = 9.352, b = 7.401, c = 16.070 Å, β = 97.76°, Z = 4. CU radiation, R = 0.051 for 1495 reflections.

In the molecular structure (Fig. 1) two of the faces of the severely distorted cubane-like structure are four-membered rings and four of the faces consist of five-membered rings. The average C-C separation in bonds involving bridging atoms C(7) and C(9) is 1.550(4) Å. The average C-C-C angles in the four- and five-membered rings are 89.2 and 102.9° respectively. The two C(NO_2) groups attached at C(9) are both essentially planar with an interplanar angle of 89.8°. The average of the lengths of the two bonds linked to C(NO_2)$_2$ is 1.507(4) Å and of the three bonds joined to C(NO_2) is 1.560(4) Å. There are no unusually short intermolecular contacts.

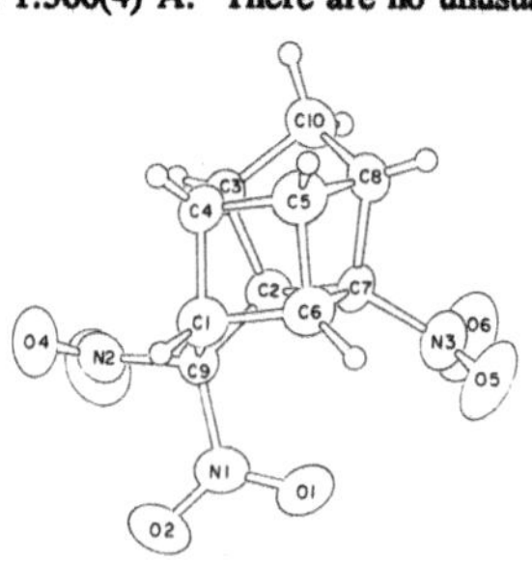

Fig. 1. $C_{10}H_9N_3O_6$: a perspective view of the molecule showing the numbering scheme.

endo-8-HYDROXYTETRACYCLO[5.3.0.0^{2,5}.0^{3,9}]DECANE

$C_{10}H_{14}O$ (I)

4-ACETOXY-6-(METHYLSULPHONYL)PENTACYCLO[5.3.0.0^{2,5}.0^{3,9}.0^{4,8}]DECANE
$C_{13}H_{16}O_5S$ (II)

A.J.H. KLUNDER, W.C.G.M. DE VALK, J.M.J. VERLAAK, J.W.M. SCHELLEKENS, J.H. NOORDIK, V. PARTHASARATHI and B. ZWANENBURG, 1985. *Tetrahedron,* **41**, 963-973.

I. Monoclinic, $P2_1/c$, a = 12.371, b = 11.382, c = 13.323 Å, β = 118.11°, Z = 8. R = 0.230.

II. Orthorhombic, $P2_12_12_1$, a = 5.939, b = 14.181, c = 15.323 Å, Z = 4. Cu radiation, R = 0.069 for 1297 reflections.

Crystals of (I) decomposed during data collection but the data were adequate to establish the connectivity. Compound (II) is shown to have the anti-structure (Fig. 1). The C(2)-C(3) bond is unusually long (1.605(9) Å).

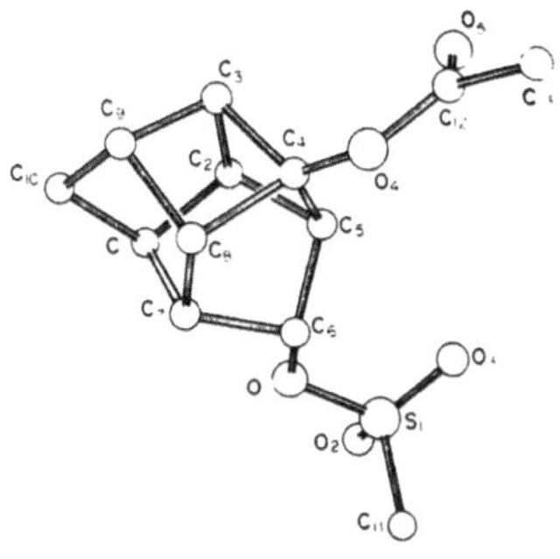

Fig. 1. A view of $C_{13}H_{16}O_5S$ (II).

exo-7-METHYLBICYCLO[3.3.1]NONAN-3-ONE
$C_{10}H_{16}O$ (I)

endo-7-METHYLBICYCLO[3.3.1]NONAN-3-ONE
$C_{10}H_{16}O$ (II)

J.F. RICHARDSON and T.S. SORENSEN, 1985. *Canad. J. Chem.*, **63**, 1166-1169.

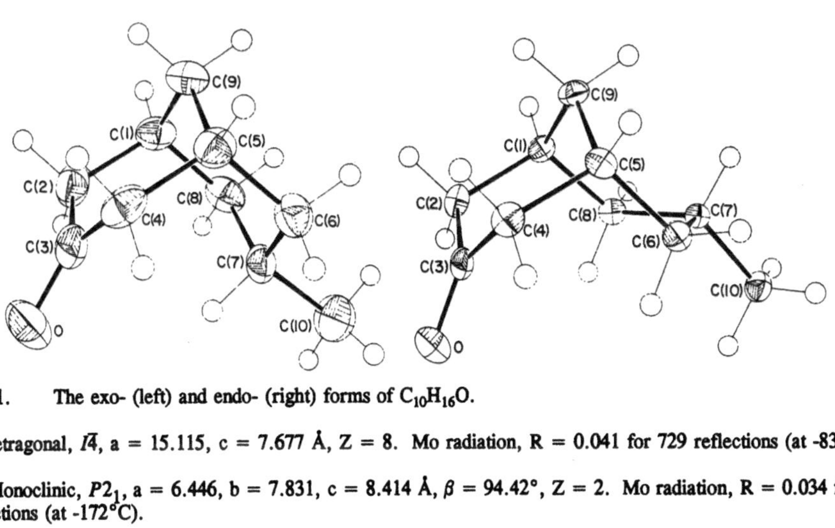

Fig. 1. The exo- (left) and endo- (right) forms of $C_{10}H_{16}O$.

I. Tetragonal, $I\bar{4}$, a = 15.115, c = 7.677 Å, Z = 8. Mo radiation, R = 0.041 for 729 reflections (at -83°C).

II. Monoclinic, $P2_1$, a = 6.446, b = 7.831, c = 8.414 Å, β = 94.42°, Z = 2. Mo radiation, R = 0.034 for 978 reflections (at -172°C).

The analyses show that molecule (I) is in a chair-chair conformation whereas the endo isomer (II) has a boat-chair conformation (Fig. 1).

2-endo-7-syn-DIMETHYLBICYCLO[2.2.1]HEPTANE-2-exo-CARBOXAMIDE
$C_{10}H_{17}NO$

H.W. THOMPSON and J.K. WONG, 1985. *J. Org. Chem.*, **50**, 4270-4276.

Monoclinic, *C2/c*, a = 15.990, b = 12.182, c = 20.789 Å, β = 100.57°, Z = 8. Cu radiation, R = 0.068 for 1377 reflections.

The crystal structure analysis was carried out to determine the stereochemistry of the apical methyl group relative to the amido group. The molecule is shown in Fig. 1.

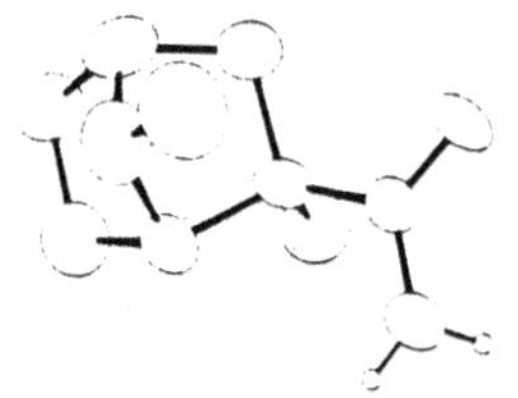

Fig. 1. The $C_{10}H_{17}NO$ molecule.

5,6-BENZOBICYCLO[2.2.1]HEPT-5-ENE-2-exo,7-anti-DIYL DINITRATE
$C_{11}H_{10}N_2O_6$

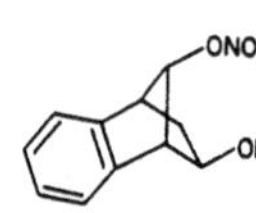

W.J. LAYTON, C.P. BROCK, P.A. CROOKS, S.L. SMITH and P. BURN, 1985. *J. Org. Chem.*, **50**, 5372-5374.

Monoclinic, *C2/c*, a = 24.100, b = 6.990, c = 17.524 Å, β = 129.07°, Z = 8. Mo radiation, R = 0.034 for 2101 reflections.

The crystal structure analysis established the stereochemistry of the molecule.

1,2,4,5,6,7,8,9-OCTAHYDRO-10,10-DIMETHOXYTRICYCLO[5.2.1.0^{2,6}]DECA-4,8-DIEN-3-ONE
$C_{12}H_6Cl_8O_3$

N. GALESIC, I. MATIJASIC and M. BRUVO, 1985. *Acta Cryst.*, **C41**, 1662-1664.

Monoclinic, $P2_1/c$, a = 11.028, b = 11.871, c = 13.231 Å, β = 93.14°, Z = 4. Mo radiation, R = 0.035 for 3518 reflections.

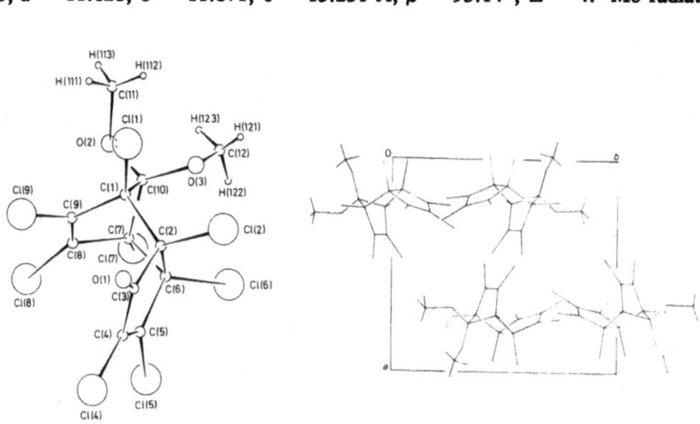

Fig. 1. Views of the $C_{12}H_6Cl_8O_3$ molecule and a projection of the crystal structure onto (001).

The $C(sp^3)$-Cl and $C(sp^2)$-Cl bond lengths in the molecule (Fig. 1) range from 1.744 to 1.767(2) and from 1.688 to 1.696(2) Å respectively. Ring C(1,2,6,7,8,9) has a boat conformation with C(1)...C(7) as keel and the dihedral angle between the two plane halves of this ring is 111.4(1)°. The five-membered ring C(2,3...6) is planar to within ±0.023 Å. The two other five-membered rings have envelope conformations in each of which C(10) is the flap. The $C(sp^3)$-$C(sp^3)$ bond lengths are somewhat elongated and range from 1.572 to 1.585(3) Å. The bridgehead C-C-C and O-C-O angles are 91.6(1) and 114.6(2)° respectively, while the O-C-O-C torsion angles about the C(10)-O(2) and C(10)-O(3) bonds are -53.5(2) and -47.6(3)° respectively. Intermolecular distances are essentially van der Waals contacts.

1,5-DIMETHYLTRICYCLO[3.3.0.0^{2,8}]OCTA-3,6-DIENE-2,6-DICARBONITRILE (2,6-DICYANO-1,5-DIMETHYLSEMIBULLVALENE)

$C_{12}H_{10}N_2$

H. QUAST, J. CHRIST, E.-M. PETERS, K. PETERS and H.G. VON SCHNERING, 1985. *Chem. Ber.*, **118**, 1154-1175.

Monoclinic, $P2_1/n$, a = 13.383, b = 9.395, c = 7.968 Å, β = 91.07°, Z = 4. Mo radiation, R = 0.057 for 2195 reflections.

The molecular conformation and details of dimensions are shown in Fig. 1.

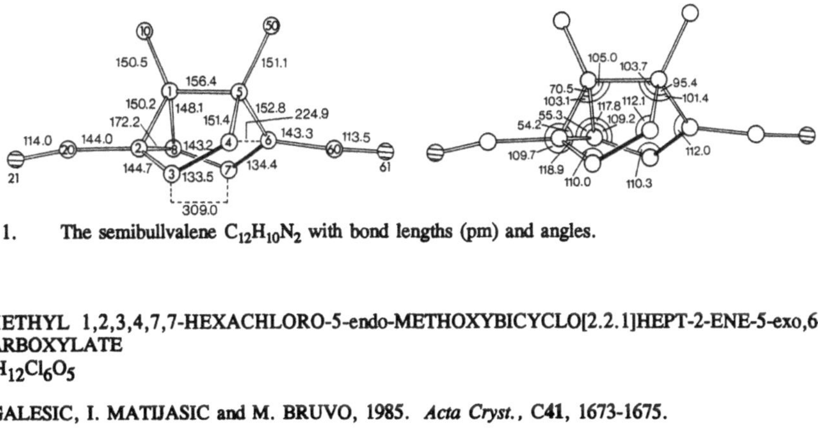

Fig. 1. The semibullvalene $C_{12}H_{10}N_2$ with bond lengths (pm) and angles.

DIMETHYL 1,2,3,4,7,7-HEXACHLORO-5-endo-METHOXYBICYCLO[2.2.1]HEPT-2-ENE-5-exo,6-endo-DICARBOXYLATE

$C_{12}H_{12}Cl_6O_5$

N. GALESIC, I. MATIJASIC and M. BRUVO, 1985. *Acta Cryst.*, **C41**, 1673-1675.

Monoclinic, $P2_1/c$, a = 16.458, b = 12.651, c = 16.863 Å, β = 96.83°, Z = 8. Cu radiation, R = 0.077 for 3948 reflections.

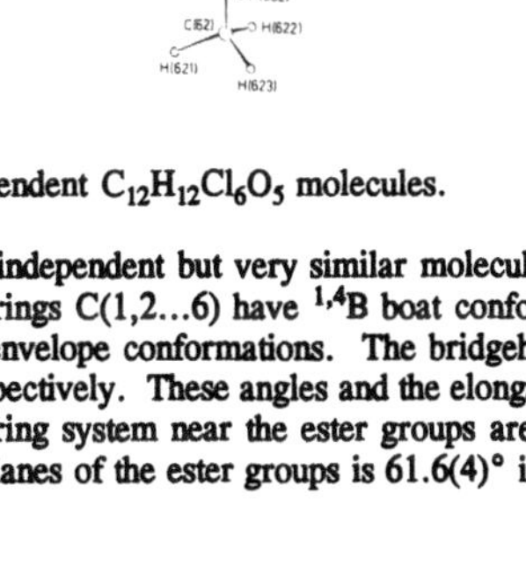

Fig. 1. A view of one of the independent $C_{12}H_{12}Cl_6O_5$ molecules.

There are two crystallographically independent but very similar molecules (Fig. 1) in the asymmetric unit. In molecules A and B the six-membered rings C(1,2...6) have $^{1,4}B$ boat conformations and the five-membered rings C(7,1,2,3,4) and C(7,1,6,5,4) exhibit envelope conformations. The bridgehead angles C(1)-C(7)-C(4) are 93.4(5) and 92.7(5)° for molecules A and B respectively. These angles and the elongation of some of the $C(sp^3)$-$C(sp^3)$ bond lengths in the parts of the norbornene ring system near the ester groups are indicative of the steric strain present. The dihedral angle between the mean planes of the ester groups is 61.6(4)° in molecule A and 50.1(3)° in molecule

B. All intermolecular distances, except for two slightly short Cl...O contacts, are consistent with essentially van der Waals interactions.

2,6-DINITROHEXACYCLO[5.4.1.$0^{2,6}$.$0^{3,10}$.$0^{5,9}$.$0^{8,11}$]DODECANE (DINITROBISHOMOPENTAPRISMANE) $C_{12}H_{12}N_2O_4$

L.A. PAQUETTE, J.W. FISCHER and P. ENGEL, 1985. *J. Org. Chem.*, **50**, 2524-2527.

Monoclinic, *P2/c*, a = 11.855, b = 7.677, c = 13.853 Å, β = 124.63°, Z = 4. Mo radiation, R = 0.049 for 2282 reflections (at -140°C).

The crystallographic asymmetric unit contains two independent molecules, A and B, which differ in the orientation of the nitro groups with respect to the bishomopentaprismane framework (Fig. 1). The five-membered rings are in half-chair conformations. The C(1)-C(5) bond length is 1.593(3) Å in A, and 1.590(3) Å in B.

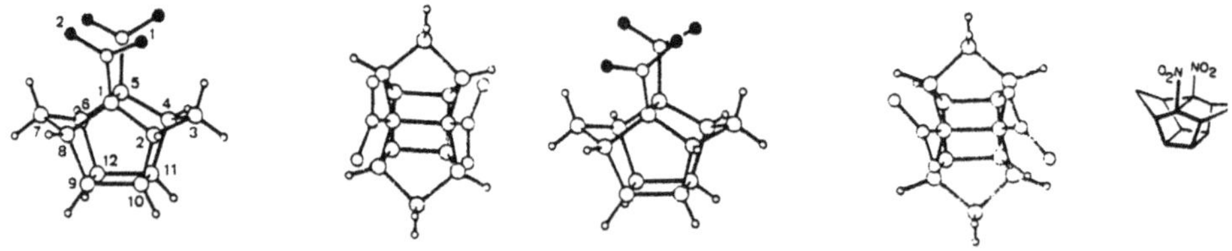

Fig. 1. Side and top views of the two independent molecules of $C_{12}H_{12}N_2O_4$ in the crystal.

TETRACYCLO[6.2.1.$1^{3,6}$.$0^{2,7}$]DODEC-2(7)-ENE (anti-SESQUINORBORNENE) $C_{12}H_{16}$

M. GAJHEDE, F.S. JORGENSEN, K.R. KOPECKY, W.H. WATSON and R.P. KASHYAP, 1985. *J. Org. Chem.*, **50**, 4395-4397.

Monoclinic, *P*2_1/*n*, a = 6.112, b = 7.395, c = 9.850 Å, β = 99.21°, Z = 2. Mo radiation, R = 0.034 for 1475 reflections (at 105 K). R = 0.060 for 641 reflections (at 243 K).

The molecule (Fig. 1) has a planar structure around the double bond; the C(1)-C(6)-C(6')-C(4) torsion angle is 0.2(1)°. The molecule is itself centrosymmetric, hence only two molecules in the unit cell.

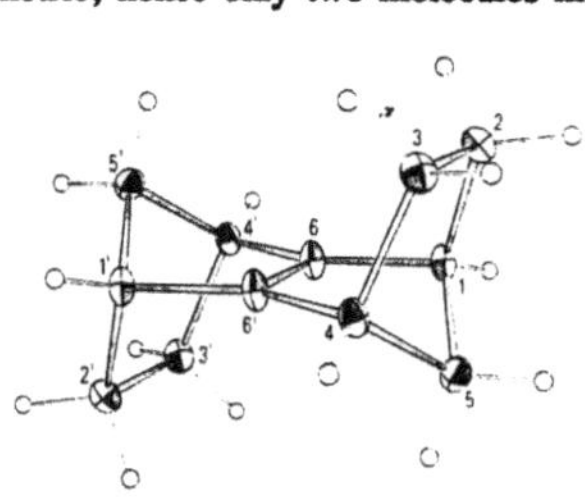

Fig. 1. ORTEP drawing of $C_{12}H_{16}$.

2-HYDROXYTETRACYCLO[6.2.1.$1^{3,6}$.$0^{2,7}$]DODECANE $C_{12}H_{18}O$

L.A. PAQUETTE, G. DELUCCA, K. OHKATA and J.C. GALLUCCI, 1985. *J. Am. Chem. Soc.*, **107**, 1015-1023.

Monoclinic, *C2/c*, a = 20.404, b = 10.519, c = 9.352 Å, β = 104.16°, Z = 8. Mo radiation, R = 0.049 for 1394 reflections.

The tetracyclodecane framework of the molecule (Fig. 1) shows almost C_{2v} symmetry with C(1)-C(2)-C(3)

(128.6(2)) and C(6)-C(7)-C(8) (129.3(2)°) angles greater than tetrahedral values. There are close non-bonded contacts involving C atoms (C(4)...C(10) 3.172(4), C(5)...C(9) 3.189(6) Å) and H atoms (H(1C4)...H(1O2) 1.70(4), H(1C9)...H(2C5) 1.76(4) Å). Bond lengths and angles are normal.

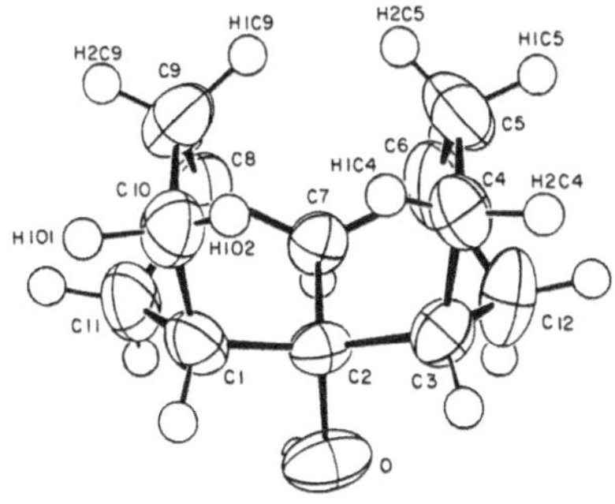

Fig. 1. $C_{12}H_{18}O$: the molecular structure.

3,7-DIMETHOXY-1,5-DIMETHYLTETRACYCLO[3.3.0.$0^{2,8}$.$0^{4,6}$]OCTANE-2,6-DICARBONITRILE
$C_{14}H_{16}N_2O_2$

H. QUAST, J. CHRIST, E.-M. PETERS, K. PETERS and H.G. VON SCHNERING, 1985. *Chem. Ber.*, **118**, 1176-1182.

Triclinic, $P\bar{1}$, a = 9.103, b = 9.731, c = 7.949 Å, α = 88.55, β = 93.80, γ = 106.95°, Z = 2. Mo radiation, R = 0.062 for 2784 reflections.

Principal dimensions of this strained molecule (Fig. 1) include C(1)-C(5) 1.536(3), C(1)-C(8) 1.497(3), C(4)-C(5) 1.495(4), C(2)-C(8) 1.517(3), C(4)-C(6) 1.518(3), C(1)-C(2) 1.537(4), C(5)-C(6) 1.537(3) Å.

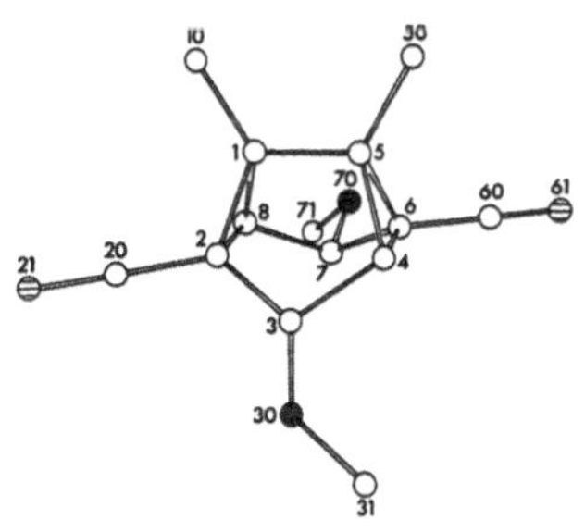

Fig. 1. The $C_{14}H_{16}N_2O_2$ molecule.

5,5-DIMETHYLTRICYCLO[6.2.2.$0^{2,7}$]DODECANE-3,9-DIONE
$C_{14}H_{20}O_2$

Z. DAUTER, J.V. GREENHILL, A. KARAULOV and C.D. REYNOLDS, 1985. *Acta Cryst.*, **C41**, 630-632.

Monoclinic, *P2/c*, a = 10.731, b = 6.411, c = 17.719 Å, β = 94.40°, Z = 4. Cu radiation, R = 0.077 for 1000 reflections.

The bridged rings of the tricyclododecane system are in the usual boat conformation (Fig. 1). The unbridged ring adopts a half-boat conformation. Bond lengths and angles are normal.

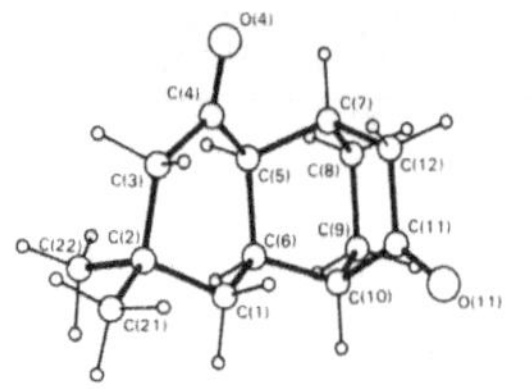

Fig. 1. $C_{14}H_{20}O_2$: the molecular structure.

TETRACYCLO[4.4.0.0^{2,8}.0^{4,7}]DEC-9-EN-5-ONE 2,4-DINITROPHENYLHYDRAZONE
$C_{16}H_{14}N_4O_4$ (I)

10-BROMOTRICYCLO[5.2.1.0^{4,8}]DEC-5-EN-2-ONE
$C_{10}H_{11}BrO$ (II)

TETRACYCLO[5.3.0.0^{2,5}.0^{4,8}]DEC-9-EN-3-YL 3,5-DINITROBENZOATE
$C_{17}H_{14}N_2O_6$ (III)

J.M.M. SMITS, J.H. NOORDIK, H.E.L. DEPRE, R.C.W. ZWANENBURG and A.J.H. KLUNDER, 1985. *Acta Cryst.*, C41, 1638-1642.

I. Monoclinic, $P2_1/n$, a = 6.745, b = 24.915, c = 8.911 Å, β = 99.0°, Z = 4. Mo radiation, R = 0.080 for 528 reflections.

II. Monoclinic, $P2_1/c$, a = 11.088, b = 6.649, c = 12.667 Å, β = 102.36°, Z = 4. Mo radiation, R = 0.048 for 987 reflections.

III. Triclinic, $P\bar{1}$, a = 6.428, b = 9.912, c = 12.455 Å, α = 94.74, β = 92.64, γ = 102.79°, Z = 2. Mo radiation, R = 0.032 for 1562 reflections.

Bond lengths and angles in these strained molecules (Fig. 1) are all in the range expected for such structures. No intermolecular distances significantly shorter than normal van der Waals contacts are observed. Only in (III) is the cage structure found to be similar to that expected in the parent compound- tetracyclo[5.3.0.0^{2,5}.0^{4,8}]dec-9-en-3-one. In the other compounds the cage structure has been modified in the chemical reaction. In (III) there are two short intramolecular contacts between C(8) and the olefinic carbon atoms C(16) and C(16) both of length 2.90 Å.

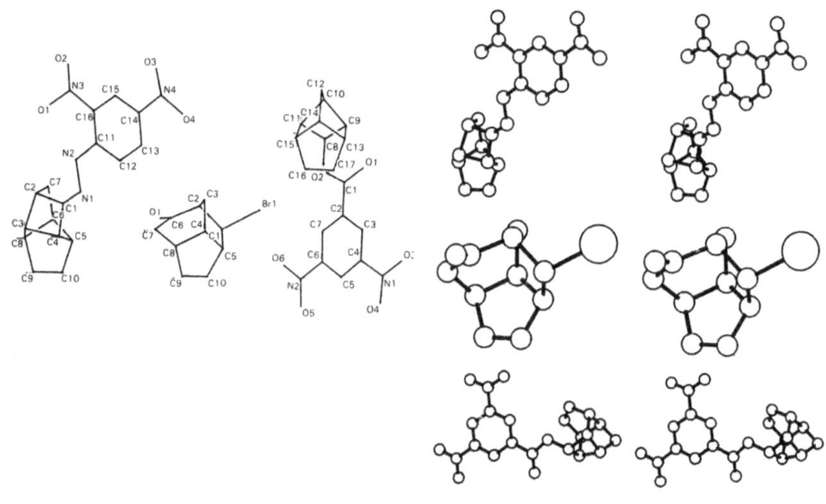

Fig. 1. Atomic numbering (left) and stereoviews (right) of $C_{16}H_{14}N_4O_4$ (top), of $C_{10}H_{11}BrO$ (centre) and of $C_{17}H_{14}N_2O_6$ (bottom).

BICYCLO[6.2.2]DODECA-8,10,12-TRIENE-9,10-DICARBOXYLIC ACID DIMETHYL ESTER ([6]PARA-CYCLOPHANE-8,9-DICARBOXYLIC ACID DIMETHYL ESTER)
$C_{16}H_{20}O_4$ (I)

cis-8,12-DIBROMOBICYCLO[6.2.2]DODECA-1(11),9-DIENE-9,10-DICARBOXYLIC ACID DIETHYL ESTER
$C_{18}H_{24}Br_2O_4$ (II)

J. LIEBE, C. WOLFF, C. KRIEGER, J. WEISS and W. TOCHTERMANN, 1985. *Chem. Ber.*, **118**, 4144-4178.

I. Monoclinic, $P2_1$, a = 8.489, b = 8.516, c = 10.467 Å, β = 105.55°, Z = 2. Mo radiation, R = 0.034 for 1680 reflections.

II. Monoclinic, $P2_1/n$, a = 9.372, b = 13.010, c = 15.948 Å, β = 91.38°, Z = 4. Mo radiation, R = 0.051 for 2444 reflections.

The benzene ring in (I) is deformed into a boat shape (Fig. 1) with deformation angles α of 19.4 and 19.5°. The stereochemistry of the dibromo compound (II) is also shown in Fig. 1. The Br-C distances are Br(1)-C(12) 2.006(7), Br(2)-C(8) 1.981(6) Å. Atoms C(1),C(8),C(9),C(10),C(11),C(12) define a plane for which the largest deviation is C(8) at 0.31 Å.

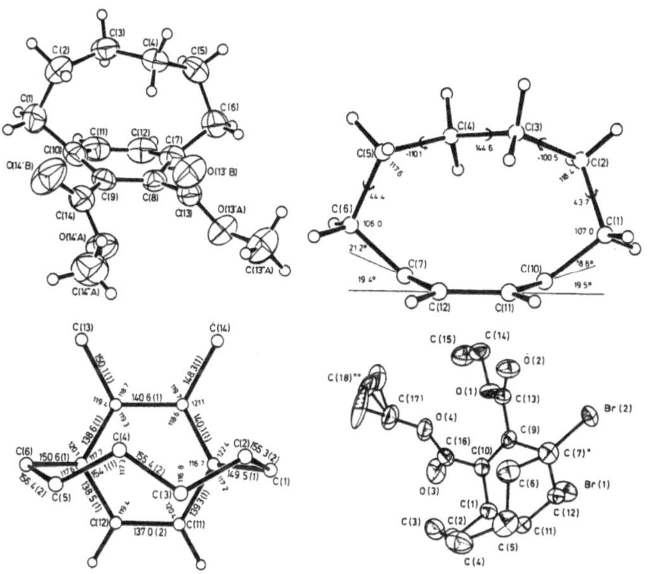

Fig. 1. $C_{16}H_{20}O_4$ (I) (left) with some dimensions (pm and °), and $C_{18}H_{24}Br_2O_4$ (II) (right) (atom C(8) is behind C(7) and the methyl group on C(17) is disordered).

12,15-DIMETHOXY[2](2,6)-p-BENZOQUINONO[2]PARACYCLOPHANE
$C_{18}H_{18}O_4$ (I)

5,8,12,15-TETRAMETHOXY[2.2]METAPARACYCLOPHANE
$C_{20}H_{24}O_4$ (II)

[2.2](2,5)-(2,6)-p-BENZOQUINONOPHANE
$C_{16}H_{12}O_4$ (III)

14,17-DIMETHOXY[3](2,6)-p-BENZOQUINONO[3]PARACYCLOPHANE
$C_{20}H_{22}O_4$ (IV)

H.A. STAAB, M. JORNS, C. KRIEGER and M. RENTZEA, 1985. *Chem. Ber.*, **118**, 796-813.

I. Orthorhombic, $P2_12_12_1$, a = 7.698, b = 8.205, c = 24.025 Å, Z = 4. Mo radiation, R = 0.036 for 1910 reflections.

II. Monoclinic, $P2_1/c$, a = 16.131, b = 8.304, c = 13.540 Å, β = 111.33°, Z = 4. Mo radiation, R = 0.040 for 3330 reflections.

III. Monoclinic, $C2/c$, a = 24.698, b = 9.077, c = 11.065 Å, β = 92.39°, Z = 8. Mo radiation, R = 0.041 for 2429 reflections.

IV. Monoclinic, $P2_1/c$, a = 8.555, b = 25.697, c = 8.370 Å, β = 114.87°, Z = 4. Mo radiation, R = 0.053 for 3050 reflections.

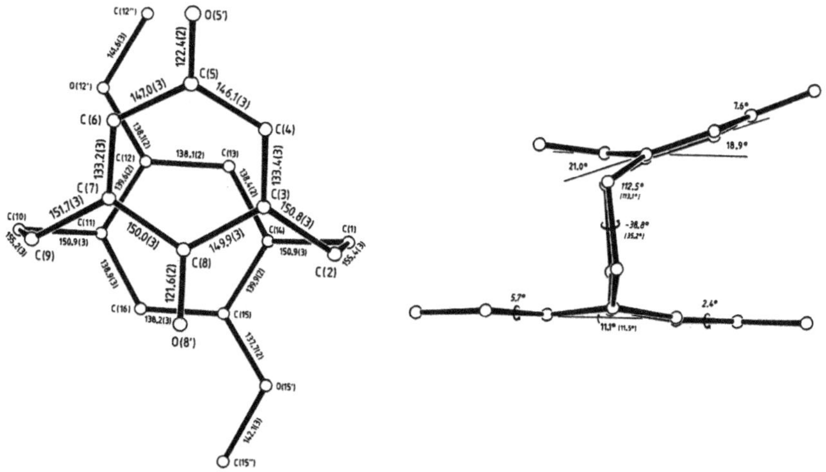

Fig. 1. Top and side views of $C_{18}H_{18}O_4$ (I) with dimensions.

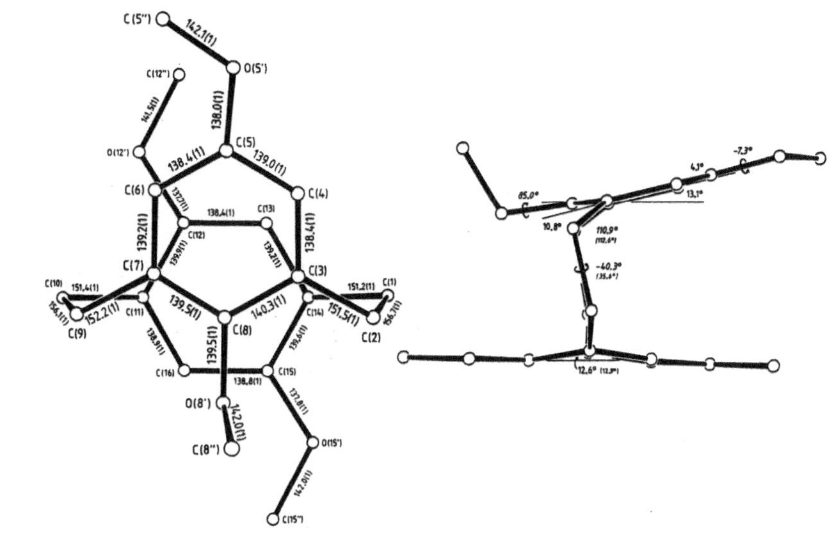

Fig. 2. Top and side views of $C_{20}H_{24}O_4$ (II) with dimensions.

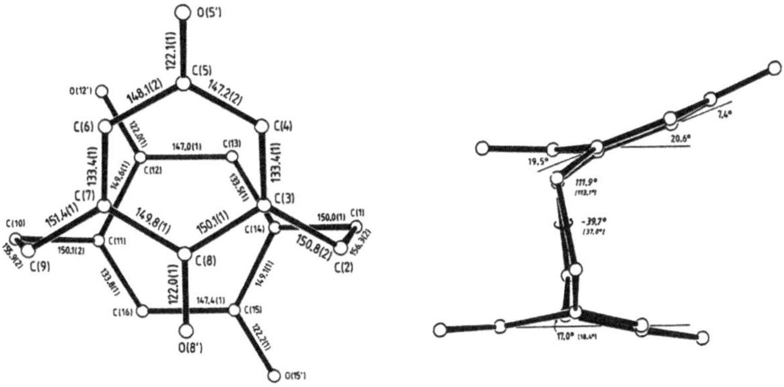

Fig. 3. Top and side views of $C_{16}H_{12}O_4$ (III) with dimensions.

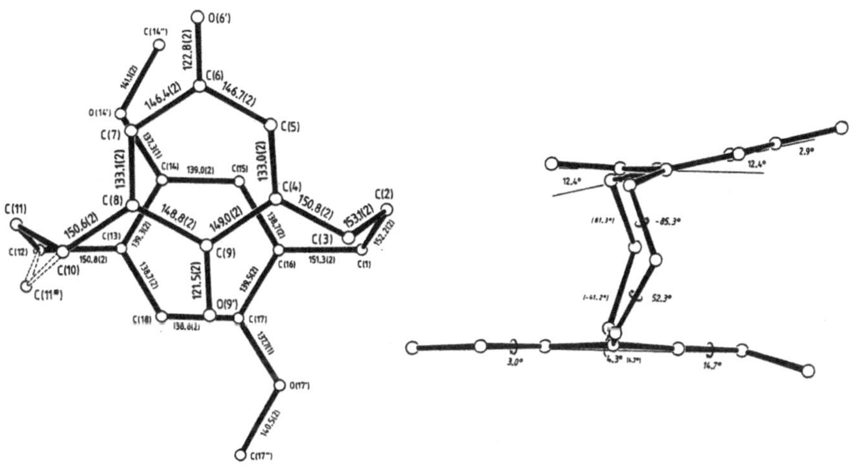

Fig. 4. Top and side views of $C_{20}H_{22}O_4$ (IV) with dimensions.

Details of dimensions and conformations for these strained molecules are shown in Figs. 1-4. In (I) the quinone ring is deformed to a boat conformation and the para-bridged aromatic ring has a typical paracyclophane boat form. In (II) both meta- and para-bridged rings are in boat form. In compound (III) both rings are quinoid and have a more pronounced boat form. Steric strain is reduced in (IV) and the para-aromatic ring deviates only slightly from planarity.

syn-13,16-DIMETHOXY[2](2,6)-p-BENZOQUINONO[2]METACYCLOPHANE
$C_{18}H_{18}O_4$ (I)

8,16-EPOXY-5,8-DIHYDRO-8,13-DIHYDROXY-5-OXO[2.2]METACYCLOPHANE
$C_{16}H_{14}O_4$ (II)

H.A. STAAB, W.R.K. REIBEL and C. KRIEGER, 1985. *Chem. Ber.*, **118**, 1230-1253.

I. Monoclinic, *C2/c*, a = 10.488, b = 11.871, c = 24.819 Å, β = 101.64°, Z = 8. Mo radiation, R = 0.047 for 3119 reflections.

II. Orthorhombic, *Pbca*, a = 10.116, b = 15.778, c = 15.672 Å, Z = 8. Mo radiation, R = 0.044 for 2183 reflections.

Details of conformation and dimensions for these strained molecules are shown in Figs. 1 and 2.

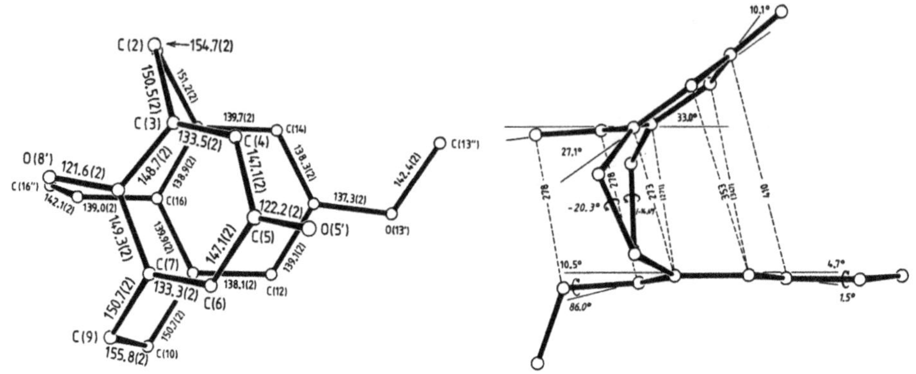

Fig. 1. Top and side views of $C_{18}H_{18}O_4$ (I) with dimensions.

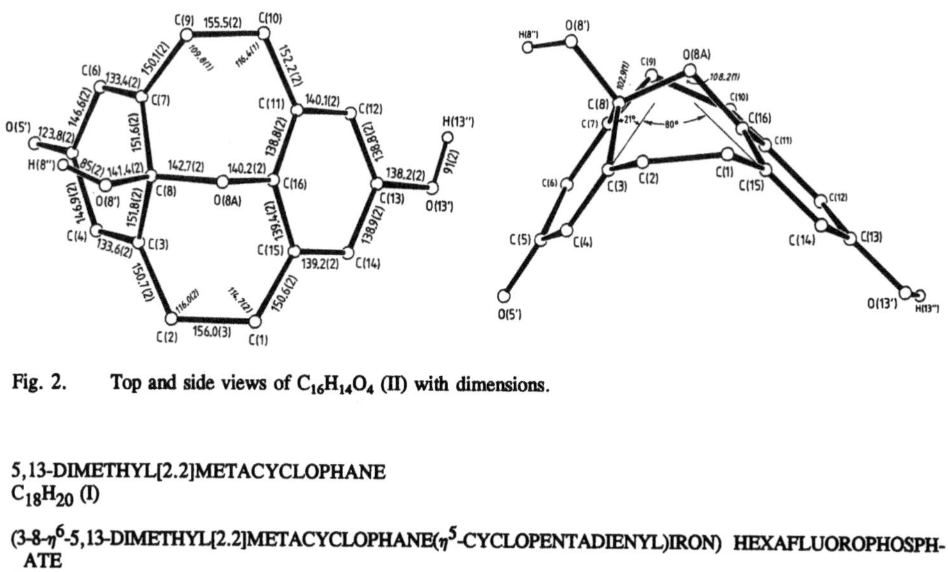

Fig. 2. Top and side views of $C_{16}H_{14}O_4$ (II) with dimensions.

5,13-DIMETHYL[2.2]METACYCLOPHANE
$C_{18}H_{20}$ (I)

(3-8-η^6-5,13-DIMETHYL[2.2]METACYCLOPHANE(η^5-CYCLOPENTADIENYL)IRON) HEXAFLUOROPHOSPHATE
$C_{23}H_{25}F_6FeP$ (II) $C_{23}H_{25}Fe^+,F_6P^-$

A.R. KORAY, T. ZAHN and M.L. ZIEGLER, 1985. *J. Organomet. Chem.*, **291**, 53-60.

I. Triclinic, $P\bar{1}$, a = 5.222, b = 8.125, c = 8.388 Å, α = 95.40, β = 107.79, γ = 94.34°, Z = 1. Mo radiation, R = 0.037 for 429 reflections.

II. Monoclinic, $P2_1/c$, a = 12.817, b = 9.674, c = 18.176 Å, β = 95.5°, Z = 4. Mo radiation, R = 0.069 for 1101 reflections.

The metacyclophane and its Fe complex are shown in Figs. 1 and 2. Molecule (I) has a transoid stepped conformation with the methyl groups directed away from one another. The geometry of the uncomplexed ring in (II) (Fig. 2) is very similar to that in (I). The Fe atom in (II) is 1.59(2) Å from the aromatic ring and 1.66(3) Å from the cyclopentadienyl ring.

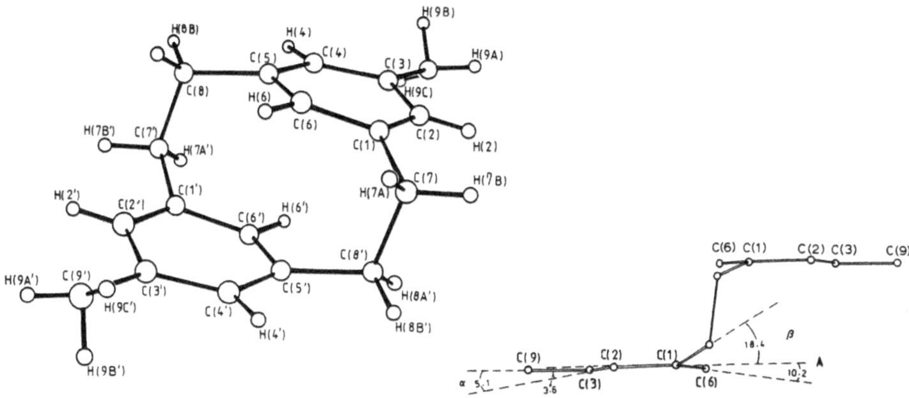

Fig. 1. The $C_{18}H_{20}$ molecule and a side view.

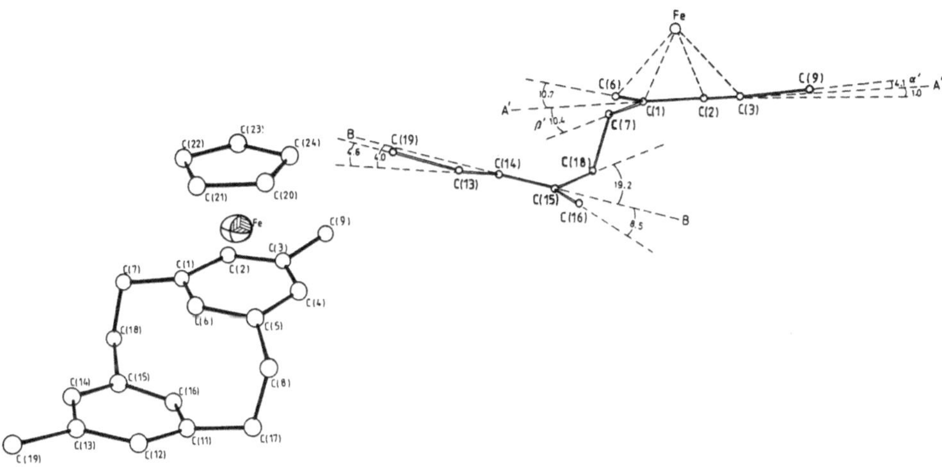

Fig. 2. The $C_{23}H_{25}Fe^+$ ion and a side view.

[3.3]METACYCLOPHANE
$C_{18}H_{20}$

M.F. SEMMELHACK, J.J. HARRISON, D.C. YOUNG, A. GUTIERREZ, S. RAFII and J. CLARDY, 1985. *J. Am. Chem. Soc.*, **107**, 7508-7514

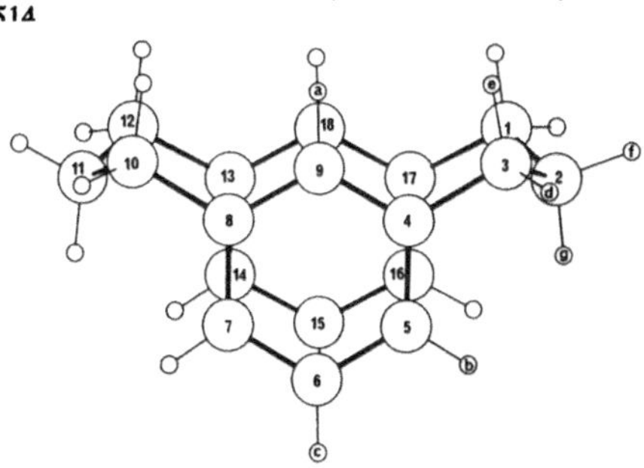

Fig. 1. A perspective drawing of $C_{18}H_{20}$.

Monoclinic, $P2_1$, a = 8.0006, b = 9.013, c = 9.401 Å, β = 9.401°, Z = 2. Mo radiation, R = 0.034 for 823 reflections (at -40°C).

The molecule assumes a syn geometry (Fig. 1) with the bridging chains in a chair-chair conformation. The two arene rings are tilted at 24° to each other; there is a second distortion, a twist of the aromatic rings by roughly 15° about an axis through the centre of each ring.

(1α,3β,6α,7α,8β,10S*)-N-(4-BROMOPHENYL)-1-METHYLTETRACYCLO[4.3.1.0^{3,10}.0^{8,10}]DECANE-7-CARBOXAMIDE
$C_{18}H_{20}BrNO$

V.B. RAO, C.F. GEORGE, S. WOLFF and W.C. AGOSTA, 1985. *J. Am. Chem. Soc.*, **107**, 5732-5739.

Triclinic, $P\bar{1}$, a = 7.619, b = 10.047, c = 21.691 Å, α = 76.89, β = 81.02, γ = 89.64°, Z = 4. Cu radiation, R = 0.055 for 3248 reflections.

There are two nearly identical molecules in the asymmetric unit, one is shown in Fig. 1. The four rings share a central quaternary C atom, C(10); bond distances involving C(10) are shortened (mean value 1.508(20) Å). One of the cyclobutane rings is essentially planar and the other two are slightly folded (fold angles 2.5-12.5°). The cyclopentane ring is in a half-chair conformation, with C(5) -0.29 Å from C(3),C(10),C(6) and C(4) +0.48 Å from the same plane. Molecules are linked by N-H...O hydrogen bonds.

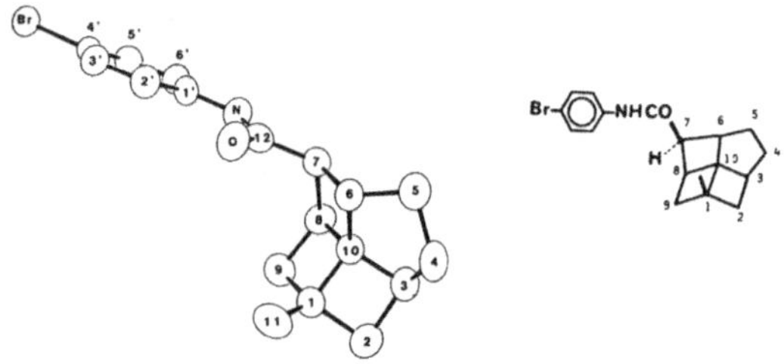

Fig. 1. The $C_{18}H_{20}BrNO$ molecule.

6β-HYDROXY-7,9α-DIMETHYL-6α-PHENYLADAMANTANE-2,4-DIONE
$C_{18}H_{20}O_3$ (I)

3-(5-BENZOYL-5-METHYL-2-OXOCYCLOHEXYL)BUTANOIC ACID
$C_{18}H_{22}O_4$ (II)

P.W. HICKMOTT, M.G. AHMED, S.A. AHMED, S. WOOD and M. KAPON, 1985. *J. Chem. Soc. Perkin Trans. 1*, 2559-2571.

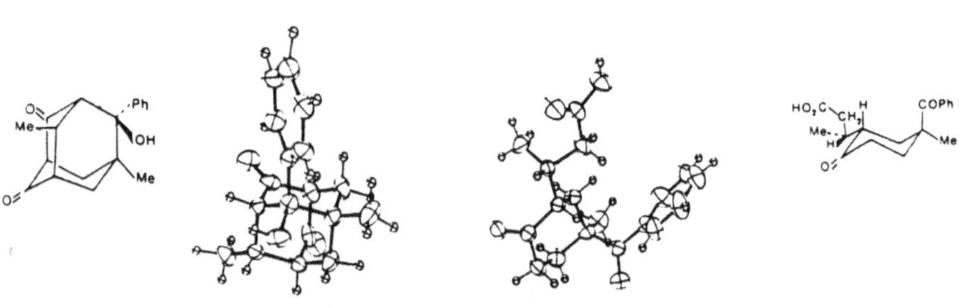

Fig. 1. ORTEP drawings of molecules of $C_{18}H_{20}O_3$ (left) and $C_{18}H_{22}O_4$ (right).

I. Orthorhombic, $P2_12_12_1$, a = 13.802, b = 12.966, c = 8.020 Å, Z = 4. Mo radiation, R = 0.049 for 927 reflections.

II. Monoclinic, $P2_1/n$, a = 16.912, b = 8.675, c = 11.435 Å, β = 106.78°, Z = 4. Mo radiation, R = 0.069

for 1662 reflections.

The crystal structure analyses established that the molecules have the configurations shown (Fig. 1).

1,8,9,10,11,11-HEXACHLOROPENTACYCLO[6.2.1.$1^{3,6}$.$0^{2,7}$.$0^{4,10}$]DODECAN-5-YL BENZOATE
$C_{19}H_{14}Cl_6O_2$

O. ERMER, S.A. MASON, F.A.L. ANET and S.S. MIURA, 1985. *J. Am. Chem. Soc.*, **107**, 2330-2334.

Orthorhombic, $Pna2_1$, a = 13.520, b = 10.639, c = 13.369 Å, Z = 4. Neutron radiation, λ = 0.8402 Å, R = 0.021 for 3290 reflections (at 113 K).

The outstanding feature of this pentacyclic half-cage compound (Fig. 1) is a short non-bonded H(5)...H(9) contact of just 1.617(3) Å, the shortest such contact yet observed. The corresponding distance from an X-ray study was found to be 1.78(5) Å. The C(5)-H(5) and C(9)-H(9) bond lengths of 1.095(3) and 1.078(3) Å respectively are considerably longer than the corresponding X-ray values of 0.90(4) and 0.97(4) Å. The short contact can be explained by an inward rocking of the H(9)-C(9)-Cl(9) group due to a short contact between Cl(9)...Cl(11a) of 3.048(1) Å. The C(5)-H(5) and C(9)-H(9) bonds are somewhat compressed by this rocking, and the angles H(9)-C(9)-Cl(9) 101.2(2) and H(5)-C(5)-O(13) 104.9(2)° are also somewhat smaller than normal.

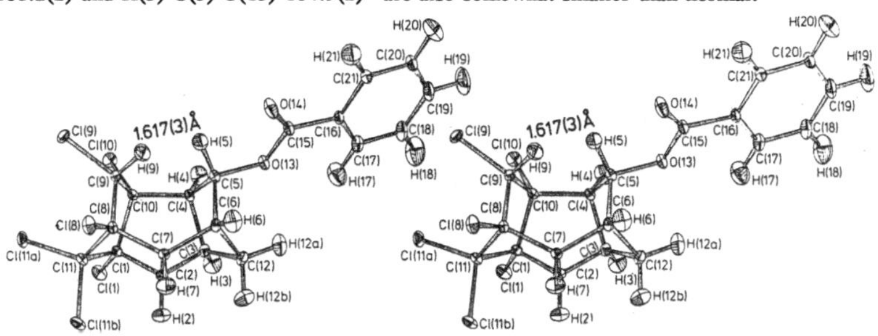

Fig. 1. View of the $C_{19}H_{14}Cl_6O_2$ structure, showing close non-bonded H...H contact.

5,5',7,7'-TETRABROMO[1]DIADAMANTANE
$C_{19}H_{24}Br_4$

J.J. SOSNOWSKI, A.L. RHEINGOLD and R.K. MURRAY Jr., 1985. *J. Org. Chem.*, **50**, 3788-3791.

Monoclinic, *C2/c*, a = 19.515, b = 12.566, c = 7.915 Å, β = 103.85°, Z = 4. Mo radiation, R = 0.039 for 1067 reflections.

The crystal structure analysis unequivocally identified the tetrabromide.

syn-[4.4.3]PROPELLA-2,4,12-TRIEN-11-OL 3,5-DINITROBENZOATE
$C_{20}H_{18}N_2O_6$

J.C. GALLUCCI, K. OHKATA and L.A. PAQUETTE, 1985. *Canad. J. Chem.*, **63**, 862-865.

Triclinic, $P\bar{1}$, a = 10.208, b = 13.355, c = 7.068 Å, α = 99.35, β = 100.63, γ = 100.79°, Z = 2. Mo radiation, R = 0.051 for 2083 reflections.

The unsaturated five-membered ring of this compound has an envelope conformation (Fig. 1) with C(1) 0.47 Å from the plane of the other four atoms (C(6),C(11),C(12),C(13)). The cyclohexadiene ring has atoms C(3),C(4),C(5),C(6) nearly planar, with C(1) +0.52 Å and C(2) +0.20 Å from this plane. The remaining cyclohexane ring is in a chair conformation.

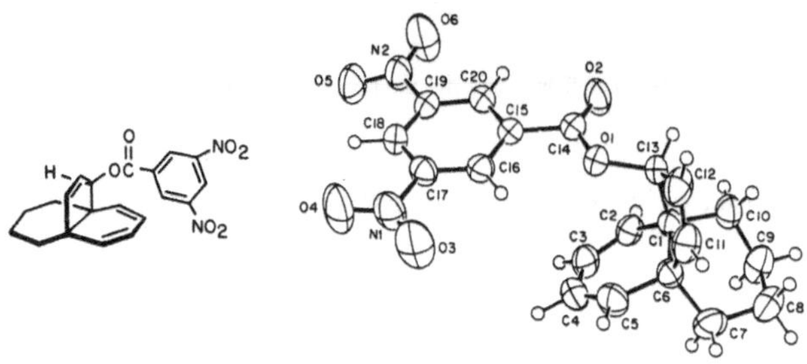

Fig. 1. The $C_{20}H_{18}N_2O_6$ molecule.

ADAMANTYLIDENE-ADAMANTANE-BROMONIUM TRIBROMIDE
$C_{20}H_{28}Br_4$ $C_{20}H_{28}Br^+,Br_3^-$

H. SLEBOCKA-TILK, R.G. BALL and R.S. BROWN, 1985. *J. Am. Chem. Soc.*, **107**, 4504-4508.

Monoclinic, $P2_1/c$, a = 7.404, b = 7.320, c = 37.354 Å, β = 95.49°, Z = 4. Mo radiation, R = 0.041 for 2097 reflections.

The cation (Fig. 1) has a slightly asymmetric three-membered ring (Br-C 2.116(6), 2.194(6), C-C 1.497(8) Å, C-Br-C 40.6(2), Br-C-C 72.5 and 66.9(3)°). The dihedral angle between the substituent planes of the carbons connecting the adamantyl rings is 32.9°. The Br_3^- anion forms a close contact with the Br^+ which may be the cause of the asymmetry in the three-membered ring.

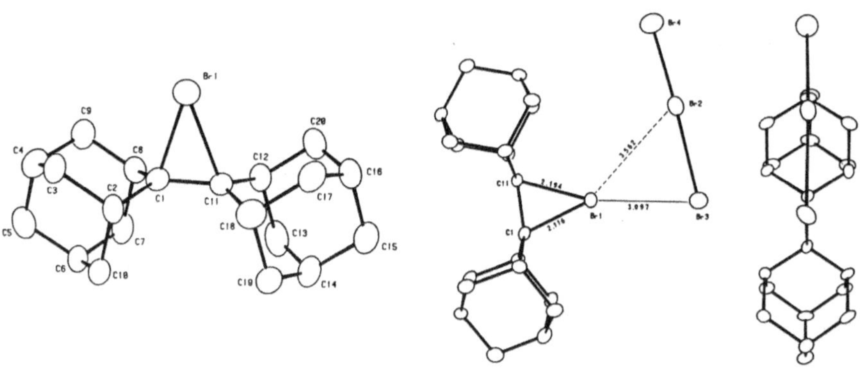

Fig. 1. Views of the $C_{20}H_{28}Br^+,Br_3^-$ system.

8,8-DIPHENYLBICYCLO[3.2.2]NONA-2,6-DIEN-9-ONE
$C_{21}H_{18}O$

C.P. FALSHAW, A. LAKOUES and G.A. TAYLOR, 1985. *J. Chem. Res.*, S106, M1536-1560.

Monoclinic, $P2_1/n$, a = 10.5866, b = 16.501, c = 8.8294 Å, β = 101.359°, Z = 4. Cu radiation, R = 0.042 for 1644 reflections.

The analysis established the structure as shown in Fig. 1.

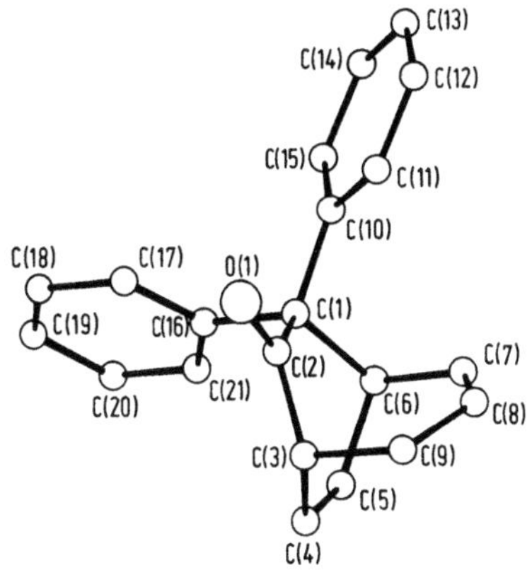

Fig. 1. The structure of $C_{21}H_{18}O$.

1,2,3,4,4a,9,9a,10-OCTAHYDRO-syn-17-METHYL-9,10[1',2']BENZENO-1,4-METHANOANTHRACENE
$C_{22}H_{22}$

H.J. LINDNER, D. LENOIR and D.N. BUTLER, 1985. *Chem. Ber.*, **118**, 363-369.

Triclinic, $P\bar{1}$, a = 8.271, b = 12.634, c = 15.725 Å, α = 73.82, β = 97.26, γ = 89.90°, Z = 4. Cu radiation, R = 0.061 for 3967 reflections.

The two independent overcrowded molecules in the asymmetric unit are very similar and have non-crystallographic mirror symmetry (Fig. 1). The C(18)-C(19) bond is long (1.604 and 1.589(5) Å) and the C-C-C angles at C(15) and C(16) are enlarged to 122.6 and 122.7(4)°. Similar angle increases are found for the C(21) bearing the C(22) methyl group (C(22)-C(21)-C 122.4-124.4(4)°).

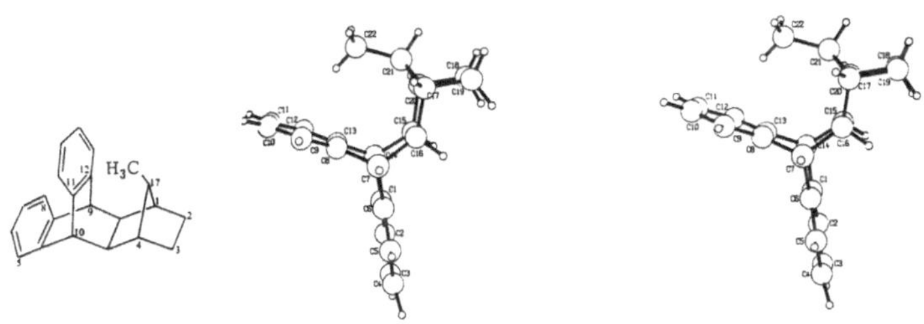

Fig. 1. The $C_{22}H_{22}$ molecule and a stereoview.

2-DIPHENYLMETHYLENE-4,5,6,7-TETRAHYDRO-4,7-METHANO-2H-INDENE
$C_{23}H_{20}$ (I)

4',5',6',7'-TETRAHYDRO-2'-ISOPROPYLIDENESPIRO(CYCLOPROPANE-1,2'-[4,7]METHANO[2H]INDENE)
$C_{15}H_{18}$ (II)

4,5,6,7-TETRAHYDRO-2,8-DIISOPROPYLIDENE-4,7-METHANO-2H-INDENE
$C_{16}H_{20}$ (III)

J.C. GALLUCCI, T.M. KRAVETZ, K.E. GREEN and L.A. PAQUETTE, 1985. *J. Am. Chem. Soc.*, **107**, 6592-6598.

I. Monoclinic, $P2_1/n$, a = 9.666, b = 17.087, c = 11.229 Å, β = 116.53°, Z = 4. Mo radiation, R = 0.070 for 1671 reflections.

II. Triclinic, $P\bar{1}$, a = 9.345, b = 10.502, c = 5.953 Å, α = 91.04, β = 95.46, γ = 84.03°, Z = 2. Mo radiation, R = 0.066 for 1974 reflections.

III. Triclinic, $P\bar{1}$, a = 8.078, b = 6.341, c = 14.003 Å, α = 89.38, β = 73.45, γ = 110.81°, Z = 2. Mo radiation, R = 0.088 for 1791 reflections.

Molecules (I) to (III) are shown in Fig. 1. The isodicyclopentafulvene core common to all three molecules is described in terms of noncrystallographic mirror symmetry. The angles about the apical methane carbon of the norbornane ring (96.2°, 96.6° and 96.7° for (I) to (III)) exhibit the usual deviation from the tetrahedral value.

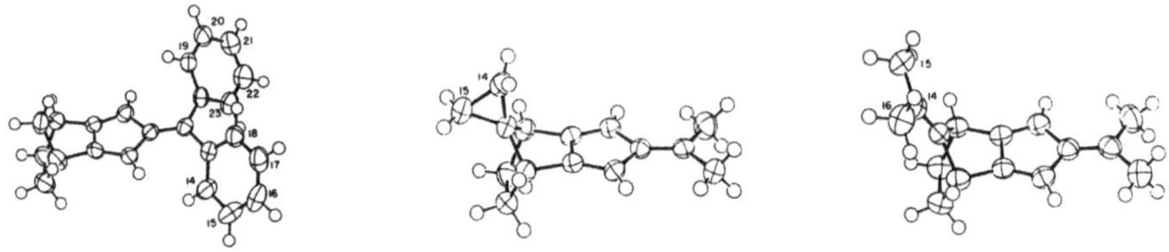

Fig. 1. Views of the methano-indenes (I)-(III) (left–right).

[2.2]PARACYCLO(4,6)[2.2]METAPARACYCLOPHANE
$C_{26}H_{26}$ (I)

[2.2]METACYCLO(4,6)[2.2]METAPARACYCLOPHANE
$C_{26}H_{26}$ (II)

Y. KOIZUMI, T. TOYODA, N. KASAI and S. MISUMI, 1985. *Bull. Chem. Soc. Jpn.*, **58**, 3158-3163.

I. Monoclinic, $P2_1/a$, a = 15.989, b = 9.932, c = 5.991 Å, β = 100.61°, Z = 2. Mo radiation, R = 0.066 for 1864 reflections.

II. Monoclinic, $P2_1/c$, a = 8.969, b = 22.744, c = 10.339 Å, β = 119.32°, Z = 4. Mo radiation, R = 0.096 for 3152 reflections.

Molecule (I) lies about an inversion centre and has the centrosymmetric stair-case structure shown in Fig. 1. The outside para-bridged ring is in a boat conformation and the central ring is deformed towards a chair (bending angle at C(12) is 4.3°). Molecule (II) has a platform shape (syn-form) (Fig. 1) and all three benzene rings adopt boat conformations.

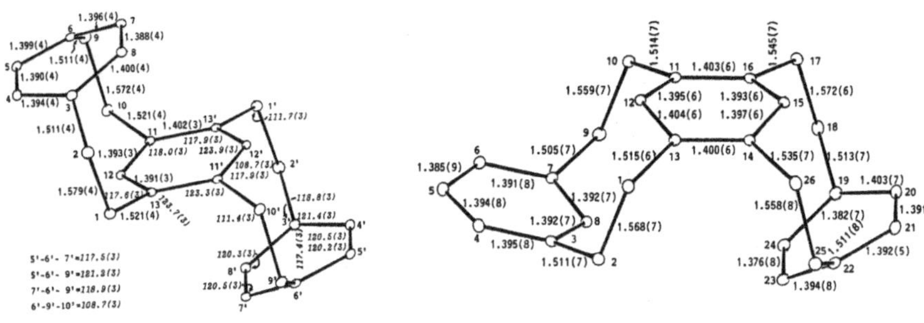

Fig. 1. Bond distances for the $C_{26}H_{26}$ paracyclophane molecules.

2,3-DI-1-ADAMANTYL-2,3-DIMETHYLBUTANE
$C_{26}H_{42}$

M.A.F. MEER, H.-D. BECKHAUS, K. PETERS, H.G. VON SCHNERING and C. RUCHARDT, 1985. *Chem. Ber.*, **118**, 4665-4673.

Monoclinic, $P2_1/n$, a = 22.82, b = 6.749, c = 13.596 Å, β = 102.12°, Z = 4. Mo radiation, R = 0.063 for 3253 reflections.

The most remarkable feature of this structure (Fig. 1) is the very long C-C bonds in the central region (C(2)-C(3) 1.639, C(2)-C(201) 1.647, C(3)-C(301) 1.640). The anti-conformation is found in the crystal. There is good agreement between the experimental results and force-field calculations.

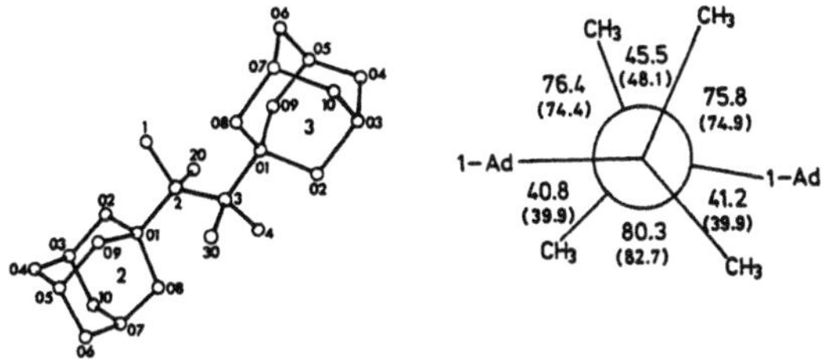

Fig. 1. The $C_{26}H_{42}$ molecule and a Newman projection along the C(2)-C(3) bond.

(4α,6aβ,8β,11aα)-METHYL 1,2,3,4,5,6,6a,7,8,11-DECAHYDRO-4,6a,9-TRIMETHYL-8,11a-METHANO-11aH-CYCLOHEPTA[a]NAPHTHYL-4-BENZOATE
$C_{27}H_{34}O_2$

G. PRECIGOUX, F. LEROY, M. TARAN and B. DELMOND, 1985. *Acta Cryst.*, **C41**, 550-552.

Monoclinic, $P2_1$, a = 11.272, b = 7.546, c = 13.164 Å, β = 96.94°, Z = 2. Mo radiation, R = 0.063 for 1230 reflections.

The molecule is shown in Fig. 1. Bond lengths and angles have normal values except for C(11a)-C(7)-C(8) which is 100.2(6)°.

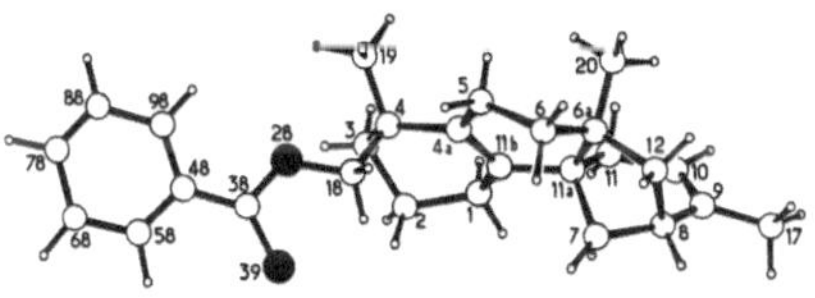

Fig. 1. The molecule $C_{27}H_{34}O_2$.

HEPTACYCLO[13.13.0^{2,7}.0^{8,13}.0^{14,28}.0^{16,21}.0^{22,27}]OCTACOSA-2(7),3,5,8(13),9,11,16(21),17,19,22(27),23,25-DODECAENE (TETRABENZOTRICYCLO[5.5.0.0^{2,8}]DODECA-3,5,9,11-TETRAENE)
$C_{28}H_{20}$

H. IRNGARTINGER, U. HUBER-PATZ, M. NIXDORF and H. RODEWALD, 1985. *Acta Cryst.*, **C41**, 1136-1138.

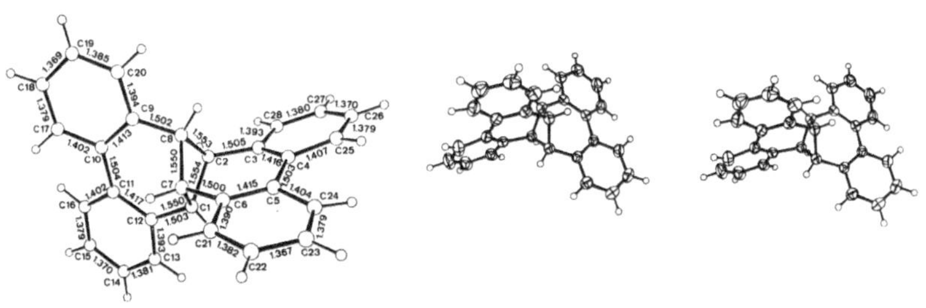

Fig. 1. $C_{28}H_{20}$: a view of the molecule showing the atomic numbering and bond lengths and a stereoview.

Orthorhombic, *Pbca*, a = 12.765, b = 12.905, c = 22.139 Å, Z = 8. Mo radiation, R = 0.042 for 2821

reflections.

The central four-membered ring in the molecule (Fig. 1) is puckered. The bond lengths and angles in this ring are not significantly different and both possible pucker angles are equal and of value 30.3(1)°. The transannular C...C distances are 2.155(2) Å. As a consequence the seven-membered ring is strained and both angles at C(3) and C(4) are significantly greater than 120° having values of 124.1(1) and 125.7(2)° respectively. The biphenyl groups deviate from a linear arrangement having twists of opposite senses about their central bonds. The torsion angles C(3)-C(4)-C(5)-C(6) and C(9)-C(10)-C(11)-C(12) are -14.9(2) and 13.3(2)° respectively. The molecule shows only small deviations from $\bar{4}$ point group symmetry.

cis-9-((BENZYLOXY)METHOXY)-4a,9,9a,10-TETRAHYDRO-9,10-o-BENZENOANTHRACENE-1,4-DIONE $C_{28}H_{22}O_4$

P. BHARADWAJ, J.A. POTENZA, R.M. ORNAF, K.E. RODRIQUES, S. KNAPP and R.A. LALANCETTE, 1985. *Acta Cryst.*, **C41**, 1520-1522.

Orthorhombic, $P2_12_12_1$, a = 13.898, b = 7.285, c = 20.704 Å, Z = 4. Cu radiation, R = 0.052 for 1285 reflections.

In the crystal, the benzyloxymethoxy group extends away from the bicyclo[2.2.2]octane unit (Fig. 1). The 2-ene-1,4-dione ring adopts a shallow boat conformation with both carbonyl groups pointed away from an underlying benzene ring (C(11-16)). The O(1) atom of a carbonyl group and O(9) of the benzyloxymethoxy group are in close non-bonded contact (2.810(6) Å). The C(9)-C(9a) and C(4a)-C(10) bond distances are 1.604(7) and 1.565(7) Å respectively, indicating the presence of strain in the molecule.

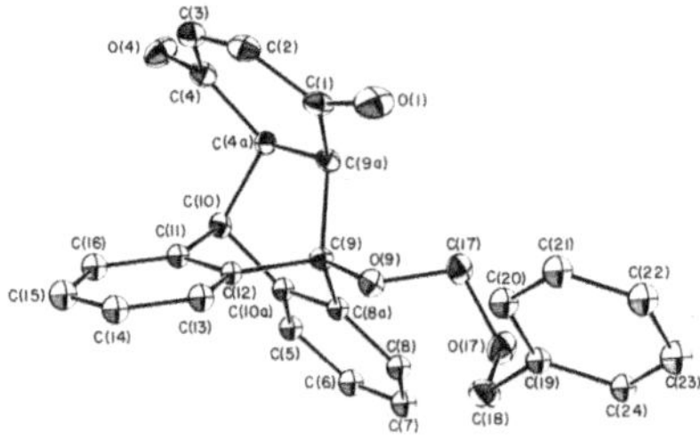

Fig. 1. $C_{28}H_{22}O_4$: the molecular structure.

9,9'-DICYANO-9,10';9',10-DIANTHRYL 9-CYANOANTHRACENE trans-DIMER $C_{30}H_{18}N_2$ (I)

9,9'-DICYANO-9,10';9',10-DIANTHRYL 9-CYANOANTHRACENE trans-DIMER $C_{30}H_{18}N_2$ (II)

C.R. THEOCHARIS and W. JONES, 1985. *J. Chem. Soc. Faraday Trans. 1*, **81**, 857-874.

I. Triclinic, $P\bar{1}$, a = 10.146, b = 17.785, c = 11.501 Å, α = 92.95, β = 95.59, γ = 89.82°, Z = 4. Cu radiation, R = 0.0436 for 2785 reflections.

II. Triclinic, $P1$, a = 10.217, b = 10.235, c = 11.594 Å, α = 95.20, β = 90.37, γ = 120.13°, Z = 2. Cu radiation, R = 0.1015 for 1172 reflections.

Crystals of (I) have one dimer in a general position and two others lying about inversion centres (Fig. 1). The three dimers differ in the dihedral angles subtended by the phenyl rings (134-139°). On heating a crystal of (I) to 420 K, a crystal of (II) was obtained; there are two independent molecules both possessing pseudo centres of symmetry. The structure did not refine well and there is a possibility that there is some 8% monomer present. The dihedral angles between the phenyl rings are in the range 132-144°.

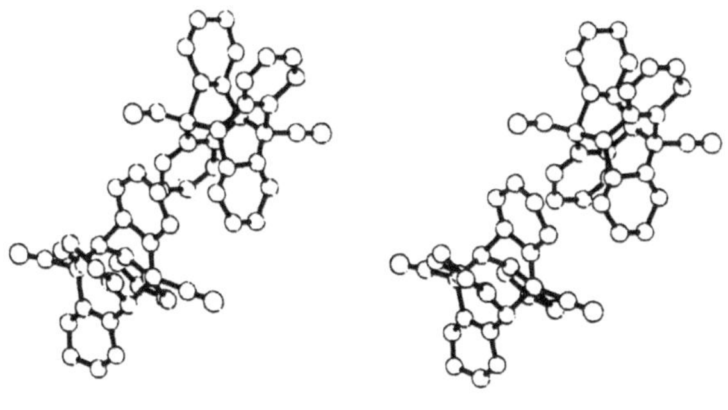

Fig. 1. A stereoview showing two $C_{30}H_{18}N_2$ dimers in (I).

anti-[2.2](2,7)FLUORENOPHANE
$C_{30}H_{24}$

M.W. HAENEL, H. IRNGARTINGER and C. KRIEGER, 1985. *Chem. Ber.*, **118**, 144-162.

Triclinic, $P\bar{1}$, a = 8.803, b = 8.839, c = 13.281 Å, α = 97.53, β = 94.27, γ = 91.94°, Z = 2. Mo radiation, R = 0.046 for 3624 reflections.

The two independent molecules in the cell lie about inversion centres and have similar geometries. Details of conformation and dimensions are in Fig. 1.

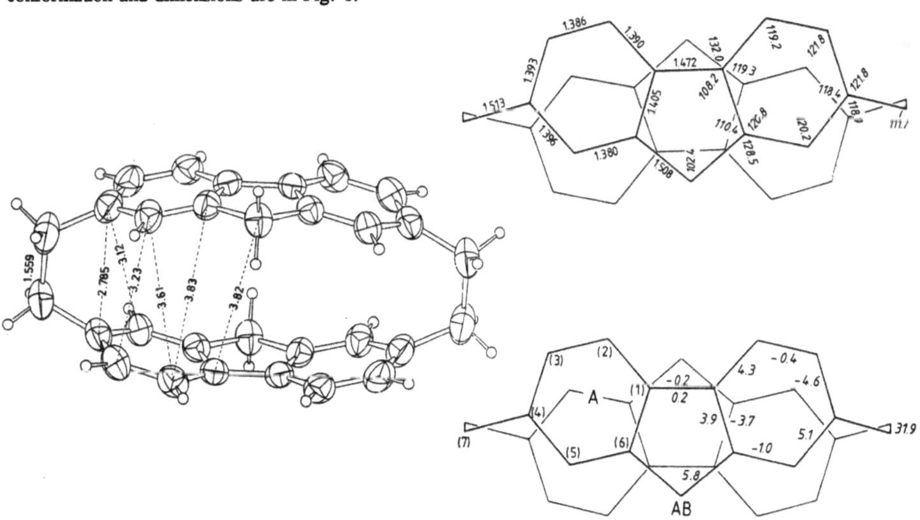

Fig. 1. The $C_{30}H_{24}$ molecule showing intramolecular distances and averaged bond lengths, angles and torsion angles.

[2_6](1,2,4,5)CYCLOPHANE (DELTAPHANE)
$C_{30}H_{30}$ (I)

14,20-DIMETHYL[2_5](1,2,4,5)(1,3,4)(1,3,4)CYCLOPHAN-17-ENE
$C_{30}H_{30}$ (II)

[2_6](1,2,4,5)CYCLOPHANESILVER TRIFLUOROMETHANESULPHONATE

$C_{31}H_{30}AgF_3O_3S$ (III) $C_{30}H_{30}Ag^+,CF_3O_3S^-$

[2_3](1,4)CYCLOPHANESILVER TRIFLUOROMETHANESULPHONATE
$C_{25}H_{24}AgF_3O_3S$ (IV) $C_{24}H_{24}Ag^+,CF_3F_3S^-$

HEECHOL KANG, A.W. HANSON, B. EATON and V. BOEKELHEIDE, 1985. *J. Am. Chem. Soc.*, **107**, 1979-1985.

I. Monoclinic, *C2/c*, a = 16.511, b = 9.3886, c = 14.133 Å, β = 108.87°, Z = 4. Cu radiation, R = 0.041 for 1317 reflections.

II. Monoclinic, $P2_1/c$, a = 15.934, b = 9.749, c = 14.066 Å, β = 96.11°, Z = 4. Cu radiation, R = 0.040 for 2951 reflections.

III. Monoclinic, $P2_1/n$, a = 18.696, b = 9.998, c = 14.022 Å, β = 93.37°, Z = 4. Cu radiation, R = 0.053 for 2500 reflections.

IV. Monoclinic, $P2_1/a$, a = 37.426, b = 10.448, c = 23.909 Å, β = 90.3°, Z = 16. Cu radiation, R = 0.058 for 6630 reflections.

The deltaphane molecule (I) (Fig. 1) shows twofold symmetry, deviating from ideal-6m2 symmetry by a twisting of the bridging methylene carbons, to relieve non-bonded H-H interactions. The carbon atoms lie in three planes tilted at 120 ° from each other. Molecule (II) (Fig. 1) also shows approximate twofold symmetry. The complex (III) (Fig. 1) has essentially the same structure as (I) with the Ag atom lying at distances in the range 2.41(1)-2.48(1) Å from C(4), C(13) and C(20), and completing tetrahedral coordination to the oxygen of the triflate group (Fig. 1), the Ag-O distance being 2.99(1) Å. The Ag atom lies 0.23 Å from the plane of the C(4)-C(13)-C(20) atoms. The cyclophane moiety in (IV) (Fig. 2) deviates considerably from $\bar{6}$m2 symmetry to almost 32 symmetry, by systematic twisting of the bridges, with C(1), C(9) and C(17) lying on one side of the bridging bonds, and C(2), C(10) and C(18) on the other. The Ag atom lies between 2.40(1) and 2.69(1) Å from the six C atoms, being on average 2.49 Å from the midpoints of the ring double bonds. The Ag-O distance is 2.41(1) Å. Remaining features of all the molecules are normal.

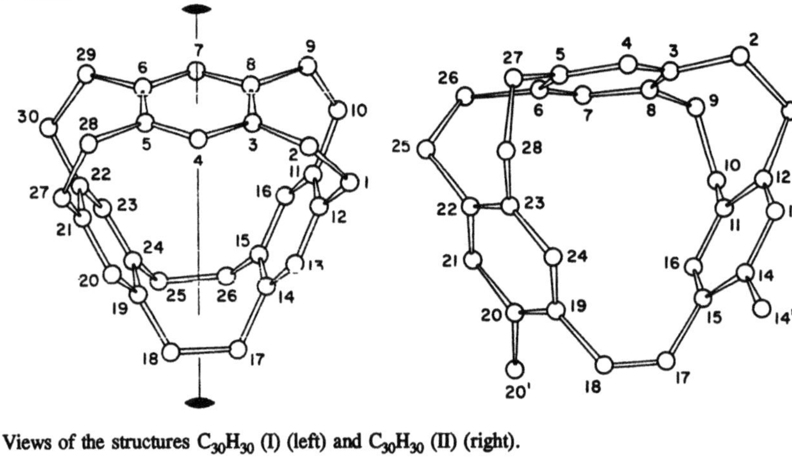

Fig. 1. Views of the structures $C_{30}H_{30}$ (I) (left) and $C_{30}H_{30}$ (II) (right).

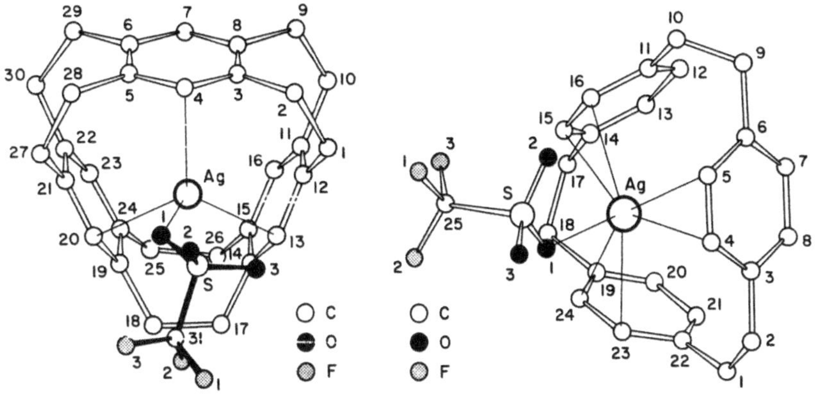

Fig. 2. Views of the complexes $C_{30}H_{30}$ (III) (left) and $C_{25}H_{24}AgF_3O_3S$ (IV) (right).

13-(METHOXYCARBONYL)-14-HYDROXY-17-METHYL-(11)-m-CYCLOPHANE (S)-α-METHOXY-2-(TRIFLUOROMETHYL)PHENYLACETATE ESTER
$C_{30}H_{37}F_3O_5$

T.H. CHAN, G.J. KANG, F. BELANGER-GARIEPY, F. BRISSE and K. STELIOU, 1985. *Canad. J. Chem.*, **63**, 3298-3304.

Orthorhombic, $P2_12_12_1$, a = 10.808, b = 13.676, c = 19.060 Å, Z = 4. Cu radiation, R = 0.048 for 2756 reflections.

The 14-membered ring has a slightly distorted rectangular conformation described as the rectangular diamond-lattice conformation [3434]. The absolute configuration was also established.

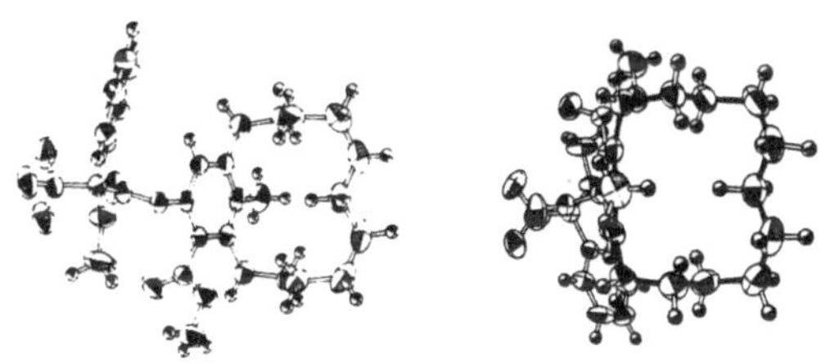

Fig. 1. Two views of $C_{30}H_{37}F_3O_5$.

syn-[2.2](1,4)ANTHRACENOPHANE
$C_{32}H_{24}$ (I)

9,10-DIHYDRO[2.2.0.0](1,4,9,10)ANTHRACENOPHANE
$C_{32}H_{24}$ (II)

T. TOYODA, N. KASAI and S. MISUMI, 1985. *Bull. Chem. Soc. Jpn.*, **58**, 2348-2356.

I. Monoclinic, $A2$, a = 25.539, b = 8.152, c = 20.561 Å, β = 106.03°, Z = 8. Mo radiation, R = 0.053 for 4413 reflections (at -149°C).

II. Monoclinic, $P2_1/c$, a = 8.065, b = 12.963, c = 19.580 Å, β = 91.46°, Z = 4. Mo radiation, R = 0.043 for 2003 reflections.

There are two independent molecules in (I). All four anthracene nuclei have normal dimensions and in each molecule the two anthracene systems are nearly parallel (Fig. 1). The ethylene-bridged benzene rings are in boat conformation. The photoisomer (II) (Fig. 2) has had additional bonds inserted to connect the upper and lower anthracene nuclei and these bonds are long (1C8-2C8 1.628(4), 1C15-2C15 1.637(5) Å).

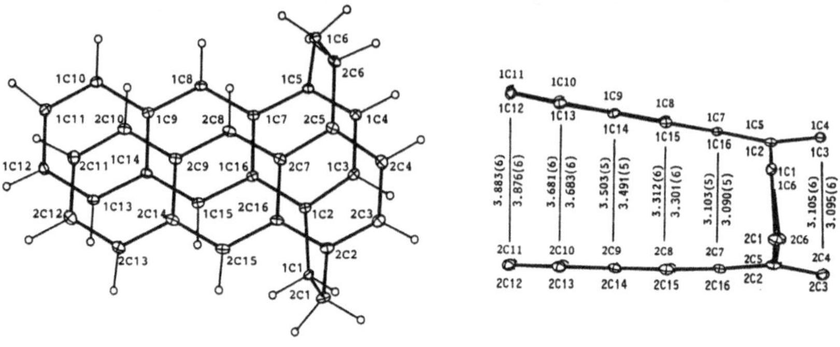

Fig. 1. One of the $C_{32}H_{24}$ (I) molecules and a side-on view.

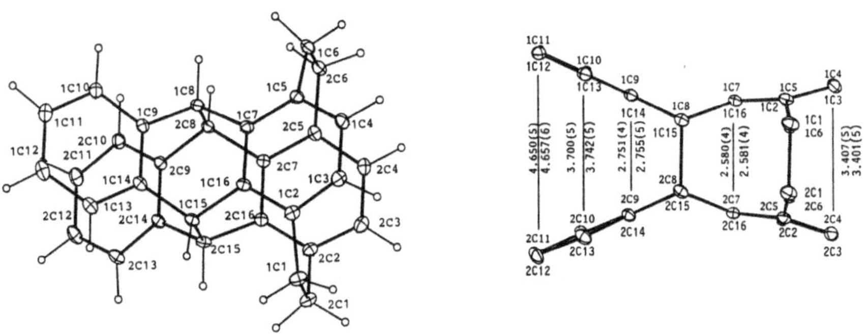

Fig. 2. The $C_{32}H_{24}$ (II) photoisomer and a side on view.

PENTACYCLO[20.2.2.2^{4,7}.2^{10,13}.2^{16,19}]DOTRIACONTA-4,6,10,12,16,18,22,24,25,27,29,31-DODECAENE ([2.2.2.2]PARACYCLOPHANE)

$C_{32}H_{32}$

A. RENAULT and C. COHEN-ADDAD, 1985. *Acta Cryst.*, C41, 773-775.

Monoclinic, $P2_1/c$, a = 12.026, b = 9.731, c = 20.978 Å, β = 92.35°, Z = 4. Cu radiation, R = 0.061 for 2763 reflections.

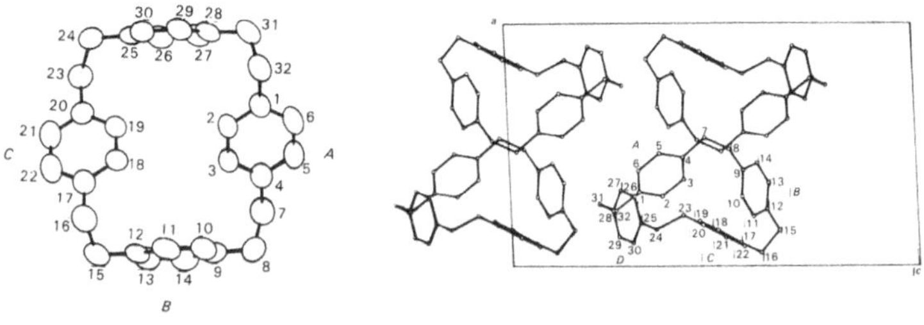

Fig. 1. $C_{32}H_{32}$: a projection of the molecule onto the plane defined by the phenyl ring centroids and a projection of the molecular packing onto (010).

In the molecule (Fig. 1) each of the four aromatic rings is planar. The dihedral angles between pairs of adjacent rings are A/B 50.6(8), B/C 64.0(8), C/D 62.3(8) and A/D 53.9(8)°. The centroids of these rings are essentially coplanar and the A and C rings make angles of -33 and 34° while the B and D rings make angles of 83 and 85° with the centroid plane. The molecule exists in the gauche conformation in agreement with NMR studies in solution.

8,9,14,15-TETRAHYDRO-4b,15(1',2'):9,14(1'',2'')DIBENZENO-4bH-DIBENZO[a,g]CYCLOPENTA[c]CYCLO-DECEN-7(7aH)-ONE
$C_{33}H_{22}O$ (I)

4-(9-ANTHRYL)-1,3a,4,5-TETRAHYDRO-5,9b(1',2')BENZENO-9bH-BENZ[e]INDEN-3(2H)-ONE
$C_{33}H_{24}O$ (II)

4-(9-ANTHRYL)-1,2,3,3a,4,5-HEXAHYDRO-5,9b(1',2')BENZENO-9bH-BENZ[e]INDEN-3-OL METHANOL SOLVATE
$C_{33}H_{26}O,CH_4O$ (III)

H.-D. BECKER, B.W. SKELTON and A.H. WHITE, 1985. *Aust. J. Chem.*, **38**, 1471-1480.

I. Monoclinic, $P2_1/c$, a = 15.233, b = 8.253, c = 19.136 Å, β = 112.95°, Z = 4. Mo radiation, R = 0.038 for 2235 reflections.

II. Monoclinic, $P2_1/n$, a = 12.844, b = 18.055, c = 10.077 Å, β = 91.68°, Z = 4. Mo radiation, R = 0.039 for 1250 reflections.

III. Monoclinic, $P2_1/n$, a = 20.728, b = 10.130, c = 12.512 Å, β = 105.97°, Z = 4. Mo radiation, R = 0.049 for 1491 reflections.

The X-ray analysis established the structures of the three compounds (Fig. 1). In (I), which is formed upon irradiation by the skeletal rearrangement of 4-(9-anthryl)-4,5-dihydro-5,9b[1',2']-benzeno-9bH-benz[e]indene-3(3aH)-one, the tetraphenyl-ethane bond, C(108)-C(208), is long (1.623(5) Å). In both (II) and (III), a long single bond (1.59, 1.61 Å) connects the 9,10 dihydroanthracene moiety with the 9-anthrylmethyl carbon. There is also a long bond (1.57, 1.60 Å) connecting the dihydroanthracene moiety with the ethano bridge. The vicinal bridge protons are trans-oriented in both (II) and (III). In (III), the methanol solvate molecule is hydrogen-bonded to the O atom (2.7 Å).

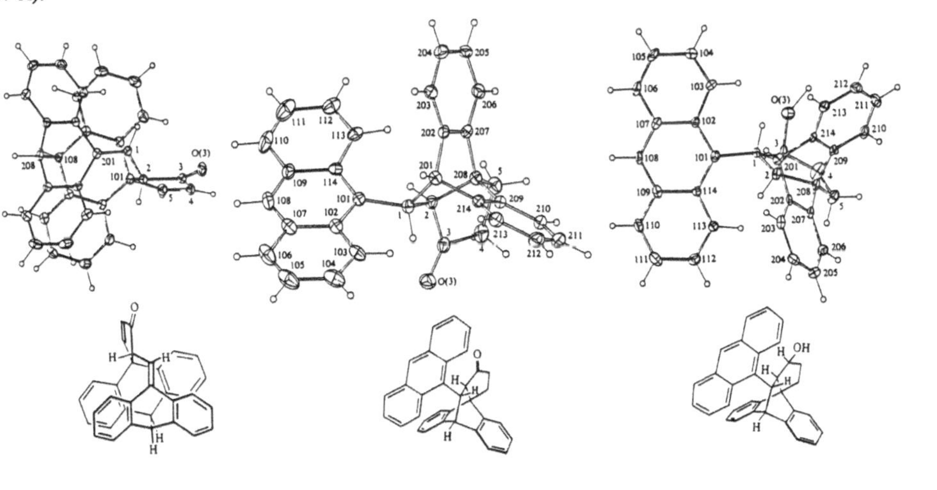

Fig. 1. The molecular skeletons and structures: $C_{33}H_{22}O$ (left), $C_{33}H_{24}O$ (centre), and $C_{33}H_{26}O,CH_4O$ (right).

exo-1,7,8,9-TETRAPHENYLTRICYCLO[5.2.1.0^{2,6}]DECA-3,8-DIEN-10-ONE
$C_{34}H_{26}O$ (I)

endo-1,7,8,9-TETRAPHENYLTRICYCLO[5.2.1.0^{2,6}]DECA-3,8-DIEN-10-ONE

$C_{34}H_{26}O$ (II)

J.M. COXON, M.J. O'CONNELL and P.J. STEEL, 1985. *Aust. J. Chem.*, **38**, 1223-1231.

I. Triclinic, $P\bar{1}$, a = 9.863, b = 10.338, c = 13.634 Å, α = 80.90, β = 68.87, γ = 66.87°, Z = 2. Mo radiation, R = 0.039 for 2215 reflections.

II. Monoclinic, $P2_1/c$, a = 11.610, b = 7.862, c = 27.675 Å, β = 100.22°, Z = 4. Mo radiation, R = 0.049 for 3134 reflections.

In the exo-isomer, the cyclopentenoid double bond is disordered over two positions with occupancies 0.62 and 0.38. The relative orientations of the four phenyl rings in the two molecules are similar (Fig. 1). The C(8)-C(9) double bonds are longer than expected in both molecules (1.354(2), 1.355(3) Å). The cyclopentene rings are planar in both (I) and (II).

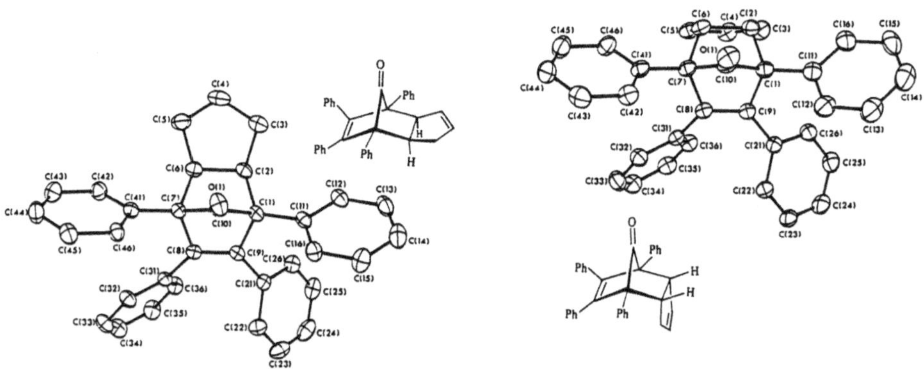

Fig. 1. $C_{34}H_{26}O$: the molecular skeleton and structure of the exo-isomer (I) (left), and of the endo-isomer (II) (right).

8,14,22,28-TETRAMETHOXY-5,11,19,25-TETRAMETHYL[2.0.2.0]METACYCLOPHANE
$C_{36}H_{40}O_4$

T. KANEDA, S. UMEDA, H. TANIGAWA, S. MISUMI, Y. KAI, H. MORII, K. MIKI and N. KASAI, 1985. *J. Am. Chem. Soc.*, **107**, 4802-4803.

Orthorhombic, *Ccca*, a = 20.080, b = 18.929, c = 8.076 Å, Z = 4. Cu radiation, R = 0.071 for 1116 reflections.

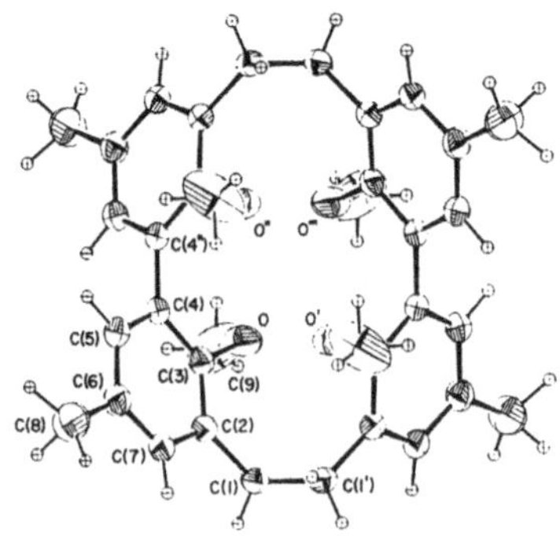

Fig. 1. The $C_{36}H_{40}O_4$ structure.

The molecule has crystallographic D_2 symmetry (Fig. 1) and the distance from the methoxy O atom to the molecule centre is 2.078 Å. The molecule is strain free with a dihedral angle of 89.9° between the benzene rings in the biphenyl moiety and a torsion angle C(2)-C(1)-C(1')-C(2') of 86.2°.

5,11,17,23,29,35-HEXA-t-BUTYL-37,38,39,40,41,42-HEXAKIS(2-METHOXYETHOXY)CALIX[6]ARENE
$C_{84}H_{120}O_{12}$

R. UNGARO, A. POCHINI, G.D. ANDREETTI and P. DOMIANO, 1985. *J. Inclusion Phenomena*, **3**, 35-42.

Monoclinic, $P2_1/c$, a = 14.581, b = 22.517, c = 11.799 Å, β = 92.13°, Z = 2. Cu radiation, R = 0.074 for 4247 reflections.

The molecule (Fig. 1) lies about an inversion centre. Three adjacent phenolic units point their ethereal chains above the mean molecular plane (which practically coincides with that found for the bridging methylene groups). Two of the three independent phenyl rings are almost normal to the molecular plane (dihedral angles 73.0(1) and 104.1(1)°); the third is tilted by 42.9(1)° to the reference plane and directs its ethereal chain towards the centre of the macroring.

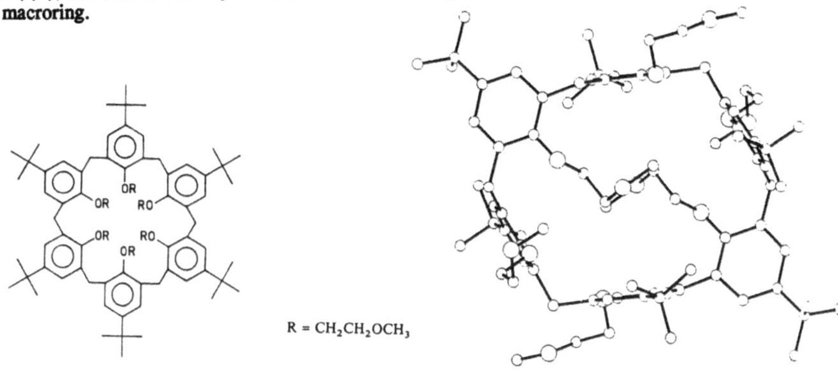

Fig. 1. The $C_{84}H_{120}O_{12}$ calix[6]arene.

p-t-BUTYLCALIX[8]ARENE
$C_{88}H_{112}O_8$

C.D. GUTSCHE, A.E. GUTSCHE and A.I. KARAULOV, 1985. *J. Inclusion Phenomena*, **3**, 447-451.

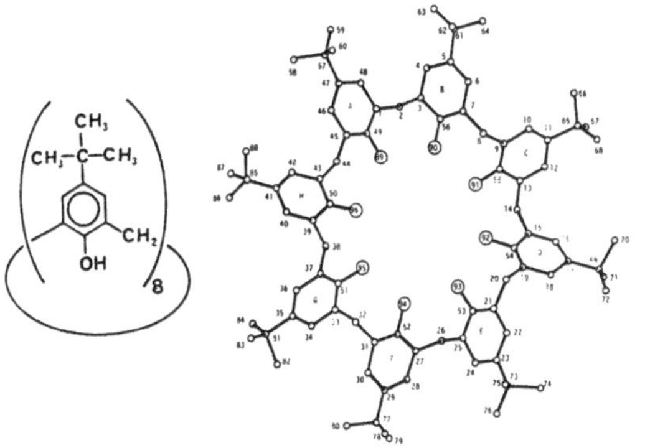

Fig. 1. The p-t-butylcalix[8]arene molecule.

Monoclinic, $P2_1/c$, a = 20.312, b = 23.020, c = 20.006 Å, β = 113.05°, Z = 4. Cu radiation, R = 0.166 for 4231 reflections.

The analysis shows that the molecule has a pleated loop conformation in which the eight hydroxyl groups are arranged in a slightly undulating, almost planar, intramolecularly hydrogen-bonded array (Fig. 1). The dihedral angles between the benzene rings and the molecular plane are 40.6, 146.3, 40.3, 48.8, 138.7, 33.5, 41.0 and 140.4°.

p-(1,1,3,3)TETRAMETHYLBUTYLCALIX[8]METHOXYETHYLOXYBENZENE OCTAPODAND
$C_{144}H_{224}O_{16}$

R. UNGARO, A. POCHINI, G.D. ANDREETTI and F. UGOZZOLI, 1985. *J. Inclusion Phenomena*, **3**, 409-420.

Triclinic, $P\bar{1}$, a = 16.647, b = 16.447, c = 14.269 Å, α = 108.56, β = 105.63, γ = 87.23°, Z = 1. Mo radiation, R = 0.150 for 2863 reflections.

The molecule lies about an inversion centre (Fig. 1) and has a roughly ellipsoidal shape with two sets of four phenolic residues following a helical arrangement. In the crystal, the polyether chains of two adjacent phenolic units partially fill the calix cavity. The dihedral angles which the four independent phenolic residues make with the mean molecular plane are 21.3(5), 119.5(5), 45.5(5) and 86.6(5)°.

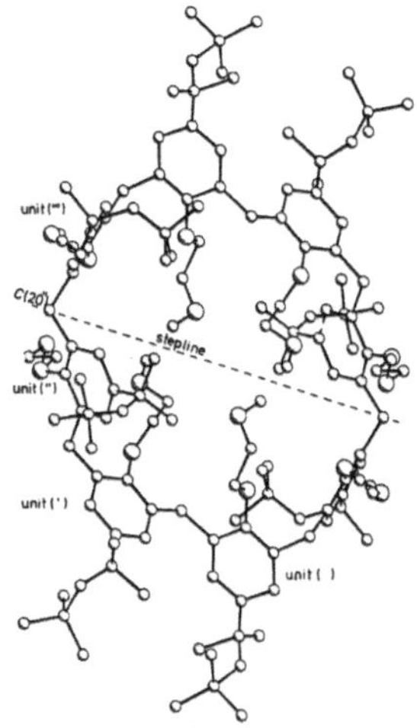

Fig. 1. A view of the calix[8] molecule $C_{144}H_{224}O_{16}$.

3,5-DIBROMO-1,2,4-TRIAZOLE
$C_2HBr_2N_3$

J. VALKONEN, I. PITKANEN and A. PAJUNEN, 1985. *Acta Chem. Scand.*, **A39**, 711-716.

Hexagonal, $P\bar{6}2c$, a = 11.697, c = 6.791 Å, Z = 6. Mo radiation, R = 0.038 for 293 reflections.

The compound forms planar molecules (Fig. 1) which are connected through hydrogen bonds. The distance between layers, parallel to the a-b plane is 3.40 Å.

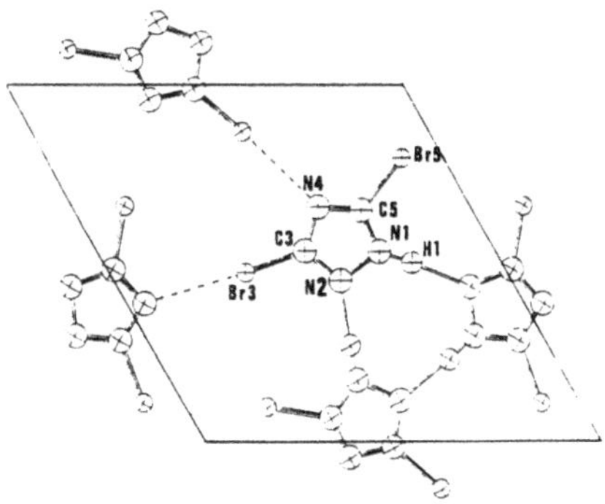

Fig. 1. Layered structure of $C_2HBr_2N_3$ projected along the c-axis.

4(5)-CARBOMETHOXYIMIDAZOLE
$C_5H_6N_2O_2$

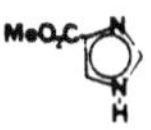

T.J. KRESS, 1985. *J. Org. Chem.*, **50**, 3073-3076.

Monoclinic, $P2_1/c$, a = 3.8421, b = 15.542, c = 9.662 Å, β = 95.32°, Z = 4. Cu radiation, R = 0.047 for 871 reflections.

The crystal structure analysis identified the molecule.

1H-3,5-DIMETHYLPYRAZOLE-4-DIAZONIUM CHLORIDE
$C_5H_7ClN_4$

R.P. BRINT, D.J. COVENEY, F.J. LALOR, G. FERGUSON, M. PARVEZ and P.Y. SIEW, 1985. *J. Chem. Soc. Perkin Trans.* 2, 139-145.

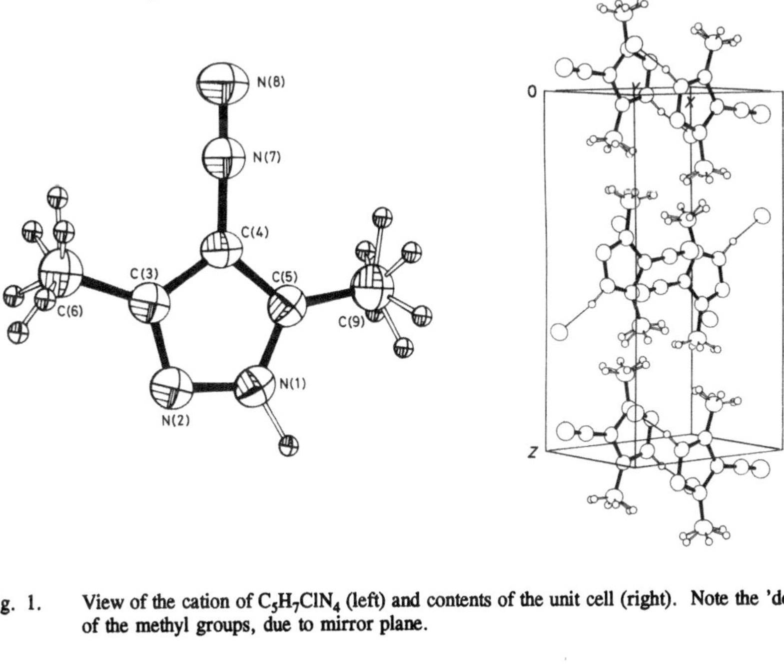

Fig. 1. View of the cation of $C_5H_7ClN_4$ (left) and contents of the unit cell (right). Note the 'double' hydrogens of the methyl groups, due to mirror plane.

Orthorhombic, *Pnma*, a = 7.964, b = 6.308, c = 15.060 Å, Z = 4. Mo radiation, R = 0.0316 for 539 reflections.

The cation and chloride ion both lie on crystallographic mirror planes; the hydrogens of each methyl group adopt two different orientations about the mirror plane. The diazonium moiety is essentially linear; the N(7)-N(8) distance is 1.096(4) Å; the C(4)-N(7) distance is 1.351(4) Å. The chloride ions lie on the same mirror-plane and are hydrogen bonded to the cations; Cl...H 1.96 and N...Cl 3.034 Å (Fig. 1).

1,1-CARBONYLBIS(AZIRIDINE)
$C_5H_8N_2O$

R. BOESE, P. RADEMACHER and L. TRESCHANKE, 1985. *J. Mol. Struct.*, **131**, 55-60.

Monoclinic, *C2/c*, a = 8.285, b = 11.148, c = 6.866 Å, β = 114.54°, Z = 4. Mo radiation, R = 0.038 for 595 reflections (at -105°C).

The molecule lies on a twofold symmetry axis (Fig. 1) and has principal dimensions C=O 1.211(2), C(1)-N(1) 1.389, N-C(ring) 1.455 and 1.460(2), C-C 1.485(2) Å, O-C-N 123.5(1), C(1)-N-C(ring) 120.2(1) and 121.6(1)°, torsion angle O-C(1)-N(1)-(C(2)-C(3) mid-point) is 83.8(1)°.

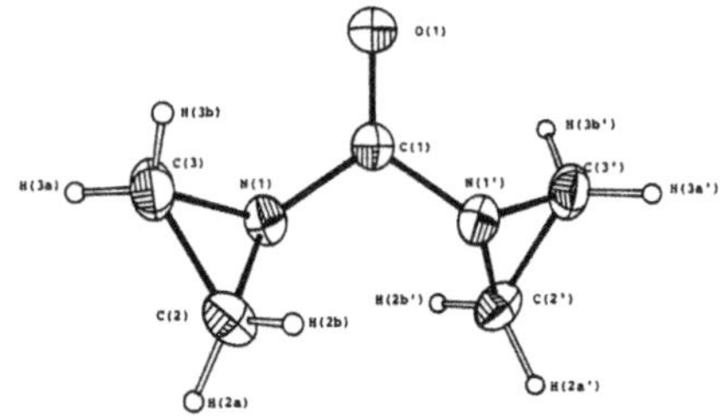

Fig. 1. The $C_5H_8N_2O$ molecule.

1-METHYL-2,2-AZIRIDINEDICARBOXAMIDE
$C_5H_9N_3O_2$

S. BRUCKNER, L. MALPEZZI, A.V. PROSYANIK and S.V. BONDARENKO, 1985. *Acta Cryst.*, **C41**, 213-215.

Monoclinic, *C2/c*, a = 23.150, b = 7.008, c = 8.137 Å, β = 91.86°, Z = 8. Mo radiation, R = 0.054 for 844 reflections.

The N atom of the three-membered ring lies 0.656 Å out of the plane of its three substituents and is pyramidal (Fig. 1). The C(5)-N(3) bond makes an angle of 122.3° with the aziridine plane. The plane defined by C(1)-C(4)-C(2) is almost perpendicular to the ring plane (88.3(2)°). The two carboxamide groups are as far apart as possible and make an angle of 64.7(3)° with each other. Molecules are associated in the crystal in layers parallel to the bc plane by an extended hydrogen-bonding network.

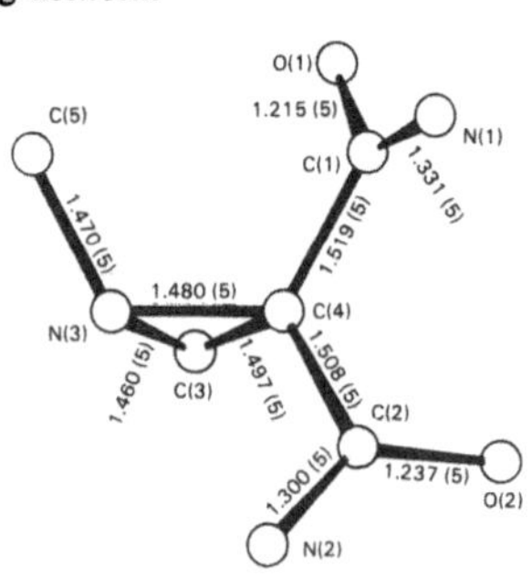

Fig. 1. $C_5H_9N_3O_2$: bond distances.

2,2,5,5-TETRAMETHYLPYRROLIDIN-3-OL
$C_8H_{17}NO$

M. CYGLER, K. DOBRYNIN, M.J. GRABOWSKI, R.B. NAZARSKI and R. SKOWRONSKI, 1985. *J. Chem. Soc. Perkin Trans.* 2, 1495-1501.

Monoclinic, $P2_1/c$, a = 11.98, b = 15.39, c = 11.26 Å, β = 118.10°, Z = 8. Cu radiation, R = 0.045 for 2000 reflections.

There are two independent molecules, A and B, in the crystallographic asymmetric unit and these have nearly identical conformations. The hydroxy group and the amine hydrogen are in a cis relationship and both assume an axial orientation to the ring. In the crystal each molecule (Fig. 1) forms four intermolecular hydrogen bonds, two of which are strong (O...N 2.82 Å) and two of which are weaker (N...O 3.40 Å).

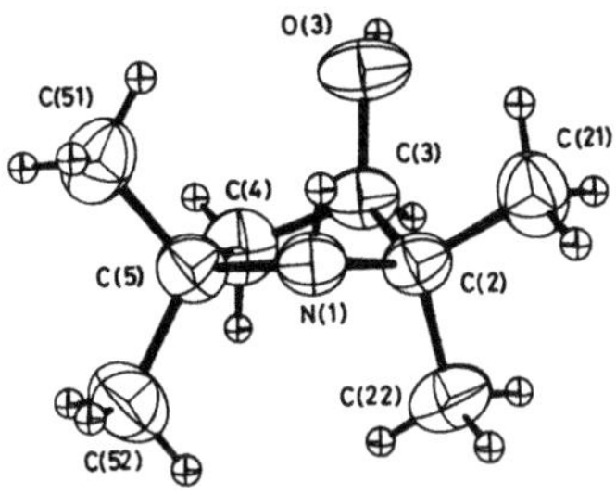

Fig. 1. An ORTEP drawing of a molecule of $C_8H_{17}NO$.

1-ETHYL-3,4-BIS(N-METHYLCARBAMOYL)PYRAZOLE (ETHIRAZOLE)
$C_9H_{14}N_4O_2$

V.S. FUNDAMENSKII, N.D. DROVTSYNA, V.L. GOL'DFARB, K.A. MOSHKOV and Yu.S. BORODKIN, 1985. *Zh. Strukt. Khim.*, **26**, 1-82-89 [J. Struct. Chem., **26**, 67-74].

Orthorhombic, *Pbca*, a = 17.237, b = 13.885, c = 8.797 Å, Z = 8. Mo radiation, R = 0.046 for 777 reflections.

The molecule (Fig. 1) exhibits normal geometry. The structure consists of chains of hydrogen-bonded molecules extending along the *b*-axis (N-H...O, N...O 2.91 Å).

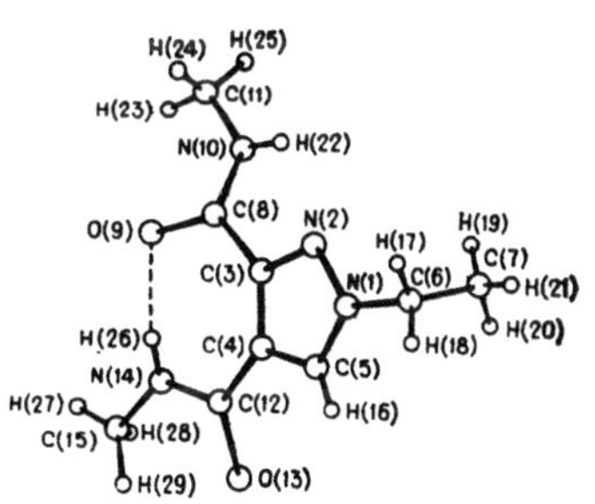

Fig. 1. The $C_9H_{14}N_4O_2$ molecule.

1-PHENYL-2,2-AZIRIDINEDICARBOXAMIDE
$C_{10}H_{11}N_3O_2$ (I) (II)

S. BRUCKNER, L. MALPEZZI, A.V. PROSYANIK and S.V. BONDARENKO, 1985. *Acta Cryst.*, **C41**, 215-219.

I. Monoclinic, $P2_1/n$, a = 6.180, b = 28.147, c = 5.919 Å, β = 105.90°, Z = 4. Mo radiation, R = 0.047 for 813 reflections.

II. Orthorhombic, $P2_12_12_1$, a = 7.236, b = 8.761, c = 15.700 Å, Z = 4. Mo radiation, R = 0.050 for 1576 reflections.

In (I) (Fig. 1) the phenyl ring plane is perpendicular to the three-membered ring, the angle between the normal to the phenyl ring and the distal C(3)-C(4) vector being 89.0(2)°. In (II) this angle is 77.5(1)°. The N atom of the aziridine ring in both (I) and (II) has a pyramidal nature with N(3) lying 0.603(3) Å out of the plane defined by its three substituents. The N(3)-C(5) bond makes an angle of 126.3(2)° with the aziridine ring. The plane defined by C(1)-C(4)-C(2) is approximately perpendicular to the aziridine plane in both (I) and (II). The molecular dimensions of the two molecules are similar. The crystal packing in both structures is governed by hydrogen bonds of the type N-H...O (2.830(6)-3.109(6)) and N-H...N (2.728(6), 2.750(3) Å).

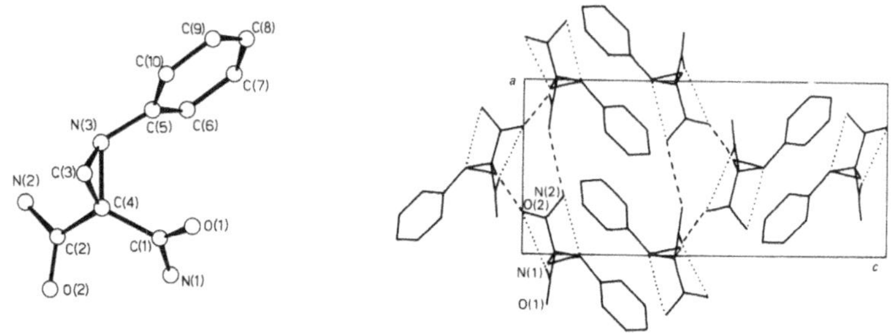

Fig. 1. $C_{10}H_{11}N_3O_2$: a molecule of (I) and packing in (II).

2-(4,5-DIMETHYLIMIDAZOL-2-YL)BENZENETHIOL HYDROCHLORIDE
$C_{11}H_{13}ClN_2S$ (I) $C_{11}H_{13}N_2S^+,Cl^-$

2-(4,5-DIMETHYLIMIDAZOL-2-YL)BENZENETHIOL
$C_{11}H_{12}N_2S$ (II)

J.P. STREET, K.I. SKOREY, R.S. BROWN and R.G. BALL, 1985. *J. Am. Chem. Soc.*, **107**, 7669-7679.

I. Monoclinic, *C2/c*, a = 20.802, b = 8.010, c = 14.497 Å, β = 99.34°, Z = 8. Mo radiation, R = 0.050 for 1564 reflections.

II. Orthorhombic, *Pbca*, a = 12.342, b = 19.869, c = 8.500 Å, Z = 8. Mo radiation, R = 0.051 for 982 reflections.

In the hydrochloride salt (I) (Fig. 1) the dihedral angle between the two rings is 57.3° and an imidazole N-H unit is hydrogen bonded to the Cl^- ion. On deprotonation (Fig. 1) the two rings become essentially coplanar (dihedral angle 2.4°). The driving force for this process likely derives from a tight S^-...H-N(2)$^+$ H-bond (H...S 1.94(7) Å).

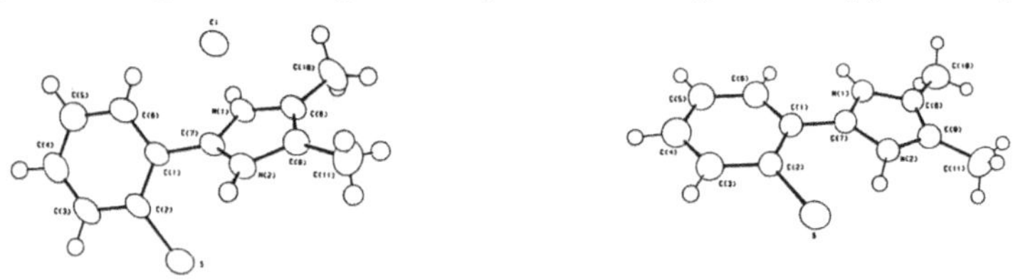

Fig. 1. Views of $C_{11}H_{13}N_2S^+,Cl^-$ (left) and $C_{11}H_{12}N_2S$.

3-ACETYL-4-AMINO-1-(p-CHLOROPHENYL)-5-IMINO-2-OXO-3-PYRROLINE
$C_{12}H_{10}ClN_3O_2$

M. BASATO, R. CAMPOSTRINI, B. CORAIN, B. LONGATO, S. SITRAN, A.C. VERONESE and G. VALLE, 1985. *J. Chem. Soc. Perkin Trans. 2*, 2019-2025.

Monoclinic, $P2_1/n$, a = 25.800, b = 7.385, c = 12.998 Å, β = 102.0°, Z = 8. Cu radiation, R = 0.092 for 1183

reflections.

The crystallographic asymmetric unit consists of two independent molecules A and B whose molecular geometries are very similar (Fig. 1). Apart from the (p-ClC_6H_4) moiety all other atoms are almost coplanar. Two different types of hydrogen bonds are observed in the structure: one involves the amide moiety of a molecule of A (or B) with the imino C=NH nitrogen of another molecule of A (or B); the other involves the amide NH group of a molecule of A (or B) and the acetyl oxygen atom of another independent molecule of B (or A) and vice versa. Atoms N(1)-O(1) are involved in an intramolecular hydrogen bond.

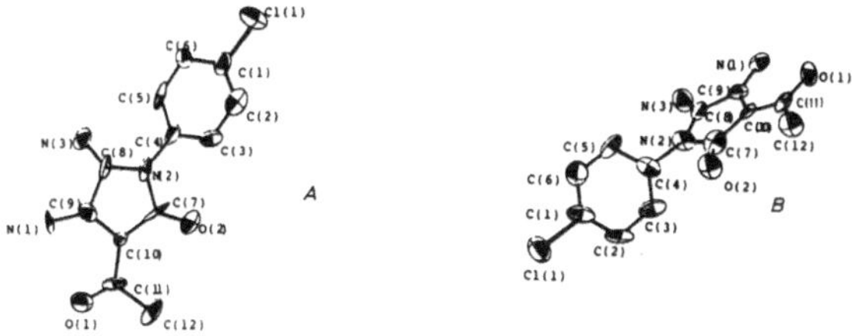

Fig. 1. ORTEP view of an asymmetric unit in the crystal of $C_{12}H_{10}ClN_3O_2$.

1-(4-BROMOBENZOYL)-3-METHYLTHIOPYRROLIDIN-2-ONE
$C_{12}H_{12}BrNO_2S$

H. TAKAHATA, T. YAMAZAKI and K. AOE, 1985. *J. Org. Chem.*, **50**, 4648-4650.

Monoclinic, $P2_1/c$, a = 15.4924, b = 5.8167, c = 7.058 Å, β = 103.774°, Z = 4. Cu radiation, R = 0.053 for 1799 reflections.

The crystal structure analysis established the stereochemistry of the molecule (Fig. 1).

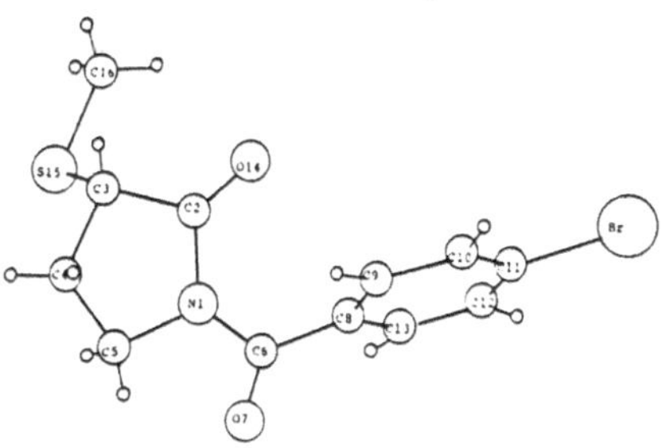

Fig. 1. The $C_{12}H_{12}BrNO_2S$ molecule.

1-ACETYL-3-METHYL-2-PHENYL-5-HYDROXYPYRAZOLIDINE
$C_{12}H_{16}N_2O_2$

A.V. DOVGILEVICH, K.N. ZELENIN, A.A. ESPENBETOV, Yu.T. STRUCHKOV, I.P. BEZHAN, L.A. SVIRIDOVA, G.A. GOLUBEVA, M.Yu. MALOV and Yu.G. BUNDEL, 1985. *Khim. Get. Soedin., SSSR*, 1242-1247.

Triclinic, $P\bar{1}$, a = 6.4936, b = 9.2607, c = 9.8671 Å, α = 87.029, β = 79.390, γ = 79.978°, Z = 2. Cu radiation, R = 0.068 for 1565 reflections.

The pyrazolidine ring is in a slightly distorted flattened envelope conformation (Fig. 1), with C(5) -0.520(3) Å from the N(1),N(2),C(3),C(4) plane. The OH group at C(5) is axial and the N(1)-N(2) bond length is 1.431(3) Å. Molecules are linked to form centrosymmetrical dimers by O(2)-H...O(1') hydrogen bonds (O(2)...O(1') 2.735(3) Å).

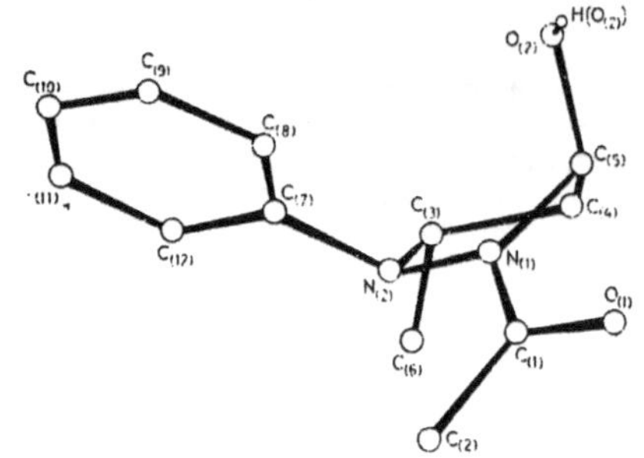

Fig. 1. A view of $C_{12}H_{16}N_2O_2$.

(RS-SR)-ETHYL 2,5-DIOXO-4-PHENYL-3-PYRROLIDINECARBOXYLATE
$C_{13}H_{13}NO_4$

H. ALPER, C.P. MAHATANTILA, F.W.B. EINSTEIN and A.C. WILLIS, 1985. *Acta Cryst.*, C41, 548-550.

Monoclinic, $P2_1/n$, a = 15.286, b = 5.377, c = 16.218 Å, β = 109.72°, Z = 4. Mo radiation, R = 0.046 for 1130 reflections.

The compound is a trans-substituted succinimide, hydrogen bonded into one dimensional chains (N...O 2.870(4) Å) running in the b direction. The five-membered ring (Fig. 1) has an envelope conformation with the phenyl-substituted C atom lying 0.224(3) Å out of the plane of the other four atoms.

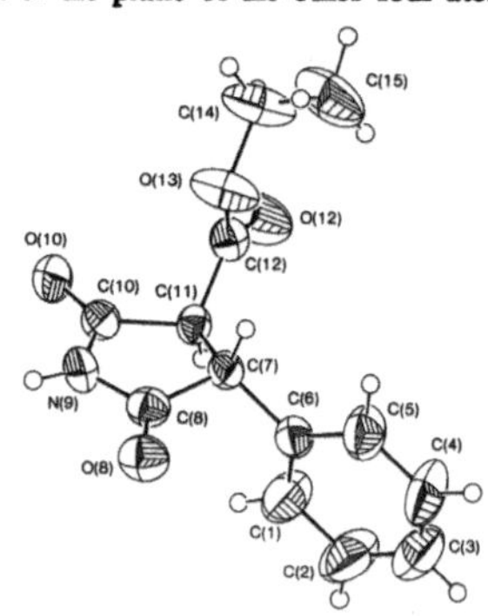

Fig. 1. $C_{13}H_{13}NO_4$: the molecular structure.

S-(-)-4-(α-METHYLBENZYL)-1-(2,3-DIBROMOPROPYL)URAZOLE
$C_{13}H_{15}Br_2N_3O_2$

S.E. MALLAKPOUR, G.B. BUTLER, H. AGHABOZORG and G.J. PALENIK, 1985. *Macromolecules*, 342-347.

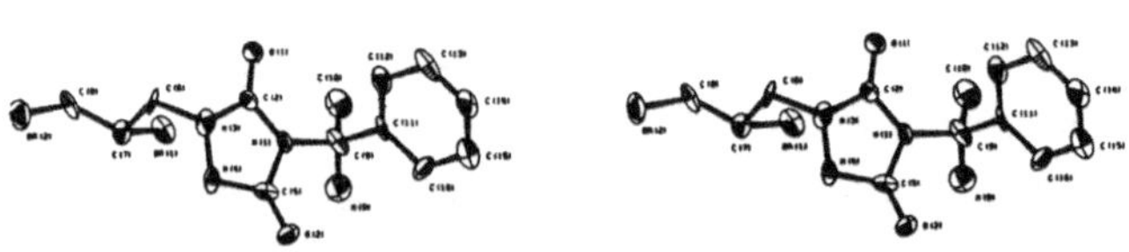

Fig. 1. A stereoview of $C_{13}H_{15}Br_2N_3O_2$.

Orthorhombic, $P2_12_12_1$, a = 5.985, b = 10.758, c = 23.063 Å, Z = 4. Mo radiation, R = 0.066 for 966 reflections.

A stereoview of the molecule showing the structure and absolute configuration is in Fig. 1. The five-membered triazoline ring is distorted from planarity towards and envelope conformation. The dihedral angle between the phenyl and the five-membered rings is 65.8°. Molecules are linked by N(4)-H)...O(1) hydrogen bonds (H...O(1) 2.01 Å).

1-(1-ADAMANTYL)PYRAZOLE
$C_{13}H_{18}N_2$ (I)

1-(3-HYDROXYADAMANTYL)-4-NITROPYRAZOLE
$C_{13}H_{17}N_3O_3$ (II)

P. CABILDO, R.M. CLARAMUNT, D. SANZ, M.C. FOCES-FOCES, F.H. CANO, J. CATALAN and J. ELGUERO, 1985. *Tetrahedron*, **41**, 473-478.

I. Monoclinic, $P2_1$, a = 7.4021, b = 10.7529, c = 6.9651 Å, β = 90.206°, Z = 2. Cu radiation, R = 0.043 for 951 reflections.

II. Monoclinic, $P2/n$, a = 31.1172, b = 6.8506, c = 12.0313 Å, β = 94.873°, Z = 8. Cu radiation, R = 0.061 for 2464 reflections.

There are two independent molecules in (II). In all three adamantane moieties of (I) and (II) the C-C-C angles are in the range 108.0-110.7(4)°; all six-membered rings (Fig. 1) have chair conformations. The pyrazole ring shows delocalization of the double bonds. Structurally, the main difference in the three molecules is that in (I) N(1)-C(5) and C(6)-C(11) bonds are eclipsed; in (II) one molecule has this same conformation and the other has N(1)-N(2) and C(6)-C(11) eclipsed.

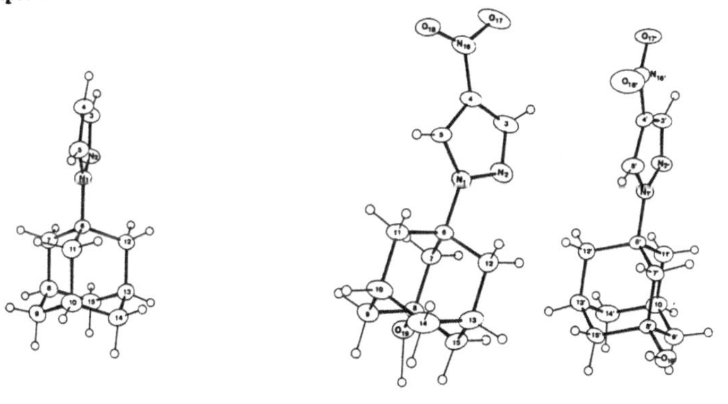

Fig. 1. The $C_{13}H_{18}N_2$ (I) molecule (left) and the two $C_{13}H_{17}N_3O_3$ (II) molecules.

4-ACETOACETYL-3-METHYL-1-PHENYLPYRAZOL-5-OL
$C_{14}H_{14}N_2O_3$ (I)

4-BENZOYL-3-METHYL-1-PHENYLPYRAZOL-5-ONE 4-BENZOYL-5-METHYL-2-PHENYLPYRAZOL-3(2H)-ONE
$C_{17}H_{14}N_2O_2$ (II)

M.J. O'CONNELL, C.G. RAMSAY and P.J. STEEL, 1985. *Aust. J. Chem.*, **38**, 401-409.

I. Triclinic, $P\bar{1}$, a = 7.142, b = 13.704, c = 14.699 Å, α = 117.36, β = 96.87, γ = 93.73°, Z = 4. Mo radiation, R = 0.049 for 2810 reflections (at 193 K).

II. Monoclinic, $P2_1/c$, a = 13.508, b = 9.124, c = 11.451 Å, β = 90.80°, Z = 4. Mo radiation, R = 0.048 for 771 reflections.

There are two crystallographically independent molecules in the asymmetric unit of (I). The structure (Fig. 1) is that of the OH tautomer. In one molecule the pyrazole ring mean plane is inclined at 12.1(4)° to the phenyl ring mean plane, while in the other this angle is 7.0(4)°. The molecules possess two intramolecular hydrogen bonds,

O(1)...O(2) 2.627(2), 2.633(2) and O(2)...O(3) 2.561(2), 2.565(2) Å. Consequently the acetoacetyl side chain lies in the same plane as the pyrazole ring. In the crystal, the molecules all lie in planes approximately perpendicular to the *a*-axis and stack in layers along the *a*-axis. Compound (II) exists as the NH tautomer. The phenyl rings are both planar, the N(1)-phenyl ring making an angle of 28.2(7)° with the pyrazole mean plane and the other phenyl ring making an angle of 59.3(6)° with this plane. The pyrazole ring is slightly puckered. Molecules related by the c-glide are strongly hydrogen bonded (O(1)...N(2) 2.656(7) Å) into infinite chains along the *c*-axis.

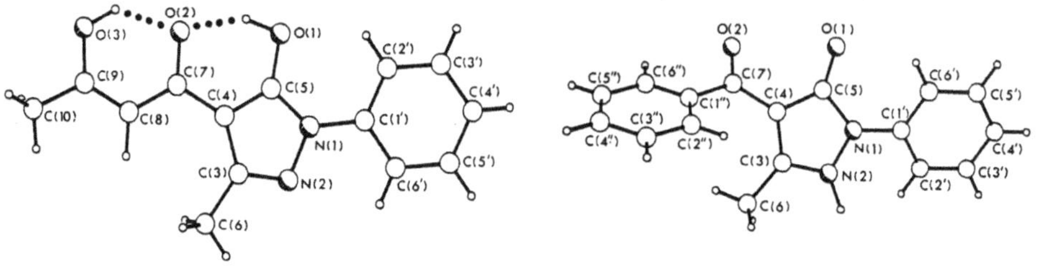

Fig. 1. The molecular structures: $C_{14}H_{14}N_2O_3$ (left) and $C_{17}H_{14}N_2O_2$ (right).

3-BENZYL-5-(METHOXYCARBONYLMETHYLTHIOMETHYL)HYDANTOIN
$C_{14}H_{16}N_2O_4S$

N.N. DHANESHWAR, S.S. TAVALE and T.N.G. ROW, 1985. *Acta Cryst.*, C41, 1320-1322.

Monoclinic, $P2_1$, a = 11.158, b = 6.010, c = 11.428 Å, β = 98.80°, Z = 2. Mo radiation, R = 0.035 for 1039 reflections.

The five-membered hydantoin ring in the molecule (Fig. 1) is planar and the substituent carbonyl O atoms lie within 0.055 Å of this plane. The aromatic ring is also planar and the interplanar dihedral angle is 89.3(7)°. Bond lengths and angles within the molecule are in good agreement with generally accepted values. There is a hydrogen bond, N(1)-H...O(2), linking the molecules along the *b*-axis and in which the N...O separation is 2.89(1) Å.

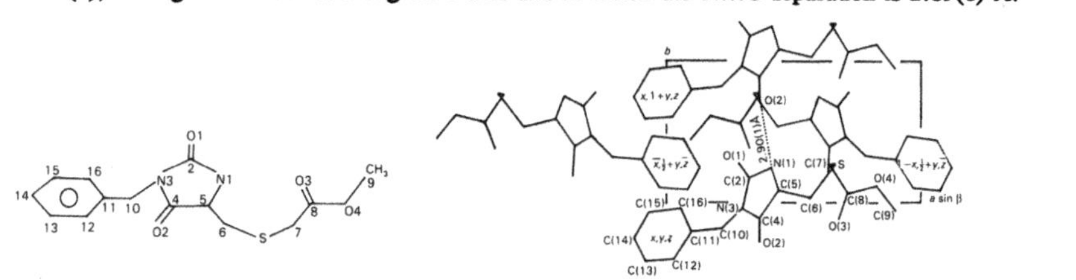

Fig. 1. $C_{14}H_{16}N_2O_4S$: structural formula and the (001) projection.

METHYL 6-(4-FORMYL-2,5-DIMETHYLPYRROL-3-YL)-3-METHOXYCARBONYL-4,5-DIAZAHEXA-3,5-DIENOATE
$C_{14}H_{17}N_3O_5$

T.H. DE LA F. GOMEZ, J.S. ARQUES, R.A. JONES, H.M. DAWES and M.B. HURSTHOUSE, 1985. *J. Chem. Soc. Perkin Trans. 1*, 899-902.

Monoclinic, $P2_1/c$, a = 14.069, b = 14.370, c = 8.018 Å, β = 97.15°, Z = 4. Mo radiation, R = 0.055 for 1544 reflections.

The crystal structure analysis established that the molecule was the title compound (Fig. 1).

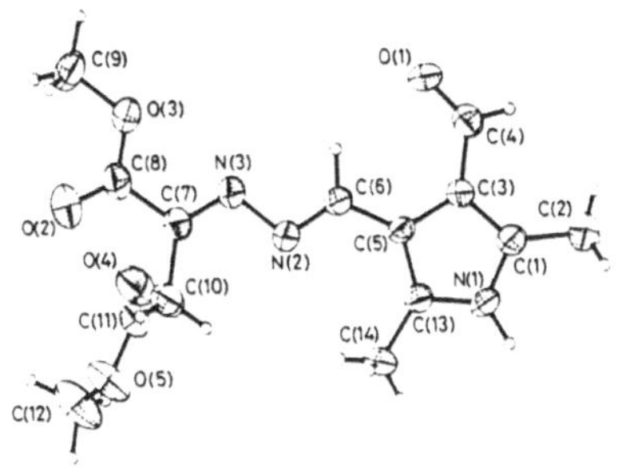

Fig. 1. Molecular structure of $C_{14}H_{17}N_3O_5$.

DIETHYL 1-(p-METHOXYPHENYLSULPHONYL)-2,2-AZIRIDINEDICARBOXYLATE
$C_{15}H_{19}NO_7S$

S. BRUCKNER, L. MALPEZZI, A.V. PROSYANIK and S.V. BONDARENKO, 1985. *Acta Cryst.*, **C41**, 219-222.

Monoclinic, $P2_1/c$, a = 17.124, b = 10.364, c = 10.144 Å, β = 106.41°, Z = 4. Mo radiation, R = 0.033 for 1347 reflections.

The three-membered heterocycle (Fig. 1) is a regular triangle with a mean side of 1.484(5) Å. The N atom is pyramidal and lies 0.700(3) Å out of the plane of its three substituents. The S-N bond is 1.675(4) Å and makes an angle of 122.4(3)° with the aziridine plane. The S-O(5) and S-O(6) bond distances are 1.434(6) and 1.426(3) Å respectively. The plane defined by C(1)-C(4)-C(2) is approximately perpendicular to the ring plane. The two ethoxycarbonyl groups are planar with an angle of 83.1(2)° between the planes. Crystal packing is governed by an extended network of C-H...O bonds.

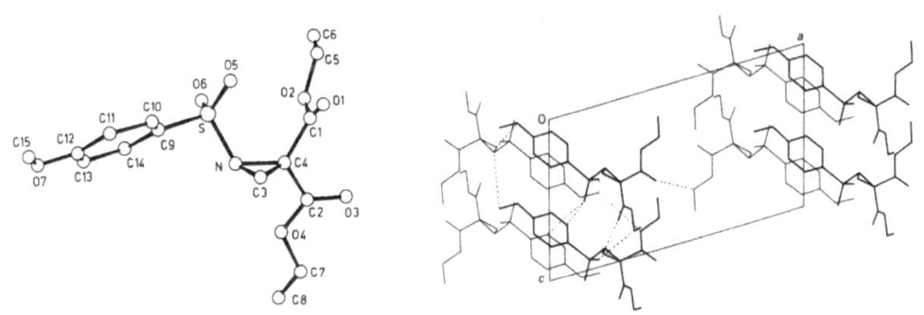

Fig. 1. $C_{15}H_{19}NO_7S$: a view of the molecule and crystal packing projected along the b-axis.

2,2',3,3',4,4',5,5'-OCTAKIS(TRIFLUOROMETHYLTHIO)-2,2'-BI-2H-PYRROLE
$C_{16}F_{24}N_2S_8$

D.J. BRAUER, M.R.C. GERSTENBERGER and A. HAAS, 1985. *Acta Cryst.*, **C41**, 579-581.

Monoclinic, $P2_1/c$, a = 8.285, b = 16.656, c = 11.431 Å, β = 98.86°, Z = 2. Mo radiation, R = 0.075 for 1162 reflections.

The crystals contain centrosymmetric dimeric molecules, the five-membered rings being joined by a C-C bond of 1.51(2) Å (Fig. 1). The average S-C(sp^2) bond length is 1.76(2) and the average S-C(sp^3) bond length is 1.86(1) Å. There is a 10° variation in S-C-F angles. The CF_3 groups are rotationally disordered. The (F_3)C-S-C fragments are approximately perpendicular to their respective rings and the SCF_3 moieties are oriented such that the hollow regions of the molecule are filled.

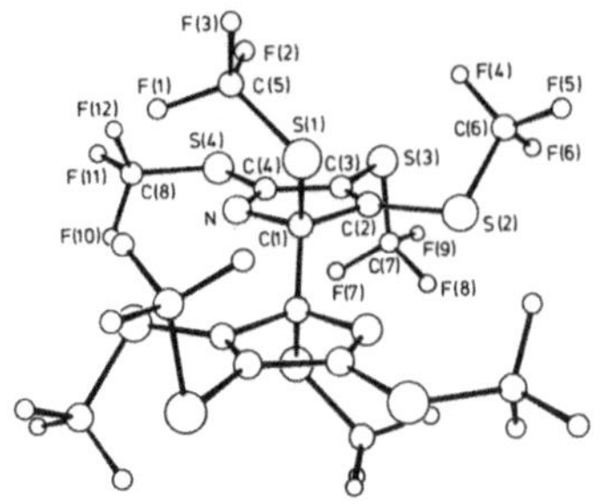

Fig. 1. $C_{16}F_{24}N_2S_8$: a drawing of the molecule.

BICYCLO[4.4.1]UNDECA-4,6,8,10-TETRAENE-2,3-DIONE-2-((2-CHLORO-3-PYRIDINYL)HYDRAZONE)
$C_{16}H_{12}ClN_3O$

R. NEIDLEIN, R. GOTTFRIED, A. GIEREN, C.-P. KAERLEIN and T. HUBNER, 1985. *Z. Naturforsch.*, **B40**, 193-198.

Orthorhombic, *F2dd*, a = 6.916, b = 23.727, c = 33.960 Å, Z = 16. Mo radiation, R = 0.031 for 1992 reflections.

The analysis determined the conformation and dimensions given in Fig. 1.

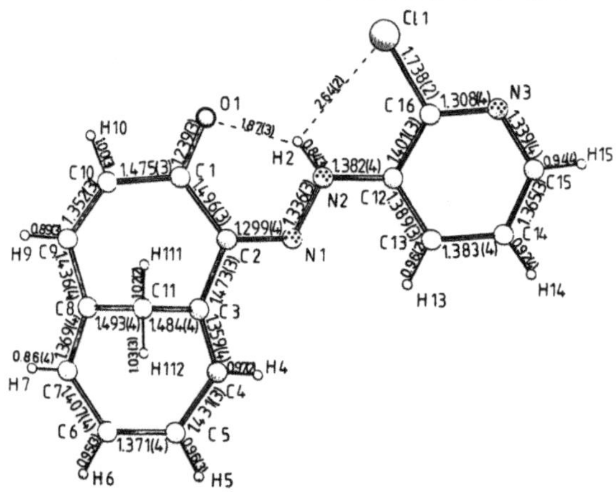

Fig. 1. The $C_{16}H_{12}ClN_3O$ molecule and bond lengths.

BENZYL 4-ETHYL-3,5-DIMETHYLPYRROLE-2-CARBOXYLATE
$C_{16}H_{19}NO_2$ (I)

METHYL E-3-(3-ETHYL-4,5-DIMETHYLPYRROL-2-YL)-2-CYANOPROPENOATE
$C_{13}H_{16}N_2O_2$ (II)

METHYL E-3-(5-(BENZYLOXYCARBONYL)-3-ETHYL-4-METHYLPYRROL-2-YL)-2-CYANOPROPENOATE
$C_{20}H_{20}N_2O_4$ (III)

J.B. PAINE III, D. DOLPHIN, J. TROTTER and T.J. GREENHOUGH, 1985. *Canad. J. Chem.*, **63**, 2683-2690.

I. Monoclinic, $P2_1/c$, a = 14.934, b = 6.674, c = 15.269 Å, β = 101.96°, Z = 4. Mo radiation, R = 0.052 for 862 reflections.

II. Monoclinic, $P2_1/n$, a = 7.303, b = 13.478, c = 12.985 Å, β = 97.48°, Z = 4. Cu radiation, R = 0.040 for 2032 reflections.

III. Monoclinic, $P2_1/c$, a = 11.157, b = 13.109, c = 14.068 Å, β = 115.47°, Z = 4. Mo radiation, R = 0.038 for 1483 reflections.

The molecules are approximately planar (Fig. 1) except for deviations of the phenyl rings and the terminal atoms of the ethyl group bound to C(3). The bond lengths (Fig. 2) show the effect of the strongly electron-withdrawing substituents.

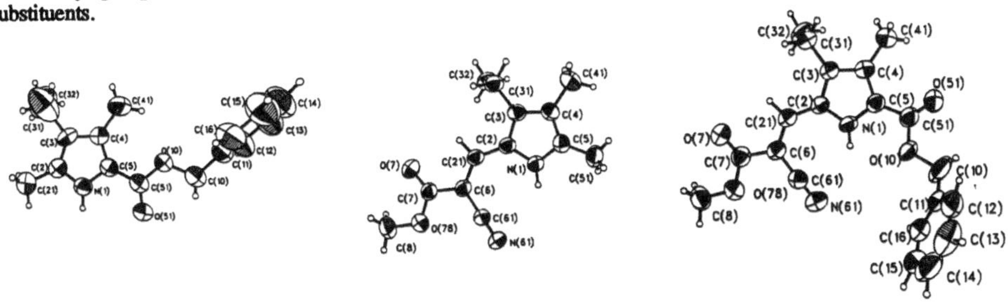

Fig. 1. The $C_{16}H_{19}NO_2$ (I), $C_{13}H_{16}N_2O_2$ (II) and $C_{20}H_{20}N_2O_4$ (III) molecules (left to right).

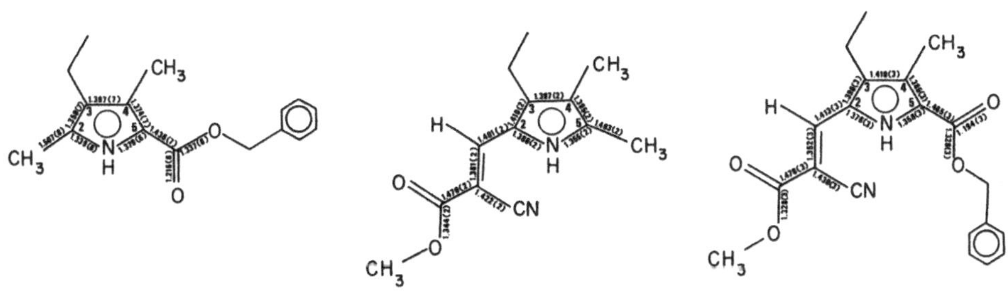

Fig. 2. Principal bond lengths in pyrroles (I)-(III) (left to right).

2-BENZOYL-5-PHENYLPYRROLE
$C_{17}H_{13}NO$

V.G. KHARCHENKO, S.N. CHALAYA, Yu.T. STRUCHKOV, A.A. ESPENBETOV, O.V. LITVINOV and N.T. KOMYAGIN, 1984. *Khim. Get. Soedin. SSSR,* 1355-1358.

Triclinic, $P\bar{1}$, a = 9.672, b = 10.061, c = 14.398 Å, α = 103.58, β = 93.31, γ = 107.24°, Z = 4. Mo radiation, R = 0.0419 for 2220 reflections.

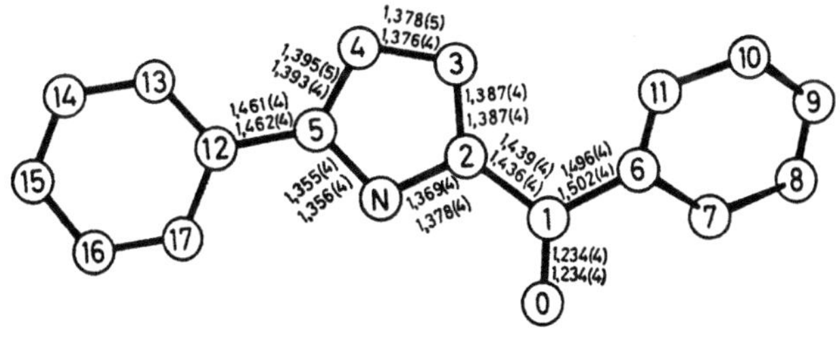

Fig. 1. A $C_{17}H_{13}NO$ molecule and bond lengths for the two independent molecules.

The asymmetric unit contains two independent molecules (Fig. 1) linked to form dimers via N-H...O hydrogen bonds (2.940(3) Å). The only significant difference between the two molecules is the phenyl ring orientation of the benzoyl group which forms a dihedral angle of 43.8° in one molecule and 50.0° in the other, with the pyrrole ring plane.

(E)-3-(2-METHOXYPHENYL)-1-PHENYL-3-(1H-1,2,4-TRIAZOL-1-YL)PROP-2-EN-1-ONE
$C_{18}H_{15}N_3O_2$

S.K. BRANCH and I.W. NOWELL, 1985. *Acta Cryst.*, **C41**, 769-771.

Orthorhombic, $P2_12_12_1$, a = 15.020, b = 14.537, c = 7.169 Å, Z = 4. Mo radiation, R = 0.064 for 1092 reflections.

The molecular conformation (Fig. 1) is such that the carbonyl group and the unsaturated C(4)=C(5) bond are cis-orientated and nearly coplanar. The dihedral angle between the planar groups of atoms, C(5,6,71),O(1) and N(1),C(31,4,5,6), is 23(1)°. The methoxyphenyl ring is twisted by 59(1)° out of the plane containing the carbon-carbon double bond while the triazolyl ring, which shows significant localization, lies almost in that plane. The unsubstituted phenyl ring and the carbonyl group are almost coplanar. The exocyclic angles of the triazolyl ring are very asymmetric with C-N-C (129.9(6)°) significantly larger than C-N-N (122.4(6)°).

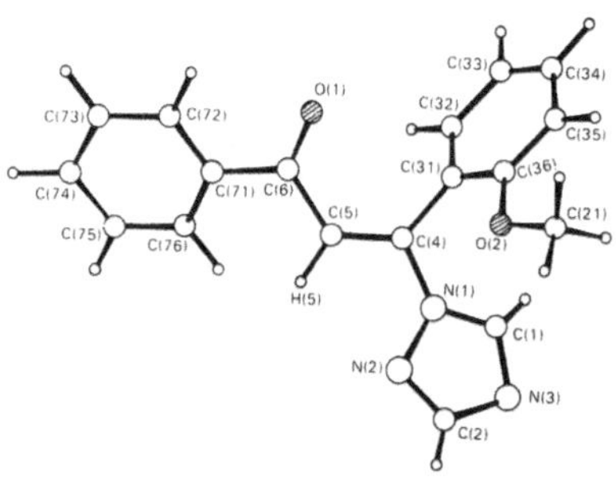

Fig. 1. $C_{18}H_{15}N_3O_2$: a view perpendicular to the N(1),C(31,4,5,6) plane showing the molecular conformation.

3-METHYL-4-(2',4'-DIMETHYLPHENYLHYDRAZONO)-1-PHENYL-5-PYRAZOLONE
$C_{18}H_{18}N_4O$

L.G. KUZ'MINA, L.P. GIGOR'EVA, Yu.T. STRUCHKOV, Z.I. EZHKOVA, B.E. ZAITSEV, V.A. ZAITSEVA and P.P. PRON'KIN, 1985. *Khim. Get. Soedin., SSSR*, 816-821.

Monoclinic, $P2_1/n$, a = 11.254, b = 19.199, c = 7.4762 Å, β = 98.21°, Z = 4. Cu radiation, R = 0.075 for 1781 reflections.

The molecule (Fig. 1) exists as the hydrazone tautomer with a planar pyrazole ring. The dimensions are consistent with substantial delocalization of the π-electron density over the whole azopyrazolone fragment. There is an intramolecular N(4)-H...O(1) hydrogen bond H...O 2.14 Å. The C(11)-C(16) and C(21)-C(26) phenyl rings are rotated (9.4 and 9.2° respectively) out of the pyrazolone plane.

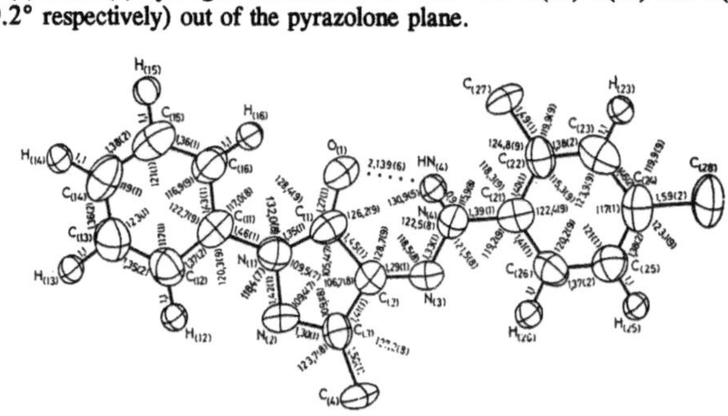

Fig. 1. The $C_{18}H_{18}N_4O$ molecule with bond lengths and angles.

4-CINNAMOYL-5-METHYL-2-PHENYLIMIDAZOLE
$C_{19}H_{16}N_2O$

A.C. CODA, G. DESIMONI and A. CODA, 1985. *Heterocycles*, **23**, 1893-1896.

Monoclinic, $P2_1/c$, a = 9.738, b = 16.497, c = 9.671 Å, β = 96.63°, Z = 4. Mo radiation, R = 0.0388 for 1970 reflections.

The vinyl imidazolyl ketone system (Fig. 1) is largely planar but both phenyl rings are rotated out of the plane. The molecules are linked by N-H...N hydrogen bonds N...N 2.976(2) Å.

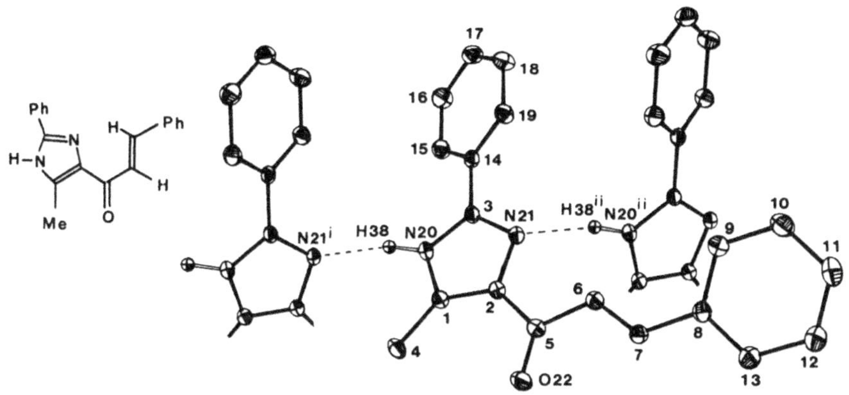

Fig. 1. The $C_{19}H_{16}N_2O$ system.

(E)-4,4-DIMETHYL-1-(1-NAPHTHYL)-1-(1H-1,2,4-TRIAZOL-1-YL)PENT-1-EN-3-ONE
$C_{19}H_{19}N_3O$

S.K. BRANCH and I.W. NOWELL, 1985. *Acta Cryst.*, **C41**, 594-597.

Monoclinic, $P2_1/c$, a = 18.971, b = 16.763, c = 10.902 Å, β = 104.64°, Z = 8. Mo radiation, R = 0.0546 for 1268 reflections.

The two crystallographically independent molecules in the asymmetric unit adopt similar conformations such that the CO group and the C(4)=C(5) bond are cis-oriented and nearly coplanar (Fig. 1). The naphthyl ring is twisted by 75.8(10) and 72.7(10)° respectively, in the two molecules, out of the plane containing the C=C double bond. The triazolyl ring lies almost in that plane. This conformation allows extensive conjugation between the α,β-unsaturated ketone and the heterocyclic group.

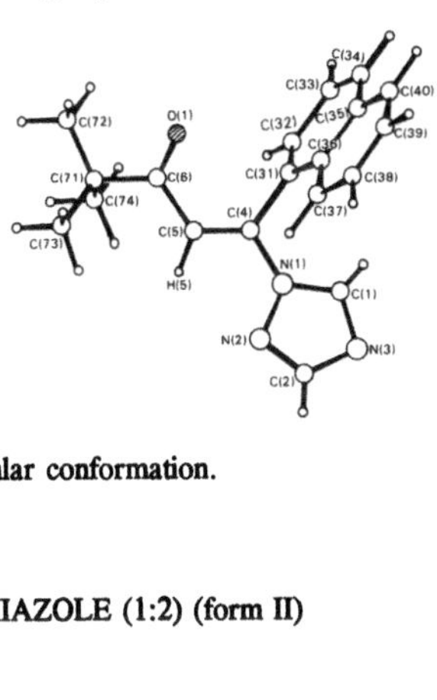

Fig. 1. $C_{19}H_{19}N_3O$: the molecular conformation.

PHENYLBUTAZONE SULPHATHIAZOLE (1:2) (form II)
$C_{19}H_{20}N_2O_2,2(C_9H_9N_3O_2S)$

B. JEANJEAN, J. RAMBAUD, S. ALBEROLA, J.P. DECLERCQ and A. DOUBOURG, 1985. *Bull. Soc. Chim. Fr.*, 611-616.

Triclinic, $P\bar{1}$, a = 19.665, b = 10.706, c = 10.413 Å, α = 77.62, β = 80.08, γ = 82.24°, Z = 2. Mo radiation, R = 0.074 for 4179 reflections.

A view of this polymorph is in Fig. 1. The n-butyl chain of the phenylbutazone is disordered over three orientations. Molecular dimensions are normal. The three molecules in the asymmetric unit are linked only by van der Waals interactions; N-H...O and N-H...N hydrogen bonds
link the molecules in the crystal lattice.

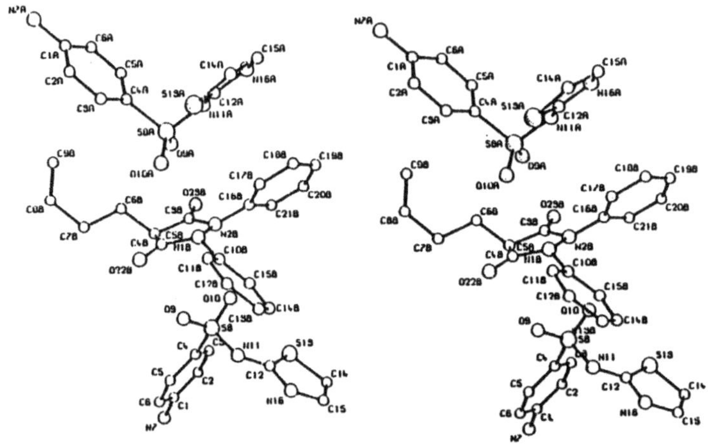

Fig. 1. A stereoview of the phenylbutazone sulphathiazole (1:2) complex.

3-ACETYL-2-ETHOXYCARBONYL-4-(3-OXOBUTYLTHIO)-5-PHENYLPYRROLE
$C_{19}H_{21}NO_4S$

C.T. GOKOU, J.-P. PRADERE, H. QUINIOU and L. TOUPET, 1985. *J. Chem. Soc. Perkin Trans. 1*, 1875-1878.

Monoclinic, $P2_1/n$, a = 10.071, b = 10.801, c = 17.001 Å, β = 93.56°, Z = 4. Mo radiation, R = 0.030 for 2525 reflections.

The crystal structure analysis established the identity of the molecule (Fig. 1). The C(2)-S bond length is 1.760(2) Å; that for S-C(10) is 1.792(4) Å and the C(2)-S-C(10) angle is 104.3(1)°.

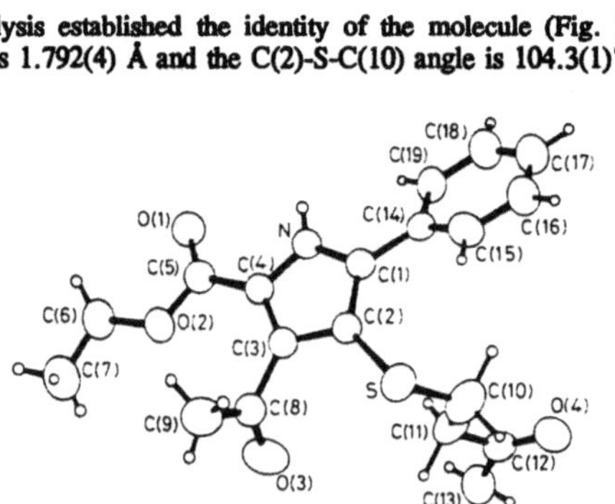

Fig. 1. General view of a molecule of $C_{19}H_{21}NO_4S$.

1-METHYL-3-((2-METHOXYETHYL)CARBAMOYL)-N-PHENYL-4-(N-PROPYLCARBAMOYL)PYRAZOLE-5-CARBOXAMIDE

$C_{19}H_{25}N_5O_4$

D. CHAMBERS, W.A. DENNY, J.S. BUCKLETON and G.R. CLARK, 1985. *J. Org. Chem.*, **50**, 4736-4738.

Monoclinic, $P2_1/n$, a = 15.271, b = 17.277, c = 15.473 Å, β = 99.23°, Z = 8. Mo radiation, R = 0.103 for 1452 reflections.

There are two independent molecules in the crystallographic asymmetric unit. The crystal structure analysis served to identify which isomer was the molecule (Fig. 1).

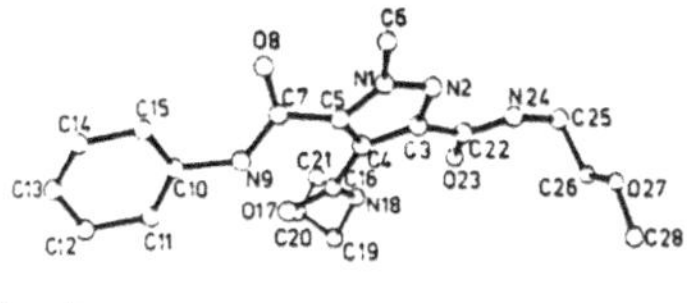

Fig. 1. The $C_{19}H_{25}N_5O_4$ molecule.

METHYL 1-(3-CHLOROBENZOYLAMINO)-4-DIETHYLAMINOCARBONYL-2,5-DIMETHYL-1H-PYRROLE-3-CARBOXYLATE
$C_{20}H_{24}ClN_3O_4$

G. GIUSEPPETTI, C. TADINI, O. ATTANASI, M. GROSSI and F. SERRA-ZANETTI, 1985. *Acta Cryst.*, **C41**, 450-453.

Monoclinic, $C2/c$, a = 11.050, b = 26.041, c = 14.413 Å, β = 90.84°, Z = 8. Mo radiation, R = 0.032 for 1477 reflections.

In the molecule (Fig. 1) the five-membered heteroring is planar as is the benzene ring. The dihedral angle between their least-squares planes is 64.7(1)°. The Cl and carbonyl substituents are displaced from the benzene mean plane by 0.024(1) and 0.034(3) Å respectively. The bond lengths and angles in the molecule are normal. In the crystal structure two homologous hydrogen bonds N(1)-H...O(4) of length 2.788(3) Å join the molecules into dimers thus forming an unusual fourteen-membered pseudo ring. There is one other slightly short intradimer possible hydrogen-bonding interaction, C(4)-H...O(4), of length 3.259(4) Å.

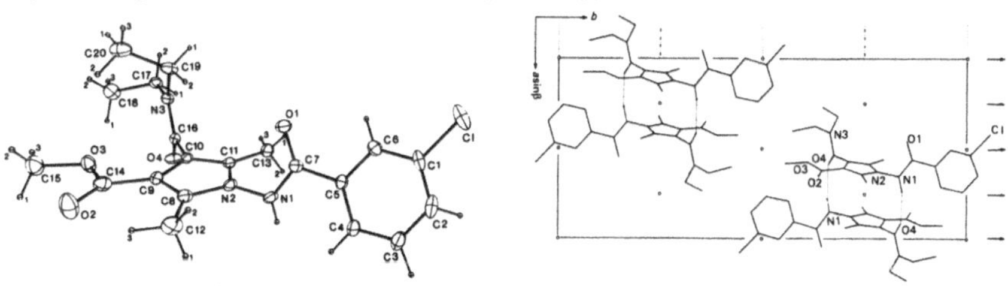

Fig. 1. $C_{20}H_{24}ClN_3O_4$: a perspective view of the molecule (left) and (right) a projection down the c-axis of four of the molecules in the unit cell showing the hydrogen bonds in the dimers as broken lines.

2-METHOXYCARBONYL-N-TRITYLAZIRIDINE
$C_{23}H_{21}NO_2$ (I)

2-METHOXYCARBONYL-N-TRITYLPYRROLIDINE (N-TRITYLPROLINE)
$C_{25}H_{25}NO_2$ (II)

A.F. MISHNEV, M.F. BUNDULE, Ya.Ya. BLEIDELIS, A.V. EREMEEV and F.D. POLYAK, 1985. *Zh. Strukt. Khim.*, **26**, 1-159-162 [J. Struct. Chem., **26**, 139-142].

I. Orthorhombic, $P2_12_12_1$, a = 9.863, b = 10.866, c = 17.749 Å, Z = 4. Cu radiation, R = 0.055 for 1471 reflections.

II. Monoclinic, $P2_1$, a = 9.614, b = 10.047, c = 10.598 Å, β = 101.56°, Z = 2. Cu radiation, R = 0.048 for

1582 reflections.

The molecules (I) and (II) are shown in Figs. 1 and 2, respectively. The relative orientations of the phenyl rings in (I) and (II) differ from one another by 5-7° in corresponding dihedral angles. The proline residue in (II) has an unusual conformation (ϕ -108.6°), the pyrrolidine ring having a C(3)-envelope conformation.

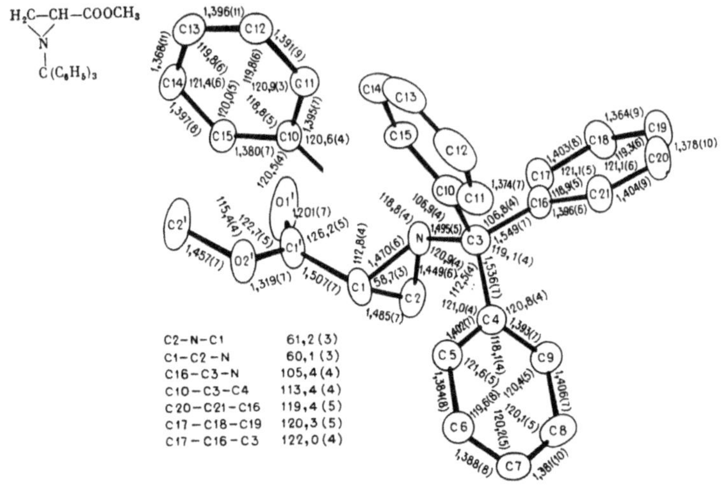

Fig. 1. Molecular geometry of $C_{23}H_{21}NO_2$ (I).

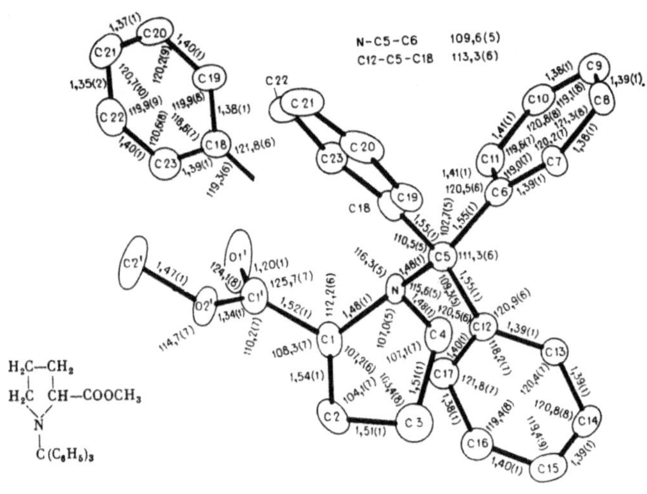

Fig. 2. Molecular geometry of $C_{25}H_{25}NO_2$ (II).

3,4,4,5-TETRAMETHYL-1-(p-METHOXYBENZOYL)-5-(p-METHOXYBENZOYLAZO)-2-PYRAZOLINE $C_{23}H_{26}N_4O_4$

J. STEPHANIDOU-STEPHANATOU, N.E. ALEXANDROU, G. SALEM and A. TERZIS, 1985. *J. Heterocycl. Chem.*, 22, 673-675.

Monoclinic, *C2/c*, a = 32.392, b = 8.927, c = 16.424 Å, β = 108.30°, Z = 8. Mo radiation, R = 0.0443 for

2590 reflections.

The analysis confirms the proposed pyrazoline structure proposed for this oxidation product. The 1-p-methoxybenzoyl group is disordered slightly over two orientations as shown in Fig. 1. The five-membered ring is in envelope form with the carbon atom bearing the p-methoxybenzoylazo group being 0.434 Å from the plane of the other four ring atoms.

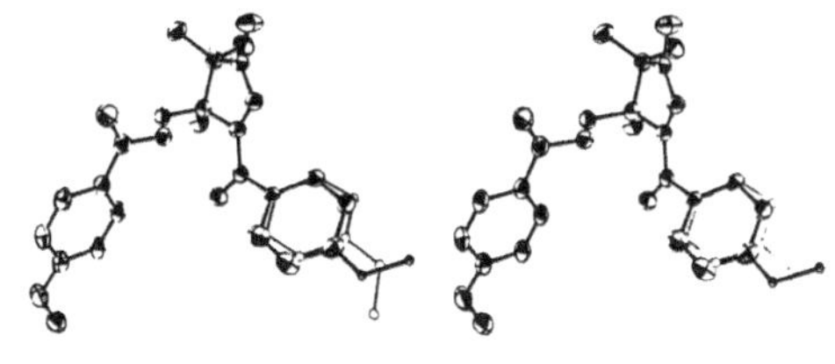

Fig. 1. A stereoview of $C_{23}H_{26}N_4O_4$ showing the disorder.

1-ISOPROPYL-2,4-DIPHENYL-5-PIVALOYL-4,5-DIHYDROIMIDAZOLE
$C_{23}H_{28}N_2O$

P. BROUANT, J. BARBE, A. MARSURA and C. LUU-DUC, 1985. *Acta Cryst.*, **C41**, 1369-1372.

Monoclinic, *P*2$_1$/*c*, a = 10.931, b = 13.989, c = 13.849 Å, β = 106.0°, Z = 4. Mo radiation, R = 0.078 for 1516 reflections.

The dihydroimidazole ring in the molecule (Fig. 1) is slightly puckered with C(3) displaced 0.18(1) Å out of the least-squares plane of the other four ring atoms. All substituents are in a trans configuration with respect to each other and have orientations such that the mutual steric hindrances are minimised. The plane of the phenyl group at C(1) is twisted by 35° with respect to that of the heteroring. The methyl groups C(22) and C(23) are in a staggered conformation with torsion angles C(22)-C(21)-N(2)-C(3) and C(23)-C(21)-N(2)-C(3) having values of 56.2(9) and 71.2(9)° respectively. The conformation of the pivaloyl group around the C(3)-C(16) bond is cisoid with respect to the N(2) lone pair.

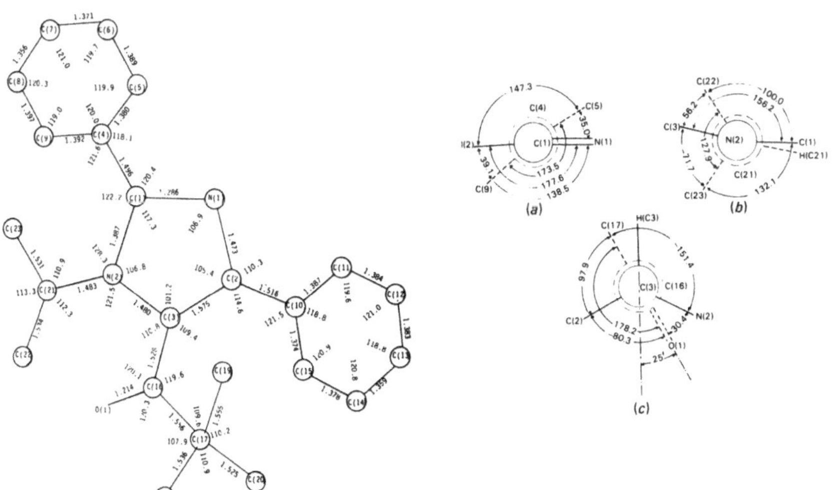

Fig. 1. $C_{23}H_{28}N_2O$: bond lengths and angles in the molecule and Newman projections along (a) the C(4)-C(1) bond, (b) the N(2)-C(21) bond and (c) the C(3)-C(16) bond.

4,5-BIS(4-METHOXYPHENYL)-2-(4-NITROPHENYL)IMIDAZOLIUM ACETATE DIHYDRATE

$C_{25}H_{23}N_3O_6,2(H_2O)$ $C_{23}H_{20}N_3O_4^+,C_2H_3O_2^-,2(H_2O)$

Y. INOUYE and Y. SAKAINO, 1985. *Acta Cryst.*, **C41**, 1106-1108.

Monoclinic, *C2/c*, a = 39.306, b = 6.974, c = 21.291 Å, β = 121.83°, Z = 8. Cu radiation, R = 0.046 for 1700 reflections.

In the molecule (Fig. 1) the imidazole ring is planar, its nitrogen atom N(11) is protonated by the acetic acid proton and it is nearly coplanar with the benzene ring with the nitro substituent. The dihedral angles between its plane and those of the other two benzene rings are 39.9 and 139.4°. Bond lengths and angles in the molecule are in agreement with generally accepted values. In the crystal two centrosymmetrically related imidazole rings are linked to each other through four water molecules at the N(14)-H sites. The N-H...O hydrogen bonds are of length 2.630(7) and 2.751(6) Å for O(34) and O(35) respectively. The O-H...O bond lengths range from 2.716(6) to 2.820(6) Å. The molecules are packed into sheets parallel to (010).

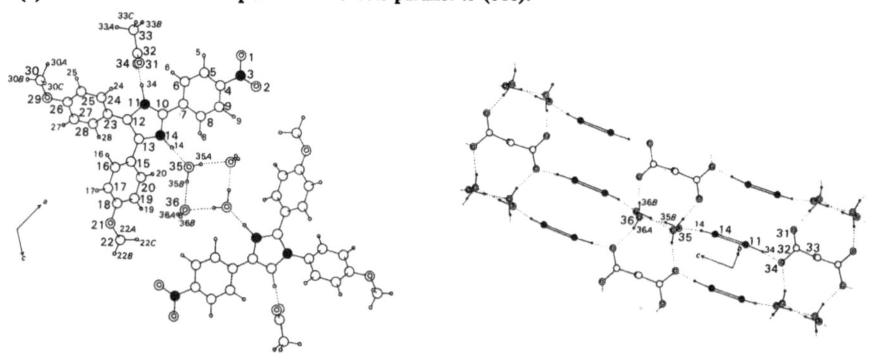

Fig. 1. $C_{25}H_{23}N_3O_6,2(H_2O)$: the molecular structure projected onto (010) and a view of the three-dimensional crystal packing with hydrogen bonds shown as dotted lines and with double circles and filled circles representing O and N atoms, respectively.

(Z)-rac-4,5-DIHYDRO-2-(2-PHENYL-2-(4-BENZYLOXYPHENYL)CYCLOPROPYL)-1H-IMIDAZOLE DIETHYL ETHER SOLVATE

$C_{25}H_{24}N_2O,0.15(C_4H_{10}O)$

J.W. TILLEY, J.W. CLADER, M. WIRKUS and J.F. BLOUNT, 1985. *J. Org. Chem.*, **50**, 2220-2224.

Monoclinic, *P2₁/a*, a = 10.203, b = 11.216, c = 17.950 Å, β = 94.30°, Z = 4. Cu radiation, R = 0.036 for 1201 reflections.

The crystal structure analysis established that the molecule (Fig. 1) was the Z isomer.

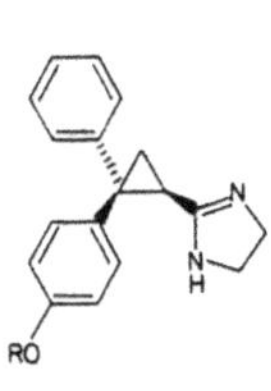

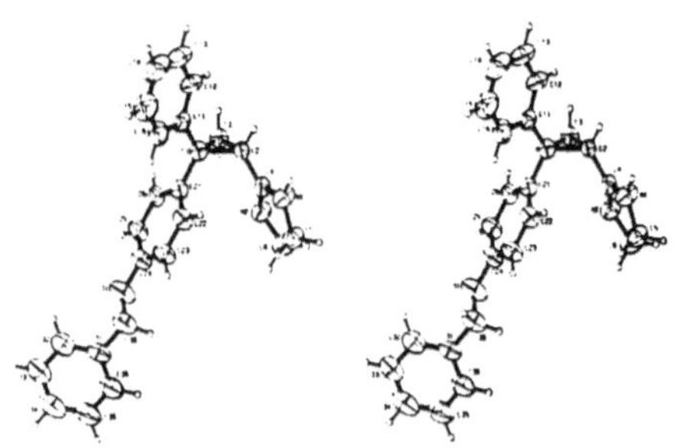

Fig. 1. A stereoview of $C_{25}H_{24}N_2O$.

1,3-DIBENZYL-2-(4,4-DIMETHYL-2,5-BISTHIOXOCYCLOHEXYLIDENE)IMIDAZOLINE
$C_{25}H_{28}N_2S_2$

J. SANDSTROM, K. STENVALL, N. SEN and K. VENKATESAN, 1985. *J. Chem. Soc. Perkin Trans. 2*, 1939-1942.

Monoclinic, $P2_1/c$, a = 11.685, b = 9.892, c = 20.864 Å, β = 102.15°, Z = 4. Mo radiation, R = 0.030 for 1067 reflections.

There is a twist of 80.8(5)° about the inter-ring bond (Fig. 1) which has a length of 1.482(6) Å.

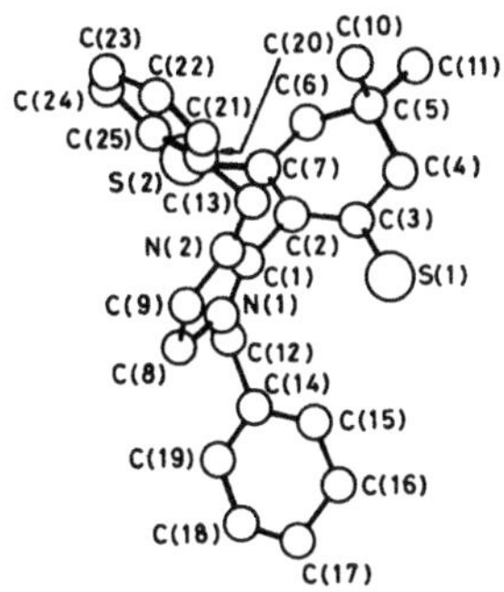

Fig. 1. View of a molecule of $C_{25}H_{28}N_2S_2$ down the crystallographic b axis.

2-(3-(4-(3-CHLOROPHENYL)-1-PIPERAZINYL)PROPYL)-5-ETHYL-2,4-DIHYDRO-4-(2-PHENOXYETHYL)-3H-1,2,4-TRIAZOLIN-3-ONE HYDROCHLORIDE
$C_{25}H_{33}Cl_2N_5O_2$ $C_{25}H_{33}ClN_5O_2^+,Cl^-$

G.D. MADDING, D.W. SMITH, R.L. SHELDON and B. LEE, 1985. *J. Heterocycl. Chem.*, **22**, 1121-1126.

Monoclinic, $P2_1/c$, a = 18.894, b = 11.202, c = 12.293 Å, β = 100.14° Z = 4. Mo radiation, R = 0.039 for 2922 reflections.

The triazole, phenoxy and chlorophenyl moieties are essentially planar (Fig. 1) while the piperazine has a chair conformation. Nitrogen N(20) carries the extra H atom and is positively charged.

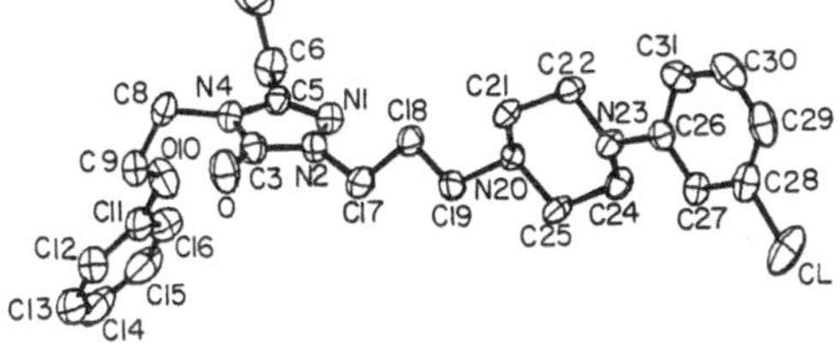

Fig. 1. A view of $C_{25}H_{33}ClN_5O_2^+$.

(3R)-1-((1'R)-(p-BENZYLOXYPHENYL)BENZYLOXYCARBONYLMETHYL)-3-ISOPROPENYLAZETIDIN-2-ONE
$C_{28}H_{27}NO_4$

S.T. HODGSON, D.M. HOLLINSHEAD, S.V. LEY, C.M.R. LOW and D.J. WILLIAMS, 1985. *J. Chem. Soc. Perkin Trans. 1*, 2375-2381.

Monoclinic, $P2_1$, a = 5.929, b = 7.985, c = 50.712 Å, β = 91.31°, Z = 4. Cu radiation, R = 0.051 for 2797 reflections.

There are two independent molecules in the crystallographic asymmetric unit and these have markedly different conformations (Fig. 1). The lactam and phenyl rings are similarly oriented with respect to each other in both molecules, but the propenyl group and the benzyl ester carbonyl have different orientations to the lactam ring in the two molecules. The terminal benzyl ring is also differently orientated.

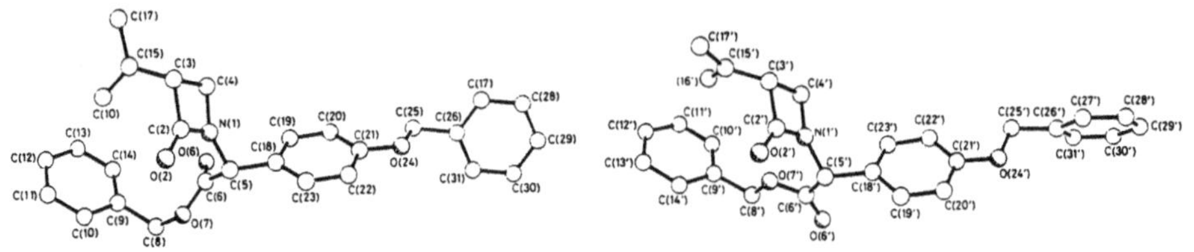

Fig. 1. The conformations of the two independent molecules in the crystal of $C_{28}H_{27}NO_4$.

9-(3-METHOXY-5-OXO-4-PHENYL-4,5-DIHYDRO-1,2,4-TRIAZOLYL)-3,7,15,18-TETRAMETHYLHEXACYCLO[9.6.1.$0^{2,10}$.$0^{3,8}$.$0^{4,17}$.$0^{14,18}$]OCTADECA-7,12,14-TRIENE-6,16-DIONE CYCLOHEXANE SOLVATE $C_{31}H_{31}N_3O_4,0.5(C_6H_{12})$

S.J. FOSTER, C.W. REES and D.J. WILLIAMS, 1985. *J. Chem. Soc. Perkin Trans. 1*, 711-717.

Triclinic, $P\bar{1}$, a = 8.927, b = 12.265, c = 15.897 Å, α = 87.57, β = 77.49, γ = 71.35°, Z = 2. Cu radiation, R = 0.086 for 2813 reflections.

The crystal structure analysis established the identity of the molecule (Fig. 1). The phenyl ring is rotated 56° from the plane of the triazole ring; the triazole ring and both its oxygen and methoxy substituents are coplanar. The folded conformation of the hexacyclic system creates a U-shaped cavity with near-parallel sides; this leads to some notable juxtapositions of atoms on opposite faces.

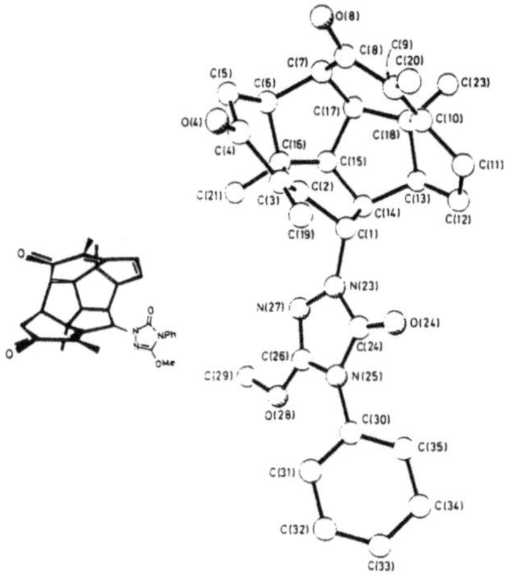

Fig. 1. Molecular structure of $C_{31}H_{31}N_3O_4$.

(1R,2S,3R,5R)-1-(1'-METHYLPYRROLIDIN-2'-ONE)-2-HYDROXY-2-PHENYL-3,5-DIBENZOYLCYCLOHEXANE $C_{31}H_{31}NO_4$ (I)

(1R,2S,3R,5S)-1-(1'-METHYLPYRROLIDIN-2'-ONE)-2-HYDROXY-2-PHENYL-3,5-DIBENZOYLCYCLOHEXANE $C_{31}H_{31}NO_4$ (II)

D. MOOTZ, B. MIDDELHAUVE, A. WILMS, H. MOHRLE and R. WILLE, 1985. *Z. Kristallogr.*, **170**, 132-133.

I. Monoclinic, $P2_1/c$, a = 11.199, b = 11.651, c = 21.986 Å, β = 118.24°, Z = 4. Mo radiation, R = 0.061 for 2674 reflections.

II. Monoclinic, $P2_1/c$, a = 17.868, b = 6.367, c = 22.809 Å, β = 102.33°, Z = 4. Mo radiation, R = 0.057 for 2635 reflections.

The analysis establishes the relative configuration as in the titles and in Fig. 1. The cyclohexane rings have chair form; all substituents are equatorial with the exception of the hydroxyl group and in (II) the 5-substituent. There is an intramolecular O-H...O hydrogen bond between the hydroxyl group and an adjacent benzoyl oxygen.

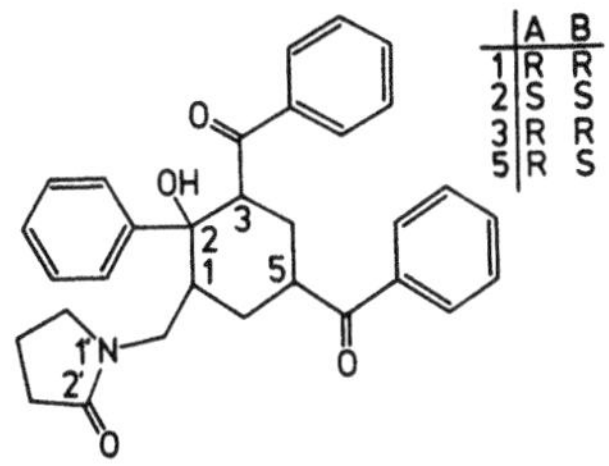

Fig. 1. The $C_{31}H_{31}NO_4$ skeleton.

2-(2,4-DINITROPHENYLTHIOAMINO)-2,3,4,5-TETRAPHENYL-2H-PYRROLE BENZENE SOLVATE
$C_{34}H_{24}N_4O_4S,C_6H_6$

R.S. ATKINSON, B.D. JUDKINS, D.R. RUSSELL and L.J.S. SHERRY, 1985. *J. Chem. Soc. Perkin Trans. 1*, - 1967-1969.

Monoclinic, $P2_1/c$, a = 11.52, b = 16.10, c = 18.67 Å, β = 95.2°, Z = 4. Mo radiation, R = 0.099 for 1070 reflections.

The crystal structure analysis established the identity of the molecule (Fig. 1). Bond lengths include S-C 1.735(19) and S-N 1.679(16) Å.

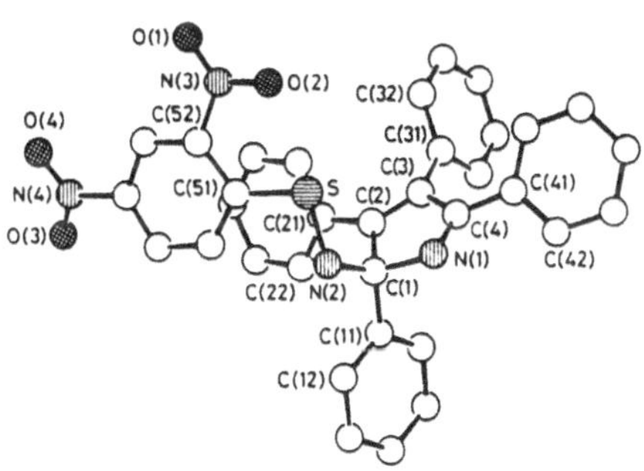

Fig. 1. Molecular structure of $C_{34}H_{24}N_4O_4S$.

1,2,3-TRIAZINE
$C_3H_3N_3$

H. NEUNHOEFFER, M. CLAUSEN, H.-D. VOTTER, H. OHL, C. KRUGER and K. ANGERMUND, 1985. *Liebigs Ann. Chem.*, 1732-1751.

Triclinic, $P\bar{1}$, a = 5.7688, b = 6.8732, c = 5.6725 Å, α = 110.080, β = 113.947, γ = 95.302°, Z = 2. Mo radiation, R = 0.0404 for 1744 reflections (at 100 K).

Principal dimensions for this planar molecule (Fig. 1) are N(1)-N(2) 1.326(1), N(2)-N(3) 1.326(2), N(3)-C(1) 1.345(1), C(1)-C(2) 1.388(1), C(2)-C(3) 1.382(1), C(3)-N(1) 1.346(1) Å, N-N-N 121.6(1), N-N-C 119.6 and

119.8(1), C-C-C 115.4(1), C-C-N 121.7 and 121.9(1)°.

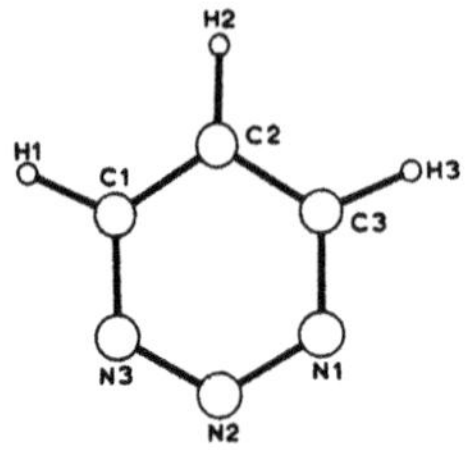

Fig. 1. 1,2,3-Triazine.

4,5-DICHLORO-6(1H)-PYRIDAZINONE
$C_4H_2Cl_2N_2O$

V.M. LYNCH, S.H. SIMONSEN, M.J. MUSMAR and G.E. MARTIN, 1985. *Acta Cryst.*, C41, 1807-1810.

I. Monoclinic, $P2_1/c$, a = 5.1967, b = 9.046, c = 12.901 Å, β = 100.090°, Z = 4. Mo radiation, R = 0.028 for 1552 reflections.

II. Monoclinic, $P2_1/c$, a = 5.3384, b = 13.827, c = 7.9107 Å, β = 94.397°, Z = 4. Mo radiation, R = 0.025 for 1597 reflections.

Molecule (I) is found as colourless block-like crystals while (II) crystallizes as pale pink plates. In each form, the compound exists in the oxo form (Fig. 1) and crystallizes as hydrogen-bonded dimers. Infinite layers of differently oriented dimers result from C-H...O (C...O 3.076(2)) and C-H...Cl (C...Cl 3.534(1) Å) type hydrogen bonds. There are only minor differences in bond lengths and angles for the two molecules. The pyridazinone rings are planar and the Cl atoms lie on opposite sides of the plane in both forms. There are intramolecular Cl...Cl and Cl...O distances (3.225(1), 2.980(1) in (I), 3.2040(4), 2.993(1) Å in II).

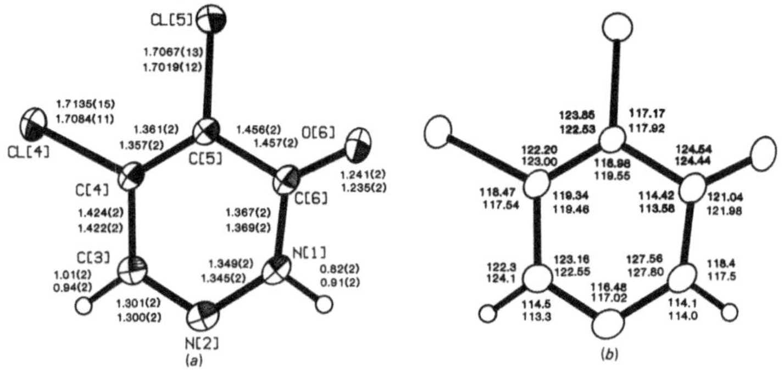

Fig. 1. Bond lengths and angles in $C_4H_2Cl_2N_2O$, (a) (I), (b) (II).

1,4-DINITROSOPIPERAZINE
$C_4H_8N_4O_2$

K. SEKIDO, K. OKAMOTO and S. HIROKAWA, 1985. *Acta Cryst.*, C41, 741-743.

Orthorhombic, *Pbca*, a = 9.495, b = 6.570, c = 10.521 Å, Z = 4. Mo radiation, R = 0.089 for 959 reflections.

The molecule (Fig. 1) is located on a crystallographic centre of symmetry. The piperazine ring has the chair conformation and the two nitroso groups are in the anti form. The mean N-C(ring) bond length is 1.450(5) Å and the C-C bond length is 1.515(5) Å. The N-N and N-O bond lengths are 1.336(4) and 1.202(4) Å respectively, and the N-N-O angle is 113.4(3)°. The molecules are packed together to form infinite ribbons along the [110] direction and are held together by van der Waals forces.

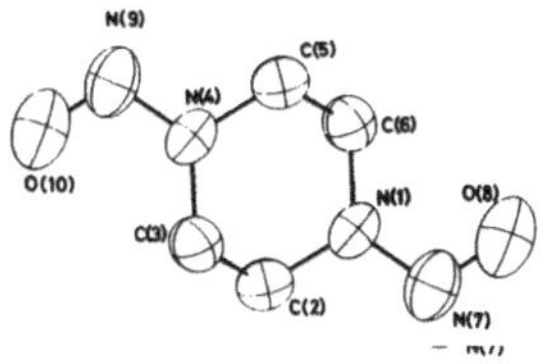

Fig. 1. A view of $C_4H_8N_4O_2$.

PYRIDINIUM PENTAIODIDE
$C_5H_6I_5N$ (I) $C_5H_6N^+,I_3^-,I_2$

β-NAPHTHYLAMMONIUM PENTAIODIDE
$C_{10}H_{10}I_5N$ (II) $C_{10}H_{10}N^+,I_3^-,I_2$

N-METHYL-γ-PICOLINIUM HEPTAIODIDE
$C_7H_{10}I_7N$ (III) $C_7H_{10}N^+,I_3^-,2(I_2)$

F.H. HERBSTEIN, G.M. REISNER and W. SCHWOTZER, 1985. *J. Inclusion Phenomena*, **3**, 173-182.

I. Monoclinic, $P2_1/m$, a = 9.221, b = 12.918, c = 6.026 Å, β = 103.60°, Z = 2. Mo radiation, R = 0.087 for 1187 reflections.

II. Triclinic, $P\bar{1}$, a = 10.390, b = 9.502, c = 4.462 Å, α = 99.19, β = 90.40, γ = 108.49°, Z = 1. Mo radiation, R = 0.051 for 1319 reflections.

III. Monoclinic, $C2/c$, a = 19.315, b = 12.714, c = 8.442 Å, β = 107.26°, Z = 4. Mo radiation, R = 0.058 for 1336 reflections.

The packing of (I) has pleated sheets of iodines pierced by channels of almost square cross-section which extend along the *c*-direction (Fig. 1). In (II) the iodine arrangement consists of zigzag chains of I_3^- and I_2 moieties both at inversion centres (Fig. 2); the cation also lies on an inversion centre with its methyl group disordered. I-I distances are 2.912(1) in I_3^-, 2.716 in I_2. The overall structure is that of a channel inclusion complex with the channel along c. In (III) the basic structural unit has the form of a twisted ladder in which I_3^- ions bridge between two zigzag chains of I_2 molecules (Fig. 3). The I_3^- ion is on an inversion centre and the two iodine molecules are each bisected by twofold axes. Channels with axes along [101] are formed and the cations are lined up within the channels (and are disordered).

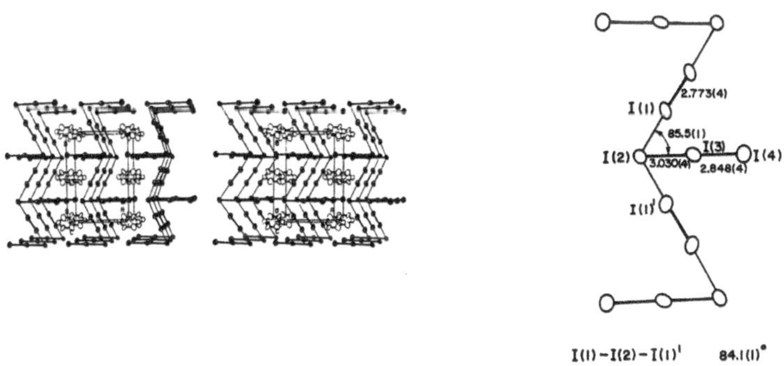

Fig. 1. A stereoview of $C_5H_6N^+,I_3^-,I_2$ and details of I-I dimensions.

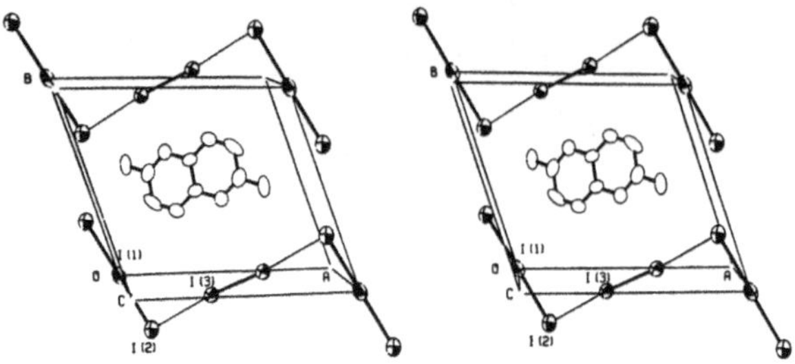

Fig. 2. A stereoview of $C_{10}H_{10}N^+,I_3^-,I_2$.

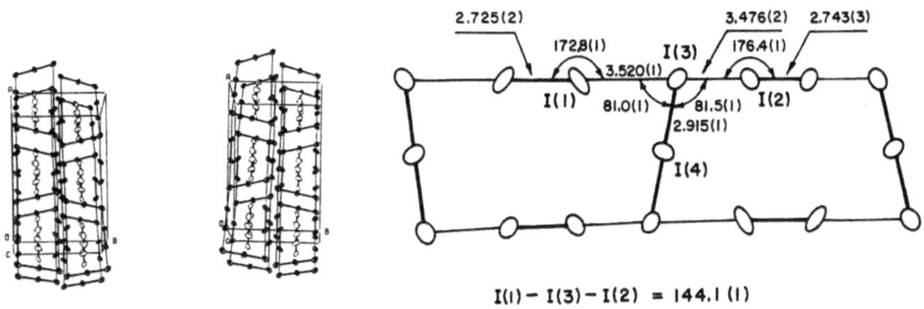

Fig. 3. A stereoview of $C_7H_{10}N^+,I_3^-,2(I_2)$ and details of I-I dimensions.

(E)-3-FLUORO-4-PYRIDINECARBALDEHYDE OXIME
$C_6H_5FN_2O$

J.M. SOROF, H.L. CARRELL, J.P. GLUSKER, J. McLICK and E. KUN, 1985. *Acta Cryst.*, **C41**, 785-788.

Monoclinic, $P2_1/a$, a = 14.225, b = 11.742, c = 3.9852 Å, β = 92.48°, Z = 4. Cu radiation, R = 0.050 for 1217 reflections.

The oxime (Fig. 1) is in the syn conformation with torsion angle C(4)-C(8)-N(9)-O(10)-178.7°; the entire molecule is approximately planar. The molecules in the crystal are held together by van der Waals forces and by hydrogen bonds from the oxime hydroxyl group to N(1) of a symmetry related molecule. The O...N separation is 2.708(2) Å. The fluorine atom is not involved in any hydrogen bonding.

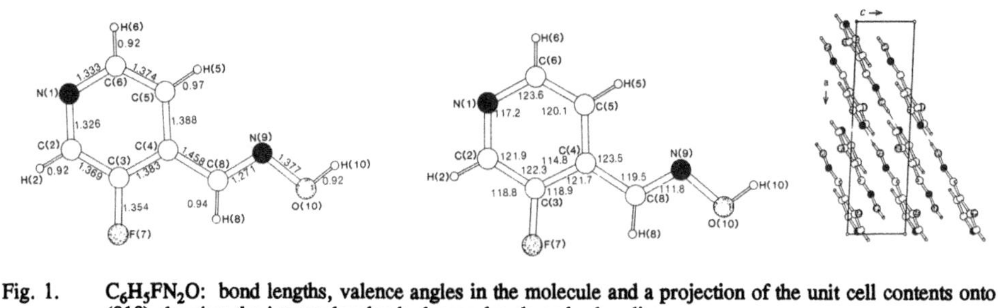

Fig. 1. $C_6H_5FN_2O$: bond lengths, valence angles in the molecule and a projection of the unit cell contents onto (010) showing the intermolecular hydrogen bonds as broken lines.

4-METHYLPYRIDINE

C_6H_7N

U. OHMS, H. GUTH, W. TREUTMANN, H. DANNOHL, A. SCHWEIG and G. HEGER, 1985. *J. Chem. Phys.*, **83**, 273-279.

I. Tetragonal, $I4_1/a$, a = 7.618, c = 18.635 Å, Z = 8. Mo radiation, R = 0.059 for 1795 reflections (at 120 K).

II. Tetragonal, $I4_1/a$, a = 7.606, c = 18.62 Å, Z = 8. Neutron radiation, R = 0.035 for 1743 reflections (at 120 K).

The molecular dimensions from the neutron and X-ray structures are in Fig. 1; the three hydrogens of the methyl group are disordered over six sites. The molecular packing has an antiparallel ordering of the molecules with respect to the *c*-axis; separations are normal van der Waals distances. An X-N synthesis agrees well with theoretical dynamic electron deformation density.

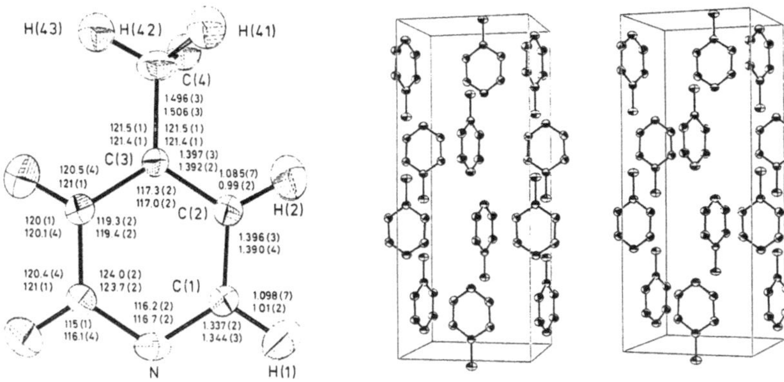

Fig. 1. 4-Methylpyridine showing dimensions (upper values, neutron; lower values, X-ray) and a stereoview of the packing.

3-(4-CHLOROPHENYL)-6-(TRIFLUOROMETHYL)DIHYDRO-1,2,4,5-TETRAZINE
$C_9H_6ClF_3N_4$

K.H. PILGRAM, L.H. GALE and R.D. SKILES, 1985. *J. Heterocycl. Chem.*, **22**, 643-647.

Triclinic, $P\bar{1}$, a = 10.294, b = 12.395, c = 13.720 Å, α = 68.21, β = 89.58, γ = 86.76°, Z = 6. Cu radiation, R = 0.052 for 2335 reflections.

There are three similar independent molecules (Fig. 1) in the asymmetric unit. The dihydro-s-tetrazine ring is in a slightly twisted boat conformation with the N atoms N(2) and N(4) at the prow and stern. The mean N(1)-C(7) and N(3)-C(8) bond lengths are 1.286 and 1.277 Å respectively consistent with localized C=N bonds. The p-chlorophenyl ring is rotated 35.6° about the C(7)-C(4) bond out of the tetrazine plane. Molecules are linked in ribbons by N-H...N hydrogen bonds.

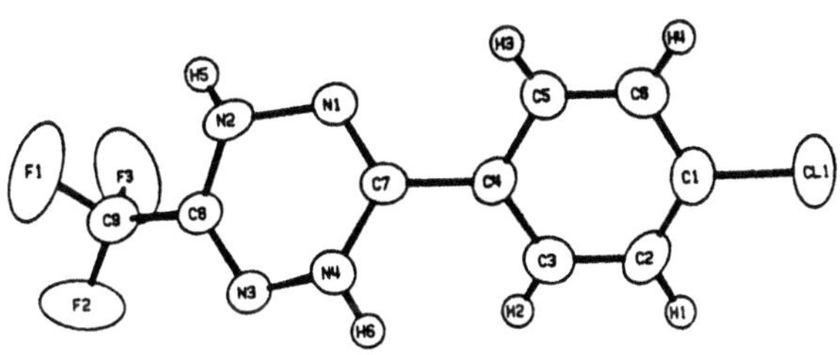

Fig. 1. One of the $C_9H_6ClF_3N_4$ molecules.

(1,4-PYRAZINIO)BIS(DICYANOMETHYLIDE)
$C_{10}H_4N_6$

G.-E. MATSUBAYASHI, Y. SAKAMOTO, T. TANAKA and K. NAKATSU, 1985. *J. Chem. Soc. Perkin Trans.* 2, 947-950.

Tetragonal, *P4₂/mnm*, a = 8.781, c = 5.9168 Å, Z = 2. Mo radiation, R = 0.047 for 341 reflections.

All the atoms lie in a plane and the molecule lies on a centre of symmetry (Fig. 1); N(1'), N(2'), C(1') and C(2') are related to N(1), N(2), C(1) and C(2) respectively; C(2''), N(2'') and C(3'') are mirror images of C(2), N(2) and C(3). The centres of the molecules are located at (0,0,0.5) and (0.5,0.5,0), the molecular planes being parallel to the c plane and perpendicular to one another. The interplanar spacing is 2.958(1) Å.

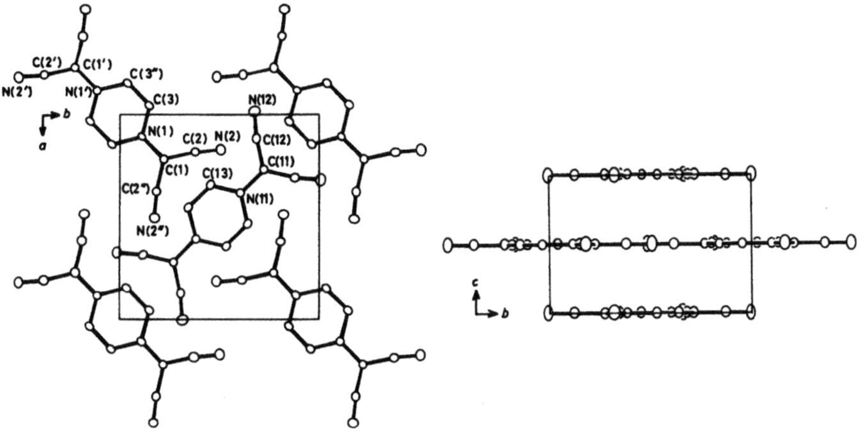

Fig. 1. $C_{10}H_4N_6$: perspective views of the unit cell along the c axis (top) and a axis (bottom).

N-(4-OXOPYRIDIN-1-YL)PYRIDINIUM TETRAFLUOROBORATE
$C_{10}H_9BF_4N_2O$ (I)

2,6-DIMETHYL-N-(4-OXOPYRIDIN-1-YL)PYRIDINIUM TETRAFLUOROBORATE
$C_{12}H_{15}BF_4N_2O_2$ (II)

M.P. SAMMES, T.-F. LAI, A.R. KATRITZKY, R. MURUGAN and H. LUCE, 1985. *J. Chem. Soc. Perkin Trans.* 2, 573-579.

I. Monoclinic, $P2_1/c$, a = 10.876, b = 8.913, c = 12.805 Å, β = 112.16°, Z = 4. Mo radiation, R = 0.057 for 637 reflections.

II. Monoclinic, *C2/m*, a = 22.390, b = 6.944, c = 9.330 Å, β = 98.96°, Z = 4. Cu radiation, R = 0.079 for 1065 reflections.

The N-N bond lengths, and the torsion angles between the ring planes are 1.417(7) Å and 74(3)° for (I), and 1.411(5) Å and 90° for (II) (crystal symmetry determined). In (I) the tetrafluoroborate ion is disordered; crystals of (II) include a water molecule. The molecules are shown in Fig. 1.

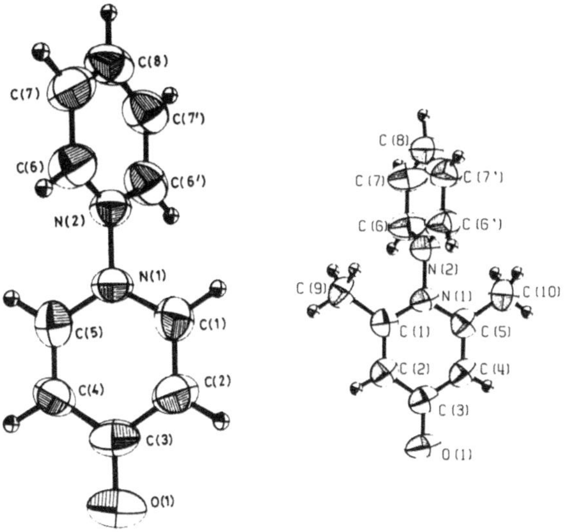

Fig. 1. Perspective drawings of molecules of $C_{10}H_9BF_4N_2O$ (left) and $C_{12}H_{15}BF_4N_2O_2$ (right) cations.

4-DIMETHYLAMINO-1-PHENYL-1,2,3,5-TETRAZIN-6-ONE
$C_{10}H_{11}N_5O$ (I)

5-DIMETHYLAMINO-2,3-DIHYDRO-3-OXO-2-p-TOLYLIMINO-1H-1,2,4-TRIAZOL-2-IUM-1-IDE MONOHYDRATE
$C_{11}H_{13}N_5O,H_2O$ (II)

1-o-CHLOROPHENYLIMINO-3,5-BIS(DIMETHYLAMINO)-1,2,4-TRIAZOLIUM TETRAFLUOROBORATE
$C_{12}H_{16}BClF_4N_6$ (III) $C_{12}H_{16}ClN_6^+,BF_4^-$

P.F. LINDLEY, A.R. WALTON, A.E. BAYDAR and G.V. BOYD, 1985. *Acta Cryst.*, **C41**, 1669-1672.

I. Orthorhombic, *Pbca*, a = 23.510, b = 11.169, c = 7.777 Å, Z = 8. Cu radiation, R = 0.050 for 645 reflections (at 278 K).

II. Triclinic, $P\bar{1}$, a = 7.930, b = 12.186, c = 6.919 Å, α = 89.088, β = 110.219, γ = 98.001°, Z = 2. Cu radiation, R = 0.057 for 1402 reflections.

III. Triclinic, $I\bar{1}$, a = 21.834, b = 6.850, c = 10.938 Å, α = 93.12, β = 87.76, γ = 91.94°, Z = 4. Cu radiation, R = 0.084 for 1718 reflections.

In molecule (I) (Fig, 1) the heteroring is planar and bond lengths in the ring indicate extensive electron delocalization. The phenyl ring is also planar and the dihedral angle between their least-squares planes is 54.0(2)°. In molecule (II), which is essentially planar, the bond N(2)-C(3) is abnormally long (at 1.546(5) Å) for a single bond. The valence angles at C(3) are unequal having values of 105.3(3), 120.4(3) and 34.3(4)°. The cation in (III) is also nearly planar. The N(1)-C(5) bond is slightly long and the difference between the exocyclic angles at C(5) is less pronounced than those at C(3) in (II). To minimise steric hindrance between the five- and six-membered rings in planar structures (II) and (III), the exocyclic angles at the bridging region are widened. The BF_4^- anion in (II) has normal geometry. Except for a hydrogen bond between the water molecule and O(31) in (II), of length 2.925(5) Å, all other intermolecular distances in (I), (II) and (III) are consistent with normal van der Waals contacts.

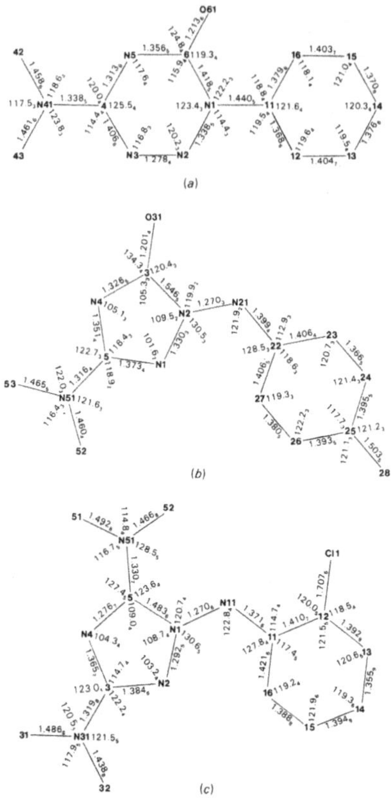

Fig. 1. Molecular dimensions of (a) the molecule $C_{10}H_{11}N_5O$, (b) the molecule $C_{11}H_{13}N_5O,H_2O$ and (c) the cation in $C_{12}H_{16}BClF_4N_6$.

4-AMINOPYRIDINIUM HEMIPERCHLORATE

$C_{10}H_{13}ClN_4O_4$ $C_5H_7N_2^+,C_5H_6N_2,ClO_4^-$

P. TEULON, R.G. DELAPLANE, I. OLOVSSON and J. ROZIERE, 1985. *Acta Cryst.*, **C41**, 479-483.

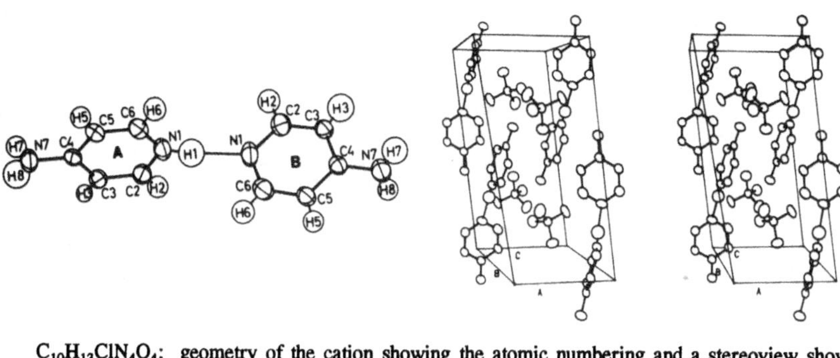

Fig. 1. $C_{10}H_{13}ClN_4O_4$: geometry of the cation showing the atomic numbering and a stereoview showing the packing in the unit cell.

Monoclinic, $P2_1/c$, a = 7.030, b = 12.924, c = 14.442 Å, β = 97.04°, Z = 4. Mo radiation, R = 0.064 for

2445 reflections (at 193 K).

The asymmetric unit of this low temperature β-phase contains well ordered tetrahedral perchlorate anions and two $NH_2C_5H_4N$ moieties asymmetrically H-bridged to form the cations (Fig. 1). Though the ring atoms within each pyridine moiety are coplanar the geometries of rings A and B have slight but significant differences. The bridging N-H...N bond is nearly linear and of length 2.697(4) Å. The dihedral angle between the planes of rings A and B is 89.8(5)°. The Cl-O bond lengths in the perchlorate anions range from 1.421(3) to 1.433(2) Å. The cations are hydrogen bonded via the amino groups to the O atoms of the perchlorate anions. This structure is compared with that

reported earlier for the α-phase (1).

1. Structure Reports, 46B, 225.

N,N'-BIS(1-METHYLETHYL)-6-METHYLTHIO-1,3,5-TRIAZINE-2,4-DIAMINE (PROMETRYN)
$C_{10}H_{19}N_5S$

H.P. AVEY, C.H.L. KENNARD and G. SMITH, 1985. *Acta Cryst.*, **C41**, 938-940.

Triclinic, $P\bar{1}$, a = 10.094, b = 11.834, c = 12.370 Å, α = 79.08, β = 79.73, γ = 77.12°, Z = 4. Mo radiation, R = 0.074 for 1740 reflections.

There are two crystallographically independent but similar molecules in the asymmetric unit. The s-triazine ring (Fig. 1) is essentially coplanar with the thiomethyl group and the first two atoms of the amine side chains. Bond lengths and angles are normal with the small C-N-C and large N-C-N angles usually found in triazines. Packing of molecules in the unit cell involves only N atoms, both amine and hetero, in weak hydrogen bonding between A and B molecules in the asymmetric unit (N(3)A...N(6)B 3.03(1) Å) and intra-unit associations (N(5)A...N(2)B 3.08(1), N(6)A...N(3)B 3.05(1) Å).

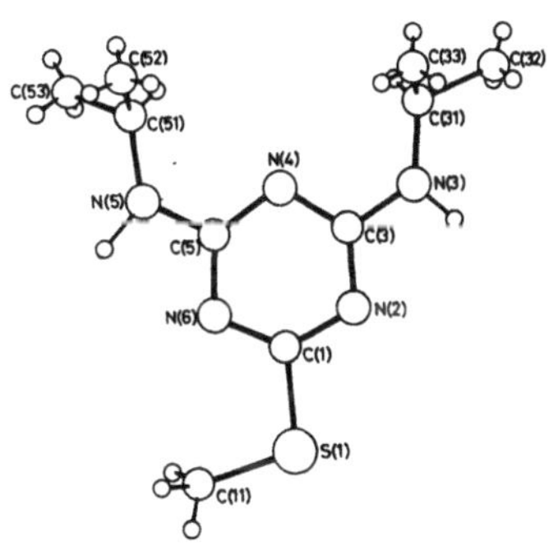

Fig. 1. The molecular structure of $C_{10}H_{19}N_5S$.

2-AMINO-5-(4-CHLOROPHENYL)-6-ETHYLPYRIMIDIN-4(3H)-ONE 4-AMINO-5-(4-CHLOROPHENYL)-6-ETHYLPYRIMIDIN-2(1H)-ONE MONOHYDRATE
$C_{12}H_{12}ClN_3O,C_{12}H_{12}ClN_3O,H_2O$

R.J. GRIFFIN, C.H. SCHWALBE, M.F.G. STEVENS and K.P. WONG, 1985. *J. Chem. Soc. Perkin Trans. 1*, - 2267-2276.

Monoclinic, $P2_1/n$, a = 24.297, b = 7.917, c = 13.312 Å, β = 98.09°, Z = 4. Mo radiation, R = 0.072 for 1486 reflections.

There are two isomers in the hydrated duplex; these are linked by three hydrogen bonds as shown in Fig. 1.

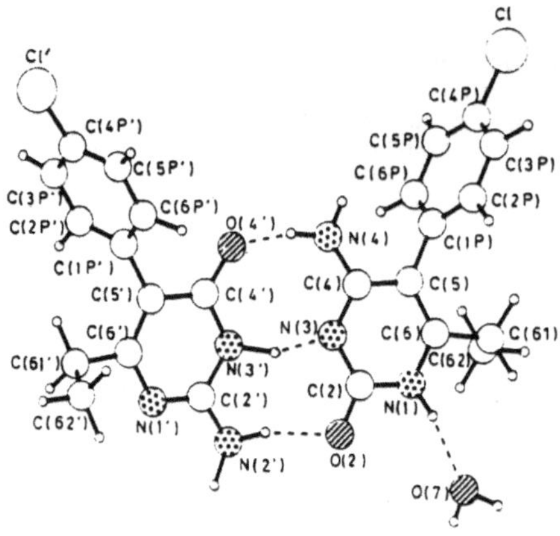

Fig. 1. Perspective drawing of the hydrated base-pair, $C_{12}H_{12}ClN_3O,C_{12}H_{12}ClN_3O,H_2O$, projected onto its least-squares plane.

N'-(4,6-DIMETHYL-2-PYRIMIDINYL)SULPHANILAMIDE METHANOL SOLVATE $C_{12}H_{14}N_4O_2S,CH_4O$

J. RAMBAUD, L. MAURY, B. PAUVERT, M. AUDRAN, Y. LASSERRE, G. BERGE and J.-P. DECLERCQ, 1985. *Acta Cryst.*, C41, 133-134.

Orthorhombic, $P2_12_12_1$, a = 19.187, b = 8.334, c = 9.749 Å, Z = 4. Mo radiation, R = 0.037 for 1002 reflections.

Bond lengths and angles in the molecule (Fig. 1) are in good agreement with other observed values. The two six-membered rings in the molecule are planar. The molecular conformation is defined by the torsion angles C(3,5)-C(4)-S(8)-N(11), C(4)-S(8)-N(11)-C(12) and S(8)-N(11)-C(12)-N(13,17) which are 95(1), -68(1) and 4(1)° respectively. The solvate molecules are in cavities between the sulphamethazine moieties and each is linked by hydrogen bonds to the sulphone oxygen atom O(9) and to the amide atom N(11) of different sulphamethazine molecules. The S(8)-O(9) and S(8)-O(10) bond lengths, 1.438(4) and 1.418 Å, are significantly different- probably due to this hydrogen bonding.

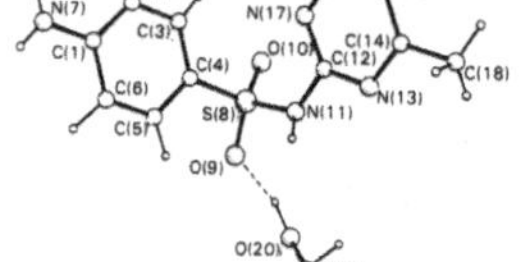

Fig. 1. $C_{12}H_{14}N_4O_2S,CH_4O$: a view of the asymmetric unit.

HYDROGENBIS(3-METHYLPYRIDINE N-OXIDE) TETRACHLOROAURATE(III) $C_{12}H_{15}AuCl_4N_2O_2$ $[H(3\text{-}MepyO)_2](AuCl_4)$

M.S. HUSSAIN and S.A.A. AL-HAMOUD, 1985. *J. Chem. Soc. Dalton Trans.*, 749-753.

Triclinic, $P\bar{1}$, a = 6.857, b = 8.011, c = 8.797 Å, α = 105.05, β = 66.89, γ = 89.39°, Z = 1. Mo radiation, R = 0.063 for 2339 reflections.

The cation (Fig. 1) lies across a crystal inversion centre. Its O atoms are linked by a short hydrogen bond (O...O

2.406(9), N-O 1.362(4) Å). The square-planar $AuCl_4^-$ anions (Au-Cl 2.275, 2.286(3) Å), also on inversion centres, are well separated from one another but there are long Au...O contacts (3.040(3) Å) to make the coordination effectively distorted octahedral.

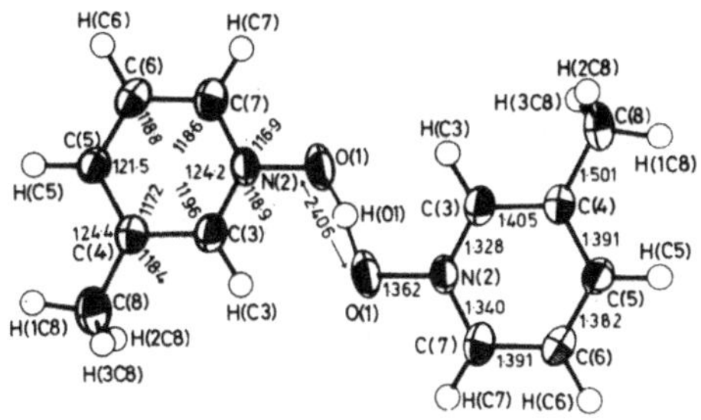

Fig. 1. The cation in $C_{12}H_{15}AuCl_4N_2O_2$.

2-((HYDROXYIMINO)(PHENYLTHIO)METHYL)-1-METHYLPYRIDINIUM CHLORIDE
$C_{13}H_{13}ClN_2OS$ $C_{13}H_{13}N_2OS^+,Cl^-$

B.J. OLEKSYN and K.M. STADNICKA, 1985. *Acta Cryst.*, C41, 743-746.

Monoclinic, $P2_1/c$, a = 9.30, b = 11.433, c = 14.02 Å, β = 114.8°, Z = 4. Mo radiation, R = 0.037 for 2513 reflections.

The molecular cation (Fig. 1) is in the (E) configuration and has a folded conformation. The cations and chloride anions are linked together through O-H(O)...Cl hydrogen bridges. In the cation the bond lengths indicate that there is no π-bond localization in the phenyl and pyridinium rings and the C(7)-N(2) bond of length 1.281(2) Å has a double bond character. The phenyl and pyridinium rings and the oxime moiety are each planar. The N(2)-C(7)-C(5)-C(4) torsion angle is 105.1(3)°.

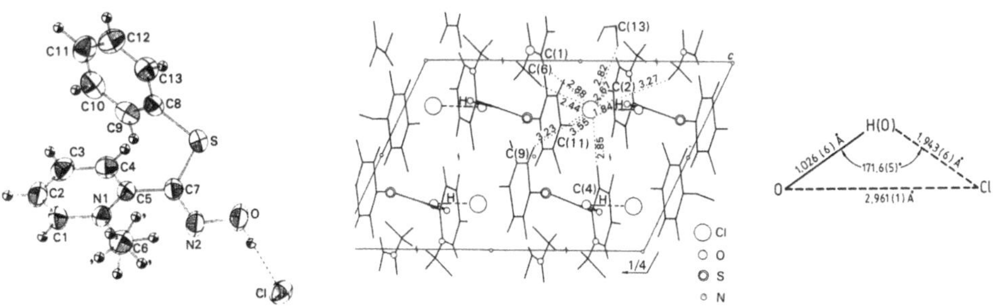

Fig. 1. $C_{13}H_{13}ClN_2OS$: a perspective view of the ion pair (left), details of the hydrogen-bonding geometry (centre) and (right) the molecular packing viewed along the b-axis.

BIS(PYRIDINIUM) OXALATE OXALIC ACID
$C_{14}H_{14}N_2O_8$ $2(C_5H_6N^+),C_2O_4^{2-},C_2H_2O_4$

G.R. NEWKOME, K.J. THERIOT and F.R. FRONCZEK, 1985. *Acta Cryst.*, C41, 1642-1644.

Triclinic, $P\bar{1}$, a = 3.999, b = 8.439, c = 11.319 Å, α = 77.96, β = 88.64, γ = 82.69°, Z = 1. Mo radiation, R = 0.041 for 1390 reflections.

The asymmetric unit in this structure (Fig. 1) consists of a planar pyridinium ion in a general position hydrogen bonded to an oxalate ion located on a crystallographic inversion centre. This oxalate ion is bonded to a centrosymmetric oxalic acid molecule forming linear chains of acid molecules and oxalate ions lying along the *c*-axis. The oxalate and oxalic acid moieties are planar with an interplanar dihedral angle of 13.0°. The lengths of the bonds C(2)-O(3) and C(2)-O(4) in the oxalate moiety are 1.261(1) and 1.230(1) Å respectively, and these do not correlate well with those of similar bonds reported previously. The C(1)-O(2) bond in the oxalic acid is rather short at 1.192(1) Å. The O(1)-H...O(3) hydrogen bond is of length 2.535(1) Å and the N...O(4) and N...O(3') separations are 2.716(1) and 2.923(1) Å suggesting that the contact between the pyridinium and oxalate ions takes the form of an asymmetrically bifurcated hydrogen bond.

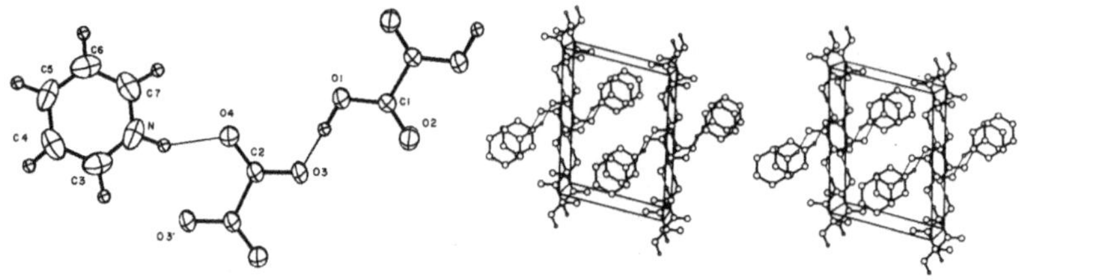

Fig. 1. $C_{14}H_{14}N_2O_8$: a view of the asymmetric unit showing the numbering scheme and a stereoview of the unit cell.

DIPIPERIDINOSQUARAINE
$C_{14}H_{20}N_2O_2$

P.H.M. BUDZELAAR, H. DIETRICH, J. MACHELEID, R. WEISS and P. VON R. SCHLEYER, 1985. *Chem. Ber.*, **118**, 2118-2126.

Orthorhombic, *Pnab*, a = 7.1798, b = 9.5362, c = 18.788 Å, Z = 4. Cu radiation, R = 0.043 for 8701 reflections.

The molecule lies about an inversion centre and consequently the central four-membered ring is exactly planar (Fig. 1). The N and O atoms are 0.003(1) and 0.007(1) Å from the ring plane and the piperidino ring is in chair form. Main dimensions are: C(1)-C(2) 1.4572(8), C(1)-C(2') 1.4546(8), C(2)-O 1.2396(7), C(1)-N 1.3112(7), N-C(3) 1.470(1), N-C(7) 1.463(1) Å, C(1)-C(2)-C(1') 88.49(5), C-C-O 135.59 and 135.93(5), (ring)C-C-N 134.12 and 134.37(5)°. The data are consistent with the presence of considerable C-N and C-O π-bonding but only a small amount of C-C π-bonding in the squaraine system.

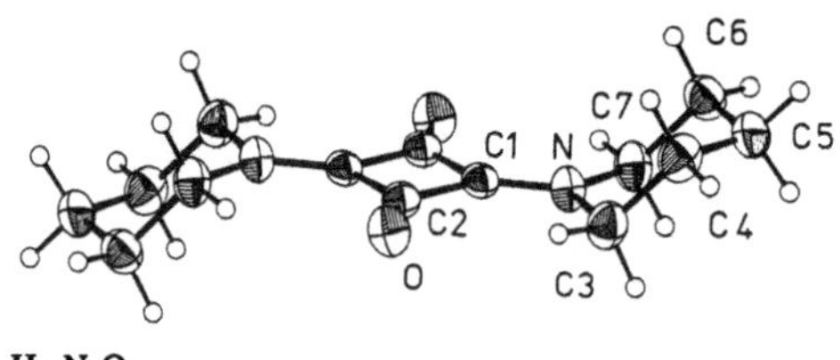

Fig. 1. A drawing of $C_{14}H_{20}N_2O_2$.

DIETHYL 1,4-DIHYDRO-2,4,6-TRIMETHYL-3,5-PYRIDINEDICARBOXYLATE
$C_{14}H_{21}NO_4$

S. FORTIER, M.E. FRASER, N.J. MOORE, Y.S. PARK, R.A. WHITNEY and G.S. MARKS, 1985. *Acta Cryst.*, **C41**, 411-413.

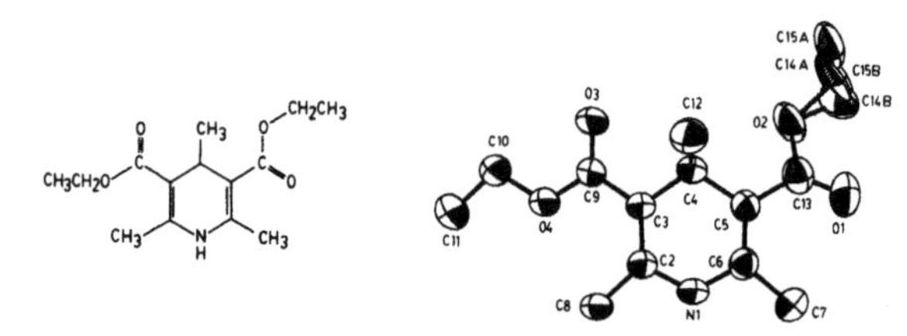

Fig. 1. $C_{14}H_{21}NO_4$: structural formula and a perspective view of the molecule.

Monoclinic, $P2_1/a$, a = 16.921, b = 7.483, c = 11.429 Å, β = 94.24°, Z = 4. Cu radiation, R = 0.056 for 1371

reflections.

The 1,4-dihydropyridine ring in the molecule (Fig. 1) adopts a flattened boat conformation. The dihedral angles between the plane of atoms C(2,3,5,6) and those of atoms C(2,6),N(1) and atoms C(3,4,5) are 8.9(5) and 17.2(5)° respectively. The methyl substituent C(12) is axial and all the others are equatorial. The ethoxycarbonyl moiety at C(5) is disordered with two equally probable positions for the terminal C_2H_5 fragment. The two ethoxycarbonyl groups are twisted in opposite directions so that the possible molecular mirror symmetry is destroyed.

(-)-1,3,3-TRIMETHYL-4-PHENYL-4-PIPERIDINOL HYDROBROMIDE

$C_{14}H_{22}BrNO$ $C_{14}H_{22}NO^+,Br^-$

F.R. AHMED, G.F. LAWS, A.E. MADANI and A.F. CASY, 1985. *J. Med. Chem.*, **28**, 1947-1949.

Orthorhombic, $P2_12_12_1$, a = 12.387, b = 12.610, c = 9.576 Å, Z = 4. Cu radiation, R = 0.039 for 1688 reflections.

The structure and absolute stereochemistry are shown in Fig. 1. The piperidine ring has the chair conformation with the phenyl ring equatorial and the hydroxyl group axial. The phenyl ring plane makes a dihedral angle of 33.4(5)° with the N(1),C(4),O plane. Each Br ion forms two hydrogen bonds N-H...Br...H'-O' that interlink the ions in chains along the *y*-axis.

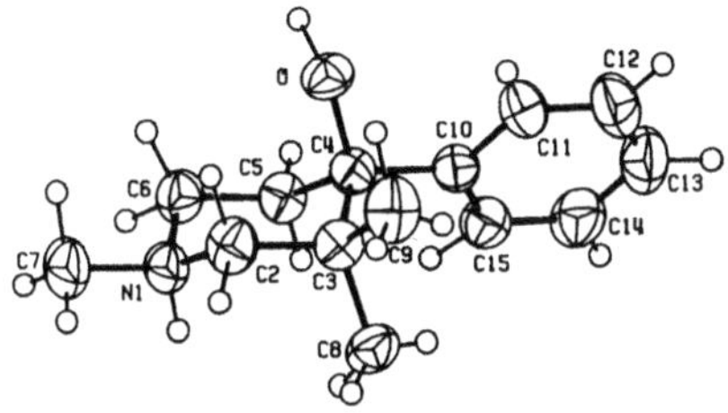

Fig. 1. The conformation and absolute configuration of the $C_{14}H_{22}NO^+$ ion.

S-(-)-3-(3-HYDROXYPHENYL)-n-PROPYLPIPERIDINE HYDROCHLORIDE

$C_{14}H_{22}ClNO$ $C_{14}H_{22}NO^+,Cl^-$

S.-O. THORBERG, L. GAWELL, I. CSOREGH and J.L.G. NILSSON, 1985. *Tetrahedron*, **41**, 129-139.

Orthorhombic, $P2_12_12_1$, a = 20.464, b = 18.367, c = 7.442 Å, Z = 8. Cu radiation, R = 0.0465 for 1949 reflections.

The two independent formula units are shown to have S absolute stereochemistry (Fig. 1). The piperidine rings have chair conformations. The molecules adopt different conformations in the solid (differing in rotation about C(4)-C(8) and N(10)-C(14)). Each Cl^- ion is an acceptor for two H-bonds from two different cations (O...Cl 3.078 and 3.066(4), N...Cl 3.125 and 3.097(5) Å).

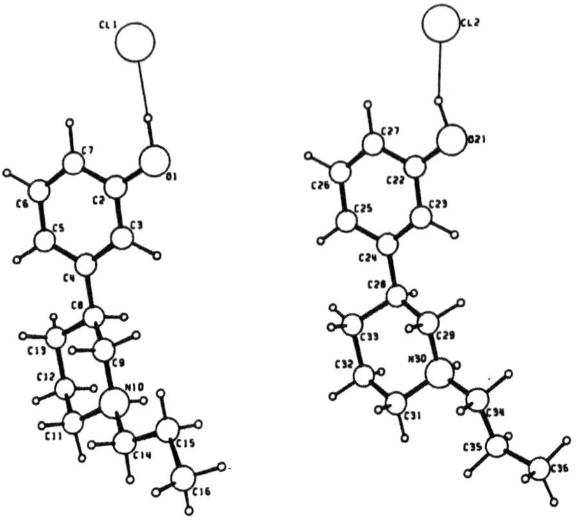

Fig. 1. The two independent $C_{14}H_{22}NO^{+},Cl^{-}$ units.

1-(4-FLUOROPHENYL)-1-(3-PYRIDYL)BUT-3-YN-1-OL
$C_{15}H_{12}FNO$

A. ALBINATI, S.V. MEILLE, A. ARNOLDI and R. GALLI, 1985. *Acta Cryst.*, C41, 97-99.

Monoclinic, $P2_1/c$, a = 12.609, b = 7.681, c = 12.645 Å, β = 94.97°, Z = 4. Mo radiation, R = 0.064 for 1232 reflections.

Distances around the tetrahedral carbon atom C(1) in the molecule (Fig. 1) are normal. The pyridine and benzene rings are planar with an interplanar dihedral angle of 84(1)°. The angles C(1)-C(5)-C(10) and C(1)-C(5)-C(6) are widened, probably due to intramolecular steric interactions, to 122.6(4) and 118.7(4)° respectively. There is one short intermolecular contact O(1)-H...N(1) of length 2.81 Å, consistent with a hydrogen-bonding interaction.

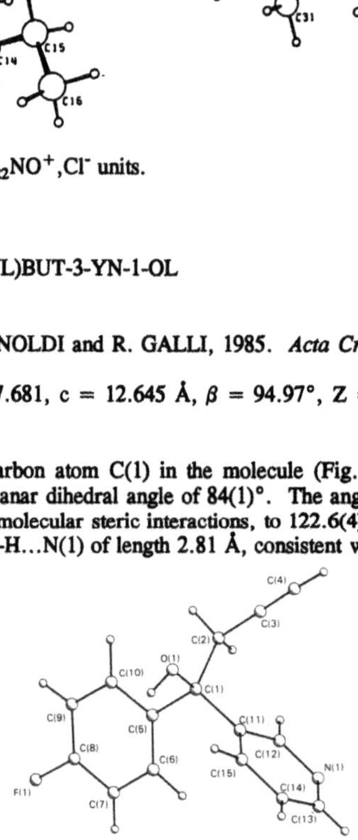

Fig. 1. $C_{15}H_{12}FNO$: a view of the molecule.

1,4-BIS(3,5-DICHLORO-2-PYRIDYLOXY)BENZENE
$C_{16}H_8Cl_4N_2O_2$ (I)

1,4-BIS(5-CHLORO-2-PYRIDYLOXY)BENZENE
$C_{16}H_{10}Cl_2N_2O_2$ (II)

1,4-BIS(3-CHLORO-2-PYRIDYLOXY)BENZENE
$C_{16}H_{10}Cl_2N_2O_2$ (III)

W.B. DRENDEL and M. SUNDARALINGAM, 1985. *Acta Cryst.*, C41, 950-953.

I. Triclinic, $P\bar{1}$, a = 3.964, b = 7.344, c = 14.546 Å, α = 87.28, β = 84.97, γ = 73.20°, Z = 1. Cu radiation, R = 0.116 for 1257 reflections.

II. Monoclinic, $P2_1/c$, a = 3.987, b = 5.587, c = 31.910 Å, β = 91.20°, Z = 2. Cu radiation, R = 0.052 for 1013 reflections.

III. Monoclinic, $P2_1/c$, a = 7.367, b = 9.611, c = 21.035 Å, β = 99.61°, Z = 4. Cu radiation, R = 0.078 for 1919 reflections.

Both (I) and (II) (Fig. 1) lie on crystallographic centres of symmetry. Both induce mouse hepatic monooxygenase activity and have very similar conformations, with the pyridyl rings anti, with respect to the phenyl rings, and phenyl-ether dihedral angles of 52.7(3) and 59.1(2)° respectively. (III), which lacks biological activity, has the pyridyl groups syn and the phenyl-ether dihedral angles average 84.0°. Bond lengths and angles lie within normal ranges. The major packing forces in all three are stacking interactions between the aromatic rings, phenyl-phenyl and pyridyl-pyridyl only, and the stacking interactions are between 3.3 and 3.7 Å.

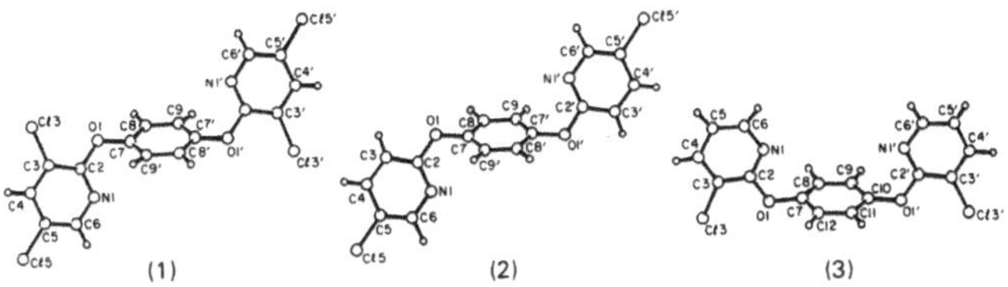

Fig. 1. The molecular conformations: $C_{16}H_8Cl_4N_2O_2$ (1), $C_{16}H_{10}Cl_2N_2O_2$ (2) and $C_{16}H_{10}Cl_2N_2O_2$ (3).

2-(2-PYRIDINYLMETHYL)AMINO-3-(2-PYRIDINYLMETHYLENE)AMINO-2-BUTENEDINITRILE PERCHLORATE
$C_{16}H_{13}ClN_6O_4$ $C_{16}H_{13}N_6^+,ClO_4^-$

V. DE FELICE, P. GANIS and G. VALLE, 1985. *Z. Kristallogr.*, **170**, 263-266.

Triclinic, $P\bar{1}$, a = 11.231, b = 10.315, c = 7.783 Å, α = 101.3, β = 97.2, γ = 92.4°, Z = 2. Cu radiation, R = 0.067 for 1122 reflections.

The organic cation (Fig. 1) has a folded conformation with an intramolecular N-H...N hydrogen bond (N(3)...N(4) 2.77 Å). The angles at C(1) and N(1) are widened (C(2)-C(1)-N(1) 129°, C(1)-N(1)-C(5) 127°). The perchlorate ion is hydrogen bonded to N(1).

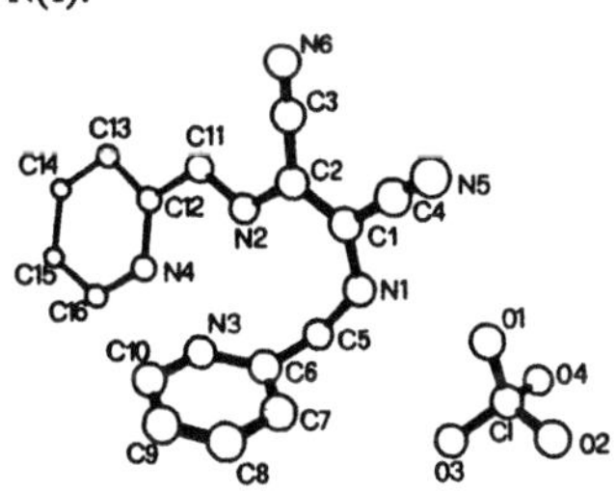

Fig. 1. A perspective drawing of $C_{16}H_{13}ClN_6O_4$.

1-BENZYL-5-CHLORO-4-PHENYL-2(1H)-PYRIMIDINONE
$C_{17}H_{13}ClN_2O$ (I)

1-BENZYL-5-CHLORO-6-PHENYL-2(1H)-PYRIMIDINONE
$C_{17}H_{13}ClN_2O$ (II)

F. RISE, C. ROMMING and K. UNDHEIM, 1985. *Acta Chem. Scand.*, **B39**, 459-468.

I. Monoclinic, $P2_1/c$, a = 6.323, b = 12.606, c = 17.973 Å, β = 97.31°, Z = 4. Mo radiation, R = 0.034 for 1979 reflections (at -130°C).

II. Orthorhombic, $P2_12_12_1$, a = 9.452, b = 10.270, c = 14.724 Å, Z = 4. Mo radiation, R = 0.039 for 1921 reflections (at -130°C).

The crystal structure analyses established the stereochemistries of the molecules (Fig. 1). In (I) the four-phenyl group makes an angle of 63.2° with the plane of the pyrimidine ring while in (II) the angle between the six-phenyl group and the pyrimidine ring is 75.5°. The pyrimidine ring in (II) is closer to planarity than that in I.

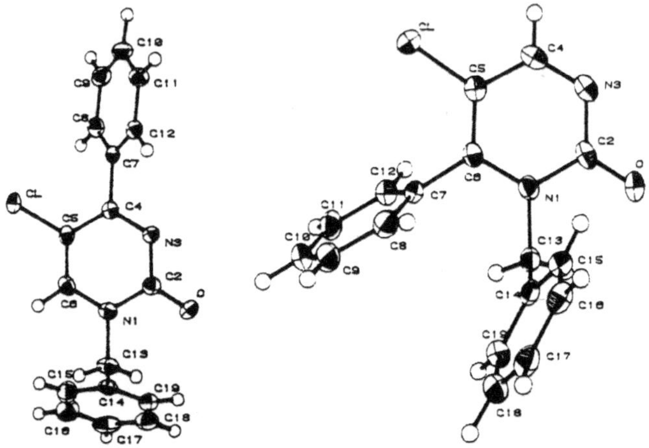

Fig. 1. ORTEP plots of molecules of $C_{17}H_{13}ClN_2O$: (I) (left) and (II) (right).

2,4,5-TRINITRO-7-PIPERIDINOFLUORENONE
$C_{18}H_{14}N_4O_7$

L.A. CHETKINA, Z.P. POVET'EVA, V.K. BEL'SKII and B.P. BESPALOV, 1985. *Kristallografiya*, **30**, 910-914.

Monoclinic, $P112_1/n$, a = 14.547, b = 21.399, c = 5.948 Å, γ = 108.76°, Z = 4. Mo radiation, R = 0.088 for 1066 reflections.

There is steric hindrance between the nitro groups at C(4) and C(5) (Fig. 1) and the nitro oxygens are not well defined. The fluorenone nucleus is not exactly planar (displacements of up to 0.077 Å). The piperidine ring at C(7) is in a chair conformation.

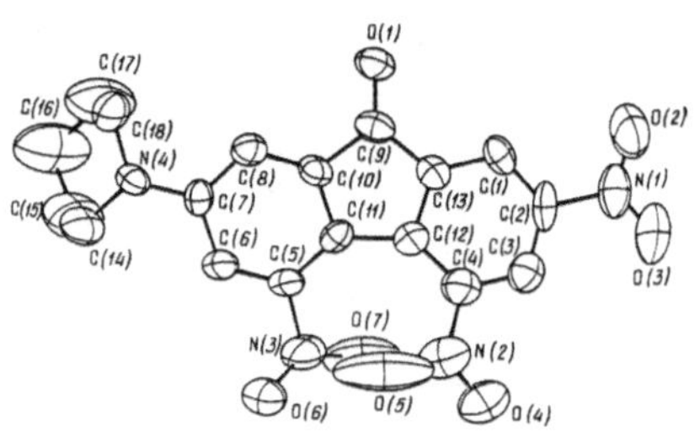

Fig. 1. The $C_{18}H_{14}N_4O_7$ molecule.

3,5-BIS(METHOXYCARBONYL)-2,6-DIMETHYL-4-(2-TRIFLUOROMETHYLPHENYL)-1,4-DIHYDRO-PYRIDINE
$C_{18}H_{18}F_3NO_4$

R. FOSSHEIM, 1985. *Acta Chem. Scand.*, **B39**, 785-790.

Monoclinic, $P2_1/a$, a = 7.775, b = 13.919, c = 15.394 Å, β = 100.84°, Z = 4. Mo radiation, R = 0.045 for 2731 reflections (at -150°C).

The dihydropyridine ring (Fig. 1) has a flat-boat conformation and the ring is approximately bisected by the plane of the pseudo-axial phenyl ring. The orientation of the phenyl is such that the ortho substituent points away from the dihydropyridine ring.

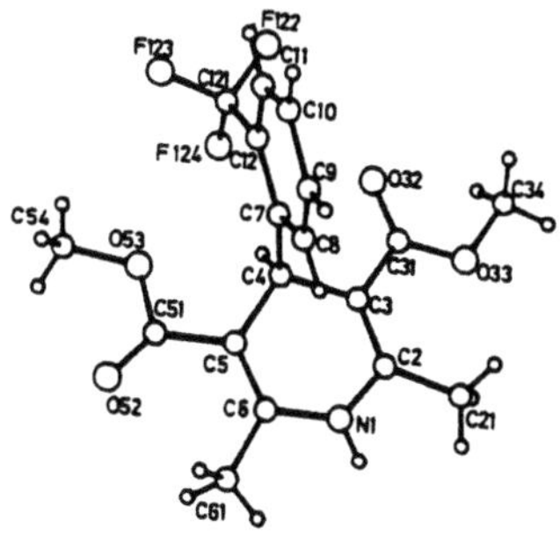

Fig. 1. Perspective drawing of a molecule of $C_{18}H_{18}F_3NO_4$.

6-DIETHYLAMINO-1-DIMETHYLTHIOCARBAMOYL-5-METHYL-3-PHENYL-1H-PYRIDAZIN-4-ONE $C_{18}H_{24}N_4OS$ (I)

5-DIETHYLTHIOCARBAMOYL-6-DIMETHYLAMINO-5-METHYL-3-PHENYL-5H-PYRIDAZIN-4-ONE $C_{18}H_{24}N_4OS$ (II)

P.F. LINDLEY, A.E. BAYDAR and G.V. BOYD, 1985. *Acta Cryst.*, **C41**, 1277-1280.

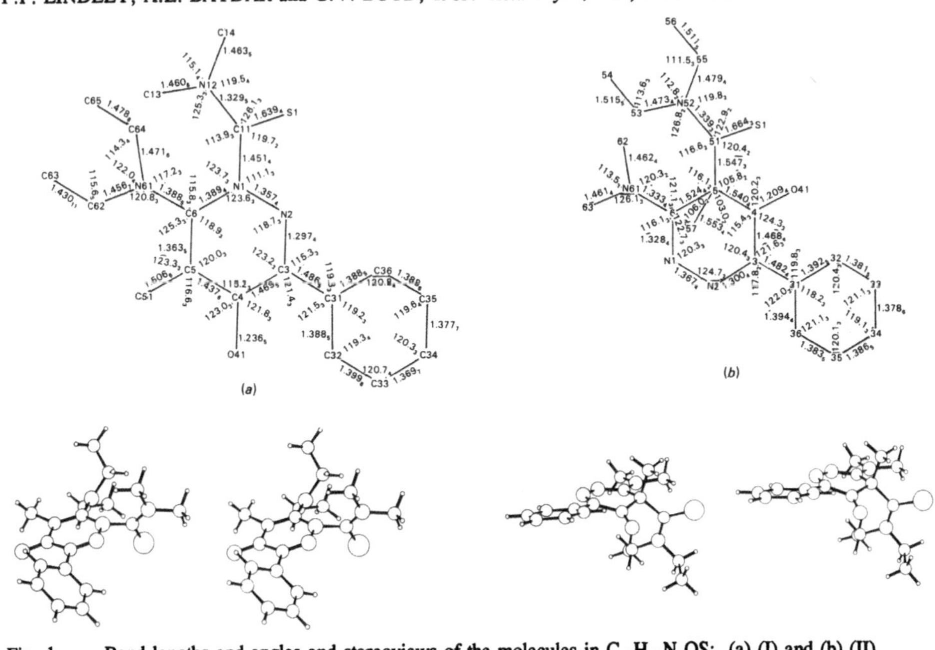

Fig. 1. Bond lengths and angles and stereoviews of the molecules in $C_{18}H_{24}N_4OS$: (a) (I) and (b) (II).

I. Monoclinic, $P2_1/c$, a = 13.210, b = 10.353, c = 14.386 Å, β = 110.586°, Z = 4. Cu radiation, R = 0.069 for 2432 reflections.

II. Monoclinic, $P2_1/n$, a = 11.154, b = 14.943, c = 11.066 Å, β = 102.239°, Z = 4. Cu radiation, R = 0.054 for 2446 reflections.

In (I) the pyridazine ring is planar while in (II) this ring shows pronounced puckering (Fig. 1) which can be described as a sofa-like conformation with a pseudo-mirror plane through N(1) and C(4). There is incomplete conjugation over the heterocyclic ring atoms in (II). The thiocarbamoyl moiety has similar geometry in (I) and (II). In (I), the dihedral angle between the plane of the thiocarbamoyl moiety and the pyridazine ring is 67.8(1)°, whereas in (II) it makes an angle of 79.9(1)° with the mean plane of atoms N(1), N(2), C(3) and C(6). The corresponding dihedral angles for the phenyl groups are 38.9(1) and 12.6(2)° respectively.

3,3'-BIS(4-HYDROXY-1-OXY-2,2,6,6-TETRAMETHYL-3-PIPERIDYL)METHANE
$C_{19}H_{36}N_2O_4$

R.P. SHIBAEVA, V.I. SUSKINA and A.B. SHAPIRO, 1985. *Kristallografiya*, **30**, 190-192.

Monoclinic, $P2_1/c$, a = 13.213, b = 13.930, c = 12.154 Å, β = 111.48°, Z = 4. Mo radiation, R = 0.045 for 1639 reflections.

Bond lengths are in Fig. 1; the hetero rings have twisted chair conformations. The nitrogen atoms are pyramidal (angle sum 355.9° at N(1), 356.7° at N(2)) and the angles between the N-O bonds and the C-N-C planes are 22.0 and 19.5° for N(1)-O(1) and N(2)-O(2) respectively.

Fig. 1. Bond lengths in $C_{19}H_{36}N_2O_4$.

trans,trans,cis-14-BENZYLOXY-2,4-DIMETHYL-5-PHENYLPIPERIDINE
$C_{20}H_{25}NO$

YEA-SHUN CHENG, E. HO, P.S. MARIANO and H.L. AMMON, 1985. *J. Org. Chem.*, **50**, 5678-5686.

Triclinic, $P\bar{1}$, a = 6.660, b = 11.268, c = 12.012 Å, α = 77.99, β = 89.68, γ = 77.19°, Z = 2. Cu radiation, R = 0.076 for 2223 reflections.

The crystal structure analysis established the stereochemistry of the molecule (Fig. 1).

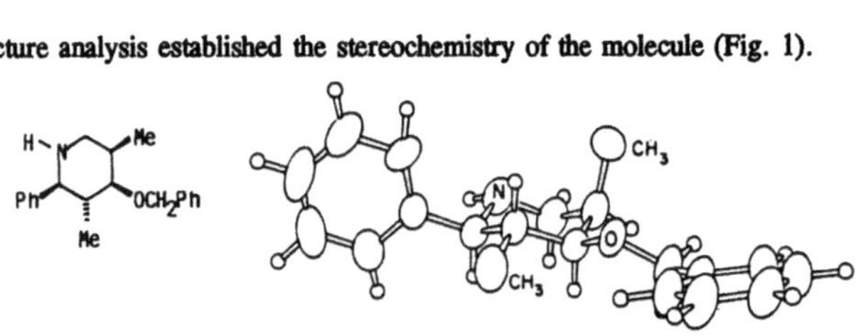

Fig. 1. ORTEP drawing of $C_{20}H_{25}NO$.

β-2-AMINO-3-BENZYL-3,5-DICYANO-6-ETHYLTHIO-4-PHENYL-3,4-DIHYDROPYRIDINE
$C_{22}H_{20}N_4S$

L. FUENTES, J.J. VAQUERO, J.C. DEL CASTILLO, M.I. ARDID and J.L. SOTO, 1985. *Heterocycles*, **23**, 93-98.

Monoclinic, $P2_1/n$, a = 11.289, b = 12.545, c = 14.153 Å, β = 90.470°, Z = 4. Mo radiation, R = 0.058.

The analysis established the structure and stereochemistry as shown in Fig. 1.

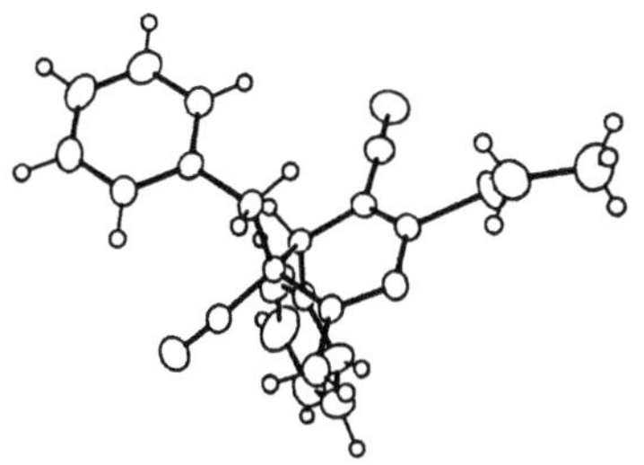

Fig. 1. A view of $C_{22}H_{20}N_4S$.

2-(HYDROXY(METHOXYCARBONYL)METHYL)-2-METHOXYCARBONYL-4,6-DIMETHYL-1-PHENYL-1,2-DIHYDROPYRIMIDINIUM PICRATE
$C_{23}H_{23}N_5O_{12}$ $C_{17}H_{21}N_2O_5^+, C_6H_2N_3O_7^-$

E. GUTIERREZ-PUEBLA, A. MONGE, C. RUIZ-VALERO and A. RODRIGUEZ-ROLDAN, 1985. *Acta Cryst.*, C41, 409-411.

Triclinic, $P\bar{1}$, a = 8.113, b = 12.289, c = 13.523 Å, α = 97.27, β = 105.67, γ = 89.11°, Z = 2. Mo radiation, R = 0.051 for 2161 reflections.

Bond lengths and angles in the unique cation and anion of the asymmetric unit (Fig. 1) are normal. These ions are interconnected by two asymmetrically bifurcated hydrogen bonds.

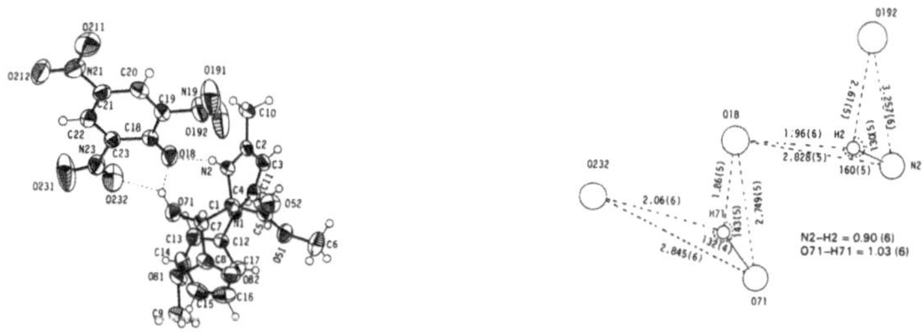

Fig. 1. $C_{23}H_{23}N_5O_{12}$: a perspective view of the cation-anion pair and geometrical details of the hydrogen-bonding network.

BIS(2-NONYLPYRIDINE 1-OXIDE)HYDROGEN TETRACHLOROAURATE(III)
$C_{28}H_{47}AuCl_4N_2O_2$

M.G.B. DREW, L.R. GLAVES and M.J. HUDSON, 1985. *J. Chem. Soc. Dalton Trans.*, 771-775.

Triclinic, $P\bar{1}$, a = 17.351, b = 10.381, c = 20.373 Å, α = 88.5, β = 101.8, γ = 99.6°, Z = 4. Mo radiation, R = 0.059 for 4846 reflections.

The asymmetric unit contains two $H(C_9H_{19}C_5H_4NO)_2^+$ cations and one $AuCl_4^-$ anion on general positions, and two anions on inversion centres. The cation (Fig. 1) contains a short hydrogen bond, O...O 2.39 and 2.41(2) Å; it is not clear whether this bond is symmetrical. The square-planar anions have dimensions Au-Cl 2.256-2.306(6) Å, Cl-Au-Cl 87.9-91.9(2)°. The Au atom on a general position is distant 3.36, 3.54(2) Å from cation O atoms, with O...Au-Cl angles within 15° of 90°.

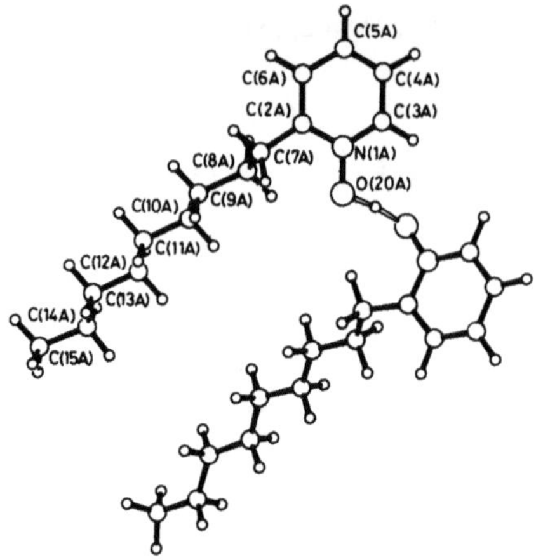

Fig. 1. The cation in $C_{28}H_{47}AuCl_4N_2O_2$.

N,N'-DITRITYL-2,5-DIKETOPIPERAZINE
$C_{42}H_{34}N_2O_2$ (I)

N,N'-DITRITYL-2,5-DIKETOPIPERAZINE METHYLENE CHLORIDE CLATHRATE
$C_{42}H_{34}N_2O_2,CH_2Cl_2$ (II)

I. GOLDBERG, Z. STEIN, L.-T.W. LIN and H. HART, 1985. *Acta Cryst.*, **C41**, 1539-1543.

I. Monoclinic, $P2_1/c$, a = 11.737, b = 16.199, c = 17.710 Å, β = 108.37°, Z = 4. Mo radiation, R = 0.047 for 2578 reflections.

II. Triclinic, $P\bar{1}$, a = 12.193, b = 12.378, c = 12.670 Å, α = 106.74, β = 104.01, γ = 90.38°, Z = 2. Mo radiation, R = 0.109 for 1940 reflections.

The N,N'-ditrityl-2,5-diketopiperazine molecule has the same overall geometry in (I) and (II) (Fig. 1). The central six-membered ring is folded about an axis passing through the two methylene C atoms. The dihedral angle between the two halves of the ring is 30.6(2) in (I) and 28(1)° in (II). The N-C(trityl) bonds are equatorial to the ring. There are several short intramolecular contacts between the central fragment and the peripheral trityl groups (2.822(4)-2.959(4) Å). In the crystals of (I), the molecules interpenetrate each other, with the trityl groups of one species lying above the central ring of an adjacent molecule. In (II), the host molecules are arranged in layers parallel to the ab plane of the unit cell, with the trityl groups in contact and the space between the diketopiperazine rings filled by methylene chloride molecules.

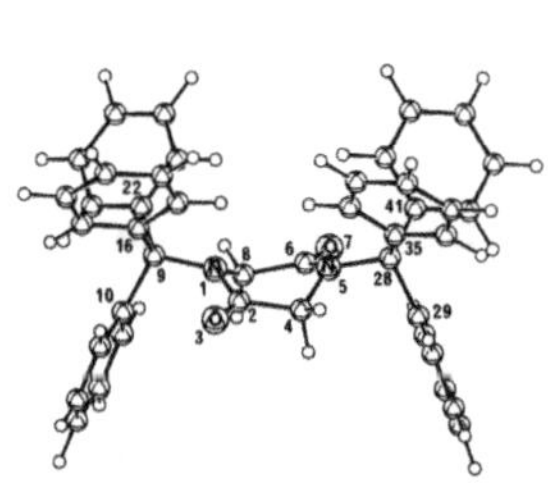

Fig. 1. The $C_{42}H_{34}N_2O_2$ molecular structure.

1,3,5,7-TETRANITRO-1,3,5,7-TETRAAZACYCLOOCTANE HYDRATE (γ-form) (OCTAHYDRO-1,3,5,7-TETRANITRO-1,3,5,7-TETRAZOCINE HYDRATE (γ-form))
$C_4H_8N_8O_8,0.25(H_2O)$

P. MAIN, R.E. COBBLEDICK and R.W.H. SMALL, 1985. *Acta Cryst.*, **C41**, 1351-1354.

Monoclinic, *Pn*, a = 13.27, b = 7.90, c = 10.95 Å, β = 106.8°, Z = 4. Mo radiation, R = 0.046 for 1267 reflections.

There are two crystallographically independent molecules in the asymmetric unit (Fig. 1). They have dimensions which are essentially identical and in good agreement with those found for the α, β and δ polymorphs of HMX (1). The γ-HMX molecules have a pseudo-diad axis of symmetry and pairs of such molecules with their axes parallel associate in the crystal. The site of the water molecule of solvation, O(9), has a 50% occupancy factor. The closest intermolecular C...O, O...N and O...O contacts are of length 3.01, 2.89 and 2.90 Å respectively. There are also several contacts with the water molecule of lengths in the range 2.91 to 3.10 Å, typical of weak hydrogen bonding.

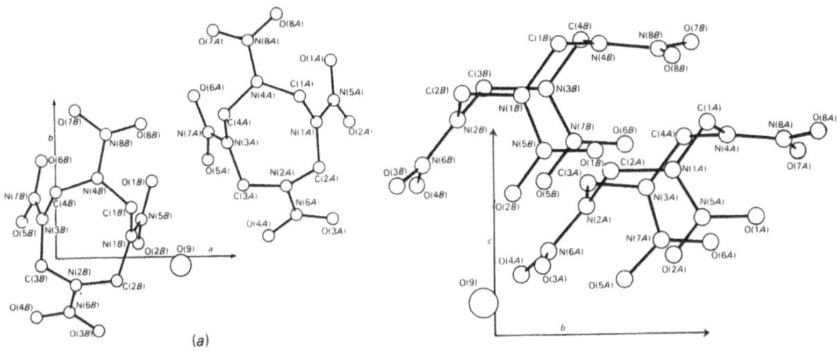

Fig. 1. $2(C_4H_8N_8O_8),0.5(H_2O)$: projection of the asymmetric unit along the *c*-axis (left) and along the *a*-axis (right).

1. Structure Reports, **28**, 467: **19**, 608: **40B**, 235.

OCTAHYDRO-1,3,5,7-TETRANITRO-1,3,5,7-TETRAZOCINE N-METHYL-2-PYRROLIDINONE $C_4H_8N_8O_8,C_5H_9NO$

T.M. HALLER, A.L. RHEINGOLD and T.B. BRILL, 1985. *Acta Cryst.*, **C41**, 963-965.

Rhombohedral, $R\bar{3}c$, a = 16.607, c = 31.506 Å, Z = 18. Mo radiation, R = 0.085 for 1287 reflections.

The structure (Fig. 1) consists of well-ordered substituted tetrazocine molecules in the chair-chair conformation and disordered N-methyl-2-pyrrolidinone molecules. The disorder in the latter is twofold rotational and the occupancies are equal. Bond distances and angles in the tetrazocine moiety are normal. The closest intermolecular contact is 3.016(7) Å (O(1)...C(2)).

Fig. 1. $C_4H_8N_8O_8,C_5H_9NO$: the tetrazocine molecule (left) and the model used for the disordered N-methyl-2-pyrrolidinone (right).

(4R,7S)-7-ISOPROPYL-4-METHYL-2-AZEPANONE ((-)-MENTHONE LACTAM) $C_{10}H_{19}NO$

H. TAKAYANAGI, E. HIROSE, K. FURUHATA and H. OGURA, 1985. *Bull. Chem. Soc. Jpn.*, **58**, 745-746.

Orthorhombic, $P2_12_12_1$, a = 8.837, b = 40.176, c = 6.028 Å, Z = 8. Cu radiation, R = 0.095 for 861

reflections.

The two independent molecules are shown in Fig. 1. The (-)-menthone lactam rings are in quasi-chair conformations with the amide chromophore being approximately coplanar. The mean value of the C-NH-CO-C torsion angle is +2°.

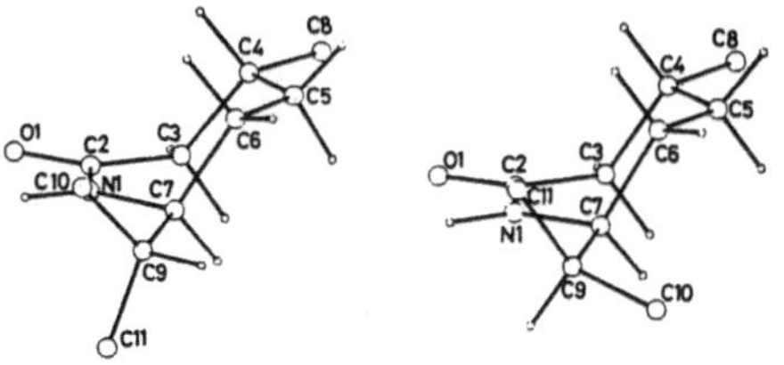

Fig. 1. Views of the two (-)-menthone lactam molecules.

4,4,7,7-TETRAMETHYLAZACYCLODECAN-2-ONE (4,4,7,7-TETRAMETHYLAZOCAN-2-ONE) $C_{13}H_{25}NO$ (I)

6,6,9,9-TETRAMETHYLAZACYCLODECAN-2-ONE (6,6,9,9-TETRAMETHYLAZOCAN-2-ONE) $C_{13}H_{25}NO$ (II)

C.L. BARNES, F.A. McGUFFEY and D. VAN DER HELM, 1985. *Acta Cryst.*, C41, 92-95.

I. Monoclinic, *C2/c*, a = 19.18(19.002), b = 16.458(16.224), c = 9.941(9.8983) Å, β = 121.33(121.14)°, Z = 8. Cu radiation, R = 0.035 for 2217 reflections (values in parentheses at 138 K).

II. Monoclinic, *P*2_1*/c*, a = 8.728(8.6282), b = 17.938(17.790), c = 9.825(9.768) Å, β = 121.15(120.855)°, Z = 4. Cu radiation, R = 0.077 for 2019 reflections (values in parentheses at 138 K).

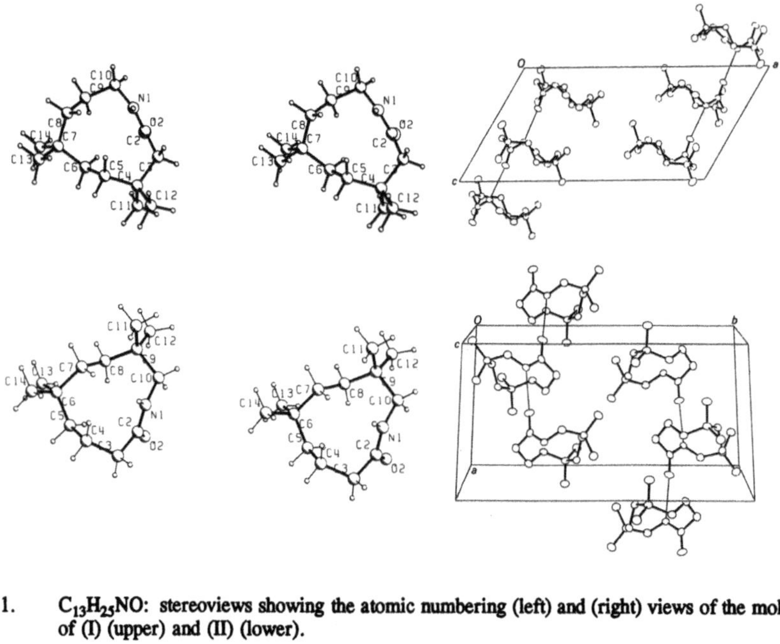

Fig. 1. $C_{13}H_{25}NO$: stereoviews showing the atomic numbering (left) and (right) views of the molecular packing of (I) (upper) and (II) (lower).

In both isomeric molecules (Fig. 1) the amide group is trans but significantly distorted from planarity. This nonplanarity is associated with its out-of-plane bonding to O(2) with N(1)...O(2) distances of 2.888(2) and 2.914(2) Å in (I) and (II) respectively. These hydrogen bonds link the molecules into chains along the c-axis in (I) and along the a-axis in (II). The conformations of the macroring in (I) and (II) are similar to each other and to that in the parent molecule, pelargolactam (1). There are several rather short transannular H...H distances in both isomers.

1. Structure Reports, 41B, 308.

2-(DIETHYLAMINO)-5-METHYL-3-(METHYLTHIO)-5-PHENYL-5H-AZEPINE
$C_{18}H_{24}N_2S$

J.W. STREEF, H.C. VAN DER PLAS, N. NIEMAN and C.H. STAM, 1985. *Rec. J. R. Neth. Chem. Soc.*, **104**, 166-170.

Monoclinic, $P2_1/a$, a = 23.4136, b = 8.1024, c = 9.5286 Å, β = 101.86°, Z = 4. Cu radiation, R = 0.035 for 2149 reflections.

The molecule structure and stereochemistry are established as shown in Fig. 1.

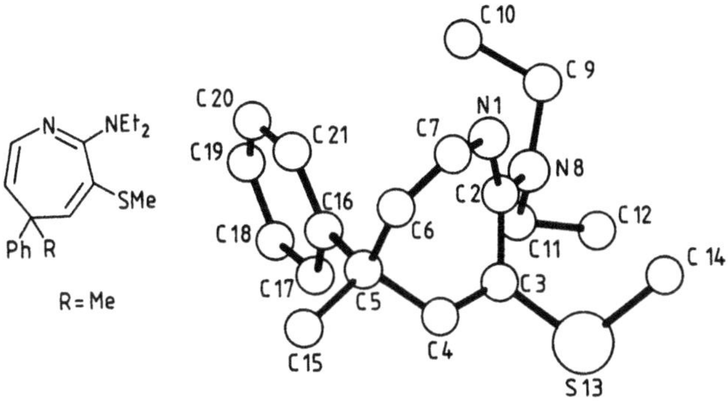

Fig. 1. The $C_{18}H_{24}N_2S$ molecule.

5,6-DIHYDRO-6-METHOXY-7H-TETRAZOLO[1,5-d](1,4)DIAZEPINE
$C_6H_9N_5O$

Y. OHBA, I. MATSUKURA, Y. FUKAZAWA and T. NISHIWAKI, 1985. *Heterocycles*, **23**, 287-290.

Monoclinic, $C2/c$, a = 19.625, b = 6.255, c = 14.053 Å, β = 105.77°, Z = 8. Mo radiation, R = 0.045 for 1501 reflections.

The analysis established the structure and stereochemistry as shown in Fig. 1.

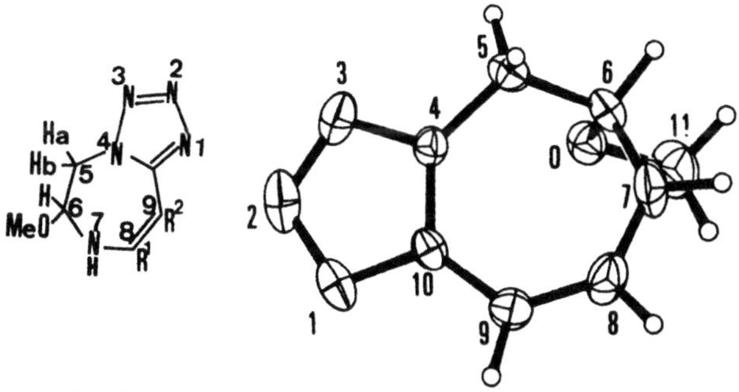

Fig. 1. The $C_6H_9N_5O$ molecule.

1,3-DIHYDRO-2H-BENZIMIDAZOL-2-ONE

$C_7H_6N_2O$

F.H. HERBSTEIN and M. KAPON, 1985. *Z. Kristallogr.*, **173**, 249-256.

Monoclinic, *P*2_1/*n*, a = 5.296, b = 5.077, c = 23.444 Å, β = 94.50°, Z = 4. Mo radiation, R = 0.0635 for 1499 reflections.

A refinement with new data and correct space group of a structure reported previously (1). Molecular dimensions and packing are shown in Fig. 1. The data establish the solid state tautomer as the keto and not the enol form.

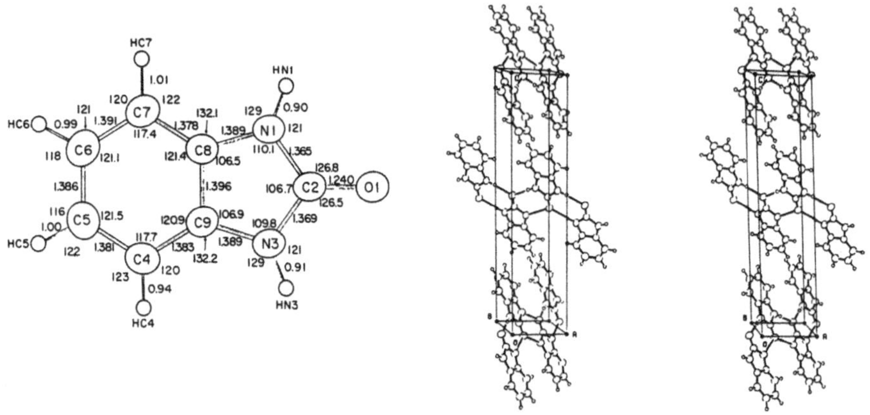

Fig. 1. Molecular dimensions and packing for $C_7H_6N_2O$.

1. Structure Reports, **48B**, 282.

8-CARBAMOYL-3-(2-CHLOROETHYL)IMIDAZO[5,1-d]-1,2,3,5-TETRAZIN-4(3H)-ONE (MITOZOLOMIDE) $C_7H_7ClN_6O_2$

P.R. LOWE, C.H. SCHWALBE and M.F.G. STEVENS, 1985. *J. Chem. Soc. Perkin Trans. 2*, 357-361.

Triclinic, *P*$\bar{1}$, a = 7.003, b = 8.680, c = 16.041 Å, α = 93.76, β = 93.99, γ = 92.08°, Z = 4. Mo radiation, R = 0.052 for 3244 reflections.

There are two independent molecules (Fig. 1) in the crystallographic asymmetric unit and these are rotamers about the C(8)-C(81) and C(8)'-C(81)' bonds, the orientation of the carbamoyl group in one rotamer facilitating an intramolecular hydrogen bond of the type N-H...N. Apart from the chloroethyl side chain, both molecules are approximately planar and intermolecular hydrogen bonds hold groups of four molecules together around the centre of symmetry.

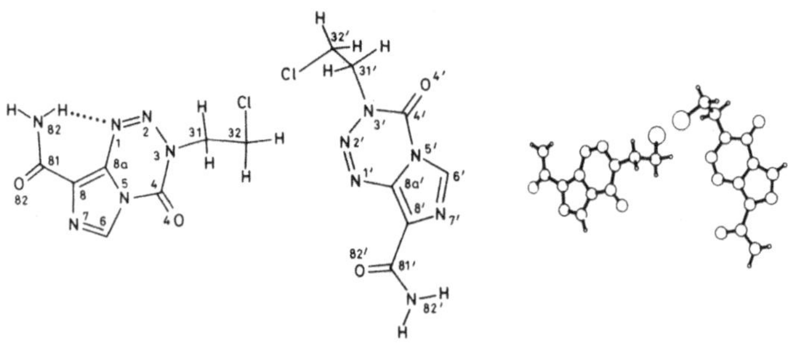

Fig. 1. Perspective drawing of the two molecules of $C_7H_7ClN_6O_2$ in the crystallographic asymmetric unit.

2-THIOXOQUINAZOL-4-ONE
$C_8H_6N_2OS$

B. TASHKHODZHAEV, S. YANGIBAEV and Kh.M. SHAKHIDAYATOV, 1985. *Zh. Strukt. Khim.*, **26**, 1-155-156 [J. Struct. Chem., **26**, 135-136].

Monoclinic, *B*112/*b*, a = 21.942, b = 17.578, c = 4.184 Å, γ = 80.14°, Z = 8. Mo radiation, R = 0.078 for 876 reflections.

The structure is shown in Fig. 1. The molecule is planar to within ±0.03 Å. Bond lengths and angles are normal. The packing is dominated by N-H...S and N-H...O hydrogen bonding (N...S 3.34, N...O 2.83 Å).

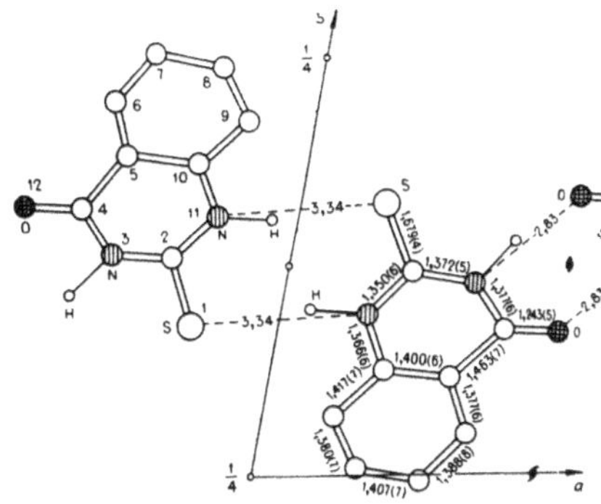

Fig. 1. Projection of the $C_8H_6N_2OS$ structure on the *ab* plane.

N-AMINOPHTHALIMIDE
$C_8H_6N_2O_2$

J.H. LOEHLIN, 1985. *Acta Cryst.*, **C41**, 210-212.

Orthorhombic, $P2_12_12_1$, a = 20.472, b = 6.672, c = 5.254 Å, Z = 4. Mo radiation, R = 0.046 for 779 reflections.

The molecule (Fig. 1), with the exception of the ammine hydrogen atoms, is planar with an N-N bond length of 1.409(3) Å. The terminal N atoms form a hydrogen-bonded chain along a 2_1 screw axis in the c direction (N(2)...N(2') 3.197(5)), with a second hydrogen bond C=O...H-C (3.349(5) Å) connecting neighbouring molecules diagonally.

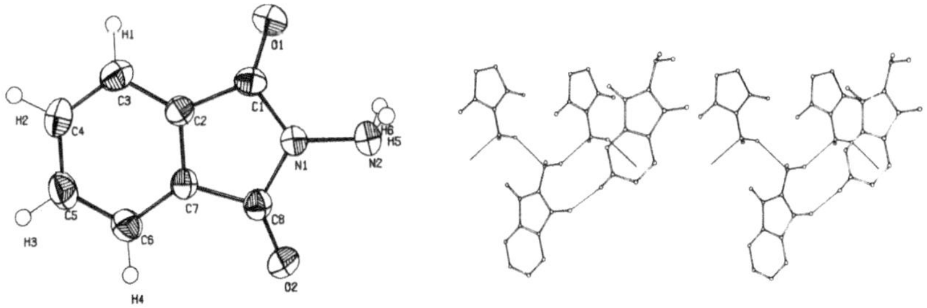

Fig. 1. $C_8H_6N_2O_2$: the molecular structure and a stereoview normal to the 2_1 axis.

4-OXOPYRIDO[1,2-a]PYRIMIDIN-1-IUM-2-OLATE
$C_8H_6N_2O_2$

N. THORUP and O. SIMONSEN, 1985. *Acta Cryst.*, **C41**, 472-474.

Monoclinic, $P2_1/c$, a = 8.358, b = 5.432, c = 15.153 Å, β = 94.11°, Z = 4. Mo radiation, R = 0.040 for 1991 reflections.

In the solid state the molecule (Fig. 1) exists in a tautomeric form with the "acidic" H atom bonded to N. The N(1)-C(10) and N(5)-C(10) bonds have partial double bond character and bonds C(2)-O(2) and C(4)-O(4) have significantly different lengths. This, it is postulated, may be due to the different strengths of hydrogen bonding in which O(2) and O(4) are involved. The N(1)-H...O(2) hydrogen bond is of length 2.698(2) Å while the C(6)-H...O(4) interaction is of length 3.180(2) Å.

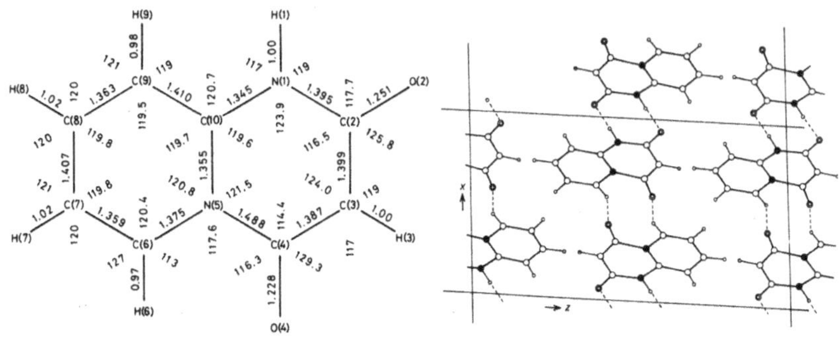

Fig. 1. Bond lengths and angles in the $C_8H_6N_2O_2$ molecule and a projection down the b-axis of the crystal structure with hydrogen bonds indicated by dotted lines.

1,4-DIACETYL-3,6-DINITROTETRAHYDROIMIDAZO[4,5-d]IMIDAZOLE-2,5(1H,3H)DIONE (3,6-DIACETYL-1,4-DINITROGLYCOLURIL)
$C_8H_8N_6O_8$

J. BOILEAU, E. WIMMER, M. PIERROT, A. BALDY and R. GALLO, 1985. *Acta Cryst.*, C41, 1680-1683.

Orthorhombic, $P2_12_12_1$, a = 6.858, b = 9.638, c = 18.551 Å, Z = 4. Mo radiation, R = 0.050 for 1633 reflections.

The two five-membered rings which are fused together along the C(7)-C(8) bond in the molecule (Fig. 1) are each planar to within 0.06 Å and the dihedral angle between their mean planes is 115(1)°. The core of the molecule has a pseudo-diad symmetry axis passing through the midpoint of the C(7)-C(8) bond and bisecting the angle between the ring planes. The two nitro substituents are in positions 1 and 4 and the acetyl substituents are in positions 3 and 6. The planes of these substituent groups are twisted with respect to those of the rings to which they are attached. Bond lengths and angles in the molecule are in good agreement with generally accepted values. Except for two short C-H...O contacts of mean length 2.87(2) Å, all other intermolecular distances are consistent with van der Waals interactions.

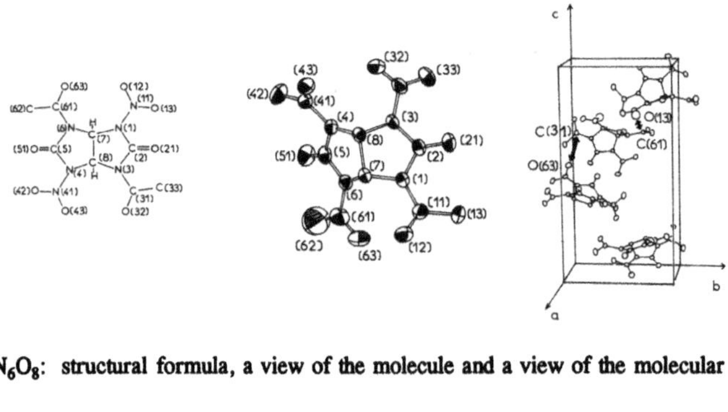

Fig. 1. $C_8H_8N_6O_8$: structural formula, a view of the molecule and a view of the molecular packing.

3-AMINO-1,3-DIHYDRO-2-METHYL-4-AZA-ISOINDOL-1-ONE
$C_8H_9N_3O$

M. WOSNIAK, H.C. VAN DER PLAS and S. HARKEMA, 1985. *J. Org. Chem.*, **50**, 3435-3437.

Triclinic, $P\bar{1}$, a = 9.984, b = 9.710, c = 9.590 Å, α = 112.51, β = 112.76, γ = 73.99°, Z = 4. Mo radiation, R = 0.059 for 1757 reflections.

There is some disorder in the crystal; this is in the amino group. The crystal structure analysis established the identity of this reaction product.

3,5,7-TRICHLORO-3-METHYL-3H-INDOL-2-ONE
$C_9H_6Cl_3NO$

D.K. CHAKRABORTY and S.K. TALAPATRA, 1985. *Acta Cryst.*, C**41**, 1365-1366.

Monoclinic, $P2_1/c$, a = 10.069, b = 10.064, c = 11.405 Å, β = 113.75°, Z = 4. Mo radiation, R = 0.035 for 1591 reflections.

The five-membered ring in the indole nucleus of the molecule (Fig. 1) is nearly planar and the dihedral angle between the least-squares planes through the two rings is only 1.2(1)°. The N(1)-C(7) bond is shortened to 1.360(4) Å due to the delocalized π-electron system of the benzenoid ring extending over atoms N(1), C(7) and O(1). The molecular dimensions agree well with values found in similar compounds. The structure is stabilized in the form of centrosymmetric hydrogen-bonded dimers linked through N(1)-H...O(1) bonds of length 2.847(3) Å. There are no unusually short inter-dimer contacts.

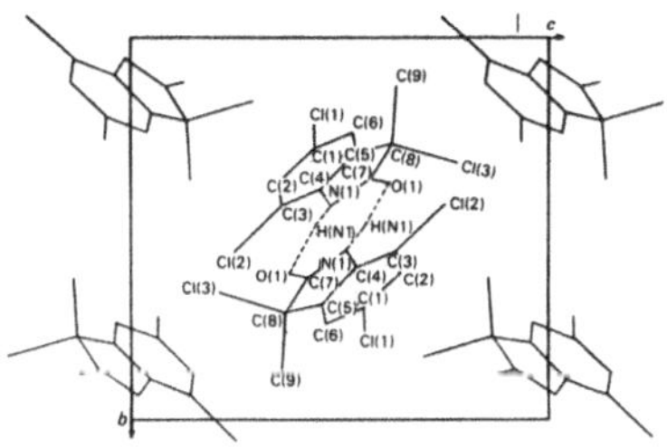

Fig. 1. A projection of the $C_9H_6Cl_3NO$ crystal structure onto (100).

5-CHLORO-3-HYDROXY-3-METHYL-3H-INDOL-2-ONE
$C_9H_8ClNO_2$

D.K. CHAKRABORTY, S.K. TALAPATRA and A. CHATTERJEE, 1985. *Acta Cryst.*, C**41**, 1363-1364.

Monoclinic, $P2_1/c$, a = 7.625, b = 9.968, c = 11.652 Å, β = 91.61°, Z = 4. Mo radiation, R = 0.029 for 1227 reflections.

The five-membered ring of the oxindole nucleus of the molecule (Fig. 1) is slightly puckered. The dihedral angle between the atomic planes N(1),C(4,5) and N(1),C(7,8) is 5.6(2)°. The N(1),C(4,7),O(1) system is planar and the N(1)-O(7) bond length of 1.351(2) Å is shortened due to resonance of the N electron pair with the carbonyl group. The benzenoid ring is planar, Cl is essentially coplanar and N(1) only deviates by 0.026(2) Å. Other bond lengths and angles in the molecule agree with expected values. The molecules are linked into centrosymmetric dimers through O(2)-H...O(1) hydrogen bonds of length 2.667(2) Å. The dimers are linked by N(1)-H...O(2) hydrogen bonds of length 2.874(3) Å. Alternating layers of oppositely oriented molecules form stacks in the crystal along the *a*-axis.

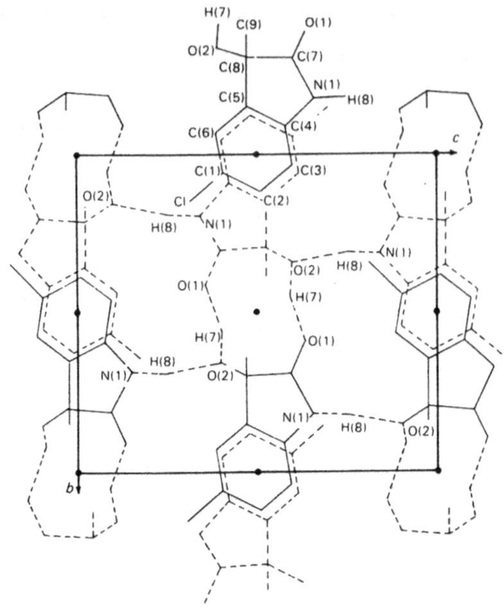

Fig. 1. $C_9H_8ClNO_2$: a projection onto (100) of the unit cell contents showing the hydrogen bonding and the stacks of oppositely oriented molecules.

3-ETHYL-4-HYDROXY-1,2,3-(4H)-BENZOTRIAZINE
$C_9H_{11}N_3O$ (I)

6-CHLORO-4-HYDROXY-3-METHYL-4-PHENYL-1,2,3-BENZOTRIAZINE
$C_{14}H_{12}ClN_3O$ (II)

KWONG-KHEE LAI, C.H. SCHWALBE, K. VAUGHAN, R.J. LAFRANCE and C.D. WHISTON, 1985. *Canad. J. Chem.*, **63**, 581-585.

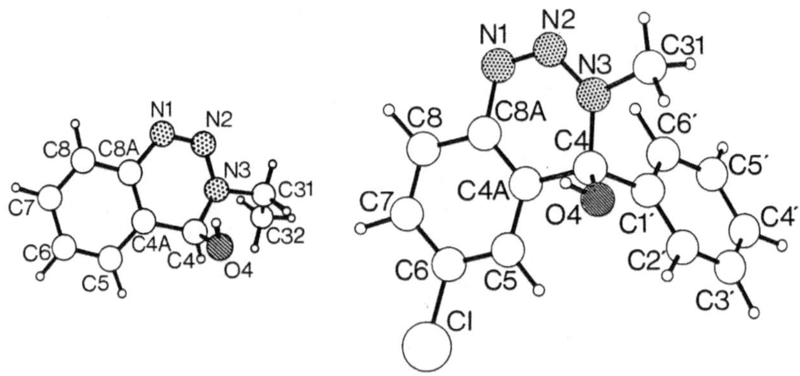

Fig. 1. Views of $C_9H_{11}N_3O$ (I) (left) and $C_{14}H_{12}ClN_3O$ (II) (right).

I. Orthorhombic, *Pbcn*, a = 14.346, b = 7.239, c = 17.276 Å, Z = 8. Mo radiation, R = 0.043 for 890 reflections.

II. Monoclinic, *P*2_1/*n*, a = 12.222, b = 7.482, c = 14.170 Å, β = 94.06°, Z = 4. Mo radiation, R = 0.060

for 2128 reflections.

In both compounds the triazine rings (Fig. 1) have short N(1)-N(2) bonds (1.268(3) (I); 1.263(3) Å (II)) with longer N(2)-N(3) bonds (1.333(3) (I); 1.316(3) Å (II)). The geometry at C(4) is tetrahedral in both (I) and (II) but the hetero ring is puckered in (I) and planar in (II). Molecules are linked in the crystals by O(4)-H...N(1) hydrogen bonds (O...N 2.842(4) Å in (I), 2.778(3) Å in (II)).

8-HYDROXY-2-METHYLAMINOQUINOLINE
$C_{10}H_{10}N_2O$ (I)

2-ANILINO-8-HYDROXYQUINOLINE
$C_{15}H_{12}N_2O$ (II)

I.N. POLYAKOVA, Z.A. STARIKOVA, B.V. PARUSNIKOV and I.A. KRASAVIN, 1985. *Kristallografiya*, **30**, 310-313.

I. Monoclinic, $P2_1/c$, a = 5.718, b = 12.164, c = 12.780 Å, β = 101.36°, Z = 4. Mo radiation, R = 0.060 for 1176 reflections.

II. Monoclinic, $P2_1/c$, a = 9.192, b = 9.261, c = 14.333 Å, β = 108.05°, Z = 4. Mo radiation, R = 0.042 for 1606 reflections.

The geometry of the 8-hydroxyquinoline fragments of (I) and (II) are similar (Fig. 1). The geometry at the exocyclic N atoms is flattened pyramidal. In (II) repulsion between quinoline and phenyl rings leads to the dihedral angle between their planes being 50.1° and the C-N-C angle being increased to 127.9(2)°. Both molecules have intramolecular O-H...N hydrogen bonds and intermolecular N(2)-H...O join up the molecules in zigzag chains.

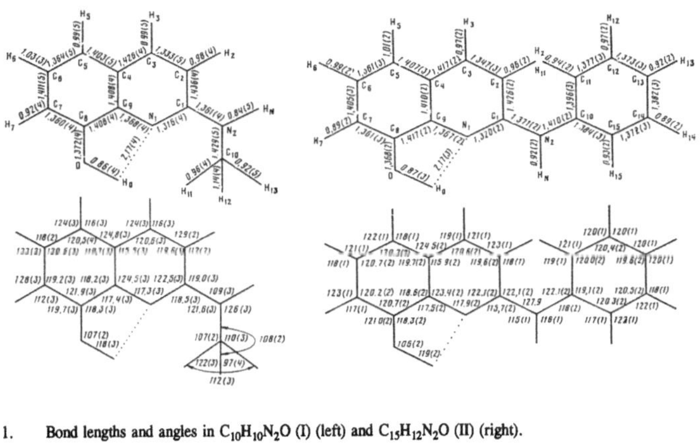

Fig. 1. Bond lengths and angles in $C_{10}H_{10}N_2O$ (I) (left) and $C_{15}H_{12}N_2O$ (II) (right).

1-METHYL-2-DIMETHYLAMINOBENZIMIDAZOLIUM PERCHLORATE
$C_{10}H_{14}ClN_3O_4$ (I)

3-METHYL-2-DIMETHYLAMINOBENZOXAZOLIUM IODIDE
$C_{10}H_{13}IN_2O,H_2O$ (II)

3-METHYL-2-DIMETHYLAMINOBENZOTHIAZOLIUM IODIDE
$C_{10}H_{13}IN_2S$ (III)

1,3-DIMETHYL-2-DIMETHYLAMINOBENZIMIDAZOLIUM PERCHLORATE
$C_{11}H_{16}ClN_3O_4$ (IV)

1,3-DIMETHYL-2-METHYLAMINOBENZIMIDAZOLIUM PERCHLORATE
$C_{10}H_{14}ClN_3O_4$ (V)

1,3-DIMETHYL-2-AMINOBENZIMIDAZOLIUM PERCHLORATE
$C_9H_{12}ClN_3O_4$ (VI)

R. BENASSI, R. GRANDI, U.M. PAGNONI, F. TADDEI, G. BOCELLI and P. SGARABOTTO, 1985. *J. Chem. Soc. Perkin Trans. 2*, 1513-1521.

I. Monoclinic, *C2*, a = 19.081, b = 6.760, c = 9.765 Å, β = 103.6°, Z = 4. Cu radiation, R = 0.074 for 1048 reflections.

II. Monoclinic, *C2/c*, a = 24.307, b = 6.891, c = 14.347 Å, β = 90.7°, Z = 8. Mo radiation, R = 0.047 for 2017 reflections.

III. Monoclinic, $P2_1/n$, a = 14.423, b = 7.434, c = 10.869 Å, β = 98.8°, Z = 4. Mo radiation, R = 0.031 for 2027 reflections.

IV. Monoclinic, $P2_1/c$, a = 6.898, b = 18.845, c = 11.127 Å, β = 109.5°, Z = 4. Cu radiation, R = 0.060 for 1377 reflections.

V. Monoclinic, *Pn*, a = 6.238, b = 8.333, c = 11.751 Å, β = 95.9°, Z = 2. Cu radiation, R = 0.059 for 871 reflections.

VI. Orthorhombic, $Pna2_1$, a = 10.069, b = 19.449, c = 5.892 Å, Z = 4. Cu radiation, R = 0.065 for 659 reflections.

Fig. 1. shows a projection of the cation of molecule (IV) onto the benzimidazolium plane and the numbering scheme. The amino group is coplanar with respect to the plane of the rings in all six molecules. The NN-dimethylamino group tends to be distorted when a methyl group is present at the heterocyclic nitrogen atom. A full discussion and comparison of the molecular geometries of the anions and cations, and the hydrogen bonding, is given in the paper.

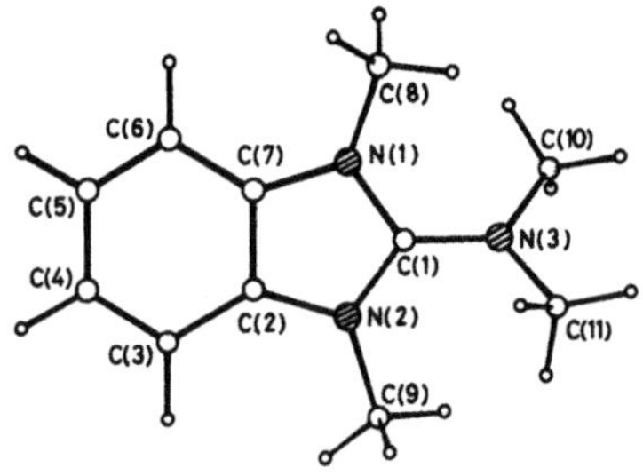

Fig. 1. A projection of the cation of $C_{11}H_{16}ClN_3O_4$ on the benzimidazolium plane.

THIOSEMICARBAZONE-8-QUINOLINALDEHYDE
$C_{11}H_{10}N_4S$

Yu.M. CHUMAKOV, V.N. BYUSHKIN and V.G. BODYU, 1985. *Zh. Strukt. Khim.*, **26**, 6-114-119 [J. Struct. Chem., **26**, 929-934].

Monoclinic, $P112_1/b$, a = 8.952, b = 9.760, c = 12.430 Å, γ = 84.18°, Z = 4. Mo radiation, R = 0.0459 for 804 reflections.

The molecule is planar to within 0.085 Å. Bond lengths are generally as expected. The thiosemicarbazone C-N bond, 1.286(6) Å, is shorter than in related molecules. Molecules are linked by N-H...S and N-H...N hydrogen bonds in the crystal (Fig. 1).

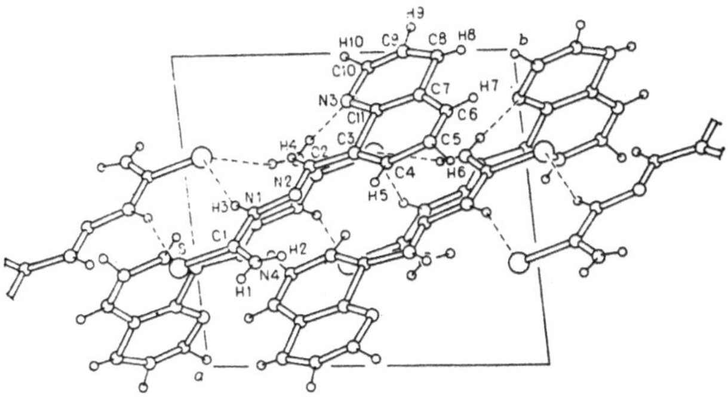

Fig. 1. Projection of the $C_{11}H_{10}N_4S$ structure on the (001) plane.

2-METHYL-3-METHYLTHIO-6-PHENYL-7H-1,2,4-TRIAZOLIDO[4,3-b](1,2,4)TRIAZOLIUM
$C_{11}H_{11}N_5S$

M. ALAJARIN, P. MOLINA, A. TARRAGA, M.J. VILAPLANA, C. FOCES-FOCES, F.H. CANO, R.M. CLARAMUNT and J. ELGUERO, 1985. *Bull. Chem. Soc. Jpn.*, **58**, 735-744.

Monoclinic, $P2_1/c$, a = 12.6939, b = 16.0936, c = 12.0239 Å, β = 107.247°, Z = 8. Cu radiation, R = 0.054 for 3560 reflections.

There are two independent molecules in the asymmetric unit (Fig. 1) and both have planar triazolo-triazole ring systems with the S-methyl group directed towards N(5) (torsion angles C(11)-S(10)-C(3)-N(4) 58.9(3) and -74.8(4)° for the two molecules). The phenyl rings are rotated -26.7(4) and -4.3(4)° out of coplanarity with the triazolo-triazole ring system.

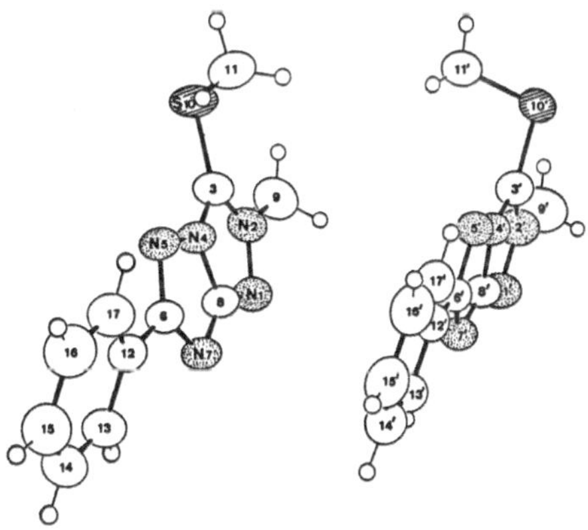

Fig. 1. The two independent $C_{11}H_{11}N_5S$ molecules.

6-HYDROXY-6-PHENYL-1-AZABICYCLO[3.2.0]HEPTAN-2-ONE
$C_{12}H_{13}NO_2$

Y. TROIN, J.-C. GRAMAIN and T. PRANGE, 1985. *Acta Cryst.*, **C41**, 1093-1095.

Monoclinic, $C2/c$, a = 19.813, b = 9.119, c = 11.642 Å, β = 104.6°, Z = 8. Cu radiation, R = 0.064 for 1769 reflections.

In the molecule (Fig. 1) the lactam ring has a flattened envelope conformation with C(2) displaced by 0.23(5)

Å from the N(1), C(3,4,5) plane. In the fused four-membered azetidine ring the dihedral angles between the N(1), C(5,6) and N(1), C(6,7) planes is 15.5(5)°. The strained ring junction forces the amidic atom, N(1), to be pyramidal. The crystal structure is stabilized by intermolecular O(6)-H...O(2) hydrogen bonds with an O...O separation of 2.71(1) Å. The pucker in the five-membered ring in the crystal is greater than that which the NMR data suggests for the molecule in solution.

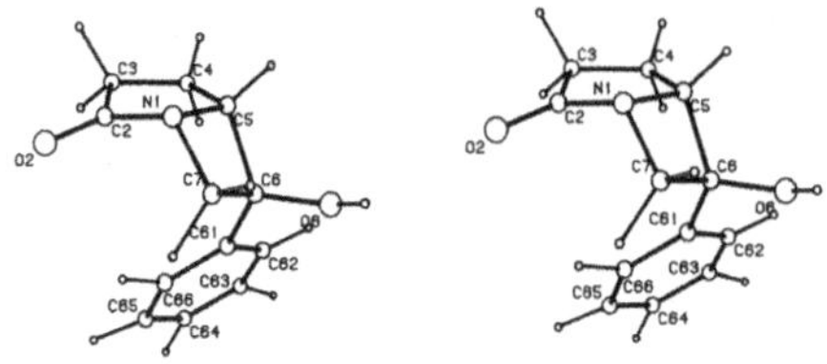

Fig. 1. A stereoview of the $C_{12}H_{13}NO_2$ molecule.

3-ETHOXY-5,6,7,8-TETRAHYDRO-1-HYDROXYISOQUINOLINE-4-CARBONITRILE
$C_{12}H_{14}N_2O_2$

S. SALISI, J.L. BRIANSO, X. SOLANS and M. FONT-ALTABA, 1985. *Acta Cryst.*, **C41**, 474-476.

Monoclinic, $P2_1/n$, a = 20.00, b = 6.219, c = 8.794 Å, β = 94.31°, Z = 4. Mo radiation, R = 0.056 for 1369 reflections.

In the molecule (Fig. 1) the pyridine ring is planar with O(16) and the carbonitrile group deviating on the same side of the ring plane. The ethoxy group and the fused six-membered ring, which has a skew-boat conformation, deviate to the other side. Bond lengths and angles agree well with generally accepted values. Enantiomorphic molecules are hydrogen bonded through the carbonyl O atom and the N atom of the isoquinoline moiety. The N(10)-H...O(16) separation is 2.821(3) Å.

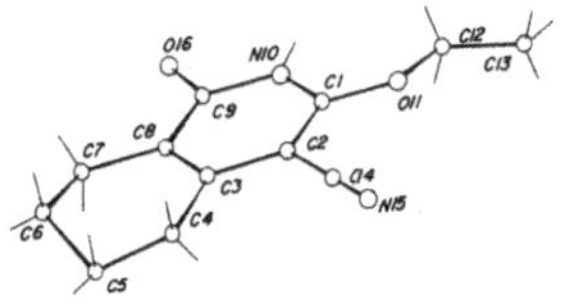

Fig. 1. A view of the $C_{12}H_{14}N_2O_2$ molecule showing the atomic numbering.

1,2-DIHYDRO-6-HYDROXY-2,2,4-TRIMETHYLQUINOLINE
$C_{12}H_{15}NO$ (I)

2,6-DIHYDRO-6-OXO-2,2,4-TRIMETHYLQUINOLINE
$C_{12}H_{13}NO$ (II)

A.E. OBODOVSKAYA, Z.A. STARIKOVA, Yu.A. IVANOV and I.E. POKROVSKAYA, 1985. *Zh. Strukt. Khim.*, **26**, 5-93-98 [J. Struct. Chem., **26**, 729-733].

I. Orthorhombic, $P2_12_12_1$, a = 12.121, b = 11.003, c = 7.797 Å, Z = 4. Mo radiation, R = 0.044 for 987 reflections.

II. Orthorhombic, *Pbcn*, a = 10.616, b = 17.156, c = 11.401 Å, Z = 8. Mo radiation, R = 0.040 for 842 reflections.

The structures are shown in Fig. 1. The dihydrogenated ring in (II) is nonplanar. The length of the C-C(=C) bond (1.470 Å) and the nonplanarity of the π-systems of the C=C double bond and of the benzene ring (dihedral angle 15.4°) indicate less conjugation in comparison with planar dienic molecules.

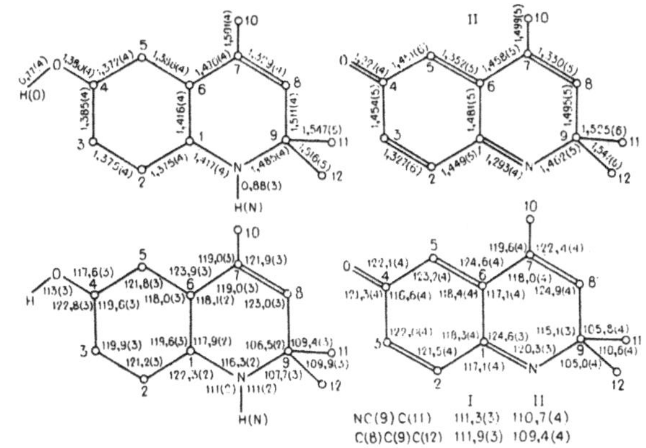

Fig. 1. Geometry of (I) (left) and (II) (right).

3aβ,4,5,7aβ-TETRAHYDRO-2,4β-DIMETHYL-4α-ETHOXY-1H-ISOINDOLE-1,3-(2H)-DIONE
$C_{12}H_{17}NO_3$

Y. KANAOKA, Y. HATANAKA, E.N. DUESLER, I.L. KARLE and B. WITKOP, 1985. *Tetrahedron,* **41**, 5365-5370.

Monoclinic, $P2_1/c$, a = 8.088, b = 22.855, c = 7.381 Å, β = 119.9°, Z = 4. Cu radiation, R = 0.090 for 1714 reflections.

The molecule has a cis ring junction with the methyl group on C(4) on the same side as the cis hydrogens on C(3a) and C(7a) (Fig. 1).

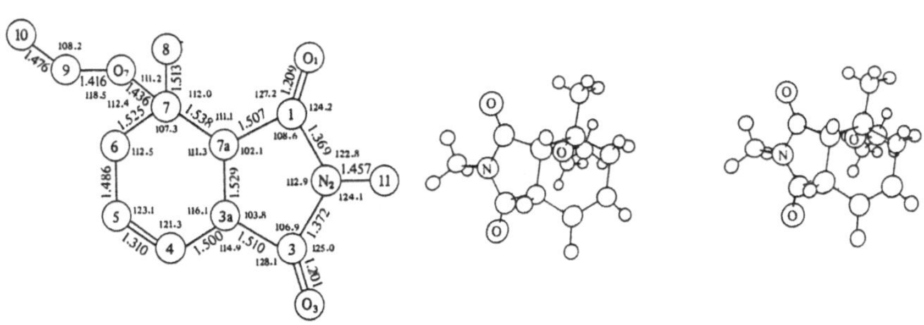

Fig. 1. Dimensions and conformation of $C_{12}H_{17}NO_3$.

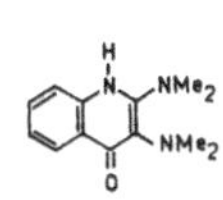

2,3-BIS(DIMETHYLAMINO)QUINOL-4-ONE
$C_{13}H_{17}N_3O$

A. BOUVY, Z. JANOUSEK, H.G. VIEHE and J.M. ARIETTA, 1985. *Bull. Soc. Chim. Belg.*, **94**, 869-871.

Tetragonal, $P4_12_12$, a = 9.746, c = 26.534 Å, Z = 8. Mo radiation, R = 0.055 for 688 reflections.

The analysis established the structure of this rearrangement product as shown above.

4-AMINO-N-(5-CHLORO-2-QUINOXALINYL)BENZENESULPHONAMIDE ACETONITRILE SOLVATE
$C_{14}H_{11}ClN_4O_2S,C_2H_3N$

H.M. DEUTSCH, D.G. VANDERVEER and L.H. ZALKOW, 1985. *Acta Cryst.*, C41, 1116-1118.

Monoclinic, $P2_1/c$, a = 7.888, b = 20.527, c = 11.068 Å, β = 98.64°, Z = 4. Mo radiation, R = 0.064 for 2107 reflections.

In the molecule (Fig. 1) the S atom is tetrahedrally coordinated to two O atoms, to the 4-aminobenzene group and through N(3) to the 5-chloro-2-quinoxalinyl moiety. Bond lengths and angles in the molecule agree with normal values. The fused ring system and its immediate attachments are coplanar and S is displaced from this plane by 0.392(4) Å. In the 4-aminobenzene fragment the ring is slightly nonplanar with a twist about the line joining the midpoints of the C(9)-C(10) and C(12)-C(13) bonds. Atoms S and N(4) are displaced from the least-squares plane of the ring by 0.040(4) and 0.070(8) Å respectively. The dihedral angle between these planes is 96.0(8)°. There appears to be a N(3)-H...N hydrogen bond to the solvate molecule with an H...N separation of 1.98(1) Å.

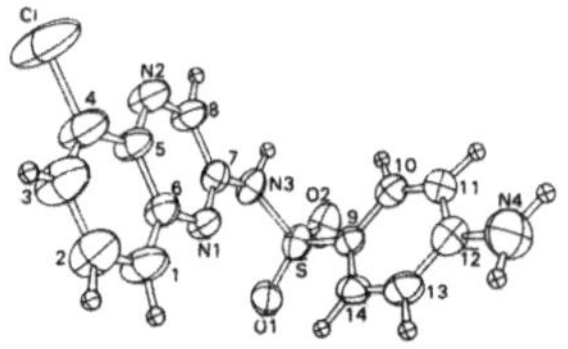

Fig. 1. A perspective view of the unsolvated $C_{14}H_{11}ClN_4O_2S$ molecule.

2-(3,6-DIMETHYLPYRAZIN-2-YL)INDOLE
$C_{14}H_{13}N_3$

Y. AKITA, A. INOUE, K. YAMAMOTO, A. OHTA, T. KURIHARA and M. SHIMIZU, 1985. *Heterocycles*, 23, 2327-2333.

Orthorhombic, *Pbnb*, a = 12.862, b = 19.692, c = 9.184 Å, Z = 8. Mo radiation, R = 0.078 for 1566 reflections.

The analysis established the structure as shown in Fig. 1.

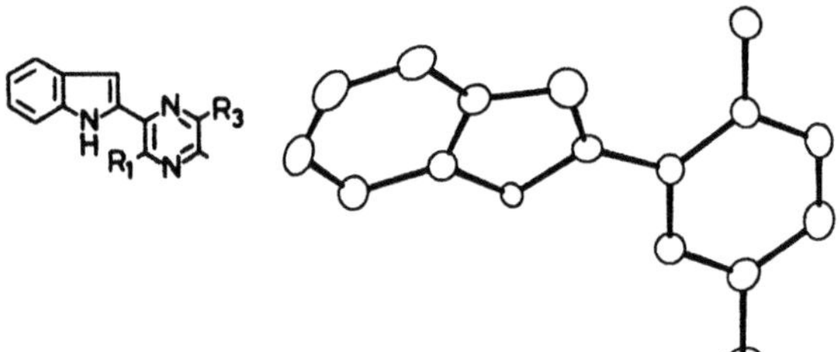

Fig. 1. The $C_{14}H_{13}N_3$ molecule (R_1 = R_3 = CH_3).

BIS(BENZIMIDAZOLIUM) BIS(μ_2-CHLORO)TETRACHLORODICUPRATE(II)
$C_{14}H_{14}Cl_6Cu_2N_2$ $2(C_7H_7N_2^+),Cl_6Cu_2^{2-}$

M. BUKOWSKA-STRZYZEWSKA, A. TOSIK, T. GLOWIAK and J. WNEK, 1985. *Acta Cryst.*, C41, 1184-1186.

Monoclinic, $P2_1/n$, a = 18.567, b = 6.443, c = 16.810 Å, β = 92.93°, Z = 4. Mo radiation, R = 0.032 for 2498 reflections.

The $Cu_2Cl_6^{2-}$ anions are located on crystallographic centres of inversion. The structure consists of polymeric $[Cu_2Cl_6^{2-}]_\infty$ chains and benzimidazolium cations, half of which form stacking columns surrounded by four parallel anionic chains (Fig. 1). The cations and anionic chains are linked by N-H(...Cl)$_2$ and N-H...Cl hydrogen bonds (N...Cl 3.175(4)-3.396(5) Å). The Cu atoms are five-coordinate with one long and four short Cu-Cl bonds (Fig. 1). Cu...Cu distances within the chains indicate the possibility of magnetic exchange between metal atoms.

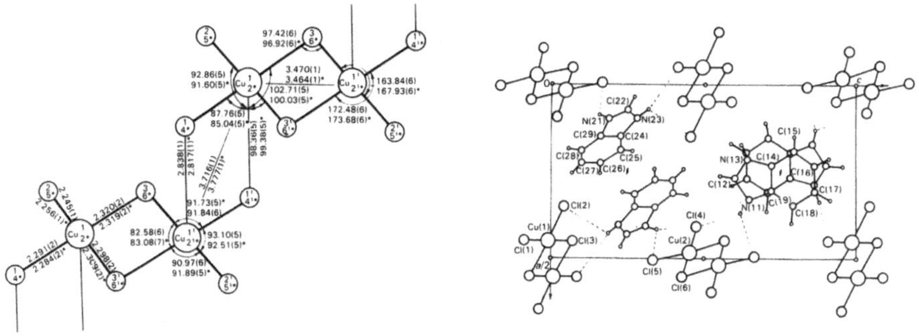

Fig. 1. $C_{14}H_{14}Cl_6Cu_2N_2$: bond lengths and angles in the two crystallographically independent $Cu_2Cl_6^{2-}$ chains and the structure viewed along the *y* axis.

2-(5'-AMINOCYCLOHEXANE-SPIRO-4'-(4'H-IMIDAZOL)-2'-YL)-1-METHYLPYRIDINIUM IODIDE

$C_{14}H_{19}IN_4$ $C_{14}H_{19}N_4^+,I^-$

P. RUBIO, F. FLORENCIO and S. GARCIA-BLANCO, 1985. *Acta Cryst.*, **C41**, 463-465.

Tetragonal, $I4_1/a$, a = 31.475, c = 6.446 Å, Z = 16. Mo radiation, R = 0.024 for 1643 reflections.

The structure (Fig. 1) contains an N-methylpyridinium moiety joined to a 4H-imidazole which has a pentamethylene group substituted at the spiranic C(1) atom. The pyridine ring is planar, the five-membered ring is approximately planar and the cyclohexane ring has a slightly-flattened chair conformation. The short C(3)-C(12) bond (1.472(7) Å) and the small torsion angle about this bond is attributed to conjugation between the pyridinium and imidazole groups. In the five-membered ring N(2)=C(3) is of length 1.281(6) Å while N(4)-C(5) and C(5)-N(17) of lengths 1.313(6) and 1.315(8) Å respectively, are somewhat short for single C-N bonds. The crystal structure is stabilized by van der Waals forces and by N(17)-H...N(4) hydrogen bonds of length 3.043(7) Å.

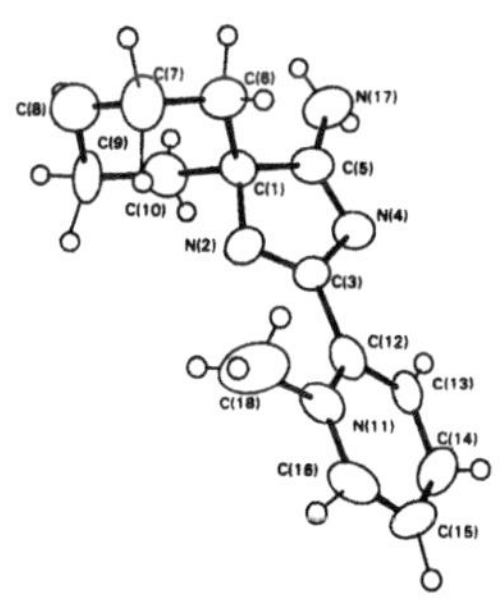

Fig. 1. A view of the $C_{14}H_{19}N_4^+$ ion showing the atom numbering.

6-ACETOXY-1,2,3,4-TETRAHYDRO-2,2,4-TRIMETHYLQUINOLINE

$C_{14}H_{19}NO_2$

A.E. OBODOVSKAYA, Z.A. STARIKOVA, Yu.A. IVANOV and I.E. POKROVSKAYA, 1985. *Zh. Strukt. Khim.*, **26**, 5-99-102 [J. Struct. Chem., **26**, 734-737].

Monoclinic, $P2_1/c$, a = 16.638, b = 6.343, c = 13.208 Å, β = 106.89°, Z = 4. Mo radiation, R = 0.052 for 664 reflections.

The molecular geometry is shown in Fig. 1. The tetrahydropyridine ring has a skewed half-chair conformation. C(10) occupies a pseudoaxial position. The CH_3CO_2 group is planar and forms a dihedral angle of 80.1° with the aromatic ring plane.

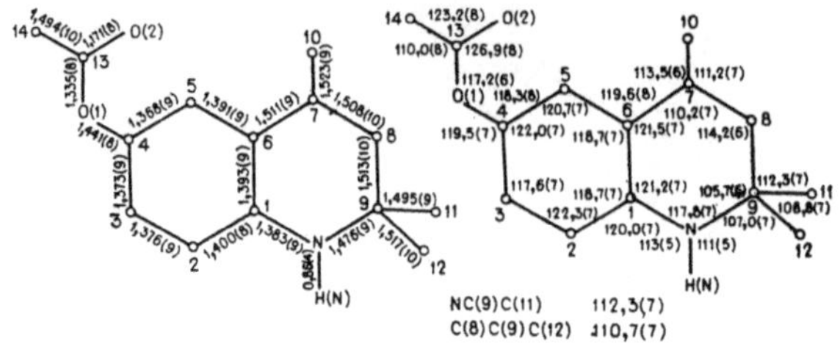

Fig. 1. Geometry of $C_{14}H_{19}NO_2$.

2-BENZYLISOINDOLE
$C_{15}H_{13}N$

R. BONNETT, M.B. HURSTHOUSE, S.A. NORTH and J. TROTTER, 1985. *J. Chem. Soc. Perkin Trans.* 2, 293-296.

Monoclinic, $P2_1/n$, a = 11.1407, b = 5.8204, c = 17.4414 Å, β = 91.419°, Z = 4. Cu radiation, R = 0.043 for 1506 reflections.

The isoindole ring and the phenyl ring are individually planar (Fig. 1) and the interplanar angle is 87.2°. The isoindole skeleton has C_{2v} symmetry. The bond lengths and angles are shown in Fig. 1 and show that the isoindole is a delocalised system.

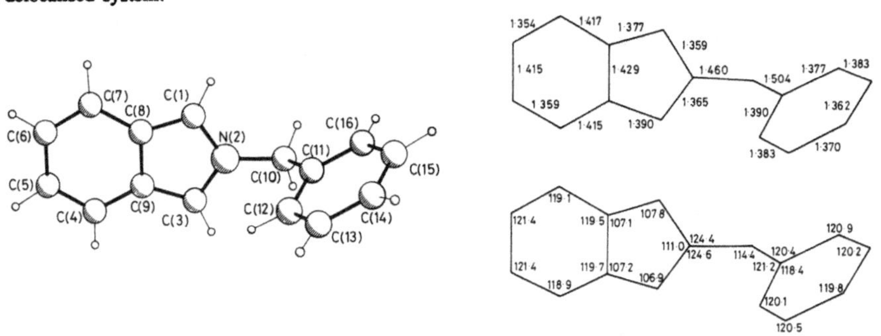

Fig. 1. Perspective drawing of a molecule of $C_{15}H_{13}N$ with bond lengths (Å) and angles (°). E.S.D.'s for bond lengths are 0.002-0.003 Å and are 0.2° for angles.

2-(N-METHYL-2-PYRIDYLAMINO)-8-HYDROXYQUINOLINE
$C_{15}H_{13}N_3O$

I.N. POLYAKOVA, Z.A. STARIKOVA, B.V. PARUSNIKOV and I.A. KRASAVIN, 1985. *Kristallografiya,* 30, 1010-1012.

Triclinic, $P\bar{1}$, a = 7.181, b = 8.31, c = 10.256 Å, α = 91.66, β = 96.44, γ = 95.7°, Z = 2. Mo radiation, R = 0.052 for 1776 reflections.

The molecule has a twisted syn-E conformation (Fig. 1) with torsion angles N(1)-C(1)-N(2)-C(15) 5.1° and N(3)-C(10)-N(2)-C(15) 46.5°. The quinoline and pyridine rings are inclined at 50.5° to one another.

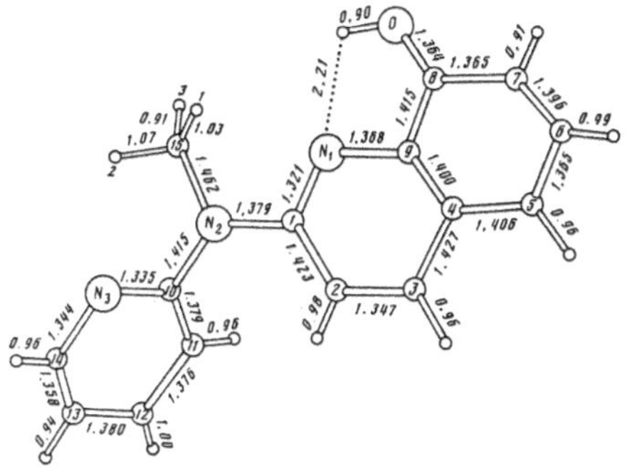

Fig. 1. The $C_{15}H_{13}N_3O$ molecule and interatomic distances.

4-PHENYL-6,7,8,9-TETRAHYDRO-1H-2,3-BENZODIAZEPINE-2-OXIDE
$C_{15}H_{16}N_2O$

M.D. WALKINSHAW, 1985. *Acta Cryst.*, **C41**, 1253-1255.

Monoclinic, $P2_1/n$, a = 8.062, b = 13.900, c = 12.007 Å, β = 108.18°, Z = 4. Mo radiation, R = 0.033 for 1194 reflections.

Bond lengths and angles are close to their expected values with C(4)-C(5), C(51)-C(91) and N(2)-N(3) only slightly longer (1.358(3), 1.341(3) and 1.279(3) Å respectively) than double bond values (Fig. 1). The N-O bond length is 1.268(2) Å. The 1,2-diazepine ring adopts a boat conformation, with C(1) as prow. Packing is governed by van der Waals interactions.

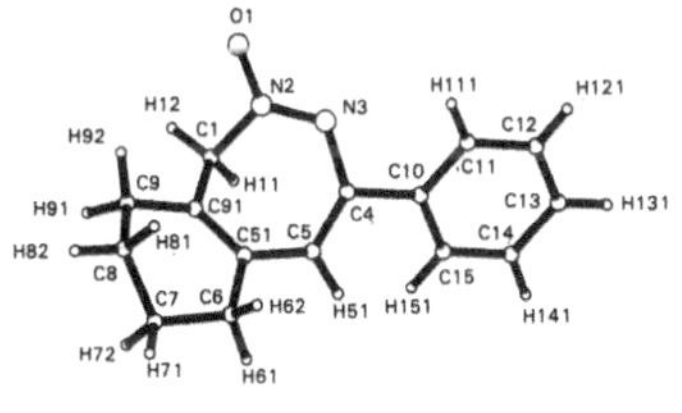

Fig. 1. $C_{15}H_{16}N_2O$: the molecular structure.

3-(1-METHYLCYCLOPENTA[b]PYRROLIDIN-2-YL)PROPAN-1-OL TARTRATE DIHYDRATE
$C_{15}H_{27}NO_7,2(H_2O)$ $C_{11}H_{22}NO^+,C_4H_5O_6^-,2(H_2O)$

A.P. KRIVEN'KO, T.G. NIKOLAEVA, A.A. ESPENBETOV, N.T. KOMYAGIN, N.N. SOROKIN, Yu.T. STRUCHKOV and V.G. KHARCHENKO, 1985. *Khim. Get. Soedin., SSSR*, 71-75.

Orthorhombic, $P2_12_12_1$, a = 7.295, b = 21.115, c = 11.762 Å, Z = 4. Mo radiation, R = 0.105 for 662 reflections.

A view of the ions with bond lengths is in Fig. 1. Both five-membered rings have envelope conformations. The H atoms were not located in the analysis, but O...O and O...N distances are indicative of hydrogen bonding. The absolute configuration of the cation is established from the known configuration of the tartrate anion.

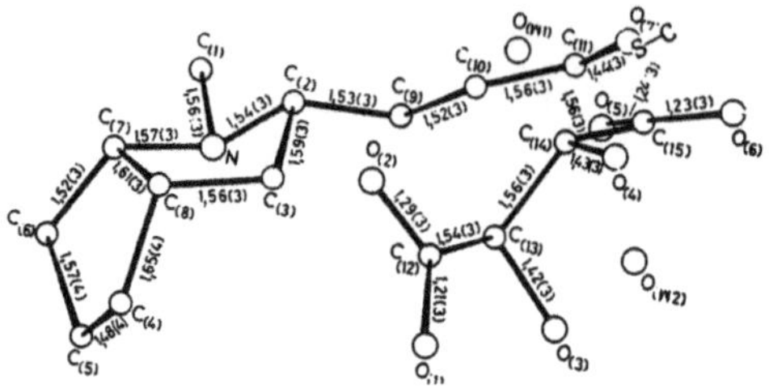

Fig. 1. Bond lengths in $C_{11}H_{22}NO^+,C_4H_5O_6^-,2(H_2O)$.

trans-2,2'-AZOQUINOXALINE
$C_{16}H_{10}N_6$

C. KRIEGER, A. KOCAK and O. BEKAROGLU, 1985. *Helv. Chim. Acta*, **68**, 581-583.

Monoclinic, $P2_1/n$, a = 469.6, b = 1224.4, c = 1205.8 Å, β = 104.10°, Z = 2. Mo radiation, R = 0.038 for 950 reflections.

The molecule (Fig. 1) is almost planar and the two chemically equivalent parts are related by a centre of symmetry; hence there is half a molecule in the asymmetric unit. The molecules are stacked along the a axis giving a 'herring-bone' pattern.

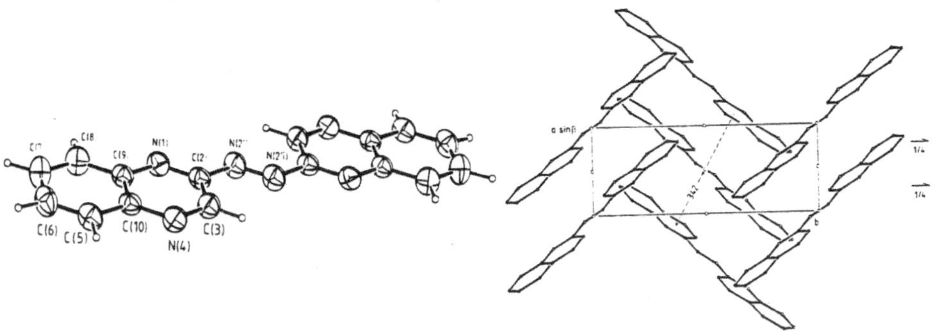

Fig. 1. Perspective drawing of a molecule of $C_{16}H_{10}N_6$ (left) with the crystal lattice projected down the c axis (right).

3-(p-CHLOROBENZOYLTHIO)-7-METHYL-2-METHYLTHIOPYRAZOLO[1,5-a]PYRIDINE
$C_{16}H_{13}ClN_2OS_2$

A. KAKEHI, S. ITO, M. ITO, T. YOTSUYA and K. NAGATA, 1985. *Bull. Chem. Soc. Jpn.*, **58**, 1432-1442.

Monoclinic, $P2_1/c$, a = 16.816, b = 6.331, c = 17.026 Å, β = 116.21°, Z = 4. Mo radiation, R = 0.091 for 1617 reflections.

The analysis established the structure as shown in Fig. 1, with N-N 1.374(13), S-C(ring) 1.741 and 1.747(13), S-C(O) 1.784(19), S-C(Me) 1.824(14) Å.

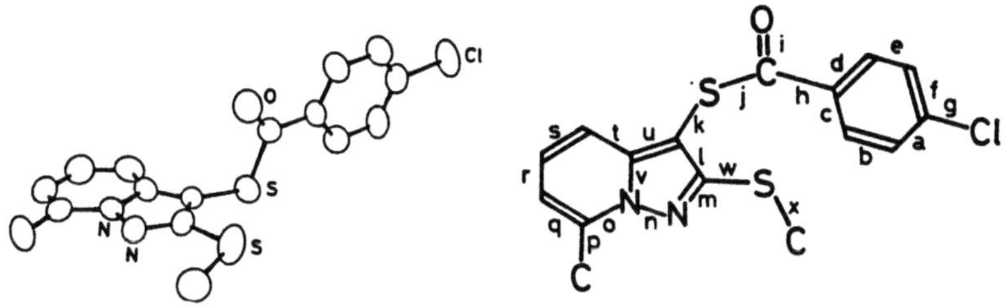

Fig. 1. The $C_{16}H_{13}ClN_2OS_2$ molecule.

7-CHLORO-1-METHYL-5-PHENYL-1H-1,5-BENZODIAZEPINE-2,4(3H,5H)DIONE (CLOBAZAM)
$C_{16}H_{13}ClN_2O_2$

H.J. BUTCHER and T.A. HAMOR, 1985. *Acta Cryst.*, C41, 1081-1083.

Rhombohedral, $R\bar{3}r$, a = 14.656, c = 111.75 Å, Z = 6. Mo radiation, R = 0.043 for 1588 reflections.

In the molecule (Fig. 1) the seven-membered ring adopts a boat conformation with C(3) as bow and C(10)-C(11) as stern. Bonds at N(1), C(2), C(4) and N(5) respectively, have nearly planar dispositions with N(1)-C(2) and C(4)-N(5) bond lengths slightly elongated to 1.361(3) Å on average. The fused chloro-substituted 'benzo' ring is planar and makes a dihedral angle of 85.2(4)° with the 5-phenyl ring. The molecular structure, except for the orientation of the 5-phenyl ring, is very similar to that of diazepam (1). There are no unusually short intermolecular contacts in the crystal structure.

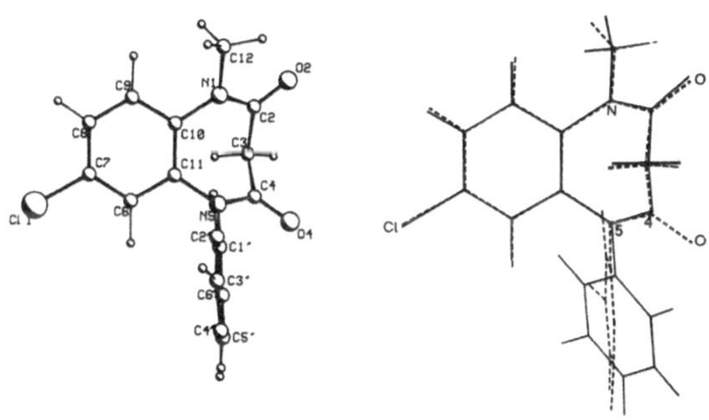

Fig. 1. $C_{16}H_{13}ClN_2O_2$: a view of the molecule perpendicular to the benzo ring C(6,7...11) and a superposition of the molecular structure with that of diazepam.

1. Structure Reports, 38B, 291.

cis-(5RS,6RS,8RS)-BENZYL 6-(1-HYDROXYETHYL)-7-OXO-1-AZABICYCLO[3.2.0]HEPT-2-ENE-2-CARB-OXYLATE
$C_{16}H_{17}NO_4$

J.H. BATESON, A.M. QUINN, T.C. SMALE and R. SOUTHGATE, 1985. *J. Chem. Soc. Perkin Trans. 1*, 2219-2234.

Triclinic, $P\bar{1}$, a = 7.581, b = 14.083, c = 7.389 Å, α = 97.8, β = 96.6, γ = 109.3°, Z = 2. Mo radiation, R = 0.051 for 1862 reflections.

The crystal structure analysis established the stereochemistry of the side-chain as (5RS,6RS,8RS) (Fig. 1).

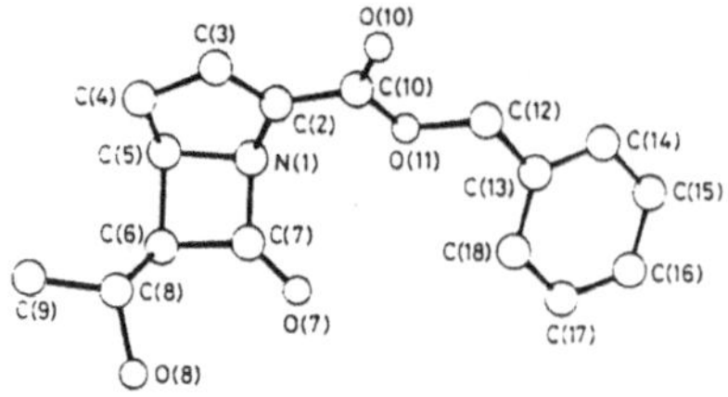

Fig. 1. X-ray structure of a molecule of $C_{16}H_{17}NO_4$.

1,4,5,6,7,8-HEXAHYDRO-7-METHYL-3-PHENYLCINNOLINE-1-CARBOXAMIDE
$C_{16}H_{19}N_3O$

G. MARONGIU, G. NAVARRA, A. MACCIONI and A. PLUMITALLO, 1985. *Acta Cryst.*, **C41**, 1059-1062.

Triclinic, $P\bar{1}$, a = 13.37, b = 15.41, c = 11.75 Å, α = 75.1, β = 69.4, γ = 94.0°, Z = 6. Cu radiation, R = 0.093 for 2943 reflections.

In the three independent molecules in the asymmetric unit (Fig. 1) the equivalent bond lengths and angles, in general, are in good agreement. The conformations of these molecules are slightly different as the angles between the least-squares planes of the phenyl and dihydropyridazine rings indicate. They are 9.8(8), 32.7(7) and 31.5(9)° in molecules 1, 2 and 3 respectively. The methyl group is bonded to C(4) in a pseudo equatorial position. The molecules in the crystal are hydrogen bonded through the amide group to form two crystallographically independent dimeric units. One dimer is centrosymmetric being composed of two molecules 1 and the other is formed from molecules 2 and 3. The N...O hydrogen bond lengths range from 2.89(1) to 2.95(1) Å. The dimers lie in planes almost parallel to (100) and their packing is regulated by normal van der Waals contacts.

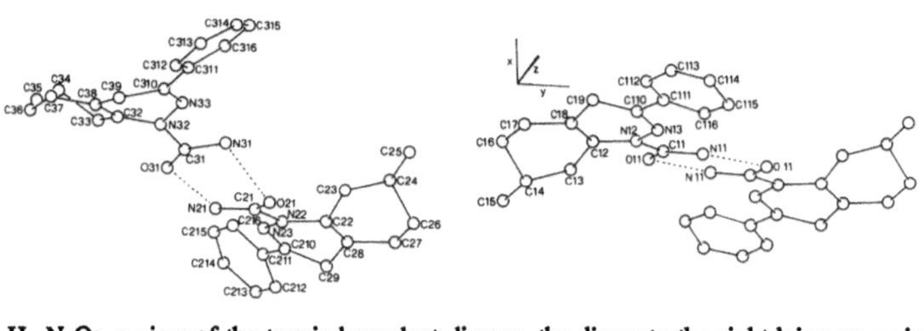

Fig. 1. $C_{16}H_{19}N_3O$: a view of the two independent dimers, the dimer to the right lying on an inversion centre.

3-(1-METHYL-2-OCTAHYDROINDOLYL)PROPAN-1-OL TARTRATE
$C_{16}H_{29}NO_7$ $C_{12}H_{24}NO^+, C_4H_5O_6^-$

A.P. KRIVEN'KO, T.G. NIKOLAEVA, A.A. ESPENBETOV, Yu.T. STRUCHKOV and V.G. KHARCHENKO, 1985. *Khim. Get. Soedin., SSSR*, 66-70.

Monoclinic, $P2_1$, a = 7.719, b = 14.440, c = 7.761 Å, β = 101.15°, Z = 2. Mo radiation, R = 0.0339 for 1478 reflections.

Views of the ions with dimensions are in Fig. 1. The five-membered ring conformation is close to half-chair and the six-membered ring is an asymmetrical chair. In the tartrate anion there is an intraionic O(4)-H...O(2) hydrogen bond (H...O 2.07 Å). The ions are linked by a hydrogen bond network involving the N-H and O-H groups. The known absolute configuration of the tartrate anion allowed the absolute configuration of the cation to be established.

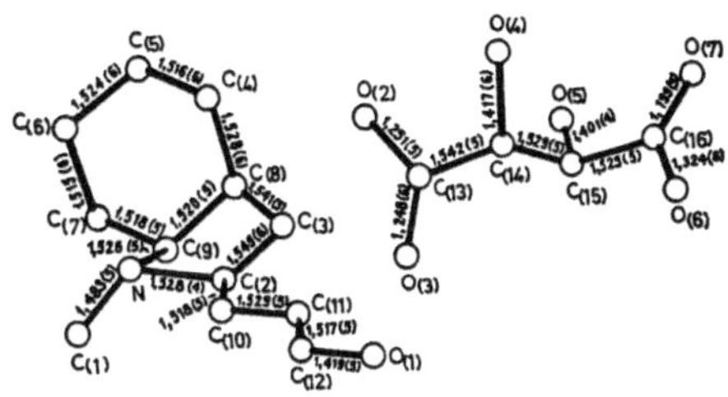

Fig. 1. Bond lengths in $C_{12}H_{24}NO^+,C_4H_5O_6^-$.

2-BENZOYL-1,2-DIHYDRO-1-ISOQUINOLINECARBONITRILE
$C_{17}H_{12}N_2O$

E. TYKARSKA, M. JASKOLSKI and Z. KOSTURKIEWICZ, 1985. *Acta Cryst.*, **C41**, 1792-1794.

Monoclinic, $P2_1/c$, a = 7.2105, b = 12.628, c = 15.953 Å, β = 112.26°, Z = 4. Cu radiation, R = 0.040 for 1579 reflections.

The heterocyclic fragment of the molecule (Fig. 1) exhibits a 1,3-diplanar conformation. The C(3)-N(2)-C(13)-O(14) torsion angle is -165.3(2)°, characterising the amide bond as anti. The angle between the phenyl group and the plane of the amide moiety is 45.0(3)°. Bond lengths in the hetero ring vary from 1.323(3) to 1.518(3) Å. The twist angle about the N(2)-C(13) bond is 7.9(2)° and the dihydroisoquinoline moiety is slightly distorted. The chiral molecules form a racemic crystal.

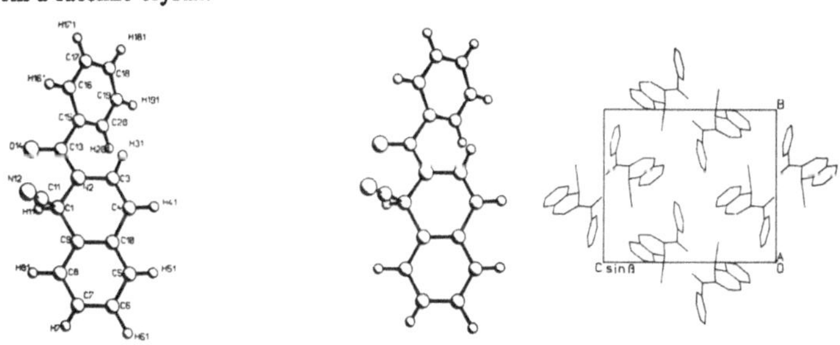

Fig. 1. $C_{17}H_{12}N_2O$: stereoview of the molecule and projection of the structure down *a*.

DIMETHYL 1-PHENYLINDAZOLE-6,7-DICARBOXYLATE
$C_{17}H_{14}N_2O_4$ (I)

DIMETHYL 4-(1,2-BIS(ETHOXYCARBONYL)ETHYL)-1-PHENYLINDAZOLE-6,7-DICARBOXYLATE
$C_{23}H_{22}N_2O_8$ (II)

B.A. GONZALEZ, R.A. JONES, M.M. SIMON, J.S. ARQUES, H.M. DAWES and M.B. HURSTHOUSE, 1985. *J. Chem. Res.*, S84-85, M1128-1172.

I. Monoclinic, $P2_1/c$, a = 8.709, b = 20.781, c = 8.245 Å, β = 98.14°, Z = 4. Cu radiation, R = 0.058 for 1672 reflections.

II. Monoclinic, $P2_1/c$, a = 9.020, b = 10.045, c = 24.318 Å, β = 92.44°, Z = 4. Mo radiation, R = 0.050 for 2829 reflections.

The crystal structure analyses established the structures of molecules (I) and (II) as shown in Fig. 1.

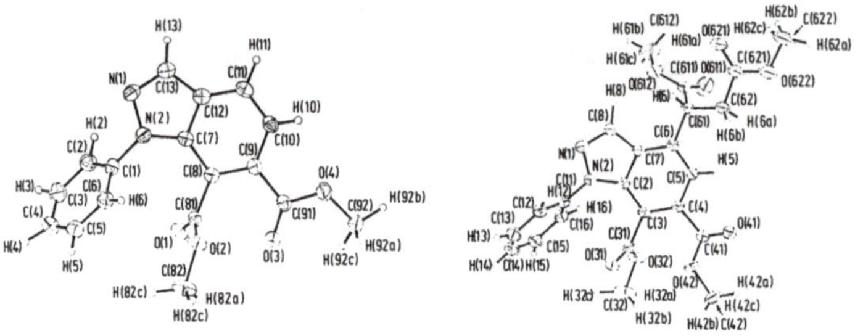

Fig. 1. Views of $C_{17}H_{14}N_2O_4$ (I) (left) and $C_{23}H_{22}N_2O_8$ (II) (right).

(±)-1-(2,3-DICHLOROPHENYL)-3,5,6,7-TETRAHYDRO-3-HYDROXY-3,6,6-TRIMETHYL-1H-INDOLE-2,4-DI-ONE
$C_{17}H_{17}Cl_2NO_3$ (I) (II)

K. GO, G. KARTHA and N. VISWANATHAN, 1985. *Acta Cryst.*, C41, 417-419.

I. Monoclinic, $P2_1/c$, a = 9.319, b = 16.098, c = 12.013 Å, β = 109.11°, Z = 4. Cu radiation, R = 0.049 for 2889 reflections.

II. Triclinic, $P\bar{1}$, a = 8.861, b = 9.855, c = 11.020 Å, α = 115.47, β = 103.05, γ = 88.13°, Z = 2. Cu radiation, R = 0.050 for 3057 reflections.

Bond lengths and angles in the diastereomers (I) and (II) (Fig. 1) are normal. In (I) the OH group at C(3) lies on the same side as the Cl atoms of the phenyl rings while in (II) it is the methyl group at C(3) which is on the same side as the Cl atoms. The dihedral angles between the least-squares planes of the phenyl and five-membered rings in (I) and (II) are 88.3(2) and 66.6(3)° respectively. In (I) there is an intramolecular hydrogen bond O(3)-H...O(4) and in (II) the bond is O(3)-H...O(2) of respective lengths 2.805(2) and 2.801(2) Å.

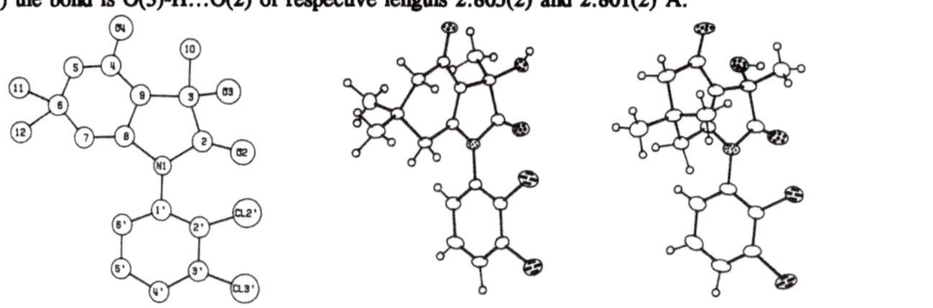

Fig. 1. $C_{17}H_{17}Cl_2NO_3$: atomic numbering and perspective views of isomer (I) (centre) and of isomer (II) (right).

3-BENZYLOXYMETHYL-2,5,6,7-TETRAHYDRO-2,6-DIMETHYL-4H-INDAZOL-4-ONE
$C_{17}H_{20}N_2O_2$

S. SUTTHIVAIYAKIT, W. SUCROW, H. WONNEMANN and C. KRUGER, 1985. *Liebigs Ann. Chem.*, 794-801.

Monoclinic, $P2_1/n$, a = 5.0378, b = 10.5565, c = 28.955 Å, β = 90.729°, Z = 4. Cu radiation, R = 0.056 for 2418 reflections.

The five-membered ring (Fig. 1) is planar, with N(1)-N(2) 1.382(3), N(2)-C(10) 1.449(4), N(2)-C(3) 1.344(3), C(3)-C(9) 1.391(3), C(9)-C(8) 1.403(3), C(8)-N(1) 1.328(3) Å.

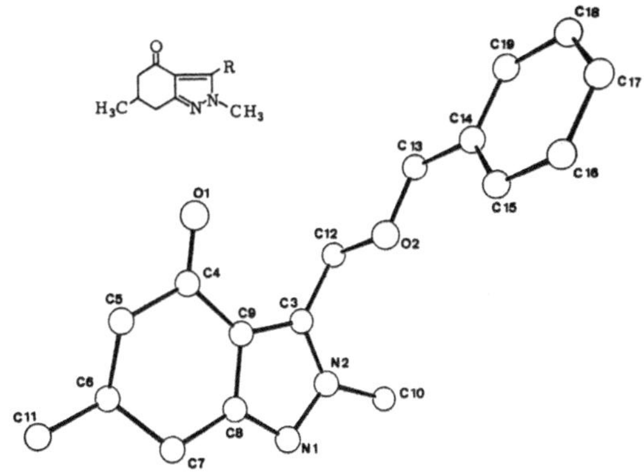

Fig. 1. The $C_{17}H_{20}N_2O_2$ molecule and conformation.

3-CARBOMETHOXY-1-METHYL-2-PHENYL-2α,3α,3aβ,4,5,7aβ-HEXAHYDRO-7(6H)-INDOLONE
$C_{17}H_{21}NO_3$

D. WENKERT, S.B. FERGUSON, B. PORTER, A. QVARNSTROM and A.T. McPHAIL, 1985. *J. Org. Chem.*, **50**, 4114-4119.

Triclinic, $P\bar{1}$, a = 9.888, b = 10.210, c = 8.413 Å, α = 84.16, β = 114.10, γ = 103.84°, Z = 2. Cu radiation, R = 0.054 for 2366 reflections.

The crystal structure analysis determined the configuration and conformation of the molecule (Fig. 1). The cyclohexanone ring has a chair conformation; the pyrrolidine ring has an approximate mirror plane of symmetry passing through C(7a) and the midpoint of the C(2)-C(3) bond and has an envelope form.

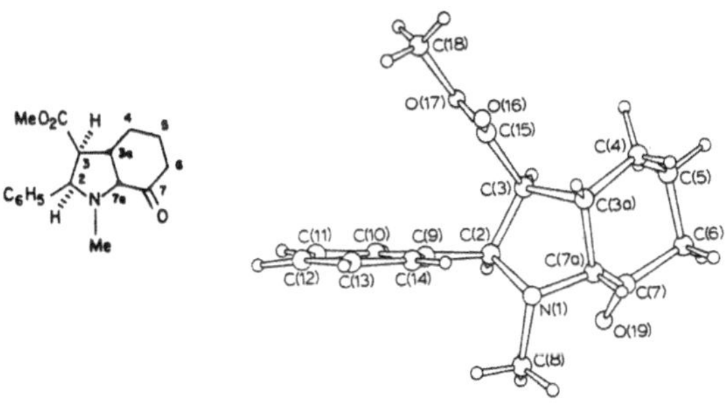

Fig. 1. Perspective drawing of a molecule of $C_{17}H_{21}NO_3$.

6-t-BUTYL-2-(3,3-DIMETHYLBUTANOYL)-2,3-DIHYDRO-3,3-DIMETHYL-1,2,4-TRIAZOLO[4,3-b]PYRIDAZINE
$C_{17}H_{28}N_4O$

R.F. ABDULLA, N.D. JONES and J.K. SWARTZENDRUBER, 1985. *Chem. Ber.*, **118**, 5009-5015.

Monoclinic, $P2_1/c$, a = 9.924, b = 18.443, c = 11.173 Å, β = 113.38°, Z = 4. Cu radiation, R = 0.0601 for 2905 reflections.

The analysis establishes that the compound is a triazolo[4,3-b]pyridazine, with the acyl group location shown to be at N(2).

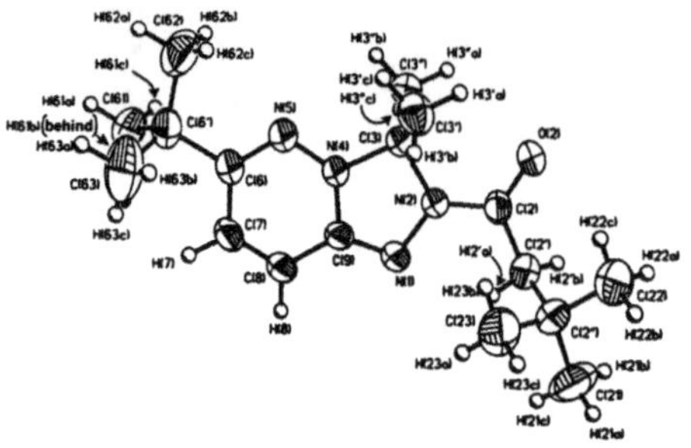

Fig. 1. A view of $C_{17}H_{28}N_4O$.

4-(4'-N-DIETHYLAMINOPHENYLAZO)PHTHALIMIDE
$C_{18}H_{18}N_4O_2$

B. GOLINSKI, 1985. *Z. Kristallogr.*, **173**, 113-117.

Monoclinic, *C2/c*, a = 20.44, b = 7.945, c = 20.808 Å, β = 100.9°, Z = 8. Cu radiation, R = 0.060 for 2584 reflections.

The molecules are almost planar (Fig. 1) and are linked by N-H...O hydrogen bonds about inversion centres to form dimers (N...O 2.890 Å). The N-N distance corresponds with a double bond.

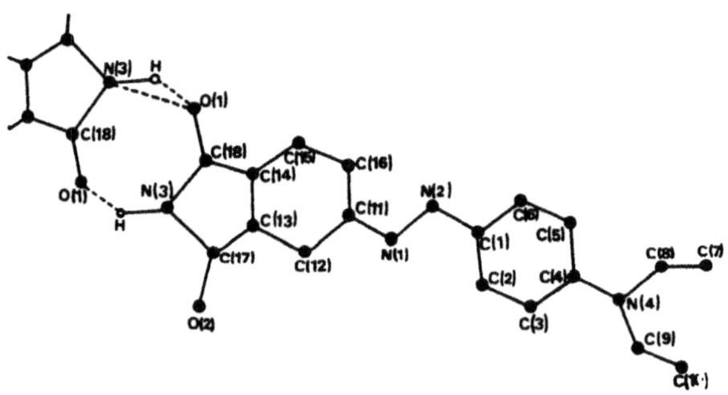

Fig. 1. A view of $C_{18}H_{18}N_4O_2$.

6-DIETHYLAMINO-2-(N,N-DIETHYLCARBAMOYL)-2,4,7-TRIMETHYL-2,2a-DIHYDROAZETO[2,3-d]-3,5-DIAZOCINE-N[1]-OXIDE
$C_{19}H_{31}N_5O_2$

A.T.M. MARCELIS, H.C. VAN DER PLAS and S. HARKEMA, 1985. *J. Org. Chem.*, **50**, 270-272.

Monoclinic, *C2/c*, a = 27.424, b = 8.270, c = 21.140 Å, β = 119.51°, Z = 8. Mo radiation, R = 0.110 for 2778 reflections.

There is some disorder in the crystal concerned with two of the diethylamino methyl groups. The molecule (Fig. 1) consists of a 1,3-diazocine ring fused to a four membered cyclic nitrone. The double bonds in the diazocine ring are all cis, and the conformation of the diazocine can be described as boat-like. The diethylcarbamoyl group and the diazocine nitrogen are on opposite faces of the dihydroazete ring.

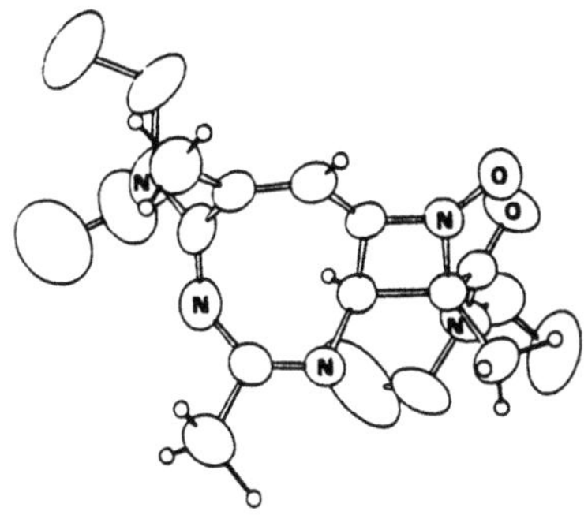

Fig. 1. ORTEP drawing of a molecule of $C_{19}H_{31}N_5O_2$.

2,2'-DIMETHYL-8,8'-DIQUINOLYL DISULPHIDE
$C_{20}H_{16}N_2S_2$ (I)

2,2',6,6'-TETRAMETHYL-8,8'-DIQUINOLYL DISULPHIDE
$C_{22}H_{20}N_2S_2$ (II)

O.G. MATYUKHINA, Ya.K. OZOLS, B.T. IBRAGIMOV, A.P. STURIS and Yu.A. BANKOVSKII, 1985. *Zh. Strukt. Khim.*, **26**, 2-119-124 [J. Struct. Chem., **26**, 246-250].

I. Monoclinic, *B*112/*b*, a = 8.146, b = 15.98, c = 12.98 Å, γ = 90.62°, Z = 4. Cu radiation, R = 0.067 for 527 reflections.

II. Orthorhombic, *Pnna*, a = 18.399, b = 24.480, c = 8.436 Å, Z = 8. Cu radiation, R = 0.059 for 1651 reflections.

The structures of (I) and (II) are shown in Figs. 1 and 2, respectively. The twist angle of the S-S bond is 81.1° in (I) and 89.9° in (II). Both compounds have equatorial conformations relative to the C-S bonds. Bond lengths: S-S 2.042(7) Å in (I), 2.036(3) Å in (II); S-C 1.755(11) Å in (I), 1.795(5) and 1.780(4) Å in (II). The planar quinoline fragments form dihedral angles of 77° in (I) and 88° in (II). (I) has exact C_2 symmetry.

Fig. 1. The $C_{20}H_{16}N_2S_2$ (I) molecule.

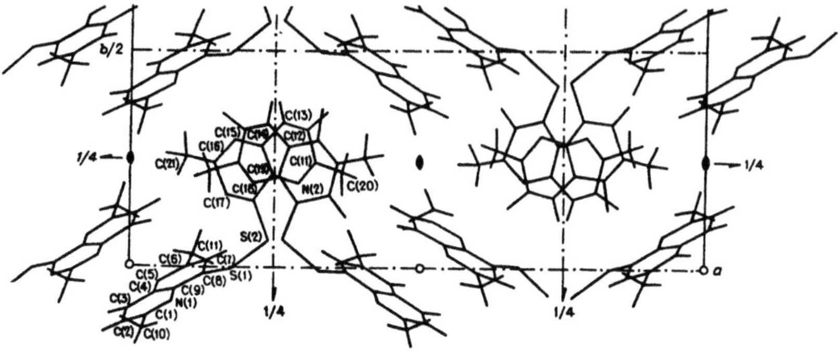

Fig. 2. The structure of $C_{22}H_{20}N_2S_2$ (II).

cis-2-CARBAMOYL-3-PHENYL-7-PHENYLMETHYLENE-3,3a,4,5,6,7-HEXAHYDRO-2H-INDAZOLE
$C_{21}H_{21}N_3O$ (I)

trans-2-CARBAMOYL-3-PHENYL-7-PHENYLMETHYLENE-3,3a,4,5,6,7-HEXAHYDRO-2H-INDAZOLE
$C_{21}H_{21}N_3O$ (II)

T. LORAND, D. SZABO, A. FOLDESI, L. PARKANYI, A. KALMAN and A. NESZMELYI, 1985. *J. Chem. Soc. Perkin Trans. 1*, 481-486.

I. Triclinic, $P\bar{1}$, a = 11.244, b = 10.808, c = 8.702 Å, α = 64.05, β = 104.57, γ = 114.45°, Z = 2. Mo radiation, R = 0.042 for 3136 reflections.

II. Triclinic, $P\bar{1}$, a = 11.413, b = 5.468, c = 14.757 Å, α = 87.54, β = 108.96, γ = 92.26°, Z = 2. Mo radiation, R = 0.044 for 3081 reflections.

In both molecules (Fig. 1) the six-membered rings have a shape which is intermediate between chair, half-chair and envelope, whilst the five-membered rings have an envelope conformation with C(5) as the flap. In the crystals of both molecules there are hydrogen bonds of the type N(8)-H...O(7) forming dimer associates around the centres of symmetry.

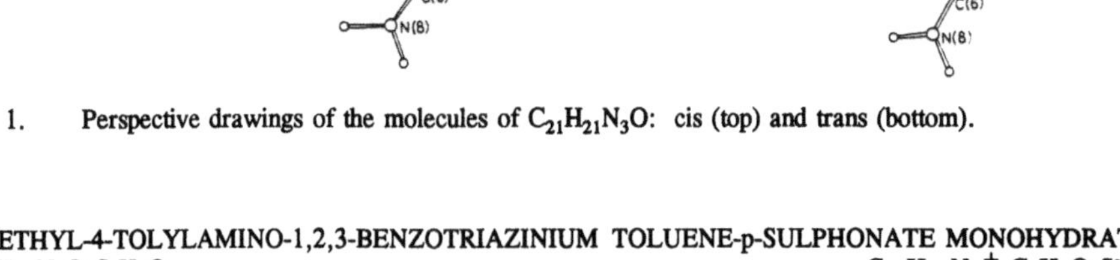

Fig. 1. Perspective drawings of the molecules of $C_{21}H_{21}N_3O$: cis (top) and trans (bottom).

2-METHYL-4-TOLYLAMINO-1,2,3-BENZOTRIAZINIUM TOLUENE-p-SULPHONATE MONOHYDRATE
$C_{22}H_{22}N_4O_3S,H_2O$ $C_{15}H_{15}N_4^+,C_7H_7O_3S^-,H_2O$

B.J. CLARK, R. GRAYSHAN and K.J. McCULLOUGH, 1985. *J. Chem. Res.*, S50-51, M749-775.

Triclinic, $P\bar{1}$, a = 7.898, b = 11.028, c = 12.325 Å, α = 89.36, β = 87.50, γ = 79.29°, Z = 2. Mo radiation, R = 0.037 for 2026 reflections.

The crystal structure is composed of discrete benzotriazinium and toluene-p-sulphonate ions together with an associated molecule of water of crystallisation (Fig. 1). The benzotriazinium ring system has been methylated at the N(2) nitrogen atom.

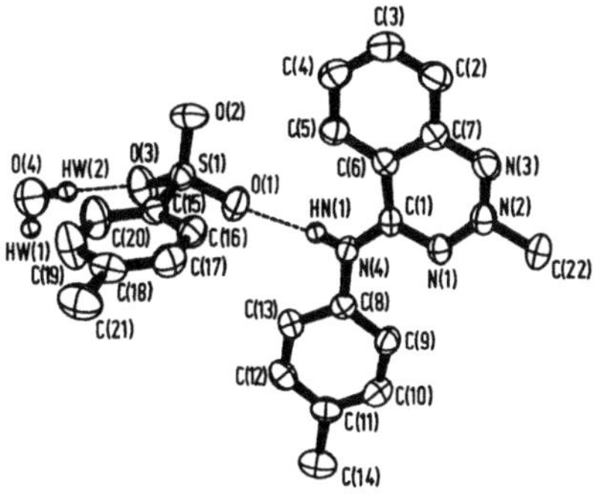

Fig. 1. ORTEP diagram of the crystal structure of the salt $C_{22}H_{22}N_4O_3S,H_2O$.

6-ISOPROPYL-2,4-DIPHENYLQUINAZOLINE
$C_{23}H_{20}N_2$

D. HUNTER, D.G. NEILSON and T.J.R. WEAKLEY, 1985. *J. Chem. Soc. Perkin Trans. 1*, 2709-2717.

Monoclinic, $P2_1/c$, a = 5.749, b = 14.762, c = 21.854 Å, β = 109.24°, Z = 4. Cu radiation, R = 0.088 for 1004 reflections.

The 2-phenyl substituent is twisted 3.8° out of the plane of the pyrimidine ring (Fig. 1) and the 4-phenyl substituent lies at an angle of 63.9° to the pyrimidine ring.

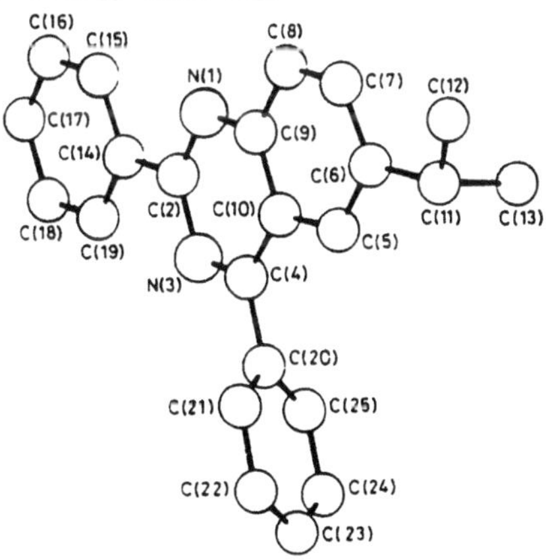

Fig. 1. Molecular structure of $C_{23}H_{20}N_2$.

N(3)-DIBENZYLAMINOMETHYL-4-HYDROXYCYCLOHEXANE-5-SPIRO-2-IMIDAZOLODINONE METHANOL SOLVATE
$C_{23}H_{29}N_3O_2,CH_4O$

L. SALAZAR, M. ESPADA, C. PEDREGAL, F. FLORENCIO and S. GARCIA-BLANCO, 1985. *J. Mol. Struct.*, **129**, 321-332.

Triclinic, $P\bar{1}$, a = 7.289, b = 11.419, c = 11.047 Å, α = 88.21, β = 75.23, γ = 79.89°, Z = 2. Mo radiation, R = 0.071 for 2617 reflections.

The five-membered ring adopts a C(3)-envelope conformation with the spiro six-membered ring in chair form (Fig. 1). Molecular dimensions are normal and N(14) is pyramidal. Molecules are linked by N-H...O and O-H...O hydrogen bonds (N...O 2.987(8), O...O 2.711(6) Å).

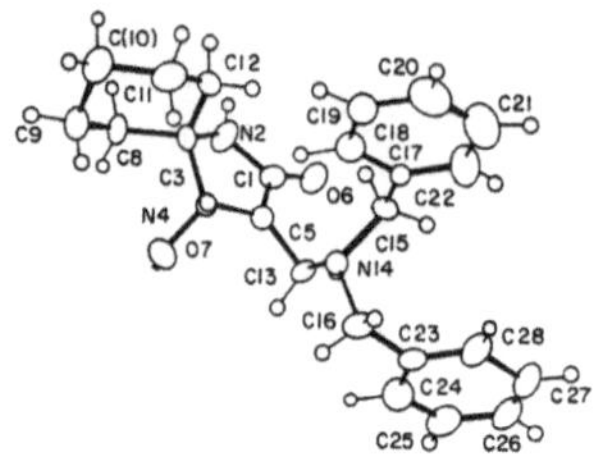

Fig. 1. A view of the $C_{23}H_{29}N_3O_2$ molecule.

DIPHTHALIMIDE PHTHALATE (BIS(1,3-DIOXO-2H-ISOINDOL-2-YL)PHTHALATE)
$C_{24}H_{12}N_2O_8$

J.W. BATS and H.-J. TEUBER, 1985. *Acta Cryst.*, **C41**, 298-300.

Triclinic, $P\bar{1}$, a = 8.025, b = 12.910, c = 20.589 Å, α = 73.90, β = 88.72, γ = 77.87°, Z = 4. Mo radiation, R = 0.040 for 3938 reflections.

There are two crystallographically independent molecules in the asymmetric unit. These differ only in the relative orientation of one of the side groups (Fig. 1). One carboxyl group is coplanar with the central phenyl group, the other forms an angle of 101.2(4)° in (I) and 111.2(4)° in (II). The phthalimide groups are almost perpendicular to the carboxyl groups. The average N-O bond length, 1.390(2) Å, corresponds to a single bond. The packing shows stacks of parallel phthalimide groups along a.

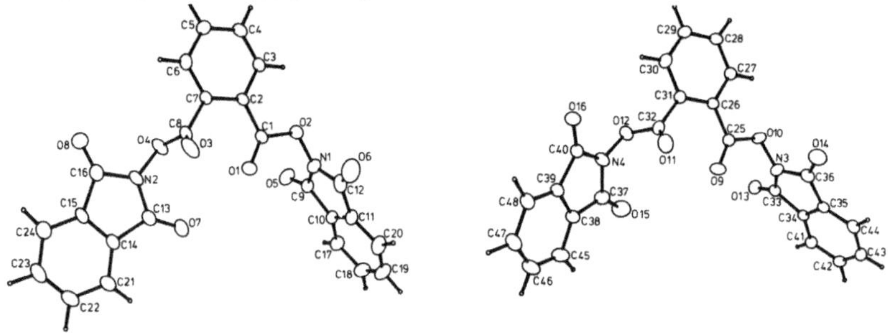

Fig. 1. $C_{24}H_{12}N_2O_8$: the geometries of the two independent molecules.

(2-PYRIDYL)-N,N'-BIS(8-QUINOLYLAMINO)METHANE
$C_{24}H_{19}N_5$

J.M. MOHEDANO, F. FLORENCIO and S. GARCIA-BLANCO, 1985. *Z. Kristallogr.*, **170**, 297-302.

Monoclinic, $P2_1/n$, a = 9.729, b = 6.709, c = 29.847 Å, β = 97.61°, Z = 4. Mo radiation, R = 0.046 for 3697 reflections.

The 8-amino-quinolyl and pyridine groups are individually planar (Fig. 1). The dihedral angle between the 8-amino-quinolyl planes is 127.4(1)° and those between the quinolyl planes and pyridine ring are 15.2(2) and 48.1(2)°. There are a number of intramolecular N-H...N hydrogen bonds (N...N 2.659-3.047(4) Å).

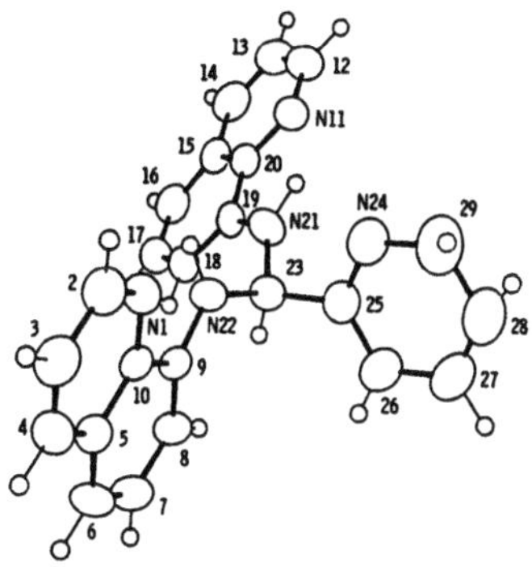

Fig. 1. A view of $C_{24}H_{19}N_5$.

1,8-DIMETHYL-4,6-DIPHENYL-1,2,3,4-TETRAHYDRO-1,5-BENZODIAZOCIN-2-ONE
$C_{24}H_{22}N_2O$

A.A. DVORKIN, Yu.A. SIMONOV, S.A. ANDRONATI and V.V. DANILIN, 1985. *Kristallografiya,* **30**, 91-95.

Monoclinic, $P2_1/b11$, a = 8.628, b = 15.430, c = 16.140 Å, α = 64.19°, Z = 4. Mo radiation, R = 0.060 for 1462 reflections.

Views of the molecule with dimensions and packing are in Fig. 1. The eight-membered ring is in boat conformation and the dihedral between the phenyl ring plane at C(6) and the plane of the condensed benzene ring is 95.7°. Atom N(1) has planar geometry (angle sum 360°).

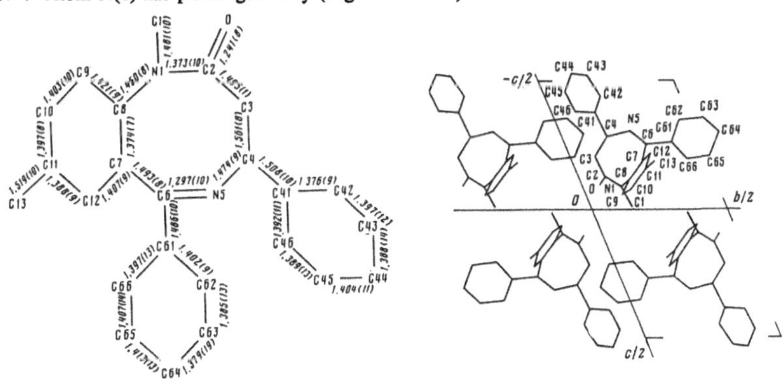

Fig. 1. Bond lengths in $C_{24}H_{22}N_2O$ and a projection of the structure down *a*.

BENZYL (±)-1-α-HYDROXY-7β-PHENOXYACETAMIDO-1-CARBACEPHALOSPORANATE
$C_{24}H_{24}N_2O_6$

G.A. SIM, 1985. *Acta Cryst.*, **C41**, 1512-1514.

Monoclinic, $I2/c$, a = 20.900, b = 10.340, c = 24.226 Å, β = 120.82°, Z = 8. Mo radiation, R = 0.045 for 2466 reflections.

The X-ray analysis established that the substituents at C(1) and C(7) are trans (Fig. 1). The β-lactam N atom is displaced by 0.20(1) Å from the plane of the three atoms to which it is bonded. The geometry at this N atom is pyramidal. The N-CO bond in the β-lactam is 1.361(4) and is longer than N(12)-C(23) (1.320(4) Å) in the side-chain. The C(8)-O(11) and C(23)-C(24) distances are 1.206(4) and 1.233(4) Å respectively. There are two hydrogen bonded contacts in the crystal (O(13)...O(24) 2.85(1), O(11)...N(12) 2.85(1) Å).

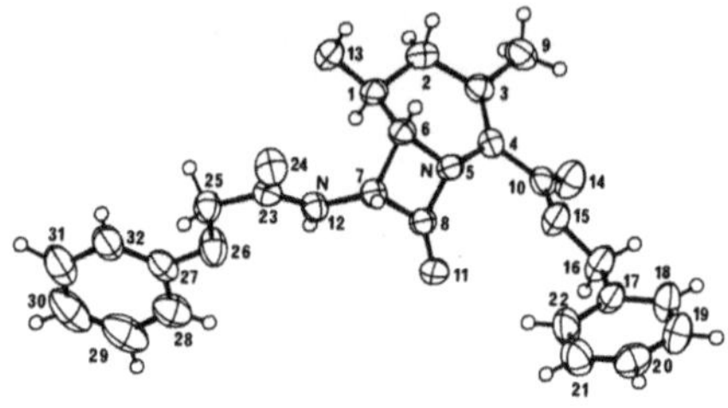

Fig. 1. $C_{24}H_{24}N_2O_6$: the molecular structure.

4,10-DIBENZYL-4,10-DIAZABICYCLO[5.5.0]DODEC-1(7)-ENE DIHYDROCHLORIDE
$C_{24}H_{32}Cl_2N_2$ $C_{24}H_{32}N_2^{2+},2(Cl^-)$

S. KASHINO, M. KIMURA, S. MOROSAWA and M. HAISA, 1985. *Acta Cryst.*, **C41**, 919-921.

Orthorhombic, *Pcab*, a = 12.737, b = 16.796, c = 10.364 Å, Z = 4. Cu radiation, R = 0.040 for 1618 reflections.

The molecule (Fig. 1) sits on a crystallographic centre of symmetry. The N atoms are protonated and hydrogen bonded to the Cl^- ions ($N^+ \ldots Cl^-$ 3.064(1) Å). Two seven-membered rings are fused through the double bond C(10)-C(10') (1.333(4) Å). These rings adopt chair conformations. The benzyl group is bonded equatorially to N. Molecules are arranged in sheets parallel to (010). These sheets are stacked along b.

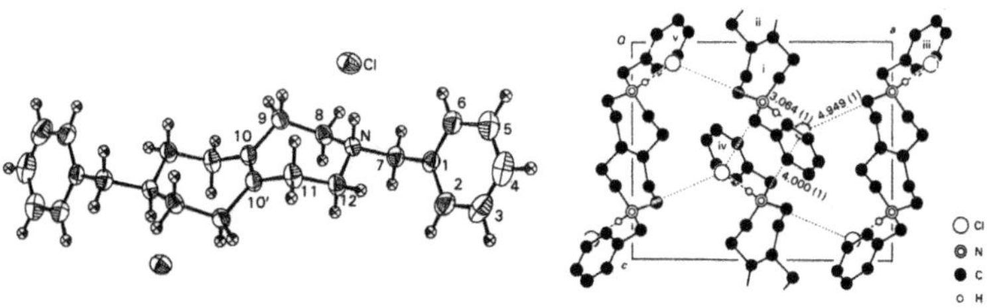

Fig. 1. The $C_{24}H_{32}Cl_2N_2$ molecule and the molecular arrangement in a sheet parallel to (010).

1,3,5-TRIBENZYL-2,4,6-TRIOXOPYRROLO[3,4-d]IMIDAZOLE
$C_{26}H_{21}N_3O_3$

D.A. CASTEEL and N.J. LEONARD, 1985. *J. Org. Chem.*, **50**, 2450-2456.

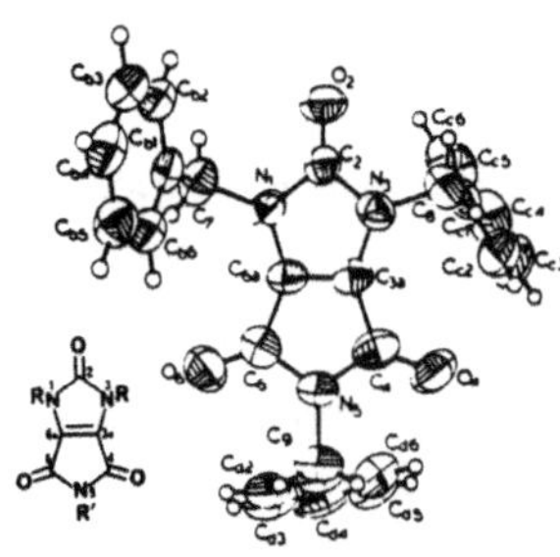

Fig. 1. ORTEP drawing of $C_{26}H_{21}N_3O_3$.

Monoclinic, $P2_1/c$, a = 9.881, b = 26.537, c = 9.048 Å, β = 110.54°, Z = 4. Mo radiation, R = 0.043 for 1481 reflections.

The heterocycle is planar (Fig. 1).

TETRAKIS(N,N'-(o-PHENYLENE)THIOUREA)SELENIUM(II) DICHLORIDE DIHYDROCHLORIDE

$C_{28}H_{26}Cl_4N_8S_4Se$ $C_{28}H_{24}N_8S_4Se^{2+}$,2($Cl^-$),2(HCl)

S. SOWRIRAJAN, G. ARAVAMUDAN, M. SESHASAYEE and G.C. ROUT, 1985. *Acta Cryst.*, **C41**, 576-579.

Monoclinic, $P2_1/a$, a = 4.830, b = 26.565, c = 13.787 Å, β = 100.07°, Z = 2. Mo radiation, R = 0.075 for 1205 reflections.

The Se atom lies on a crystallographic centre of symmetry and is bonded to four S atoms of the ligand (in the thione form) in an almost square-planar arrangement (Fig. 1). The Se-S distances are 2.525(6) and 2.543(6) Å, and S-Se-S is 89.1(2)°. The Cl^- ions are not bonded to Se but provide lattice stabilization through extensive hydrogen bonding. The phenylenethiourea ligands are planar with a mean C-S distance of 1.72(2) Å.

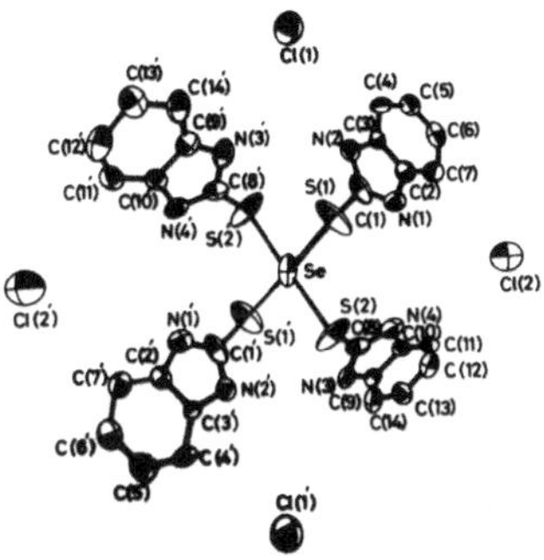

Fig. 1. A view of the $C_{28}H_{26}Cl_4N_8S_4Se$ molecule.

2-(o-AMMONIOBENZYL)-3-(3-INDOLYL)-1H-QUINOLINIUM BIS(ETHANESULPHONATE)

$C_{28}H_{31}N_3O_6S_2$ $C_{24}H_{21}N_3^{2+}$,2($C_2H_5O_3S^-$)

V. CODY, 1985. *Acta Cryst.*, **C41**, 413-415.

Triclinic, $P\bar{1}$, a = 13.492, b = 14.789, c = 8.109 Å, α = 90.17, β = 113.38, γ = 108.85°, Z = 2. Cu radiation, R = 0.052 for 3665 reflections.

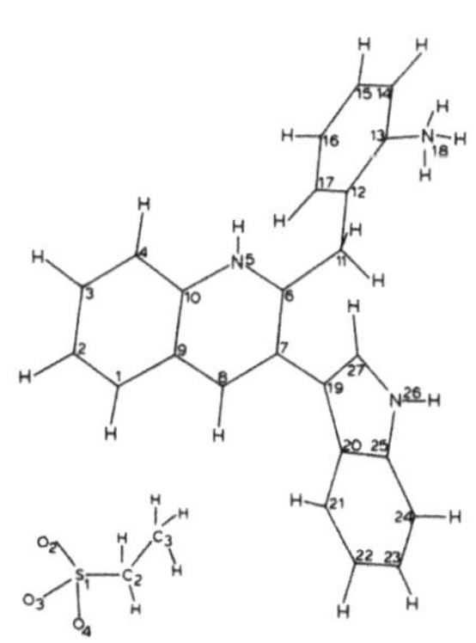

Fig. 1. $C_{28}H_{31}N_3O_6S_2$: numbering scheme for the complex molecule and for one of the ethanesulphonate anions.

In the cation (Fig. 1) the ammoniobenzyl ring is twist-skewed with respect to the quinoline ring having C(7)-C(6)-C(11)-C(12) and C(6)-C(11)-C(12)-C(17) torsion angles of 113.4(3) and -26.4(4)° respectively. The o-amino group is distal to the quinoline ring and is protonated, as is the quinoline nitrogen atom. The indolyl ring is twisted from the quinoline ring with a torsion angle of C(6)-C(7)-C(19)-C(27) of -42.6(4)°. The N(26)-C(27) bond length of

1.340(4) Å is somewhat shorter than that observed in other indole and ethanesulphonate structures. There is a three-dimensional network of hydrogen bonds linking the nitrogen atoms of the molecule to the oxygen atoms of the two crystallographically independent ethanesulphonate anions.

3-(N$^{\alpha}$-TRITYLMETHIONYL)BENZOTRIAZOLE-1-OXIDE

$C_{30}H_{28}N_4O_2S$

K. BARLOS, D. PAPAIOANNOU, S. VOLIOTIS, R. PREWO and J.H. BIERI, 1985. *J. Org. Chem.*, **50**, 696-697.

Orthorhombic, $P2_12_12_1$, a = 8.7008, b = 14.3777, c = 20.846 Å, Z = 4. Mo radiation, R = 0.064 for 4838 reflections (at 130 K).

The crystal structure analysis proved that the methionyl moiety is attached at N(3) (Fig. 1). The H(9)...N(2) distance is 2.38 Å.

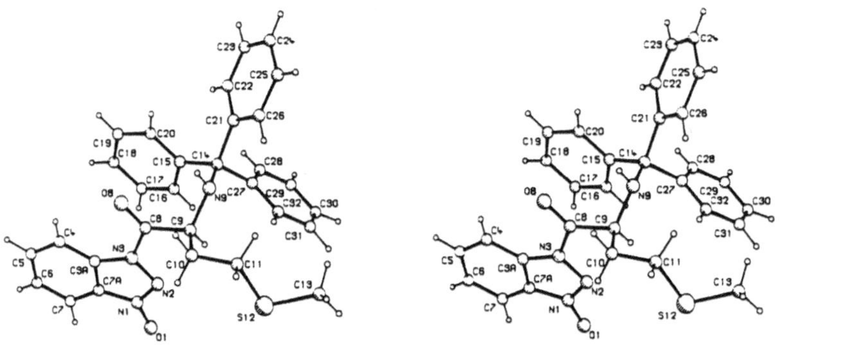

Fig. 1. A stereoview of $C_{30}H_{28}N_4O_2S$.

N-2-(1,3-DICHLOROINDOLIZINYL)PYRIDINIUM TETRAPHENYLBORATE

$C_{37}H_{29}BCl_2N_2$ $C_{13}H_9Cl_2N_2^+, C_{24}H_{20}B^-$

K.A. SMITH, K.C. WATERMAN and A. STREITWEISER Jr., 1985. *J. Org. Chem.*, **50**, 3360-3365.

Monoclinic, $P2_1/c$, a = 16.3668, b = 10.0669, c = 18.5781 Å, β = 98.170°, Z = 4. Mo radiation, R = 0.0327 for 3940 reflections.

The structure of the molecule, without the tetraphenylborate anion, is shown in Fig. 1.

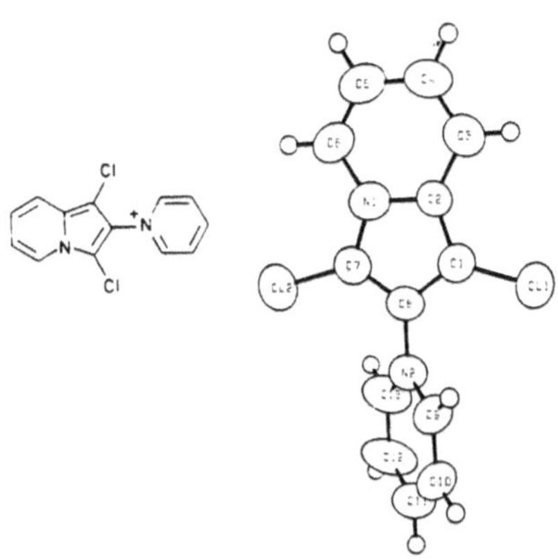

Fig. 1. The cation of the salt $C_{37}H_{29}BCl_2N_2$.

1,5,7-TRIS(1,3,3-TRIMETHYL-2-INDOLINYL)[2.2.2]HEPTAMETHINEDIIUM BIS(TETRAFLUOROBORATE) DICHLOROMETHANE SOLVATE

$C_{40}H_{45}B_2F_8N_3,CH_2Cl_2$ (I) $C_{40}H_{45}N_3^{2+},2(BF_4^-),CH_2Cl_2$

1,5,7-TRIS(1,3,3-TRIMETHYL-2-INDOLINYL)[2.2.2]HEPTAMETHINEDIIUM DIIODIDE DICHLOROMETHANE SOLVATE

$C_{40}H_{45}I_2N_3,CH_2Cl_2$ (II) $C_{40}H_{45}N_3^{2+},2(I^-),CH_2Cl_2$

R. ALLMANN, W. GRAHN, J. KNECHT, D. KUCHARCZYK and C. REICHARDT, 1985. *Chem. Ber.*, **118**, 1295-1303.

I. Orthorhombic, $P2_12_12_1$, a = 27.712, b = 12.024, c = 12.429 Å, Z = 4. Mo radiation, R = 0.084 for 1424 reflections.

II. Orthorhombic, $P2_12_12_1$, a = 27.544, b = 11.922, c = 12.320 Å, Z = 4. Mo radiation, R = 0.077 for 1190 reflections.

The structures are isomorphous and the central portion of the cations is approximately planar and in all-trans configuration (Fig. 1). The compounds exist as the vinylogous amidinium ions and not (as formerly supposed) as methinylogous guanidinium ions.

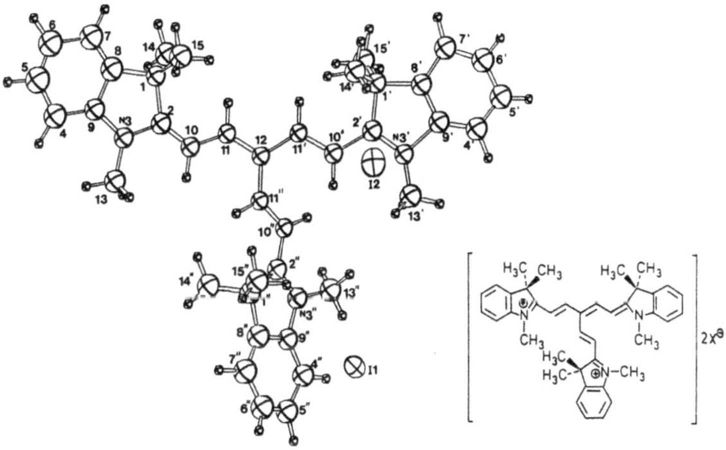

Fig. 1. The $C_{40}H_{45}N_3^{2+},2I^-$ system.

8,8-BIS(METHOXYCARBONYL)-2,3,4,5,9 PENTAPHENYL-1,7-DIAZABICYCLO[5.3.0]DECA-2,4-DIEN-6-ONE n-HEPTANE SOLVATE

$C_{42}H_{34}N_2O_5,C_7H_{16}$

R. MILLINI, H. KISCH and C. KRUGER, 1985. *Z. Naturforsch.*, **B40**, 187-192.

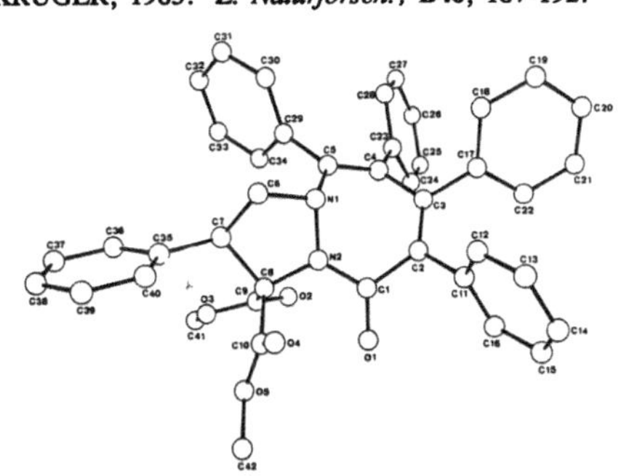

Fig. 1. A view of $C_{42}H_{34}N_2O_5$.

Triclinic, $P\bar{1}$, a = 9.9194, b = 10.1234, c = 19.250 Å, α = 97.575, β = 90.197, γ = 106.109°, Z = 2. Cu radiation, R = 0.062 for 5277 reflections.

The molecule is shown to have the structure and conformation given in Fig. 1. The N-N (1.430(3) Å) and N(1)-C(5) (1.436(3) Å) are consistent with a delocalized bonding in the N(1)-N(2)-C(5) fragment. There is no appreciable π-conjugation in the diazepinone ring.

6,6'-BIS(2-DIMETHOXYPHOSPHORYL-3,4,6-TRIPHENYL-6H-PYRROLINO[1,2-b]PYRAZOLE) CHLOROFORM SOLVATE MONOHYDRATE
$C_{52}H_{44}N_4O_6P_2,CHCl_3,H_2O$

S.G. KHBEIS, G. MAAS and M. REGITZ, 1985. *Tetrahedron*, **41**, 811-817.

Monoclinic, $P2_1/c$, a = 12.754, b = 10.917, c = 20.413 Å, β = 97.86°, Z = 2. Mo radiation, R = 0.074 for 1730 reflections.

The structure and stereochemistry are established as shown in Fig. 1. The molecule lies about an inversion centre and the C(6)-C(6') bond is long (1.574(14) Å). The P=O bond length is 1.466(6) Å with P-O 1.554 and 1.561(6) Å.

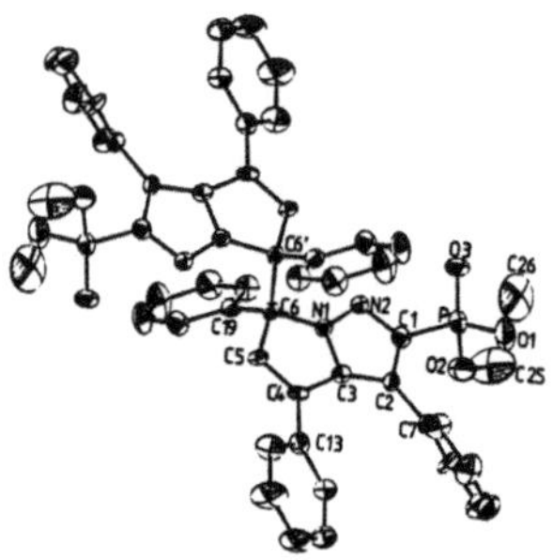

Fig. 1. The $C_{52}H_{44}N_4O_6P_2$ molecule.

6-AZIDOTETRAZOLO[5,1-a]PHTHALAZINE
$C_8H_4N_8$

M.B. HOSSAIN, D. VAN DER HELM, R. SANDUJA and M. ALAM, 1985. *Acta Cryst.*, **C41**, 1199-1202.

Orthorhombic, $P2_12_12_1$, a = 12.458, b = 14.669, c = 4.980 Å, Z = 4. Mo radiation, R = 0.034 for 892 reflections.

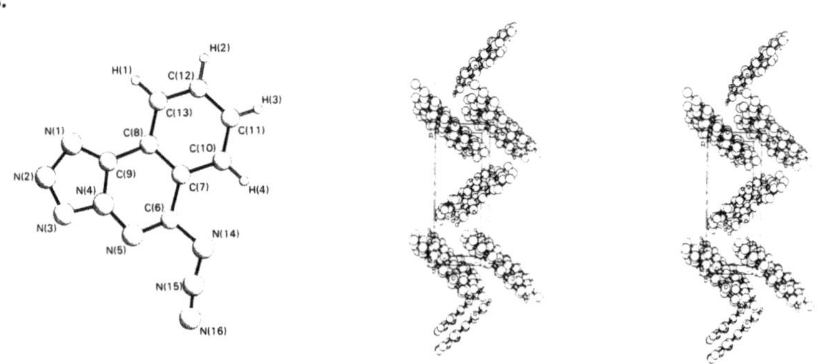

Fig. 1. $C_8H_4N_8$: the molecular structure and packing viewed down the *b*-axis.

The molecule (Fig. 1) is planar with respect to the ring system and the azido group is bent (9.1°) out of this plane. In the tetrazole ring, the N(1)-N(2) and N(3)-N(4) bonds are longer than expected at 1.360(3) and 1.349(3) Å respectively. The N(2)-N(3) distance is 1.301(3) Å. In the pyridazine ring, bonds C(6)-C(7) (1.455(3)), C(7)-C(8) (1.400(3)) and C(8)-C(9) (1.436(3) Å) are significantly longer than expected. The N-N distances in the azide group are 1.262(3) and 1.113(4) Å. In the crystal, planar molecules are arranged in such a way that they lie approximately on two mutually perpendicular lattice planes forming a see-saw pattern.

4-CYANO-1,3,6-TRIAZACYCLO[3.3.3]AZINE
$C_{10}H_5N_5$

G.J. ASHWELL, L.A. MARCH, I.W. NOWELL and J.T. SHAW, 1985. *J. Chem. Soc. Perkin Trans. 2*, 153-156.

Monoclinic, $P2_1/c$, a = 6.411, b = 15.366, c = 9.211 Å, β = 104.30°, Z = 4. Mo radiation, R = 0.045 for 737 reflections.

In the crystal the molecules are arranged in distorted columns (Fig. 1), parallel to the c axis with two types of overlap and interplanar spacings of 3.09 and 3.31 Å which alternate.

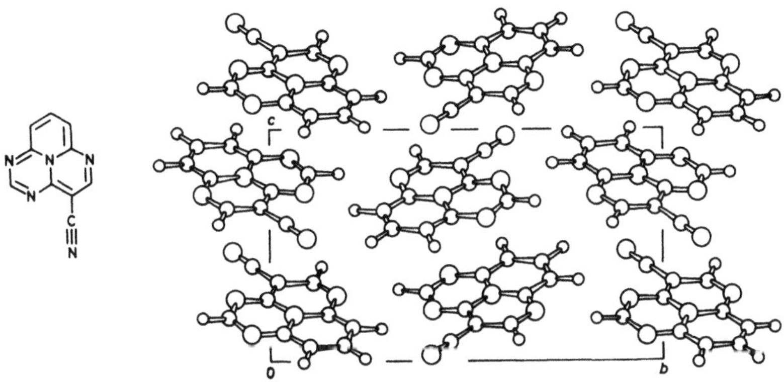

Fig. 1. Crystal structure of $C_{10}H_5N_5$ projected along a and the molecular skeleton.

1-AZACARBAZOLE
$C_{11}H_8N_2$

K. SUWINSKA, 1985. *Acta Cryst.*, C41, 973-975.

Monoclinic, $P2_1/n$, a = 11.318, b = 5.547, c = 13.516 Å, β = 95.71°, Z = 4. Cu radiation, R = 0.055 for 1160 reflections.

The molecule as a whole (Fig. 1) deviates significantly from planarity with N(1) and N(9) deviating in opposite directions by 0.028(2) and -0.039(2) Å respectively from the mean plane. The azacarbazole molecules are assembled pairwise in the crystal, the members of each pair being related to each other by a centre of symmetry and linked by two hydrogen bonds (N(9)...N(1') and N(9')...N(1) 2.946(2) Å).

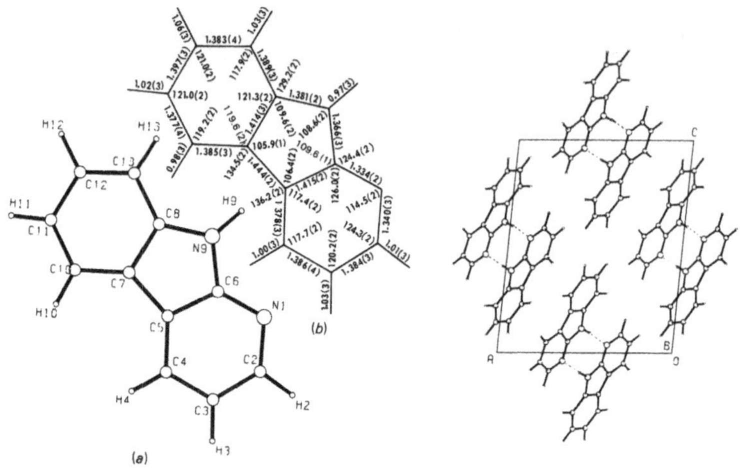

Fig. 1. The $C_{11}H_8N_2$ molecular skeleton, bond lengths and angles, and the molecular packing.

CARBAZOLE
$C_{12}H_9N$

V.K. BEL'SKII, 1985. *Kristallografiya*, **30**, 193-194.

Orthorhombic, *Pbnm*, a = 5.713, b = 7.775, c = 19.111 Å, Z = 4. Mo radiation, R = 0.035 for 363 reflections.

The molecule lies across a mirror plane with the N-H bond (Fig. 1) on the mirror plane. The angle between the benzene ring planes is 1.9°. There are no hydrogen bonds and intermolecular contacts correspond with van der Waals interactions. For previous photographic work see (1).

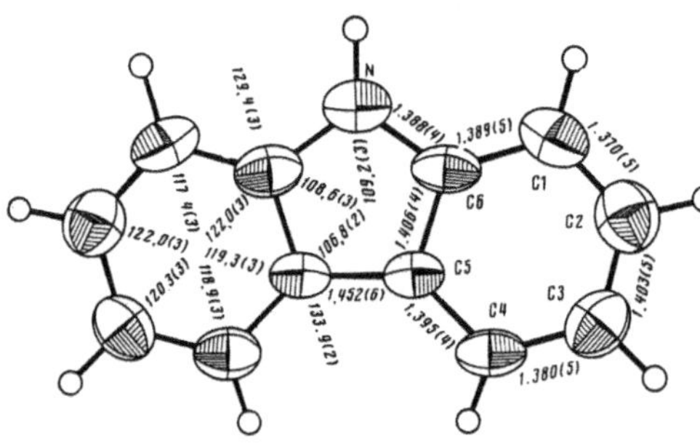

Fig. 1. The carbazole molecule with dimensions.

1. Structure Reports, 34B, 157.

6,12-DIHYDRODIPYRIDO[1,2-a:1',2'-d]PYRAZINIUM DIBROMIDE
$C_{12}H_{12}Br_2N_2,CH_4O$ $C_{12}H_{12}N_2^{2+}.2Br^-.CH_3OH$

M.R. BRYCE, J.G. EAVES, D. PARKER, J.A.K. HOWARD and O. JOHNSON, 1985. *J. Chem. Soc. Perkin Trans. 2*, 433-436.

Orthorhombic, *Cmc*2_1, a = 8.113, b = 12.214, c = 13.700 Å, Z = 4. Mo radiation, R = 0.048 for 634 reflections.

The central unsaturated ring of the cation (Fig. 1) has an eclipsed boat conformation; the dihedral angle between the outer rings is 141°. The molecules lie on crystallographic mirror planes and there is some atom-type disorder present among the quaternary sites in the cation (C(13) and N(11)).

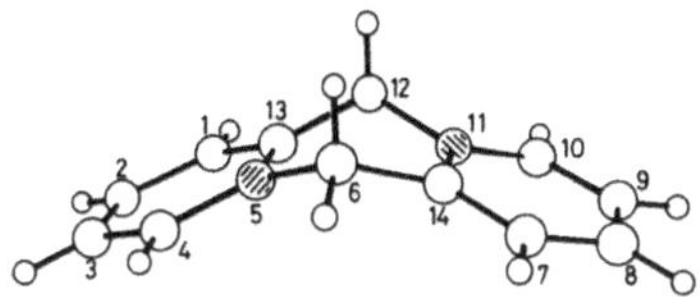

Fig. 1. Perspective drawing of a cation in the crystal structure of $C_{12}H_{12}Br_2N_2$.

1,2-PENTAMETHYLENE-5-NITRO-1,2-DIHYDRO-3H-INDAZOL-3-ONE
$C_{12}H_{13}N_3O_3$

V.J. ARAN, M. STUD, C. FOCES-FOCES, F.H. CANO and M. MARTINEZ-RIPOLL, 1985. *J. Heterocycl. Chem.*, **22**, 1743-1745.

Monoclinic, $P2_1/c$, a = 7.2465, b = 20.0005, c = 8.3068 Å, β = 105.543°, Z = 4. Cu radiation, R = 0.053 for 1350 reflections.

In the molecule (Fig. 1) the five-membered ring has a twist conformation with the binary axis through N(1)-N(2) distorted to an envelope at N(2). The seven-membered ring has a chair form with a quasi mirror plane through N(1)-N(2), distorted towards a twist-chair with a quasi binary axis through N(2). The geometry at N(1) is not planar (angle sum 354.9(3)°), while at N(2) the bond angle sum is 359.2(3)°. The nitro group plane makes an angle of 1.7° with the phenyl ring.

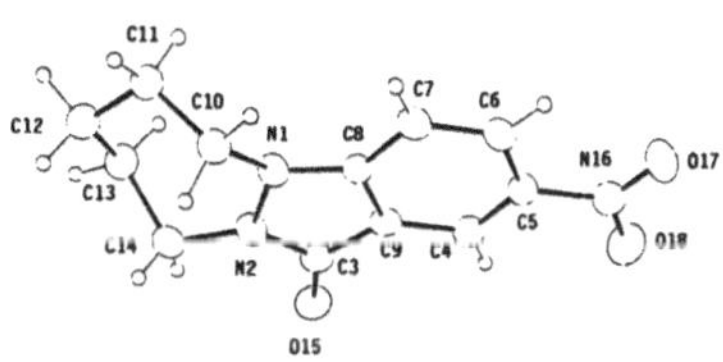

Fig. 1. The $C_{12}H_{13}N_3O_3$ molecule.

3-METHYL-2(1H)-BENZO[g]QUINOXALINONE
$C_{13}H_{10}N_2O$

M.J. BERTRAND, L. MALTAIS, F. BRISSE and M.-J. OLIVIER, 1985. *Canad. J. Chem.*, **63**, 3386-3389.

Monoclinic, $C2/c$, a = 19.834, b = 4.636, c = 22.360 Å, β = 96.30°, Z = 8. Cu radiation, R = 0.050 for 781 reflections.

The analysis shows that the amide form is present in the crystal lattice and that nearly planar molecules form centrosymmetric dimers through N-H...O hydrogen bonds (Fig. 1).

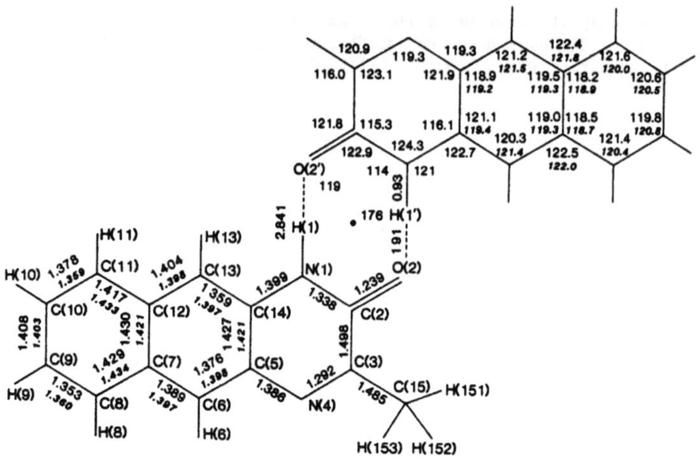

Fig. 1. Dimensions for $C_{13}H_{10}N_2O$. Values in italics refer to corresponding data from anthracene and anthracene complexes.

1-CARBAMOYLBENZ[c,d]INDOL-2-ONE
$C_{13}H_{10}N_2O_2$

L.G. KUZ'MINA, Yu.T. STRUCHKOV, L.P. GRIGOR'EVA, Z.I. EZHKOVA, B.E. ZAITSEV, S.P. TRUNOV, N.I. VASIL'EVA and K.M. DYUMAEV, 1984. *Khim. Get. Soedin. SSSR*, 1110-1114.

Monoclinic, *P*2_1/*a*, a = 19.685, b = 9.845, c = 5.3289 Å, β = 96.71°, Z = 4. Cu radiation, R = 0.085 for 980 reflections.

The analysis establishes the benz[c,d]indol-2-one skeleton for the molecule as shown in Fig. 1. There is an intramolecular N(2)-H...O(1) hydrogen bond H...O 1.76 Å; there is also a weak intermolecular H...O contact (H...O 2.245 Å).

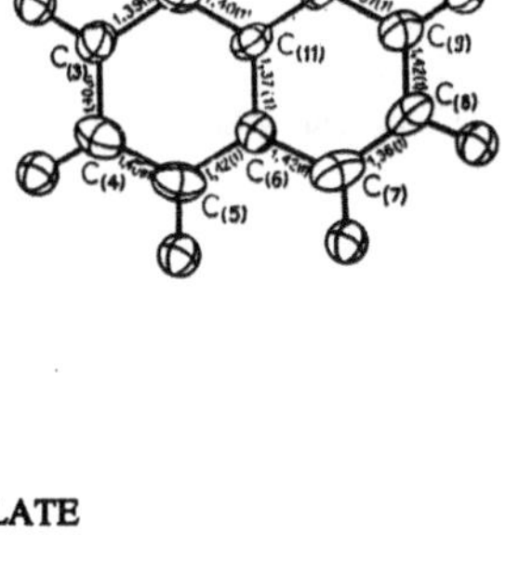

Fig. 1. The $C_{13}H_{10}N_2O_2$ molecule.

METHYL β-CARBOLINE-2-CARBOXYLATE
$C_{13}H_{10}N_2O_2$

A.K.S. MUIR and P.W. CODDING, 1985. *Canad. J. Chem.*, 63, 2752-2756.

Monoclinic, $P2_1/c$, a = 11.4866, b = 5.8091, c = 32.417 Å, β = 97.111°, Z = 8. Cu radiation, R = 0.049 for 3290 reflections.

In both molecules in the asymmetric unit the ester side chain has an extended conformation and is coplanar with the β-carboline moiety (Fig. 1). The carbonyl O atom (O(34)) and aromatic N atom (N(2)) are cis and form a three-centre hydrogen bond to the amine N atom of the other molecule.

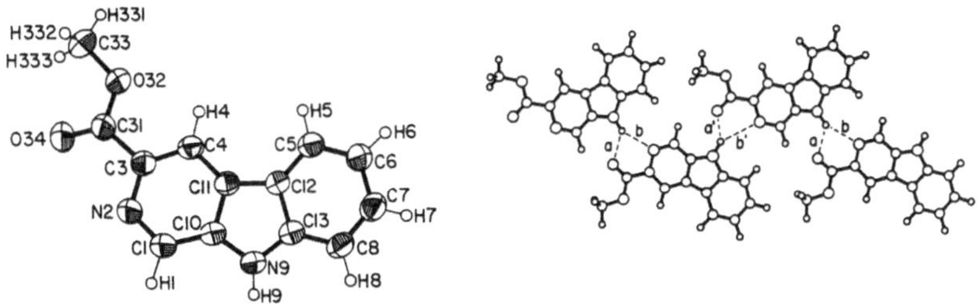

Fig. 1. One of the $C_{13}H_{10}N_2O_2$ molecules and a view showing the three-centre hydrogen bonds.

DIMETHYL PYRIDAZINO[2,3-b]BENZO-1,2,4-TRIAZINE-3,4-DICARBOXYLATE
$C_{14}H_{11}N_4O_4$

B. BOVIO, 1985. *J. Cryst. Spect. Res.*, **15**, 147-156.

Monoclinic, $P2_1/c$, a = 9.222, b = 15.335, c = 9.544 Å, β = 102.18°, Z = 4. Mo radiation, R = 0.040 for 590 reflections.

All the bonds in the fused rings have partial double-bond character consistent with an extensive conjugation within the molecule (with the exclusion of the 4-carbomethoxy group); average dimensions are: C-C 1.387, C-N 1.358, N-N 1.356 Å, C-C-C 121.5, C-C-N 120.4, C-N-C 117.8, N-C-N 121.4, N-N-C 119.8°.

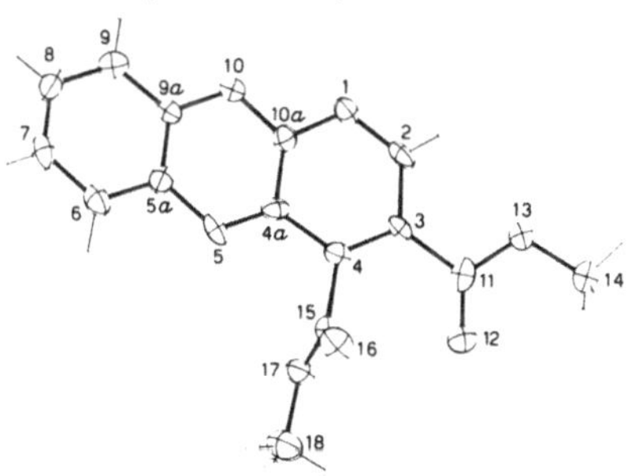

Fig. 1. A plot of $C_{14}H_{11}N_4O_4$.

N-ETHYLCARBAZOLE
$C_{14}H_{13}N$

T. KIMURA, Y. KAI, N. YASUOKA and N. KASAI, 1985. *Bull. Chem. Soc. Jpn.*, **58**, 2268-2271.

Orthorhombic, *Pbca*, a = 7.971, b = 24.615, c = 56.587 Å, Z = 40. Cu radiation, R = 0.062 for 5134 reflections.

The five crystallographically independent molecules in the asymmetric unit are not significantly different, one is shown in Fig. 1. The carbazole group is planar and the ethyl group bonded to N is almost perpendicular to the carbazole ring (average 87.8°). The average C-N bond length in the carbazole rings is 1.372(8) Å and the C-C bond linking the two aromatic rings has a mean value 1.437(9) Å; the C-N-C intraring angle is 109.0(2)°.

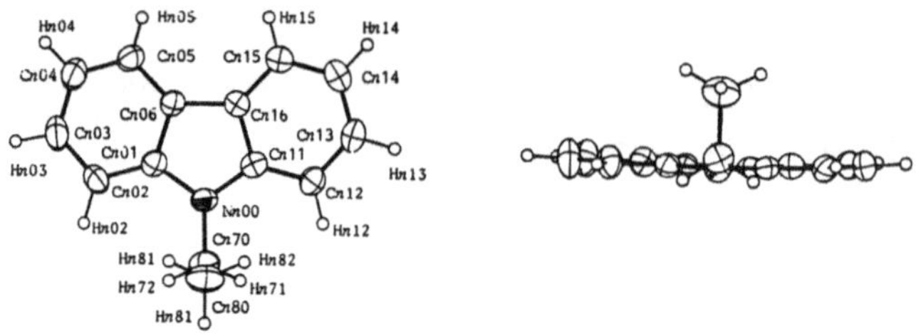

Fig. 1. Two views of an N-ethylcarbazole molecule.

4-(2-CARBOXYETHYL)-1,2,3,4-TETRAHYDROCYCLOPENT[b]INDOLE $C_{14}H_{15}NO_2$

J.R. RODRIGUEZ, F. TEMPRANO, C. ESTEBAN-CALDERON, M. MARTINEZ-RIPOLL and S. GARCIA-BLANCO, 1985. *Tetrahedron,* **41**, 3813-3823.

Triclinic, $P\bar{1}$, a = 11.642, b = 9.950, c = 11.455 Å, α = 107.11, β = 110.13, γ = 82.53°, Z = 4. Cu radiation, R = 0.056 for 2324 reflections.

There are two independent molecules which form centrosymmetric dimers by O-H...O hydrogen bonding (Fig. 1). In the packing, there is overlap between pairs of centrosymmetrically related molecules forming an apparent charge transfer complex with interplanar separations of 3.55 and 3.68Å.

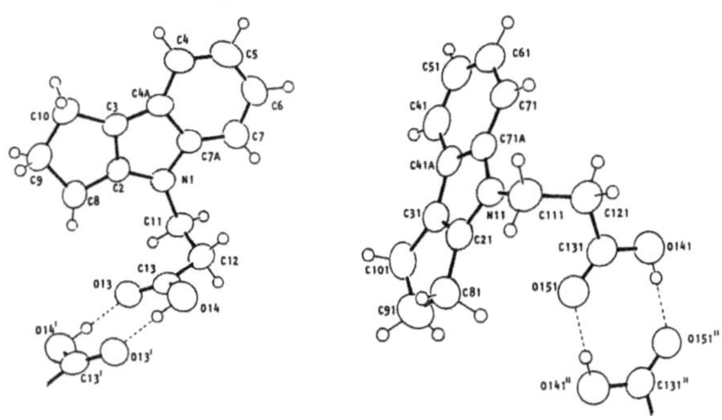

Fig. 1. Views of the two independent $C_{14}H_{15}NO_2$ molecules.

9-ACETOXYMETHYL-9a-METHOXY-3-OXO-1,2,9,9a-TETRAHYDRO-3H-PYRROLO[1,2-a]INDOLE $C_{15}H_{17}NO_4$

M. DARTMANN, W. FLITSCH, B. KREBS and P. RUSSKAMP, 1985. *Liebigs Ann. Chem.*, 1437-1447.

Triclinic, $P\bar{1}$, a = 7.731, b = 9.359, c = 10.560 Å, α = 99.48, β = 109.78, γ = 102.08°, Z = 2. Mo radiation, R = 0.047 for 2707 reflections (at -133°C).

The analysis established the structure and stereochemistry as shown in Fig. 1.

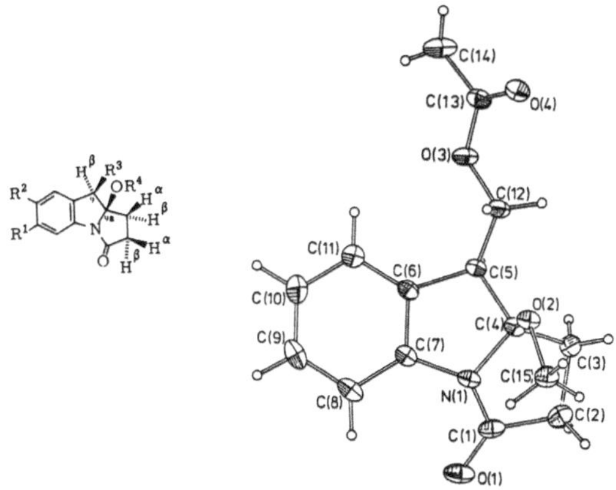

Fig. 1. The $C_{15}H_{17}NO_4$ molecule ($R^1=R^2=H$, $R^3=CH_2OAc$, $R^4=CH_3$).

2-PHENYL-2,5-DIHYDROPYRAZOLO[4,3-c]QUINOLIN-3(3H)-ONE $C_{16}H_{11}N_3O$ (I)

2-(4-CHLOROPHENYL)-2,5-DIHYDROPYRAZOLO[4,3-c]QUINOLIN-3(3H)-ONE $C_{16}H_{10}ClN_3O$ (II)

V. FERRETTI, V. BERTOLASI, G. GILLI and P.A. BOREA, 1985. *Acta Cryst.*, C41, 107-110.

I. Monoclinic, $P2_1/c$, a = 8.147, b = 12.855, c = 12.521 Å, β = 103.62°, Z = 4. Mo radiation, R = 0.037 for 1147 reflections.

II. Orthorhombic, *Pbca*, a = 14.790, b = 9.515, c = 18.326 Å, Z = 8. Mo radiation, R = 0.046 for 1117 reflections.

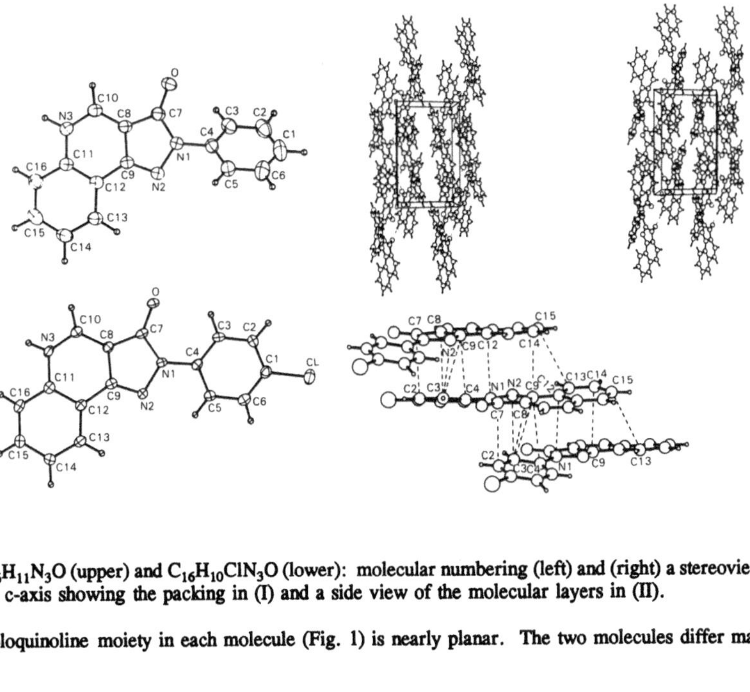

Fig. 1. $C_{16}H_{11}N_3O$ (upper) and $C_{16}H_{10}ClN_3O$ (lower): molecular numbering (left) and (right) a stereoview along the c-axis showing the packing in (I) and a side view of the molecular layers in (II).

The pyrazoloquinoline moiety in each molecule (Fig. 1) is nearly planar. The two molecules differ mainly in the conformation of the C(1,2...6) phenyl ring. The torsion angle N(2)-N(1)-C(4)-C(5) is -41.2(4) and 10.2(5)° in (I) and (II) respectively. The average of the lengths of the N(3)-C(10) 'single' and the C(8)-C(10) 'double' bonds are 1.323(4) and 1.363(5) Å respectively. These values were ascribed to the effects of the infinite chains of C=O...H-N hydrogen bonds affecting the charge distributions on the polar canonical forms contributing to the bond lengths. In (I) these infinite chains of molecules, linked head to tail, run along the b axis. The packing of these columns is governed by weak van der Waals forces. In (II) a similar but more efficient packing of the chains occur due to the smaller values of the N(2)-N(1)-C(4)-C(5) torsion angle. Several short intermolecular C-H...O contacts were observed in both (I) and (II).

7,12-DIHYDROPHTHALAZINO[2,3-b]PHTHALAZINE-5,14-DITHIONE
$C_{16}H_{12}N_2S_2$

V.N. NESTEROV, V.E. SHKLOVER, Yu.T. STRUCHKOV, K.A. POTEKHIN, V.N. LISITSIN and A.G. BOROVKOV, 1985. *Acta Cryst.*, C41, 1771-1773.

Orthorhombic, $P2_12_12_1$, a = 13.064, b = 13.1891, c = 7.756 Å, Z = 4. Mo radiation, R = 0.032 for 1602 reflections.

The tetracyclic molecule (Fig. 1) has a non-planar twisted conformation bent along the C(5)...C(14) vector by 45.0(1)°. Ring B has a chair conformation and ring C a twisted boat conformation. Ring A is planar and coplanar with the flat part of ring B. Ring D is also planar but forms a dihedral angle of 22.9(1)° with the plane of the bottom of the boat formed by ring C. The average C=S length is 1.676(2) Å. There is conjugation in the C(7a)C(7)S(1)N(6) and C(11a)C(12)S(2)N(13) fragments of the molecule but not in the N(6)-N(13) bond (1.406(2) Å). In the crystal, there are intermolecular contacts between the benzene rings of molecules related by twofold screw axes parallel to [100]. The molecules thus form infinite chains along the *a*-axis.

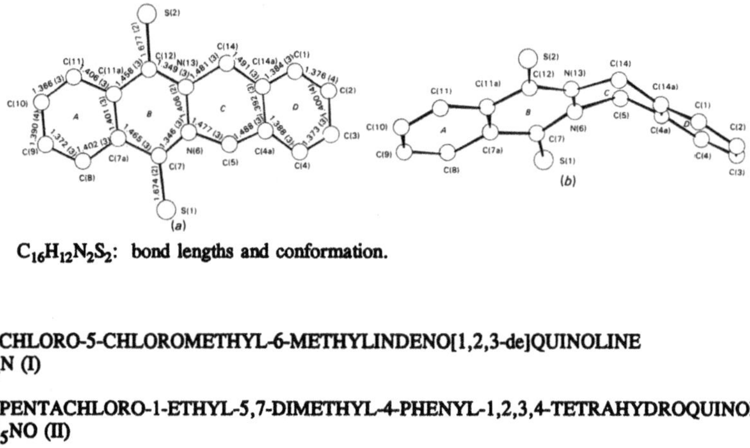

Fig. 1. $C_{16}H_{12}N_2S_2$: bond lengths and conformation.

1,2,4-TRICHLORO-5-CHLOROMETHYL-6-METHYLINDENO[1,2,3-de]QUINOLINE
$C_{17}H_9Cl_4N$ (I)

3,3,4,6,8-PENTACHLORO-1-ETHYL-5,7-DIMETHYL-4-PHENYL-1,2,3,4-TETRAHYDROQUINOLIN-2-ONE
$C_{19}H_{16}Cl_5NO$ (II)

L. DENNER, J.L.C. MARAIS and B. STASKUN, 1985. *Tetrahedron*, **41**, 5615-5618.

I. Triclinic, $P\bar{1}$, a = 10.417, b = 9.455, c = 8.555 Å, α = 106.64, β = 90.68, γ = 111.51°, Z = 2. Cu radiation, R = 0.0640 for 2165 reflections.

II. Monoclinic, $P2_1/n$, a = 19.818, b = 10.714, c = 9.214 Å, β = 99.43°, Z = 4. Mo radiation, R = 0.065 for 2472 reflections.

The analyses established the structures and connectivity as shown in Fig. 1.

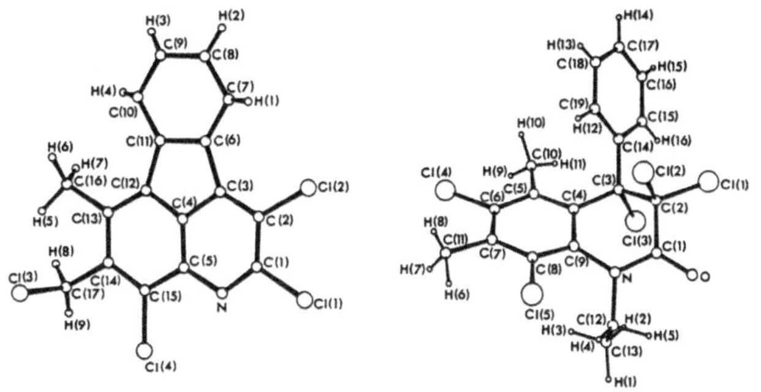

Fig. 1. Views of $C_{17}H_9Cl_4N$ (I) (left) and $C_{19}H_{16}Cl_5NO$ (II) (right).

(1'RS,2SR,3RS,4'SR)-2,3-(CYCLOHEX-2'-EN-1',4'-YLENE)-N-(4''-METHYLPHENYL)SUCCINIMIDE
$C_{17}H_{17}NO_2$

T.W. HAMBLEY, I.A. NEWSOM and S. STERNHELL, 1985. *Aust. J. Chem.*, **38**, 513-517.

Monoclinic, $P2_1/n$, a = 11.442, b = 6.509, c = 18.660 Å, β = 80.29°, Z = 4. Mo radiation, R = 0.038 for 1319 reflections.

The structure (Fig. 1) consists of discrete molecules held together by van der Waals interactions. The phenyl ring is essentially planar with C(1) and N(8) 0.050(3) and 0.046(2) Å respectively, above this plane. The five-membered ring and its associated carbonyl groups is approximately planar as is the imide moiety, O(10),C(9),N(8),C(13),O(14),C(5). The dihedral angles between the planes through the phenyl and the imide groups, C(6),C(5),N(8),C(13) and C(4),C(5),N(8),C(9), are 66(1) and 64(1)° respectively. Bond lengths and angles lie within the expected ranges.

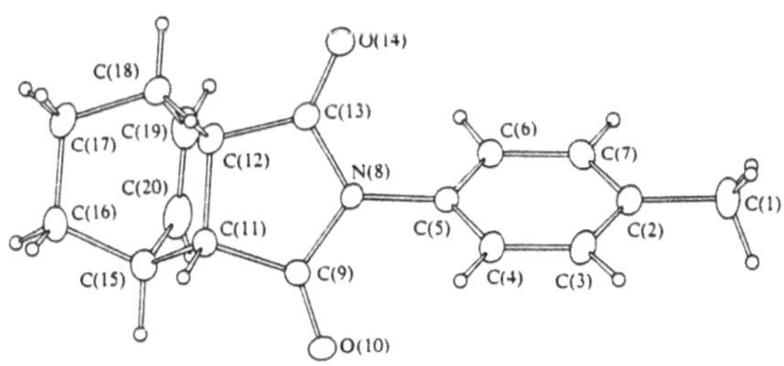

Fig. 1. $C_{17}H_{17}NO_2$: the molecular structure.

(-)-(3S,4aR,10aR)-9-BROMO-1,2,3,4,4a,5,10,10a-OCTAHYDRO-1,6-DIMETHOXY-3-(METHOXYCARBONYL)-BENZO[g]QUINOLINE
$C_{17}H_{22}BrNO_4$

R. NORDMANN and A. WIDMER, 1985. *J. Med. Chem.*, **28**, 1540-1542.

Triclinic, $P1$, a = 9.246, b = 9.898, c = 12.749 Å, α = 129.56, β = 98.75, γ = 95.13°, Z = 2. Cu radiation, R = 0.031 for 2183 reflections.

The structure and absolute stereochemistry are shown in Fig. 1. C(10a) has R absolute configuration.

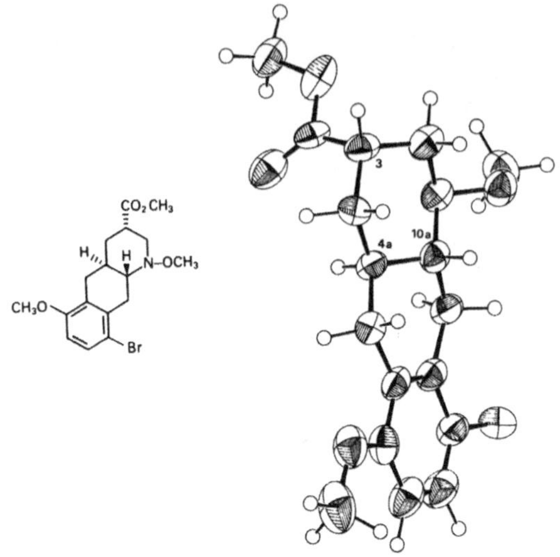

Fig. 1. The $C_{17}H_{22}BrNO_4$ structure.

2-CHLORO-1-METHYL-4-PHENYL-α-CARBOLINE HYDROCHLORIDE MONOHYDRATE
$C_{18}H_{14}Cl_2N_2,H_2O$ $C_{18}H_{14}ClN_2^+,Cl^-,H_2O$

E.V. SLAVYANOV, V.N. BYUSHKIN, V.V. KONONOVA and A.A. SEMENOV, 1984. *Khim. Get. Soedin. SSSR*, 779-782.

Orthorhombic, $Pna2_1$, a = 8.872, b = 15.315, c = 12.323 Å, Z = 4. Cu radiation, R = 0.062 for 1040 reflections.

The six-membered rings of the carboline system are inclined at 3° to the central pyrrole ring and 6° to one another (Fig. 1). The phenyl ring is rotated 52.7° out of the carboline plane.

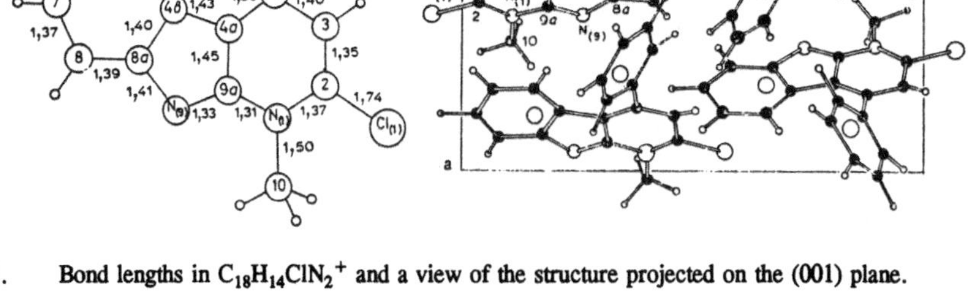

Fig. 1. Bond lengths in $C_{18}H_{14}ClN_2^+$ and a view of the structure projected on the (001) plane.

(Z)-3-(α-HYDROXYBENZYLIDENE)-1,2,3,9-TETRAHYDROPYRROLO[2,1-b]QUINAZOLIN-9-ONE
$C_{18}H_{14}N_2O_2$

J.C. BARNES, J.N. LOW, J.D. PATON, A.D. DUNN and K.I. KINNEAR, 1985. *Acta Cryst.*, **C41**, 282-284.

Monoclinic, $P2_1/c$, a = 11.309, b = 11.121, c = 11.672 Å, β = 110.23°, Z = 4. Mo radiation, R = 0.065 for 1115 reflections.

The phenyl group of the molecule (Fig. 1) is approximately coplanar with the pyrroloquinazolinone fragment. The molecule is the (Z)-isomer and is stabilised by a hydrogen bond between O(22) and N(4) 2.571(6) Å.

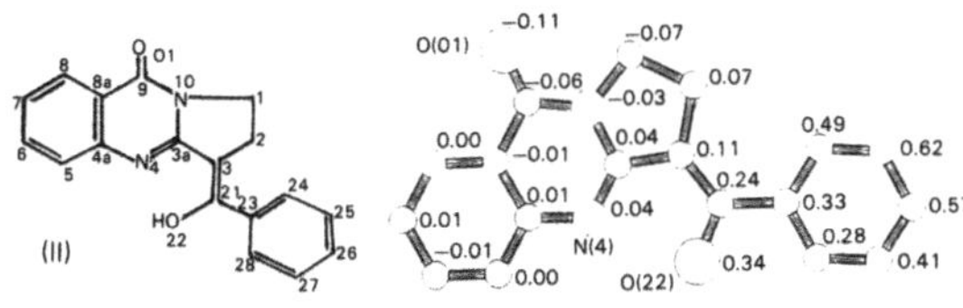

Fig. 1. $C_{18}H_{14}N_2O_2$: the molecular skeleton and structure.

3a-AZONIA(4-PHENYL-6-METHYL-4,9-DIHYDROQUINAZOLINO[3a,9a]-3,10-DIHYDRO-1H-PYRAZIN-2-ONE) BROMIDE

$C_{18}H_{18}BrN_3O$ $C_{18}H_{18}N_3O^+, Br^-$

O.P. RUDENKO, M.M. BOTOSHANSKII, Yu.A. SIMONOV, A.V. BOGATSKII and O.P. POVOLOTSKAYA, 1984. *Khim. Get. Soedin. SSSR*, 983-988.

Monoclinic, $B112/b$, a = 24.192, b = 13.678, c = 10.748 Å, γ = 104.2°, Z = 8. Mo radiation, R = 0.050 for 1335 reflections.

The cation dimensions are given in Fig. 1. Nitrogen atoms N(2) and N(3) are protonated. The A and B rings have flattened boat conformations and the C and D rings are individually planar. The ions are linked by N-H...Br hydrogen bonds (N(2)...Br 3.24 Å, N(3)...Br 3.39 Å).

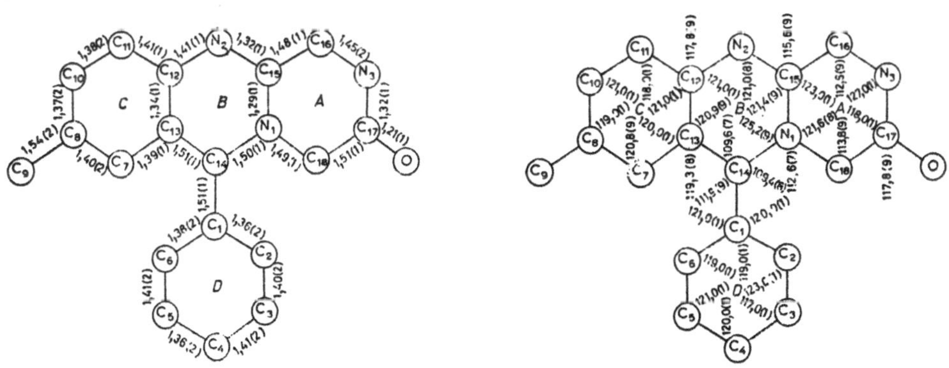

Fig. 1. Bond lengths and angles in $C_{18}H_{18}N_3O^+$.

4,5-DIAZA-N-PHENYL-7,8,8-TRIMETHYLTRICYCLO[4.2.1.0^{3,7}]NONANE-4,5-DICARBOXIMIDE

$C_{18}H_{21}N_3O_2$

W. ADAM, W.D. GILLASPEY, E.-M. PETERS, K. PETERS, R.J. ROSENTHAL and H.G. VON SCHNERING, 1985. *J. Org. Chem.*, **50**, 580-586.

Triclinic, $P\bar{1}$, a = 10.206, b = 22.799, c = 7.092 Å, α = 94.05, β = 98.54, γ = 99.85°, Z = 4. Mo radiation, R = 0.071 for 4313 reflections.

The crystallographic asymmetric unit contains two molecules of the urazole.

2,3,4,5,6,7,8,9-OCTAHYDRO-2,2,5,5,8,8-HEXAMETHYL-1H-BENZO[1,2-c:3,4-c':5,6-c'']TRIPYRROLIUM TRIBROMIDE DIHYDRATE

$C_{18}H_{30}Br_3N_3,2(H_2O)$ $C_{18}H_{30}N_3{}^{3+},3(Br^-),2(H_2O)$

S.L. LAWTON, J. CIRIC, G.T. KOKOTAILO and G.W. GRIFFIN, 1985. *Acta Cryst.*, C41, 1683-1685.

Triclinic, $P\bar{1}$, a = 11.097, b = 13.502, c = 9.124 Å, α = 92.87, β = 111.37, γ = 66.08°, Z = 2. Mo radiation, R = 0.070 for 2525 reflections.

The structure contains discrete trications (Fig. 1) which have C_3 pseudo-symmetry. The central ring in the cation is planar. The pyrrolium rings are puckered with the N atoms displaced, on average, by 0.5(1) Å above the mean planes of the four other atoms in the rings. One of the methyl groups on the N atoms is in an axial position and the other is early equatorial with respect to the plane of the cation. Two cations are held together electrostatically in the form of a sandwich by a central network of four water molecules and two bromide anions. These form a planar centrosymmetric six-membered ring held together by hydrogen bonds O-H...Br^- with lengths ranging from 3.20 to 3.28 Å and O-H...O of length 2.75 Å. Each member of this ring is located either above or below a pyrrolium ring in the sandwich.

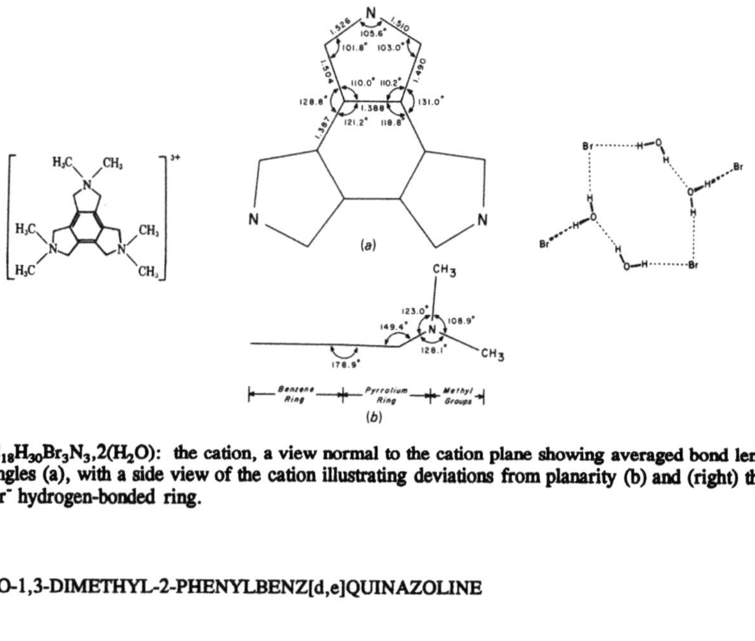

Fig. 1. $C_{18}H_{30}Br_3N_3,2(H_2O)$: the cation, a view normal to the cation plane showing averaged bond lengths and angles (a), with a side view of the cation illustrating deviations from planarity (b) and (right) the water-Br^- hydrogen-bonded ring.

2,3-DIHYDRO-1,3-DIMETHYL-2-PHENYLBENZ[d,e]QUINAZOLINE

$C_{19}H_{18}N_2$

A.A. ESPENBETOV, A.F. POZHARSKII, Yu.T. STRUCHKOV and A.N. SUSLOV, 1985. *Khim. Get. Soedin., SSSR*, 977-985.

Monoclinic, $P2_1/n$, a = 7.3136, b = 11.2316, c = 18.352 Å, β = 97.308°, Z = 4. Cu radiation, R = 0.0446 for 1319 reflections.

A view of the molecule with dimensions is in Fig. 1. The naphthalene moiety shows a slight departure from planarity. The dihydropyrimidine ring has an asymmetrically compressed half-chair conformation with C(11) 0.401 Å and N(2) -0.143 Å from the N(1),C(1),C(9),C(8) plane. The phenyl substituent at C(11) is axial.

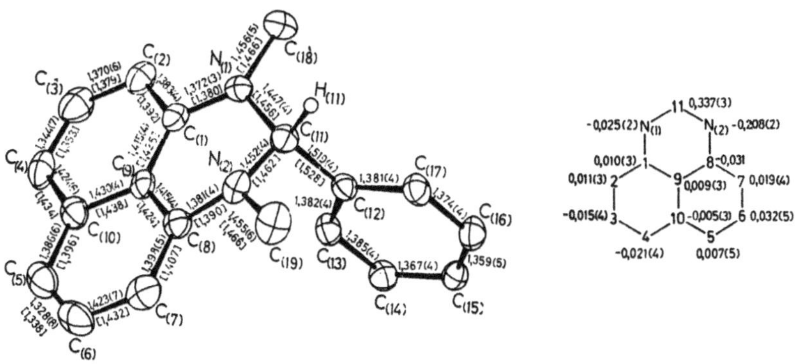

Fig. 1. The $C_{19}H_{18}N_2$ molecule showing bond lengths (values in [ ] corrected for libration) and displacements of the atoms from the naphthalene plane.

DIMETHYL 1-p-CHLOROPHENYLPYRIDAZINO[3,4-b]INDOLE-3,4-DICARBOXYLATE
$C_{20}H_{14}ClN_3O_4$

A.C. CODA, G. DESIMONI, M. PAPPALARDO, P.P. RIGHETTI, P.F. SENECI, G. TACCONI and R. OBERTI, 1985. *Tetrahedron*, **41**, 2545-2555.

Triclinic, $P\bar{1}$, a = 12.118, b = 10.788, c = 7.601 Å, α = 86.30, β = 98.26, γ = 113.60°, Z = 2. Cu radiation, R = not stated, for 1726 reflections.

The structure of this unexpected reaction product was shown by the analysis to be the pyridazino[3,4-b]indole derivative (Fig. 1).

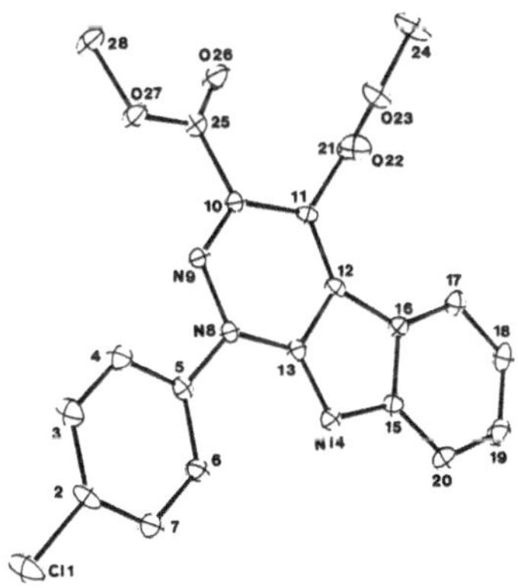

Fig. 1. A view of $C_{20}H_{14}ClN_3O_4$.

TRIMETHYL 1,2,3,6α-TETRAHYDRO-2,4-DIMETHYL-1,2,6-METHENO-3-BENZAZOCINE-3,5,11α-TRICARBOXYLATE
$C_{20}H_{21}NO_6$ (I)

DIMETHYL 1,4,5,8-TETRAHYDRO-11-METHOXY-4,6-DIMETHYL-1,4,8-METHENO-2,3,5-BENZOTRIAZECINE-7,13α-DICARBOXYLATE
$C_{19}H_{21}N_3O_5$ (II)

D.C. REMY, S.W. KING, D. COCHRAN, J.P. SPRINGER and J. HIRSHFIELD, 1985. *J. Org. Chem.*, **50**, 4120-4125.

I. Orthorhombic, *Pbca*, a = 13.854, b = 14.237, c = 18.575 Å, Z = 8. Cu radiation, R = 0.062 for 2039 reflections.

II. Triclinic, $P\bar{1}$, a = 9.699, b = 12.688, c = 8.122 Å, α = 96.41, β = 110.88, γ = 77.74°, Z = 2. Cu radiation, R = 0.057 for 2138 reflections.

The crystal structure analyses established the stereochemistries of the molecules (Fig. 1).

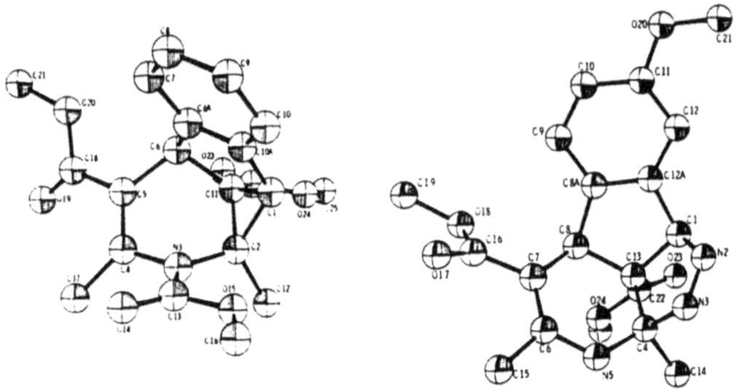

Fig. 1. Drawings of $C_{20}H_{21}NO_6$ (I) (left) and $C_{19}H_{21}N_3O_5$ (II) (right), hydrogen atoms omitted.

9-((5-(DIMETHYLAMINO)PENTYL)IMINO)-1-NITRO-9,10-DIHYDROACRIDINE MONOHYDRATE
$C_{20}H_{24}N_4O_2,H_2O$ (I)

9-((3-(METHYLAMINO)PROPYL)IMINO)-1-NITRO-9,10-DIHYDROACRIDINE MONOHYDRATE
$C_{17}H_{18}N_4O_2,H_2O$ (II)

9-((2-HYDROXYETHYL)AMINO)-1-NITRO-9,10-DIHYDROACRIDINIUM CHLORIDE
$C_{15}H_{14}ClN_3O_3$ (III) $C_{15}H_{14}N_3O_3^+,Cl^-$

9-((3-CARBOXYPROPYL)AMINO)-1-NITRO-9,10-DIHYDROACRIDINIUM CHLORIDE
$C_{16}H_{14}ClN_3O_4$ (IV) $C_{16}H_{14}N_3O_4^+,Cl^-$

J.J. STEZOWSKI, P. KOLLAT, M. BOGUCKA-LEDOCHOWSKA and J.P. GLUSKER, 1985. *J. Am. Chem. Soc.*, **107**, 2067-2077.

I. Monoclinic, $P2_1/c$, a = 8.446, b = 11.475, c = 20.702 Å, β = 97.475°, Z = 4. Cu radiation, R = 0.052 for 3120 reflections.

II. Triclinic, $P\bar{1}$, a = 11.0684, b = 11.4679, c = 14.9562 Å, α = 90.228, β = 115.806, γ = 77.750°, Z = 4. Cu radiation, R = 0.036 for 4515 reflections.

III. Monoclinic, $P2_1/n$, a = 8.3393, b = 20.7131, c = 8.9282 Å, β = 112.862°, Z = 4. Cu radiation, R = 0.049 for 2434 reflections.

IV. Orthorhombic, *Pbca*, a = 8.1847, b = 16.834, c = 22.667 Å, Z = 8. Cu radiation, R = 0.060 for 2425 reflections.

Molecules (I) and (II) (Fig. 1) are crystallised as free bases but (III) and (IV) (Fig. 1) as salts. Among bond lengths, the most significant differences are in the bonds C(9)-N(9), C(9)-C(1a), C(9)-C(8a) and C(2)-C(3), with free base averages and salt averages as follows: C(9)-N(9) 1.287, 1.316, C(9)-C(1a) 1.479, 1.456, C(9)-C(8a) 1.480, 1.444, C(2)-C(3) 1.386, 1.406 Å. The main difference in bond angles is due to steric overcrowding with N(9) pushed further from the nitro groups in the salts (III, IV) than in the free bases (I, II). The torsion angles about the nitro group are less in the cations than in the free bases, with C(1a)-C(1)-N(1)-O(1) 33° in (I), 35° in (II), 53° in (III), 61° in (IV), and C(8a)-C(1)-N(1)-O(2) 26, 26, 53 and 58°. The side chain direction is opposite in cations and free bases, with C(1a)-C(9)-N(9)-C 3, 8° in (I), (II) and 174, 172° in (III), (IV), probably due to the proton on N(9) in the salts. The angles between the outer ring of the acridine moiety are in the range 13-21°, and the central ring is the least planar. There is substantial hydrogen bonding in all four structures, involving N(10) protons binding to H_2O in the free bases and to Cl^- in the cations, and in the salts the N(9) proton binds to Cl^-. In addition, the functional group at the end of the side chain is involved if possible.

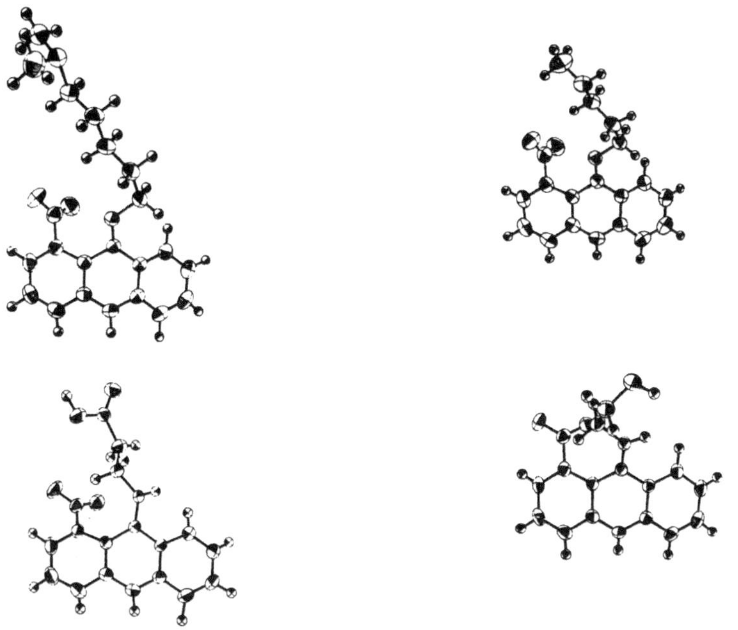

Fig. 1. Views of the four structures $C_{20}H_{24}N_4O_2,H_2O$ (I), $C_{17}H_{18}N_4O_2,H_2O$ (II), $C_{15}H_{14}ClN_3O_3$ (III) and $C_{16}H_{14}ClN_3O_4$ (IV) (clockwise from top left).

3-CHLORO-5-PHENYL-5,11-DIHYDROISOINDOLO[2,1-a]QUINAZOLINE
$C_{21}H_{15}ClN_2$

J. TROSCHUTZ, A. KIRFEL and G. WILL, 1985. *Arch. Pharm.*, **318**, 690-694.

Monoclinic, $P2_1/c$, a = 9.608, b = 12.075, c = 13.876 Å, β = 101.4°, Z = 4. Mo radiation, R = 0.067 for 1203 reflections.

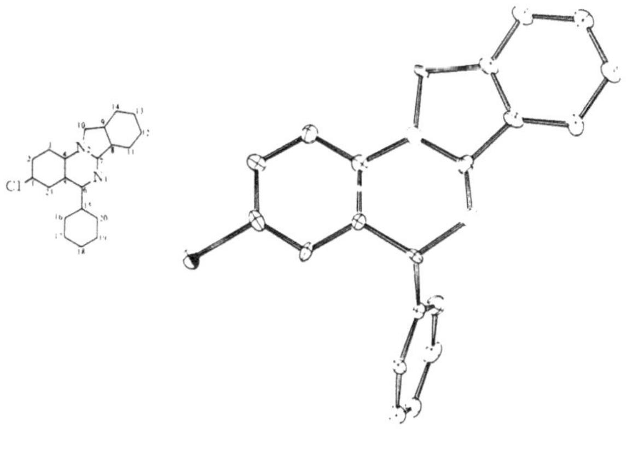

Fig. 1. Molecular structure of $C_{21}H_{15}ClN_2$ with numbering scheme.

The molecule is shown to be a dihydroisoindolo[2,1-a]quinazoline (Fig. 1). The bond distance C(7)-N(1), at 1.270(11) Å, indicates the presence of a double bond. The other distances are normal with the exception of C(9)

C(10) at 1.547(12) Å, and C(10)-N(2) at 1.500(11) Å, indicating the strain present in the five-membered ring C(7),C(8),C(9),C(10),N(2).

4'-(ACRIDIN-9-YLAMINO)-3'-METHOXYMETHANESULPHONANILIDE
$C_{21}H_{19}N_3O_3S$ (I)

4'-(ACRIDIN-9-YLAMINO)-N,N-BISMETHANESULPHONYL-3'-METHOXYANILINE
$C_{22}H_{21}N_3O_5S_2$ (II)

Z.H.L. ABRAHAM, S.D. CUTBUSH, R. KURODA, S. NEIDLE, R.M. ACHESON and G.N. TAYLOR, 1985. *J. Chem. Soc. Perkin Trans.* 2, 461-465.

I. Monoclinic, $P2_1/c$, a = 11.179, b = 21.117, c = 8.795 Å, β = 112.27°, Z = 4. Cu radiation, R = 0.041 for 2243 reflections.

II. Triclinic, $P\bar{1}$, a = 7.862, b = 9.875, c = 15.382 Å, α = 70.22, β = 78.94, γ = 77.79°, Z = 2. Cu radiation, R = 0.039 for 2838 reflections.

The molecules of (I) and (II) are shown in Fig. 1. They have a distinct orientation of the substituted phenyl ring with respect to the acridine group.

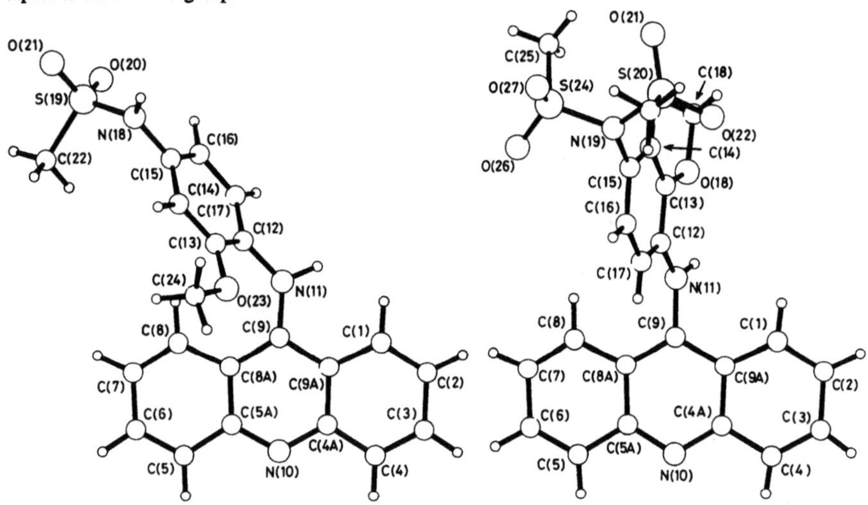

Fig. 1. Perspective diagrams of molecules of $C_{21}H_{19}N_3O_3S$ (left) and $C_{22}H_{21}N_3O_5S_2$ (right).

DIMETHYL (5S,6aR,7R,11bS,12R,13R)-5,6,6a,11b-TETRAHYDRO-5,7-ETHANO-7H-INDOLO[2,3-c]ISOQUINOLINE-12,13-DICARBOXYLATE METHANOL SOLVATE HYDRATE
$C_{21}H_{20}N_2O_4$,0.083(CH_4O),0.167(H_2O) (I)

(S)-6-(CARBOXYMETHYL)-2,3-DIHYDRO-3-PHENYLPYRAZOLO[5,1-a]ISOQUINOLINE-1,2-DICARBOXYLIC ACID TRIMETHYL ESTER
$C_{24}H_{22}N_2O_6$ (II)

I.L. KARLE, J.L. FLIPPEN-ANDERSON and R. HUISGEN, 1985. *Acta Cryst.*, C41, 1095-1100.

I. Triclinic, $P\bar{1}$, a = 12.121, b = 17.280, c = 11.612 Å, α = 91.2, β = 129.8, γ = 94.6°, Z = 4. Cu radiation, R = 0.108 for 5270 reflections.

II. Triclinic, $P\bar{1}$, a = 10.728, b = 13.126, c = 8.788 Å, α = 96.34, β = 113.03, γ = 103.45°, Z = 2. Cu radiation, R = 0.062 for 2816 reflections.

Both crystallographically independent molecules in the asymmetric unit of (I) (Fig. 1) have very similar geometries and conformations. The six-membered rings A and E are planar while ring B has an envelope conformation with N(18) displaced by 0.70(4) Å out of the plane of the other five ring atoms. Ring C has a boat conformation and there is a cis junction at the fusion of ring B with the non-planar five-membered ring D. The two methyl esters attached to ring C are gauche with respect to each other. The tricyclic moiety in (II) and the ester on C(1) are essentially coplanar. The carboxyl methyl ester on C(2) and the phenyl ring on N(3) are mutually anti gauche with a torsion angle C(13)-C(2)-N(3)-C(15) of -133(1)°. The carboxyl methyl ester on C(6) lies on the same side of the molecule as the phenyl group. In the crystals of (I) the two independent molecules have different environments. The water and methanol solvate sites are only partially occupied and appear to be involved in a network of weak hydrogen bonds. In the crystal structure of (II) the isoquinolyl moieties of centrosymmetrically related molecules stack in parallel planes of spacing 3.35 Å. No hydrogen bonding is reported.

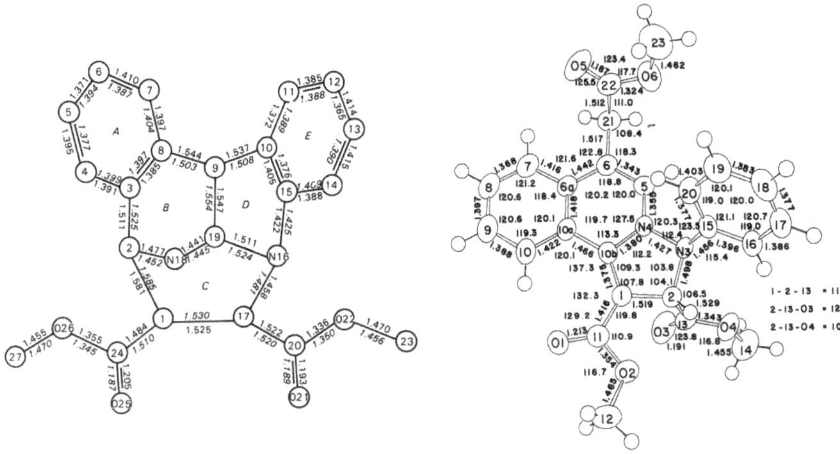

Fig. 1. Some geometrical parameters for $C_{21}H_{20}N_2O_4$ (I) (left) and in $C_{24}H_{22}N_2O_6$ (II) (right).

(1α,2α,9β,9aα)-2,3,9,9a-TETRAHYDRO-1,2-DIHYDROXY-6,7-DIMETHOXY-9-(PHENYLMETHYL)-1H-PYRROLO[1,2-a]INDOLE-9-CARBONITRILE
$C_{21}H_{22}N_2O_4$

W. VERBOOM, B.H.M. LAMMERINK, R.J.M. EGBERINK, D.N. REINHOUDT and S. HARKEMA, 1985. *J. Org. Chem.*, **50**, 3797-3806.

Monoclinic, $P2_1/c$, a = 10.489, b = 8.923, c = 20.158 Å, β = 98.82°, Z = 4. Mo radiation, R = 0.039 for 2052 reflections.

The crystal structure analysis established the stereochemistry of the molecule (Fig. 1).

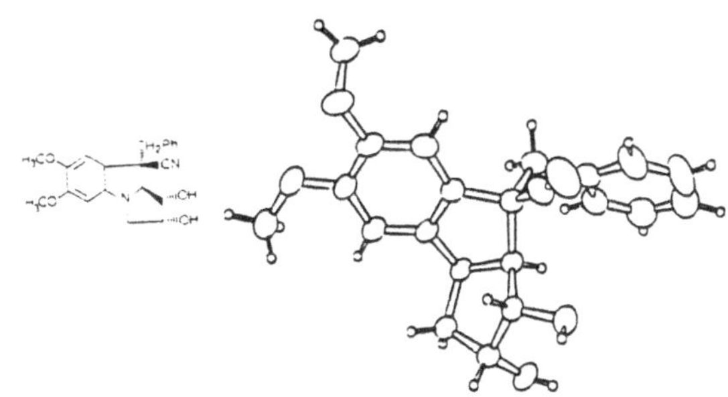

Fig. 1. The $C_{21}H_{22}N_2O_4$ molecule.

11-BENZOYL-1,2,3,4-TETRADEUTERO-10H-INDENO[2,1-b]INDOLIZIN-10-ONE
$C_{22}H_9D_4NO_2$

A. CAROTTI, G. CASINI, E. GAVUZZO and F. MAZZA, 1985. *Heterocycles*, **23**, 1659-1669.

Monoclinic, *Cc*, a = 4.238, b = 22.830, c = 15.839 Å, β = 90.76°, Z = 4. Mo radiation, R = 0.088 for 903 reflections.

The analysis determined the structure and stereochemistry to be as shown in Fig. 1. The four fused ring system is essentially planar (maximum deviation C(10a), 0.04 Å). The conformation of the benzoyl group is defined by the torsion angles C(10a)-C(11)-C(12)-C(13) -32(3)° and C(11)-C(12)-C(13)-C(18) -30(3)°.

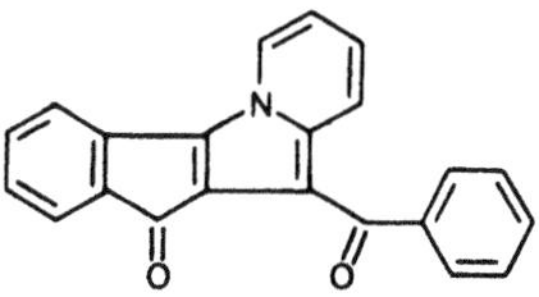

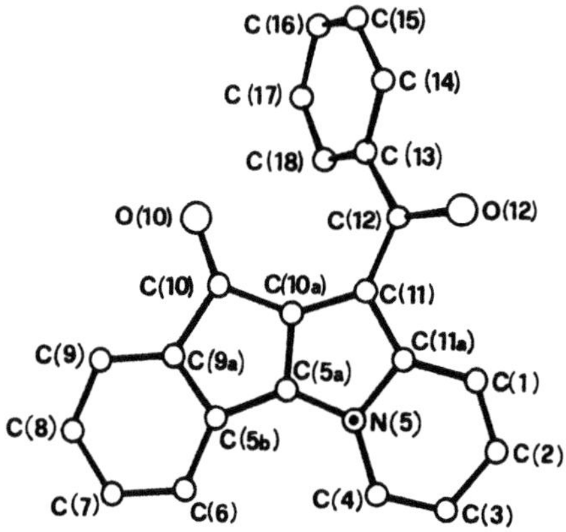

Fig. 1. The $C_{22}H_9D_4NO_2$ molecule (hydrogens omitted).

DIMETHYL 1-ETHYL-INDOLO[2,3-a]PERHYDROQUINOLIZINE-2-MALONATE BROMIDE MONOHYDRATE
$C_{22}H_{29}BrN_2O_4,H_2O$ $C_{22}H_{29}N_2O_4{}^+,Br^-,H_2O$

A. CHIARONI, C. RICHE, D.S. GRIERSON and H.-P. HUSSON, 1985. *Canad. J. Chem.*, **63**, 2979-2982.

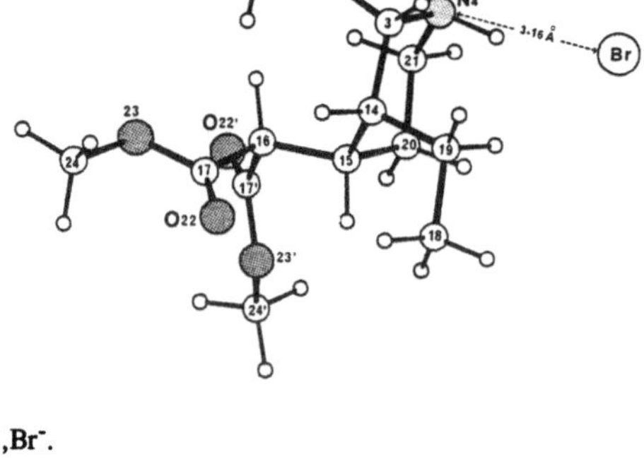

Fig. 1. A view of $C_{22}H_{29}N_2O_4{}^+,Br^-$.

Orthorhombic, *Pbca*, a = 11.181, b = 15.130, c = 26.74 Å, Z = 8. Cu radiation, R = 0.055 for 1902 reflections.

The analysis shows that the molecule has the cis indoloquinolizidine conformation with C(3), C(14) and C(15) substituents axial (Fig. 1). Ring D is in a slightly flattened chair conformation. The bromide ion is hydrogen bonded to N(4).

6,7-BENZO-3-BENZYL-5-HYDROXY-4-PHENYL-1,3-DIAZABICYCLO[3.3.0]OCTAN-8-ONE $C_{23}H_{20}N_2O_2$

J.D. COYLE, L.E. SMART, J.F. CHALLINER and E.J. HAWS, 1985. *J. Chem. Soc. Perkin Trans. 1*, 121-129.

Orthorhombic, *Pbca*, a = 10.1082, b = 13.7636, c = 27.0813 Å, Z = 8. Mo radiation, R = 0.042 for 2885 reflections.

The X-ray crystal structure analysis established the structure of one of the two diastereoisomers of the molecule (Fig. 1). The OH group (C(8), O(1)) and the phenyl substituent (C(9)-C(18)) are in a trans arrangement. The dihedral angle between these groups is 113°.

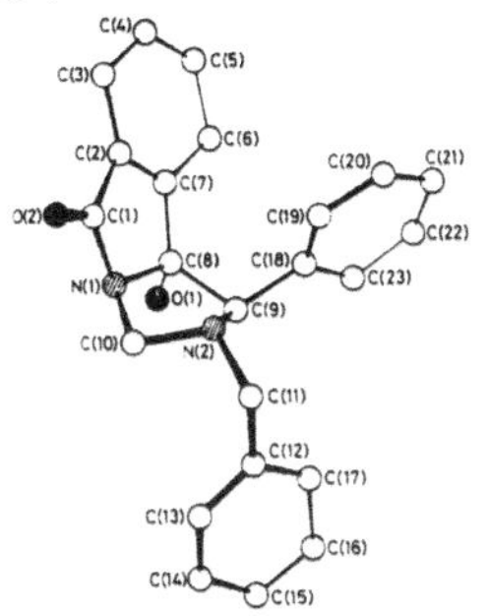

Fig. 1. Perspective view of a molecule of $C_{23}H_{20}N_2O_2$.

2,4a-DIMETHYL-4-PHENYL-3-ETHOXYCARBONYL-5-OXO-4a,5-DIHYDRO-4H-INDENO[1,2-b]PYRIDINE $C_{23}H_{21}NO_3$

M.F. BUNDULE, A.F. MISHNEV, Ya.Ya. BLEIDELIS, V.K. LUSIS, D.KH. MUTSENIETSE and G.Ya. DUBAR, 1984. *Khim. Get. Soedin. SSSR*, 1388-1392.

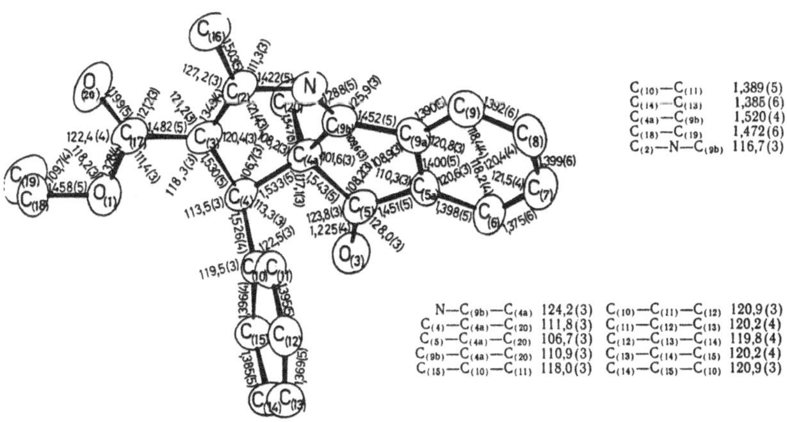

Fig. 1. The $C_{23}H_{21}NO_3$ molecule with dimensions.

Triclinic, $P\bar{1}$, a = 8.4856, b = 9.7660, c = 12.5959 Å, α = 88.445, β = 72.910, γ = 68.308°, Z = 2. Cu radiation, R = 0.070 for 2880 reflections.

The molecular conformation and dimensions are given in Fig. 1. The 3,4-dihydro pyridine ring has a distorted eclipsed boat conformation. Packing of the molecules is controlled by van der Waals interactions.

10,12-ETHANO-11-ISOPROPYLIDENE-2,8-METHANO-5-PHENYL-5-AZAPENTACYCLO[7.3.0.0 ^{3,7}]PENTADEC-1(9)-ENE-4,6-DIONE
$C_{23}H_{23}NO_2$ (I)

11,13-ETHANO-12-ISOPROPYLIDENE-2,8-METHANOPENTACYCLO[8.3.0.0 ^{3,8}]HEXADECA-1(10),5-DIENE-4,7-DIONE (isomer 1)
$C_{19}H_{20}O_2$ (II)

11,13-ETHANO-12-ISOPROPYLIDENE-2,8-METHANOPENTACYCLO[8.3.0.0 ^{3,8}]HEXADECA-1(10),5-DIENE-4,7-DIONE (isomer 2)
$C_{19}H_{20}O_2$ (III)

L.A. PAQUETTE, T.M. KRAVETZ and LEH-YEH HSU, 1985. *J. Am. Chem. Soc.*, **107**, 6598-6603.

I. Monoclinic, *C2/c*, a = 28.470, b = 11.473, c = 11.597 Å, β = 99.95°, Z = 8. Mo radiation, R = 0.045 for 2008 reflections.

II. Monoclinic, $P2_1/c$, a = 15.210, b = 6.260, c = 15.159 Å, β = 95.06°, Z = 4. Mo radiation, R = 0.057 for 1634 reflections.

III. Monoclinic, $P2_1/m$, a = 8.1369, b = 11.386, c = 8.835 Å, β = 113.85°, Z = 2. Mo radiation, R = 0.044 for 699 reflections.

The analyses established the stereochemistries as shown in Fig. 1. There is folding at C8-C9 in (I) (15.4°), at C7-C12 in (II) (16.3°). Molecule (III) has crystallographic mirror symmetry.

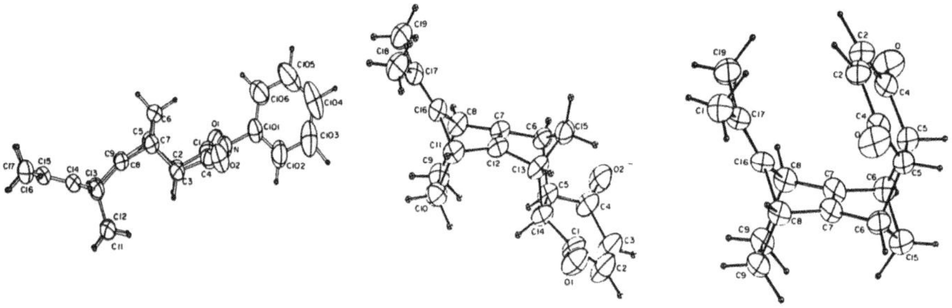

Fig. 1. Views of diene-diones (I)-(III) (left to right).

1,4-DI-t-BUTYL-3-(t-BUTYLIMINO)-3,4,5,6,7,8-HEXAHYDRO-5,8-METHANOPYRROLO[3,2-b]INDOL-2(1H)-ONE
$C_{23}H_{35}N_3O$

A. NONNENMACHER, H. PLIENINGER and M.L. ZIEGLER, 1985. *Chem. Ber.*, **118**, 3275-3286.

Monoclinic, $P2_1/a$, a = 11.867, b = 12.810, c = 14.799 Å, β = 102.31°, Z = 4. Mo radiation, R = 0.057 for 883 reflections.

The molecule is shown in Fig. 1 along with bond lengths which lie in the expected ranges. The angle sum at each of C(1), C(12), N(2) and N(1) is 360°.

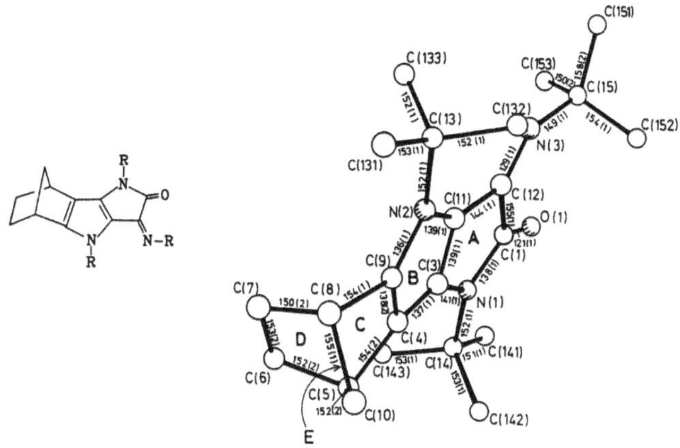

Fig. 1. The $C_{23}H_{35}N_3O$ molecule and bond lengths (pm).

2-ETHYL-7,9-DIMETHYL-4,5-DIHYDRO-5-(2,4-DINITROPHENYLHYDRAZONO)-4-OXO-1H-PYRROLO[2,3-f]QUINOLINE-2,7,9-TRICARBOXYLATE
$C_{24}H_{18}N_6O_{11}$

H. VAN KONINGSVELD, J.C. JANSEN, J.A. JONGEJAN, J.F. JZN and J.A. DUINE, 1985. *Acta Cryst.*, **C41**, 89-92.

Triclinic, $P\bar{1}$, a = 11.018, b = 15.451, c = 7.507 Å, α = 92.68, β = 104.69, γ = 86.63°, Z = 2. Mo radiation, R = 0.053 for 2238 reflections.

The molecule is planar, to within 0.15 Å, and this conformation is stabilized by an intramolecular bifurcated hydrogen bond between H(N3) and O(4) and O(41) and by an intramolecular hydrogen bond between H(N1) and O(91). The NO_2 group involved in the bifurcated bond is tilted 13.2(9)° out of the plane of the phenyl ring. The carboxylate substituents at C(2), C(7) and C(9) are inclined at 5.7, 11.0 and 33.8(9)° respectively, to the plane of the associated aromatic ring.

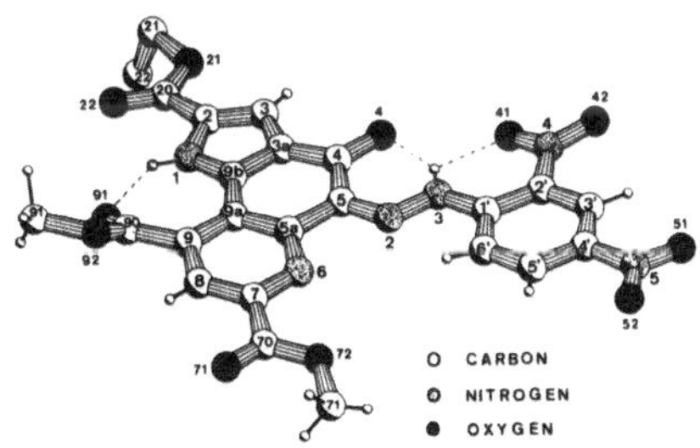

Fig. 1. $C_{24}H_{18}N_6O_{11}$: a view of the molecule showing the atomic numbering.

3-(p-CHLOROANILINO)-10-(p-CHLOROPHENYL)-2,10-DIHYDRO-2-(ISOPROPYLIMINO)PHENAZINE (CLOFAZIMINE)
$C_{27}H_{22}Cl_2N_4$ (I), (II)

2-((p-CHLOROPHENYL)IMINO)-10-(p-CHLOROPHENYL)-2,10-DIHYDRO-3-(ISOPROPYLAMINO)PHENAZINE (ISOCLOFAZIMINE)
$C_{27}H_{22}Cl_2N_4$ (III)

U. RYCHLEWSKA, M.B.H. BROOM, D.S. EGGLESTON and D.J. HODGSON, 1985. *J. Am. Chem. Soc.*, **107**,

4768-4772.

I. Monoclinic, $P2_1/a$, a = 7.788, b = 22.960, c = 13.362 Å, β = 98.58°, Z = 4. Mo radiation, R = 0.036 for 2180 reflections.

II. Triclinic, $P\bar{1}$, a = 10.507, b = 12.852, c = 9.601 Å, α = 95.96, β = 97.22, γ = 69.73°, Z = 2. Mo radiation, R = 0.062 for 1462 reflections.

III. Triclinic, $P\bar{1}$, a = 12.849, b = 13.252, c = 7.598 Å, α = 96.20, β = 104.80, γ = 105.44°, Z = 2. Mo radiation, R = 0.043 for 1597 reflections.

Isoclofazimine and monoclinic clofazimine are shown in Fig. 1; the triclinic form of clofazimine has a very similar conformation to that of (I) but has disordered isopropyl groups. In all three structures the dihydrophenazine moieties are not aromatic; the 14-atom ring systems in (II) and (III) are approximately planar. The monoclinic form (I) has a small (6.7°) fold about N(5)-N(10). The angles at N(3) (thought to be indicative of activity) are 131.6(2), 129.9(5) and 125.4(3)° in (I)-(III) respectively; angles 125.5±1 are inactive while angles of 131±1 are active.

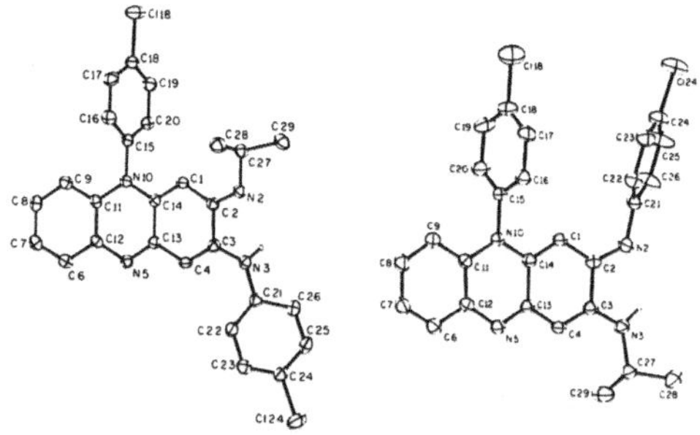

Fig. 1. The monoclinic form of clofazimine (I) (left) and isoclofazimine (II).

TETRAMETHYL 5-METHYL-7,8,9,10,11,11a-HEXAHYDRO-5H-CYCLOOCTA[b]INDOLE-6,11a-DIYLDI-MALEATE
$C_{27}H_{31}NO_8$ (I)

(Z,Z)-DIMETHYL 7,8-BIS(METHOXYCARBONYL)-4,5,11,12,13,14,15,15a-OCTAHYDRO-11H-CYCLOUN-DECA[4,5]PYRROLO[3,2,1-hi]INDOL-15a-YLMALEATE
$C_{29}H_{33}NO_8$ (II)

(Z,E)-METHYL 3,4,5-TRIS(METHOXYCARBONYL)-5-(5-METHYL-5,7,8,9,10,11,12,13-OCTAHYDROCYCLO-DECA[b]INDOL-6-YLIDENE)PENT-3-ENOATE
$C_{29}H_{35}NO_8$ (III)

R.M. LETCHER, TING-FONG LAI and T.C.W. MAK, 1985. *J. Chem. Res.*, S282-283, M3001-3088.

I. Triclinic, $P\bar{1}$, a = 9.159, b = 11.920, c = 12.643 Å, α = 107.78, β = 91.15, γ = 90.24°, Z = 2. Mo radiation, R = 0.062 for 4246 reflections.

II. Monoclinic, $P2_1/n$, a = 11.500, b = 20.112, c = 12.116 Å, β = 109.70°, Z = 4. Mo radiation, R = 0.065 for 3045 reflections.

III. Monoclinic, $P2_1/c$, a = 9.519, b = 20.630, c = 28.520 Å, β = 98.15°, Z = 8. Mo radiation, R = 0.092 for 4324 reflections.

In (I) (Fig. 1) one methyl group is disordered. In (III) there are two independent molecules in the crystallographic asymmetric unit. The crystal structure analyses unambiguously identified the molecules.

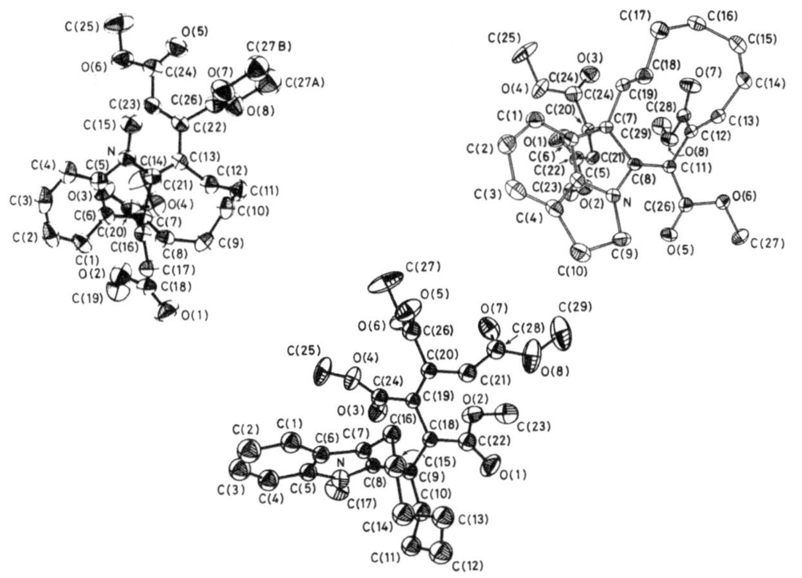

Fig. 1. Views of the molecules of $C_{27}H_{31}NO_8$ (I) (top left), $C_{29}H_{33}NO_8$ (II) (top right) and $C_{29}H_{35}NO_8$ (III) (bottom).

9,9'-OXYBIS(10-METHYLACRIDINIUM) BIS(TETRAFLUOROBORATE)
$C_{28}H_{22}B_2F_8N_2O$ (I) $(C_{28}H_{22}N_2O)^{2+}.2BF_4^-$

9,9'-OXYBISACRIDINE
$C_{26}H_{16}N_2O$ (II)

G. MAAS, 1985. *J. Chem. Soc. Perkin Trans. 2*, 1985-1989.

I. Monoclinic, *C2/c*, a = 23.251, b = 5.761, c = 18.748 Å, β = 100.22°, Z = 4. Mo radiation, R = 0.044 for 1014 reflections.

II. Monoclinic, *P*2$_1$/*c*, a = 12.137, b = 13.170, c = 11.554 Å, β = 96.36°, Z = 4. Mo radiation, R = 0.0426 for 1960 reflections.

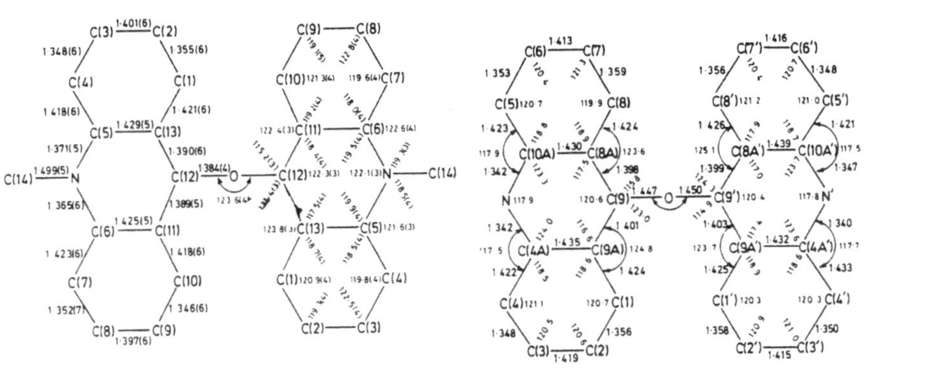

Fig. 1. Bond lengths (Å) and angles (°) in the molecules of $C_{28}H_{22}B_2F_8N_2O$ (left) and $C_{26}H_{16}N_2O$ (right).

Both compounds have twist conformations around the ether structural fragment. The angles between the acridine rings and the C-O-C plane are 46.3, 46.3° in (I) and 57.4, 51.1° in (II). The most notable difference between (I) and (II) is that the length of the ether bond is 0.06 Å shorter in (I) than in (II). The molecular geometries are shown in Fig. 1. Molecules of (I) have C_2 symmetry.

4,15-DIPHENYL-8,10,19,21-TETRAMETHYL-2,4,6,8,10,13,15,17,19,21-DECAAZAHEXACYCLO[10.10.0.-$0^{1,18}.0^{2,6}.0^{7,12}.0^{13,17}$]DOCOSAN-3,5,9,11,14,16,20,22-OCTAONE

$C_{28}H_{24}N_{10}O_8$

H. WAMHOFF, K. WALD, A. KIRFEL, L. FARKAS, N. SAMIMI and G. WILL, 1985. *Chem. Ber.*, **118**, 436-443.

Monoclinic, $P2_1/a$, a = 11.031, b = 21.683, c = 13.803 Å, β = 95.29°, Z = 4. Mo radiation, R = 0.132 for 1341 reflections.

The analysis established the constitution of this tetrazocane as shown in Fig. 1.

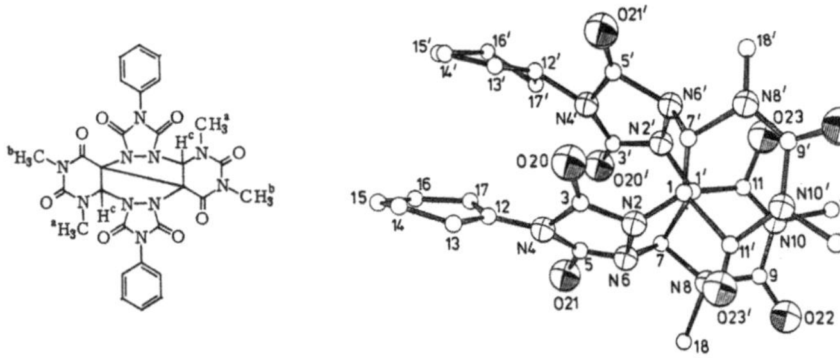

Fig. 1. The $C_{28}H_{24}N_{10}O_8$ molecule.

trans-N-(3-METHYL-2-p-TOLUENESULPHONIMIDO-3-AZATRICYCLO[8.5.0.0^{4,9}]PENTADECA-4(9),5,7-TRI-EN-10-YL)-p-TOLUENESULPHONAMIDE

$C_{29}H_{33}N_3O_4S_2$

J.L. TOMLIN, T.S. CAMERON, C.J. CARDIN and A.S. BAILEY, 1985. *Acta Cryst.*, C41, 1194-1196.

Triclinic, $P\bar{1}$, a = 14.67, b = 13.00, c = 8.25 Å, α = 99.9, β = 109.0, γ = 101.9°, Z = 2. Cu radiation, R = 0.0754 for 1375 reflections.

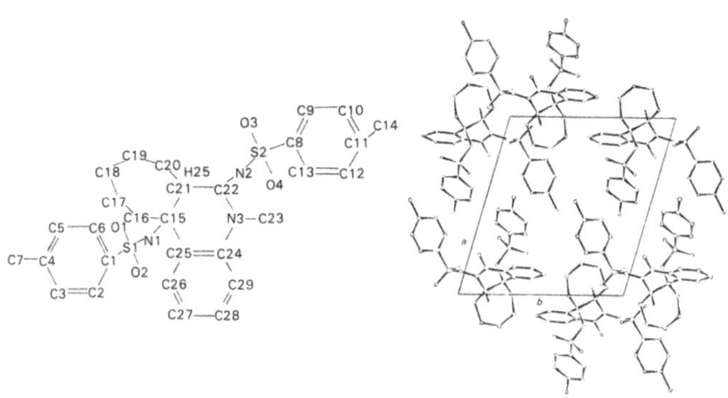

Fig. 1. $C_{29}H_{33}N_3O_4S_2$: the molecular skeleton and packing.

The compound (Fig. 1) has trans stereochemistry with the seven-membered ring in a chair-like conformation. The phenyl rings make angles of 97 and 129° with their respective C,N,S planes. The sulphonyl S-N(imino) distance (1.60(1)) is shorter than the sulphonyl S-N(amino) distance (1.64(1) Å). The S(1)-O(2) and S(1)-O(1) distances are 1.45(1) and 1.43(1) Å respectively, while the S(2)-O(3) and S(2)-O(4) distances are 1.46(1) and 1.47(1) Å respectively. The individual molecules are held together by a hydrogen bond linking N(1) of one molecule to O(3) of another (N...O 2.960 Å).

1,8-DIHYDRO-6,13-BIS(2-HYDROXYBENZOYL)DIBENZO[b,i]-1,4,8,11-TETRAAZACYCLOTETRADECA-4,6,-11,13-TETRAENE
$C_{32}H_{24}N_4O_4$ (I)

3-(2-BENZIMIDAZOLYL)-4H-(1)-BENZOPYRAN-4-ONE
$C_{16}H_{10}N_2O_2$ (II)

G. RIHS, I. SIGG, G. HAAS and T. WINKLER, 1985. *Helv. Chim. Acta*, **68**, 1933-1935.

I. Monoclinic, $P2_1/c$, a = 4.302, b = 25.835, c = 11.340 Å, β = 101.63°, Z = 2. Mo radiation, R = 0.072 for 1998 reflections.

II. Orthorhombic, *Pbca*, a = 16.692, b = 23.617, c = 6.095 Å, Z = 8. Mo radiation, R = 0.078 for 1833 reflections.

In (I) the molecules are centrosymmetric and sit at a crystallographic centre of symmetry. In (I) the hydrogens attached to the N-atoms in the cavity seem to be evenly distributed amongst all four N-atoms (ie. 0.5 H per N). There is in (I) an intramolecular hydrogen bond between the O-atom of the keto group and the O-atom of the Oh-group with an O...H distance of 1.58 Å; the keto group forms an angle of 27° with the macrocycle and of 15° with the o-hydroxyphenyl ring. Molecule (II) is planar with an intramolecular N-H...O hydrogen bond (N...O 2.790 Å). The molecules are shown in Fig. 1.

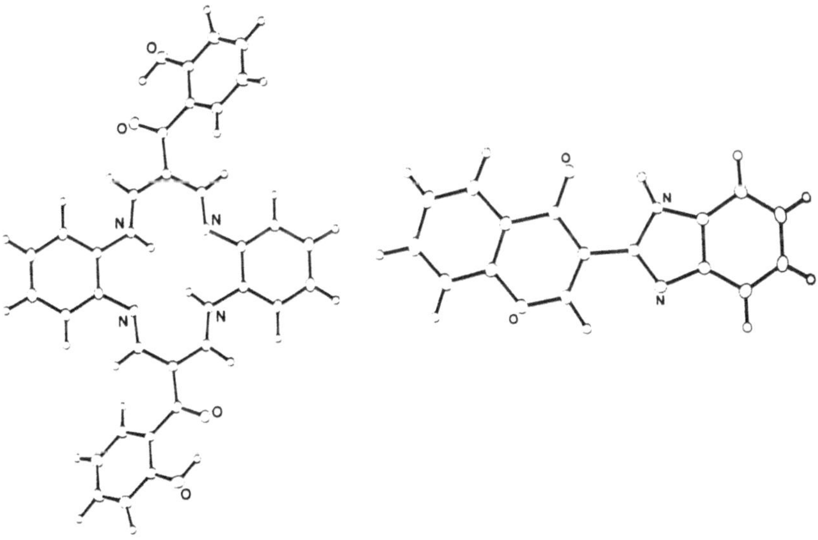

Fig. 1. ORTEP drawings of molecules of $C_{32}H_{24}N_4O_4$ (top left) and $C_{16}H_{10}N_2O_2$ (bottom right).

1,8,9,20-TETRAHYDRO-3,6-DIMETHYL-1,8-DIPHENYLDIBENZO[c,m]DIPYRAZOLO[3,4-f:4',3'-j](1,2,5,8,9,-12)HEXAAZACYCLOTETRADECINE
$C_{32}H_{26}N_{10}$

I.A. KROL, V.M. AGRE, V.K. VIGRAN and V.M. DZIOMKO, 1985. *Khim. Get. Soedin., SSSR*, 1564-1569.

Monoclinic, $P2_1/c$, a = 10.058, b = 26.272, c = 11.007 Å, β = 104.15°, Z = 4. Mo radiation, R = 0.072 for 2234 reflections.

The 14-membered ring of the molecule has a non-planar conformation as shown in Fig. 1. The N(3)-N(4) and N(8)-N(9) bond lengths (1.257(7) and 1.26(1) Å respectively) correspond to localized N=N double bonds. The conformation is stabilized by intramolecular N-H...N hydrogen bonds (Fig. 1).

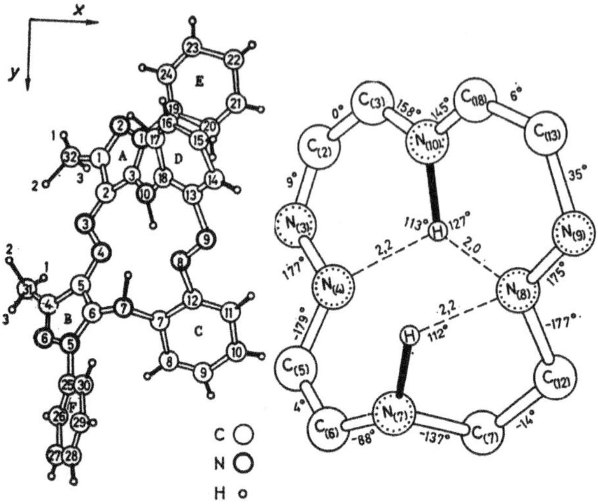

Fig. 1. The $C_{32}H_{26}N_{10}$ molecule and the conformation of the 14-membered ring.

6-METHYL-4-(METHYL(PHENYL)AMINO)-3-PHENYL-1-(p-TOLYL)-1H-PYRROLO[2,3-b]QUINOLINE
$C_{32}H_{27}N_3$

G. HIMBERT, W. SCHWICKERATH and G. MAAS, 1985. *Liebigs Ann. Chem.*, 1389-1397.

Monoclinic, *C2/c*, a = 25.519, b = 14.552, c = 15.149 Å, β = 117.07°, Z = 8. Mo radiation, R = 0.048 for 2488 reflections.

The analysis confirmed the pyrrolo[2,3-b]quinoline structure as shown in Fig. 1. Molecular dimensions appear normal.

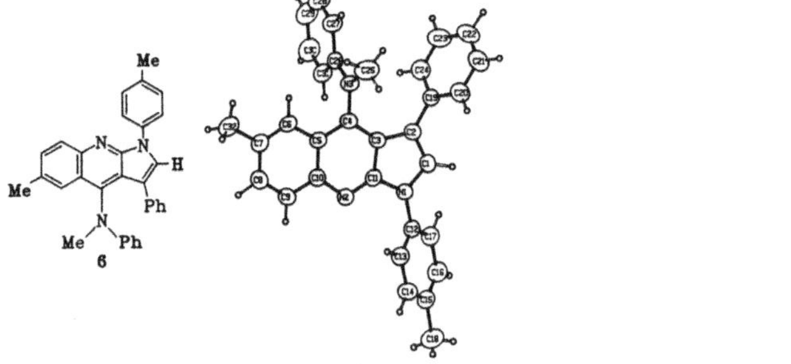

Fig. 1. The molecule $C_{32}H_{27}N_3$.

7,8,17,18-TETRAHYDRO-10,20-DIPHENYL-5,9,15,19-TETRAAZADIBENZO[a,i]CYCLOHEXADECENE-6,16-(5H,15H)DIONE
$C_{32}H_{28}N_4O_2$

A.A. DVORKIN, T.I. MALINOVSKY, Yu.A. SIMONOV, S.A. ANDRONATI, A.V. KUZMINA and A.S. YAVORSKY, 1985. *Acta Cryst.*, C41, 796-798.

Orthorhombic, *Pcan*, a = 7.331, b = 18.945, c = 18.420 Å, Z = 4. Mo radiation, R = 0.046 for 1044

reflections.

The molecule (Fig. 1) possesses crystallographic symmetry C_2 with the diad axis passing through the centre of the macrocycle. The 16-membered macrocycle has a deep boat conformation and this is stabilized by strong intramolecular N-H...N hydrogen bonds of length 2.650(4) Å. The amide groups are in the trans form. There is also a shortened intramolecular C(2)-H...O interaction of length 2.883(4) Å. The conformation of the phenyl substituents is defined by the torsion angle C(6)-C(7)-C(11)-C(12) of value 81.6(6)°. All intermolecular distances are essentially van der Waals contacts.

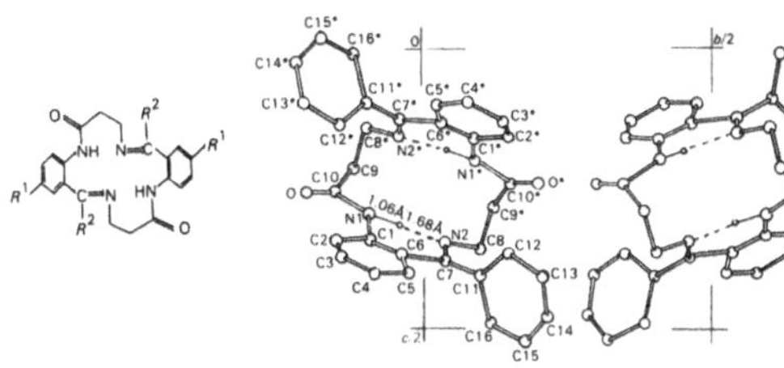

Fig. 1. $C_{32}H_{28}N_4O_2$: molecular formula and a projection of the crystal structure onto (100).

3'-PHENYLDISPIRO(FLUORENE-9,4'-[1,2]THIAZOLE-5',9''-FLUORENE)-1',1'-DIOXIDE $C_{33}H_{21}NO_2S$

R.C. HALTIWANGER, W.K.L. VAN HAVERE, P.T. BEURSKENS, B.F. BONINI and B. ZWANENBURG, 1985. *J. Cryst. Spect. Res.*, **15**, 89-94.

Monoclinic, $P2_1/n$, a = 15.542, b = 10.116, c = 16.661 Å, β = 112.47°, Z = 4. Mo radiation, R = 0.024 for 2221 reflections.

The fluorene ring systems are found to be perpendicular to the five-membered ring (Fig. 1); the dihedral angle between the phenyl and five-membered rings is 23.4(5)°. The five-membered ring approximates an envelope form with C(5) at the flap; this folding reduces strong π-π interactions between the two fluorene systems. All bond lengths and angles are normal.

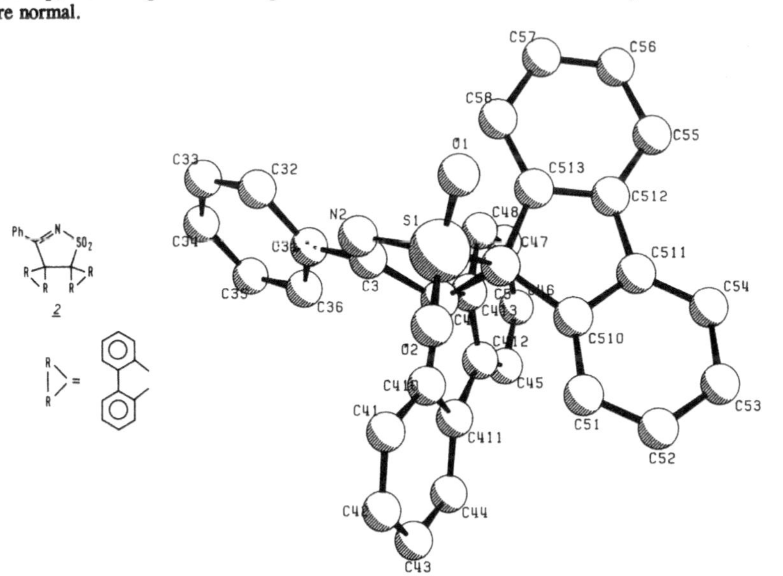

Fig. 1. The molecular structure of $C_{33}H_{21}NO_2S$.

1-((9-(N,N-DIALLYL)AMINO)PHENANTHREN-10-YL)-1a,9b-DIHYDROPHENANTHRO[9,10-b]AZIRINE
$C_{34}H_{30}N_2$

M. WEITZBERG, Z. AIZENSHTAT and J. BLUM, 1985. *J. Heterocycl. Chem.*, **22**, 865-868.

Triclinic, *P*1, a = 14.532, b = 14.588, c = 12.688 Å, α = 93.33, β = 97.52, γ = 79.51°, Z = 4. Mo radiation, R = 0.098 for 4283 reflections.

The analysis established the structure of this reaction product to be as shown in Fig. 1; the two nitrogen atoms are in trans-configuration.

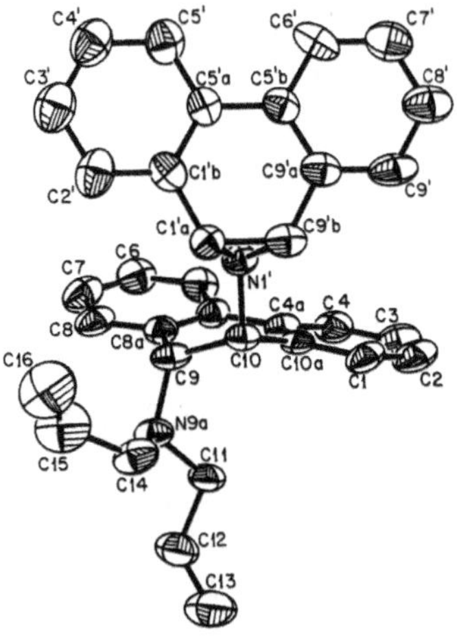

Fig. 1. A view of one of the $C_{34}H_{30}N_2$ molecules.

2,2-(BIPHENYL-2,2'-DIYL)-2,3-DIHYDRO-1,3-DIISOPROPYL-4-(4-METHOXYPHENYL)PHENYLMETHYLENEAMINO-1,3-DIAZETIUM HEXACHLOROANTIMONATE
$C_{34}H_{34}Cl_6N_3OSb$ $C_{34}H_{34}N_3O^+,Cl_6Sb^-$

M. AL-TALIB, I. JIBRIL, J.C. JOCHIMS and G. HUTTNER, 1985. *Tetrahedron*, **41**, 527-536.

Triclinic, *P*$\bar{1}$, a = 10.744, b = 12.312, c = 14.919 Å, α = 112.63, β = 84.57, γ = 101.94°, Z = 2. Mo radiation, R = 0.050 for 3544 reflections (at 223 K).

A view of the cation with principal dimensions is in Fig. 1. The four-membered ring is planar but the bonds to the isopropyl substituents do not lie in this plane. The C(1)-N(1)-C(2) bond angle is 127.3(7)°.

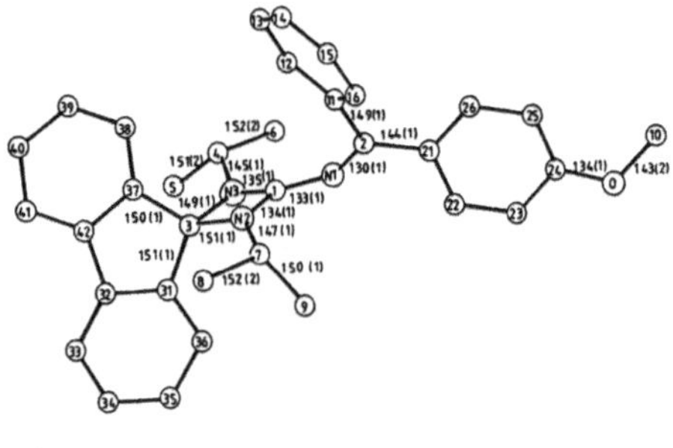

Fig. 1. A plot of $C_{34}H_{34}N_3O^+$ with principal dimensions (pm).

N,N'-(4,4'-OXYDIBENZYL)BIS(PHENANTHRIDINIUM) DICHLORIDE MONOHYDRATE
$C_{40}H_{30}Cl_2N_2,H_2O$ (I) $C_{40}H_{30}N_2O^{2+},2(Cl^-),H_2O$

N,N'-DECAMETHYLENEBIS(PHENANTHRIDINIUM) DIBROMIDE DIHYDRATE
$C_{36}H_{38}Br_2N_2,2(H_2O)$ (II) $C_{36}H_{38}N_2^{2+},2(Br^-),2(H_2O)$

M. CORY, D.D. McKEE, J. KAGAN, D.W. HENRY and J.A. MILLER, 1985. *J. Am. Chem. Soc.*, **107**, 2528-2536.

I. Monoclinic, $P2_1/n$, a = 11.021, b = 22.810, c = 12.920 Å, β = 106.71°, Z = 4. Cu radiation, R = 0.123 for 1133 reflections (at -110°C).

II. Monoclinic, $P2_1/c$, a = 8.898, b = 18.190, c = 10.611 Å, β = 110.23°, Z = 2. Cu radiation, R = 0.085 for 2033 reflections.

Compound (I) (Fig. 1) has a structure corresponding to the model of a bifunctional intercalator. In the crystal each molecule intercalates with its neighbours, in a way such that two rings of the three in the phenanthridium group superimpose over an adjacent ring system. The separation of the rings is in good agreement with that required for bifunctional intercalation into the DNA helix. The structure of (II) shows instead stacking of the phenanthridium ring above the aliphatic chain of a neighbouring molecule. The structure of (II) does not show the classical "C" conformation of (I), instead being extended as shown (Fig. 1).

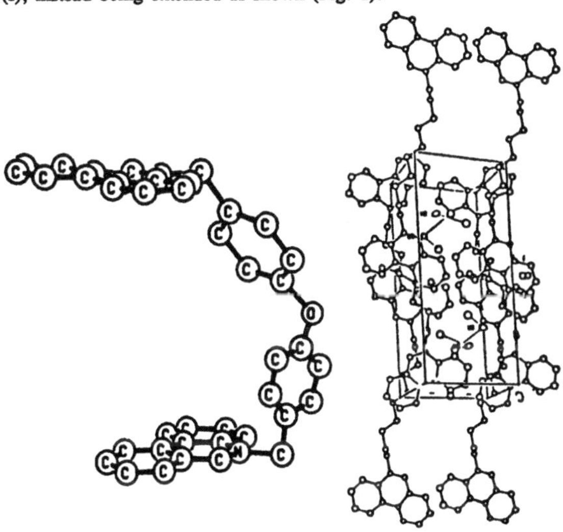

Fig. 1. View of $C_{40}H_{30}Cl_2N_2,H_2O$ (I) showing coplanarity of phenanthridium rings (left) and of the unit cell of $C_{36}H_{38}Br_2N_2,2(H_2O)$ (II) showing the extended conformation of the molecule (right).

15,26-ETHENO-3,4,5,6,21,22,23,34-HEXAKIS(METHOXYCARBONYL)-7,20-DIAZAHEXACYCLO[12.12.0.-$0^{2,7}.0^{2,12}.0^{15,25}.0^{20,25}$]HEXACOSA-3,5,8,10,12,16,18,21,23-NONAENE
$C_{42}H_{38}N_2O_{16}$

O. METH-COHN, C. CUORE and P.H. VAN ROOYEN, 1985. *J. Chem. Soc. Perkin Trans. 1*, 1793-1801.

Monoclinic, $P2_1/c$, a = 18.288, b = 12.498, c = 17.237 Å, β = 97.81°, Z = 4. Cu radiation, R = 0.049 for 4527 reflections.

The crystal structure analysis showed the stereochemistry of the adduct to be that shown in Fig. 1.

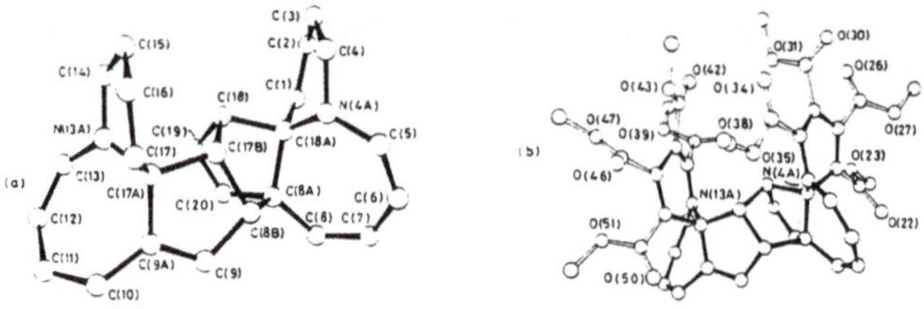

Fig. 1. Perspective drawings of a molecule of the adduct $C_{42}H_{38}N_2O_{16}$: without methoxycarbonyl groups (left) and full molecule (right).

BIS((-)-cis-2,8-DIMETHYL-1,2,3,4,4a,9b-HEXAHYDRO-2H-PYRIDO[4,3-b]INDOLIUM) (S,S)-2,3-DI-o-BENZOYLTARTRATE

$C_{44}H_{50}N_4O_8$ $2(C_{13}H_{19}N_2{}^+),C_{18}H_{12}O_8{}^{2-}$

V. KETTMANN, L. BENES and M. TICHY, 1985. *Acta Cryst.*, C41, 208-210.

Monoclinic, *C*2, a = 17.470, b = 7.975, c = 14.406 Å, β = 94.10°, Z = 2. Cu radiation, R = 0.034 for 1252 reflections.

The structure of this antiarrhythmic drug consists of groups of three ions, two cations and one anion linked by hydrogen bonds (Fig. 1). The distances between the protonated N(1) and O(1) and O(2) of the deprotonated carboxyl group of (S,S)-2,3-di-O-benzoyltartrate are 3.077(3) and 2.754(2) Å respectively. The latter is definitely a hydrogen bonding distance while the former may be part of a bifurcated hydrogen bond or just a long contact. Each group of three ions possesses crystallographic C_2 symmetry and is held together by van der Waals forces only. The saturated six-membered ring in the cation exists in a flattened chair conformation while the five-membered ring adopts a symmetric C_s envelope conformation. The two chiral centres are C(3)-R, C(4)-R. Bond lengths and angles are unexceptional.

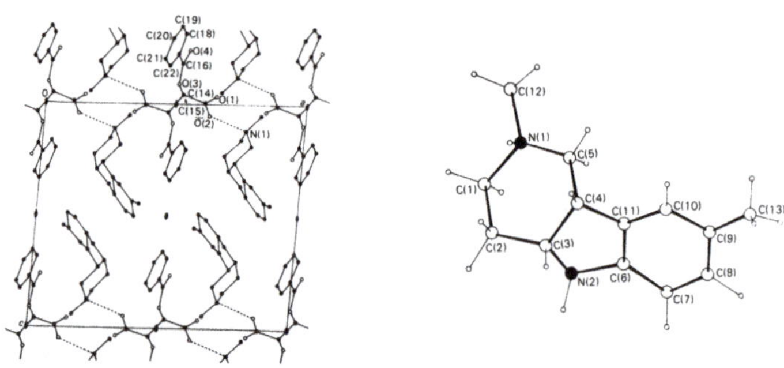

Fig. 1. $C_{44}H_{50}N_4O_8$: b-axis projection and the cation.

3,7-DIMETHYL-3,7-DIAZABICYCLO[3.3.1]NONAN-9-OL ETHANOL SOLVATE DIHYDRATE
$C_9H_{18}N_2O,C_2H_6O,2(H_2O)$

E. GALVEZ, M.S. ARIAS, J. BELLANATO, J.V. GARCIA-RAMOS, F. FLORENCIO, P. SMITH-VERDIER and S. GARCIA-BLANCO, 1985. *J. Mol. Struct.*, 127, 185-201.

Monoclinic, $P2_1/c$, a = 12.036, b = 14.404, c = 14.777 Å, β = 108.76°, Z = 8. Mo radiation, R = 0.066 for 1887 reflections.

There are two very similar independent molecules in the asymmetric unit with no significant differences between them (Fig. 1). The chair-chair conformation is found and is stabilized by hydrogen bonding with the water and ethanol molecules. Molecular dimensions are as anticipated. The N(3)...N(7) intramolecular distance is 2.89 Å.

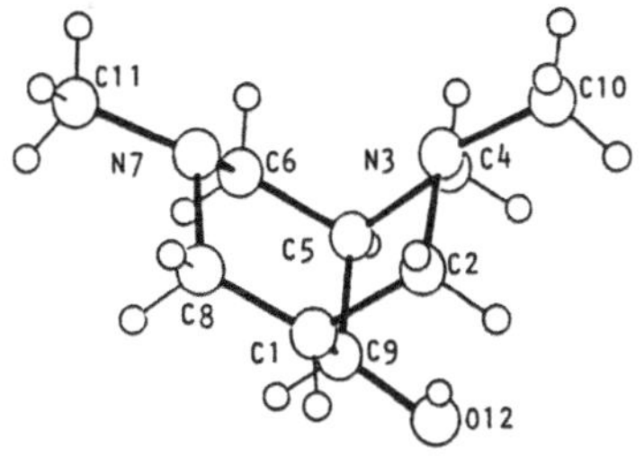

Fig. 1. A view of $C_9H_{18}N_2O$.

8-OXO-N-METHYL-4,5-DIAZATRICYCLO[4.3.0.0^{3,7}]NONANE-4,5-DICARBOXIMIDE
$C_{10}H_{11}N_3O_3$

W. ADAM, O. DE LUCCHI, K. HILL, E.-M. PETERS, K. PETERS and H.G. VON SCHNERING, 1985. *Chem. Ber.*, **118**, 3070-3088.

Orthorhombic, *Pbca*, a = 18.609, b = 18.182, c = 5.944 Å, Z = 8. Mo radiation, R = 0.081 for 1522 reflections.

The analysis provided rigorous proof of the structure and stereochemistry as shown in Fig. 1. The N(2)-N(6) bond length is 1.447(5) Å.

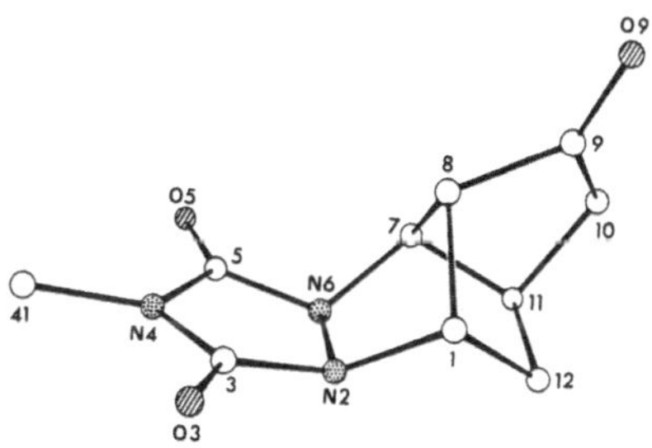

Fig. 1. A view of $C_{10}H_{11}N_3O_3$.

meso-3,7-DIAZONIATRICYCLO[4.2.2.2^{2,5}]DODECA-3,7,9,11-TETRAENE-4,8-DIAMINIUM TETRACHLOROPALLADATE(II)
$C_{10}H_{14}Cl_4N_4Pd$ $C_{10}H_{14}N_4^{2+}, Cl_4Pd^{2-}$

M.C. NAVARRO-RANNINGER, S. MARTINEZ-CARRERA and S. GARCIA-BLANCO, 1985. *Polyhedron*, **4**, 1379-1381.

Monoclinic, $P2_1/a$, a = 12.442, b = 8.810, c = 6.687 Å, β = 101.94°, Z = 2. Mo radiation, R = 0.027 for 1717 reflections.

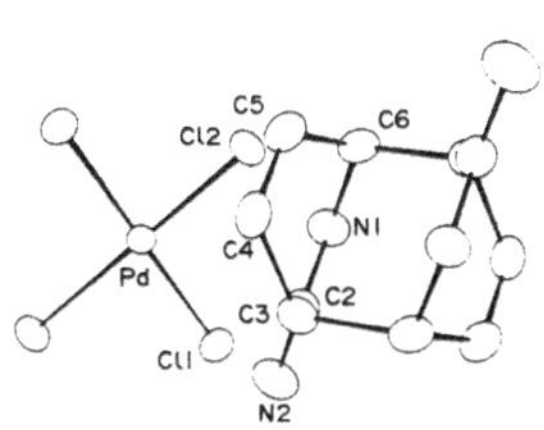

Fig. 1. Cation and anion in $C_{10}H_{14}Cl_4N_4Pd$.

The dication (Fig. 1) consists of two moieties derived from 2-aminopyridine, joined across a crystal inversion centre by long C(3)-C(6') and C(3')-C(6) bonds (1.610(5) Å). Bond lengths around each six-membered ring are (starting with N(1)-C(2)): 1.310, 1.512, 1.517, 1.320, 1.509, 1.466(5) Å. The Pd-Cl bond lengths are 2.302, 2.303(2) Å.

5,6,7,8-TETRAFLUORO-9-METHYL-1,4-DIHYDRO-1,4-IMINONAPHTHALENE
$C_{11}H_7F_4N$

M.P. BYRN, C.E. STROUSE, G.W. GRIBBLE and C.S. LEHOULLIER, 1985. *Acta Cryst.*, **C41**, 238-239.

Monoclinic, $P2_1/n$, a = 6.258, b = 31.85, c = 9.481 Å, β = 99.71°, Z = 8. Mo radiation, R = 0.057 for 1687 reflections (at 115 K).

There are two crystallographically independent molecules in the asymmetric unit. There is also disorder in the imine bridge and ethylene fragments of both molecules. The disorder is created by a twofold rotation of the molecule, which superimposes the fluorinated rings (Fig. 1). The C-F bonds are normal (average 1.358(6)) and the C-N distances range from 1.468(11) to 1.515(11) Å. The distances from the N-methyl C atom to the centre of the benzene ring in the two molecules are 3.608(8) and 3.616(11) Å. The corresponding imine bridge bond angles are 95.0(7) and 96.0(4)°.

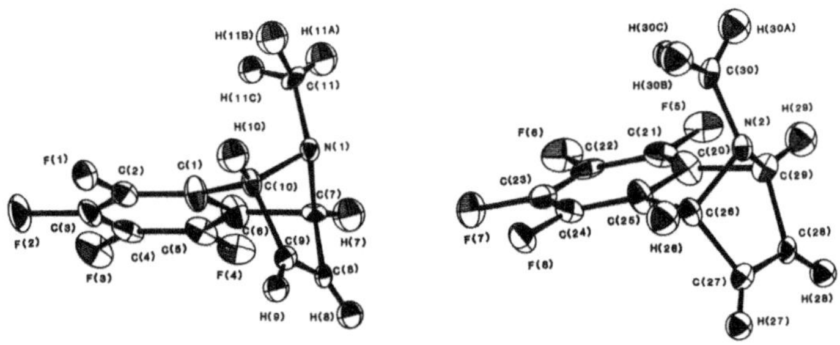

Fig. 1. $C_{11}H_7F_4N$: the high occupancy fragments for both molecules in the asymmetric unit.

14-METHYL-4,6,10,12-TETRAAZATETRACYCLO[8.4.0.0^{2,6}.0^{3,14}]TETRADEC-4-ENE-11,13-DIONE
$C_{11}H_{14}N_4O_2$

M. GDANIEC, G. WENSKA and A.E. KOZIOL, 1985. *Acta Cryst.*, **C41**, 1230-1232.

Triclinic, $P\bar{1}$, a = 5.793, b = 9.431, c = 10.902 Å, α = 106.21, β = 81.11, γ = 109.74°, Z = 2. Cu radiation, R = 0.036 for 1337 reflections.

The molecule (Fig. 1) has a cis-syn configuration, with the dihydrothymine ring significantly non-planar and the diazacycloheptane ring in a chair conformation. The dihydroimidazole ring adopts a flattened envelope conformation with C(5P) as flap and the cyclobutane ring is puckered. Molecules in the crystal are linked by a pair of intermolecular N(3)...O(1) hydrogen bonds (2.894(3) Å) to form dimers about a centre of symmetry.

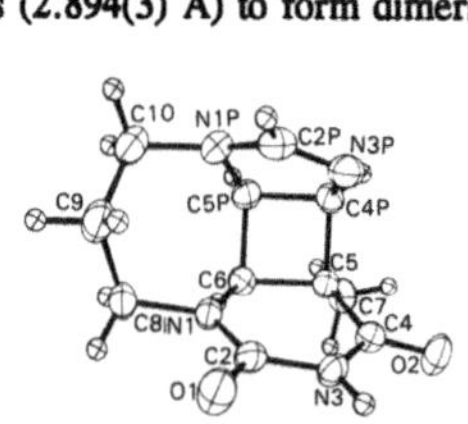

Fig. 1. $C_{11}H_{14}N_4O_2$: the molecular structure.

9-METHOXY-7,8-DIAZABICYCLO[4.2.2]DECA-2,4,9-TRIENE-N-METHYL-7,8-DICARBOXIMIDE
$C_{12}H_{13}N_3O_3$ (I)

2-METHOXY-7,8-DIAZABICYCLO[4.2.2]DECA-2,4,9-TRIENE-N-METHYL-7,8-DICARBOXIMIDE
$C_{12}H_{13}N_3O_3$ (II)

W. ADAM, G. KLUG, E.-M. PETERS, K. PETERS and H.G. VON SCHNERING, 1985. *Tetrahedron*, **41**, 2045-2056.

I. Triclinic, $P\bar{1}$, a = 10.118, b = 13.891, c = 9.765 Å, α = 99.10, β = 116.72, γ = 78.18°, Z = 4. Mo radiation, R = 0.050 for 3186 reflections.

II. Monoclinic, *C2/c*, a = 22.866, b = 8.427, c = 14.073 Å, β = 120.55°, Z = 8. Mo radiation, R = 0.044 for 1862 reflections.

The conformations of these urazole derivatives are as shown in Fig. 1.

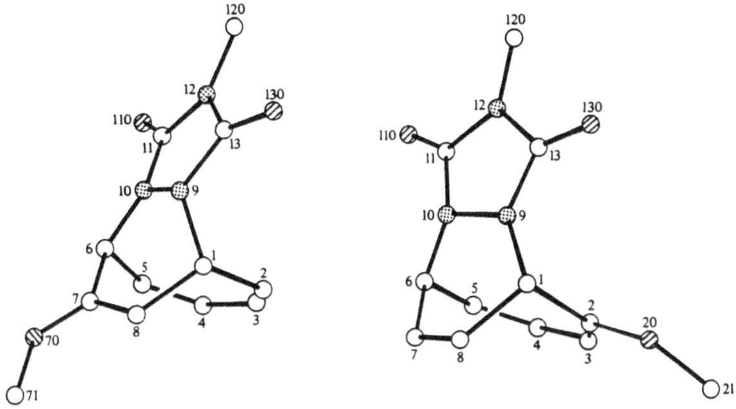

Fig. 1. Views of $C_{12}H_{13}N_3O_3$, (I) (left) and (II) (right).

2-BENZYL-7-anti-BROMO-endo,endo-5,6-DIHYDROXY-2-AZABICYCLO[2.2.1]HEPTAN-3-ONE DEUTERO-CHLOROFORM SOLVATE
$C_{13}H_{14}BrNO_3,0.5(CDCl_3)$

W.C. FAITH, C.A. BOOTH, B.M. FOXMAN and B.B. SNIDER, 1985. *J. Org. Chem.*, **50**, 1983-1985.

Monoclinic, *C2/c*, a = 23.851, b = 6.288, c = 24.942 Å, β = 127.59°, Z = 8. Mo radiation, R = 0.054 for 940 reflections.

The crystal structure analysis established the stereochemistry of the diol molecule (Fig. 1).

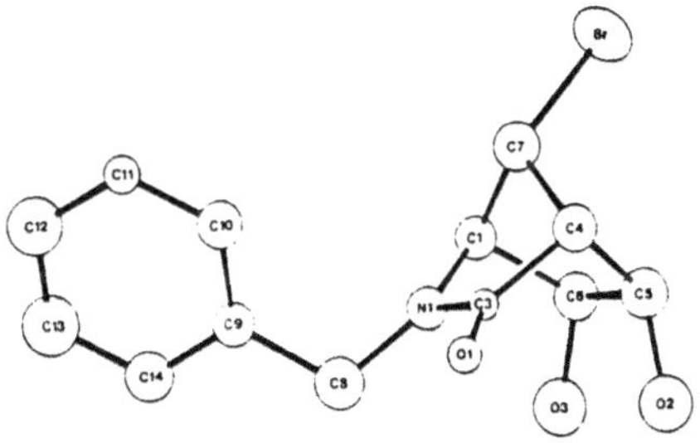

Fig. 1. Perspective drawing of a molecule of $C_{13}H_{14}BrNO_3,0.5(CDCl_3)$.

13-METHYL-[7](2,6)PYRIDINOPHANIUM PERCHLORATE
$C_{13}H_{20}ClNO_4$ $C_{13}H_{20}N^+,ClO_4^-$

H. WEBER, J. PANT and H. WUNDERLICH, 1985. *Chem. Ber.*, **118**, 4259-4270.

Monoclinic, $P2_1/n$, a = 15.090, b = 11.411, c = 16.913 Å, β = 91.20°, Z = 8. Mo radiation, R = 0.140 for 3538 reflections.

The cation is severely crowded and there are two independent cations in the asymmetric unit (Fig. 1). In each, the pyridinium ring has an approximate envelope conformation, with the N atom 0.20 Å out of the plane of the five C atoms. The heptamethylene bridge is strained and has torsion angles some 30° from usual values.

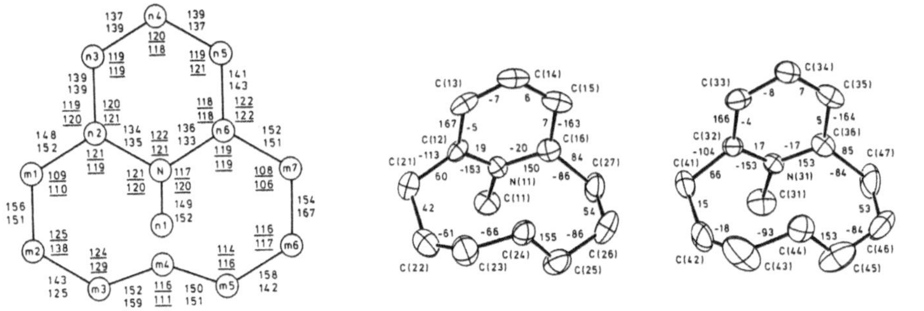

Fig. 1. Details of the $C_{13}H_{20}N^+$ bond lengths (pm) and angles (°, and underlined) (left) and views of the two independent cations with torsion angles (°).

syn-1,6:8,13-DIIMINO[14]ANNULENE
$C_{14}H_{12}N_2$ (I)

syn-1,6-(METHYLIMINO)-8,13-IMINO[14]ANNULENE
$C_{15}H_{14}N_2$ (II)

syn-1,6:8,13-BIS(METHYLIMINO)[14]ANNULENE
$C_{16}H_{16}N_2$ (III)

syn-1,6-IMINO-8,13-IMINIUM[14]ANNULENE PERCHLORATE
$C_{14}H_{13}ClN_2O_4$ (IV) $C_{14}H_{13}N_2^+,ClO_4^-$

R. DESTRO, T. PILATI, M. SIMONETTA and E. VOGEL, 1985. *J. Am. Chem. Soc.*, **107**, 3185-3191.

I. Orthorhombic, *Pbcn*, a = 15.977, b = 7.329, c = 18.613 Å, Z = 8. Mo radiation, R = 0.038 for 1393 reflections.

II. Monoclinic, $P2_1/c$, a = 9.490, b = 18.777, c = 6.532 Å, β = 96.89°, Z = 4. Mo radiation, R = 0.043 for 1400 reflections.

III. Monoclinic, $P2_1/c$, a = 9.227, b = 19.463, c = 7.320 Å, β = 101.03°, Z = 4. Mo radiation, R = 0.040 for 1980 reflections.

IV. Monoclinic, $P2_1/c$, a = 8.477, b = 9.735, c = 16.841 Å, β = 101.61°, Z = 4. Mo radiation, R = 0.059 for 1940 reflections.

The structure of (I) (Fig. 1) is disordered, there being two equivalent positions for each of the two imino hydrogens H(N1) and H(N2). When both adopt the endo position they are separated by 1.19(4) Å and the C(1)...C(6) and C(8)...C(13) transannular distances are equal at 2.294(2) Å. The crystal structure consists of parallel ribbons of molecules along <001> linked by weak N-H...N hydrogen bonds. The geometries of (I), (II), (III) and (IV) as side views show their basic similarity (Fig. 2). In (I) and (II) there is an intramolecular H-bond with N-H...N parameters as shown. In (III) the repulsion of the bridging groups is the sole interaction in this region and leads to small but significant changes: C(1)...C(6) is 2.290(3) in (II), 2.279(2) Å in (III) and the two inner dihedral angles in (II) are 111.1(2), 106.2(1) *cf.* 107.6(1), 108.1(1)° in (III). In (IV) there is severe disorder in the perchlorate anion. However the following features can be noted: The N...N separation is very short at 2.464(4) Å, the ~10° dihedral angle between the planes of the bridges observed in (I), (II), (III) is ~0° in (IV), the N-C(ring) bond length is 1.460 in (IV), 1.437 Å in (I) and the transannular distances C(1)...C(6) 2.325(4) and

C(8)...C(13) 2.356(4) Å are increased.

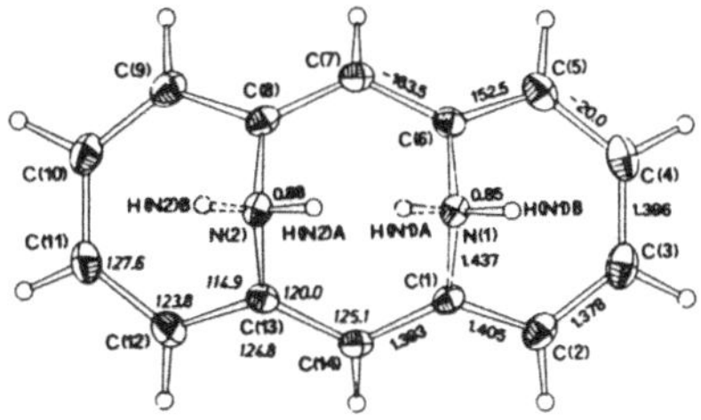

Fig. 1. View of $C_{14}H_{12}N_2$ (I) showing bond parameters.

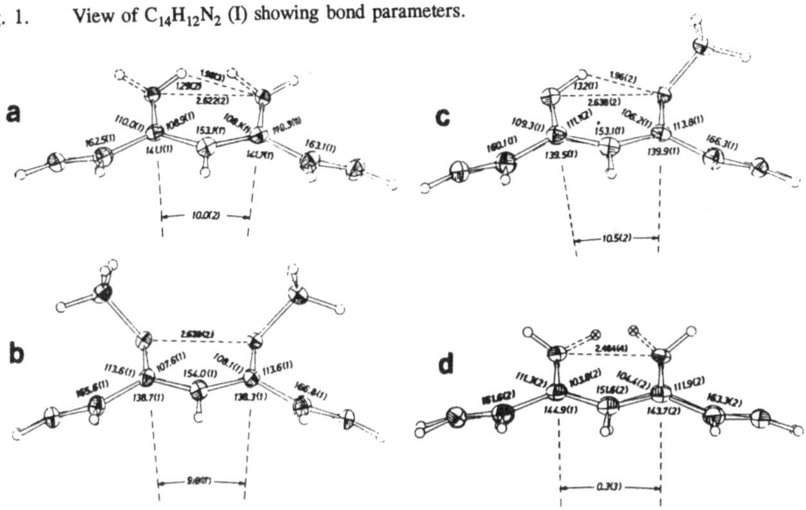

Fig. 2. Side views of all four annulenes (I) (a), (II) (b), (III) (c) and (IV) (b).

1,3-DIMETHYL-1,2,3,4,5,6-HEXAHYDRO-1,5-METHANO-3-BENZAZOCIN-9-OL HYDROBROMIDE
$C_{14}H_{20}BrNO$ $C_{14}H_{20}NO^+,Br^-$

A. ITAI, Y. IITAKA, T. KOMETANI and S. SHIOTANI, 1985. *Acta Cryst.*, **C41**, 222-224.

Orthorhombic, $P2_12_12_1$, a = 12.371, b = 14.394, c = 7.571 Å, Z = 4. Cu radiation, R = 0.032 for 1512 reflections.

The molecule (Fig. 1), a synthetic analgesic, is L-shaped. The C ring adopts a chair conformation and the H atom at the protonated N atom takes a β-axial position, oriented towards ring A. The distance between the N atom and the centre of ring A is 4.11 Å. A Br atom connects two adjacent molecules by two hydrogen bonds (O...Br 3.23, N...Br 3.35 Å) resulting in an infinite chain in the crystal packing.

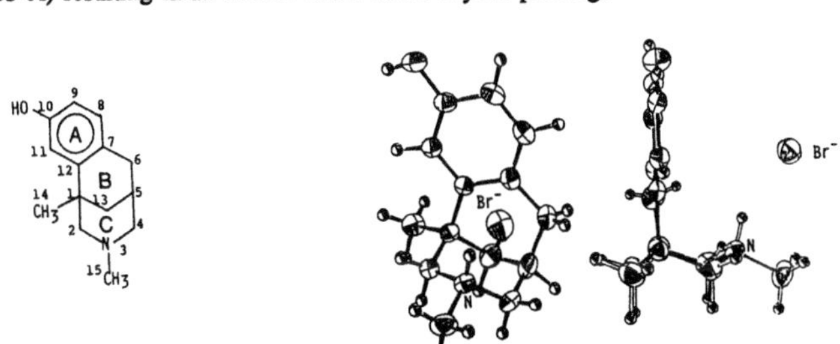

Fig. 1. $C_{14}H_{20}BrNO$: the molecular skeleton and two views of the molecule.

8,8'-BI(8-AZABICYCLO[3.2.1]OCTANE)
$C_{14}H_{24}N_2$ (I)

8,8'-BI(8-AZABICYCLO[3.2.1]OCTENIUM) NITRATE MONOHYDRATE
$C_{14}H_{24}N_3O_3,H_2O$ (II) $C_{14}H_{24}N_2^+,NO_3^-,H_2O$

S.F. NELSEN, G.T. CUNKLE, D.H. EVANS, K.J. HALLER, M. KAFTORY, B. KIRSTE, H. KURRECK and T. CLARK, 1985. *J. Am. Chem. Soc.*, **107**, 3829-3839.

I. Orthorhombic, *Pbca*, a = 10.698, b = 9.204, c = 12.438 Å, Z = 4. Mo radiation, R = 0.064 for 1063 reflections.

II. Monoclinic, *C2/m*, a = 12.889, b = 10.521, c = 6.514 Å, β = 114.06°, Z = 2. Mo radiation, R = 0.063 for 863 reflections (at 198 K).

Both (I) and its salt in (II) are in the anti-AA conformation (Fig. 1). Molecule (I) lies about an inversion centre and (II) has 2/m crystallographic symmetry. The change in geometry at N on going from (I) to (II) is substantial as shown in the dimensions (Fig. 1).

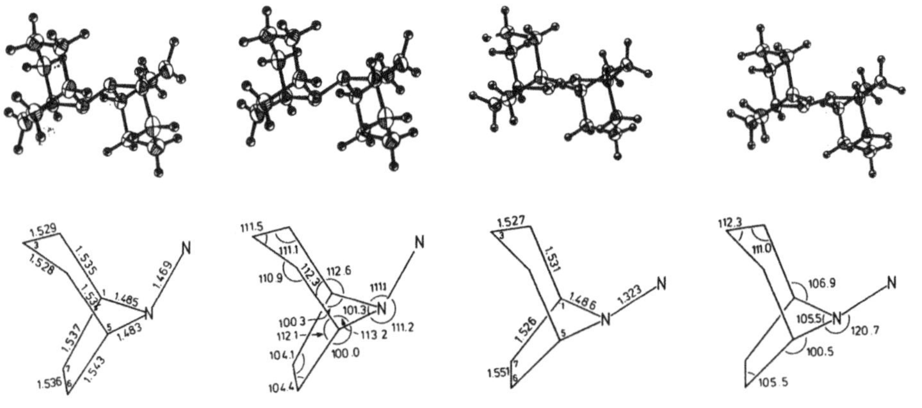

Fig. 1. Stereoviews and dimensions for $C_{14}H_{24}N_2$ (I) (left) and for $C_{14}H_{24}N_2^+$ (right).

1,6:8,13-DIIMINOMETHANO[14]ANNULENE
$C_{15}H_{12}N_2$ (I)

1,6:8,13-DIIMINOETHANO[14]ANNULENE
$C_{16}H_{14}N_2$ (II)

1,6:8,13-DIIMINOPROPANO[14]ANNULENE
$C_{17}H_{16}N_2$ (III)

R. DESTRO, T. PILATI, M. SIMONETTA and E. VOGEL, 1985. *J. Am. Chem. Soc.*, **107**, 3192-3196.

I. Monoclinic, $P2_1/n$, a = 8.685, b = 6.687, c = 18.985 Å, β = 95.67°, Z = 4. Mo radiation, R = 0.051 for 1111 reflections.

II. Orthorhombic, *Fmm2*, a = 17.391, b = 11.561, c = 5.995 Å, Z = 4. Mo radiation, R = 0.041 for 383 reflections.

III. Monoclinic, *C2*, a = 17.407, b = 6.310, c = 11.975 Å, β = 93.13°, Z = 4. Mo radiation, R = 0.042 for 960 reflections.

The structure of (I) shows approximate mm symmetry, as does the annulene nucleus of the ~m symmetric molecule (III), and (II) has crystallographic mm symmetry. The increase in ring deformation going from (I) to (II) is considerably less than expected, as measured by the parameters τ (misalignment angles in $2p_z$ orbitals) and D (r.m.s. distances of 14 carbon atoms from the mean ring plane). For (I) these parameters are 24.5(3)°, 0.24 Å, for

(II) 28.9(3)°, 0.34 Å and for (III) 30.4(4), 0.36 Å. The similarity of the latter two is remarkable and is in spite of the different numbers of $-CH_2-$ groups in their bridges. In the current series of compounds relief from steric compression is achieved by flattening at the N atoms. This trend is shown by the increase in C(ring)-N-C(bridge) angles (109.8 in (I), 118.8 in (II), 122.8° in (III)) and a decrease of 0.033 Å in N-C(ring) distance in going from (I) to (III). C(15) and C(17) lie only ~0.3 Å from the C(ring)-N-C(ring) planes in (III). The dihedral angle between the bridges is 17.3(4)° in (III), surprisingly smaller than in (II) (18.5(3)°). Remaining features are normal.

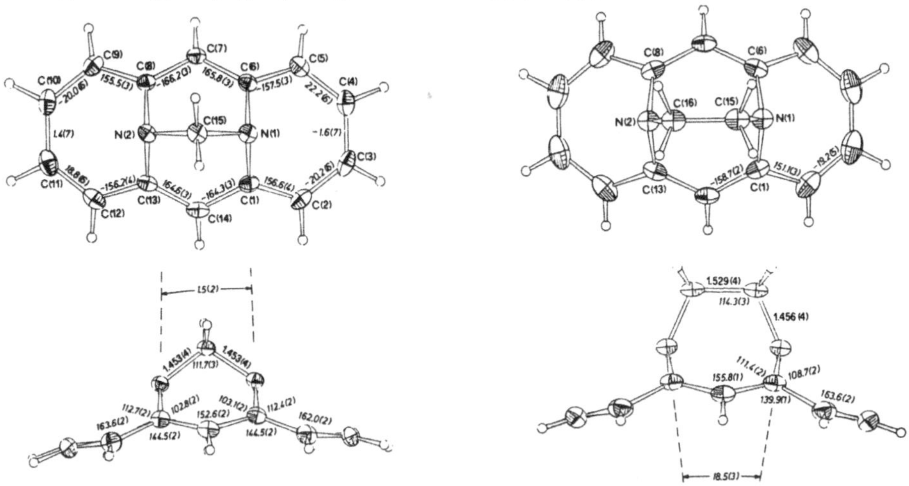

Fig. 1. $C_{15}H_{12}N_2$ (I) (left), $C_{16}H_{14}N_2$ (II) (right): views and molecular geometry of the structures.

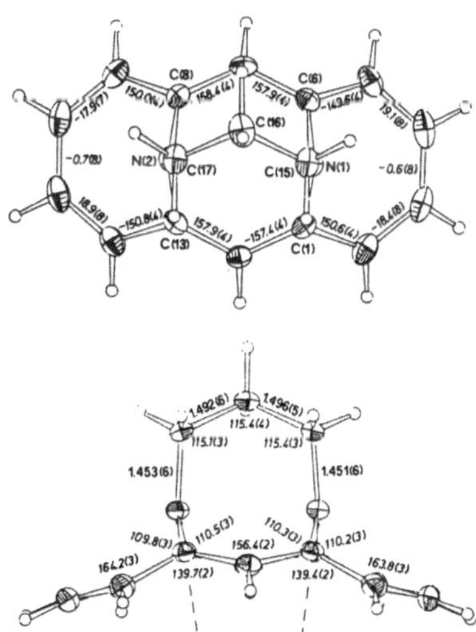

Fig. 2. View of $C_{17}H_{16}N_2$ (III) showing details of geometry.

2,2-DIMETHYL-5-HYDROXY-4-KETO-7-PYRROLIDINYLBENZOPYRAN
$C_{15}H_{19}NO_3$

J.W. ApSIMON, L.W. HERMAN and C. HUBER, 1985. *Canad. J. Chem.*, **63**, 2589-2596.

Monoclinic, $P2_1/c$, a = 11.319, b = 6.376, c = 19.563 Å, β = 104.59°, Z = 4. Mo radiation, R = 0.040 for

1498 reflections.

With the exception of C(12) (an axial Me group) (Fig. 1) all atoms are within 0.5 Å of planarity. The pyrrolidine ring has envelope conformation (C(14) 0.395(6) Å from the plane of the other four atoms).

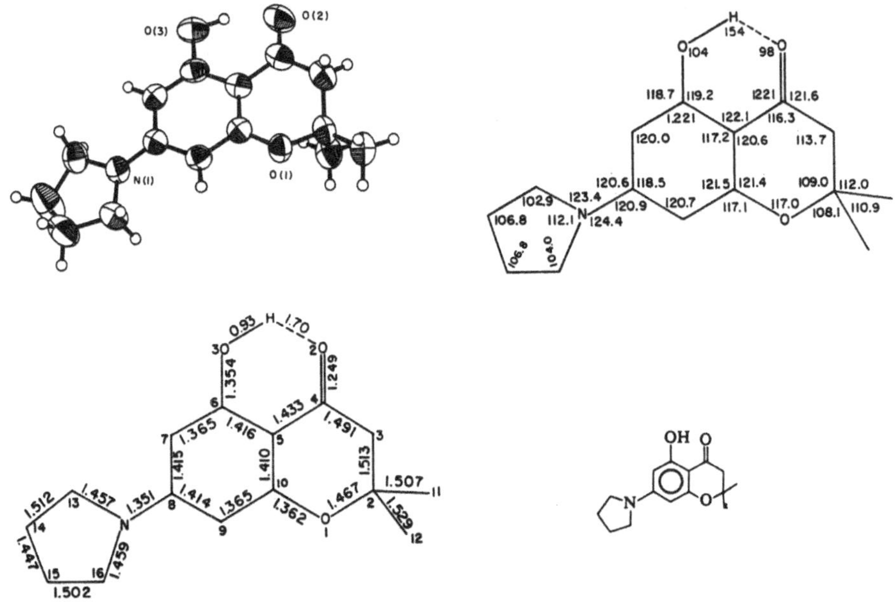

Fig. 1. The $C_{15}H_{19}NO_3$ molecule with bond lengths (esd's 0.003-0.006 Å) and angles (esd's 0.2-0.4°).

9-HYDROXY-3,7-DIMETHYL-2,3,4,5,6,7-HEXAHYDRO-2,7-METHANO-1H-3-BENZAZONINE HYDROBROMIDE

$C_{15}H_{22}BrNO$ $C_{15}H_{22}NO^+,Br^-$

A. ITAI, Y. IITAKA, T. KOMETANI and S. SHIOTANI, 1985. *Acta Cryst.*, C41, 1057-1059.

Orthorhombic, $P2_12_12_1$, a = 10.766, b = 16.137, c = 8.291 Å, Z = 4. Cu radiation, R = 0.052 for 1589 reflections.

This synthetic analgesic has a benzomorphan structure in which ring C is enlarged (Fig. 1). The molecule is L-shaped and the N-containing ring C protrudes from the plane of the rest of the molecule. The seven-membered ring C adopts a chair conformation in which a pseudo mirror plane passes through N(2) and the mid-point of the N-C(6) bond. Bond lengths and angles in the molecule are normal. The crystal structure is stabilized by two N...Br hydrogen bonds of lengths 3.32(1) and 3.28(1) Å and by normal van der Waals interactions.

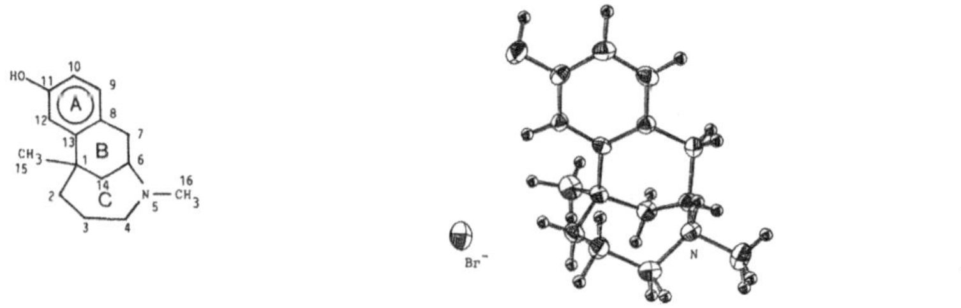

Fig. 1. $C_{15}H_{22}BrNO$: structural formula and a view of the asymmetric unit.

13,14-DIMETHOXY-3,10-DIAZABICYCLO[10.2.2]HEXADECA-12,14,15-TRIENE-2,11-DIONE MONOHYDRATE
$C_{16}H_{22}N_2O_4,H_2O$

R.E. PATTERSON, S.J. RODGERS and S.D. STULTS, 1985. *Acta Cryst.*, **C41**, 1322-1324.

Orthorhombic, $P2_12_12_1$, a = 10.610, b = 12.614, c = 12.771 Å, Z = 4. Mo radiation, R = 0.0291 for 2102 reflections.

In the molecule (Fig. 1) the hexamethylene bridge is situated above and is asymmetrically oriented with respect to the benzene ring. It is tilted towards the methoxy groups. This induces a slight distortion in the ring at C(1) and C(4) which are both displaced by about 0.05 Å above the least-squares plane of the benzene ring. The dihedral angle between this plane and that of the amides is about 70°. Bond lengths and angles in the molecule are in good agreement with generally accepted values. The solvate water molecule forms an intermolecular bridge between amide and carbonyl groups of adjacent molecules through hydrogen bonds in which the O...H separations are 1.950(1) and 1.763(1) Å respectively.

Fig. 1. $C_{16}H_{22}N_2O_4,H_2O$: views of the molecule showing the atomic numbering and side-on to the benzene ring.

2'-(1-(2-PYRIDYL)ETHYLIDENE)-3-AZABICYCLO[3.2.2]NONANE-3-CARBOTHIOHYDRAZIDE
$C_{16}H_{22}N_4S$ (I)

2'-(1-(2-PYRIDYL)ETHYLIDENE)-3-AZABICYCLO[3.2.2]NONANE-3-CARBOSELENOHYDRAZIDE
$C_{16}H_{22}N_4Se$ (II)

N.N. DHANESHWAR, S.S. TAVALE, T.N.G. ROW and Y.K. BHOON, 1985. *Acta Cryst.*, **C41**, 1188-1191.

I. Monoclinic, *Cc*, a = 6.433, b = 21.796, c = 11.533 Å, β = 90.48°, Z = 4. Mo radiation, R = 0.046 for 991 reflections.

II. Triclinic, $P\bar{1}$, a = 8.462, b = 9.178, c = 11.128 Å, α = 107.56, β = 95.07, γ = 86.86°, Z = 2. Mo radiation, R = 0.038 for 1821 reflections.

The azabicyclo[3.2.2]nonane ring adopts the boat-boat conformation in both compounds (Fig. 1). The pyridine rings are planar. The C-S bond distance in (I) is 1.71(1) and the C-Se bond distance in (II) is 1.883(5) Å, indicating partial double bond character in both. Likewise the C(1)-N(2) bond distances of 1.37(1) (I) and 1.331(7) Å (II) indicate partial double bonds. The C(2)-N(3) bonds are double (1.31(1) (I), 1.297(6) Å (II)). There is an intramolecular hydrogen bond between N(3) and S in (I) (2.78(1)) and N(3) and Se in (II) (2.933(5) Å).

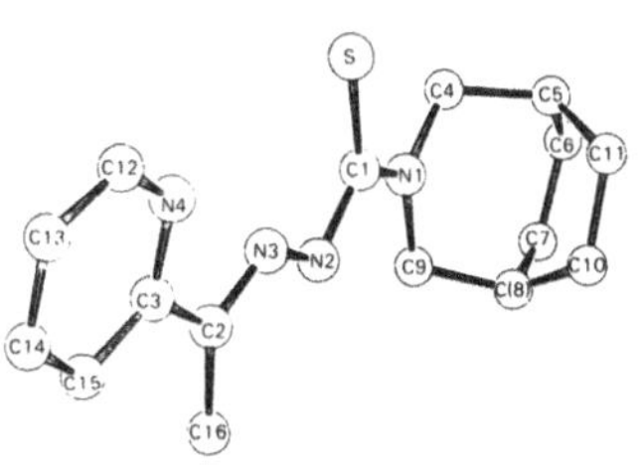

Fig. 1. $C_{16}H_{22}N_4S$: the molecular structure.

5,9:15,19-DIMETHANO-1,3,5,9,11,13,15,19-OCTAAZATETRACYCLO[9.9.2.0^{3,22}.0^{13,21}]DOCOSAN-2-ONE-12-THIONE
$C_{16}H_{26}N_8OS$

W.L. MOCK, T. MANIMARAN, W.A. FREEMAN, R.M. KUKSUK, J.E. MAGGIO and D.H. WILLIAMS, 1985. *J. Org. Chem.*, **50**, 60-62.

Monoclinic, $P2_1$, a = 7.275, b = 13.052, c = 9.726 Å, β = 104.92°, Z = 2. Mo radiation, R = 0.0787 for 1562 reflections.

In the crystal the molecule half-surrounds a point equidistant from the S-atom, the O-atom and the nominally equivalent N-atom pairs N(3), N(3A) and N(4), N(4A) (Fig. 1.) The molecule is disordered about this point. The molecule is the diastereoisomer in which the methano bridges C(4) and C(9) are syn to the methine hydrogens of the thioglycoluril moiety.

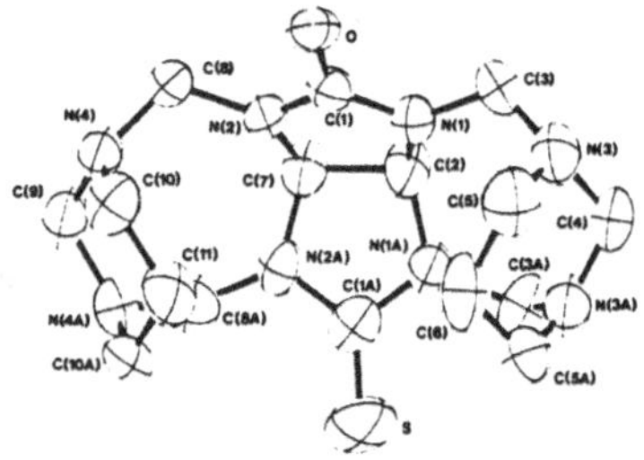

Fig. 1. A view of a molecule of $C_{16}H_{26}N_8OS$.

8,10-DIMETHYL-4-PHENYL-2,4,6-TRIAZATRICYCLO[5.3.2.0^{2,6}]DODEC-11-ENE-3,5,9-TRIONE
$C_{17}H_{17}N_3O_3$

D.I. SCHUSTER, LI WANG and J.M. VAN DER VEEN, 1985. *J. Am. Chem. Soc.*, **107**, 7045-7053.

Orthorhombic, $P2_12_12_1$, a = 7.230, b = 9.301, c = 21.854 Å, Z = 4. Mo radiation, R = 0.072 for 1666 reflections.

The analysis establishes that the molecule is the cis-dienone shown in Fig. 1.

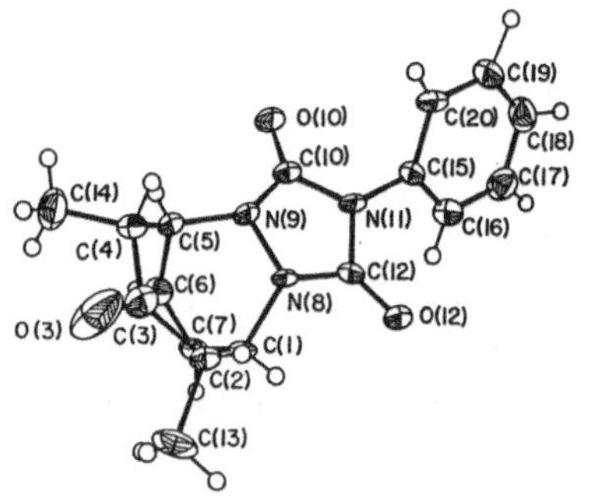

Fig. 1. A view of $C_{17}H_{17}N_3O_3$.

9,10-DIMETHOXY-1,2,4,5-TETRAHYDRO-6,8-ETHENO-3H-3-BENZAZONINE-3-CARBONITRILE
$C_{17}H_{18}N_2O_2$ (I)

PHENYL 10,11-DIMETHOXY-1,4,5,6-TETRAHYDRO-7,9-ETHENO-3-BENZAZECINE-3(2H)-CARBOXYLATE
$C_{24}H_{25}NO_4$ (II)

J.B. BREMNER, L.M. ENGELHARDT, A.H. WHITE and K.N. WINZENBERG, 1985. *J. Am. Chem. Soc.*, **107**, 3910-3915.

I. Triclinic, $P\bar{1}$, a = 11.468, b = 9.579, c = 7.133 Å, α = 94.70, β = 104.59, γ = 102.47°, Z = 2. Mo radiation, R = 0.041 for 1812 reflections.

II. Triclinic, $P\bar{1}$, a = 14.531, b = 8.800, c = 8.251 Å, α = 87.90, β = 77.95, γ = 76.68°, Z = 2. Mo radiation, R = 0.044 for 1999 reflections.

Distortion of the molecular geometries in these strained compounds is manifested in deviation of bond and torsion angles from normal values (e.g. the C8-C16-C15-C1 torsion angles in (I) and (II) (Fig. 1) deviate by 40° and 16.4°, respectively, from 180°).

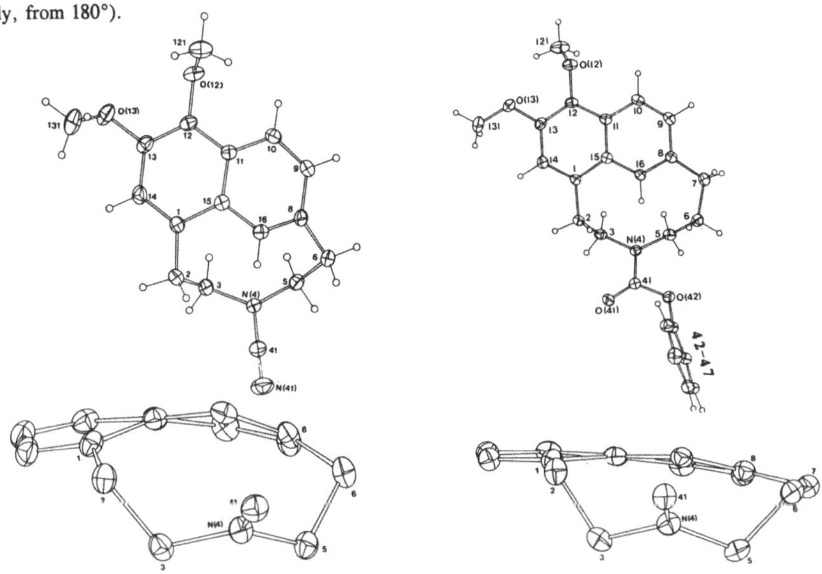

Fig. 1. Projections normal to the naphthalene plane and along the central bond of the naphthalene ring in $C_{17}H_{18}N_2O_2$ (I) (left) and $C_{24}H_{25}NO_4$ (II) (right).

4-ACETOXY-N-PHENYL-7,8-DIAZABICYCLO[4.2.2]DECA-2,9-DIENE-7,8-DICARBOXIMIDE
$C_{18}H_{17}N_3O_4$

W. ADAM, N. GRETZKE, L. HASEMANN, G. KLUG, E.-M. PETERS, K. PETERS, H.G. VON SCHNERING and B. WILL, 1985. *Chem. Ber.*, **118**, 3357-3379.

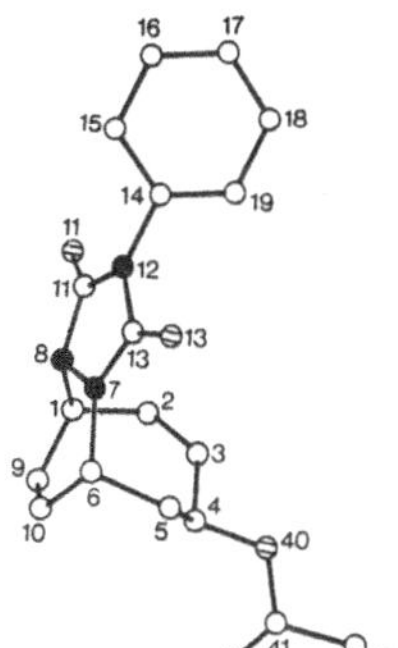

Fig. 1. A view of the urazole $C_{18}H_{17}N_3O_4$.

Triclinic, $P\bar{1}$, a = 8.262, b = 13.303, c = 8.123 Å, α = 101.03, β = 97.67, γ = 104.16°, Z = 2. Mo radiation, R = 0.046 for 2967 reflections.

The X-ray analysis establishes the *endo*-acetate substituent at the C(4) position (Fig. 1) and played a key role in the spectral assignments of endoperoxides.

[2](1,5)NAPHTHALINO[2](2,6)PYRIDINOPHANE-1,11-DIENE
$C_{19}H_{13}N$ (I)

[2](1,5)NAPHTHALINO[2](2,6)PYRIDINOPHANE
$C_{19}H_{17}N$ (II)

[2](1,4)NAPHTHALINO[2](2,6)PYRIDINOPHANE
$C_{19}H_{17}N$ (III)

M.W. HAENEL, B. LINTNER, R. BENN, A. RUFINSKA, G. SCHROTH, C. KRUGER, S. HIRSCH, H. IRNGARTINGER and D. SCHWEITZER, 1985. *Chem. Ber.*, **118**, 4884-4906.

I. Triclinic, $P\bar{1}$, a = 8.384, b = 12.187, c = 13.347 Å, α = 99.56, β = 95.19, γ = 94.21°, Z = 4. Mo radiation, R = 0.047 for 1902 reflections.

II. Orthorhombic, $Pna2_1$, a = 16.295, b = 8.271, c = 20.929 Å, Z = 8. Mo radiation, R = 0.037 for 1376 reflections.

III. Orthorhombic, $P2_12_12_1$, a = 8.202, b = 11.983, c = 14.571 Å, Z = 4. Mo radiation, R = 0.0492 for 1412 reflections.

There are two independent molecules in each of (I) and (II) and one independent molecule in (III); details of dimensions and conformation are in Figs. 1-3. In (I) the two molecules (M and N) have different orientations of the pyridine ring relative to the naphthalene moiety. Molecule (III) is the exo-isomer and has an interplanar angle of 14.3° between the pyridine and naphthalene planes; the N atom is 2.74 Å from the C(2),C(3),C(9),C(10) plane.

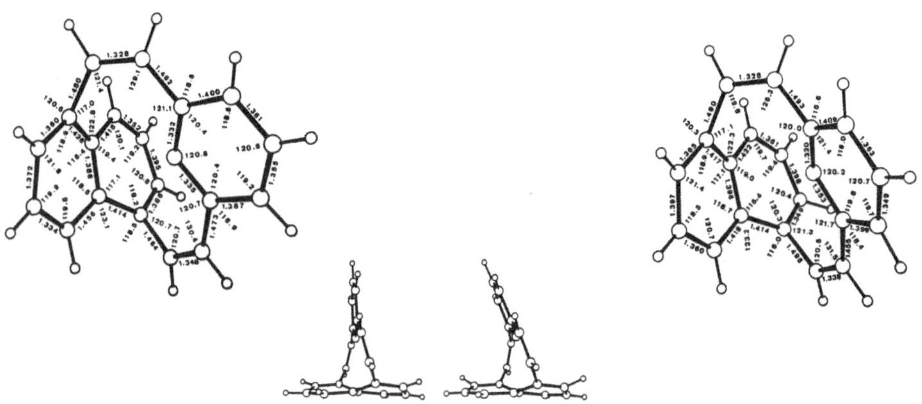

Fig. 1. The two $C_{19}H_{13}N$ molecules and the relative orientation of the pyridine rings.

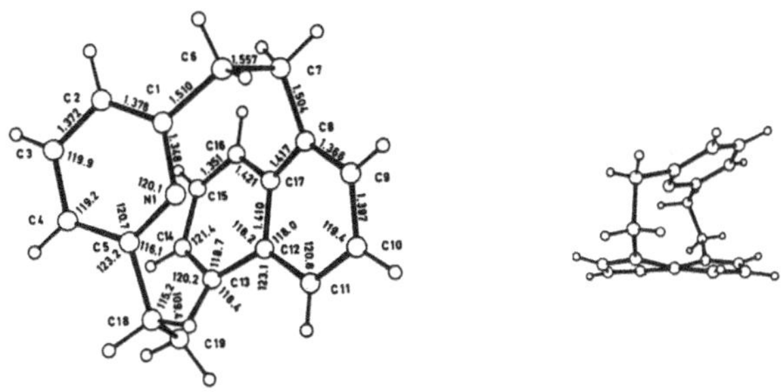

Fig. 2. One of the $C_{19}H_{17}N$ molecules in (II) with mean dimensions and a view showing the pyridine ring orientation relative to the naphthalene plane.

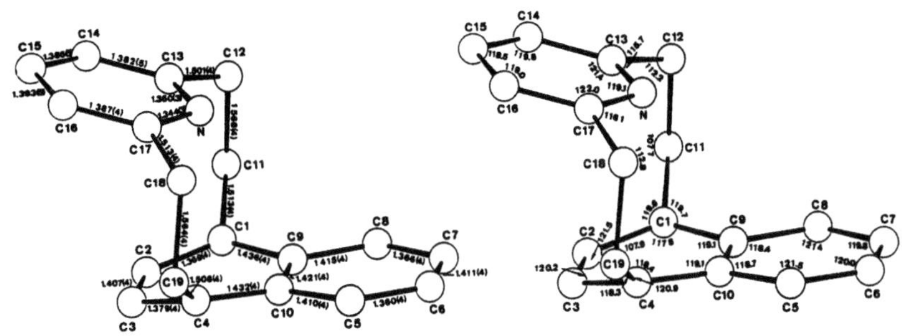

Fig. 3. The $C_{19}H_{17}N$ (III) structure with bond lengths and angles.

6-CYANO-7-ETHYL-1-AZABICYCLO[5.3.1]CYCLOUNDECANO[4:5][3,2-b]INDOLE HYDROBROMIDE METHANOL SOLVATE

$C_{19}H_{24}BrN_3,0.5(CH_4O)$ $C_{19}H_{24}N_3^+,Br^-,0.5(CH_4O)$

G. KALAUS, N. MALKIEH, I. KATONA, M. KAJTAR-PEREDY, T. KORITSANSZKY, A. KALMAN, L. SZABO and C. SZANTAY, 1985. *J. Org. Chem.*, **50**, 3760-3767.

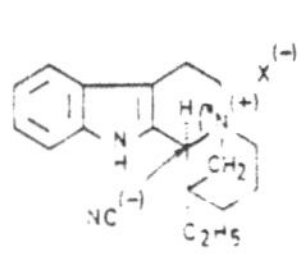

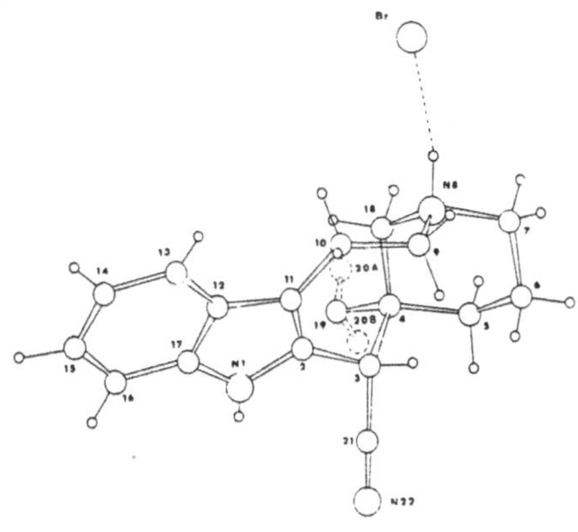

Fig. 1. A view of the molecular structure of $C_{19}H_{24}BrN_3,0.5(CH_4O)$. Two positions are shown for the disordered carbon atom (C(20)). The hydrogen bond between N(8) and the bromine atom is shown (dashed).

Monoclinic, $P2_1/c$, a = 7.831, b = 13.685, c = 20.139 Å, β = 123.01°, Z = 4. Cu radiation, R = 0.043 for

3377 reflections.

The molecule (Fig. 1) has a distorted eight-membered heteroring fused to a coplanar aromatic indole moiety at the C(2)-C(11) (1.361(3) Å) double bond and sharing three atoms, including the protonated N(8) atom, with a piperidine ring. The C(20) atom is disordered. The nitrile group has an equatorial orientation with respect to the eight-membered ring. The piperidine ring has a chair conformation.

(1α,5α,6α,7β,8α,9α)-2,7-BIS(ACETYLAMINO)-3-ACETYLIMINO-2,4-DIAZABICYCLO[3.3.1]NONANE-6,8,9-TRIOL TRIACETATE
$C_{19}H_{27}N_5O_9$ (I)

(1α,8α,9α,10β,11α,12α)-7-ACETYL-10-ACETYLAMINO-4-METHYL-2,3,5,7-TETRAAZATRICYCLO[6.3.1.-$0^{2,6}$]DODECA-3,5-DIENE-9,11,12-TRIOL TRIACETATE
$C_{19}H_{25}N_5O_8$ (II)

(1α,5α,6α,7β,9β,10α)-8-OXA-2,4-DIAZATRICYCLO[5.2.1.$0^{5,9}$]DEC-2-ENE-3,6,10-TRIOL TRIACETATE
$C_{13}H_{16}N_2O_7$ (III)

(1α,5α,6α,7β,8α,9α)-4-ACETYL-7-CHLORO-3-DIACETYLAMINO-2,4-DIAZABICYCLO[3.3.1]NON-2-ENE-6,-8,9-TRIOL TRIACETATE DIETHYL ETHER SOLVATE
$C_{19}H_{24}ClN_3O_9,C_4H_{10}O$ (IV)

W. FRITSCHE-LANG, P. WILHARM, E. HADICKE, H. FRITZ and H. PRINZBACH, 1985. *Chem. Ber.*, **118**, 2044-2078.

I. Monoclinic, $P2_1/c$, a = 11.562, b = 13.124, c = 14.944 Å, β = 99.78°, Z = 4. Cu radiation, R = 0.046 for 2124 reflections.

II. Monoclinic, $C2/c$, a = 21.370, b = 8.940, c = 26.441 Å, β = 102.86°, Z = 8. Cu radiation, R = 0.057 for 2620 reflections.

III. Monoclinic, $P2_1/c$, a = 11.294, b = 10.572, c = 13.328 Å, β = 118.27°, Z = 4. Cu radiation, R = 0.043 for 1667 reflections.

IV. Tetragonal, $P4_12_12$, a = 11.846, c = 38.114 Å, Z = 8. Cu radiation, R = 0.080 for 2278 reflections.

The analyses established the structures and stereochemistries as shown in Figs. 1 and 2. In (I) there are two intra- and one inter-molecular N-H...O hydrogen bonds; the hexahydropyrimidine ring is in a sofa conformation and the cyclohexane ring in a slightly distorted chair form. In (II) there is one intermolecular N-H...O hydrogen bond; the hexahydropyrimidine ring is in sofa form and the cyclohexane ring has a distorted boat shape. Molecule (III) has both intra- and inter-molecular N-H...O hydrogen bonds and the N-containing ring is in sofa form.

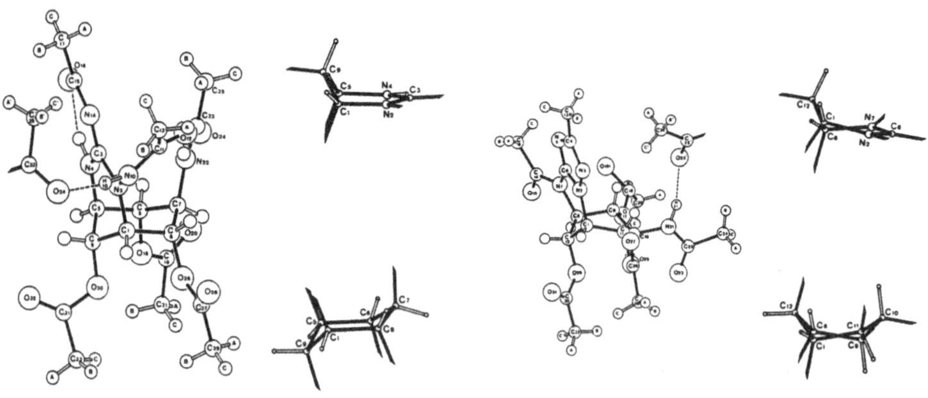

Fig. 1. Views of $C_{19}H_{27}N_5O_9$ (I) (left) and $C_{19}H_{25}N_5O_8$ (II) (right) with details of ring conformation.

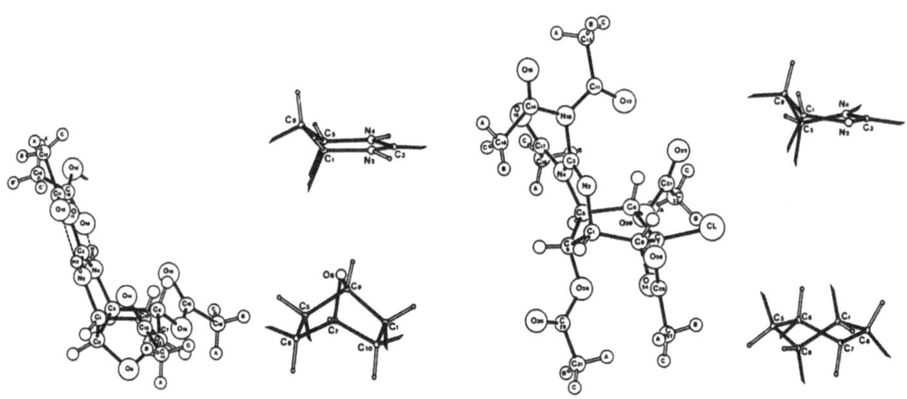

Fig. 2. Views of $C_{13}H_{16}N_2O_7$ (III) (left) and $C_{19}H_{24}ClN_3O_9$ (IV) right with details of ring conformation.

2,4-DIPHENYL-3-AZABICYCLO[3.3.1]NON-2-EN-9-OL
$C_{20}H_{21}NO$

A.A. ESPENBETOV, A.I. YANOVSKII, Yu.T. STRUCHKOV, T.T. OMAROV and I.A. AMANZHOLOV, 1985. *Zh. Strukt. Khim.*, **26**, 6-144-147 [J. Struct. Chem., **26**, 961-964].

Monoclinic, $P2_1/c$, a = 7.907, b = 22.521, c = 8.801 Å, β = 100.06°, Z = 4. Mo radiation, R = 0.0386 for 2386 reflections (at -120°C).

The molecular structure is shown in Fig. 1. Bond lengths are as expected. The azacyclohexane and cyclohexane rings have half-chair and chair conformations, respectively.

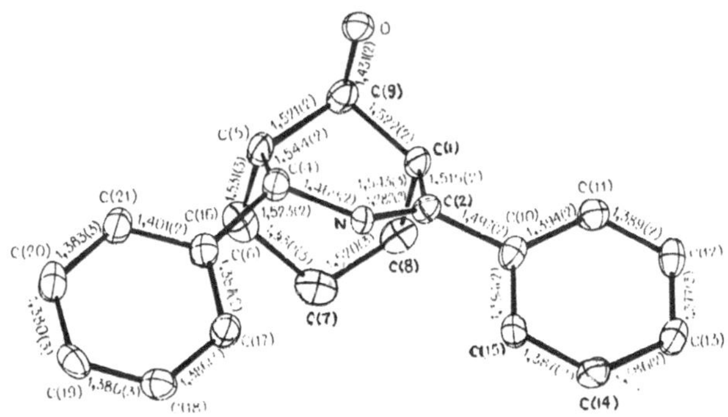

Fig. 1. The $C_{20}H_{21}NO$ molecule.

METHYL 4,5,7,8,9,10,11,12,13,14-DECAHYDRO-7,9-DIOXO-8,15-METHENOPYRROLO[3,2,1-op](1)-BENZAZACYCLODODECINE-16-CARBOXYLATE
$C_{20}H_{21}NO_4$ (I)

METHYL (Z)-1',2',4',5'-TETRAHYDRO-2,4'-DIOXOSPIRO(CYCLOHEPTANE-1,6'-(6H)-PYRROLO[3,2,1-ij]-QUINOLIN)-5'-YLIDENE-ACETATE
$C_{20}H_{21}NO_4$ (II)

endo-METHYL (Z)-1,1a,2,3,4,8,9,9a-OCTAHYDRO-1-METHOXYCARBONYLMETHYL-11-METHOXYCARBONYLMETHYLENE-12-OXO-1b,4a-EPOXYETHANOINDOLIZINO[2,3,4,5,6-jklm]CARBAZOLE-1

CARBOXYLATE
$C_{26}H_{27}NO_8$ (III)

R.M. LETCHER, TING-FONG LAI and T.C.W. MAK, 1985. *J. Chem. Soc. Perkin Trans. 1*, 1921-1926.

I. Orthorhombic, $P2_12_12_1$, a = 8.587, b = 11.131, c = 17.118 Å, Z = 4. Mo radiation, R = 0.040 for 1712 reflections.

II. Monoclinic, $P2_1/c$, a = 7.239, b = 16.486, c = 14.679 Å, β = 109.37°, Z = 4. Mo radiation, R = 0.046 for 1126 reflections.

III. Orthorhombic, $Pca2_1$, a = 13.080, b = 12.710, c = 27.34 Å, Z = 8. Mo radiation, R = 0.108 for 1651 reflections.

In (III) there are two independent molecules in the crystallographic asymmetric unit. The analyses served to establish the identities of the three molecules (Fig. 1).

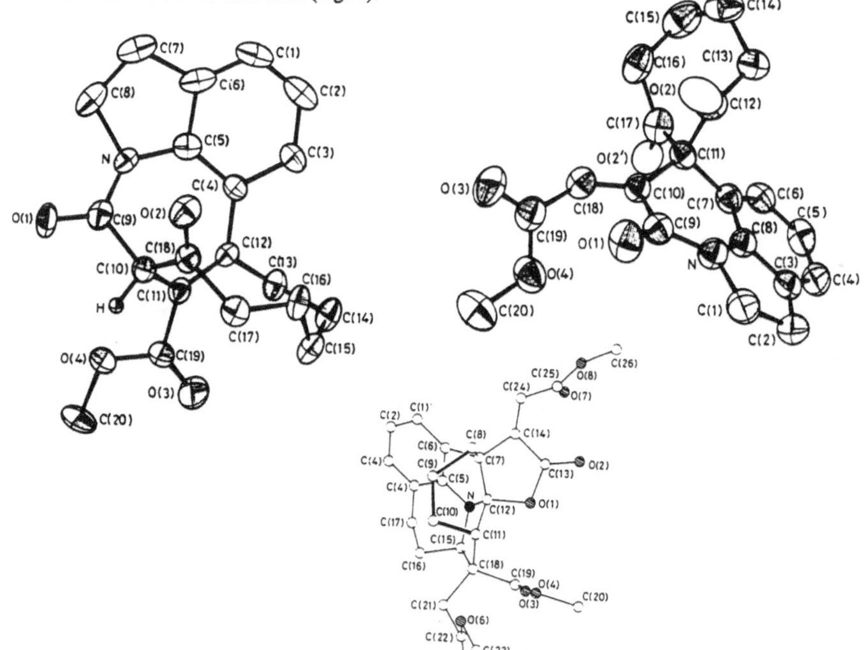

Fig. 1. Perspective views of the molecules of $C_{20}H_{21}NO_4$ (I) (top left), $C_{20}H_{21}NO_4$ (II) (top right) and $C_{26}H_{27}NO_8$ (bottom).

endo-10-BENZOYL-4-PHENYL-4,10-DIAZATRICYCLO[5.2.1.$0^{2,6}$]DEC-8-ENE-3,5-DIONE
$C_{21}H_{16}N_2O_3$ (I)

exo-10-ACETYL-4-PHENYL-4,10-DIAZATRICYCLO[5.2.1.$0^{2,6}$]DECANE-3,5-DIONE
$C_{16}H_{16}N_2O_3$ (II)

M.G.B. DREW, A.V. GEORGE, N.S. ISAACS and H.S. RZEPA, 1985. *J. Chem. Soc. Perkin Trans. 1*, 1277-1284.

I. Monoclinic, $P2_1/n$, a = 17.189, b = 6.836, c = 15.182 Å, β = 106.2°, Z = 4. Mo radiation, R = 0.061 for 1101 reflections.

II. Orthorhombic, *Pcab*, a = 10.95, b = 16.94, c = 15.24 Å, Z = 8. Mo radiation, R = 0.069 for 1089

reflections.

The angle between the plane of atoms C(1,9,8,7) and that of C(2,3,5,6),N(4) is 55.1° in (I) and 4.4° in (II). Apart from this major difference several aspects of the molecular shape are common to both molecules: N(4) is closely trigonal planar; N(10) is approximately pyramidal; the conformation of the phenyl ring attached to the maleimide nitrogen is similar and makes angles of 55.3° (I) and 55.8° (II) with the adjacent five-membered ring (Fig. 1).

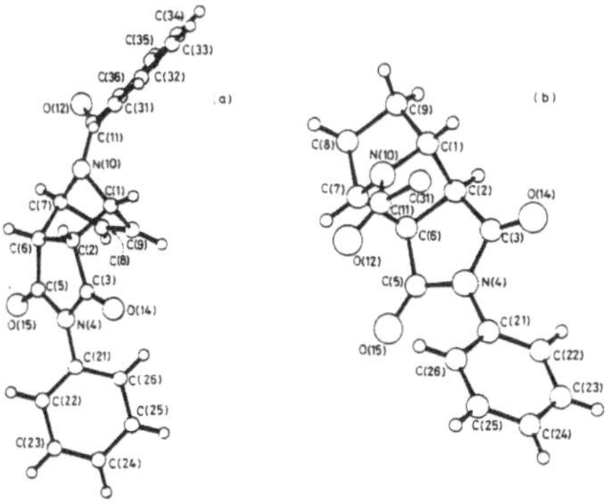

Fig. 1. Perspective drawings of molecules of $C_{21}H_{16}N_2O_3$ (left) and $C_{16}H_{16}N_2O_3$ (right).

N-(8-BENZYL-1αH,5αH-NORTROPAN-3α-YL)-2,6-DIMETHOXYBENZAMIDE
$C_{23}H_{28}N_2O_3$

F. DURANT, P. RENARD, G. EVRARD and A. MICHEL, 1985. *Acta Cryst.*, **C41**, 1361-1362.

Monoclinic, $P2_1/c$, a = 10.341, b = 22.478, c = 9.070 Å, β = 94.66°, Z = 4. Cu radiation, R = 0.036 for 1916 reflections.

In the benzamide moiety of the molecule (Fig. 1) the dimethoxyphenyl ring and the amide group are not coplanar. The dihedral angle between their least-squares planes is 65.2(4)°. This occurs due to steric hindrance between O(27) of one of the ortho methoxy substituents and the exocyclic carbonyl atom O(18). Their intramolecular O...O separation is 2.895(3) Å and due to their noncoplanarity no hydrogen bond is formed. The N-benzyl group is in an axial position. Bond lengths and angles in the molecule are in good agreement with generally accepted values. The crystal structure is stabilized by N(16)-H...O(18) intermolecular hydrogen bonds of length 2.998(4) Å.

Fig. 1. $C_{23}H_{28}N_2O_3$: structural formula and bond lengths and angles in the molecule.

BIS(μ_2-ETHYLENE(N^4-$N^{10'}$))BIS(6,7,8-TRIFLUORO-1,4,10-TRIAZATRICYCLO[7.3.1.$0^{5,13}$]TRIDECA-3,6,8-TRIENE)
$C_{24}H_{24}F_6N_6$

M.C. ETTER and J.E. RICHMAN, 1985. *J. Org. Chem.*, **50**, 112-114.

Orthorhombic, *Pccn*, a = 14.514, b = 15.205, c = 9.403 Å, Z = 4. Cu radiation, R = 0.034 for 1751 reflections.

The molecule is illustrated in Fig. 1. The bridging C(11)-C(12) bond length is 1.537(4) Å, N(1) is nearly planar, N(2) is nearly tetrahedral and N(3) is intermediate between these extremes. The aromatic rings are coplanar with a van der Waals separation of 3.4 Å, and they pack in a head-to-tail arrangement within the crown cavity such that the fluorines extend outside of the cavity.

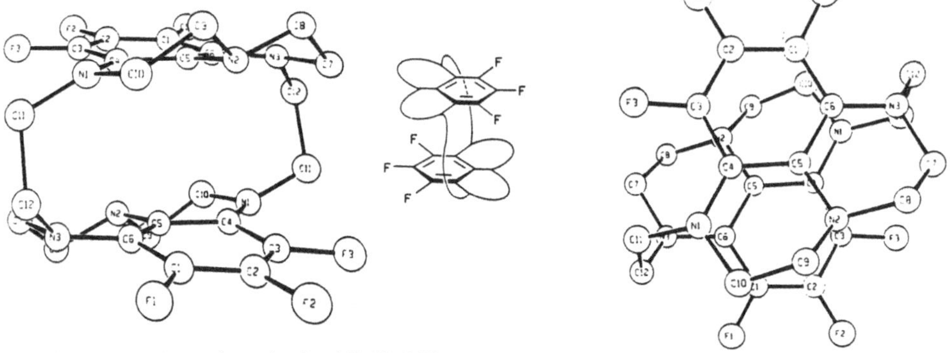

Fig. 1. Two views of a molecule of $C_{24}H_{24}F_6N_6$.

9,10-BENZO-1-CHLORO-7-DIPHENYLPHOSPHORYL-4-PHENYL-2,4,6-TRIAZATETRACYCLO[5.5.1.0^{2,6}.-0^{8,13}]TRIDEC-11-ENE-3,5-DIONE DICHLOROMETHANE SOLVATE
$C_{32}H_{23}ClN_3O_3P,CH_2Cl_2$

M. BOHSHAR, G. MAAS, H. HEYDT and M. REGITZ, 1985. *Tetrahedron*, **41**, 825-835.

Monoclinic, $P2_1/c$, a = 10.512, b = 20.972, c = 14.869 Å, β = 110.76°, Z = 4. Mo radiation, R = 0.0439 for 2374 reflections.

The seven-membered ring has a boat conformation and the $POPh_2$ moiety is exo at C(12) (Fig. 1). The triazole ring bridges C(2) and C(12) and is endo, with N(1) and N(2) pyramidal.

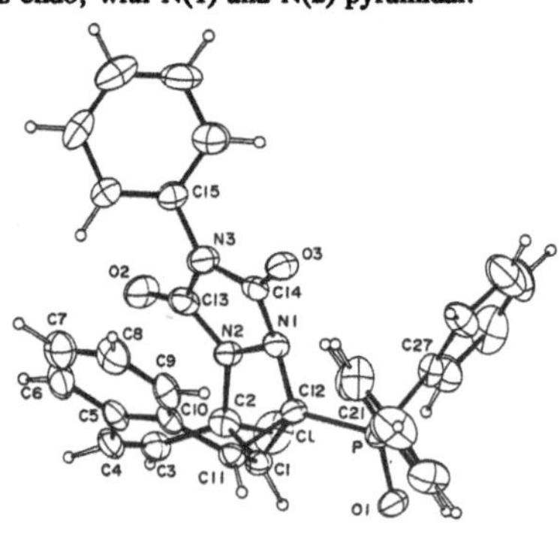

Fig. 1. The $C_{32}H_{23}ClN_3O_3P$ molecule.

2,7:10,15-BIS(4-PHENYL-1,2,4-TRIAZOLINE-3,5-DIONE-1,2-DIYL)TRICYCLO[8.6.0.0^{2,9}]HEXADECA-3,8,11,-16-TETRAENE DICHLOROMETHANE SOLVATE
$C_{32}H_{26}N_6O_4,CH_2Cl_2$

H. MEIER, T. ECHTER, K. BLECKMANN and W. WINTER, 1985. *Chem. Ber.*, **118**, 221-226.

Monoclinic, $P2_1/c$, a = 8.686, b = 23.210, c = 15.200 Å, β = 103.02°, Z = 4. Mo radiation, R = 0.047 for 2581 reflections.

The X-ray analysis shows that the skeleton of the molecule has two boat-shaped eight-membered rings and a central slightly folded four-membered ring (Fig. 1).

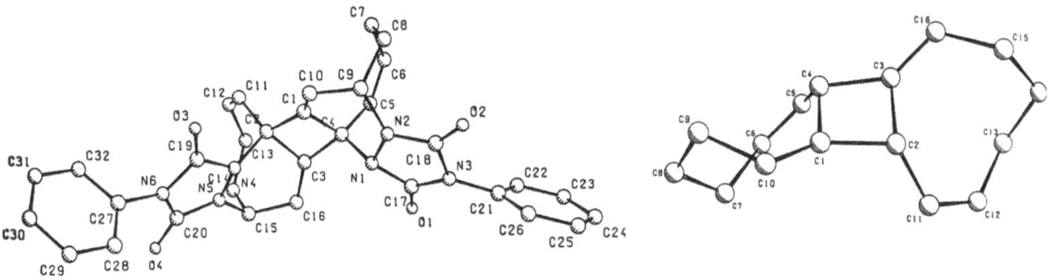

Fig. 1. The $C_{32}H_{26}N_6O_4$ molecule and a view of the three fused rings.

13-(2-AMINOPHENYL)-5,9b-DIHYDRO-5,9b[1',2']-BENZENODIBENZO[c,d-g]DIBENZ[3,4:6,7]AZEPINO[1,2-a]INDOLE
$C_{40}H_{26}N_2$

H. QUAST, P. ECKERT, B. SEIFERLING, E.-M. PETERS, K. PETERS and H.G. VON SCHNERING, 1985. *Chem. Ber.*, **118**, 3058-3069.

Monoclinic, $P2_1/n$, a = 15.276, b = 16.513, c = 11.716 Å, β = 112.55°, Z = 4. Mo radiation, R = 0.063 for 2409 reflections.

The molecule is obtained by photolysis of 9-azidotriptycene, is highly strained and has structure and stereochemistry as shown in Fig. 1 with the NH_2 group of the aminophenyl moiety directed below the molecule.

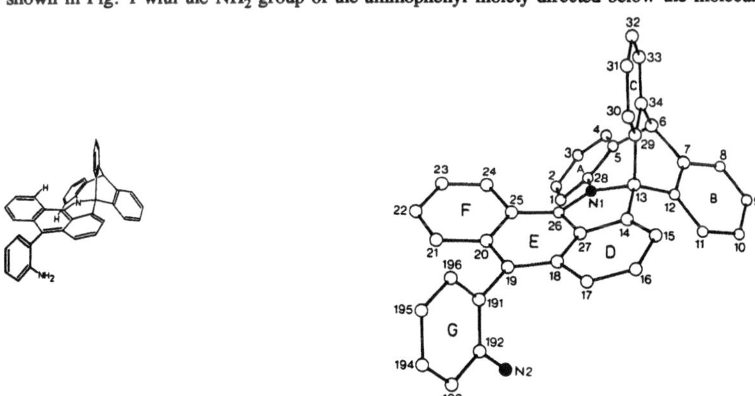

Fig. 1. A view of $C_{40}H_{26}N_2$.

2,2-DIMETHYL-1,3-DIOXANE-4,6-DIONE (MELDRUM'S ACID)
$C_6H_8O_4$

C.E. PFLUGER and P.D. BOYLE, 1985. *J. Chem. Soc. Perkin Trans. 2*, 1547-1549.

Orthorhombic, *Pbca*, a = 6.055, b = 13.746, c = 16.629 Å, Z = 8. Mo radiation, R = 0.074 for 1100

reflections.

The molecule (Fig. 1) has a boat conformation; the lactone groups (C(2), O(1), C(6), O(6), C(5) and C(2), O(3), C(4), O(4), C(5)) are both essentially planar. There is a short intermolecular distance in the crystal (C-H...O, H...O 2.43(3) Å).

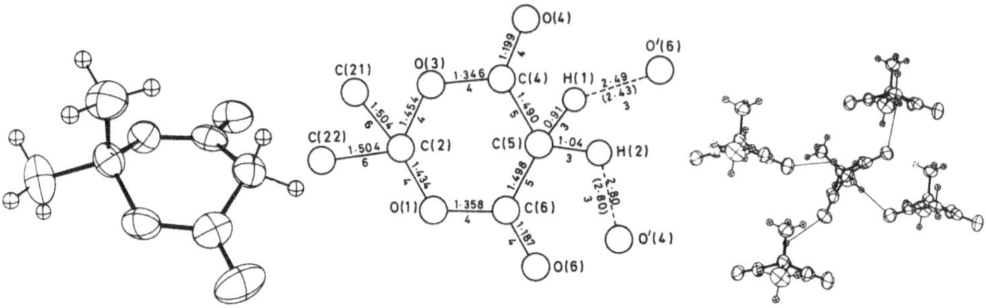

Fig. 1. $C_6H_8O_4$: perspective drawing (left), bond lengths (Å) (centre) and crystal packing (right).

(3S)-(+)-2-exo-BROMO-4,8-DIOXATRICYCLO[4.2.1.0^{3,7}]NONAN-5-ONE
$C_7H_7BrO_3$

S. OGAWA, Y. IWASAWA, T. NOSE, T. SUAMI, S. OHBA, M. ITO and Y. SAITO, 1985. *J. Chem. Soc. Perkin Trans. 1*, 903-906.

Monoclinic, $P2_1$, a = 10.137, b = 5.447, c = 6.770 Å, β = 90.88°, Z = 2. Mo radiation, R = 0.043 for 1241 reflections.

The absolute configuration was determined from anomalous dispersion. The molecule is a bromolactone (Fig. 1).

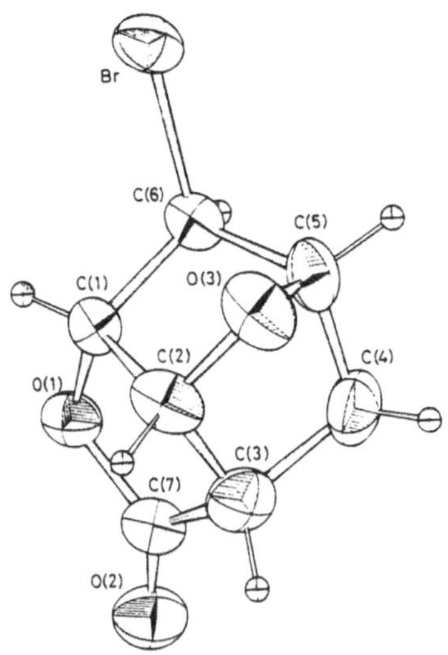

Fig. 1. Molecular structure of $C_7H_7BrO_3$.

PYRIDOXAL
$C_8H_9NO_3$ (I)

PYRIDOXAMINE DIHYDRATE
$C_8H_{12}N_2O_2,2(H_2O)$ (II)

C.L. MacLAURIN and M.F. RICHARDSON, 1985. *Acta Cryst.*, C**41**, 261-263.

I. Monoclinic, *Cc*, a = 12.093, b = 9.1067, c = 7.317 Å, β = 113.99°, Z = 4. Mo radiation, R = 0.034 for 1693 reflections.

II. Monoclinic, $P2_1/c$, a = 10.8819, b = 4.6033, c = 20.9791 Å, β = 96.563°, Z = 4. Mo radiation, R = 0.039 for 1394 reflections.

Pyridoxal exists as the hemiacetal in the zwitterionic form in which the pyridine ring is protonated [coordinates not published previously or deposited (**1**)]. Pyridoxamine (Fig. 1) is a zwitterion in the crystal, where a proton has been transferred from the phenolic to the aminomethyl group. The C(1)-O(1) distances in (I) and (II) are 1.290(1) and 1.327(2) Å respectively. In (II) the pyridine ring is essentially planar. Each of the eight potential hydrogen-bond donors is involved in hydrogen bonding. In the crystal, translationally related molecules are stacked along the b-axis (4.6033 Å).

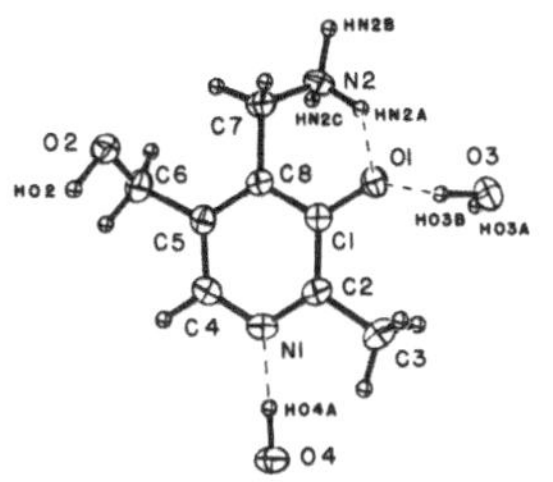

Fig. 1. $C_8H_{12}N_2O_2,2(H_2O)$: the molecular structure.

1. S.P.S. RAO, M.D. POOJARY and H. MANOHAR, 1982. Curr. Sci., **51**, 410-411.

BICYCLO[2.2.1]HEPTA-2,5-DIENE-2,3-DICARBOXYLIC ANHYDRIDE
$C_9H_6O_3$ (I)

BICYCLO[2.2.1]HEPT-2-ENE-2,3-DICARBOXYLIC ANHYDRIDE
$C_9H_8O_3$ (II)

W.H. WATSON and R.P. KASHYAP, 1985. *Acta Cryst.*, C**41**, 1226-1229.

I. Monoclinic, $P2_1/n$, a = 12.642, b = 11.412, c = 11.773 Å, β = 117.77°, Z = 8. Cu radiation, R = 0.046 for 1498 reflections.

II. Orthorhombic, *Pnma*, a = 6.233, b = 10.421, c = 11.843 Å, Z = 4. Cu radiation, R = 0.049 for 435 reflections.

There are two crystallographically independent molecules in the asymmetric unit of (I). The anhydride ring and the associated carbonyl O atoms (Fig. 1) are coplanar in (I) and (II). The C(2) and C(3) atoms are pyramidalised and the π-electron system in (II) deviates from planarity by 10.6(6)°. The corresponding deviations in the two molecules of (I) are 1.9(7) and 4.1(7)°. There is consistency in the molecular dimensions observed for (I) and (II).

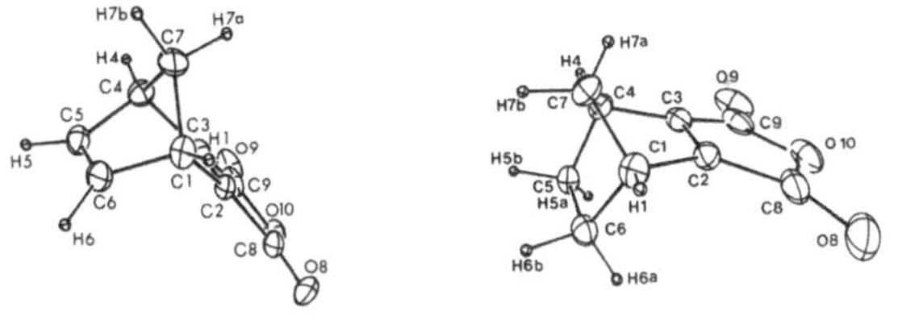

Fig. 1. The molecular structures of $C_9H_6O_3$ (left) and $C_9H_8O_3$ (right).

UMBELLIFERONE
$C_9H_6O_3$ (I)

7-ETHOXYCOUMARIN (7-O-ETHYLUMBELLIFERONE)
$C_{11}H_{10}O_3$ (II)

K. UENO, 1985. *Acta Cryst.*, **C41**, 1786-1789.

I. Monoclinic, $P2_1/c$, a = 3.892, b = 11.022, c = 16.722 Å, β = 90.58°, Z = 4. Cu radiation, R = 0.042 for 1255 reflections.

II. Monoclinic, $P2_1/n$, a = 4.001, b = 9.770, c = 24.213 Å, β = 93.12°, Z = 4. Cu radiation, R = 0.041 for 1691 reflections.

Both molecules (Fig. 1) are approximately planar. In (II), the torsion angle around the C(ring)-O(ethoxy) bond is 2.6(3)° resulting in a C-O-C ethoxy angle of 117.7(1)°. In (I), the C(2)-O(11), C(5)-C(10) and C(6)-C(7) bond lengths are longer than expected at 1.224(2), 1.411(2) and 1.396(3) Å respectively, whereas the C(2)-C(3) (1.433(3)), C(4)-C(10) (1.424(2)), and C(8)-C(9) (1.376(2) Å) bonds are shorter. In (II), the C(7)-O(12), C(8)-C(9), C(4)-C(10) and C(2)-O(11) bond lengths are 1.364(1), 1.390(1), 1.432(1) and 1.213(1) Å respectively. In both crystals, the coumarin moieties are stacked along the *a* axes. In (I), O(11)...O(12) hydrogen bonds (2.746(1) Å) form ribbons of molecules along the *c*-axis.

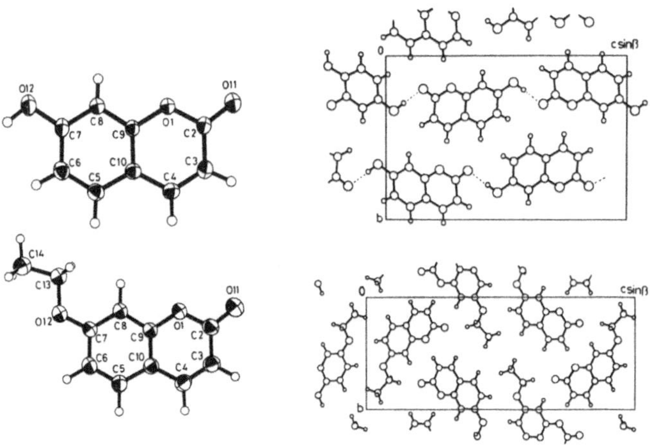

Fig. 1. Molecular structures and packing in $C_9H_6O_3$ (top) and $C_{11}H_{10}O_3$ (bottom).

erythro-2-METHOXYMETHYL-2,4-DIMETHYLGLUTARIC ANHYDRIDE (erythro-3-METHOXYMETHYL-3,5-DIMETHYL-1-OXACYCLOHEXANE-2,6-DIONE)
$C_9H_{14}O_4$

J. HASEK, 1985. *Acta Cryst.*, **C41**, 583-586.

Monoclinic, $P2_1/c$, a = 6.554, b = 7.297, c = 20.983 Å, β = 95.69°, Z = 4. Mo radiation, R = 0.064 for 2227 reflections.

The oxacyclohexane ring of the molecule (Fig. 1) has a half-boat conformation with C(4) at the tip. The dihedral angle between the planes of the sp^2 bonds of C(2) and C(6) is 7.9(5)°. Packing is governed by van der Waals interactions.

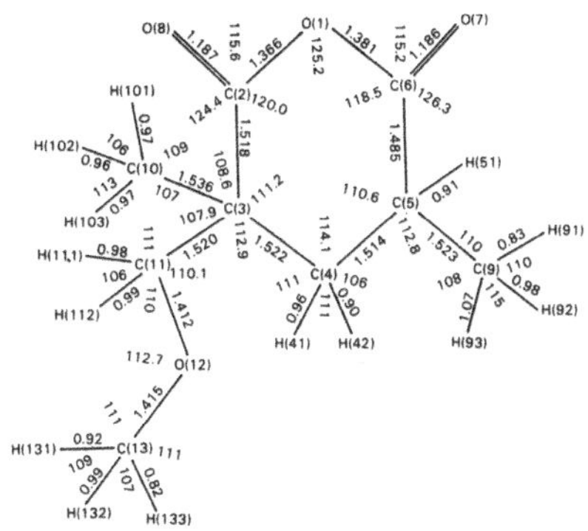

Fig. 1. $C_9H_{14}O_4$: bond lengths and angles.

1,4,7,10-TETRAOXACYCLOTRIDECANE-11,13-DIONE
$C_9H_{14}O_6$

HUANG JINSHUN, WANG MANFANG, HUANG ZAIFU, YANG XINGSHUI, LI ZHONGDE, LI ZHUNG and YOU XIAOZENG, 1985. *Jiegou Huaxue (J. Struct. Chem.)*, **4**, 24-26.

Monoclinic, $P2_1/a$, a = 8.267, b = 16.233, c = 8.524 Å, β = 114.20°, Z = 4. Mo radiation, R = 0.045 for 944 reflections.

The molecular conformation is shown in Fig. 1; the oxygen atoms in the ring are nearly planar. Mean bond lengths and angles are: C-O 1.437, C-C 1.493 Å, O-C-C 110.44, C-O-C 115.21°.

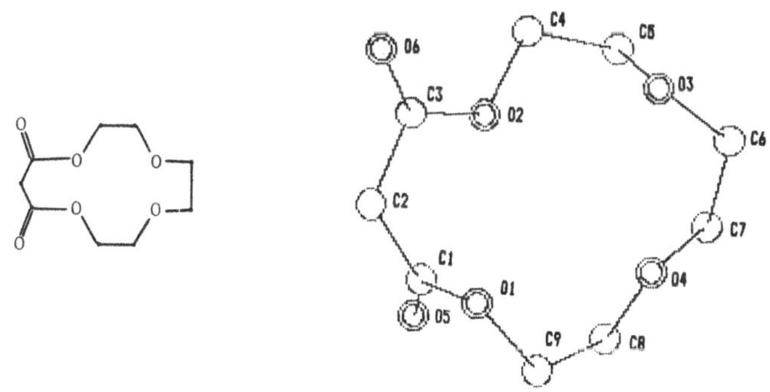

Fig. 1. The $C_9H_{14}O_6$ molecule.

(4S,6S,10S)-(+)-4,10-DIHYDROXY-1,7-DIOXASPIRO[5.5]UNDECANE
$C_9H_{16}O_4$ (I)

(4R,6S,10R)-4,10-DIHYDROXY-1,7-DIOXASPIRO[5.5]UNDECANE
$C_9H_{16}O_4$ (II)

K. MORI, T. UEMATSU, K. YANAGI and M. MINOBE, 1985. *Tetrahedron*, **41**, 2751-2758.

I. Orthorhombic, $P2_12_12_1$, a = 8.800, b = 15.367, c = 7.496 Å, Z = 4. Cu radiation, R = 0.033 for 923 reflections.

II. Orthorhombic, $P2_12_12_1$, a = 9.138, b = 13.557, c = 7.731 Å, Z = 4. Mo radiation, R = 0.045 for 740 reflections.

The structures and stereochemistries of these spiroundecanes are shown in Fig. 1. There is intermolecular O-H...O hydrogen bonding in (I) and intramolecular O-H...O hydrogen bonding in (II) (which stabilizes the axial orientation of the OH groups).

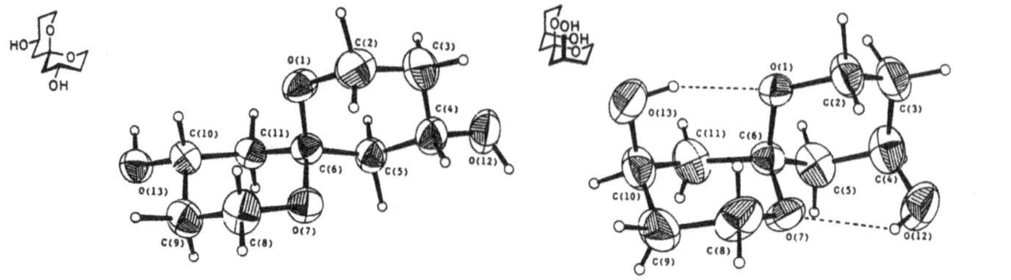

Fig. 1. Views of the $C_9H_{16}O_4$ isomer, (I) (left) and (II) (right).

3,9-DIMETHYL-2,4,8,10-TETRAOXASPIRO[5.5]UNDECANE
$C_9H_{16}O_4$ (I)

3,3,9,9-TETRAMETHYL-2,4,8,10-TETRAOXASPIRO[5.5]UNDECANE
$C_{11}H_{20}O_4$ (II)

A.E. OBODOVSKAYA, M.A. KHUSAINOV, Z.A. STARIKOVA, R.S. MUSAVIROV, E.A. KANTOR and D.L. RAKHMANKULOV, 1985. *Zh. Strukt. Khim.*, **26**, 4-119-124 [J. Struct. Chem., **26**, 595-600].

I. Orthorhombic, $P2_12_12_1$, a = 5.95, b = 9.335, c = 18.083 Å, Z = 4. Mo radiation, R = 0.058 for 731 reflections.

II. Monoclinic, $C2/c$, a = 8.339, b = 9.472, c = 14.890 Å, β = 92.08°, Z = 4. Mo radiation, R = 0.031 for 836 reflections.

The molecules (I) and (II) have similar structures; that for (I) is shown in Fig. 1. In (II) there are two independent molecules having twofold symmetry. The six-membered rings have chair conformations.

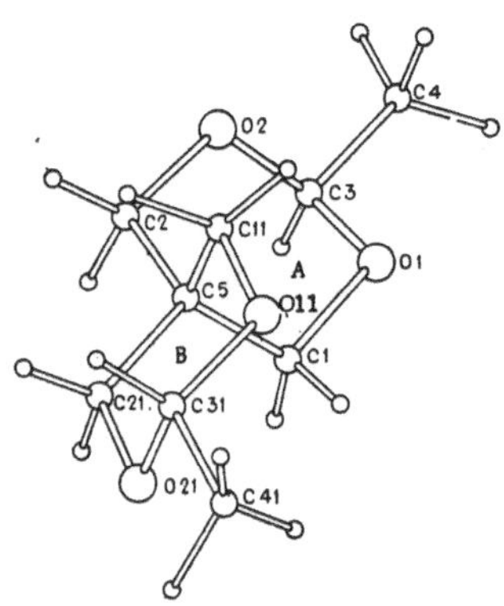

Fig. 1. The $C_9H_{16}O_4$ molecule.

3,8-DIHYDROXY-3-METHYL-3,4-DIHYDROISOCOUMARIN
$C_{10}H_{10}O_4$ (I)

2-ACETONYL-6-METHOXYBENZOIC ACID
$C_{11}H_{12}O_4$ (II)

K.-I. KAWAI, H. ITO, H. NAGASE, R. YAMAGUCHI and S. NAKAJIMA, 1985. *Acta Cryst.*, **C41**, 415-417.

I. Monoclinic, $P2_1/c$, a = 7.011, b = 9.493, c = 13.943 Å, β = 93.87°, Z = 4. Mo radiation, R = 0.0657 for 1041 reflections.

II. Orthorhombic, $P2_12_12_1$, a = 12.954, b = 15.043, c = 5.218 Å, Z = 4. Mo radiation, R = 0.0635 for 788 reflections.

The structures of molecules (I) and (II) in the crystals take, respectively, the lactol and the keto acid forms (Fig. 1). Bond lengths and angles in the molecules are normal. The dihydropyran ring in (I) adopts a half-chair form with the hydroxyl at C(3) in an axial position. There is an intramolecular O(8)-H...O(1) hydrogen bond of length 2.607 Å. The aromatic rings in both molecules are planar. In (II) there is a short intramolecular O(72)...C(9) contact of length 3.017 Å. The crystal structures of both compounds are stabilized mainly by intermolecular hydrogen bonds.

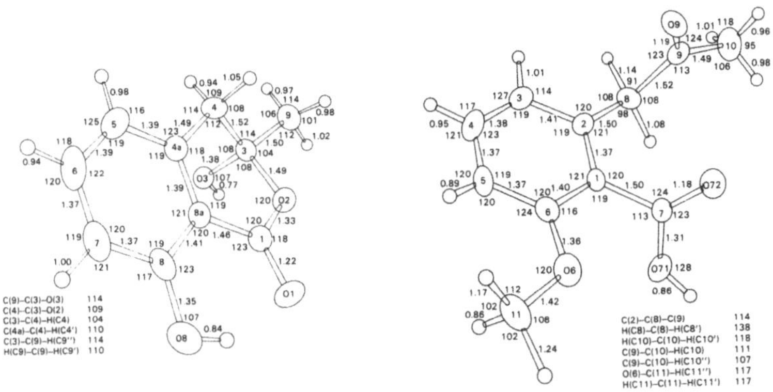

Fig. 1. Bond lengths and angles in $C_{10}H_{10}O_4$ (left) and in $C_{11}H_{12}O_4$ (right).

s-cis-2,6-DIMETHYL-4-(2-OXOPROPYLIDENE)-4H-PYRAN
$C_{10}H_{12}O_2$

A.T. BALABAN, V. WRAY, N.G. FURMANOVA, V.I. MINKIN, L.S. MINKINA, Yu.E. CZERNYSCH and G.S. BORODKIN, 1985. *Liebigs Ann. Chem.*, 1587-1595.

Orthorhombic, $P2_12_12_1$, a = 12.386, b = 14.254, c = 5.177 Å, Z = 4. Mo radiation, R = 0.052 for 472 reflections.

The crystal decayed by 20% during data collection, but the data were adequate to show that the molecule is practically planar and that it has the s-cis conformation (Fig. 1). The sterically crowded conformation of the molecule is due to strong π-electron delocalization.

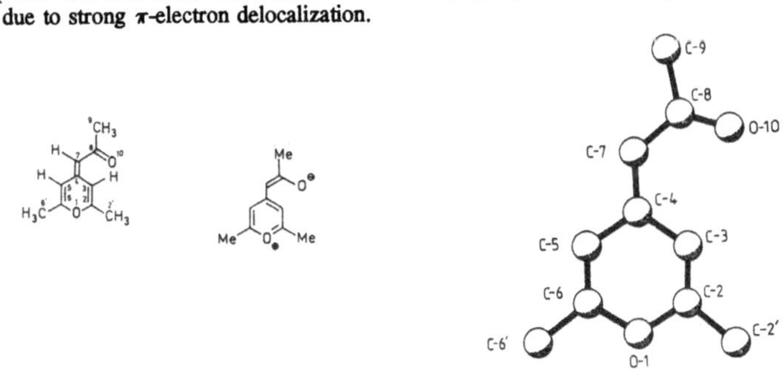

Fig. 1. The $C_{10}H_{12}O_2$ molecule.

7-exo-SUCCINIMIDO-2-OXABICYCLO[3.2.0]HEPTAN-6-exo-OL
$C_{10}H_{13}NO_4$

B. SHELDRICK, D. AKRIGG, M.I. PAGE and D. AGATHOCLEOUS, 1985. *Acta Cryst.*, C41, 1816-1817.

Orthorhombic, $P2_12_12_1$, a = 9.748, b = 9.918, c = 10.136 Å, Z = 4. Cu radiation, R = 0.0685 for 894 reflections.

The cyclobutanone and furan rings of the oxaheptan-6-exo-ol moiety (Fig. 1) make an angle of 118.2(7)°. Bond lengths and angles lie within acceptable ranges. The OH group forms a hydrogen bond to N with a length of 3.06(1) Å.

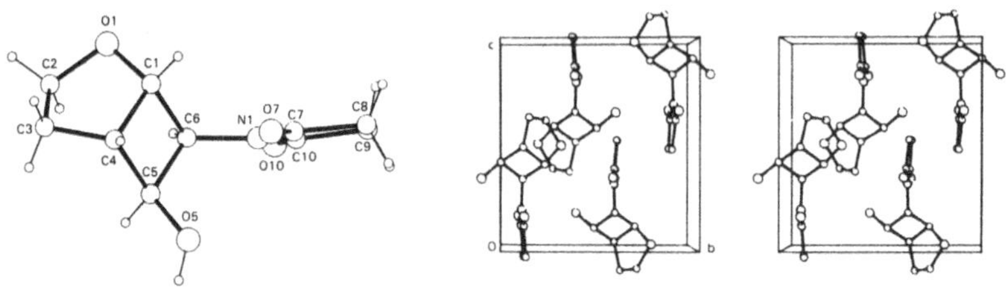

Fig. 1. $C_{10}H_{13}NO_4$: the molecular structure and a stereoview of the packing.

ETHYL ((1S,5R,8R)-8-HYDROXY-3-OXO-2-OXA-6-AZABICYCLO[3.3.0]OCT-6-YL)ACETATE HYDROBROMIDE

$C_{10}H_{16}BrNO_5$ $C_{10}H_{16}NO_5^+,Br^-$

J.F. RICHARDSON and V.K. YADAV, 1985. *Acta Cryst.*, C41, 1687-1689.

Orthorhombic, $P2_12_12_1$, a = 5.5495, b = 12.392, c = 18.825 Å, Z = 4. Mo radiation, R = 0.038 for 1100 reflections.

In the fused ring system in the molecule (Fig. 1), the five-membered ring O(5),C(6,1,2,3) has a conformation intermediate between a flattened half-chair and a flattened envelope. The ring N(10),C(1,6,7,9) has an almost ideal half-chair conformation with a pseudo-diad axis passing through C(1) and the mid-point of the C(7)-C(9) bond. The ethyl acetate side chain of this ring is in an extended roughly planar conformation and the hydroxyl O(8) is axial. The carbonyl C(3)=O(4) is equatorial. Bond lengths and angles in the molecule are normal and no unusually short intermolecular contacts are reported.

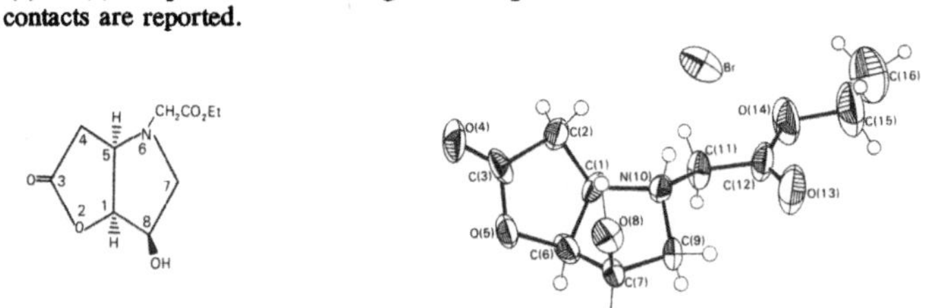

Fig. 1. $C_{10}H_{16}BrNO_5$: chemical numbering scheme in the cation and a view of the ion-pair.

1,7-DIOXACYCLODODECANE-2,8-DIONE

$C_{10}H_{16}O_4$ (I)

1,7,13-TRIOXACYCLOOCTADECANE-2,8,14-TRIONE

$C_{15}H_{24}O_6$ (II)

P. GROTH, 1985. *Acta Chem. Scand.*, A39, 749-759.

I. Monoclinic, $P2_1/n$, a = 15.179, b = 17.378, c = 15.356 Å, β = 96.28°, Z = 16. Mo radiation, R = 0.067 for 4014 reflections (at -150°C).

II. Orthorhombic, $Pca2_1$, a = 33.412, b = 4.732, c = 10.359 Å, Z = 4. Mo radiation, R = 0.05 for 1279 reflections (at -150°C).

In (I) there are four independent molecules in the crystallographic asymmetric unit (Fig. 1). In (II) there is some crystallographic disorder.

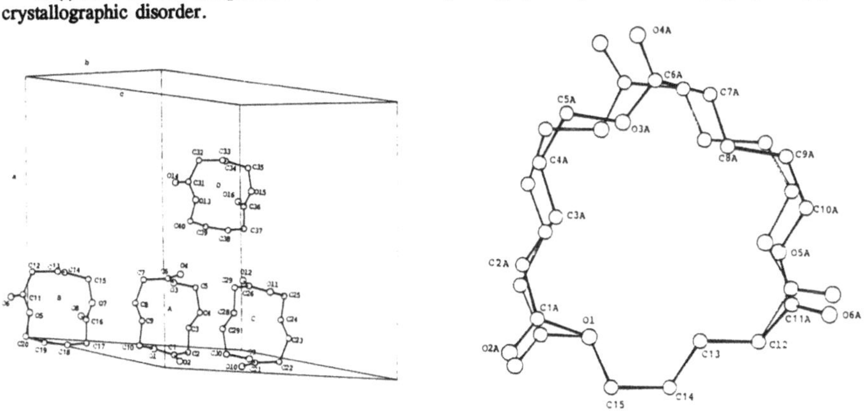

Fig. 1. Perspective drawing of the four independent molecules of $C_{10}H_{16}O_4$ (left) and of a molecule of $C_{15}H_{24}O_6$ (right) showing the disordered atoms.

(RS)-3-METHYL-r-3,t-3a-DIHYDROXY-trans-OCTAHYDROCYCLOHEPTA[6]FURAN-2-ONE
$C_{10}H_{16}O_4$

O. PEDERSEN, C. CHRISTOPHERSEN, L. BREHM and S.B. CHRISTENSEN, 1985. *Acta Chem. Scand.*, **B39**, 375-390.

Orthorhombic, *Pbca*, a = 18.182, b = 10.305, c = 10.745 Å, Z = 8. Mo radiation, R = 0.043 for 1084 reflections (at 110 K). (a = 18.431, b = 10.422, c = 10.872 Å. R = 0.039 for 1052 reflections at room temperature.)

The crystal structure analysis shows that the two rings are trans-fused and that the hydroxy groups at C(3) and C(3a) are trans to each other; consequently, the relative configuration at the asymmetric centres is 3RS, 3aSR and 8aRS. The crystal is partially disordered at atoms C(5), C(6) and C(7) (Fig. 1). The conformation of the γ-lactone ring is a distorted envelope with C(3a) at the flap of the envelope. The cycloheptane ring is in a twist-chair conformation.

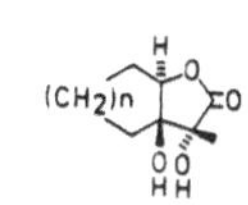

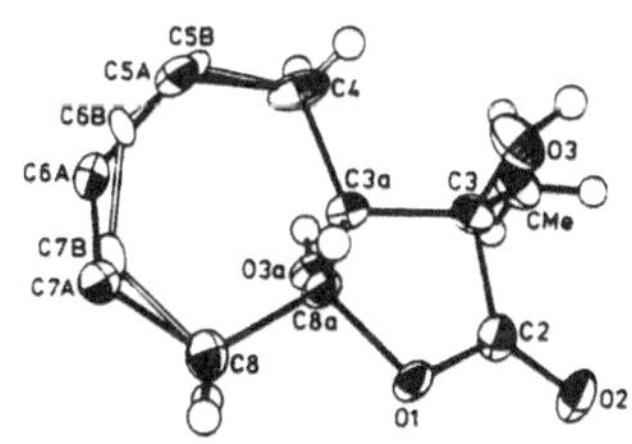

Fig. 1. Perspective drawing of a molecule of $C_{10}H_{16}O_4$ (at 110 K) showing the alternative positions of the disordered atoms.

(4S)-2,3-DIHYDRO-6-FLUOROSPIRO(4H-1-BENZOPYRAN-4,4'-IMIDAZOLIDINE)-2',5'-DIONE (SORBINIL)
$C_{11}H_9FN_2O_3$ (I)

(R)-N1',N3'-BIS(3-BROMOBENZYL)-2,3-DIHYDRO-6-FLUOROSPIRO(4H-1-BENZOPYRAN-4,4'-IMIDAZOLINE)-2',5'-DIONE
$C_{25}H_{19}Br_2FN_2O_3$ (II)

R. SARGES, J. BORDNER, B.W. DOMINY, M.J. PETERSON and E.B. WHIPPLE, 1985. *J. Med. Chem.*, **28**, 1716-1720.

I. Orthorhombic, $P2_12_12_1$, a = 6.397, b = 7.507, c = 21.028 Å, Z = 4. Mo radiation, R = 0.040 for 1258 reflections [see also the following Report].

II. Monoclinic, $P2_1$, a = 8.821, b = 8.092, c = 16.037 Å, β = 93.92°, Z = 2. Mo radiation, R = 0.055 for 3667 reflections.

The analyses determined the structures and absolute stereochemistries as shown in Figs. 1 and 2.

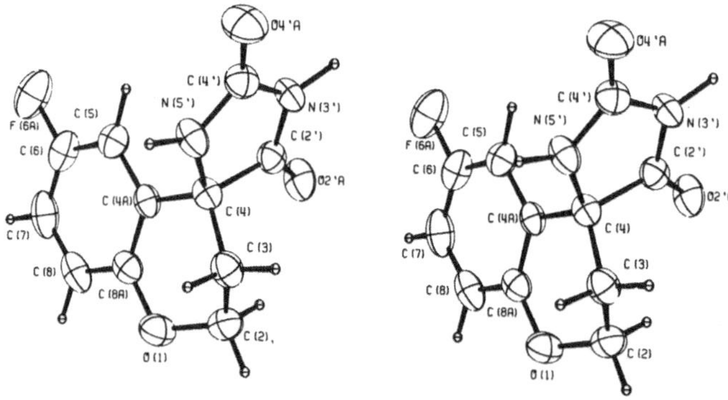

Fig. 1. A stereoview of sorbinil (I).

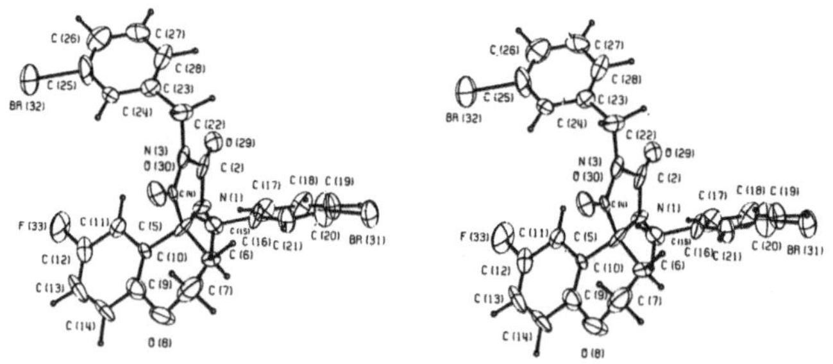

Fig. 2. A stereoview of $C_{25}H_{19}Br_2FN_2O_3$ (II).

(4S)-2,3-DIHYDRO-6-FLUOROSPIRO(4H-1-BENZOPYRAN-4,4'-IMIDAZOLIDINE)-2',5'-DIONE (SORBINIL) $C_{11}H_9FN_2O_3$

C.R. KISSINGER, E.T. ADMAN, J.I. CLARK and R.E. STENKAMP, 1985. *Acta Cryst.*, **C41**, 988-990.

Orthorhombic, $P2_12_12_1$, a = 7.5096, b = 6.397, c = 21.018 Å, Z = 4. Cu radiation, R = 0.042 for 749 reflections [see also the preceding Report].

The molecule, a potent inhibitor of the enzyme aldose reductase, consists of fluorobenzene and hydantoin rings connected via a nonplanar pyran ring (Fig. 1). The obtuse angle between the planes of the fluorobenzene and hydantoin rings is 98°. Bond distances and angles are normal except for the CH_2-CH_2 of the pyran ring which is somewhat short, 1.484(7) Å. In the crystal, the molecules form intercalated stacks with the benzopyran ring systems parallel to the ac face. Molecules in each stack are connected by hydrogen bonds along the b direction (N(14)...O(15) 2.986(6) Å). There are also hydrogen bonds between adjacent stacks (N(12)...O(16) 2.787(6) Å).

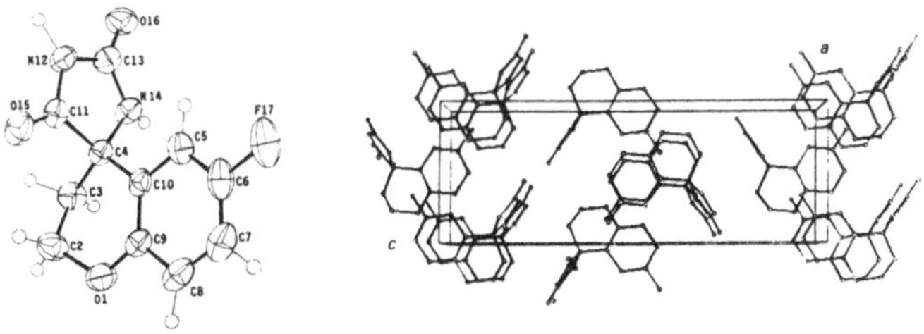

Fig. 1. $C_{11}H_9FN_2O_3$: the molecular structure and packing.

(+)-DIBROMOPHAKELLINE DIHYDRATE FORMIC ACID SOLVATE
$C_{11}H_{11}Br_2N_5O,2(H_2O),2(CH_2O_2)$ (I)

DIBROMOCANTHARELLINE DIHYDRATE METHANOL SOLVATE
$C_{11}H_{11}Br_2N_5O,CH_4O,2(H_2O)$ (II)

G. DE NANTEUIL, A. AHOND, J. GUILHEM, C. POUPAT, E. TRAN HUU DAU, P. POTIER, M. PUSSET, J. PUSSET and P. LABOUTE, 1985. *Tetrahedron*, **41**, 6019-6023.

I. Triclinic, *P*1, a = 11.168, b = 10.242, c = 10.431 Å, α = 133.87, β = 92.01, γ = 103.09°, Z = 2. Cu radiation, R = 0.073 for 2164 reflections.

II. Monoclinic, $P2_1$, a = 16.281, b = 5.782, c = 9.507 Å, β = 99.53°, Z = 2. Cu radiation, R = 0.087 for 1369 reflections.

These new pyrrolic metabolites are shown to have structure and stereochemistry indicated in Fig. 1.

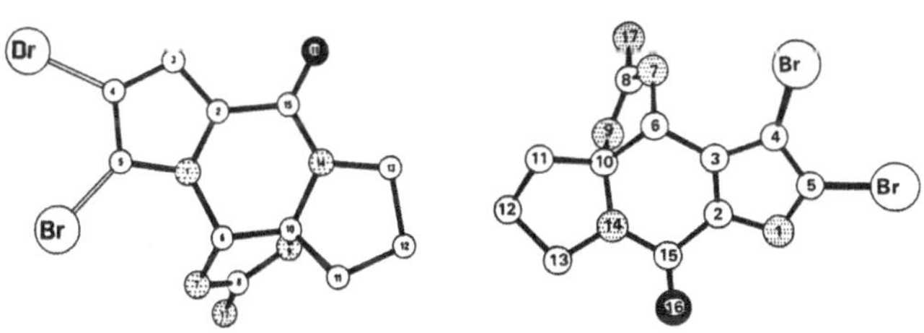

Fig. 1. Views of (+)-dibromophakelline (left) and dibromocantharelline (right).

1-METHYL-11-OXATRICYCLO[5.3.1.$0^{2,6}$]UNDEC-4-ENE-3,9-DIONE
$C_{11}H_{12}O_3$ (I)

1-ACETYL-11-OXATRICYCLO[5.3.1.$0^{2,6}$]UNDEC-4-ENE-3,9-DIONE
$C_{12}H_{12}O_4$ (II)

B.E. LA BELLE, M.J. KNUDSEN, M.M. OLMSTEAD, H. HOPE, M.D. YANUCK and N.E. SCHORE, 1985. *J. Org. Chem.*, **50**, 5215-5222.

I. Monoclinic, $P2_1/c$, a = 11.598, b = 8.290, c = 9.819 Å, β = 78.73°, Z = 4. Mo radiation, R = 0.067 for 1737 reflections (at 140 K).

II. Monoclinic, $P2_1/n$, a = 6.002, b = 23.169, c = 7.667 Å, β = 108.20°, Z = 4. Cu radiation, R = 0.039 for 1630 reflections (at 140 K).

The crystal structure analyses unambiguously established the structures of the two molecules (Fig. 1).

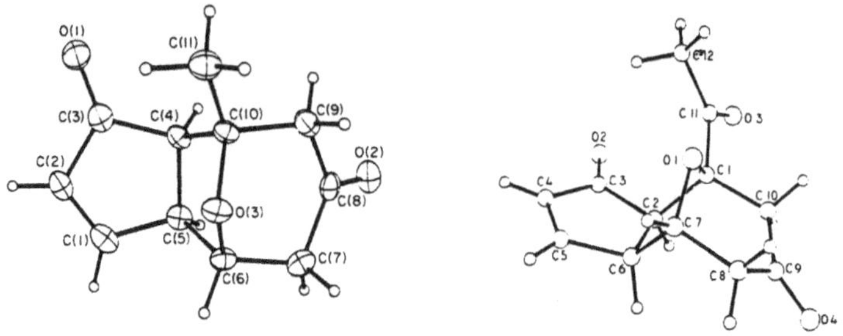

Fig. 1. Molecules of $C_{11}H_{12}O_3$ (I) (left) and $C_{12}H_{12}O_4$ (II) (right).

9-METHYL-3-OXO-2-OXABICYCLO[4.4.0]DEC-4-ENE-6,9-CARBOLACTONE
$C_{11}H_{12}O_4$

M. NAKAYAMA, J. KURAMOTO, T. MATSUI and J.D. WHITE, 1985. *Bull. Chem. Soc. Jpn.*, **58**, 3051-3052.

Triclinic, $P\bar{1}$, a = 7.601, b = 11.179, c = 12.147 Å, α = 85.26, β = 80.18, γ = 80.57°, Z = 4. Mo radiation, R = 0.071 for 3707 reflections.

The two molecules in the asymmetric unit are shown to have the same stereochemistry as indicated in Fig. 1.

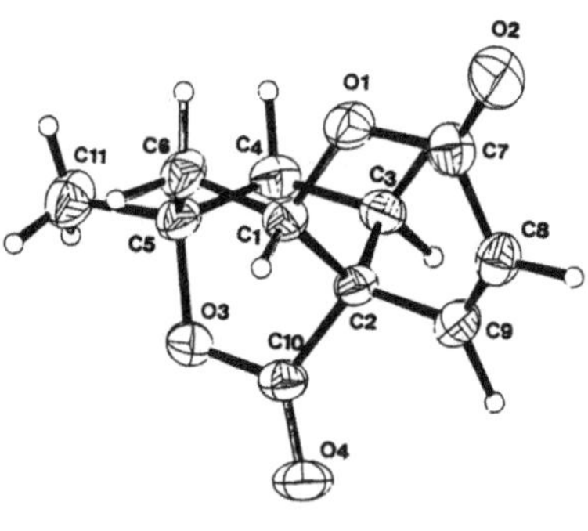

Fig. 1. A view of one of the $C_{11}H_{12}O_4$ molecules.

(1RS,3RS,5RS,6RS,7SR)-3,4:5,6-DIEPOXY-1,5,7-TRIMETHYLBICYCLO[4.2.0]OCTAN-7-OL
$C_{11}H_{16}O_3$ (I)

(1RS,3SR,4SR,5SR,6SR,7SR)-3,4:5,6-DIEPOXY-1,5,7-TRIMETHYLBICYCLO[4.2.0]OCTAN-7-YL p-NITRO-BENZOATE
$C_{18}H_{19}NO_6$ (II)

R. PHAFF, N. BISCHOFBERGER, P. MATHIES, W. PETTER, B. FREI and O. JEGER, 1985. *Helv. Chim. Acta*, **68**, 1204-1216.

I. Monoclinic, $P2_1/c$, a = 7.85, b = 11.63, c = 11.66 Å, β = 98.99°, Z = 4. Mo radiation, R = 0.073 for 1197 reflections.

II. Monoclinic, $P2_1/n$, a = 7.86, b = 15.59, c = 13.78 Å, β = 97.12°, Z = 4. Mo radiation, R = 0.052 for 1166 reflections.

The crystal structure analyses established the stereochemistries of the molecules (Fig. 1).

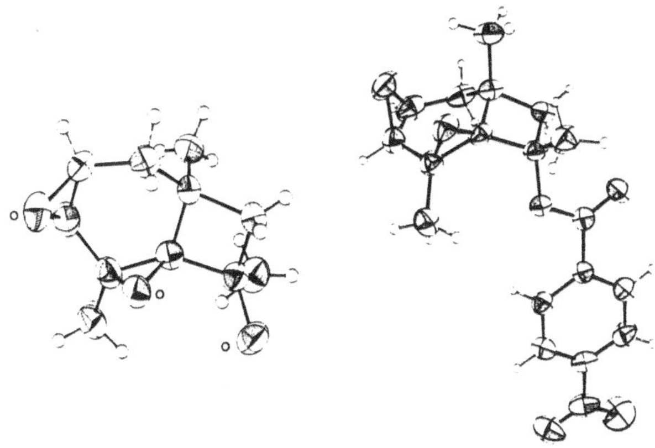

Fig. 1. ORTEP drawings of molecules of $C_{11}H_{16}O_3$ (left) and $C_{18}H_{19}NO_6$ (right).

9,12-DIHYDROXY-9-METHYL-10,11-DIOXATRICYCLO[6.3.1]DODECANE
$C_{11}H_{18}O_4$

G.R. CLARK, M.M. NIKAIDO, C.K. FAIR and J. LIN, 1985. *J. Org. Chem.*, **50**, 1994-1996.

Monoclinic, $P2_1/n$, a = 11.444, b = 5.813, c = 16.351 Å, β = 107.49°, Z = 4. Mo radiation, R = 0.035 for 595 reflections.

The crystal structure analysis established the stereochemistry of the molecule (Fig. 1).

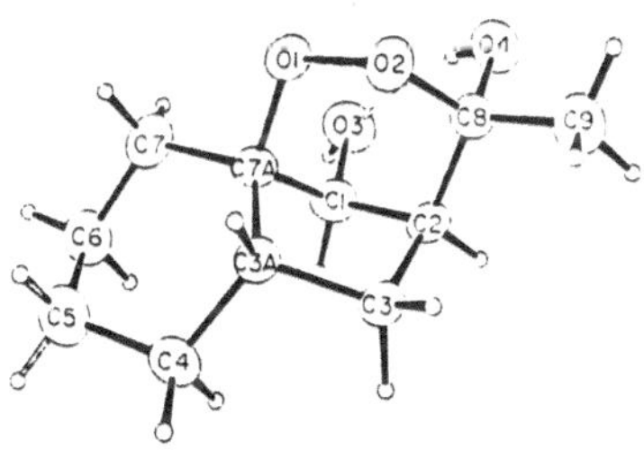

Fig. 1. ORTEP plot of a molecule of $C_{11}H_{18}O_4$.

(2R,3R,3aR,4aR,8aS,8bR)-PERHYDRO-2-(HYDROXYMETHYL)FURO[3,2-b]BENZOFURAN-3-OL
$C_{11}H_{18}O_4$

F. MICHEEL, A. KLEMER, M. KOHLA, H. PESCHKE and R. MATTES, 1985. *Liebigs Ann. Chem.*, 383-395.

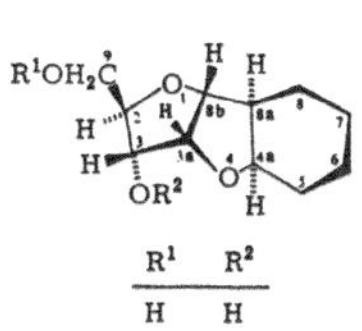

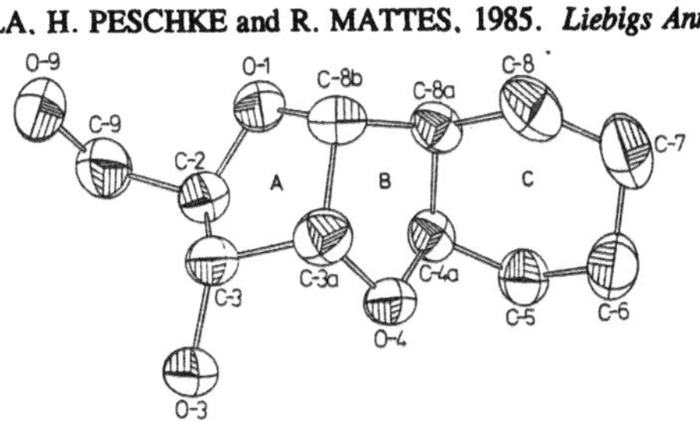

Fig. 1. An ORTEP plot of the $C_{11}H_{18}O_4$ molecule with H atoms omitted.

Monoclinic, $P2_1$, a = 5.721, b = 10.689, c = 8.968 Å, β = 92.07°, Z = 2. Mo radiation, R = 0.037 for 1098 reflections (at -140°C).

The analysis shows that the molecule has structure and conformation as in Fig. 1; the cyclohexane ring is in a distorted chair conformation.

cis-PERHYDRO-1,8a-EPOXY-2-NAPHTHYL METHANESULPHONATE
$C_{11}H_{18}O_4S$ (I)

trans-PERHYDRO-1,8a-EPOXY-2-NAPHTHYL METHANESULPHONATE
$C_{11}H_{18}O_4S$ (II)

E.O. SCHLEMPER and G.R. CLARK, 1985. *Acta Cryst.*, C41, 130-132.

I. Monoclinic, $P2_1/n$, a = 12.289, b = 5.741, c = 17.293 Å, β = 93.83°, Z = 4. Mo radiation, R = 0.051 for 1243 reflections.

II. Monoclinic, $P2_1/c$, a = 6.283, b = 21.429, c = 9.080 Å, β = 108.30°, Z = 4. Mo radiation, R = 0.040 for 1166 reflections (at 145 K).

In both isomers (I) and (II) (Fig. 1) the saturated six-membered rings are in chair conformations while the epoxide rings have half-chair conformations. Bond lengths and angles in both molecules agree well with generally accepted values. The dihedral between the plane containing the midpoint of C(10)...O(4), C(1) and C(2) and the plane containing atoms C(1), C(2) and O(2) is 79.2 in (I) and -166.9° in (II). The relationship between the epoxide and methanesulphonate units in (I) is "bisected" and in (II) is "perpendicular"

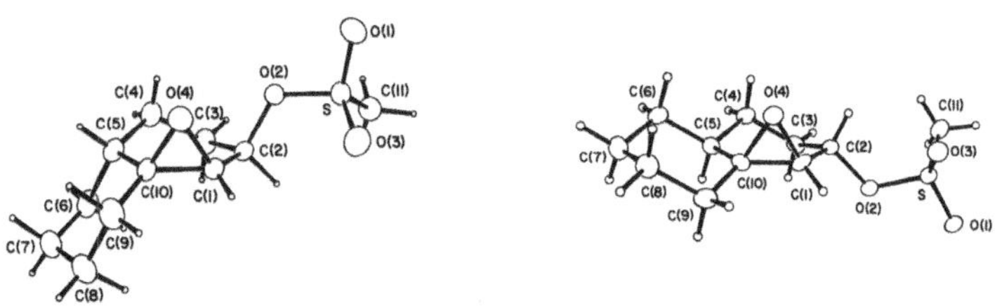

Fig. 1. $C_{11}H_{18}O_4S$: views and atomic numbering in the cis-fused isomer, (I) (left) and in the trans-fused isomer, (II) (right).

2-NITRONAPHTHO[1,2-b]FURAN
$C_{12}H_7NO_3$

J.-P. BIDEAU, G. BRAVIC and M. COTRAIT, 1985. *Acta Cryst.*, C41, 558-560.

Monoclinic, $P2_1/c$, a = 6.500, b = 19.866, c = 7.375 Å, β = 90.77°, Z = 4. Cu radiation, R = 0.047 for 1620 reflections.

The molecule is a weakly active mutagenic agent. It is planar and has normal bond lengths and angles (Fig. 1). Packing is governed by van der Waals interactions.

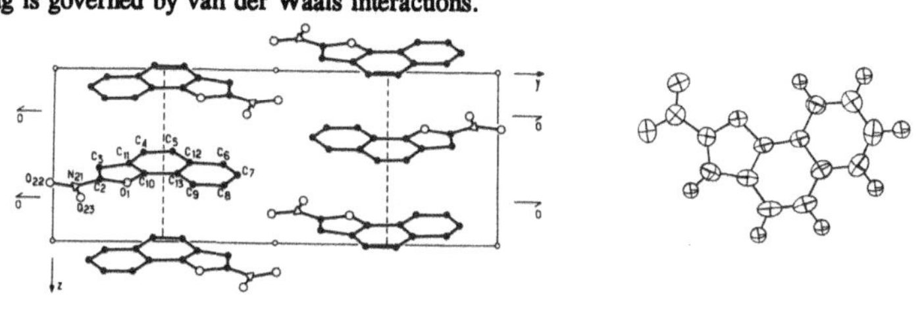

Fig. 1. $C_{12}H_7NO_3$: the molecular structure and projection of the structure along x.

1-ETHYL-1,4-DIHYDRO-4-OXO[1,3]DIOXOLO[4,5g]CINNOLINE-3-CARBOXYLIC ACID (CINOXACIN)
$C_{12}H_{10}N_2O_5$

M.J. ROSALES, R.A. TOSCANO, N. BARBABEHRENS and J. GARCIA, 1985. *Acta Cryst.*, C41, 1825-1826.

Triclinic, $P\bar{1}$, a = 6.946, b = 9.119, c = 9.129 Å, α = 80.42, β = 83.46, γ = 80.66°, Z = 2. Cu radiation, R = 0.055 for 1345 reflections.

The three-ring system of the molecule (Fig. 1) is practically planar with a maximum angle of 1.8° between the end rings. The maximum deviations from planarity are found in the O-containing ring. The C(6)-O(4), O(4)-C(7), C(7)-O(5) and O(5)-C(8) bond lengths are 1.358(2), 1.422(3), 1.429(3) and 1.357(3) Å respectively. Other bond lengths and angles agree with expected values. There is a hydrogen bond between H(1) of the carboxyl group and the O atom of the ketone (H(1)...O(3) 1.732 Å). The C(3)-O(3) distance is 1.248(3) Å.

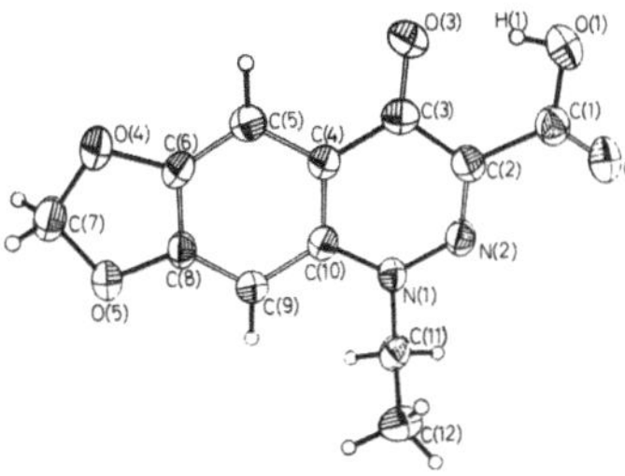

Fig. 1. $C_{12}H_{10}N_2O_5$: the molecular structure.

2,2-DIMETHYL-1,3-DIOXANE-4,5,6-TRIONE 5-(2,4-DINITROPHENYLHYDRAZONE)
$C_{12}H_{10}N_4O_8$

B. VICKERY, G.R. WILLEY and M.G.B. DREW, 1985. *Acta Cryst.*, C41, 586-589.

Monoclinic, $P2_1/c$, a = 14.236, b = 5.220, c = 20.111 Å, β = 99.4°, Z = 4. Mo radiation, R = 0.072 for 1089 reflections.

The two nitro groups and the phenyl ring of the molecule (Fig, 1) are closely planar while the dioxane ring is in an envelope conformation. The dimensions of the phenylhydrazone moiety are: C(4)-N(7) 1.386(9), N(7)-N(8) 1.330(8) and N(8)-C(9) 1.315(8) Å. The amine hydrogen H(7) forms a bifurcated intramolecular hydrogen bond to the ortho nitro group and to a carbonyl O atom (H(7)...O(11) 1.88(5), H(7)...O(1) 2.03(5) Å).

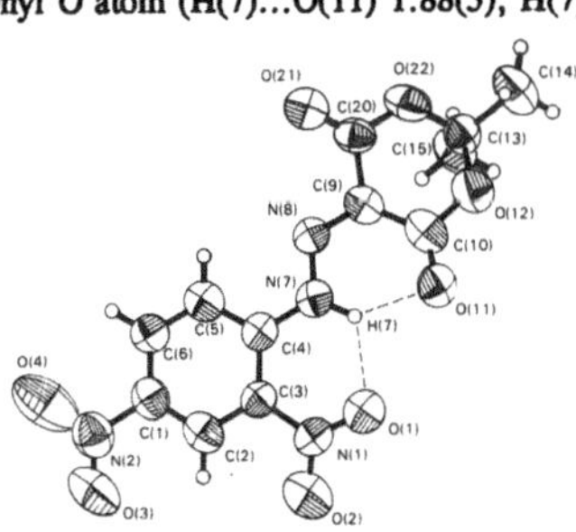

Fig. 1. $C_{12}H_{10}N_4O_8$: the molecular structure.

2-NITRO-6,7,8,9-TETRAHYDRONAPHTHO[2,1-b]FURAN
$C_{12}H_{11}NO_3$

J.-P. BIDEAU, G. BRAVIC and M. COTRAIT, 1985. *Acta Cryst.*, C41, 1799-1801.

Triclinic, $P\bar{1}$, a = 10.968, b = 7.291, c = 7.395 Å, α = 92.57, β = 94.89, γ = 119.56°, Z = 2. Cu radiation, R = 0.043 for 1513 reflections.

The nitro group in the molecule (Fig. 1) is essentially coplanar with the benzofuran system. Bond lengths and angles are normal. Stacking forces resulting from π-π interactions between parallel molecules (separation 3.37(1) Å) play an important role in crystal cohesion.

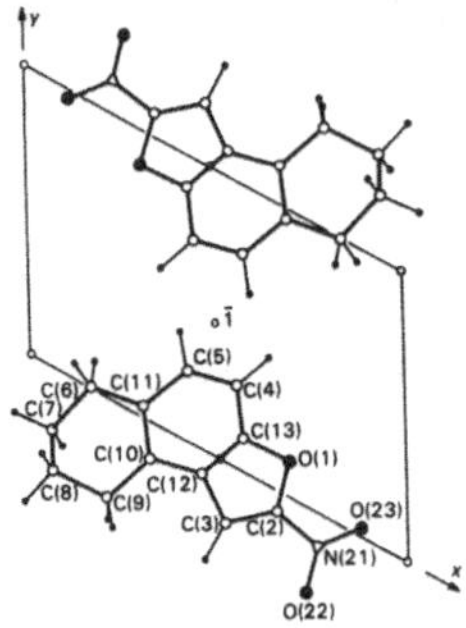

Fig. 1. $C_{12}H_{11}NO_3$: projection of the structure along the z-axis.

2-ACETYL-4,6-DIMETHOXYBENZOFURAN (CALEBERTIN A)
$C_{12}H_{12}O_4$ (I)

CALETEUCRIN
$C_{13}H_{14}O_5$ (II)

A.G. OBER, F.R. FRONCZEK and N.H. FISCHER, 1985. *J. Nat. Prod.*, **48**, 242-248.

I. Orthorhombic, *Pnma*, a = 10.992, b = 6.767, c = 14.643 Å, Z = 4. Mo radiation, R = 0.043 for 624 reflections.

II. Triclinic, $P\bar{1}$, a = 7.786, b = 11.072, c = 14.623 Å, α = 85.79, β = 83.49, γ = 84.46°, Z = 4. Mo radiation, R = 0.051 for 1848 reflections.

Calebertin (I) lies on a crystallographic mirror plane (Fig. 1) with the methyl keto substituent s-trans to the benzofuran double bond. There are two independent molecules of caleteucrin (II) in the asymmetric unit with similar but non-identical conformations. In both molecules, the benzo ring is planar while the dioxan ring has a twist conformation which is more severe for one molecule over the other.

Fig. 1. Views of calebertin A (left) and one of the caleteucrin molecules (right).

2,4,6-TRIS(1',1',2'-TRICHLOROPROPYL)-1,3,5-TRIOXANE (α-PARABUTYLCHLORAL)
$C_{12}H_{15}Cl_9O_3$ (I)

2,4,6-TRIS(1',1',2'-TRICHLOROPROPYL)-1,3,5-TRIOXANE (β-PARABUTYLCHLORAL)
$C_{12}H_{15}Cl_9O_3$ (II)

S.J. ARCHER, H.M.N.H. IRVING, K.R. KOCH and L.R. NASSIMBENI, 1985. *J. Cryst. Spect. Res.*, **15**, 333-349.

I. Orthorhombic, $P2_12_12_1$, a = 12.165, b = 9.964, c = 17.433 Å, Z = 4. Mo radiation, R = 0.043 for 1667 reflections.

II. Orthorhombic, $Pna2_1$, a = 12.387, b = 10.488, c = 16.605 Å, Z = 4. Mo radiation, R = 0.047 for 1417 reflections.

Both forms exist in cis,cis-chair conformations and the isomerism arises from the chirality of the side chains at the β-carbon atoms. In the α-form only two of the side chains have the same chirality but in the β-form the chirality at all the β-carbon atoms is the same. Bond lengths are shown in Fig. 1.

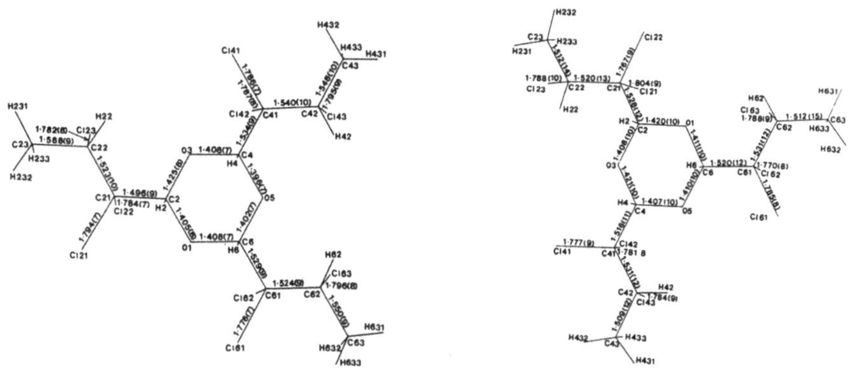

Fig. 1. Bond lengths in the α (left) and β- (right) forms of $C_{12}H_{15}Cl_9O_3$.

exo-3,endo-7-DIACETYL-1-HYDROXY-endo-3,5-exo-7-TRIMETHYL-cis-2,4,6-TRIOXABICYCLO[3.3.0]OCTANE
$C_{12}H_{18}O_6$ (I)

11-HYDROXY-1,3,5,7,11-PENTAMETHYL-cis-2,4,6,8-TETRAOXATRICYCLO[3.3.3.0 ^{3,7}]UNDECAN-9-ONE
$C_{12}H_{18}O_6$ (II)

M. POJE, I. PERINA, I. VICKOVIC and M. BRUVO, 1985. *Tetrahedron,* **41**, 1985-1987.

I. Monoclinic, $P2_1/n$, a = 14.902, b = 15.858, c = 5.566 Å, β = 93.37°, Z = 4. Mo radiation, R = 0.052 for 1355 reflections.

II. Monoclinic, $P2_1/n$, a = 10.831, b = 15.724, c = 7.591 Å, β = 110.13°, Z = 4. Mo radiation, R = 0.053 for 1131 reflections.

The structures and stereochemistries of these oxo-bridged species are as shown in Fig. 1.

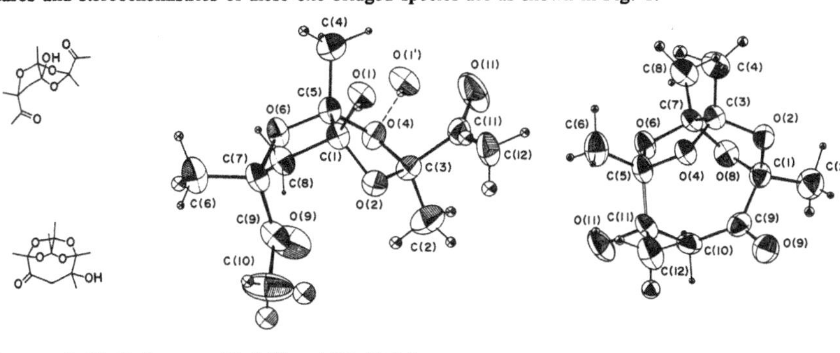

Fig. 1. The two $C_{12}H_{18}O_6$ isomers, (I) (left) and (II) (right).

1,8-DIOXACYCLOTETRADECANE-2,9-DIONE
$C_{12}H_{20}O_4$

P. GROTH, 1985. *Acta Chem. Scand.*, A39, 659-661.

Monoclinic, $P2_1/c$, a = 8.552, b = 10.111, c = 7.533 Å, β = 113.83°, Z = 2. Mo radiation, R = 0.032 for 946 reflections (at -150°C).

The molecules are centrosymmetric and lie about a crystallographic inversion centre (Fig. 1). The ring conformation may be described as [3 4 3 4].

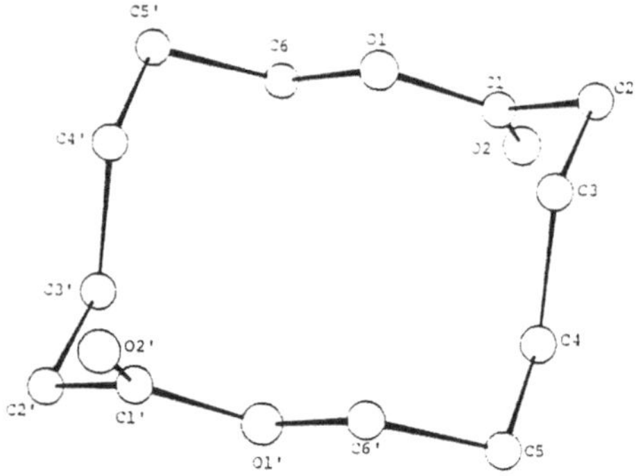

Fig. 1. Perspective drawing of a molecule of $C_{12}H_{20}O_4$.

18-CROWN-6 AMIDOSULPHURIC ACID ETHANOL SOLVATE
$C_{12}H_{27}NO_9S,0.5(C_2H_6O)$ $C_{12}H_{24}O_6,H_3NO_3S,0.5(C_2H_6O)$

F. SEEL, N. KLEIN, B. KREBS, M. DARTMANN and G. HENKEL, 1985. *Z. Anorg. Allg. Chem.*, **524**, 95-100.

Orthorhombic, $Pna2_1$, a = 16.705, b = 13.748, c = 8.371 Å, Z = 4. Mo radiation, R = 0.053 for 1640 reflections (at -120°C).

The amidosulphuric acid molecule is linked to the 18-crown-6 molecule by three N-H...O hydrogen bridges (Fig. 1) (2.87-3.10 Å).

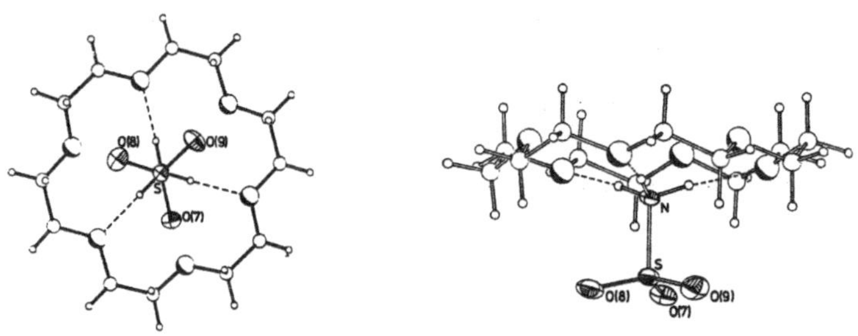

Fig. 1. Two views of $C_{12}H_{24}O_6,H_3NO_3S$.

DIAQUAOXODIPEROXOMOLYBDENUM(VI) 1,4,7,10,13,16-HEXAOXACYCLOOCTADECANE MONOHYDRATE (DIAQUAOXODIPEROXOMOLYBDENUM(VI) 18-CROWN-6 MONOHYDRATE)
$C_{12}H_{28}MoO_{13},H_2O$

C.B. SHOEMAKER, D.P. SHOEMAKER, L.V. McAFEE and C.W. DeKOCK, 1985. *Acta Cryst.*, **C41**, 347-350.

Orthorhombic, $A2_1ma$, a = 12.882, b = 13.683, c = 12.020 Å, Z = 4. Mo radiation, R = 0.028 for 1252 reflections.

The Mo atom (Fig. 1) lies on a crystallographic mirror plane of symmetry and has pentagonal bipyramidal seven-coordinate geometry with two peroxo groups and one water molecule in the equatorial plane and the double-bonded O and another water molecule at the apices. Two oxygen atoms O(1) and O(10) of the 18-crown-6 entity lie on the mirror plane but the ether molecule has pseudo-D_{3d} symmetry with gauche configurations at C-C bonds and trans at C-O bonds. The six O atoms of the macro-ring are coplanar to within 0.25 Å and five of these O atoms are hydrogen-bonded to water molecules to form infinite columns, with mirror plane symmetry, extending in the direction of the a-axis. O(1) is the only oxygen atom in the macro-ring not involved in hydrogen bonding. The average C-C and C-O bond lengths in the ether molecule are 1.48(2) and 1.41(1) Å respectively, and the O...O separations in the hydrogen-bonding contacts are in the range 2.489 to 2.891 Å

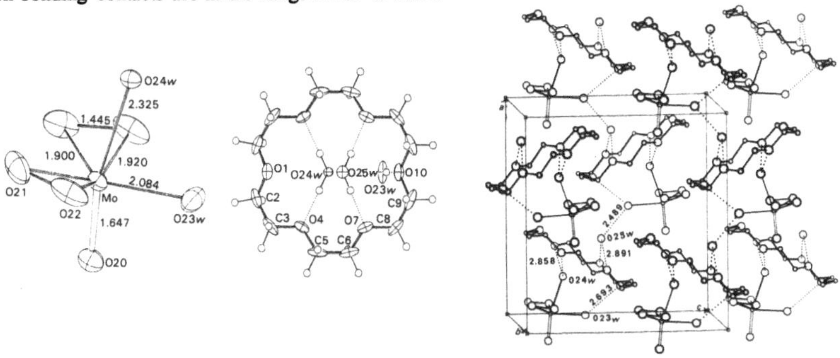

Fig. 1. $C_{12}H_{28}MoO_{13},H_2O$: the Mo coordination, a view showing the numbering in and the hydrogen-bonding contacts to the 18-crown-6 molecule, and the packing diagram showing the unit cell contents with dotted lines representing hydrogen bonds.

4-METHYL-7-(PROP-2-YN-1-YLOXY)CHROMEN-2-ONE (GIPARMENE)
$C_{13}H_{10}O_3$

P. TOFFOLI, P. KHODADAD, N. RODIER, R. CEOLIN and J. ASTOIN, 1985. *Acta Cryst.*, **C41**, 933-935.

Orthorhombic, $P2_12_12_1$, a = 10.99, b = 23.46, c = 4.001 Å, Z = 4. Mo radiation, R = 0.024 for 831 reflections.

The two rings in the molecule are planar with a dihedral angle of 1° between their mean planes (Fig. 1). Bond distances and angles in the coumarinic group are normal. The acetylenic bond is 1.157(5) Å in length. Packing is governed by van der Waals interactions.

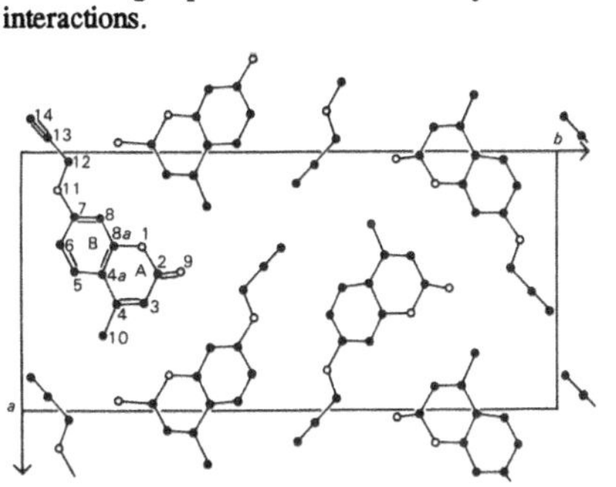

Fig. 1. $C_{13}H_{10}O_3$: projection onto the (001) plane.

3,4-DIMETHYL-1-(2-PYRIDYL)PYRANO[2,3-c]PYRAZOL-6(1H)-ONE
$C_{13}H_{11}N_3O_2$

C.G. RAMSAY and P.J. STEEL, 1985. *Acta Cryst.*, **C41**, 135-136.

Monoclinic, *C2/c*, a = 14.081, b = 17.781, c = 9.828 Å, β = 112.91°, Z = 8. Mo radiation, R = 0.043 for 1235 reflections.

The pyranopyrazole ring system in the molecule (Fig. 1) is essentially planar. The 4-methyl carbon atom is displaced from this plane by 0.110(4) Å. The dihedral angle between the least-squares plane of the pyranopyrazole ring system and that of the pyridine ring is 19.0(4)°. The molecules pack in layers along the a-axis. All intermolecular distances are essentially van der Waals contacts.

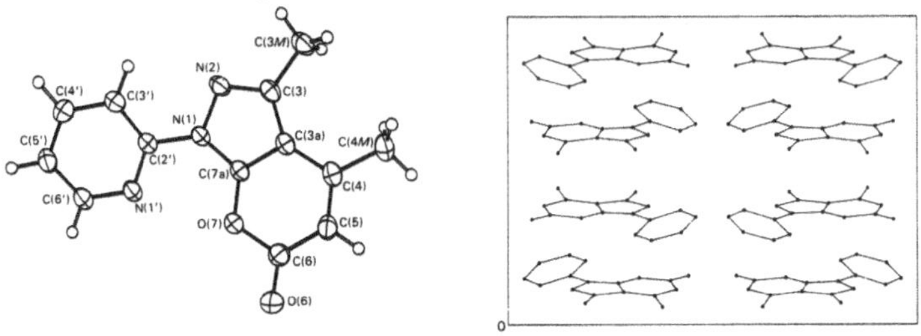

Fig. 1. $C_{13}H_{11}N_3O_2$: a view of the molecule and the unit cell contents projected down the c-axis (b-axis horizontal).

1-ETHYL-1,4-DIHYDRO-6,7-METHYLENEDIOXY-4-OXO-3-QUINOLINECARBOXYLIC ACID (OXOLINIC ACID)
$C_{13}H_{11}NO_5$

M. CYGLER and C.P. HUBER, 1985. *Acta Cryst.*, **C41**, 1052-1055.

Monoclinic, $P2_1/c$, a = 7.182, b = 10.575, c = 14.758 Å, β = 94.26°, Z = 4. Cu radiation, R = 0.038 for 1565 reflections.

The molecule (Fig. 1) is roughly planar except for the terminal carbon atom of the N-ethyl group which is displaced by 1.36 Å from the mean plane of the rest of the molecule. In the benzene ring there is considerable double-bond localization and in the pyridine ring a significant difference between the two C-N bond lengths- the bond adjacent to the benzene ring being longer by 0.058(2) Å. The molecular conformation is stabilized by a strong intramolecular bond O(16)-H...O(17) with the formation of a quasi six-membered ring. This hydrogen atom also appears to be involved in a weak intermolecular bond to O(9) of the dioxole ring with an O(16)...O(9) separation of 2.986(2) Å. The molecules form zigzag chains along the *c*-axis and all other intermolecular contacts are governed by van der Waals forces.

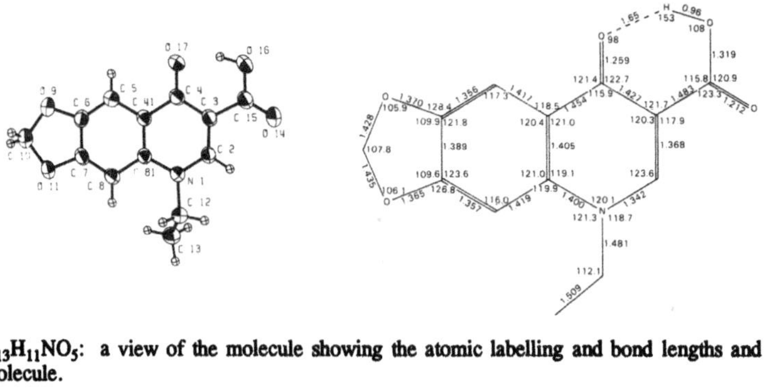

Fig. 1. $C_{13}H_{11}NO_5$: a view of the molecule showing the atomic labelling and bond lengths and angles in the molecule.

2-(4-CHLOROANILINO)-4-OXO-4,5-DIHYDRO-3-FUROIC ACID ETHYL ESTER
$C_{13}H_{12}ClNO_4$

K. EICHHORN and K.F. FISCHER, 1985. *Acta Cryst.*, **C41**, 1316-1318.

Monoclinic, $P2_1/c$, a = 4.508, b = 13.547, c = 20.784 Å, β = 97.61°, Z = 4. Mo radiation, R = 0.039 for 1571 reflections.

In the molecule (Fig. 1) the five-membered heterocyclic ring is nearly planar with C(8) tilted 1.8(4)° out of the plane of the other four atoms. There is an intramolecular N-H(5)...O(4) hydrogen bridge between the amino and carboxyl groups which closes a planar six-membered ring N, C(7,10,11), O(4) and H(5). The ethyl group is in a staggered conformation and is nearly coplanar with the dihydrofuran ring. The phenyl ring is planar but due to steric hindrance is twisted out of the plane of the hetero ring by 23.9(2)°. The Cl and N atoms are significantly displaced from the phenyl ring plane by 0.074(4) and 0.105(4) Å respectively. Bond lengths and angles in the molecule are normal and intermolecular separations are consistent with van der Waals packing forces.

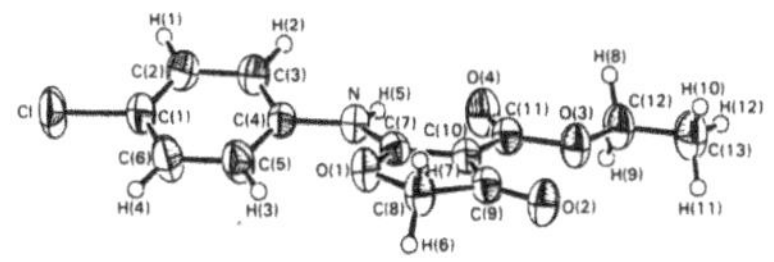

Fig. 1. A view of the $C_{13}H_{12}ClNO_4$ molecule.

11-ACETOXY-3-OXABICYCLO[6.3.1]DODECA-1(12),8,10-TRIENE
$C_{13}H_{16}O_3$

K.J. SHEA, L.D. BURKE and R.J. DOEDENS, 1985. *J. Am. Chem. Soc.*, **107**, 5305-5306.

Triclinic, $P\bar{1}$, a = 6.743, b = 12.221, c = 7.605 Å, α = 91.27, β = 109.91, γ = 100.27°, Z = 2. Mo radiation, R = 0.067 for 1982 reflections.

The benzene ring exists in an unsymmetrical boat conformation as shown in Fig. 1. Overcrowding causes several bond angles to have non-normal values, e.g. C(3)-C(4)-C(11) 126.1, C(5)-C(4)-C(11) 113.2, C(8)-C(9)-C(10) 120.6, C(9)-C(10)-C(11) 119.2°.

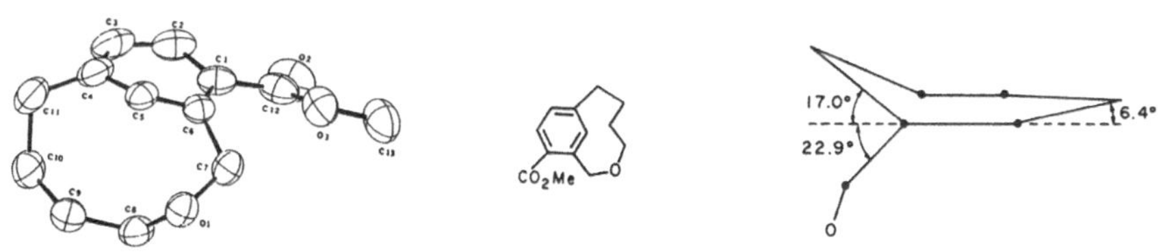

Fig. 1. The $C_{13}H_{16}O_3$ molecule and details of the benzene distortion.

cis-BICYCLO[3.3.0]OCT-3-ENE-2,7-DIONE-7-(2,2-DIMETHYLTRIMETHYLENE ACETAL)
$C_{13}H_{18}O_3$

X. SOLANS, M. FONT-ALTABA, A. MOYANO, E. CARCELLER and F. SERRATOSA, 1985. *Acta Cryst.*, **C41**, 1628-1629.

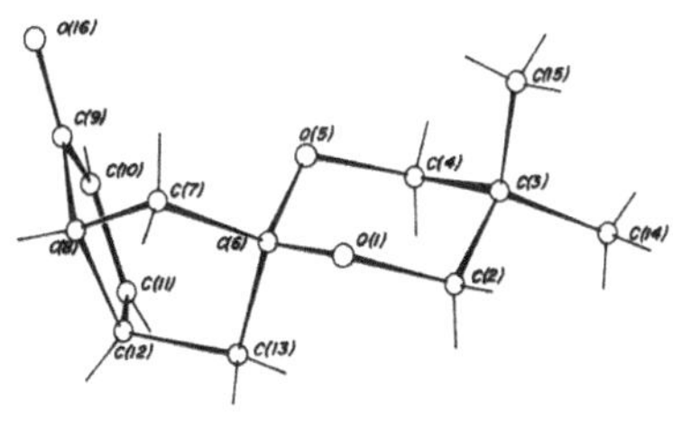

Fig. 1. A view of the $C_{13}H_{18}O_3$ molecule.

Orthorhombic, *Pbca*, a = 19.633, b = 12.211, c = 10.147 Å, Z = 8. Mo radiation, R = 0.048 for 1045 reflections.

In the molecule (Fig. 1) the six-membered heteroring has a chair conformation with C(3) and C(6) deviating by 0.679(5) and -0.624(5) Å respectively, from the mean plane of the four other ring atoms. The equatorial C(6)-C(7) and axial C(6)-C(13) bond lengths are 1.507(5) and 1.548(5) Å respectively. The two five-membered rings C(6,7,8,12,13) and C(8,9,10,11,12) have skew-envelope and envelope conformations respectively. Most of the bond lengths and angles are normal and deviations are ascribed to steric and to electron conjugation effects. The intermolecular distances are consistent with essentially van der Waals contacts.

1,3,7,7-TETRAMETHYL-2,11-DIOXABICYCLO[4.4.1]UNDECA-3,5-DIEN-10-ONE $C_{13}H_{18}O_3$

F.S. JORGENSEN, M. GAJHEDE and B. FREI, 1985. *Helv. Chim. Acta*, **68**, 2148-2157.

Orthorhombic, $P2_12_12_1$, a = 7.872, b = 12.015, c = 12.250 Å, Z = 4. Mo radiation, R = 0.045 for 2510 reflections (at 105 K).

The diene unit is twisted by 21° and the acetal conformations can be considered as being a distorted (+95°, -55°) gauche (+), gauche (-) conformation (Fig. 1). The acetal bond lengths differ by 0.010(2) Å.

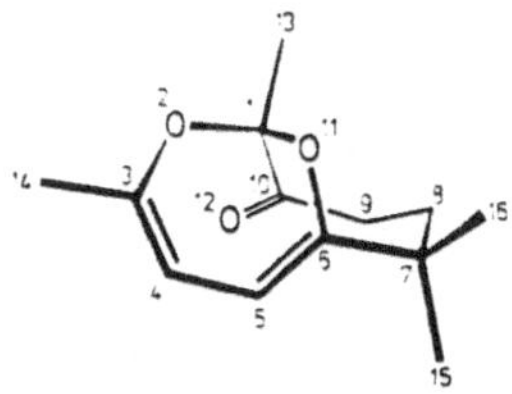

Fig. 1. A molecule of $C_{13}H_{18}O_3$.

N-(6-CHLORO-4-HYDROXY-2-OXO-2H-PYRAN-3-CARBONYL)-p-TOLYLCARBAMATE $C_{14}H_{10}ClNO_6$

W. RIED, J. NENNINGER and J.W. BATS, 1985. *Chem. Ber.*, **118**, 1371-1382.

Monoclinic, $P2_1/n$, a = 10.373, b = 8.412, c = 16.947 Å, β = 107.32°, Z = 4. Mo radiation, R = 0.040 for 2232 reflections.

The pyran ring (Fig. 1) is almost coplanar with the plane of the amido-ketone and is stabilised by intramolecular H-bonding (H(2)...O(4) 1.67(2) Å, H(3)...O(2) 1.93(2) Å). The p-tolyl ring is inclined at 102° to the carboxyl group plane.

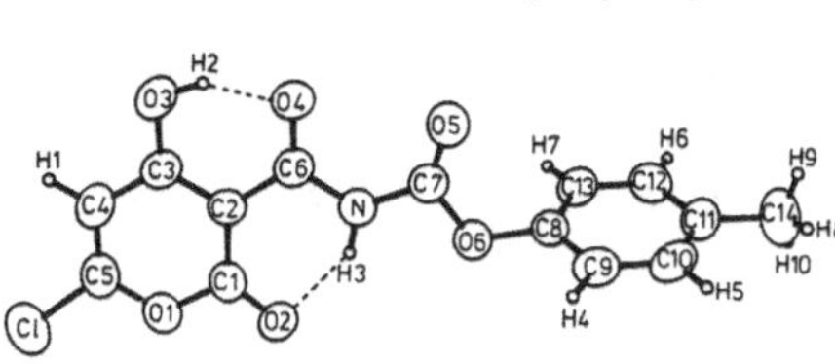

Fig. 1. A view of $C_{14}H_{10}ClNO_6$.

5-(α-PHENYLETHYL)-2,2-DIMETHYL-1,3-DIOXANE-4,6-DIONE $C_{14}H_{16}O_4$

HU SHENGZHI, DONG NAN, HUANG XIAN and T.C.W. MAK, 1985. *Jiegou Huaxue (J. Struct. Chem.)*, **4**, 107-110.

Monoclinic, $P2_1/n$, a = 11.394, b = 5.770, c = 20.534 Å, β = 106.09°, Z = 4. Mo radiation, R = 0.057 for 1984 reflections.

The molecule has a twist-boat conformation (Fig. 1) with the α-phenylethyl group occupying a bow position. All dimensions are normal.

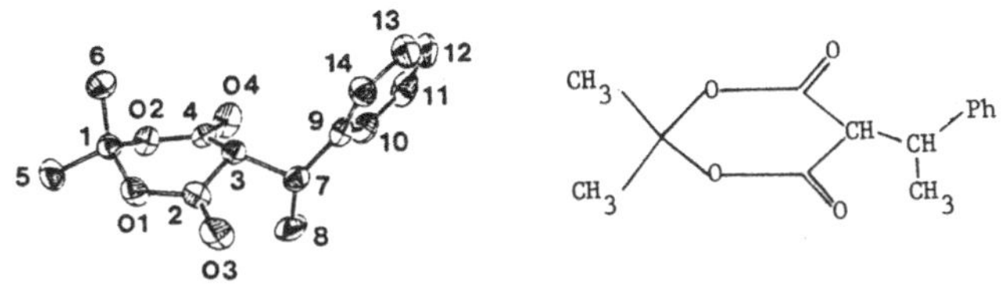

Fig. 1. A view of $C_{14}H_{16}O_4$.

METHYL (2R)-3-CAMPHANOYLAMINO-2-FLUOROPROPANOATE
$C_{14}H_{20}FNO_5$

D. GANI, P.B. HITCHCOCK and D.W. YOUNG, 1985. *J. Chem. Soc. Perkin Trans. 1*, 1363-1372.

Orthorhombic, $P2_12_12_1$, a = 7.375, b = 9.522, c = 21.793 Å, Z = 4. Cu radiation, R = 0.101 for 1132 reflections.

The crystal structure analysis together with other evidence showed that the molecule has the absolute configuration shown (Fig. 1).

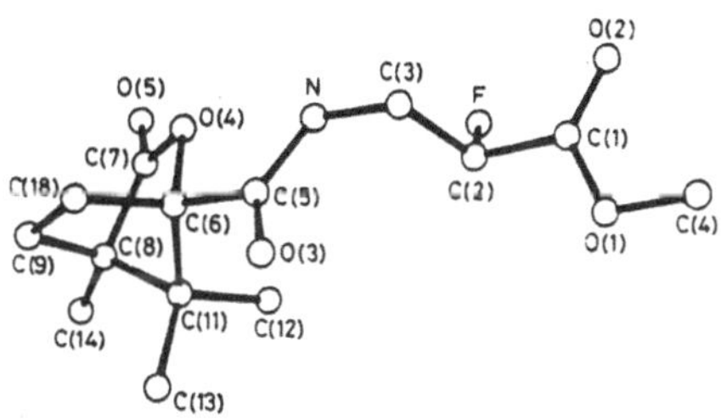

Fig. 1. Perspective drawing of a molecule of $C_{14}H_{20}FNO_5$.

19-METHYL-3,6,9,12,15-PENTAOXA-19,20-DIAZABICYCLO[15.2.1]EICOSA-1(20),17-DIENE-2,16-DIONE
$C_{14}H_{20}N_2O_7$

J. ELGUERO, P. NAVARRO, M.I. RODRIGUEZ-FRANCO, F.H. CANO, C. FOCES-FOCES and A. SAMAT, 1985. *J. Chem. Res.*, S312-313, M3401-3418.

Monoclinic, $P2_1/c$, a = 10.6687, b = 18.7967, c = 8.7977 Å, β = 114.247°, Z = 4. Cu radiation, R = 0.067 for 2253 reflections.

The molecule (Fig. 1) has a pseudosymmetry plane through C(18) and O(9). The cavity is a hole of ellipsoidal shape which is flattened in the C(18)...O(9) direction. The ellipse has axes of 2.375 and 2.950 Å. The molecule could be expected to interact with ions such as Li^+ or Na^+.

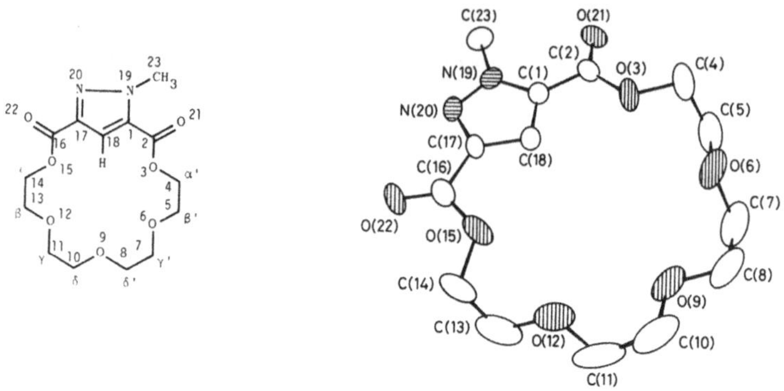

Fig. 1. The molecular skeleton and an ORTEP plot of $C_{14}H_{20}N_2O_7$.

METHYL 3-((3aα,5β,7aα)OCTAHYDRO-5,7a-DIMETHYL-2,4-DIOXOBENZOFURAN-3a-YL)PROPIONATE $C_{14}H_{20}O_5$

J.W. MUSKOPF and R.M. COATES, 1985. *J. Org. Chem.*, **50**, 69-76.

Monoclinic, $P2_1/c$, a = 14.072, b = 13.395, c = 7.803 Å, β = 106.41°, Z = 4. Cu radiation, R = 0.053 for 1250 reflections.

The γ-lactone is fused cis to the cyclohexanone ring (Fig. 1) and the methyl group adjacent to the ketone is trans to the angular propionate substituent.

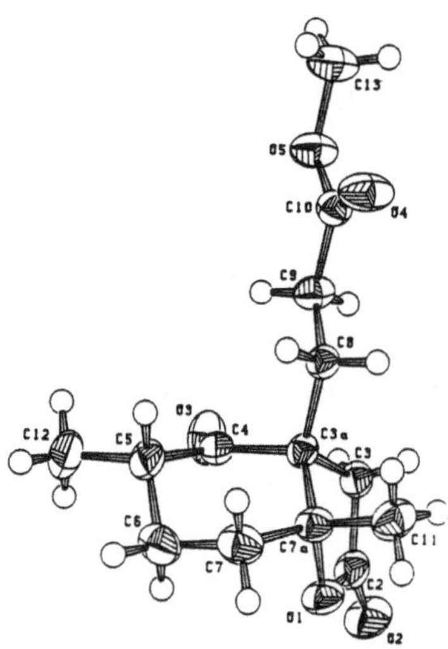

Fig. 1. An ORTEP drawing of a molecule of $C_{14}H_{20}O_5$.

cis-6-t-BUTYL-2-METHYL-2,3,5,6,7,8-HEXAHYDRO-4H-BENZO[b]PYRAN-4-ONE $C_{14}H_{22}O_2$

R. DOLMAZON, S. GELIN, R. FAURE and H. LOISELEUR, 1985. *Acta Cryst.*, **C41**, 1367-1369.

Monoclinic, $P2_1/c$, a = 11.120, b = 6.087, c = 22.670 Å, β = 121.16°, Z = 4. Cu radiation, R = 0.063 for 1734 reflections.

In the molecule (Fig. 1) the heterocyclic dihydropyranic ring has an approximate sofa conformation. The fused cyclohexenyl ring has a half-chair form which is severely flattened towards the sofa form. The bond lengths and angles in the molecule are normal. No unusually short intermolecular distances are reported.

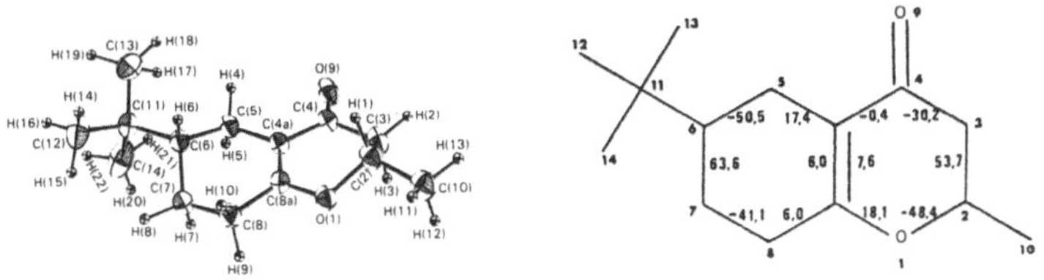

Fig. 1. $C_{14}H_{22}O_2$: views of the molecule showing the atomic numbering and the endocyclic torsion angles.

3-BENZYLIDENEPHTHALIDE
$C_{15}H_{10}O_2$

Y. WANG, M.C. CHENG, L.C. LIN, V.C. KOH and M. YANG, 1985. *Acta Cryst.*, C**41**, 924-926.

Monoclinic, $P2_1/n$, a = 15.166, b = 8.002, c = 19.013 Å, β = 103.64°, Z = 8. Mo radiation, R = 0.030 for 1890 reflections.

There are two molecules in the asymmetric unit. Both are in the Z form with the phenyl ring trans to the benzene ring of the phthalide (Fig. 1). The molecule is approximately planar with an average C-C distance of 1.389(6) Å for the benzene rings. The C=C and C=O distances average 1.328 and 1.202 Å respectively. The dihedral angle between the phthalide plane and the benzyl group is 7.1(1)°. The angle between the two molecular planes is 60.9(1)°.

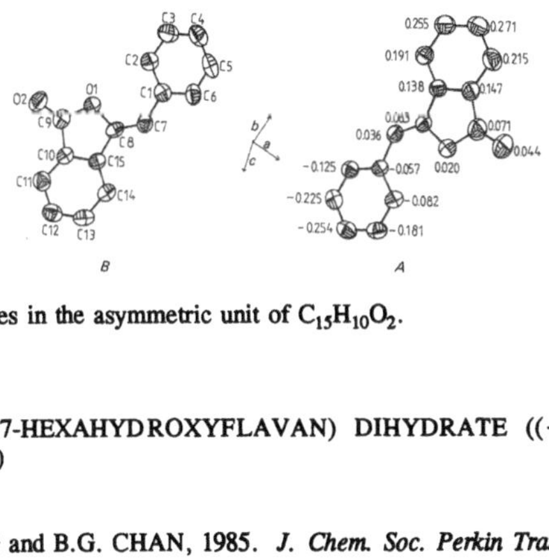

Fig. 1. The two molecules in the asymmetric unit of $C_{15}H_{10}O_2$.

((2R,3S,4R)-(+)-3,3',4,4',5,7-HEXAHYDROXYFLAVAN) DIHYDRATE ((+)-2,3-trans-3,4-trans-LEUCO-CYANIDIN DIHYDRATE)
$C_{15}H_{14}O_7,2(H_2O)$

L.J. PORTER, R.Y. WONG and B.G. CHAN, 1985. *J. Chem. Soc. Perkin Trans. 1*, 1413-1418.

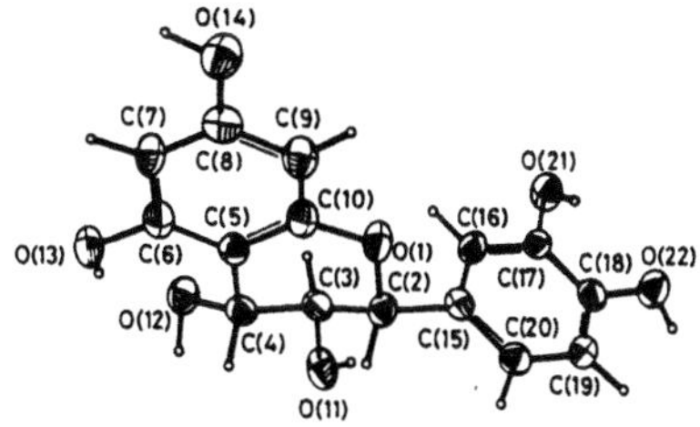

Fig. 1. Perspective drawing of a molecule of $C_{15}H_{14}O_7$.

Orthorhombic, $P2_12_12_1$, a = 4.862, b = 15.796, c = 19.362 Å, Z = 4. Cu radiation, R = 0.031 for 1247 reflections.

The water molecules are involved in a hydrogen bonding system in the crystal. A full discussion of the structure is given in the paper. The molecule is shown in Fig. 1.

8aβ-BROMO-5aα,5,6,7,8,8a-HEXAHYDRO-1,7α-DIHYDROXY-8α-METHOXYCARBONYLXANTHONE MONOHYDRATE
$C_{15}H_{15}BrO_6,H_2O$

G. FERGUSON, B. KAITNER, J. GILMORE, V.O.T. OMUARU and W.B. WHALLEY, 1985. *J. Chem. Soc. Perkin Trans. 1*, 1343-1347.

Triclinic, $P\bar{1}$, a = 8.648, b = 17.597, c = 20.309 Å, α = 94.36, β = 91.39, γ = 93.41°, Z = 8. Mo radiation, R = 0.050 for 2939 reflections.

The crystals are centrosymmetric and the molecule (Fig. 1) and its enantiomer are present equally in the crystal lattice. Aromatic ring A is planar; ring B has a C(9) envelope conformation; ring C has a chair conformation. In the crystal the water molecule takes part in three O-H...O hydrogen bonds and is firmly anchored in the lattice. There is an intramolecular hydrogen bond between hydroxy O(1)-H and adjacent carbonyl oxygen O(2) (O...O 2.604 Å).

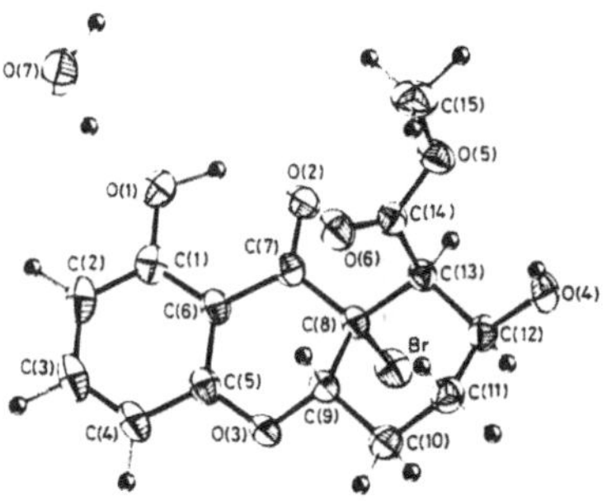

Fig. 1. Perspective drawing of a molecule of $C_{15}H_{15}BrO_6,H_2O$ in the crystal.

1,3-DIHYDRO-4-((2S)-2,5-DIHYDRO-2-FURYL)-3-METHYL-1-p-TOLYL-2H-IMIDAZOLE-2-THIONE
$C_{15}H_{16}N_2OS$

E. MORENO, A. LOPEZ-CASTRO and R. MARQUEZ, 1985. *Acta Cryst.*, **C41**, 767-769.

Monoclinic, $P2_1$, a = 6.033, b = 7.488, c = 15.563 Å, β = 99.55°, Z = 2. Mo radiation, R = 0.047 for 1695 reflections.

In the molecule (Fig. 1) the phenyl group (A) is planar with a mean C-C bond length of 1.384(5) Å. The dihydroimidazole ring (B) is also planar with substituents S(1), C(11), C(12) and C(4) displaced from its least-squares plane by 0.037(2), -0.069(5), 0.050(5) and -0.079(4) Å respectively. The furanose ring (C) is also nearly planar owing to the presence of the C(13)=C(14) double bond. The pyranose ring pucker corresponds to a flattened $^3T_2$ conformation. The dihedral angles between the rings A/B and C/B are 55.2(2) and 99.1(2)° respectively. Intermolecular distances are consistent with van der Waals contacts.

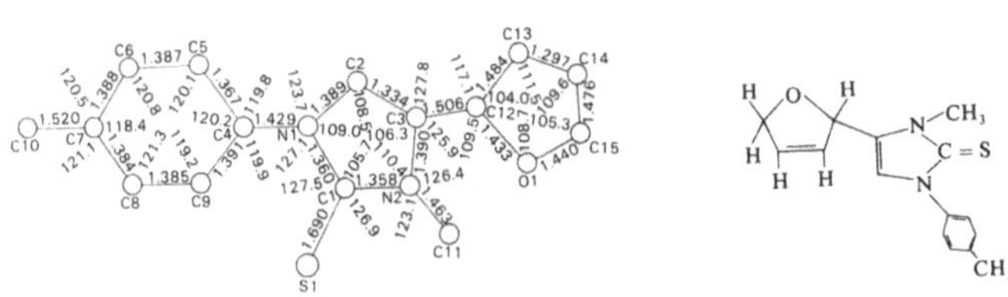

Fig. 1. Bond lengths and angles in the $C_{15}H_{16}N_2OS$ molecule.

4,6-DIMETHYL-5-(N-METHYLPHENYLSULPHONYLAMINO-N)-4-VINYL-1,3-DIOXAN-2-ONE
$C_{15}H_{19}NO_5S$

P.J. MAURER, C.G. KNUDSEN, A.D. PALKOWITZ and H. RAPOPORT, 1985. *J. Org. Chem.*, **50**, 325-332.

Orthorhombic, $P2_12_12_1$, a = 8.2136, b = 11.1000, c = 18.2287 Å, Z = 4. Mo radiation, R = 0.0281 for 1269 reflections.

The crystal structure analysis established the stereochemistry of this molecule (Fig. 1) and a group of related molecules.

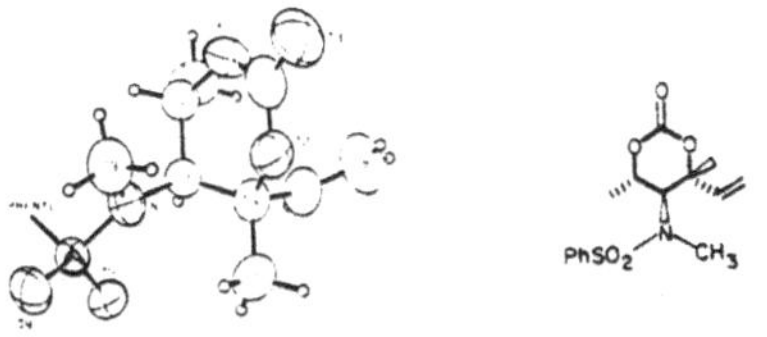

Fig. 1. ORTEP drawing of a molecule of $C_{15}H_{19}NO_5S$, with the phenyl group omitted for clarity.

2-(1,3,4,5,6,7-HEXAHYDRO-1-ISOBENZOFURANYL)CYCLOHEXANECARBOXYLIC ACID
$C_{15}H_{20}O_4$

H. BAYER, M. SCHREYER, A. GIEREN and V. LAMM, 1985. *Chem. Ber.*, **118**, 3413-3418.

Triclinic, $P\bar{1}$, a = 6.679, b = 9.932, c = 11.470 Å, α = 108.61, β = 106.20, γ = 95.03°, Z = 2. Mo radiation, R = 0.046.

The molecule is formed from two molecules of hexahydrophthalic anhydride with loss of one molecule of Co_2. It crystallizes as centrosymmetric hydrogen-bonded dimers (O...O 2.623(2) Å) (Fig. 1). Chiral centres C(1), C(9), C(14) have SSS or RRR configuration. The five-membered ring and its substituents C(5) and C(8) are planar and the attached cyclohexene ring has a half-chair form. The lone cyclohexane ring is in chair form.

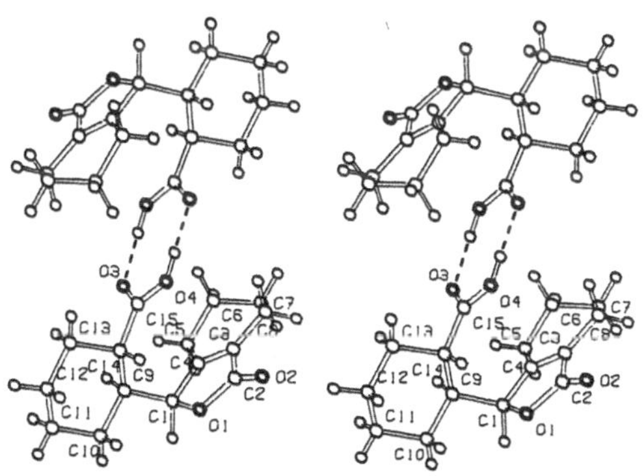

Fig. 1. A stereoview of $C_{15}H_{20}O_4$.

8-ACETOXY-7-AZA-6-ETHOXYCARBONYL-7-FORMYL-3,4-DIHYDROXY-3,4-O-ISOPROPYLIDENE-2-OXA-BICYCLO[3.3.0]OCTANE
$C_{15}H_{21}NO_8$

C.W. HOLZAPFEL, J.M. KOEKEMOER, G.J. KRUGER and M.S. VAN DYK, 1985. *S. Afr. J. Chem.*, **38**, 207-211.

Monoclinic, $P2_1$, a = 12.65, b = 7.36, c = 9.55 Å, β = 102.0°, Z = 2. Mo radiation, R = 0.084 for 1657 reflections.

The molecule, a 11-deoxy-11-azaprostaglandin synthon, is shown to have the dioxolane and aza-rings cis-fused on the same side to the central furanose ring (Fig. 1). The conformations of the aza-ring, the central furanose ring and the dioxolane ring are $^{2}T_{1}/^{2}E$, E_1 and $^{4}T_5$ respectively.

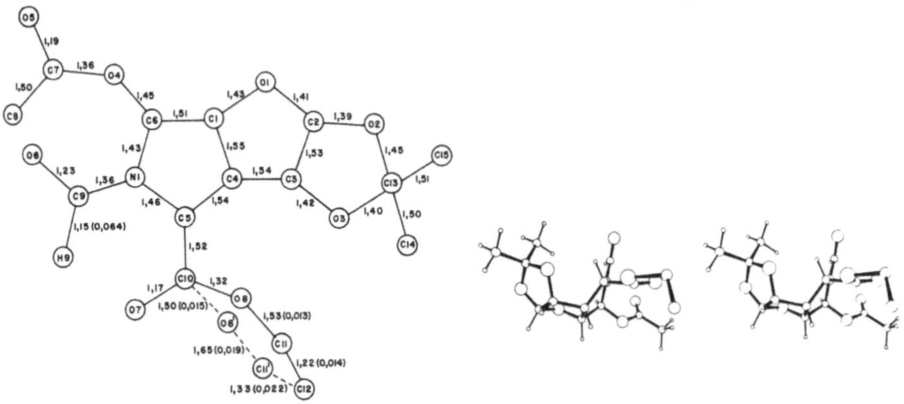

Fig. 1. The $C_{15}H_{21}NO_8$ molecule and a stereoview; the ethoxycarbonyl fragment is disordered.

(2R,3S)-1-(1-(t-BUTYLOXYCARBONYLAMINO)-2-PHENYLETHYL)OXIRANE
$C_{15}H_{21}NO_3$

B.E. EVANS, K.E. RITTLE, C.F. HOMNICK, J.P. SPRINGER, J. HIRSHFIELD and D.F. VEBER, 1985. *J. Org. Chem.*, **50**, 4615-4625.

Monoclinic, $P2_1/n$, a = 6.205, b = 11.932, c = 20.039 Å, β = 94.14°, Z = 4. R = 0.046 for 1827 reflections.

The crystal structure analysis showed that the molecule (Fig. 1) was the 2R,3S isomer.

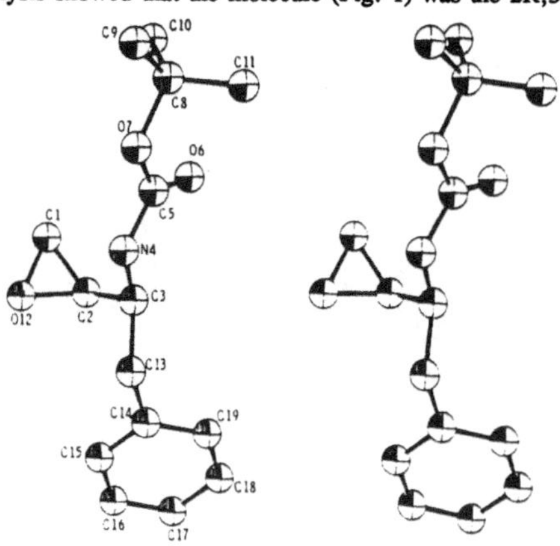

Fig. 1. A stereoview of $C_{15}H_{21}NO_3$ with hydrogen atoms omitted.

10-ACETOXY-4,7-DIMETHYL-8-OXOTETRAHYDROFURAN[2,3-i]-2-OXA-PERHYDRONAPHTHALENE (13-ACETOXY-6,10-DIMETHYL-2,12-DIOXATRICYCLO[7.4.0.0^{1,5}]TRIDECAN-11-ONE)
$C_{15}H_{22}O_5$

XUAN-DE LUO, H.J.C. YEH, A. BROSSI, J.L. FLIPPEN-ANDERSON and R. GILARDI, 1985. *Heterocycles*, 23, 881-887.

Monoclinic, *C*2, a = 29.181, b = 8.152, c = 12.648 Å, β = 98.65°, Z = 8. Cu radiation, R = 0.040 for 2187 reflections [see also the following Report].

There are two independent similar molecules in the asymmetric unit; a view of one of them is in Fig. 1. The five-membered ring is a flattened envelope with C(1) being the out-of-plane atom. The C_6 ring has a normal chair conformation. The six-membered lactone ring is flattened in the vicinity of O(2), but the flattening is more pronounced in one molecule than the other (compare torsion angles C(10)-O(2)-C(11)-C(4) -4.4° and C(9)-C(10)-O(2)-C(11) -2.3°, O(2)-C(11)-C(4)-C(5) -23.1° in one molecule with the corresponding values (27.4, -11.8 and 50.9° respectively) in the other).

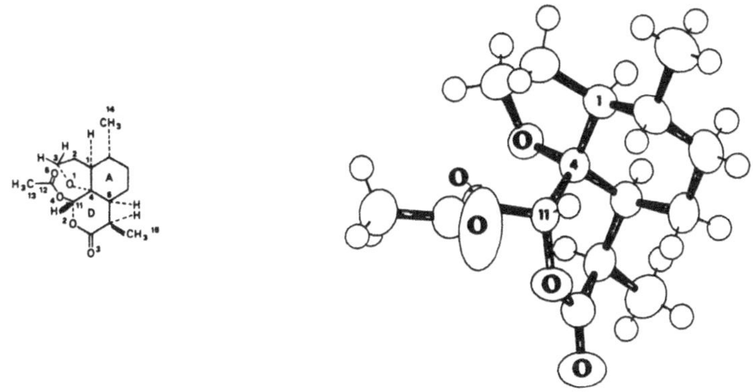

Fig. 1. The $C_{15}H_{22}O_5$ structure.

13-ACETOXY-6,10-DIMETHYL-2,12-DIOXATRICYCLO[7.4.0.0^{1,5}]TRIDECAN-11-ONE
$C_{15}H_{22}O_5$ (I)

12,14-EPOXY-11-HYDROXY-4,8,12-TRIMETHYL-2,13-DIOXATRICYCLO[7.4.1.0^{5,14}]TETRADECAN-3-ONE
$C_{15}H_{22}O_5$ (II)

A.J. LIN, D.L. KLAYMAN, J.M. HOCH, J.V. SILVERTON and C.F. GEORGE, 1985. *J. Org. Chem.*, **50**, 4504-4508.

I. Monoclinic, *A*2, a = 12.626, b = 8.142, c = 29.177 Å, β = 98.65°, Z = 8. Cu radiation, R = 0.040 for 3228 reflections [see also the preceding Report].

II. Monoclinic, *P*2$_1$, a = 5.465, b = 13.752, c = 9.831 Å, β = 101.71°, Z = 2. Cu radiation, R = 0.065 for 966 reflections.

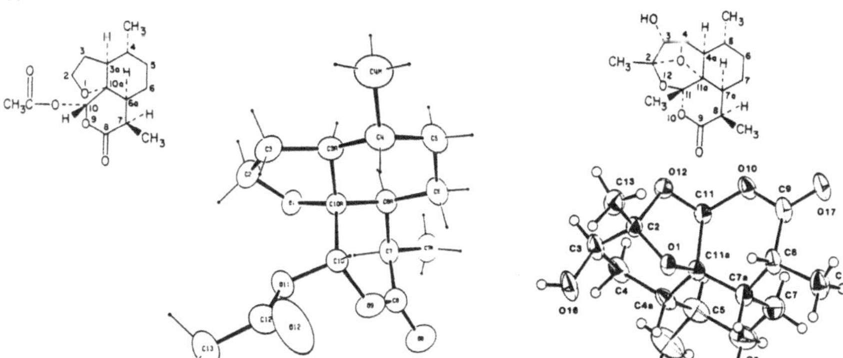

Fig. 1. Drawings of one of the independent molecules in the crystals of $C_{15}H_{22}O_5$ (I) and of $C_{15}H_{22}O_5$ (II) (right).

In (I) there are two independent molecules in the asymmetric unit. The crystal structure analyses identified the decomposition products as (I) and (II) (Fig. 1).

(2-METHOXY-1,3-XYLYL)-15-CROWN-4-TRIAQUA(TRINITRATO)DYSPROSIUM(III)
$C_{15}H_{22}O_5,H_6DyN_3O_{12}$

G. TOMAT, G. VALLE, P. DI BERNARDO and P.L. ZANONATO, 1985. *Inorg. Chim. Acta*, **110**, 113-117.

Triclinic, $P\bar{1}$, a = 12.312, b = 10.193, c = 10.264 Å, α = 95.1, β = 94.5, γ = 92.5°, Z = 2. Mo radiation, R = 0.042 for 3829 reflections.

The Dy atom is bonded to three water ligands and to three bidentate NO_3 ligands (Fig. 1). The coordination geometry is tricapped trigonal prismatic ($\bar{6}2m$) with one oxygen of each NO_3 group at an equatorial position. Bond lengths about Dy are: OH_2 2.319-2.374(7), O(N)(equatorial) 2.455-2.495(6), (other) 2.420-2.436(6) Å. The crown-ether is not bonded to Dy. Four of the five ether O atoms are hydrogen-bonded to H_2O (2.68-2.76 Å) and there are also inter-complex hydrogen bonds (2.74, 2.86 Å).

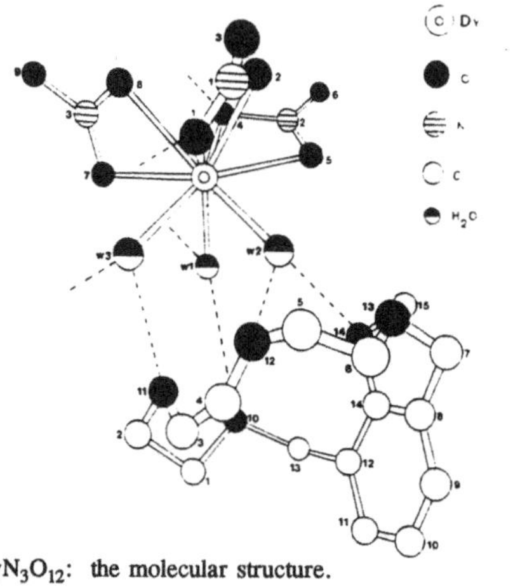

Fig. 1. $C_{15}H_{22}O_5,H_6DyN_3O_{12}$: the molecular structure.

[1]BENZOPYRANO[4,3-c][1]BENZOPYRAN-5,11-DIONE (DIBENZONAPHTHYRONE)
$C_{16}H_8O_4$ (I)

1,3,7,9-TETRA-t-BUTYL[1]BENZOPYRANO[4,3-c][1]BENZOPYRAN-5,11-DIONE
$C_{32}H_{40}O_4$ (II)

1,7-DI-t-BUTYL-3,9-DIMETHYL[1]BENZOPYRANO[4,3-c][1]BENZOPYRAN-5,11-DIONE
$C_{26}H_{28}O_4$ (III)

1,7-DI-t-BUTYL-3,9-DIMETHYL[1]BENZOPYRANO[4,3-c][1]BENZOPYRAN-5,11-DIONE
$C_{26}H_{28}O_4$ (IV)

1,7-DI-t-BUTYL-3,9-DIMETHOXY[1]BENZOPYRANO[4,3-c][1]BENZOPYRAN-5,11-DIONE
$C_{26}H_{28}O_6$ (V)

1,7-DI-t-BUTYL-3,9-DIMETHOXY[1]BENZOPYRANO[4,3-c][1]BENZOPYRAN-5,11-DIONE
$C_{26}H_{28}O_6$ (VI)

H.-D. BECKER, C.L. RASTON, B.W. SKELTON and A.H. WHITE, 1985. *Aust. J. Chem.*, **38**, 97-110.

I. Monoclinic, $P2_1/c$, a = 4.621, b = 9.569, c = 12.776 Å, β = 95.29°, Z = 2. Mo radiation, R = 0.071 for 375 reflections.

II. Monoclinic, $P2_1/a$, a = 18.824, b = 11.677, c = 14.977 Å, β = 120.92°, Z = 4. Mo radiation, R = 0.060

for 1580 reflections.

III. Triclinic, $P\bar{1}$, a = 10.442, b = 8.307, c = 6.532 Å, α = 102.29, β = 104.49, γ = 91.78°, Z = 1. Mo radiation, R = 0.043 for 994 reflections.

IV. Triclinic, $P\bar{1}$, a = 10.927, b = 7.383, c = 7.309 Å, α = 110.44, β = 90.19, γ = 105.19°, Z = 1. Cu radiation, R = 0.071 for 813 reflections.

V. Orthorhombic, *Pcab*, a = 10.923, b = 12.127, c = 17.308 Å, Z = 4. Mo radiation, R = 0.080 for 527 reflections.

VI. Triclinic, $P\bar{1}$, a = 9.885, b = 9.150, c = 6.822 Å, α = 79.20, β = 67.38, γ = 80.32°, Z = 1. Cu radiation, R = 0.045 for 703 reflections.

In the parent compound, (I), the asymmetric unit consists of one half of the molecule and the structure is centrosymmetric (Fig. 1). The two carbonyl groups are folded out of the plane of the central C=C double bond in opposite directions and, consequently, the two aromatic rings are parallel. The same mode of 'anti'-oriented folding of the CO groups is observed in centrosymmetric structures (III) to (V), although the degree of folding varies considerably. Molecules of (II), however, are syn-oriented with respect to the plane of the ethylenic double bond. Consequently, the two aromatic rings are twisted out of the plane of the ethylene double bond in opposite directions, the dihedral angle between the planes of the two benzo rings being 25°. In structures (I)-(IV) and (VI) the H(2)...O(7') distances are 2.11(1), 2.10(7) and 2.15(8), 2.17(3), 2.16(6) and 2.08(5) Å respectively. The intermolecular packing in (I)-(VI) is such that adjacent molecules are aligned parallel along the central C=C double bond. Only for compounds (I) and (VI) is intermolecular overlap substantial and suggestive of possible charge-transfer interaction.

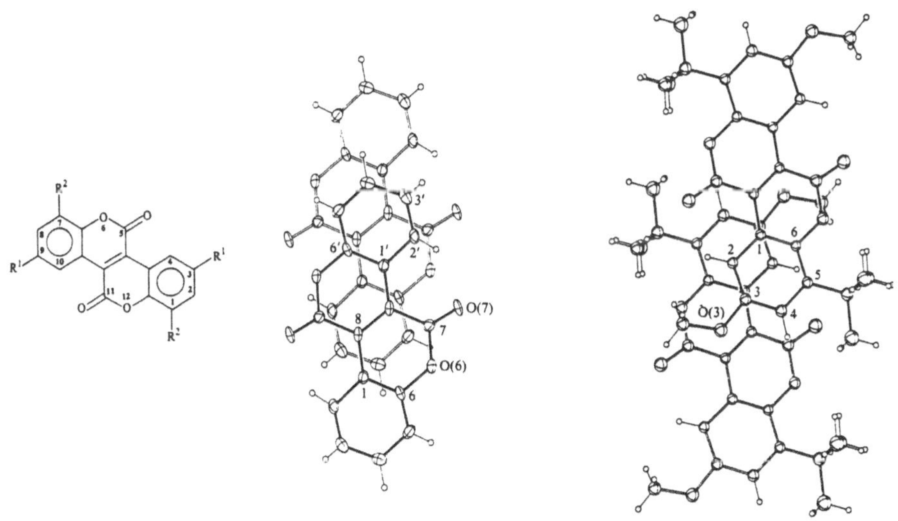

Fig. 1. The molecular skeleton (left), and molecular overlap in $C_{16}H_8O_4$ (centre) and $C_{26}H_{28}O_6$ (VI) (right).

(E)-3-(2-OXOBENZOFURAN-3(2H)-YLIDENE)BENZOFURAN-2(3H)-ONE (ISOXINDIGO)
$C_{16}H_8O_4$ (I)

(E)-5,5',7,7'-TETRA-t-BUTYL-3-(2-OXOBENZOFURAN-3(2H)-YLIDENE)BENZOFURAN-2(3H)-ONE(5,5',7,-7'-TETRA-t-BUTYLISOXINDIGO)
$C_{32}H_{40}O_4$ (II)

(E)-7,7'-DI-t-BUTYL-5,5'-DIMETHOXY-3-(2-OXOBENZOFURAN-3(2H)-YLIDENE)BENZOFURAN-2(3H)-ONE (7,7'-DI-t-BUTYL-5,5'-DIMETHOXYISOXINDIGO) (α-form)
$C_{26}H_{28}O_6$ (III)

(E)-7,7'-DI-t-BUTYL-5,5'-DIMETHOXY-3-(2-OXOBENZOFURAN-3(2H)YLIDENE)BENZOFURAN-2(3H)-ONE (7,7'-DI-t-BUTYL-5,5'-DIMETHOXYISOXINDIGO) (β-form)
$C_{26}H_{28}O_6$ (IV)

H.-D. BECKER, C.L. RASTON, B.W. SKELTON and A.H. WHITE, 1985. *Aust. J. Chem.*, **38**, 85-96.

I. Monoclinic, $P2_1/n$, a = 17.692, b = 5.690, c = 5.693 Å, β = 100.52°, Z = 2. Mo radiation, R = 0.067 for 385 reflections.

II. Monoclinic, $P2_1/n$, a = 16.402, b = 12.313, c = 13.913 Å, β = 96.75°, Z = 4. Mo radiation, R = 0.088 for 1261 reflections.

III. Orthorhombic, *Pcab*, a = 10.449, b = 12.477, c = 17.825 Å, Z = 4. Mo radiation, R = 0.058 for 959 reflections.

IV. Monoclinic, $P2_1/n$, a = 6.828, b = 15.464, c = 10.770 Å, β = 98.58°, Z = 2. Mo radiation, R = 0.065 for 598 reflections.

In (I), (III) and (IV) the molecules are disposed about crystallographic symmetry centres (Fig. 1). The fused ring skeletons are essentially planar in each of these. In (I) (Fig. 1) the central C=C bond is very short (1.27(2) Å) and there is a short intramolecular contact distance of 1.78(9) Å between H(2) and the carbonyl O atom, O(7'). Molecules in the crystal are arranged such that adjacent molecules have the central C=C double bonds aligned parallel. Molecules of (II) (Fig. 1) are twisted about the central C=C bond (1.34(2) Å) by 16.8° and the H(2')...O(7) intramolecular distance is 1.9(1) Å. The crystal structure is similar to that of (I). Polymorphs (III) and (IV) (Fig. 2) have the trans configuration. In both, the central C=C double bond is 1.36 Å and the intramolecular distances between carbonyl oxygen O(7) and aromatic hydrogen H(2') are 2.05(4) and 2.01(4) Å. Also, in both (III) and (IV) the methoxy group is coplanar with the aromatic ring and is oriented towards H(2). Packing is similar to that observed in (I) and (II)

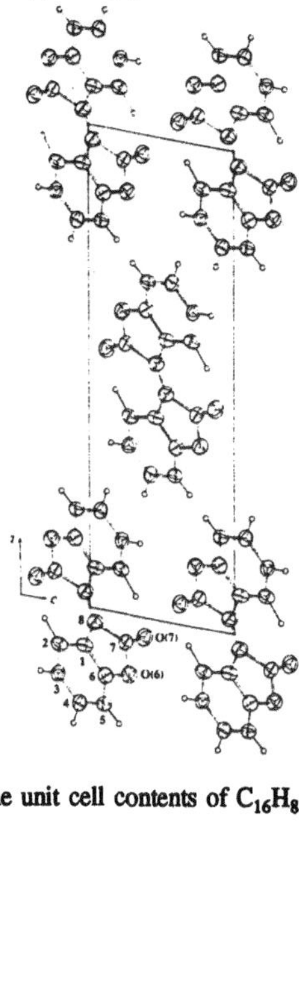

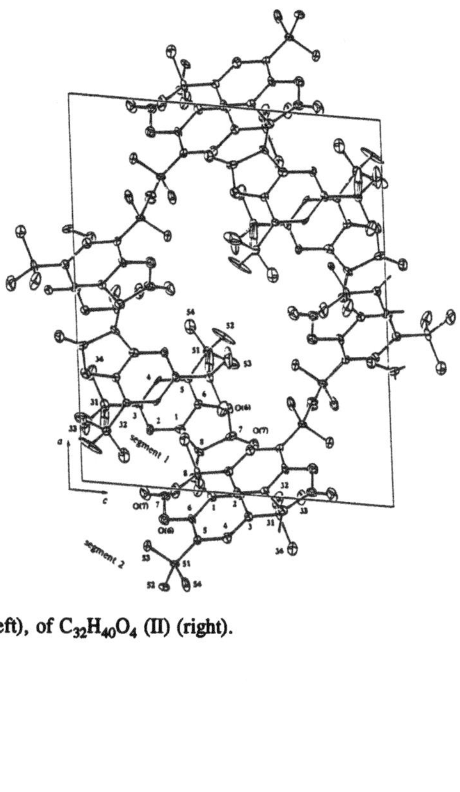

Fig. 1. The unit cell contents of $C_{16}H_8O_4$ (I) (left), of $C_{32}H_{40}O_4$ (II) (right).

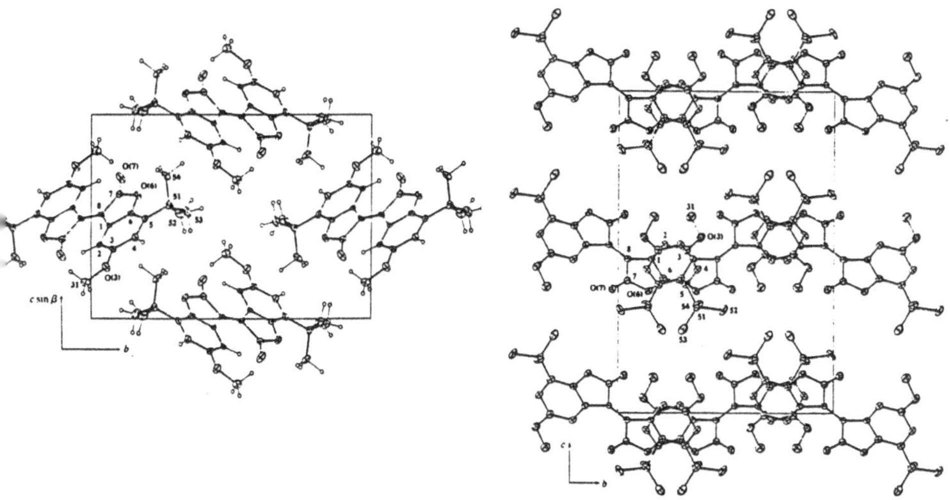

Fig. 2. The unit cell contents of $C_{26}H_{28}O_6$ (III) (left) and $C_{26}H_{28}O_6$ (IV) (right).

DISPIRO(6-ETHYL-7-OXABICYCLO[4.3.0]NONAN-8-ONE-2,2':4,2''-BIS(2,5-DIHYDROFURAN-5-ONE))
$C_{16}H_{16}O_6$

J. BIGORRA, J. FONT, R.M. ORTUNO, F. SANCHEZ-FERRANDO, F. FLORENCIO, S. MARTINEZ-CARRERA and S. GARCIA-BLANCO, 1985. *Tetrahedron*, **41**, 5589-5594.

Monoclinic, $P2_1/n$, a = 10.740, b = 13.293, c = 10.094 Å, β = 91.518°, Z = 4. Cu radiation, R = 0.045 for 1966 reflections.

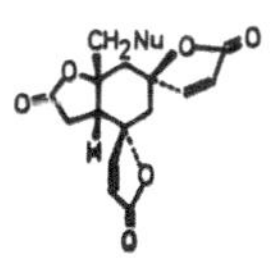

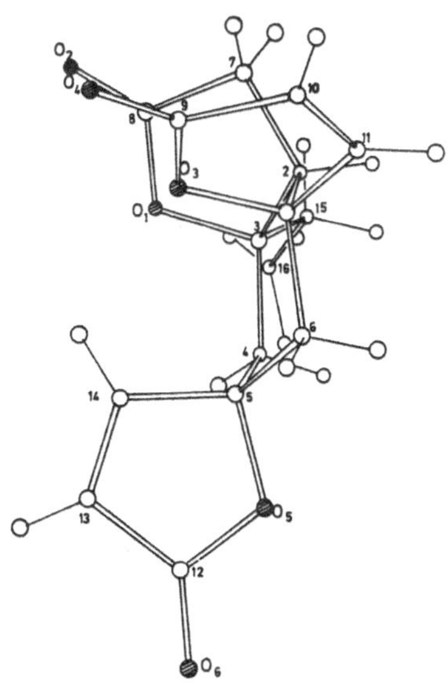

Fig. 1. The $C_{16}H_{16}O_6$ molecule.

This product of reaction of protoanemonin with lithium dimethylcuprate is shown in Fig. 1. The six-membered ring has a distorted flattened-chair conformation, the cis-fused butyrolactone has a half-chair conformation, while both spiranic butenolide rings are planar.

2-HYDROXY-1,3-XYLYL-18-CROWN-5
$C_{16}H_{24}O_6$ (I)

5-NITRO-2-HYDROXY-1,3-XYLYL-18-CROWN-5
$C_{16}H_{23}NO_8$ (II)

AMMONIUM 5-NITRO-1,3-XYLYL-18-CROWN-5-2-OLATE
$C_{16}H_{26}N_2O_8$ (III) $C_{16}H_{22}NO_8^-,H_4N^+$

C.M. BROWNE, G. FERGUSON, M.A. McKERVEY, D.L. MULHOLLAND, T. O'CONNOR and T. PARVEZ, 1985. *J. Am. Chem. Soc.*, **107**, 2703-2712.

I. Orthorhombic, $Pc2_1b$, a = 9.779, b = 18.943, c = 8.915 Å, Z = 4. Mo radiation, R = 0.053 for 759 reflections.

II. Monoclinic, $P2_1/c$, a = 10.503, b = 9.003, c = 18.924 Å, β = 101.98°, Z = 4. Mo radiation, R = 0.032 for 1189 reflections.

III. Monoclinic, $I2/a$, a = 20.476, b = 9.092, c = 21.211 Å, β = 108.92°, Z = 8. Mo radiation, R = 0.053 for 2635 reflections.

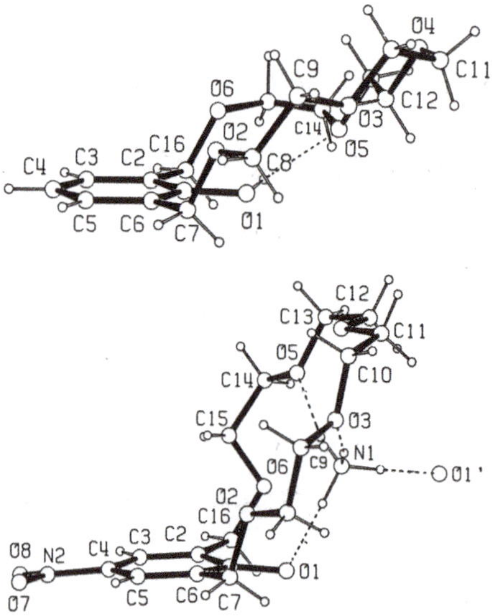

Fig. 1. Side view of the structures of $C_{16}H_{24}O_6$ (I) (left), $C_{16}H_{23}NO_8$ (II) (right) and $C_{16}H_{26}N_2O_8$ (III) (bottom).

The structures of (I) and (II) have an intra-annular relationship between the aromatic ring substituent and the macrocycle cavity but the orientation of the phenolic OH group differs. In (I) the angle between the aromatic ring plane and the mean plane of the macrocycle is 28°, in (II) this angle is 47° (Fig. 1). The hydrogen bond O(1)...O(5) 2.856(15), H(O1)...O(5) 2.05 Å in (I) is replaced by an intramolecular hydrogen bond O(1)...O(2) 2.856(15), H(O1)...O(2) 1.91 Å in (II). The macrocycle in (I) has an "opened-out" conformation, with torsion angles O(4)-C(12)-C(13)-O(5) 178°, O(5)-C(14)-C(15)-O(6) 1°. In both compounds the C-C torsion angles are removed from the usual coordinated 18-crown-6 gauche values, likewise the C-O torsion angles are not the usual anti. In the substituted compound (III) the C-C torsion angles are close to gauche, C-O close to anti, apart from C(2)-C(16)-O(6)-C(15)-70°. This conformation is a consequence of NH_4^+ being incorporated into the cavity with three hydrogen bonds, to phenoxide oxygen (N...O(1) 2.690(4)) and two ethereal oxygens (N...O(3) 2.884(4), N...O(5) 2.883(5) Å). The dihedral angle between aromatic and macrocycle planes is opened further than in (I) and (II), being 58° (Fig. 1). The fourth H of NH_4^+ is involved in an intermolecular H-bond to the phenoxy O of an adjacent molecule (N...O 2.781(4) Å), forming centrosymmetrically related dimers. In (III) the phenoxide rings stack with an interplanar spacing of 3.39 Å. Bond lengths and angles in all three structures are unexceptional. There is large thermal motion/disorder in (I), especially in the aromatic moiety.

(3RS,13SR)-3-BROMOACETOXY-13-TETRADECANOLIDE(4RS,14SR)-4-BROMOACETOXY-14-METHYL-1-OXACYCLOTETRADECAN-2-ONE
$C_{16}H_{27}BrO_4$ (I)

(3RS,13RS)-3-BROMOACETOXY-13-TETRADECANOLIDE(4RS,14RS)-4-BROMOACETOXY-14-METHYL-1-OXACYCLOTETRADECAN-2-ONE
$C_{16}H_{27}BrO_4$ (II)

M.N. PONNUSWAMY and J. TROTTER, 1985. *Acta Cryst.*, C**41**, 1109-1111.

I. Triclinic, $P\bar{1}$, a = 11.546, b = 14.478, c = 5.283 Å, α = 92.42, β = 96.93, γ = 92.81°, Z = 2. Mo radiation, R = 0.054 for 1101 reflections.

II. Triclinic, $P\bar{1}$, a = 11.868, b = 13.904, c = 5.482 Å, α = 90.51, β = 98.82, γ = 98.48°, Z = 2. Cu radiation, R = 0.057 for 2394 reflections.

Both isomeric molecules contain 14-membered rings (Fig. 1). In the cis or S,R isomer, (I), the molecule is disordered and the ring exhibits equal amounts of 3344 and 3434 conformations. In the trans isomer or S,S isomer, (II), the ring exhibits a 3355 conformation. In both molecules the 13-methyl and bulky 3-bromoacetoxy substituents lie in equatorial positions. In the cis isomer C=O(2) and C=O(4) are nearly parallel and the conformational change giving the trans isomer results in these bonds being nearly antiparallel. Bond lengths and angles in the molecule are generally close to normal values. No unusually short intermolecular contacts are reported.

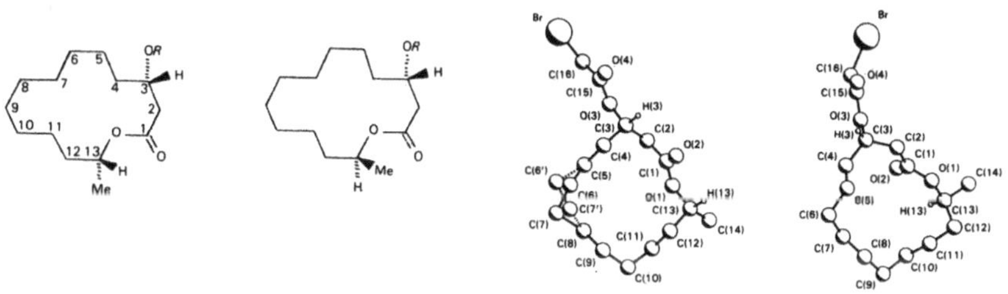

Fig. 1. $C_{16}H_{27}BrO_4$: structural formulae of (I) and (II) and views of the cis and trans isomers showing the nature of the disorder in the ring of I.

7-BENZYLOXYMETHYLBICYCLO[2.2.1]HEPT-5-ENE-2,3-DICARBOXYLIC ANHYDRIDE
$C_{17}H_{16}O_4$

L.M. TOLBERT, J.C. GREGORY and C.P. BROCK, 1985. *J. Org. Chem.*, **50**, 548.

Orthorhombic, $P2_12_12_1$, a = 6.2161, b = 9.0987, c = 26.0744 Å, Z = 4. Mo radiation, R = 0.043 for 1013 reflections.

The crystal structure analysis confirmed the identity of this adduct (Fig. 1).

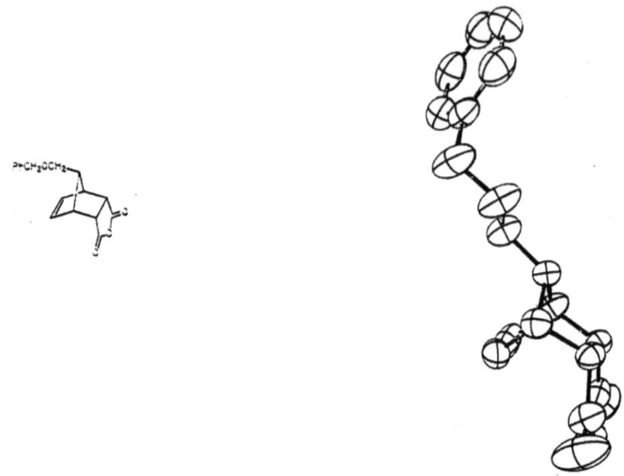

Fig. 1. A molecule of $C_{17}H_{16}O_4$.

cis-2-(4'-CHLOROPHENYL)-2-METHYL-5-PHENYL-1,3-DIOXANE
$C_{17}H_{17}ClO_2$ (I)

trans-2-(4'-CHLOROPHENYL)-2-METHYL-5-PHENYL-1,3-DIOXANE
$C_{17}H_{17}ClO_2$ (II)

K.J. SHEA and T.K. DOUGHERTY, 1985. *J. Org. Chem.*, **50**, 4439-4442.

I. Monoclinic, $P2_1/c$, a = 14.758, b = 7.761, c = 14.396 Å, β = 112.43°, Z = 4. Mo radiation, R = 0.043 for 2130 reflections.

II. Monoclinic, $P2_1/n$, a = 8.164, b = 10.433, c = 17.56 Å, β = 91.35°, Z = 4. Mo radiation, R = 0.045 for 2034 reflections.

The crystal structure analyses established the cis and trans configurations of the two isomers (Fig. 1). In (I) the planes of the two aryl groups are orthogonal but this is quite different to the arrangement in (II).

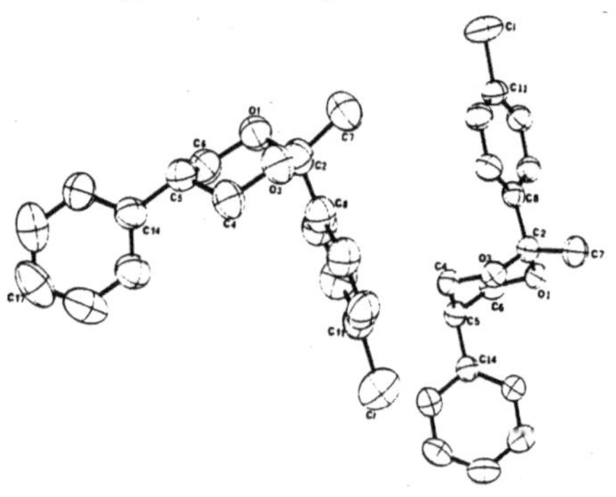

Fig. 1. ORTEP drawings of molecules of $C_{17}H_{17}ClO_2$: cis (I) on the left with trans (II) on the right.

1,9a-EPOXY-1,4,4a,6,7,8,8a,9,9a,10-DECAHYDRO-4a,5,8a-TRIMETHYL-6-OXO-2(3H)-ANTHRACENONE
$C_{17}H_{22}O_3$

L.A. PAQUETTE, D.T. BELMONT and YEH-LEH HSU, 1985. *J. Org. Chem.*, **50**, 4667-4672.

Monoclinic, $P2_1/n$, a = 13.191, b = 9.107, c = 13.438 Å, β = 108.63°, Z = 4. Mo radiation, R = 0.029 for 1175 reflections.

The molecule (Fig. 1) is highly concave; the R dihedral angle between plane O(1)-C(1)-C(2)-C(14) in the first ring and plane O(2)-C(7)-C(8)-C(9) in the third ring is 70.2°. Both angular methyl groups lie on the convex side of the molecule.

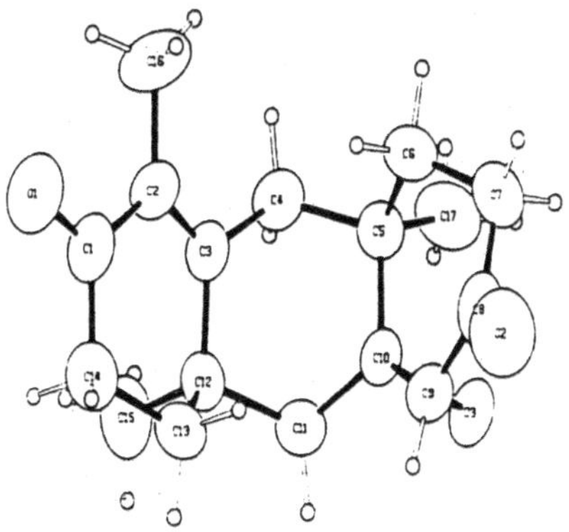

Fig. 1. ORTEP diagram of $C_{17}H_{22}O_3$.

ADAMANTANE-2-SPIRO-3'-(6'-t-BUTYL-4'-METHOXY-1',2',5'-TRIOXANE)
$C_{17}H_{28}O_4$

C.W. JEFFORD, J. BOUKOUVALAS, S. KOHMOTO and G. BERNARDINELLI, 1985. *Tetrahedron,* **41**, 2081-2088.

Orthorhombic, $Pna2_1$, a = 22.033, b = 6.3679, c = 11.502 Å, Z = 4. Mo radiation, R = 0.038 for 605 reflections.

It is shown that the trioxane ring has a chair conformation with the OMe substituent axial at C(2); the t-butyl group is equatorial.

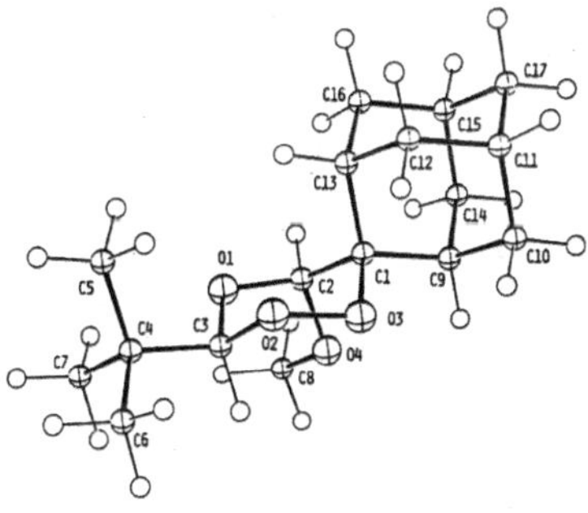

Fig. 1. The structure and stereochemistry of $C_{17}H_{28}O_4$.

cis-transoid-cis-6,11-DIOXATRICYCLO[7.3.0.$0^{4,8}$]DODEC-2-EN-5,5,7,7,10,10,12,12-OCTACARBONITRILE
$C_{18}H_6N_8O_2$

A. RODRIGUEZ-ROLDAN, A. MONGE and E. GUTIERREZ-PUEBLA, 1985. *Acta Cryst.*, **C41**, 1128-1130.

Monoclinic, $P2_1/c$, a = 6.771, b = 16.149, c = 16.407 Å, β = 100.08°, Z = 4. Mo radiation, R = 0.050 for 824 reflections.

In this adduct (Fig. 1) formed from one benzene molecule and two molecules of tetracyanoethylene oxide, the benzene ring has lost its aromatic character. It is nonplanar with deviations of up to 0.243(9) Å from the least-squares mean ring plane and the bond lengths in the ring range from 1.316(16) to 1.571(12) Å. The fused five-membered rings are also nonplanar and both have distorted half-chair conformations. The three bonds in these rings at the fusion with the six-membered ring show evidence of steric strain present in the adduct and are elongated with lengths ranging from 1.558(12) to 1.598(12) Å.

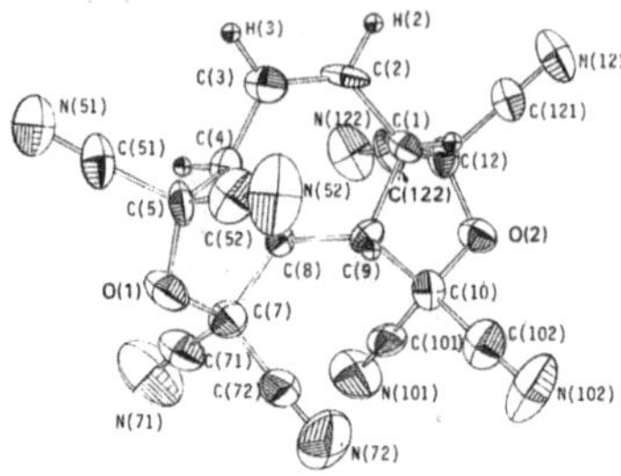

Fig. 1. A view of the $C_{18}H_6N_8O_2$ molecule showing the atomic numbering.

(-)-5,6:9,10-DIBENZO-4,11-DIOXATRICYCLO[6.4.0.0^{2,7}]DODECADIEN-3,12-DIONE (anti-(3S,3'S,4S,4'S)-COUMARIN DIMER (HEAD TO HEAD))
$C_{18}H_{12}O_4$

K. SAIGO, N. YONEZAWA, K. SEKIMOTO, M. HASEGAWA, K. UENO and H. NAKANISHI, 1985. *Bull. Chem. Soc. Jpn.*, **58**, 1000-1005.

Orthorhombic, $P2_12_12_1$, a = 8.711, b = 12.493, c = 12.728 Å, Z = 4. Mo radiation, R = 0.055 for 1593 reflections.

The structure and absolute stereochemistry are shown in Fig. 1. The cyclobutane ring is puckered with a 32.5° fold angle.

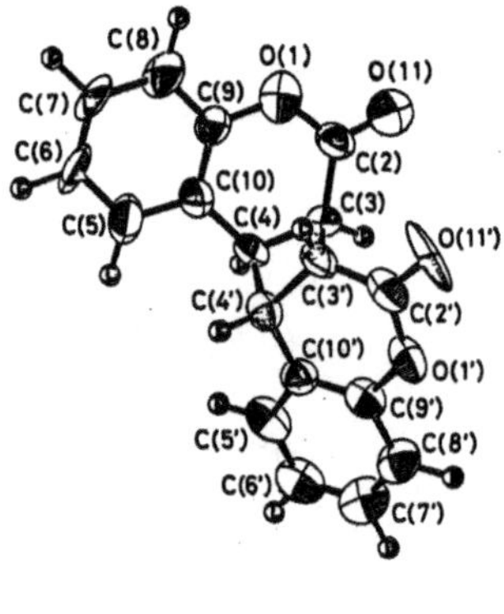

Fig. 1. A view of $C_{18}H_{12}O_4$.

5-BENZYLIDENE-4,5-DIHYDRO-4-OXO-3-PHENYLFURAN-2-CARBOXYLIC ACID METHANOL SOLVATE
$C_{18}H_{12}O_4,CH_4O$

G. KOLLENZ, A. BIEDERMANN, K. PETERS, E.-M. PETERS and H.G. VON SCHNERING, 1985. *J. Chem. Res.*, S288-289, M2976-2994.

Triclinic, $P\bar{1}$, a = 1035.9, b = 1405.2, c = 608.5 Å, α = 98.69, β = 101.32, γ = 108.33°, Z = 2. Mo radiation, R = 0.037 for 1885 reflections.

The structure and identity of this hydrolysis product was established by the crystal structure analysis to be that shown (Fig. 1). The furanone ring is nearly planar.

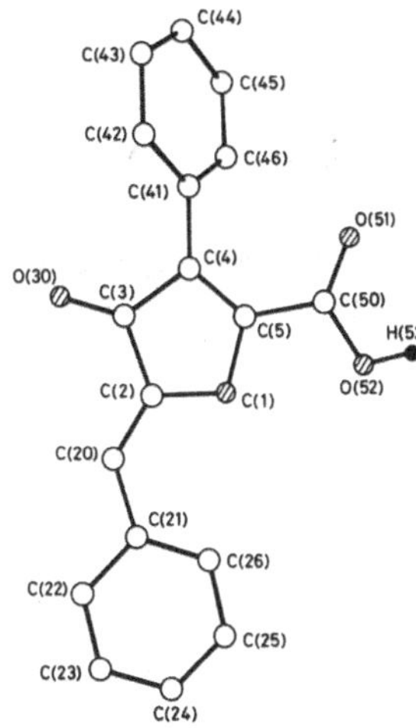

Fig. 1. Perspective drawing of a molecule of $C_{18}H_{12}O_4,CH_4O$.

(1R)-1-(2-HYDROXYPHENYL)-2,2,2-TRICHLOROETHYL-(1S)-CAMPHANATE
$C_{18}H_{19}Cl_3O_5$ (I)

(1R)-1-(2-HYDROXY-3-METHYLPHENYL)-2,2,2-TRICHLOROETHYL-(1S)-CAMPHANATE
$C_{19}H_{21}Cl_3O_5$ (II)

F. BIGI, G. CASIRAGHI, G. CASNATI, G. SARTORI, G.G. FAVA and M.F. BELICCHI, 1985. *J. Org. Chem.*, **50**, 5018-5022.

I. Trigonal, $P3_2$, a = 11.635, c = 12.662 Å, Z = 3. Cu radiation, R = 0.0406 for 1745 reflections.

II. Orthorhombic, $P2_12_12_1$, a = 21.491, b = 15.277, c = 6.476 Å, Z = 4. Cu radiation, R = 0.0562 for 1431 reflections.

The stereochemistries of the molecules (Fig. 1) were established by the crystal structure analyses. In particular, the stereodispositions of C(2) and C(10) in (I), and of C(2) and C(11) in (II) were established.

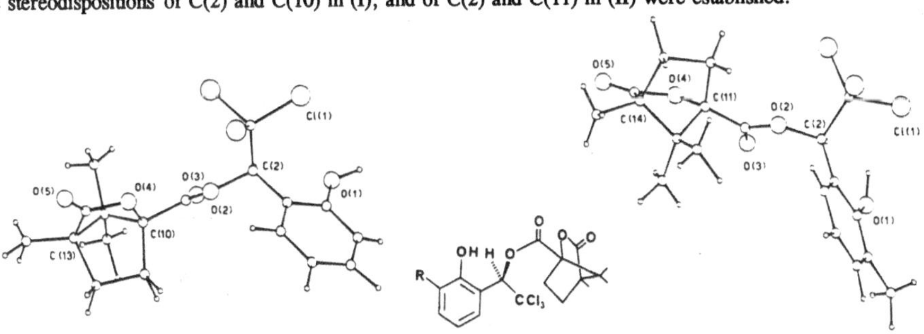

Fig. 1. Perspective drawings of the molecules of $C_{18}H_{19}Cl_3O_5$ (left) and $C_{19}H_{21}Cl_3O_5$ (right). Molecular skeleton of (II) inset.

2-(2',5'-DIHYDRO-5'-ETHYLIDENE-4'-HYDROXY-2'-OXO-3'-FURYL)-7-ETHYLIDENE-5-OXO-2,4,4-TRIMETHYL(2,5-DIHYDROFURO[3,4-b]3,4-DIHYDRO-(2H)-PYRAN) (3-(2',5'-DIHYDRO-5'-ETHYLIDENE-4'-HYDROXY-2'-OXO-3'-FURYL)-9-ETHYLIDENE-3,5,5-TRIMETHYL-2,8-DIOXABICYCLO[4.3.0]NON-1(6)-EN-7-ONE)

$C_{18}H_{20}O_6$

M.J. BEGLEY, N.G. CLEMO and G. PATTENDEN, 1985. *J. Chem. Soc. Perkin Trans. 1*, 2393-2397.

Monoclinic, $P2_1/n$, a = 13.107, b = 11.592, c = 11.721 Å, β = 104.74°, Z = 4. Mo radiation, R = 0.0417 for 2130 reflections.

In the lactone rings there are exocyclic bond angles in excess of 130°; both lactone rings are planar while the dihydropyran has a half-chair conformation (Fig. 1).

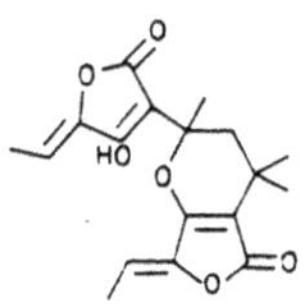

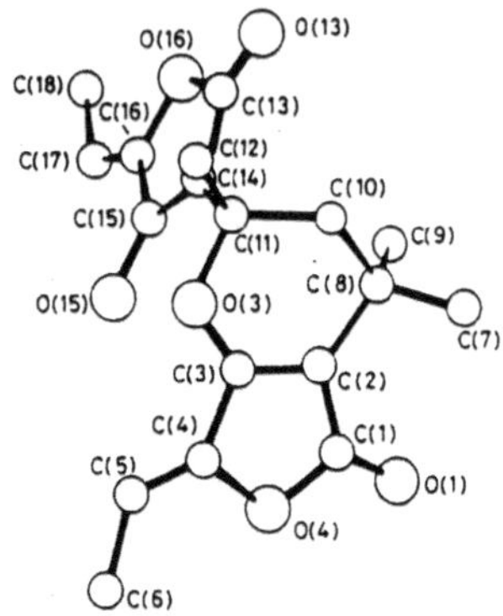

Fig. 1. Perspective drawing of a molecule of $C_{18}H_{20}O_6$.

15-(N,N-DIMETHYLCARBAMOYL)-14-HYDROXY-11,11-DIMETHYL-4,6-DIOXATETRACYCLO[7.6.0.-
[illegible].$0^{10,14}$]PENTADECA-1,3(7),8-TRIEN-13-ONE
$C_{18}H_{21}NO_5$

S. RAUCHER, D.S. JONES and R.E. STENKAMP, 1985. *J. Org. Chem.*, **50**, 4523-4526.

Monoclinic, $P2_1/c$, a = 6.944, b = 21.59, c = 12.04 Å, β = 67.27°, Z = 4. Cu radiation, R = 0.083 for 901 reflections.

The crystal structure analysis confirmed that the molecule (Fig. 1) was the postulated ketone.

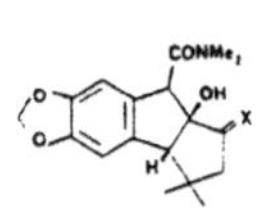

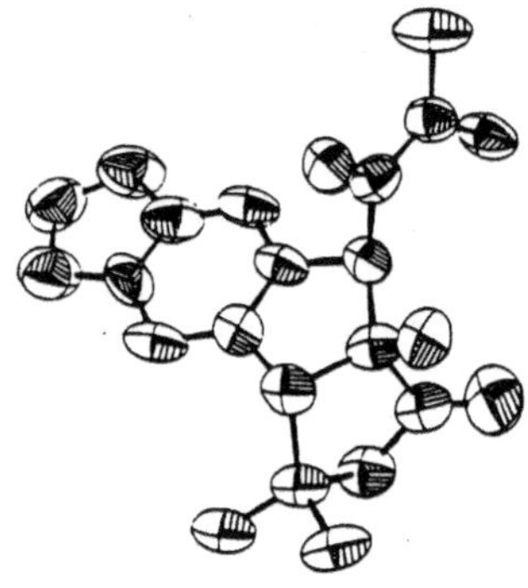

Fig. 1. A $C_{18}H_{21}NO_5$ molecule with hydrogen atoms omitted.

2α-(2-(4-CARBOXYPHENYL)ETHOXY)-trans-1-OXADECALIN
$C_{18}H_{24}O_4$ (I)

2α-(2,2-BIS(4-CHLOROPHENYL)ETHOXY)-trans-1-OXADECALIN
$C_{23}H_{26}Cl_2O_2$ (II)

P.G. JONES, G.M. SHELDRICK, A.J. KIRBY and A.J. BRIGGS, 1985. *Acta Cryst.*, **C41**, 1380-1383.

I. Monoclinic, $P2_1/n$, a = 7.558, b = 8.894, c = 25.256 Å, β = 93.86°, Z = 4. Mo radiation, R = 0.063 for 1518 reflections.

II. Monoclinic, $P2_1/c$, a = 19.269, b = 8.995, c = 13.170 Å, β = 108.42°, Z = 4. Mo radiation, R = 0.055 for 2237 reflections.

The geometry at the acetal centre is almost identical in both molecules (Fig. 1). Both adopt the (+sc,+sc) conformation with torsion angles C(9)-O(1)-C(2)-O(10) and O(1)-C(2)-O(10)-C(11) of values 67.3(5) and 65.7(3)° in (I) and 67.1(5) and 64.6(4)° for (II). Equivalent bonds in both molecules are essentially identical. The exocyclic C-OR bond lengths are 1.411(6) and 1.406(6) Å for (I) and (II), respectively, and are significantly shorter than the endocyclic C-O bonds which are of lengths 1.428(6) and 1.416(6) Å. Molecules of (I) pack in the crystal as hydrogen-bonded centrosymmetric carboxylic acid dimers with an O(22)...O(21) separation of 2.62(1) Å. The crystal structure of (II) contains no unusually short intermolecular contacts all of which are consistent with essentially van der Waals packing forces.

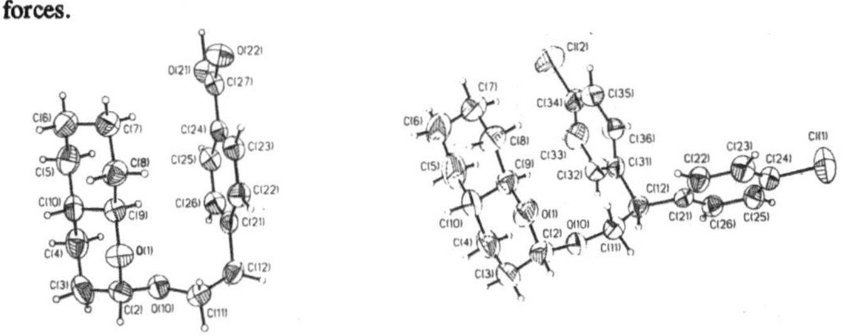

Fig. 1. Views of $C_{18}H_{24}O_4$ (left) and of $C_{23}H_{26}Cl_2O_2$ (right).

3,4:5,6-BIS(ISOPROPYLIDENEDIOXY)-2-(3,5-DIMETHYL-4-OXO-2H-3,4-DIHYDROPYRAN-2-YL)-2H-3,4,5,6-TETRAHYDROPYRAN

$C_{18}H_{26}O_7$

S. DANISHEFSKY, K.-H. CHAO and G. SCHULTE, 1985. *J. Org. Chem.*, **50**, 4650-4652.

Monoclinic, $P2_1$, a = 6.004, b = 16.051, c = 9.755 Å, β = 100.17°, Z = 2. Mo radiation, R = 0.062 for 1712 reflections.

The crystal structure analysis confirmed the stereochemistry of the pyrone molecule (Fig. 1).

Fig. 1. The $C_{18}H_{26}O_7$ molecule.

3-(2-ISOPROPYL-5-METHYLCYCLOHEXYLOXY)-4,10-DIOXATRICYCLO[5.2.1.0^{2,6}]DECANE

$C_{18}H_{30}O_3$ (I)

5-(7,7-DIMETHYL-2-OXOBICYCLO[2.2.1]HEPTANE-1-CARBOXYLATO-O)-4,10-DIOXATRICYCLO[5.2.1.-0^{2,6}]DECANE

$C_{18}H_{24}O_5$ (II)

P.W. SPRAGUE, J.E. HEIKES, J.Z. GOUGOUTAS, M.F. MALLEY, D.N. HARRIS and R. GREENBERG, 1985. *J. Med. Chem.*, **28**, 1580-1590.

I. Monoclinic, $P2_1$, a = 5.761, b = 8.266, c = 17.838 Å, β = 96.72°, Z = 2. Cu radiation, R = 0.05 for 848 reflections.

II. Monoclinic, $P2_1$, a = 7.220, b = 9.219, c = 12.406 Å, β = 100.34°, Z = 2. Cu radiation, R = 0.06 for 826 reflections.

The structures and absolute stereochemistries are shown in Fig. 1. The endo-acetal ring of (I) is in an O(1) envelope conformation and the exo-acetal ring in (II) is in a half-chair conformation. The exocyclic O(2)-C(11) bond in both structures is pseudo-axial.

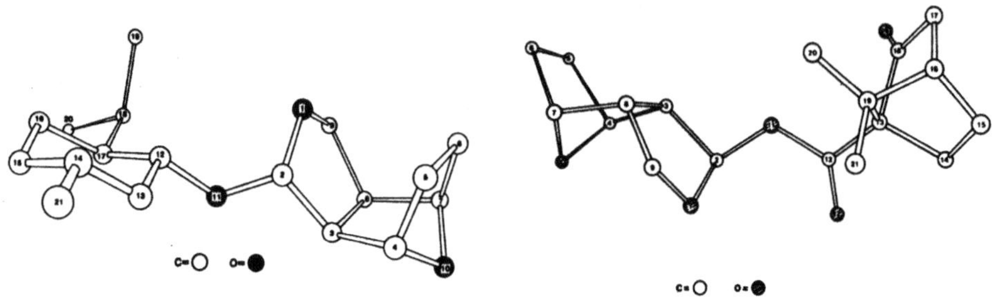

Fig. 1. The solid-state conformations of $C_{18}H_{30}O_3$ (I) (left) and $C_{18}H_{24}O_5$ (II) (right).

(5aSR,6aRS,7SR,10aRS,11aRS)-5a,11a-EPOXY-5a,6a,7,10,10a,11a-HEXAHYDRONAPHTHACENE-7-METHOXY-5,6,9,11,12(8H)-PENTAONE
$C_{19}H_{14}O_7$

R.C. GUPTA, D.A. JACKSON, R.J. STOODLEY and D.J. WILLIAMS, 1985. *J. Chem. Soc. Perkin Trans. 1*, - 525-533.

Orthorhombic, $P2_12_12_1$, a = 10.330, b = 11.800, c = 13.294 Å, Z = 4. Cu radiation, R = 0.030 for 1234 reflections.

The crystal structure analysis established the identity and stereochemistry of the molecule (Fig. 1). The A ring has a chair-like conformation in the crystal. The B ring has a boat-like geometry.

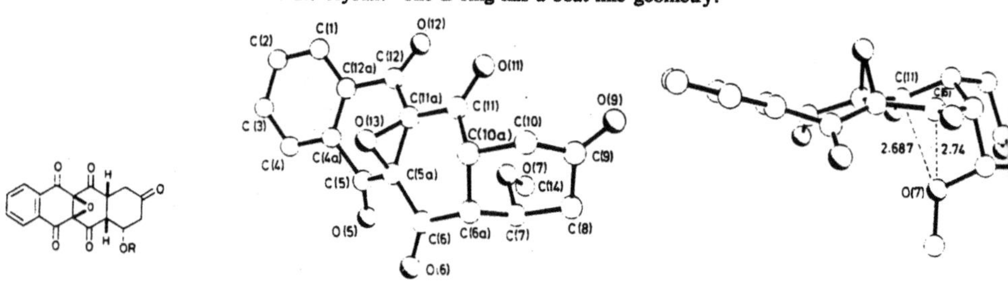

Fig. 1. Perspective drawing of a molecule of $C_{19}H_{14}O_7$ (left) with a view showing the A-ring and the close approach of the O(7)-C(11) and O(7)-C(6) atoms (right); distances in Å.

(±)-BOSTRYCIN ACETONIDE
$C_{19}H_{20}O_8$

T.R. KELLY, J.K. SAHA and R.R. WHITTLE, 1985. *J. Org. Chem.*, **50**, 3679-3685.

Triclinic, $P\bar{1}$, a = 8.047, b = 11.126, c = 11.442 Å, α = 113.60, β = 106.52, γ = 99.67°, Z = 2. Mo radiation, R = 0.036 for 2080 reflections.

The crystal structure analysis established that the molecule (Fig. 1) has the C(4) hydroxyl in a β configuration. This contrasts with previous work (1) which places it in an α configuration.

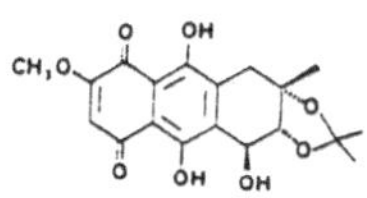

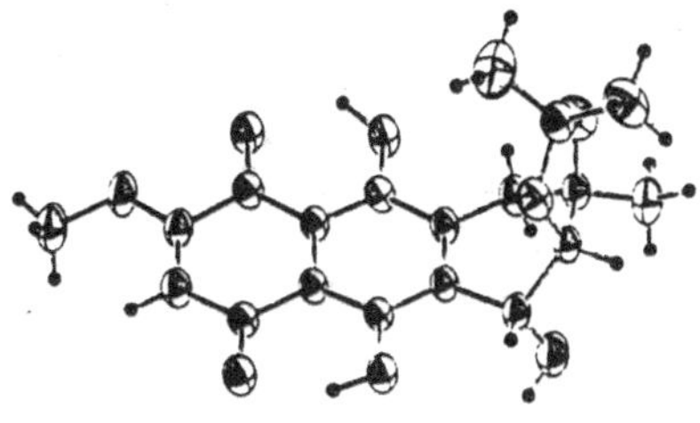

Fig. 1. ORTEP drawing of $C_{19}H_{20}O_8$.

1. W. KREISER, L. JANITSCHKE and W.S. SHELDRICK, 1977. *J. Chem. Soc., Chem. Commun.*, 269.

4a-CYANO-3,3-ETHYLENEDIOXY-6,7-DIMETHOXY-1,2,3,4,4a,10a-HEXAHYDROPHENANTHRENE $C_{19}H_{21}NO_4$

T. KAMETANI, Y. SUZUKI and T. HONDA, 1985. *Heterocycles*, **23**, 563-566.

Monoclinic, $P2_1/c$, a = 12.951, b = 10.279, c = 20.871 Å, β = 142.2°, Z = 4. R = 0.093 for 835 reflections.

The molecule (Fig. 1) is shown to have a B/C-trans ring junction.

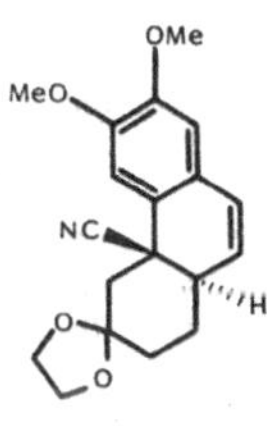

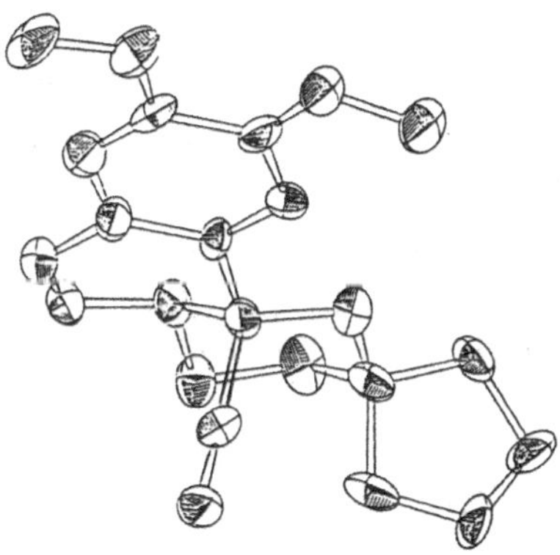

Fig. 1. The $C_{19}H_{21}NO_4$ structure.

2',5,7'-TRICHLOROSPIRO(BENZOFURAN-3(2H),9'-[9H]-XANTHEN)-2-ONE $C_{20}H_9Cl_3O_3$

M. KIMURA, 1985. *Bull. Chem. Soc. Jpn.*, **58**, 905-910.

Monoclinic, $C2/c$, a = 21.316, b = 17.267, c = 9.158 Å, β = 94.05°, Z = 8. Mo radiation, R = 0.088 for 1993 reflections.

The molecule has three phenyl rings fused to a spirolactone. The conformation and selected dimensions are shown in Fig. 1. Lactone ring B is coplanar with ring A and its orientation is almost perpendicular to the xanthene ring. The A/C. A/E and C/E dihedral angles are 73.6, 83.6 and 22.81° respectively.

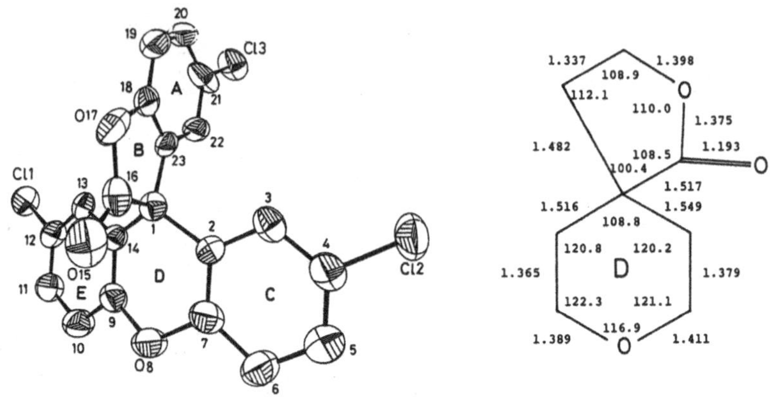

Fig. 1. A view of $C_{20}H_9Cl_3O_3$ and selected dimensions.

DINAPHTHO[1,2-b:2',1'-d]FURAN
$C_{20}H_{12}O$

R.E. GERKIN and W.J. REPPART, 1985. *Acta Cryst.*, C41, 961-963.

Monoclinic, $P2_1/c$, a = 14.309, b = 4.665, c = 20.827 Å, β = 109.30°, Z = 4. Mo radiation, R = 0.043 for 2672 reflections.

Molecular dimensions are shown in Fig. 1. The molecule is not planar, the mean planes of the naphthalenic moieties being inclined at 1.0°. All atomic positions, except C(4-6) and C(14-16), exhibit near twofold symmetry with respect to an axis which passes through the O atom and the midpoint of the C(10)-C(20) bond. Packing is governed by van der Waals interactions.

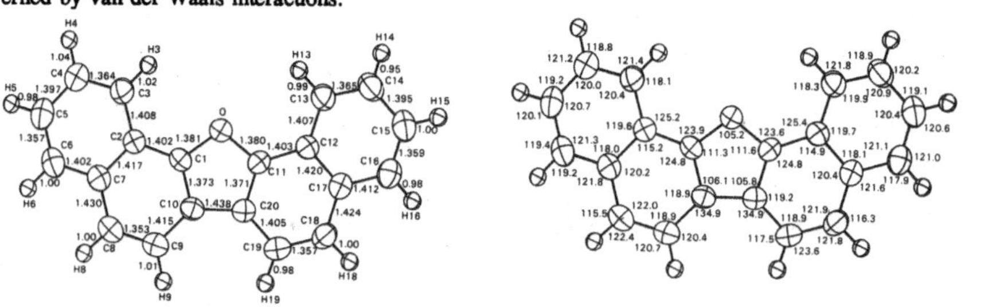

Fig. 1. $C_{20}H_{12}O$: bond lengths and angles (σ 0.003 Å, 0.2°).

5-FORMYL-4-PHENYL-α-(PHENYLMETHYLENE)-4H-1,3-DIOXIN-2-ACETALDEHYDE
$C_{20}H_{16}O_4$

C. REICHARDT, KYEONG-YEOL YUN, W. MASSA and R.E. SCHMIDT, 1985. *Liebigs Ann. Chem.*, 1987-1996.

Monoclinic, $P2_1/c$, a = 9.827, b = 18.145, c = 9.700 Å, β = 112.85°, Z = 4. Mo radiation, R = 0.033 for 1943 reflections.

The molecule has structure and stereochemistry shown in Fig. 1. The atoms C(2),O(1),C(6),C(5),C(4) of the dioxin ring and the formyl group C(51)-O(51) are approximately planar; atom O(3) is bent 51.5° from this plane. The propenal residue is also approximately planar and is inclined at 63.2° to the dioxin plane. The orientation of the phenyl ring at C22 is defined by the C(21)-C(22)-C(221)-C(226) torsion angle of -21.6°. The phenyl rings at

C(4) and C(22) are almost normal to one another (dihedral angle 92.0°).

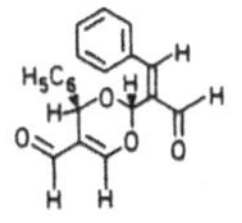

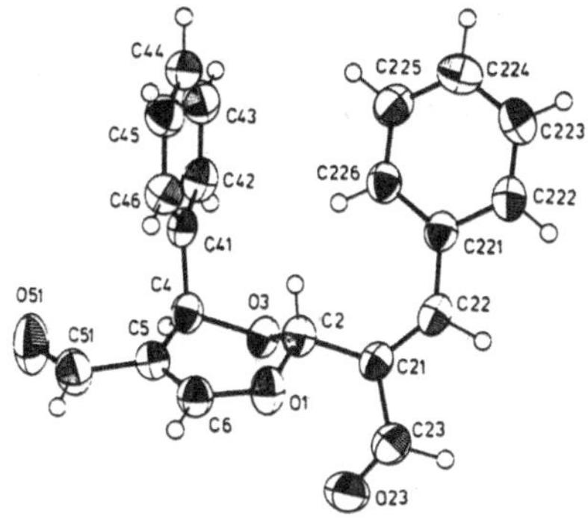

Fig. 1. The $C_{20}H_{16}O_4$ molecular structure.

1,3-DIHYDRO-4-((2R)-2,5-DIHYDRO-2-FURYL)-3-PHENYL-1-(p-TOLYL)-2H-IMIDAZOLE-2-THIONE
$C_{20}H_{18}N_2OS$

A. CRIADO, A. CONDE and R. MARQUEZ, 1985. *Acta Cryst.*, **C41**, 1215-1217.

Orthorhombic, $P2_12_12_1$, a = 9.366, b = 20.616, c = 9.137 Å, Z = 4. Mo radiation, R = 0.052 for 1979 reflections.

In the molecule (Fig. 1), the furyl ring is approximately planar due to the presence of the C(6)=C(7) double bond (1.289(9) Å). The dihedral angle between the imidazole and furyl mean planes is 69.9(2)° and the glycosidic torsion angle is 105.6(6)°, i.e. syn. The C-S bond distance is 1.677(5) Å. The N(1)- and N(2)-phenyl rings are both essentially planar and the dihedral angles between these mean planes and the imidazole ring are 64.6(2) and 69.7(2)° respectively. Packing is governed by a C(6)-H...O hydrogen bond which gives rise to infinite helical chains parallel to [001] (C(6)...O 3.454(8) Å).

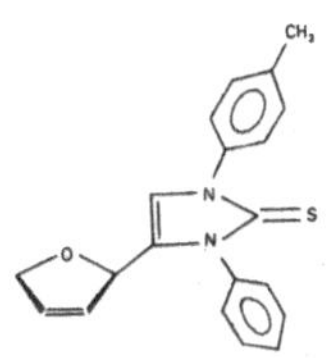

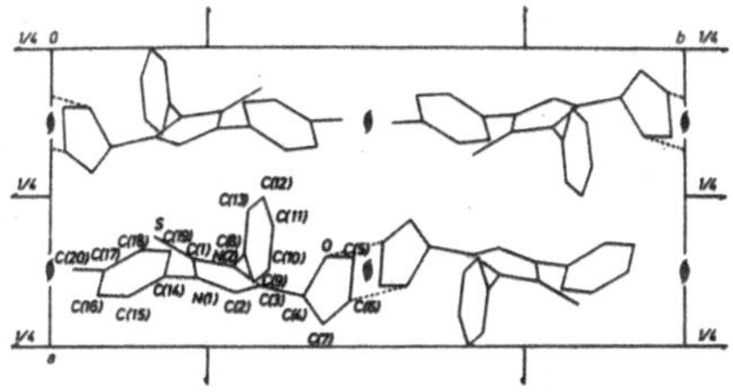

Fig. 1. $C_{20}H_{18}N_2OS$: the molecular skeleton and packing.

6-BROMO-3,3',4',5,7-PENTA-O-METHYLCATECHIN
$C_{20}H_{23}BrO_6$

F.W.B. EINSTEIN, E. KIEHLMANN and E.K. WOLOWIDNYK, 1985. *Canad. J. Chem.*, **63**, 2176-2180.

Triclinic, $P1$, a = 9.589, b = 11.576, c = 11.326 Å, α = 118.80, β = 93.23, γ = 111.44°, Z = 2. Mo radiation, R = 0.0268 for 2213 reflections.

The unit cell contains two molecules which differ in the conformation of the heterocyclic ring (Fig. 1). Atom C(2) is above and C(3) is below the ring plane in both molecules; atoms O and C(11) are above this plane in molecule B (sofa) and below in molecule (A) (half-chair). The analysis shows that the Br atom is at C(6) and establishes the trans diaxial arrangement of H(2)/H(3) and the quasi-equatorial orientation of the 3,4-dimethoxyphenyl group.

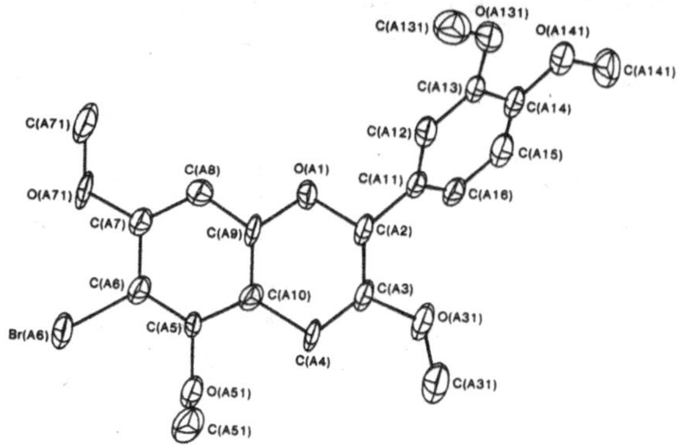

Fig. 1. A view of molecule A of $C_{20}H_{23}BrO_6$.

(1aRS,3aSR,4aRS,5SR,8RS,8aSR,9aSR)-3a-METHOXY-2,2-DIMETHYL-8a-(PROP-2-ENYL)-1a,2,5,8-TETRAHYDRO-1H-5,8-METHANOCYCLOPROPA[c]NAPHTHO[2,3-b]FURAN-4,9(3aH,8aH)-DIONE
$C_{20}H_{24}O_4$

R.F.C. BROWN, K.B. CALDWELL, B.M. GATEHOUSE and P.Y.T. TEO, 1985. *Aust. J. Chem.*, **38**, 1339-1349.

Monoclinic, $P2_1/n$, a = 11.522, b = 13.506, c = 11.052 Å, β = 92.66°, Z = 4. Cu radiation, R = 0.093 for 1174 reflections.

The X-ray analysis established the identity of the title compound which has an 8a-allyl group syn to the methoxyl group and the fused cyclopropane ring (Fig. 1). Bond lengths and angles have acceptable values.

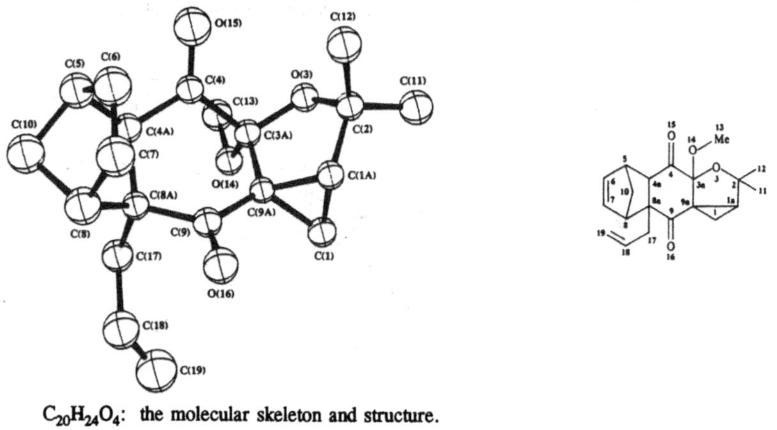

Fig. 1. $C_{20}H_{24}O_4$: the molecular skeleton and structure.

1,4,6,9-TETRAISOPROPYL-1,4,6,9-TETRAAZA-5,10-DIOXAPERHYDROANTHRACENE
$C_{20}H_{40}N_4O_2$

R.L. WILLER, D.W. MOORE, C.K. LOWE-MA and D.J. VANDERAH, 1985. *J. Org. Chem.*, **50**, 2368-2372.

Triclinic, $P\bar{1}$, a = 5.981, b = 9.529, c = 10.021 Å, α = 97.53, β = 101.07, γ = 99.66°, Z = 1. Mo radiation, R = 0.080 for 1653 reflections.

The crystal structure analysis confirmed the ring juncture stereochemistry of the molecule (Fig. 1). The molecule lies on an inversion centre. All three rings have chair conformations; the angles of the plane of the isopropyl groups to the C-N-C plane are 70.4 and 69.8°. The dihedral angle between H(1) and H(4) is 52.6°. The four hydrogen atoms at the ring junctures are in a cis-transoid-cis configuration.

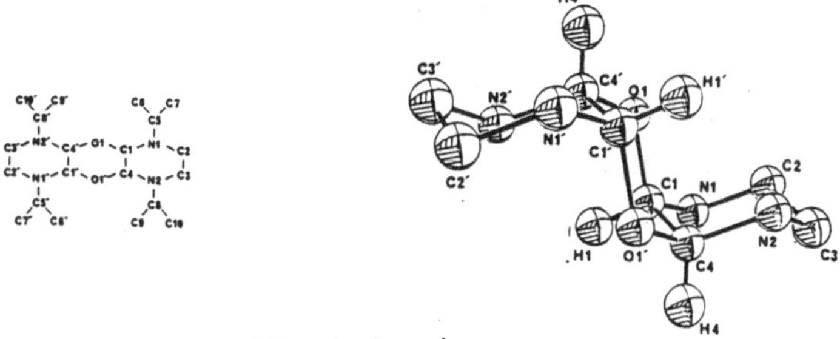

Fig. 1. The centrosymmetric $C_{20}H_{40}N_4O_2$ molecule.

4a(p-BROMOBENZOATOMETHYL)-8-METHOXY-6-OXO-OCTAHYDROBENZO[b,e]COUMARIN
$C_{21}H_{19}BrO_6$

R.V. STEVENS and A.P. VINOGRADOFF, 1985. *J. Org. Chem.*, **50**, 4056-4062.

Triclinic, *P*1, a = 7.383, b = 10.280, c = 13.658 Å, α = 109.89, β = 94.58, γ = 108.10°, Z = 1. Mo radiation, R = 0.043 for 2608 reflections (at 115 K).

The crystal structure analysis established the cis nature of all ring junctions (Fig. 1).

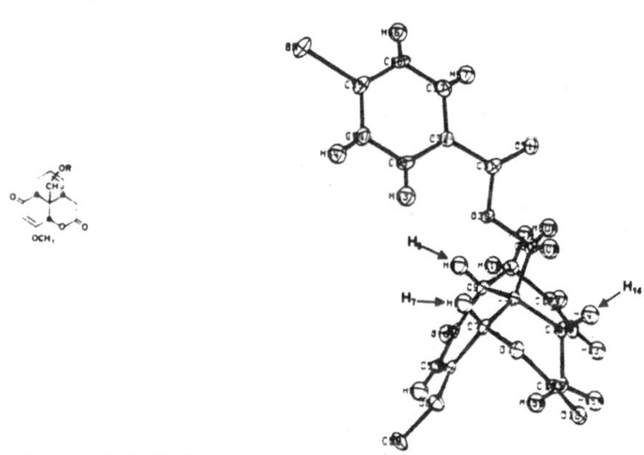

Fig. 1. ORTEP diagram of $C_{21}H_{19}BrO_6$.

2,2,8,8-TETRAMETHYL-5-PHENYL-1,3,4,7,9,10-HEXAHYDRODIPYRANO[4,3-b;4,3-d]PYRIDINE
$C_{21}H_{25}NO_2$

A.A. ESPENBETOV, A.I. YANOVSKII, Yu.T. STRUCHKOV, T.T. OMAROV and A.Sh. GUBASHEVA, 1985. *Zh. Strukt. Khim.*, **26**, 6-147-150 [J. Struct. Chem., **26**, 964-966].

Triclinic, $P\bar{1}$, a = 9.7025, b = 10.394, c = 10.643 Å, α = 112.72, β = 112.70, γ = 94.38°, Z = 2. Cu radiation, R = 0.044 for 2353 reflections.

The molecule is shown in Fig. 1. The pyridine is slightly non-planar due to steric repulsion between the benzene ring and the nearest CH_2 groups of the dihydropyran rings. Both dihydropyran rings have half-chair conformations. Bond lengths are as expected.

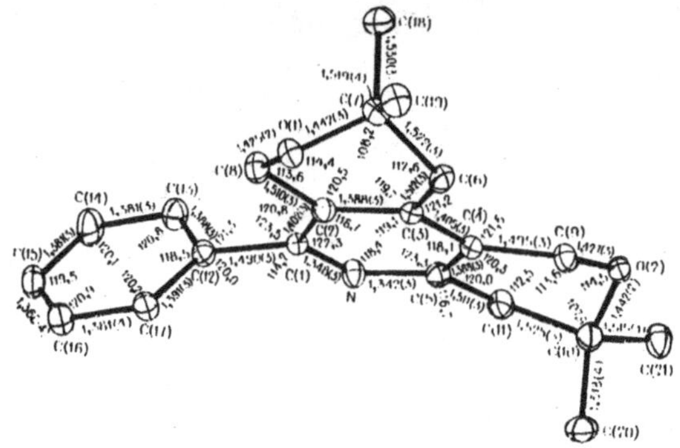

Fig. 1. The structure of $C_{21}H_{25}NO_2$.

4-(3α-HYDROXY-4α-METHOXYCARBONYL-8aβ-METHYL-1,6-DIOXO-2a,3,4,5,6,7,8,8a-OCTAHYDROACENAPHTH-3-YL)BUTAN-2-ONE ETHYLENE ACETAL

$C_{21}H_{28}O_7$

C.K. FAIR, E.O. SCHLEMPER, B. NASSIM and P. CRABBE, 1985. *Acta Cryst.*, **C41**, 1665-1667.

Monoclinic, $P2_1$, a = 12.190, b = 6.488, c = 12.702 Å, β = 102.33°, Z = 2. Mo radiation, R = 0.038 for 858 reflections.

There are no unusual bond lengths or angles in the molecule (Fig. 1). The angles at the C(sp^3) atoms in the tricyclic ring system range from 101.8(5) to 114.4(4)°. The largest deviations from the tetrahedral value occur at C(8a) and C(2a) and strains at these sites are ascribed to the two exo double bonds of the five-membered ring to which these atoms belong. This ring has an envelope conformation with C(8b) as flap. Both unsaturated six-membered rings have nearly ideal sofa conformations with atoms C(3) and C(8) deviating from the mean planes defined by the five other atoms in the rings. The other five-membered ring has a half-chair conformation. Except for an intermolecular hydrogen bond O(3)-H...O(6) of length 2.909(5) Å the intermolecular contacts are consistent with van der Waals interactions.

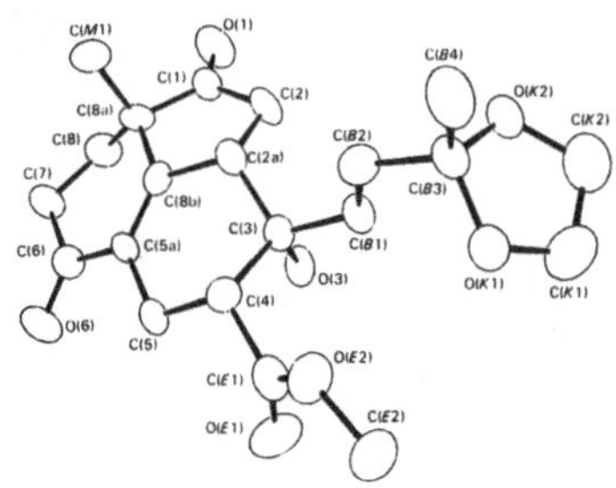

Fig. 1. A view of the $C_{21}H_{28}O_7$ molecule.

4,4-BIS(4-AMINO-5,6-DIHYDRO-6,6-DIMETHYLFURO[2,3-d]PYRIMIDIN-2-YL)-2-METHYL-2-BUTANOL DIHYDRATE

$C_{21}H_{30}N_6O_3,2(H_2O)$

H. WAMHOFF, H.-A. THIEMIG, H. PUFF and E. FRIEDRICHS, 1985. *Chem. Ber.*, **118**, 4782-4793.

Triclinic, $P\bar{1}$, a = 12.927, b = 13.513, c = 15.201 Å, α = 71.19, β = 85.76, γ = 83.35°, Z = 4. Cu radiation, R = 0.107 for 2686 reflections.

The analysis establishes the constitution as shown in Fig. 1. The two independent molecules in the asymmetric unit are linked by N-H...N hydrogen bonds.

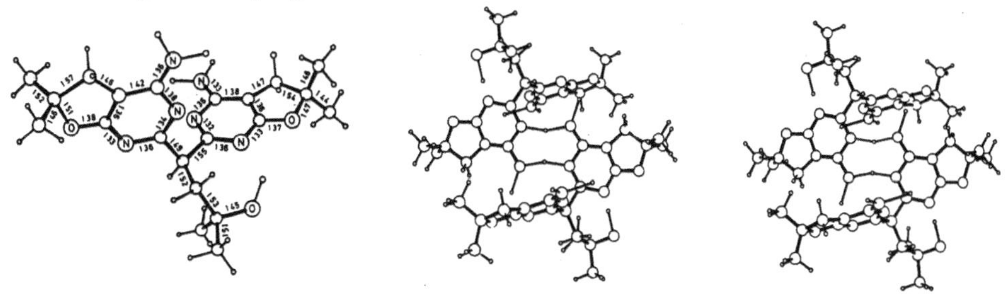

Fig. 1. Bond lengths in $C_{21}H_{30}N_6O_3$ and a stereoview of a pair of molecules.

(2S-(2α(R*),4aα,5α,8α,8aα))-OCTAHYDRO-8,9,9-TRIMETHYL-2-(1-PHENYLETHOXY)-5,8-METHANO-2H-1-BENZOPYRAN

$C_{21}H_{30}O_2$

C.R. NOE, M. KNOLLMULLER, E. WAGNER and H. VOLLENKLE, 1985. *Chem. Ber.*, **118**, 1733-1745.

Orthorhombic, $P2_12_12_1$, a = 19.336, b = 14.541, c = 6.5074 Å, Z = 4. Mo radiation, R = 0.092 for 935 reflections.

The pyran ring is in a slightly deformed boat conformation (Fig. 1). A general rule is given to allow prediction of absolute configuration of acyclic alkyl-aryl carbinols from the course of acetal formation between enantiomorphically pure lactal and racemic alcohol.

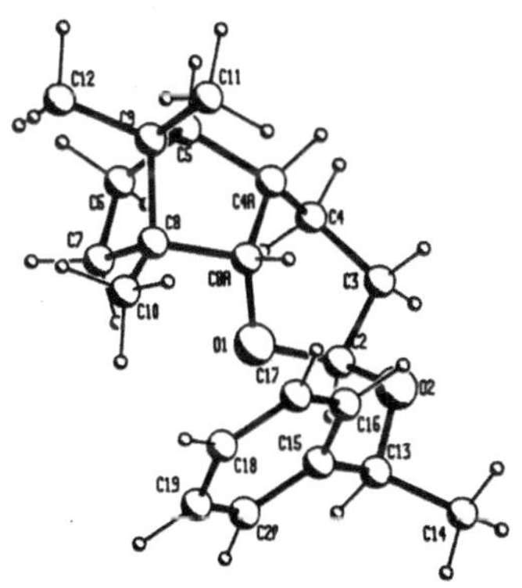

Fig. 1. A view of the $C_{21}H_{30}O_2$ molecule.

rel-METHYL (3R,3aR,4aS,5S,7R,8R,8aS,9aS)-5-BROMO-DODECAHYDRO-3-ISOPROPYL-8,9a-DIMETHYL-10-OXO-4a,8-(EPOXYMETHANO)-4aH-BENZ[f]INDENE-7-CARBOXYLATE

$C_{21}H_{31}BrO_4$ (I)

rel-METHYL (3R,3aS,4aS,5S,7R,8R,8aS,9aS)-5-BROMO-DODECAHYDRO-3-ISOPROPYL-8,9a-DIMETHYL-10-OXO-4a,8-(EPOXYMETHANO)-4aH-BENZ[f]INDENE-7-CARBOXYLATE

$C_{21}H_{31}BrO_4$ (II)

E.J. COREY, T.A. ENGLER and J.M. BERG, 1985. *Acta Cryst.*, C41, 289-292.

I. Triclinic, $P\bar{1}$, a = 7.042, b = 11.760, c = 13.160 Å, α = 102.21, β = 96.25, γ = 95.20°, Z = 2. Mo radiation, R = 0.0338 for 211 reflections.

II. Monoclinic, $P2_1/c$, a = 11.855, b = 7.190, c = 24.571 Å, β = 97.59°, Z = 4. Mo radiation, R = 0.0362 for 1922 reflections.

Compounds (I) and (II) (Fig. 1) differ in their configuration at C(3a). There is a trans A/B ring fusion in (I) and a cis fusion in (II). As a consequence ring B has a chair conformation in (I) and a twist-boat conformation in (II). Also quite large differences are seen in bond angles at the ring-junction C atoms 3a (118.3(3), 113.4(3)°), 4a (117.5(3), 114.3(3)°), 8a (113.4(3), 108.7(3)°) and 9a (99.7(3), 104.5(3)°).

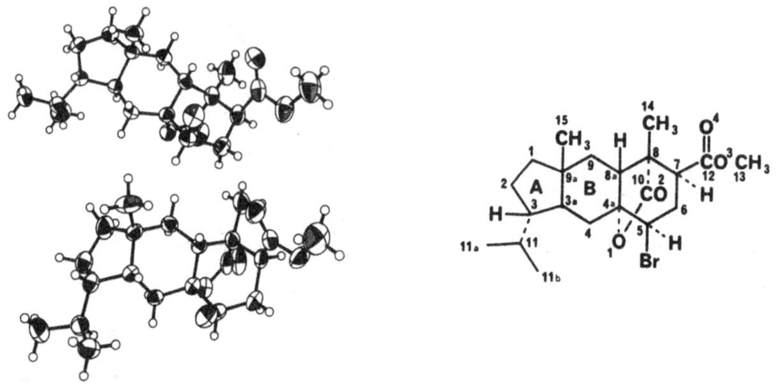

Fig. 1. $C_{21}H_{31}BrO_4$: the molecular structures of (I) (top left), (II) (bottom left) and the molecular skeleton (right).

BIS(2,2,2-TRICHLOROETHYL) exo,exo-2,5-DIPHENYL-7-OXA-3,4-DIAZABICYCLO[4.1.0]HEPTANE-3,4-DICARBOXYLATE
$C_{22}H_{18}Cl_6N_2O_5$

W. ADAM, A. BERKESSEL, K. HILDENBRAND, E.-M. PETERS, K. PETERS and H.G. VON SCHNERING, 1985. *J. Org. Chem.*, **50**, 4899-4909.

Triclinic, $P\bar{1}$, a = 9.962, b = 15.331, c = 9.205 Å, α = 94.04, β = 102.40, γ = 108.50°, Z = 2. Mo radiation, R = 0.042 for 4233 reflections.

The crystal structure analysis established the stereochemistry of the molecule (Fig. 1).

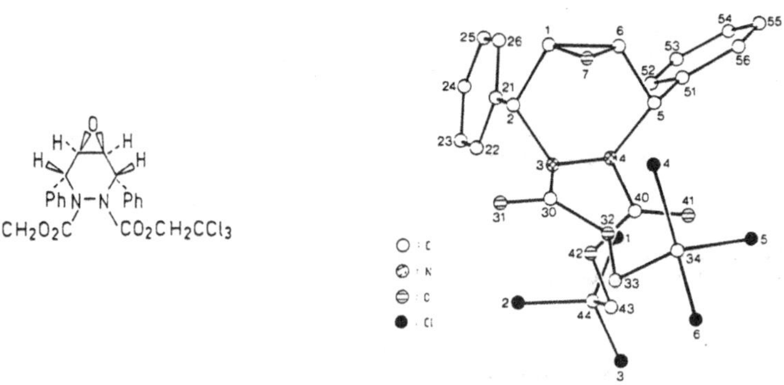

Fig. 1. The $C_{22}H_{18}Cl_6N_2O_5$ molecule.

8,12-EPOXY-12-ETHOXY-3,11-BIS(5-NITRO-2-PYRIDYL)-3,11-DIAZATRICYCLO[5.3.1.1^{2,6}]DODEC-4-EN-10-ONE
$C_{22}H_{20}N_6O_7$

G. FERGUSON, K.J. FISHER, B.E. IBRAHIM, C.Y. ISHAG, G.M. ISKANDER and M. PARVEZ, 1985. *J. Chem. Res.*, S212-213, M2473-2487.

Triclinic, $P\bar{1}$, a = 10.021, b = 13.809, c = 8.614 Å, α = 98.31, β = 113.63, γ = 80.17°, Z = 2. Mo radiation, R = 0.0365 for 1357 reflections.

This reaction product of 1-(5-nitro-2-pyridyl)-3-oxidopyridinium betaine with ethanol in triethylamine was shown to have the structure (Fig. 1). The molecule has three six-membered rings and one saturated substituted tetrahydrofuran-type ring. The central ring (C(17,27,28,29,111) and N(12)) has a chair confirmation; ring (N(12), C(17,18,19,110,111)) has a twisted boat form; and ring (N(22), C(27,28,210,211)) has a deformed envelope conformation with C(28) at the flap. The five-membered ring (O(1), C(28,29,111,110)) has an envelope conformation with C(29) at the flap.

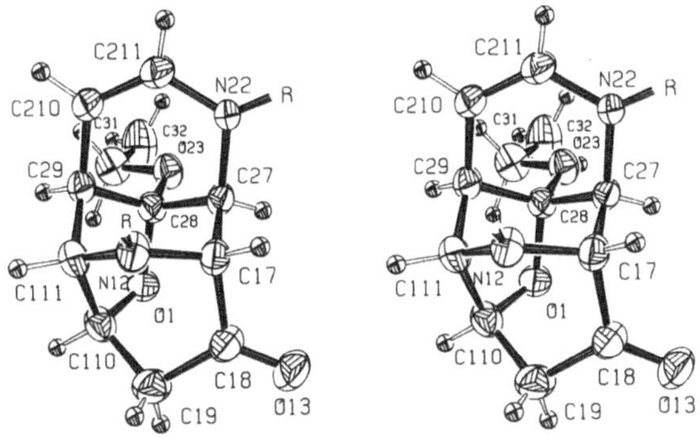

Fig. 1. A stereoview of $C_{22}H_{20}N_6O_7$.

3-exo-6-DIPHENYL-4-OXA-exo-TRICYCLO[6.2.1.0^{2,7}]UNDEC-2-EN-5-ONE
$C_{22}H_{20}O_2$ (I)

4,4a,5,6,7,8,9,10-OCTAHYDRO-1,exo-4-DIPHENYL-3H-CYCLOOCTA[c]PYRAN-3-ONE
$C_{23}H_{24}O_2$ (II)

endo-7-BENZOYL-1-PHENYLTRICYCLO[4.4.0.0^{3,10}]DECAN-2-ONE
$C_{23}H_{22}O_2$ (III)

M. CHRISTL, U. LANZENDORFER, J. HEGMANN, K. PETERS, E.-M. PETERS and H.G. VON SCHNERING, 1985. *Chem. Ber.*, **118**, 2940-2973.

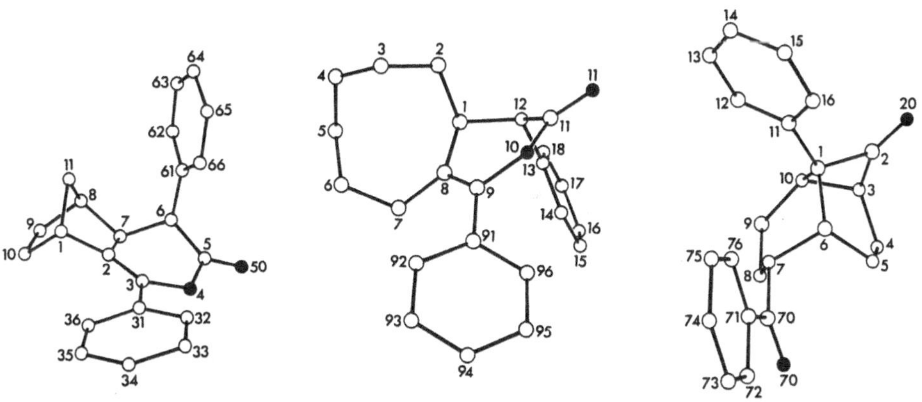

Fig. 1. Views (left to right) of $C_{22}H_{20}O_2$ (I), $C_{23}H_{24}O_2$ (II) and $C_{23}H_{22}O_2$ (III).

I. Monoclinic, $P2_1/a$, a = 18.253, b = 8.721, c = 10.506 Å, β = 104.22°, Z = 4. Mo radiation, R = 0.038.

II. Triclinic, $P\bar{1}$, a = 9.663, b = 11.074, c = 9.114 Å, α = 92.20, β = 110.35, γ = 102.74°, Z = 2. Mo radiation, R = 0.041.

III. Monoclinic, $P2_1/a$, a = 20.380, b = 11.768, c = 15.285 Å, β = 100.97°, Z = 8. Mo radiation, R = 0.060.

The analyses established the structures and stereochemistries as shown in Fig. 1.

anti-1,4,7,10-TETRAMETHYL-1,4:7,10-DIEPOXY-1,4,7,10-TETRAHYDRONAPHTHACENE $C_{22}H_{20}O_2$

G.W. GRIBBLE, R.B. PERNI and K.D. ONAN, 1985. *J. Org. Chem.*, **50**, 2934-2939.

Monoclinic, $P2_1/c$, a = 8.393, b = 8.019, c = 13.656 Å, β = 112.45°, Z = 2. Cu radiation, R = 0.056 for 1023 reflections.

Molecules (Fig. 1) of the anti isomer lie on crystallographic centres of symmetry.

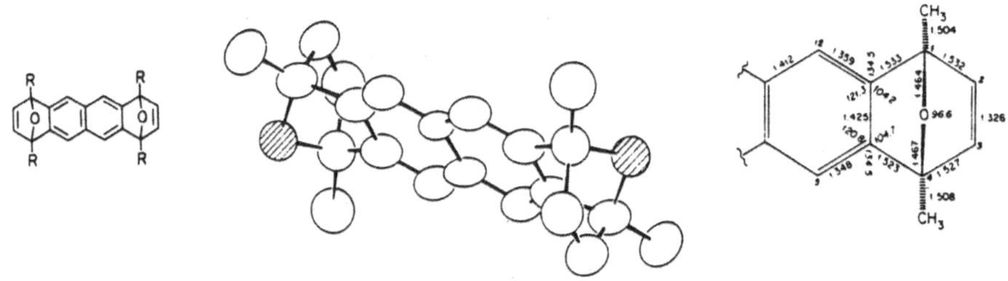

Fig. 1. A molecule of the anti isomer of $C_{22}H_{20}O_2$ with selected bond lengths (Å) and angles (°).

cis-6,11,17,18-TETRAHYDRO-5,12-DIOXATRIBENZO[a,e,i]CYCLODODECENE-17,18-DIOL $C_{22}H_{20}O_4$

F.R. FRONCZEK, S.J. JUNGK and R.D. GANDOUR, 1985. *Acta Cryst.*, **C41**, 426-427.

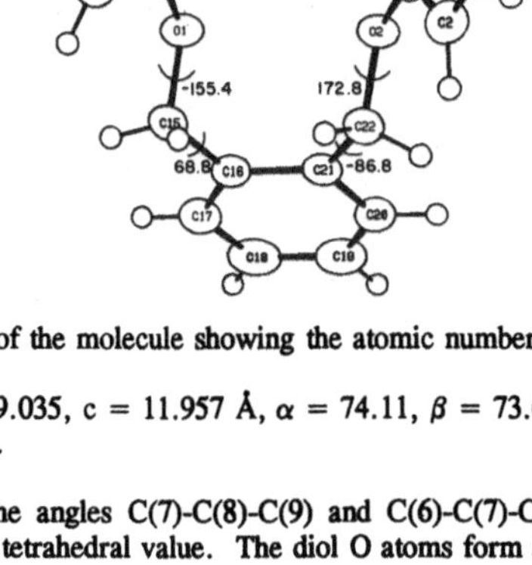

Fig. 1. $C_{22}H_{20}O_4$: a view of the molecule showing the atomic numbering and selected torsion angles.

Triclinic, $P\bar{1}$, a = 8.633, b = 9.035, c = 11.957 Å, α = 74.11, β = 73.61, γ = 83.00°, Z = 2. Mo radiation, R = 0.031 for 2016 reflections.

In the molecule (Fig. 1) the angles C(7)-C(8)-C(9) and C(6)-C(7)-C(8), of mean value 115.5(2)°, deviate considerably from the expected tetrahedral value. The diol O atoms form a gauche interaction with a O-C-C-O torsion angle of 42.5(3)°- smaller by some 20° than in related compounds. There is an intramolecular hydrogen bond O(3)-H...O(4) with an H...O separation of 2.14(2) Å. The phenylene ether bond angles at O(1) and O(2) are 119.4(1) and 114.4(1)° respectively. The aromatic rings are all planar but substituent atoms deviate significantly from their least-squares planes. There is a weak intermolecular hydrogen bond O(4)-H...O(2)of length 3.019(1) Å.

DL-rel-(1R,5S,6S)-6-METHYL-10-OXO-11-OXATRICYCLO[4.3.2.0^{1,5}]UNDEC-8-EN-4-YLIDENEHYDRAZINE
$C_{22}H_{24}N_2O_4$ (I)

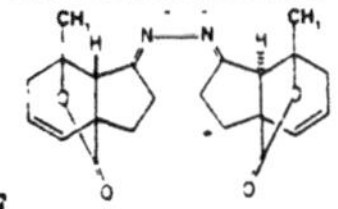

rel-(1S,4R,5R,8S)-4-CYANO-8-METHYL-9-OXATRICYCLO[6.2.1.0^{1,5}]UNDECAN-10-ONE
$C_{12}H_{15}NO_2$ (II)

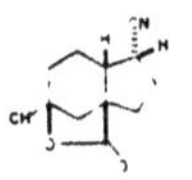

B. CHENERA, CHE-PING CHUANG, D.J. HART and LEH-YEH HSU, 1985. *J. Org. Chem.*, **50**, 5409-5410.

I. Monoclinic, $P2_1/c$, a = 10.965, b = 11.556, c = 15.221 Å, β = 95.80°, Z = 4. Mo radiation, R = 0.037 for 1352 reflections.

II. Triclinic, $P\bar{1}$, a = 7.760, b = 19.993, c = 7.125 Å, α = 91.18, β = 98.17, γ = 83.45°, Z = 4. Mo radiation, R = 0.030 for 2245 reflections.

The crystal structure analyses established the stereochemistries of the molecules.

2-METHOXYCARBONYL-2-PHENYL-4H-2,3,3a,11c-TETRAHYDROPYRRO[2,3-d]NAPHTHO[2,1-b]PYRAN
$C_{23}H_{21}NO_3$

P. ARMSTRONG, R. GRIGG, M.W. JORDAN and J.F. MALONE, 1985. *Tetrahedron*, **41**, 3547-3558.

Monoclinic, $P2_1/n$, a = 15.12, b = 10.88, c = 11.24 Å, β = 104.9°, Z = 4. Cu radiation, R = not stated, for 2257 reflections.

The molecule is shown to have the stereochemistry depicted in Fig. 1.

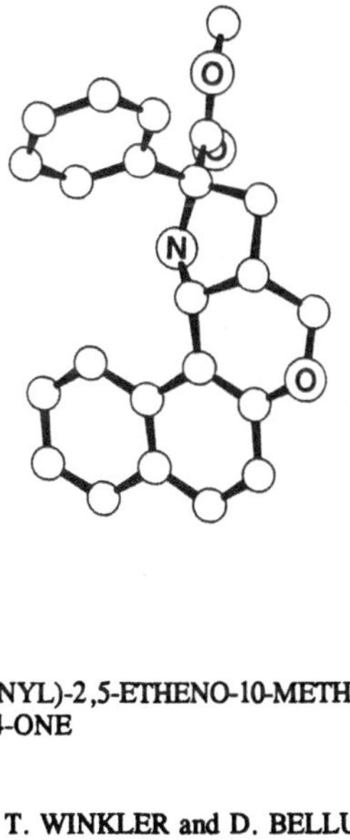

Fig. 1. The $C_{23}H_{21}NO_3$ molecule.

1,6-DICHLORO-7,8-BIS(p-CHLOROPHENYL)-2,5-ETHENO-10-METHOXYCARBONYL-2-TRIFLUOROMETHYL-3-OXABICYCLO[4.2.0]OCT-7-EN-4-ONE
$C_{24}H_{13}Cl_4F_3O_4$

P. MARTIN, E. STEINER, J. STREITH, T. WINKLER and D. BELLUS, 1985. *Tetrahedron*, **41**, 4057-4078.

Orthorhombic, $P2_12_12_1$, a = 9.120, b = 15.573, c = 16.676 Å, Z = 4. Mo radiation, R = 0.062 for 3366 reflections.

The structure of this tricyclic cyclobutene derivative was established as shown in Fig. 1.

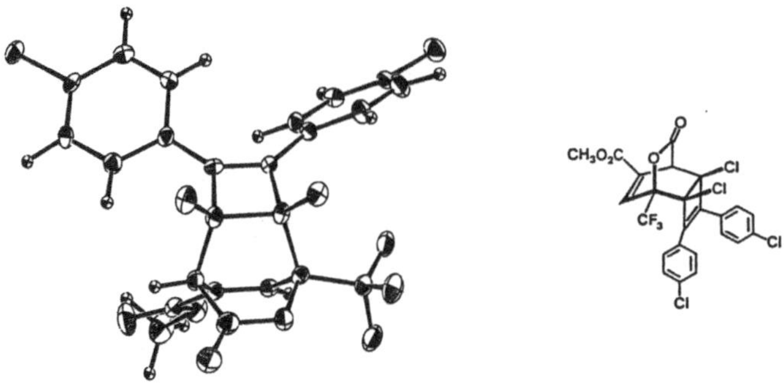

Fig. 1. A view of $C_{24}H_{13}Cl_4F_3O_4$

10,15,20-TRIHYDRO-5H-5,10,15,20-TETRAOXATETRABENZO[a,d,g,j]CYCLODODECINE
$C_{24}H_{16}O_4$ (I)

2,3,7,8,12,13,17,18-OCTAMETHOXY-10,15,20-TRIHYDRO-5H-5,10,15,20-TETRATHIATETRABENZO[a,d,g,j]-CYCLODODECINE CHLOROFORM SOLVATE
$C_{32}H_{32}O_8S_4,2(CHCl_3)$ (II)

K. VON DEUTEN, W. HINRICHS, T. WEISS and G. KLAR, 1985. *J. Chem. Res.*, S52-53, M501-557.

I. Rhombohedral, $R\bar{3}$, a = 2433.4, c = 905.6 Å, Z = 9. Mo radiation, R = 0.050 for 1559 reflections.

II. Triclinic, $P\bar{1}$, a = 855.3, b = 1079.3, c = 1261.5 Å, α = 68.65, β = 70.42, γ = 87.58°, Z = 1. Mo radiation, R = 0.046 for 2207 reflections.

Both compounds (Fig. 1) lie about crystallographic inversion centres. The four chalcogen atoms and two non-adjacent phenylene rings are nearly coplanar, while the other two rings extend almost perpendicularly above and below the plane. Bond lengths (C-O 1.392, C-S 1.779 Å) are in the normal range.

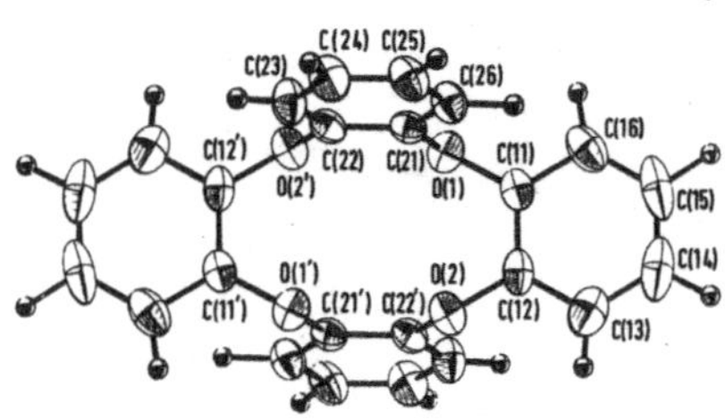

Fig. 1. Molecular structure of $C_{24}H_{16}O_4$.

3a,5,6a-TRIPHENYL-3,3a-DIHYDRO-2H-FURO[3,2-b]PYRROLE-2,6(6aH)-DIONE
$C_{24}H_{17}NO_3$

S. MATAKA, K. UEHARA, K. TAKAHASHI, M. TASHIRO, K. YOSHIHIRA, K. KAWAZOE, S. SATO and C. TAMURA, 1985. *Bull. Chem. Soc. Jpn.*, **58**, 3043-3044.

Monoclinic, $P2/a$, a = 16.606, b = 7.314, c = 16.295 Å, β = 111.67°, Z = 4. Cu radiation, R = 0.067 for 2453 reflections.

The 1-pyrrolene ring is planar (Fig. 1) and the tetrahydrofuran ring is close to envelope conformation (C(3)

0.33(2) Å from the plane of the other four atoms). The dihedral angle between the cis-fused rings is 104.1(2)°.

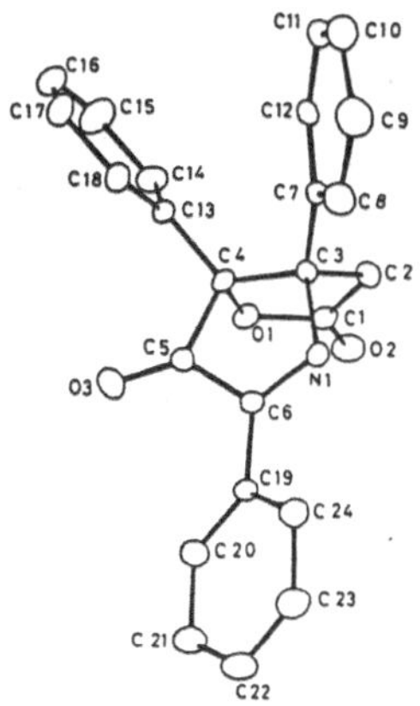

Fig. 1. A view of $C_{24}H_{17}NO_3$.

8-BENZYL-11,17-ETHENO-14-OXA-8-AZATETRACYCLO[8.7.0.0^{2,7}.0^{12,16}]HEPTADECA-9,13,15-TRIONE
$C_{24}H_{25}NO_4$

T. NAITO, I. NINOMIYA, M. DOI and M. INOUE, 1985. *Heterocycles*, **23**, 1215-1217.

Monoclinic, $P2_1/c$, a = 7.085, b = 10.941, c = 26.592 Å, β = 102.82°, Z = 4. Cu radiation, R = 0.0967 for 2971 reflections.

The analysis establishes the structure and stereochemistry as shown in Fig. 1.

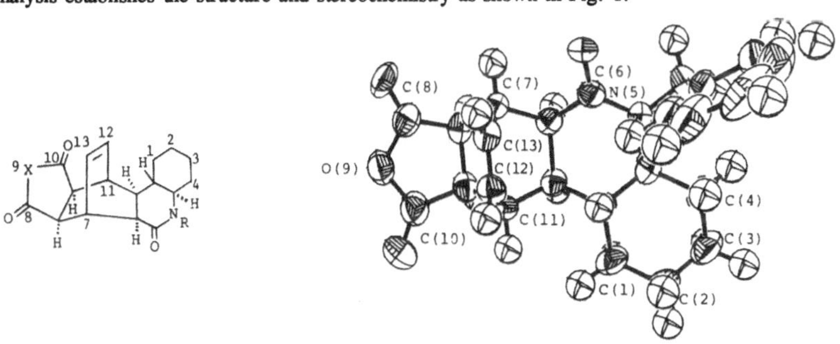

Fig. 1. The $C_{24}H_{25}NO_4$ molecule (X = O, R = CH_2Ph).

2,6-BIS((2-PHENYL-1,3-DIOXOLAN-2-YL)METHYL)PYRIDINE
$C_{25}H_{25}NO_4$

D.E. BENNETT, B.A. DIXON, A.C. GREENE, J.E. MORNINGSTAR, G.R. NEWKOME, F.R. FRONCZEK and S.F. WATKINS, 1985. *Acta Cryst.*, **C41**, 1473-1474.

Monoclinic, $C2/c$, a = 21.92, b = 6.276, c = 15.845 Å, β = 95.29°, Z = 4. Mo radiation, R = 0.041 for 1273 reflections.

The molecule (Fig. 1) has crystallographic symmetry 2. Both pyridine and phenyl rings are planar and they form a dihedral angle of 55.0(2)°. The torsion angles N-C(3)-C(4)-C(5), C(3)-C(4)-C(5)-C(6) and C(4)-C(5)-C(6)-C(7) are -113.8(3), 58.9(3) and -87.3(3)°, respectively. The acetal ring has an envelope conformation with O(2) as the flap. The crystal packing is dominated by the parallel stacking of aromatic rings.

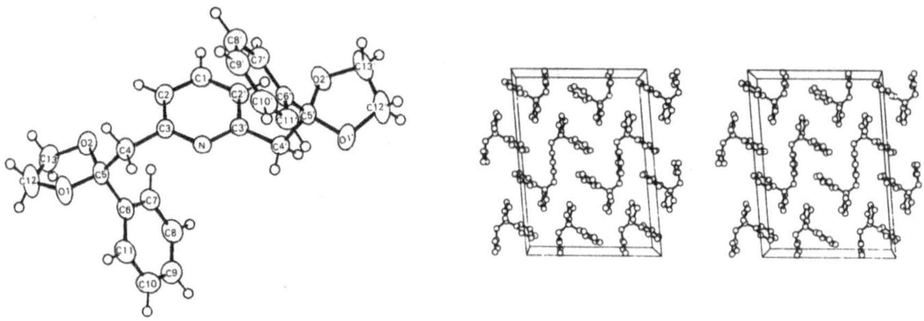

Fig. 1. $C_{25}H_{25}NO_4$: the molecular structure (left) and a stereoview of the unit cell down the symmetry axis.

6,6'-BIS(2-(6-BROMO-2-PYRIDYL)-1,3-DIOXOLAN-2-YL)-2,2'-BIPYRIDINE
$C_{26}H_{20}Br_2N_4O_4$

F.R. FRONCZEK, H.C.R. TAYLOR, V.K. GUPTA and G.R. NEWKOME, 1985. *Acta Cryst.*, **C41**, 1548-1549.

Triclinic, $P\bar{1}$, a = 6.231, b = 9.859, c = 10.299 Å, α = 88.45, β = 84.65, γ = 77.09°, Z = 1. Mo radiation, R = 0.037 for 1795 reflections.

The molecule (Fig. 1) lies on a crystallographic centre of inversion. The bipyridine moiety is in the anti-conformation, the dioxolane rings are in the half-chair conformation and the pyridine N atoms are oriented anti to the dioxolane O atoms. Bond lengths and angles are normal.

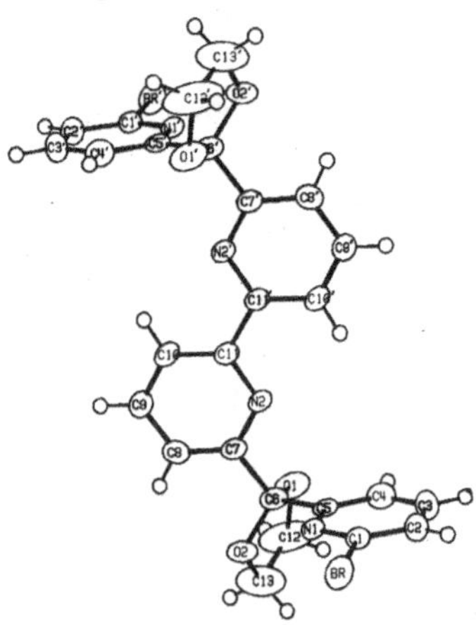

Fig. 1. $C_{26}H_{20}Br_2N_4O_4$: the conformation.

7-BENZYLOXY-6,2',3',4'-TETRAMETHOXYISOFLAV-3-ENE
$C_{26}H_{26}O_6$ (I)

2,6,7,2',3',4'-HEXAMETHOXYISOFLAV-3-ENE
$C_{21}H_{24}O_7$ (II)

T.C.A. AFONYA, F.B.M. EPELLE, W.B. WHALLEY, G. FERGUSON and M. PARVEZ, 1985. *J. Cryst. Spect. Res.*, **15**, 289-299.

I. Monoclinic, $P2_1/a$, a = 8.528, b = 13.294, c = 19.629 Å, β = 99.92°, Z = 4. Mo radiation, R = 0.053 for 1673 reflections.

II. Monoclinic, $C2/c$, a = 28.088, b = 10.662, c = 12.869 Å, β = 97.32°, Z = 8. Mo radiation, R = 0.041 for 1424 reflections.

The analyses establish the structures as shown in Fig. 1. Molecular dimensions are normal. In (I), ring C(1),C(6)-C(9),O(6) has an O(6) envelope conformation with O(6) 0.297 Å from the C(1),C(6)-C(9) plane. In (II) the six-membered ring O(1),C(1),C(6)-C(9) has a twist conformation with O(1) 0.207 Å above and C(9) 0.234 Å below the best plane through the ring.

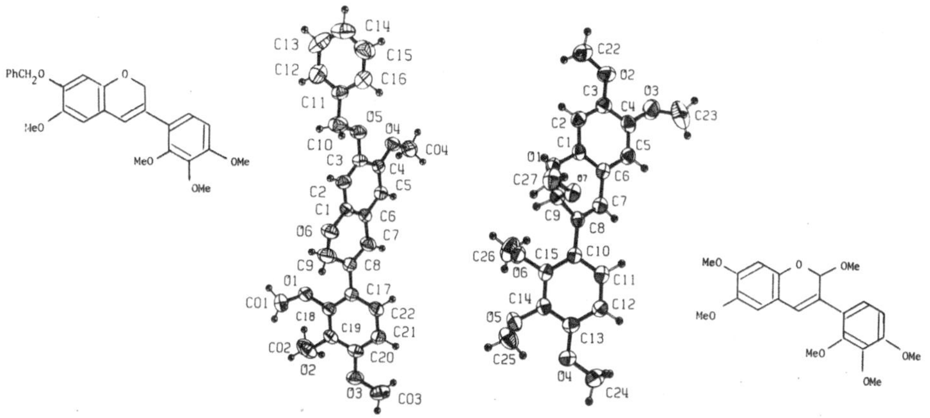

Fig. 1. The $C_{26}H_{26}O_6$ (I) (left) and $C_{21}H_{24}O_7$ (II) (right) molecules.

3-(p-ETHOXYPHENYL)-7-NONYLOXYCOUMARIN
$C_{26}H_{32}O_4$

O.S. FILIPENKO, V.I. PONOMAREV, L.O. ATOVMYAN, T.N. ANISIMOVA and B.M. BOLOTIN, 1985. *Kristallografiya*, **30**, 904-909.

Triclinic, $P\bar{1}$, a = 14.814, b = 16.628, c = 10.766 Å, α = 95.78, β = 102.10, γ = 117.02°, Z = 4. Cu radiation, R = 0.055 for 3844 reflections.

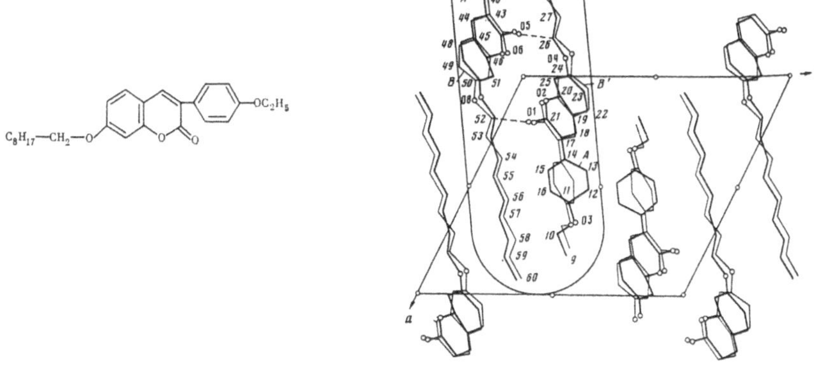

Fig. 1. The $C_{26}H_{32}O_4$ molecule and a projection of the crystal structure onto the *ab* plane.

The asymmetric unit contains two independent molecules A and B which are contact conformers and which differ in their dihedral angles between the planes of the phenyl and coumarin rings (33° for A, 23° for B). The A and B molecules form stacks (Fig. 1) with the flat coumarin fragments parallel to one another with the nonyloxy chains maximally extended. The stacks are joined in blocks parallel to ($1\bar{2}0$) about an inversion centre with short O...C contacts of 3.28-3.30 Å.

6a,7a,13b,13c-TETRAHYDRO-6a-HYDROXY-2,3,11,12-TETRAMETHOXY-6,6,8,8-TETRAMETHYL-6H,8H-BISCHROMENO[3,4-b:4',3'-d]FURAN
$C_{26}H_{32}O_8$

F. CAMPS, J. COLL, A. CONCHILLO, A. MESSEGUER, E. MOLINS and C. MIRAVITLLES, 1985. *J. Cryst. Spect. Res.*, **15**, 483-492.

Orthorhombic, *Pcab*, a = 9.857, b = 15.274, c = 32.601 Å, Z = 8. Mo radiation, R = 0.071 for 932 reflections.

The analysis shows that the central tetrahydrofuran ring has a half-chair conformation; the adjacent six-membered rings have distorted boat conformations. The angle between the tetrahydrofuran ring and the chroman fragments is 100° and the angle between both chroman ring systems is 41°.

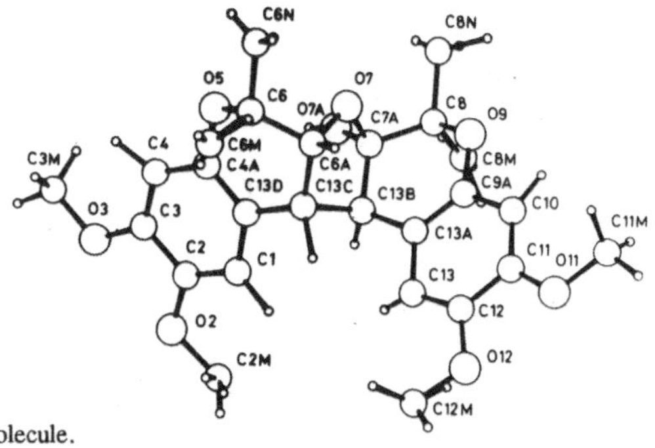

Fig. 1. The $C_{26}H_{32}O_8$ molecule.

1,4-DIBENZYL-3-(2'-THIOPYRIDYL)-6-(2''-γ-BUTYROLACTONYL)PIPERAZINE-2,5-DIONE
$C_{27}H_{25}N_3O_4S$

R.M. WILLIAMS, R.W. ARMSTRONG, L.K. MARUYAMA, JEN-SEN DUNG and O.P. ANDERSON, 1985. *J. Am. Chem. Soc.*, **107**, 3246-3253.

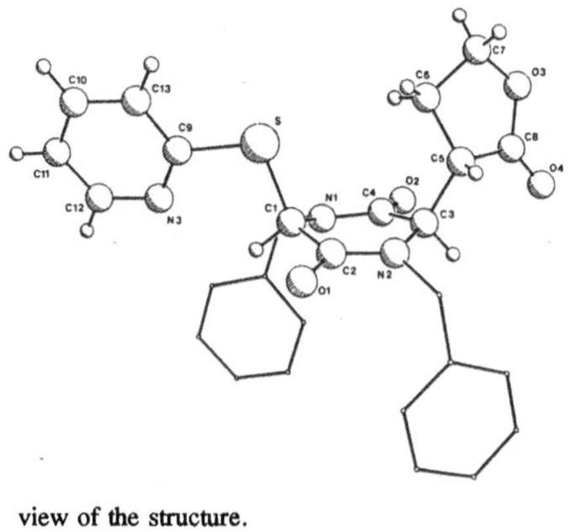

Fig. 1. $C_{27}H_{25}N_3O_4S$: view of the structure.

Orthorhombic, $Pna2_1$, a = 7.929, b = 16.094, c = 18.891 Å, Z = 4. Mo radiation, R = 0.038 for 1801 reflections.

In the structure (Fig. 1) all bond lengths and angles are unremarkable. The C-3 methine proton of the compound is confirmed in the syn conformation.

(6-DIETHYLAMINO-9-(2-FLUOROSULPHONYL-4-SULPHONATOPHENYL)-3H-XANTHENE-3-YLIDENE)DI-ETHYLAMMONIUM ETHANOL SOLVATE
$C_{27}H_{29}FN_2O_6S_2,2.5(C_2H_6O)$

I.N. POLYAKOVA, Z.A. STARIKOVA, B.V. PARUSNIKOV and I.A. KRASAVIN, 1985. *Zh. Strukt. Khim.*, **26**, 4-125-133 [J. Struct. Chem., **26**, 600-605].

Triclinic, $P\bar{1}$, a = 11.265, b = 12.870, c = 12.957 Å, α = 100.36, β = 88.94, γ = 109.82°, Z = 2. Mo radiation, R = 0.104 for 2308 reflections.

The rhodamine dye molecule is shown in Fig. 1. The planes of the benzene ring and the xanthene moiety are perpendicular to each other. The structure of the fragment may be described as a combination (with differing contributions) of two p-quinoid forms. The S-O bonds in the SO_3^- group (1.388-1.420 Å) and S-F in SO_2F (1.443 Å) are unusually short. The xanthene moieties of neighbouring molecules are situated side by side at a distance of 3.78 Å, and the packing of the "dimers" is loose. In the resulting "channels", ethanol molecules are situated.

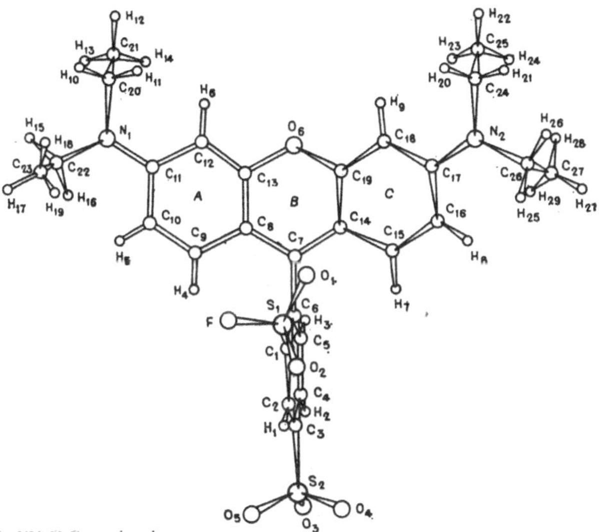

Fig. 1. The $C_{27}H_{29}FN_2O_6S_2$ molecule.

rel-(1S,3S,6R,8S,11R,13R,16S,18R)-2,2,7,7,12,12,17,17-OCTAMETHYL-21,22,23,24-TETRAOXAPENTA-CYCLO[16.2.1.$1^{3,6}$.$1^{8,11}$.$1^{13,16}$]TETRACOSANE
$C_{28}H_{48}O_4$ (I)

rel-(1S,3R,6S,8R,11S,13R,16S,18R)-2,2,7,7,12,12,17,17-OCTAMETHYL-21,22,23,24-TETRAOXAPENTA-CYCLO[16.2.1.$1^{3,6}$.$1^{8,11}$.$1^{13,16}$]TETRACOSANE
$C_{28}H_{48}O_4$ (II)

rel-(1S,3R,6S,8R,11S,13R,16S,18R)-2,2,7,7,12,12,17,17-OCTAMETHYL-21,22,23,24-TETRAOXAPENTA-CYCLO[16.2.1.$1^{3,6}$.$1^{8,11}$.$1^{13,16}$]TETRACOSANELITHIUM PICRATE
$C_{34}H_{50}LiN_3O_{11}$ (III)

M. VAN BEYLEN, B. ROLAND, G.S.D. KING and J. AERTS, 1985. *J. Chem. Res.*, S388-389, M4201-4289.

I. Triclinic, $P\bar{1}$, a = 10.966, b = 13.366, c = 10.958 Å, α = 98.54, β = 112.84, γ = 110.87°, Z = 2. Mo radiation, R = 0.066 for 2840 reflections.

II. Triclinic, $P\bar{1}$, a = 11.028, b = 12.882, c = 10.379 Å, α = 70.19, β = 71.89, γ = 80.06°, Z = 2. Mo radiation, R = 0.072 for 2054 reflections.

III. Orthorhombic, *Pcab*, a = 21.387, b = 22.169, c = 15.253 Å, Z = 8. Mo radiation, R = 0.072 for 2360 reflections.

The crystal structure analyses established the stereochemistries of the two isomers, (I) and (II), and of the complex, (III). The molecules are shown in Fig. 1. The complexing of isomer-2 has no effect on the bond lengths, but changes the conformation.

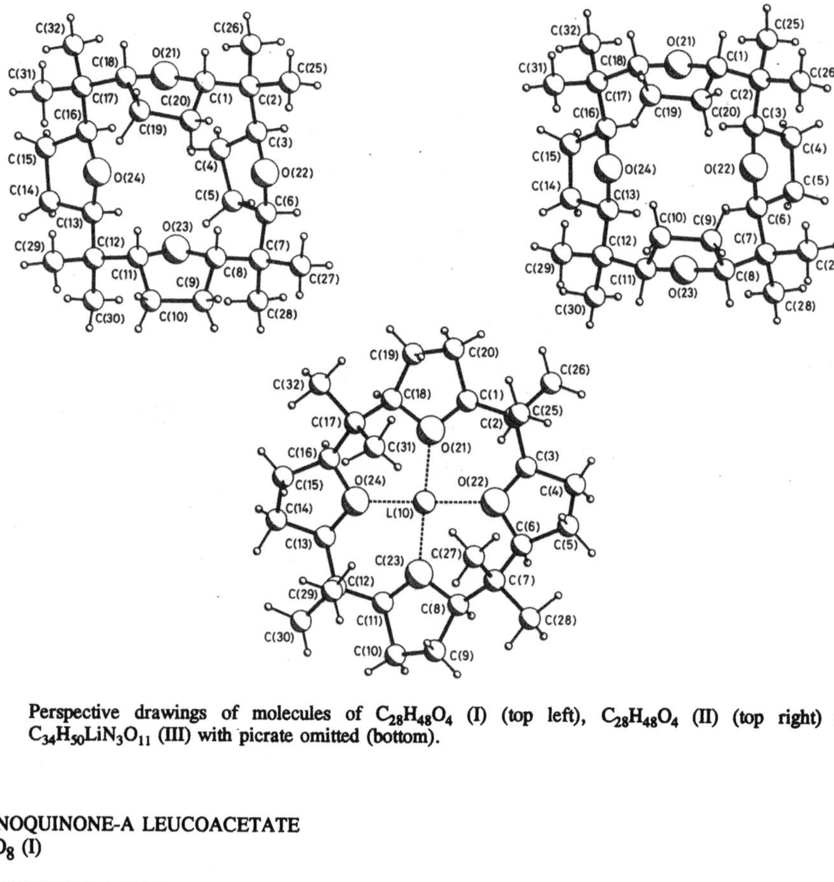

Fig. 1. Perspective drawings of molecules of $C_{28}H_{48}O_4$ (I) (top left), $C_{28}H_{48}O_4$ (II) (top right) and $C_{34}H_{50}LiN_3O_{11}$ (III) with picrate omitted (bottom).

ECKLONOQUINONE-A LEUCOACETATE
$C_{29}H_{36}O_8$ (I)

(1,9-DIMETHYL-4,6-BIS(1-METHYLETHYL)-7,8-DIOXO-7,8-DIHYDRODIBENZO[b,e](1,4)DIOXIN-2-YL)-METHANESULPHONATE
$C_{21}H_{24}O_7S$ (II)

W.L. LUTOLF, R. PREWO, J.H. BIERI, P. RUEDI and C.H. EUGSTER, 1985. *Helv. Chim. Acta*, **68**, 860-881.

I. Monoclinic, $P2_1/c$, a = 9.221, b = 15.283, c = 21.933 Å, β = 116.03°, Z = 4. Mo radiation, R = 0.055 for 4888 reflections (at -140°C).

II. Monoclinic, $P2_1/c$, a = 14.068, b = 9.657, c = 19.460 Å, β = 128.39°, Z = 4. Mo radiation, R = 0.088 for 1779 reflections (at -80°C).

The crystal structure analyses of (I) and (II) established the stereochemistries of the related molecules ecklonoquinone A, ecklonoquinone B, isoecklonoquinone A and isoecklonoquinone B. The molecules are shown in Fig. 1.

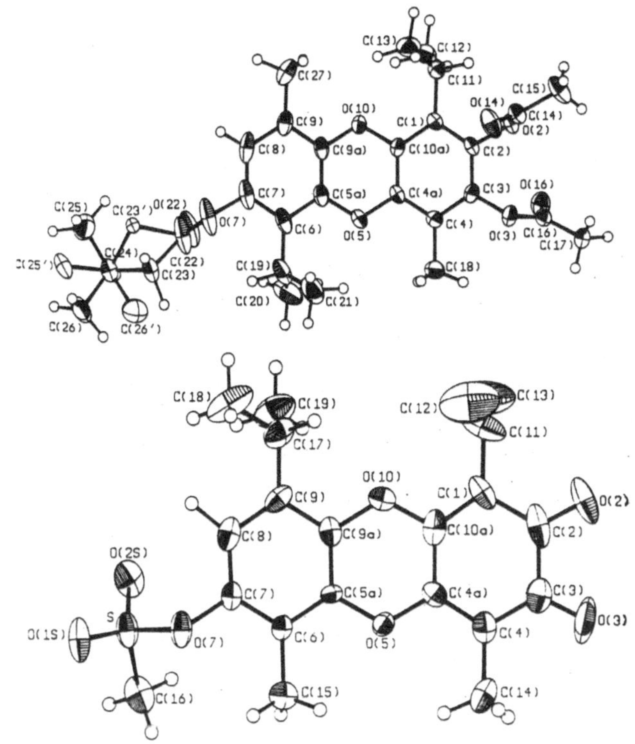

Fig. 1. ORTEP drawings of molecules of $C_{29}H_{36}O_8$ (top) and $C_{21}H_{24}O_7S$ (bottom).

METHYL (1RS,2SR,5SR,6SR,9SR,10SR)(2-ACETYL-9,10-EPOXY-1,2,3,4,5,6,7,8,9,10-DECAHYDRO-1,5-BIS(2-NITROPHENYL)SULPHENYL-2-ANTHRACENE-6-CARBOXYLATE)
$C_{30}H_{28}N_2O_8S_2$

J.-M. TORNARE, P. VOGEL, A.A. PINKERTON and D. SCHWARZENBACH, 1985. *Helv. Chim. Acta*, **68**, 2195-2215.

Triclinic, $P\bar{1}$, a = 7.7382, b = 9.381, c = 20.346 Å, α = 83.680, β = 82.951, γ = 78.200°, Z = 2. Cu radiation, R = 0.059 for 3342 reflections.

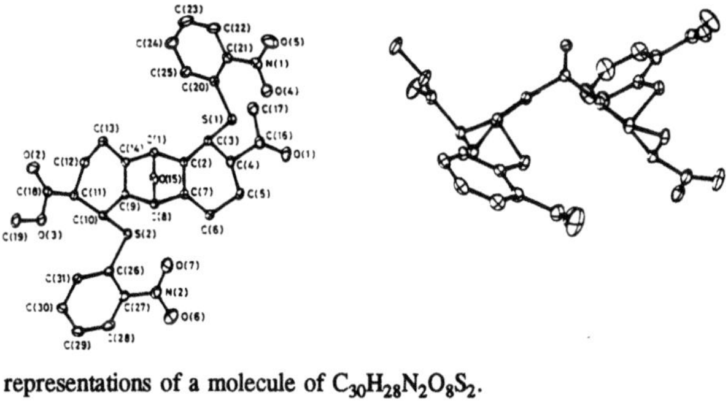

Fig. 1. ORTEP representations of a molecule of $C_{30}H_{28}N_2O_8S_2$.

The crystal structure analysis established the stereochemistry of this adduct (Fig. 1); a 7-oxabicyclo[2.2.1]hepta

2,5-diene system annellated to two cyclohexene rings.

DINAPHTHO[2,1-q:1,2-t](1,4,7,10,13,16-HEXAOXACYCLOEICOSANE)
$C_{31}H_{34}O_6$ (I)

19-(3,4,5-TRIMETHOXYPHENYL)DINAPHTHO[2,1-q:1,2-t](1,4,7,10,13,16-HEXACYCLOEICOSANE)
$C_{40}H_{44}O_9$ (II)

W. CLEGG, J.C. LOCKHART and M.B. McDONNELL, 1985. *J. Chem. Soc. Perkin Trans. 1,* 1019-1023.

I. Monoclinic, *P21/c*, a = 10.011, b = 12.736, c = 20.986 Å, β = 98.56°, Z = 4. Mo radiation, R = 0.075 for 2650 reflections.

II. Monoclinic, *C2/c*, a = 19.389, b = 11.704, c = 31.556 Å, β = 90.95°, Z = 8. Mo radiation, R = 0.050 for 4424 reflections.

The cyclic ether parts of the two crowns, (I) and (II), are very similar; the torsion angles are shown in Fig. 1.

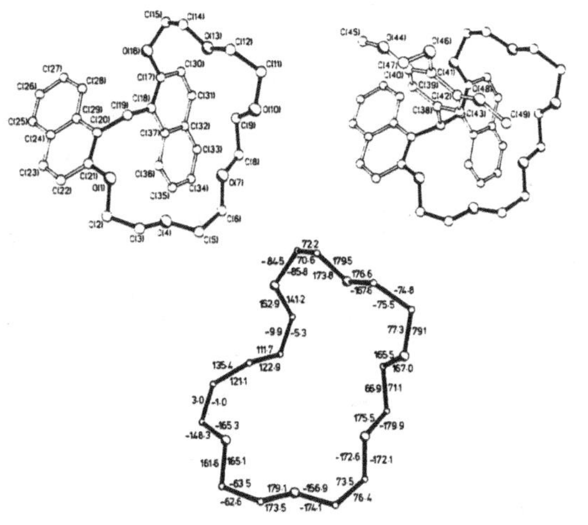

Fig. 1. Structures of $C_{31}H_{34}O_6$ (top left) and $C_{40}H_{44}O_9$ (top right): only the additional atoms are labelled for (II), other labelling is as for (I). Torsion angles for each on the bottom (values for (I) are shown inside the ring and for (II) outside).

1',3',4',5',6b,12,13,13a-OCTAHYDRO-1',7-DIMETHOXY-1',4'-EPOXY-7H-ACENAPHTHO[1',2'-3,4]FURANO-[2,1,5-a,b]-(2)-BENZPYRAN-12-SPIRO-3'-(2)-BENZOXEPIN-5'-ONE
$C_{32}H_{24}O_6$

J. TOYODA, T. IBATA, H. TAMURA, K. OGAWA, T. NISHINO and M. TAKEBAYASHI, 1985. *Bull. Chem. Soc. Jpn.*, **58**, 2212-2216.

Triclinic, $P\bar{1}$, a = 12.104, b = 12.477, c = 9.112 Å, α = 102.98, β = 110.16, γ = 70.65°, Z = 2. Cu radiation, R = 0.057 for 3681 reflections.

The molecule was shown to have the RRS structure (Fig. 1). The C(2)-C(13) bond of the acenaphthene ring is long (1.571(3) Å).

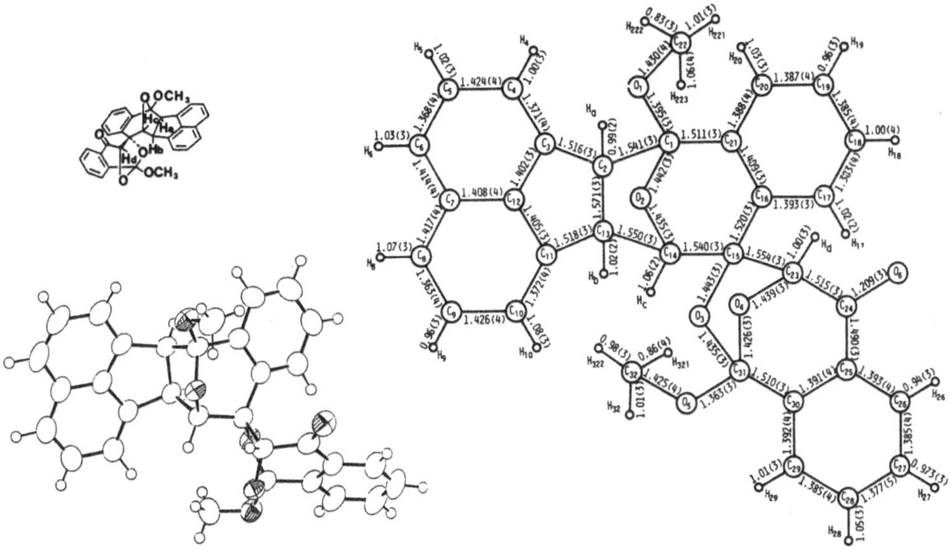

Fig. 1. The $C_{32}H_{24}O_6$ molecule and bond lengths.

(+)-STRIGOL-N-((R)-1-(1-NAPHTHYL)ETHYL)CARBAMATE
$C_{32}H_{33}NO_7$

D.W. BROOKS, H.S. BEVINAKATTI and D.R. POWELL, 1985. *J. Org. Chem.*, **50**, 3779-3781.

Monoclinic, $P2_1$, a = 13.215, b = 9.548, c = 12.705 Å, β = 118.50°, Z = 2. Cu radiation (at -140°C).

The crystal structure analysis determined the stereochemistry of this molecule (Fig. 1) and hence of a number of related molecules.

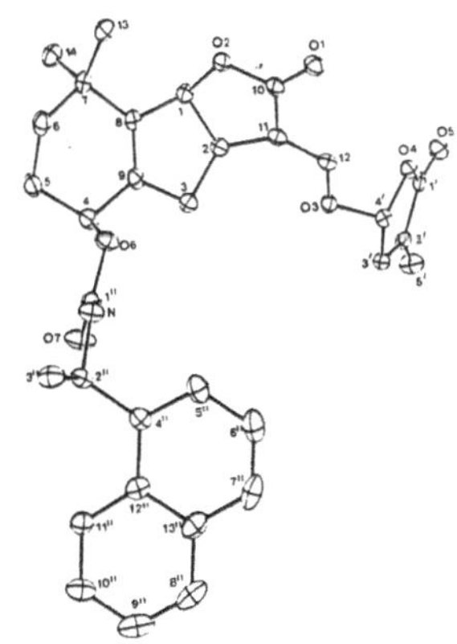

Fig. 1. Perspective drawing of a molecule of $C_{32}H_{33}NO_7$.

DISPIRO(TRICYCLO[3.3.1.1^{3,7}]DECANE-2,6'-6'H,12'H-DIBENZO[b,f](1,5)DIOXOCIN-12',2''-TRICYCLO-[3.3.1.1^{3,7}]DECANE)
$C_{32}H_{36}O_2$

M.F. GARBAUSKAS, I.A. EVANS and J.J. TALLEY, 1985. *Acta Cryst.*, **C41**, 140-142.

Monoclinic, $P2_1/a$, a = 13.961, b = 11.189, c = 16.208 Å, β = 107.94°, Z = 4. Mo radiation, R = 0.0488 for 1901 reflections.

In the molecule (Fig. 1) the central 1,5-dioxocin ring adopts a tub conformation which effectively minimises the steric hindrance between the two bulky adamantane moieties. One bridgehead C atom is above and the other is below the tub. Bond lengths and angles in the molecule are in agreement with generally accepted values. There are no unusually short intermolecular contacts. The correlation between the structure and its ^{13}C-NMR spectrum is briefly discussed.

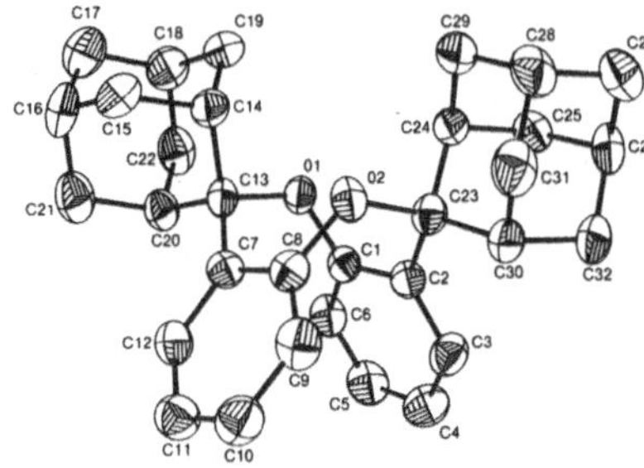

Fig. 1. $C_{32}H_{36}O_2$: a view of the molecule.

t-BUTYL 1,6,8-TRI-t-BUTYL-3,4-(9,10-PHENANTHRO)-2,5-DIOXABICYCLO[4.2.0]OCTA-3,7-DIENE-7-CARBOXYLATE
$C_{35}H_{44}O_4$

J. FINK, G. MAAS and M. REGITZ, 1985. *J. Org. Chem.*, **50**, 3839-3843.

Orthorhombic, *Pbca*, a = 15.412, b = 19.045, c = 20.835 Å, Z = 8. Mo radiation, R = 0.054 for 3159 reflections.

The crystal structure analysis established the stereochemistry of the molecule (Fig. 1) and hence the stereochemistries of related systems.

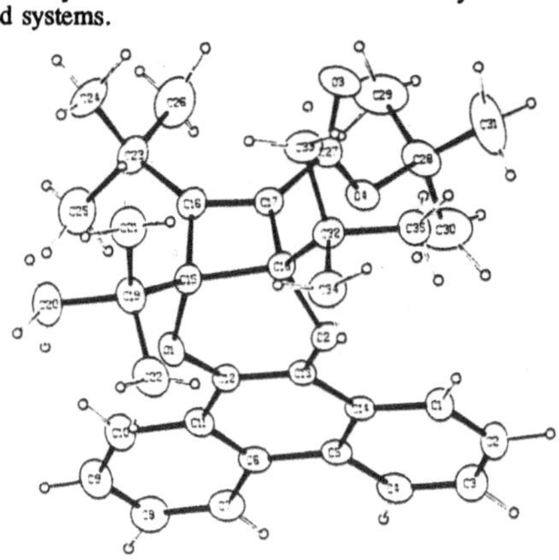

Fig. 1. The $C_{35}H_{44}O_4$ molecule.

4,5-BENZO-3,7,8,9,10-PENTAPHENYL-1-OXASPIRO[4.5]DECA-4,7,9-TRIEN-2-ONE
$C_{43}H_{30}O_2$

M. REGITZ, K. URGAST and G. MAAS, 1985. *Z. Naturforsch.*, **B40**, 67-76.

Triclinic, $P\bar{1}$, a = 9.014, b = 12.000, c = 14.891 Å, α = 82.96, β = 97.12, γ = 100.33°, Z = 2. Mo radiation, R = 0.059 for 5532 reflections.

The structure was established to be as shown in Fig. 1. The dihedral angle between the C(10)-C(1)-C(13) and O(1)-C(1)-C(9) planes at the spiro junction is exactly 90°.

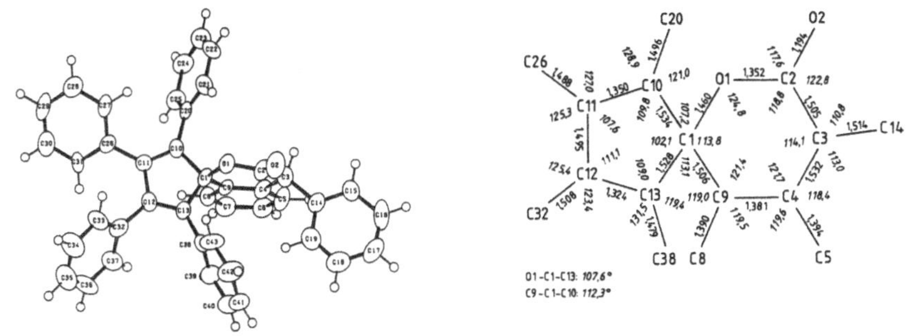

Fig. 1. The $C_{43}H_{30}O_2$ molecule and details of dimensions in the spiro ring systems.

TETRA-1-NAPHTHOIDE
$C_{44}H_{24}O_8$

R. GERDIL and G. BERNARDINELLI, 1985. *Acta Cryst.*, C**41**, 1523-1525.

Tetragonal, $P4_2/n$, a = 13.1416, c = 9.719 Å, Z = 2. Mo radiation, R = 0.034 for 562 reflections.

The molecules (Fig. 1) lie in special positions on crystallographic inversion tetrads with four planar $C_{11}H_6O_2$ units arranged up and down about the central symmetry axis. These four symmetry-related fragments are bonded together via four trans ester linkages giving rise to a sixteen-membered ring fused to four naphthalene moieties. Two naphthalene rings related by a twofold axis are inclined to one another at an angle of 66.8°. The O(1)...O(1) separations in the large internal ring are 2.626(3) Å. The sixteen-membered ring is locked in a rigid conformation mainly by the short contacts involving the carboxylate O atoms. There are no short intermolecular contacts. Bond lengths and angles are normal.

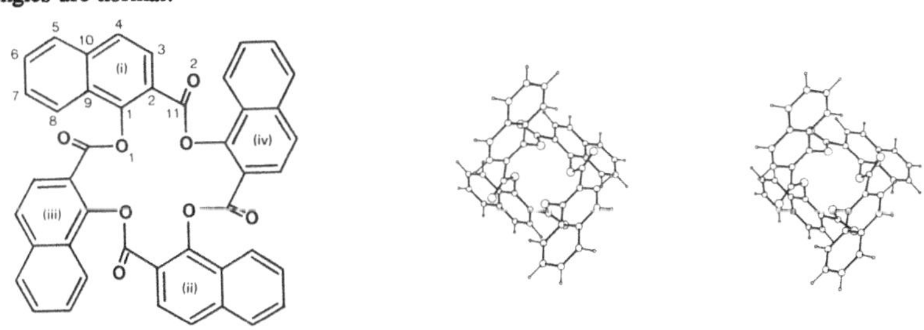

Fig. 1. $C_{44}H_{24}O_8$: the molecular skeleton and a stereoview of the molecule on a special position on the $\bar{4}$ axis, as viewed down *c*.

DISTEGYL ETHER
$C_{44}H_{42}O_{15}$

N. HOULBERT, E. BROWN, J.-P. ROBIN, D. DAVOUST, A. CHIARONI, T. PRANGE and C. RICHE, 1985. *J. Nat. Prod.*, **48**, 345-356.

Monoclinic, $P2_1$, a = 12.353, b = 12.871, c = 13.609 Å, β = 114.64°, Z = 2. Cu radiation, R = 0.051 for 2674 reflections.

The molecule is shown to have structure and stereochemistry as illustrated in Fig. 1.

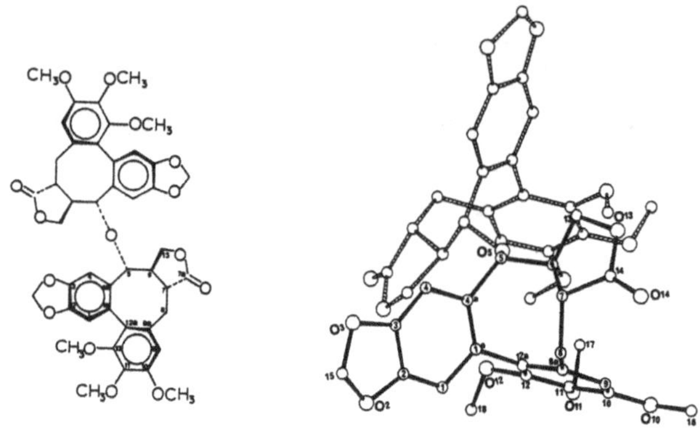

Fig. 1. Distegyl ether and a perspective view.

2,4,4-TRIFLUORO-1,3-DITHIETAN-2-YLIUM HEXAFLUOROARSENATE
$C_2AsF_9S_2$ $C_2F_3S_2^+,AsF_6^-$

J. ANTEL, K. HARMS, P.G. JONES, R. MEWS, G.M. SHELDRICK and A. WATERFELD, 1985. *Chem. Ber.*, **118**, 5006-5008.

Orthorhombic, *Pmmn*, a = 7.649, b = 8.317, c = 6.903 Å, Z = 2. Mo radiation, R = 0.045 for 445 reflections.

The cation has mm symmetry (Fig. 1) and the C-F and C-S bonds to the cationic carbon C(1) are significantly shorter than the other C-F and C-S bonds (C(1)-F(1) 1.248(12), C(1)-S 1.681(6), C(2)-F 1.294(7), C(2)-S 1.848(6) Å, S-C(1)-S' 108.3(6), S-C(2)-S' 95.0(4), C(1)-S-C(2) 78.4(4)°). The AsF_6^- ion has twofold disorder.

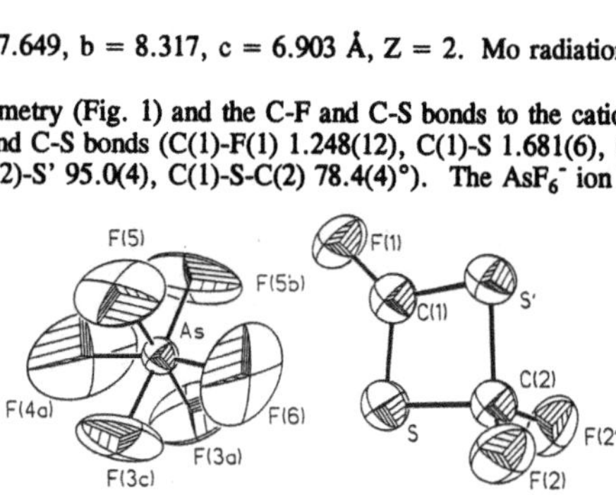

Fig. 1. The anion and cation in $C_2AsF_9S_2$.

DIPOTASSIUM 1,2-DITHIOLE-3-THIONE-4,5-DITHIOLATE
$C_3K_2S_5$ $K_2C_3S_5$

J. SIELER, F. BEYER, E. HOYER, L. ANDERSEN and O. LINDQVIST, 1985. *Acta Chem. Scand.*, **A39**, 153-156.

Orthorhombic, *Cmcm*, a = 14.436, b = 7.616, c = 8.383 Å, Z = 4. Mo radiation, R = 0.058 for 816 reflections.

The $C_3S_5^{2-}$ anion has exact C_{2v} symmetry. The C-S distance is 1.721(6) Å and the C-C distance is 1.420(7) Å. The S-S distance is 2.067(3) Å. Each potassium atom is coordinated by eight sulphur atoms (Fig. 1); the shortest K-S distance is 3.182(1) Å.

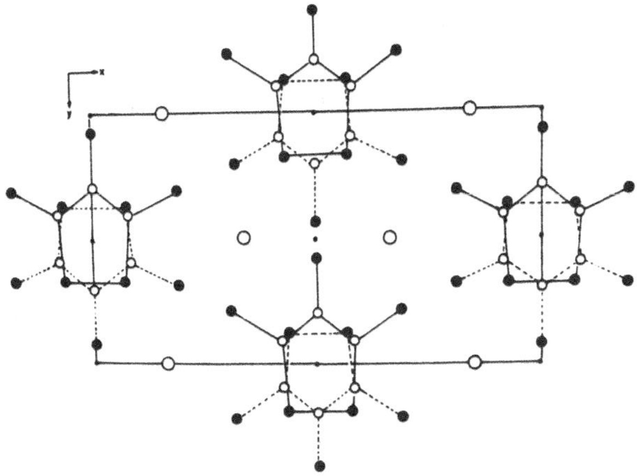

Fig. 1. The packing of the $C_3K_2S_5$ structure.

3H-1,2-DITHIOLE-3-THIONE
$C_3H_2S_3$

CHIN HSUAN WEI, 1985. *Acta Cryst.*, C41, 1768-1770.

Tetragonal $P4_2/mbc$, a = 10.654, c = 9.304 Å, Z = 8. Mo radiation, R = 0.049 for 571 reflections.

The crystal structure (Fig. 1) is disordered in that half of the units in the crystal are related to the other half by the crystallographic mirror plane through z = 0. With S(3) and C(3) atoms constrained to the mirror plane, the other atoms are required to take alternative positions with 50% occupancy on opposite sides of the mirror plane. A consequence of the disorder is that the molecular parameters are not as precise as might be expected otherwise. Molecules are approximately planar and the plane normal of the five-membered ring makes an angle of 91.2(1)° with the z-axis. The C(4)-C(5) bond length (1.34(1) Å) indicates some double bond character. The S(1)-S(2) bond length is 2.050(3) Å.

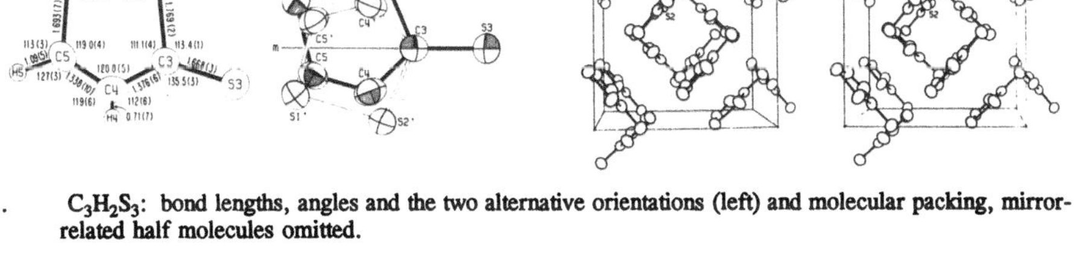

Fig. 1. $C_3H_2S_3$: bond lengths, angles and the two alternative orientations (left) and molecular packing, mirror-related half molecules omitted.

8-CYANOISOTHIAZOLOPENTATHIEPIN
$C_4N_2S_6$ (I)

THIENO[3,4-f]-1,2,3,4,5-PENTATHIEPIN
$C_4H_2S_6$ (II)

7-PHENYLPYRAZOLOPENTATHIEPIN
$C_9H_6N_2S_5$ (III)

6-METHYLBENZOPENTATHIEPIN

$C_7H_6S_5$ (IV)

6-TRIFLUOROMETHYL-4-NITROBENZOTRITHIOLE
$C_7H_2F_3NO_2S_3$ (V)

B.L. CHENARD, R.L. HARLOW, A.L. JOHNSON and S.A. VLADUCHICK, 1985. *J. Am. Chem. Soc.*, **107**, 3871-3879.

I. Monoclinic, *P*2_1*/c*, a = 16.722, b = 6.240, c = 9.249 Å, β = 104.49°, Z = 4. Mo radiation, R = 0.051 for 1706 reflections.

II. Orthorhombic, *Pbca*, a = 7.811, b = 24.395, c = 8.867 Å, Z = 8. Mo radiation, R = 0.025 for 1382 reflections (at -100°C).

III. Monoclinic, *P*2_1*/c*, a = 14.852, b = 7.367, c = 12.305 Å, β = 115.86°, Z = 4. Mo radiation, R = 0.034 for 2242 reflections.

IV. Triclinic, *P*$\bar{1}$, a = 7.829, b = 9.354, c = 7.760 Å, α = 97.84, β = 103.03, γ = 112.23°, Z = 2. Mo radiation, R = 0.036 for 1877 reflections (at -100°C).

V. Monoclinic, *P*2_1*/c*, a = 15.376, b = 8.184, c = 7.816 Å, β = 96.79°, Z = 4. Mo radiation, R = 0.045 for 1359 reflections (at -100°C).

Molecules (I)-(IV) have the sulphur containing rings in chair conformation (Fig. 1). There is a remarkable similarity in S-S distances in the compounds (2.051-2.063(1) (I); 2.052-2.057(1) (II); 2.043-2.053(1) (III); 2.047-2.058(1) (IV)). The C_2S_3 ring in (V) has an envelope conformation with S-S 2.065 and 2.071(1) Å. (The paper also reports a redetermination of orthorhombic S_8 (1) which yielded S-S 2.051 Å).

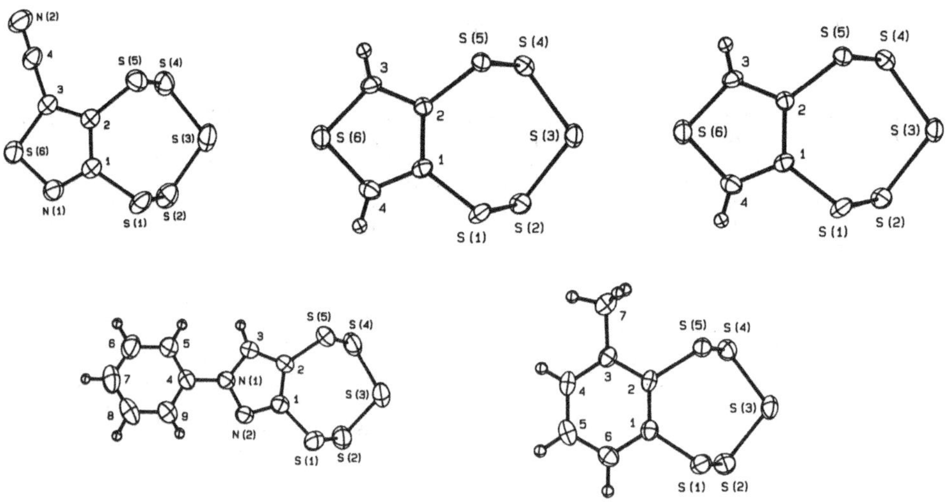

Fig. 1. The pentathiepins (I)-(IV) and benzotrithiole (V) (clockwise from top left).

1. P. COPPENS, Y.W. YANG, R.H. BLESSING, W.F. COOPER and S.F.K. LARSEN, 1977. J. Am. Chem. Soc., **99**, 760-766.

4-SELENANONE
C_5H_8OSe

M.D. THOMPSON, E.M. HOLT, K.D. BERLIN and B.J. SCHERLAG, 1985. *J. Org. Chem.*, **50**, 2580-2581.

Monoclinic, *P*2_1*/n*, a = 6.949, b = 5.620, c = 16.926 Å, β = 112.03°, Z = 4. Mo radiation, R = 0.081.

The six-membered selanone ring (Fig. 1) has a chair conformation.

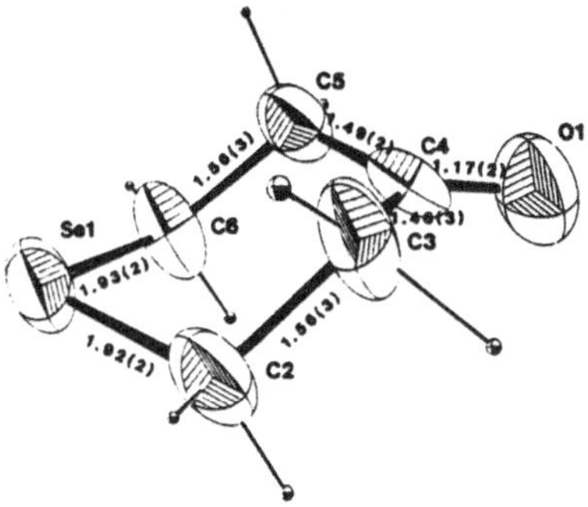

Fig. 1. Projection view of a molecule of C_5H_8OSe showing some bond lengths (Å).

1,4,6,9-TETRATHIASPIRO[4.4]NONANE
$C_5H_8S_4$ (I)

1,5,7,11-TETRATHIASPIRO[5.5]UNDECANE
$C_7H_{12}S_4$ (II)

J.D. KORP and I. BERNAL, 1985. *J. Cryst. Spect. Res.*, **15**, 201-209.

I. Tetragonal, *I*4_1*cd*, a = 9.521, c = 19.052 Å, Z = 8. Mo radiation, R = 0.064 for 189 reflections.

II. Orthorhombic, *Pbca*, a = 8.955, b = 12.356, c = 18.425 Å, Z = 8. Mo radiation, R = 0.025 for 1429 reflections.

Molecule (I) has twofold crystallographic symmetry and the unique five-membered ring has a half-chair conformation (Fig. 1). There is probably some disorder, either positional or thermal; the S-C distances are in the range 1.780(9)-1.793(7) Å with C-C 1.476(9) Å. Compound (II) has both rings in normal chair conformations with mean S-C 1.815(2), C-C 1.506(4) Å. Bond angles at C(1) of (II) show an unusually large range from 101.3(1)° (S(1)-C(1)-C(3)) to 113.7(1)° (S(2)-C(1)-S(4)).

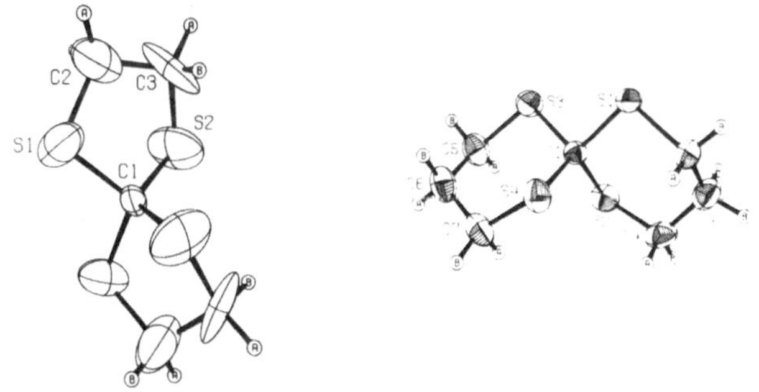

Fig. 1. Views of $C_5H_8S_4$ (I) (left) and $C_7H_{12}S_4$ (II) (right).

2,6-DIMETHYL-4H-1-THIAPYRAN-4-ONE (2,6-DIMETHYL-4H-THIIN-4-ONE)
C_7H_8OS

A. BANERJEE, C.J. BROWN and P.C. JAIN, 1985. *Acta Cryst.*, **C41**, 1505-1507.

Monoclinic, *I*2/*a*, a = 9.036, b = 11.348, c = 7.360 Å, *β* = 104.4°, Z = 4. Mo radiation, R = 0.028 for 456 reflections.

Molecules in the crystal are essentially planar and stack in a close-packed arrangement in layers perpendicular to *c*, with the S, O and one C atom lying on a twofold axis (Fig. 1). The S-C distance is 1.737(2) Å. Packing is governed by van der Waals interactions.

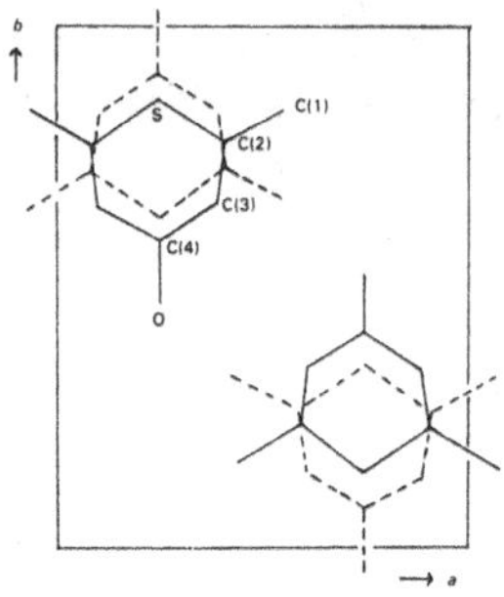

Fig. 1. Arrangement of C_7H_8OS molecules in projection of the unit cell.

4-METHYL-5-(2-PYRAZINYL)-3H-1,2-DITHIOL-3-ONE
$C_8H_6N_2OS_2$

CHIN HSUAN WEI, 1985. *Acta Cryst.*, C41, 1525-1528.

Monoclinic, $P2_1/a$, a = 11.863, b = 19.873, c = 7.552 Å, β = 96.01°, Z = 8. Mo radiation, R = 0.048 for 2664 reflections.

There are two independent molecules of this Oltipraz analogue in the asymmetric unit. Corresponding molecular parameters closely resemble one another (Fig. 1). Individual six- and five-membered rings are nearly planar, the angle between the plane normals being 4.6(1)° in one molecule and 9.6(1)° in the other. Packing is governed by van der Waals interactions.

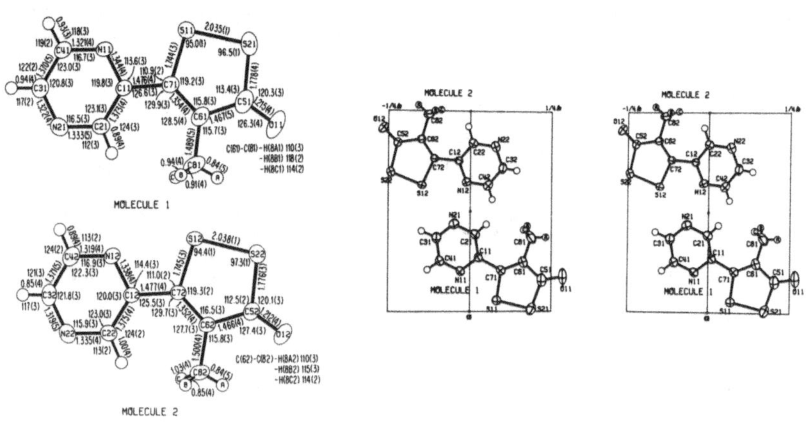

Fig. 1. $C_8H_6N_2OS_2$: bond lengths and angles and a stereoview of the two independent molecules in the asymmetric unit.

3-AMINO-4-CYANO-2-(2,2-DICYANOVINYL)THIOPHENE METHANOL SOLVATE
$C_9H_4N_4S,CH_4O$

S.A. MARINUZZI-BROSEMER, D.C. DITTMER, M.H.M. CHEN and J. CLARDY, 1985. *J. Org. Chem.*, 50, 799-802.

Triclinic, $P\bar{1}$, a = 12.205, b = 14.018, c = 3.836 Å, α = 94.61, β = 81.76, γ = 118.12°, Z = 2. Cu radiation, R = 0.070 for 1050 reflections.

The three nitrile bonds in the molecule (Fig. 1) have lengths of 1.124(10), 1.131(14) and 1.126(12) Å respectively; S(1)-C(5) is 1.742 but S(1)-C(2) is 1.698 Å; the two α-β carbon bonds C(4)-C(5) and C(2)-C(3) measure 1.429 and 1.363 Å respectively.

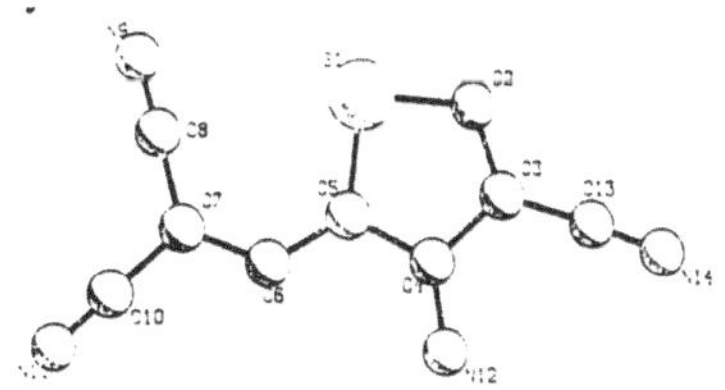

Fig. 1. A molecule of $C_9H_4N_4S$ with hydrogen atoms omitted.

2-METHYL-1-BENZOTHIOPHENE
C_9H_8S

H.H. SUTHERLAND and A. RAWAS, 1985. *Acta Cryst.*, C41, 929-931.

Orthorhombic, $Pna2_1$, a = 11.71, b = 11.03, c = 5.96 Å, Z = 4. Cu radiation, R = 0.079 for 553 reflections - (at 283 K).

The individual five- and six-membered rings are planar but the molecule as a whole (Fig. 1) is non-planar, the rings being inclined at 0.6(4)° to one another. The average C-C bond length in the phenyl ring is 1.398(20) Å. The average C-S bond length is 1.756(16) Å.

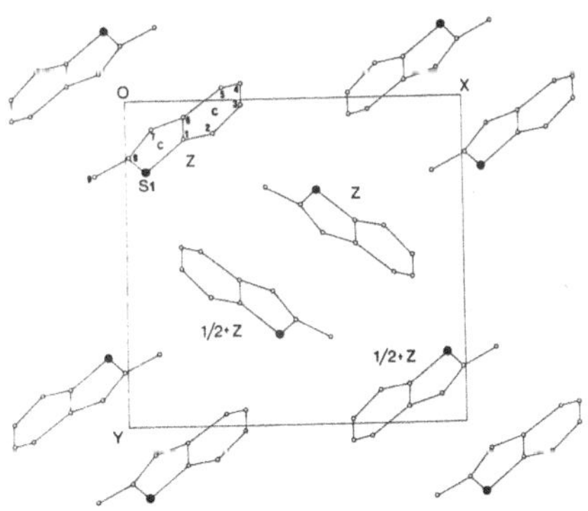

Fig. 1. The arrangement of the C_9H_8S molecules in the unit cell viewed along **c**.

1,4-DITHIANIUM-1-BIS(METHOXYCARBONYL)METHYLIDE (1,4-DITHIANYLIUM-1-(BIS(METHOXYCARBONYL))METHANIDE)
$C_9H_{15}O_4S_2$

W.H. WATSON, I. VICKOVIC and A.L. TERNAY Jr., 1985. *Acta Cryst.*, C41, 1090-1092.

Monoclinic, $P2_1/n$, a = 12.942, b = 10.344, c = 8.974 Å, β = 108.87°, Z = 4. Cu radiation, R = 0.036 for 1306 reflections.

The S(1)-C(7) bond length in the molecule (Fig. 1) is 1.722(3) Å and the four S-O(sp^3) bond distances are statistically equivalent with an average value of 1.802(4) Å. The 1,4-dithiane ring adopts a chair conformation with the ylide S(1) atom displaced by 0.688(3) Å out of the plane of the three attached C atoms. The malonylide moiety shows significant deviations from planarity with O(9) displaced from the least squares plane by 0.23(1) Å and with torsion angles within the group deviating by up to 15° from the planar values of 0 and 180°. The dithiane ring is almost perpendicular to the carbanion moiety with short intramolecular contacts O(9)...O(13), O(10)...S(1) and O(12)...S(1) of lengths 2.693(4), 2.739(3) and 2.996(4) Å respectively. There are also slightly short C-H...O(9) intermolecular contacts.

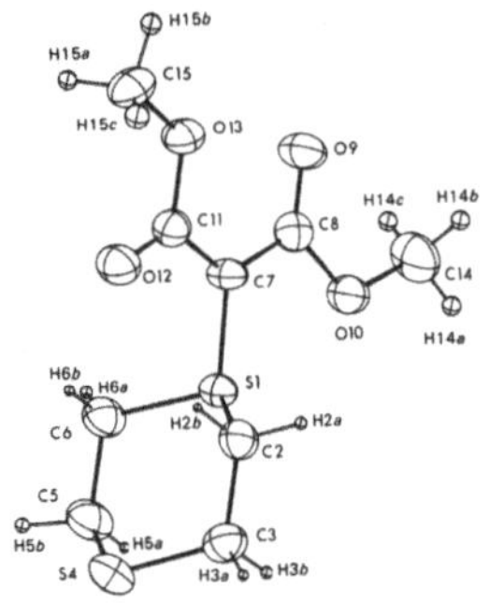

Fig. 1. $C_9H_{15}O_4S_2$: a view of the molecule showing the atomic numbering.

1,4-DIAZATHIANTHRENE
$C_{10}H_6N_2S_2$

S.B. LARSON, S.H. SIMONSEN, W.W. LAM, G.E. MARTIN, C.M. LINDSAY and K. SMITH, 1985. *Acta Cryst.*, **C41**, 1784-1786.

Triclinic, $P\bar{1}$, a = 8.3751, b = 10.7988, c = 11.1715 Å, α = 83.947, β = 74.611, γ = 75.073°, Z = 4. Mo radiation, R = 0.0387 for 4176 reflections (at 163 K).

Bond lengths and angles in the two crystallographically independent molecules (A,B) in the asymmetric unit are in good agreement (Fig. 1). The benzene and pyrazine rings are planar with dihedral angles of 138.93(7) and 137.00(7)°. In the crystal, one molecule appears to fit into the fold of the other independent molecule.

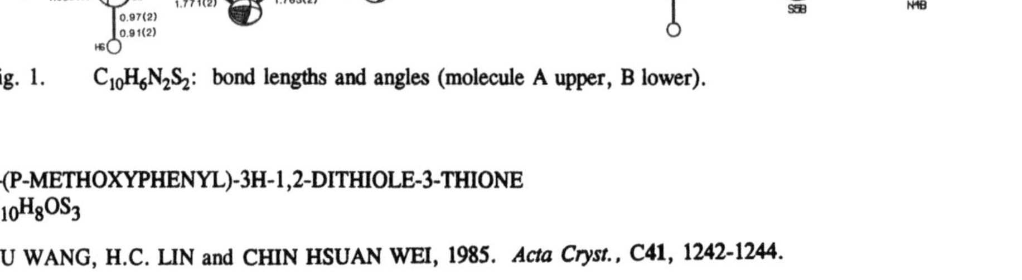

Fig. 1. $C_{10}H_6N_2S_2$: bond lengths and angles (molecule A upper, B lower).

5-(P-METHOXYPHENYL)-3H-1,2-DITHIOLE-3-THIONE
$C_{10}H_8OS_3$

YU WANG, H.C. LIN and CHIN HSUAN WEI, 1985. *Acta Cryst.*, **C41**, 1242-1244.

Monoclinic, $P2_1/n$, a = 10.834, b = 13.446, c = 7.517 Å, β = 104.96°, Z = 4. Mo radiation, R = 0.029 for 1652 reflections.

The molecule (Fig. 1) consists of planar methoxyphenyl and 1,2-dithiole-3-thione rings with a dihedral angle between the normals to the mean planes of 7.7(1)°. The S(2)-S(3) distance is 2.047(1) and the exocyclic C(1)=S(1) distance is 1.666(3) Å.

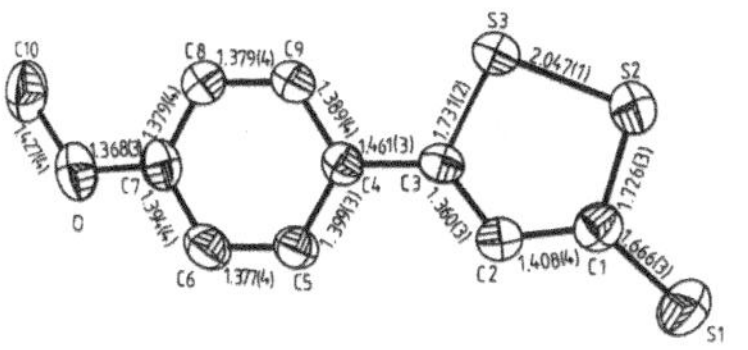

Fig. 1. $C_{10}H_8OS_3$: the molecular dimensions.

3,4'-DIMETHYL-3',4-BIS(METHYLTHIO)-2,2',5,5'-TETRATHIAFULVALENE
$C_{10}H_{12}S_6$

L. OUAHAB and P. BATAIL, 1985. *Acta Cryst.*, **C41**, 928-929.

Triclinic, $P\bar{1}$, a = 7.883, b = 9.101, c = 5.193 Å, α = 96.32, β = 106.63, γ = 99.73°, Z = 1. Mo radiation, R = 0.062 for 1171 reflections.

The methylthio substituents are trans to each other on a planar tetrathiafulvalene (TTF) unit located at a centre of symmetry in the crystal (Fig. 1). The S(5) atom is in the TTF plane while C(7) is 1.29 Å out of this plane.

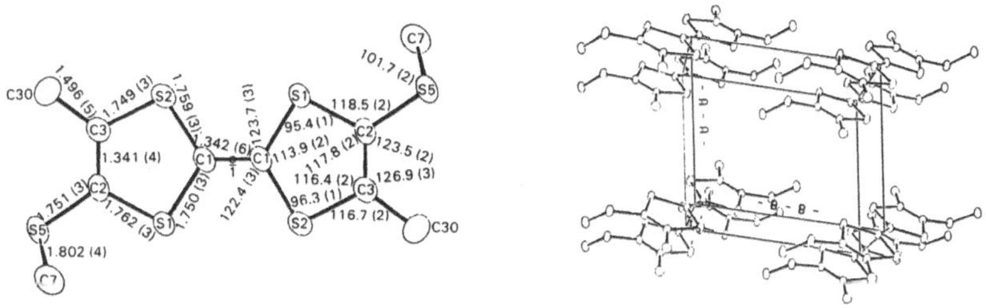

Fig. 1. $C_{10}H_{12}S_6$: the molecular dimensions and the crystal structure.

2-METHYL-2-(4-NITROPHENYL)-1,3-DITHIANE
$C_{11}H_{13}NO_2S_2$

H. FINK and I. STAHL, 1985. *Acta Cryst.*, **C41**, 151-153.

Monoclinic, $P2_1/n$, a = 11.039, b = 16.942, c = 6.381 Å, β = 91.35°, Z = 4. Mo radiation, R = 0.036 for 659 reflections.

In the molecule (Fig. 1) the 1,3-dithiane ring has a chair conformation. The planar phenyl substituent is in an axial position and has an orientation perpendicular to the pseudo-mirror plane of the hetero ring. The C-S distances are normal and range from 1.809(6) to 1.820(6) Å. There are no unusually short intermolecular distances.

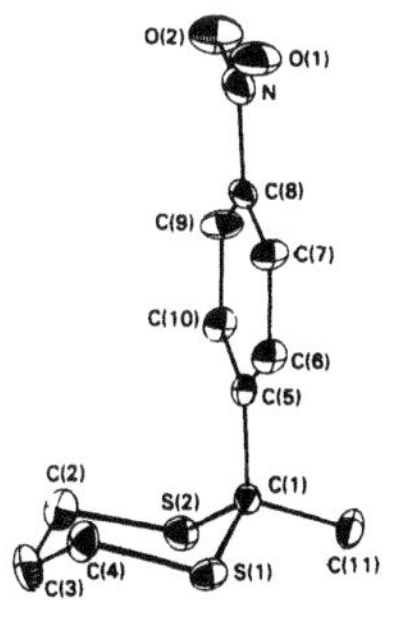

Fig. 1. $C_{11}H_{13}NO_2S_2$: a view of the molecule.

1-METHYL-4-THIATRICYCLO[5.4.0.$0^{2,6}$]UNDECAN-8-ONE-4,4-DIOXIDE
$C_{11}H_{16}O_3S$ (I)

1-METHYL-4-THIATRICYCLO[5.4.0.$0^{2,6}$]UNDECAN-8-OL-4,4-DIOXIDE
$C_{11}H_{18}O_3S$ (II)

J.R. WILLIAMS, C. LIN and D.F. CHODOSH, 1985. *J. Org. Chem.*, **50**, 5815-5822.

I. Monoclinic, $P2_1/n$, a = 8.874, b = 13.034, c = 9.378 Å, β = 96.33°, Z = 4. Mo radiation, R = 0.036 for 2790 reflections.

II. Monoclinic, $P2_1/n$, a = 6.643, b = 22.695, c = 7.449 Å, β = 99.99°, Z = 4. Mo radiation, R = 0.036 for 2374 reflections.

The crystal structure analyses established that (I) is the cis-anti-cis photoadduct and (II) is the cis-trans alcohol (Fig. 1).

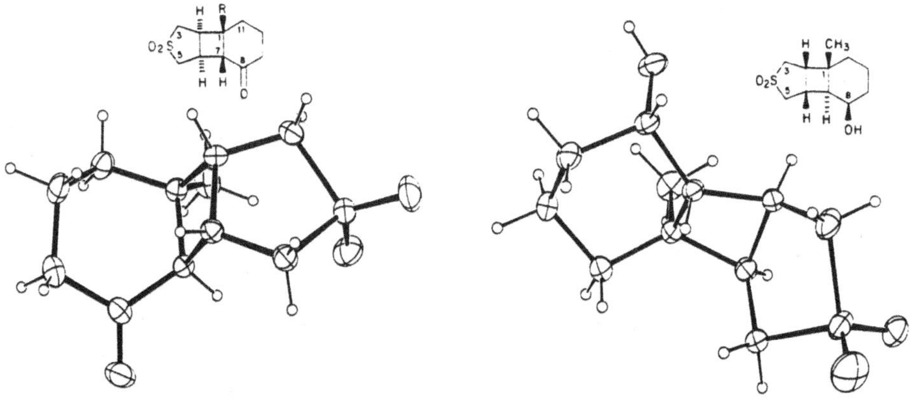

Fig. 1. Drawings of $C_{11}H_{16}O_3S$ (I) (left) and $C_{11}H_{18}O_3S$ (II) (right).

2-ISOBUTYL-1,3-DIHYDROBENZO[c]THIOPHENIUM PERCHLORATE
$C_{12}H_{17}ClO_4S$ $C_{12}H_{17}S^+,ClO_4^-$

J.F. KING, G.T.Y. TSANG, M.M. ABDEL-MALIK and N.C. PAYNE, 1985. *J. Am. Chem. Soc.*, **107**, 3224-3232.

Monoclinic, $P2_1/n$, a = 17.382, b = 9.1731, c = 8.7828 Å, β = 91.7°, Z = 4. Cu radiation, R = 0.044 for 2054 reflections.

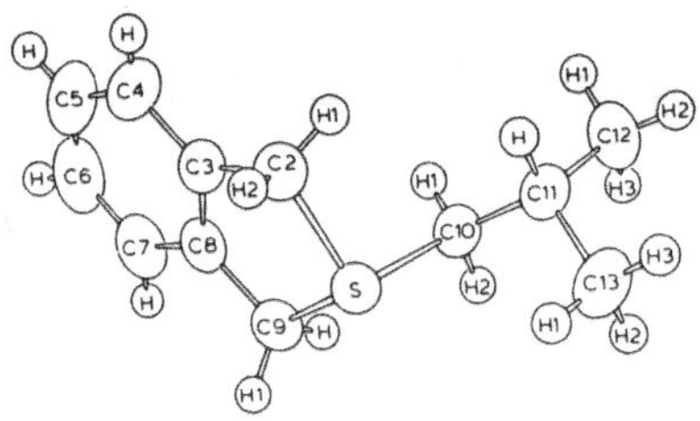

Fig. 1. $C_{12}H_{17}ClO_4S$: view of the $C_{12}H_{17}S^+$ cation.

The structure (Fig. 1) consists of discrete cations and anions. The ordered perchlorate anions are unexceptional. In the cation, the S atom adopts a normal trigonal geometry but with the C(2)-S-C(9) angle being small at 95.7(1)° and the S atom being 0.85 Å displaced from the C3 plane. The isopropyl group on C(10) is anti (178°) to C(9) whereas the C(10)-C(11) bond is orthogonal (87°) to the C(2)-S bond. The S...O(3) (perchlorate) distance of 3.080(3) Å suggests a weak interaction. The C(2)-C(9)-S plane is at 3.8° to the arene fragment plane which makes the C=S$^+$ bond almost coplanar with the benzene ring. The five-membered ring is strained as shown by the small

C(2)-S-C(9) angle. The only remaining short contact is one of 2.77(4) Å between S and H(1,C(13)).

BIS(DIMETHYLTRIMETHYLENETETRATHIAFULVALENE) HEXAFLUOROARSENATE
$C_{12}H_{24}AsF_6S_8$ (I) $C_{11}H_{12}S_4^+, C_{11}H_{12}S_4, AsF_6^-$

BIS(DIMETHYLTRIMETHYLENETETRASELENAFULVALENE) HEXAFLUOROARSENATE
$C_{12}H_{24}AsF_6Se_8$ (II) $C_{11}H_{12}Se_4^+, C_{11}H_{12}Se_4, AsF_6^-$

D. CHASSEAU, 1985. *Nouv. J. Chim.*, **9**, 347-350.

I. Triclinic, $P\bar{1}$, a = 7.183, b = 7.831, c = 13.455 Å, α = 99.26, β = 89.88, γ = 112.31°, Z = 1. Mo radiation, R = 0.100 for 1718 reflections.

II. Triclinic, $P\bar{1}$, a = 7.334, b = 7.795, c = 13.682 Å, α = 100.14, β = 88.32, γ = 109.87°, Z = 1. Mo radiation, R = 0.090 for 990 reflections.

The structures are isomorphous and have the AsF_6^- ions on inversion centres with the thiafulvalenes in general positions related by inversion centres (Fig. 1). Inter-cation spacings are 3.575 and 3.606 Å in (I) and 3.655 and 3.675 Å in (II).

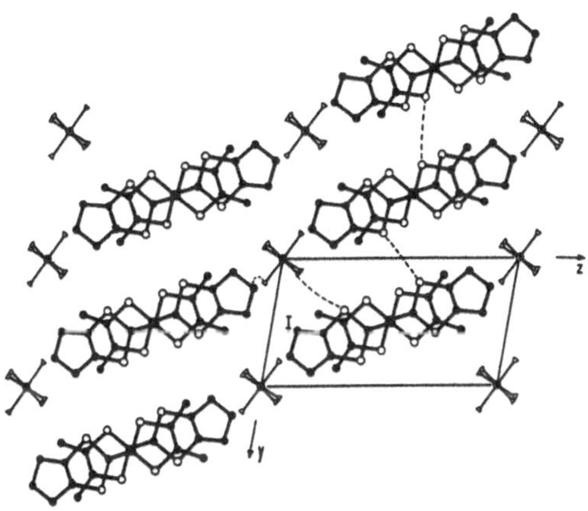

Fig. 1. A projection of the $C_{12}H_{24}AsF_6S_4$ structure along the *a*-direction.

3-CHLORO-2-METHYL-3-(2-THIENYL)ACROLEIN 2,4-DINITROPHENYLHYDRAZONE
$C_{14}H_{11}ClN_4O_4S$

A. CARPITA, A. LEZZI, R. ROSSI, F. MARCHETTI and S. MERLINO, 1985. *Tetrahedron*, **41**, 621-625.

Monoclinic, $P2_1/c$, a = 12.352, b = 11.287, c = 11.214 Å, β = 93.56°, Z = 4. Mo radiation, R = 0.051 for 2316 reflections.

The structure and stereochemistry were established as shown in Fig. 1. The thienyl ring plane forms a dihedral angle of 133.6° with the Cl,C(4)-C(8) plane, which is itself almost coplanar (dihedral angle 7.7°) with the plane of the 2,4-DNP moiety.

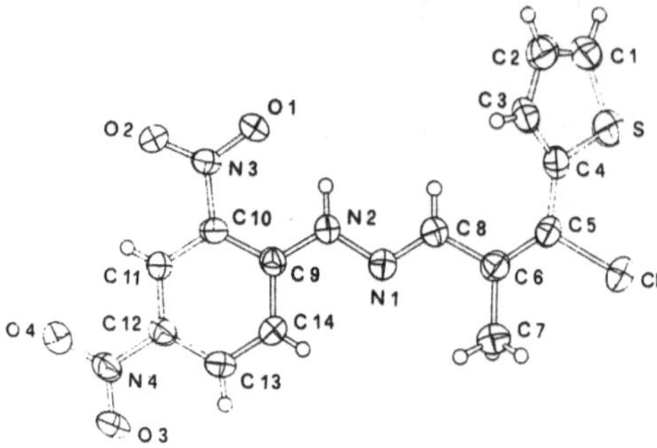

Fig. 1. The $C_{14}H_{11}ClN_4O_4S$ molecule.

2-BROMO-4-(2-CHLOROPHENYL)-9-METHYL-6H-THIENO[3,2-f](1,2,4)TRIAZOLO[4,3-a](1,4)DIAZEPINE (BROTIZOLAM)
$C_{15}H_{10}BrClN_4S$

H.J. BUTCHER and T.A. HAMOR, 1985. *Acta Cryst.*, **C41**, 265-266.

Monoclinic, $P2_1/c$, a = 15.749, b = 12.928, c = 7.346 Å, β = 95.44°, Z = 4. Mo radiation, R = 0.047 for 1730 reflections.

The angle between the mean planes of the phenyl and thieno rings of this hypnotic agent (Fig. 1) is 71.7(5)°. The seven-membered ring adopts a boat conformation with bow and stern angles of 56.8(6) and 29.2(6)°. The phenyl, triazolo and thieno rings are each planar. The Cl atom points to the same side of the molecule as C(4). Molecular dimensions and intermolecular contacts are normal.

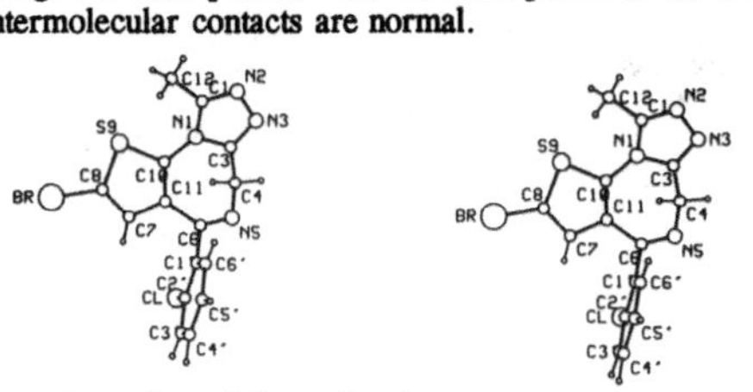

Fig. 1. $C_{15}H_{10}BrClN_4S$: a stereoview of the molecule.

3a,6-DIHYDRO-1-METHYLCYCLOPENTA[6,7][1,2]DITHIEPINO[5,4-b][1]BENZOTHIOPHEN-2(3H)-ONE
$C_{15}H_{12}OS_3$

G. ALEX, S. SRINIVASAN, S.R. RAMADAS and J.A. RAO, 1985. *Acta Cryst.*, **C41**, 1463-1465.

Orthorhombic, $P2_12_12_1$, a = 8.747, b = 8.787, c = 18.173 Å, Z = 4. Mo radiation, R = 0.040 for 767 reflections.

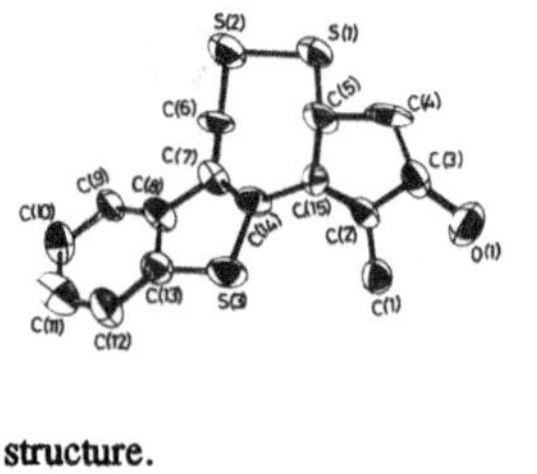

Fig. 1. $C_{15}H_{12}OS_3$: the molecular structure.

The S atoms of the dithiepin ring are disordered such that the ring conformation in the major component (75%)

is a chair and that in the minor component is twist-boat (Fig. 1). The C(5)-S(1) bond length of 1.959(13) Å and the C-S-S angles (96.5(10)-102.1(5)°) indicate internal strain in the dithiepine480- ring. The thiophene ring adopts an envelope conformation with C(8) as flap.

1,3-DIHYDRO-3-METHYL-4-(4-THIO-β-D-ERYTHROFURANOSYL)-1-(p-TOLYL)-2H-IMIDAZOLE-2-THIONE
$C_{15}H_{18}N_2O_2S_2$

E. MORENO, S. PEREZ, A. LOPEZ-CASTRO and R. MARQUEZ, 1985. *Acta Cryst.*, **C41**, 1465-1467.

Orthorhombic, $P2_12_12_1$, a = 11.408, b = 17.502, c = 7.846 Å, Z = 4. Mo radiation, R = 0.038 for 2172 reflections.

The thiotetrafuranose ring in the molecule (Fig. 1) has a conformation intermediate between twist $^2{}_3T$ and envelope E_3. The phenyl/imidazole and thiofuranose/imidazole dihedral angles are 46.7(1) and 52.9(1)°. Packing is governed by van der Waals interactions.

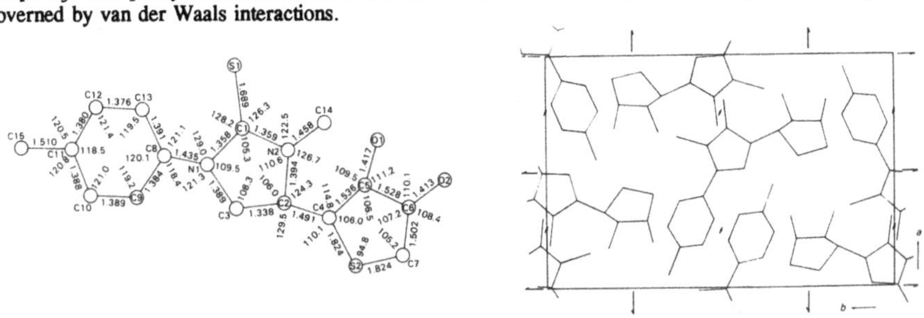

Fig. 1. $C_{15}H_{18}N_2O_2S_2$: bond lengths and angles (σ 0.004 Å, 0.3°) and the unit-cell contents viewed along [001].

CYCLOOCTA[1,2-c:3,4-c':5,6-c'':7,8-c''']TETRATHIOPHENE (TETRA(3,4-THIENYLENE))
$C_{16}H_8S_4$

H. IRNGARTINGER, U. HUBER-PATZ and H. RODEWALD, 1985. *Acta Cryst.*, **C41**, 1088-1089.

Monoclinic, *P2/c*, a = 9.956, b = 15.670, c = 13.700 Å, β = 91.80°, Z = 6. Mo radiation, R = 0.037 for 3341 reflections.

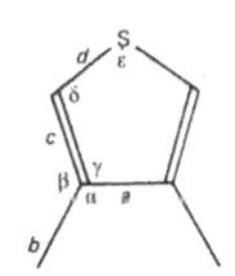

	(1)	(2)	(1)	(2)
a	1·436 (3)	1·400 (2)	α 126·0 (4)	122·5 (3)
b	1·469 (5)	1·494 (4)	β 122·3 (4)	118·6 (3)
c	1·364 (6)	—	γ 111·6 (3)	118·8 (2)
d	1·704 (5)	—	δ 112·6 (3)	—
			ε 91·5 (2)	—
b–a–b'	1·4 (6)	1·0 (6)		
a–b–a'	58·1 (15)	66·1 (17)		
c–b–c'	49·4 (15)	—		

Fig. 1. $C_{16}H_8S_4$: a view of the molecule with the numbering of the symmetric molecule shown in square brackets. Also shown are average values of the indicated bond lengths and angles with a comparison between them (1) and those of tetrabenzo[a,c,e,g]cyclooctatetraene **(1)** (shown in columns headed (2)).

The asymmetric unit consists of one molecule (Fig. 1) and a second on a crystallographic diad axis. The eight-membered ring in each molecule has a tub conformation with average valence and torsion angles of 126.0(4) and 58(2)° respectively. The bond lengths in this ring range from 1.436(3) to 1.469(5) Å and have little alternating nature. This suggests that this ring is more like a octamethylene-cyclooctane rather than a cyclooctatetraene system. The thiophene rings are tilted outwards from the eight-membered ring and by 3.7(1)° from a plane passing through both C atoms common to the central ring and thiophene and the corresponding adjacent C atoms within the eight membered ring. The dihedral angle between the planes of neighbouring thiophene rings is 53.7(14)°. No unusually short intermolecular contacts are reported.

1. Structure Reports, **48B**, 182.

4,4',5,5'-TETRAMETHOXY-2,2'-DITHIOBIPHENYL IODINE
$C_{16}H_{16}I_{1.17}O_4S_2$

W. HINRICHS, J. KOPF, K.-W. STENDER and G. KLAR, 1985. *Z. Naturforsch.*, **B40**, 39-44.

Monoclinic, *C2/c*, a = 23.123, b = 6.985, c = 25.558 Å, β = 123.52°, Z = 8. Mo radiation, R = 0.062.

The crystal structure (Fig. 1) consists of stacks of partially oxidised DTB units with alternating orientations, with disordered polyiodide chains between these stacks. In the non-planar DTB molecule, the dihedral angle between the two phenyl rings is 25.1° and the dithio bridge is inclined at 41.7° to the longitudinal axis. The C(ar)-S-S-C(ar) torsion angle is 55.6°, S-S 2.055(2) Å.

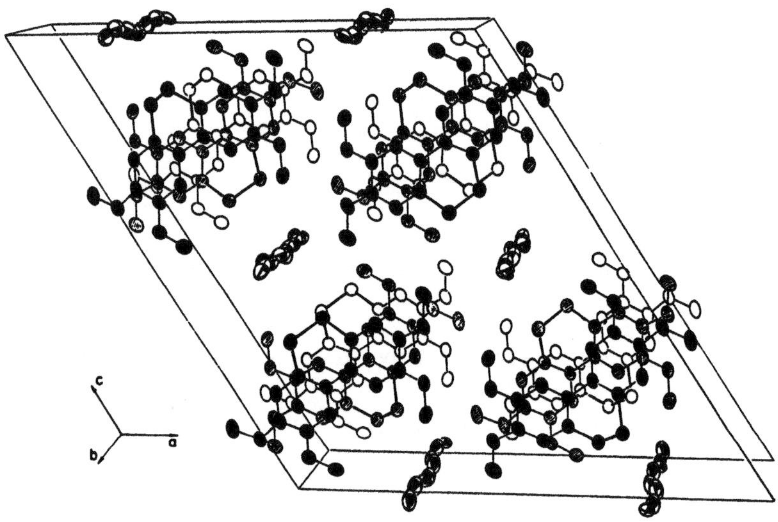

Fig. 1. The unit cell of $C_{16}H_{16}I_{1.17}O_4S_2$.

N-(4-DIETHYLAMINO-5-METHYL-1,3-DITHIOL-2-YLIDENE)-p-METHOXYTHIOBENZAMIDE
$C_{16}H_{20}N_2OS_3$

A. DIBO, M. STAVAUX, N. LOZAC'H and A. HORDVIK, 1985. *Acta Chem. Scand.*, B39, 103-112.

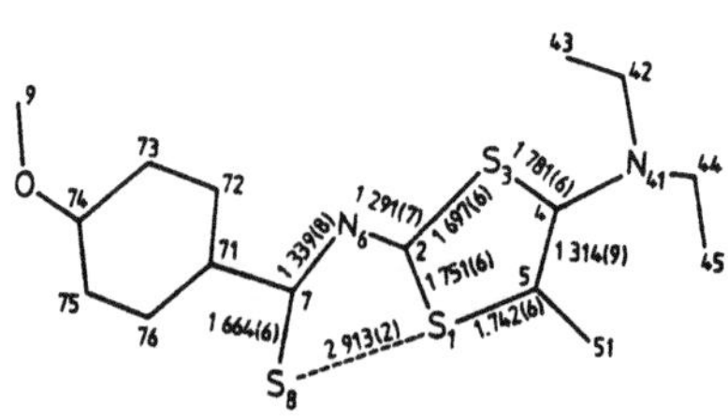

Fig. 1. Molecular structure of $C_{16}H_{20}N_2OS_3$ showing bond lengths (Å).

Monoclinic, $P2_1/c$, a = 10.967, b = 20.88, c = 7.941 Å, β = 106.03°, Z = 4. Mo radiation, R = 0.050 for 1788 reflections.

All atoms in the molecule (Fig. 1) with the exception of C(42,43,44,45) lie in the same plane, and the plane through C(4,42,44) forms a dihedral angle of 88° with it. The amino group is pyramidal. The carbon atoms of the diethylamino group are probably disordered.

2,6-DIBUTOXYCARBONYLTETRATHIAFULVALENE
$C_{16}H_{20}O_4S_4$

R.M. LOBKOVSKAYA, R.P. SHIBAEVA, Ya.N. KREITSBERGA and O.Ya. NEILAND, 1985. *Kristallografiya*, **30**, 806-808.

Monoclinic, $P112_1/a$, a = 15.97, b = 8.222, c = 8.007 Å, γ = 112.61°, Z = 2. Mo radiation, R = 0.034 for 1109 reflections.

The molecule lies about an inversion centre (Fig. 1) and the five-membered ring is folded 3.9° about the S...S line. The carboxyl group plane is rotated 8.6° out of the S(1),S(2),C(2),C(3) plane. The butoxy fragment is maximally extended but not quite planar.

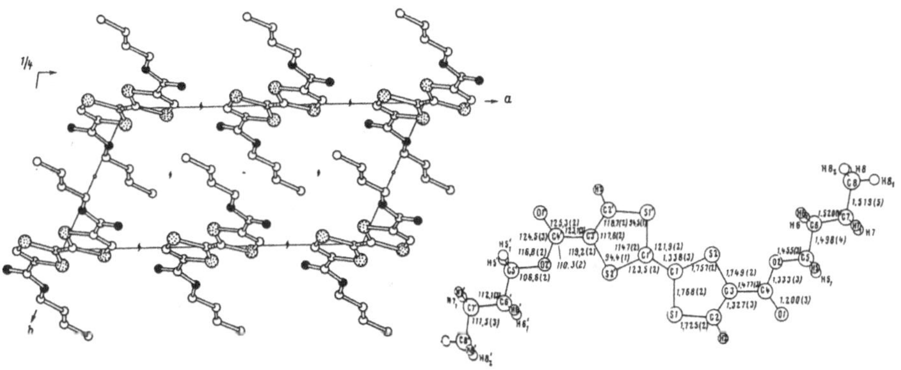

Fig. 1. A projection of the $C_{16}H_{20}O_4S_4$ structure along *c* and bond lengths and angles.

(R)-3-ETHYL-3-PHENYL-2,2-(TRIMETHYLENEDITHIO)PENT-4-ENAL
$C_{16}H_{20}OS_2$

C.G. YOUNG, B.R. JAMES and S.J. RETTIG, 1985. *Canad. J. Chem.*, **63**, 1035-1040.

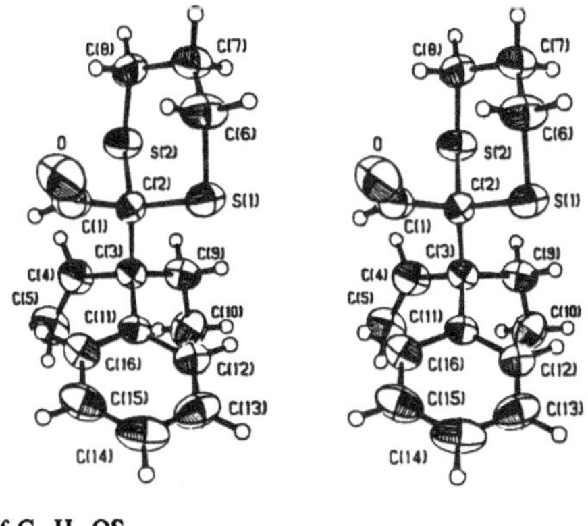

Fig. 1. A stereoview of $C_{16}H_{20}OS_2$.

Monoclinic, $P2_1$, a = 6.7224, b = 7.0172, c = 16.269 Å, β = 98.443°, Z = 2. Cu radiation, R = 0.071 for

1536 reflections.

The structure and absolute configuration are established as shown in Fig. 1. On the (R)-C(3) atom the vinyl and ethyl substituents were clearly differentiated, with bond lengths 1.319(9) and 1.490(9) Å respectively. The trimethylenedithio ring has a chair conformation; S-C distances are 1.811(6)-1.848(5) Å.

1-METHYLPHENANTHRO[3,4b]THIOPHENE
$C_{17}H_{12}S$

M.J. MUSMAR, G.E. MARTIN, R.T. GAMPE Jr., V.M. LYNCH, S.H. SIMONSEN, M.L. LEE, M.L. TEDJAMULIA and R.N. CASTLE, 1985. *J. Heterocycl. Chem.*, **22**, 545-553.

Monoclinic, $P2_1/c$, a = 9.201, b = 7.6143, c = 17.812 Å, β = 103.602°, Z = 4. Mo radiation, R = 0.049 for 2522 reflections (at 163 K).

Intramolecular overcrowding causes the molecule to be helical; dimensions are in Fig. 1. Each individual ring is non-planar, with the phenyl rings approaching twist-boat conformations. Methyl carbon C(12) is 2.980(3) Å from C(11) and H(123) is 2.12(3) Å from H(C11).

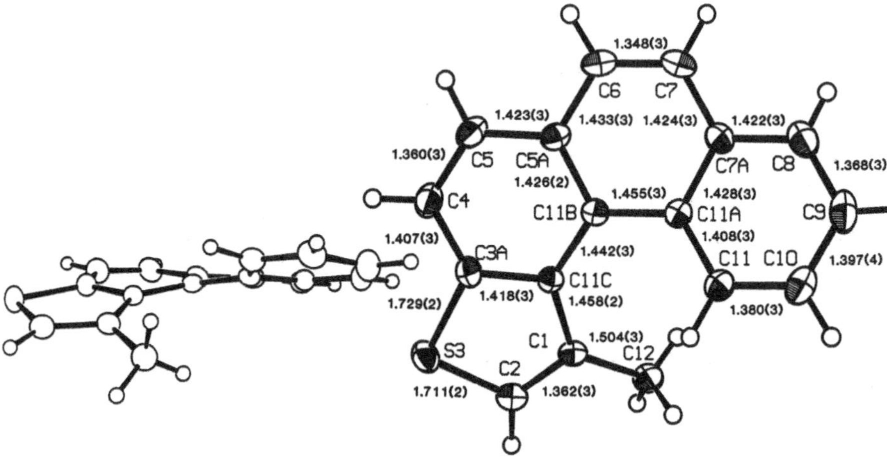

Fig. 1. An edge-on view of $C_{17}H_{12}S$ and an ORTEP plot showing bond lengths (Å).

(4S,5S)-(+)-2,2-DIMETHYL-4-PHENYL-5-((2-THIENYLMETHYLENE)AMINO)-1,3-DIOXANE
$C_{17}H_{19}NO_2S$ (I)

(4S,5S)-(+)-2,2-DIMETHYL-4-PHENYL-5-(BENZYLIDENEAMINO)-1,3-DIOXANE
$C_{19}H_{21}NO_2$ (II)

K. WEINGES, H. BRACHMANN, P. STAHNECKER, H. RODEWALD, M. NIXDORF and H. IRNGARTINGER, 1985. *Liebigs Ann. Chem.*, 566-578.

I. Orthorhombic, $P2_12_12_1$, a = 8.474, b = 9.591, c = 20.686 Å, Z = 4. Mo radiation, R = 0.045 for 1031 reflections (at -13°C).

II. Orthorhombic, $P2_12_12_1$, a = 8.851, b = 9.777, c = 19.945 Å, Z = 4. Mo radiation, R = 0.041 for 1115 reflections.

The analyses establish the previously postulated E-configuration of the C(8)=N(7) bond and its chiral shielding by the phenyl substituent at C(4); details of bond lengths are in Figs. 1 and 2. There are short non-bonded contacts N(7)...C(16) (3.223(7) Å in (I), 3.209(5) Å in (II)). The dioxane ring has a chair conformation in both compounds with ring torsion angles ±55°.

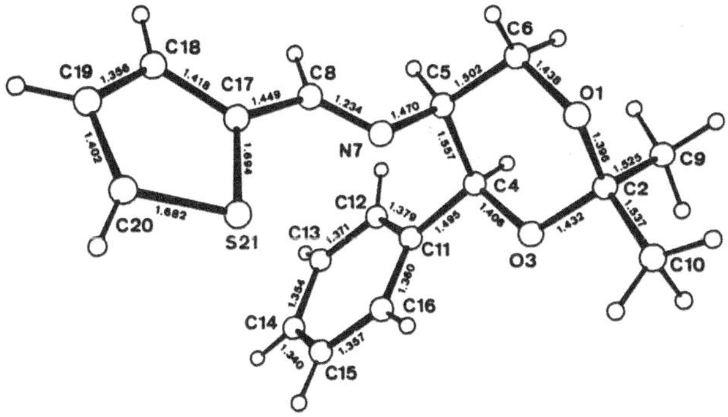

Fig. 1. A view of $C_{17}H_{19}NO_2S$ (I) with bond lengths (σ 0.005-0.011 Å).

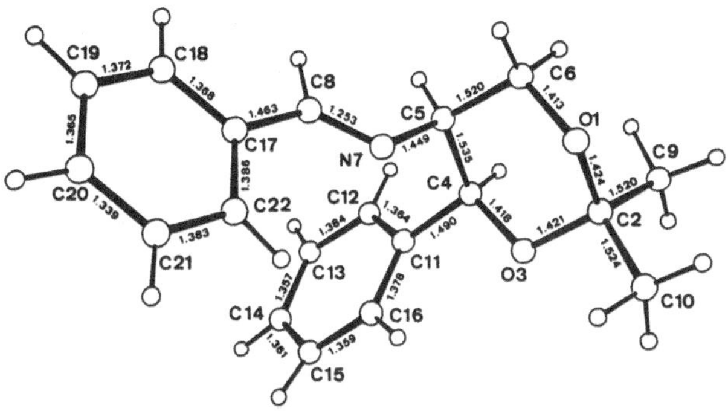

Fig. 2. A view of $C_{19}H_{21}NO_2$ (II) with bond lengths (σ 0.004-0.008 Å).

17-DEAZA-17-METHYLTHIONIUM ISOMORPHINAN PERCHLORATE (S-METHYL ISOSULFORPHANOL PERCHLORATE)

$C_{17}H_{23}ClO_5S$ (I) $C_{17}H_{23}OS^+,ClO_4^-$

10-METHOXYBENZO[1,2-b]DECAHYDRO-9-THIAPHENALENE

$C_{17}H_{22}OS$ (II)

B. BELLEAU, U. GULINI, B. GOUR-SALIN and F.R. AHMED, 1985. *Canad. J. Chem.*, **63**, 1268-1274.

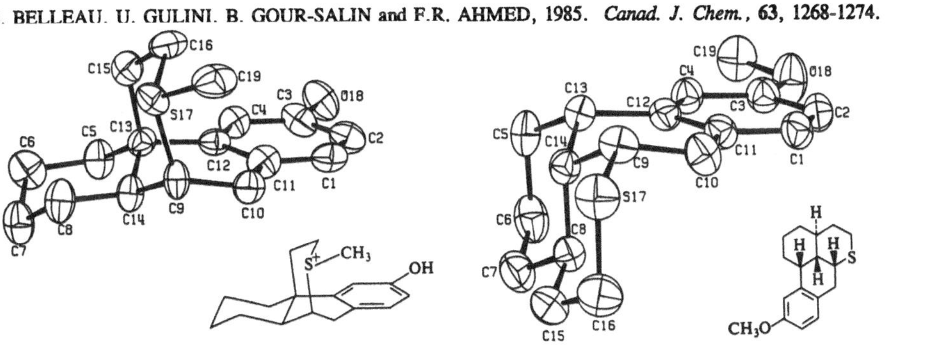

Fig. 1. The $C_{17}H_{23}OS^+$ cation in (I) (left) and the $C_{17}H_{22}OS$ (II) (right).

I. Orthorhombic, $Pna2_1$, a = 10.934, b = 9.219, c = 17.131 Å, Z = 4. Cu radiation, R = 0.053 for 1018 reflections.

II. Monoclinic, $P2_1/a$, a = 17.080, b = 9.372, c = 9.327 Å, β = 108.67°, Z = 4. Cu radiation, R = 0.034 for 2545 reflections.

These structure analyses establish that substitution of a basic N by S in a Grewe-like process for morphinan synthesis unexpectedly leads to a complete reversal of the B/C ring junction stereochemistry of the product. The molecules are shown in Fig. 1. The cation in (I) is shown to be an S-methyl isosulforphanol.

2-CHLORO-9-OXOTHIOXANTHENIUM-10-BIS(METHOXYCARBONYL)METHYLIDE $C_{18}H_{13}ClO_5S$

S.S.C. CHU, V. NAPOLEONE, P. DE MEESTER and T.L. CHU, 1985. *Acta Cryst.*, **C41**, 985-987.

Triclinic, $P\bar{1}$, a = 5.562, b = 12.176, c = 12.763 Å, α = 99.01, β = 99.80, γ = 78.35°, Z = 2. Mo radiation, R = 0.042 for 1720 reflections.

Bond lengths and angles (Fig. 1) in the thioxanthone ring have acceptable values though the two C-S bond lengths in the central ring (1.777(4) Å) are longer than expected. The ring system is slightly folded with the central ring adopting a flat boat conformation (fold angle 173.7(2)°). The bis(methoxycarbonyl)methylide group is in a pseudoequatorial conformation with respect to the central ring of the thioxanthone moiety. It is planar. Both of the ester groups have the endo conformation with one carbonyl O anti and the other syn to the non-bonding electron pair on S.

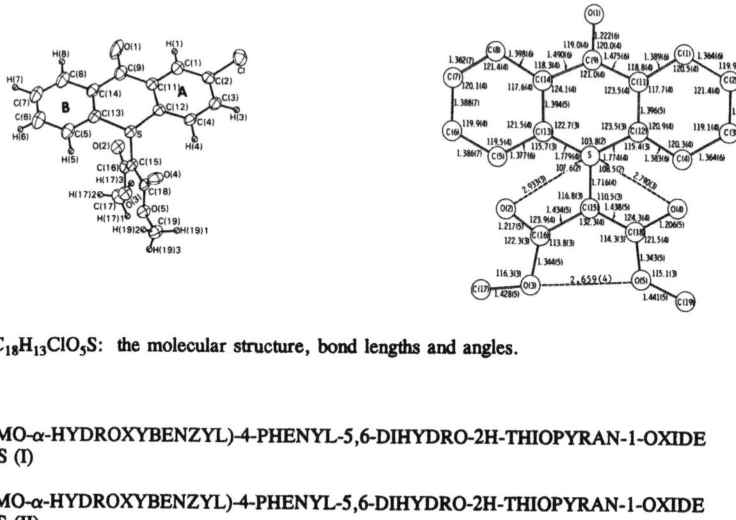

Fig. 1. $C_{18}H_{13}ClO_5S$: the molecular structure, bond lengths and angles.

α-2-(4'-BROMO-α-HYDROXYBENZYL)-4-PHENYL-5,6-DIHYDRO-2H-THIOPYRAN-1-OXIDE $C_{18}H_{17}BrO_2S$ (I)

γ-2-(4'-BROMO-α-HYDROXYBENZYL)-4-PHENYL-5,6-DIHYDRO-2H-THIOPYRAN-1-OXIDE $C_{18}H_{17}BrO_2S$ (II)

R.L. CRUMBIE, D.D. RIDLEY and P.J. STEEL, 1985. *Aust. J. Chem.*, **38**, 119-132.

I. Triclinic, $P\bar{1}$, a = 12.715, b = 13.351, c = 12.622 Å, α = 116.70, β = 97.78, γ = 111.62°, Z = 4. Mo radiation, R = 0.057 for 2247 reflections.

II. Monoclinic, $P2_1/n$, a = 12.060, b = 7.612, c = 18.606 Å, β = 107.24°, Z = 4. Mo radiation, R = 0.052 for 1805 reflections.

There are two crystallographically independent molecules of (I) in the asymmetric unit. These have similar dimensions but differ in conformation (Fig. 1). The mean planes through the C(4) phenyl rings are inclined to the plane of the C(3)-C(4) double bond at 22.8° in one molecule and 6.5° in the other. The corresponding angle in (II) (Fig. 1) is 33.2°. In the crystal of (I), the two independent molecules are hydrogen bonded to one another and this pair is further hydrogen-bonded to another pair related by a crystallographic centre of inversion. In (II), two molecules are mutually hydrogen-bonded about a centre of inversion.

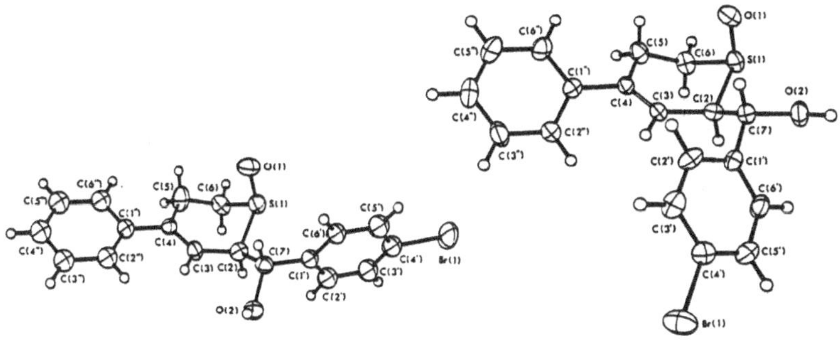

Fig. 1. The molecular structures of the α-form of $C_{18}H_{17}BrO_2S$ (left) and in the γ-form (right).

6,9-DIMETHOXY-15,18-DINITRO-2,11-DITHIA[3.3]METACYCLOPHANE
$C_{18}H_{18}N_2O_6S_2$ (I)

anti-5,8-DIMETHOXY-13-NITRO[2.2]METACYCLOPHANE
$C_{18}H_{19}NO_4$ (II)

6,9-DIMETHOXY-15-NITRO-2,11-DITHIA[3.3]METACYCLOPHANE
$C_{18}H_{19}NO_4S_2$ (III)

H.A. STAAB, L. SCHANNE, C. KRIEGER and V. TAGLIEBER, 1985. *Chem. Ber.*, **118**, 1204-1229.

I. Monoclinic, $P2_1/n$, a = 8.891, b = 16.769, c = 12.929 Å, β = 101.05°, Z = 4. Mo radiation, R = 0.040 for 3808 reflections.

II. Monoclinic, $C2/c$, a = 23.886, b = 7.981, c = 18.531 Å, β = 115.63°, Z = 8. Mo radiation, R = 0.044 for 2834 reflections.

III. Monoclinic, $P2_1/n$, a = 7.747, b = 17.130, c = 13.481 Å, β = 91.53°, Z = 4. Mo radiation, R = 0.040 for 3708 reflections.

Details of conformation and dimensions for these metacyclophanes are in Figs. 1-3. The angles between the aromatic ring planes are 12.5° in (I) and 19.9° in (III). In (II), the angle between the C(3),C(4),C(6),C(7) and C(11),C(12),C(14),C(15) planes is 15.1°.

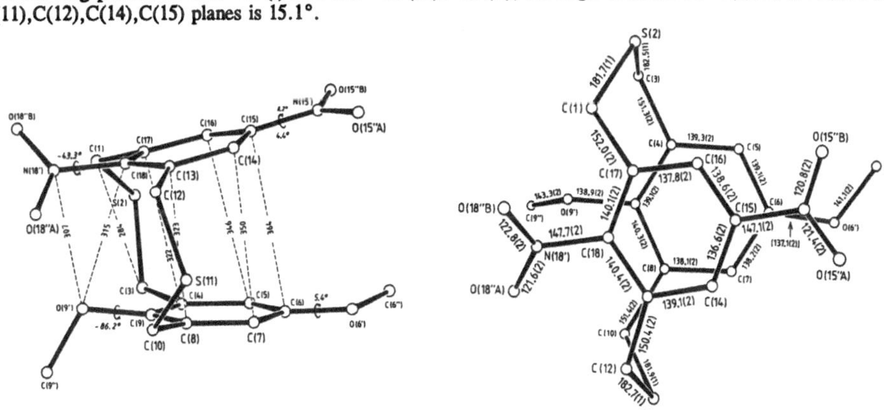

Fig. 1. Top and side views of $C_{18}H_{18}N_2O_6S_2$ (I) with dimensions.

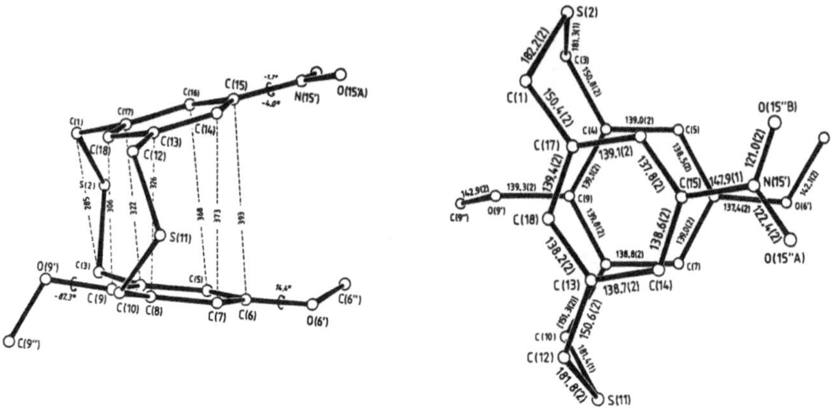

Fig. 2. Top and side views of $C_{18}H_{19}NO_4S_2$ (III).

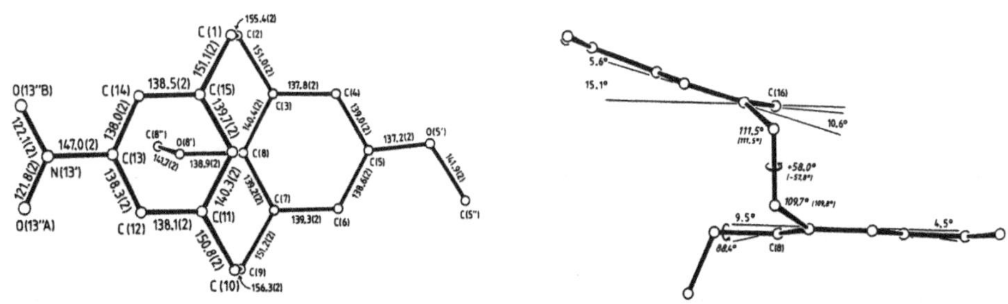

Fig. 3. Top and side views of $C_{18}H_{19}NO_4$ (II).

1,4-BIS(t-BUTYLTHIO)-3,6-BIS(ETHYLTHIO)THIENO[3,4-c]THIOPHENE
$C_{18}H_{28}S_6$

S. YONEDA, K. OZAKI, T. INOUE, A. SUGIMOTO, K. YANAGI and M. MINOBE, 1985. *J. Am. Chem. Soc.*, **107**, 5801-5802.

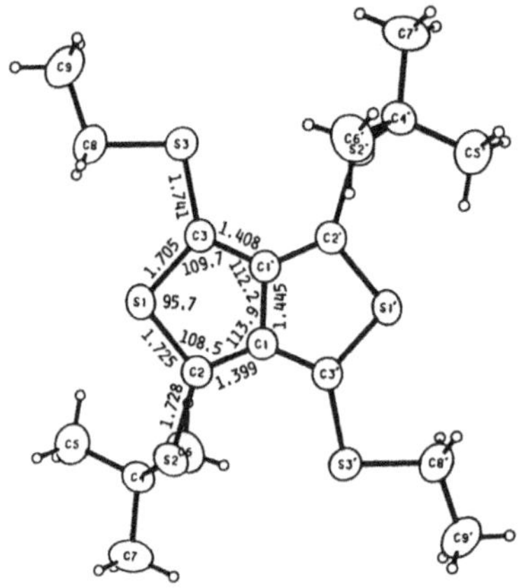

Fig. 1. The $C_{18}H_{28}S_6$ molecule with ring dimensions.

Triclinic, $P\bar{1}$, a = 10.229, b = 10.870, c = 5.790 Å, α = 102.00, β = 103.20, γ = 60.74°, Z = 1. Mo radiation, R = 0.026.

The molecule lies about an inversion centre. Ring dimensions are shown in Fig. 1.

(1RS)-15-THIABICYCLO[10.7.0]NONADEC-1(12)-ENE-15,15-DIOXIDE
$C_{18}H_{32}O_2S$

S.K. RAMASESHA, S.F. WATKINS and F.R. FRONCZEK, 1985. *Acta Cryst.*, **C41**, 232-234.

Triclinic, $P\bar{1}$, a = 8.618, b = 12.403, c = 16.792 Å, α = 88.24, β = 79.82, γ = 87.29°, Z = 4. Mo radiation, R = 0.067 for 3033 reflections.

There are two crystallographically independent molecules in the asymmetric unit. Each molecule (Fig. 1) consists of a 5-thiacyclononene ring trans fused to a cyclododecene ring. The flexibility of the latter results in an apparent disorder such that three distinct conformers can be resolved in the unit cell. All the molecules have identical nine-membered rings but there are three different 12-membered rings.

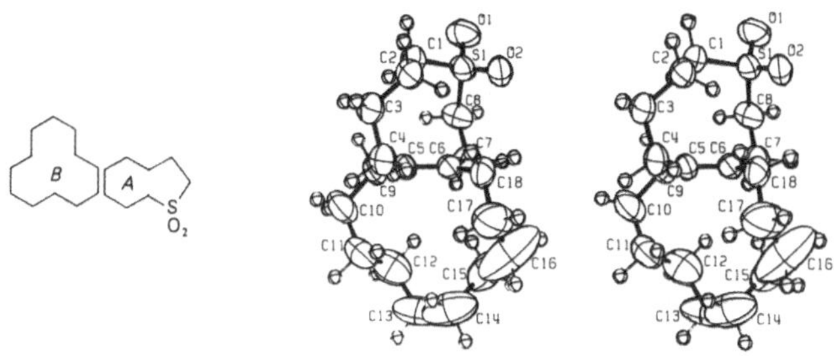

Fig. 1. $C_{18}H_{32}O_2S$: stereoview of the molecule showing one of the conformers.

rac-4,8-DI-1,3-DITHIAN-2-YLIDENE-4,4a,8,8a-TETRAHYDRO-4a,8a-METHANO-1,5-NAPHTHOQUINONE
$C_{19}H_{18}O_2S_4$

R. NEIDLEIN, G. HARTZ, A. GIEREN, H. BETZ and T. HUBNER, 1985. *Chem. Ber.*, **118**, 1455-1462.

Triclinic, $P\bar{1}$, a = 7.180, b = 10.321, c = 12.927 Å, α = 95.63, β = 97.97, γ = 102.23°, Z = 2. Mo radiation, R = 0.036 for 3396 reflections.

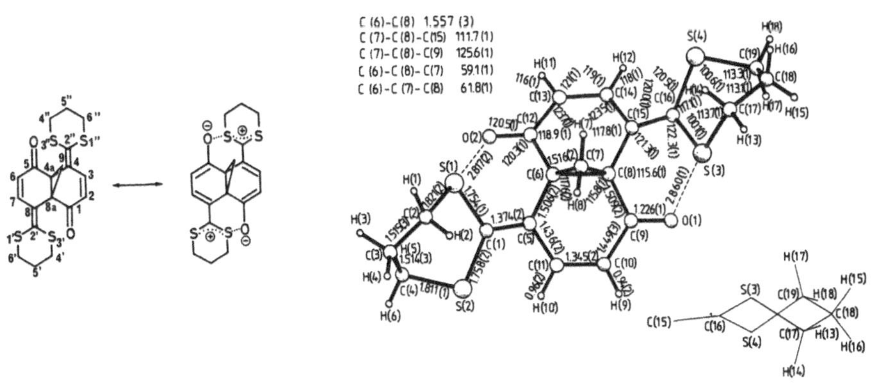

Fig. 1. The $C_{19}H_{18}O_2S_4$ molecule and the twist conformation of the 1,3-dithian rings.

The molecule has effective non-crystallographic C_2 symmetry with details of averaged dimensions and conformation shown in Fig. 1. The six-membered dithian rings are in twist conformations.

3,6-DIHYDRO-4,5-DIMETHYL-2-(2-(METHOXYMETHYL)PYRROLIDINESULPHONYL)-2-PHENYL-2H-THIA-PYRAN-1-OXIDE
$C_{19}H_{27}NO_4S_2$

R.C. HALTIWANGER, P.T. BEURSKENS, P.A.T.W. PORSKAMP, L.A.G.M. VANDEN BROEK and B. ZWANENBURG, 1985. *J. Cryst. Spect. Res.*, **15**, 109-113.

Orthorhombic, $P2_12_12_1$, a = 8.9604, b = 13.5928, c = 16.5621 Å, Z = 4. Mo radiation, R = 0.036 for 1813 reflections.

The molecule (Fig. 1) is shown to have E-geometry and the absolute configuration is R at C(1) and S at S(1). The dihydrothiapyran ring features are normal.

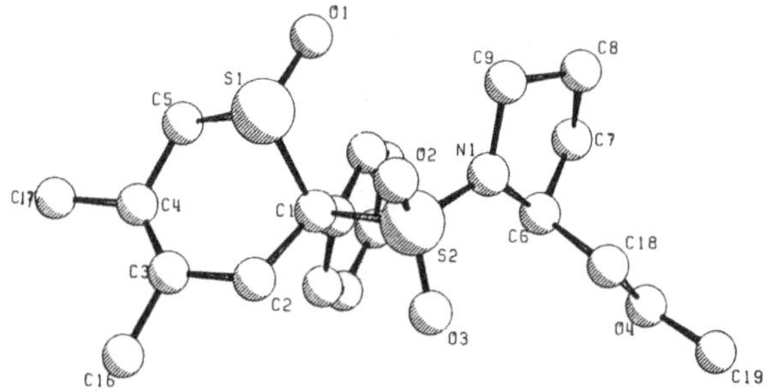

Fig. 1. The $C_{19}H_{27}NO_4S_2$ molecule.

cis-4a,4b-trans-4b-2-(α-1,3-DITHIOLAN-2-YL)-1,2,3,4,4a,4b,5,6,7,8,8a,9-DODECAHYDRO-2,4b-DIMETHYL-PHENANTHREN-3-ONE
$C_{19}H_{28}OS_2$ (I)

cis-4a,4b-trans-4b-2-(β-1,3-DITHIOLAN-2-YL)-1,2,3,4,4a,4b,5,6,7,8,8a,9-DODECAHYDRO-2,4b-DIMETHYL-PHENANTHREN-3-ONE
$C_{19}H_{28}OS_2$ (II)

J.H. NOORDIK, J.M.M. SMITS, A. SICHERER-ROETMAN, B.J.M. JANSEN and A. DE GROOT, 1985. *Acta Cryst.*, **C41**, 570-573.

I. Orthorhombic, $P2_12_12_1$, a = 8.096, b = 3.338, c = 26.465 Å, Z = 4. Mo radiation, R = 0.028 for 1962 reflections.

II. Monoclinic, $P2_1/n$, a = 11.340, b = 6.386, c = 25.159 Å, β = 101.20°, Z = 4. Mo radiation, R = 0.043 for 1121 reflections.

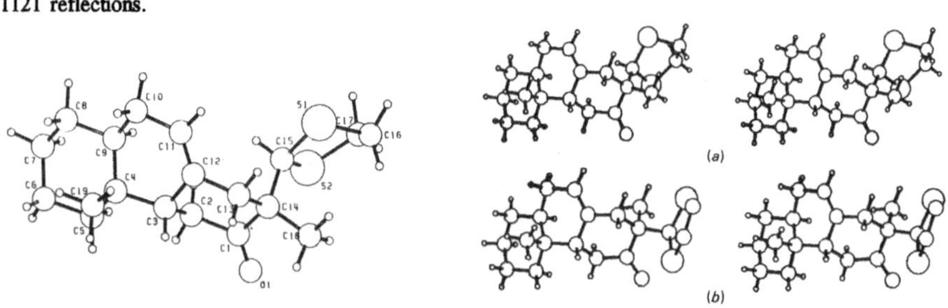

Fig. 1. $C_{19}H_{28}OS_2$: molecule (I) and a stereoview of (I) (top right) and of (II) (bottom right).

Both epimers (Fig. 1) have a conformation with trans-syn coupled rings, in which the ring containing the dithiolane-ring substituted C atom has a chair conformation. The geometries of the epimers are very similar apart from the difference in stereochemistry at C(14).

3-PHENYL-4-(1-PHENYLETHYL)-5-(2-THIENYL)-4,5-DIHYDRO-1,2,4-OXADIAZOLE
$C_{20}H_{18}N_2OS$

C. RUIZ-VALERO, A. MONGE and E. GUTIERREZ-PUEBLA, 1985. *Acta Cryst.*, **C41**, 1789-1790.

Monoclinic, $P2_1/n$, a = 8.84, b = 12.235, c = 15.903 Å, β = 91.90°, Z = 4. Mo radiation, R = 0.052 for 1750 reflections.

The structure (Fig. 1) consists of individual molecules held together by van der Waals forces. C(13) is anticlinal to C(6) and synperiplanar to H(2). The dihydrooxadiazole ring is not planar. The thienyl and phenyl rings are essentially planar. The molecular dimensions are normal.

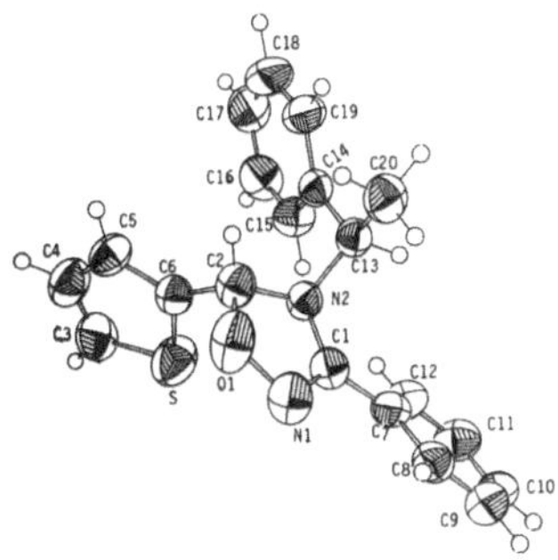

Fig. 1. $C_{20}H_{18}N_2OS$: the molecular structure.

2,3,4,4a,6,10b-HEXAHYDRO-6-PHENYL-1,4-METHANO-1H-DIBENZO[b,d]THIOPYRAN
$C_{20}H_{20}S$

L.W. SPRUCE, K. RAMALINGAM, K.D. BERLIN and E.M. HOLT, 1985. *Phosphorus and Sulfur*, **22**, 337-348.

Monoclinic, $P2_1/n$, a = 6.413, b = 15.727, c = 15.841 Å, β = 99.36°, Z = 4. Mo radiation, R = 0.048 for 1636 reflections.

The hydrogens on the C atoms adjacent to the S atom are shown to be syn (Fig. 1). The six-membered hetero ring has a boat conformation with S(5)-C(6) 1.824(4), S(5)-C(4a) 1.810(4) Å.

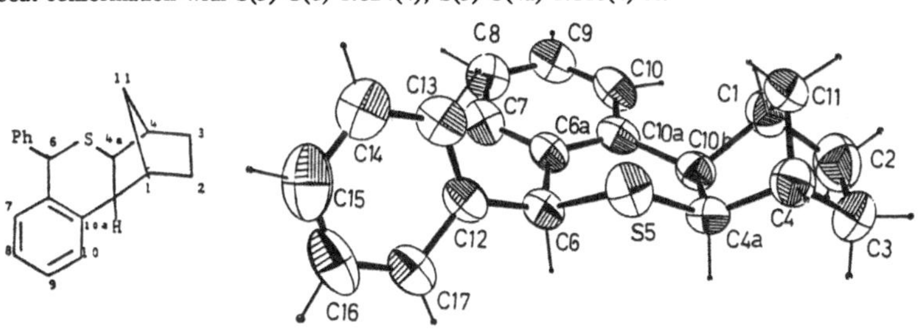

Fig. 1. The $C_{20}H_{20}S$ molecule.

(2S-(2αS*,3aα,4α,7α,7aα))-OCTAHYDRO-7,8,8-TRIMETHYL-2-(1-PHENYLETHOXY)-4,7-METHANOBENZO-[b]THIOPHENE
$C_{20}H_{28}OS$

C.R. NOE, M. KNOLLMULLER, E. WAGNER and H. VOLLENKLE, 1985. *Chem. Ber.*, **118**, 3299-3310.

Orthorhombic, $P2_12_12_1$, a = 19.461, b = 12.746, c = 7.3476 Å, Z = 4. Mo radiation, R = 0.097 for 2056 reflections.

The thiophene ring is shown to have a pronounced envelope conformation (Fig. 1) with S-C(2) 1.825(5), S-C(7A) 1.828(5), C(2)-O 1.417(6) Å, torsion angles C(3)-C(3A)-C(7A)-S 5.1, C(7A)-S-C(2)-C(3) -35.1, S-C(2)-C(3)-C(3A) 43.5, C(2)-C(3)-C(3a)-C(7a) -31.4°.

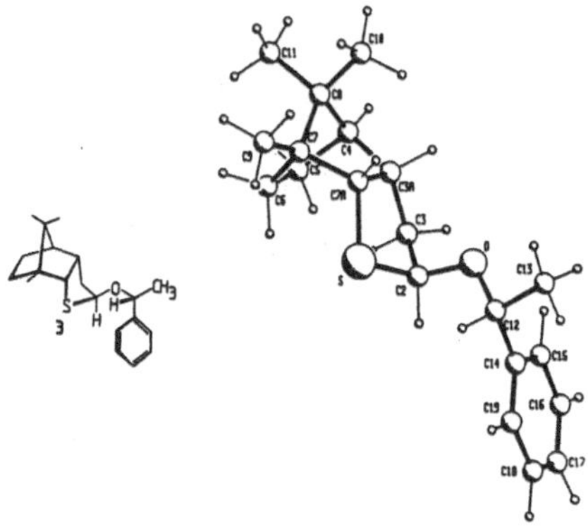

Fig. 1. The $C_{20}H_{28}OS$ molecule.

(+)-(S)-5,10-DISPIRO(2'-(4',5'-DIMETHYL-1',3'-DITHIACYCLOPENTANE))-2,7-DIHYDROXY-2,7-ETHANO-BICYCLO[4.4.0]DECANE
$C_{20}H_{32}O_2S_4$

H. BUDING, B. DEPPISCH, H. MUSSO and G. SNATZKE, 1985. *Chem. Ber.*, **118**, 4597-4612.

Monoclinic, $P2_1$, a = 10.855, b = 7.687, c = 13.779 Å, β = 108.76°, Z = 2. Cu radiation, R = 0.046 for 1947 reflections (at 160 K).

The molecule is shown to have the S-configuration (Fig. 1). The six-membered rings all have somewhat deformed chair conformations with ring torsion angle values 34.6 to 68.2°. The longest C-C bond is between C(4) and C(9) (1.562(9) Å). The C-S distances are in the range 1.808-1.859(7) Å.

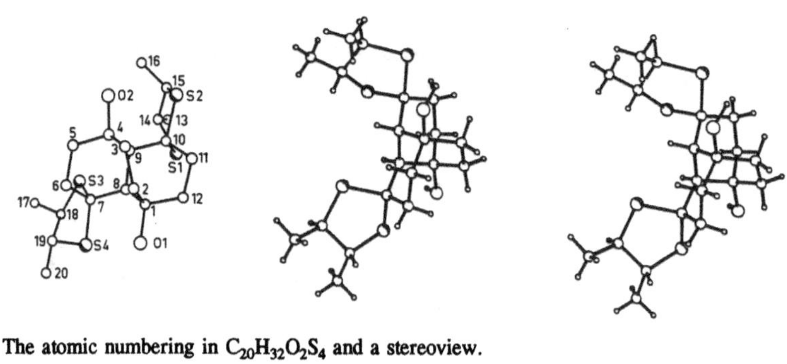

Fig. 1. The atomic numbering in $C_{20}H_{32}O_2S_4$ and a stereoview.

2,4,6-TRIPHENYL-1,3,5-TRITHIANE
$C_{21}H_{18}S_3$

K. SEKIDO and S. HIROKAWA, 1985. *Acta Cryst.*, C41, 397-400.

Monoclinic, *C2/c*, a = 19.711, b = 9.578, c = 19.459 Å, β = 92.09°, Z = 8. Mo radiation, R = 0.069 for 4395 reflections.

The trithiane ring in the molecule (Fig. 1) has a chair conformation and is nearly parallel to (010). The S-C bond lengths lie in the range 1.806(3) to 1.829(3) Å with an average value of 1.821 Å and the C-S-C and S-C-S average valence angles are 99.5 and 114.1° respectively. All three planar phenyl rings are in equatorial positions- two are nearly perpendicular to the trithiane ring and the third is inclined at about 45°. The trithiane rings stack along the screw axes and close van der Waals contacts are observed between centrosymmetrically related phenyl groups with an interplanar spacing of 3.41 Å.

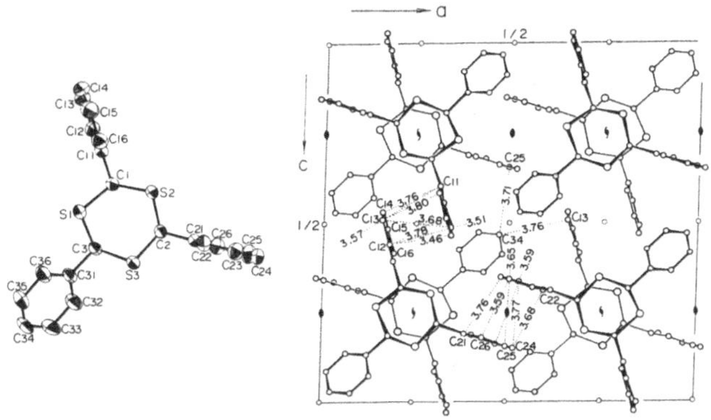

Fig. 1. $C_{21}H_{18}S_3$: a perspective view of the molecule and a projection of the crystal structure down the b-axis showing intermolecular distances shorter than 3.81 Å.

6,7-DIHYDRO-5-(((cis-2-HYDROXY-trans-3-PHENOXYCYCLOPENTYL)AMINO)METHYL)-2-METHYL-BENZO[b]THIOPHEN-4(5H)-ONE
$C_{21}H_{25}NO_3S$

J.R. McCARTHY, M.B. ZIMMERMAN, D.L. TREPANIER, M.E. LE TOURNEAU, P.E. WIEDEMAN, J.P. WHITTEN, R.J. BROERSMA, P.J. SHEA, N.L. WIECH and J.C. HUFFMAN, 1985. *J. Med. Chem.*, **28**, 1142-1145.

Triclinic, *P*$\bar{1}$, a = 12.161, b = 12.469, c = 6.327 Å, α = 87.84, β = 95.89, γ = 98.04°, Z = 2. Mo radiation, R = 0.050 for 2136 reflections (at -160°C).

The analysis confirmed the structure as shown in Fig. 1.

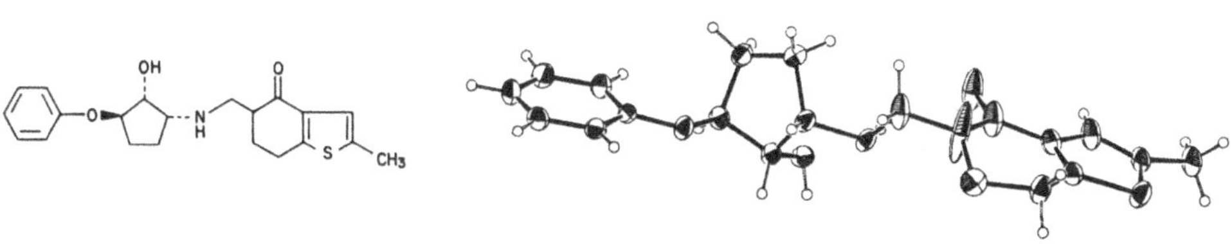

Fig. 1. The $C_{21}H_{25}NO_3S$ molecule.

TETRATHIA[7]HETEROHELICENE (racemate)
$C_{22}H_{10}S_4$ (I)

TETRATHIA[7]HETEROHELICENE (enantiomer)
$C_{22}H_{10}S_4$ (II)

H. NAKAGAWA, A. OBATA, K.-I. YAMADA, H. KAWAZURA, M. KONNO and H. MIYAMAE, 1985. *J. Chem. Soc. Perkin Trans. 2*, 1899-1903.

I. Orthorhombic, *Pbcn*, a = 15.383, b = 14.195, c = 7.767 Å, Z = 4. Cu radiation, R = 0.048 for 1227 reflections.

II. Orthorhombic, $P2_12_12_1$, a = 16.975, b = 11.236, c = 9.186 Å, Z = 4. Cu radiation, R = 0.045 for 2145 reflections.

The aromatic rings in the molecules are flat with terminal rings A and A' (Fig. 1) being almost planar; the planarity is poorest in the central D ring. The racemate molecule has C_2 symmetry in the crystal but the enantiomer lacks this symmetry because of differences in the dihedral angles between the successive aromatic rings. In the racemate the molecules are stacked in a column with alternating chirality, whilst in the enantiomer the molecules are confined orthogonally to each other, resulting in strain.

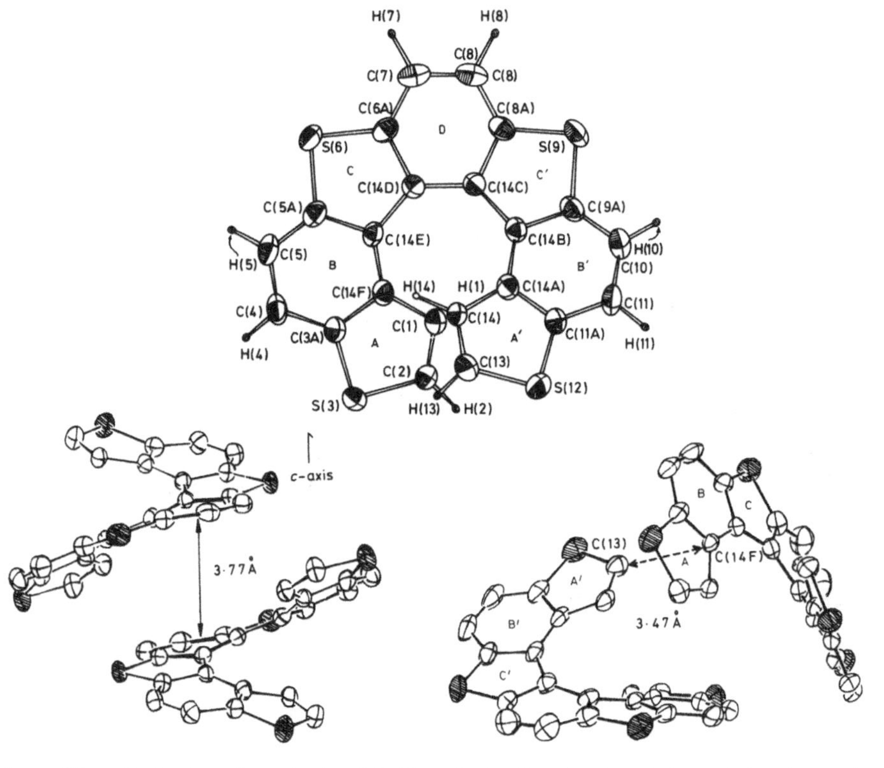

Fig. 1. Schematic drawing of a molecule of $C_{22}H_{10}S_4$ (top) with a drawing of the stacked molecules in the racemate (bottom left) and the orthogonal packing of the enantiomer (bottom right).

7-BENZYL-3-FORMYL-6-PHENYLBENZO[b]THIOPHENE
$C_{22}H_{16}OS$

B. HANQUET, R. GUILARD and Y. DUSAUSOY, 1985. *J. Chem. Soc. Perkin Trans. 1*, 983-989.

Triclinic, $P\bar{1}$, a = 7.067, b = 9.266, c = 13.396 Å, α = 102.95, β = 99.56, γ = 101.06°, Z = 2. Cu radiation, R = 0.045 for 2382 reflections.

Each molecule consists of two ortho-condensed and coplanar rings (a thiophene ring and a six-membered ring) (Fig. 1). The aldehyde function is on C(3) of the thiophene ring. The angle between the two phenyl groups is 94° and the angles between the thiophene and each of these planes are respectively 126 and 98°.

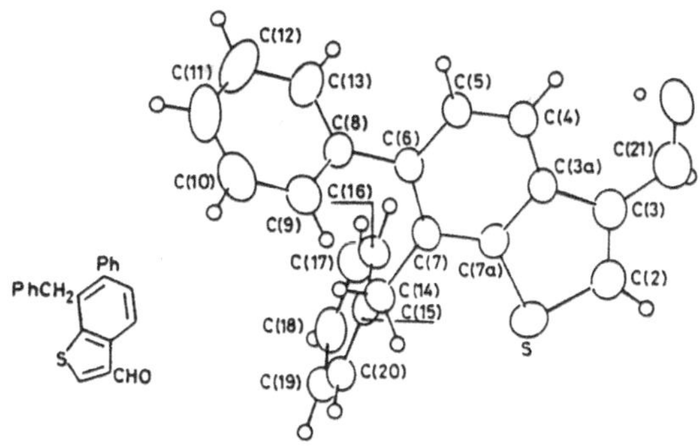

Fig. 1. ORTEP view of a molecule of $C_{22}H_{16}OS$.

5-(2-FLUOROPHENYL)-1-METHYL-2-(3-THENOYLAMINOMETHYL)-2,3-DIHYDRO-1H-1,4-BENZODIAZEPINE HYDROCHLORIDE (TIFLUADOM HYDROCHLORIDE)
$C_{22}H_{21}ClFN_3OS$ (I) $C_{22}H_{21}FN_3OS^+,Cl^-$

(+)-5-(2-FLUOROPHENYL)-1-METHYL-2-(3-THENOYLAMINOMETHYL)-2,3-DIHYDRO-1H-1,4-BENZODIAZEPINE p-TOLUENESULPHONATE ((+)-TIFLUADOM p-TOLUENESULPHONATE)
$C_{29}H_{28}FN_3O_4S_2$ (II) $C_{22}H_{21}FN_3OS^+,C_7H_7O_3S^-$

T.J. PETCHER, A. WIDMER, U. MAETZEL and H. ZEUGNER, 1985. *Acta Cryst.*, **C41**, 909-912.

I. Orthorhombic, $P2_12_12_1$, a = 21.362, b = 14.919, c = 6.484 Å, Z = 4. Cu radiation, R = 0.051 for 2146 reflections.

II. Orthorhombic, $P2_12_12_1$, a = 10.319, b = 10.824, c = 24.537 Å, Z = 4. Cu radiation, R = 0.041 for 2428 reflections.

The absolute configuration of (II) is 2S. In both structures (Fig. 1), the 2-substituent is axially placed on the diazepine ring. In (I), this ring has a boat conformation with a symmetry plane passing through C(3) while in (II) this ring has a twist-boat conformation with a symmetry axis passing through C(5a). The conformations are otherwise very similar in the two compounds, including the presence of two intermolecular hydrogen bonds from N(4) and the amide group of the 2-substituent to the anion (N(4)...Cl 3.013(2), N(12)...Cl 3.267(2), N(4)...F(2') 2.868(3), I; N(4)...O(8'') 2.687(4), N(12)...O(9'') 2.941(4) Å, II). Bond distances and angles are normal.

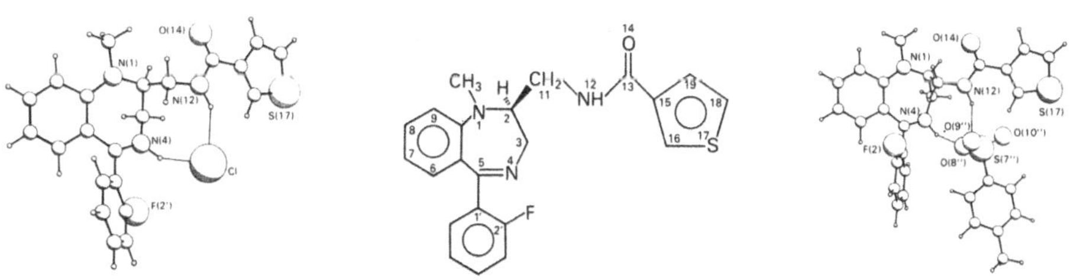

Fig. 1. The cation in $C_{22}H_{21}ClFN_3OS$ (I) (left), in $C_{29}H_{28}FN_3O_4S_2$ (II) (right) and the skeleton of tifluadom (centre).

trans-2,3-BIS(MESITYLTHIO)-4-NITRO-2,3-DIHYDROTHIOPHENE
$C_{22}H_{25}NO_2S_3$

G. PETRILLO, M. NOVI, G. GARBARINO, C. DELL'ERBA and A. MUGNOLI, 1985. *J. Chem. Soc. Perkin Trans. 2*, 1291-1296.

Monoclinic, $P2_1/c$, a = 7.722, b = 16.665, c = 17.318 Å, β = 92.13°, Z = 4. Mo radiation, R = 0.037 for

2988 reflections.

The trans-configuration of the two mesitylthio groups is confirmed by the S(2)-C(2)-C(3)-S(3) torsion angle (-157.5(1)°). The dihydrothiophene ring is puckered (Fig. 1); its conformation can be described as a half-chair with a pseudo-twofold axis running through atom C(5) and the midpoint of the C(2)-C(3) bond. The whole nitroethylene system is almost planar.

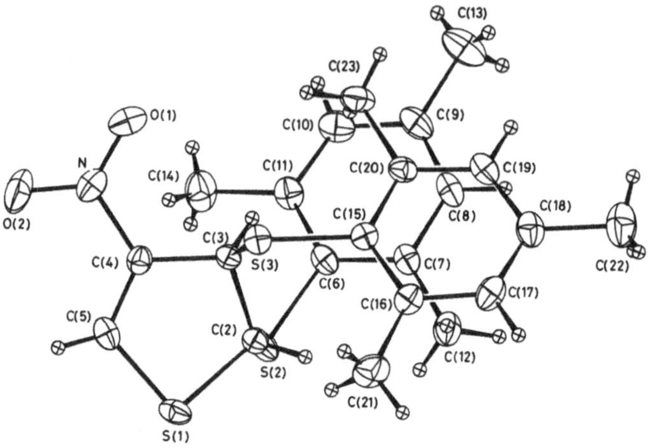

Fig. 1. Molecular structure of $C_{22}H_{25}NO_2S_3$.

5,6-DIPHENYLNAPHTHO[1,8-bc]THIOPHENE
$C_{23}H_{16}S$

B. HANQUET, M. FARNIER, R. GUILARD, C. LECOMTE and Y. DUSAUSOY, 1985. *Canad. J. Chem.*, **63**, 2089-2099.

Monoclinic, $P2_1/c$, a = 11.711, b = 9.707, c = 15.452 Å, β = 109.41°, Z = 4. Cu radiation, R = 0.059 for 2118 reflections.

The molecular conformation is shown in Fig. 1. The three fused rings are planar to within 0.07 Å, with C(9) -0.26 Å and C(18) +0.28 Å from this plane. The phenyl ring planes are inclined at 55.3 and 55.8° to the plane of the three fused rings and the C(18)-C(15)-C(8) and C(9)-C(7)-C(8) bond angles are 124.2 and 123.7(4)° respectively.

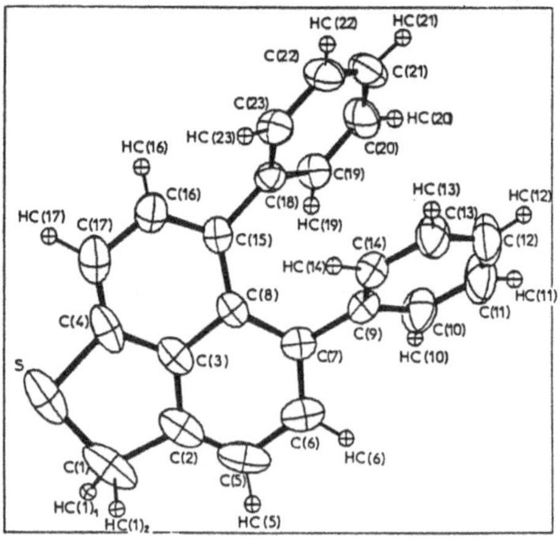

Fig. 1. A view of $C_{23}H_{16}S$.

ETHYL (E)-p-(2-(4,4-DIMETHYLTHIOCHROMAN-6-YL)PROPENYL)BENZOATE
$C_{23}H_{26}O_2S$ (I)

6-ACETYL-4,4-DIMETHYLTHIOCHROMAN-1,1-DIOXIDE
$C_{13}H_{16}O_3S$ (II)

K.M. WAUGH, K.D. BERLIN, W.T. FORD, E.M. HOLT, J.P. CARROL, P.R. SCHOMBER, M.D. THOMPSON and L.J. SCHIFF, 1985. *J. Med. Chem.*, **28**, 116-124.

I. Triclinic, $P\bar{1}$, a = 20.568, b = 14.760, c = 7.679 Å, α = 113.33, β = 79.45, γ = 79.98°, Z = 4. Mo radiation, R = 0.058 for 3463 reflections.

II. Triclinic, $P\bar{1}$, a = 9.292, b = 9.219, c = 7.951 Å, α = 102.16, β = 77.49, γ = 79.60°, Z = 2. Mo radiation, R = 0.064 for 5373 reflections.

There are two similar independent molecules in (I) and there is also some disorder. The stereochemistry at the C(11)-C(13) double bond was established (Fig. 1). In each molecule the two aromatic rings are nearly perpendicular to each other (86.4° and 84.2°). The sulphur containing ring has a half-chair conformation. The hetero ring in (II) adopts a C(2)/C(3) half-chair conformation, with S-C(8') 1.767(3) and S-C(2) 1.761(4) Å, and S-O 1.443 and 1.447(3) Å.

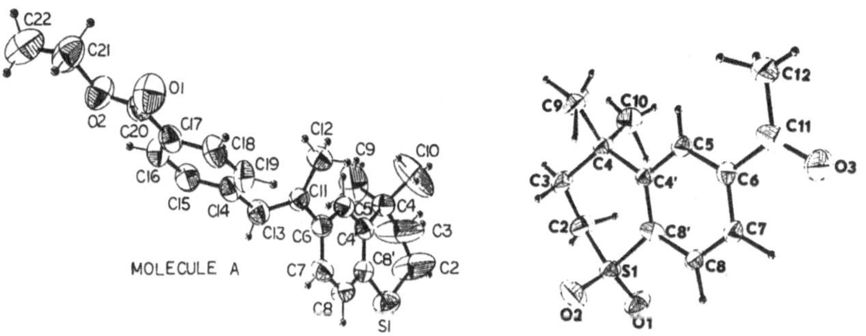

Fig. 1. One of the $C_{23}H_{26}O_2S$ molecules (left) and the $C_{13}H_{16}O_3S$ structure (right).

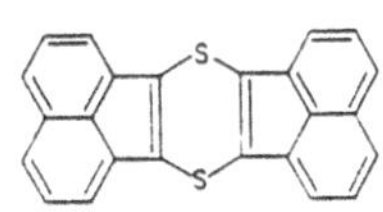

BIS(ACENAPHTHO)[1,2-b:1,2-e]-1,4-DITHIIN
$C_{24}H_{12}S_2$

A. ORAHOVATZ, M.I. LEVINSON, P.J. CARROLL, M.V. LAKSHMIKANTHAM and M.P. CAVA, 1985. *J. Org. Chem.*, **50**, 1550-1552.

Orthorhombic, *Pbca*, a = 8.285, b = 13.651, c = 14.758 Å, Z = 4. Mo radiation, R = 0.075 for 848 reflections.

The crystal structure analysis established the identity of the compound.

6,6-DIPHENYL-4-MESITYLIMINO-5-THIASPIRO[2.3]HEXANE
$C_{26}H_{25}NS$

A. BATTAGLIA, G. BARBARO, P. GIORGIANNI, E. FORESTI, P. SABATINO and A. DONDONI, 1985. *J. Org. Chem.*, **50**, 5368-5370.

Monoclinic, $P2_1/n$, a = 9.210, b = 12.716, c = 18.102 Å, β = 99.52°, Z = 4. Mo radiation, R = 0.00 for 2517 reflections.

The crystal structure analysis established the stereochemistry of the molecule (Fig. 1). The S(1)-C(2) distance is 1.887(3) Å; S(1)-C(4) 1.780(4) Å. The dihedral angle between the three-membered ring and the least-squares mean plane through the four-membered ring is 86.5°.

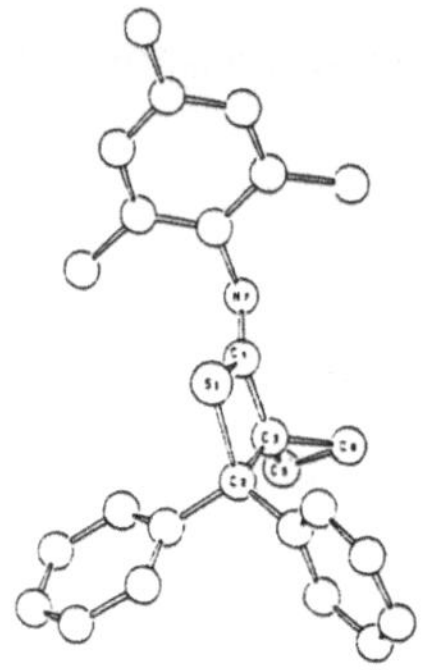

Fig. 1. The molecule $C_{26}H_{25}NS$, with hydrogen atoms omitted.

THIENO[2,3-l:5,4-l']BIS(BENZO[d]BENZO[1,2-b:4,3-b']DITHIOPHENE) PENTATHIA[9]HETEROHELICENE $C_{28}H_{12}S_5$

H. NAKAGAWA, H. NAGASHIMA, K. YAMADA and H. KAWAZURA, 1985. *Acta Cryst.*, **C41**, 778-780.

Monoclinic, *C2/c*, a = 14.685, b = 10.657, c = 13.964 Å, β = 100.31°, Z = 4. Mo radiation, R = 0.062 for 1054 reflections.

The molecule (Fig. 1) possesses crystallographic symmetry C_2, the diad axis passing through S(1), and displays a helical geometry with more than one turn. The region of overlap near the A and B rings gives rise to steric repulsion with closest non-bonding contacts of 3.02 and 3.08 Å. The C-C bonds in the inner core are elongated while those in the outer core are shortened. The dihedral angles between consecutive rings average about 10° while that between facing planes of maximum overlap is 22.6(3)°. In the crystal P and M enantiomers are alternately stacked along the c-axis with a greater part of the molecular planes overlapping each other. Molecules of the same chirality form layers parallel to (001).

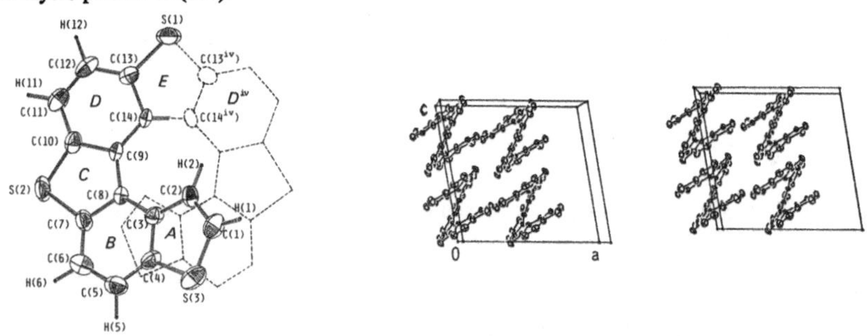

Fig. 1 A view of the $C_{28}H_{12}S_5$ molecule in which the broken lines represent the twofold related part of the molecule and a stereoview of the molecular packing viewed down the b-axis.

10-((-)-1-MENTHYLOXYCARBONYL)-(P)-(+)-[2](1,3)BENZENO[0](3,4)THIOPHENO[0](1,3)BENZENOPHANE $C_{29}H_{32}O_2S$

A. AIGNER, F. VOGTLE, S. FRANKEN and H. PUFF, 1985. *Chem. Ber.*, **118**, 3643-3651.

Monoclinic, $P2_1$, a = 7.117, b = 12.097, c = 14.157 Å, β = 91.41°, Z = 2. Mo radiation, R = 0.041 for 3318 reflections.

The analysis establishes the P-(+) configuration for the highly strained molecule (Fig. 1), and large torsion angles of 19.8° between the A, B and C rings.

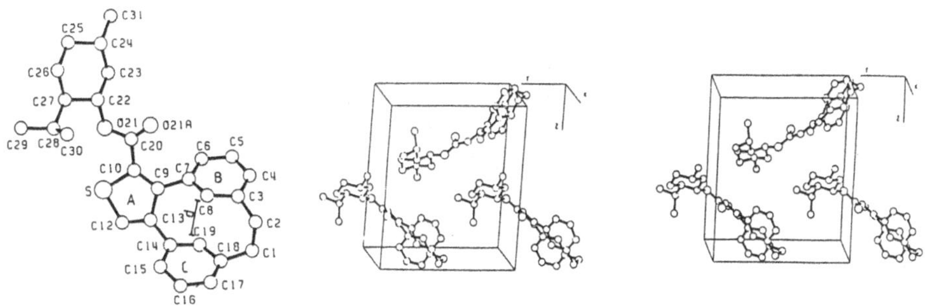

Fig. 1. The $C_{29}H_{32}O_2S$ molecule and a stereoview of the unit cell.

2,4,10-TRIPHENYLBENZO[e]THIOPYRANO[3,4-b]THIOPYRAN
$C_{30}H_{20}S_2$

C.H. CHEN, L.W. KELTS and H.R. LUSS, 1985. *J. Org. Chem.*, **50**, 2727-2729.

Monoclinic, $P2_1/c$, a = 21.107, b = 8.484, c = 13.616 Å, β = 106.57°, Z = 4. Mo radiation, R = 0.042 for 1919 reflections.

The important details of the molecule are shown in Fig. 1.

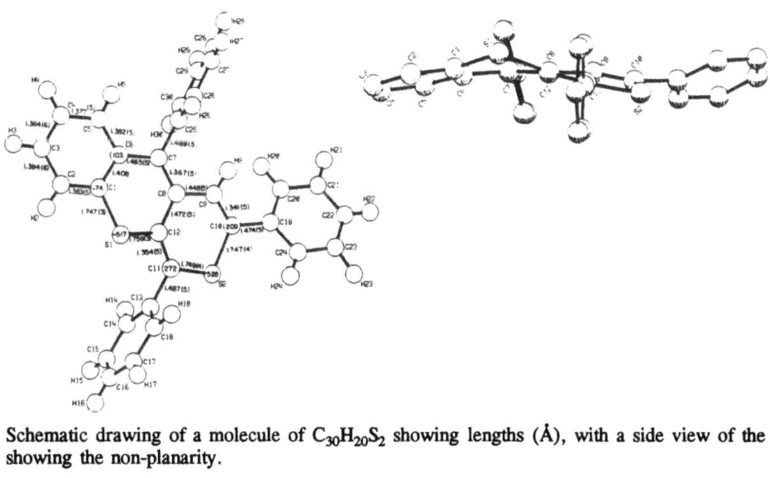

Fig. 1. Schematic drawing of a molecule of $C_{30}H_{20}S_2$ showing lengths (Å), with a side view of the molecule showing the non-planarity.

TRIS(BIS(ETHYLENEDITHIO)TETRATHIAFULVALENE) NITRATE (α-phase)
$C_{30}H_{24}NO_3S_{24}$ $C_{10}H_8S_8{}^+,2(C_{10}H_8S_8),NO_3{}^-$

A. WEBER, H. ENDRES, H.J. KELLER, E. GOGU, I. HEINEN, K. BENDER and D. SCHWEITZER, 1985. *Z. Naturforsch.*, **B40**, 1658-1663.

Monoclinic, $P2_1/n$, a = 5.890, b = 31.125, c = 12.915 Å, β = 103.73°, Z = 2. Mo radiation, R = 0.057 for 1888 reflections.

There are two independent BEDT-TTF moieties, one in a general position and one on an inversion centre. In the crystal, sheets of BEDT-TTF radical cations, with inter-ion S...S contacts as short as 3.315(4) Å, are separated by anion layers. The C-S bond lengths are in the range 1.707(15)-1.762(12) Å. Crystal data are also given for the β- and γ-phases.

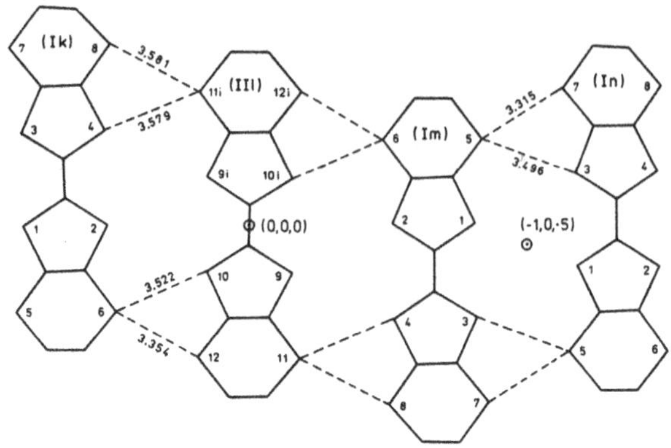

Fig. 1. A section of a BEDT-TTF ribbon showing inter-ion S...S contacts.

TETRAKIS(4,5-BIS(CARBOMETHOXY)-1,3-DITHIOL-2-YLIDENE)CYCLOBUTANE
$C_{32}H_{24}O_{16}S_8$

T. SUGIMOTO, H. AWAJI, Y. MISAKI, Z.-I. YOSHIDA, Y. KAI, H. NAKAGAWA and N. KASAI, 1985. *J. Am. Chem. Soc.*, **107**, 5792-5793.

Orthorhombic, $P2_12_12_1$, a = 17.9317, b = 25.757, c = 8.539 Å, Z = 4. Mo radiation, R = 0.052 for 3073 reflections.

The molecule has approximate D_{2d} symmetry (excluding eight carbomethoxy groups) and has a puckered conformation for the central four-membered ring. Columns of molecules are formed in the crystal lattice (Fig. 1).

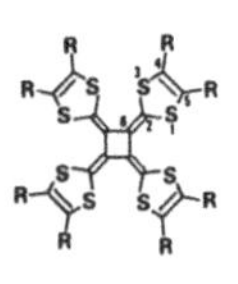

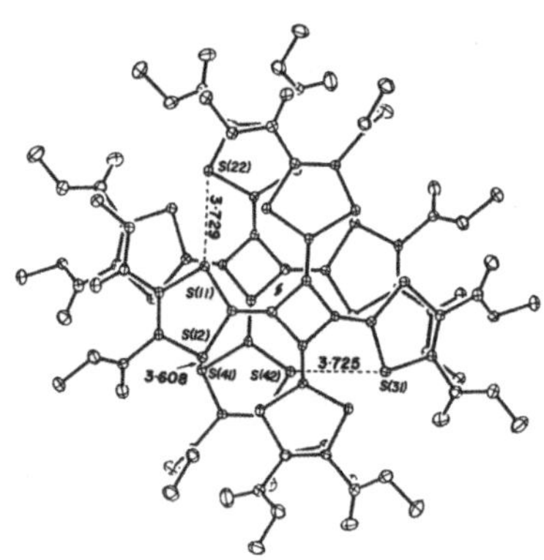

Fig. 1. The stacking arrangement of $C_{32}H_{24}O_{16}S_8$.

2-BENZYLIDENE-3-HYDROXY-4-(2-METHOXYBENZOYL)-3-(2-METHOXYPHENYL)-5-PHENYLTETRA-HYDROTHIOPHENE
$C_{32}H_{28}O_4S$

R.T. LALONDE, R.A. FLORENCE, B.A. HORENSTEIN, R.C. FRITZ, L. SILVEIRA, J. CLARDY and B.S. KRISHNAN, 1985. *J. Org. Chem.*, **50**, 85-91.

Triclinic, $P\bar{1}$, a = 9.973, b = 11.561, c = 13.063 Å, α = 77.65, β = 77.83, γ = 64.49°, Z = 2. Cu radiation, R = 0.059 for 3254 reflections.

The crystal structure analysis established the identity of the molecule (Fig. 1).

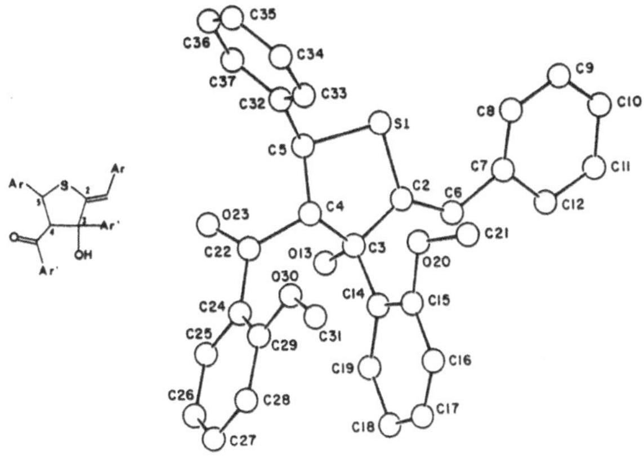

Fig. 1. Schematic drawing of a molecule of $C_{32}H_{28}O_4S$ with hydrogen atoms omitted.

2,3,6,8,9-PENTAPHENYL-1-THIA-6-AZASPIRO[4.4]NONA-2,8-DIENE-4,7-DIONE
$C_{37}H_{25}NO_2S$

N. YASUOKA, Y. KAI and N. KASAI, 1985. *Bull. Chem. Soc. Jpn.*, **58**, 2943-2946.

Monoclinic, $P2_1/c$, a = 20.810, b = 16.000, c = 18.131 Å, β = 107.77°, Z = 8. Cu radiation, R = 0.062 for 4609 reflections.

The two independent molecules in the asymmetric unit have the same overall structure (Fig. 1); they differ in conformation, with four out of five phenyl rings taking different orientations in the two molecules. The dihedral angles between the planes of the heterocyclic rings are 84.6 and 85.3°.

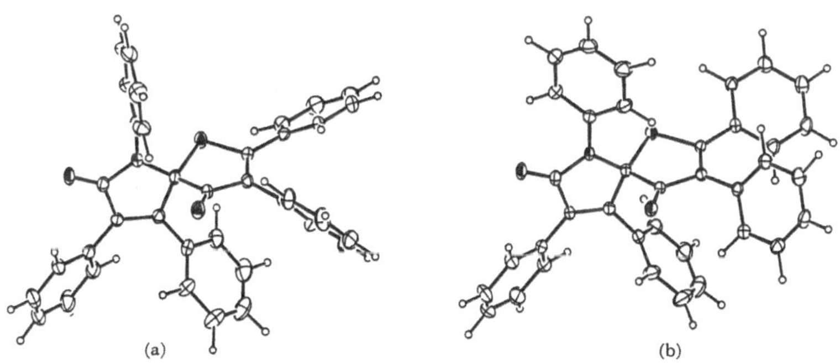

Fig. 1. The two independent $C_{37}H_{25}NO_2S$ molecules.

34,35,36,37,38,39-HEXAMETHOXY-28-THIAHEPTACYCLO[28.3.1.1^{2,6}.1^{7,11}.1^{12,16}.1^{17,21}.1^{22,26}]NONATRI-ACONTA-1(34),2,4,6(39),7,9,11(38),12,14,16(37),17,19,21(36),22,24,26(35),30,32-OCTADECAENE
$C_{44}H_{40}O_6S$

C.B. KNOBLER, K.N. TRUEBLOOD, M. DEGRANDPRE and D.J. CRAM, 1985. *Acta Cryst.*, **C41**, 1337-1340.

Monoclinic, Cc, a = 12.693, b = 18.382, c = 16.009 Å, β = 99.50°, Z = 4. Mo radiation, R = 0.092 for 3155 reflections.

The molecule (Fig. 1) is chiral and has a pseudo twofold axis of symmetry. Two of the six methyl groups, those adjacent to the methylene groups, point into the cavity in the molecule and the others turn away from the cavity in an alternating up-down-up-down arrangement. Averaged values of bond lengths of different types are in good agreement with expected values but in individual bonds there are some large deviations from these values. The average CH_2-S-CH_2-C(aromatic) torsion angle is 66(2)°. The cross-cavity C...O and O...O distances are 4.67, 5.08 and 4.91 Å. The tilts of the aromatic rings with respect to the median plane of the molecule vary systematically being least at the C(23)-C(27) bond and greatest in the more flexible part of the molecule near the-CH_2-S-CH_2-linkage.

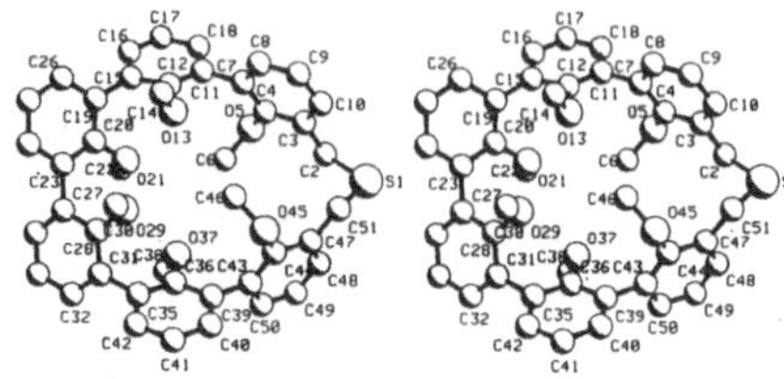

Fig. 1. A stereoview of the $C_{44}H_{40}O_6S$ molecule.

HEPTAKIS(DIBENZOTETRATHIAFULVALENE) TRIS(HEXACHLORONIOBATE)

$C_{98}H_{56}Cl_{18}Nb_3S_{28}$ $3(C_{14}H_8S_4^{+}),4(C_{14}H_8S_4),3(Cl_6Nb^{-})$

R.P. SHIBAEVA and R.M. LOBKOVSKAYA, 1985. *Kristallografiya*, **30**, 693-697.

Tetragonal, *P*$\bar{4}$*n*2, a = 15.152, c = 24.99 Å, Z = 2. Mo radiation, R = 0.098 for 514 reflections.

In the crystal structure (Fig. 1) two independent $NbCl_6^-$ ions occupy special positions with symmetries $\bar{4}$ and 2 and have octahedral geometry with Nb-Cl 2.34-2.36 Å. One of the four crystallographically independent cation radicals occupies a special position of symmetry 22 and the others have twofold crystallographic symmetry. Owing to the limited accuracy of the coordinates for the organic moieties they were not analyzed in detail. The crystal structure has isolated cation and anion stacks running along the *c*-axis. The plane-to-plane separation in the cation stacks are 3.50-3.58 Å.

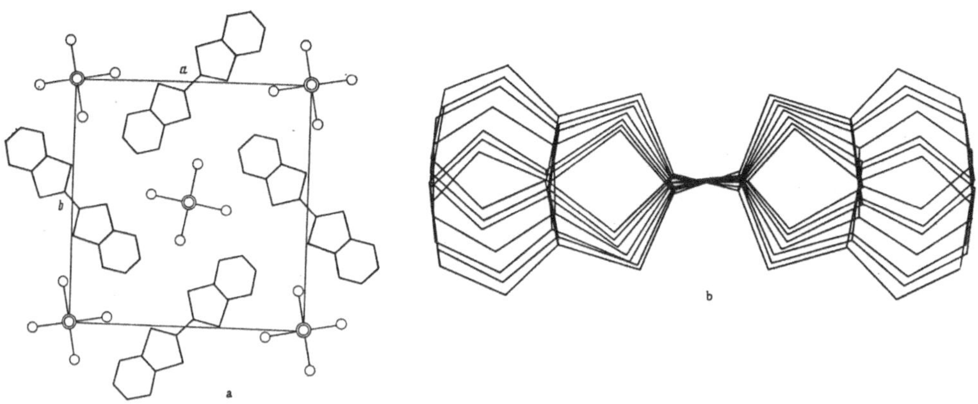

Fig. 1. A projection of the $(C_{14}H_8S_4)_7(NbCl_6)_3$ structure on the ab plane; for the DBTTF cation only the mean position is shown but there is really a rosette of 7 DBTTF moieties as shown on the right.

3-AMINO-4-NITROFURAZANE

$C_2H_2N_4O_3$ (I)

3,4-DIAMINO-FURAZANE

$C_2H_4N_4O$ (II)

A.S. BATSANOV and Yu.T. STRUCHKOV, 1985. *Zh. Strukt. Khim.*, **26**, 1-65-69 [J. Struct. Chem., **26**, 52-56].

I. Orthorhombic, $Pna2_1$, a = 14.864, b = 5.385, c = 11.630 Å, Z = 8. Cu radiation, R = 0.044 for 929 reflections (at -120°C).

II. Monoclinic, $C2/c$, a = 3.6035, b = 11.141, c = 10.329 Å, β = 94.80°, Z = 4. Cu radiation, R = 0.072 for 332 reflections.

The structures are shown in Fig. 1. The diamine (II) has exact twofold symmetry while for (I) there are two independent molecules having slightly different conformations. The hydrogen bonds in (II) connect the molecules into a three-dimensional network and those in (I) result in dimer chains. The substituents strongly affect the intra-annular N-O distances.

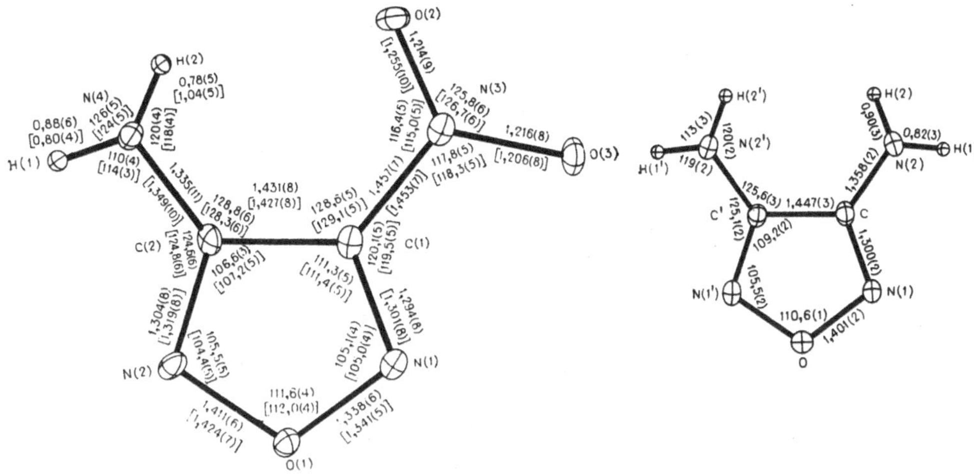

Fig. 1. Geometry of $C_2H_2N_4O_3$ (I) (left) and $C_2H_4N_4O$ (II) (right). Bracketed values for (I) refer to the second molecule in the asymmetric unit.

2-AMINO-4-METHYLTHIO-6-OXO-6H-1,3-OXAZINE-5-CARBONITRILE
$C_6H_5N_3O_2S$

H. KRISTINSSON, T. WINKLER, G. RIHS and H. FRITZ, 1985. *Helv. Chim. Acta,* **68**, 1155-1159.

Triclinic, $P\bar{1}$, a = 7.429, b = 8.682, c = 13.399 Å, α = 88.25, β = 94.05, γ = 115.28°, Z = 4. Mo radiation, R = 0.041 for 2940 reflections.

There are two independent molecules in the asymmetric unit but there are no significant differences between them. The crystal structure analysis confirmed the structure of the molecule (Fig. 1).

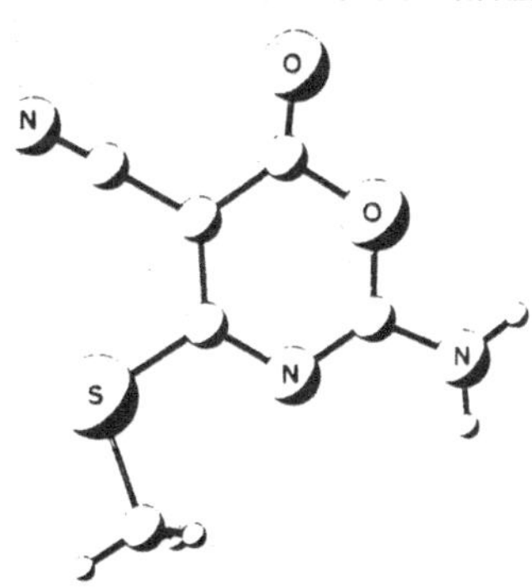

Fig. 1. Perspective view of a molecule of $C_6H_5N_3O_2S$.

1,6-DIAZA-3,4,8,9,12,13-HEXAOXABICYCLO[4.4.4]TETRADECANE HEXAMETHYLENE TRIPEROXIDE-DIAMINE
$C_6H_{12}N_2O_6$

W.P. SCHAEFER, J.T. FOURKAS and B.G. TIEMANN, 1985. *J. Am. Chem. Soc.*, **107**, 2461-2463.

Rhombohedral, *R*3*m*, a = 10.417, c = 6.975 Å, Z = 3. Mo radiation, R = 0.034 for 1620 reflections.

There is unusual exactly planar threefold coordination about the bridgehead N atoms (Fig. 1). N(1) is 0.006(7) Å from the C(1) plane and N(2) is 0.025(8) from the C(2) plane. All C-N-C bonds are 120.0(5)°. Bond lengths in the title compound tend to be somewhat shorter than in comparable structures, at 1.426(8) and 1.416(8) Å. The C-O-O-C torsion angle of 130° is ~30° larger than in related compounds. Remaining features are quite normal. The N atoms reside on the threefold axis.

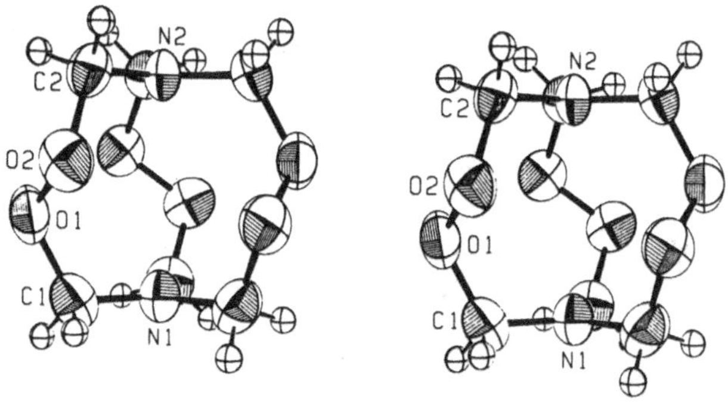

Fig. 1. Stereoview of $C_6H_{12}N_2O_6$, showing planar N atoms.

(RS)-3-HYDROXY-4,5,6,7-TETRAHYDROISOXAZOLO[5,4-c]PYRIDINE-7-CARBOXYLIC ACID MONOHYDRATE
$C_7H_8N_2O_4,H_2O$

P. KROGSGAARD-LARSEN, L. BREHM, J.S. JOHANSEN, P. VINZENTS, J. LAURIDSEN and D.R. CURTIS, 1985. *J. Med. Chem.*, **28**, 673-679.

I. Monoclinic, $P2_1/c$, a = 7.614, b = 11.515, c = 9.729 Å, β = 90.87°, Z = 4. Mo radiation, R = 0.046 for 1885 reflections.

II. Monoclinic, $P2_1/c$, a = 7.5053, b = 11.445, c = 9.686 Å, β = 90.08°, Z = 4. Mo radiation, R = 0.041 for 2560 reflections (at 105 K).

The analyses show that the isoxazole ring is planar and that the six-membered ring adopts a half-chair conformation with the carboxylate group equatorial (Fig. 1). All available hydrogen atoms take part in a hydrogen bond network.

Fig. 1. The $C_7H_8N_2O_4,H_2O$ structure.

METHYLAMMONIUM 1-METHYLCAFFOLIDE
$C_7H_{10}N_4O_5$ $C_6H_4N_3O_5^-,CH_6N^+$

M. POJE, A. PALKOVIC, I. PERINA, I. VICKOVIC and M. BRUVO, 1985. *Tetrahedron*, **41**, 4681-4684.

Monoclinic, $P2_1/n$, a = 12.385, b = 10.190, c = 8.104 Å, β = 96.89°, Z = 4. Cu radiation, R = 0.049 for 1064

reflections.

A view of methylammonium caffolide is shown in Fig. 1. Anions are linked by N(3)-H...N(7) hydrogen bonds (H...N 1.81(1) Å).

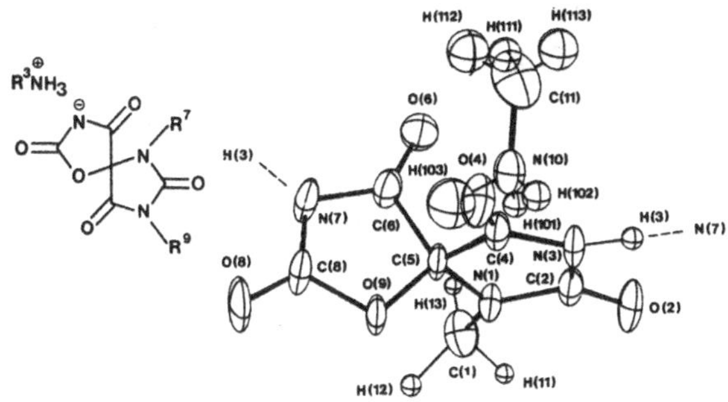

Fig. 1. The methylammonium caffolide structure.

cis-5,6-TRIMETHYLENE-5,6-DIHYDRO-4H-1,3-OXAZIN-2(3H)-ONE
$C_7H_{11}NO_2$ (I)

cis-5,6-TRIMETHYLENE-5,6-DIHYDRO-4H-1,3-OXAZINE-2(3H)-THIONE
$C_7H_{11}NOS$ (II)

3-METHYL-cis-5,6-TETRAMETHYLENE-3,4,5,6-TETRAHYDRO-1,3-OXAZIN-2-ONE
$C_9H_{15}NO_2$ (III)

S. STANKOVIC, A. KAPOR, B. RIBAR, A. KALMAN, G. ARGAY, L. KARANOVIC, G. STAJER and G. BERNATH, 1985. *J. Mol. Struct.*, **131**, 45-53.

I. Monoclinic, $P2_1/c$, a = 12.013, b = 6.133, c = 9.899 Å, β = 100.83°, Z = 4. Mo radiation, R = 0.078 for 909 reflections.

II. Monoclinic, $P2_1/n$, a = 6.291, b = 21.316, c = 6.058 Å, β = 102.95°, Z = 4. Mo radiation, R = 0.049 for 1372 reflections.

III. Monoclinic, $P2_1/c$, a = 6.207, b = 14.365, c = 10.474 Å, β = 104.30°, Z = 4. Mo radiation, R = 0.074 for 953 reflections.

The analyses show that the six-membered oxazin-2-one and thione rings try to assume envelope conformation (Fig. 1). The five-membered carbocyclic rings in (I) and (II) have a marked influence on the puckering of the 1,3-oxazine moiety; the endo bond angles are reduced to a mean value of 105.2° from 111.4° in (III), which results in bond angle opening at C(5) in (I) and (II) (C(4)-C(5)-C(6) 112.2(6), 111.5(4), 108.4(7)° in (I)-(III) respectively).

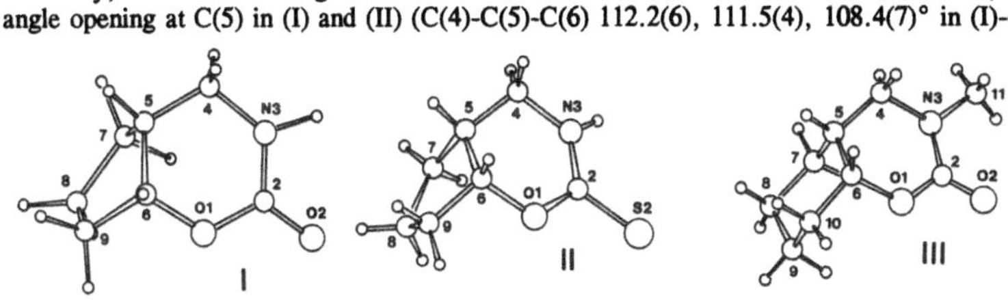

Fig. 1. Perspective views (left to right) of $C_7H_{11}NO_2$ (I), $C_7H_{11}NOS$ (II), and $C_9H_{15}NO_2$ (III).

2,2-BIS(HYDROXYMETHYL)-1-AZA-4-OXABICYCLO[4.1.0]HEPTAN-5-ONE
$C_7H_{11}NO_4$

A.V. EREMEEV, O.N. KRUTIUS, A.F. MISHNEV, Ya.Ya. BLEIDELIS, E.E. LIEPIN'SH, A.G. ODINETS,

D.A. BERZINYA and A.A. KIMENIS, 1984. *Khim. Get. Soedin. SSSR*, 1349-1354.

Orthorhombic, *Pcab*, a = 18.161, b = 8.515, c = 9.843 Å, Z = 8. Cu radiation, R = 0.053 for 1178 reflections.

The morpholine ring of the molecule (Fig. 1) has a distorted boat conformation; the C(1),N(7),C(5),C(4) fragment is planar. The aziridine plane is inclined at 96° to the mean morpholine plane. The packing is stabilized by O(12)-H...N(7) (2.81 Å) and O(11)-H...O(8) (2.78 Å) intermolecular hydrogen bonds.

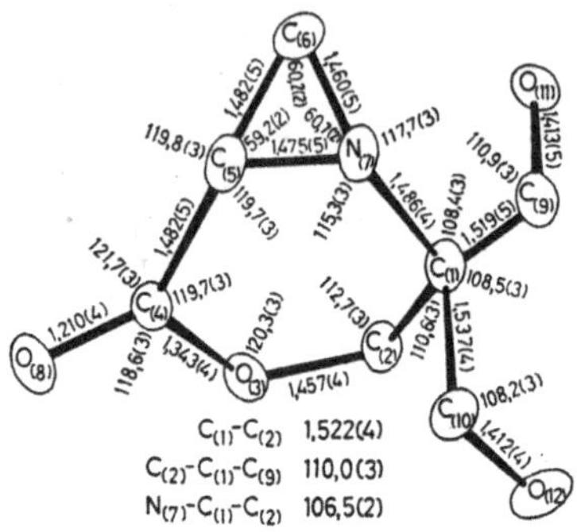

Fig. 1. The $C_7H_{11}NO_4$ molecule with bond lengths.

5-(1',2'-DIHYDROXY-2'-METHOXYCARBONYLETHYL)ISOXAZOLINE
$C_7H_{11}NO_5$

K.B.G. TORSSELL, A.C. HAZELL and R.G. HAZELL, 1985. *Tetrahedron*, **41**, 5569-5575.

Monoclinic, $P2_1/n$, a = 22.552, b = 9.026, c = 4.708 Å, β = 93.11°, Z = 4. Mo radiation, R = 0.046 for 644 reflections.

The molecule (Fig. 1) has three almost planar moieties; the isoxazoline ring, carbon atoms C(4)-C(8) and atoms O(3),C(7),C(8),O(4),O(5). In the five-membered ring, atom C(5) is 0.150(5) Å from the plane of remaining four atoms, with C(6) axial.

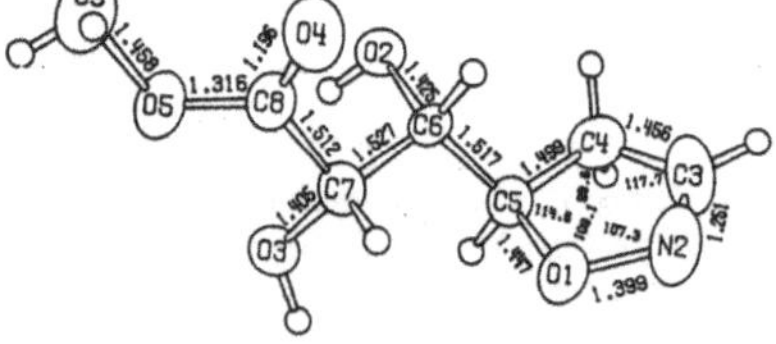

Fig. 1. The $C_7H_{11}NO_5$ molecule with dimensions (σ 0.005-0.009 Å, 0.4-0.6°).

RUBIDIUM 2-PHENYL-1,3,4-OXADIAZOL-5(4H)-ONATE
$C_8H_5N_2O_2Rb$ (I) $C_8H_7N_2O_2^-,Rb^+$

SILVER 2-PHENYL-1,3,4-OXADIAZOL-5(4H)-ONATE DIHYDRATE
$C_8H_5N_2O_2Ag$ (II) $C_8H_7N_2O_2^-,Ag^+,2(H_2O)$

M. DESSOLIN, M. GOLFIER and T. PRANGE, 1985. *J. Org. Chem.*, **50**, 4461-4465.

I. Orthorhombic, *Pbca*, a = 35.310, b = 7.466, c = 6.744 Å, Z = 8. Cu radiation, R = 0.063 for 955 reflections.

II. Monoclinic, $P2_1/c$, a = 8.395, b = 18.256, c = 7.058 Å, β = 111.7°, Z = 4. Cu radiation, R = 0.041 for

1192 reflections; Mo radiation, R = 0.053 for 1084 reflections..

The orientation of the phenyl group with respect to the oxadiazolone ring is the same in the two structures. The dihedral angles are 11 and 8.1° for the Rb^+ and Ag^+ salts respectively. The bond distances in the oxadiazolone rings are shown in Fig. 1.

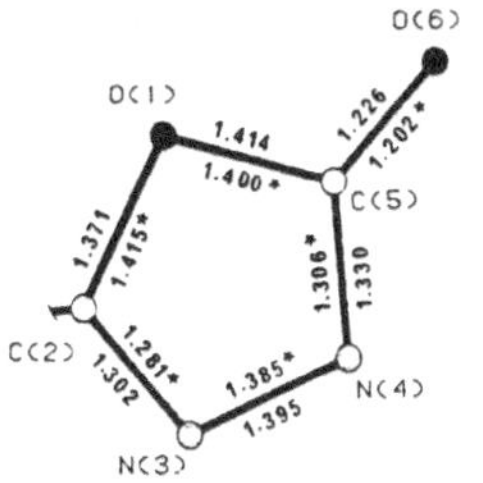

Fig. 1. Bond distances (Å) in the oxadiazolone rings of $C_8H_5N_2O_2Rb$ (starred) and $C_8H_5N_2O_2Ag$.

cis-4a,5,8,8a-TETRAHYDRO-4H-3,1-BENZOXAZIN-2(1H)-ONE
$C_8H_{11}NO_2$ (I)

cis-4a,5,8,8a-TETRAHYDRO-4H-3,1-BENZOXAZINE-2(1H)-THIONE
$C_8H_{11}NOS$ (II)

cis-4a,5,6,7,8,8a-HEXAHYDRO-4H-3,1-BENZOXAZINE-2(1H)-THIONE
$C_8H_{13}NOS$ (III)

G. ARGAY, A. KALMAN, A. KAPOR, G. STAJER and G. BERNATH, 1985. *J. Mol. Struct.*, **131**, 31-43.

I. Monoclinic, $P2_1/n$, a = 11.266, b = 9.638, c = 7.500 Å, β = 109.16°, Z = 4. Mo radiation, R = 0.037 for 1343 reflections.

II. Monoclinic, $P2_1/n$, a = 7.681, b = 10.007, c = 11.207 Å, β = 106.06°, Z = 4. Mo radiation, R = 0.038 for 1605 reflections.

III. Orthorhombic, *Pbca*, a = 11.862, b = 15.751, c = 18.375 Å, Z = 16. Mo radiation, R = 0.042 for 2595 reflections.

There are two independent similar molecules in the asymmetric unit of (III); views of (I), (II) and the two molecules of (III) are in Fig. 1. The hetero ring in all four molecules has a conformation close to envelope ^5E slightly distorted towards half-chair $^5H_6$, which may be attributed to the intrinsic properties of these rings, such as rather short endo C(2)-O bonds.

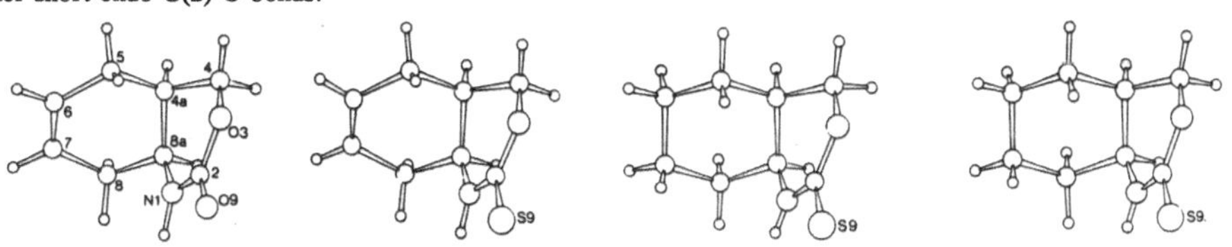

Fig. 1. Solid state conformations for benzoxazines (I), (II) and the two molecules of (III) (left to right).

3,4,5,6-TETRAMETHYL-1-AZAPYRYLIUM TETRACHLOROALUMINATE
$C_8H_{12}AlCl_4NO$ $C_8H_{12}NO^+,AlCl_4^-$

Q.B. BROXTERMAN, H. HOGEVEEN, R.F. KINGMA and F. VAN BOLHUIS, 1985. *J. Am. Chem. Soc.*, **107**, 5722-5729.

Orthorhombic, *Pnma*, a = 17.871, b = 14.243, c = 10.648 Å, Z = 8. Mo radiation, R = 0.054 for 1830 reflections (at -110°C).

There are two $AlCl_4^-$ ions on mirror planes and an essentially planar azapyrylium ion in a general position (Fig. 1).

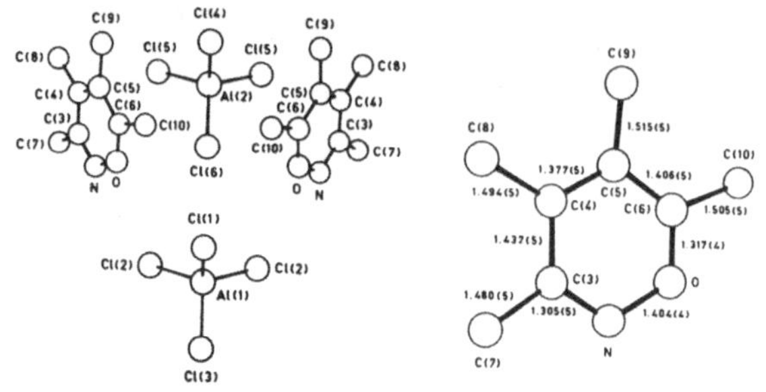

Fig. 1. The arrangement of cations and anions in $C_8H_{12}AlCl_4NO$ and bond lengths in the cation.

4,8-DIMETHYL-4,8-DIAZA-1,5-DIOXADECALIN
$C_8H_{16}N_2O_2$ (I)

PERHYDRO-4,8-DIETHYL-4,8-DIAZA-1,5,9,10-TETRAOXAANTHRACENE
$C_{12}H_{22}N_2O_4$ (II)

A. LE ROUZIC, M. MAUNAYE and P. L'HARIDON, 1985. *J. Chem. Res.*, S35, M601-628.

I. Orthorhombic, *Pbcn*, a = 14.39, b = 4.733, c = 13.34 Å, Z = 4. Mo radiation, R = 0.053 for 323 reflections.

II. Monoclinic, $P2_1/c$, a = 10.559, b = 7.864, c = 8.555 Å, β = 110.5°, Z = 2. Mo radiation, R = 0.044 for 1105 reflections.

The analyses establishes the structures as shown above.

β-(2-(PERHYDRO-1,3,2-DIOXAZINYL))ISOVALERIANIC ACID METHYLAMIDE
$C_9H_{18}N_2O_3$

A.B. ZOLOTOI, O.A. D'YACHENKO, I.I. CHERVIN, V.F. RUDCHENKO, L.O. ATOVMYAN and R.G. KOSTYANOVSKII, 1985. *Khim. Get. Soedin., SSSR*, 1341-1347.

Orthorhombic, *Pbca*, a = 23.829, b = 9.555, c = 9.834 Å, Z = 8. Cu radiation, R = 0.046 for 1388 reflections.

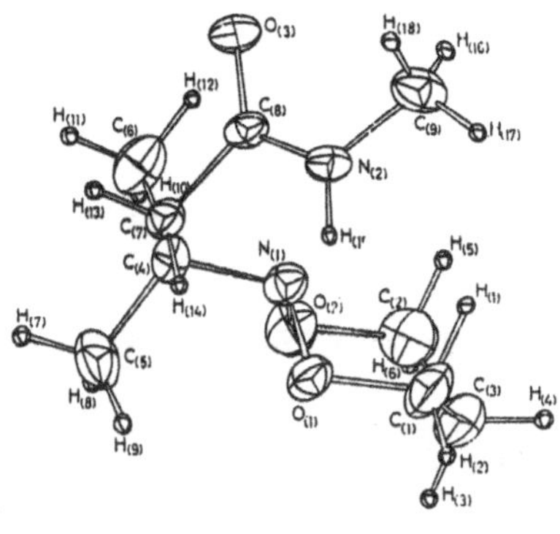

Fig. 1. The $C_9H_{18}N_2O_3$ molecule.

The molecule is shown to be in the chair conformation with equatorial orientation of the substituent on N (Fig. 1). Some dimensions are: mean N-O 1.437, mean C-O 1.437 Å, mean C-O-N-O torsion angle 76.2°. In the crystals, molecules are linked by N(2)-H...O(3') hydrogen bonds (H...O 2.03 Å).

N-(5-METHYL-3-ISOXAZOLYL)SULPHANILAMIDE (form II) (SULPHAMETHOXAZOLE (form II))
$C_{10}H_{11}N_3O_3S$

L. MAURY, J. RAMBAUD, B. PAUVERT, Y. LASSERRE, G. BERGE and M. AUDRAN, 1985. *Canad. J. Chem.*, **63**, 3012-3018.

Monoclinic, *C2/c*, a = 25.106, b = 7.229, c = 14.847 Å, β = 117.9°, Z = 8. Mo radiation, R = 0.0408 for 1246 reflections.

The crystal structure of form (II) is shown in Fig. 1. Molecules are linked by NH...O and N-H...N hydrogen bonds. For details of form (I) see **1**.

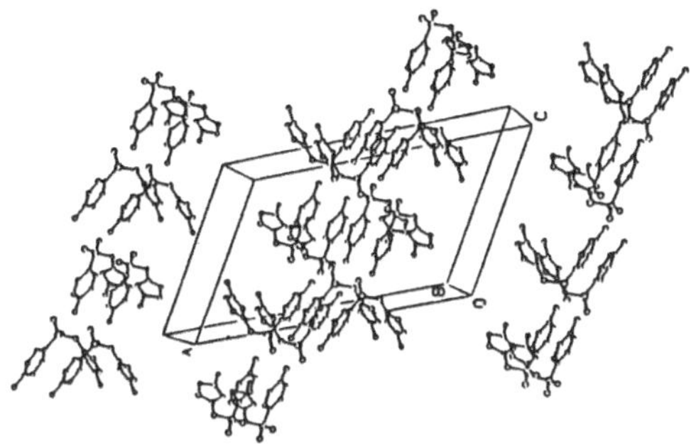

Fig. 1. A view of the $C_{10}H_{11}N_3O_3S$ crystal structure.

1. J. RAMBAUD, R. ROQUES, S. ALBIOLA and F. SABON, 1980. Bull. Soc. Chim. Fr. **1**, 56.

1,3-DITHIO-meso-OXALIC ACID DIMORPHOLIDE
$C_{11}H_{16}N_2O_3S_2$

G. MOTTE-COPPE, F. DUTRON-WOITRIN, T.G.C. BIRD, H.G. VIEHE, J.P. DECLERCQ, G. GERMAIN and M. VAN MEERSSCHE, 1985. *Tetrahedron*, **41**, 693-697.

Orthorhombic, *Pnab*, a = 7.688, b = 9.047, c = 19.770 Å, Z = 4. Cu radiation, R = 0.045 for 855 reflections.

The analysis established the structure and stereochemistry as shown in Fig. 1. A crystallographic twofold axis lies along the central C=O bond. Main bond lengths are: N(6)-C(5) 1.474(4), C(7)-N(6) 1.317(5), C(8)-C(7) 1.513(4), S(10)-C(7) 1.661(4), O(9)-C(8) 1.214(6) Å. The S(10)-C(7)-N(6) angle is 128.7(3)°. The heterocyclic rings have chair conformations.

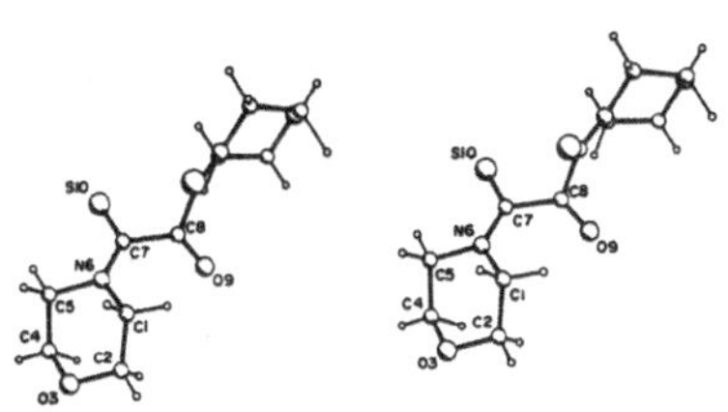

Fig. 1. A stereoview of $C_{11}H_{16}N_2O_3S_2$.

6-(6(e)-CARBAMOYLTETRAHYDRO-2(e)-PYRANYL)-8-OXA-6-AZABICYCLO[3.2.1]OCTAN-7-ONE
$C_{12}H_{18}N_2O_4$

T. YAMANE, H. NANAYAMA, T. ASHIDA, K. HASHIMOTO and H. SUMITOMO, 1985. *Bull. Chem. Soc. Jpn.*, **58**, 2304-2306.

Monoclinic, $P2_1$, a = 10.814, b = 9.964, c = 11.975 Å, β = 101.44°, Z = 4. Cu radiation, R = 0.062 for 2009 reflections.

The two independent molecules in the asymmetric unit have fairly similar dimensions (Fig. 1) except for C(2)-C(3). The tetrahydropyran ring has chair form with the substituents equatorial. The conformation around N(6)-C(66) is anticlinal (*ac*+) with torsion angle C(7)-N(6)-C(66)-O(61) 118.4(3) in one molecule and 104.8(4)° in the other.

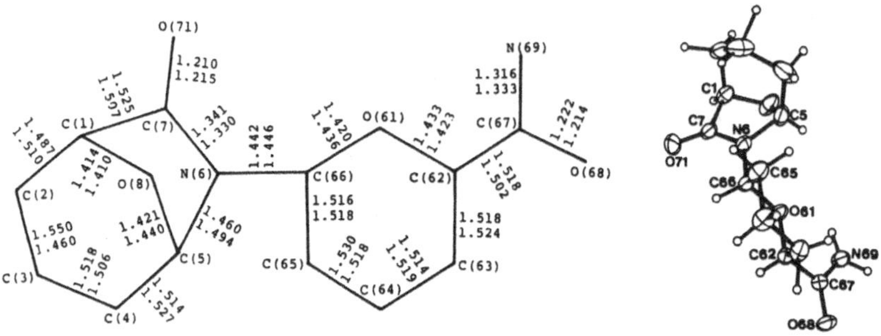

Fig. 1. Bond lengths in $C_{12}H_{18}N_2O_4$ and a view of one of the molecules.

4-((Z)-1-ETHOXYETHYLIDENE)-2-PHENYL-5-OXAZOLONE
$C_{13}H_{13}NO_3$

R.A. BELL, R. FAGGIANI, C.J.L. LOCK and R.A. McLEOD, 1985. *Canad. J. Chem.*, **63**, 3618-3630.

Monoclinic, $P2_1/n$, a = 9.219, b = 19.899, c = 7.459 Å, β = 118.01°, Z = 4. Mo radiation, R = 0.0709 for 1419 reflections.

The molecule is essentially planar (Fig. 1) and most bond lengths and angles are normal. Exceptions are C(6)-O(3) 1.339(4) Å, C(6)-O(3)-C(8) 122.1(3), C(7)-C(6)-O(3) 108.1(4)°. These are attributed to delocalization in the O(ether)CCCO(carbonyl) system and strong intramolecular repulsive interactions.

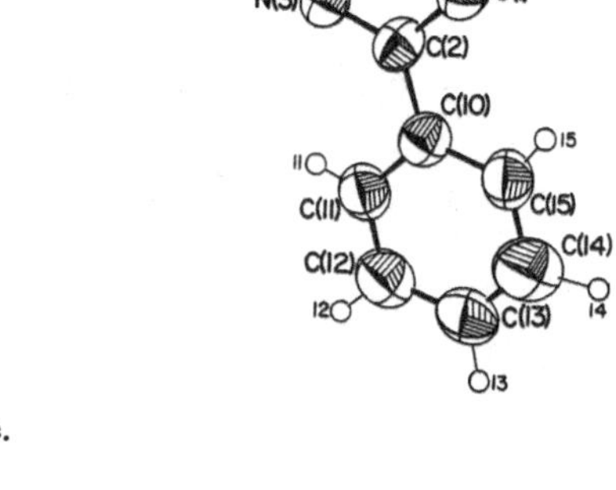

Fig. 1. The $C_{13}H_{13}NO_3$ molecule.

(RS)-ETHYL 4,6,7,9-TETRAHYDRO-3-METHOXY-5,7-DIOXOISOXAZOLO[5,4-f]INDOLIZINE-4a(5H)

CARBOXYLATE
$C_{13}H_{14}N_2O_6$

L. BREHM, J.S. JOHANSEN and P. KROGSGAARD-LARSEN, 1985. *Acta Chem. Scand.*, **B39**, 807-811.

Monoclinic, $P2_1/c$, a = 13.059, b = 14.014, c = 7.225 Å, β = 95.05°, Z = 4. Mo radiation, R = 0.040 for 3252 reflections (at 110 K).

The crystal structure analysis confirmed the stereochemistry of the molecule (Fig. 1). The isoxazole ring is planar; the six-membered ring has a half-chair conformation with C(4a) and N(8) at 0.35 and -0.26 Å respectively from the best plane of the other four atoms; the five-membered ring of the tetrahydroindolizine moiety has a flattened envelope conformation with C(4a) at the flap and 0.21 Å from the plane of the other four atoms.

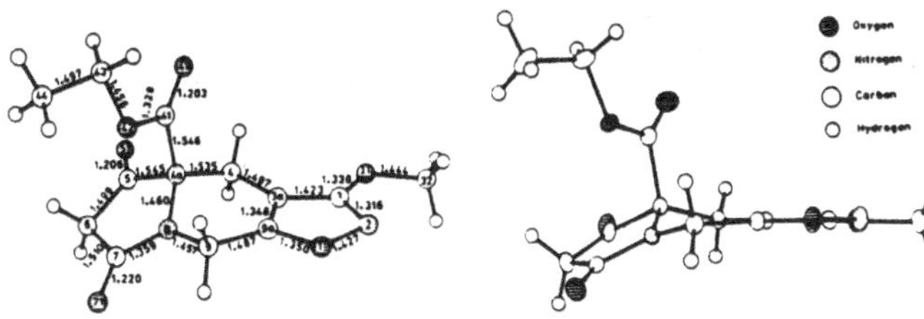

Fig. 1. ORTEP drawings of a molecule of $C_{13}H_{14}N_2O_6$ with bond lengths in Å.

METHYL 3-PHENYL-4a,7,8,8a-TETRAHYDROPYRIDO[4,3-e][1,4,2]DIOXAZINE-7-CARBOXYLATE
$C_{14}H_{14}N_2O_4$

R.G. BALL, E.E. KNAUS and S.K. DUBEY, 1985. *Acta Cryst.*, **C41**, 456-458.

Monoclinic, $P2_1/c$, a = 14.304, b = 10.297, c = 9.092 Å, β = 102.09°, Z = 4. Mo radiation, R = 0.061 for 769 reflections.

In the molecule (Fig. 1) the phenyl ring is planar and nearly coplanar with the N(1), C(6), O(1) plane. The pyrido and dioxazine rings are fused in an endo conformation. The methoxycarbonyl group is approximately coplanar with N(2), C(3,4). Bond lengths and angles within the molecule all have normal values and the intermolecular distances are consistent with van der Waals interactions.

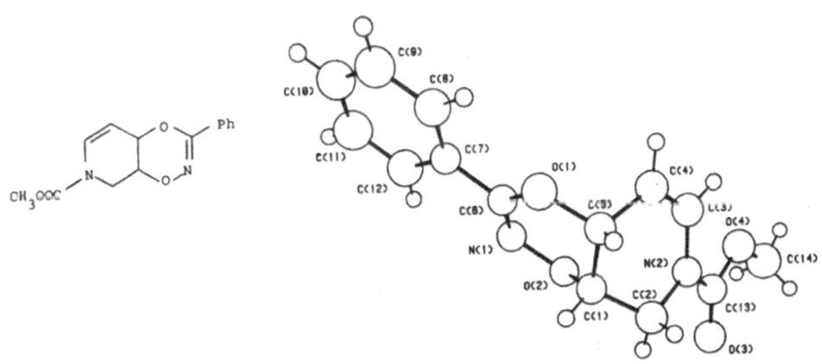

Fig. 1. Structural formula and a view of the $C_{14}H_{14}N_2O_4$ molecule.

N-(2-METHOXYACETYL)-N-(2,6-XYLYL)-3-AMINO-1,3-OXAZOLIDIN-2-ONE
$C_{14}H_{18}N_2O_4$

J.C.J. BART, R. SCORDAMAGLIA and M. CALCATERRA, 1985. *J. Cryst. Spect. Res.*, **15**, 433-444.

Monoclinic, $P2_1/n$, a = 9.363, b = 12.715, c = 12.360 Å, β = 102.64°, Z = 4. Mo radiation, R = 0.062 for 1175 reflections.

The analysis shows that the dimethylphenyl group plane is inclined at 76.0° to the amidic plane (Fig. 1). The oxazolidinone ring has a C(11) envelope conformation (C(11) 0.43 Å from plane of remaining ring atoms) and makes a 66.5° dihedral angle with the amidic plane. The N(1)-N(2) distance is 1.395(7) Å.

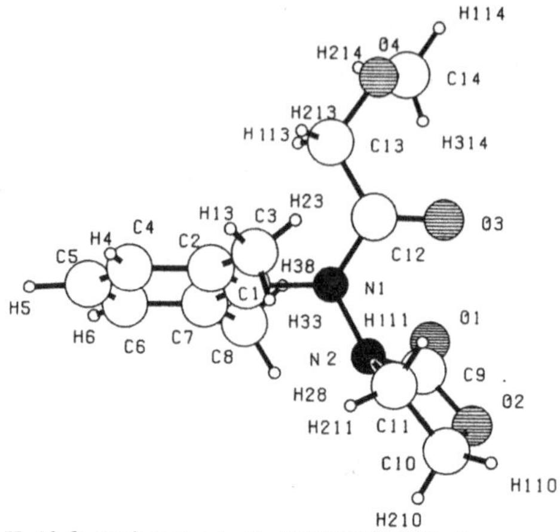

Fig. 1. A view of $C_{14}H_{18}N_2O_4$ projected onto the C(1),C(12),N(2) plane.

4,7,13,16-TETRAOXA-1,10-DIAZABICYCLO[8.8.2]EICOSANE
$C_{14}H_{28}N_2O_4$

P. GROTH, 1985. *Acta Chem. Scand.*, **A39**, 59-62.

Orthorhombic, *Pbcn*, a = 7.841, b = 14.072, c = 14.210 Å, Z = 4. Mo radiation, R = 0.034 for 1063 reflections (at -130°).

This molecule has two 12-membered rings, one of which is 1,4-condensed on the other (Fig. 1).

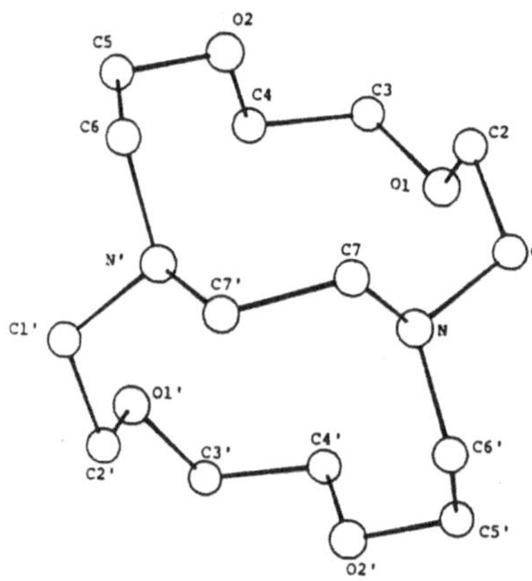

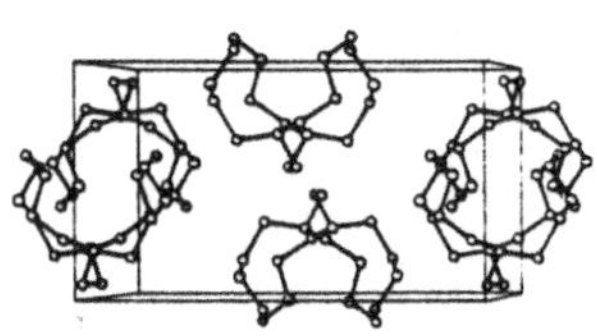

Fig. 1. $C_{14}H_{28}N_2O_4$: schematic drawing of molecule (left) and unit cell contents (right).

3-(p-BENZOYLOXYPHENYL)SYDNONE
$C_{15}H_{12}N_2O_3$ (I)

3-(p-ISOPROPOXYPHENYL)SYDNONE
$C_{11}H_{12}N_2O_3$ (II)

CHUEN-HER UENG, P.L. LEE, Y. WANG and MOU-YUNG YEH, 1985. *Acta Cryst.*, **C41**, 1776-1779.

I. Monoclinic, $P2_1/c$, a = 6.582, b = 6.190, c = 20.848 Å, β = 96.77°, Z = 4. Mo radiation, R = 0.048 for 2222 reflections.

II. Triclinic, $P\bar{1}$, a = 6.916, b = 8.089, c = 10.820 Å, α = 108.49, β = 82.89, γ = 108.42°, Z = 2. Mo radiation, R = 0.041 for 1609 reflections.

The sydnone ring of the molecule (Fig. 1) has a cyclic azomethine-imide type of bonding. The dihedral angles between the sydnone ring and the phenyl ring of (I) and (II) are 39.1(1) and 24.4(1)° respectively. Compound (I) is more planar than (II).

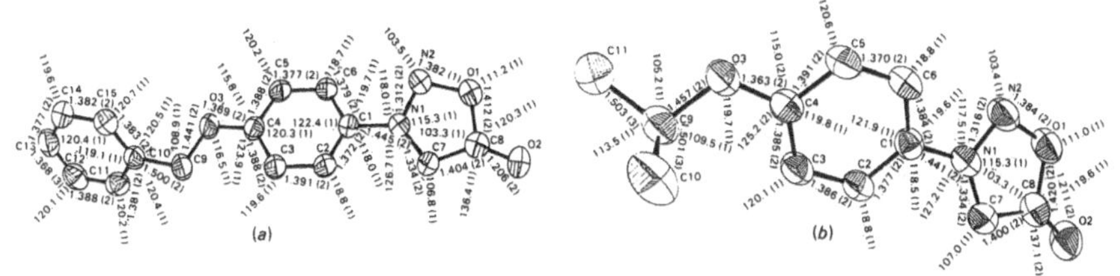

Fig. 1. Bond lengths and angles in $C_{15}H_{12}N_2O_3$ (left) and $C_{11}H_{12}N_2O_3$ (right).

(6R,5R,2S)-6-(3-CHLOROPHENOXY)-2-METHYL-1-OXA-4-AZA-SPIRO[4.5]DECAN-3-ONE
$C_{15}H_{18}ClNO_3$ (I)

(6S,5R,2S)-6-(3-CHLOROPHENOXY)-2-METHYL-1-OXA-4-AZA-SPIRO[4.5]DECAN-3-ONE
$C_{15}H_{18}ClNO_3$ (II)

(6S,5S,2S)-6-(3-CHLOROPHENOXY)-2-METHYL-1-OXA-4-AZA-SPIRO[4.5]DECAN-3-ONE
$C_{15}H_{18}ClNO_3$ (III)

L. DUPONT, O. DIDEBERG and A. MONTEIL, 1985. *Acta Cryst.*, C41, 627-630.

I. Orthorhombic, $P2_12_12_1$, a = 7.913, b = 28.843, c = 6.285 Å, Z = 4. Cu radiation, R = 0.071 for 1550 reflections.

II. Monoclinic, $P2_1$, a = 11.933, b = 7.883, c = 9.094 Å, β = 119.02°, Z = 2. Cu radiation, R = 0.074 for 1474 reflections.

III. Monoclinic, $P2_1$, a = 11.225, b = 16.861, c = 8.473 Å, β = 108.88°, Z = 4. Cu radiation, R = 0.089 for 2754 reflections.

There are two crystallographically independent molecules in the asymmetric unit of (III) (Fig. 1). These both have the same configuration C(7)S, C(12)S, C(14)S. In each of the molecules the cyclohexyl ring has a chair conformation with the heterocyclic ring more or less planar. The angle made by the C(1)O(1)C(7) plane and the benzene ring is 11.7(3) in (I), 13.4(3) in (II) and 3.3(7) and 184.6(7)° in (III). Packing is characterised by N-H...O hydrogen bonds (N(1)...O(3) 2.947(6) (I), N(1)...O(3) 2.846(7) (II) and N(1)...O(3) 2.873(9), 2.880(9) Å (III)).

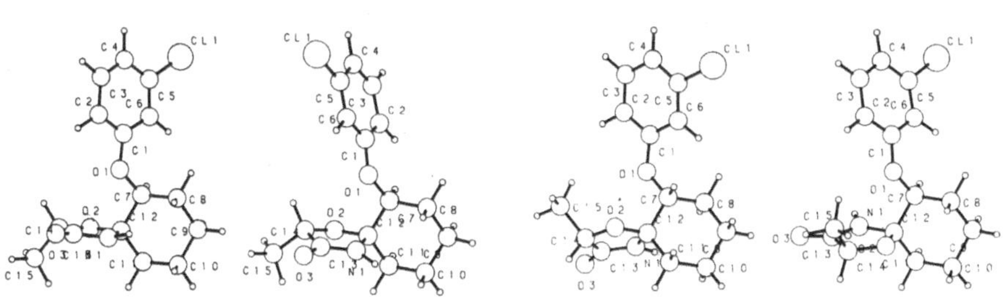

Fig. 1. $C_{15}H_{18}ClNO_3$: the molecular structures of (I) (left), (II) (centre left) and (III) two molecules (right).

2-(4-METHOXYBENZYL)-5-PHENYL-1,3,4-OXADIAZOLE
$C_{16}H_{14}N_2O_2$ (I)

5-(4-METHOXYPHENYL)-2-(2-NITROPHENYL)-6H-1,3,4-OXADIAZINE
$C_{16}H_{13}N_3O_4$ (II)

2'-(2-PHENYL-4H-1,3,4-OXADIAZIN-5(6H)-YLIDENE)BENZOHYDRAZIDE
$C_{16}H_{14}N_4O_2$ (III)

R.W. SAALFRANK, B. WEISS, K. PETERS and H.G. VON SCHNERING, 1985. *Chem. Ber.*, **118**, 4026-4034.

I. Triclinic, $P\bar{1}$, a = 7.427, b = 16.696, c = 5.756 Å, α = 95.31, β = 99.81, γ = 76.62°, Z = 2. Mo radiation, R = 0.043 for 1678 reflections.

II. Triclinic, $P\bar{1}$, a = 10.144, b = 12.613, c = 6.184 Å, α = 100.82, β = 95.71, γ = 109.60°, Z = 2. Mo radiation, R = 0.040 for 1510 reflections.

III. Orthorhombic, *Iba*2, a = 13.206, b = 28.614, c = 7.769 Å, Z = 8. Mo radiation, R = 0.040 for 461 reflections.

The constitution of the three molecules was established as shown in Fig. 1.

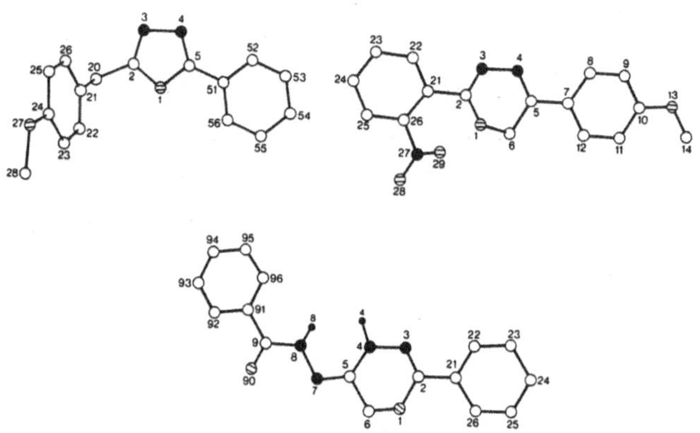

Fig. 1. The $C_{16}H_{14}N_2O_2$ (I) (top left), $C_{16}H_{13}N_3O_4$ (II) (top right) and $C_{16}H_{14}N_4O_2$ (III) (bottom) molecules.

N,N-DIMETHYL-1,10,11-TRIMETHOXY-2-OXA-5-AZANIUMBICYCLO[7.4.0]TRIDECA-8,10,13(8)TRIENE IODIDE
$C_{16}H_{26}INO_4$ (I) $C_{16}H_{26}NO_4^+,I^-$

2,6,6-TRIMETHYL-2,11,12-TRIMETHOXY-3-OXA-6-AZONIABICYCLO[7.4.0]TRIDECA-1(9),10,12-TRIENE IODIDE
$C_{17}H_{28}INO_4$ (II) $C_{17}H_{28}NO_4^+,I^-$

M.F.C. LADD, J. MOHYLA, D.C. POVEY and H.P.G. WILLIAMS, 1985. *J. Cryst. Spect. Res.*, **15**, 501-511.

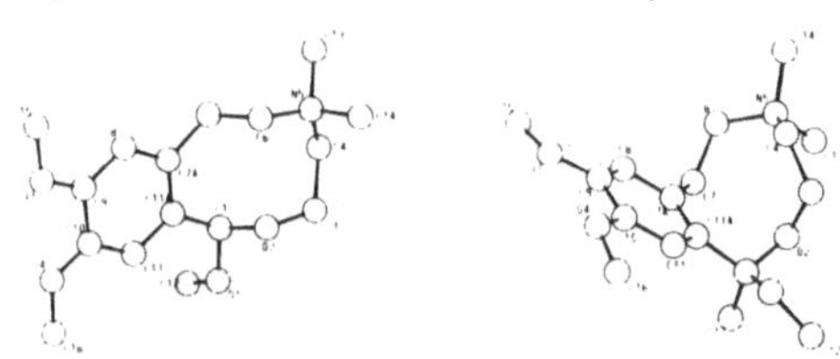

Fig. 1. Views of the $C_{16}H_{26}NO_4^+$ (left) and $C_{17}H_{28}NO_4^+$ (right) cations.

I. Orthorhombic, *Pbcn*, a = 30.774, b = 9.314, c = 13.200 Å, Z = 8. Cu radiation, R = 0.089 for 2287 reflections.

II. Monoclinic, $P2_1/c$, a = 11.361, b = 14.516, c = 11.913 Å, β = 107.35°, Z = 4. Cu radiation, R = 0.084 for 3086 reflections.

Both molecules have similar stereochemistry, differing chemically only at C(1) (Fig. 1) with a CH_3 group replacing a hydrogen. The nine-membered rings have a puckered conformation.

(±)-5-METHYL-1-PHENYL-1,3,4,6-TETRAHYDRO-5H-2,5-BENZOXAZOCINE HYDROCHLORIDE
$C_{17}H_{20}ClNO$ $C_{17}H_{20}NO^+,Cl^-$

P. KLUFERS and A. VON PETERSENN, 1985. *Z. Kristallogr.*, **170**, 99-101.

Monoclinic, $P2_1/c$, a = 11.761, b = 7.739, c = 16.897 Å, β = 97.43°, Z = 4. Mo radiation, R = 0.045 for 2673 reflections.

The molecular conformation is shown in Fig. 1; the tetrahydrooxazocine ring is under strain with ring bond angles some 5° greater than normal. The Cl^- ion is hydrogen bonded to the N atom.

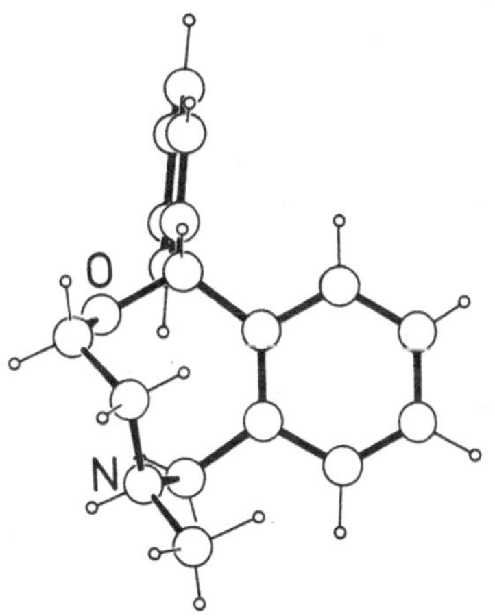

Fig. 1. The $C_{17}H_{20}NO^+$ cation.

3-MORPHOLINOETHYLTHIO-4-METHYL-6-PHENYLPYRIDAZINE
$C_{17}H_{21}N_3OS$

C. VAN DER BREMPT, G. EVRARD and F. DURANT, 1985. *Bull. Soc. Chim. Belg.*, **94**, 261-265.

Triclinic, $P\bar{1}$, a = 11.162, b = 10.793, c = 9.142 Å, α = 103.05, β = 98.72, γ = 126.42°, Z = 2. Cu radiation, R = 0.038 for 2559 reflections.

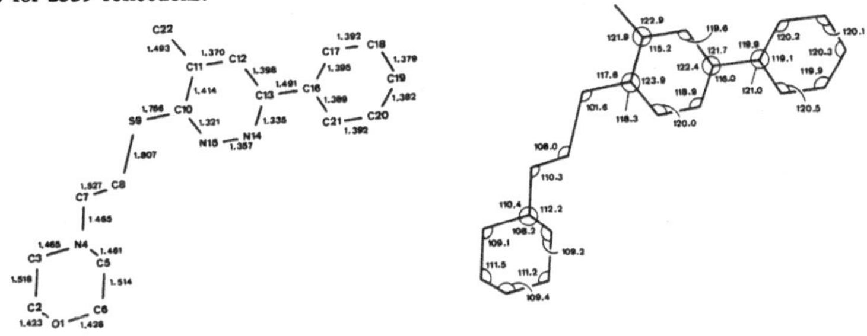

Fig. 1. Bond lengths and angles in $C_{17}H_{21}N_3OS$.

Molecular dimensions are in Fig. 1. The S(9) atom is sp^2 hybridised and the torsion angle C(8)-S(9)-C(10)-C(11) is 175.6(2)°. The morpholino ring has chair form, with N(4)-C(7)-C(8)-S(9) -172.1(2)°.

(E)-N,N'-DICYCLOHEXYL-4-MORPHOLINECARBOXAMIDINE
$C_{17}H_{31}N_3O$

F. BELLUCCI, V. BERTOLASI, V. FERRETTI and G. GILLI, 1985. *Acta Cryst.*, **C41**, 544-546.

Monoclinic, *Cc*, a = 14.043, b = 9.695, c = 12.727 Å, β = 90.32°, Z = 4. Mo radiation, R = 0.031 for 1515 reflections.

The crystal consists of discrete monomeric units held together by van der Waals forces. The compound has a centre of syn-anti isomerism related to the presence of the C(5)-N(2) double bond. The conformation about C(5)-N(3) is synclinal and the morpholine and cyclohexane rings have chair conformations (Fig. 1). A complete range of C-N distances is found ranging from standard $N(sp^3)$-$C(sp^3)$ for N(3)-C(12) (1.471(2)) to $N(sp^2)$-$C(sp^2)$ for N(2)-C(5) (1.276(3) Å). C(5)-N(1) (1.414(2)) and C(5)-N(3) (1.381(2) Å) are intermediate in length. The average C-C distance in morpholine is 1.505(3) and in cyclohexane is 1.520(3) Å.

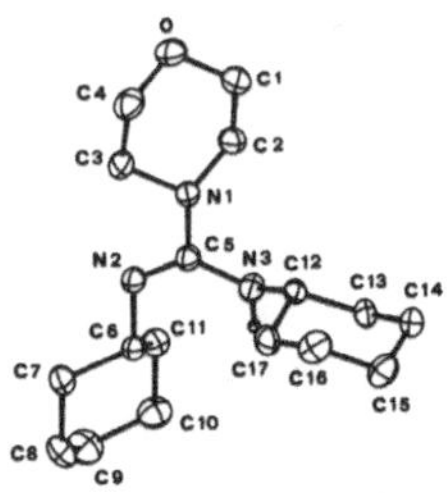

Fig. 1. $C_{17}H_{31}N_3O$: the molecular structure.

(4R*)-3-(4-((3-CHLOROPHENYL)METHOXY)PHENYL)-5-(METHYLAMINOMETHYL)-1,3-OXAZOLIDIN-2-ONE METHANESULPHONIC ACID
$C_{18}H_{19}ClN_2O_3,CH_4O_3S$

F. DURANT, B. NORBERG, F. BUFKENS and G. EVRARD, 1985. *Acta Cryst.*, **C41**, 247-249.

Orthorhombic, $P2_12_12_1$, a = 7.396, b = 45.969, c = 5.822 Å, Z = 4. Cu radiation, R = 0.050 for 1937 reflections.

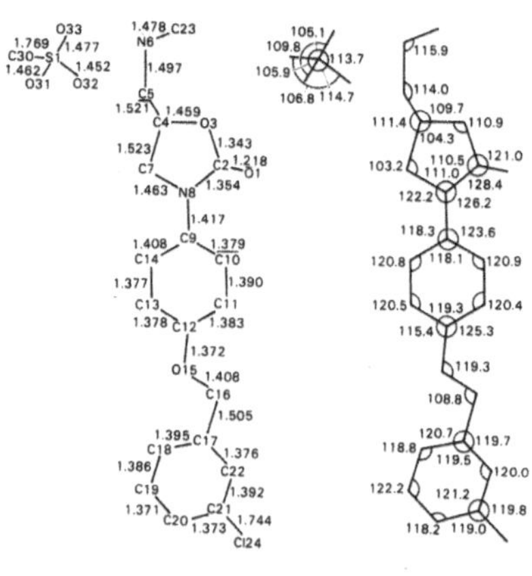

Fig. 1. $C_{18}H_{19}ClN_2O_3,CH_4O_3S$: bond lengths and angles (σ = 0.011 Å, 0.6°).

The oxazolidinone moiety shows little distortion and is coplanar with the adjacent phenyl ring (Fig. 1). The N(8) atom is trigonal and the dihedral angles along the N(8) side chain show that the molecule is stretched and flattened. The N atom of the methylaminomethyl moiety is bonded to methanesulphonic acid molecules by two hydrogen bridges.

MORPHOLINIUM 5-ACETYL-3-CYANO-1,4-DIHYDRO-6-METHYL-4-(2-NITROPHENYL)-2-PYRIDINETHIOLATE

$C_{19}H_{22}N_4O_4S$ $C_4H_{10}NO^+,C_{15}H_{12}N_3O_3S^-$

V.N. NESTEROV, V.E. SHKLOVER, Yu.T. STRUCHKOV, Yu.A. SHARANIN, A.M. SHESTOPALOV and L.A. RODINOVSKAYA, 1985. *Acta Cryst.*, **C41**, 1191-1194.

Monoclinic, *C2/c*, a = 21.221, b = 11.665, c = 16.598 Å, β = 108.75°, Z = 8. Mo radiation, R = 0.045 for 2913 reflections.

The crystal structure consists of discrete morpholinium cations and heterocyclic anions (Fig. 1). In the anion, the six-membered 1-azacyclohexa-2,5-diene moiety has a flattened boat conformation with a fold angle, along N(1)...C(4), of 12.1(1)°. There is conjugation in the N(2)≡C(6)-C(5)=C(1)-S chain and also in the O(1)=C(14)-C(3)=C(2)-N(1) chain. The C(1)-S distance is 1.725(3) Å. The pseudo-axial 2-nitrophenyl substituent is rotated by 75.6(1)° with respect to the bottom of the boat (C(1),C(2),C(3),C(5)). The cation has a chair conformation and mean values of 1.498(6), 1.492(2) and 1.422(1) Å for the C-C, C-N and C-O bond lengths. Intermolecular N-H...O, NH...S and N-H...N hydrogen bonds of 2.729(4), 3.407(3) and 2.990(4) Å are present.

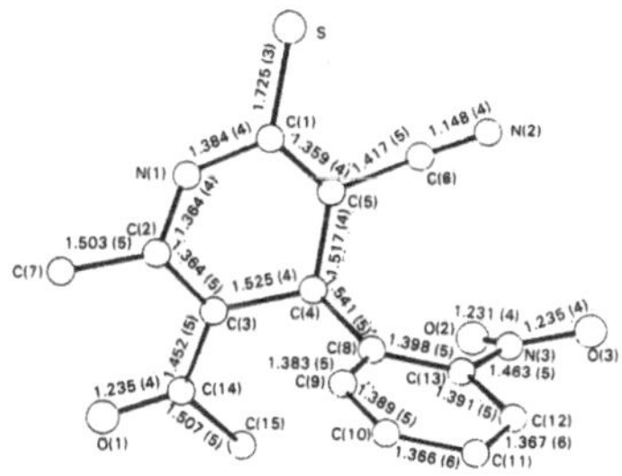

Fig. 1. $C_{19}H_{22}N_4O_4S$: the anion.

3-MORPHOLINIUM-ETHYLAMINO-5-METHYL-6-PHENYLPYRIDAZINIUM OXALATE OXALIC ACID

$C_{19}H_{24}N_4O_5,C_2H_2O_4$ $C_{17}H_{24}N_4O^{2+},C_2O_4^{2-},C_2H_2O_4$

C. VAN DER BREMPT, F. DURANT, B. NORBERG and G. EVRARD, 1985. *Acta Cryst.*, **C41**, 1532-1535.

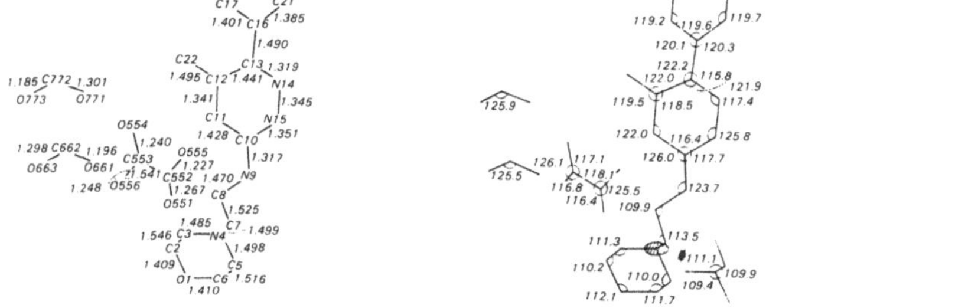

Fig. 1. $C_{19}H_{24}N_4O_5,C_2H_2O_4$: bond lengths and angles.

Triclinic, *P*$\bar{1}$, a = 9.949, b = 16.477, c = 8.345 Å, α = 87.75, β = 109.88, γ = 116.59°, Z = 2. Mo radiation, R = 0.037 for 1268 reflections.

The dihedral angle between the pyridazine and phenyl rings in this 5-methyl analogue of minaprine is -50.8(7)°

as a result of steric hindrance between the methyl group and the phenyl ring (Fig. 1). The bond between the pyridazine and phenyl moieties (C(13)-C(16)) is 1.490(9) Å. The morpholino ring adopts a chair conformation and the side-chain methylene groups 7 and 8 are syn-clinal, with N(4)-C(7)-C(8)-N(9)-80.7(4)°. Crystal cohesion is achieved by a dense network of hydrogen bonds between two oxalate moieties, the protonated N(4) and N(15) and the exocyclic N(9) atoms. Hydrogen bonds are formed by alternation of oxalate ions and oxalic acid molecules to form zigzag chains. The terminal ions are linked to two 5-methyl analogue molecules via four different N-H...O bonds. The oxalic acid molecules are planar, the oxalate ion slightly distorted.

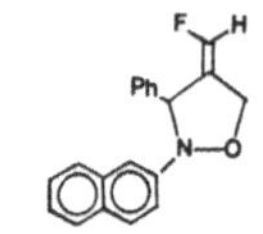

4-FLUOROMETHYLENE-3-PHENYL-2-(2-NAPHTHYL)ISOOXAZOLIDINE
$C_{20}H_{16}FNO$

W.R. DOLBIER Jr., C.R. BURKHOLDER, G.E. WICKS, G.J. PALENIK and M. GAWRON, 1985. *J. Am. Chem. Soc.*, **107**, 7183-7184.

Monoclinic, $P2_1/c$, a = 5.940, b = 23.685, c = 22.672 Å, β = 91.69°, Z = 8. Mo radiation, R = 0.073 for 3545 reflections.

The analysis established the stereochemistry as shown.

cis-3-(4-(5-METHOXYMETHYL-4-METHYL-2-OXO-1,3-OXAZOLIDIN-3-YL)PHENOXYMETHYL)BENZO-NITRILE
$C_{20}H_{20}N_2O_4$ (I)

trans-3-(4-(5-METHOXYMETHYL-4-METHYL-2-OXO-1,3-OXAZOLIDIN-3-YL)PHENOXYMETHYL)BENZO-NITRILE
$C_{20}H_{20}N_2O_4$ (II)

F. DURANT, F. BUFKENS, F. LEFEVRE, G. EVRARD and A. MICHEL, 1985. *Acta Cryst.*, **C41**, 243-246.

I. Monoclinic, $P2_1/c$, a = 13.491, b = 14.597, c = 9.187 Å, β = 99.85°, Z = 4. Cu radiation, R = 0.050 for 2010 reflections.

II. Monoclinic, $P2_1/c$, a = 16.961, b = 14.656, c = 7.525 Å, β = 97.40°, Z = 4. Cu radiation, R = 0.030 for 1137 reflections.

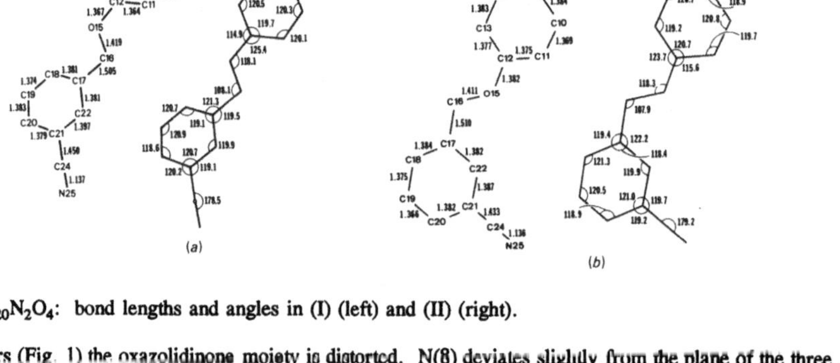

Fig. 1. $C_{20}H_{20}N_2O_4$: bond lengths and angles in (I) (left) and (II) (right).

In both isomers (Fig. 1) the oxazolidinone moiety is distorted. N(8) deviates slightly from the plane of the three adjacent C atoms, C(2), C(7) and C(9) (-0.083 Å in (I) and 0.048 Å in II). The dihedral angles between the mean planes of the oxazolidinone and phenyl rings are 25.3° and 43.9° in (I) and (II) respectively. The two O atoms O(3) and O(6) have the gauche conformation in both.

3-(p-FLUOROPHENYLIMINO)METHYL-4,5-DIHYDRO-5-MORPHOLINO-1-PHENYL-1H-PYRAZOLE
$C_{20}H_{21}FN_4O$ (I)

3-(p-FLUOROPHENYLIMINO)METHYL-1-(p-METHOXYPHENYL)-1H-PYRAZOLE
$C_{17}H_{14}FN_3O$ (II)

R. DESTRO, E. ERBA, L. FORTI, D. POCAR and D. SCARCELLA, 1985. *Liebigs Ann. Chem.*, 1377-1388.

I. Orthorhombic, $Pna2_1$, a = 13.490, b = 22.290, c = 6.119 Å, Z = 4. Mo radiation, R = 0.041 for 1189 reflections.

II. Monoclinic, $P2_1/c$, a = 3.918, b = 21.722, c = 16.730 Å, β = 91.07°, Z = 4. Mo radiation, R = 0.037 for 1627 reflections.

In (I) the 4,5-dihydropyrazoline ring is almost planar (Fig. 1); very high thermal motion affects the phenyl ring bonded to N(3). The five-membered ring in (II) is also planar (Fig. 1) and the dimensions are consistent with extensive delocalization. The phenyl ring bonded to N(1) is twisted 21° to the base plane to relieve interactions between the H atoms on C(3) and C(6).

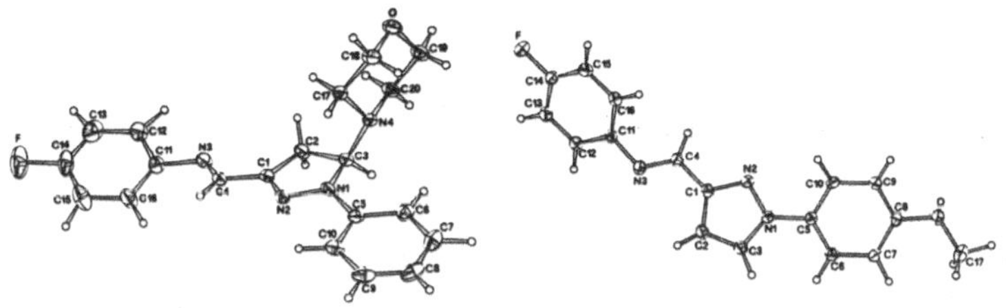

Fig. 1. Views of $C_{20}H_{21}FN_4O$ (I) (left) and $C_{17}H_{14}FN_3O$ (II) (right).

2,4-BIS(DIMETHYLAMINO)-6-((DIPHENYLMETHYLENE)AMINO)-1,3,5-OXADIAZINIUM HEXACHLORO-ANTIMONATE
$C_{20}H_{22}Cl_6N_5OSb$ (I) $C_{20}H_{22}N_5O^+,Cl_6Sb^-$

1-(DIPHENYLMETHYL)-2,3,4,5-TETRAHYDRO-3,5-DIISOPROPYL-6-(ISOPROPYLIDENEIMINO)-4-(ISO-PROPYLIMINO)-2-OXO-1,3,5-TRIAZINIUM HEXACHLOROANTIMONATE DICHLOROMETHANE SOLV-ATE
$C_{28}H_{38}Cl_6N_5OSb,CH_2Cl_2$ (II) $C_{28}H_{38}N_5O^+,Cl_6Sb^-,CH_2Cl_2$

M. AL-TALIB, J.C. JOCHIMS, L. ZSOLNAI and G. HUTTNER, 1985. *Chem. Ber.*, **118**, 1887-1902.

I. Orthorhombic, $P2_12_12_1$, a = 9.65, b = 14.73, c = 19.13 Å, Z = 4. Mo radiation, R = 0.032 for 1783 reflections (at 213 K).

II. Monoclinic, $P2_1/c$, a = 8.900, b = 15.04, c = 29.22 Å, β = 92.52°, Z = 4. Mo radiation, R = 0.067 for 3288 reflections (at 238 K).

Details of conformation and dimensions for (I) and (II) are in Fig. 1. In both compounds the 2-azaallenium units are bent ((I): C-N(4)-C 129.2(8)°, (II): C-N(5)-C 129(1)°) and show unequal C=N bond lengths. The observed dimensions suggest that these compounds are azomethines rather than 2-azaallenium salts. The positive charge in the cation of (I) is mainly delocalized over the polymethine unit (N(3)-C(1)-N(1)-C(2)-N(5)); in (II) the positive charge is delocalized over the guanidinium unit N(4)-C(3)-N(5)-N(2). The triazinium ring of (II) is not planar (C(3) and N(2) below the C(1),N(6),C(5),N(4) plane).

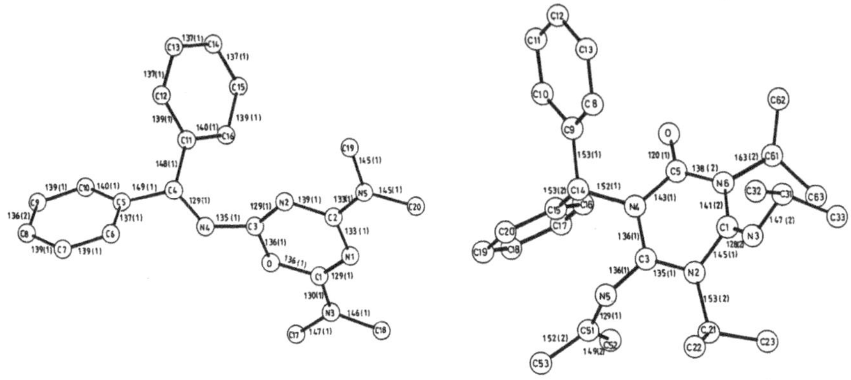

Fig. 1. The $C_{20}H_{22}N_5O^+$ (left) and $C_{28}H_{38}N_5O^+$ (right) cations.

DIMETHYL rel-(2S,3S,3aR)-7-t-BUTYL-2,3,3a,4-TETRAHYDRO-4,4-DIMETHYLISOXAZOLO[2,3-a]INDOLE-2,3-DICARBOXYLATE
$C_{20}H_{27}NO_5$ (I)

DIMETHYL rel-(2R,3aS,9S,10S)-5,7-DI-t-BUTYL-3,3-DIMETHYL-3a,2-EPOXYETHANO-3H-INDOLE-9,10-DICARBOXYLATE
$C_{24}H_{35}NO_5$ (II)

D. DOPP, C. KRUGER, G. MAKEDAKIS and A.M. NOUR-EL-DIN, 1985. *Chem. Ber.*, **118**, 510-525.

I. Monoclinic, $P2_1/c$, a = 9.829, b = 19.653, c = 10.456 Å, β = 100.74°, Z = 4. Mo radiation, R = 0.0503 for 2144 reflections.

II. Triclinic, $P\bar{1}$, a = 9.5173, b = 10.1415, c = 14.3345 Å, α = 82.291, β = 73.103, γ = 66.925°, Z = 2. Cu radiation, R = 0.0406 for 3249 reflections.

The analyses established the structures and stereochemistries as shown in Fig. 1.

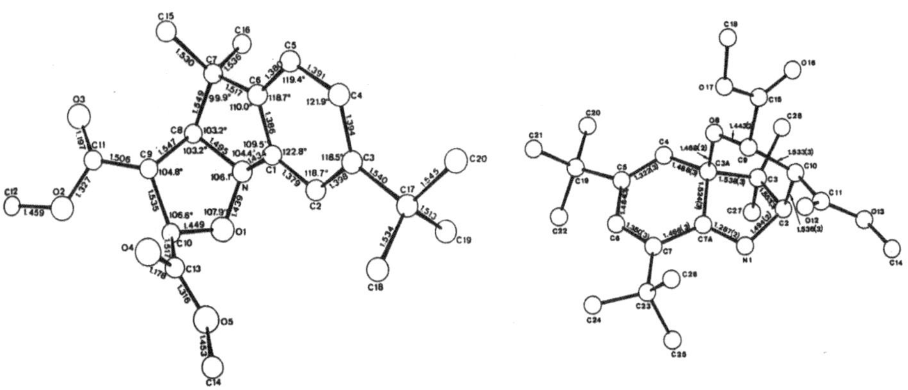

Fig. 1. Views of $C_{20}H_{27}NO_5$ (I) (left) and $C_{24}H_{35}NO_5$ (II) (right) with dimensions.

(3RS,4SR,5RS)-2-BENZYL-4-METHOXYCARBONYL-5-METHYL-3-((4RS)-2,2,5,5-TETRAMETHYL-1,3-DIOXOLAN-4-YL)ISOXAZOLIDINE
$C_{20}H_{29}NO_5$

M.J. FRAY, R.H. JONES and E.J. THOMAS, 1985. *J. Chem. Soc. Perkin Trans. 1*, 2753-2761.

Triclinic, $P\bar{1}$, a = 11.549, b = 6.563, c = 14.146 Å, α = 101.44, β = 90.38, γ = 103.89°, Z = 2. Cu radiation, R = 0.050 for 3371 reflections.

The crystal structure analysis established the stereochemistry of the molecule (Fig. 1).

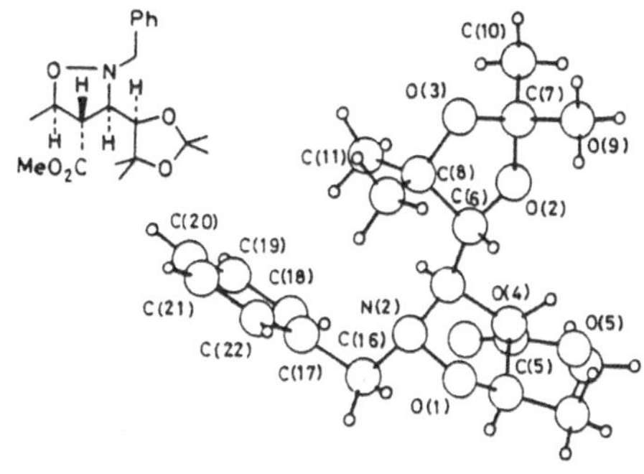

Fig. 1. Schematic drawing of a molecule of $C_{20}H_{29}NO_5$.

BIS[3,5-DI(N-MORPHOLINIO)-1,2,4-TRITHIOLANE(2+)] HEXADECAIODIDE(4-)
$C_{20}H_{32}I_{16}N_4O_4S_6$ $[S_2\{C(S)NC_4H_8O\}_2]_2(I_{16})$

F. BIGOLI, M.A. PELLINGHELLI, G. CRISPONI, P. DEPLANO and E.F. TROGU, 1985. *J. Chem. Soc. Dalton Trans.*, 1349-1343.

Triclinic, $P\bar{1}$, a = 7.834, b = 7.959, c = 24.142 Å, α = 86.80, β = 88.55, γ = 70.86°, Z = 2. Mo radiation, R = 0.038 for 1954 reflections.

The organic di-cation contains $N^+=C(S)$ moieties (N-C 1.27, 1.29(2), C-S 1.73-1.78(2), S-S 2.04(1) Å). The OC_4N rings have distorted chair conformations, and the C_2S_3 ring has a twisted conformation. The novel I_{16}^{4-} anion lies across a crystal inversion centre (Fig. 1); successive I-I bond lengths and I-I-I angles from the end are: 2.838, 2.976, 3.358, 2.741, 3.471, 3.018, 2.827, 3.550(3) Å, 177.2, 144.9, 172.3, 170.9, 161.4, 172.4, 147.2(1)° i.e. $I_3 \ldots I_2 \ldots I_3^- \ldots I_3^- \ldots I_2 \ldots I_3^-$.

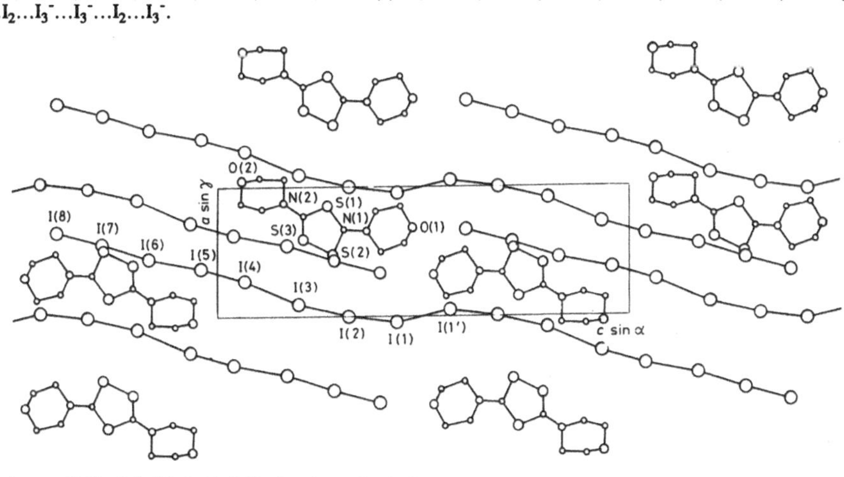

Fig. 1. Cell of $C_{20}H_{32}I_{16}N_4O_4S_6$ (b-axis projection).

3,9-DIMETHYL-6,14,17,22,25-PENTAOXA-1,3,9,11-TETRAAZABICYCLO[9.8.8]HEPTACOSANE-2,10-DIONE

$C_{20}H_{38}N_4O_7$

A.A. DVORKIN, Yu.A. SIMONOV, N.G. LUK'YANENKO, T.I. KIRICHENKO, S.V. SHCHERBAKOV, T.I. MALINOVSKII and A.V. BOGATSKII, 1985. *Kristallografiya*, **30**, 915-918.

Monoclinic, *B*11*b*, a = 21.040, b = 14.873, c = 9.172 Å, γ = 54.95°, Z = 4. Cu radiation, R = 0.052 for 1381 reflections.

The crystal structure contains discrete molecules which have approximate twofold symmetry (Fig. 1). A cup-shaped molecule is formed with the base of the cup being the corrugated oxyethylene chain. Torsion angles are close to gauche for the C-C and C-N bonds and close to trans for the C-O bonds.

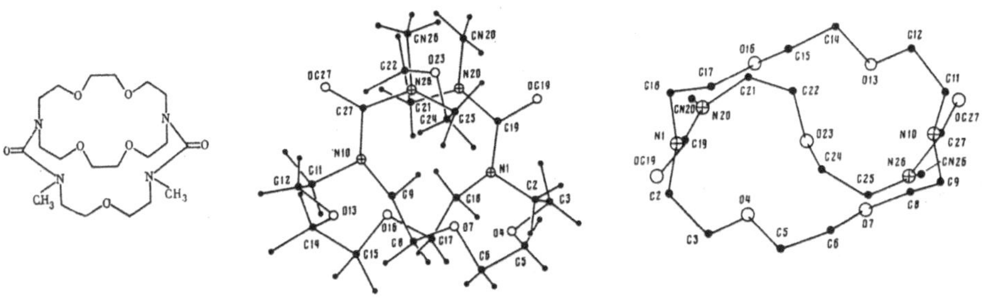

Fig. 1. The $C_{20}H_{38}N_4O_7$ molecule and views of the conformation.

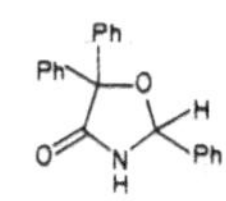

2,5,5-TRIPHENYLOXAZOLIDIN-4-ONE
$C_{21}H_{17}NO_2$

M. ABOU-GHARBIA, D.M. KETCHA, D.E. ZACHARIAS and D. SWERN, 1985. *J. Org. Chem.*, **50**, 2224-2228.

Orthorhombic, $P2_12_12_1$, a = 12.185, b = 24.481, c = 11.002 Å, Z = 8. Cu radiation, R = 0.038 for 3398 reflections.

There are two molecules in the crystallographic asymmetric unit. The structure analysis established the cis form in which the substituents are pseudoequatorial. The ring has an envelope conformation with the oxygen as the out-of-plane atom.

5,7-DICHLORO-2,3-DIHYDRO-2-(p-METHOXYPHENYL)-4-(p-NITROBENZOYL)-4H-1,4-BENZOXAZINE
$C_{22}H_{16}Cl_2N_2O_5$ (I)

5,7-DICHLORO-2,3-DIHYDRO-2-METHOXY-4-(p-NITROBENZOYL)-3-PHENYL-4H-1,4-BENZOXAZINE
$C_{22}H_{16}Cl_2N_2O_5$ (II)

5,7-DICHLORO-2-ETHOXY-2,3-DIHYDRO-4-(p-NITROBENZOYL)-4H-1,4-BENZOXAZINE
$C_{17}H_{14}Cl_2N_2O_5$ (III)

6,8-DICHLORO-3a,9a-DIHYDRO-9-(p-NITROBENZOYL)-9H-FURO[3,2-b](1,4)BENZOXAZINE
$C_{17}H_{10}Cl_2N_2O_5$ (IV)

M.F. GARBAUSKAS, E.A. WILLIAMS and H.W. HEINE, 1985. *Acta Cryst.*, **C41**, 1217-1222.

I. Monoclinic, $P2_1/a$, a = 11.696, b = 9.151, c = 19.72 Å, β = 93.33°, Z = 4. Mo radiation, R = 0.0948 for 1289 reflections.

II. Monoclinic, $P2_1/n$, a = 11.269, b = 13.585, c = 13.965 Å, β = 90.98°, Z = 4. Mo radiation, R = 0.0368 for 2043 reflections.

III. Triclinic, $P\bar{1}$, a = 7.174, b = 7.859, c = 16.965 Å, α = 85.50, β = 79.27, γ = 71.62°, Z = 2. Mo

radiation, R = 0.0681 for 1404 reflections.

IV. Monoclinic, $P2_1/n$, a = 11.967, b = 8.663, c = 15.996 Å, β = 91.19°, Z = 4. Mo radiation, R = 0.0507 for 1555 reflections.

All four compounds (Fig. 1) exhibit slight shortening of the N-carbonyl bond (1.378-1.397 Å), indicating partial double bond character. Compounds (I)-(III) exist in a half-chair conformation in which the bulky substituents are axial. (IV) contains a cis-fused dihydrofuran ring and is constrained to a boat conformation. Bond lengths and angles lie within normal ranges. There is some evidence for disorder in the NO_2 groups but this was not quantified.

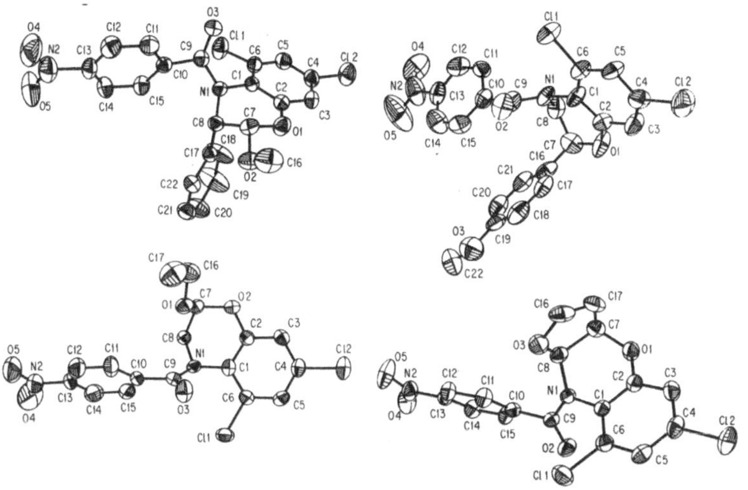

Fig. 1. The molecular structures: $C_{22}H_{16}Cl_2N_2O_5$, (I) (top left), $C_{22}H_{16}Cl_2N_2O_5$, (II) (top right), $C_{17}H_{14}Cl_2N_2O_5$ (bottom left) and $C_{17}H_{10}Cl_2N_2O_5$ (bottom right).

2-N-PHENYLIMINO-3-PHENYL-5-PHENOXYMETHYL-1,3-OXAZOLIDINE $C_{22}H_{20}N_2O_2$

G.V. GRIDUNOVA, V.E. SHKLOVER, Yu.T. STRUCHKOV, A.M. FAINLEIB, Ts.M. FRENKEL and V.A. PANKRATOV, 1984. *Khim. Get. Soedin. SSSR,* 1026-1031.

Orthorhombic, *Pbca*, a = 36.535, b = 10.2431, c = 9.7945 Å, Z = 8. Cu radiation, R = 0.057 for 1679 reflections.

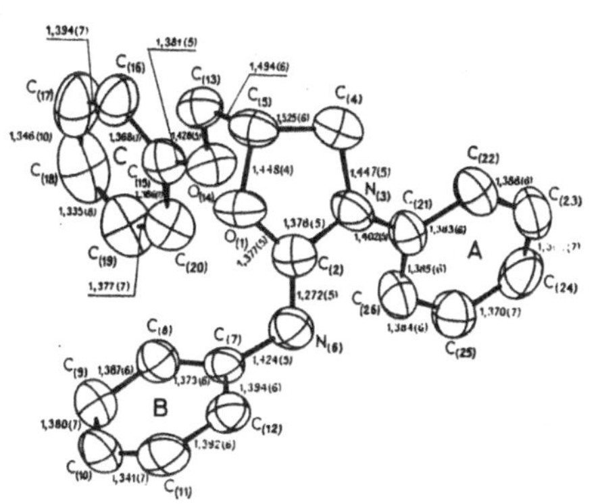

Fig. 1. The $C_{22}H_{20}N_2O_2$ molecule with bond lengths.

The central oxazolidine ring of the molecule has a conformation intermediate between an envelope and a half chair (Fig. 1). Intermolecular contacts correspond to van der Waals interactions.

(1RS,4RS,5RS,6RS,9SR)-8-BENZYL-4-ETHYL-1,5-DIMETHYL-9-PHENYL-3,7-DIOXA-8-AZABICYCLO[4.3.0]-NONAN-2-ONE
$C_{23}H_{27}NO_3$

M.J. FRAY, E.J. THOMAS and D.J. WILLIAMS, 1985. *J. Chem. Soc. Perkin Trans. 1*, 2763-2767.

Monoclinic, $P2_1/c$, a = 11.501, b = 9.439, c = 19.587 Å, β = 106.67°, Z = 4. Cu radiation, R = 0.041 for 1922 reflections.

The crystal structure analysis established the stereochemistry of the molecule (Fig. 1).

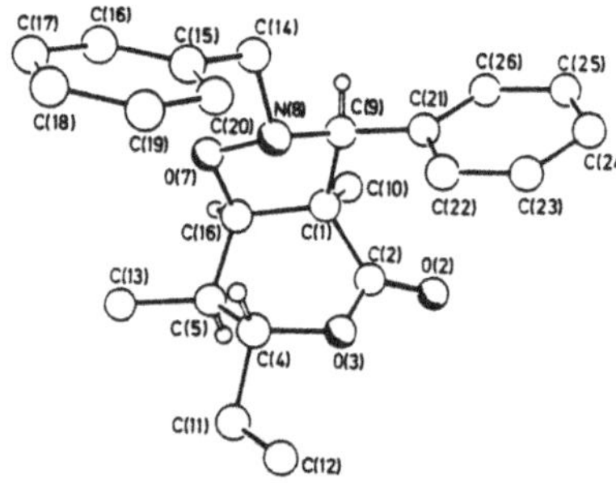

Fig. 1. Perspective diagram of a molecule of $C_{23}H_{27}NO_3$.

(2R,5R,6S,2S,5S,6R)-2-ETHYL-5,6-DIHYDRO-5-METHYL-2,4,6-TRIPHENYL-2H-1,3-OXAZINE
$C_{25}H_{25}NO$

J. BARLUENGA, J. JOGLAR, S. FUSTERO, V. GOTOR, C. KRUGER and M.J. ROMAO, 1985. *Chem. Ber.*, **118**, 3652-3663.

Triclinic, $P\bar{1}$, a = 10.1459, b = 11.2989, c = 10.091 Å, α = 99.744, β = 116.419, γ = 71.767°, Z = 2. Cu radiation, R = 0.053 for 3378 reflections.

The oxazine ring (Fig. 1) is puckered into a half-chair conformation (atoms C(2),C(1),N,C(4) planar) with O and C(3) 0.293 Å and 0.478 Å respectively displaced to opposite sides of the four-atom plane. Molecular dimensions are as anticipated; N-C(1) 1.274(2), N-C(4) 1.452(2) Å.

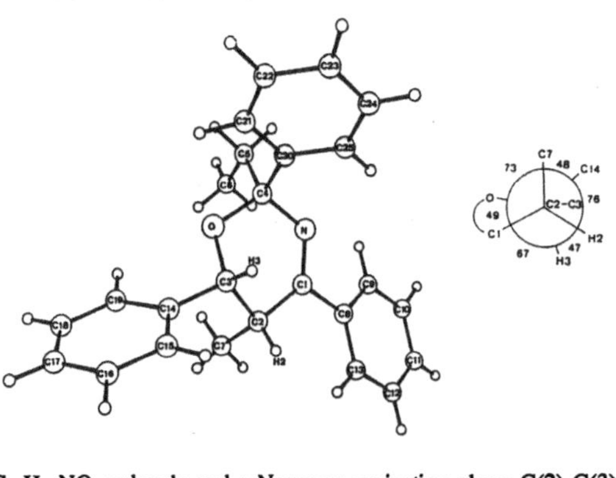

Fig. 1. The $C_{25}H_{25}NO$ molecule and a Newman projection along C(2)-C(3) with torsion angles (°).

((2,2-DIMETHYL-4-PHENYL-Δ^3-1,3-OXAZOLIN-5-YL)PHENYL)TRIMETHYLAMMONIUM 2-OXO-10-BORNANESULPHONATE ACETONITRILE SOLVATE

$C_{30}H_{40}N_2O_5S,C_2H_3N$ $C_{20}H_{25}N_2O^+,C_{10}H_{15}O_4S^-,C_2H_3N$

J. FORICHER, F. MONTAVON, K.-H. PFOERTNER and P. SCHONHOLZER, 1985. *Helv. Chim. Acta*, **68**, 592-599.

Monoclinic, $P2_1$, a = 15.774, b = 8.447, c = 12.418 Å, β = 101.17°, Z = 2. Mo radiation, R = 0.072.

The crystal structure analysis served to establish the absolute configurations of a group of related compounds. The cation is shown in Fig. 1.

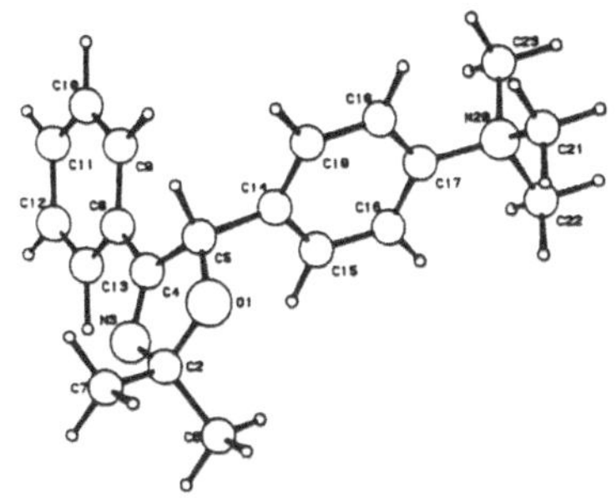

Fig. 1. Perspective drawing of a cation of the molecule of $C_{30}H_{40}N_2O_5S$.

N-t-BUTYL-N-((5-t-BUTYL-1,3,7,8-TETRAPHENYL-2,4,9-TRIOXA-5-AZATRICYCLO[4.2.1.0 ^{3,7}]NON-8-YL)-METHYLENE)AMINE-N-OXIDE

$C_{38}H_{40}N_2O_4$

H.G. AURICH, G. BAUM, W. MASSA and K.-D. MOGENDORF, 1985. *Acta Cryst.*, C**41**, 573-576.

Monoclinic, $P2_1/c$, a = 10.566, b = 15.629, c = 38.579 Å, β = 90.34°, Z = 8. Mo radiation, R = 0.055 for 5608 reflections (at 210 K).

The structure is shown to be the strained tricyclic system (Fig. 1) consisting of condensed tetrahydroisoxazole and tetrahydrofuran rings which both have envelope conformations and are linked by an oxygen (O(2)) bridge. The two independent molecules in the asymmetric unit differ in the relative orientations of the phenyl rings. The longest bonds in the structure are between fully-substituted carbon atoms (mean C(2)-C(2') 1.5772(4), mean C(2)-C(3) 1.569(4) Å). The mean N(1')-O(1') distance is 1.279(4) Å.

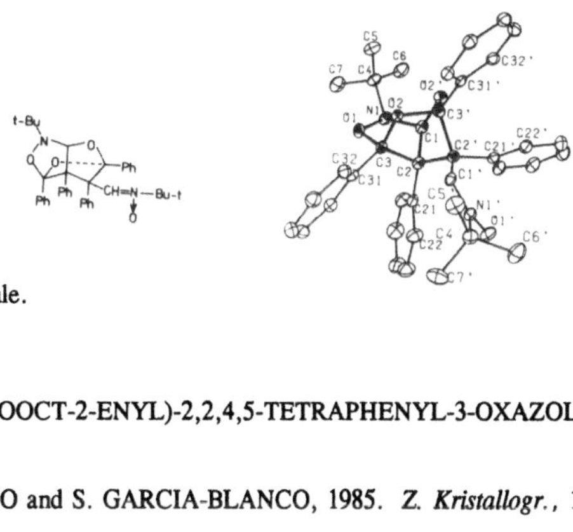

Fig. 1. The $C_{38}H_{40}N_2O_4$ molecule.

5-(4-(p-CHLOROBENZOYL)CYCLOOCT-2-ENYL)-2,2,4,5-TETRAPHENYL-3-OXAZOLINE

$C_{42}H_{36}ClNO_2$

J.M. MOHEDANO, F. FLORENCIO and S. GARCIA-BLANCO, 1985. *Z. Kristallogr.*, **172**, 239-244.

Triclinic, $P\bar{1}$, a = 16.766, b = 10.284, c = 10.887 Å, α = 94.02, β = 116.51, γ = 82.73°, Z = 2. Cu radiation, R = 0.077 for 3837 reflections.

The cyclooctene ring (Fig. 1) is distorted and the two groups bonded to it are in a cis arrangement. Bond lengths are normal. The eight-membered ring has asymmetric parameters Q_2 0.904(1), Q_3 0.715(2), Q_4 -0.273(1), ϕ_2 13.19(4), ϕ_3 10.18(4), total puckering amplitude Q_T 1.185(2).

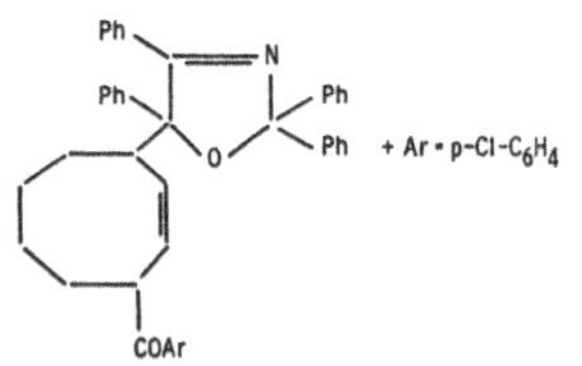

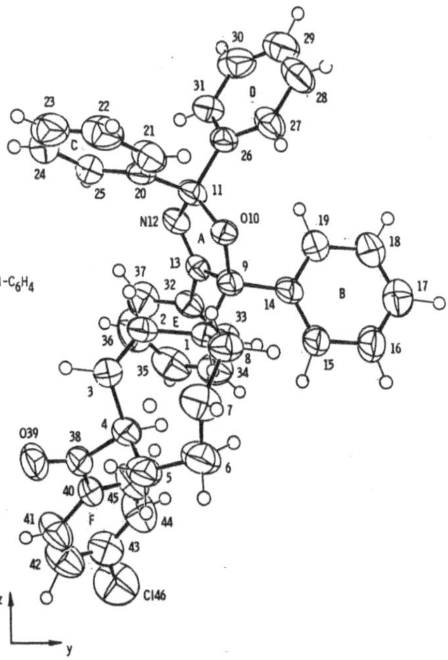

Fig. 1. The $C_{42}H_{36}ClNO_2$ molecule.

4-TRIFLUOROMETHYL-1,2,3,5-DITHIADIAZOLE
$C_2F_3N_2S_2$ (I)

4-TRIFLUOROMETHYL-1,2,3,5-DITHIADIAZOLIUM BIS(4-TRIFLUOROMETHYL-1,2,3,5-DITHIADIAZOLE) CHLORIDE
$C_6F_9ClN_6S_6$ (II) $C_2F_3N_2S_2^+$,2($C_2F_3N_2S_2$),Cl^-

H.-U. HOFS, J.W. BATS, R. GLEITER, G. HARTMANN, R. MEWS, M. ECKERT-MAKSIC, H. OBERHAMMER and G.M. SHELDRICK, 1985. *Chem. Ber.*, **118**, 3781-3804.

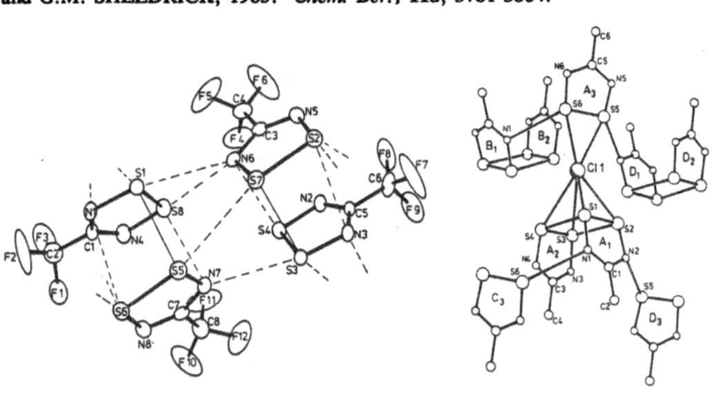

Fig. 1. The four $C_2F_3N_2S_2$ molecules (left) and part of the $C_2F_3N_2S_2^+$,2 ($C_2F_3N_2S_2$),Cl^- system (without F atoms).

I. Triclinic, $P\bar{1}$, a = 8.814, b = 10.795, c = 12.505 Å, α = 97.80, β = 90.02, γ = 102.36°, Z = 8. Mo radiation, R = 0.073 for 2138 reflections.

II. Monoclinic, $C2/c$, a = 31.828, b = 7.863, c = 14.556 Å, β = 99.99°, Z = 8. Mo radiation, R = 0.066 for 2629 reflections.

The asymmetric unit of (I) has four independent molecules which have planar five-membered rings and nearly C_{2v} symmetry. There are short S...S contacts between pairs of molecules (2.997 and 2.978(2) Å) and S...N contacts (3.188 and 3.241(5) Å) between dimers. The S-S distances in the rings are 2.082-2.091(2) Å and S-N distances are 1.616-1.639(5) Å. In (II), in contrast to (I), *endo*-dimers are formed and there are short contacts to the chloride ion (S...Cl 2.844-3.167 Å). Further N...S contacts lead to an infinite chain structure.

5-METHYL-1,3,2,4-DITHIADIAZOLIUM HEXAFLUOROARSENATE(V)

$C_2H_3AsF_6N_2S_2$ (I) [MeCSNSN](AsF_6)

4-METHYL-1,3,2-DITHIAAZOLIUM HEXAFLUOROARSENATE(V)

$C_3H_4AsF_6NS_2$ (II) [MeCSNSCH](AsF_6)

G.K. MacLEAN, J. PASSMORE, M.N.S. RAO, M.J. SCHRIVER, P.S. WHITE, D. BETHELL, R.S. PILKINGTON and L.H. SUTCLIFFE, 1985. *J. Chem. Soc. Dalton Trans.*, 1405-1416.

I. Monoclinic, $P2_1/c$, a = 8.182, b = 9.822, c = 11.515 Å, β = 110.91°, Z = 4. Mo radiation, R = 0.12 for 893 reflections.

II. Monoclinic, $P2_1/c$, a = 8.559, b = 9.941, c = 12.053 Å, β = 116.69°, Z = 4. Mo radiation, R = 0.049 for 1077 reflections.

The crystals contain discrete planar $MeCS_2N_2^+$ (Fig. 1) or $MeC_2HS_2N^+$ cations and AsF_6^- anions. The ring C and S atoms in each case make contacts with anion F atoms either in, or roughly perpendicular to, the ring plane ((I), (II): C...F 2.98-3.22(3), 3.04-3.18(4); S...F 2.87-3.13(2), 2.96-3.29(5) Å). Bond lengths in the rings are, in order: (I), C-N 1.32(3), N-S 1.57(2), S-N 1.59(2), N-S 1.65(2), S-C 1.77(2); (II), C-C(Me) 1.399(10), C(Me)-S 1.670(8), S-N 1.589(7), N-S 1.612(7), S-C 1.679(7) Å. The anion in (II) has alternative orientations.

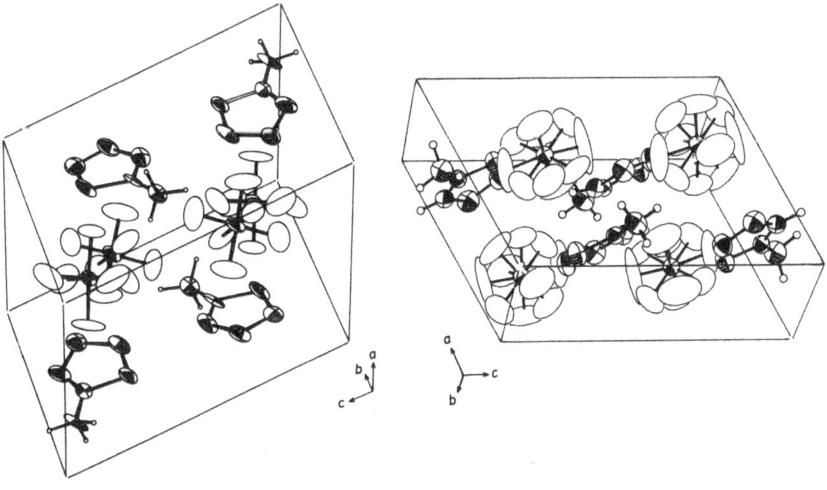

Fig. 1. Unit cell of $C_2H_3AsF_6N_2S_2$ (left) and of $C_3H_4AsF_6NS_2$ (right).

2-METHYL-5-METHOXY-2H-1,2,4,6-THIATRIAZIN-3(4H)-ONE-1,1-DIOXIDE

$C_4H_7N_3O_4S$

G. HAMPRECHT, R.-D. ACKER and E. HADICKE, 1985. *Liebigs Ann. Chem.*, 2363-2370.

Triclinic, $P\bar{1}$, a = 6.581, b = 8.009, c = 8.454 Å, α = 101.49, β = 110.28, γ = 105.70°, Z = 2. Mo radiation, R = 0.041.

All atoms (except S(1)) of the thiatriazine ring are coplanar, and S(1) is 0.259 Å from this plane. Molecular dimensions are in Fig. 1, and are indicative of some delocalization.

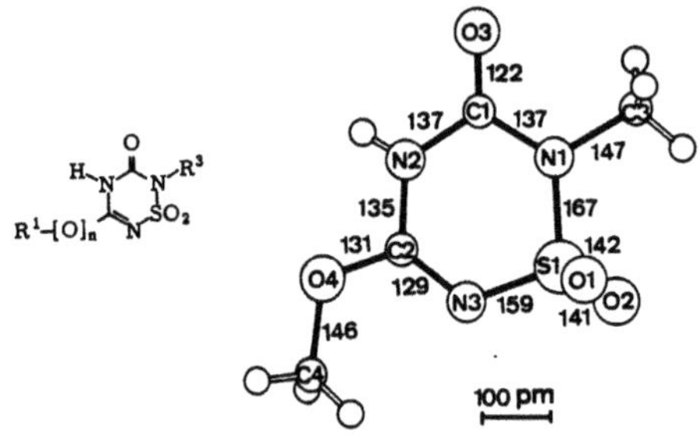

Fig. 1. The $C_4H_7N_3O_4S$ molecule with bond lengths (pm).

5-IMINO-4-METHYL-3-METHYLAMINO-4,5-DIHYDRO-1,2,4-THIADIAZOLIUM CHLORIDE MONOHYDRATE
$C_4H_9ClN_4S,H_2O$ $C_4H_9N_4S^+,Cl^-,H_2O$

H. SENDA and J. MARUHA, 1985. *Acta Cryst.*, C41, 1626-1628.

Monoclinic, $P2_1/n$, a = 12.153, b = 11.338, c = 6.421 Å, β = 91.69°, Z = 4. Mo radiation, R = 0.056 for 1368 reflections.

The crystal structure (Fig. 1) consists of a protonated heterocyclic cation, a chloride ion and a water molecule. The cation is almost planar with a maximum deviation from the mean plane of the five-membered ring of 0.068(3) Å by N(3) of the methylamino side chain. Bond lengths and angles in the ring agree well with those found in other related compounds and are indicative of considerable π-electron conjugation. The ions and water of crystallization are held together by a network of hydrogen bonds.

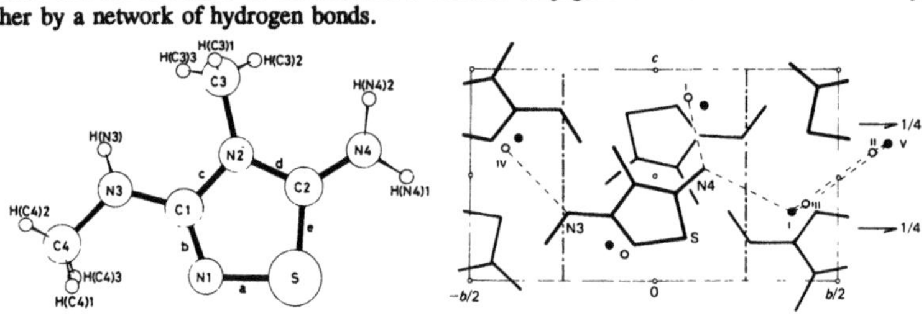

Fig. 1. $C_4H_9ClN_4S,H_2O$: views of the cation projected onto the mean plane of the five-membered ring and a projection onto (100) of the crystal structure (in which water molecules and Cl^- ions are represented by empty and filled circles, respectively, and with hydrogen bonds shown as broken lines).

N-(5-NITRO-2-THIAZOLYL)ACETAMIDE (AMINITROZOLE)
$C_5H_5N_3O_3S$

O.M. PEETERS, N.M. BLATON and C.J. DE RANTER, 1985. *Acta Cryst.*, C41, 965-967.

Monoclinic, $P2_1/c$, a = 9.815, b = 14.699, c = 5.098 Å, β = 90.41°, Z = 4. Cu radiation, R = 0.051 for 1056 reflections.

The thiazole ring of the molecule (Fig. 1) is exactly planar and the carbonyl moiety is cis to the S atom of the thiazolyl ring. The C(4)-C(5), C(5)-N(6), C(2)-N(9) and N(9)-C(10) bond distances are 1.340(2), 1.430(5), 1.368(4) and 1.378(4) Å respectively, indicating little interaction of the nitro and the amido group with the thiazole ring. In the crystal the molecules form a dimer around the inversion centre through N(9)...N(3) hydrogen bonds (N(9)...N(3) 2.972(5) Å). An intermolecular distance between O(8) and H(4'') (2.445 Å) stabilizes the crystal structure in the a direction.

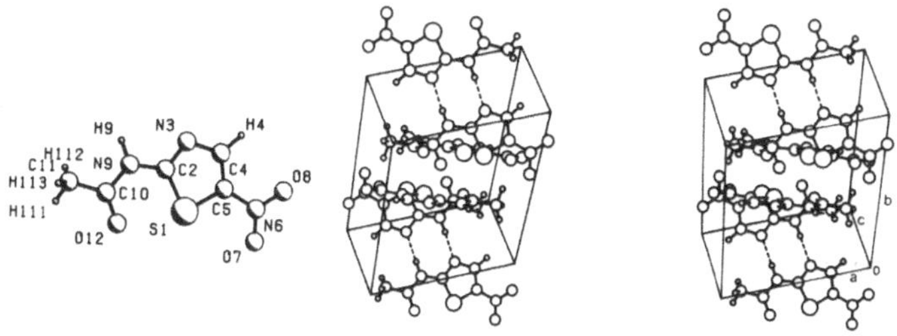

Fig. 1. $C_5H_5N_3O_3S$: the molecular structure and a stereoview of the packing.

7-t-BUTYL-3-CHLORO-1,3,5,2,4,6,8-TRITHIATETRAZOCINE
$C_5H_9ClN_4S_3$

R.T. BOERE, R.T. OAKLEY and A.W. CORDES, 1985. *Acta Cryst.*, **C41**, 1686-1687.

Tetragonal, $P4_2/n$, a = 17.856, c = 6.646 Å, Z = 8. Mo radiation, R = 0.049 for 1201 reflections.

The molecule (Fig. 1) has a folded butterfly structure with almost planar SNCNS and SNSNS fragments. The chlorine atom is endo to this folded eight-membered ring and the transannular S(2)...S(3) separation of only 2.378(2) Å is the shortest such distance reported for a thiazyl compound. The S(1)-Cl bond length is 2.218(2) Å and the bond lengths and angles in the molecule are in agreement with expected values. No unusually short intermolecular contacts are reported.

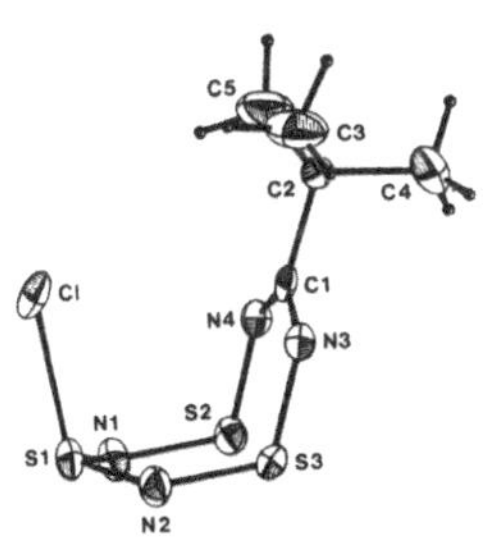

Fig. 1. A view of the $C_5H_9ClN_4S_3$ molecule.

5,6-DIHYDRO-3-DIMETHYLAMINO-2H-1,4-THIAZIN-2-ONE-OXIME HYDROCHLORIDE
$C_5H_{10}ClN_3OS$ (I) $C_5H_{10}N_3OS^+,Cl^-$

N'-(2-(((5-((DIMETHYLAMINO)METHYL)-2-FURANYL)METHYL)THIO)ETHYL)-2-HYDROXYIMINO-N-METHYL-2-NITRO-N-NITROSO-ETHANIMIDAMIDE
$C_{13}H_{20}N_6O_5S$ (II)

T.J. CHOLERTON, J. CLITHEROW, J.H. HUNT, J.W.M. MACKINNON, M. MARTIN-SMITH, B.J. PRICE, J. MURRAY-RUST and P. MURRAY-RUST, 1985. *J. Chem. Res.*, S250-251, M2818-2846.

I. Monoclinic, $P2_1/c$, a = 9.01, b = 7.81, c = 14.44 Å, β = 119.65°, Z = 4. Mo radiation, R = 0.044 for 841 reflections.

II. Monoclinic, $P2_1/c$, a = 11.17, b = 9.85, c = 16.68 Å, β = 80.15°, Z = 4. Mo radiation, R = 0.043 for 1112 reflections.

The crystal structure analyses established the stereochemistries of the two molecules (Fig. 1). In the crystal of (I) the cations are hydrogen bonded to the chloride ion. In the crystal of (II) the molecule has a zwitterionic form and there is strong hydrogen bonding (O(2)...H...N(6) 2.55(1) Å) which causes the molecule to adopt a U-shape.

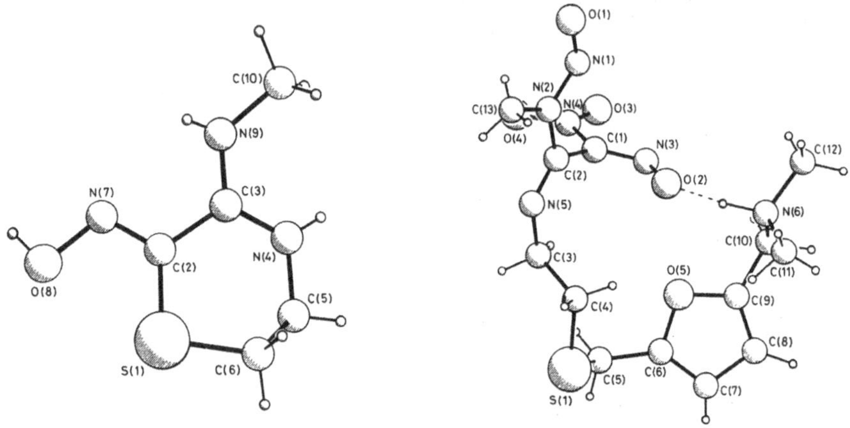

Fig. 1. Views of $C_5H_{10}ClN_3OS$ (I) (left) and $C_{13}H_{20}N_6O_5S$ (II) (right).

3-METHYL-4-OXO-7-THIA-1,3-DIAZABICYCLO[3.3.0]OCTANE-2-THIONE
$C_6H_8N_2OS_2$ (I)

8-METHYL-7-OXO-4-THIA-1,8-DIAZABICYCLO[4.3.0]NONANE-9-THIONE
$C_7H_{10}N_2OS_2$ (II)

A.V. EREMEEV, R. NURDINOV, F.D. POLYAK, R.M. ZOLOTOYABKO, A.F. MISHNEV, M.F. BUNDULE and Ya.Ya. BLEIDELIS, 1985. *Khim. Get. Soedin., SSSR,* 1327-1332.

I. Orthorhombic, $P2_12_12_1$, a = 7.359, b = 8.049, c = 13.792 Å, Z = 4. Cu radiation, R = 0.047 for 696 reflections.

II. Orthorhombic, $Pca2_1$, a = 16.539, b = 5.692, c = 19.850 Å, Z = 8. Cu radiation, R = 0.102 for 1122 reflections.

Views of (I) and the two independent molecules of (II) are in Fig. 1. The imidazolidine ring of (I) is planar and the thiazolidine ring has a somewhat distorted envelope conformation (C(5) 0.542 Å from the NC_2S plane). Molecules in (II) have similar geometry, with planar imidazolidine rings and chair conformations for the tetrahydrothiazine rings.

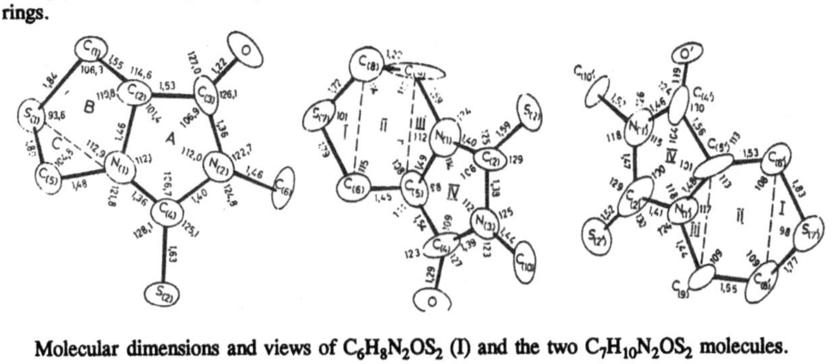

Fig. 1. Molecular dimensions and views of $C_6H_8N_2OS_2$ (I) and the two $C_7H_{10}N_2OS_2$ molecules.

5-AMINOTHIAZOLE-4-CARBOXYLIC ACID ETHYL ESTER
$C_6H_8N_2O_2S$

B. GOLANKIEWICZ, P. JANUSZCZYK, M. GDANIEC and Z. KOSTURKIEWICZ, 1985. *Tetrahedron*, **41**, 5989-5994.

Monoclinic, $P2_1/n$, a = 9.1239, b = 11.903, c = 7.4043 Å, β = 94.04°, Z = 4. Cu radiation, R = 0.048 for 1201 reflections.

The thiazole ring (Fig. 1) is planar and there is an intramolecular N(6)-H...O(8) hydrogen bond (O(8)...H 2.30(4) Å). The C(4)-C(7)-O(9)-C(10) torsion angle is 179.5(3)°. C-S distances are 1.726 and 1.732(3) Å with C(2)-N(3) 1.292(4), N(3)-C(4) 1.381(3), C(4)-C(5) 1.389(3), C(5)-N(6) 1.338(3) Å. Molecules are linked by a hydrogen-bond network as shown in Fig. 1.

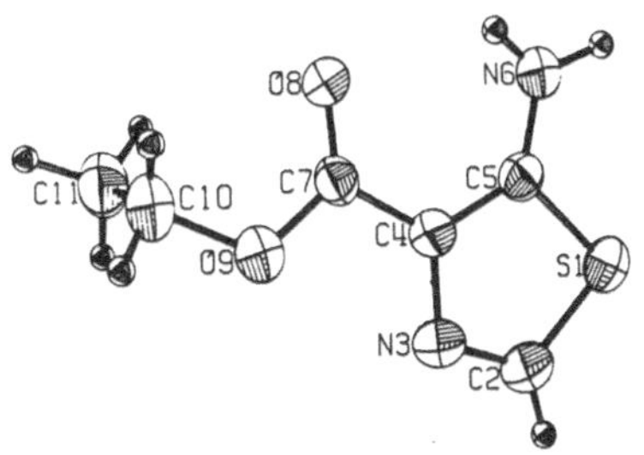

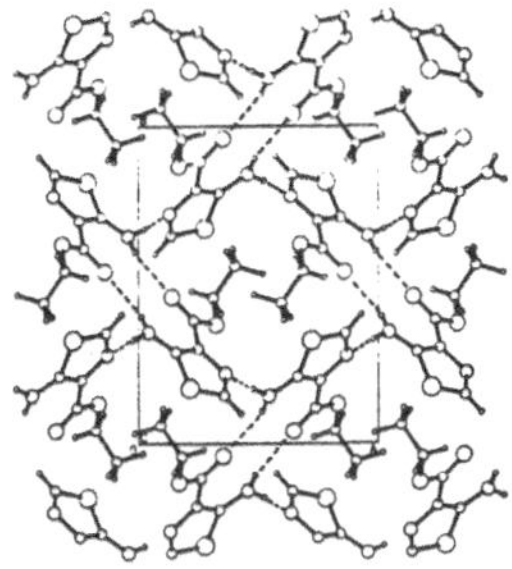

Fig. 1. The $C_6H_8N_2O_2S$ molecule and a projection of the crystal structure along *c*.

1,3-DICHLORO-5-PHENYL-1λ^4,3λ^4,2,4,6-DITHIATRIAZINE
$C_7H_5Cl_2N_3S_2$

J.B. GRAHAM III, A.W. CORDES, R.T. OAKLEY and R.T. BOERE, 1985. *Acta Cryst.*, C**41**, 1835-1836.

Triclinic, $P\bar{1}$, a = 6.028, b = 9.985, c = 10.157 Å, α = 117.33, β = 106.73, γ = 90.31°, Z = 2. Mo radiation, R = 0.032 for 1547 reflections.

The S(1)N(1)C(1)N(2)S(2) fragment of the heterocyclic ring is planar with the N(3) atom displaced 0.336(2) Å from this plane on the side opposite the S-bonded Cl atoms (Fig. 1). The S-N distances range from 1.581(1) to 1.615(2) and the N-C distances are 1.327(2) and 1.344(2) Å. The S-Cl distances are 2.180(1) and 2.129(1) Å.

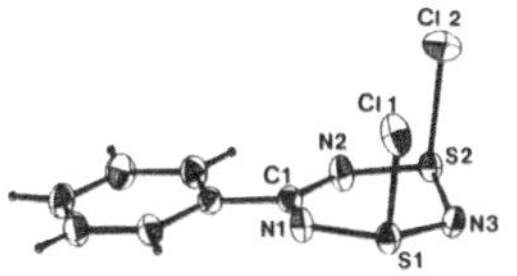

Fig. 1. The $C_7H_5Cl_2N_3S_2$ molecular structure.

7-PHENYL-1λ^4,3λ^4,Δ^2,5λ^4-TRITHIA-2,4,6,8,9-PENTAAZABICYCLO[3.3.1]NONANE
$C_7H_5N_5S_3$

A.W. CORDES, R.T. OAKLEY and R.T. BOERE, 1985. *Acta Cryst.*, C**41**, 1833-1834.

Monoclinic, $P2_1/n$, a = 5.958, b = 22.954, c = 7.428 Å, β = 106.25°, Z = 4. Mo radiation, R = 0.030 for 1363 reflections.

The planar S(1)N(1)C(1)N(2)S(2) and S(1)N(4)S(3)N(5)S(2) moieties of the bicyclo fragment make angles of 137.1 and 118.3° with the SNS bridging unit (Fig, 1). The S-N bonds connecting the NSN moiety to the CN_3S_2 ring are much longer (1.728(2), 1.728(2) Å) than those in the remainder of the molecule (1.546-1.630 Å).

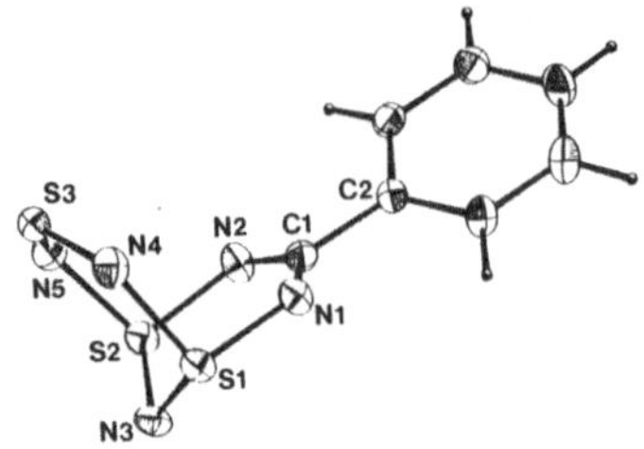

Fig. 1. The $C_7H_5N_5S_3$ molecular structure.

2-MERCAPTOBENZOTHIAZOLE
$C_7H_5NS_2$

A. RADHA, 1985. *Z. Kristallogr.*, **171**, 225-228.

Monoclinic, $P2_1/n$, a = 8.014, b = 6.004, c = 15.393 Å, β = 100.72°, Z = 0. Mo radiation, R = 0.035 for 1017 reflections.

For previous work see (1). Bond lengths and angles in this determination are normal. The five- and six-membered rings are individually planar (to within 0.007(3) Å and 0.012(3) Å respectively); the dihedral angle between the two planes is 1.2°.

1. Structure Reports, 37B, 200.

2-(5-AMINO-1,3,4-THIADIAZOL-2-YL)BENZENESULPHONAMIDE
$C_8H_8N_4O_2S_2$ (I)

2-(5-AMINO-1,3,4-THIADIAZOL-2-YL)-N-METHYLBENZENESULPHONAMIDE
$C_9H_{10}N_4O_2S_2$ (II)

E. FORESTI, L. RIVA DI SANSEVERINO and P. SABATINO, 1985. *Acta Cryst.*, **C41**, 240-243.

I. Monoclinic, $P2_1/n$, a = 14.604, b = 14.605, c = 5.206 Å, β = 94.78°, Z = 4. Mo radiation, R = 0.038 for 1781 reflections.

II. Monoclinic, $P2_1/a$, a = 21.308, b = 13.420, c = 8.155 Å, β = 94.49°, Z = 8. Mo radiation, R = 0.053 for 3900 reflections.

There are two crystallographically independent molecules in the asymmetric unit of (II), one being an approximate mirror image of the other. Compound (I) exhibits only one of the conformers (Fig. 1). The thiadiazole ring is planar in both (I) and (II) while the S endocyclic angles are 86.7° in (I) and 87.0(1) and 87.2(1)° in (II). The dihedral angle between the thiadiazole and phenyl ring is 55.7(1)° in (I) and 52.3(1) and 33.8(1)° in (II). The two compounds exhibit a different hydrogen bonding pattern.

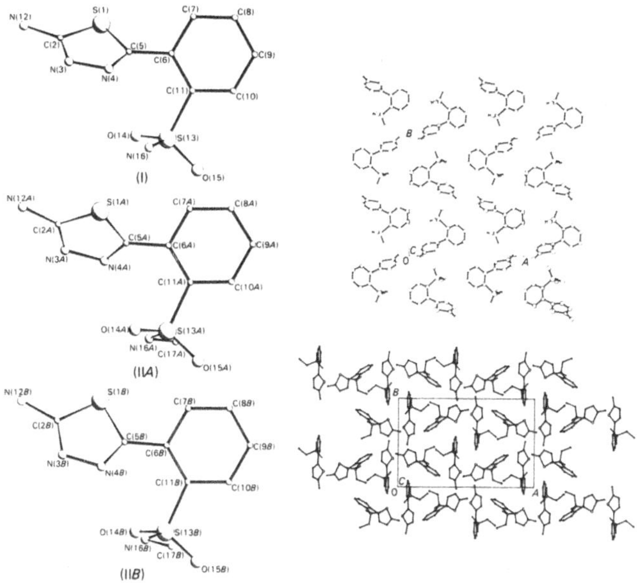

Fig. 1. The molecular structures (left) and packing (right): $C_8H_8N_4O_2S_2$ (top) and $C_9H_{10}N_4O_2S_2$ (bottom).

4-ACETYL-trans-3-METHYLCARBAMOYL-1,4-THIAZINANE 1-OXIDE
$C_8H_{14}N_2O_3S$

F. VANHOUTEGHEM, A.T.H. LENSTRA, H.J. GEISE, C. VAN der AUWERA and M.J. ANTEUNIS, 1985. *Acta Cryst.*, **C41**, 1818-1821.

Monoclinic, $P2_1/n$, a = 10.363, b = 12.613, c = 8.307 Å, β = 97.95°, Z = 4. Mo radiation, R = 0.038 for 2033 reflections.

In the molecule (Fig. 1), the six-membered thiazinane ring adopts a distorted chair conformation with the methylcarbamoyl group axial and the orientation of the S=O and the acetyl group inclinal. The C(6)O(2)C(5)N(1)C(3)C(7)O(3)N(2)C(8) fragment constitutes a dipeptide in the α-helical conformation. Bond distances include: S=O(1) 1.504(2), S-C(1) 1.795(2), S-C(4) 1.804(2), N(1)-C(5) 1.362(2), N(1)-C(3) 1.457(2), N(1)-C(2) 1.466(2), C(3)-C(7) 1.535(2), N(2)-C(7) 1.323(2) and N(2)-C(8) 1.449(2) Å.

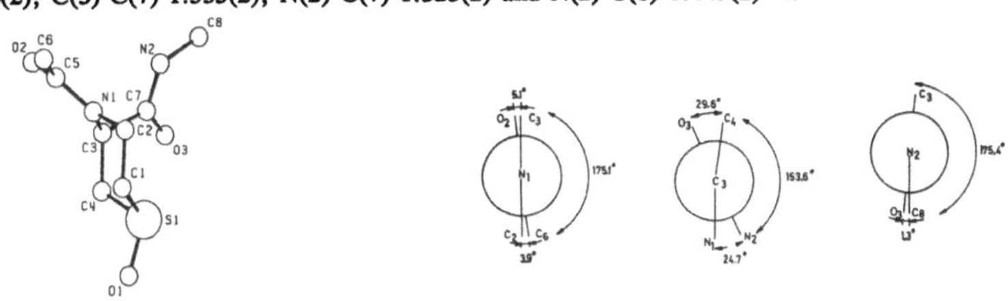

Fig. 1. $C_8H_{14}N_2O_3S$: molecular structure and Newman projections along C(5)-N(1), C(7)-C(3) and N(2)-C(7).

(S)-2,2,5,5-TETRAMETHYLTHIAZOLIDINE-4-CARBOXYLIC ACID
$C_8H_{15}NO_2S$

H.E. HOWARD-LOCK, C.J.L. LOCK and P.S. SMALLEY, 1985. *Canad. J. Chem.*, **63**, 2411-2419.

Monoclinic, $P2_1$, a = 11.351, b = 8.303, c = 11.969 Å, β = 116.69°, Z = 4. Mo radiation, R = 0.0774 for

2388 reflections.

There are two very similar independent molecules in the asymmetric unit and these are shown to exist in the amino acid form (rather than the zwitterion) with two distinctly different C-O bond lengths (C=O 1.183(8), 1.206(7) Å, C-OH 1.331(8), 1.307(7) Å). Molecules are linked by hydrogen bonds (O(2)...N 2.666 and 2.687(7) Å). The mean C-S distance is 1.850. The five-membered ring has atoms C(5),S,C(2) defining a plane with N (0.382(4) Å) and C(4) (0.269(5) Å) asymmetrically twisted out of the plane on opposite sides in one molecule and similarly (N 0.346(4) Å, C(4) 0.296(5) Å) in the other.

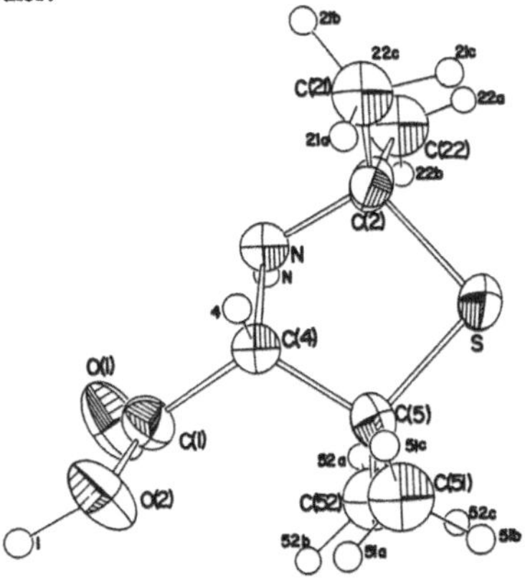

Fig. 1. A view of one of the $C_8H_{15}NO_2S$ molecules.

4-CHLORO-1,2,3-DITHIAZOLIUM-5-(4'-METHYLPHENOLATE)
$C_9H_6ClNOS_2$

R. APPEL, H. JANSSEN, M. SIRAY and F. KNOCH, 1985. *Chem. Ber.*, **118**, 1632-1643.

Orthorhombic, $Pna2_1$, a = 15.971, b = 13.229, c = 4.676 Å, Z = 4. Mo radiation, R = 0.034 for 1477 reflections.

The analysis shows that the molecule is planar (Fig. 1) with a short O-S distance of 2.16 Å and a linear O-S-S moiety.

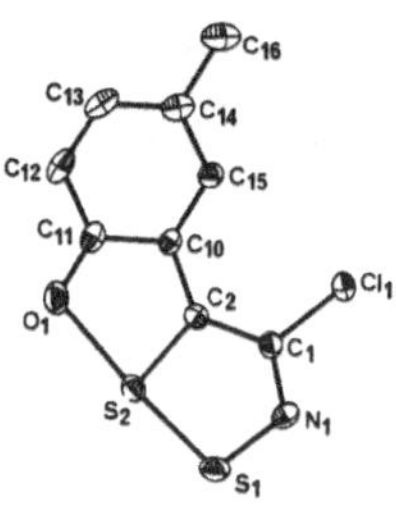

Fig. 1. The $C_9H_6ClNOS_2$ molecule.

5,6-DIHYDRO-5-HYDROXY-4-PHENYL-1,3,4-THIADIAZINE
$C_9H_{10}N_2OS$

Y. MATSUBARA, S. YAMADA, M. YOSHIHARA and T. MAESHIMA, 1985. *Phosphorus and Sulfur*, **22**, 41-47.

Monoclinic, $P2_1/c$, a = 19.256, b = 10.076, c = 19.486 Å, β = 150.76°, Z = 8. Mo radiation, R = 0.109 for 3536 reflections.

There are two independent molecules in the asymmetric unit and the asymmetric unit is chosen to contain one S and one R form (Fig. 1). The groups of atoms S(1)-C(1)-N(1)-N(2) and S(2)-C(10)-N(3)-N(4) are each coplanar; the esd's of bond lengths are large (0.07-0.83 Å).

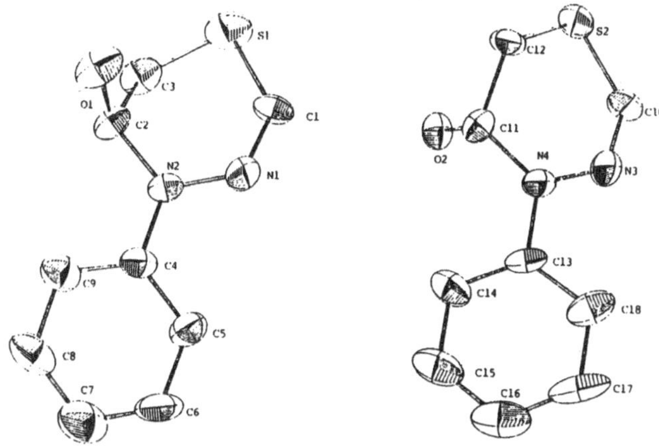

Fig. 1. Drawings of the S and R forms of $C_9H_{10}N_2OS$.

ETHYL 1,3-BIS(TRICHLOROMETHYL)-5-THIA-2,7-DIAZA-8-OXOBICYCLO[2.2.2]OCT-2-ENE-6-CARBOXYLATE
$C_{10}H_8Cl_6N_2O_3S$

L.F. LEE, M.G. DOLSON, R.K. HOWE and B.R. STULTS, 1985. *J. Org. Chem.*, **50**, 3216-3218.

Triclinic, $P\bar{1}$, a = 8.584, b = 9.343, c = 11.957 Å, α = 107.86, β = 109.45, γ = 97.381°, Z = 2. Mo radiation, R = 0.0274 for 1981 reflections.

The crystal structure analyses established the identity of the molecule (Fig. 1).

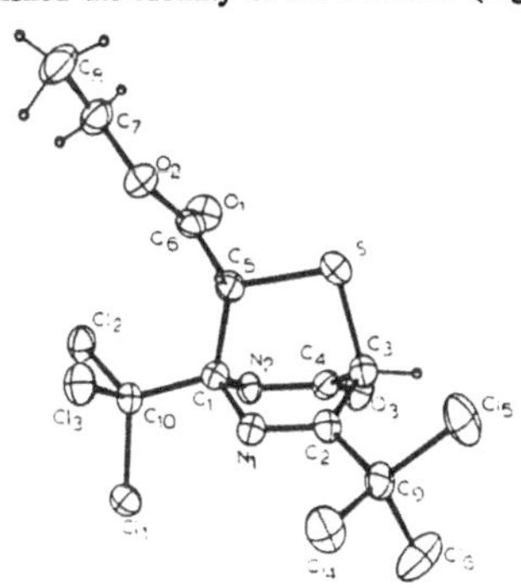

Fig. 1. ORTEP drawing of $C_{10}H_8Cl_6N_2O_3S$.

6-(N-ETHYLAMINO)-2,3-DIHYDROIMIDAZO[2,1-b]BENZOTHIAZOLE MONOHYDRATE
$C_{11}H_{13}N_3S,H_2O$

C. VAN DER BREMPT, A. SOREE, B. NORBERG, G. EVRARD and F. DURANT, 1985. *Bull. Soc. Chim. Belg.*, **94**, 159-162.

Triclinic, $P\bar{1}$, a = 8.939, b = 10.796, c = 7.551 Å, α = 100.11, β = 96.29, γ = 122.49°, Z = 2. Mo radiation, R = 0.045 for 1188 reflections.

The molecule is flattened (Fig. 1), with the C(5) side chain lying in a stretched "anti" conformation (C(4)-C(5)-N(13)-C(14) -15.0(9)°, C(5)-N(13)-C(14)-C(15) -172.1(6)°). The dimensions (Fig. 1) are consistent with the electron delocalisation shown. The water molecule links benzothiazoles via hydrogen bonding.

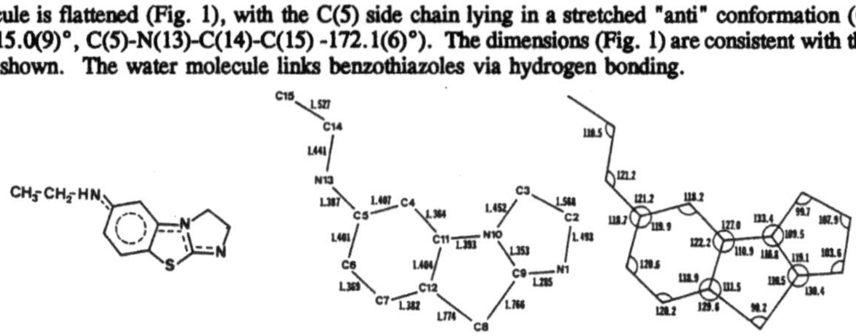

Fig. 1. The $C_{11}H_{13}N_3S$ molecule and dimensions.

2-ISOPROPYL-3-PHENYL-1,2-THIAZETIDINE-1,1-DIOXIDE
$C_{11}H_{15}NO_2S$

E. MEYLE, E. KELLER and H.-H. OTTO, 1985. *Liebigs Ann. Chem.*, 802-812.

Triclinic, $P\bar{1}$, a = 9.363, b = 11.584, c = 6.265 Å, α = 105.19, β = 100.49, γ = 67.08°, Z = 2. Mo radiation, R = 0.084 for 1403 reflections.

The four-membered β-sultam ring is puckered (fold angle 23 to 24°) and the isopropyl group at N is disordered (as shown in Fig. 1) over two orientations. Principal dimensions include S-O 1.420 and 1.428(7), S-N 1.661(6), S-C(2) 1.761(9) Å, O-S-O 116.4(4), C(2)-S-N 80.0(3), C(1)-N-S 93.6(4)°.

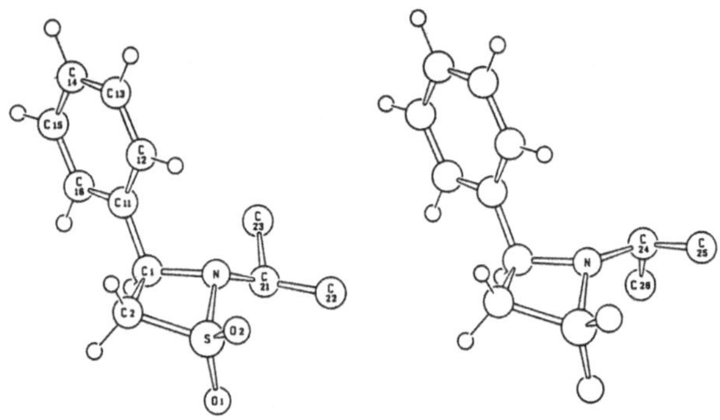

Fig. 1. The $C_{11}H_{15}NO_2S$ molecule showing the different isopropyl orientations.

1,3-DIMETHYL-1-(4-(3-FLUOROBENZYL)-5-THIOXO-1,3,4-THIADIAZOL-2-YL)UREA
$C_{12}H_{13}FN_4OS_2$

A.R. MOORMAN, D.C. FINDAK and HAN SAN KU, 1985. *J. Heterocycl. Chem.*, 22, 915-920.

Monoclinic, $P2_1/n$, a = 13.057, b = 8.449, c = 13.703 Å, β = 102.23°, Z = 4. Mo radiation, R = 0.036 for 1409 reflections.

X-ray analysis was used to establish the structure of this photoproduct as shown in Fig. 1.

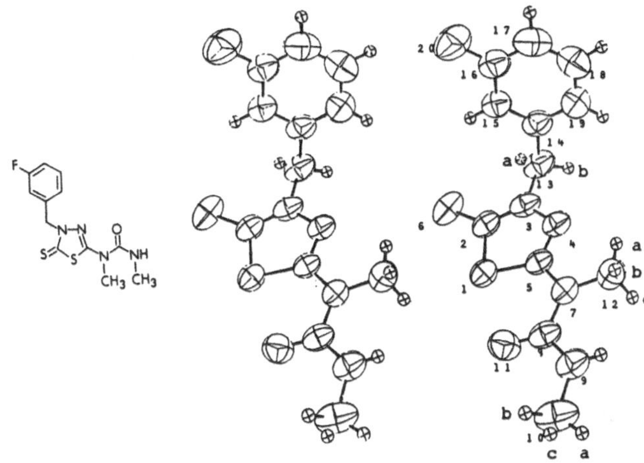

Fig. 1. The $C_{12}H_{13}FN_4OS_2$ molecule and a stereoview.

3-ISOPROPYL-5-PHENYLTHIAZOLIDIN-4-ONE-2-THIONE (3-ISOPROPYL-5-PHENYLRHODANINE) $C_{12}H_{13}NOS_2$

K. RANG, J. SANDSTROM, L. THELL and Q. YANG, 1985. *Acta Chem. Scand.*, **B39**, 123-129.

Monoclinic, $P2_1$, a = 12.273, b = 5.245, c = 9.888 Å, β = 111.40°, Z = 2. Mo radiation, R = 0.022 for 1136 reflections.

The 3-isopropyl group is bisected by the thiazolidine ring plane, with the methyl groups turned towards the carbonyl oxygen (Fig. 1). The thiazolidine and phenyl rings are practically planar with an 86.2° angle between the planes.

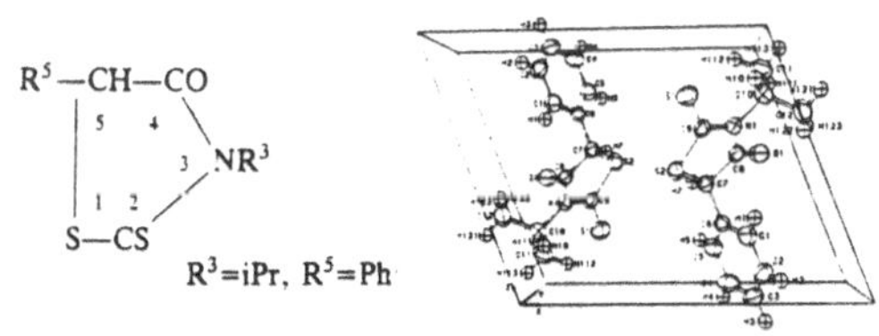

Fig. 1. The $C_{12}H_{13}NOS_2$ molecule and the contents of a unit cell in the crystal.

1-((3-METHYL-4-ISOTHIAZOLYL)METHYL)GUANIDINIUM HEMISULPHATE $C_{12}H_{22}N_8O_4S_3$ $2(C_6H_{11}N_4S^+),O_4S^{2-}$

J.-M. LEGER, J.-C. COLLETER and A. CARPY, 1985. *Acta Cryst.*, **C41**, 622-624.

Monoclinic, $P2_1/c$, a = 8.653, b = 7.473, c = 32.042 Å, β = 92.36°, Z = 4. Cu radiation, R = 0.059 for 2136 reflections.

There are two independent molecules in the asymmetric unit. These differ in conformation due to the position of the guanidine group relative to the ring (Fig. 1). The angle between the plane of the ring and that of the guanidine group is 97(1)° in one molecule and 65(1)° in the other. The mean bond length difference between the two independent molecules is 0.008 Å. Protonation occurs on double-bonded N(2) which makes the terminal N(1) and N(3) atoms equivalent. Crystalline cohesion is ensured by strong hydrogen bonds involving the sulphate ions.

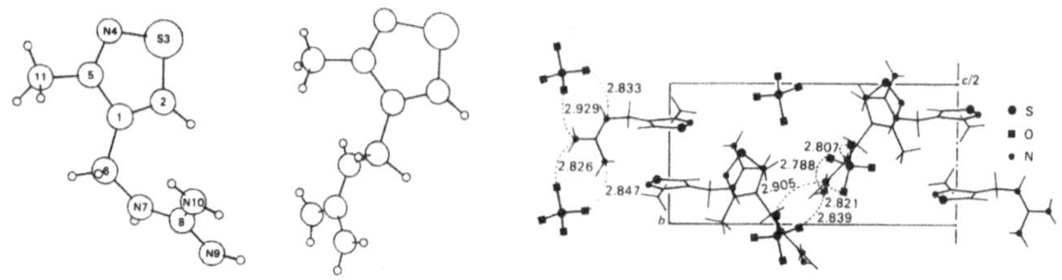

Fig. 1. $C_{12}H_{22}N_8O_4S_3$: the two molecules and the packing.

10-METHYLPHENOTHIAZINE-5-OXIDE
$C_{13}H_{11}NOS$

S.S.C. CHU, P. DE MEESTER, M.V. JOVANOVIC and E.R. BIEHL, 1985. *Acta Cryst.*, C41, 1111-1114.

Monoclinic, $P2_1/n$, a = 11.289, b = 10.831, c = 8.818 Å, β = 91.80°, Z = 4. Mo radiation, R = 0.044 for 1980 reflections.

The benzene rings in the molecule (Fig. 1) are planar with an interplanar dihedral angle of 151.2(8)°. The central ring is in a boat conformation with the 10-methyl substituent in an equatorial and the sulphinyl oxygen O(5) in an axial conformation. Bond lengths and angles in the molecule are in good agreement with those observed in other phenothiazines. Intermolecular packing in the crystal is governed by van der Waals forces.

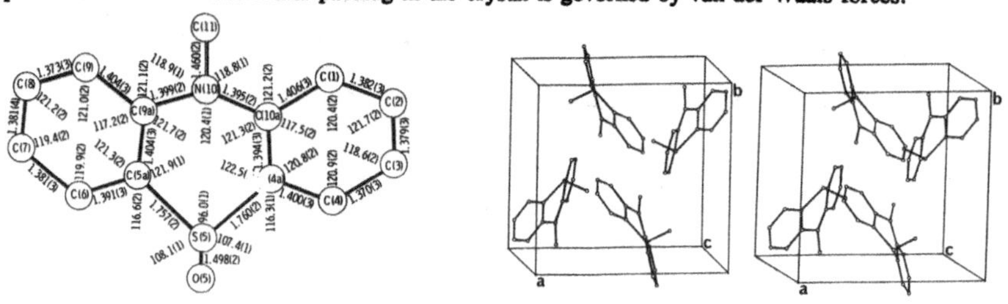

Fig. 1. $C_{13}H_{11}NOS$: bond lengths and angles in the molecule and a stereoview of the molecular packing in the unit cell.

8-BENZYL-4-THIA-1,8-DIAZABICYCLO[4.3.0]NONANE-2,7,9-TRIONE
$C_{13}H_{12}N_2O_3S$

N.N. DHANESHWAR, S.S. TAVALE and T.N.G. ROW, 1985. *Acta Cryst.*, C41, 1528-1530.

Orthorhombic, *Pcab*, a = 8.851, b = 12.035, c = 23.506 Å, Z = 8. Mo radiation, R = 0.062 for 1491 reflections.

The hydantoin ring in the molecule (Fig. 1) is planar and the thiazine ring adopts a boat conformation. The dihedral angle between the hydantoin ring plane and the aromatic ring plane is 113.9(4)°. Bond lengths and angles in the hydantoin ring are normal and the C(8)-S and C(6)-S bond distances in the thiazine ring are 1.817(4) and 1.806(5) Å respectively, close to a single-bond value for C-S.

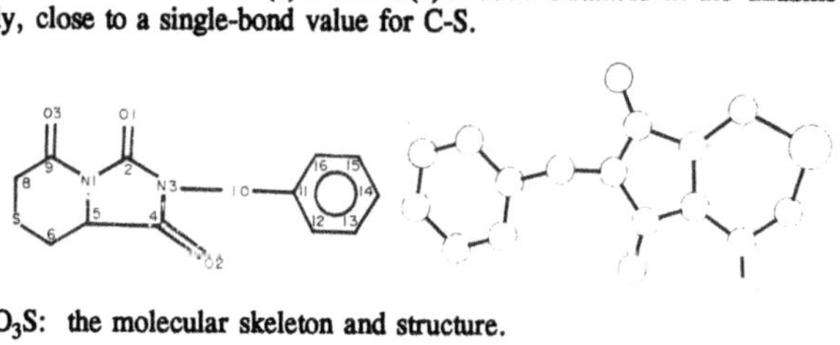

Fig. 1. $C_{13}H_{12}N_2O_3S$: the molecular skeleton and structure.

9-ACETYL-2,4-DIMETHYL-3a,4,9,9a-TETRAHYDROTHIAZOLO[4,5-b]QUINOXALINE
$C_{13}H_{15}N_3OS$

D.S. YUFIT, Yu.T. STRUCHKOV, V.N. DROZD, V.G. BAKLIKOV, V.N. CHARUSHIN and O.N. CHUPAKHIN, 1985. *Khim. Get. Soedin., SSSR,* 1417-1421.

Monoclinic, $P2_1/n$, a = 9.448, b = 10.074, c = 14.233 Å, β = 106.562°, Z = 4. Cu radiation, R = 0.053 for 1741 reflections.

The tetrahydropyrazine ring has a twisted boat conformation (Fig. 1). Atom N(4) has pyramidal and N(9) planar geometry (with N(4)-C(4a) 1.405(5), N(9)-C(8a) 1.425(5) Å). The thiazoline ring has an envelope conformation, with C(9a) 0.198(4) Å from the plane of the remaining four ring atoms. The dihedral angle between the best planes of the thiazoline and quinoxaline systems is 90.8°.

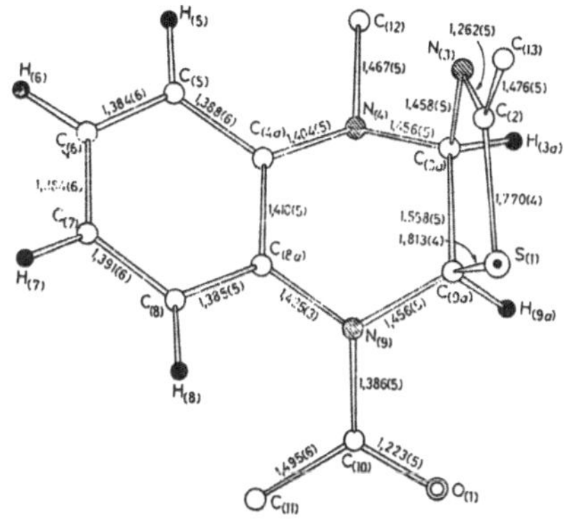

Fig. 1. The $C_{13}H_{15}N_3OS$ molecule with bond lengths.

BIS(5-PHENYL-1,3,2,4,6-DITHIATRIAZINE)
$C_{14}H_{10}N_6S_4$

R.T. BOERE, C.L. FRENCH, R.T. OAKLEY, A.W. CORDES, J.A.J. PRIVETT, S.L. CRAIG and J.B. GRAHAM, 1985. *J. Am. Chem. Soc.*, **107**, 7710-7717.

Triclinic, $P\bar{1}$, a = 6.093, b = 10.874, c = 12.458 Å, α = 109.13, β = 100.59, γ = 92.52°, Z = 2. Mo radiation, R = 0.051 for 1327 reflections.

The dimer consists of a cofacial arrangement of two dithiatriazine units linked by two long S...S interactions (2.527(3) Å) (Fig. 1). The two planar SNCNS segments within each dimer are mutually inclined with a dihedral angle of 12.5°. Dimeric pairs are stacked in a head-to-tail fashion with a 3.65 Å interplanar separation between each phenyl ring and a neighbouring CN_3S_2 ring.

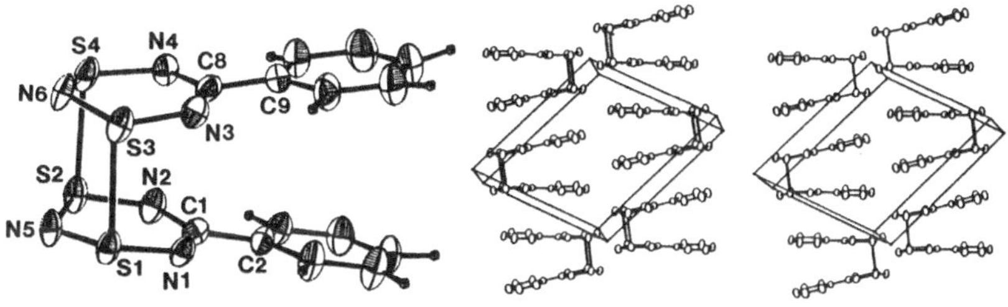

Fig. 1. The $C_{14}H_{10}N_6S_4$ molecule and crystal packing.

5-IMINO-4-PHENYL-3-PHENYLAMINO-4H-1,2,4-THIADIAZOLINE (HECTOR'S BASE)
$C_{14}H_{12}N_4S$ (I)

5-ACETYLIMINO-4-PHENYL-3-PHENYLAMINO-4H-1,2,4-THIADIAZOLINE
$C_{16}H_{14}N_4OS$ (II)

N-(1,3-BENZOTHIAZOL-2-YL)-N-(ANILINOPHENYLIMINO)METHYLANILINE (HUGERHOFF'S BASE)
$C_{26}H_{20}N_4S$ (III)

I.I. KANDROR, V.K. BEL'SKII, B.V. KOPYLOVA and L.V. YASHKINA, 1985. *Khim. Get. Soedin., SSSR*, 206-210.

I. Orthorhombic, $P2_12_12_1$, a = 11.021, b = 9.513, c = 12.198 Å, Z = 4. Mo radiation, R = 0.028 for 803 reflections.

II. Monoclinic, $P112_1/a$, a = 19.677, b = 10.504, c = 7.420 Å, γ = 102.05°, Z = 4. Mo radiation, R = 0.032 for 1413 reflections.

III. Monoclinic, $P112_1/n$, a = 9.770, b = 18.447, c = 12.116 Å, γ = 98.03°, Z = 4. Mo radiation, R = 0.030 for 1068 reflections.

Views of the three heterocyclic oxidation products of N-arylthioureas are in Figs. 1-3 with main dimensions. For previous work in (I) see (1). There is a short intramolecular S...O contact of 2.471 Å in (II).

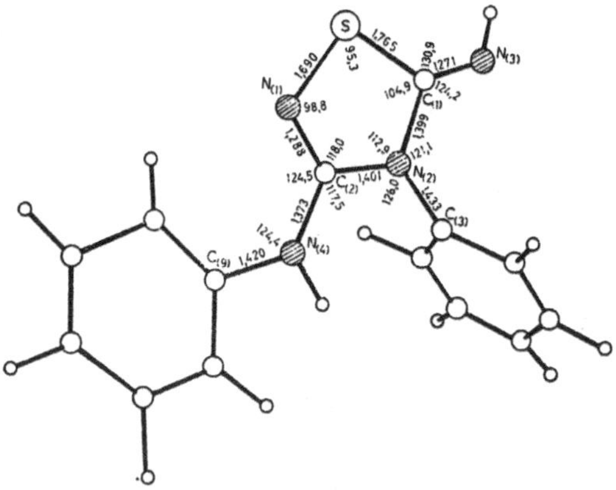

Fig. 1. Hector's base with main bond lengths.

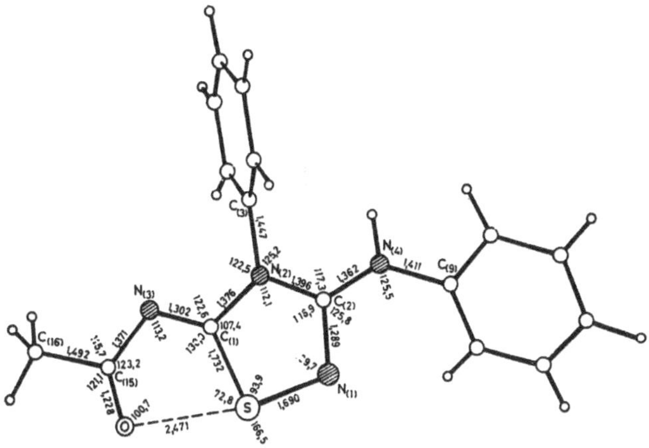

Fig. 2. A view of $C_{16}H_{14}N_4OS$ (II) with main bond lengths.

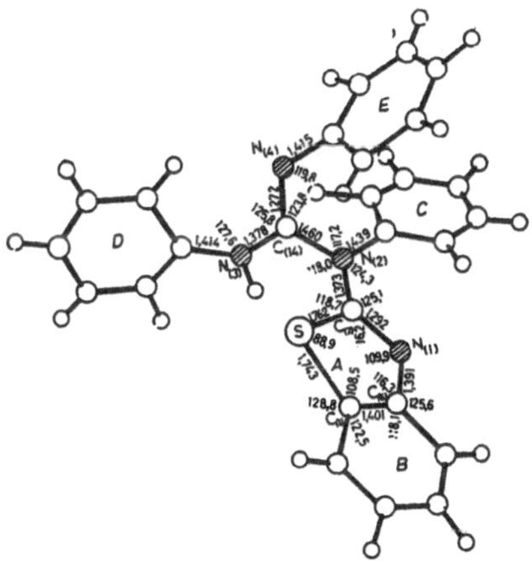

Fig. 3. Hugerhoff's base with main bond lengths.

1. Structure Reports, **44B**, 363.

(S,S)-METHYL 7-OXO-8-PHENYL-9-THIOXO-4,1,8-THIADIAZABICYCLO[4.3.0]NONANE-2-CARBOXYLATE $C_{14}H_{14}N_2O_3S_2$

A.V. EREMEEV, R. NURDINOV, F.D. POLYAK, E.E. LIEPIN'SH, A.V. MISHNEV, M.F. BUNDULE and Ya.Ya. BLEIDELIS, 1985. *Khim. Get. Soedin., SSSR,* 1056-1059.

Orthorhombic, $P2_12_12_1$, a = 6.644, b = 9.727, c = 23.660 Å, Z = 4. Cu radiation, R = 0.057 for 1525 reflections.

The thiomorpholine ring (Fig. 1) has a chair conformation with S(1) 0.942 and N(1) -0.543 Å from the C_4 plane. The imidazoline ring is planar and both N atoms have planar coordination. The mean C-S bond length is 1.812(10) Å with C=S 1.633(7) Å. The phenyl ring is rotated 68.3° out of the imidazoline plane.

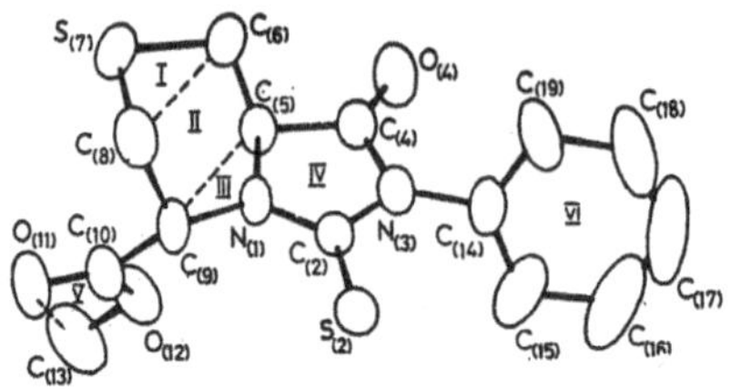

Fig. 1. The conformation of $C_{14}H_{14}N_2O_3S_2$.

2-p-TOLUIMIDO-4,4-DIMETHYL-(4H)-1,3-THIAZINE
$C_{14}H_{16}N_2OS$

L.G. KUZ'MINA, A.A. ESPENBETOV, Yu.T. STRUCHKOV, S.T. CHOLPANKULOVA and L.A. TSOI, 1984. *Khim. Get. Soedin. SSSR,* 1638-1643.

Monoclinic, $P2_1/c$, a = 10.970, b = 13.4591, c = 11.339 Å, β = 114.82°, Z = 4. Mo radiation, R = 0.086 for 959 reflections.

The molecule is shown to exist in the imino form (Fig. 1) and the thiazine ring has a conformation intermediate between a flattened twisted boat and semi-chair.

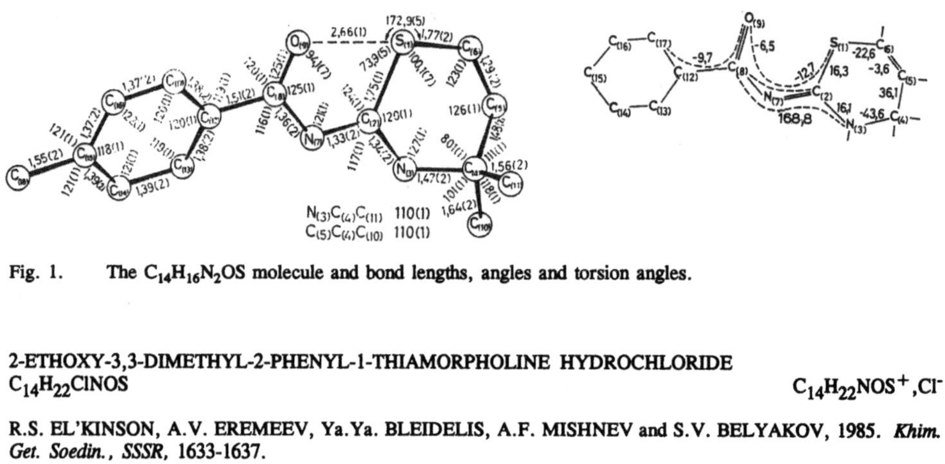

Fig. 1. The $C_{14}H_{16}N_2OS$ molecule and bond lengths, angles and torsion angles.

2-ETHOXY-3,3-DIMETHYL-2-PHENYL-1-THIAMORPHOLINE HYDROCHLORIDE
$C_{14}H_{22}ClNOS$ $C_{14}H_{22}NOS^+,Cl^-$

R.S. EL'KINSON, A.V. EREMEEV, Ya.Ya. BLEIDELIS, A.F. MISHNEV and S.V. BELYAKOV, 1985. *Khim. Get. Soedin., SSSR,* 1633-1637.

Monoclinic, $P2_1/c$, a = 13.724, b = 6.508, c = 16.727 Å, β = 90.25°, Z = 4. Cu radiation, R = 0.030 for 1814 reflections.

A view of the cation with dimensions is in Fig. 1. The thiomorpholine ring has a chair conformation, with S 0.956(1) and N -0.636(4) Å from the C_4 plane. The dihedral angle between the phenyl and thiomorpholine best planes is 64.7°. N(4) is protonated and tetrahedral; cations are linked into centrosymmetric pairs via N-H...Cl hydrogen bonds (N...Cl 3.109(4) Å).

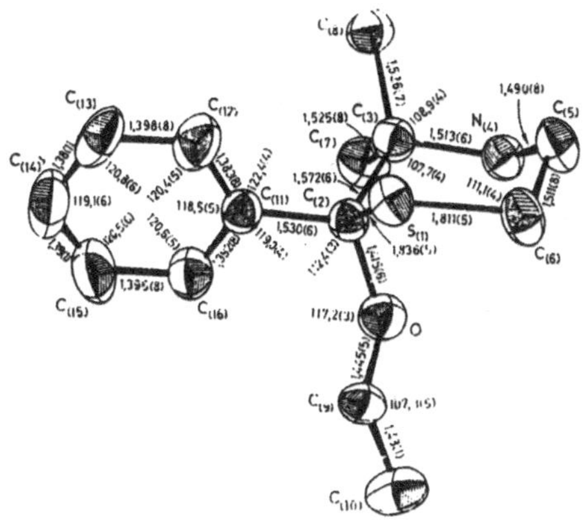

Fig. 1. The $C_{14}H_{22}NOS^+$ cation with dimensions.

1,3,4,6,9,10a-HEXAMETHYL-5,10-DIHYDRO-1,2,6-THIADIAZINO[4,3-g](2,1,3)BENZOTHIADIAZINE-2,2,8,8-TETRAOXIDE
$C_{14}H_{22}N_4O_4S_2$

P. GOYA, A. MARTINEZ, C. OCHOA, M. STUD, M.L. JIMENO, C. FOCES-FOCES, F.H. CANO and M. MARTINEZ-RIPOLL, 1985. *Tetrahedron*, **41**, 3105-3116.

Orthorhombic, *Pbca*, a = 25.2814, b = 12.1977, c = 11.0033 Å, Z = 8. Cu radiation, R = 0.082 for 833 reflections.

The structure and stereochemistry are shown in Fig. 1. The rings containing S atoms have envelope conformations with the S atoms at the flaps. The central ring has a half-chair conformation. The S-O distances are in the range 1.404-1.436(12) Å with S(1)-N(2) 1.622(18), S(1)-N(13) 1.688(15), S(2)-N(8) 1.642(15), S(2)-N(9) 1.656(14) Å.

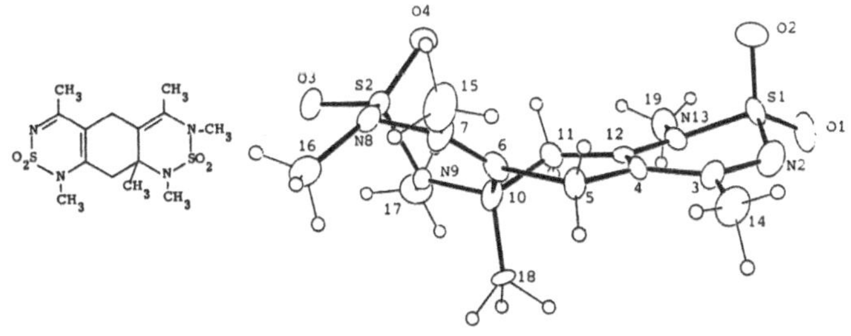

Fig. 1. The $C_{14}H_{22}N_4O_4S_2$ molecule and conformation.

SPIRO((2,3-DIHYDRO-2-OXO-IMIDAZO[1,2-a]PYRIDINE)-3,3'-(3,4-DIHYDRO-2-METHYL-4-OXO-2H-1,2-BENZOTHIAZINE 1,1-DIOXIDE))
$C_{15}H_{11}N_3O_4S$

C.P. NORRIS, H. BERKE, J.G. LOMBARDINO and J. BORDNER, 1985. *J. Heterocycl. Chem.*, **22**, 837-840.

Monoclinic, $P2_1/c$, a = 9.677, b = 15.996, c = 10.010 Å, β = 111.51°, Z = 4. Mo radiation, R = 0.055 for 1454 reflections.

The analysis established the structure of this unexpected product to be as shown in Fig. 1.

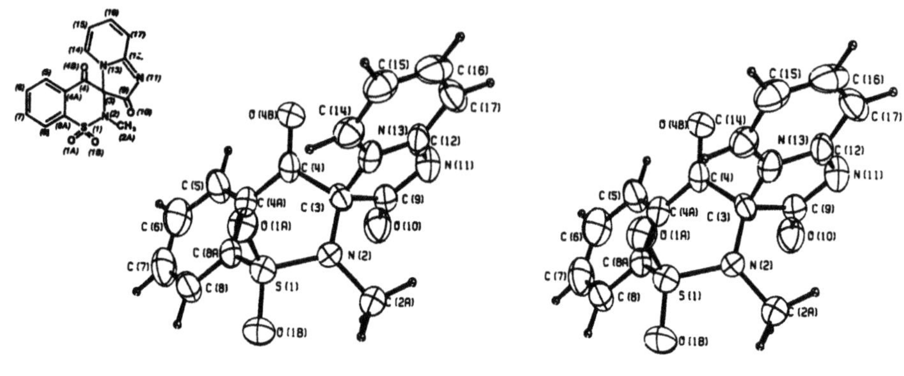

Fig. 1. The $C_{15}H_{11}N_3O_4S$ molecule.

12-CHLORO-6-METHYL-6,7-DIHYDRO-5H,12H-DIBENZO[b,g][1,5]THIAZOCINIUM HEXAFLUOROPHOSPHATE
$C_{15}H_{15}ClF_6NPS$ (I) $C_{15}H_{15}ClNS^+,F_6P^-$

12-METHOXY-6-METHYL-6,7-DIHYDRO-5H,12H-DIBENZO[b,g][1,5]THIAZOCINIUM HEXACHLOROANTIMOATE
$C_{16}H_{18}Cl_6NOSSb$ (II) $C_{16}H_{18}NOS^+,Cl_6Sb^-$

6,12-DIMETHYL-6,7-DIHYDRO-5H,12H-DIBENZO[b,g][1,5]THIAZOCINIUM HEXAFLUOROPHOSPHATE
$C_{16}H_{18}F_6NPS$ (III) $C_{16}H_{18}NS^+,F_6P^-$

6-METHYL-6,7-DIHYDRO-5H-DIBENZO[b,g][1,5]THIAZOCINE-S-OXIDE
$C_{15}H_{15}NOS$ (IV)

F. IWASAKI and K.-Y. AKIBA, 1985. *Acta Cryst.*, **B41**, 445-452.

I. Orthorhombic, *Pbca*, a = 17.101, b = 15.327, c = 13.195 Å, Z = 8. Cu radiation, R = 0.078 for 1973 reflections.

II. Monoclinic, $P2_1/c$, a = 11.791, b = 10.593, c = 18.860 Å, β = 98.25°, Z = 4. Mo radiation, R = 0.053 for 3221 reflections.

III. Orthorhombic, *Pbca*, a = 16.702, b = 16.352, c = 12.770 Å, Z = 8. Cu radiation, R = 0.064 for 2272 reflections.

IV. Triclinic, $P\bar{1}$, a = 8.5145, b = 9.4444, c = 16.831 Å, α = 99.23, β = 99.12, γ = 91.69°, Z = 4. Cu radiation, R = 0.043 for 4260 reflections.

Views of the molecules are shown in Fig. 1. There are two independent molecules (A and B) in the asymmetric unit of (IV). The conformation of the thiazocine (or thiazocinium) rings of these compounds is a distorted boat form without C_s or C_2 symmetry. Very strong transannular interactions are found between S and N atoms, which show the hypervalency of S. The S...N distances are 2.090, 2.207, 2.466, 2.593 and 2.624 Å for (I), (II), (III), (IV) A and (IV) B, respectively. The X_{ap}-S...N angles are 176.9, 175.0, 176.9, 179.5 and 179.1° for (I), (II), (III), (IV)A and (IV)B respectively, where the X_{ap}'s are Cl, OMe, Me and O for (I), (II), (III) and (IV) respectively. The configuration about the S atom is a distorted trigonal bipyramid with two apical S...N and S-X_{ap} bonds, two equatorial S-C bonds and the lone pair considered to occupy the third equatorial position. Therefore these compounds are concluded to be a new type of σ-ammoniosulfurane.

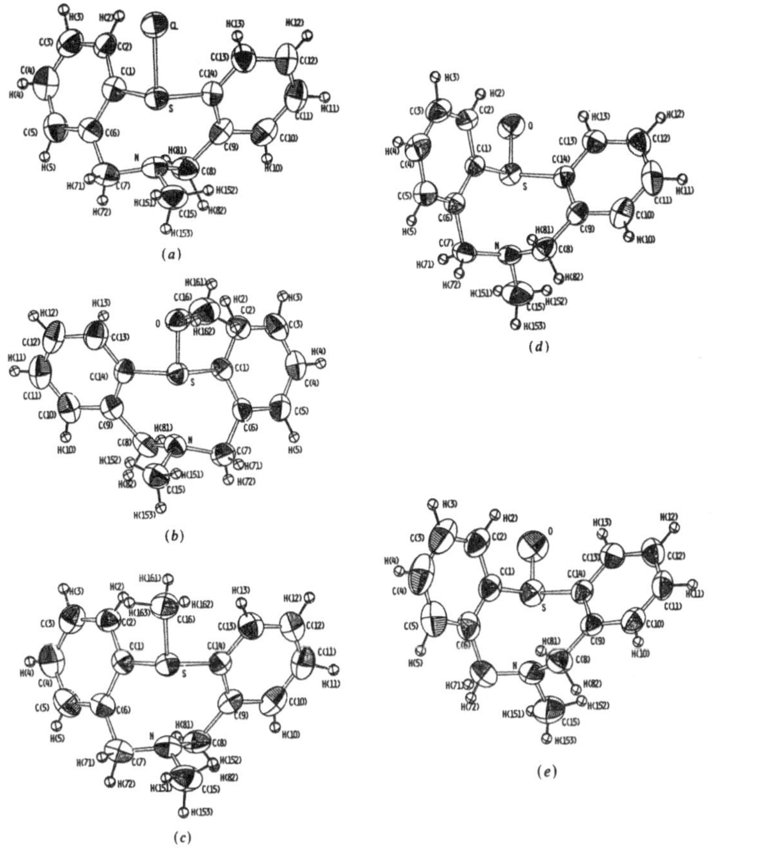

Fig. 1. ORTEP drawings of the dibenzo[b,g][1,5]thiazocine compounds (I) → (IV) ((a) (I), (b) (II), (c) (III), (d) (IVA), (e) (IVB)).

cis-2-METHYL-3,4-DIPHENYL-1,2-THIAZETIDINE-1,1-DIOXIDE
$C_{15}H_{15}NO_2S$

A. CHIARONI, C. RICHE, P. LOISEAU, M. BONNAFOUS and Y. ADAM, 1985. *Acta Cryst.*, C41, 1265-1267.

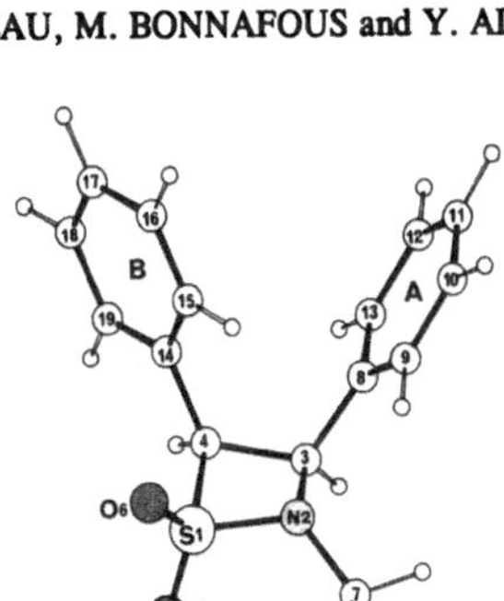

Fig. 1. $C_{15}H_{15}NO_2S$: the molecule.

Monoclinic, $P2_1/c$, a = 10.136, b = 13.016, c = 10.192 Å, β = 108.93°, Z = 4. Cu radiation, R = 0.054 for

2181 reflections.

The thiazetidine ring (Fig. 1) is puckered by 14.7°. The S-C(4) and S-N(2) distances are 1.806(2) and 1.643(2) Å respectively. The N atom is pyramidal and the C(7) methyl trans to phenyl ring A. The aromatic ring planes make angles of 53 and 87° respectively with the thiazetidine ring plane and an angle of 54° with each other.

2,3,4,5-TETRAHYDRO-2,2,5-TRIMETHYL-4-OXO-1H-PHENOTHIONIAAZIN-10-IDE $C_{15}H_{17}NOS$ (I)

5-ETHYL-2,3,4,5-TETRAHYDRO-4-OXO-1H-PHENOTHIONIAAZIN-10-IDE $C_{14}H_{15}NOS$ (II)

G.M. ISKANDER, I.E. KHAWAD, G. YOUSIF, K. FISHER, C.K. FAIR and E.O. SCHLEMPER, 1985. *Helv. Chim. Acta*, **68**, 2216-2225. G.M. ISKANDER, I.E.S.E. KHAWAD, G. YOUSIF, K.J. FISHER, K.O. FAIR and E.O. SCHLEMPER, 1985. *Gazz. Chim. Ital.*, **115**, 669-675.

I. Monoclinic, $P2_1/n$, a = 11.987, b = 9.919, c = 12.099 Å, β = 110.03°, Z = 4. Mo radiation, R = 0.050 for 1630 reflections.

II. Monoclinic, $P2_1/n$, a = 13.690, b = 7.469, c = 13.748 Å, β = 119.37°, Z = 4. Mo radiation, R = 0.045 for 2015 reflections.

The molecules of (I) and (II) are both non-planar and have boat conformations. The S-atom is pyramidal with the C-S-C angles less than tetrahedral. The S-C(6) bonds are 1.735 and 1.73 Å. Bond lengths and angles are shown in Fig. 1.

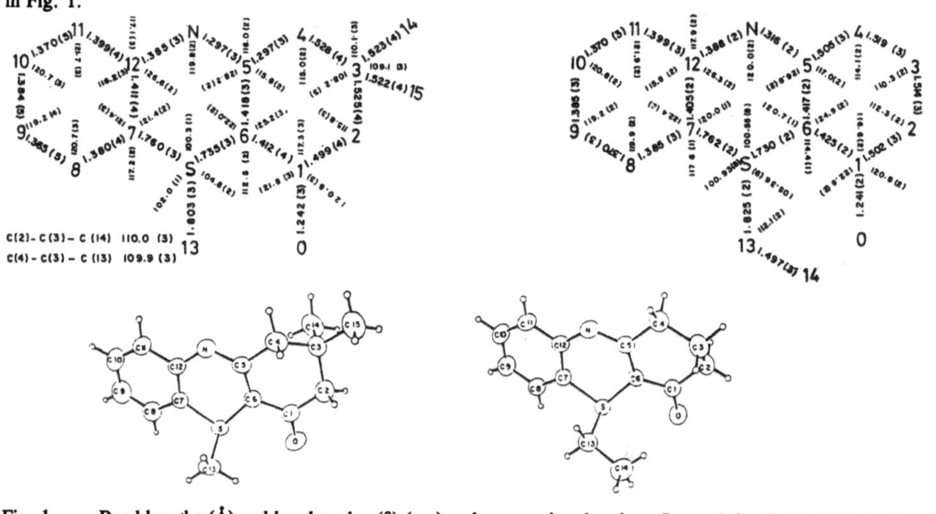

Fig. 1. Bond lengths (Å) and bond angles (°) (top) and perspective drawings (bottom) for $C_{15}H_{17}NOS$ (left) and $C_{14}H_{15}NOS$ (right).

2-(1,2,3-BENZOTHIADIAZOL-5-YL)-7-CHLORO-4-OXO-1,2,3,4-TETRAHYDROPHTHALAZIN-1-YLACETIC ACID METHANOL SOLVATE $C_{16}H_{11}ClN_4O_3S,CH_4O$

L. DUPONT and O. DIDEBERG, 1985. *Acta Cryst.*, **C41**, 1263-1265.

Monoclinic, $P2_1/n$, a = 11.560, b = 13.249, c = 12.317 Å, β = 101.16°, Z = 4. Mo radiation, R = 0.041 for 2502 reflections.

The benzothiadiazole and phthazinyl moieties in the molecule (Fig. 1) are approximately planar with a dihedral angle between these mean planes of 95.8(5)°. The C(1)C(2)C(7)C(8)N(2)N(1) ring adopts a half-chair conformation.

Cohesion in the crystal is achieved through van der Waals contacts and N-H...O (N...O 2.797(3)) and O-H...O (O...O 2.592(3) Å) hydrogen bonds.

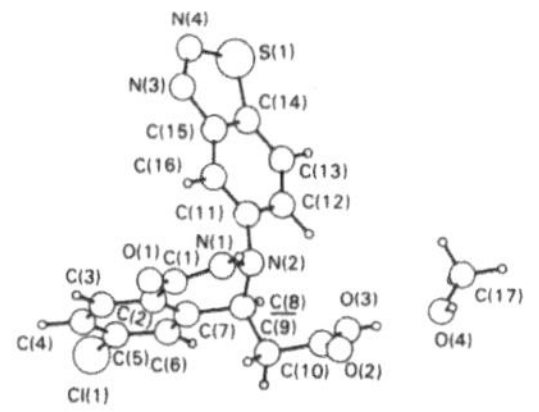

Fig. 1. $C_{16}H_{11}ClN_4O_3S,CH_4O$: the molecular structure.

10-(2'-PYRIDYL)PYRIDO[3,2-b][1,4]BENZOTHIAZINE
$C_{16}H_{11}N_3S$

M.V. JOVANOVIC, E.R. BIEHL, P. DE MEESTER and S.S.C. CHU, 1985. *J. Heterocycl. Chem.*, **22**, 777-783.

Monoclinic, *C2/c*, a = 27.887, b = 7.982, c = 14.252 Å, β = 124.04°, Z = 8. Cu radiation, R = 0.060 for 1860 reflections.

The central ring of the molecule (Fig. 1) is in a flattened boat form, with a folding angle of 166.1(1)° between the benzo and pryido rings. The plane of the 10-pyridyl ring nearly bisects the pyridobenzothiazine ring with dihedral angles of 95.3(1) and 97.5(1)° for the pyridyl/benzo rings and for the pyridyl/pyrido rings respectively.

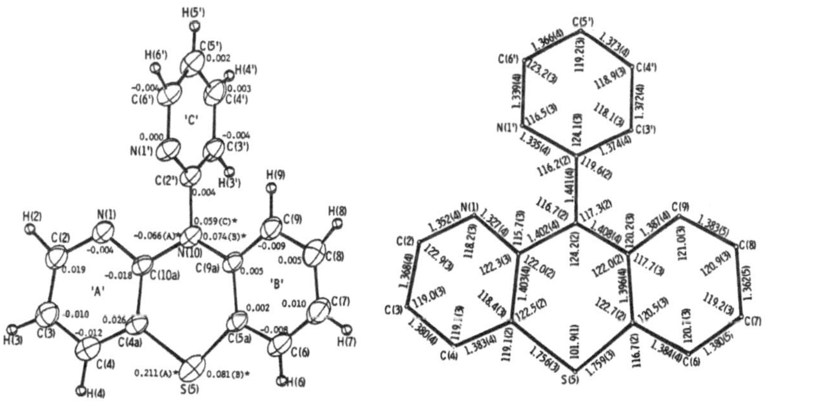

Fig. 1. The $C_{16}H_{11}N_3S$ molecule showing bond lengths (Å), angles (°) and deviations of atoms from the ring planes.

ETHYL 2-(2-PHENYLTHIAZOLO[3,2-b](1,2,4)TRIAZOL-6-YL)THIAZOLE-4-CARBOXYLATE
$C_{16}H_{12}N_4O_2S_2$

V. WARIN, F. BAERT, R. HOUSSIN, J.L. BERNIER and J.P. HENICHART, 1985. *Acta Cryst.*, **C41**, 1238-1239.

Orthorhombic, $P2_12_12_1$, a = 30.148, b = 10.579, c = 4.982 Å, Z = 4. Mo radiation, R = 0.044 for 1282 reflections.

The molecule (Fig. 1) is essentially planar, the angle between the thiazolotriazole mean plane and the phenyl ring mean plane being 10(1)°. Bond lengths and angles are normal. There is a short intramolecular contact between S(1) and N(13) of the molecule (2.92(1) Å). The geometry of the thiatriazole moiety fits well with the stacking of a guanine-cytosine base pair of DNA.

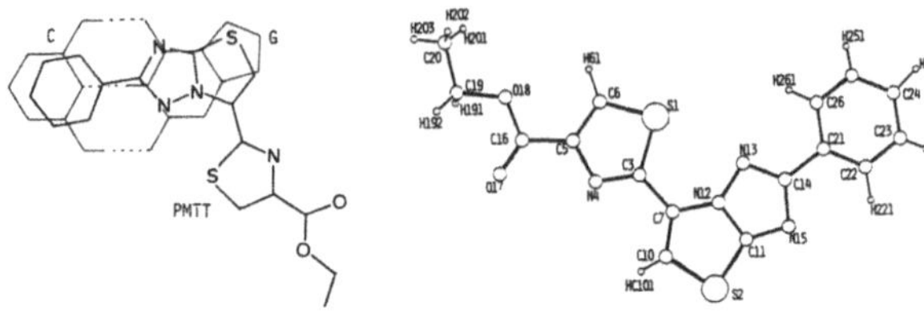

Fig. 1. $C_{16}H_{12}N_4O_2S_2$: the stacking of the molecules and the molecular structure.

((3-METHYLBENZO[4,5]THIAZOLIN-2-YLIDENE)METHYL)PHENYLKETONE
$C_{16}H_{13}NOS$

M. REGITZ, G. WEISE, B. LENZ, U. FORSTER, K. URGAST and G. MAAS, 1985. *Bull. Soc. Chim. Belg.*, **94**, 499-520.

Orthorhombic, *Pbca*, a = 26.971, b = 10.426, c = 9.545 Å, Z = 8. Mo radiation, R = 0.0402 for 1262 reflections.

The analysis established the structure as shown in Fig. 1. The C(8)-C(9)-C(10)-C(11) torsion angle is 171.3(3)°.

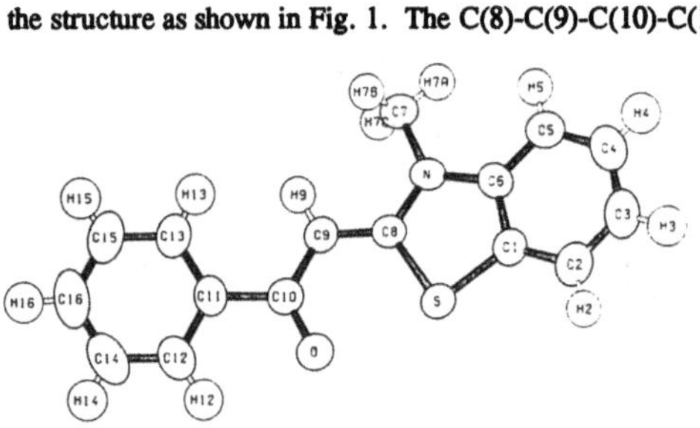

Fig. 1. A view of $C_{16}H_{13}NOS$.

3,6-BIS(4-METHOXYPHENYL)-1,4,2,5-DITHIADIAZINE
$C_{16}H_{14}N_2O_2S_2$

J.M.M. SMITS, P.T. BEURSKENS, B.G. LENZ and B. ZWANENBURG, 1985. *Acta Cryst.*, **C41**, 1086-1087.

Orthorhombic, $P2_12_12$, a = 11.260, b = 17.086, c = 3.948 Å, Z = 2. Mo radiation, R = 0.024 for 839 reflections.

The molecule (Fig. 1) is not planar; the heteroring system, through which a crystallographic diad axis passes, is in a boat conformation. The C(1)-N(1A), C(1)-S(1) and N(1)-S(1) bond lengths are 1.272(4), 1.756(3) and 1.707(3) Å respectively. The benzene ring plane and that of the C(2), C(1), S(1), N(1A) moiety have a dihedral angle of 31.3(1)°. This non-coplanarity probably arises due to steric hindrance suggested by the short intramolecular S(1)...C(3) contact of length 3.125(3) Å. No unusually short intermolecular distances are reported.

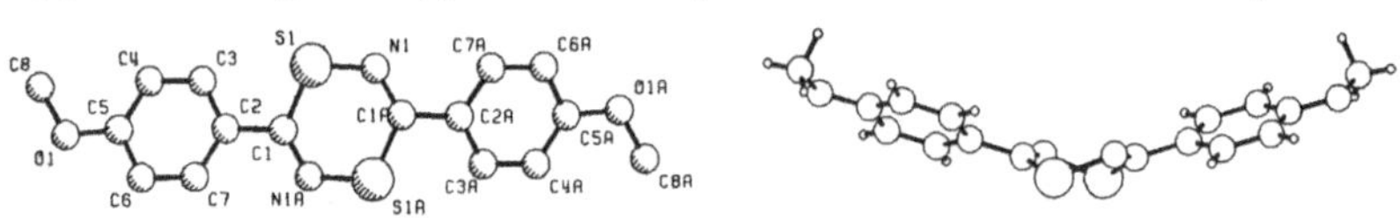

Fig. 1. $C_{16}H_{14}N_2O_2S_2$: atomic numbering and a side-on view of the molecule.

2-CHLORO-10-(3-METHYLAMINOPROPYL)PHENOTHIAZINE 5-OXIDE (MONO-DE-N-METHYLCHLORPROMAZINE SULPHOXIDE)

$C_{16}H_{17}ClN_2OS$

E. HOUGH, E. WOLD and S.G. DAHL, 1985. *Acta Cryst.*, **C41**, 386-389.

Orthorhombic, *Pbca*, a = 8.088, b = 19.051, c = 19.604 Å, Z = 8. Mo radiation, R = 0.039 for 2140 reflections (at 120 K).

The central thiazine ring in the molecule (Fig. 1) has a boat conformation with substituent O(15) axial. The N(10)-C(17) bond to the 3-methylaminopropyl fragment is equatorial. Atoms N(10) and S(5) are displaced from the least-squares plane of atoms C(11,12,13,14) by 0.19(4) and 0.57(4) Å respectively. Bond lengths in the aromatic rings are normal varying between 1.379(6) and 1.416(5) Å and the C(2)-Cl(6) bond distance is 1.740(4) Å. The dihedral angle between the least-squares planes of the aromatic rings is 155.3(3)°. Except for one possible intermolecular hydrogen bond between N(20) and O(15) of length 3.024(4) Å all other intermolecular distances are van der Waals contacts.

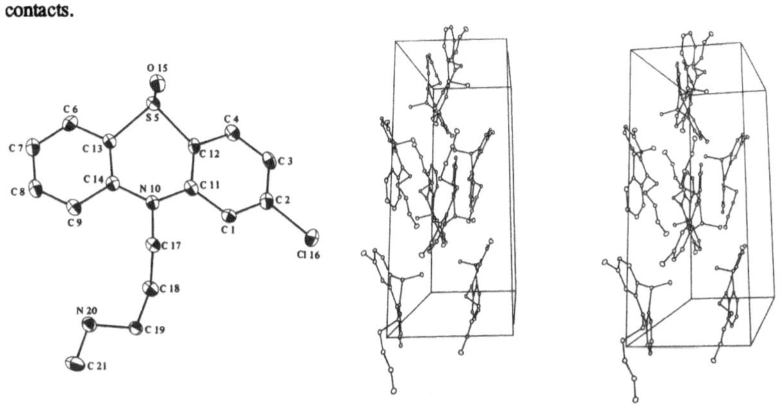

Fig. 1. $C_{16}H_{17}ClN_2OS$: a perspective view of the molecule (left) and a stereoview approximately down the b-axis of the molecular packing (right).

5-(N,N-DIISOPROPYLTHIOCARBAMOYLIMINO)-4-METHYL-3-PHENYLTHIO-(4H)-1,2,4-THIADIAZOLE
$C_{16}H_{22}N_4S_3$

J. GOERDELER, H.W. LINDEN, H. PUFF and R. HUNDT, 1985. *Chem. Ber.*, **118**, 3241-3247.

Orthorhombic, $P2_12_12_1$, a = 7.404, b = 11.623, c = 21.513 Å, Z = 4. Mo radiation, R = 0.038 for 1229 reflections (at 185 K).

The molecule and bond lengths are shown in Fig. 1. The central portion of the molecule is close to being planar.

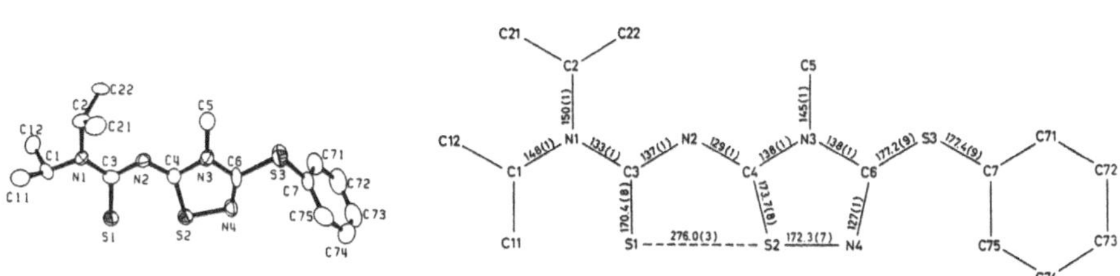

Fig. 1. A view of $C_{16}H_{22}N_4S_3$ and bond lengths (pm).

2,3-DIHYDRO-6,6-DIPHENYLIMIDAZO[2,1-b]THIAZOL-5(6H)-ONE
$C_{17}H_{14}N_2OS$ (I)

2,3-DIHYDRO-5,5-DIPHENYLIMIDAZO[2,1-b]THIAZOL-6(5H)-ONE BENZENE SOLVATE $C_{17}H_{14}N_2OS,C_6H_6$ (II)

J. KAROLAK-WOJCIECHOWSKA, M. MIKOLAJCZYK, A. ZATORSKI, K. KIEC-KONONOWICZ and A. ZEJC, 1985. *Tetrahedron*, **41**, 4593-4602.

I. Orthorhombic, $P2_12_12_1$, a = 11.376, b = 12.255, c = 8.434 Å, Z = 4. Cu radiation, R = 0.105 for 1256 reflections.

II. Monoclinic, $P2_1/c$, a = 11.539, b = 10.242, c = 16.353 Å, β = 95.45°, Z = 4. Mo radiation, R = 0.048 for 2069 reflections.

In both compounds the two fused five-membered ring systems are essentially coplanar (Fig. 1). The phenyl rings bonded to C(3) are inclined at 61° and 65° (I), 72.2° and 68.5° (II) to the plane of the thiohydantoin ring. The angle sum at N(1) in (II) is 358.0°. In the non-solvated form (1) the value reported is 346.0°.

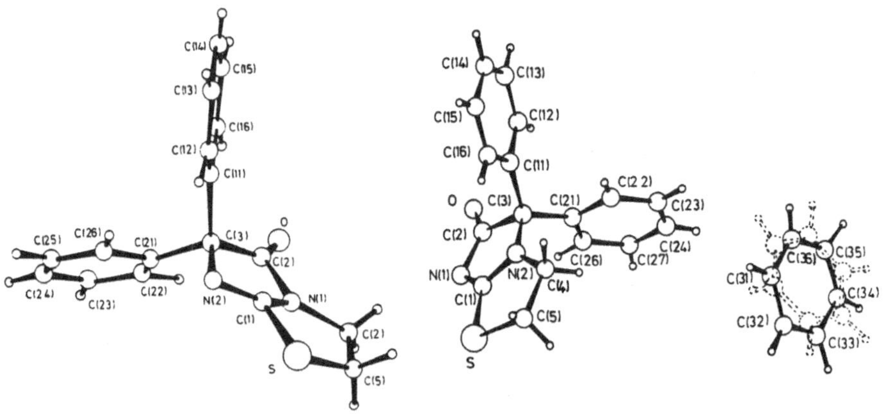

Fig. 1. Views of $C_{17}H_{14}N_2OS$; (I) (left), (II) (right).

1. Structure Reports, **48B**, 458.

erthyro-3,4-DIHYDRO-2(α-HYDROXYBENZYL)-2,4-DIMETHYL-3-OXO-2H-1,4-BENZOTHIAZINE $C_{17}H_{17}NO_2S$

F. BABUDRI, S. FLORIO, L. ZUCCARO, G. CASCARANO and F. STASI, 1985. *Tetrahedron*, **41**, 569-573.

Monoclinic, $P2_1/c$, a = 12.609, b = 12.054, c = 10.218 Å, β = 106.15°, Z = 4. Mo radiation, R = 0.050 for 2579 reflections.

The analysis establishes the stereochemistry as shown in Fig. 1. The six-membered ring has a half-chair conformation, with C(10) -1.053 and C(8) -0.336 Å out of the quasi-planar S(1)-C(13)-C(12)-N(1) moiety. The hydroxybenzyl group is (-)-synclinal with respect to the C=O group (O(2)-C(10)-C(8)-C(7) -15.0°); the methyl group is axial. There is a hydrogen bond between carbonyl and OH groups (O...O 2.764(3) Å).

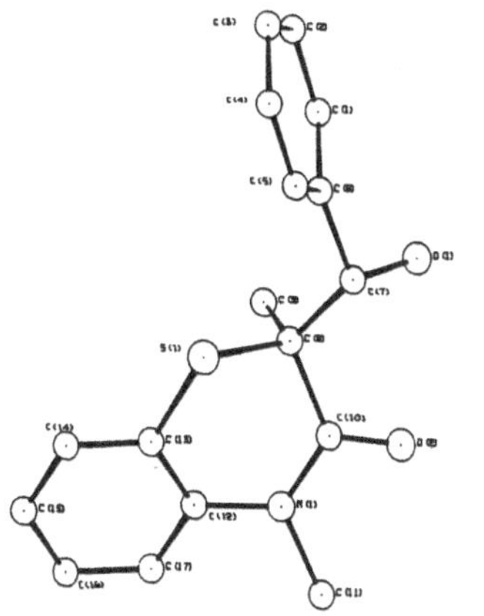

Fig. 1. A view of $C_{17}H_{17}NO_2S$.

CHLORPROMAZINE PERCHLORATE PERCHLORIC ACID
$C_{17}H_{19}Cl_2N_2O_4$,0.5($HClO_4$) $C_{17}H_{19}ClN_2S^+$,ClO_4^-,0.5($HClO_4$)

F.L. RUPEREZ, J.C. CONESA, J. SORIA, M.C. APREDA, F.H. CANO and C. FOCES-FOCES, 1985. *J. Phys. Chem.*, **89**, 1178-1182.

Monoclinic, *C2/c*, a = 24.522, b = 6.9100, c = 25.260 Å, β = 107.46°, Z = 8. Cu radiation, R = 0.068.

Views of the CPZ^+ radical are in Fig. 1; the chlorine atom is disordered over two sites (77/23) at C(1) and C(10). The perchloric acid molecule is disordered about a crystallographic twofold axis. The radical is folded about an axis close to the N...S line (168.7(3)°) and the plane of the side chain is essentially normal to the plane of the ring system. The C-S distances are 1.776(13) and 1.733(22) Å. For details of CPZ see (1).

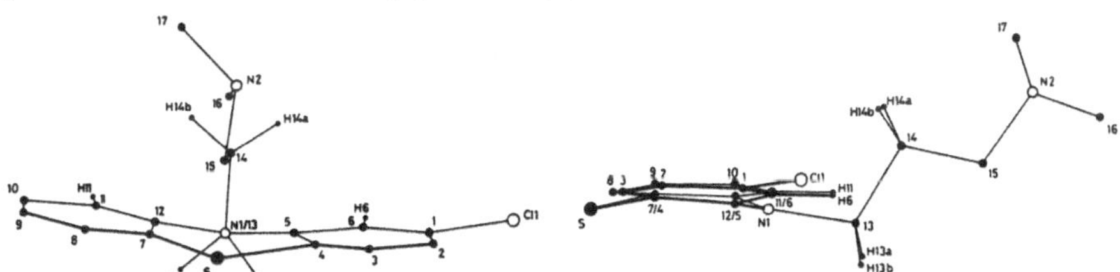

Fig. 1. A view of the CPZ^+ radical along N(1)-C(13), and a side view.

1. Structure Reports, **34B**, 213.

2-CHLORO-10-(3-DIMETHYLAMINOPROPYL)PHENOTHIAZINE 5-OXIDE (CHLORPROMAZINE SULPHOXIDE)
$C_{17}H_{19}ClN_2OS$

E. HOUGH, M. HJORTH and S.G. DAHL, 1985. *Acta Cryst.*, **C41**, 383-386.

Orthorhombic, *Pbca*, a = 10.357(10.210), b = 14.090(13.961), c = 23.585(23.441) Å, Z = 8. Cu radiation, R = 0.067(0.051) for 2144(2634) reflections (values in parentheses are at 120 K).

The mean bond lengths at 120 K in the planar aromatic rings of the molecule (Fig. 1) are normal, 1.399(6) Å in the substituted ring and 1.402(5) Å in the unsubstituted aromatic ring. The C-Cl bond length is 1.734(3) Å. The angle between the planes of the phenyl rings is 159.5(3)°. The central thiazine ring has a boat conformation with O(15) in an axial position. Atoms N(10) and S(5) are displaced by 0.18(4) and 0.44(4) Å respectively, from the least-squares plane of atoms C(11,12,13,14). The molecules are packed into two nearly mutually perpendicular layers parallel to the c-axis. Intermolecular distances are van der Waals contacts.

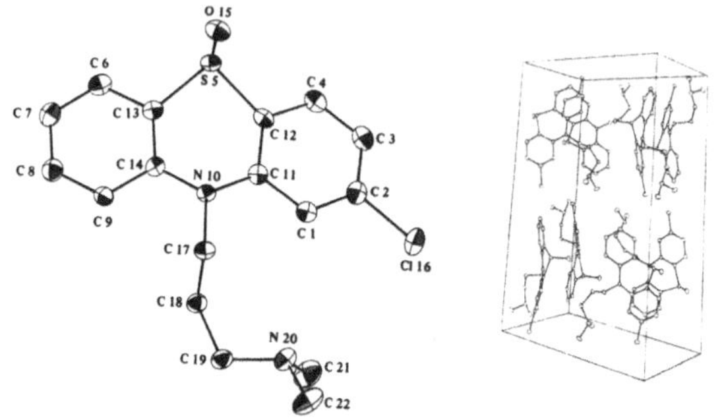

Fig. 1. $C_{17}H_{19}ClN_2OS$: a perspective view of the molecule showing the atomic numbering, and a stereoview of the molecular packing.

4-METHYL-3,5-DIPHENYL-1,4-THIAZINANE-1,1-DIOXIDE
$C_{17}H_{19}NO_2S$

D. KLEFFEL, C. KRATKY and H.-H. OTTO, 1985. *Liebigs Ann. Chem.*, 1012-1020.

Monoclinic, $P2_1/n$, a = 12.122, b = 20.824, c = 12.942 Å, β = 98.53°, Z = 8. Mo radiation, R = 0.043 for 2337 reflections.

There are two independent similar molecules in the asymmetric unit (Fig. 1). The six-membered rings are in chair conformation with equatorial phenyl and methyl groups. The mean S-C distance is 1.75 Å and mean S-O 1.44 Å; other dimensions are normal.

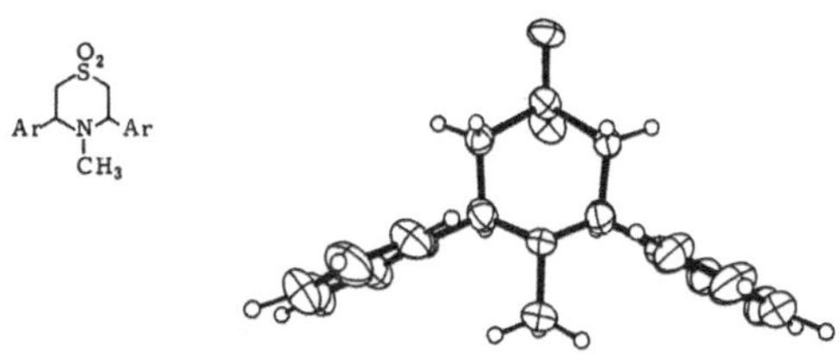

Fig. 1. One of the $C_{17}H_{19}NO_2S$ molecules.

(1,4)BENZOTHIAZINO[3,2-i]PHENOTHIAZINE
$C_{18}H_{10}N_2S_2$

J. GARBARCZYK, 1985. *Acta Cryst.*, C41, 1062-1064.

Monoclinic, $P2_1/c$, a = 10.472, b = 5.430, c = 12.356 Å, β = 107.6°, Z = 2. Cu radiation, R = 0.071 for 669 reflections.

The molecule (Fig. 1) is located on a crystallographic centre of symmetry and is quasi-planar. The external benzene ring, R1, has normal geometry while the central benzene ring, R3, has a quinonoid structure with unequal C-C bond lengths with C(1B)-C(3'B) of length 1.350(12) Å- shorter by some 0.10 Å than the other ring unique bond lengths. In the essentially planar phenothiazine ring, R2, the S and N atoms deviate from the R1 ring plane by -0.082(2) and 0.032(2) Å respectively. The molecules are arranged in the unit cell in alternate parallel layers with the planes of neighbouring layers almost perpendicular. Intermolecular distances are consistent with essentially van der Waals contacts.

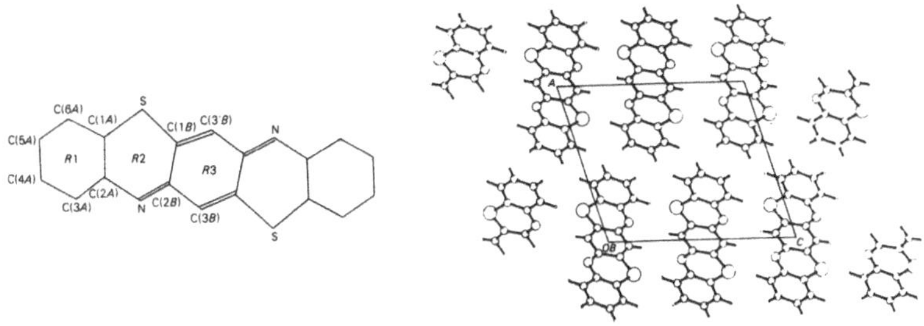

Fig. 1. $C_{18}H_{10}N_2S_2$: atom numbering in the molecule and the unit cell contents projected onto (010) with the large circles representing the S atoms.

N-PHENYLPHENOTHIAZINE
$C_{18}H_{13}NS$

C.L. KLEIN, J.M. CONRAD III and S.A. MORRIS, 1985. *Acta Cryst.*, **C41**, 1202-1204.

Triclinic, $P\bar{1}$, a = 10.369, b = 10.596, c = 15.468 Å, α = 68.26, β = 65.49, γ = 68.04°, Z = 4. Mo radiation, R = 0.055 for 2921 reflections (at 110 K).

There are two crystallographically independent molecules in the asymmetric unit. The phenyl ring is bonded equatorially to the phenothiazine skeleton at N (Fig. 1). The fold angles between the two least-squares planes of the two benzo rings are 162.6(5) and 150.7(5)°. Bond lengths and angles lie within normal ranges. There are no hydrogen bonds and only one intermolecular contact (2.80(4) Å).

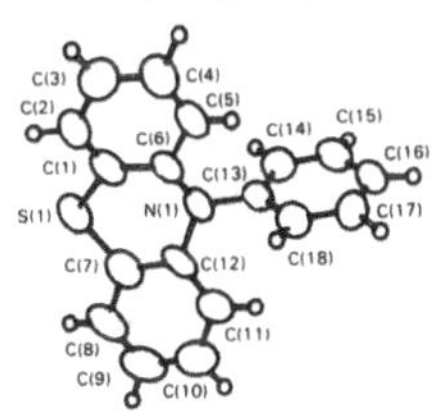

Fig. 1. $C_{18}H_{13}NS$: the molecular structure.

10-(4-METHOXYPHENYL)PYRIDO[3,2-b](1,4)BENZOTHIAZINE
$C_{18}H_{14}N_2OS$

P. DE MEESTER, S.S.C. CHU, M.V. JOVANOVIC and E.R. BIEHL, 1985. *Acta Cryst.*, **C41**, 1246-1249.

Triclinic, $P\bar{1}$, a = 13.680, b = 10.982, c = 11.782 Å, α = 101.63, β = 116.24, γ = 70.97°, Z = 4. Mo radiation, R = 0.051 for 3529 reflections.

There are two crystallographically independent molecules in the asymmetric unit (Fig. 1). The methoxy carbon, C(7'), of molecule B is disordered and the C(7')-O(4') bond lengths of 1.33(1) and 1.31(1) Å are much shorter than the corresponding distance in molecule A (1.415(8) Å). Otherwise the molecular dimensions of A and B show good agreement. The tricyclic moiety is folded with the central ring in a boat conformation. The fold angles are 157.1(1) and 155.2(1)° for A and B respectively. The 10-(4-methoxyphenyl) group adopts a boat-equatorial conformation with respect to the central ring. Packing is governed by van der Waals interactions.

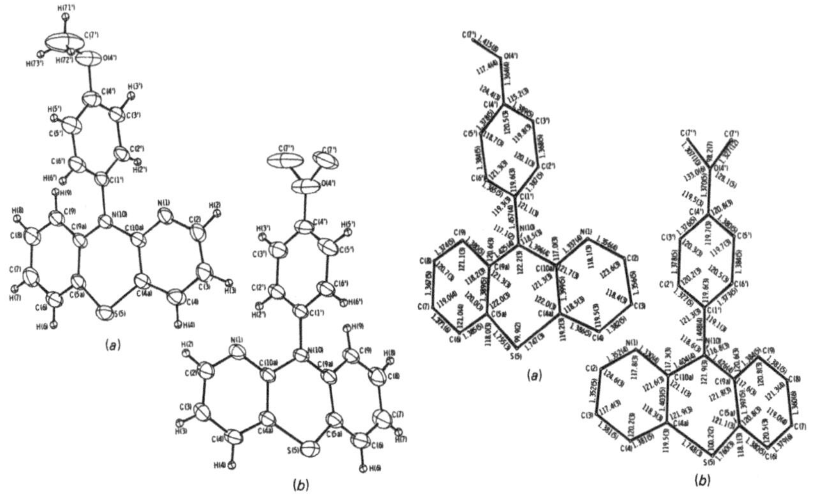

Fig. 1. $C_{18}H_{14}N_2OS$: the molecular structures, bond lengths and angles ((a) A, (b) B).

(8aRS,9RS,10aSR)-10-PHENYL-1,2,3,4,5,6,7,8,8a,10a-DECAHYDRO-10H-9-THIA-10-AZAPHENANTHRENE-9-OXIDE
$C_{18}H_{23}NOS$

Z. DAUTER, P. HANSON, C.D. REYNOLDS, W.A. STOCKBURN and T.W. STONE, 1985. *Acta Cryst.*, **C41**, 1514-1516.

Monoclinic, $P2_1$, a = 5.661, b = 18.071, c = 8.178 Å, β = 103.74°, Z = 2. Cu radiation, R = 0.062 for 1112 reflections.

The S-O bond of the molecule (Fig. 1) is trans to the C-H bonds of the ring junctions, C(1) and C(12). The heterocyclic ring adopts a conformation such that S(13) and N(15) are displaced by -0.65(1) and 0.18(1) Å out of the mean plane through C(1), C(6), C(7) and C(12). The S(13)-O, S(13)-N(15) and S(13)-C(12) bond lengths are 1.475(5), 1.680(6) and 1.779(3) Å respectively. Ring A adopts a chair conformation and ring B a distorted chair conformation. Ring C is essentially planar.

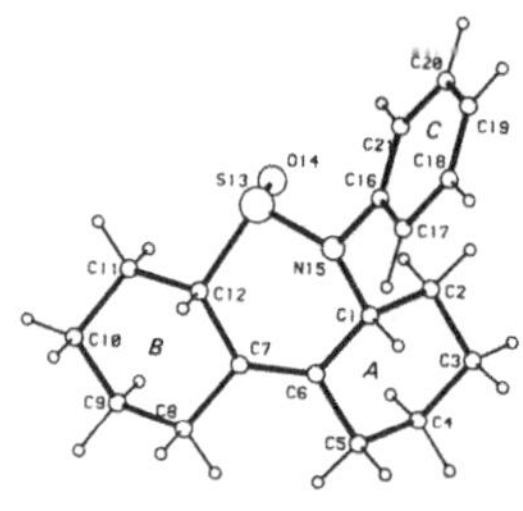

Fig. 1. $C_{18}H_{23}NOS$: the molecular structure.

DIPHENYL-N-(p-TOLUENESULPHONYL)SELENIMIDE TETRAHYDROFURAN
$C_{19}H_{17}NO_2SSe,0.5(C_4H_8O)$ (I) $MeC_6H_4SO_2NSePh_2,0.5C_4H_80$

TRICHLORO(TRIPHENYLPHOSPHORANYLIDENEAMINO)SELANE DICHLOROMETHANE
$C_{18}H_{15}Cl_3NPSe,CH_2Cl_2$ (II) $Ph_3PNSeCl_3,CH_2Cl_2$

DICHLOROBIS(TRIPHENYLPHOSPHORANYLIDENEAMINO)SELANE

$C_{36}H_{30}Cl_2N_2P_2Se$ (III) $(Ph_2PN)_2SeCl_2$

CHLORO-N-(4,6-DI-t-BUTYL-2-OXOCYCLOHEXA-3,5-DIEN-1-YLIDENE)SELENAMIDE
$C_{14}H_{20}ClNOSe$ (IV) $Bu^t_2C_6H_2ONSeCl$

H.W. ROESKY, K.-L. WEBER, U. SESEKE, W. PINKERT, M. NOLTEMEYER, W. CLEGG and G.M. SHELDRICK, 1985. *J. Chem. Soc. Dalton Trans.*, 565-571.

I. Monoclinic, $P2_1/n$, a = 19.423, b = 10.935, c = 21.211 Å, β = 114.62°, Z = 8. Mo radiation, R = 0.062 for 4485 reflections.

II. Triclinic, $P\bar{1}$, a = 9.771, b = 10.452, c = 11.919 Å, α = 91.91, β = 92.83, γ = 110.71°, Z = 2. Mo radiation, R = 0.050 for 2696 reflections.

III. Orthorhombic, $Pna2_1$, a = 21.646, b = 8.898, c = 16.737 Å, Z = 4. Mo radiation, R = 0.031 for 2746 reflections.

IV. Monoclinic, $C2/c$, a = 15.059, b = 13.777, c = 15.522 Å, β = 100.63°, Z = 8. Mo radiation, R = 0.041 for 1616 reflections.

Two independent but very similar molecules are present in (I). The molecules of (II) dimerize across a crystal inversion centre. All four compounds ((I)-(IV), Fig. 1) contain Se-N bonds which are shorter than expected for single bonds (although the formally double Se-N bond in (I) is the longest). Bond lengths and angles round the Se atoms include: (I), Se-N 1.787(6), Se-C 1.933(7) Å, N-Se-C 99.5(3), C-Se-C 98.2(3)°: (II), Se-N 1.680(4), Se-Cl 2.634, 2.983(μ), 2.248, 2.291(2) Å, Se-Cl-Se 96.1(1), N-Se-Cl 93.4-104.6(1)°: (III), Se-N 1.735(4), Se-Cl 2.595(2) Å, N-Se-Cl 99.4(1), N-Se-N 102.5(2), Cl-Se-Cl 149.9(1)°: (IV), Se-N 1.766(4), Se-Cl 2.302(2), Se-O 2.079(3) Å, N-Se-Cl 91.1(1), N-Se-O 83.7(1), O-Se-Cl 174.7(1)°.

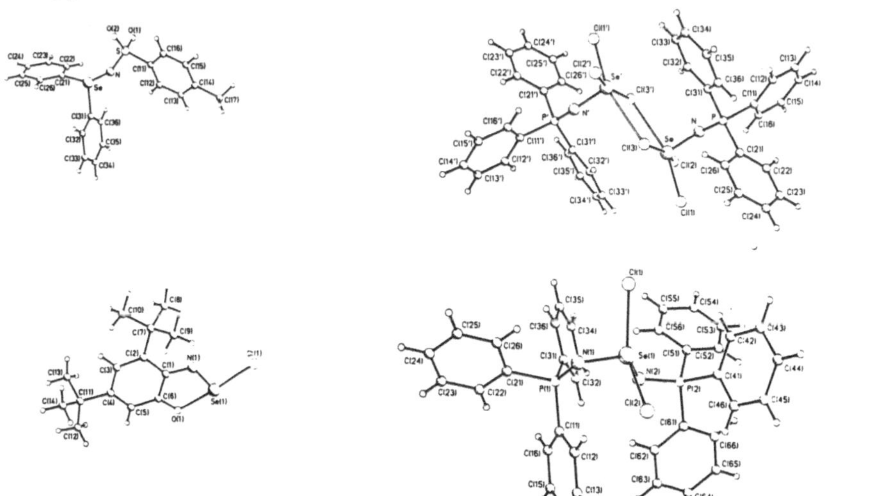

Fig. 1. Views (clockwise from top left) of (I), $C_{19}H_{17}NO_2SSe$, (II), $C_{18}H_{15}Cl_3NPSe$, (III), $C_{36}H_{30}Cl_2N_2P_2Se$, and (IV), $C_{14}H_{20}ClNOSe$.

DI-t-BUTYL-2-METHYL-2-((2α,3aβ,7aβ)-OCTAHYDROBENZOTHIAZOL-2'-YL)MALONATE
$C_{19}H_{33}NO_4S$ (I)

rel-(1R,3R,7S)-2-THIA-6-AZATRICYCLO[5.4.0.$0^{3,6}$]UNDECAN-5-ONE
$C_9H_{13}NOS$ (II)

R.C. CAMBIE, G.R. CLARK, T.C. JONES, P.S. RUTLEDGE, G.A. STRANGE and P.D. WOODGATE, 1985. *Aust. J. Chem.*, **38**, 745-764.

I. Monoclinic, $P2_1/c$, a = 10.743, b = 11.198, c = 18.644 Å, β = 105.76°, Z = 4. Mo radiation, R = 0.046 for 2418 reflections.

II. Orthorhombic, *Pbca*, a = 8.088, b = 22.031, c = 10.383 Å, Z = 8. Mo radiation, R = 0.056 for 749 reflections.

Both unit cells contain a racemic mixture of enantiomers. In (I) and (II) (Fig. 1), the cyclohexane rings have similar dimensions. The similarity extends to the S-C (cyclohexane) bonds of 1.833(3) and 1.832(4) Å of (I) and (II) respectively. In (I), the S(1)-C(2') bond in the thiazolidine moiety is unusually long (1.863(3) Å). The S(2)-C(3) distance in (II) is 1.810(4) Å. There is no intermolecular hydrogen bonding in either molecule.

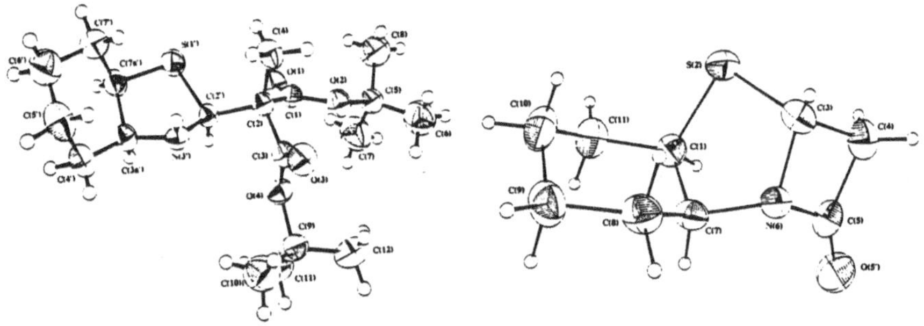

Fig. 1. $C_{19}H_{33}NO_4S$ (left) and $C_9H_{13}NOS$ (right).

1-(4-NITROPHENYL)-3-(4-CHLOROPHENYL)-8,9-DIMETHOXY-10b-METHYL-S-TRIAZOLOBENZOTHIAZINE DIHYDRATE
$C_{24}H_{21}ClN_4O_4S,2(H_2O)$ (I)

1-PHENYL-3-(4-CHLOROPHENYL)-8,9-DIMETHOXY-10b-METHYL-S-TRIAZOLOBENZOTHIAZINE
$C_{24}H_{22}ClN_3O_2S$ (II)

1-(4-CHLOROPHENYL)-3-CARBETHOXY-8,9-DIMETHOXY-10b-METHYL-S-TRIAZOLOBENZOTHIAZINE
$C_{21}H_{22}ClN_3O_4S$ (III)

A. KALMAN, G. ARGAY, P. SOHAR, L. FODOR, J. SZABO and G. BERNATH, 1985. *J. Mol. Struct.*, **129**, 271-276.

I. Triclinic, $P\bar{1}$, a = 9.292, b = 11.521, c = 12.077 Å, α = 102.83, β = 95.59, γ = 86.98°, Z = 2. Mo radiation, R = 0.083 for 3282 reflections.

II. Triclinic, $P\bar{1}$, a = 8.688, b = 10.248, c = 13.707 Å, α = 104.52, β = 94.30, γ = 65.88°, Z = 2. Mo radiation, R = 0.034 for 2590 reflections.

III. Triclinic, $P\bar{1}$, a = 7.344, b = 12.186, c = 13.440 Å, α = 76.80, β = 78.67, γ = 70.87°, Z = 2. Mo radiation, R = 0.044 for 2488 reflections.

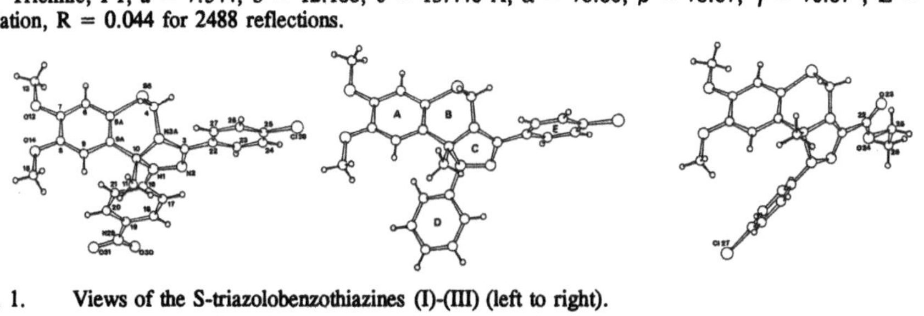

Fig. 1. Views of the S-triazolobenzothiazines (I)-(III) (left to right).

The analyses determined the molecular conformations as shown in Fig. 1, and all have cis B/C ring junctions. The thiazine ring in (I) has a skew-boat conformation shifted towards a twist-boat, while in (II) an almost perfect skew-boat conformation is found. In (III) the flexible thiazine ring has a conformation intermediate between a half-chair and envelope. The triazole ring has an envelope conformation in (I) and (III) but a half-chair conformation in

(II).

3,5-BIS(DIPHENYLAMINO)-1,2,4-THIADIAZOLE
$C_{26}H_{20}N_4S$

H. SENDA and J. MARUHA, 1985. *Acta Cryst.*, **C41**, 1329-1331.

Triclinic, $P\bar{1}$, a = 9.741, b = 13.232, c = 9.631 Å, α = 110.27, β = 109.36, γ = 71.96°, Z = 2. Mo radiation, R = 0.047 for 4043 reflections.

The heterocyclic ring, H, in the molecule (Fig. 1) is planar and its exocyclic atoms N(3) and N(4) deviate from this plane by -0.030 and 0.033 Å respectively. The four phenyl groups are nearly planar and the A/H, B/H, C/H and D/H interplanar dihedral angles are 20.9, 100.3, 47.6 and 76.0° respectively. The S-N(1) bond (1.651(1) Å) is intermediate between the values for a single and a double bond and the C(2)-N(2) bond (1.302(2) Å) exhibits considerable double-bond character. All bond lengths and angles in the phenyl groups are normal. The intermolecular distances are consistent with van der Waals contacts.

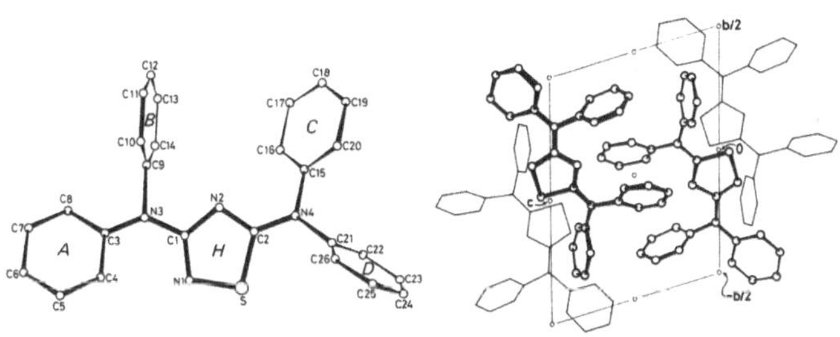

Fig. 1. $C_{26}H_{20}N_4S$: view of the molecule showing the atomic numbering scheme and the molecular packing projected onto (100).

TETRAPHENYLPHOSPHONIUM 6,6,7,7-TETRAFLUORO-1,3,5,2,4-TRITHIADIAZEPANE-3,3-DIOXIDE CHLORIDE
$C_{26}H_{22}ClF_4N_2O_2PS_3$ $C_2H_2F_4N_2O_2S_3,C_{24}H_{20}P^+,Cl^-$

H.W. ROESKY, A. THIEL, M. NOLTEMEYER and G.M. SHELDRICK, 1985. *Chem. Ber.*, **118**, 2811-2821.

Monoclinic, $P2_1/c$, a = 14.234, b = 14.225, c = 15.359 Å, β = 109.57°, Z = 4. Mo radiation, R = 0.054 for 2700 reflections.

The seven-membered ring conformation is shown in Fig. 1 and the heterocyclic rings are linked by H...Cl...H hydrogen bonds to form infinite chains (H...Cl 2.14 Å). In the ring, the S-N distances are S(1)-N 1.611 and 1.616(5), S(2,3)-N 1.649 and 1.664(5) Å with S-C 1.800 and 1.809(6) Å.

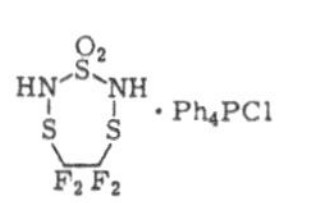

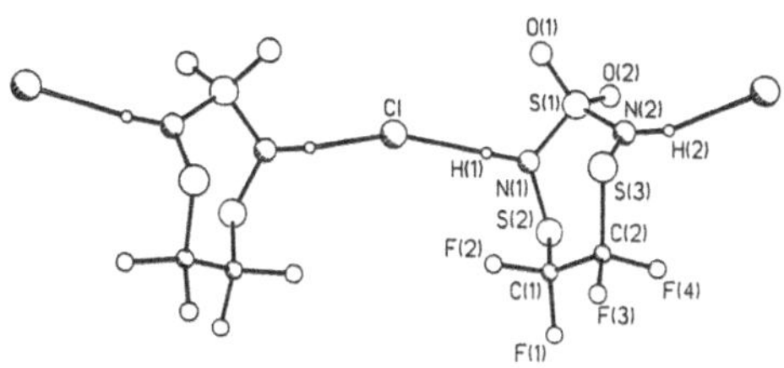

Fig. 1. The seven-membered ring in the trithiadiazepane system linked to form infinite chains by H...Cl...H hydrogen bridges.

BIS(3,5-DIPHENYL-1,2,4,6-THIATRIAZINE(S-S))

$C_{28}H_{20}N_6S_2$

P.J. HAYES, R.T. OAKLEY, A.W. CORDES and W.T. PENNINGTON, 1985. *J. Am. Chem. Soc.*, **107**, 1346-1351.

Triclinic, $P\bar{1}$, a = 8.894, b = 11.058, c = 12.814 Å, α = 81.76, β = 86.75, γ = 76.00°, Z = 2. Mo radiation, R = 0.050 for 1125 reflections.

In the structure (Fig. 1), discrete pairs of thiatriazine rings are linked in a cofacial fashion. The closest contact between the two ring systems is S(1)...S(2) (2.666(3) Å). The C_2N_3S rings are approximately coplanar (dihedral angle 14°) and are in eclipsed conformation. The phenyl rings make dihedral angles of 3.8-11.4° with the planes of the C_2N_3S rings.

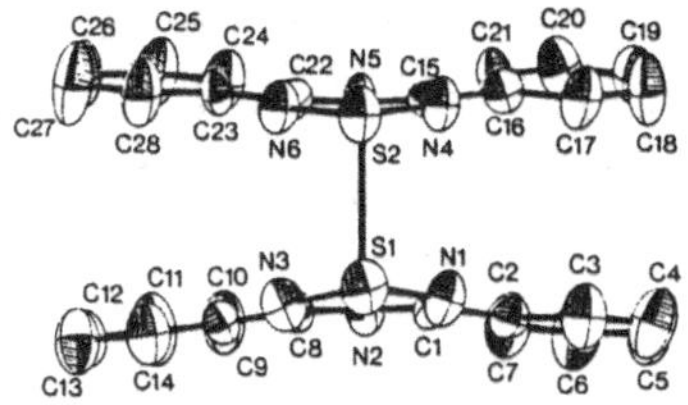

Fig. 1. $C_{28}H_{20}N_6S_2$: a side-on view of the dimeric structure.

2,4-BIS(DIMETHYLAMINO)-6-PHENYL-1,3,5-THIADIAZINIUM TETRAPHENYLBORATE
$C_{37}H_{37}BN_4S$ $C_{13}H_{17}N_4S^+,C_{24}H_{20}B^-$

P.F. LINDLEY, W.E. BOTTOMLEY and G.V. BOYD, 1985. *Acta Cryst.*, **C41**, 1260-1262.

Monoclinic, $P2_1/a$, a = 26.076, b = 9.517, c = 13.046 Å, β = 102.11°, Z = 4. Cu radiation, R = 0.057 for 3899 reflections.

In the planar thiadiazinium cation (Fig. 1) the S(1)-C(2), S(1)-C(6) and N(5)-C(6) distances are 1.751(3), 1.738(3) and 1.281(4) Å respectively. The latter indicates a double bond. The bond lengths N(21)-C(2) 1.321(5), C(2)-N(3) 1.314(3), N(3)-C(4) 1.331(4) and C(4)-N(41) 1.321(4) Å indicate delocalisation of electron density in this region. The tetraphenylborate ion is normal.

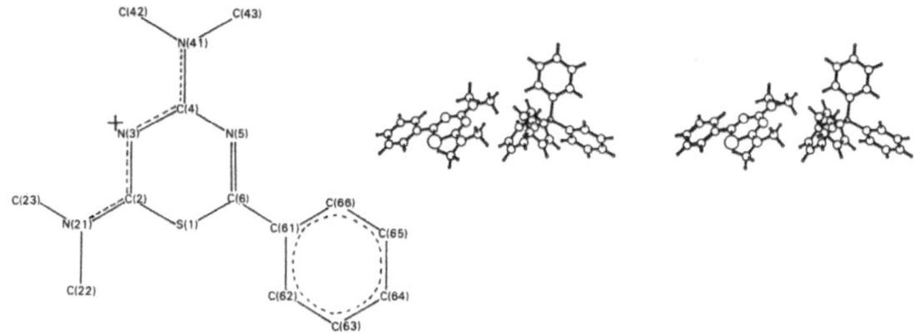

Fig. 1. $C_{37}H_{37}BN_4S$: the cation (left) and a stereoview of the asymmetric unit (right).

5-NITROBENZ[1,6-d]-3H-1,2-OXATHIOLE-S,S-DIOXIDE
$C_7H_5NO_5S$ (I)

6-NITRONAPHTHOL[1,8-cd]-1,2-OXATHIOLE-S,S-DIOXIDE
$C_{10}H_5NO_5S$ (II)

E. TRAN HUU-DAU, J. GUILHEM, F. ACHER and M. WAKSELMAN, 1985. *J. Chem. Res.*, S126-127, M1501-1535.

I. Triclinic, $P\bar{1}$, a = 8.862, b = 9.355, c = 12.054 Å, α = 94.97, β = 95.42, γ = 121.89°, Z = 4. Cu

radiation, R = 0.063 for 2482 reflections.

II. Monoclinic, *C2/c*, a = 17.655, b = 8.063, c = 15.093 Å, β = 115.42°, Z = 8. Cu radiation, R = 0.065 for 1443 reflections.

The sulphur atom of the benzosultone (I) (Fig. 1) is 0.204 Å above the mean plane of the molecule whereas the naphthosultone (II) is quite planar. In (I) there are two molecules in the crystallographic asymmetric unit.

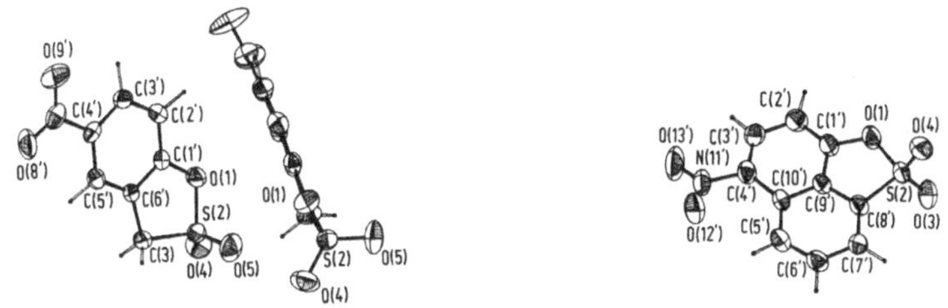

Fig. 1. ORTEP drawings of the molecules of $C_7H_5NO_5S$ (left) (both independent molecules) and of $C_{10}H_5NO_5S$ (right).

1,4,9-TRIAZAPHENOXATHIIN
$C_9H_5N_3OS$

S.B. LARSON, S.H. SIMONSEN, G.E. MARTIN and K. SMITH, 1985. *Acta Cryst.*, **C41**, 1781-1783.

Orthorhombic, *Pbca*, a = 14.892, b = 15.778, c = 7.1688 Å, Z = 8. Mo radiation, R = 0.0451 for 1842 reflections (at 163 K).

The molecule (Fig. 1) is composed of a planar pyridine ring and a planar pyrazine ring. The dihedral angle between the rings is 172.47(8)°. Both N atoms in the pyrazine ring have unsymmetrical bond lengths as do S and O in the central ring. In the crystal, molecules pack in undulating layers stacked in the *c* direction.

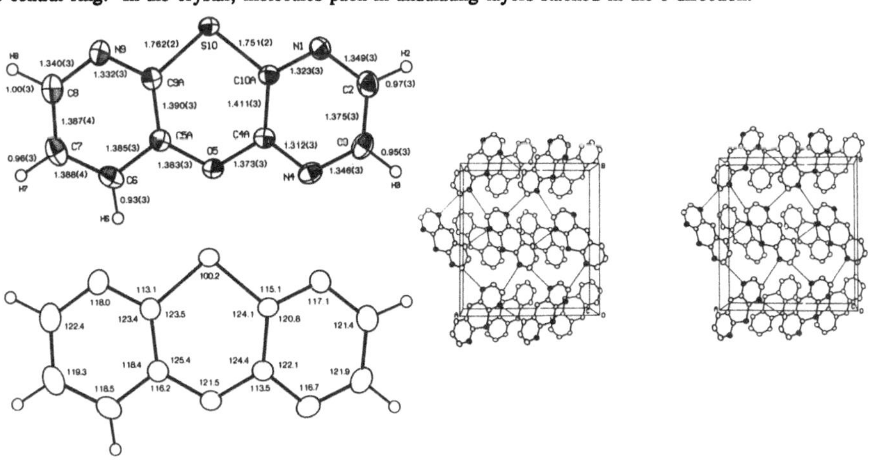

Fig. 1. $C_9H_5N_3OS$: bond lengths, angles and a stereoview of molecular packing viewed down the *c*-axis (S, N dark).

PIASELENOLE PIASELENOLIUM PENTAIODIDE
$C_{12}H_9I_5N_4Se_2$ $C_6H_5N_2Se^+,C_6H_4N_2Se,I_3^-,I_2$

A. GIEREN, T. HUBNER, V. LAMM, R. NEIDLEIN and D. DROSTE, 1985. *Z. Anorg. Allg. Chem.*, **523**, 33

44.

Monoclinic, $P2_1/n$, a = 9.320, b = 13.812, c = 17.159 Å, β = 96.11°, Z = 4. Mo radiation, R = 0.031 for 4156 reflections.

The structure contains layer shaped polyiodide aggregates built up by linear asymmetric I_3^- anions and I_2 molecules (Fig. 1). Almost linear I_3^- anions are connected by I_2 molecules to form slightly puckered layers. The piaselenole and its conjugated acid form a ribbon like array, connected as shown in Fig. 1 by N-H...N hydrogen bonds and by N...Se contacts.

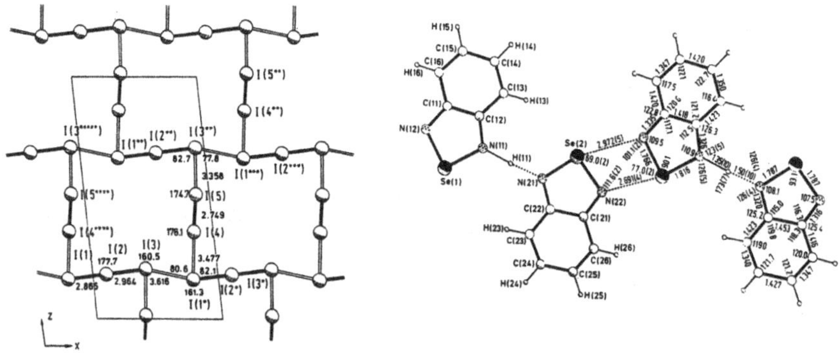

Fig. 1. The polyiodide system in $C_6H_5N_2Se^+,C_6H_4N_2Se,I_3^-,I_2$ projected onto the (010) plane and the piaselenole-piaselenolium ribbon.

15-HYDROXY-12-METHYL-5-OXA-2,8-DITHIA[9]METACYCLOPHANE
$C_{13}H_{18}O_2S_2$ (I)

15,30-DIHYDROXY-12,27-DIMETHYL-5,20-DIOXA-2,8,17,23-TETRATHIA[9.9]METACYCLOPHANE (β-form)
$C_{26}H_{36}O_4S_4$ (II)

M. CANNAS, G. MARONGIU and G. NAVARRA, 1985. *Acta Cryst.*, C41, 458-460.

I. Monoclinic, $P2_1/c$, a = 12.61, b = 7.72, c = 14.14 Å, β = 102.8°, Z = 4. Cu radiation, R = 0.083 for 1721 reflections.

II. Triclinic, $P\bar{1}$, a = 14.85, b = 9.18, c = 5.17 Å, α = 103.4, β = 98.7, γ = 90.3°, Z = 1. Cu radiation, R = 0.079 for 1427 reflections.

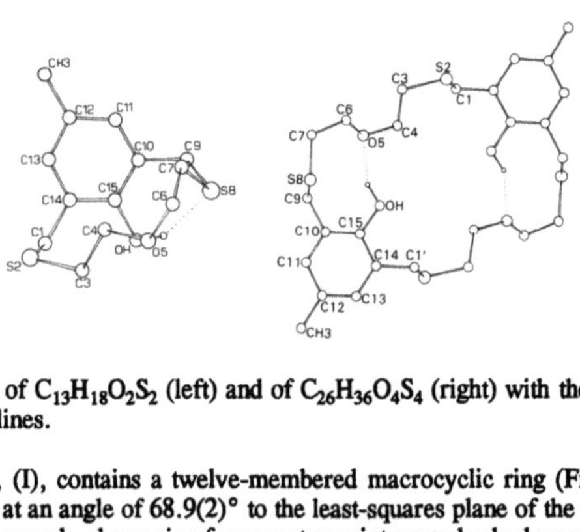

Fig. 1. Perspective views of $C_{13}H_{18}O_2S_2$ (left) and of $C_{26}H_{36}O_4S_4$ (right) with the transannular hydrogen bonds shown as broken lines.

The monomeric molecule, (I), contains a twelve-membered macrocyclic ring (Fig. 1). The plane through the 4-methylphenol atoms is tilted at an angle of 68.9(2)° to the least-squares plane of the atoms of the macrocyclic ring. The hydroxyl substituent on the nearly planar ring forms a strong intraannular hydrogen bond O(H)-H...S(8) of length 3.15(1) Å. The S(2)...O(H) separation is 3.846(9) Å and as a consequence the torsion angles around the bonds of the two O(5)-C-C-S-C-C chains in the macroring are markedly different. (II), the centrosymmetric dimer of (I), contains a 24-membered crown macrocycle in which hydroxyl groups bond to the oxa ring atoms O(5). The

O(H)...O(5) separations are 2.82(1) Å. Bond lengths and angles in both molecules are in good agreement with accepted values. Packing in both crystal structures is governed by van der Waals interactions. The structure of the β-form of the dimer is compared with that of the α-form reported in 1.

1. Structure Reports, 51B, 422.

10-(2-DIMETHYLAMINOETHYL)PHENOTHIAZINE HYDROCHLORIDE MONOHYDRATE (FENETHAZINE HYDROCHLORIDE MONOHYDRATE)

$C_{16}H_{19}ClN_2S,H_2O$ (I) $C_{16}H_{19}N_2S^+,Cl^-,H_2O$

10-(2-DIMETHYLAMINOETHYL)PHENOTHIAZINE TETRACHLOROCUPRATE(II) (FENETHAZINE TETRACHLOROCUPRATE(II))

$C_{16}H_{19}Cl_4CuN_2S$ (II) $C_{16}H_{19}N_2S^{2+},Cl_4Cu^{2-}$

A. OBATA, M. YOSHIMORI, K.-I. YAMADA and H. KAWAZURA, 1985. *Bull. Chem. Soc. Jpn.*, **58**, 437-441.

I. Monoclinic, $P2_1/a$, a = 8.686, b = 31.660, c = 6.140 Å, β = 100.44°, Z = 4. Cu radiation, R = 0.052 for 2297 reflections.

II. Triclinic, $P\bar{1}$, a = 20.433, b = 12.721, c = 7.526 Å, α = 91.31, β = 94.56, γ = 99.69°, Z = 4. Cu radiation, R = 0.050 for 4478 reflections.

The phenothiazine ring in the cation in (I) is folded (Fig. 1) with a 140.3° dihedral angle between the benzene ring planes. Formation of the cation radical in (II) results in the phenothiazine ring being flattened (dihedral angles between the benzene ring planes being 167.5 and 174.7° (for the two independent molecules of (II))). The 2-aminoethyl chain has a trans conformation in (II) and a near gauche form in (I).

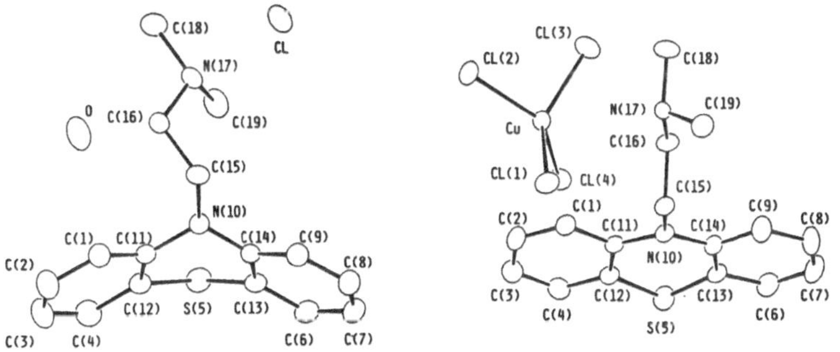

Fig. 1. Views of $C_{16}H_{19}ClN_2S,H_2O$ (I) (left) and $C_{16}H_{19}Cl_4CuN_2S$ (II) (right).

2-exo-(5,5-DIMETHYL-1,3-DIOXOCYCLOHEX-2-YLIDENE)-6,6-DIMETHYL-4,5,6,7-TETRAHYDRO-1,3-BENZOXATHIOL-4-ONE

$C_{17}H_{20}O_4S$

S.C. KOKKOU, G.A. STERGIOUDIS, A.P. BOZOPOULOS and P.J. RENTZEPERIS, 1985. *Acta Cryst.*, **C41**, 1676-1677.

Monoclinic, $P2_1/n$, a = 26.855, b = 10.515, c = 5.757 Å, β = 93.43°, Z = 4. Mo radiation, R = 0.085 for 532 reflections.

The fused five- and six-membered rings in the molecule (Fig. 1) are roughly coplanar. The dihedral angle between the least-squares planes of the oxathiole and cycloenone rings is 3.8(7)°. The oxathiole ring is planar and the cycloenone ring has an envelope conformation with C(5) as flap. The 5,5-dimethyl-1,3-dioxo-2-cyclohexanedione ring also has an envelope conformation with C(13) as flap. The dihedral angle between the mean planes of the oxathiole and the 5,5-dimethyl-1,3-dioxo-2-cyclohexanedione rings is 13.4(7)°. Atoms O(3) and O(4) are displaced on either side of the oxathiole ring by 0.20 and 0.14(4) Å respectively. Bond lengths and angles in the molecule appear to be normal and except for four slightly short C-H...O intermolecular distances, all such contacts are consistent with essentially van der Waals interactions.

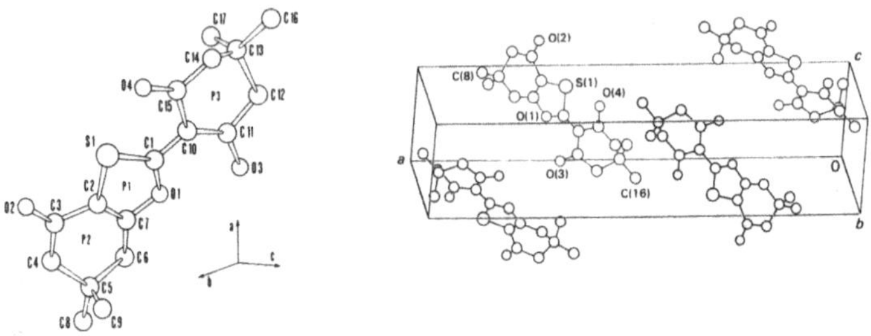

Fig. 1. View of the $C_{17}H_{20}O_4S$ molecule and a clinographic projection showing the molecular packing.

4,13-DIOXA-1,16-DITHIACYCLOTETRACOSANE-5,12,17,24-TETRADIONE
$C_{20}H_{32}O_6S_2$ (I)

catena(μ_2-4,13-DIOXA-1,16-DITHIACYCLOTETRACOSA-5,12,17,24-TETRADIONE-S,S')(μ_2-NITRATO-O,O,-O')SILVER
$(C_{16}H_{24}AgNO_9S_2)_n$ (II)

Y. TOR, J. LIBMAN, F. FROLOW, H.E. GOTTLIEB, R. LAZAR and A. SHANZER, 1985. *J. Org. Chem.*, **50**, 5476-5480.

I. Orthorhombic, $P2_12_12_1$, a = 54.041, b = 7.175, c = 5.602 Å, Z = 4. Mo radiation, R = 0.0369 for 2331 reflections (at 83 K).

II. Triclinic, $P\bar{1}$, a = 11.746, b = 8.931, c = 20.041 Å, α = 90.90, β = 97.38, γ = 88.67°, Z = 4. Mo radiation, R = 0.073 for 4449 reflections (at 83 K).

The X-ray crystal structure analyses established the shapes of the molecules (Fig. 1).

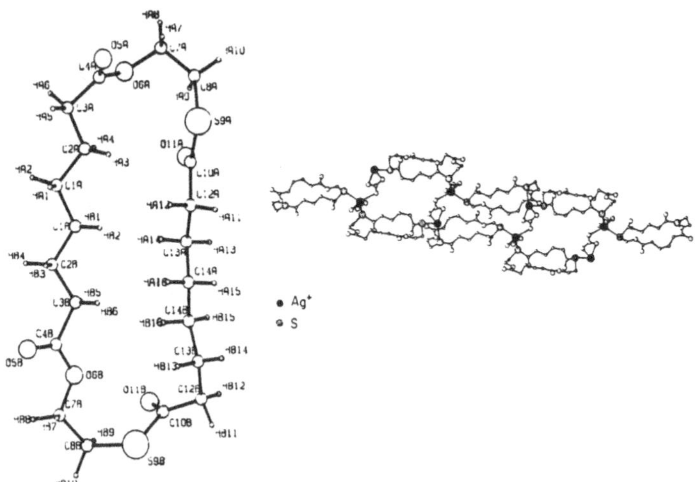

Fig. 1. Drawings of $C_{20}H_{32}O_6S_2$ (I) (left) and the crystal structure of $(C_{16}H_{24}AgNO_9S_2)_n$ (II) (right).

TETRAPHENYLPHOSPHONIUM PENTABORATETRATHIAADAMANTANE
$C_{25}H_{32}B_5PS_4$ $C_{24}H_{20}P^+,CH_{12}B_5S_4^-$

K. WOLFER, H.-D. HAUSEN and H. BINDER, 1985. *Z. Naturforsch.*, **B40**, 235-239.

Tetragonal, $P4/n$, a = 13.517, c = 7.900 Å, Z = 2. Mo radiation, R = 0.043 for 812 reflections.

Views of the cation and anion are in Fig. 1. The P atom lies on a site with $\bar{4}$ symmetry, and in the anion, C(7) and B(2) lie on a twofold axis. The mean B-S distance is 1.899 Å, with C-S 1.807(4) Å.

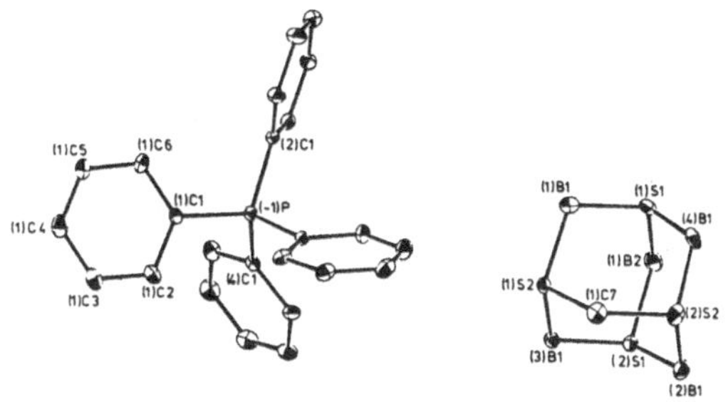

Fig. 1. The $C_{24}H_{20}P^+$ and $CH_{12}B_5S_4^-$ ions.

FLUORENE-9-SPIRO-6'-(3',6'-DIHYDRO-3',3',5'-TRIPHENYL-1',2',4'-OXATHIAZINE)
$C_{33}H_{23}NOS$

J.M.M. SMITS, P.T. BEURSKENS, B.F. BONINI and B. ZWANENBURG, 1985. *Acta Cryst.*, **C41**, 1084-1086.

Monoclinic, $P2_1$, a = 7.799, b = 10.556, c = 15.291 Å, β = 103.59°, Z = 2. Mo radiation, R = 0.099 for 2006 reflections.

The molecule (Fig. 1) is a δ-sultene- an unusual cyclic sulphenate. The S-O bond length in the six-membered ring is 1.66(1) Å and is in good agreement with that found in the five-membered ring in γ-sultene (1). Owing to the poor quality of the data due to the effect of x-radiation on the crystals, idealized phenyl rings were incorporated into the refinement process. The other bond lengths and angles appear normal and no unusually short intermolecular contacts are reported.

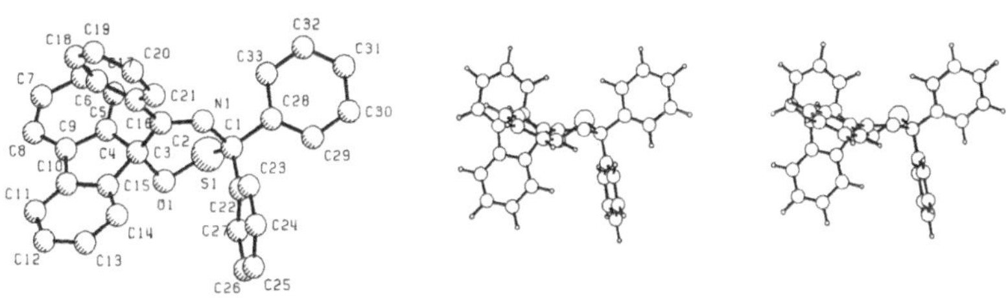

Fig. 1. $C_{33}H_{23}NOS$: a view of the molecule showing the atomic numbering and a stereoview of the molecule.

1. E. SCHAUMANN, J. EHLERS and U. BEHRENS, 1978. Angew. Chem. **90**, 480.

2,4,5,6-(1H,3H)-PYRIMIDINETETRONE (ALLOXAN)
$C_4H_2N_2O_4$

S. SWAMINATHAN, B.M. CRAVEN and R.K. McMULLAN, 1985. *Acta Cryst.*, **B41**, 113-122.

I. Tetragonal, $P4_12_12$, a = 5.841, c = 13.853 Å, Z = 4. Neutron radiation, R = 0.029 for 609 reflections (at 42 K).

II. Tetragonal, $P4_12_12$, a = 5.850, c = 13.913 Å, Z = 4. Neutron radiation, R = 0.024 for 614 reflections (at 123 K).

III. Tetragonal, $P4_12_12$, a = 5.850, c = 13.915 Å, Z = 4. Mo radiation, R = 0.043 for 1097 reflections (at 123 K).

Alloxan (Fig. 1) has an unusually high crystal density (1.93 g cm^{-3} at 295 K), apparently due to short intermolecular O=C...O distances (2.73 Å at 123 K). There is no H-bonding except for a weak bifurcated interaction with long H...O distances (2.32, 2.35 Å) at 123 K. Bond lengths and angles from neutron diffraction (corrected for thermal motion) have e.s.d.'s less than 0.002 Å and 0.1°. They indicate that the effects of conjugation in alloxan are small. With fixed values for all nuclear positional and thermal parameters as determined by neutron diffraction, the X-ray data were used to determine the charge-density distribution. The resulting electron population parameters were also used to map the electrostatic potential for an isolated molecule and for a selected group of molecules removed from the lattice. Electrostatic properties show the bifurcated H bond to be weakly attractive. In the carbonyl groups, electron density at the C atom is depleted from both sides of the molecular plane, thus deshielding the atomic nucleus. Although the shortest intermolecular C...O distances are formed with O atoms carrying a small negative charge, -0.11(3) e, these interactions are weak. It appears that the van der Waals radius for carbonyl C atoms should be smaller than for aromatic C atoms by at least 0.2 Å.

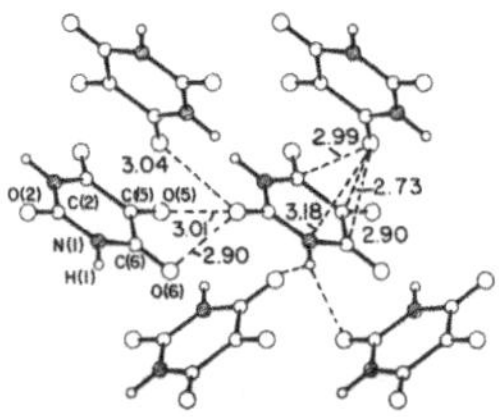

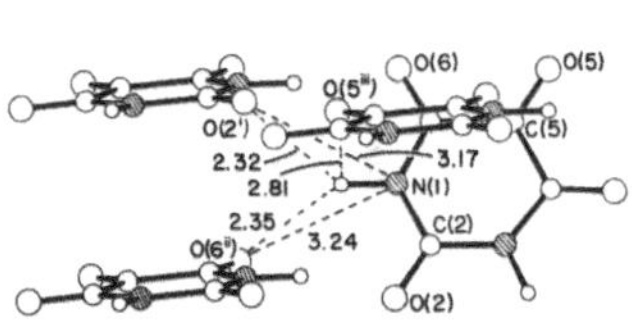

Fig. 1. Alloxan seen in its b-axis projection (left) and the crystal environment of the N-H group (right).

AMMONIUM 5-NITROBARBITURATE

$C_4H_6N_4O_5$ $C_4H_2N_3O_5^-,H_4N^+$

O. SIMONSEN, 1985. *Acta Cryst.*, C**41**, 1258-1260.

Triclinic, $P\bar{1}$, a = 4.918, b = 7.694, c = 9.819 Å, α = 104.91, β = 92.66, γ = 105.08°, Z = 2. Mo radiation, R = 0.038 for 1596 reflections.

The anion (Fig. 1) is approximately planar with only small deviations from a non-crystallographic mirror plane through O(2),C(2),C(5),N(5) and normal to the molecular plane. Bond distances and angles (σ 0.002 Å, 0.2°) have acceptable values. The anions are connected in infinite chains by pairs of N-H...O hydrogen bonds grouped around centres of symmetry (N...O 2.825(2)-2.831(2) Å). The distances from the ammonium N atom to the surrounding O atoms range from 2.935(2) to 3.068(2) Å.

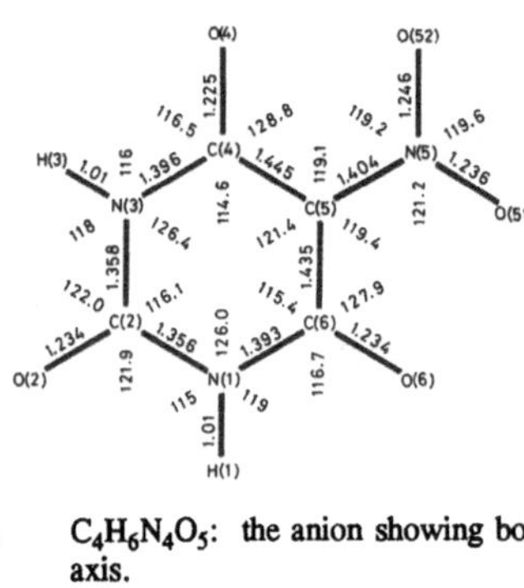

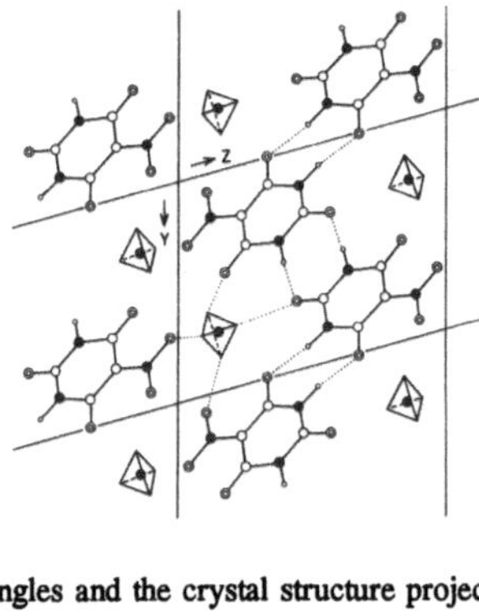

Fig. 1. $C_4H_6N_4O_5$: the anion showing bond lengths and angles and the crystal structure projected along the *x*-axis.

4-AMINO-2-THIOXO-6-PYRIMIDINONE MONOHYDRATE (6-AMINO-2-THIOXO-2,3-DIHYDRO-4(1H)--PYRIMIDINONE MONOHYDRATE)

$C_4H_5N_3OS,H_2O$

E.S. RAPER, R.E. OUGHTRED and I.W. NOWELL, 1985. *Acta Cryst.*, **C41**, 758-760.

Monoclinic, *C2/c*, a = 14.892, b = 7.593, c = 12.991 Å, β = 113.51°, Z = 8. Mo radiation, R = 0.037 for 1117 reflections.

In the molecule (Fig. 1) the heterocyclic ring is nearly planar with a maximum deviation of 0.015(2) Å for atom N(3). The average C-C and C-N bond lengths in the ring are 1.348 and 1.356 Å respectively. The C(2)-S(2) and C(6)-O(6) bond lengths are 1.671(1) and 1.252(2) Å respectively. The asymmetric units are linked by hydrogen bonds involving the water molecule, the ketonic O atom and the imido-N atom forming ribbons parallel to the c-axis. The O-H...O and N-H...O bond lengths lie in the range 2.730 to 2.850(2) Å.

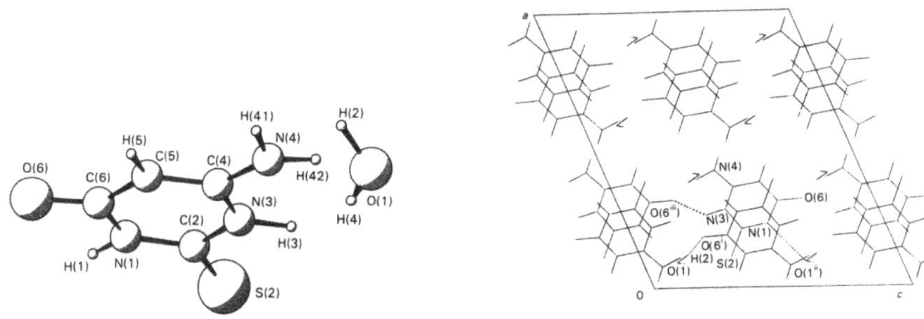

Fig. 1 $C_4H_5N_3OS,H_2O$: a view of the asymmetric unit and the unit cell contents viewed down the b-axis in which the unique set on hydrogen bonds is shown as dotted lines.

1,3,5,5-TETRANITROHEXAHYDROPYRIMIDINE
$C_4H_6N_6O_8$

Y. OYUMI, T.B. BRILL, A.L. RHEINGOLD and T.M. HALLER, 1985. *J. Phys. Chem.*, **89**, 4317-4324.

Orthorhombic, $P2_12_12_1$, a = 11.130, b = 11.227, c = 15.705 Å, Z = 8. Mo radiation, R = 0.0564 for 1685 reflections.

There are two independent molecules and both have some nitro group disorder (Fig. 1). The pyrimidine rings adopt chair conformations and the NNO_2 groups are axial. The N-N bond distances (1.37-1.40(1) Å) are indicative of partial π-electron delocalization occurring over the entire NNO_2 unit.

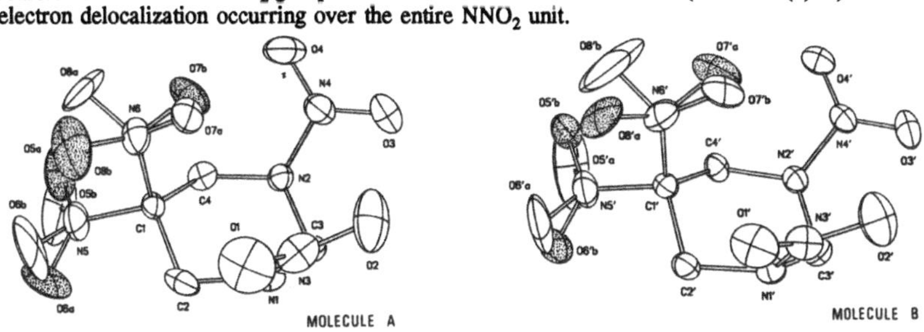

Fig. 1. The two $C_4H_6N_6O_8$ molecules showing the disordered nitro groups.

5-METHYL-2-THIOCYTOSINE HEMIHYDRATE
$C_5H_7N_3S,0.5(H_2O)$

V. RAVICHANDRAN, K.K. CHACKO, M.N. PONNUSWAMY and J. TROTTER, 1985. *Acta Cryst.*, **C41**, 1470-1473.

Orthorhombic, $P2_12_12$, a = 9.923, b = 12.969, c = 5.409 Å, Z = 4. Mo radiation, R = 0.049 for 1565

reflections.

Bond lengths and angles in the molecule have acceptable values. The S atom acts as an acceptor of N-H...S and OW-H...S hydrogen bonds (Fig. 1). The water O atom sits on a twofold axis along *c* and acts as a donor of two symmetry-related OW-H...S hydrogen bonds and at the same time acts as an acceptor of two symmetry-related N-H...OW hydrogen bonds. The environment around OW is tetrahedral.

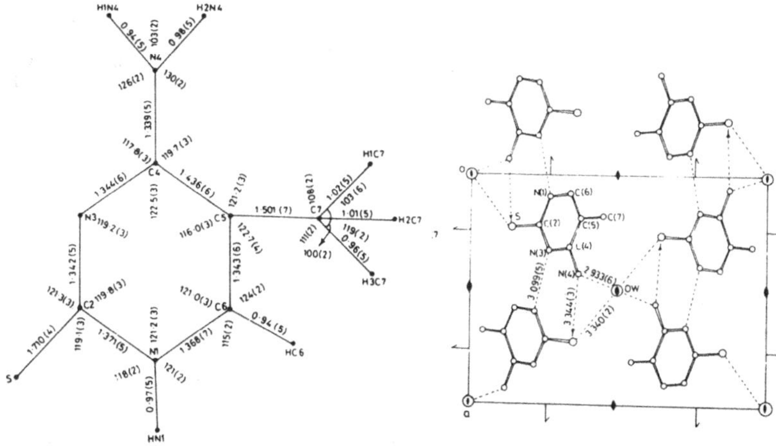

Fig. 1. $C_5H_7N_3S,0.5(H_2O)$: bond lengths and angles and the unit cell viewed down *c*.

1-(2-HYDROXYETHYL)URACIL
$C_6H_8N_2O_3$

M. SHIBATA, A. TAKENAKA and Y. SASADA, 1985. *Acta Cryst.*, C41, 1499-1500.

Orthorhombic, *Pccn*, a = 10.993, b = 19.032, c = 6.625 Å, Z = 8. Cu radiation, R = 0.068 for 962 reflections.

The uracil moieties (Fig. 1) are hydrogen bonded (N(3)...O(2)) in pairs around the twofold axis. The base pairs are stacked along the *c*-axis, the stacking being stabilised by the hydrogen bonds between O(4) and the hydroxyl group. The pyrimidine ring is planar and bond lengths and angles have acceptable values.

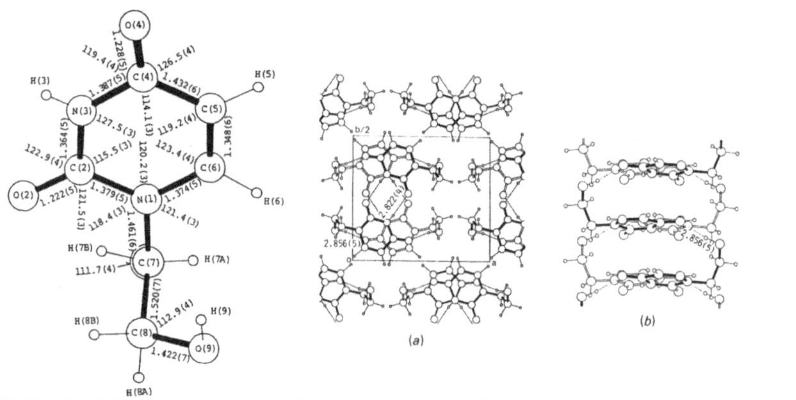

Fig. 1. $C_6H_8N_2O_3$: bond distances and angles, the crystal structure projected along *c* (centre) and the stacking of uracil pairs.

1-(2-HYDROXYETHYL)CYTOSINE

$C_6H_9N_3O_2$

M. SHIBATA, A. TAKENAKA, Y. SASADA and M. OHKI, 1985. *Acta Cryst.*, **C41**, 1354-1356.

Orthorhombic, *Pbca*, a = 12.376, b = 7.554, c = 15.407 Å, Z = 8. Cu radiation, R = 0.049 for 1109 reflections.

The pyrimidine ring in the molecule (Fig. 1) is planar and the bond lengths and angles in the cytosine moiety are in good agreement with those found in similar compounds. The torsion angles C(2)-N(1)-C(7)-C(8) and N(1)-C(7)-C(8)-O(9) are 74.5(2) and 68.3(2)° respectively. The molecules form centrosymmetrically hydrogen-bonded dimers through N(4)-H...N(3) bonds of length 3.006(2) Å. The N(4) amino group is also hydrogen bonded to the hydroxyl group O(9) of a neighbouring molecule thus forming sheets parallel to (010). Atom O(9) also forms a donor hydrogen bond with O(2). The N(4)...O(9) and O(9)...O(2) separations are 2.939(2) and 2.754(2) Å respectively.

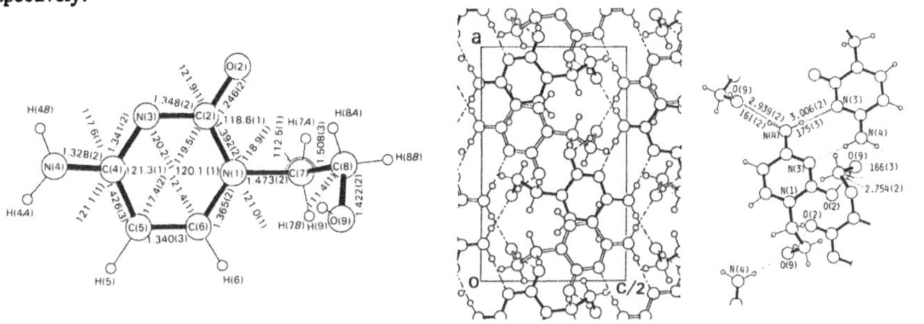

Fig. 1. $C_6H_9N_3O_2$: bond lengths and angles in the molecule, a projection of the crystal structure along the *b*-axis and geometrical details of the intermolecular hydrogen-bonding scheme.

9-(2-HYDROXYETHYL)GUANINE
$C_7H_9N_5O_2$

M. SHIBATA, A. TAKENAKA and Y. SASADA, 1985. *Acta Cryst.*, **C41**, 1501-1503.

Triclinic, $P\bar{1}$, a = 12.060, b = 13.832, c = 4.993 Å, α = 89.86, β = 92.66, γ = 75.11°, Z = 4. Cu radiation, R = 0.051 for 1633 reflections.

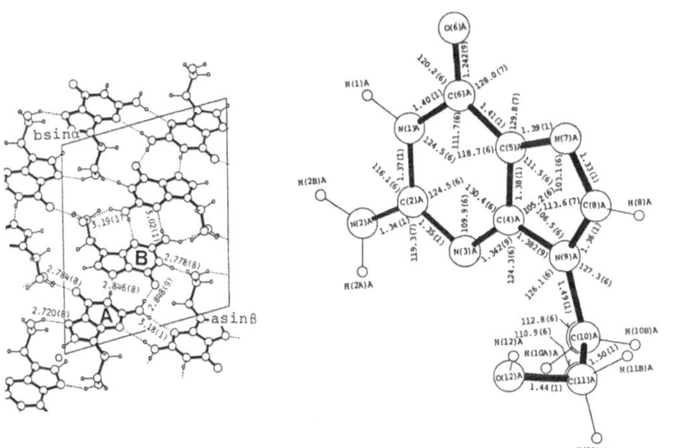

Fig. 1. $C_7H_9N_5O_2$: the crystal structure projected down *c* and bond distances and angles for molecule A.

The two independent molecules (A, B) in the asymmetric unit (Fig. 1) are self-associated into base pairs through N(2)-H...N(3) hydrogen bonds around inversion centres. These base-pairs are linked by N(1)-H...N(7) and N(2)

H...O(6) hydrogen bonds to form ribbons along *b*. The hydroxyl groups interact with guanine moieties at N(7), O(6), N(1) and N(2). These hydrogen bonds join the molecular ribbons to form a sheet parallel to the (013) plane. There are no significant differences in the dimensions of the two molecules. The purine ring is planar in both.

1-(2-HYDROXYETHYL)THYMINE MONOHYDRATE
$C_7H_{10}N_2O_3,H_2O$

M. SHIBATA, A. TAKENAKA, Y. SASADA and M. OHKI, 1985. *Acta Cryst.*, C41, 1356-1358.

Monoclinic, *C2/c*, a = 12.210, b = 9.266, c = 16.338 Å, β = 105.39°, Z = 8. Cu radiation, R = 0.045 for 1328 reflections.

The pyrimidine ring in the molecule (Fig. 1) is planar and the dimensions in the thymine moiety agree well with expected values. Torsion angles are 273.9(2) and 296.7(2)° for C(2)-N(1)-C(7)-C(8) and N(1)-C(7)-C(8)-O(9) respectively. The hydroxyl group at O(9) is hydrogen-bonded to O(4) of an adjacent thymine to form centrosymmetric dimers. These dimers stack to form a column along [110]. There is no hydrogen bond between the thymine moieties. Each is surrounded by three water molecules which form hydrogen bonds to O(2), N(3) and O(4) of lengths 2.804(2), 2.801(2) and 2.853(2) Å respectively.

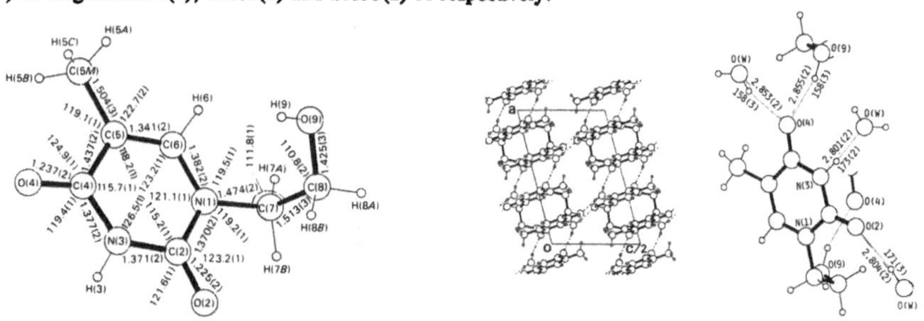

Fig. 1. $C_7H_{10}N_2O_3,H_2O$: molecular geometry, a projection of the crystal structure along the *b*-axis and geometric details of the hydrogen-bonding scheme.

3,7-DIHYDRO-3-PROPYL-1H-PURINE-2,6-DIONE (ENPROFYLLINE)
$C_8H_{10}N_4O_2$

C. STALHANDSKE, H.J.B. SLOT and P.T. BEURSKENS, 1985. *Acta Cryst.*, C41, 1517-1520.

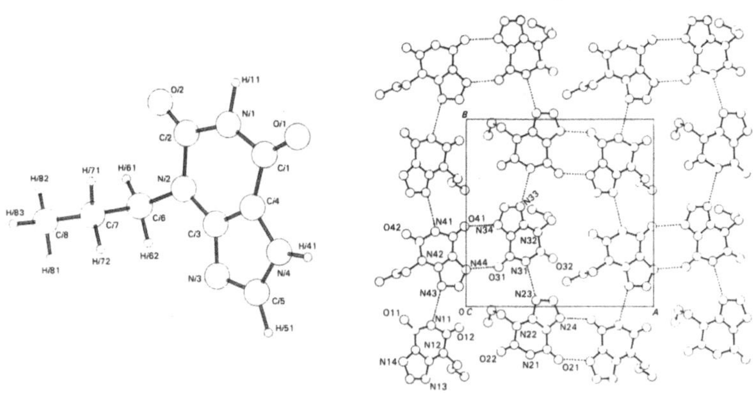

Fig. 1. $C_8H_{10}N_4O_2$: the molecular structure (i = 1-4) and one molecular layer parallel to the ab plane showing the hydrogen-bonding pattern.

Monoclinic, *P*2_1*/c*, a = 15.11, b = 13.606, c = 19.877 Å, β = 111.92°, Z = 16. Cu radiation, R = 0.079 for 1671 reflections.

The four crystallographically independent molecules in the asymmetric unit show no significant differences. The molecules (Fig. 1) are planar except for the-CH_2CH_3 moieties of the propyl groups which are disordered. In the crystal, the molecules are stacked in layers, two adjacent layers being separated by disordered ethyl groups. Molecules within a layer are linked by hydrogen bonds (N...N 2.94(1)-2.96(1), N...O 2.69(1)-2.71(1) Å).

N-FORMYL-1,2,3,4-TETRAHYDRO-2-OXOPYRROLO[1,2-a]PYRIMIDINE-8-CARBOXAMIDE
$C_9H_9N_3O_3$

H. ROSEMEYER, K. KAISER, F. SEELA, V. ZABEL and W. SAENGER, 1985. *Helv. Chim. Acta,* **68**, 534-544.

Orthorhombic, *Pbca*, a = 13.715, b = 18.717, c = 6.956 Å, Z = 8. Mo radiation, R = 0.034 for 965 reflections.

The crystal structure analysis established the stereochemistry of the molecule (Fig. 1).

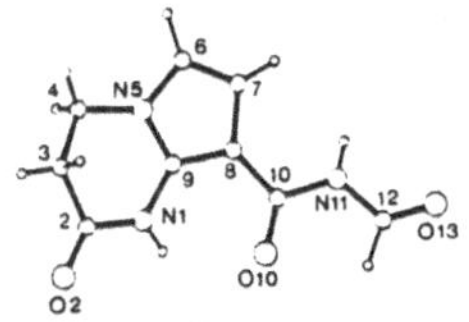

Fig. 1. ORTEP drawing of a molecule of $C_9H_9N_3O_3$.

1,9-DIMETHYL-6-ACETYLIMINO-8-OXOPURINE
$C_9H_{11}N_5O_2$

G. CIMINO, A. DE GIULIO, S. DE ROSA, S. DE STEFANO, R. PULITI, C.A. MATTIA and L. MAZZARELLA, 1985. *J. Nat. Prod.*, **48**, 523-528.

Monoclinic, $P2_1/c$, a = 9.797, b = 29.026, c = 7.781 Å, β = 106.22°, Z = 8. Cu radiation, R = 0.088 for 1421 reflections.

The two independent molecules in the asymmetric unit have similar geometries (Fig. 1) with an unusual oxo-function at position 8. Only N(7) is protonated. The C(2)-N(3) (1.310, 1.294(7) Å) and C(6)-N(10) (1.328, 1.344(6) Å) are consistent with the formulation shown in Fig. 1.

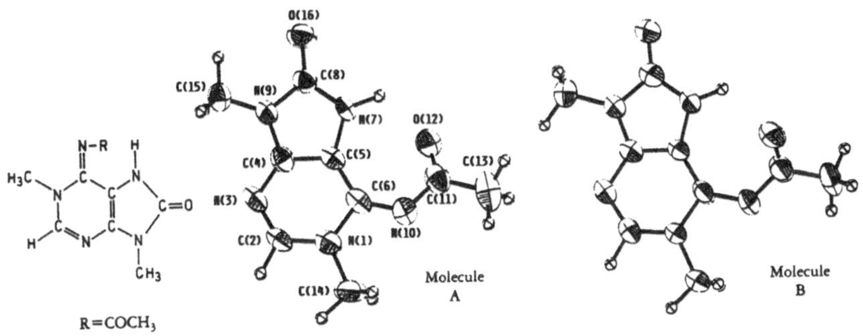

Fig. 1. The $C_9H_{11}N_5O_2$ structure and the two independent molecules.

6-N,9(N)-DIACETYL-1(N)-METHYLISOGUANINE
$C_{10}H_{11}N_5O_3$

R.J. NACHMAN, R.Y. WONG, W.F. HADDON and R.E. LUNDIN, 1985. *J. Chem. Soc. Perkin Trans. 1,* 1315-1321.

Triclinic, $P\bar{1}$, a = 5.403, b = 10.292, c = 10.729 Å, α = 74.27, β = 75.11, γ = 84.45°, Z = 2. Cu radiation,

R = 0.081 for 1229 reflections.

The isoguanine moiety is significantly non-planar; the six-membered ring is puckered so that C(12) and O(17) are appreciably out-of-plane (-1.416(7) and 0.919(6) Å respectively) (Fig. 1). There is folding along the C(4)-C(5) bond with a dihedral angle of 2.0(1)°. The acetyl group on the imidazole ring assumes a position in which its carbonyl oxygen O(18) is directed towards the hydrogen atom of the imino nitrogen N(3) on the pyrimidine ring, thus forming a strong hydrogen bond between them (N(3)...O(18) 2.777(5) Å). The imino nitrogen group, N(3)-H, is also involved in intramolecular hydrogen bonding.

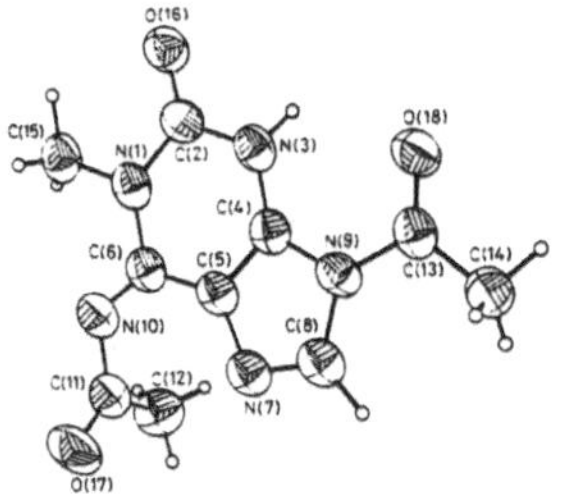

Fig. 1. Perspective view of a molecule of $C_{10}H_{11}N_5O_3$.

5-HYDROXY-2-METHYL-2-(2-PYRIDINYL)HEXAHYDRO-1,3-PYRIMIDINE
$C_{10}H_{15}N_3O$

D.E. FENTON, R. MOODY, U. CASELLATO, P.A. VIGATO and R. GRAZIANI, 1985. *J. Heterocycl. Chem.*, 22, 1639-1642.

Monoclinic, $P2_1/c$, a = 10.385, b = 11.171, c = 17.415 Å, β = 93.05°, Z = 8. Mo radiation, R = 0.055 for 2070 reflections.

There are two similar independent molecules in the asymmetric unit. The pyrimidine rings are in chair form and have equatorial OH and CH_3 groups and axial pyridine substituents. The packing of the molecules is controlled by two independent hydrogen-bond sequences. Molecular dimensions are as anticipated.

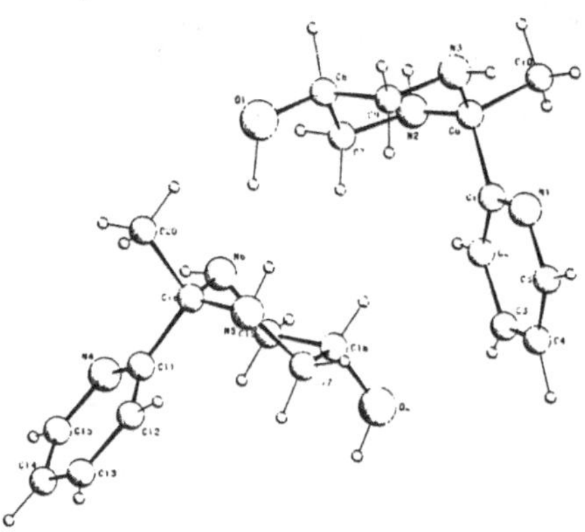

Fig. 1. The two $C_{10}H_{15}N_3O$ molecules in the asymmetric unit.

N[1]-(2,6-DIMETHYL-4-PYRIMIDINYL)SULPHANILAMIDE (SULPHISOMIDINE)
$C_{12}H_{14}N_4O_2S$

T.P. SINGH, R.K. TIWARI and M. SINGH, 1985. *Acta Cryst.*, C41, 752-755.

Orthorhombic, $Pca2_1$, a = 12.614, b = 11.278, c = 9.374 Å, Z = 4. Mo radiation, R = 0.036 for 2006 reflections.

The benzene ring in the molecule (Fig. 1) is planar while the pyrimidine ring is slightly folded about the C(12)-N(15) axis. The dihedral angle between the least-squares planes of these two rings is 78.8(2)°. Bond lengths in the benzene ring are normal and those in the hetero-ring lie in the range 1.323(4) to 1.414(4) Å. The molecule has a gauche conformation with a torsion angle of 59.6(2)° about the S(8)-N(11) bond. The bond distances around the S atom are S(8)-C(1) 1.755(3), S(8)-O(9) 1.430(2), S(8)-O(10) 1.541(2) and S(8)-N(11) 1.630(3) Å. The amino nitrogen atom N(7) is a proton donor in a hydrogen bond to O(10) of a symmetry related molecule with a N...O separation of 3.131(5) Å. The other hydrogen bond is N(11)-H...N(15) of length 2.948(3) Å. Other contacts are van der Waals interactions though the O(9)...N(15) separation is very short (2.939(4) Å).

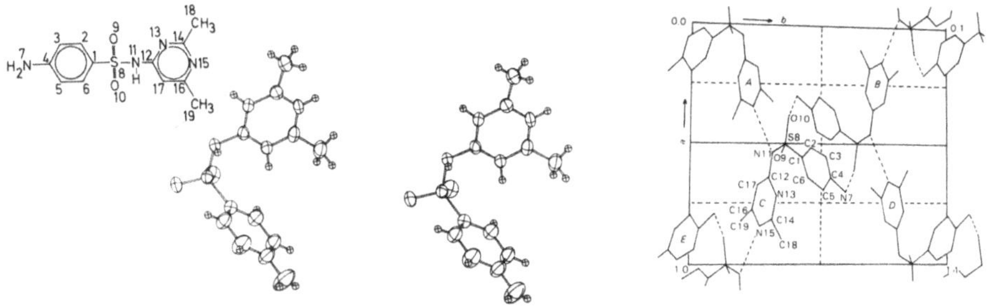

Fig. 1. $C_{12}H_{14}N_4O_2S$: a view of the molecule and the molecular packing viewed along the c-axis with hydrogen bonds shown as broken lines.

N[1]-(6-METHOXY-2-METHYL-4-PYRIMIDINYL)SULPHANILAMIDE (SULPHAMETHOMIDINE)
$C_{12}H_{14}N_4O_3S$ (I)

N[1]-(6-METHOXY-2-METHYL-4-PYRIMIDINIUMYL)SULPHANILAMIDE MONOHYDRATE (SULPHAMETHOMIDINE MONOHYDRATE)
$C_{12}H_{14}N_4O_3S,H_2O$ (II)

J. RAMBAUD, L. MAURY, B. PAUVERT, G. BERGE, M. AUDRAN, Y. LASSERRE and J.-P. DECLERCQ, 1985. *Acta Cryst.*, C41, 775-778.

I. Triclinic, $P\bar{1}$, a = 14.180, b = 11.643, c = 8.619 Å, α = 91.08, β = 101.39, γ = 90.62°, Z = 4. Mo radiation, R = 0.0514 for 3221 reflections.

II. Orthorhombic, $P2_12_12_1$, a = 7.525, b = 16.924, c = 11.785 Å, Z = 4. Mo radiation, R = 0.0405 for 1108 reflections.

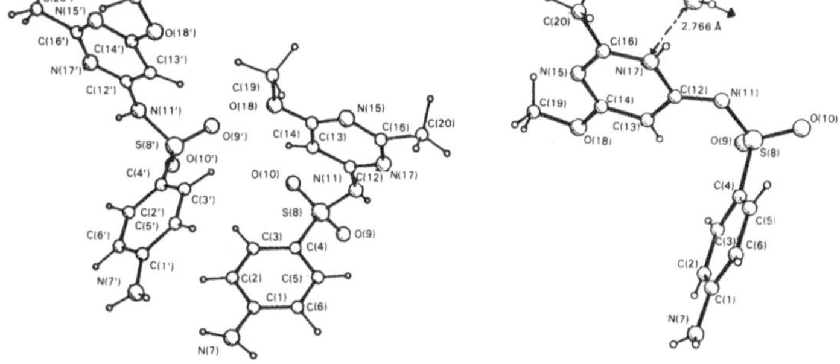

Fig. 1. Views of the independent $C_{12}H_{14}N_4O_3S$ molecules and (right) a view of the asymmetric unit in $C_{12}H_{14}N_4O_3S,H_2O$.

The asymmetric unit of (I) contains two crystallographically independent molecules having very similar bond lengths and angles. There are significant differences in some of the geometrical parameters of (II) compared with the average values in the unsolvated molecule. The respective bond lengths in (I) and (II) are: S(8)-N(11) 1.644(3), 1.582(4), C(12)-N(11) 1.400(6), 1.339(6) and C(12)-N(17) 1.342(5), 1.375(6) Å. Angles S(8)-N(11)-C(12) and

C(12)-N(17)-C(16) are 125.9(3), 122.1(3) and 116.1(3), 124.3(4)° in (I) and (II) respectively. Intermolecular distances in (I) are van der Waals contacts. In (II) there is a strong hydrogen bond N(17)-H...O(water) of length 2.766 Å.

9-(2-CARBAMOYLETHYL)GUANINE 1-METHYLCYTOSINE

$C_{13}H_{17}N_9O_4$ $C_8H_{10}N_6O_2,C_5H_7N_3O$

S. FUJITA, A. TAKENAKA and Y. SASADA, 1985. *Biochemistry*, **24**, 508-512.

Monoclinic, $P2_1/c$, a = 8.540, b = 12.693, c = 14.249 Å, β = 94.02°, Z = 4. Cu radiation, R = 0.100 for 1035 reflections.

Molecular dimensions are in Fig. 1. The bases form a Watson-Crick pair with N...O 2.79-2.92(1) Å.

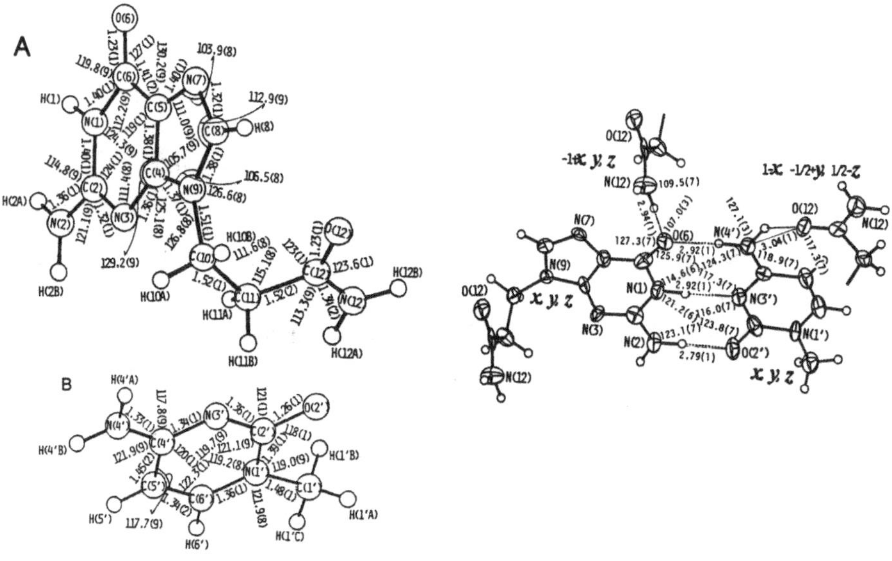

Fig. 1. The $C_8H_{10}N_6O_2$ and $C_5H_7N_3O$ molecules and their bending mode.

6-FURFURYLPURINE PICRATE (KINETIN PICRATE)

$C_{16}H_{12}N_8O_8$ $C_{10}H_{10}N_5O^+,C_6H_2N_3O_7^-$

M. SORIANO-GARCIA, R.A. TOSCANO and G. ESPINOSA, 1985. *J. Cryst. Spect. Res.*, **15**, 651-662.

Monoclinic, $P2_1/n$, a = 4.995, b = 13.931, c = 26.065 Å, β = 90.99°, Z = 4. Cu radiation, R = 0.051 for 2040 reflections.

The adenine moiety is shown to be protonated at N(3) and N(7). Ion dimensions are consistent with a major contribution of the resonance form shown in Fig. 1. The adenine moiety is planar; the furfuryl ring has atoms C(12) (0.023 Å) and C(13) (-0.020 Å) from the mean plane. The angle between the adenine and furfuryl planes is 66.8(7)°. The stacking interactions of the ions involve charge-transfer and dipole-dipole interactions between parallel and adjacent adenine moieties of kinetin and picrate ions. For previous work on kinetin see (1).

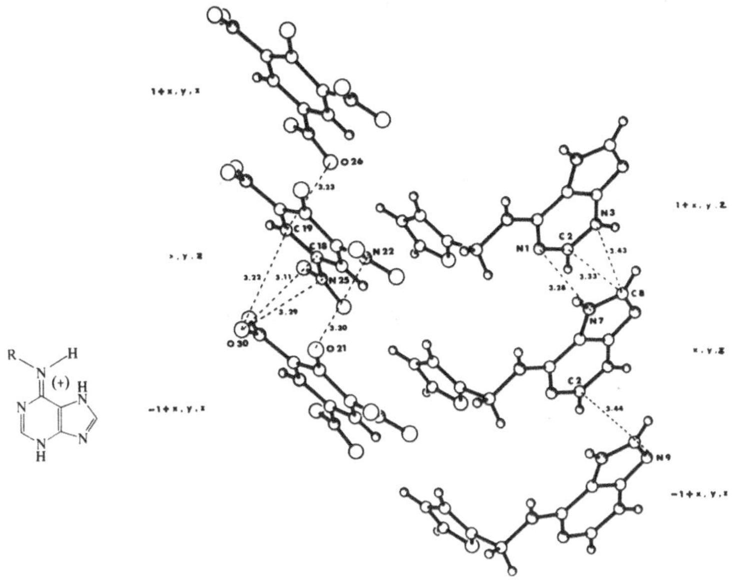

Fig. 1. The predominant resonance form for the kinetin ion and a view of the ion stacking.

1. Structure Reports, 43B, 520.

2,4-DIAMINO-5-(3,4,5-TRIMETHOXYBENZYL)PYRIMIDINIUM BENZOATE BENZOIC ACID SOLVATE (TRIMETHOPRIM BENZOATE BENZOIC ACID SOLVATE)

$C_{21}H_{24}N_4O_5,C_7H_6O_2$ $C_{14}H_{19}N_4O_3^+,C_7H_5O_2^-,C_7H_6O_2$

G.P. BETTINETTI, F. GIORDANO, A.LA MANNA, G. GIUSEPPETTI and C. TADINI, 1985. *Acta Cryst.*, **C41**, 1249-1253.

Triclinic, $P\bar{1}$, a = 10.192, b = 14.595, c = 9.457 Å, α = 97.41, β = 89.69, γ = 104.57°, Z = 2. Mo radiation, R = 0.037 for 2881 reflections.

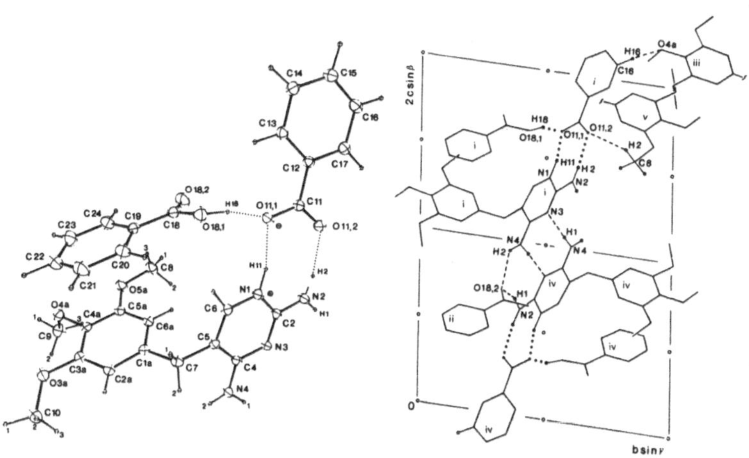

Fig. 1. $C_{21}H_{24}N_4O_5,C_7H_6O_2$: the molecular structure and packing.

The trimethoprim cation (Fig. 1) is in a twist conformation and is linked to one benzoate anion via two N-H...O hydrogen bonds (N...O 2.673(2), 2.870(2) Å). The other benzoic acid molecule is linked through a short O-H...O hydrogen bond (2.531(2) Å). The phenyl ring of this molecule is almost superimposed on the phenyl moiety of the cation, the respective planes being almost parallel. Cyclic centrosymmetrical N-H...N and linear N-H...O hydrogen bonds are responsible for molecular packing.

THIAMINE-INDOLE-3-PROPIONATE PERCHLORATE HYDROPERCHLORATE METHANOL SOLVATE

$C_{23}H_{25}ClN_5O_3S,HClO_4.CH_4O$ (I) $C_{23}H_{27}N_5O_2S^+,ClO_4^-,HClO_4,CH_4O$

5-(2-(INDOL-3-YL-PROPIONYLOXY)ETHYL)-3,4-DIMETHYLTHIAZOLIUM IODIDE

$C_{18}H_{21}IN_2O_2S$ (II) $C_{18}H_{21}N_2O_2S^+,I^-$

4-AMINO-5-((INDOL-3-YL-PROPIONYLAMINO)METHYL)-2-METHYLPYRIMIDINE

$C_{17}H_{19}N_5O$ (III)

T. ISHIDA, M. MATSUI, M. INOUE, H. HIRANO, M. YAMASHITA, K. SUGIYAMA, M. SUGIURA and K.-I. TOMITA, 1985. *J. Am. Chem. Soc.*, **107**, 3305-3314.

I. Triclinic, $P\bar{1}$, a = 14.960, b = 14.170, c = 7.436 Å, α = 101.79, β = 102.36, γ = 94.15°, Z = 2. Cu radiation, R = 0.100 for 4477 reflections.

II. Monoclinic, $P2_1/c$, a = 7.663, b = 10.870, c = 23.100 Å, β = 105.20°, Z = 4. Cu radiation, R = 0.065 for 2825 reflections.

III. Monoclinic, $P2_1/c$, a = 4.945, b = 22.416, c = 14.372 Å, β = 95.47°, Z = 4. Cu radiation, R = 0.092 for 2045 reflections.

In all three structures (Fig. 1) the thiamin moiety takes the F conformation. In (I) there is a cis conformation at the C(11)I-C(12)I bond of two trans zigzag chains. Both (II) and (III) have open conformations with nearly extended chains connecting two aromatic rings, (III) being stabilised by an intramolecular hydrogen bond between N(4'α) and O(12)I atoms as follows: N(4'α)...O(12)I 3.132(4), H(4'α)...O(12)I 2.27(3) Å, N(4'α)-H(4'α)...O(12)I 153(3)°. In (I) and (II) the $\chi^{2,1}$ torsion angle defining the indole side chain shows a very unusual cis conformation. In all three compounds crystal packing is stabilised by H-bonding, involving in (I) the perchlorate ion and solvent methanol, and by stacking interactions between indole and pyridinium rings in (I), indole and thiazolium rings in (II) and indole and pyridinium rings in (III). In (I) these interactions lead to the formation of a dimer via stacking of rings and H-bonding between centrosymmetrically related molecules. This stacking is at distances of 3.338-3.759 Å, with dihedral angles between the aromatic rings of 3.4(3) to 11.7(1)°. The stacking pair of (I) has the shortest and most parallel stacking, at less than the van der Waals distance, and a π-π charge-transfer interaction is likely to exist. Bond distances and angles in all three compounds are unexceptional and the ring planes are close to planar but with significant deviations, especially the C(2'α) atom in the pyrimidine or pyrimidinium group which deviates by 0.109(9) Å in (I) and 0.116(5) Å in (III).

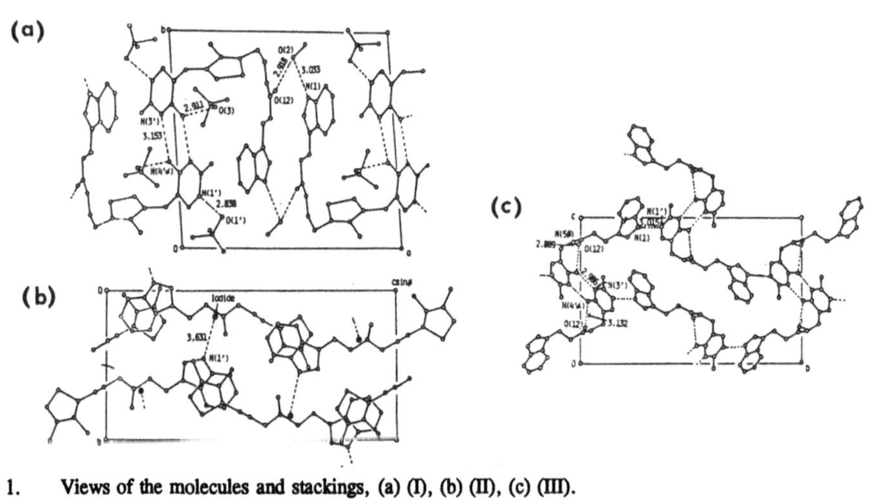

Fig. 1. Views of the molecules and stackings, (a) (I), (b) (II), (c) (III).

β-DL-ARABINOSE
$C_5H_{10}O_5$

F. LONGCHAMBON, H. GILLIER-PANDRAUD, R. WIEST, B. REES, A. MITSCHLER, R. FELD, M. LEHMANN and P. BECKER, 1985. *Acta Cryst.*, **B41**, 47-56.

I. Monoclinic, $P2_1/c$, a = 5.8732, b = 7.7704, c = 13.2478 Å, β = 99.96°, Z = 4. Mo radiation, R = 0.039 for 2165 reflections (at 75 K).

II. Monoclinic, $P2_1/c$, a = 5.873, b = 7.762, c = 13.243 Å, β = 100.03°, Z = 4. Neutron radiation, λ = 0.8400, R = 0.043 for 2944 reflections (at 75 K).

III. Monoclinic, $P2_1/c$, a = 5.880, b = 7.764, c = 13.264 Å, β = 99.90°, Z = 4. Mo radiation, R = 0.089 for 1620 reflections (at 175 K).

For previous work see 1. Electron deformation density maps around O atoms and the anomeric C atom have been computed by X- N methods. Experimental data concerning C-O bonds and O lone pairs are compared to theoretical *ab initio* studies on anomeric and exoanomeric effects. The most important points are: (i) lone pairs of O atoms O(5) and O(1) (Fig. 1) seem to be quite identical except d_2 of O(1) which interacts with the C(1)-O(5) bond; (ii) electron deformation density of anomeric C(5)-O(5) and O(1)-H bonds seems to be smaller than means in C-O and O-H bonds. These phenomena are consistent with predictions based on theoretical studies of anomeric and exoanomeric effects.

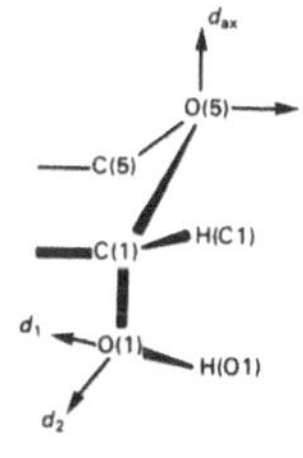

Fig. 1. The anomeric region in β-DL-arabinose

1. Structure Reports, 32B, 211; 45B, 457

CALCIUM D-GLUCARATE TETRAHYDRATE
$C_6H_8CaO_8,4(H_2O)$ $C_6H_8O_8^{2-},Ca^{2+},4(H_2O)$

C. BURDEN, W. MACKIE and B. SHELDRICK, 1985. *Acta Cryst.*, **C41**, 693-695.

Orthorhombic, $P2_12_12_1$, a = 16.220, b = 11.347, c = 6.558 Å, Z = 4. Cu radiation, R = 0.0514 for 1012 reflections.

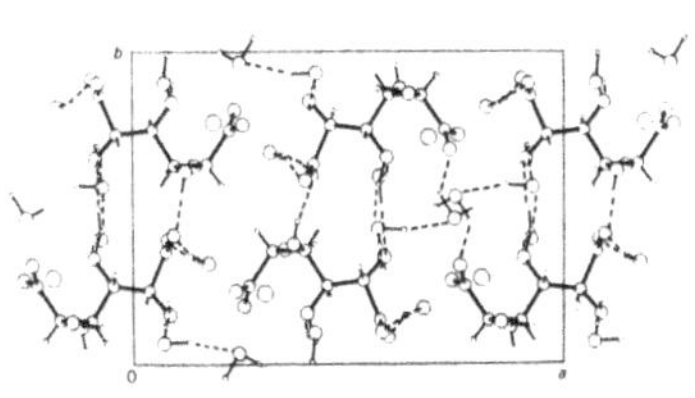

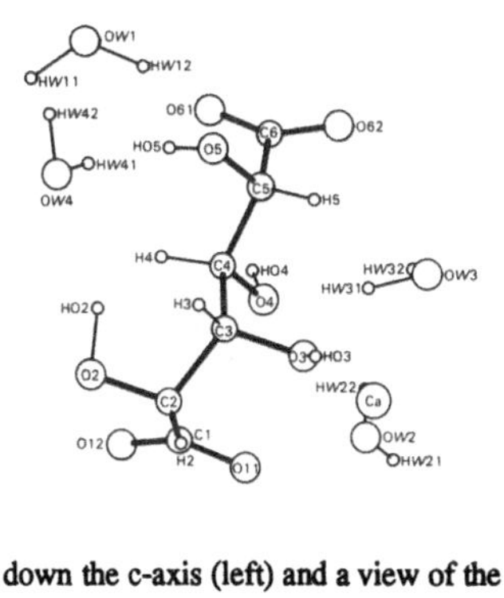

Fig. 1. $C_6H_8CaO_8,4(H_2O)$: a projection of the structure down the c-axis (left) and a view of the asymmetric unit (right).

The structure is in good agreement with that reported earlier (1). The Ca^{2+} ion has eight-coordination to O atoms in a distorted square antiprism arrangement. The Ca-O distances lie in the range 2.37(1) to 2.55(1) Å. Except for O(4) all the water molecules are involved in a hydrogen-bonding network.

1. Structure Reports, 42B, 566.

SODIUM α-L-GULURONATE DIHYDRATE

$C_6H_9NaO_7,2(H_2O)$ $C_6H_9O_7^-,Na^+,2(H_2O)$

F. MO, T.J. BROBAK and I.R. SIDDIQUI, 1985. *Carbohydr. Res.*, **145**, 13-24.

Orthorhombic, $P2_12_12_1$, a = 7.591, b = 18.884, c = 6.842 Å, Z = 4. Mo radiation, R = 0.043 for 1587 reflections (at 87 K).

The analysis indicates that there is partial anomeric disorder in the structure with $\alpha:\beta \sim 90:10$. The guluronate ring has the $^1C_4$(L) conformation. The Na^+ ion is surrounded by seven O atoms (Fig. 1) with two contacts (Na-O(5) 2.725(2), Na-O(2) 2.849(2) Å) distinctly longer than the other five which are in the range 2.344-2.412(2) Å. In the crystal structure there are stacks of sodium ions and guluronate residues in alternating sequence along *c*. The stacks are linked by a system of hydrogen bonds involving all oxygen atoms in the structure.

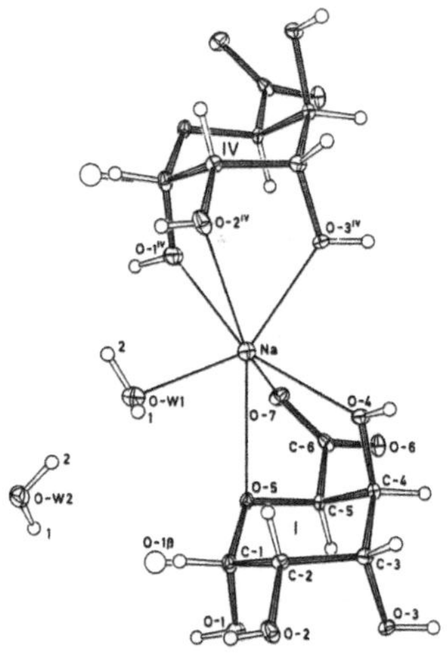

Fig. 1. Part of the sodium α-L-guluronate dihydrate structure.

β-D-GALACTOPYRANOSYLAMINE

$C_6H_{13}NO_5$

C.T. SAMUDZI, J.R. RUBLE and G.A. JEFFREY, 1985. *Carbohydr. Res.*, **142**, 39-45.

Orthorhombic, $P2_12_12_1$, a = 7.703, b = 7.788, c = 12.645 Å, Z = 4. Mo radiation, R = 0.074 for 2117 reflections.

The six-membered ring conformation is $^4C_1$(D) and molecular dimensions are in Fig. 1. The primary alcohol group is oriented gauche/trans with O(5)-C(5)-C(6)-O(6) +57°. An unexpected feature of the hydrogen bonding is that the amino group accepts a strong O-H...N bond but has no donor functionality in the crystal structure.

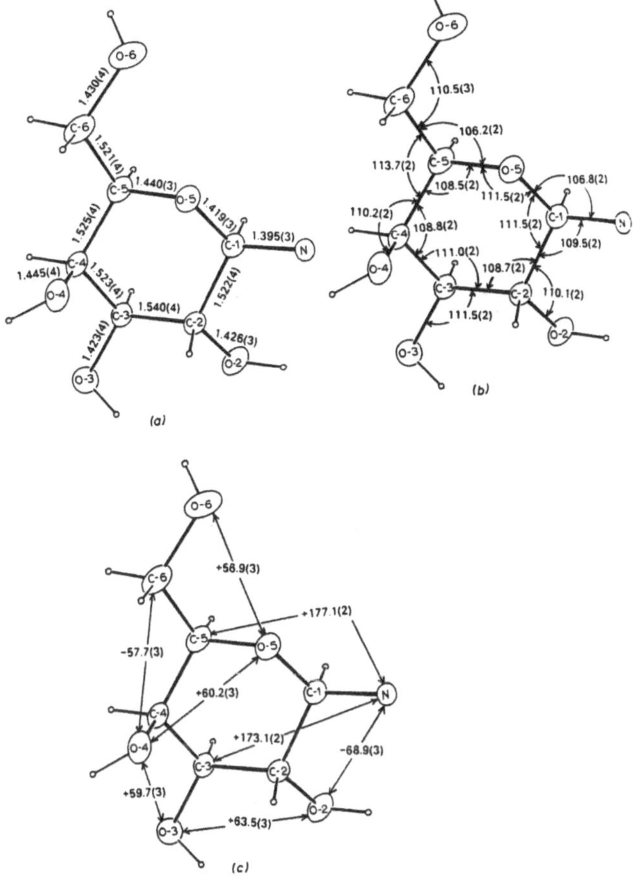

Fig. 1. Bond lengths, angles and torsion angles in $C_6H_{13}NO_5$.

2,3-ANHYDRO-4-DEOXY-α-DL-ribo-HEXOPYRANOSIDE
$C_7H_{12}O_4$ (I)

2,3-ANHYDRO-4-DEOXY-α-DL-lyxo-HEXOPYRANOSIDE
$C_7H_{12}O_4$ (II)

J.W. KRAJEWSKI, P. GLUZINSKI, Z. URBANCZYK-LIPKOWSKA, A. BANASZEK, L. PARKANYI and A. KALMAN, 1985. *Carbohydr. Res.*, **144**, 13-22.

I. Monoclinic, $P2_1/c$, a = 7.694, b = 12.954, c = 8.415 Å, β = 110.34°, Z = 4. Cu radiation, R = 0.049 for 1456 reflections.

II. Tetragonal, $I4_1/a$, a = 22.995, c = 6.086 Å, Z = 16. Cu radiation, R = 0.042 for 1109 reflections.

A hybrid conformation ^5E + $^5H_0$ was found for the anhydrohexopyranose ring in (I), whereas a slightly deformed $^5H_0$ conformation was found in (II) (Fig. 1). In (I) centrosymmetric hydrogen-bonded dimers are formed and in (II) the molecules form helices around the 4_1 axes.

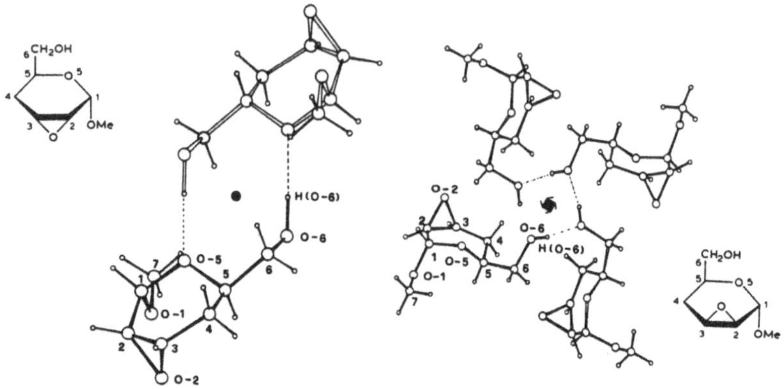

Fig. 1. The $C_7H_{12}O_4$ molecules and projections showing their intermolecular hydrogen bonding.

METHYL 2-AZIDO-2-DESOXY-β-D-GALACTOPYRANOSIDE
$C_7H_{13}N_3O_5$ (I)

ETHYL 2-AZIDO-2-DESOXY-β-D-GALACTOPYRANNOSIDE
$C_8H_{15}N_3O_5$ (II)

N. DARBON, Y. ODDON, E. GUY, B. FERRARI, A.A. PAVIA, G. PEPE and J.P. REBOUL, 1985. *Acta Cryst.*, C41, 1100-1104.

I. Monoclinic, *C*2, a = 16.630, b = 8.866, c = 6.924 Å, β = 97.57°, Z = 4. Mo radiation, R = 0.036 for 1002 reflections.

II. Orthorhombic, $P22_12_1$, a = 4.670, b = 6.933, c = 34.941 Å, Z = 4. Mo radiation, R = 0.039 for 699 reflections.

In both compounds (Fig. 1) the pyranose is in the normal $^4C_1$ chair conformation and the conformation about the C(5)-C(6) bond of the hydroxyl group at O(6) is trans-gauche. Bond lengths and angles in the molecules are in good agreement with generally accepted values except for the C-OCH_3 and C-OCH_2 bonds which are slightly shorter than in similar compounds. The conformation of the azide group is determined by the torsion angles N(3)-N(2)-N(1)-C(2) and N(2)-N(1)-C(2)-C(1) which have values of 176.7(5), 85.6(4)° for (I) and 180.0(7), 68.1(5)° for (II). The molecules in the crystals are linked into columns along the b-axis by O-H...O hydrogen bonds with O...O separations of 2.772(6) and 2.805(6) Å in (I) and 2.682(6) and 2.785(5) Å in (II).

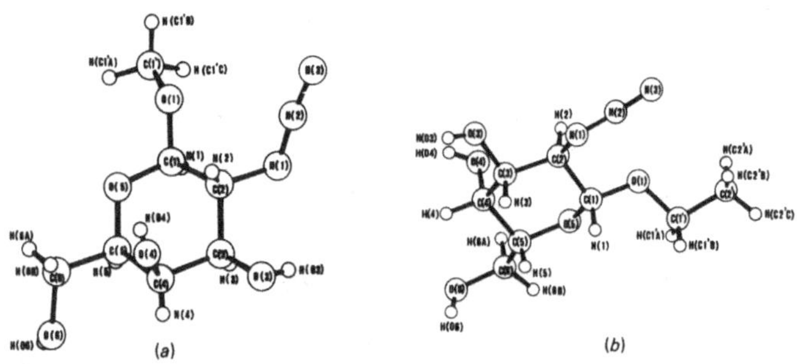

Fig. 1. Views of the molecules ($C_7H_{13}N_3O_5$ (left) and $C_8H_{15}N_3O_5$ (right)).

METHYL 3,6-DIDEOXY α D-arabino-HEXOPYRANOSIDE (METHYL TYVELOSIDE)
$C_7H_{14}O_4$

G.I. BIRNBAUM and D.R. BUNDLE, 1985. *Canad. J. Chem.*, **63**, 739-744.

Orthorhombic, $P2_12_12_1$, a = 7.478, b = 7.933, c = 14.064 Å, Z = 4. Cu radiation, R = 0.038 for 998 reflections.

The pyranose ring has an almost perfect $^4C_1$ chair conformation and the conformation adopted by the glycosidic methyl group is in agreement with the requirements of the exo-anomeric effect. Both acetal O atoms act as hydrogen bond acceptors.

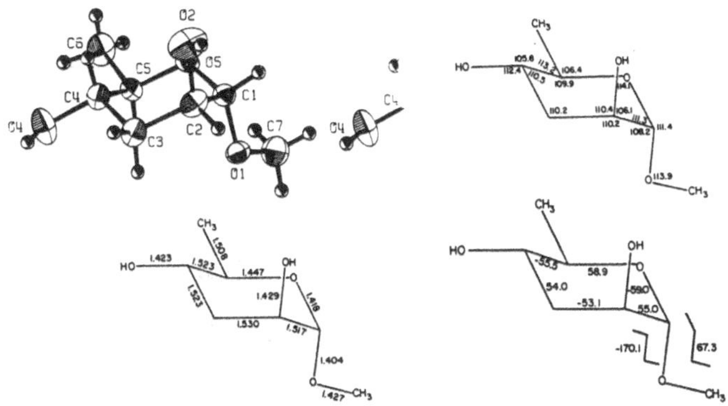

Fig. 1. The methyl tyveloside molecule and bond lengths, angles and torsion angles.

2-ACETAMIDO-3-AMINO-2,3-DIDEOXY-β-D-GLUCOFURANURONO-6,3-LACTAM $C_8H_{12}N_2O_5$

Y. SUGAWARA, H. IWASAKI, N. SUZUKI and S. OKUDA, 1985. *Carbohydr. Res.*, **142**, 195-201.

Monoclinic, $P2_1$, a = 11.704, b = 5.333, c = 7.399 Å, β = 91.63°, Z = 2. Cu radiation, R = 0.046 for 796 reflections.

The molecule and dimensions are shown in Fig. 1. The dihedral angle between the mean plane of the furanose and that of the γ-lactam ring is 106.2(2)°. The furanose ring conformation is $^1T_2$ and the C(3),C(4),C(6),N(3) atoms of the γ-lactam ring are nearly coplanar (C(5) -0.33(1) Å from this plane).

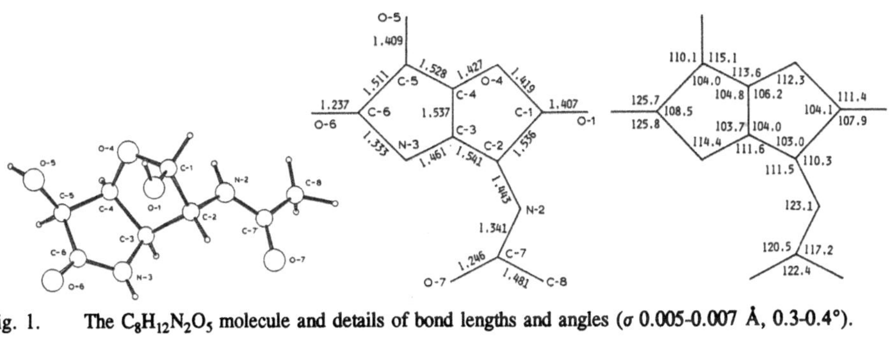

Fig. 1. The $C_8H_{12}N_2O_5$ molecule and details of bond lengths and angles (σ 0.005-0.007 Å, 0.3-0.4°).

2-β-D-RIBOFURANOSYLSELENAZOLE-4-CARBOXAMIDE (β-SELENAZOFURIN) $C_9H_{12}N_2O_5Se$ (I)

2-α-D-RIBOFURANOSYLSELENAZOLE-4-CARBOXAMIDE (α-SELENAZOFURIN) $C_9H_{12}N_2O_5Se$ (II)

B.M. GOLDSTEIN, F. TAKUSAGAWA, H.M. BERMAN, P.C. SRIVASTAVA and R.K. ROBINS, 1985. *J. Am. Chem. Soc.*, **107**, 1394-1400.

I. Orthorhombic, $P2_12_12_1$, a = 5.128, b = 13.083, c = 16.536 Å, Z = 4. R = 0.0342.

II. Orthorhombic, $P2_12_12_1$, a = 5.454, b = 9.396, c = 21.656 Å, Z = 4. R = 0.0285.

Both structures contain selenazole rings (Fig. 1) in which the Se-C bond lengths range from 1.848(3) to 1.878(5) Å. The C(2)-Se-C(5) bond angle is 84.3(2) in (I) and 84.5(1)° in (II). The sugar moiety in (I) has the C(2')-endo (2E) pucker, while (II) has the C(4')-C(5') conformation. The conformation about the C-glycosyl bond is such that the O(1') atom of the ribose is cis to the Se of the ring, with torsion angles x (O(1')-C(1')-C(2)-Se) of 30.5(5) in (I) and -13.6(3)° in (II). The O(1')-Se contact in each case is shorter than the sum of van der Waals radii. In both structures, there is substantial intermolecular hydrogen bonding.

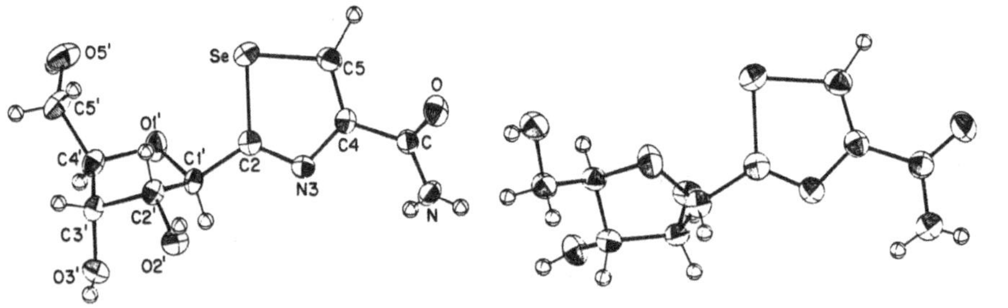

Fig. 1. Views of the two anomeric forms of $C_9H_{12}N_2O_5Se$.

5-MERCAPTO-1-β-D-RIBOFURANOSYLIMIDAZOLE-4-CARBOXAMIDE HEMIHYDRATE (5-THIOBREDININ HEMIHYDRATE)
$C_9H_{13}N_3O_5S,0.5(H_2O)$

S.G. WOOD, K.G. UPADHYA, N.K. DALLEY, P.A. McKERNAN, P.G. CANONICO, R.K. ROBINS and G.R. REVANKAR, 1985. *J. Med. Chem.*, **28**, 1198-1203.

Monoclinic, $P2_1$, a = 4.991, b = 12.857, c = 9.555 Å, β = 93.73°, Z = 2. Mo radiation, R = 0.034 for 1433 reflections.

Thiobredinin is shown to have a zwitterionic structure (Fig. 1). The C(5)-S bond length (1.708(3) Å) is consistent with double bond character; the positive charge is on N(3) which is protonated.

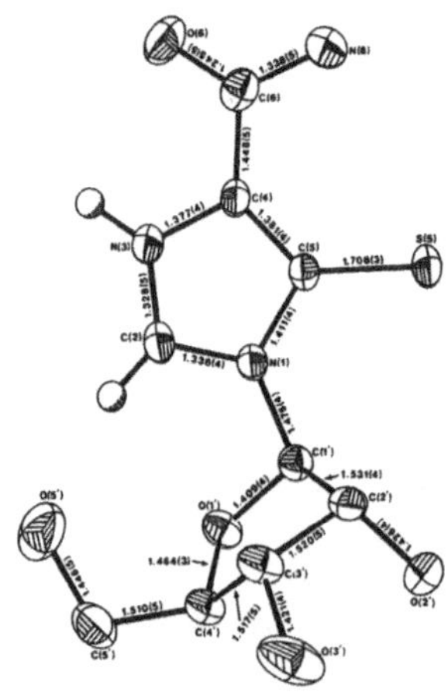

Fig. 1. A view of 5-thiobredinin.

1,2-O-(S)-(1-AMINOMETHYLETHYLIDENE)-α-D-GLUCOPYRANOSE
$C_9H_{17}NO_6$ (I)

1,2-O-(R)-(1-t-BUTOXYETHYLIDENE)-α-D-GLUCOPYRANOSE
$C_{12}H_{22}O_7$ (II)

F.H. CANO, C. FOCES-FOCES, A. ALEMANY, M. BERNABE, M.L. JIMENO and M. MARTIN-LOMAS, 1985. *Carbohydr. Res.*, **139**, 65-73.

I. Monoclinic, $P2_1$, a = 9.6926, b = 8.2376, c = 7.0159 Å, β = 105.997°, Z = 2. Cu radiation, R = 0.033 for 1026 reflections.

II. Orthorhombic, $P2_12_12_1$, a = 19.3634, b = 12.0554, c = 5.8866 Å, Z = 4. Cu radiation, R = 0.039 for 559 reflections.

The pyranoid rings in (I) and (II) are shown to have flattened $^4C_1$ conformations flattened around the dioxolane ring fusion (Fig. 1). The dioxolane ring in (I) has a half-chair $^1T_{0\text{-}1}$ conformation and that in (II) a $^2T_{0\text{-}2}$ distorted towards ^2E.

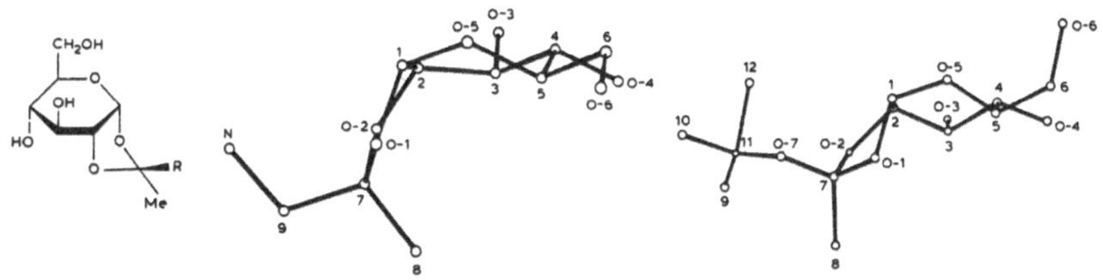

Fig. 1. The molecular structures of $C_9H_{17}NO_6$ (I) and $C_{12}H_{22}O_7$ (II).

N-ISOPROPYL-D-GLUCONAMIDE
$C_9H_{19}NO_6$ (I)

N,N-DIETHYL-D-GLUCONAMIDE
$C_{10}H_{21}NO_6$ (II)

N. DARBON-MEYSSONNIER, Y. ODDON, E. DECOSTER, A.A. PAVIA, G. PEPE and J.P. REBOUL, 1985. *Acta Cryst.*, **C41**, 1324-1327.

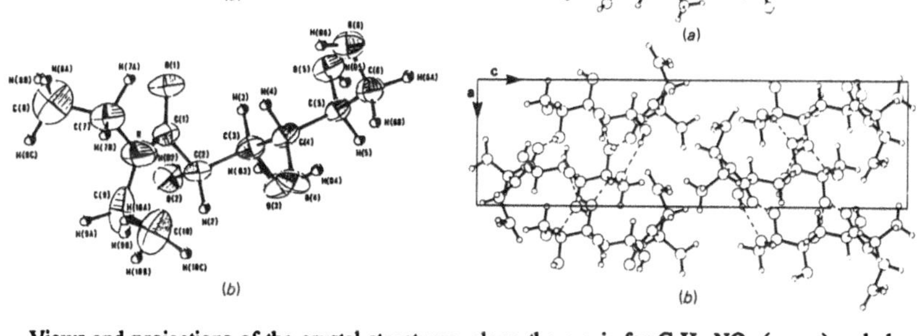

Fig. 1. Views and projections of the crystal structures, along the *c*-axis for $C_9H_{19}NO_6$ (upper) and along the *b*-axis for $C_{10}H_{21}NO_6$.

I. Orthorhombic, $P2_12_12_1$, a = 25.153, b = 8.615, c = 5.350 Å, Z = 4. Cu radiation, R = 0.049 for 585 reflections.

II. Orthorhombic, $P2_12_12_1$, a = 6.870, b = 7.655, c = 24.029 Å, Z = 4. Cu radiation, R = 0.059 for 982 reflections.

The asymmetric unit in these structures is composed of individual molecules (Fig. 1) and in (II) the ethyl group at C(9)-C(10) occurs in two possible sites with occupancy factors of 0.75 and 0.25. The gluconamide group in both compounds is in an extended conformation and atoms C(2,3,4,5,6) and atoms C(7), N, C(1,2), O(1) define two planar moieties in both molecules. The dihedral angles between these planes are 62.7 and 76.3° for (I) and (II) respectively. Bond lengths and angles in each molecule are normal and the configuration of (I) is stabilized by an intramolecular N-H...O(2) hydrogen bond. In both crystal structures the molecules are involved in hydrogen bonding. In (I) each molecule is bonded to two vicinal ones by eight hydrogen bonds and to four neighbours by ten hydrogen bonds in (II).

4-(R)-(1-HYDROXYETHYL)-5-METHYL-1-β-D-RIBOFURANOSYLIMIDAZOLE
$C_{11}H_{18}N_2O_5$

J.A. MONTGOMERY, H.J. THOMAS, A.L. ZELL, H.M. EINSPAHR and C.E. BUGG, 1985. *J. Med. Chem.*, 28, 1751-1753.

Orthorhombic, $P2_12_12_1$, a = 7.156, b = 8.379, c = 19.98 Å, Z = 4. Cu radiation, R = 0.051 for 1148 reflections.

The compound was shown by the analysis to be the R-diastereomer (Fig. 1).

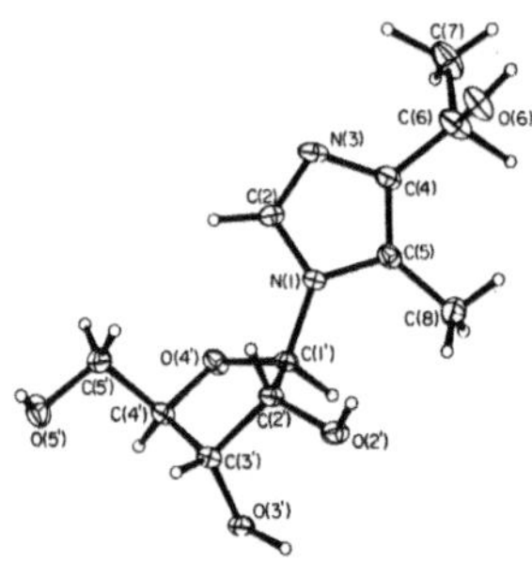

Fig. 1. The $C_{11}H_{18}N_2O_5$ molecule.

5-β-D-GALACTOPYRANOSYL-1,3-DIMETHYLBARBITURIC ACID MONOHYDRATE
$C_{12}H_{18}N_2O_8,H_2O$

M. MILLAN, C.F. CONDE, A. CONDE and R. MARQUEZ, 1985. *Acta Cryst.*, C41, 274-277.

Orthorhombic, $P2_12_12_1$, a = 9.147, b = 23.054, c = 6.921 Å, Z = 4. Mo radiation, R = 0.055 for 1623 reflections.

The galactopyranose ring (Fig. 1) adopts a $^4C_1$ conformation and the dihedral angle between the two rings in the molecule is 89.1(1)°. The mean value of the endocyclic torsion angle is 56.7(4)°. In the galactopyranose ring the mean values for C-C and C-O are 1.532(6) and 1.431(6) Å respectively. In the barbituric acid moiety the mean values for the endocyclic N-C and C-C bond lengths are 1.386(5) and 1.498(6) Å respectively. The ring adopts a twist-boat conformation. The molecules are linked by an extensive three-dimensional hydrogen-bond network involving the water molecule (O...O 2.690(5)-3.024(5), C...O 3.052(6)-3.498(6) Å).

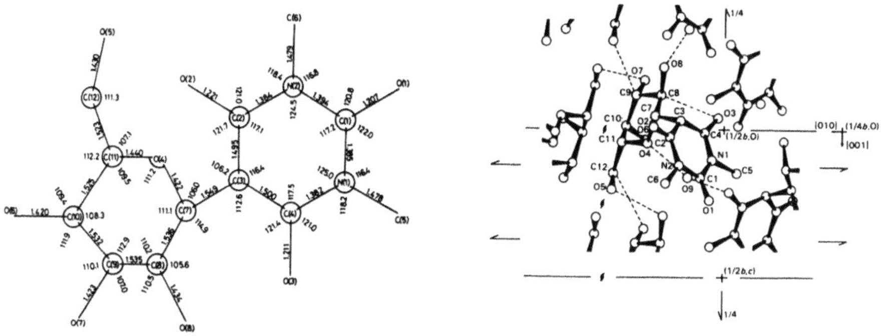

Fig. 1. $C_{12}H_{18}N_2O_8,H_2O$: molecular dimensions and the asymmetric unit with its surroundings.

2-((2,2-DIACETYLVINYL)AMINO)-2-DEOXY-α-D-GLUCOPYRANOSE
$C_{12}H_{19}NO_7$

M.J. DIANEZ, A. LOPEZ-CASTRO and R. MARQUEZ, 1985. *Acta Cryst.*, **C41**, 149-151.

Monoclinic, $P2_1$, a = 12.495, b = 4.632, c = 12.391 Å, β = 93.9°, Z = 2. Mo radiation, R = 0.051 for 1317 reflections.

In the glucopyranose ring in the molecule (Fig. 1) the mean C-C and C-O bond lengths are 1.529(9) and 1.42(1) Å respectively. The pyranose ring is in a chair conformation in which the only axial substituent is O(11). In the (2,2-diacetylvinyl)amino group the C(3)=C(6) bond length is elongated to 1.407(9) Å and the twist angle around this bond is -179.0(7)°. There is an intramolecular hydrogen bond N(1)-H...O(2) of length 2.58(1) Å and a network of O-H...O intermolecular hydrogen bonds, with lengths in the range 2.683(6) to 2.74(1) Å, stabilizing the structure.

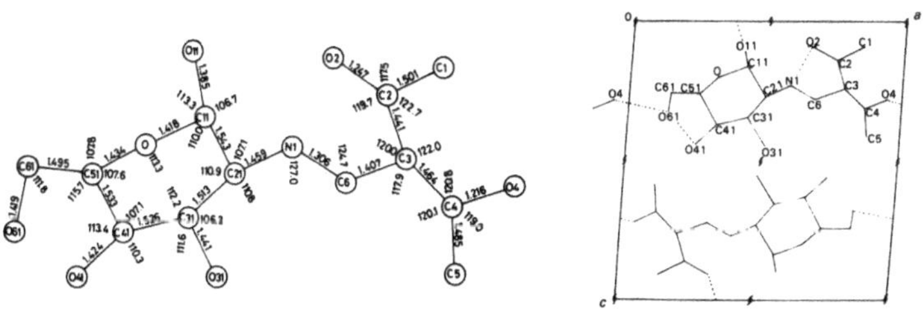

Fig. 1. $C_{12}H_{19}NO_7$: a view of the molecule showing bond lengths and angles and a projection of the unit cell contents onto (010) with hydrogen bonds shown as broken lines.

trans-(6R,9S:6S,9R)-1,7,10,15-TETRAOXADISPIRO[5.2.5.2]HEXADECANE
$C_{12}H_{20}O_4$

W.A. SZAREK, O.R. MARTIN, R.J. RAFKA and T.S. CAMERON, 1985. *Canad. J. Chem.*, **63**, 1222-1227.

Monoclinic, $P2_1/c$, a = 6.532, b = 8.856, c = 10.268 Å, β = 102.30°, Z = 2. Mo radiation, R = 0.041 for 578 reflections.

The molecules lie about inversion centres and hence have trans-stereochemistry (Fig. 1). All rings have chair conformations with mean torsion angles ±53.5 (central) and ±54.1 (external). The C-C bond lengths are in the range 1.484-1.519(4) Å. C-O bond lengths are in the range 1.417-1.431(3) Å.

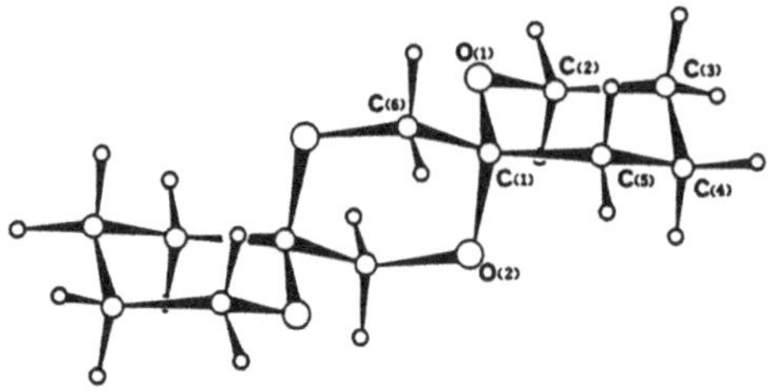

Fig. 1. A view of $C_{12}H_{20}O_4$.

2,3:5,6-DI-O-ISOPROPYLIDENE-α-D-MANNOFURANOSE
$C_{12}H_{20}O_6$

B. SHELDRICK, W. MACKIE and D. AKRIGG, 1985. *Acta Cryst.*, C41, 431-433.

Orthorhombic, $P2_12_12_1$, a = 6.665, b = 10.816, c = 18.891 Å, Z = 4. Cu radiation, R = 0.0437 for 1353 reflections.

In the molecule (Fig. 1) the fused five-membered rings C(2,3,7),O(2,3) and C(1,2,3,4),O(4) both have envelope conformations with C(7) and O(4) as flaps which deviate by -0.205(4) and 0.247(2) Å respectively, from the planes of the four other ring atoms. The angle between these planes is 115.9(4)°. The third five-membered ring is also in the envelope conformation with C(6) as flap. The C-O bond distances in the isopropylidene rings are nearly equal. There is a weak intermolecular hydrogen bond between the hydroxyl group O(1) and O(5) with an O...O separation of 3.014(7) Å.

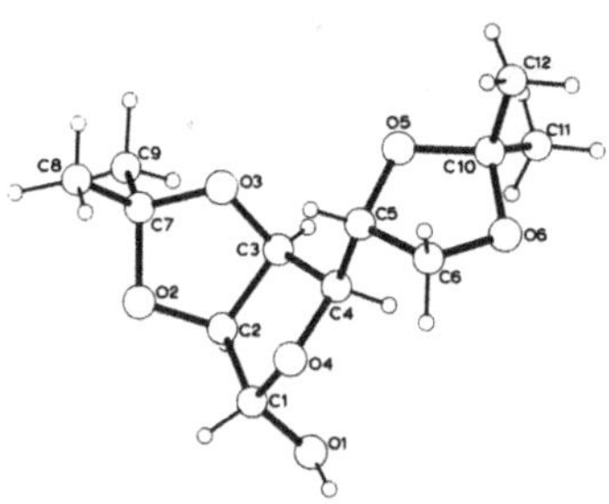

Fig. 1. A view of the $C_{12}H_{20}O_6$ molecule.

α-D-GLUCOPYRANOSYL-α-D-GLUCOPYRANOSIDE (α,α-TREHALOSE)
$C_{12}H_{22}O_{11}$

G.A. JEFFREY and R. NANNI, 1985. *Carbohydr. Res.*, 137, 21-30.

Monoclinic, $P2_1$, a = 12.971, b = 8.229, c = 6.789 Å, β = 98.12°, Z = 2. Mo radiation, R = 0.059 for 1564 reflections (at -150°C).

The molecular structure (Fig. 1) is very similar to that found in the dihydrate (1). It has approximate twofold symmetry, with both rings in $^4C_1$ conformations and linkage torsion angles O(5)-C(1)-O(1)-C(1') 60.8(6), O(5')-C(1')-O(1)-C(1) 60.1(6)°. The primary alcohol groups are oriented gauche/gauche and gauche/trans as in the dihydrate. Molecules are linked by an extensive O-H...O hydrogen bond network as shown in Fig. 1.

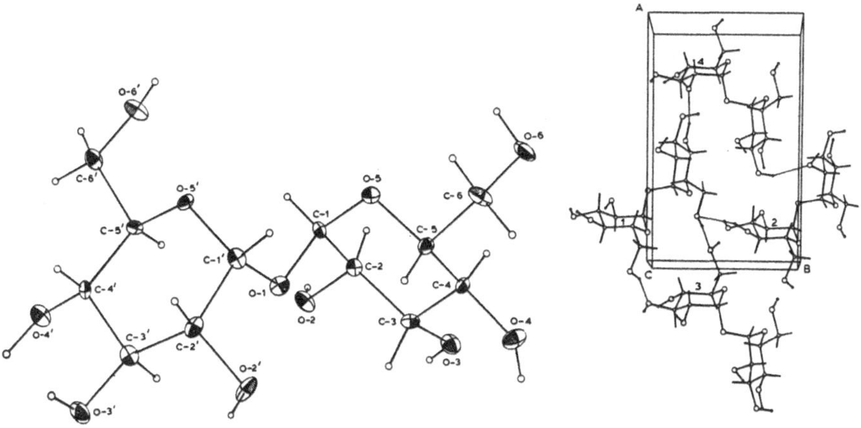

Fig. 1. The α,α-trehalose molecule and a view of the crystal structure.

1. Structure Reports, **38B**, 441.

(1R,4S)-2,3-endo-1-(β-D-RIBOFURANOSYL)-7-OXABICYCLO[2.2.1]HEPT-5-ENE-2,3-DICARBOXIMIDE
$C_{13}H_{15}NO_7$ (I)

(1S,4R)-2,3-exo-1-(2,3-O-ISOPROPYLIDENE-β-D-RIBOFURANOSYL)-7-OXABICYCLO[2.2.1]HEPT-5-ENE-2,3-DICARBOXIMIDE
$C_{16}H_{19}NO_7$ (II)

I. MAEBA, F. USAMI, T. ISHIKAWA, H. FURUKAWA, T. ISHIDA and M. INOUE, 1985. *Carbohydr. Res.*, **141**, 1-12.

I. Orthorhombic, $P2_12_12_1$, a = 5.609, b = 10.316, c = 21.805 Å, Z = 4. Cu radiation, R = 0.053 for 1278 reflections.

II. Monoclinic, $P2_1$, a = 11.166, b = 5.736, c = 12.506 Å, β = 108.04°, Z = 2. Cu radiation, R = 0.059 for 1432 reflections.

The analyses establish the endo-configuration for (I) and the exo-configuration for (II).

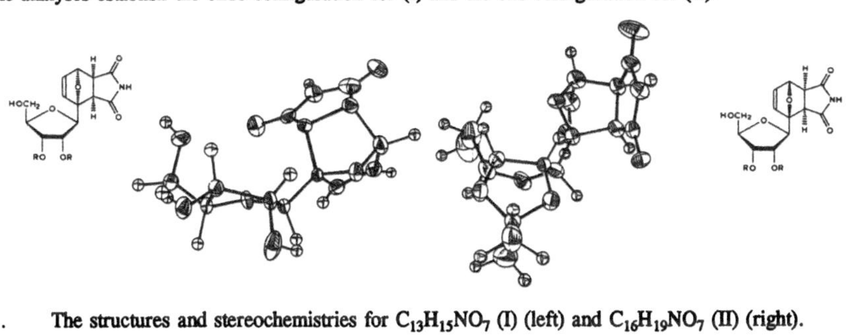

Fig. 1. The structures and stereochemistries for $C_{13}H_{15}NO_7$ (I) (left) and $C_{16}H_{19}NO_7$ (II) (right).

METHYL 3-C-(ACETOXYMETHYL)-1,4-DI-O-ACETYL-α-erythro-PENTOFURANOSIDE
$C_{13}H_{20}O_9$

M. IRIE, K. FUKUYAMA, T. TSUKIHARA, Y. KATSUBE, Y. MORIMOTO, Y. SHIGEMASA and R. NAKASHIMA, 1985. *Acta Cryst.*, **C41**, 1814-1816.

Orthorhombic, *Pbca*, a = 21.614, b = 17.239, c = 8.518 Å, Z = 8. Cu radiation, R = 0.076 for 1668 reflections.

The five-membered ring in the molecule (Fig. 1) is puckered at C(3) and has the endo conformation. The hydroxyl and methoxy groups occupy quasi-axial positions. There is an intermolecular hydrogen bond of 3.022(6) Å between the hydroxyl group and one of the acetyl groups of a molecule related by the c-glide plane.

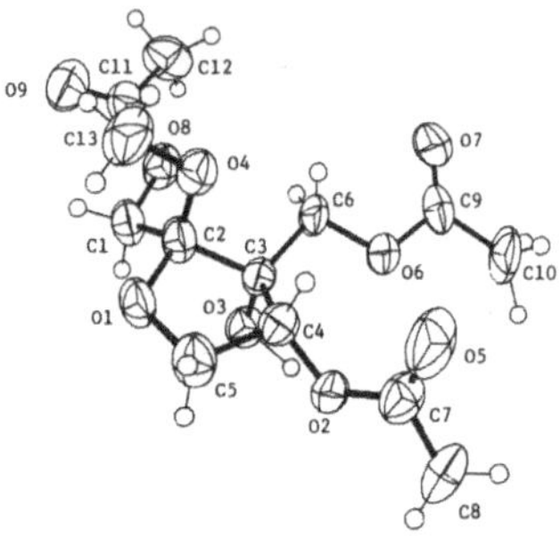

Fig. 1. $C_{13}H_{20}O_9$: the molecular structure.

(2R,3R,4S)-3,4-DIHYDROXY-2-HYDROXYMETHYL-8-METHYL-6-PHENYL-1-OXA-6,8-DIAZASPIRO[4.4]-NONANE-7-THIONE
$C_{14}H_{18}N_2O_4S$

E. MORENO, A. LOPEZ-CASTRO and R. MARQUEZ, 1985. *Acta Cryst.*, **C41**, 602-604.

Tetragonal, $P4_12_12$, a = 11.255, c = 24.829 Å, Z = 8. Mo radiation, R = 0.072 for 2023 reflections.

The molecule (Fig. 1) consists of three rings, a tetrahydrofuran and an imidazolidine joined by a common spiro C atom (C(2)) and a phenyl group. The tetrahydrofuran ring of the lyxo-tetritol group adopts a conformation intermediate between twist and envelope. The N(1), C(1), N(2), C(3) and C(2) sequence of the imidazolidine group shows a slight half-chair conformation with the pseudo twofold axis through C(1). The phenyl ring is planar. The dihedral angle between the mean planes through the spiro-connected rings is 90.7(2)° and the phenyl ring makes angles of 86.3(2) and 92.2(2)° with the imidazolidine and tetrahydrofuran respectively. There are three intermolecular hydrogen bonds, O(3)...O(4) 2.673(7), O(2)...O(1) 2.988(6) and O(4)...O(3) 2.697(7) Å.

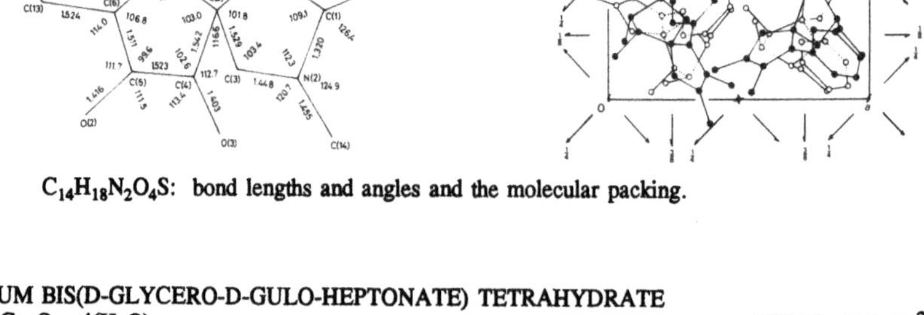

Fig. 1. $C_{14}H_{18}N_2O_4S$: bond lengths and angles and the molecular packing.

CALCIUM BIS(D-GLYCERO-D-GULO-HEPTONATE) TETRAHYDRATE
$C_{14}H_{26}Ca_2O_{16},4(H_2O)$ $2(C_7H_{13}O_8^-),Ca^{2+},4(H_2O)$

T. TAGA, T. KAJI and K. OSAKI, 1985. *Bull. Chem. Soc. Jpn.*, **58**, 30-35.

Monoclinic, $P2_1$, a = 10.064, b = 10.540, c = 11.340 Å, β = 108.92°, Z = 2. Mo radiation, R = 0.030 for 1754 reflections.

The two heptonate ions have different conformations (as shown in Fig. 1), they also coordinate to the Ca^{++} ion in different ways. The Ca^{++} ion is surrounded by eight O atoms in a deformed dodecahedral form (Fig. 1). Ca-O distances are in the range 2.40-2.48 Å.

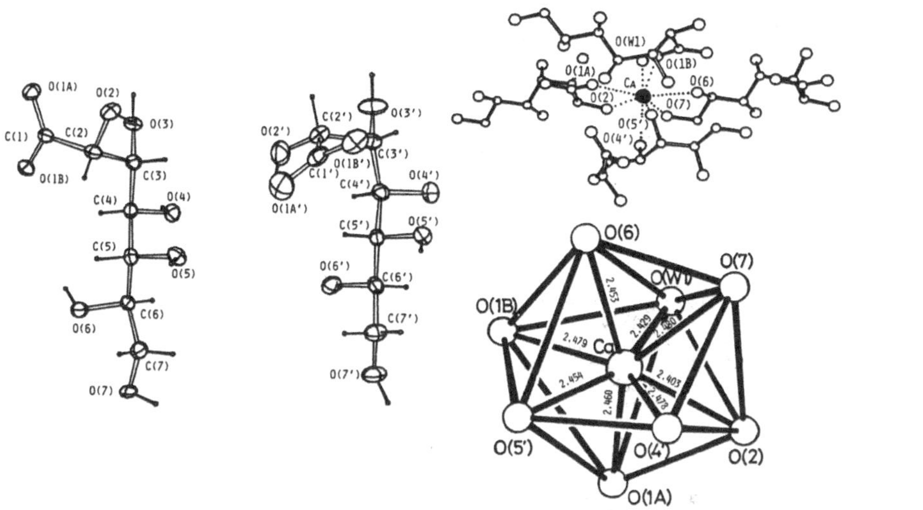

Fig. 1. The heptonate ions and Ca coordination in $C_{14}H_{26}Ca_2O_{16},4(H_2O)$.

(2R,3R)-3-BENZOYLOXY-2,3-DIHYDRO-2,3-DIMETHOXY-6-METHYL-4H-PYRAN-4-ONE $C_{15}H_{16}O_6$

F.W. LICHTENTHALER, S. NISHIYAMA, P. KOHLER and H.J. LINDNER, 1985. *Carbohydr. Res.*, **136**, 13-26.

Monoclinic, $P2_1$, a = 9.461, b = 7.117, c = 11.677 Å, β = 86.50°, Z = 2. Cu radiation, R = 0.069 for 973 reflections.

The analysis shows that the vicinal methoxy groups in the molecule (Fig. 1) are trans with a O(21)-C(2)-C(3)-O(31) -172.3°. Bond lengths and angles are in the expected ranges. The pyranoid adopts a C(2) sofa conformation ^2S somewhat distorted towards a $^2H_3$ half-chair.

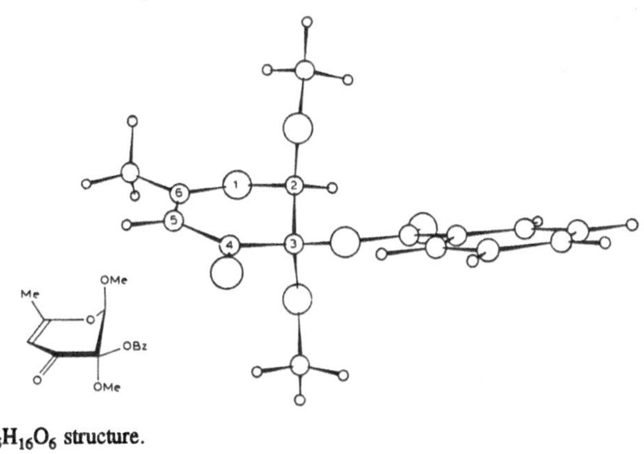

Fig. 1. The $C_{15}H_{16}O_6$ structure.

6-O-(β-D-GLUCOSYLOXY)-7-HYDROXYCOUMARIN SESQUIHYDRATE (ESCULETIN 6-GLUCOSIDE SESQUIHYDRATE)
$C_{15}H_{16}O_9,1.5(H_2O)$ (I)

7-O-(β-D-GLUCOSYLOXY)-6-HYDROXYCOUMARIN DIHYDRATE (ESCULETIN 7-GLUCOSIDE DIHYDRATE)
$C_{15}H_{16}O_9,2(H_2O)$ (II)

K. UENO, M. SHIRAKI, M. SATO and N. SAITO, 1985. *Bull. Chem. Soc. Jpn.*, **58**, 230-235.

I. Orthorhombic, $P2_12_12_1$, a = 11.358, b = 36.850, c = 7.663 Å, Z = 8. Mo radiation, R = 0.073 for 2036 reflections.

II. Orthorhombic, $P2_12_12_1$, a = 4.507, b = 13.077, c = 27.748 Å, Z = 4. Mo radiation, R = 0.069 for 1037 reflections.

There are two independent molecules in (I) in the asymmetric unit; these are shown in Fig. 1 along with the molecule of (II). The conformations of the glucosyl linkages in the three molecules are different, with C(5)-C(6)-O(1')-C(1') being -58.2, 14.3 and 9.9°. The coumarin moieties are nearly planar. The angles C(6)-O(1')-C(1') and C(5)-C(6)-O(1') are 118.5(8) and 126.2(8)° in molecule (B) of (I) compared with values of 113.7(7) and 121.4(9) for molecule (A) and 119.5(6) and 125.4(7)° for (II). The glucose molecules are in C_1 chair conformations. All hydroxyl groups and water molecules form a three-dimensional hydrogen-bonded network in the structures.

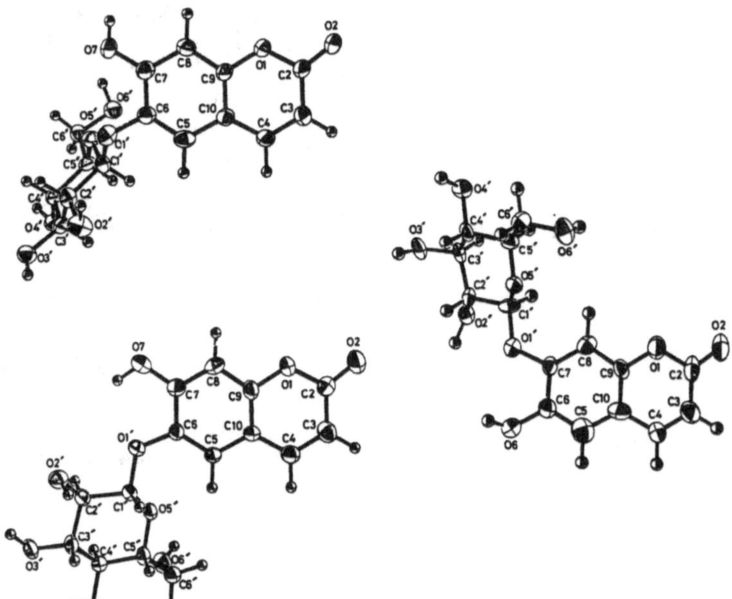

Fig. 1. The $C_{15}H_{16}O_9$ molecules, (I) (left), (II) (right).

4-(α-D-erythro-FURANOSYL)-1,3-DIHYDRO-3-METHYL-1-(p-TOLYL)-2H-IMIDAZOLE-2-THIONE
$C_{15}H_{18}N_2O_3S$

A. CRIADO, A. CONDE and R. MARQUEZ, 1985. *Acta Cryst.*, **C41**, 1212-1214.

Monoclinic, $P2_1$, a = 14.686, b = 5.359, c = 9.439 Å, β = 98.68°, Z = 2. Mo radiation, R = 0.037 for 1987 reflections.

The sugar ring (Fig. 1) has a conformation intermediate between $^3T_2$ and ^3E and is anti to the imidazole ring, glycosidic angle -9.3(4)°. The imidazole and phenyl rings are both essentially planar with a dihedral angle between these planes of 42.2(1)°. Bond lengths and angles lie within normal ranges. The crystal packing is governed by hydrogen bonds involving OH groups and S atoms to form chains along [001] (O(2)...S 3.244(3), O(3)...O(2) 2.781(2) Å).

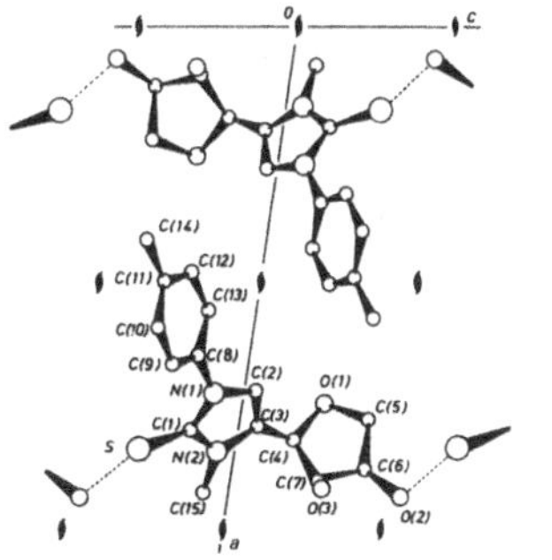

Fig. 1. $C_{15}H_{18}N_2O_3S$: the unit cell viewed along *b*.

1'-(p-BROMOPHENYL)-3'-ETHYL-1',3',4',5'-TETRAHYDRO-1,2-DIDEOXY-α-L-GLUCOFURANOSO[2,1-d]-IMIDAZOLE-2'-THIONE MONOHYDRATE (6-(p-BROMOPHENYL)-2-(1,2-DIHYDROXYETHYL)-4-ETHYL-3-HYDROXY-2,3,3a,5,6,6a-HEXAHYDROFURO[2,3-d]IMIDAZOLE-5(4H)-THIONE MONOHYDRATE)
$C_{15}H_{19}BrN_2O_4S,H_2O$

C.F. CONDE, M. MILLAN, A. CONDE and R. MARQUEZ, 1985. *Acta Cryst.*, C41, 277-280.

Monoclinic, $P2_1$, a = 7.531, b = 8.148, c = 14.625 Å, β = 97.14°, Z = 2. Mo radiation, R = 0.065 for 1672 reflections.

The imidazolidine ring is not planar and neither is the glucofuranose ring which adopts a $^4T_3$ conformation (Fig. 1). The two rings have a cis junction and the dihedral angle between the mean planes of the rings is 72.9(4)°. The phenyl ring is planar and forms a dihedral angle of 47.2(4)° with the imidazolidine ring. A three dimensional network of hydrogen bonds links the molecules to build up the crystal structure.

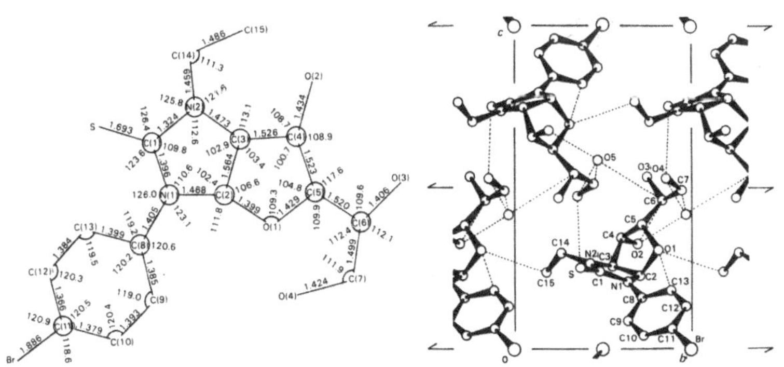

Fig. 1. $C_{15}H_{19}BrN_2O_4S,H_2O$: the molecular dimensions and a view along [100] of the unit cell contents.

3,4,6-TRI-O-ACETYL-1,2-O-(R)-(1-(endo-CYANO)ETHYLIDENE)-α-D-GLUCOPYRANOSE
$C_{15}H_{19}NO_9$ (I)

3,4,6-TRI-O-ACETYL-1,2-O-(R)-(1-CYANOETHYLIDENE)-α-D-GALACTOPYRANOSE
$C_{15}H_{19}NO_9$ (II)

F.H. CANO, C. FOCES-FOCES, M. BERNABE, J. JIMENEZ-BARBERO, M. MARTIN-LOMAS and S.

PENADES-ULLATE, 1985. *Tetrahedron*, **41**, 3875-3876.

I. Monoclinic, $P2_1$, a = 20.6021, b = 8.0438, c = 5.5541 Å, β = 95.588°, Z = 2. Cu radiation, R = 0.050 for 1560 reflections.

II. Monoclinic, $C2$, a = 20.3361, b = 10.0907, c = 18.9115 Å, β = 112.399°, Z = 8. Cu radiation, R = 0.065 for 2553 reflections.

Views of (I) and the two independent molecules of (II) are in Fig. 1. Dimensions are as anticipated; the pyranoid rings are $^4C_1$ chairs and the dioxolane rings have distorted E_1 envelope conformations.

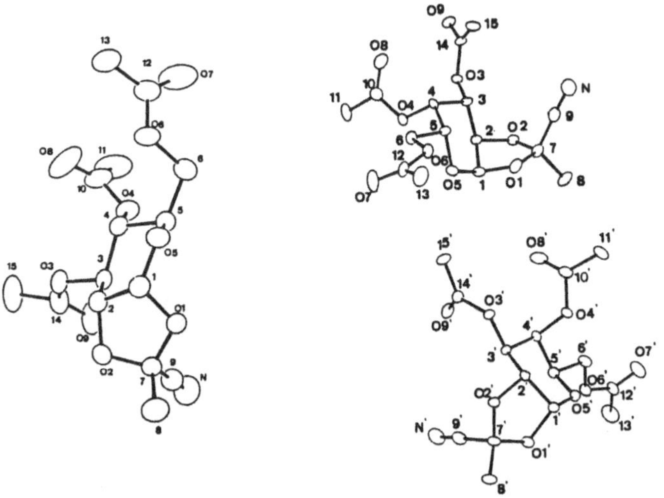

Fig. 1. Views of $C_{15}H_{19}NO_9$, (I) (left) and the two molecules in (II) (right).

BAUHININ DIHYDRATE ((Z)-6β-(β-D-GLUCOSYLOXY)-4β-METHOXY-5α-HYDROXYCYCLOHEX-2-ENE-C$\Delta^{1,\alpha}$-ACETONITRILE DIHYDRATE)
$C_{15}H_{21}NO_8,2(H_2O)$

CHIEN-CHIH CHEN, YUH-PAN CHEN, HONG-YEN HSU, KUO-HSIUNG LEE, SHOHEI TANI and A.T. McPHAIL, 1985. *J. Nat. Prod.*, **48**, 933-937.

Monoclinic, $P2_1$, a = 11.729, b = 14.205, c = 5.784 Å, β = 105.92°, Z = 2. Cu radiation, R = 0.042 for 1659 reflections.

The analysis confirms the proposed structure as shown in Fig. 1. The cyclohexene ring has approximate C_2 symmetry and the ring has a half-chair conformation. In the lattice there is an extensive network of hydrogen bonds.

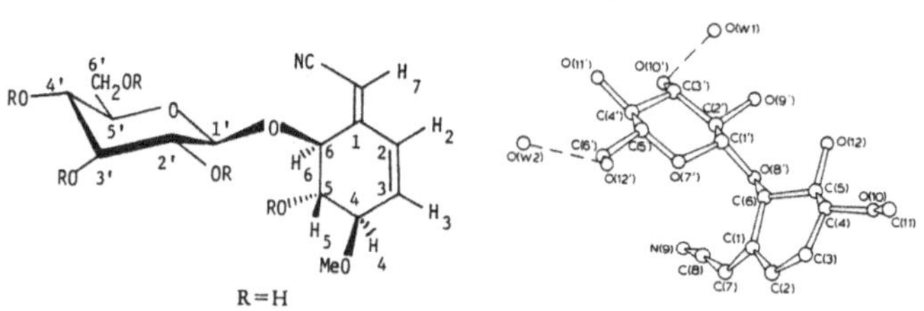

Fig. 1. The bauhinin structure and conformation in the crystal.

ISOBUTYL 2,3,4-TRI-O-ACETYL-1-THIO-β-D-XYLOPYRANOSIDE

$C_{15}H_{24}O_7S$

G.A. JEFFREY, J.R. RUBLE and B. SEPEHRNIA, 1985. *Carbohydr. Res.*, **144**, 197-203.

Monoclinic, $P2_1$, a = 10.134, b = 7.748, c = 11.726 Å, β = 96.63°, Z = 2. Mo radiation, R = 0.08 for 1311 reflections.

The molecule has a $^4C_1(D)$ conformation (Fig. 1) with substituents equatorial. The acetyl groups have the planar (S)-cis configuration with the acetyl planes within ±30° of the C-H bond at the ring carbon to which they are attached. There is some evidence for C-H...O hydrogen bonding in the crystal lattice.

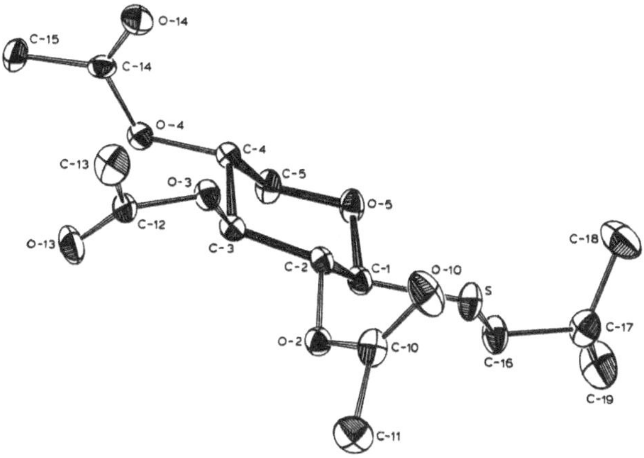

Fig. 1. A view of $C_{15}H_{24}O_7S$.

(-)-(5R,6S)-5-ETHENYL-6-(β-D-GLUCOPYRANOSYLOXY)-5,6-DIHYDRO-1H,3H-PYRANO[3,4-c]PYRAN-1-ONE HEMIHYDRATE (GENTIOPICRIN HEMIHYDRATE)
$C_{16}H_{20}O_9,0.5(H_2O)$

B. KOJIC-PRODIC, A.L. SPEK, W.G. VAN DER SLUIS and R.P. LABADIE, 1985. *Acta Cryst.*, **C41**, 798-802.

Orthorhombic, $P2_12_12_1$, a = 8.175, b = 12.810, c = 31.99 Å, Z = 8. Mo radiation, R = 0.0417 for 1789 reflections.

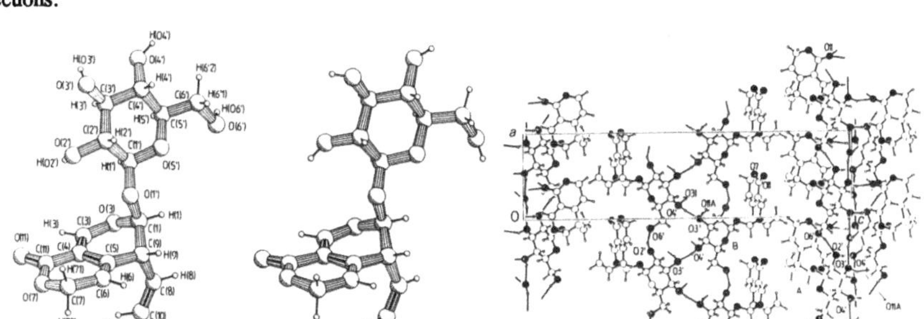

Fig. 1. $C_{16}H_{20}O_9,0.5(H_2O)$: views of molecules A and B and a projection of the structure onto (010) showing the A-A and B-B hydrogen-bonded spirals.

There are two crystallographically independent molecules A and B in the asymmetric unit (Fig. 1). The δ-lactone ring of the secoiridoid moieties in these molecules have different conformations. In molecule A it is a half-chair and in B it is nearly planar with a mean ring torsion angle of 4(1)°. In both molecules the pyran ring exhibits a skew-boat conformation and the β-D-glucose moieties are in the $^4C_1$ chair conformation. Bond lengths and angles in the A and B molecules are similar and in reasonable agreement with generally accepted values. Intermolecular hydrogen bonds between the sugar residues in A and those of B form spirals in the a-axis direction. These A-A and B-B spirals are interconnected by hydrogen bonds between the sugar residue in B and the secoiridoid residue of A. Intermolecular hydrogen bonds exist between the water molecule and both sugar moieties forming layers parallel to (001) separated by ethenyl residues.

6-(p-BROMOPHENYL)-4-ETHYL-3-HYDROXY-2-(1,2,3-TRIHYDROXYPROPYL)-2,3,3a,5,6,6a-HEXAHYDRO-FURO[2,3-d]IMIDAZOLE-5(4H)-THIONE MONOHYDRATE

$C_{16}H_{21}BrN_2O_5S,H_2O$

C.F. CONDE, M. MILLAN, A. CONDE and R. MARQUEZ, 1985. *Acta Cryst.*, **C41**, 1658-1662.

Orthorhombic, $P2_12_12_1$, a = 6.073, b = 15.977, c = 19.213 Å, Z = 4. Mo radiation, R = 0.080 for 2147 reflections.

In the molecule (Fig. 1) the furanose ring adopts the ^4E conformation. The imidazoline ring is slightly nonplanar and the dihedral angle between the least-squares planes of the cis-fused rings in the bicyclic system is 115.0(3)°. The ethyl group is twisted with respect to the imidazoline mean plane: the torsion angle C(1)-N(2)-C(15)-C(16) is 98.9(9)°. Bond lengths and angles in the molecule are in reasonable agreement with those observed in related compounds. The crystal structure is held together by a three-dimensional network of O-H...O, O-H...Br and O-H...S hydrogen bonds.

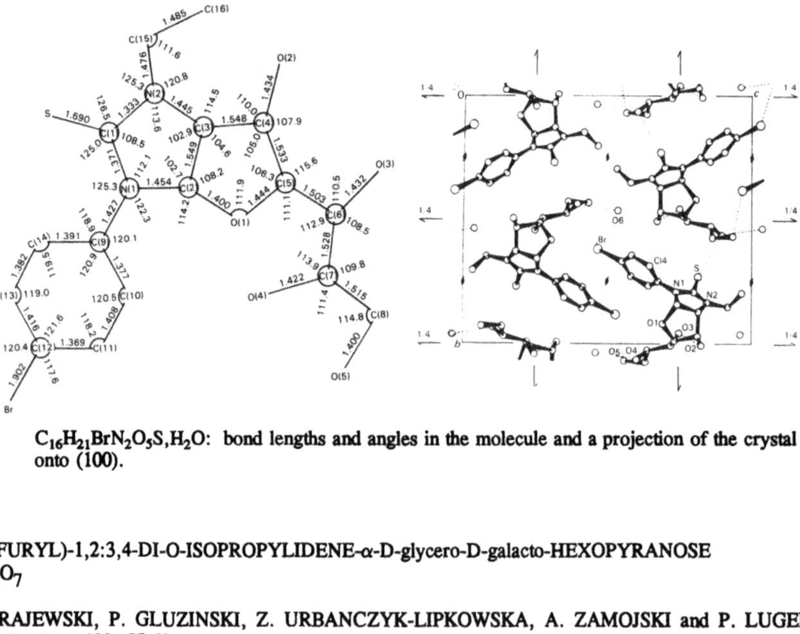

Fig. 1. $C_{16}H_{21}BrN_2O_5S,H_2O$: bond lengths and angles in the molecule and a projection of the crystal structure onto (100).

6-C-(2-FURYL)-1,2:3,4-DI-O-ISOPROPYLIDENE-α-D-glycero-D-galacto-HEXOPYRANOSE

$C_{16}H_{22}O_7$

J.W. KRAJEWSKI, P. GLUZINSKI, Z. URBANCZYK-LIPKOWSKA, A. ZAMOJSKI and P. LUGER, 1985. *Carbohydr. Res.*, **139**, 55-63.

Monoclinic, $P2_1$, a = 13.796, b = 7.887, c = 8.035 Å, β = 106.68°, Z = 2. Cu radiation, R = 0.056 for 1390 reflections.

The pyranose ring is shown to adopt a hybrid twist-boat conformation ($^{O}T_2 + B_{2,5}$). The conformations of the 1,3-dioxolane rings differ significantly; the 1,2 ring has an envelope ^3E conformation whereas the 3,4 ring has a hybrid ($^2T_1 + {}^2E$) conformation. Molecules are linked in the crystal by O(6)-H...O(3) intermolecular hydrogen bonds (O(6)...O(3) 2.804 Å) to form infinite chains parallel to [010].

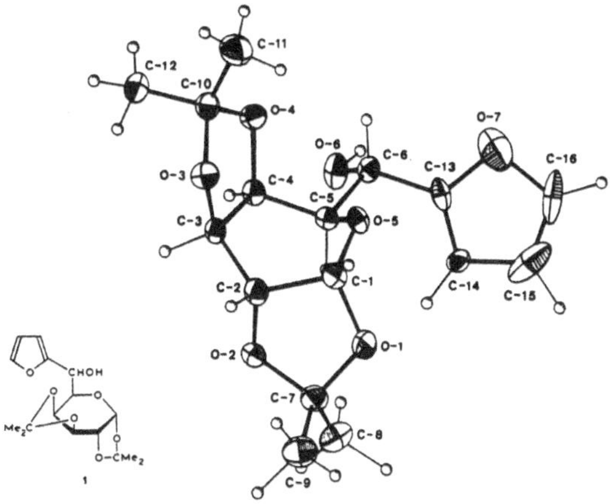

Fig. 1. The $C_{16}H_{22}O_7$ molecule and an ORTEP plot.

METHYL 4-ACETAMIDO-3-p-TOSYL-2,4,6-TRIDEOXY-arabino-D,L-HEXOPYRANOSE
$C_{16}H_{23}NO_6S$

J.T. WELCH and B.-M. SVAHN, 1985. *J. Carbohydr. Chem.*, **4**, 421-427.

Triclinic, $P\bar{1}$, a = 8.050, b = 10.852, c = 10.800 Å, α = 87.40, β = 76.23, γ = 77.57°, Z = 2. Mo radiation, R = 0.069 for 1516 reflections.

The analysis shows that the acetamide is at C(4) and the tosylate at C(3) (Fig. 1).

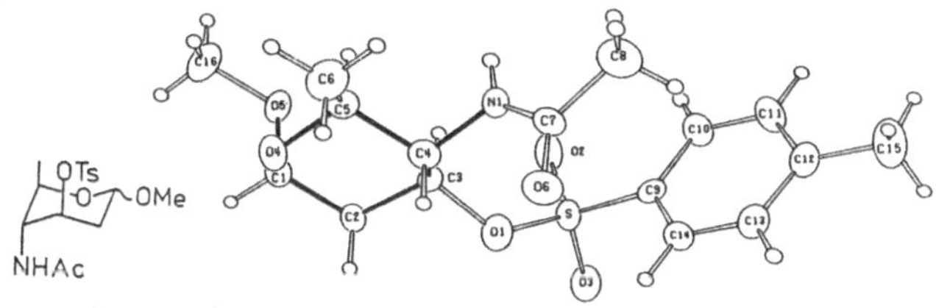

Fig. 1. A view of $C_{16}H_{23}NO_6S$.

3,4,6-TRI-O-ACETYL-2-DEOXY-α-D-threo-HEXENOPYRANOSYLBENZENE
$C_{18}H_{20}O_7$

H. GILLIER-PANDRAUD, R. BRAHMI, V. BELLOSTA-DECHAVANNE and S. CZERNECKI, 1985. *Canad. J. Chem.*, **63**, 491-494.

Orthorhombic, $P2_12_12_1$, a = 22.946, b = 10.319, c = 7.684 Å, Z = 4. Cu radiation, R = 0.080 for 1786 reflections.

The analysis shows that the sugar moiety has a half-chair $^0H_5$ conformation (Fig. 1) with the aromatic ring quasi-axial. The anomeric carbon atom has the α-configuration. NMR data show that the solution conformation is very similar to that in the solid.

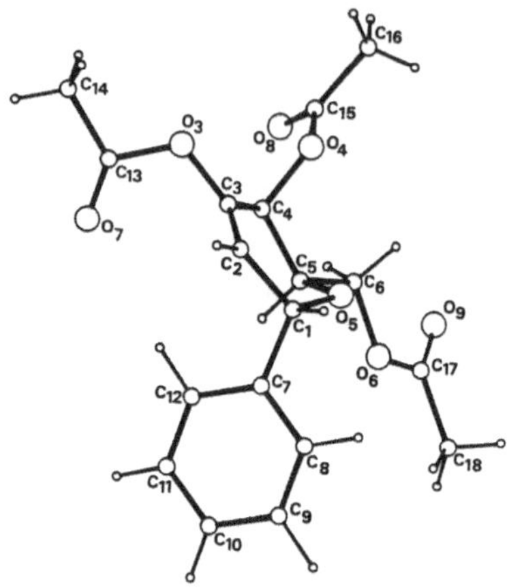

Fig. 1. A view of $C_{18}H_{20}O_7$.

(R)-1-(TETRA-O-ACETYL-β-D-GLUCOPYRANOSYLOXY)-2-CYCLOPENTENECARBONITRILE (DEIDACLIN TETRAACETATE)
$C_{20}H_{25}NO_{10}$

J.W. JAROSZEWSKI and B. JENSEN, 1985. *Acta Chem. Scand.*, B39, 867-875.

Monoclinic, *C*2, a = 21.057, b = 6.827, c = 15.321 Å, β = 101.11°, Z = 4. Mo radiation, R = 0.046 for 3353 reflections (at 100 K).

The hydrogen atoms attached to the 6-O-acetyl (C6'2) group are disordered. The crystal structure analysis unambiguously determined the absolute configuration at the cyanohydrin centre of the glucoside (Fig. 1).

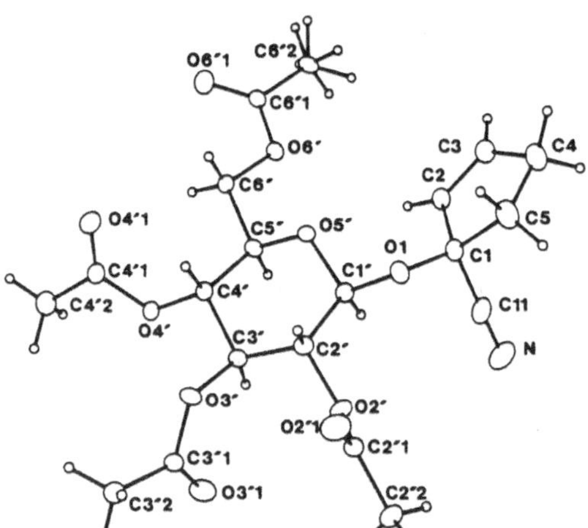

Fig. 1. Schematic drawing of a molecule of $C_{20}H_{25}NO_{10}$ showing the disorder at C(6'2).

N-(2,3:5,6-DI-O-ISOPROPYLIDENE-α-D-MANNOFURANOSYL)-2-BENZYLOXYETHYLIMINE N-OXIDE
$C_{21}H_{29}NO_7$ (I)

(+)-(1S,2'S)-DIETHYL (1-(3,3,3-TRIFLUORO-2-METHOXY-2-PHENYLPROPIONAMIDO)ETHYL)PHOSPHONATE
$C_{16}H_{23}F_3NO_6P$ (II)

R. HUBER, A. KNIERZINGER, J.-P. OBRECHT and A. VASELLA, 1985. *Helv. Chim. Acta*, **68**, 1730-1747.

I. Orthorhombic, $P2_12_12_1$, a = 5.1320, b = 15.541, c = 29.409 Å, Z = 4. Mo radiation, R = 0.074 for 2370 reflections (at -100°C).

II. Monoclinic, $P2_1$, a = 11.571, b = 8.468, c = 19.812 Å, β = 98.83°, Z = 4. Mo radiation, R = 0.170 (at 170 K).

Crystals of (II) decomposed during intensity measurements and the parameters were refined with isotropic thermal parameters. However, there seems to be no reason to doubt the relative configuration of C(1) and C(2') (Fig. 1). The absolute configuration of the molecules of (I) was determined.

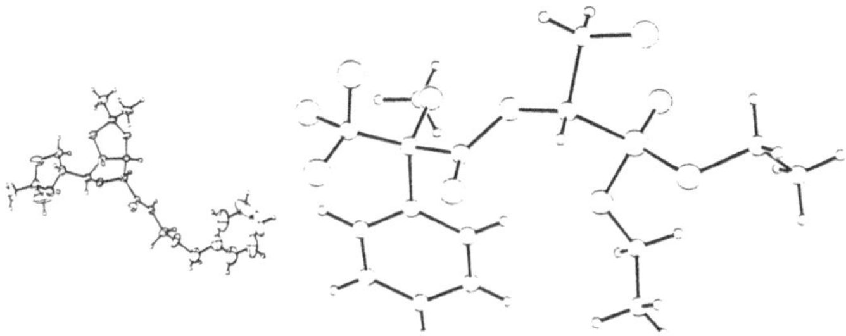

Fig. 1. Perspective drawings of molecules of $C_{21}H_{29}NO_7$ (left) and of $C_{16}H_{23}F_3NO_6P$ (right).

METHYL (BENZYL 2-((BENZYLOXYCARBONYL)AMINO)-2,3,4-TRIDEOXY-5-FLUORO-α-D-erythro-HEX-3-ENOPYRANOSIDE)URONATE
$C_{22}H_{22}FNO_6$

J. KISS, P.C. WYSS, G. FLESCH, W. ARNOLD, K. NOACK and P. SCHONHOLZER, 1985. *J. Carbohydr. Chem.*, **4**, 347-361.

Orthorhombic, $P2_12_12_1$, a = 4.929, b = 9.897, c = 41.483 Å, Z = 4. Cu radiation, R = 0.0451 for 1046 reflections (at 170 K).

The analysis established the structure and stereochemistry as shown in Fig. 1.

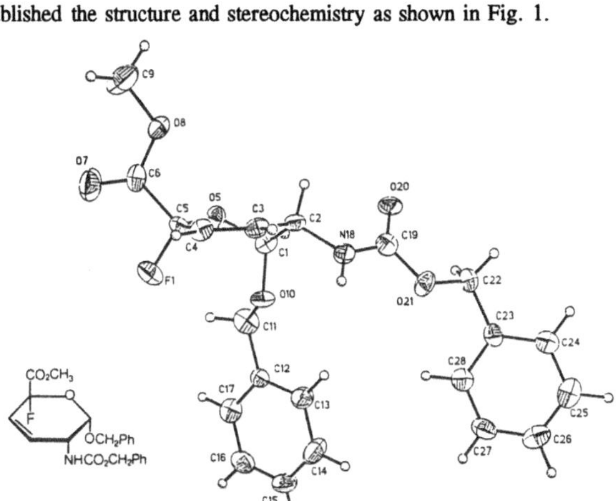

Fig. 1. The $C_{22}H_{22}FNO_6$ molecule.

(-)-(4aS)-6-(3'-O-(3-HYDROXYBENZOYL)-β-D-GLUCOPYRANOSYLOXY)-5-VINYL-4,4aα,5β,6α-TETRAHYDRO-1H,3H-PYRANO[3,4-c]PYRAN-1-ONE (DECENTAPICRIN A)
$C_{23}H_{26}O_{11}$

B. KOJIC-PRODIC, A.L. SPEK, W.G. VAN DER SLUIS and R.P. LABADIE, 1985. *Acta Cryst.*, **C41**, 379-383.

Orthorhombic, $P2_12_12_1$, a = 5.8234, b = 12.512, c = 31.693 Å, Z = 4. Mo radiation, R = 0.052 for 1088 reflections.

In the molecule (Fig. 1) the fused δ-lactone and pyran rings of the secoiridoid moiety have approximate envelope conformations with C(6) and C(9) as their respective flaps. The vinyl group attached at C(9) has an extended conformation. The β-glucose moiety has the usual $^4C_1$ chair conformation and the conformation about the glycosidic bond at O(1') is defined by the torsion angle 169.9(7)° for the C(1')-O(1')-C(1)-C(9) atom sequence. The m-hydroxybenzoyl moiety has normal geometry with the hydroxyl O(4'') coplanar with the benzene ring. The molecular packing is dominated by a three-dimensional network of intermolecular O-H...O hydrogen bonds of lengths in the range 2.65(1) to 3.19(1) Å.

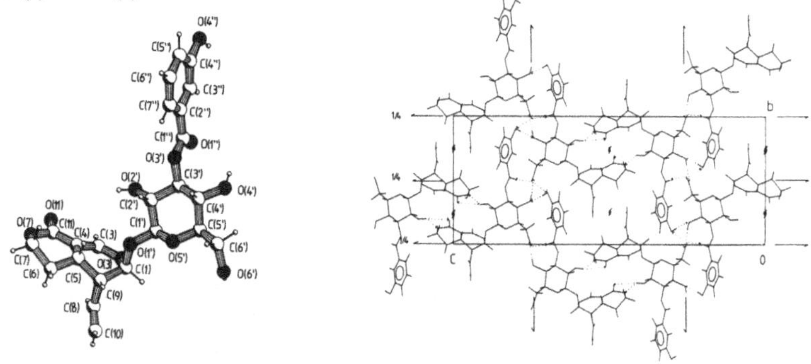

Fig. 1. $C_{23}H_{26}O_{11}$: a perspective view of the molecule and a projection of the crystal structure onto (100) with hydrogen bonds shown as dotted lines.

O-6-DEOXY-β-D-GLUCOPYRANOSYL(1-2)-6-DEOXY-α-D-GLUCOPYRANOSE HEXAACETATE (PHYLLANTHOSE HEXAACETATE)
$C_{24}H_{34}O_{15}$

L.R. NASSIMBENI, M.L. NIVEN, G.M. CRAGG and G.R. PETTIT, 1985. *Acta Cryst.*, **C41**, 728-731.

Monoclinic, $P2_1$, a = 13.077, b = 9.306, c = 12.009 Å, β = 92.33°, Z = 2. Cu radiation, R = 0.042 for 1930 reflections.

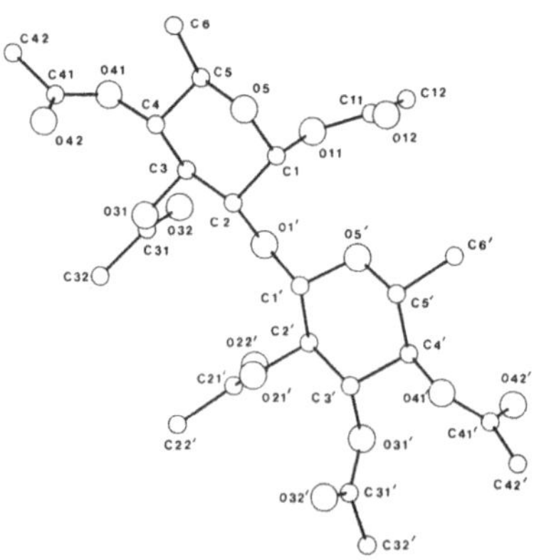

Fig. 1. A view of the $C_{24}H_{34}O_{15}$ molecule.

The C-C bond distances in the two pyranose rings of the molecule (Fig. 1) are in the range 1.516(7) to 1.536(8) Å. The C(2)-O(1') glycosidic bond of length 1.394(6) Å is significantly shorter than the glycosidic C(1')-O(1') bond of length 1.427(5) Å. The averages of the ring and other exocyclic C-O bond lengths are 1.429(10) and 1.446(5) Å respectively. Both rings are puckered and have $^4C_1$ conformations with a C(1')-O(1')-C(2) bridge angle of 115.4(3)°. Each acetate group is planar and the H-C(ring)-O-C(acetate) torsion angles lie in the range 12.8(8) to -25.7(7)°. Intermolecular distances are essentially van der Waals contacts.

3,5-DI-o-BENZOYL-1,2-DIDEOXY-1-PHENYL-β-D-RIBOFURANOSE
$C_{25}H_{22}O_5$

J. GUNNING, S. NEIDLE, T.A. MILLICAN, M.A.W. EATON, G.A. MOCK and J. MANN, 1985. *Acta Cryst.*, C**41**, 720-722.

Orthorhombic, $P2_12_12_1$, a = 4.946, b = 15.887, c = 26.555 Å, Z = 4. Cu radiation, R = 0.054 for 648 reflections.

The molecule (Fig. 1) has a propeller shape with benzoyl substituents attached to the hydroxyl positions of the deoxyribose sugar. Phenyl rings B2 and B3 act as shielding groups and B1 as a substitute for a base. The ribofuranose ring has an unsymmetrical $^2T_3$ conformation with a C(2')-endo, C(3')-exo pucker. Bond lengths and angles in the structure are normal and the molecules pack with no intermolecular stacking interactions between the phenyl groups. Intermolecular distances are van der Waals contacts.

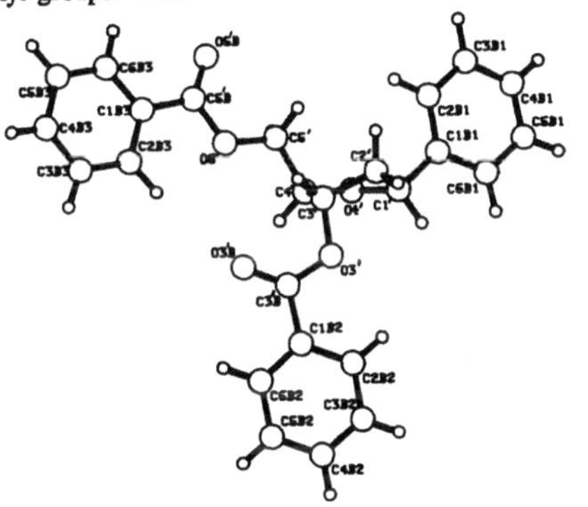

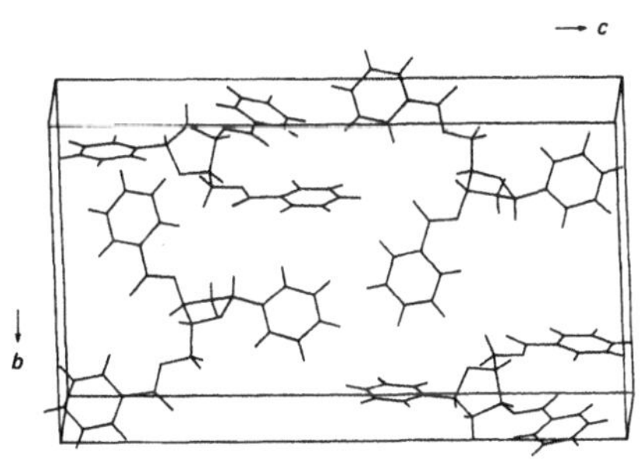

Fig. 1. $C_{25}H_{22}O_5$: the molecular structure showing the atomic numbering and molecular packing in the unit cell viewed down the a-axis.

METHYL 2,3-O-D-GLUCOPYRANOSYLIDENE-α-D-MANNOPYRANOSIDE HEXAACETATE
$C_{25}H_{34}O_{17}$

K. ASANO, S. HORITO, A. SAITO, J. YOSHIMURA and K. UENO, 1985. *Carbohydr. Res.*, **136**, 1-11.

Monoclinic, $P2_1$, a = 9.568, b = 8.893, c = 18.175 Å, β = 94.66°, Z = 2. Cu radiation, R = [not quoted] for 2620 reflections.

From the known D-mannopyranoside and D-glucopyranoside moieties the absolute configuration of the orthoester carbon was determined to be S (Fig. 1).

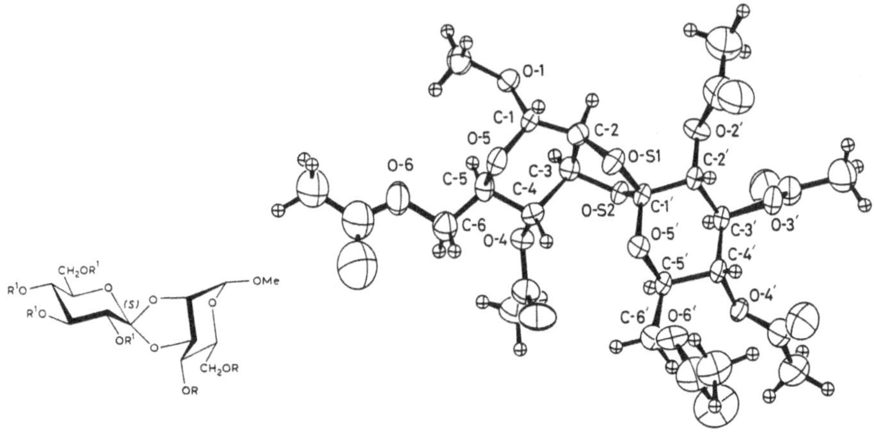

Fig. 1. The $C_{25}H_{34}O_{17}$ absolute configuration.

HEXAACETYLLEONURIDE
$C_{27}H_{36}O_{15}$

K. WEINGES, M. ZOURARI, H. SMUDA, H. RODEWALD, M. NIXDORF and H. IRNGARTINGER, 1985. *Liebigs Ann. Chem.*, 1063-1081.

Monoclinic, $P2_1$, a = 13.979, b = 6.709, c = 16.382 Å, β = 92.58°, Z = 2. Mo radiation, R = 0.044 for 2669 reflections (at -35°C).

The analysis establishes the absolute configuration of the leonuride moiety (Fig. 1) and shows that atoms C(1), C(8), C(9) have S configuration and C(5), C(6) have R-configuration. Oxygen atoms O(17) and O(21) are cis.

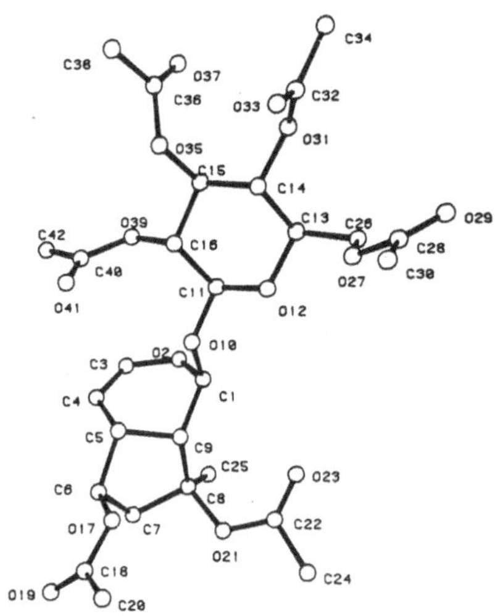

Fig. 1. The absolute stereochemistry of $C_{27}H_{36}O_{15}$.

3,3,5-TRIMETHYL-4-(3-OXO-1-BUTENYL) 4,5-EPOXYCYCLOHEXYL-2,3,4,6-TETRA-O-ACETYL-β-D-GLUCOPYRANOSIDE
$C_{27}H_{38}O_{12}$

F. PAVELČIK, K. HAVETTA and V. SUCHY, 1985. *Acta Cryst.*, **C41**, 1272-1274.

Orthorhombic, $P2_12_12_1$, a = 12.702, b = 13.432, c = 18.263 Å, Z = 4. Cu radiation, R = 0.061 for 2261 reflections.

The epoxycyclohexane ring has an unsymmetrical sofa conformation, distorted towards half-chair (Fig. 1). Bond lengths and angles are normal and show the usual variation associated with the C(15)-O(4)-C(14)-O(1)-C(1) sequence, 1.420(5), 1.415(5), 1.397(5) and 1.441(5) Å. The O(4)-C(14)-O(1)-C(1) torsion angle is -86.6°. The ester group at C(19) is disordered. In the crystal, discrete molecules are held together by van der Waals interactions.

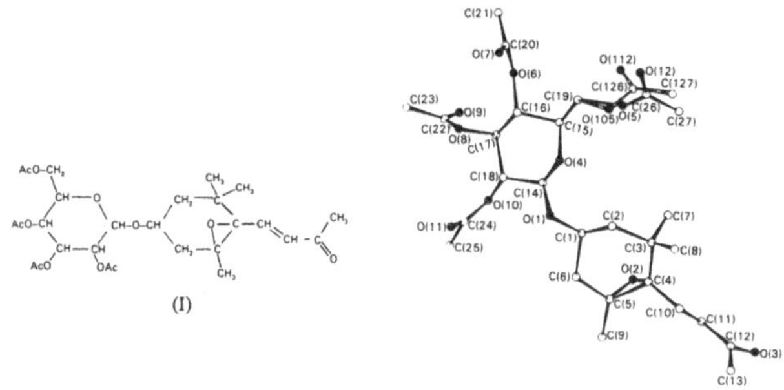

(I)

Fig. 1. $C_{27}H_{38}O_{12}$: the molecular skeleton and structure.

1,2,4,6-TETRA-O-ACETYL-3-O-(2,3,4,6-TETRA-O-ACETYL-β-D-GLUCOPYRANOSYL)-β-D-GLUCOPYRANOSE (LAMINARABIOSE OCTAACETATE)

$C_{28}H_{38}O_{19}$

S. PEREZ, C. VERGELATI and V.H. TRAN, 1985. *Acta Cryst.*, **B41**, 262-267.

Orthorhombic, $P2_12_12_1$, a = 10.379, b = 22.943, c = 14.599 Å, Z = 4. Cu radiation, R = 0.095 for 1660 reflections.

The two D-glucose residues (Fig. 1) have the $^4C_1$ pyranose conformation and are β-(1→3) linked. The conformational angles φ and ψ at the glucoside linkage are -81.1 and 134.8° respectively.

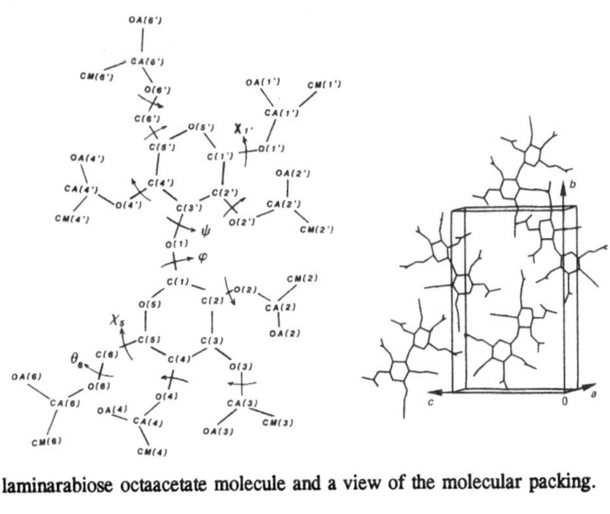

Fig. 1. The laminarabiose octaacetate molecule and a view of the molecular packing.

2-ACETAMIDO-1,3,6-TRI-O-ACETYL-2-DESOXY-4-O-(2,3,4,6-TETRA-O-ACETYL-α-L-IDOPYRANOSYL)-α-D-GLUCOPYRANOSE

$C_{28}H_{39}NO_{18}$

A. NEUMAN, J. BECQUART, D. AVENEL, H. GILLIER-PANDRAUD and P. SINAY, 1985. *Carbohydr. Res.*, **139**, 23-34.

Monoclinic, $P2_1$, a = 9.331, b = 15.385, c = 11.989 Å, β = 97.07°, Z = 2. Cu radiation, R = 0.045 for 2412 reflections.

The analysis shows that the L-idosyl residue has a $^4C_1$(L) conformation (Fig. 1) which results in the unusual axial position of the acetoxymethyl group at C(5). The glucopyranose ring has the normal $^4C_1$(D) chair conformation. Molecular dimensions are generally in accord with literature values.

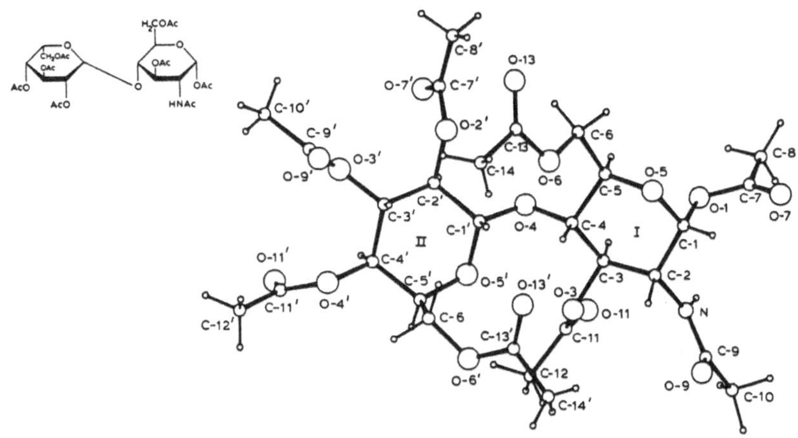

Fig. 1. The $C_{28}H_{39}NO_{18}$ molecule and an ORTEP plot.

1-(S)-ACETOXY-3-(6-(R)-O-BENZYL-1,2:3,4-DI-O-ISOPROPYLIDENE-α-D-GALACTOPYRANOS-6-YL)-1-(METHYL 2,3,4-TRI-O-BENZYL-6-DEOXY-β-D-GALACTOPYRANOSID-6-YL)PROP-2-YNE
$C_{52}H_{60}O_{13}$

J.W. KRAJEWSKI, P. GLUZINSKI, S. JAROSZ, A. ZAMOJSKI, J. BLEIDELIS, A. MISHNYOV and A. KEMME, 1985. *Carbohydr. Res.*, **144**, 183-195.

Monoclinic, $P2_1$, a = 14.896, b = 8.295, c = 20.547 Å, β = 102.66°, Z = 2. Cu radiation, R = 0.041 for 3839 reflections.

In this condensation product (Fig. 1) the configuration at the new chiral centre (C(1)) was established as S(D). The galactopyranose rings have conformations $^4C_1$ (tri-O-benzylated moiety), and $^0S_5$ + $^0T_2$ (di-O-isopropylenated moiety). The 1,2- and 3,4-O-isopropylidene rings have $^3T_2$ and ^2E conformations respectively.

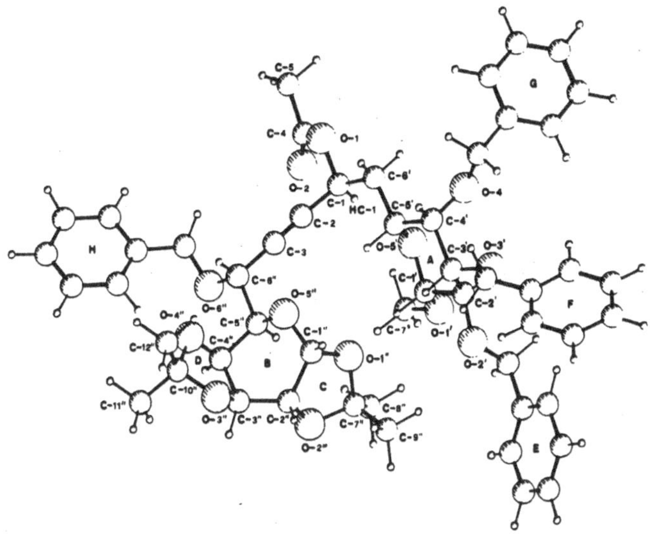

Fig. 1. A view of the condensation product $C_{52}H_{60}O_{13}$.

TRISODIUM D-2-PHOSPHOGLYCERATE HEXAHYDRATE

$C_3H_4Na_3O_7P,6(H_2O)$ $C_3H_4O_7P^{3-},3(Na^+),6(H_2O)$

T. LIS, 1985. *Acta Cryst.*, C41, 1578-1580.

Orthorhombic, $P2_12_12$, a = 34.82, b = 6.334, c = 6.146 Å, Z = 4. Cu radiation, R = 0.032 for 2500 reflections.

In the glycerate residue of the trianion (Fig. 1) all the C-O and C-C bond lengths are normal and the carboxyl group is oriented perpendicular to the C-atom chain. Atom O(3) is antiperiplanar to C(1). The phosphate ester bond P-O(2) is of length 1.625(2) Å and the orientation of the phosphate group is similar to that found in 1. The phosphate oxygen atom O(5) is pentacoordinate as it acts as an acceptor in hydrogen bonds with four water molecules. These bond lengths range from 2.762(3) to 2.833(2) Å. There are three different kinds of Na^+ ions in the structure. Na(1) is five-coordinate, Na(2) is five-coordinate but in addition has a slightly weaker link to O(2) and Na(3) has distorted octahedral coordination. The crystal structure is stabilized by an extensive three-dimensional network of hydrogen bonds.

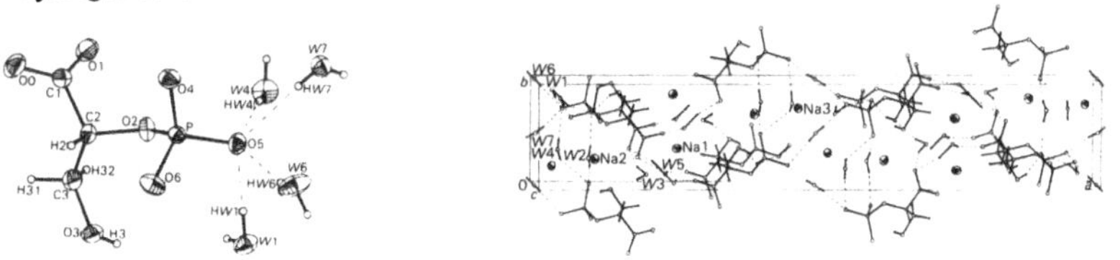

Fig. 1. $C_3H_4Na_3O_7P,6(H_2O)$: a view of the trianion and the hydrogen bonds, shown as dashed lines, formed with O(5) and a view of the crystal packing.

1. T. LIS and P. STARYNOWICZ, 1983. Abstracts of XXVI Konservatorium Kristalograficzne, p. 30.

DIPOTASSIUM FRUCTOSE-6-PHOSPHATE HEXAHYDRATE

$C_6H_{11}K_2O_9P,6(H_2O)$ $C_6H_{11}O_9P^{2-},2(K^+),6(H_2O)$

N. NARENDRA, T.P. SESHADRI and M.A. VISWAMITRA, 1985. *Acta Cryst.*, C41, 1612-1614.

Orthorhombic, $P2_12_12_1$, a = 9.855, b = 10.593, c = 17.057 Å, Z = 4. Cu radiation, R = 0.080 for 1358 reflections.

In the dianion (Fig. 1) the conformation about the C(5)-C(6) bond is gauche-trans. In the tetrahedral phosphate group two negative charges are distributed over atoms O(7,8,9). The average of these P-O bond lengths is 1.517(8) Å while the P-O(6) bond length is 1.609(7) Å. The furanose ring adopts the ^3E envelope conformation in which C(3) is displaced by 0.55(1) Å from the plane of the four other ring atoms. The two endocyclic C-O bonds, C(2)-O(2) and C(5)-O(5), have respective lengths of 1.40(1) and 1.44(1) Å. The two crystallographically independent K^+ ions bind to the phosphate group indirectly through water bridges and to the exocyclic hydroxyl groups O(2,3,4). K(1) is seven- and K(2) is six-coordinate. The crystal structure is stabilized by hydrogen bonds between the molecular O atoms and the water molecules.

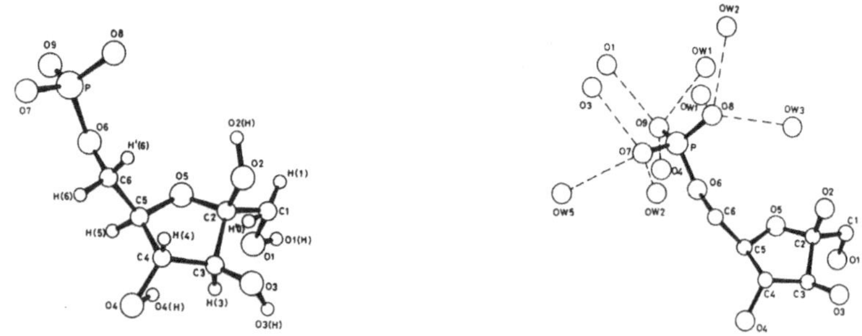

Fig. 1. $C_6H_{11}K_2O_9P,6(H_2O)$: views of the dianion and the environment of the phosphate group (with broken lines representing hydrogen bonds).

TRISODIUM FRUCTOSE 1,6-DIPHOSPHATE OCTAHYDRATE

$C_6H_{11}Na_3O_{12}P_2,8(H_2O)$ $C_6H_{11}O_{12}P_2^{3-},3(Na^+),8(H_2O)$

N. NARENDRA, T.P. SESHADRI and M.A. VISWAMITRA, 1985. *Acta Cryst.*, **C41**, 31-34.

Monoclinic, $P2_1$, a = 7.085, b = 11.645, c = 13.290 Å, β = 102.36°, Z = 2. Cu radiation, R = 0.055 for 2303 reflections.

The phosphate residues at the 1 and 6 positions in the molecule (Fig. 1) have different phosphorus ester bond lengths. The P(1)-O(1) and P(2)-O(2) distances are 1.591(3) and 1.622(3) Å respectively. The conformation about the C(5)-C(6) bond is gauche-trans and that about the C(1)-C(2) bond is gauche-gauche. The furanose ring is in the ^3E conformation with C(3) displaced from the O(5)-C(2)-C(4)-C(5) plane by 0.580(4) Å. The endocyclic bonds C(2)-O(5) and C(5)-O(5) are of lengths 1.434(5) and 1.448(5) Å respectively. The Na^+ ions bond to the phosphate groups indirectly through water bridges and the crystal structure is stabilized by an extensive network of hydrogen bonds.

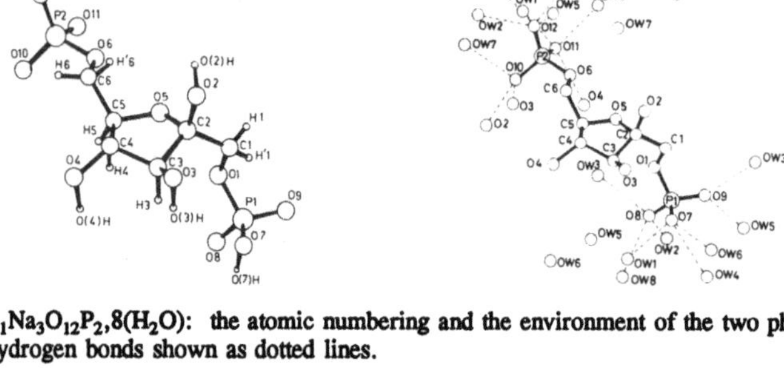

Fig. 1. $C_6H_{11}Na_3O_{12}P_2,8(H_2O)$: the atomic numbering and the environment of the two phosphate groups with the hydrogen bonds shown as dotted lines.

SODIUM D-GLUCOSE-6-(HYDROGENPHOSPHATE)

$C_6H_{12}NaO_9P$ $C_6H_{12}O_9P^-,Na^+$

T. LIS, 1985. *Carbohydr. Res.*, **135**, 187-194.

Monoclinic, $P2_1$, a = 5.774, b = 7.189, c = 12.69 Å, β = 106.66°, Z = 2. Cu radiation, R = 0.033 for 977 reflections [see also the following Report].

The pyranose ring adopts the $^4C_1$ conformation (Fig. 1) with torsion angles O(6)-C(6)-C(5)-O(5) 64.2(8) and O(6)-C(6)-C(5)-C(4) -175.6(7)°. The phosphate ester bond P-O(6) is 1.584(3) Å. All bond distances and angles are normal. The sodium ion is coordinated to five mono anions with Na-O 2.38-2.60 Å and two slightly longer distances of 2.87 Å. All OH groups take part in an inter-ion hydrogen bonded network.

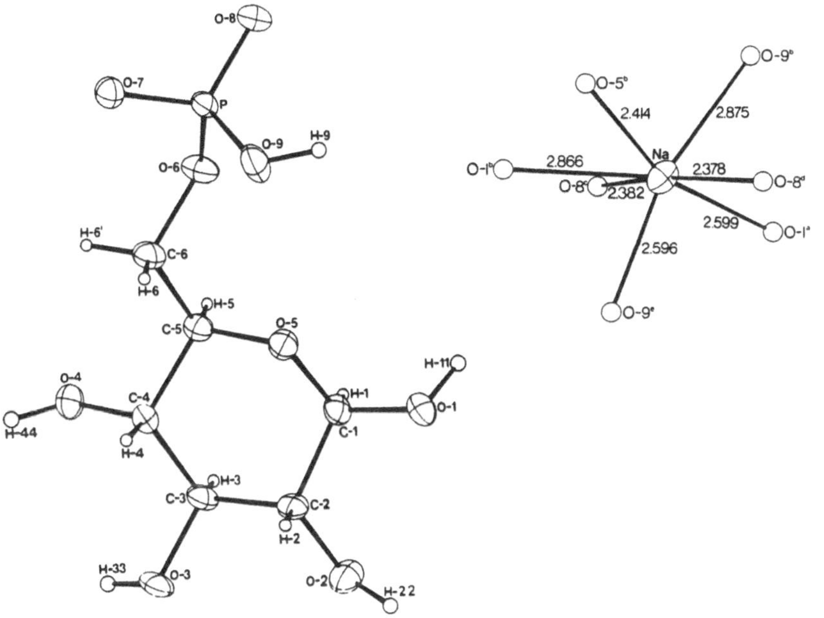

Fig. 1. The D-glucose phosphate ion and the Na^+ coordination.

SODIUM D-GLUCOSE-6-(HYDROGENPHOSPHATE)

$C_6H_{12}NaO_9P$ $C_6H_{12}O_9P^-,Na^+$

N. NARENDRA and M.A. VISWAMITRA, 1985. *Acta Cryst.*, **C41**, 1621-1624.

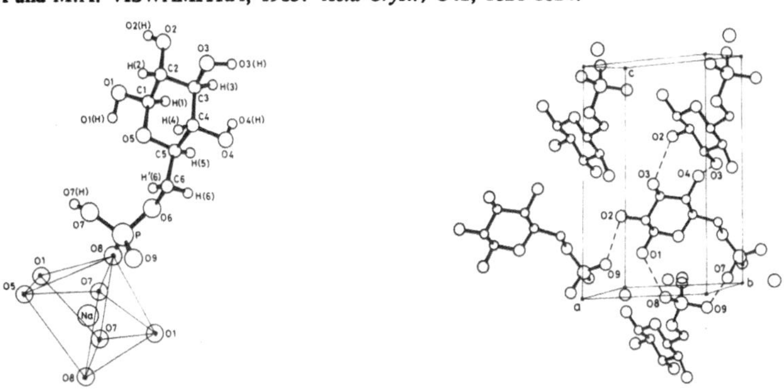

Fig. 1. $C_6H_{12}NaO_9P$: views showing the sodium coordination polyhedron and the packing arrangement (with hydrogen bonds shown as dashed lines).

Monoclinic, $P2_1$, a = 5.762, b = 7.163, c = 12.313 Å, β = 99.97°, Z = 2. Cu radiation, R = 0.042 for 922 reflections [see also the preceding Report].

In the phosphate moiety in the structure (Fig. 1) the P-O(6) ester bond is of length 1.575(5) Å and the single negative charge carried by the group is distributed between O(8) and O(9). The pyranose ring adopts a $^4C_1$ chair conformation and the endocyclic C-O bonds are of nearly equal length (1.436(9) Å). The conformation about the exocyclic C(5)-C(6) bond is gauche-trans. The Na^+ ion is surrounded by seven O atoms, four are phosphate O atoms and the other three are sugar O atoms. The crystal structure is stabilized by hydrogen bonds between O atoms of symmetry-related molecules. These O-H...O separations lie in the range of 2.774(9) to 3.184(9) Å.

OXONIUM BIS(p-NITROPHENYL)PHOSPHATE
$C_{12}H_{11}N_2O_9P$ (I) $C_{12}H_8N_2O_8P^-,H_3O^+$

TRIETHYLAMMONIUM BIS(p-NITROPHENYL)PHOSPHATE
$C_{18}H_{24}N_3O_8P$ (II) $C_{12}H_8N_2O_8P^-,C_6H_{16}N^+$

D.R. BOND, T.A. MODRO, M.L. NIVEN and L.R. NASSIMBENI, 1985. *S. Afr. J. Chem.*, **38**, 78-82.

I. Monoclinic, $P2_1/c$, a = 9.101, b = 6.316, c = 25.450 Å, β = 96.28°, Z = 4. Mo radiation, R = 0.034 for 1902 reflections.

II. Monoclinic, Pc, a = 9.069, b = 8.616, c = 13.471 Å, β = 96.23°, Z = 2. Mo radiation, R = 0.058 for 1660 reflections.

In the oxonium salt (I), the oxonium hydrogens were clearly located and three strong O-H...O hydrogen bonds cross-link the phosphoryl oxygens (O...O 2.44-2.64 Å) (Fig. 1). In the triethylammonium salt (II) a single N-H...O hydrogen bond (N...O 2.651 Å) groups the ions in pairs. Anion dimensions in (I) include P-O(1,2) 1.471 and 1.482(2), P-O(3,4) 1.589 and 1.595(2) Å, O(1)-P-O(2) 119.4, O(3)-P-O(4) 99.6(1)°; the corresponding values for (II) are 1.470 and 1.477(5), 1.616 and 1.628(4) Å, 122.6(3) and 100.7(2)°. There is a correlation between the pK_a of the cations and the O(1)-P-O(2) angle.

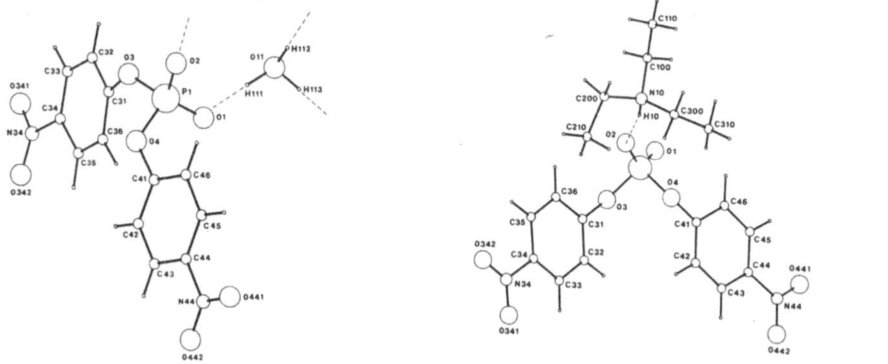

Fig. 1. Views of $C_{12}H_8N_2O_8P^-,H_3O^+$ (I) (left) and $C_{12}H_8N_2O_8P^-,C_6H_{16}N^+$ (II) (right).

QUINOLIN-8-YL BIS(p-NITROPHENYL)PHOSPHATE
$C_{21}H_{14}N_3O_8P$ (I)

QUINOLIN-8-YL (p-NITROPHENYL)PHOSPHATE
$C_{15}H_{11}N_2O_6P$ (II)

1-NAPHTHYL BIS(p-NITROPHENYL)PHOSPHATE
$C_{22}H_{15}N_2O_8P$ (III)

D.R. BOND, T.A. MODRO and L.R. NASSIMBENI, 1985. *J. Org. Chem.*, **50**, 2281-2287.

I. Triclinic, $P\bar{1}$, a = 9.452, b = 8.751, c = 13.000 Å, α = 100.59, β = 107.35, γ = 85.21°, Z = 2. Mo

radiation, R = 0.0449 for 1869 reflections.

II. Triclinic, $P\bar{1}$, a = 7.106, b = 9.559, c = 10.521 Å, α = 84.32, β = 82.89, γ = 78.60°, Z = 2. Mo radiation, R = 0.0507 for 1212 reflections.

III. Monoclinic, $P2_1/n$, a = 7.282, b = 12.683, c = 22.40 Å, β = 97.59°, Z = 4. Cu radiation, R = 0.0523 for 2743 reflections.

In all three molecules the phosphorus atom has a geometry which is reasonably close to tetrahedral. The O-P-O angle in (II) is significantly larger than in (I) and (III). (II) exists as a zwitterion. A molecule of (I) is shown in Fig. 1.

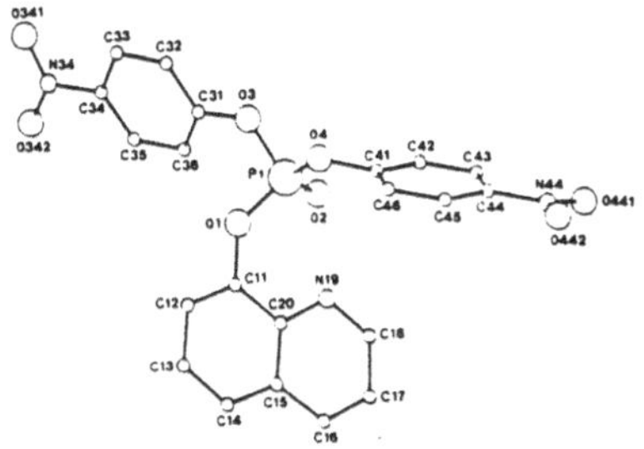

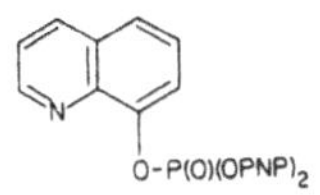

Fig. 1. Perspective view of a molecule of $C_{21}H_{14}N_3O_8P$.

2,3-O-ISOPROPYLIDENE-4-O-DIMETHOXYPHOSPHORYL-5-O-TRITYL-D-RIBONITRILE $C_{29}H_{32}NO_7P$

R. MEUWLY and A. VASELLA, 1985. *Helv. Chim. Acta*, **68**, 997-1009.

Orthorhombic, $P2_12_12_1$, a = 8.634, b = 13.335, c = 23.254 Å, Z = 4. Mo radiation, R = 0.042 for 5826 reflections.

The crystal structure analysis established the identity of the molecule (Fig. 1).

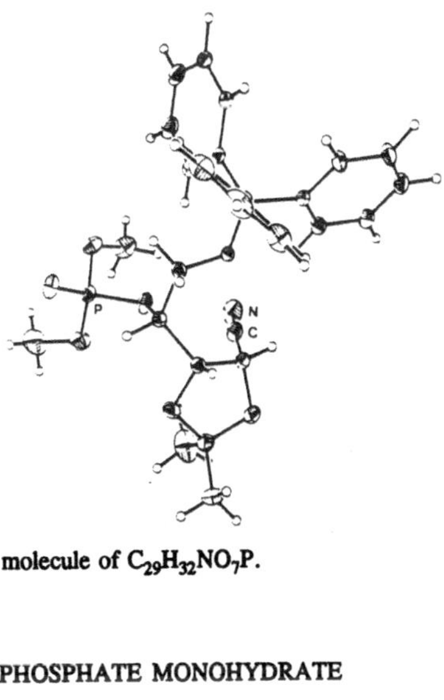

Fig. 1. ORTEP drawing of a molecule of $C_{29}H_{32}NO_7P$.

2'-DEOXYCYTIDINE-5'-MONOPHOSPHATE MONOHYDRATE $C_9H_{14}N_3O_7P,H_2O$

D.A. PEARLMAN and SUNG-HOU KIM, 1985. *Biopolymers*, **24**, 327-357.

Orthorhombic, $P2_12_12_1$, a = 6.7482, b = 11.3307, c = 16.5692 Å, Z = 4. Mo radiation, R = 0.027 for 5122 reflections (at -150°C).

For previous work see (1). This new study was carried out to determine the partial charges of the atoms in the structure and the results, shown in Fig. 1, are in general qualitative agreement with theoretical values, but significant differences exist.

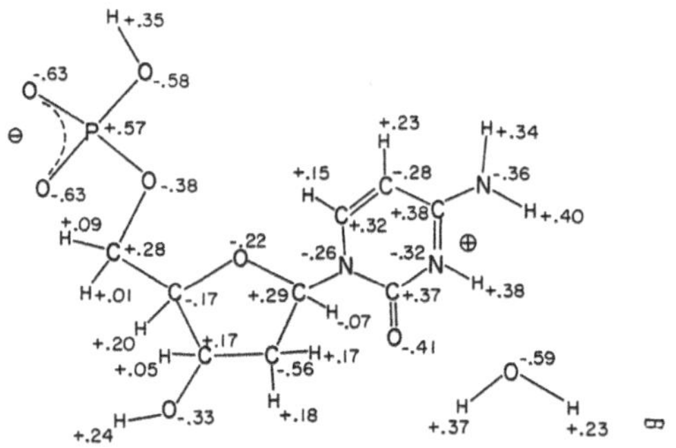

Fig. 1. The $C_9H_{14}N_3O_7P,H_2O$ zwitterion structure with atomic partial charges (in units of an electron charge).

1. Structure Reports, 37B, 254.

2-AZACOFORMYCIN
$C_{10}H_{15}N_5O_5$

G.I. BIRNBAUM, J.R. BRISSON, S.H. KRAWCZYK and L.B. TOWNSEND, 1985. *J. Am. Chem. Soc.*, **107**, 7635-7640.

Orthorhombic, $P2_12_12_1$, a = 7.022, b = 8.491, c = 20.773 Å, Z = 4. Mo radiation, R = 0.043 for 1502 reflections.

The molecule and dimensions are shown in Fig. 1. The diazepine ring is in a somewhat distorted and flattened sofa conformation, C(7) being displaced 0.751 Å from the mean plane. The glycosyl torsion angle is in the high-anti range (χ_{CN} 72.4°), the furanose ring has a C(2') endo/C(3')exo ($^2T_3$) pucker, and the conformation of the-CH_2OH side chain is gauche.

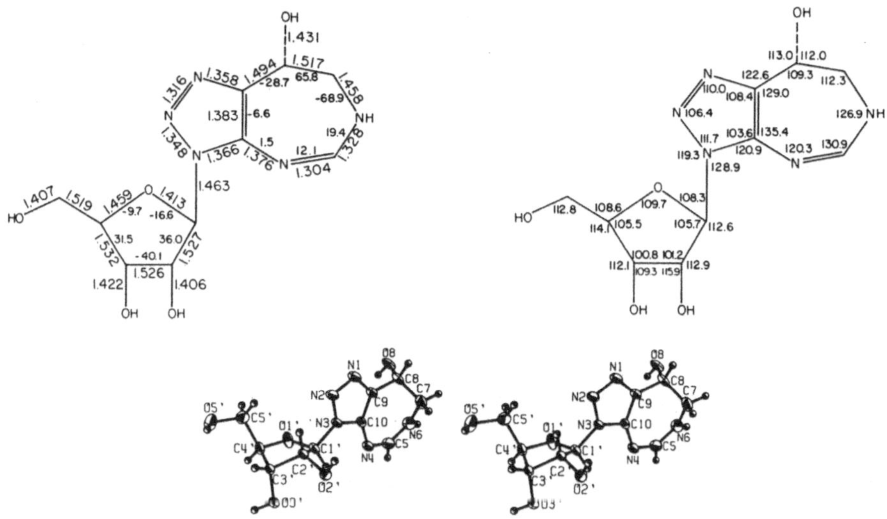

Fig. 1. The 2-azacoformycin molecule and dimensions.

3-METHYLCYTIDINIUM CHLORIDE
$C_{10}H_{16}ClN_3O_5$ $C_{10}H_{16}N_3O_5{}^+,Cl^-$

M. JASKOLSKI and M. ALEJSKA, 1985. *Acta Cryst.*, **C41**, 599-602.

Orthorhombic, $P2_12_12_1$, a = 7.225, b = 11.639, c = 14.921 Å, Z = 4. Cu radiation, R = 0.030 for 1959 reflections.

The dimensions (Fig. 1) and conformation of the cation (χ 45.2, P 163.0, τ_m 40.3, ψ 55.2°) are very similar to those of the cytidinium cation in other molecules. The pyrimidine ring is significantly non-planar (maximum deviation 0.030(2) Å) and N(4) is 0.118 Å from the pyrimidine plane. There is a two-dimensional network of hydrogen bonds in which all available proton donors are utilized: NH...O(2) 2.89, NH...Cl^- 3.20, OH...Cl^- 3.07, 3.25 and 3.27 Å.

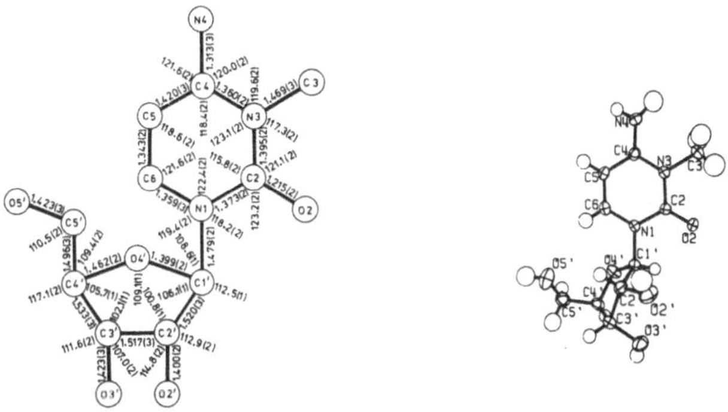

Fig. 1. $C_{10}H_{16}ClN_3O_5$: bond distances and angles, and a view of the cation.

6,2'-ANHYDRO-6-ETHYL-1β-D-ARABINOFURANOSYLURACIL
$C_{11}H_{14}N_2O_5$

Y. YAMAGATA, N. OKABE, K.-I. TOMITA, S. SHUTO, H. INOUE and T. UEDA, 1985. *Acta Cryst.*, **C41**, 1653-1656.

Orthorhombic, $P2_12_12_1$, a = 6.400, b = 9.734, c = 17.797 Å, Z = 4. Cu radiation, R = 0.057 for 1030 reflections.

The glycosidic torsion angle in the molecule (Fig. 1), O(4')-C(1')-N(1)-C(2), is 268.1(3)° implying an anti conformation. The sugar ring has the unusual $^0T_4$ conformation. The six-membered ring formed between the sugar and the base has an envelope conformation such that steric hindrance between the C(5') atom and the atoms of the fused ring moiety is minimised. The O(5') atom is disordered between two sites with occupancy factors of 0.6 and 0.4. The two possible conformations corresponding to these positions are gauche-trans and trans-gauche, respectively. Bond lengths and angles in the molecule agree well with generally accepted values. In the crystal no base stacking is observed and the structure is stabilized by a network of N-H...O and O-H...O hydrogen bonds.

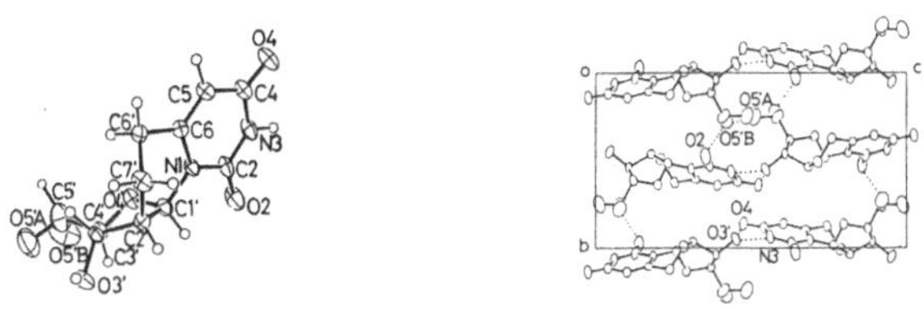

Fig. 1. View of the $C_{11}H_{14}N_2O_5$ molecule showing the disorder at O(5') and of the molecular packing.

2'-CHLORO-2'-DEOXY-1,3-DIMETHYL-ψ-URIDINE
$C_{11}H_{15}ClN_2O_5$

K.W. PANKIEWICZ, K.A. WATANABE, H. TAKAYANAGI, T. ITOH and H. OGURA, 1985. *J. Heterocycl. Chem.*, **22**, 1703-1710.

Monoclinic, $P2_1$, a = 14.677, b = 16.655, c = 5.420 Å, β = 95.41°, Z = 4. Cu radiation, R = 0.070 for 2334 reflections.

There are two independent molecules in the asymmetric unit which differ primarily in the conformation of the 5'-OH group. The analysis established the configuration of the molecules, with the 2'-Cl atom in the "down" ribo configuration. The conformation of the sugar rings is C3'-*endo*-C2'-*exo*-($^3T_2$) and of the base to sugar is *anti*.

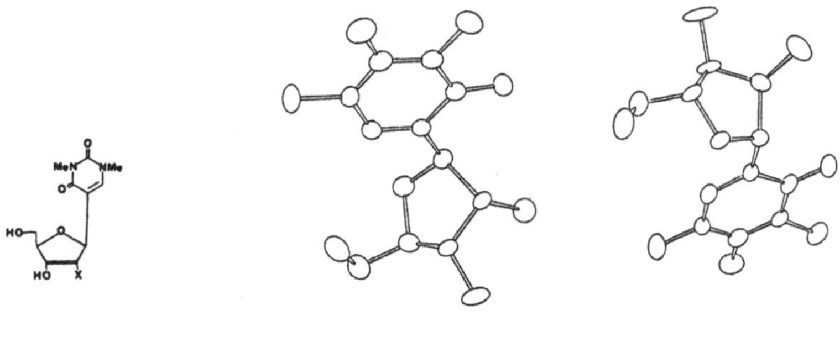

Fig. 1. The $C_{11}H_{15}ClN_2O_5$ skeleton (X=Cl) and views of the two independent molecules.

N^6-METHYL-2'-DEOXYADENOSINE
$C_{11}H_{15}N_5O_3$

T. SATO, 1985. *Acta Cryst.*, **C41**, 1624-1625.

Orthorhombic, $P2_12_12_1$, a = 10.246, b = 22.686, c = 5.233 Å, Z = 4. Mo radiation, R = 0.037 for 1490 reflections.

The deoxyribose ring in the molecule (Fig. 1) is puckered in the $_1{}^2T$ twist form with C(2') endo and C(1') exo. The base is in the anti conformation with torsion angle O(4')-C(1')-N(9)-C(8) of value 67.2(3)°. The conformation about the C(4')-C(5') bond is gauche-trans. The purine ring is slightly buckled, the largest displacement from the mean plane of the imidazole ring being 0.075(2) Å for C(6). Bond lengths and angles in the base are very similar to those in the unsubstituted adenine (1) except for the exocyclic bond angles involving N(6). The methyl substituent at N(6) is anti with respect to the purine ring. Atom C(1') is displaced by 0.254(2) Å from the imidazole ring plane. No appreciable ring stacking is observed in the crystal structure which is fully hydrogen bonded. There is also a short intermolecular C(8)-H...N(3) contact of length 3.208(3) Å.

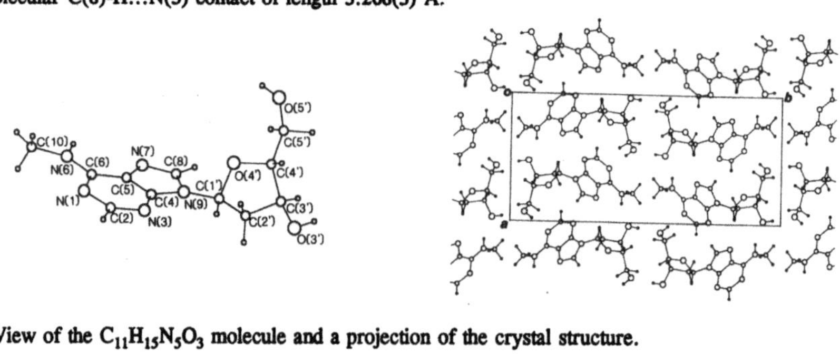

Fig. 1. View of the $C_{11}H_{15}N_5O_3$ molecule and a projection of the crystal structure.

1. R. TAYLOR and O. KENNARD, 1982. J. Mol. Struct., 78, 1-28.

8-METHYLGUANOSINE TRIHYDRATE

$C_{11}H_{15}N_5O_5,3(H_2O)$

K. HAMADA, I. HONDA, S. FUJII, T. FUJIWARA and K.-I. TOMITA, 1985. *Acta Cryst.*, **C41**, 1486-1488.

Orthorhombic, *C*222$_1$, a = 8.464, b = 30.848, c = 12.233 Å, Z = 8. Cu radiation, R = 0.053 for 1448 reflections.

In the molecule (Fig. 1) the ribose ring has a C(2')-endo pucker and the conformation around the C(5')-C(4') bond is gauche-gauche (g,g). The glycosidic torsion angle is 52.7(5)°. An intramolecular hydrogen bond connects the O(5') hydroxyl group with the N(3) atom of the guanine base (O(5')...N(3) 2.839(5) Å). The crystal structure is stabilised by intermolecular face-to-face base stackings and extensive hydrogen-bond networks involving guanine-guanine base-pairing.

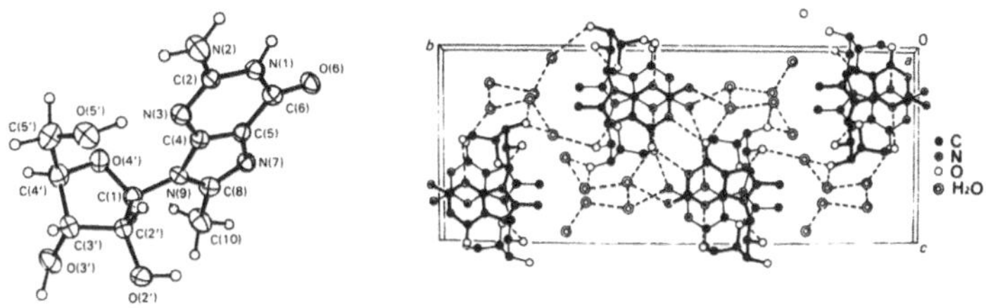

Fig. 1. $C_{11}H_{15}N_5O_5,3(H_2O)$: the molecular structure and packing in the unit cell.

1-(β-D-2',3'-seco-RIBOFURANOSYL)-5,6-DICHLOROBENZIMIDAZOLE
$C_{12}H_{14}Cl_2N_2O_4$

G.I. BIRNBAUM, R. STOLARSKI, Z. KAZIMIERCZUK and D. SHUGAR, 1985. *Canad. J. Chem.*, **63**, 1215-1221.

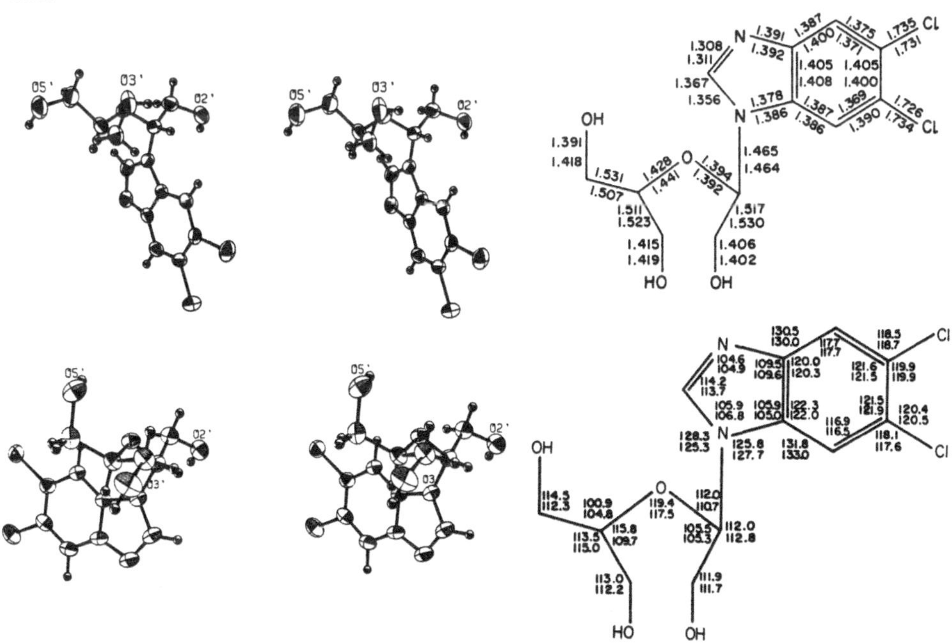

Fig. 1. Views of the two $C_{12}H_{14}Cl_2N_2O_4$ molecules and bond lengths (σ 0.003-0.006 Å) and angles (σ 0.2-0.3°).

Monoclinic, *P*2$_1$, a = 11.861, b = 7.897, c = 14.527 Å, β = 91.28°, Z = 4. Mo radiation, R = 0.036 for 2980

reflections.

There are two independent molecules in the asymmetric unit (Fig. 1) with different conformations. The glycosidic torsion angle [C(2)-N(1)-C(1')-O(1')] is 53.9° in molecule A and 129.8° in B. In both molecules the C(1')-O(1') to C(2')-O(2') orientation is trans.

5'-O-ACETYL-6,3'-ANHYDRO-2'-DEOXY-6-METHYL-1β-D-XYLOFURANOSYLURACIL $C_{12}H_{14}N_2O_5$

Y. YAMAGATA, N. OKABE, K.-I. TOMITA, S. SHUTO and T. UEDA, 1985. *Acta Cryst.*, C41, 1656-1658.

Monoclinic, $P2_1$, a = 8.312, b = 5.559, c = 13.048 Å, β = 98.50°, Z = 2. Cu radiation, R = 0.049 for 1055 reflections.

In the molecule (Fig. 1) the glycosidic torsion angle O(4')-C(1')-N(1)-C(2) is 265.3(5)° and lies in the anti range. The furanose ring has an E_2 envelope conformation. The six-membered fused ring, formed by cyclization between the sugar C(3') and atom C(6) of the uracil base, is in an envelope conformation with C(2') as flap. Bond lengths and angles in the uracil moiety agree well with accepted values except for the angle C(1')-C(2')-C(3') which, at 98.5(4)°, is significantly smaller than the average value. This is ascribed to the strained sugar pucker due to cyclization. Atom O(5') is trans-gauche with respect to the C(4')-C(5') bond. No base stacking is observed in the crystal structure which is stabilized by an intermolecular N(3)-H...O(4) hydrogen bond of length 2.834(5) Å which links screw-related molecules into columns along the *b*-axis.

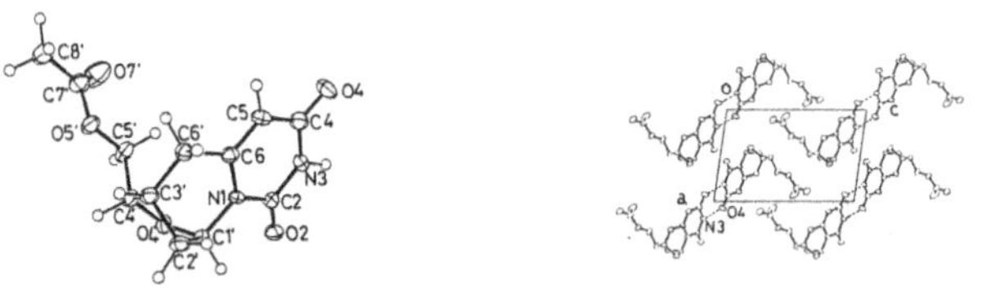

Fig. 1. Views showing the $C_{12}H_{14}N_2O_5$ molecular conformation and the molecular packing (the dotted lines indicate hydrogen bonds).

3',5'-DI-O-ACETYL-5-BROMO-2'-DEOXYURIDINE 3',5'-DI-O-ACETYL-5-IODO-2'-DEOXYURIDINE $C_{13}H_{15}Br_{0.5}I_{0.5}N_2O_7$ $0.5(C_{13}H_{15}BrN_2O_7),0.5(C_{13}H_{15}IN_2O_7)$

C.C. WILSON and P. TOLLIN, 1985. *Acta Cryst.*, C41, 465-468.

Orthorhombic, $P2_12_12_1$, a = 7.708, b = 26.835, c = 7.805 Å, Z = 4. Mo radiation, R = 0.062 for 2400 reflections.

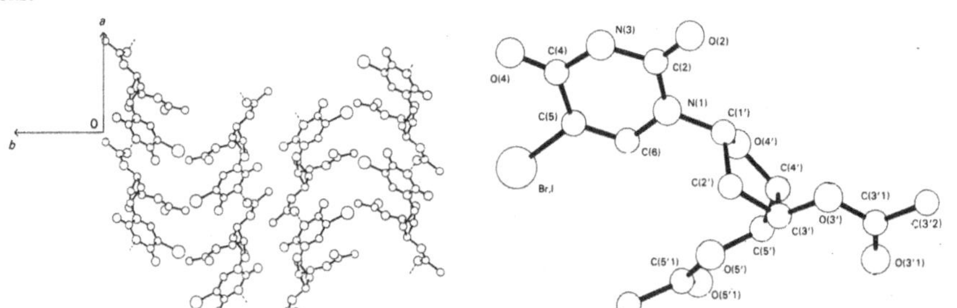

Fig. 1. $C_{13}H_{15}Br_{0.5}I_{0.5}N_2O_7$: a perspective view of the molecule showing the atomic numbering with an averaged position for the halogen atom and a view along the c-axis showing the molecular stacking.

In the structure of this mixed crystal the halogen five-position is occupied with equal probability by either Br or I atoms (Fig. 1). The molecule is anti about the glycosidic bond with torsion angle C(2)-N(1)-C(1')-O(4') of value-133(1)°. The sugar ring has $^2T_1$ pucker with C(1')-exo and C(2')-endo. The conformation about the C(4')-C(5') bond is gauche-gauche. The pyrimidine ring is planar as are the acetyl groups at the 3'- and 5'-positions. In the molecular stacking there is one hydrogen bond N(3)...O(3'1) of length 2.91(2) Å. There are also close contacts between halogen atoms and O(5'1), of average length 3.21(2) Å, and between the acetyl O(5'1) and a neighbouring base ring, 2.93(3) Å.

2',3',5'-TRI-O-ACETYLGUANOSINE
$C_{16}H_{19}N_5O_8$

C.C. WILSON, J.N. LOW and P. TOLLIN, 1985. *Acta Cryst.*, **C41**, 1123-1125.

Monoclinic, $P2_1$, a = 7.414, b = 11.491, c = 11.790 Å, β = 99.47°, Z = 2. Mo radiation, R = 0.053 for 1241 reflections.

The purine base in the molecule (Fig. 1) and the three acetyl groups are each essentially planar. The glycosidic torsion angle C(4)-N(9)-C(1')-O(4') is -160.6(9)° and lies in the high anti region. The sugar pucker is C(2')-endo and the C(4')-C(5') conformation is gauche-gauche with torsion angles C(3')-C(4')-C(5')-O(5') and O(4')-C(4')-C(5')-O(5') having values of 50.8(9) and -68.1(9)° respectively. Bond lengths and angles in the molecule are unexceptional. The molecules are base-paired through N(7)...N(1) and O(6)...N(2) hydrogen bonds of length 2.86(1) and 2.94(1) Å respectively. An additional O(5')-N(2) hydrogen bond of length 3.03(1) Å links the base-paired molecules into chains which stack along the *b*-axis. In these stacks an interaction between the 2'-acetyl O atom and two base rings is reported.

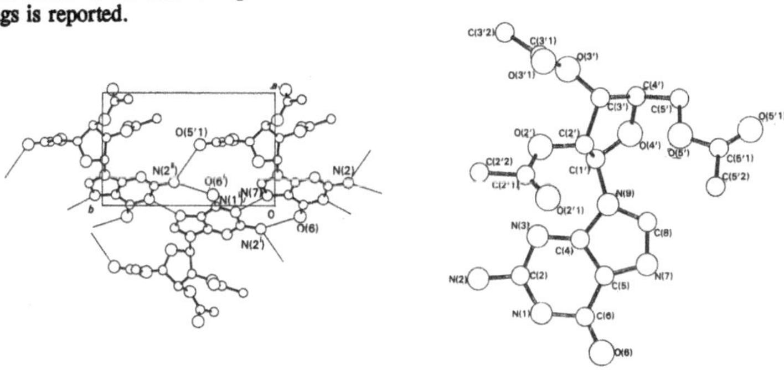

Fig. 1. $C_{16}H_{19}N_5O_8$: views of the molecule showing the atomic numbering and of the molecular packing with the hydrogen bonds represented by thin lines.

1(S)-(α-D-ARABINOFURANOSYL)-1,2,3,4-TETRAHYDRO-β-CARBOLINE
$C_{16}H_{20}N_2O_4$ (I)

4-(α-D-ARABINOFURANOSYL)IMIDAZO[4,5-c]-4,5,6,7-TETRAHYDROPYRIDINE
$C_{11}H_{17}N_3O_4$ (II)

I.M. PIPER, D.B. McLEAN, R. FAGGIANI and C.J.L. LOCK. 1985. *Canad. J. Chem.*, **63**, 2915-2921.

I. Monoclinic, $P2_1$, a = 13.091, b = 5.365, c = 11.323 Å, β = 115.78°, Z = 2. Mo radiation, R = 0.0514 for 3210 reflections.

II. Orthorhombic, $P2_12_12_1$, a = 8.118, b = 13.715, c = 10.963 Å, Z = 4. Mo radiation, R = 0.0312 for 1569 reflections.

Molecules (I) and (II) (Fig. 1) have normal bond lengths and angles. In (II) the base and sugar adopt a syn arrangement which is maintained by an internal O(2')-H...N(3) hydrogen bond (2.714(2) Å). The sugar adopts a $^2T_3$ twist conformation. In (I) the sugar has the opposite anti arrangement and is close to an envelope conformation with O(41) out of the plane. Both molecules are held in the crystals by extensive hydrogen bond networks.

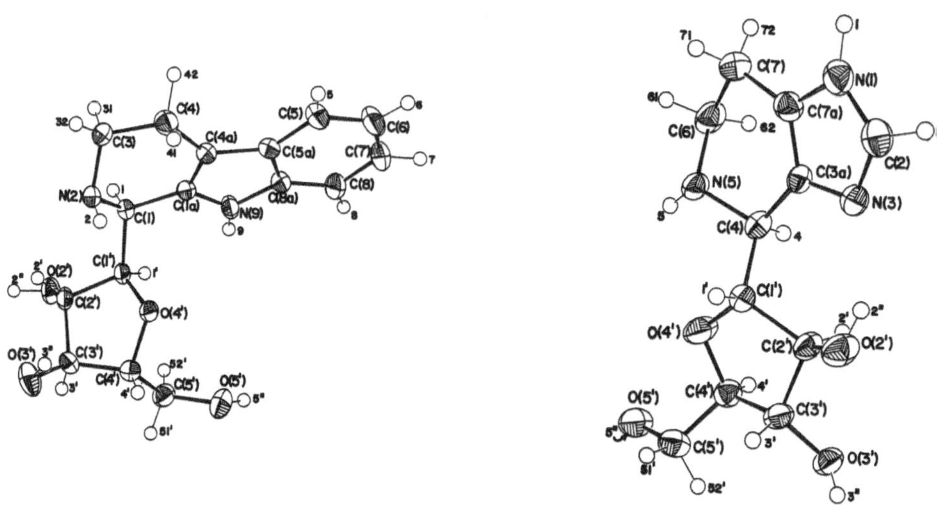

Fig. 1. Views of $C_{16}H_{20}N_2O_4$ (I) (left) and $C_{11}H_{17}N_3O_4$ (II) (right).

6-BENZYL-3-METHYL-1-(β-D-RIBOFURANOSYL)-7H-PYRAZOLO[4,3-d]PYRIMIDIN-7-ONE
$C_{18}H_{20}N_4O_5$

V. BERTOLASI, V. FERRETTI, F. BELLUCCI, G. GILLI and P.G. BARALDI, 1985. *Acta Cryst.*, **C41**, 1449-1452.

Orthorhombic, $P2_12_12_1$, a = 5.442, b = 13.387, c = 23.759 Å, Z = 4. Mo radiation, R = 0.039 for 1154 reflections.

The pyrazolopyrimidinone moiety in this analogue of formycin A and B is essentially planar (Fig. 1). The conformational parameters of this compound are a C(3')-endo ($^3E/^3T_2$) ribose conformation, a mid-anti glycosidic bond rotation (χ 66.4(4)°) and a gauche-trans conformation around the ribose C(4')-C(5') bond. There is extensive hydrogen bonding in the crystal (O...N 2.766(4), O...O 2.789(4)-3.323(4) Å).

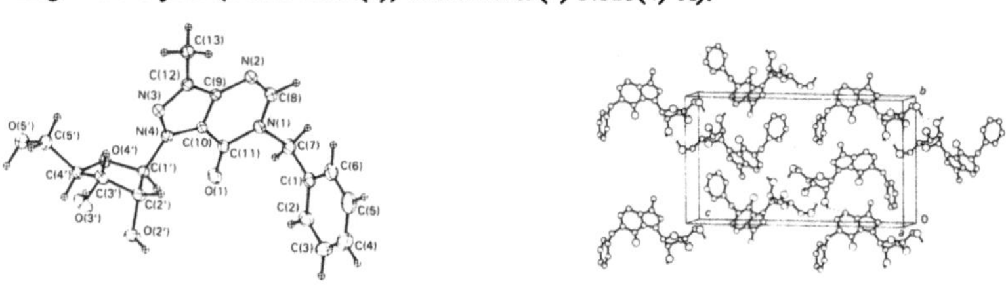

Fig. 1. $C_{18}H_{20}N_4O_5$: the molecular structure and packing along *a*.

N-(1'-(9-ADENYL)-β-D-RIBOFURANURONOSYL)-L-PHENYLALANINE HEMIHYDRATE ETHANOL SOLVATE
$C_{19}H_{20}N_6O_6,0.5(H_2O),0.25(C_2H_6O)$

F. TAKUSAGAWA, H.M. BERMAN, M. DABROW and R.K. ROBINS, 1985. *Acta Cryst.*, **C41**, 400-404.

Monoclinic, $P2_1$, a = 9.207, b = 10.503, c = 22.208 Å, β = 101.4°, Z = 4. Cu radiation, R = 0.053 for 4132 reflections.

The two crystallographically independent molecules in the asymmetric unit (Fig. 1) are both zwitterions of very different conformations with protonated bases and with negative charges on the carboxylate groups. Molecule A has a closed conformation with a glycosidic torsion angle of -129.4(3)° and so the base is syn. The sugar ring has C(3') endo and the conformation about the C(4')-C(5') bond is gauche$^-$. Molecule B has an extended conformation and the base is in the anti conformation with a glycosidic torsion angle of 77.8(3)°. The sugar conformation is C(2') endo and the conformation about C(4')-C(5') is gauche$^+$. The amide linkages C(5')-N(6') are trans in both molecules. The ethanol solvate is disordered about the C-OH bond with a site occupancy factor of 0.55(8). There is a complex three-dimensional network of N-H...O and O-H...O hydrogen bonds linking the molecules to each other and to the solvate molecules.

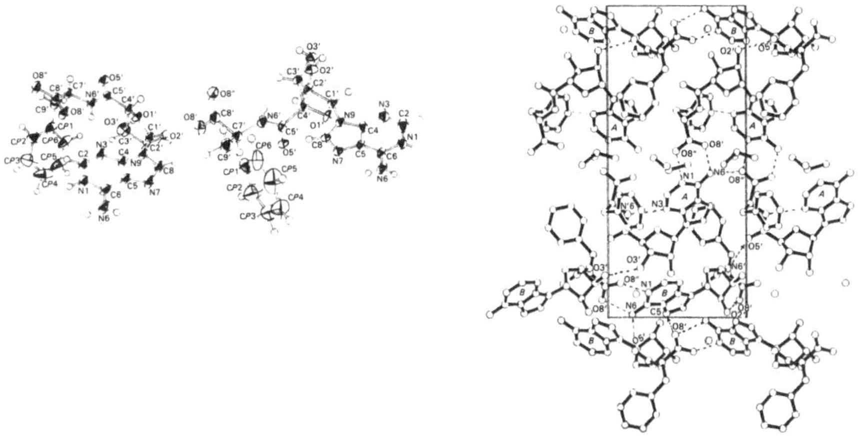

Fig. 1. $C_{19}H_{20}N_6O_6$,0.5(H_2O),0.25(C_2H_6O): perspective views (a) of molecule A, (b) of molecule B and a view down the a-axis of the molecular packing with hydrogen bonds shown as broken lines.

trans-5-ISOPROPYL-2'-DEOXYURIDINE-(3'-5')-CYCLIC-N-BENZYLPHOSPHORAMIDATE $C_{19}H_{24}N_3O_6P$

J. BERES, W.G. BENTRUDE, L. PARKANJI, A. KALMAN and A.E. SOPCHIK, 1985. *J. Org. Chem.*, **50**, 1271-1278.

Triclinic, $P1$, a = 6.524, b = 8.564, c = 9.996 Å, α = 72.94, β = 78.09, γ = 85.44°, Z = 1. Mo radiation, R = 0.064 for 2432 reflections.

The $PhCH_2NH$ is trans to the thymin-1-yl ring and is attached equatorially to the chair-form 1,3,2-dioxaphosphorinane ring, which has a distorted chair conformation (Fig. 1). Molecules are bound together in the crystal by two infinite intermolecular hydrogen bond chains along the [110] direction.

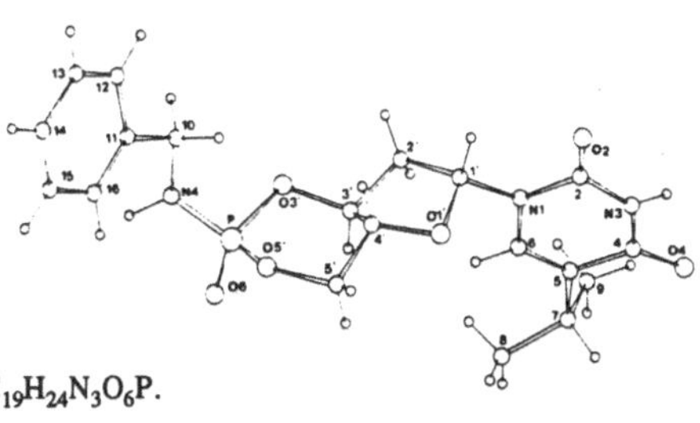

Fig. 1. A molecule of $C_{19}H_{24}N_3O_6P$.

THYMIDYLYL-(3'-5')-THYMIDINE CYANOETHYL ESTER cis-syn PHOTODIMER DIHYDRATE $C_{23}H_{30}N_5O_{12}P$,2(H_2O)

J. CADET, L. VOITURIEZ, F.E. HRUSKA and A. GRAND, 1985. *Biopolymers*, **24**, 897-903.

Orthorhombic, $P2_12_12_1$, a = 11.596, b = 14.834, c = 15.946 Å, Z = 4. Mo radiation, R = 0.042 for 4388 reflections.

This photodimer has a cis-syn cyclobutane ring (Fig. 1) which is puckered with a dihedral angle of 151° between C(5A),C(5B),C(6B) and C(5A),C(6A),C(6B) planes. The C(5A)-C(5B) bond length is 1.595(6) and C(6A)-C(6B) is 1.591(6) Å; the C(5)-C(6) type bond lengths are 1.538 and 1.545(6) Å. Saturation of the C(5)-C(6) bonds leads to a half-chair puckering of the pyrimidines.

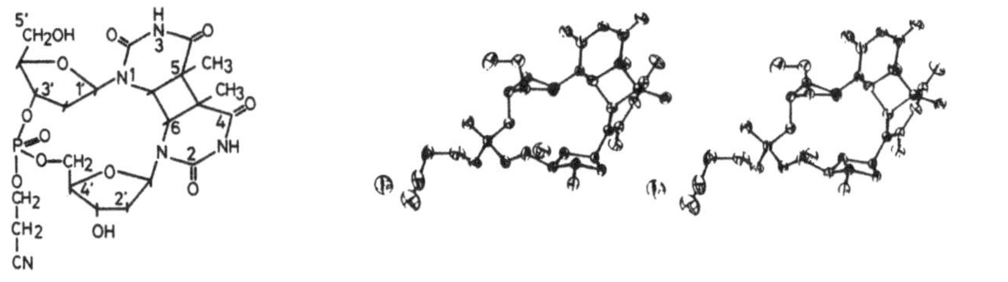

Fig. 1. The $C_{23}H_{30}N_5O_{12}P$ molecule and a stereoview.

GLYCINEHYDROXAMIC ACID
$C_2H_6N_2O_2$

T. GLOWIAK and M. KORALEWICZ, 1985. *J. Cryst. Spect. Res.*, **15**, 621-628.

Monoclinic, $P2_1/c$, a = 5.682, b = 7.133, c = 9.540 Å, β = 104.38°, Z = 4. Cu radiation, R = 0.042 for 633 reflections.

The molecule is a zwitterion with conformation shown in Fig. 1. Main bond lengths are: N(1)-O(1) 1.434(2), N(1)-C(1) 1.294(3), N(2)-C(2) 1.491(3), C(1)-O(2) 1.296(2), C(1)-C(2) 1.520(3) Å; the O(1)-N(1)-C(1)-C(2) and N(1)-C(1)-C(2)-N(2) torsion angles are -0.6(2) and -103.1(3)° respectively. Molecules are held together by a network of N^+-H...O, N^+-H...N and O-H...O hydrogen bonds (N...O 2.777 and 2.877(2), N...N 2.887(2), O...O 2.776(2) Å.

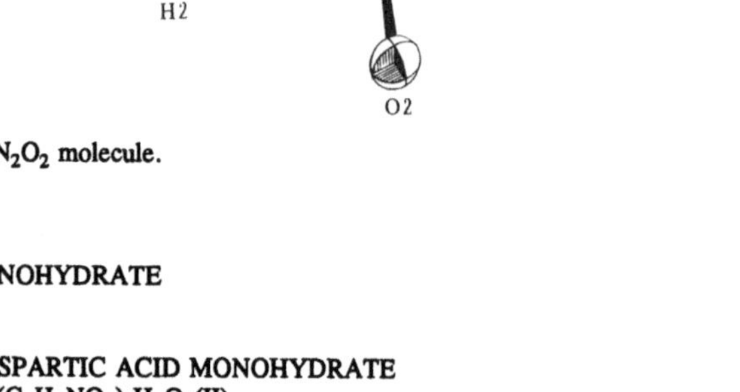

Fig. 1. The $C_2H_6N_2O_2$ molecule.

L-ASPARAGINE MONOHYDRATE
$C_4H_8N_2O_3,H_2O$ (I)

L-ASPARAGINE L-ASPARTIC ACID MONOHYDRATE
$0.85(C_4H_8N_2O_3),0.15(C_4H_7NO_4),H_2O$ (II)

J.L. WANG, Z. BERKOVITCH-YELLIN and L. LEISEROWITZ, 1985. *Acta Cryst.*, **B41**, 341-348.

I. Orthorhombic, $P2_12_12_1$, a = 5.584, b = 9.735, c = 11.701 Å, Z = 4. Mo radiation, R = 0.038 for 5339 reflections (at 100 K).

II. Orthorhombic, $P2_12_12_1$, a = 5.565, b = 9.774, c = 11.698 Å, Z = 4. Mo radiation, R = 0.0349 for 2672 reflections (at 100 K).

For previous work on (I) see (1). Stereoviews of both structures are in Fig. 1. The aspartic acid carboxyl group adopts the (normal) synplanar conformation rather than the alternative antiplanar conformation.

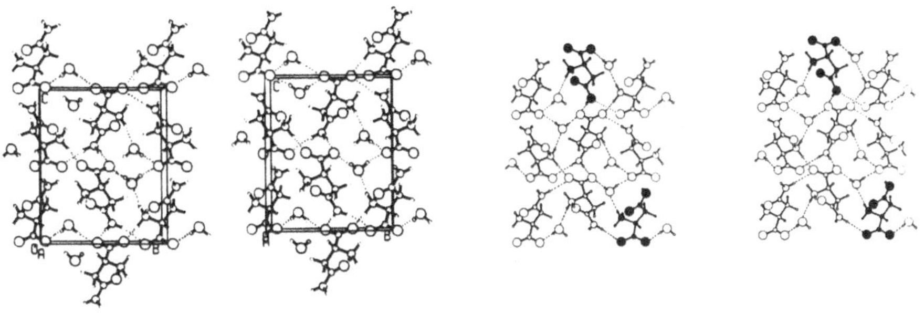

Fig. 1. Stereoviews of the molecular packing in L-asparagine monohydrate (I) (left) and in asparagine/aspartic acid monohydrate (II) (right).

1. Structure Reports, **38B**, 476.

L-(+)-GLUTAMIC ACID HYDROIODIDE
$C_5H_9NO_4,HI$

A. KIRFEL and F. WALLRAFEN, 1985. *Z. Kristallogr.*, **171**, 121-128.

Orthorhombic, $P2_12_12_1$, a = 5.892, b = 11.461, c = 13.405 Å, Z = 4. Mo radiation, R = 0.022 for 925 reflections.

The material is isomorphous with L-glutamic acid hydrochloride (1). The compound is not a halogen acid salt but a 1:1 complex of L-glutamic acid and undissociated HI connected by hydrogen bonds (N...I 3.525 and 3.527 Å) (Fig. 1). The HI hydrogen is disordered over two sites.

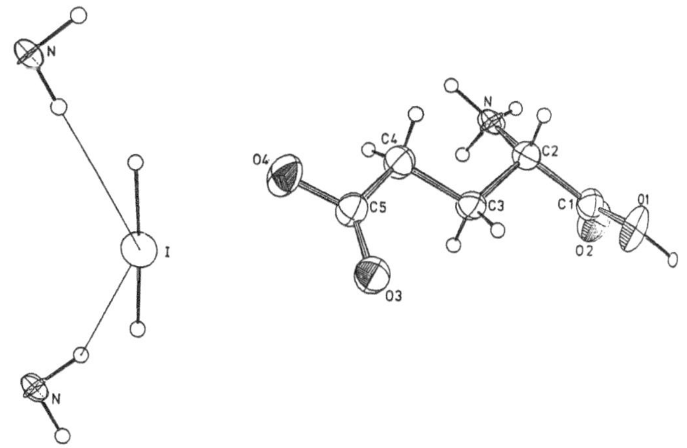

Fig. 1. The asymmetric unit of $C_5H_9NO_4,HI$.

1. Structure Reports, **38B**, 482.

TRIMETHYLAMMONIOACETATE CALCIUM DICHLORIDE DIHYDRATE

$C_5H_{11}CaCl_2NO_2,2(H_2O)$ $C_5H_{11}NO_2,Ca^{2+},2(Cl^-)2(H_2O)$

W. BRILL, W. SCHILDKAMP and J. SPILKER, 1985. *Z. Kristallogr.*, **172**, 281-289.

Orthorhombic, *Pnma*, a = 10.97, b = 10.15, c = 10.82 Å, Z = 4. Mo radiation, R = 0.033 for 1496 reflections.

The calcium and trimethylammonioacetate moieties lie on a crystallographic mirror plane. The Ca ion has distorted octahedral coordination (Fig. 1) with Ca-O(1) 2.28(3), Ca-O(2) 2.249(3), Ca-O(3) 2.391(3), Ca-Cl 2.767(2) Å. A three-dimensional network is formed by O-H...Cl hydrogen bonds (O...Cl 3.195(3) and 3.239(4) Å).

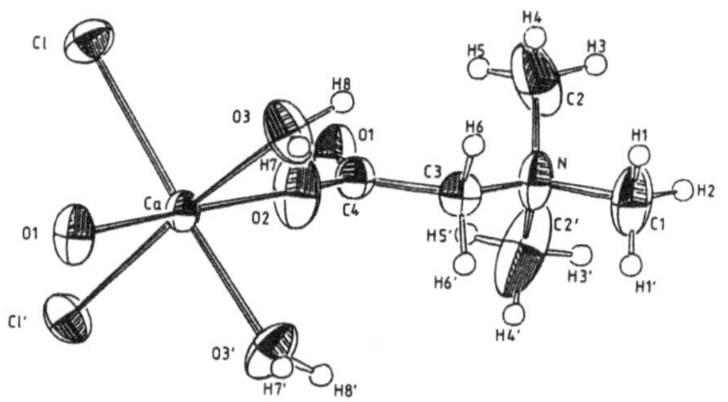

Fig. 1. The Ca coordination in $C_5H_{11}CaCl_2NO_2,2(H_2O)$.

DL-METHIONINE NITRATE

$C_5H_{12}N_2O_5S$ $C_5H_{12}NO_2S^+,NO_3^-$

A. MOSTAD and S. NATARAJAN, 1985. *Z. Kristallogr.*, **172**, 175-182.

Monoclinic, $P2_1/n$, a = 13.663, b = 5.556, c = 13.174 Å, β = 103.90°, Z = 4. Mo radiation, R = 0.036 for 2087 reflections (at 121 K).

The molecule and its packing are shown in Fig. 1. Bond C(2)-C(3) (1.545(2) Å) appears to be longer than the other C-C bonds (1.522(2) and 1.532(2) Å). There is extensive hydrogen bonding involving all possible donors.

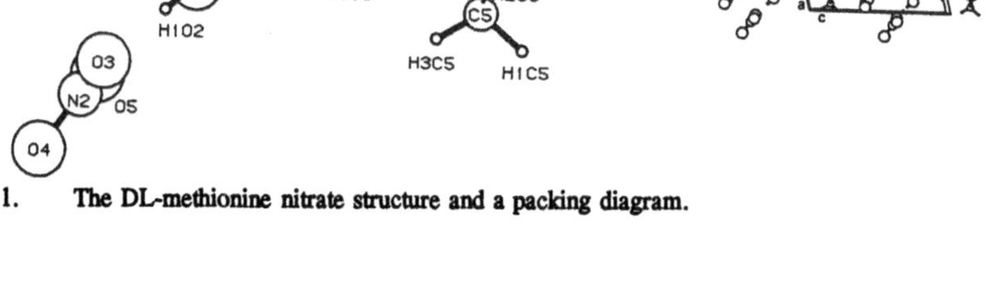

Fig. 1. The DL-methionine nitrate structure and a packing diagram.

3,6-DIMETHYLENEPIPERAZINE-2,5-DIONE (cyclo-BIS(DEHYDROALANYL))

$C_6H_6N_2O_2$

K.H. ONGANIA, G. GRANOZZI, V. BUSETTI, M. CASARIN and D. AJO, 1985. *Tetrahedron*, **41**, 2015-2018.

Monoclinic, $P2_1$, a = 5.261, b = 13.617, c = 4.570 Å, β = 104.58°, Z = 2. Mo radiation, R = 0.071 for 385

reflections.

The molecules are planar and are associated in planar ribbons by N-H...O hydrogen bonds (Fig. 1).

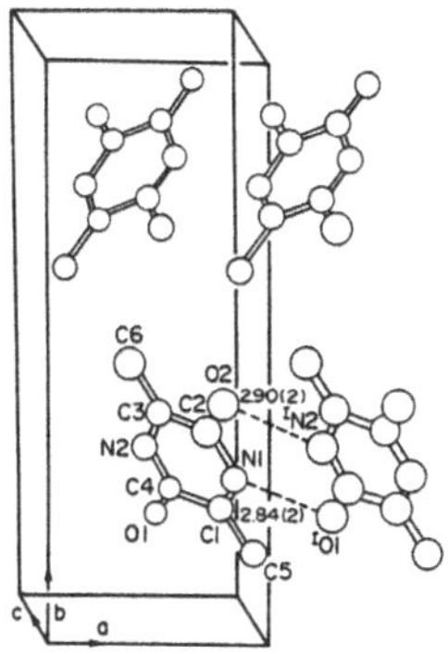

Fig. 1. The $C_6H_6N_2O_2$ crystal packing.

1,4,5,6-TETRAHYDRO-2-METHYL-4-PYRIMIDINE CARBOXYLIC ACID METHANOL SOLVATE (ECTOINE METHANOL SOLVATE)
$C_6H_{10}N_2O_2,CH_4O$

W. SCHUH, H. PUFF, E.A. GALINSKI and H.G. TRUPER, 1985. *Z. Naturforsch.*, C40, 780-784.

Orthorhombic, $P2_12_12_1$, a = 7.563, b = 9.242, c = 13.319 Å, Z = 4. Mo radiation, R = 0.063 for 808 reflections.

This cyclic amino acid isolated from a halophilic phototropic bacterium forms zwitterions with delocalized π-bonding in the N-C-N group of the hydrated pyrimidine ring (Fig. 1). Atoms N(1),C(2),N(3),C(4),C(6) are planar and C(5) is 0.64 Å off this plane producing a half-chair conformation. In the crystal, molecules are connected by N-H...O hydrogen bonds.

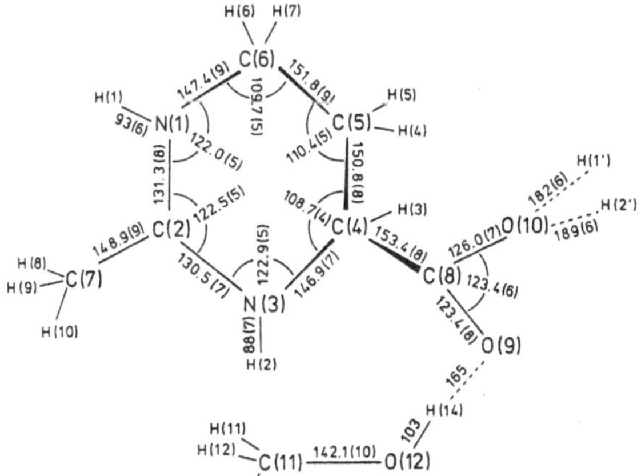

Fig. 1. Molecular dimensions (pm and °) for $C_6H_{10}N_2O_2,CH_4O$.

N-ACETYL-α-AMINOISOBUTYRIC ACID
$C_6H_{11}NO_3$ (I)

N-TRIFLUOROACETYL-α-AMINOISOBUTYRIC ACID
$C_6H_8F_3NO_3$ (II)

G. VALLE, C. TONIOLO and G.M. BONORA, 1985. *Acta Cryst.*, C41, 235-238.

I. Monoclinic, $P2_1/n$, a = 10.165, b = 8.030, c = 9.712 Å, β = 111.9°, Z = 4. Mo radiation, R = 0.042 for 1053 reflections.

II. Orthorhombic, *Pcab*, a = 15.174, b = 11.773, c = 10.092 Å, Z = 8. Mo radiation, R = 0.085 for 1121 reflections.

Rotational disorder is found in (II), there being three conformations of the molecule. In both compounds (Fig. 1) the conformation of the amide bond is trans and the aminoisobutyric residue shows a set of ϕ, ψ_T torsion values which fall in the region of the Ramachandran map where α- and 3_{10}- helices are found. The dihedral angle between the amide and carboxylic acid groups is 101.1(3) and 73.0(6)° for (I) and (II) respectively. Bond lengths and angles are normal. Molecules of (I) are held together in the crystal by a network of O-H...O=C(amide) and N-H...O(carboxylic acid) hydrogen bonds (2.606(3), 3.01(1) Å). In (II), pairs of molecules are held together across a centre of symmetry by two OH...C=O(carboxylic acid) hydrogen bonds (2.615(6) Å). There are additional hydrogen bonds of the type NH...O=C(amide) (2.966(7) Å).

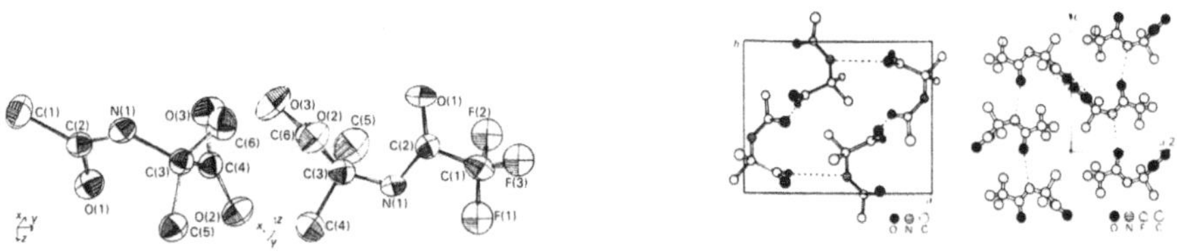

Fig. 1. Molecular structures of $C_6H_{11}NO_3$ and $C_6H_8F_3NO_3$ (left) and the packing (right).

L-2-AMINO-4,5-DITHIAOCTANEDIOIC ACID 2-DEAMINO-L-CYSTINE
$C_6H_{11}NO_4S_2$

M. RAJESWARAN and R. PARTHASARATHY, 1985. *Acta Cryst.*, C41, 726-728.

Triclinic, $P1$, a = 5.101, b = 10.195, c = 5.020 Å, α = 96.72, β = 111.27, γ = 95.63°, Z = 1. Cu radiation, R = 0.040 for 1944 reflections.

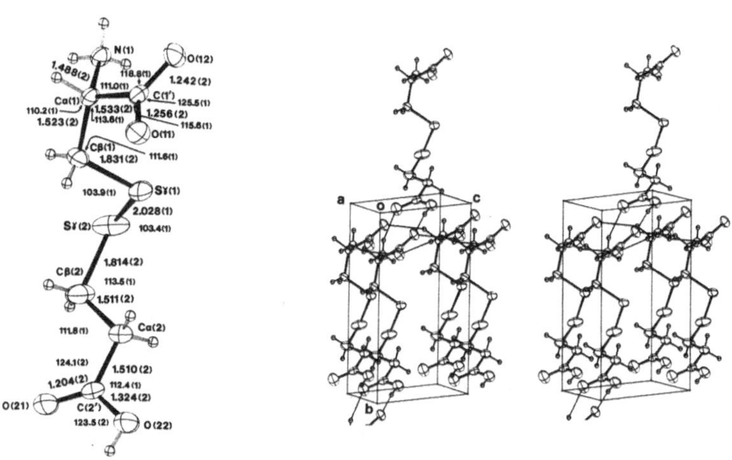

Fig. 1. $C_6H_{11}NO_4S_2$: a view of the molecule with bond lengths and angles and a stereoview of the molecular packing (the thinner lines represent hydrogen bonds).

In the zwitterionic molecule (Fig. 1) bond lengths in the cysteine moiety are significantly longer than those of the β-mercaptopropionic acid (MPA) moiety. Each carboxylate group with its immediate attachment is planar. The

C'-Cα-Cβ-Sγ backbone is zigzag planar in the MPA half but twisted in the cysteine half of the molecule. There is a short intramolecular 1,4-Sγ(1)...Cα(2) non-bonding interaction of length 3.341(3) Å in the structure. All three amino group hydrogen atoms are involved in a network of intermolecular hydrogen bonds and the molecules in the crystal are linked in a head-to-tail fashion.

N-ACETYL-DL-METHIONINE
$C_7H_{13}NO_3S$

M.N. PONNUSWAMY and J. TROTTER, 1985. *Acta Cryst.*, **C41**, 917-919.

Monoclinic, $P2_1/c$, a = 5.882, b = 9.285, c = 21.934 Å, β = 124.88°, Z = 4. Mo radiation, R = 0.041 for 2131 reflections.

The peptide group of the molecule (Fig. 1) is planar within 0.03 Å and the carboxyl group exists in the un-ionized form, with the H atom bonded to O(3). The group is approximately planar with the N atom displaced from the plane by 0.536(1) Å. Bond lengths and angles are normal. Molecules are linked by N-H...O (2.953(2)) and O-H...O (2.550(2) Å) hydrogen bonds.

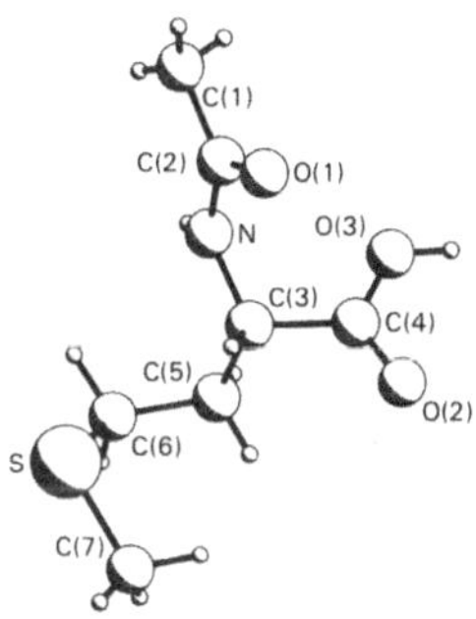

Fig. 1. The $C_7H_{13}NO_3S$ molecule.

N-SALICYLOYLGLYCINE
$C_9H_9NO_4$

W.J.J. SMEETS, J.A. KANTERS and K. VENKATASUBRAMANIAN, 1985. *Acta Cryst.*, **C41**, 272-274.

Monoclinic, $P2_1/c$, a = 13.977, b = 5.064, c = 13.127 Å, β = 105.31°, Z = 4. Mo radiation, R = 0.045 for 601 reflections.

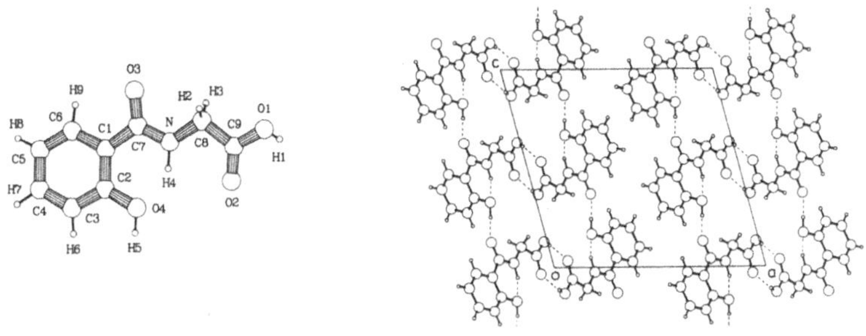

Fig. 1. $C_9H_9NO_4$: the molecule (left) and projection of the crystal structure along b (right).

The non-hydrogen atoms of the molecule are moderately coplanar (Fig. 1). The peptide-like link has the trans configuration with torsion angles of -179.3(6) for C(1)-C(7)-N-C(8) and -175.5(5)° for C(7)-N-C(8)-C(9). Bond lengths in the benzene ring range from 1.363(10) to 1.408(8) Å. There are three hydrogen bonds in the structure: one intramolecular (N...O(4) 2.632(7)) and two intermolecular (O(1)...O(2) 2.689(7), O(4)...O(3) 2.612(6) Å). The carboxyl groups form dimer-type cyclic H bonds across centres of symmetry.

(+)-3,6-EPIDITHIO-L-PROLYL-L-PROLINE ANHYDRIDE
$C_{10}H_{12}N_2O_2S_2$

J.D. KORP and I. BERNAL, 1985. *J. Cryst. Spect. Res.*, **15**, 181-190.

Orthorhombic, $P2_12_12$, a = 9.573, b = 10.143, c = 5.761 Å, Z = 2. Mo radiation, R = 0.058 for 714 reflections.

The molecule has crystallographic twofold symmetry (Fig. 1) and the C-S-S-C fragment has S-S 2.078(3) Å and torsion angle -8.52°. The C-S distance is 1.899(7) Å. The central 2,5-piperazinedione moiety is in boat form and prolyl rings are very nearly envelopes.

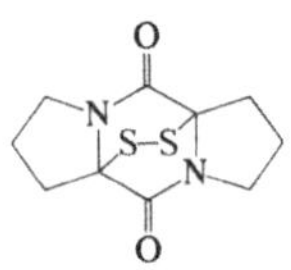

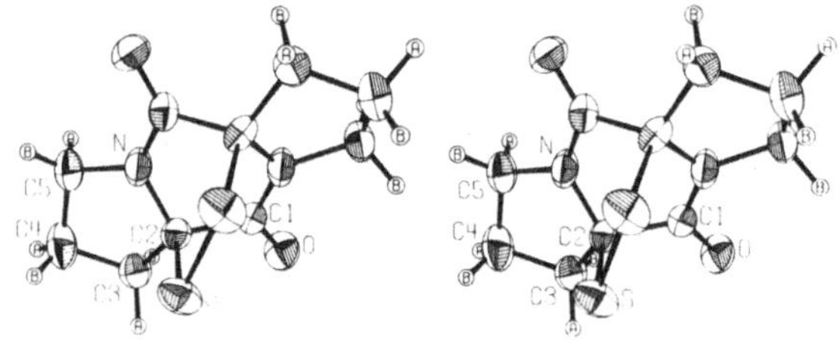

Fig. 1. The $C_{10}H_{12}N_2O_2S_2$ molecule and a stereoview.

CALCIUM BIS(HYDROXY-L-PROLINE) PENTAHYDRATE
$C_{10}H_{16}CaN_2O_6,5(H_2O)$ $2(C_5H_8NO_3^-),Ca^{2+},5(H_2O)$

E.E. KIM, A. SICIGNANO and K. ERIKS, 1985. *J. Am. Chem. Soc.*, **107**, 6042-6046.

Monoclinic, $P2_1$, a = 6.213, b = 14.905, c = 9.483 Å, β = 96.41°, Z = 2. Mo radiation, R = 0.033 for 1222 reflections.

The calcium is seven-coordinated by a carboxyl O atom (Ca-O 2.367, 2.330(3) Å) and a N atom (Ca-N 2.613, 2.595(3) Å) from each of two hydroxy-L-proline molecules and three water molecules (Ca-O 2.376, 2.452, 2.345(3) Å) leading to a distorted pentagonal-bipyramidal geometry (Fig. 1). The units are linked by an extensive hydrogen-bonded network (Fig. 1).

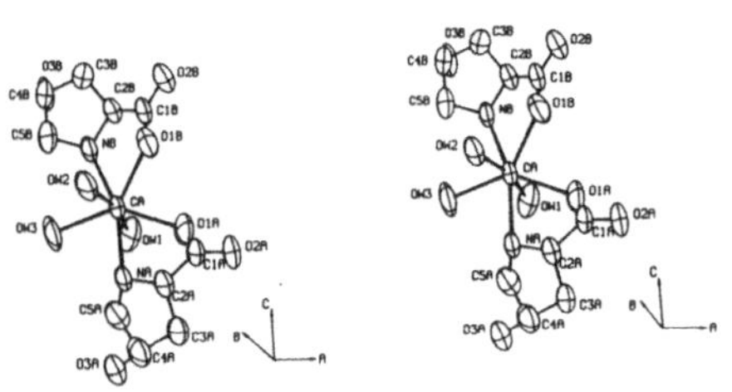

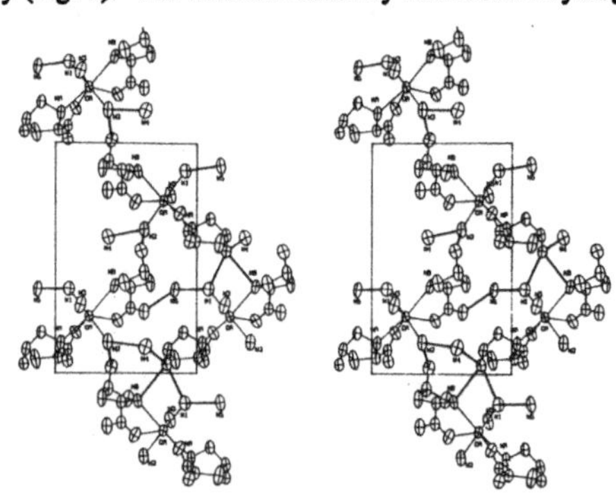

Fig. 1. Stereoviews of the Ca^{++} coordination in $C_{10}H_{16}CaN_2O_6,5(H_2O)$ and of the crystal structure (hydrogen bonds are shown by heavy lines).

(Z)-N-ACETYL-α,β-DEHYDROPHENYLALANINE METHYLAMIDE
$C_{12}H_{14}N_2O_2$

A. AUBRY, F. ALLIER, G. BOUSSARD and M. MARRAUD, 1985. *Biopolymers*, **24**, 639-646.

Monoclinic, *Cc*, a = 10.241, b = 15.252, c = 8.643 Å, β = 129.98°, Z = 4. Cu radiation, R = 0.048 for 1070 reflections.

The molecule is compared with the homologous saturated derivative, N-acetyl-L-phenylalanine methylamide (1) in Fig. 1. The geometrical parameters are not drastically altered by the α,β-unsaturation, but the N-C$^{\alpha}$-C' angle is increased by nearly 5° and the dimensions (and probably the electronic conjugation) of the N-terminal amide bond appear to be affected by the presence of the vicinal C$^{\alpha}$=C$^{\beta}$ double bond. Molecules are hydrogen bonded intermolecularly to four neighbouring molecules in a three-dimensional network, with alternating layers of interacting amide bonds and orthogonally arranged phenyl rings.

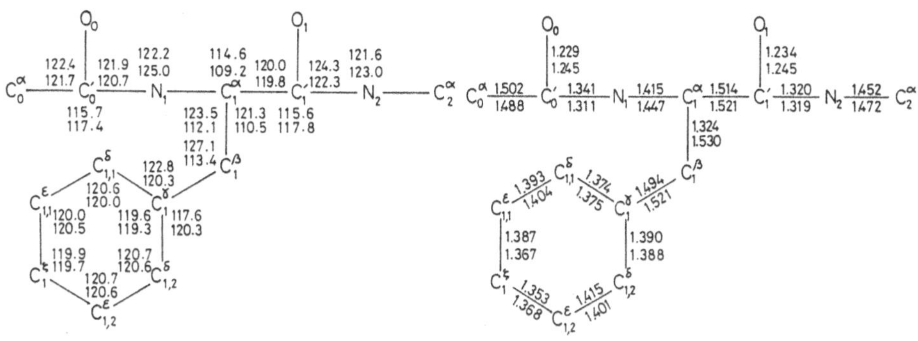

Fig. 1. Bond lengths and angles in $C_{12}H_{14}N_2O_2$ (upper values) and in N-acetyl-L-phenylalanine methylamide (lower values).

1. Structure Reports, 43B, 608.

cyclo-(L-HISTIDYL-L-HISTIDYL) HYDRATE
$C_{12}H_{14}N_6O_2,2.5(H_2O)$

Y. KOJIMA, T. YAMASHITA, S. NISHIDE, K. HIROTSU and T. HIGUCHI, 1985. *Bull. Chem. Soc. Jpn.*, **58**, 409-413.

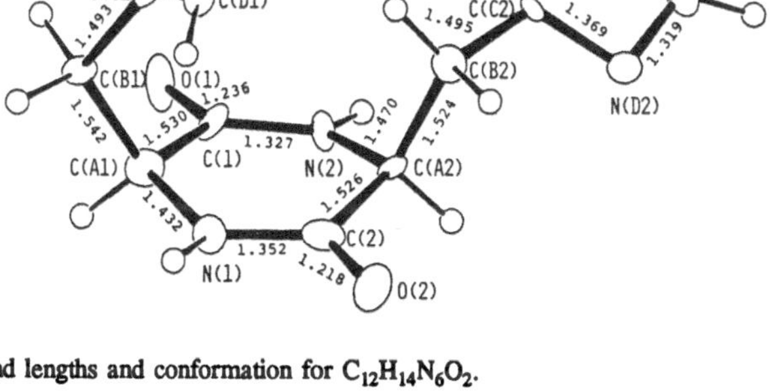

Fig. 1. Bond lengths and conformation for $C_{12}H_{14}N_6O_2$.

Orthorhombic, *P*2_12_12, a = 19.91, b = 12.66, c = 6.163 Å, Z = 4. Mo radiation, R = 0.054 for 680 reflections.

The peptide geometry and conformation are shown in Fig. 1. The piperazine ring is in a flagpole boat form with C(A1) and C(A2) 0.16(1) and 0.19(1) Å respectively from the plane of the other four ring atoms. Both imidazole rings are individually planar; one is folded back over the piperazine ring. All available N-H and O-H hydrogens take part in a three-dimensional network.

N-BENZOYL-DL-ALANINE ETHYL DITHIOESTER
$C_{12}H_{15}NOS_2$

R.H. ANGUS, P.R. CAREY, H. LEE, A.C. STORER and K.I. VARUGHESE, 1985. *Canad. J. Chem.*, **63**, 2169-2175.

Orthorhombic, *Pca*2_1, a = 9.522, b = 6.587, c = 21.025 Å, Z = 4. Mo radiation, R = 0.046 for 1641 reflections.

The molecule takes up the "B-type" conformation (Fig. 1) characterised by a small N-C-C-S(thiol) torsion angle (-23.0°) resulting from an N...S interaction (2.830 Å).

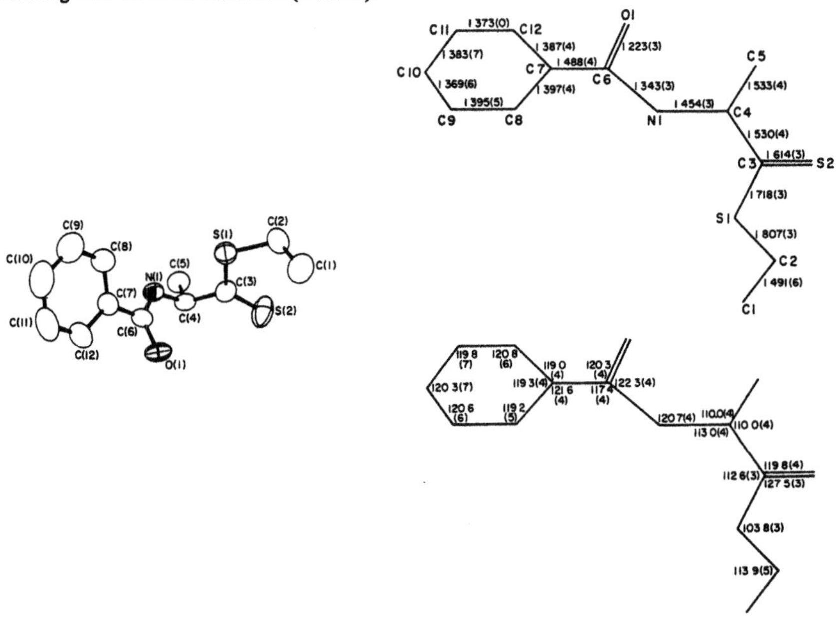

Fig. 1. The $C_{12}H_{15}NOS_2$ molecule with bond lengths and angles.

cyclo-[α-AMINOISOBUTYRYL-L-PHENYLALANYL]
$C_{13}H_{16}N_2O_2$

K. SUGUNA, S. RAMAKUMAR, R. NAGARAJ and P. BALARAM, 1985. *Acta Cryst.*, C41, 284-286.

Monoclinic, *P*2_1, a = 9.646, b = 6.133, c = 10.757 Å, β = 101.97°, Z = 2. Cu radiation, R = 0.065 for 820 reflections.

In the molecule (Fig. 1) the peptide units are planar, the diketopiperazine ring assumes a flat boat conformation and the phenylalanyl ring has folded geometry with χ^1 = 64.6(7) and χ^2 = 96.4(7)°. The crystal structure is stabilised by N-H...O hydrogen bonds. Each molecule forms a pair of hydrogen bonds to adjacent b-translated molecules. The hydrophobic and hydrophillic groups are neatly segregated into adjacent layers parallel to the c-axis.

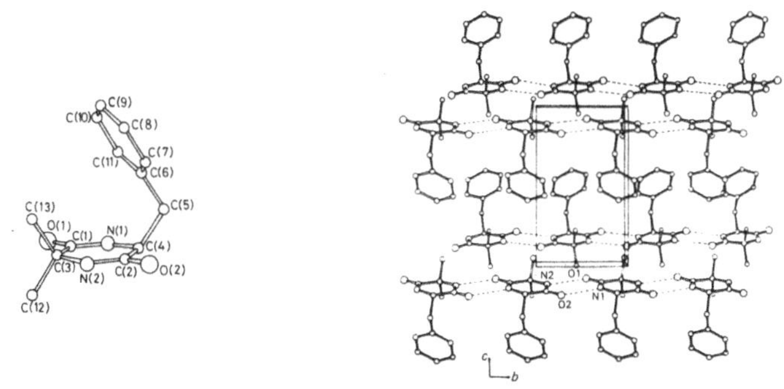

Fig. 1. $C_{13}H_{16}N_2O_2$: the molecule and the packing viewed down the a* axis.

GLYCINIUM DIPHENYLPHOSPHATE
$C_{14}H_{16}NO_6P_4$ $C_2H_6NO_2{}^+,C_{12}H_{10}O_4P^-$

T. GLOWIAK and I. WNEK, 1985. *J. Cryst. Spect. Res.*, **15**, 157-171.

Orthorhombic, *Pbca*, a = 11.412, b = 24.624, c = 10.775 Å, Z = 8. Mo radiation, R = 0.043 for 1289 reflections.

The structure exists as $^+NH_3CH_2CO_2H,(C_6H_5O)_2PO_2{}^-$ with the two ions (Fig. 1) linked by an O(5)-H...O(1) hydrogen bond (2.569(4) Å). There are three N-H...O hydrogen bonds (N...O 2.683-2.915(5) Å). The diphenylphosphate conformation is gauche with respect to both P-O ester bonds; P-O bond lengths are P-O(1,2) 1.493, 1.468(3), P-O(3,4) 1.606, 1.592(3) Å.

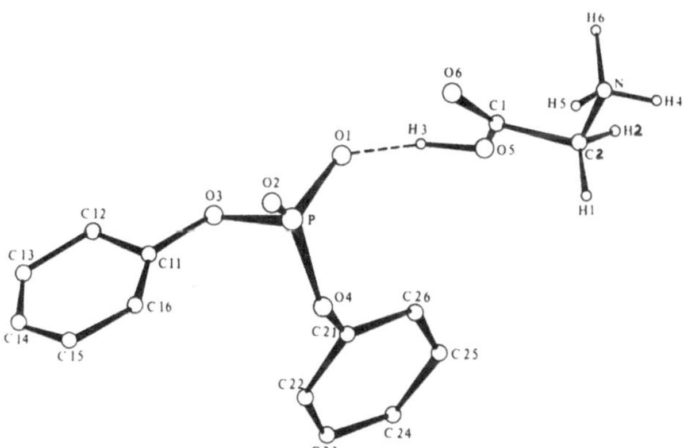

Fig. 1. The glycinium diphenylphosphate structure.

PHENYLACETYL-D-ALANYL-D-ALANINE
$C_{14}H_{18}N_2O_4$

H. LABISCHINSKI, G. BARNICKEL, D. NAUMANN, W. RONSPECK and H. BRADACZEK, 1985. *Biopolymers*, **24**, 2087-2112.

Orthorhombic, $P2_12_12_1$, a = 16.191, b = 9.605, c = 9.403 Å, Z = 4. Cu radiation, R = 0.036 for 1416 reflections.

The molecule adopts a partially folded S-shaped molecule (Fig. 1). Both CONH groups adopt the trans-planar conformation as defined by ω_0 and ω_1 values of 178°. There are no intramolecular hydrogen bonds, only intermolecular ones (two N-H...O=C, N...O 2.406(4) and 2.861(4) Å and one O-H...O, O...O 2.627(4) Å).

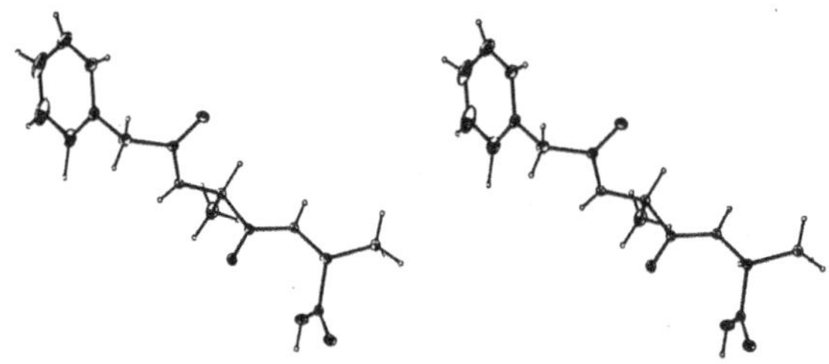

Fig. 1. A stereoview of $C_{14}H_{18}N_2O_4$.

L-ASPARTYL-L-PHENYLALANINE METHYL ESTER HEMIHYDRATE (ASPARTAME HEMIHYDRATE)
$C_{14}H_{18}N_2O_5,0.5(H_2O)$

M. HATADA, J. JANCARIK, B. GRAVES and SUNG-HOU KIM, 1985. *J. Am. Chem. Soc.*, **107**, 4279-4282.

Tetragonal, $P4_1$, a = 17.685, c = 4.919 Å, Z = 4. Mo radiation, R = 0.045 for 899 reflections.

A view of this peptide sweetener is shown in Fig. 1; also shown are bond lengths (which are normal) and torsion angles. The conformations of the bonds in the dipeptide are all close to trans. The N-terminus is positively charged and the aspartate side chain is negatively charged.

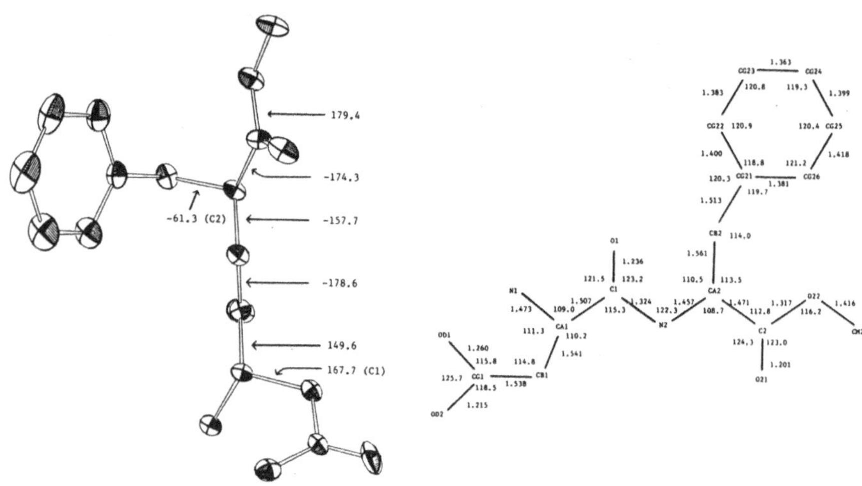

Fig. 1. The aspartame molecule with torsion angles and bond lengths.

3-IODO-L-THYRONINE HYDROCHLORIDE DIHYDRATE
$C_{15}H_{15}ClINO_4,2(H_2O)$ $C_{15}H_{15}INO_4^+,Cl^-,2(H_2O)$

N. OKABE, E. TAKIMOTO, T. FUJIWARA, Y. YAMAGATA and K.-I. TOMITA, 1985. *Acta Cryst.*, **C41**, 87-89.

Monoclinic, $P2_1$, a = 10.296, b = 11.883, c = 14.869 Å, β = 91.20°, Z = 4. Cu radiation, R = 0.098 for 1267 reflections.

There are two crystallographically independent conformers (Fig. 1) in the asymmetric unit. Both show overall transoid conformations in which the alanine moiety and the outer ring lie on opposite sides of the inner ring plane. In the conformers the 3-iodo atoms are also located on opposite sides. The torsion angles C(5)-C(4)-O(4)-C(1') and C(4)-O(4)-C(1')-C(6') are respectively 50 and 54° for conformer (I) and -63 and 33° for conformer (II). The crystal structure is stabilized by an extensive hydrogen-bonding network involving the cations, the chloride anions and the water molecules.

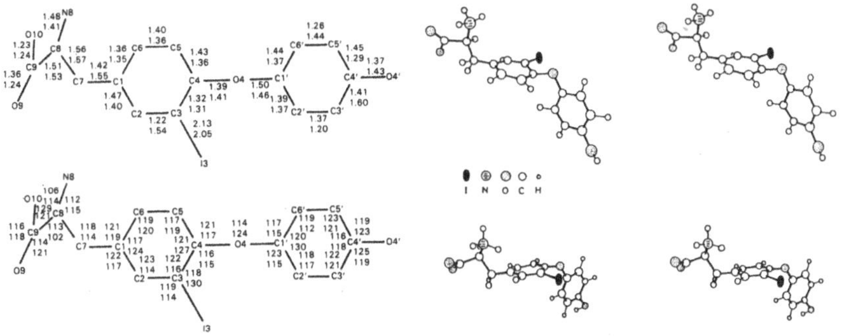

Fig. 1. $C_{15}H_{15}ClINO_4,2(H_2O)$: (left) bond lengths and angles in the conformers (I) (upper) and (II) (lower) and (right) stereoviews (a) of conformer (I) and (b) of conformer (II).

DL-METHYL-N-(2,6-DIMETHYLPHENYL)-N-(2-METHOXYACETYL)ALANINATE (METALAXYL) $C_{15}H_{21}NO_4$

J.C.J. BART, R. SCORDAMAGLIA and M. CALCATERRA, 1985. *J. Cryst. Spect. Res.*, **15**, 61-71.

Monoclinic, $P2_1/c$, a = 7.851, b = 13.119, c = 15.107 Å, β = 101.71°, Z = 4. Mo radiation, R = 0.079 for 1203 reflections.

The conformation found in the crystal has the phenyl ring plane almost perpendicular (87.2°) to the amidic plane (Fig. 1). The ester function has the extended conformation. The N atom is sp^2 hybridized, with C(1)-N-C(5) 115.7(3)°.

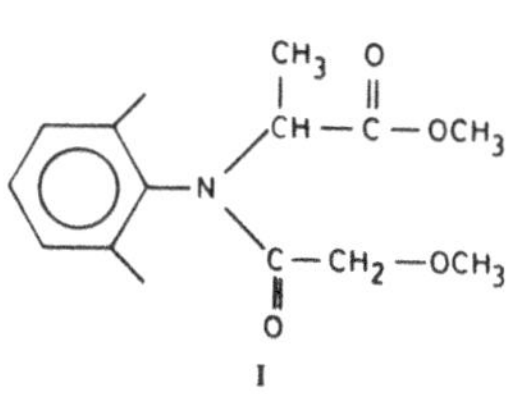

I

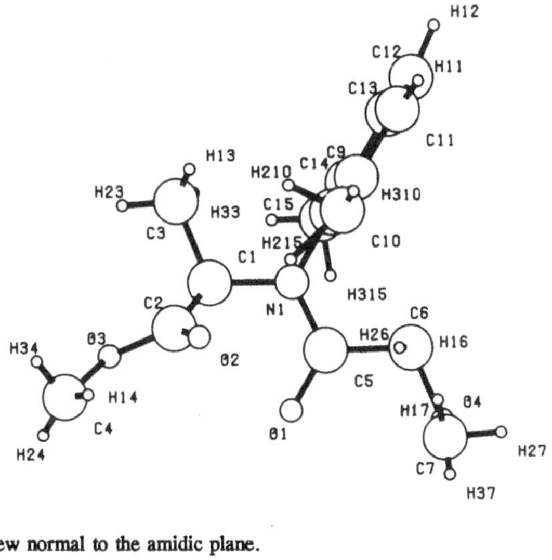

Fig. 1. The $C_{15}H_{21}NO_4$ molecule and a view normal to the amidic plane.

cyclo-(-L-2-HYDROXYISOVALERYL-L-PROLYL-L-PROLYL-)
$C_{15}H_{22}N_2O_4$ (I)

cyclo-(-D-2-HYDROXYISOVALERYL-L-PROLYL-L-PROLYL-)
$C_{15}H_{22}N_2O_4$ (II)

F. PINNEN, G. ZANOTTI, G. LUCENTE, S. CERRINI, W. FEDELI and E. GAVUZZO, 1985. *J. Chem. Soc. Perkin Trans.* 2, 1931-1937.

I. Orthorhombic, $P2_12_12_1$, a = 9.642, b = 10.177, c = 15.626 Å, Z = 4. Mo radiation, R = 0.043 for 1915 reflections.

II. Orthorhombic, $P2_12_12_1$, a = 6.138, b = 15.695, c = 15.794 Å, Z = 4. Cu radiation, R = 0.049 for 1256 reflections.

The prolyl residues have approximately C_2 symmetry in (I) and C_s symmetry in (II). All peptide bonds have the cis conformation and the two lactone bonds have the trans-conformation. Perspective views of the molecules are shown in Fig. 1 and a detailed discussion of the geometries is given in the paper.

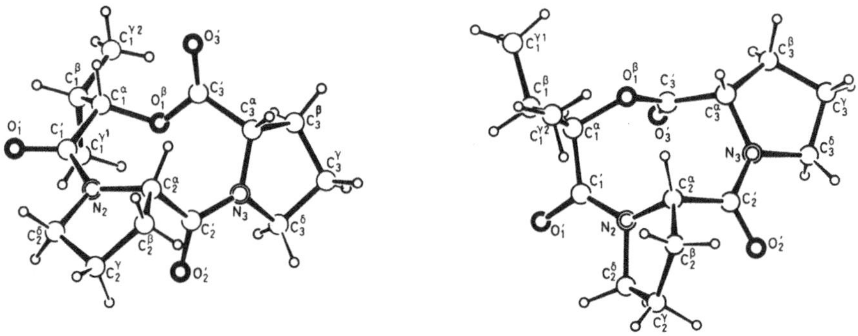

Fig. 1. Views of the molecules of $C_{15}H_{22}N_2O_4$: (I) (left) and (II) (right).

PIVALOYL-L-PROLYL-N-METHYL-L-THREONINAMIDE
$C_{15}H_{27}N_3O_4$

A. AUBRY and M. MARRAUD, 1985. *Acta Cryst.*, C41, 65-67.

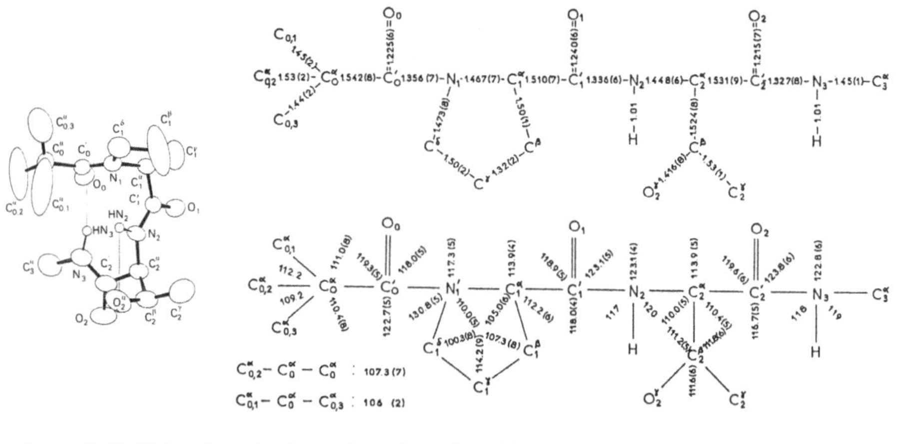

Fig. 1. $C_{15}H_{27}N_3O_4$: the molecular conformation and bond lengths and valence angles.

Hexagonal, $P6_5$, a = 10.308, c = 28.117 Å, Z = 6. Cu radiation, R = 0.052 for 1240 reflections.

The molecule adopts a folded conformation (Fig. 1) stabilized by intramolecular N(3)-H...O and N(2)-H...O hydrogen bonds of lengths 2.956(8) and 2.929(6) Å respectively. The pyrrolidine ring of the proline has an envelope conformation with C(1ν) as flap. The threonine hydroxyl group is intermolecularly hydrogen bonded to the proline carbonyl group. The O(2ν)-H...O(1b) separation is 2.810(8) Å.

t-BUTYLOXYCARBONYLGLYCYL-S-THIOALANYL-2-METHYLALANINE METHYL ESTER $C_{15}H_{27}N_3O_5S$ (I)

t-BUTYLOXYCARBONYLTHIOGLYCYL-S-ALANYL-2-METHYLALANINE METHYL ESTER $C_{15}H_{27}N_3O_5S$ (II)

O.E. JENSEN, S.-O. LAWESSON, R. BARDI, A.M. PIAZZESI and C. TONIOLO, 1985. *Tetrahedron*, **41**, 5595-5606.

I. Trigonal, $P3_1$, a = 12.052, c = 11.756 Å, Z = 3. Mo radiation, R = 0.040 for 1677 reflections.

II. Orthorhombic, $P2_12_12_1$, a = 16.341, b = 11.376, c = 10.739 Å, Z = 4. Mo radiation, R = 0.049 for 1423 reflections.

General views of the two molecules are in Fig. 1. Molecular dimensions are as anticipated. The C(4)-O(1)-C(5) angle is 121.1° for (I) and 120.3° for (II). The thiopeptide group is planar (ω_2 180° in (I), ω_1 175.5° in (II)). Molecule (I) has a folded β-turn stabilized by a 4-1 intramolecular N-H...O hydrogen bond (3.02 Å). Molecule (II) has no intramolecular H-bond and adopts an extended conformation. In the crystal structures molecules are linked by N-H...O (2.99-3.09 Å) and N-H...S (3.33 and 3.36 Å) hydrogen bonds.

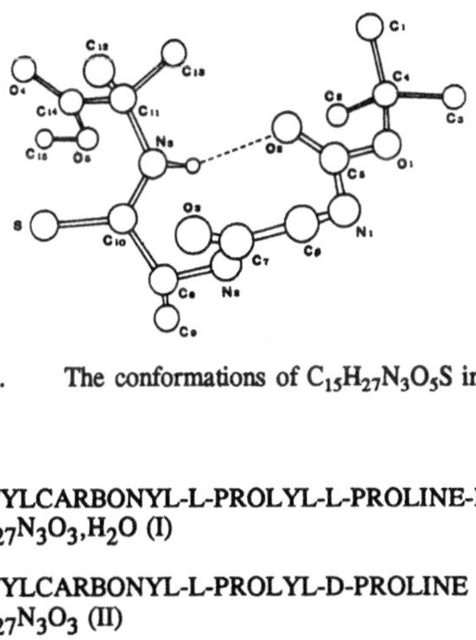

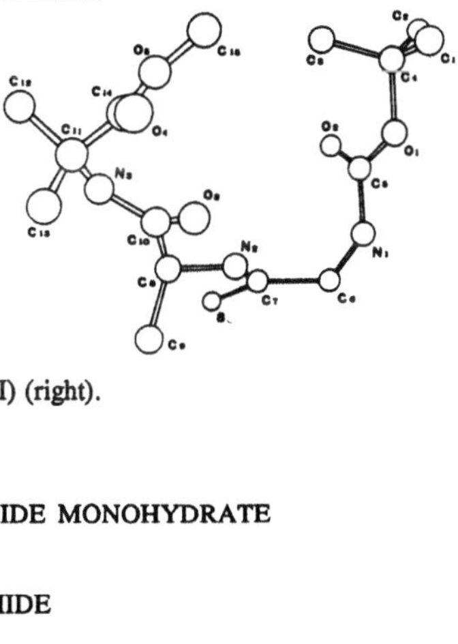

Fig. 1. The conformations of $C_{15}H_{27}N_3O_5S$ in (I) (left) and (II) (right).

t-BUTYLCARBONYL-L-PROLYL-L-PROLINE-N-METHYLAMIDE MONOHYDRATE $C_{16}H_{27}N_3O_3,H_2O$ (I)

t-BUTYLCARBONYL-L-PROLYL-D-PROLINE N-METHYLAMIDE $C_{16}H_{27}N_3O_3$ (II)

A. AUBRY, B. VITOUX and M. MARRAUD, 1985. *Biopolymers*, **24**, 1089-1100.

I. Monoclinic, $P2_1$, a = 6.662, b = 11.067, c = 12.205 Å, β = 96.28°, Z = 2. Cu radiation, R = 0.0435 for 1734 reflections.

II. Monoclinic, $P2_1$, a = 10.770, b = 15.039, c = 11.325 Å, β = 110.00°, Z = 4. Cu radiation, R = 0.0551 for 2847 reflections.

Views of these diastereoisomeric derivatives are shown in Fig. 1. The homochiral sequence in (I) accommodates an open conformation with an intermolecular hydrogen bond to a water molecule, whereas the heterochiral sequence in (II) is βII folded by a classical i+3→i interaction. Comparison with crystal structures of unstrained oligopeptides shows that the conformational properties of L-Pro-L-Pro sequences are probably a simple combination of those found for isolated L-Pro residues.

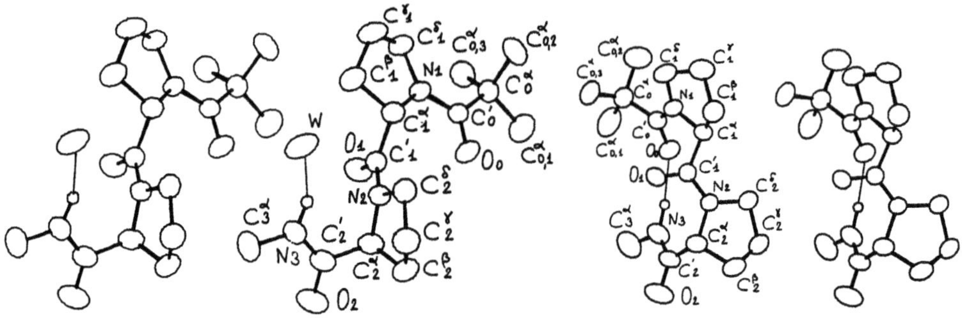

Fig. 1. Views of $C_{16}H_{27}N_3O_3,H_2O$ (I) (left) and $C_{16}H_{27}N_3O_3$ (II) (right).

DL-METHYL-N-(2,6-DIMETHYLPHENYL)-N-(2-FUROYL)ALANINATE (FURALAXYL)
$C_{17}H_{19}NO_4$

J.C.J. BART, R. SCORDAMAGLIA and M. CALCATERRA, 1985. *J. Cryst. Spect. Res.*, **15**, 387-397.

Monoclinic, *Cc*, a = 15.997, b = 7.929, c = 14.664 Å, β = 118.70°, Z = 4. Mo radiation, R = 0.068 for 852 reflections.

The molecule (Fig. 1) is the active ingredient of a systemic fungicide; the dimethylphenyl ring plane is twisted 85.1° out of the amidic plane and the planar furan ring is only rotated 6.9° out of this plane. The N atom is sp^2 hybridized (N-C(1) 1.340(14), N-C(6) 1.464(11), N-C(14) 1.477(12) Å).

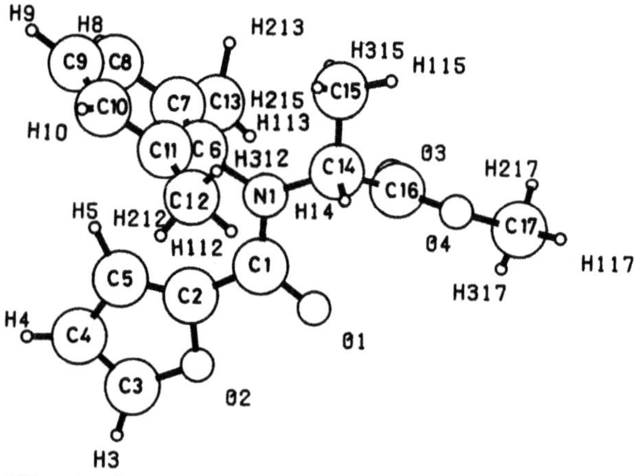

Fig. 1. The $C_{17}H_{19}NO_4$ molecule.

(2R,5S,11S,12S)-cyclo-α-MERCAPTOPROPIONYL-(S)-PHENYLALANYL-(S)-PROLYL-LACTAM
$C_{17}H_{20}N_2O_3S$

T. DEBAERDEMAEKER, T. SCHOEN and M. ROTHE, 1985. *Acta Cryst.*, **C41**, 267-268.

Monoclinic, $P2_1$, a = 9.331, b = 10.597, c = 8.922 Å, β = 108.99°, Z = 2. Mo radiation, R = 0.068 for 1290 reflections.

The compound (Fig. 1) is the 1-thio analogue of the peptide part of ergotamine. The thiazolidine ring has an envelope conformation while the pyrrolidine and piperazine rings adopt half-chair and boat conformations respectively. The molecules in the unit cell are separated by the only short distance of 2.63(1) Å between O(12) and

O(6).

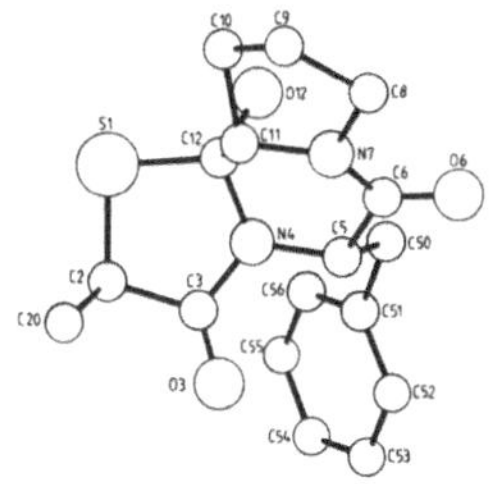

Fig. 1. $C_{17}H_{20}N_2O_3S$: the molecular structure.

t-BUTYLCARBONYL-L-PROLYL-L-LEUCYLMETHYLAMINE
$C_{17}H_{31}N_3O_3$ (I)

t-BUTYLCARBONYL-L-PROLYL-L-METHYLCYSTEINYLMETHYLAMINE
$C_{15}H_{27}N_3O_3S$ (II)

t-BUTYLCARBONYL-L-PROLYL-L-PHENYLALANYLMETHYLAMINE
$C_{20}H_{29}N_3O_3$ (III)

t-BUTYLCARBONYL-L-PROLYL-L-TYROSYLMETHYLAMINE
$C_{20}H_{29}N_3O_4$ (IV)

t-BUTYLCARBONYL-L-PROLYL-D-TYROSYLMETHYLAMINE
$C_{20}H_{29}N_3O_4$ (V)

A. AUBRY, M.T. CUNG and M. MARRAUD, 1985. *J. Am. Chem. Soc.*, **107**, 7640-7647.

I. Monoclinic, $P2_1$, a = 9.883, b = 9.518, c = 10.678 Å, β = 100.17°, Z = 2. Mo radiation, R = 0.048 for 1808 reflections.

II. Monoclinic, $P2_1$, a = 9.642, b = 9.842, c = 9.646 Å, β = 99.76°, Z = 2. Mo radiation, R = 0.052 for 1312 reflections.

III. Orthorhombic, $P2_12_12_1$, a = 20.951, b = 17.331, c = 5.440 Å, Z = 4. Mo radiation, R = 0.058 for 1589 reflections.

IV. Triclinic, $P1$, a = 5.715, b = 7.736, c = 11.450 Å, α = 86.53, β = 82.39, γ = 84.96°, Z = 1. Mo radiation, R = 0.031 for 1810 reflections.

V. Monoclinic, $P2_1$, a = 5.228, b = 15.043, c = 11.540 Å, β = 96.84°, Z = 2. Mo radiation, R = 0.045 for 1892 reflections.

In these dipeptides the middle and C-terminal amide groups have standard dimensions but steric hindrance between the t-butyl group and the Pro C^{δ} atom induces some angular variation around C' and N (Fig. 1). All five dipeptides are folded by a (i+3)→i interaction typical of a β-turn which is of the βII type whatever the LL or LD sequence. In the crystals the molecules are linked by hydrogen bonds.

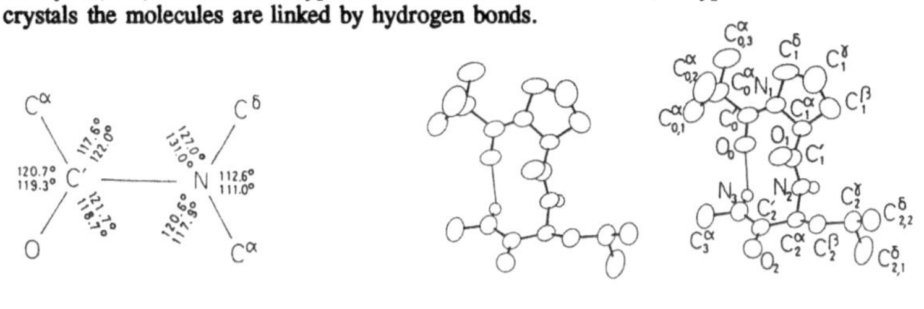

Fig. 1. Compared bond angles for the t-BuCO-Pro (lower values) and standard X-Pro (upper values) amide bonds; a stereoview of $C_{17}H_{31}N_3O_3$ (I) is also shown.

3,6-BIS(PHENYLMETHYLENE)PIPERAZINE-2,5-DIONE
$C_{18}H_{14}N_2O_2$

D. AJO, M. CASARIN, R. BERTONCELLO, V. BUSETTI, H.C.J. OTTENHEIJM and R. PLATE, 1985. *Tetrahedron,* **41**, 5543-5552.

Monoclinic, *C2/c*, a = 40.774, b = 6.237, c = 17.731 Å, β = 107.76°, Z = 12. Mo radiation, R = 0.061 for 1486 reflections.

One molecule lies in a general position in the cell and another lies about an inversion centre. The molecules have approximately planar piperazinedione rings and are linked in two series of hydrogen bonded ribbons (Fig. 1). The N-C_α bond lengths range from 1.389(8) to 1.423(9) Å.

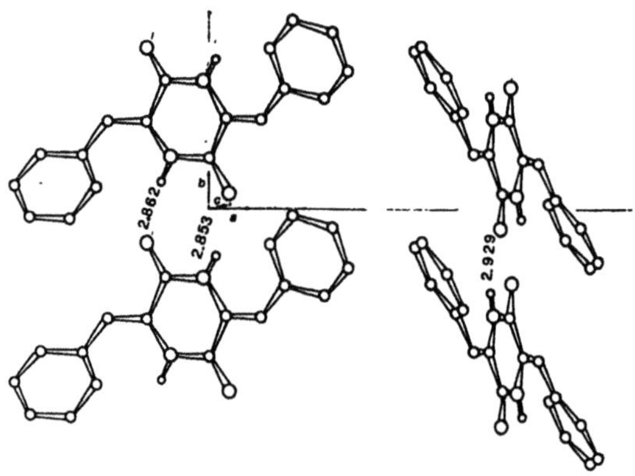

Fig. 1. The hydrogen-bond network in $C_{18}H_{14}N_2O_2$.

2-(R)-AMINO-4-(R)-HYDROXY-4-(p-HYDROXYPHENYL)-3-(R)-METHYLBUTANOIC ACID (S)-α-PHENYL-ETHYLAMIDE
$C_{19}H_{24}N_2O_3$

G. ZIMMERMANN, W. HASS, H. FAASCH, H. SCHMALLE and W.A. KONIG, 1985. *Liebigs Ann. Chem.*, 2165-2177.

Monoclinic, $P2_1$, a = 18.696, b = 6.024, c = 7.796 Å, β = 91.15°, Z = 2. Cu radiation, R = 0.0873 for 1494 reflections.

The analysis establishes the absolute configuration 2R,3R,4R as shown in Fig. 1.

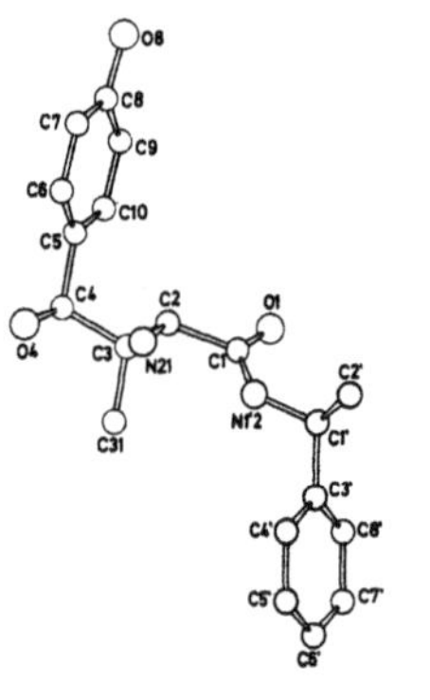

Fig. 1. A view of $C_{19}H_{24}N_2O_3$.

trans-N-PIVALOYL-D-PHENYLALANYL-L-THIAZOLIDINE-4-(N'-ETHYL)CARBOXAMIDE
$C_{20}H_{29}N_3O_3S$

S. WU, B. TINANT, J.P. DECLERCQ, M. VAN MEERSSCHE, F.A.M. BORREMANS and M.J.O. ANTEUNIS, 1985. *Bull. Soc. Chim. Belg.*, **94**, 791-796.

Orthorhombic, $P2_12_12_1$, a = 19.466, b = 16.768, c = 6.384 Å, Z = 4. Mo radiation, R = 0.051 for 1233 reflections.

The peptide chain (Fig. 1) forms a bend of the type II' β, with ϕ_1, ψ_1, ϕ_2, ψ_2, 62.5, -125.1, -92.5, 10.1° respectively. The three amides are arranged trans and thus conformation is stabilized by an intramolecular N-H...O hydrogen bond (N(3)-H...O(1), N...O 2.93 Å).

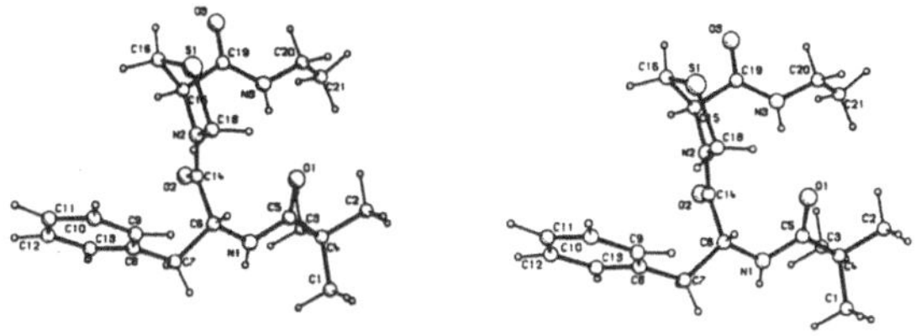

Fig. 1. A stereoview of $C_{20}H_{29}N_3O_3S$.

N-PIVALOYL-D-PHENYLALANYL-L-PROLINE-N-ETHYLAMIDE
$C_{21}H_{31}N_3O_3$

S. WU, B. TINANT, J.P. DECLERCQ, M. VAN MEERSSCHE, F.A.M. BORREMANS and M.J.O. ANTEUNIS, 1985. *Bull. Soc. Chim. Belg.*, **94**, 785-789.

Orthorhombic, $P2_12_12_1$, a = 19.336, b = 16.818, c = 6.488 Å, Z = 4. Mo radiation, R = 0.084 for 1307 reflections.

The peptide backbone is in the all-transoid form and takes the type II' β-bend conformation (Fig. 1) characterised by ϕ_1,ψ_1,ϕ_2 and ψ_2 values of 61.8, -123.2, -84.8 and 0.8° respectively, and by an intramolecular hydrogen bond involving the ethylamide NH and pivaloyl carbonyl oxygen (2.94 Å). The phenyl group stacks over the central amide bond in the "extended-to-oxygen" rotamer.

Fig. 1. A stereoview of $C_{21}H_{31}N_3O_3$.

cyclo-BIS(N-METHYLVALYL-α-HYDROXYISOVALERYL)
$C_{22}H_{38}N_2O_6$

V.I. SMIRNOVA and G.N. TISHCHENKO, 1985. *Kristallografiya*, **30**, 497-503.

Orthorhombic, $P2_12_12_1$, a = 18.30, b = 12.14, c = 10.93 Å, Z = 4. Cu radiation, R = 0.055 for 1446 reflections.

The depsipeptide ring of this LDDD isomer has an asymmetric form (Fig. 1) with one of the C=O groups (C(2)-O(2)) on the opposite side of the ring from the other three C=O groups. Bond lengths and angles are in agreement with those found in related structures. A comparison is made between the geometrical characteristics of the amide and ester groups in cyclic depsipeptides.

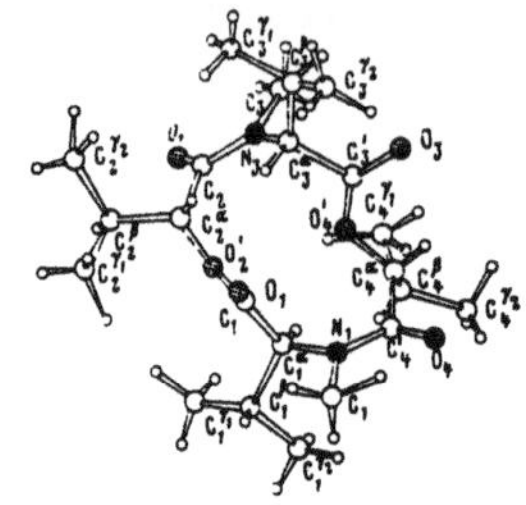

Fig. 1. The conformation of LDDD $C_{22}H_{38}N_2O_6$.

cyclo-[L-ALANYL-L-PROLYLGLYCYL-D-PHENYLALANYL-L-PROLYL]
$C_{24}H_{31}N_5O_5$

L.M. GIERASCH, I.L. KARLE, A.L. ROCKWELL and K. YENAL, 1985. *J. Am. Chem. Soc.*, **107**, 3321-3327.

Orthorhombic, $P2_12_12_1$, a = 9.142, b = 11.000, c = 23.885 Å, Z = 4. Cu radiation, R = 0.070 for 1544 reflections.

In the crystal structure (Fig. 1) the presence of L-Ala as opposed to D-Ala introduces a cis peptide bond between Ala^1 and Pro^2, directing O(1) to the exterior of the backbone ring, thus being unable to form an intramolecular hydrogen bond with N_4H. Also there is no γ-turn (3→1 H-bond) due to a change of 57° in $\psi^5(C_i^\alpha\text{-}C_i')$ and N_1 is buried within the molecule. The Pro^5 residue is trans in the present structure, the cis conformation being sterically untenable with L-Ala^1. There is some indication of puckering disorder in the two pyrrolidine rings and rotational disorder in the phenyl rings, possible since these groups lie close to a 4 Å void in the crystal structure. There are pairs of H bonds (N3...O5 3.007 and N4...O1 2.907 Å) linking adjacent molecules into zigzag ribbons (Fig. 1) with weak van der Waals interactions between hydrophobic groups linking ribbons together.

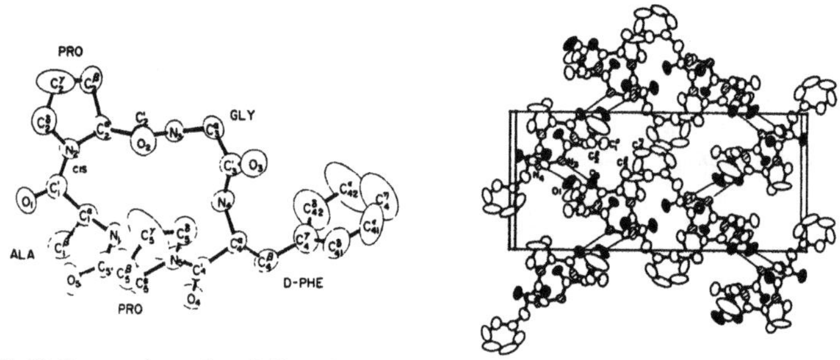

Fig. 1. $C_{24}H_{31}N_5O_5$: conformation (left) and a view of the crystal stacking.

N-t-BUTOXYCARBONYL-L-PROLYL-L-METHIONYLGLYCINE BENZYL ESTER
$C_{24}H_{35}N_3O_6S$

T. YAMANE, Y. SHIRAISHI and T. ASHIDA, 1985. *Acta Cryst.*, C41, 946-950.

Monoclinic, $P2_1$, a = 9.683, b = 35.038, c = 9.113 Å, β = 118.6°, Z = 4. Cu radiation, R = 0.061 for 3328 reflections.

There are two crystallographically independent molecules in the asymmetric unit. These molecules (Fig. 1) are related by a pseudo-twofold rotation axis and are conformationally similar apart from the orientations of the phenyl groups. Bond distances and angles are not significantly different for the two molecules except for S-C(14) (1.710, 1.507 Å). The mean torsion angles ϕ, ψ of the central Met are -136 and 138°. Molecules are alternately stacked to form a ribbon of an antiparallel β sheet infinitely extended along the a-axis. A pseudo twofold rotation axis exists at each centre of two neighbouring peptide molecules which are linked by a pair of hydrogen bonds of length 2.888 Å.

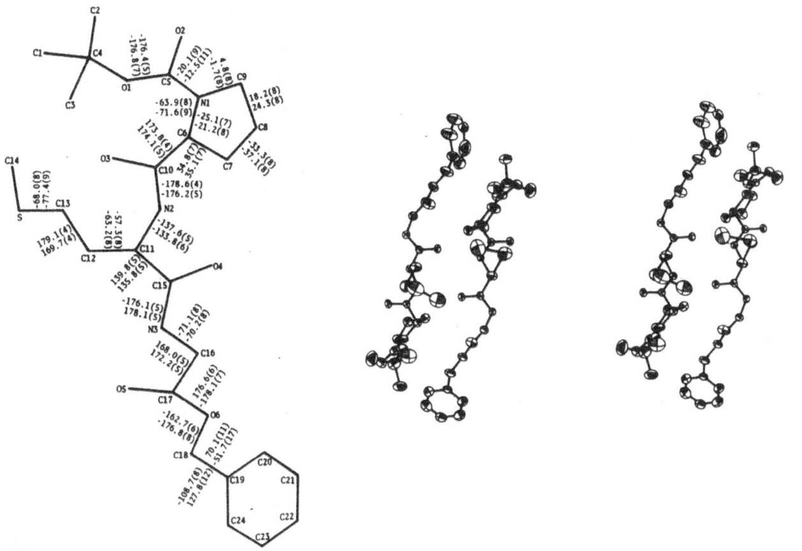

Fig. 1. The $C_{24}H_{35}N_3O_6S$ molecule showing the principal torsion angles, and a stereoview of a pair of independent molecules.

t-BUTYLOXY-L-PROLYL-α-AMINOISOBUTYRYL-L-ALANYL-α-AMINOISOBUTYRYL-L-ALANINE $C_{24}H_{41}N_5O_8$

R. BOSCH, G. JUNG, H. SCHMITT and W. WINTER, 1985. *Acta Cryst.*, C41, 1821-1825.

Orthorhombic, $P2_12_12$, a = 18.839, b = 18.776, c = 9.179 Å, Z = 4. Cu radiation, R = 0.056 for 2429 reflections.

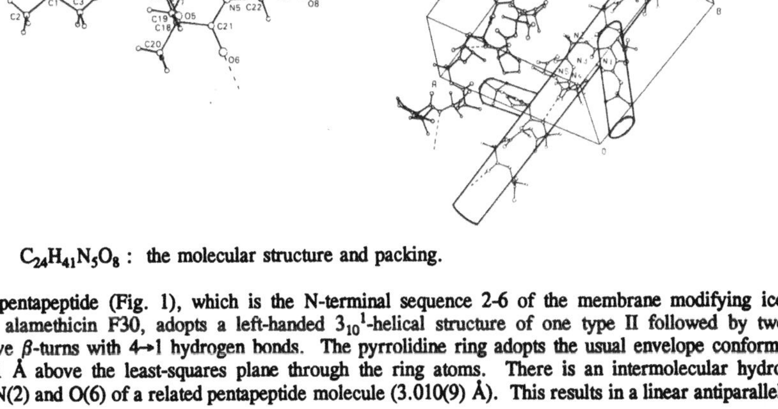

Fig. 1. $C_{24}H_{41}N_5O_8$: the molecular structure and packing.

This pentapeptide (Fig. 1), which is the N-terminal sequence 2-6 of the membrane modifying icosapeptide antibiotic alamethicin F30, adopts a left-handed $3_{10}{}^1$-helical structure of one type II followed by two type III consecutive β-turns with 4→1 hydrogen bonds. The pyrrolidine ring adopts the usual envelope conformation with C(8) 0.21 Å above the least-squares plane through the ring atoms. There is an intermolecular hydrogen bond between N(2) and O(6) of a related pentapeptide molecule (3.010(9) Å). This results in a linear antiparallel chainlike head-to-tail arrangement parallel to [010]. The unit cell also contains four undefined and disordered hydrocarbon solvent molecules.

t-BUTOXYCARBONYL-L-PROLYL-D-PROLYL-L-PROLYL-D-PROLINE
$C_{25}H_{38}N_4O_7$

M. COLAPIETRO, P. DE SANTIS, F. NOCILLI, A. PALLESCHI and R. SPAGNA, 1985. *Acta Cryst.*, C41, 126-130.

Monoclinic, $P2_1$, a = 14.123, b = 9.048, c = 11.900 Å, β = 110.58°, Z = 2. Mo radiation, R = 0.054 for 2513 reflections.

The molecule (Fig. 1) has two conformationally quasi-equivalent dipeptide halves with an overall left handed screw trend. Torsion angles around the C(2)-N(1) and C(7)-N(2) bonds are 51.9(6) and -80.9(7)° respectively. For the other two proline residues the absolute values of the equivalent torsion angles are both 64.9(5)°. All the pyrrolidine rings have envelope conformations with atoms C(4), C(8), C(14) and C(18) as flaps. The crystal structure is stabilized by hydrogen bonds involving O(1) and O(4) which bridge pairs of screw-axis related molecules. The structure is compared with and nearly identical to that predicted by minimization of the conformational energy.

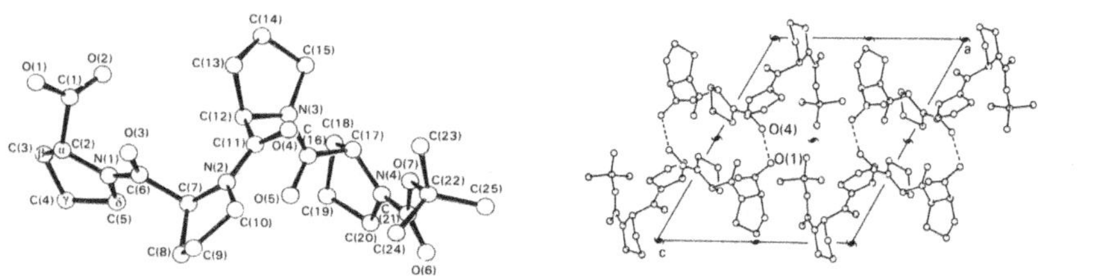

Fig. 1. $C_{25}H_{38}N_4O_7$: a view of the molecule showing the atomic numbering (left) and (right) a projection down the b-axis of the unit cell contents in which the broken lines represent the hydrogen bonds.

N-(2-(N-CARBOBENZYLOXY-α-ALANYL)AMINO-5-BROMOBENZHYDRYL)GLYCINE METHYL ESTER
$C_{27}H_{28}BrN_3O_5$

M.M. BOTOSHANSKII, Yu.A. SIMONOV, T.I. MALINOVSKII, O.P. RUDENKO and G.N. TISHCHENKO, 1985. *Kristallografiya*, 30, 96-100.

Monoclinic, $P112_1/b$, a = 13.906, b = 22.102, c = 8.938 Å, γ = 74.68°, Z = 4. Mo radiation, R = 0.066 for 1908 reflections.

The molecule has a folded chain conformation (Fig. 1) which is stabilized by an N(8)-H...N(12) intramolecular hydrogen bond (2.77 Å). Molecules are further linked by intermolecular N(5)-H...O(41) hydrogen bonds to form dimers about an inversion centre (N5...O41 2.87 Å).

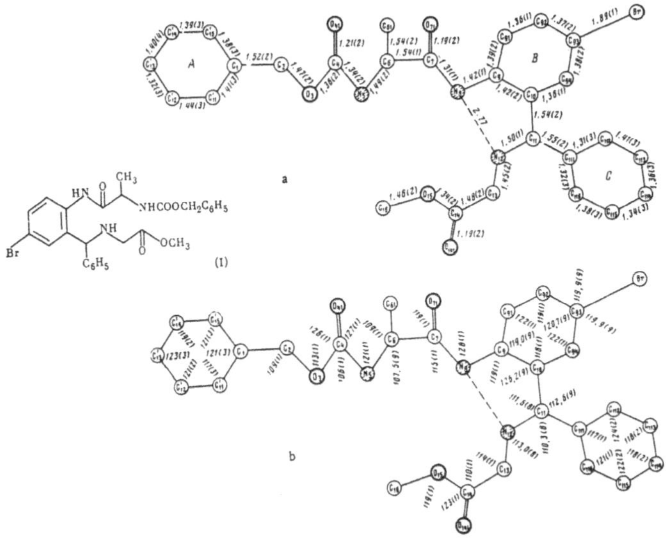

Fig. 1. The $C_{27}H_{28}BrN_3O_5$ molecule and dimensions.

(3-HYDROXY-4-METHYL-2-NITROBENZYL)L-THREONYL-D-VALYL-L-PROLYLSARCOSYL-N-METHYL-L-ALANINE LACTONE ETHYL ACETATE SOLVATE HEMIHYDRATE
$C_{29}H_{40}N_6O_{10},0.5(C_4H_8O_2),0.5(H_2O)$

A.B. MAUGER, O.A. STUART, J.A. FERRETTI and J.V. SILVERTON, 1985. *J. Am. Chem. Soc.*, **107**, 7154-7163.

Orthorhombic, $P2_12_12_1$, a = 12.891, b = 21.740, c = 24.909 Å, Z = 8. Cu radiation, R = 0.0636 for 7747 reflections.

There are two independent molecules in the asymmetric unit; their conformation angles are shown in Fig. 1. There is extensive hydrogen bonding and pairs of conformers have a juxtapositional geometry remarkably similar to that known for actinomycin.

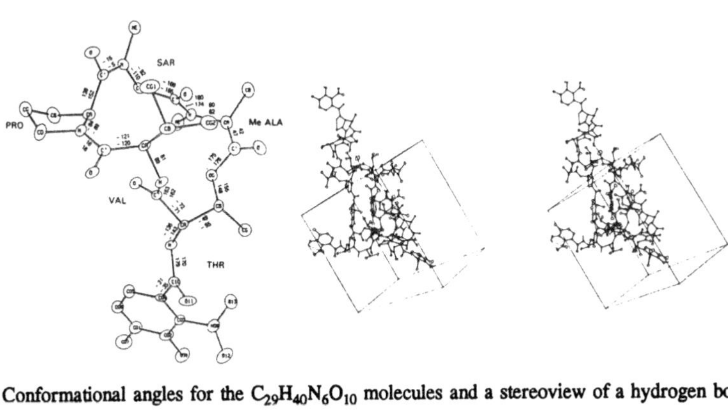

Fig. 1. Conformational angles for the $C_{29}H_{40}N_6O_{10}$ molecules and a stereoview of a hydrogen bonded dimer.

cyclo-BIS(D-ALANYLGLYCYL-PROLYL-D-PHENYLALANYL) METHANOL SOLVATE DIHYDRATE

$C_{38}H_{48}N_8O_8,CH_4O,2(H_2O)$

K.D. KOPPLE, G. KARTHA, K.K. BHANDARY and K. ROMANOWSKA, 1985. *J. Am. Chem. Soc.*, **107**, 4893-4897.

Monoclinic, $P2_1$, a = 14.656, b = 12.585, c = 15.657 Å, β = 134.07°, Z = 2. Mo radiation, R = 0.081 for 4219 reflections.

The conformation of the peptide is shown in Fig. 1. The backbone shows approximate C_2 symmetry and all peptide links are trans. The backbone has type II β-turns at L-Pro-D-Phe, followed by type III' turns at D-Phe-D-Ala. It differs from a solution conformation derived from NMR data primarily in an approximately 40° rotation of the plane of the Phe-Ala peptide bond relative to the local plane of α-carbons.

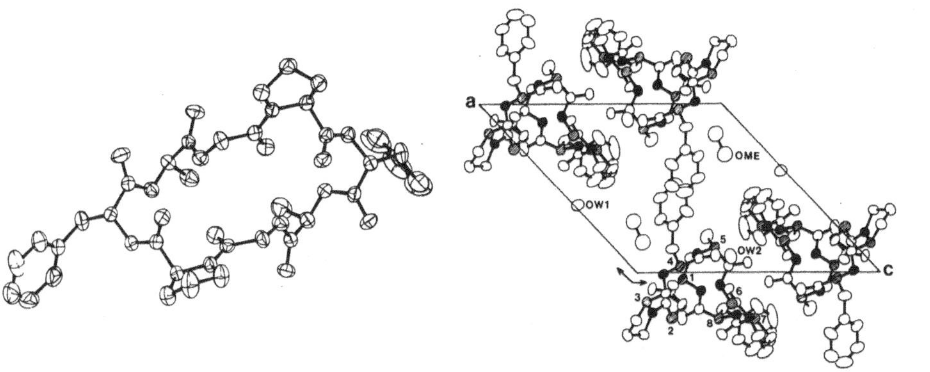

Fig. 1. The conformation of $C_{38}H_{48}N_8O_8$ and the packing viewed along the *b*-axis.

t-BUTYLOXYCARBONYL-L-ALANYL-α-AMINOISOBUTYRYLALANYL-α-AMINOISOBUTYRYLALANYL-BENZYLOXYGLUTAMOYLALANYL-α-AMINOISOBUTYRYLALANYL-α-AMINOISOBUTYRYLALANINE METHYL ESTER DICHLOROMETHANE SOLVATE

$C_{52}H_{83}N_{11}O_{16},3(CH_2Cl_2)$

R. BOSCH, G. JUNG, H. SCHMITT and W. WINTER, 1985. *Biopolymers*, **24**, 961-978.

Monoclinic, $P2_1$, a = 10.602, b = 23.884, c = 13.622 Å, β = 95.61°, Z = 2. Cu radiation, R = 0.093 for 4325 reflections.

This undecapeptide adopts a right-handed α-helical conformation in the solid state with intramolecular 5→1 hydrogen bonds. An additional intramolecular hydrogen bond is bifurcated, forming a stronger 4→1 interaction and a weaker 5→1 interaction, thus prolonging the α-helical part up to nine residues. The α-helix radius is 2.1 Å, height per residue 1.53 Å. The length of the α-helical part of the undecapeptide containing nine residues is 13.8 Å. In the crystal the α-helices are linked head-to-tail via two hydrogen bridges forming continuous chains. Adjacent helices are oriented in an antiparallel fashion with their helix axes, and have only van der Waals contacts.

t-BUTYLOXYCARBONYL-L-LEUCYL-α-AMINOISOBUTYRYLPROLYLVALYLBIS(α-AMINOISOBUTYRYL)-BENZYLOXYGLUTAMOYLGLUTAMINYLPHENYLALANINOL MONOHYDRATE

$C_{59}H_{90}N_{10}O_{14},H_2O$

R. BOSCH, G. JUNG, H. SCHMITT and W. WINTER, 1985. *Biopolymers*, **24**, 979-999.

Orthorhombic, $P2_12_12_1$, a = 15.666, b = 16.192, c = 26.876 Å, Z = 4. Cu radiation, R = 0.053 for 5859 reflections.

The molecular conformation of this nonapeptide is right-handed helical (Fig. 1) with three α-(5→1 hydrogen bonds) and three β-turns (4→1 hydrogen bonds). All but two of the hydrogen bonds are significantly longer than usual and show some bifurcation. The molecules are arranged head-to-tail in the *a*-direction and the resulting linear antiparallel chains are linked by a weak intermolecular hydrogen bridge, forming a two-dimensional layer structure in the *ab*-plane.

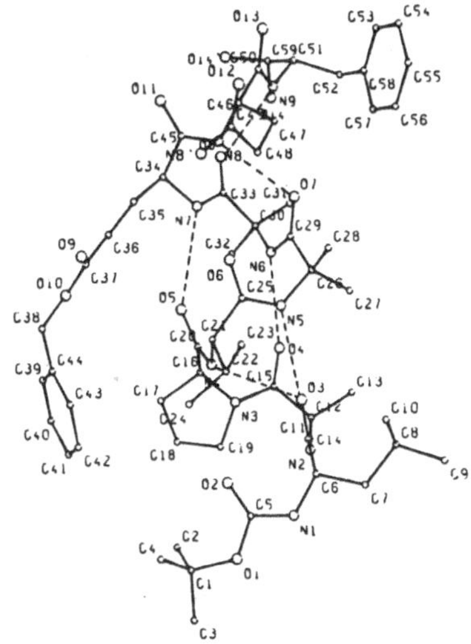

Fig. 1. A view of the $C_{59}H_{90}N_{10}O_{14}$.

CYCLOSPORIN A
$C_{62}H_{111}N_{11}O_{12}$

H.-R. LOOSLI, H. KESSLER, H. OSCHKINAT, H.-P. WEBER, T.J. PETCHER and A. WIDMER, 1985. *Helv. Chim. Acta*, **68**, 682-704.

Tetragonal, $P4_1$, a = 13.837, c = 41.242 Å, Z = 4. Cu radiation, R = 0.114 for 5057 reflections.

A detailed discussion of the molecular structure is given in the paper. A perspective drawing of the molecule is shown in Fig. 1, but this does not emphasise the important points of the structure.

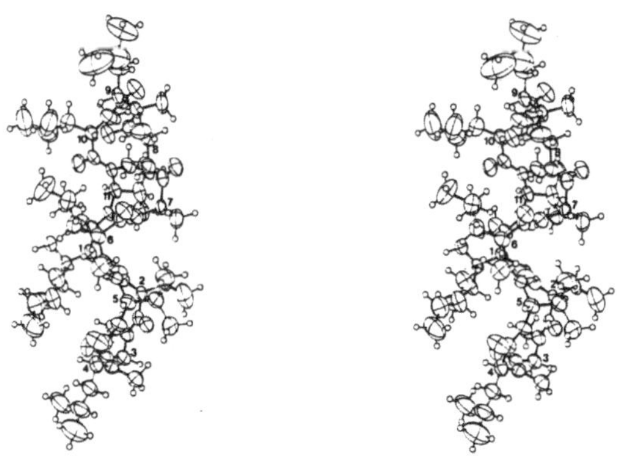

Fig. 1. An ORTEP drawing of a molecule of $C_{62}H_{111}N_{11}O_{12}$.

5-AZA-13,17-DIETHYL-2,3,7,8,12,18-HEXAMETHYLPORPHYRINATOCOBALT(II)
$C_{29}H_{31}CoN_5$

A.M. ABEYSEKERA, R. GRIGG, J.F. MALONE, T.J. KING and J.O. MORLEY, 1985. *J. Chem. Soc. Perkin*

Trans. 2, 395-402.

Orthorhombic, *Pbca*, a = 8.622, b = 37.788, c = 14.781 Å, Z = 8. Mo radiation, R = 0.069 for 966 reflections.

The molecules pack in columns (Fig. 1) with their mean planes parallel to c and inclined to b by 38°. In the columns adjacent molecules are rotated with respect to each other, through an axis perpendicular to the ring by 90°. Some average bond lengths are Co-N 1.962(10), N-C(1) 1.380(11), C(1)-C(2) 1.446(11), C(2)-C(3) 1.383(15), C(meso)-C 1.382(12) and N(meso)-C 1.321(20) Å.

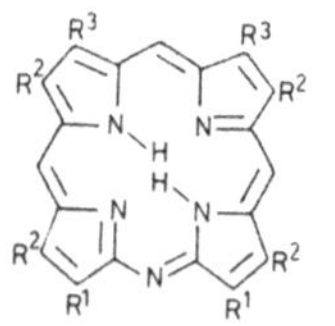

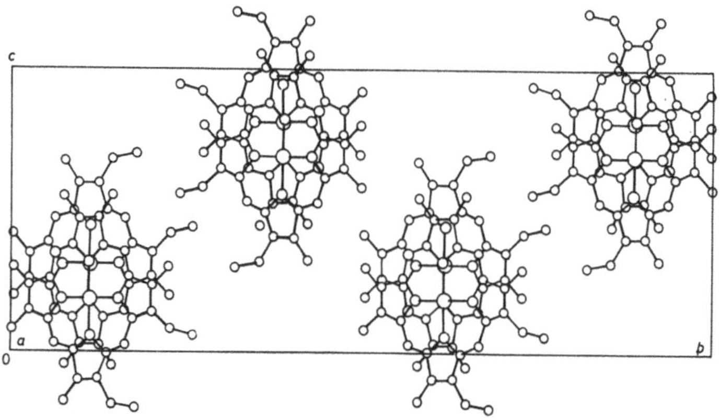

Fig. 1. Projection of the molecules of $C_{29}H_{31}CoN_5$ down the b axis.

DICHLOROPHTHALOCYANINATOTITANIUM(IV)
$C_{32}H_{16}Cl_2N_8Ti$

V.L. GOEDKEN, G. DESSY, C. ERCOLANI, V. FARES and L. GASTALDI, 1985. *Inorg. Chem.*, **24**, 991-995.

Triclinic, $P\bar{1}$, a = 8.744, b = 12.244, c = 12.835 Å, α = 79.73, β = 95.32, γ = 103.27°, Z = 2. Mo radiation, R = 0.0785 for 2158 reflections.

The Ti atom is six-coordinate, with the Cl atoms in cis positions (Fig. 1), and lies 0.84 Å above the N_4 plane. The orientation of the Ti-Cl bonds with respect to the Ti-N bonds is staggered but rotated 10° towards eclipsed. Bond lengths and angles include: Ti-Cl 2.309, 2.324(6), Ti-N 2.060 to 2.105(12) Å, Cl-Ti-Cl 82.4(2), N-Ti-N 78.7-82.7(4), N-Ti-Cl 79.4-88.0(3)°.

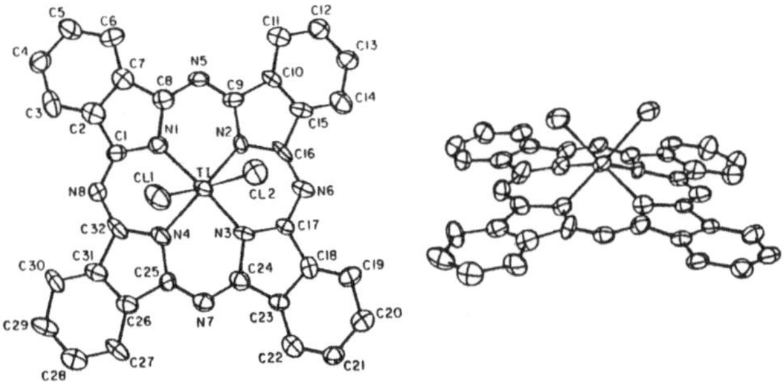

Fig. 1. Two views of the $C_{32}H_{16}Cl_2N_8Ti$ molecule.

BIS(CHLOROPHTHALOCYANINATOIRON(III)) IODINE
$2(C_{32}H_{16}ClFeN_8),I_2$ (I)

PHTHALOCYANINATOIRON(III) IODIDE
$C_{32}H_{16}FeIN_8$ (II) $C_{32}H_{16}FeN_8^+,I^-$

S.M. PALMER, J.L. STANTON, N.K. JAGGI, B.M. HOFFMAN, J.A. IBERS and L.H. SCHWARTZ, 1985. *Inorg. Chem.*, **24**, 2040-2046.

I. Triclinic, $P\bar{1}$, a = 12.270, b = 13.202, c = 8.335 Å, α = 92.80, β = 97.17, γ = 101.97°, Z = 1. Mo radiation, R = 0.042 for 4790 reflections (at 113 K).

II. Tetragonal, *P4/mcc*, a = 13.84, c = 6.77 Å, Z = 2. Mo radiation, R = 0.070 for 185 reflections (at 97 K).

Molecule (I) (Fig. 1) consists of two Fe(pc)Cl units, with Fe in square-pyramidal coordination, linked across a crystal inversion centre by Cl-I-I-Cl bonds. The Fe atom is 0.30 Å above the N_4 plane. Bond lengths and angles include: Fe-N 1.941-1.949(3), Fe-Cl 2.320(2), Cl-I 3.184(2), I-I 2.722(1) Å, Cl-I-I 179.4(1), Fe-Cl-I 108.9(1)°. The structure of (II) is similar to that of Ni(pc)I (**1**); planar Fe(pc) units on 4/m sites are stacked parallel to **c**, successive units being staggered by 40°, and I_3^- units lie in channels between the stacks.

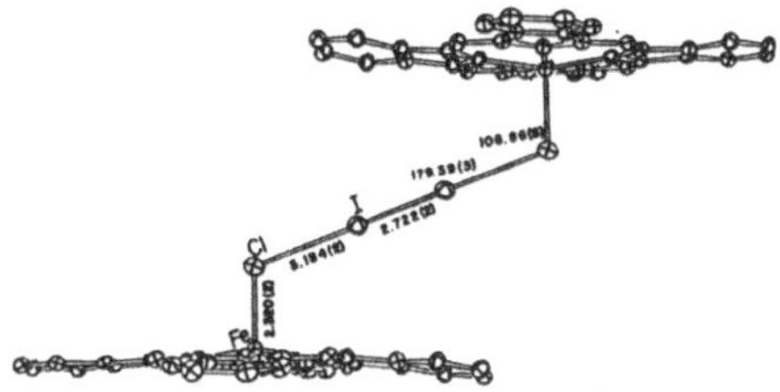

Fig. 1. $2(C_{32}H_{16}ClFeN_8),I_2$: the dimer.

1. Structure Reports, **46B**, 495.

CHLORO(PHTHALOCYANINATO)ZINC
$C_{32}H_{16}ClN_8Zn$ Zn(pc)Cl

M. MOSSOYAN-DENEUX, D. BENLIAN, M. PIERROT, A. FOURNEL and J.P. SORBIER, 1985. *Inorg. Chem.*, **24**, 1878-1882.

Tetragonal, *P4/nnc*, a = 13.85, c = 13.09 Å, Z = 4. Mo radiation, R = 0.098 for 665 reflections.

The molecule has fourfold imposed symmetry. The pc ring (Fig. 1) is planar, and the Zn atom is 0.59 Å above it in the direction of the Cl atom (Zn-Cl 2.35, Zn-N 1.98 Å). Pairs of pc rings on the same fourfold axis are separated by 3.16 Å and mutually rotated 39°; overlapping rings on different fourfold axes are separated by 3.38 Å.

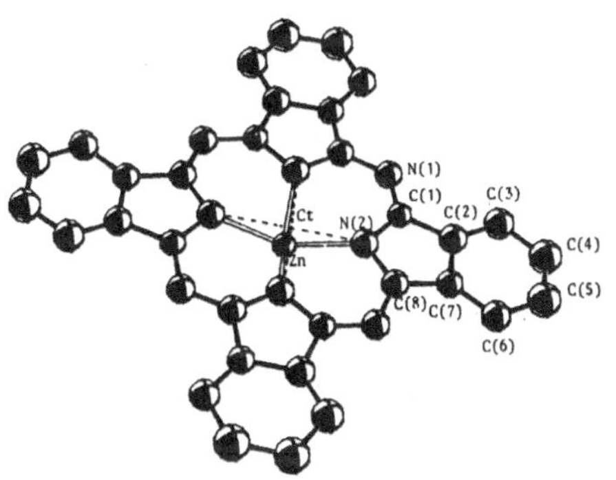

Fig. 1. $C_{32}H_{16}ClN_8Zn$: the molecule structure (Cl omitted).

PHTHALOCYANINATOCOBALT(II) IODIDE
$C_{32}H_{16}CoIN_8$ Co(pc)I

J. MARTINSEN, J.L. STANTON, R.L. GREENE, J. TANAKA, B.M. HOFFMAN and J.A. IBERS, 1985. *J. Am. Chem. Soc.*, **107**, 6915-6920.

Tetragonal, *P4/mcc*, a = 13.9266, c = 6.2466 Å, Z = 2. Cu radiation, R = 0.055 for 451 reflections (at 116 K).

The crystal packing of Co(pc)I consists of linear chains of I atoms segregated from macrocycle stacks (Fig. 1) both of which run parallel to the *c*-axis. The two Co(pc) molecules in the cell are staggered by 40°. Each Co(pc) unit is on a site of 4/m symmetry. The I atom is on a site of 422 symmetry. The Co-N(1) distance is 1.918(4) Å.

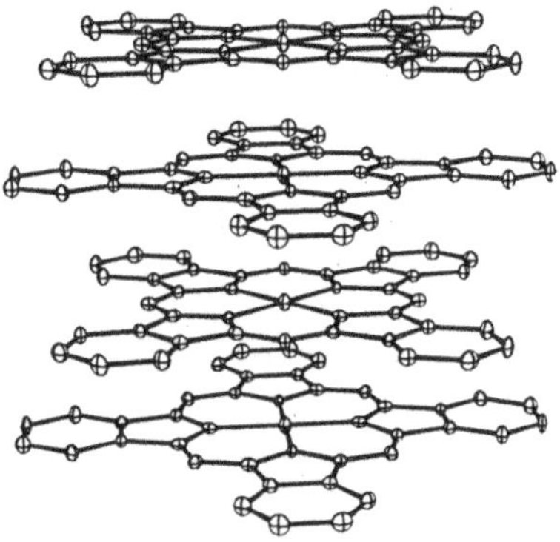

Fig. 1. A view of into a Co(pc) stack.

catena-(FLUOROPHTHALOCYANINATOGALLIUM(III))
$(C_{32}H_{16}FGaN_8)_n$ (I)

(μ_2-OXO)BIS(PHTHALOCYANINATOALUMINIUM(III))
$C_{64}H_{32}Al_2N_{16}O$ (II)

K.J. WYNNE, 1985. *Inorg. Chem.*, **24**, 1339-1343.

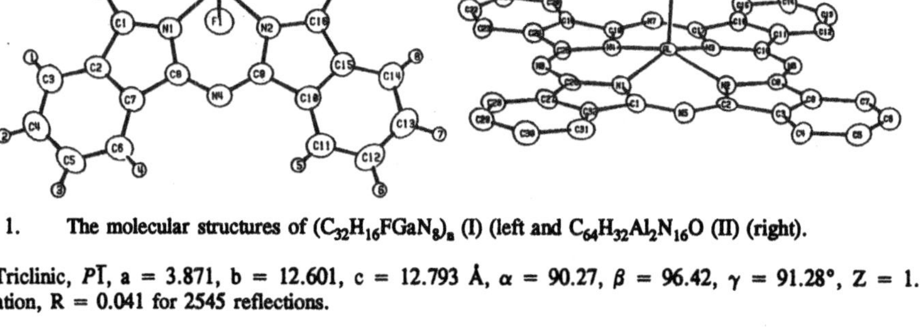

Fig. 1. The molecular structures of $(C_{32}H_{16}FGaN_8)_n$ (I) (left and $C_{64}H_{32}Al_2N_{16}O$ (II) (right).

I. Triclinic, $P\bar{1}$, a = 3.871, b = 12.601, c = 12.793 Å, α = 90.27, β = 96.42, γ = 91.28°, Z = 1. Cu radiation, R = 0.041 for 2545 reflections.

II. Triclinic, $P1$, a = 7.694, b = 12.520, c = 12.705 Å, α = 91.03, β = 94.54, γ = 90.37°, Z = 1. Cu radiation, R = 0.090 for 5018 reflections (at -110°C).

The Ga and F atoms in (I) lie on alternate crystal inversion centres along a. Octahedral coordination geometry for Ga is completed by the four inner N atoms of a virtually planar pc ring (Fig. 1). The rings are stacked in eclipsed fashion with interplanar spacing 3.872 Å. Bond lengths include: Ga-F 1.936(1), Ga-N 1.969, 1.970(2) Å. The oxygen atom which links two Al(pc) units in (II) (Fig. 1) lies on a crystal inversion centre. The pc rings are again eclipsed; the interplanar distances are 4.27 Å (intramolecular) and 3.60(2) Å (intermolecular). The Al atom has square-pyramidal geometry with Al-O 1.679(2), Al-N 1.968-1.994(5) Å, and lies 0.46 Å above the N_4 basal plane.

(2,3,7,8,12,13,17,18-OCTAETHYLPORPHYRINATO)IRON(II)
$C_{36}H_{44}FeN_4$ (I)

(trans-7,8-DIHYDRO-2,3,7,8,12,13,17,18-OCTAETHYLPORPHYRINATO)IRON(II)
$C_{36}H_{46}FeN_4$ (II)

S.H. STRAUSS, M.E. SILVER, K.M. LONG, R.G. THOMPSON, R.A. HUDGENS, K. SPARTALIAN and J.A. IBERS, 1985. *J. Am. Chem. Soc.*, **107**, 4207-4215.

I. Triclinic, $P\bar{1}$, a = 4.812, b = 13.252, c = 13.369 Å, α = 113.13, β = 92.20, γ = 93.20°, Z = 1. Cu radiation, R = 0.072 for 1273 reflections.

II. Orthorhombic, *Pbcn*, a = 21.880, b = 15.795, c = 8.554 Å, Z = 4. Mo radiation, R = 0.060 for 1929 reflections (at -150°C).

Molecule (I) lies on an inversion centre and molecule (II) has a twofold crystallographic symmetry axis along the N3-Fe-N2 axis (Fig. 1). The Fe and four N atoms are rigorously planar in both compounds. The porphyrin macrocycle in (I) is essentially planar while the chlorine macrocycle in (II) is significantly S_4 ruffled.

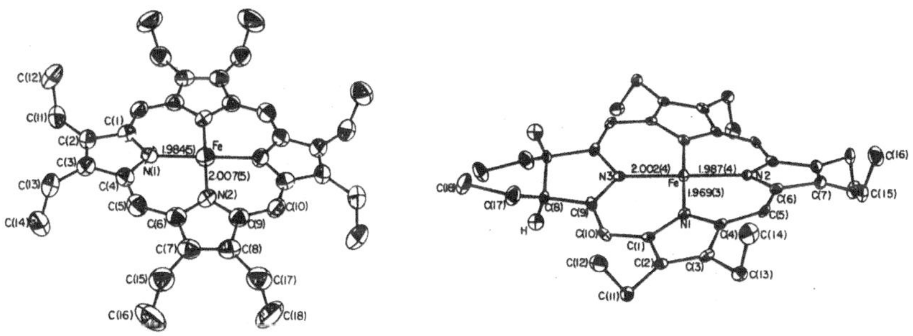

Fig. 1. Views (with Fe-N distances) for $C_{36}H_{44}FeN_4$ (I) (left) and $C_{36}H_{46}FeN_4$ (II) (right).

(OCTAETHYLPORPHYRIN N-OXIDE-N,N',N'',O)NICKEL(II)
$C_{36}H_{44}N_4NiO$

A.L. BALCH, YEE-WAI CHAN and M.M. OLMSTEAD, 1985. *J. Am. Chem. Soc.*, **107**, 6510-6514.

Triclinic, $P\bar{1}$, a = 10.610, b = 13.203, c = 13.199 Å, α = 115.49, β = 102.33, γ = 102.85°, Z = 2. Mo radiation, R = 0.068 for 3639 reflections (at 140 K).

The molecule has a nickel ion bound to three of the four N atoms and the oxygen in a somewhat distorted planar array (Fig. 1) with Ni-N 1.922(4), 1.929(5), 1.900(4), Ni-O 1.788(4) Å. The N-O distance is 1.363(6) Å, the oxygen lies 0.651(1) Å above the porphyrin and the pyrrole ring to which it is attached is tilted by 38.3(4)° out of the porphyrin plane. The structure is disordered as a result of packing of pairs of major (82.6%) and minor (17.4%) form.

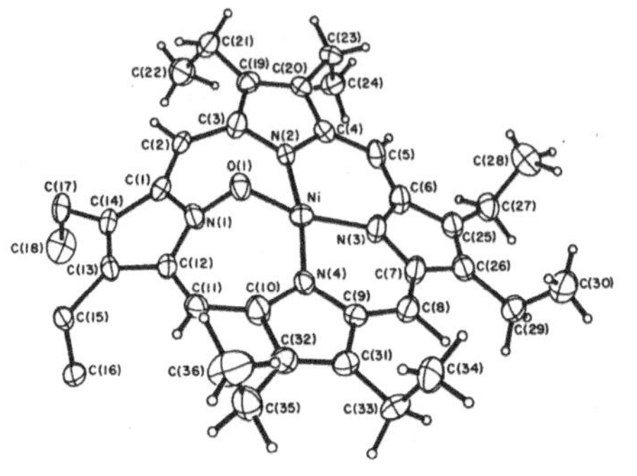

Fig. 1. A view of the major form of $C_{36}H_{44}N_4NiO$.

OCTAETHYLPORPHYRIN N-OXIDE
$C_{36}H_{46}N_4O$

A.L. BALCH, YEEWAI CHAN, M. OLMSTEAD and M.W. RENNER, 1985. *J. Am. Chem. Soc.*, **107**, 2393-2398.

Triclinic, $P\bar{1}$, a = 7.612, b = 9.740, c = 10.566 Å, α = 91.61, β = 98.49, γ = 108.43°, Z = 1. Mo radiation, R = 0.062 for 1588 reflections (at 140 K).

The oxygen atom disorders about four sites, allowing for a centre of symmetry to exist in the molecule. The O atom has 42% occupancy attached to N(1) and N(1') related by the symmetry centre, and 8% on N(2) and N(2'). The molecule is planar to within 0.153(4) Å (N(1) deviation) apart from the O atom and methyl groups. The O atom is out of plane by 0.972(6) Å in the major (42%) and by -0.986(24) Å in the minor site. The N-O distance of 1.398(7) Å is normal. The pyrrole ring containing N(1) tilts by 6.1° out of the porphyrin plane. There is some hydrogen bonding within the cavity, with H atoms attached to N(2) and N(2') contacting O(1) at distances of 1.88 and 1.82 Å respectively. The N(2)...O(1) distance is 2.564 Å while that of N(2')...O(1) is 2.517 Å. Remaining features of the structure are normal.

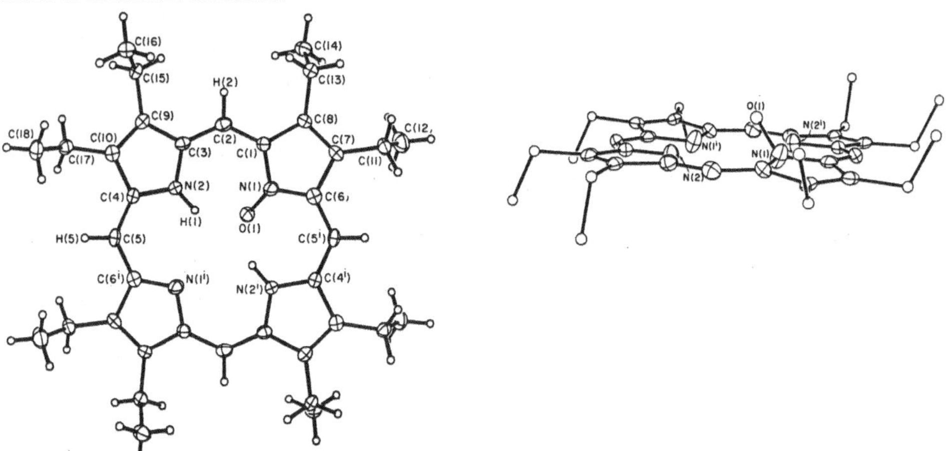

Fig. 1. $C_{36}H_{46}N_4O$: view normal to the N_4 plane showing one of the four possible O positions and a side view.

cis-DICHLORO(25-ETHOXYCARBONYL-2,3,7,8,12,13,17,18-OCTAETHYL-23,24-METHANO-23H,24H-PORPHYRIN-N^{21},N^{22},H)PALLADIUM(II) CHLOROFORM SOLVATE
$C_{40}H_{50}Cl_2N_4O_2Pd,CHCl_3$

T.J. BARTCZAK, 1985. *Acta Cryst.*, **C41**, 865-869.

Orthorhombic, *Pbca*, a = 16.577, b = 26.325, c = 19.951 Å, Z = 8. Mo radiation, R = 0.053 for 2442 reflections.

The $PdCl_2$ and ethyl acetate groups are bonded to the two opposite pairs of cis-N pyrrole atoms (Fig. 1). The porphyrin macrocycle is very folded, the dihedral angles of the pyrrole rings with the mean plane defined by atoms N(21), N(22), N(23) and N(24) being 25.4, 19.3, -14.1 and -16.9° for N(21) to N(24) pyrrole rings respectively. Coordination about Pd is best described as trigonally bipyramidally distorted square-planar. The fifth coordination site is occupied by H(37). Coordination distances are: Pd-H(37) 2.47, Pd-N(21) 1.974(10), Pd-N(22) 2.029(9), Pd-Cl(1) 2.292(3) and Pd-Cl(2) 2.312(3) Å. The solvent molecule is well defined.

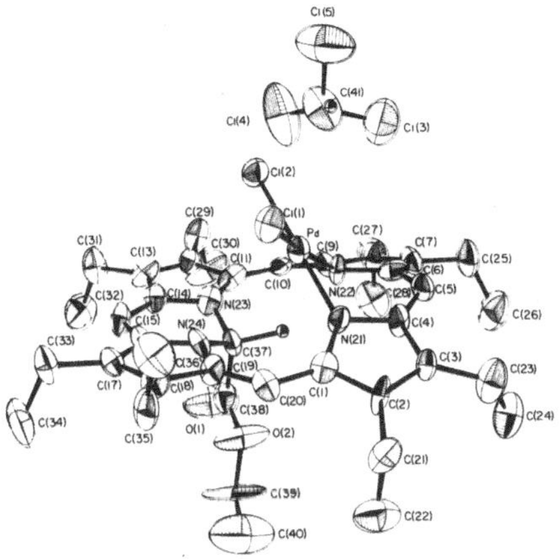

Fig. 1. $C_{40}H_{50}Cl_2N_4O_2Pd,CHCl_3$: the molecular structure.

(2-METHYLIMIDAZOL-3-YL)OCTAETHYLPORPHINATOIRON(III) PERCHLORATE CHLOROFORM SOLVATE

$C_{40}H_{50}ClFeN_6O_4,CHCl_3$ (I) $C_{40}H_{50}FeN_6^+,ClO_4^-,CHCl_3$

(2-METHYLIMIDAZOL-3-YL)OCTAETHYLPORPHINATOIRON(III) PERCHLORATE DIBROMOMETHANE SOLVATE

$C_{40}H_{50}ClFeN_6O_4,CH_2Br_2$ (II) $C_{40}H_{50}FeN_6^+,ClO^{4-},CH_2Br_2$

W.R. SCHEIDT, D.K. GEIGER, YOUNGJA LEE, C.A. REED and G. LANG, 1985. *J. Am. Chem. Soc.*, **107**, 5693-5699.

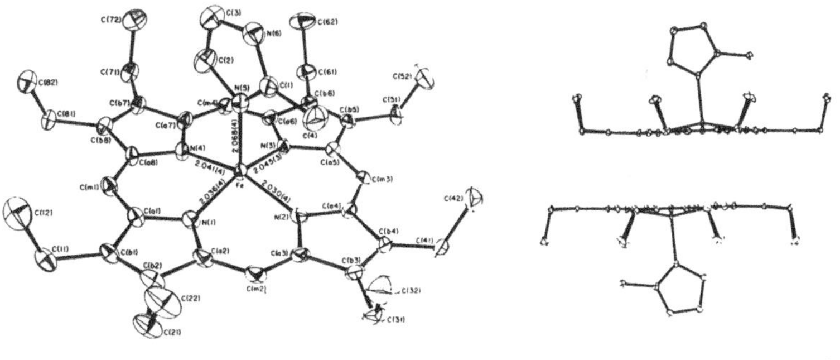

Fig. 1. The $C_{40}H_{50}FeN_6^+$ cation with dimensions and a view showing a pair of inversion related cations.

I. Monoclinic, *C2/c*, a = 22.786, b = 15.028, c = 49.974 Å, β = 101.70°, Z = 16. Mo radiation, R = 0.174 for 3933 reflections (at 96 K).

II. Monoclinic, $P2_1/n$, a = 15.059, b = 19.040, c = 15.093 Å, β = 92.40°, Z = 4. Mo radiation, R = 0.087 for 8658 reflections (at 100 K).

Molecular dimensions for (II) are shown in Fig. 1 with mean Fe-N_p 2.038(6) and the Fe atom displaced 0.36 Å from the porphinato ligand plane. Both structures have an intermolecular π-π interaction between two cations as shown in Fig. 1.

25-ETHOXYCARBONYL-2,3,7,8,12,13,17,18-OCTAETHYL-21,22-METHANO-21H,22H-PORPHYRIN HYDROBROMIDE

$C_{40}H_{51}BrN_4O_2$ $C_{40}H_{51}N_4O_2^+,Br^-$

T.J. BARTCZAK, 1985. *Acta Cryst.*, C41, 604-607.

Monoclinic, *C2/c*, a = 20.324, b = 19.947, c = 18.976 Å, β = 98.31°, Z = 8. Cu and Mo radiation, R = 0.073 for 1811 reflections.

The porphyrin ring (Fig. 1) is considerably folded as seen from the angles the pyrrole rings make with the mean N(21-24) plane (-9.8, -3.9, 12.8 and 6.1° respectively). The crystal structure consists of discrete molecules of N(21),N(22)-bridged octaethylporphyrin. The Br atom is disordered, half-occupying two sites, one a centre of symmetry the other a general position. The ethoxycarbonylmethylene group is also disordered but no useful model was obtained.

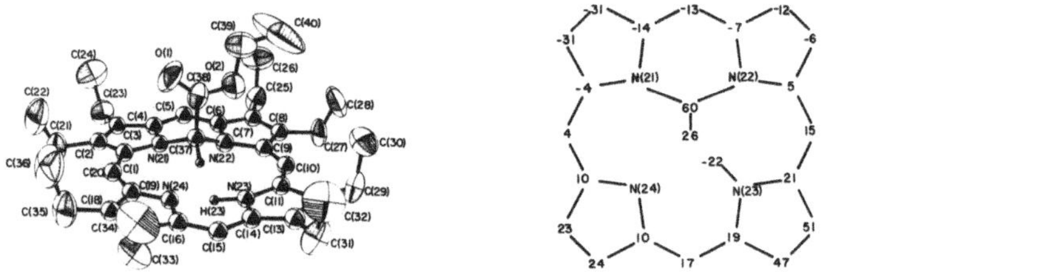

Fig. 1. $C_{40}H_{51}BrN_4O_2$: the molecular structure and deviations from the mean plane defined by N(21),N(22),N(23),N24).

(HEXAFLUOROANTIMONY)(meso-TETRAPHENYLPORPHINATO)IRON(III) FLUORORBENZENE

$C_{44}H_{28}F_6FeN_4Sb,C_6H_5F$

K. SHELLY, T. BARTCZAK, W.R. SCHEIDT and C.A. REED, 1985. *Inorg. Chem.*, 24, 4325-4330.

Orthorhombic, $Pna2_1$, a = 25.754, b = 10.748, c = 15.707 Å, Z = 4. Mo radiation, R = 0.049 for 6126 reflections.

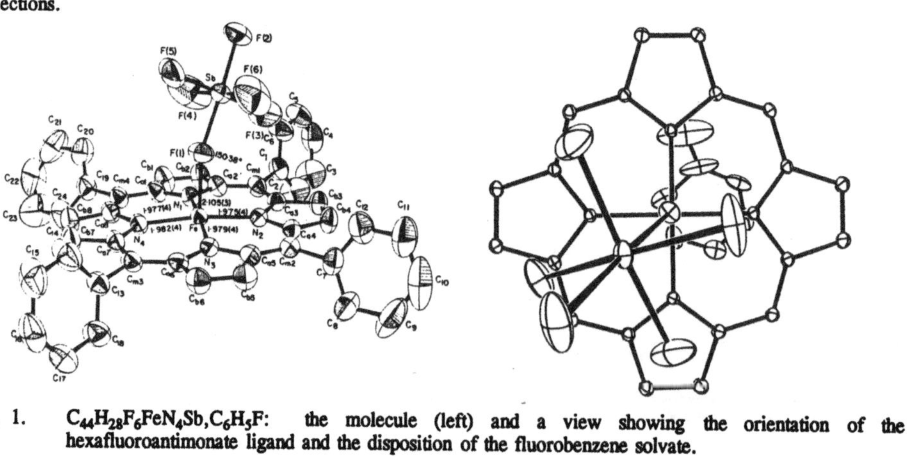

Fig. 1. $C_{44}H_{28}F_6FeN_4Sb,C_6H_5F$: the molecule (left) and a view showing the orientation of the hexafluoroantimonate ligand and the disposition of the fluorobenzene solvate.

The coordination geometry about Fe is square pyramidal, with the tpp N atoms in basal sites and an F atom of SbF_6^- in the apical site (Fig. 1). The 24-atom core shows deviations of up to 0.12 Å from planarity, and the Fe atom is 0.15 Å above the mean plane. The ttp macrocycle and the C_6H_5F molecule (on the side of the ring away from FE) form a π-complex (maximum separation between planes 3.30 Å). Bond lengths include: Fe-N 1.975-1.982(4), Fe-F 2.105(3), Sb-F 1.877-1.929(6) Å.

meso-TETRATOLYLPORPHYRIN
$C_{48}H_{38}N_4$

R.J. BUTCHER, G.B. JAMESON and C.B. STORM, 1985. *J. Am. Chem. Soc.*, **107**, 2978-2980.

Monoclinic, $P2_1/n$, a = 9.878, b = 9.286, c = 20.991 Å, β = 99.91°, Z = 2. Mo radiation, R = 0.043 for 1535 reflections.

In the structure (Fig. 1) the amino hydrogens are disordered and the molecule has 1- symmetry. The two amino proton sites are equally populated, as shown by the Ca-N-Ca angles of 107.9(3) and 107.7(3)°, midway between the amino (105.8(4)) and imino (109.8(6)°) values, and by N-Ca-Cb angles (average 108.5(4) *cf.* amino 110.5(6) and imino 107.2(4)°). Again the N...N transannular distances of 4.079(6) and 4.154(6) Å are midway between amino (4.19(1)) and imino (4.06(1) Å) values and the N...N separations between adjacent N atoms are unequal (2.894(4), 2.940(4) Å). Remaining features are unexceptional. Relevant bond parameters are given in Fig. 1.

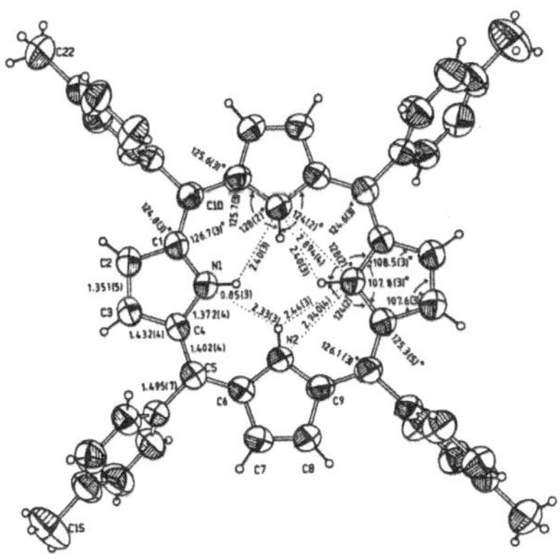

Fig. 1. View of the structure of $C_{48}H_{38}N_4$ showing selected bonding parameters.

PIPERIDINE-5,10,15,20-TETRAPHENYLPORPHYRINATOCADMIUM(II) o-XYLENE SOLVATE
$C_{49}H_{39}CdN_5,C_8H_{10}$

P.F. RODESILER, E.A.H. GRIFFITH, N.G. CHARLES, L. LEBIODA and E.L. AMMA, 1985. *Inorg. Chem.*, 24, 4595-4600.

Triclinic, $P\bar{1}$, a = 13.504, b = 14.753, c = 11.863 Å, α = 102.94, β = 94.76, γ = 91.04°, Z = 2. Mo radiation, R = 0.048 for 10672 reflections.

The Cd atom has square-pyramidal bond geometry (Fig. 1) and lies 0.66 Å from the basal (tpp) plane, towards the piperidine N atom. Bond lengths and angles include: Cd-N(pip) 2.323, (tpp) 2.190-2.222(3), average 2.203(3) Å, N(tpp)-Cd-N(tpp) 84.7-85.1(1), N(pip)-Cd-N(tpp) 101.0, 101.8, 112.1, 114.1(1)°.

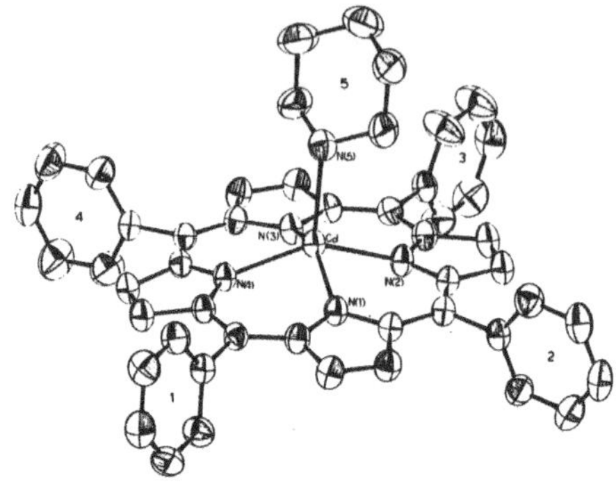

Fig. 1. $C_{49}H_{39}CdN_5$: view of an isolated molecule.

CHLOROPHENYLTETRAPHENYLPORPHINATOMOLYBDENUM BENZENE SOLVATE

$C_{50}H_{33}ClMoN_4{,}0.5(C_6H_6)$ Mo(TPP)(C_6H_5)(Cl)

J. COLIN and B. CHEVRIER, 1985. *Organometallics*, 4, 1090-1093.

Monoclinic, $P2_1/n$, a = 14.778, b = 23.513, c = 13.163 Å, β = 101.70°, Z = 4. Mo radiation, R = 0.054 for 4467 reflections.

The four pyrrole nitrogens of the molecule are coplanar and the Mo atom is 0.089 Å out of this plane towards the Cl atom (Fig. 1). The Mo-C(σ-phenyl) distance is 2.241(4) Å and the phenyl ring is staggered with respect to the pyrrole N atoms. The Mo-N distances are 2.061(3)-2.078(3) Å and Mo-Cl 2.382(1) Å.

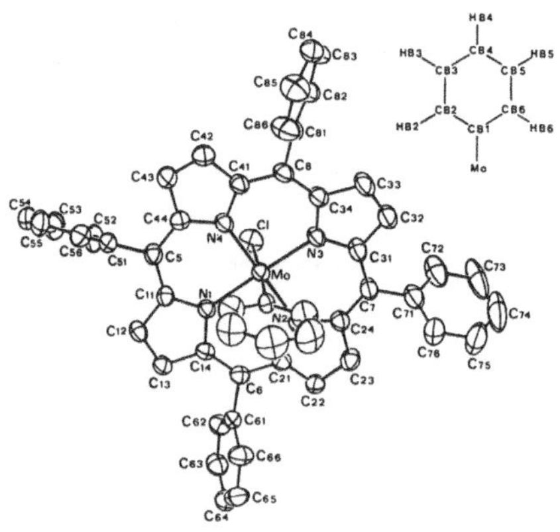

Fig. 1. A view of Mo(TPP)(C_6H_5)(Cl).

ISOTHIOCYANATOPYRIDINE(meso-TETRAPHENYLPORPHINATO)IRON(III)

$C_{50}H_{33}FeN_6S$

D.K. GEIGER, V. CHUNPLANG and W.R. SCHEIDT, 1985. *Inorg. Chem.*, 24, 4736-4741.

I. Triclinic, $P\bar{1}$, a = 11.244, b = 17.132, c = 11.167 Å, α = 100.01, β = 108.84, γ = 72.49°, Z = 2. Mo radiation, R = 0.118 for 8938 reflections.

II. Triclinic, $P\bar{1}$, a = 11.423, b = 17.297, c = 11.324 Å, α = 99.98, β = 109.43, γ = 72.43°, Z = 2. Mo radiation, R = 0.130 for 4655 reflections (at 96 K).

There are two independent molecules, each lying on a crystal inversion centre with each axial site occupied by an averaged $(py)_{0.5}(NCS)_{0.5}$ ligand. Molecule 1 (Fig. 1) is low-spin at 96 K and in high-spin/low-spin equilibrium at 293 K, whereas molecule 2 is high-spin at both temperatures. Bond lengths are (molecule 1, 2): 293 K, Fe-N(tpp) 2.023, 2.034; 2.052, 2.072; (py) 2.216, 2.402; (NCS) 1.908, 1.970(4); 96 K, Fe-N(tpp) 1.983, 1.993; 2.040, 2.073; (py) 2.027, 2.328; (NCS) 1.887, 2.030(4) Å. The displacement of Fe from the ttp N_4 plane towards NCS is 0.12, 0 Å for molecule 1, and 0.19, 0.17 Å for molecule 2, at 96, 293 K.

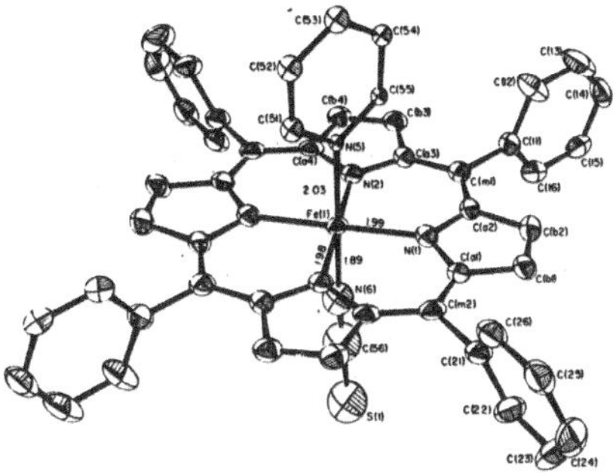

Fig. 1. $C_{50}H_{33}FeN_6S$: molecule 1 at 96 K.

BIS(N,N-DIMETHYLFORMAMIDE)TETRAPHENYLPORPHINATOMANGANESE(III) PERCHLORATE
$C_{50}H_{42}ClMnN_6O_6$

C.L. HILL and M.M. WILLIAMSON, 1985. *Inorg. Chem.*, 24, 2836-2841.

Monoclinic, *C2/c*, a = 17.739, b = 17.718, c = 15.948 Å, β = 118.69°, Z = 4. Mo radiation, R = 0.084 for 2671 reflections.

The Mn and Cl atoms lie respectively on a crystal inversion centre and on a diad axis. The O-bonded dmf ligands lie axially to the planar MnN_4 group (Fig. 1). Bond lengths and angles include: Mn-O 2.217(4), Mn-N 2.001, 2.018(5) Å, N-Mn-N 89.3, 90.5(2)°.

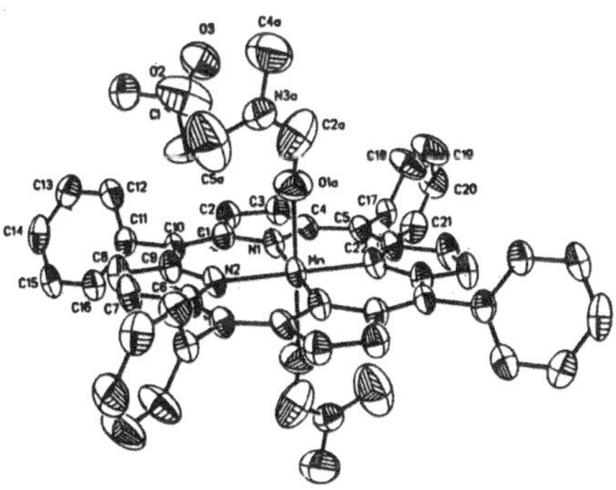

Fig. 1. The $C_{50}H_{42}ClMnN_6O_6$ molecule.

BIS(TETRAHYDROFURAN)(5,10,15,20-TETRAPHENYLPORPHINATO)ZINC(II)
$C_{52}H_{44}N_4O_2Zn$

C.K. SCHAUER, O.P. ANDERSON, S.S. EATON and G.R. EATON, 1985. *Inorg. Chem.*, 24, 4082-4086.

Triclinic, $P\bar{1}$, a = 9.572, b = 11.115, c = 11.720 Å, α = 102.71, β = 103.78, γ = 115.01°, Z = 1. Mo radiation, R = 0.040 for 3178 reflections (at -130°C).

The Zn atom is six-coordinate (Fig. 1) and lies on a crystal inversion centre. The Zn-O bonds are inclined at

9.2° to the normal to the ZnN_4 plane, and the phenyl substituents make angles of 63.2 and 80.8° with this plane. Bond lengths include: Zn-O 2.380(2), Zn-N 2.056, 2.058(2) Å.

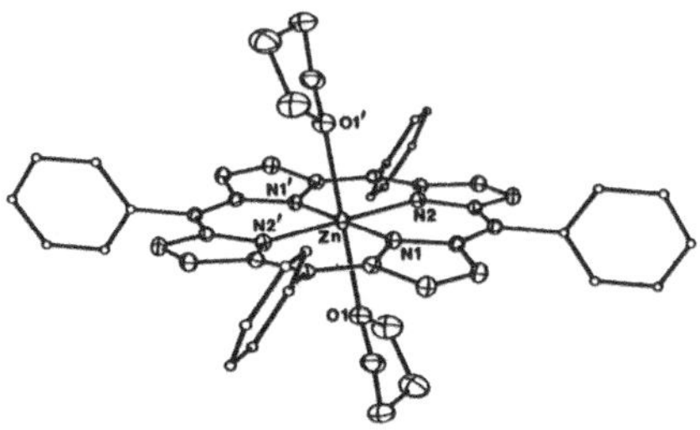

Fig. 1. $C_{52}H_{44}N_4O_2Zn$: the molecule.

((PENTACARBONYLMANGANESE)MERCURY(TETRACARBONYLMANGANESE)TETRAPHENYLPORPHINATOTIN) DICHLOROMETHANE SOLVATE
$C_{53}H_{28}HgMn_2N_4O_9Sn,0.5(CH_2Cl_2)$

S. ONAKA, Y. KONDO, M. YAMASHITA, Y. TATEMATSU, Y. KATO, M. GOTO and T. ITO, 1985. *Inorg. Chem.*, 24, 1070-1076.

I. Triclinic, $P\bar{1}$, a = 14.79, b = 14.87, c = 13.49 Å, α = 92.3, β = 111.3, γ = 109.3°, Z = 2. Mo radiation, R = 0.094 for 5419 reflections.

The molecule (Fig. 1) contains a Sn-Mn-Hg-Mn sequence with Sn-Mn 2.554(3), Mn-Hg 2.582, 2.659(3) Å, Sn-Mn-Hg 94.5(1), Mn-Hg-Mn 164.4(1)°. The Sn atom is bonded to the four N atoms of a saddle-shaped tetraphenylporphyrin macrocycle (Sn-N 2.165-2.195(15) Å), and lies 0.85 Å above their mean plane. The Mn atoms carry CO groups to achieve overall octahedral coordination.

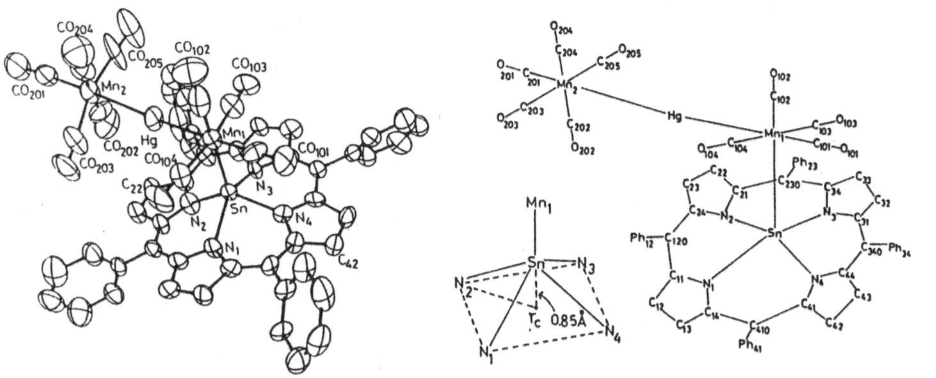

Fig. 1. The $C_{53}H_{28}HgMn_2N_4O_9Sn$ molecule.

BIS(PYRIDINE)(5,10,15,20-TETRAPHENYLPORPHYRINATO)IRON(II) PYRIDINE SOLVATE
$C_{54}H_{38}FeN_6,2(C_5H_5N)$

NAIYIN LI, V. PETRICEK, P. COPPENS and J. LANDRUM, 1985. *Acta Cryst.*, C41, 902-905.

Triclinic, $P\bar{1}$, a = 9.472, b = 10.325, c = 13.545 Å, α = 80.25, β = 77.79, γ = 77.79°, Z = 1. Mo radiation, R = 0.034 for 3577 reflections.

The Fe atom lies on a crystallographic centre of symmetry (Fig. 1) and consequently is located exactly in the plane of the porphyrinato ring. It is displaced 0.064 Å from each of the pyridine planes. The average Fe-N (pyrrole) distance is 1.993 Å while the axial Fe-N(pyridine) distance is 2.039(1) Å. The pyridine ligand is planar. The angles between the plane of the porphyrinato core and the planes of the two independent phenyl groups are 58.8 and 87.2° respectively. There are large voids in the crystal structure.

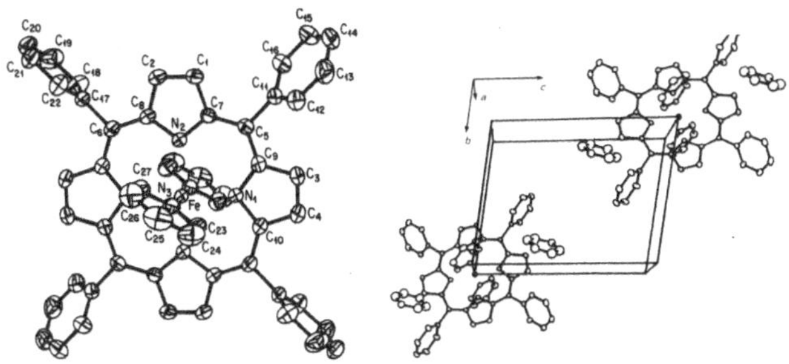

Fig. 1. The $C_{54}H_{38}FeN_6$ molecule and a packing diagram.

BIS(2,6-LUTIDINE N-OXIDE)(TETRAPHENYLPORPHINATO)MANGANESE(III) PERCHLORATE HEPTANE SOLVATE

$C_{58}H_{46}ClMnN_6O_6,C_7H_{16}$ $[Mn(C_7H_9NO)_2(C_{44}H_8N_4)](ClO_4),C_7H_{16}$

C.L. HILL and M.M. WILLIAMSON, 1985. *Inorg. Chem.*, **24**, 3024-3030.

Monoclinic, *C2/c*, a = 36.644, b = 12.316, c = 24.096 Å, β = 95.81°, Z = 8. Mo radiation, R = 0.091 for 6134 reflections.

The molecule (Fig. 1) contains a trans-octahedral MnN_4O_2 group, with long axial Mn-O bonds. The Mn atom is within 0.003 Å of the N_4 plane, and the lutidine rings are tilted at 39.6, 44.3° to this plane. Bond lengths and angles include: Mn-O 2.263, 2.264, Mn-N 1.991-2.004(4) Å, O-Mn-O 173.2(2), O-Mn-N 85.2-94.0(2)°. The porphyrin core has a flattened saddle-shape.

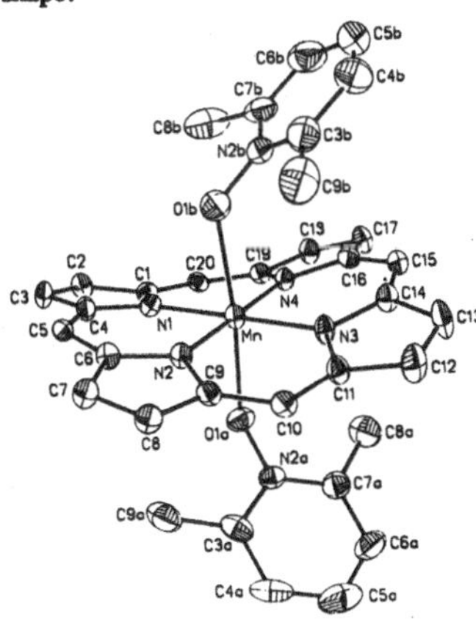

Fig. 1. The cation in $C_{58}H_{46}ClMnN_6O_6,C_7H_{16}$ (phenyl groups omitted).

BIS(PHTHALOCYANINATO)LUTETIUM DICHLOROMETHANE

$C_{64}H_{32}LuN_{16},CH_2Cl_2$ (I) $[Lu(pc)_2],CH_2Cl_2$

(PHTHALOCYANINATO)(ACETATO)DIAQUALUTETIUM HYDRATE METHANOL

$C_{34}H_{23}LuN_8O_4,H_2O,2(CH_4O)$ (II) $[Lu(pc)(OAc)(H_2O)_2],H_2O,2MeOH$

A. DE CIAN, M. MOUSSAVI, J. FISCHER and R. WEISS, 1985. *Inorg. Chem.*, **24**, 3162-3167.

I. Orthorhombic, *Pnma*, a = 28.242, b = 22.877, c = 8.050 Å, Z = 4. Cu radiation, R = 0.044 for 2663 reflections.

II. Monoclinic, *P*2_1/*n*, a = 12.033, b = 20.807, c = 14.011 Å, β = 95.34°, Z = 4. Cu radiation, R = 0.027 for 4044 reflections.

Both molecules (Fig. 1) contain eight-coordinate Lu atoms with square-antiprismatic coordination geometry. The sandwich molecule (I) contains one pc^{2-} and one pc^{-} group (not distinguishable). The Lu atom lies on a crystal mirror-plane which bisects both pc ligands. The latter are severely distorted from planarity in both compounds. The pyrrole rings make dihedral angles of 4.6-8.7 and 8.4-17.4° with the central N_4 plane in (I) and (II) respectively. Bond lengths include: (I), Lu-N 2.369-2.386(5), (II), Lu-N 2.333-2.344(3), Lu-OAc 2.366, 2.426(3), Lu-OH_2 2.306, 2.356(3) Å.

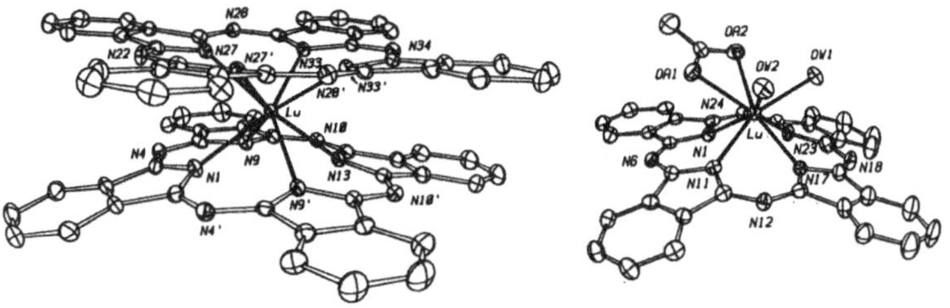

Fig. 1. The $C_{64}H_{32}LuN_{16}$ molecule (left) and the $C_{34}H_{23}LuN_8O_4$ molecule (right).

DIPHTHALOCYANINETHORIUM(IV)

$C_{64}H_{32}N_{16}Th$ Pc_2Th

A.N. DAROVSKIKH, O.V. FRANK-KAMENETSKAYA, V.S. FUNDAMENSKII and O.A. GOLYNSKAYA, 1985. *Kristallografiya*, **30**, 1085-1089.

Monoclinic, *C*2/*c*, a = 18.768, b = 18.767, c = 15.815 Å, β = 113.2°, Z = 4. Mo radiation, R = 0.052 for 2693 reflections.

The complex has twofold crystallographic symmetry (Fig. 1) with the Th atom on the twofold axis. The phthalocyanine ligands are mutually rotated through 37° which leads to a distorted antiprism coordination for the Th atom, with Th-N 2.48-2.49 Å. The phthalocyanin ligand is saucer shaped with bending away from the Th atom.

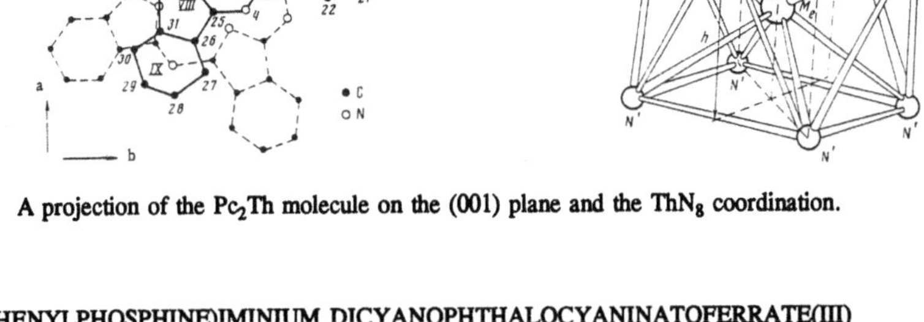

Fig. 1. A projection of the Pc_2Th molecule on the (001) plane and the ThN_8 coordination.

BIS(TRIPHENYLPHOSPHINE)IMINIUM DICYANOPHTHALOCYANINATOFERRATE(III)

$C_{70}H_{46}FeN_{11}P_2$ $C_{34}H_{16}FeN_{10}^{-}$,$C_{36}H_{30}NP_2^{+}$

H. KUPPERS, W. KALZ and H. HOMBORG, 1985. *Acta Cryst.*, **C41**, 1420-1423.

Triclinic, $P\bar{1}$, a = 12.940, b = 12.320, c = 10.322 Å, α = 65.22, β = 68.24, γ = 76.34°, Z = 1. Mo radiation, R = 0.034 for 3023 reflections (at 180 K).

The cation and anion (Fig. 1) lie in special positions, site symmetry $\bar{1}$. The cation is thus unusual in that it has a linear P-N-P moiety with a PNP angle of 180°. It also displays approximate $\bar{3}$ symmetry. The N-P distance (1.555(2) Å) is shorter than expected. In the anion, the axial cyano group at Fe also shows a linear arrangement with Fe-C-N 176.7(3)°. In the crystal, there is a slight overlap of peripheral phenylene groups of neighbouring phthalocyanine ligands (mean separation 3874(14) Å).

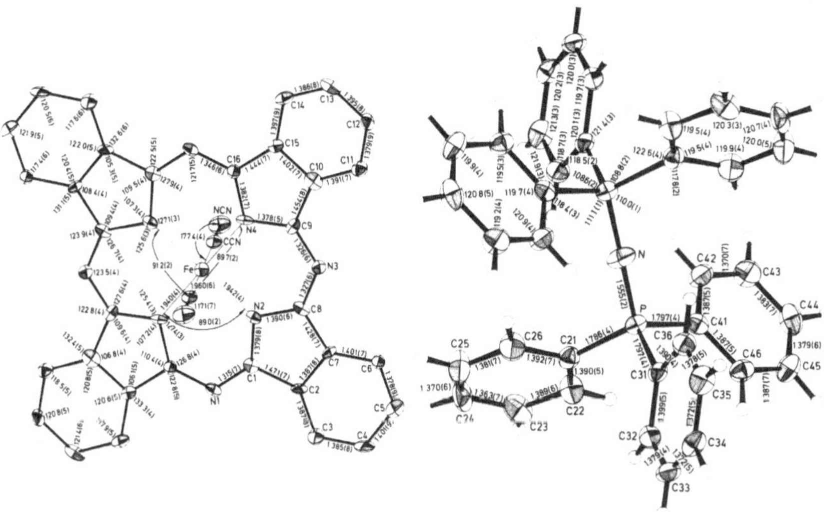

Fig. 1. $C_{70}H_{46}FeN_{11}P_2$: the anion (left) and the cation (right) showing the molecular dimensions.

(μ_2-OXO)BIS((5,10,15,20-TETRAPHENYLPORPHYRINATO)IRON(III))
$C_{88}H_{56}Fe_2N_8O$

P.N. SWEPSTON and J.A. IBERS, 1985. *Acta Cryst.*, C41, 671-673.

Orthorhombic, *C2cb*, a = 15.094, b = 24.938, c = 17.825 Å, Z = 4. Mo radiation, R = 0.093 for 6574 reflections (at 122 K).

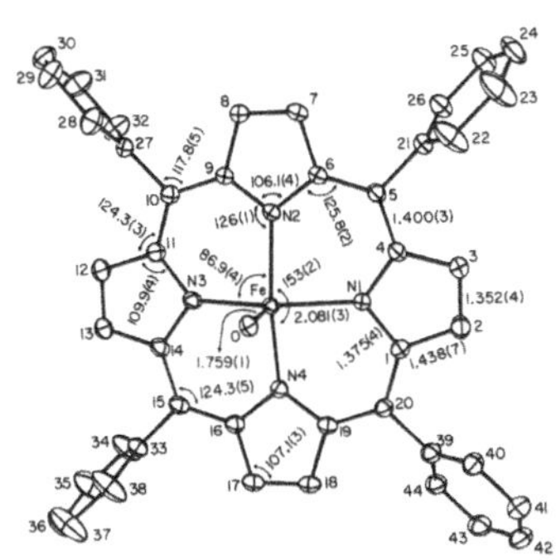

Fig. 1. $C_{88}H_{56}Fe_2N_8O$: the atomic numbering and averaged bond lengths and angles in the molecule.

The molecule (Fig. 1) has crystallographic symmetry C_2 and this low-temperature redetermination confirms the previously reported room-temperature structure (1). The absolute structure was determined and there are no significant differences between the room- and low-temperature geometries.

1. Structure Reports, 38B, 576.

7-HYDROXYGUANINE DIHYDRATE
$C_5H_5N_5O_2,2(H_2O)$

M. KITAHARA, K. ISHII, H. KAWAHARADA, K. WATANABE, T. SUGA, T. HIRATA and S. NAKAMURA, 1985. *J. Antibiot.*, **38**, 977-980.

Orthorhombic, *Pcab*, a = 6.723, b = 9.552, c = 25.935 Å, Z = 8. Mo radiation, R = 0.059 for 1216 reflections.

The analysis establishes that this antimetabolite is 7-hydroxyguanine (Fig. 1).

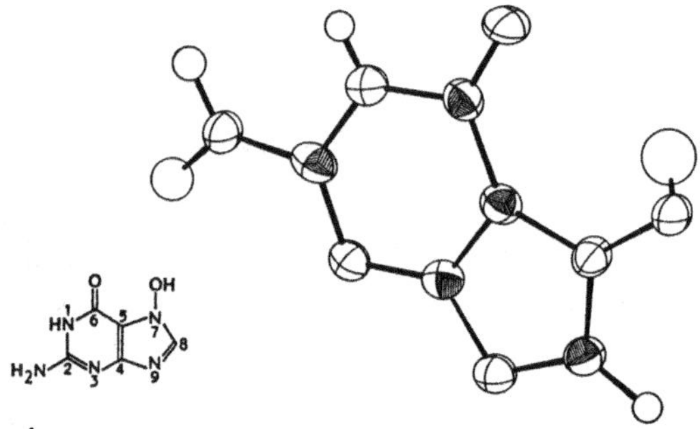

Fig. 1. 7-Hydroxyguanine.

NOJIRIMYCIN BISULPHITE ADDUCT MONOHYDRATE
$C_6H_{13}NO_7S,H_2O$

Y. KODAMA, T. TSURUOKA, T. NIWA and S. INOUYE, 1985. *J. Antibiot.*, **38**, 116-118.

Triclinic, *P*1, a = 8.997, b = 9.322, c = 6.045 Å, α = 88.5, β = 88.6, γ = 77.0°, Z = 2. Cu radiation, R = 0.057 for 1805 reflections.

The analysis establishes the structure and conformation as shown in Fig. 1 with the six-membered ring in chair form with equatorial substituents.

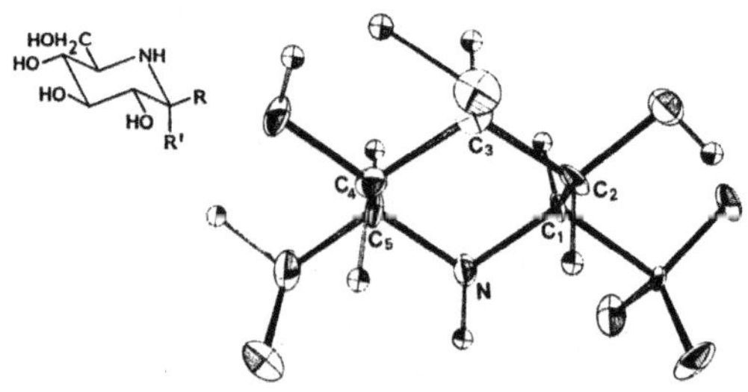

Fig. 1. The nojirimycin bisulphite adduct

METHYL 6α-BROMO-6β-IODOPENICILLANATE-1-(R)-OXIDE
$C_9H_{11}BrINO_4S$

K. SIMON, J.C. JASZBERENYI and L. PARKANYI, 1985. *J. Mol. Struct.*, **127**, 369-374.

Monoclinic, $P2_1$, a = 11.943, b = 6.153, c = 9.795 Å, β = 100.30°, Z = 2. Mo radiation, R = 0.040 for 1207 reflections.

The analysis shows that the geometry at (α)-sulphoxide is 1R, with the Br at C(6) in the α and the iodine in the β-position (Fig. 1). The four-membered ring has a maximum deviation of 0.04 Å from planarity and the five-membered thiazolidine ring has an envelope conformation with the S atom 0.85 Å from the four-atom plane.

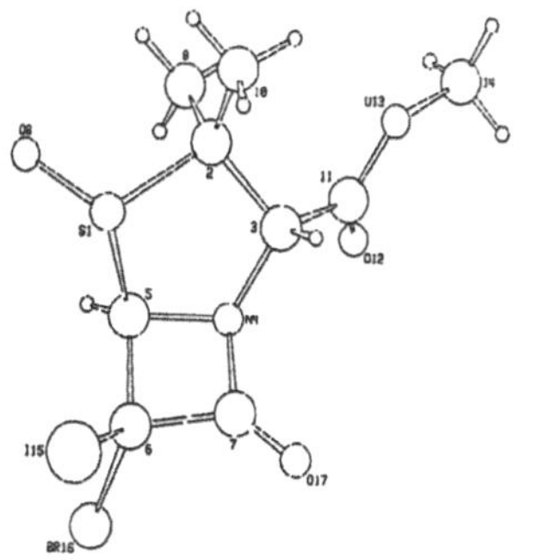

Fig. 1. The $C_9H_{11}BrINO_4S$ structure.

(3S,5R,6R)-PIVALOYLOXYMETHYL-6-BROMO-6-CHLOROPENICILLANATE
$C_{14}H_{19}BrClNO_5S$ (I)

(3S,5R,6R)-PIVALOYLOXYMETHYL-6-IODO-6-BROMOPENICILLANATE
$C_{14}H_{19}BrINO_5S$ (II)

D.U. BELINZONI, O.A. MASCARETTI, P.M. ALZARI, G. PUNTE, C. FAERMAN and A. PODJARNY, 1985. *Canad. J. Chem.*, **63**, 3177-3181.

I. Orthorhombic, $P2_12_12_1$, a = 10.198, b = 12.956, c = 13.720 Å, Z = 4. Mo radiation, R = 0.066 for 1661 reflections (at -70°C).

II. Orthorhombic, $P2_12_12_1$, a = 10.337, b = 12.988, c = 13.961 Å, Z = 4. Mo radiation, R = 0.066 for 2028 reflections (at -70°C).

The analyses determined the stereochemistry to be as shown in Fig. 1. The thiazolidine ring adopts the α-COOR axial, α-CH_3 equatorial, β-CH_3 axial conformation with C(3) out of the plane of the other four ring atoms. In the β-lactam ring, N(4) is pyramidal.

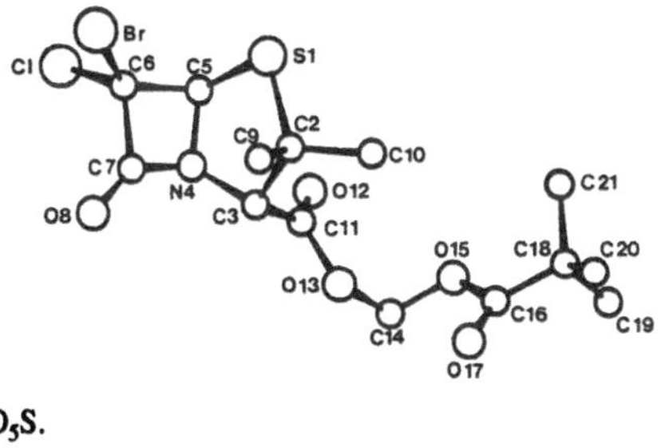

Fig. 1. A view of $C_{14}H_{19}BrINO_5S$.

DIPOTASSIUM 2'-METHOXYGRISEOLIC ACID TRIHYDRATE
$C_{15}H_{13}K_2N_5O_8,3(H_2O)$ $C_{15}H_{13}N_5O_8^{2-},2(K^+),3(H_2O)$

S. TAKAHASHI, F. NAKAGAWA, K. KAWAZOE, Y. FURUKAWA, S. SATO, C. TAMURA and A. NAITO, 1985. *J. Antibiot.*, **38**, 830-834.

Orthorhombic, $P2_12_12_1$, a = 35.749, b = 8.315, c = 6.945 Å, Z = 4. Cu radiation, R = 0.056 for 1923 reflections.

The structure and absolute stereochemistry are established as shown in Fig. 1. The adenine base is anti with respect to the pseudo-sugar moiety, with a glycosidic torsion angle χ_{CN} of 6.8°. The adenine bases are linked in pairs by N-H...N hydrogen bonds (2.895 and 3.143 Å). All available O-H and N-H groups participate in hydrogen bonding.

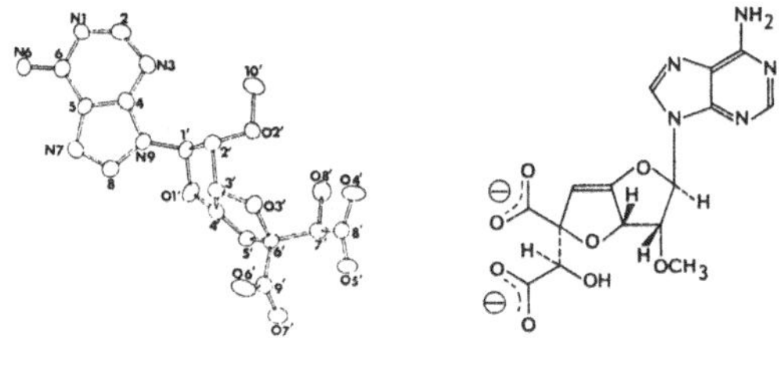

Fig. 1. The $C_{15}H_{13}N_5O_8^{2-}$ ion.

4-(2-(2-HYDROXY-3,5-DIMETHYLPHENYL)-2-OXOETHYL)-2,6-PIPERIDINEDIONE
$C_{15}H_{17}NO_4$

K.D. ONAN, H.S.P. RAO and R.J. PARRY, 1985. *Acta Cryst.*, C41, 428-431.

Monoclinic, $P2_1/n$, a = 10.772, b = 6.594, c = 20.185 Å, β = 100.60°, Z = 4. Cu radiation, R = 0.068 for 1443 reflections.

In the actiphenol molecule (Fig. 1) the benzene ring is planar with substituents C(7),C(16) coplanar and O(15),C(17) displaced on opposite sides of the ring by 0.039(3) and 0.049(4) Å respectively. The carbonyl moiety is nearly coplanar with the aromatic ring and there is an intramolecular O(15)-H...O(18) hydrogen bond of length 2.573(4) Å stabilizing this orientation. The glutarimide ring has a 1,2-diplanar conformation and bond lengths and angles agree with those in similar compounds. Centrosymmetrically related molecules form hydrogen bonded dimers through N(12)-H...O(20) bonds of length 2.887(4) Å. Contacts between dimers correspond to normal van der Waals interactions.

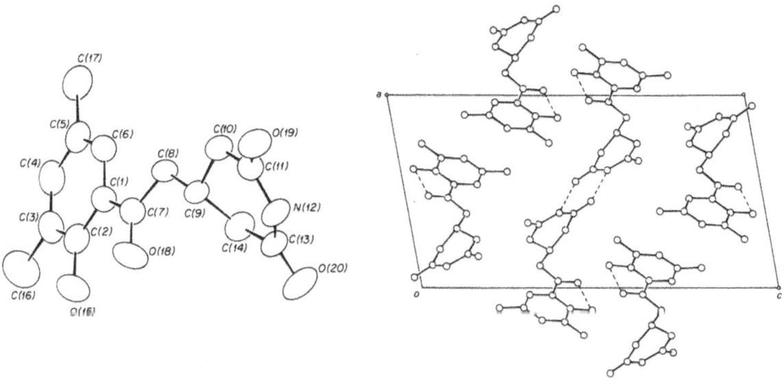

Fig. 1. $C_{15}H_{17}NO_4$: a perspective view of the molecule showing the atomic numbering and (right) the unit cell contents projected on to (010) with hydrogen bonds shown as broken lines.

OBAFLUORIN ACETONITRILE SOLVATE
$C_{17}H_{14}N_2O_7,C_2H_3N$

A.A. TYMIAK, C.A. CULVER, M.F. MALLEY and J.Z. GOUGOUTAS, 1985. *J. Org. Chem.*, 50, 5491-5495.

Monoclinic, $P2_1$, a = 7.519, b = 11.539, c = 10.805 Å, β = 91.23°, Z = 2. Cu radiation, R = 0.034 for 1198

reflections.

The crystal structure analysis confirmed the stereochemistry of the molecule (Fig. 1). A comparison of the molecular geometry with those of other β-lactones and β-lactams is given in the paper.

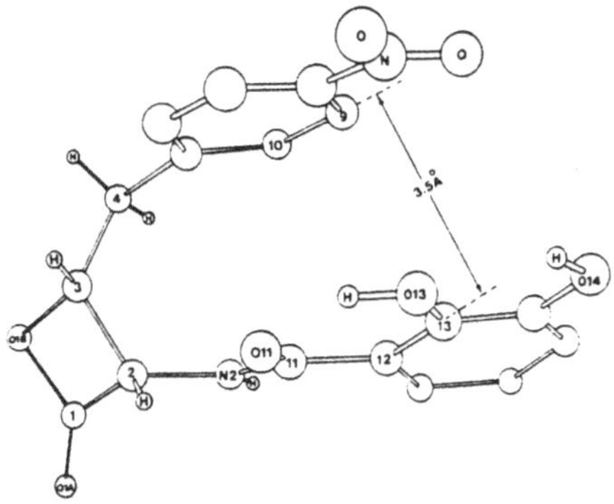

Fig. 1. Drawing of obafluorin.

1,5,6-TRIHYDROXY-3-HYDROXYMETHYL-8-METHOXY-5,6-DIHYDROBENZ[a]ANTHRAQUINONE METHANOL SOLVATE
$C_{20}H_{16}O_7,CH_4O$

J.H. WILTON, D.C. CHENEY, G.C. HOKANSON, J.C. FRENCH, HE CUN-HENG and J. CLARDY, 1985. *J. Org. Chem.*, **50**, 3936-3938.

Monoclinic, $P2_1$, a = 9.157, b = 7.243, c = 14.064 Å, β = 82.78°, Z = 2. Cu radiation, R = 0.048 for 1250 reflections.

The crystal structure analysis established the stereochemistry of the molecule (Fig. 1) and in particular that the dihydrodiol moiety is located at the 5,6-position, and that the resulting 5,6-dihydrobenz[a]anthraquinone is not planar.

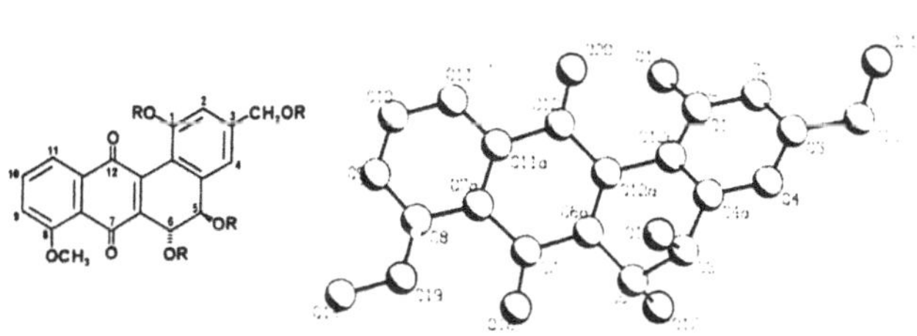

Fig. 1. A drawing of $C_{20}H_{16}O_7$ with the hydrogen atoms omitted.

HELIOMYCIN MONOHYDRATE
$C_{22}H_{16}O_6,H_2O$

S.K. ARORA, 1985. *J. Antibiot.*, **38**, 113-115.

Monoclinic, $P2_1/c$, a = 8.513, b = 19.047, c = 11.065 Å, β = 105.6°, Z = 4. Mo radiation, R = 0.0495 for 1566 reflections (at -11°C).

The molecule is planar except for the ring with the gem-dimethyl group (Fig. 1) which adopts a half-chair conformation. All oxygen atoms are involved in hydrogen bonding.

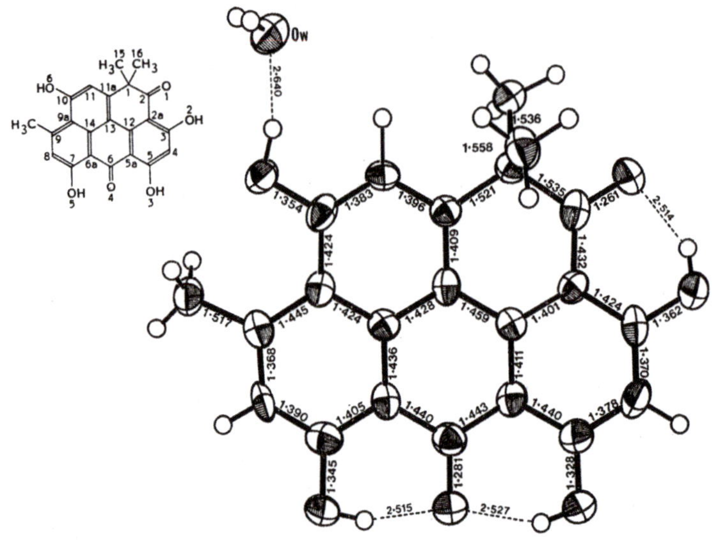

Fig. 1. The heliomycin structure and bond lengths.

ELLORAMYCIN MONOHYDRATE
$C_{32}H_{36}O_{15},H_2O$

P.G. JONES, G.M. SHELDRICK, J. ROHR and A. ZEECK, 1985. *Acta Cryst.*, **C41**, 255-257.

Monoclinic, $P2_1$, a = 10.801, b = 9.277, c = 16.660 Å, β = 105.71°, Z = 2. Mo radiation, R = 0.041 for 3566 reflections.

The molecule (Fig. 1) is an anthracycline analogue containing a permethylated α-L-rhamnose moiety. All the ring systems have the expected conformations, the non-aromatic fused rings both being half-chairs. The sugar ring adopts a chair conformation. The extended structure is linked by hydrogen bonds.

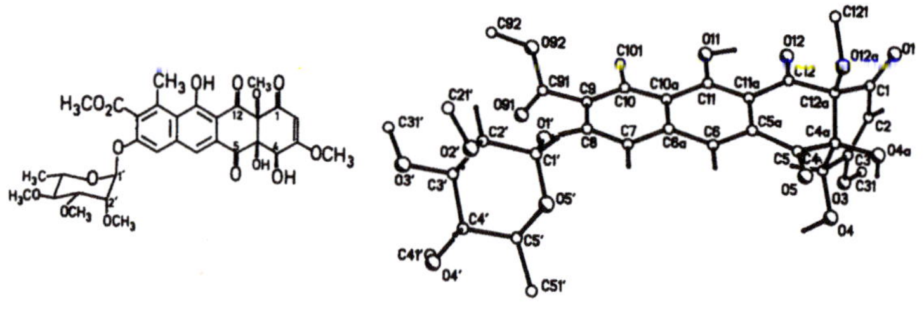

Fig. 1. $C_{32}H_{36}O_{15},H_2O$: the molecular skeleton and structure.

5-DEDESOSAMINYL-5,4''-O,O-DIACETYLMYCINAMICIN I METHANOL SOLVATE
$C_{33}H_{50}O_{12},CH_4O$

K. KINOSHITA, S. SATOI, K.-I. HARADA, M. SUZUKI and K. NAKATSU, 1985. *J. Antibiot.*, **38**, 522-526.

Orthorhombic, $P2_12_12_1$, a = 13.630, b = 23.231, c = 11.987 Å, Z = 4. Mo radiation, R = 0.149 for 2051 reflections.

The structure and stereochemistry are determined to be as shown in Fig. 1. See also **1** for related work on mycinolide. Based on this study it was possible to establish the absolute configuration of the mycinamicins.

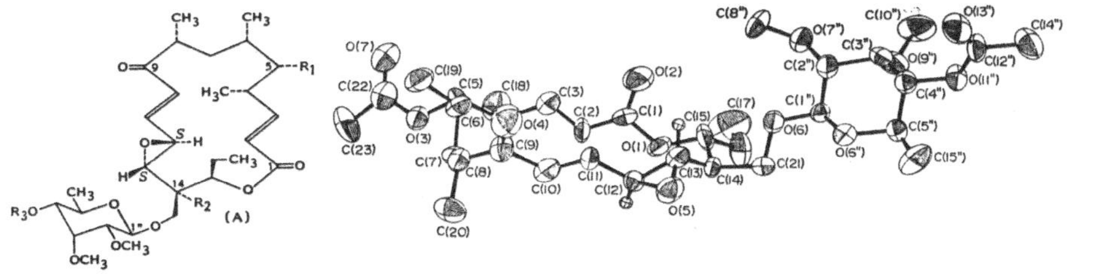

Fig. 1. The $C_{33}H_{50}O_{12}$ molecule (R_1 = OAc, R_2 = H, R_3 = Ac).

1. Structure Reports, 49B, 597.

ENNIATIN B
$C_{33}H_{57}N_3O_9$

C. KRATKY and M. DOBLER, 1985. *Helv. Chim. Acta*, **68**, 1798-1803.

Rhombohedral, *R*3, a = 14.626, c = 16.309 Å, Z = 3. Mo radiation, R = 0.113 for 575 reflections.

There appeared to be an included solvent molecule, either H_2O or MeOH and this was refined as an O-atom. The crystal structure of a hydrated form has been reported (1). The ester group is tilted by 23° with respect to the hexagonal axis and the amide group by 17°. The molecules are arranged in hexagonal layers with the centres of molecules at (0,0,0); (2/3,1/3,1/3) and (1/3,2/3,1/3) (Fig. 1).

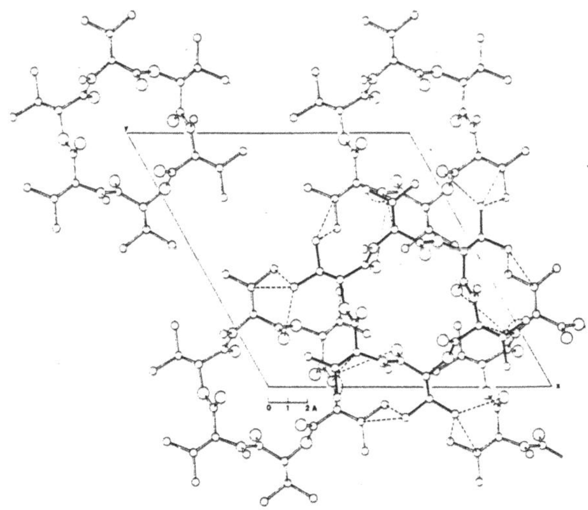

Fig. 1. Packing of molecules of $C_{33}H_{57}N_3O_9$ in the unit cell: molecules with open bonds are in the layer Z = 0, the molecule with solid bonds is at (2/3,1/3,1/3).

1. Structure Reports, 49B, 740.

ANGIOLAM A
$C_{34}H_{53}NO_7$

W. KOHL, B. WITTE, B. KUNZE, V. WRAY, D. SCHOMBURG, H. REICHENBACH and G. HOFLE, 1985. *Liebigs Ann. Chem.*, 2088-2097.

Orthorhombic, $P2_12_12_1$, a = 11.9691, b = 10.1313, c = 28.5184 Å, Z = 4. Cu radiation, R = 0.040 for 3485 reflections.

The analysis establishes the structure and stereochemistry as shown in Fig. 1. The 19-membered ring has an overall boat conformation. Molecules are linked by O-H...O and N-H...O hydrogen bonds.

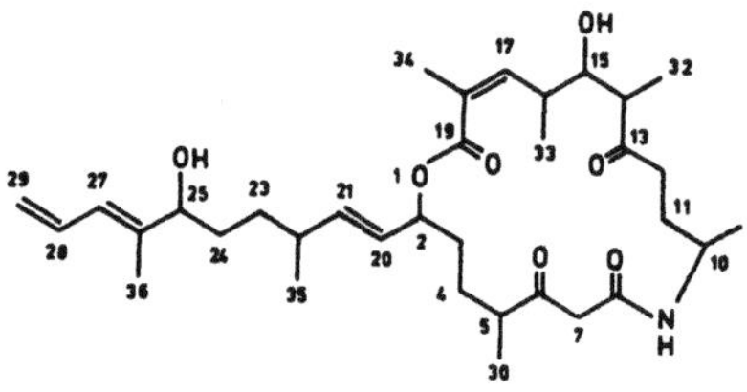

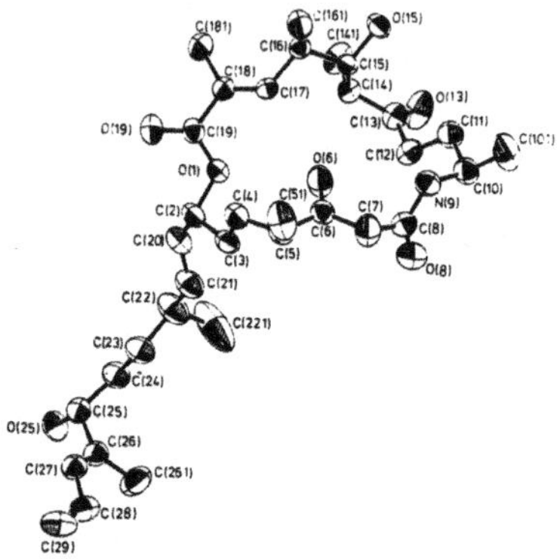

Fig. 1. The antibiotic Angiolam A.

MYXOVIRESCIN-A_1-BISACETAL
$C_{41}H_{69}NO_8$

W. TROWITZSCH, K. BORGSCHULTE, V. WRAY, D. SCHOMBURG and G. HOFLE, 1985. *Liebigs Ann. Chem.*, 1629-1652.

Orthorhombic, $P2_12_12_1$, a = 10.1846, b = 29.9769, c = 13.7690 Å, Z = 4. Cu radiation, R = 0.055 for 5662 reflections (at 150K).

The structure and conformation determined by the analysis are shown in Fig. 1. Both five-membered rings have envelope conformations.

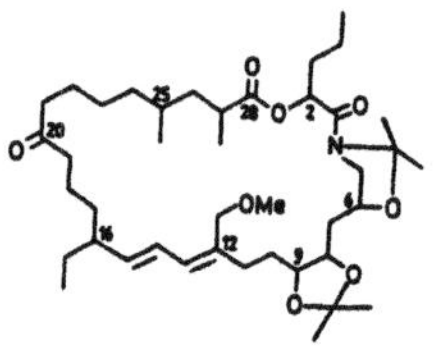

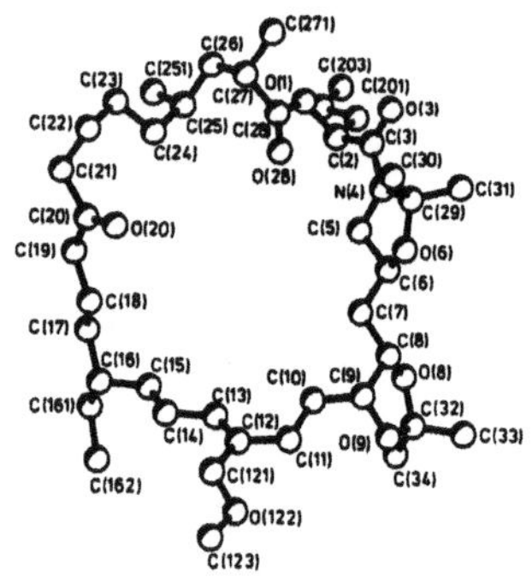

Fig. 1. The structure and stereochemistry of myxovirescin-A_1-bisacetal.

10β-HYDROXY-ESTRA-1,4-DIENE-3,17-DIONE
$C_{18}H_{22}O_3$

J.H. BIERI, R. PREWO, J.L. BRIANSO and J.F. PINIELLA, 1985. *Acta Cryst.*, C41, 1530-1532.

Orthorhombic, $P2_12_12_1$, a = 7.014, b = 12.176, c = 17.648 Å, Z = 4. Mo radiation, R = 0.061 for 2506 reflections (at around 130K).

As in other 1,4-dien-3-one steroids the A ring (Fig. 1) is almost planar, B and C have chair conformations and

D has a C(14)-envelope conformation. There is an indication of disorder at C(16) in this ring. O(19) is in the β-position. There is a hydrogen bond from H(19) to O(3) of a neighbouring molecule.

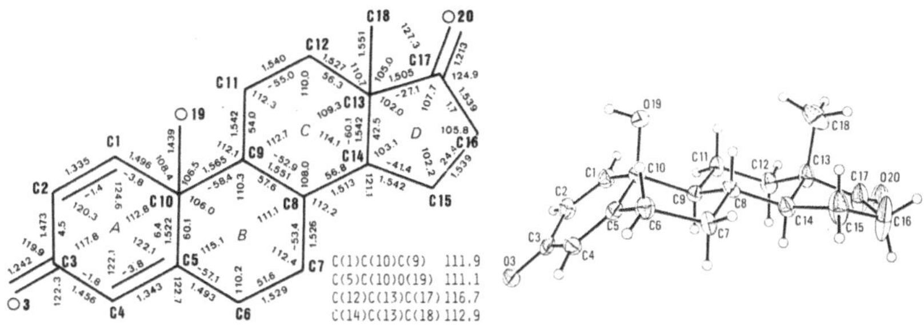

Fig. 1. $C_{18}H_{22}O_3$: bond lengths, angles and torsion angles and a view of the molecule.

3-METHOXY-19-NORPREGNA-1,3,5(10),20-TETRAEN-4-OL
$C_{21}H_{28}O_2$

A.J. BLACKMAN, A. HEATON, B.W. SKELTON and A.H. WHITE, 1985. *Aust. J. Chem.*, **38**, 565-573.

Orthorhombic, $P2_12_12_1$, a = 65.01, b = 7.527, c = 7.265 Å, Z = 8. Cu radiation, R = 0.066 for 1435 reflections.

The title compound is one of two pregnane derivatives isolated from the soft coral 'Capnella' sp. nov. There are two crystallographically, independent but similar, molecules in the asymmetric unit (Fig. 1). Although the C=CH$_2$ side chain is disordered in both molecules, the skeletal geometry is in accord with the expectations for such a system.

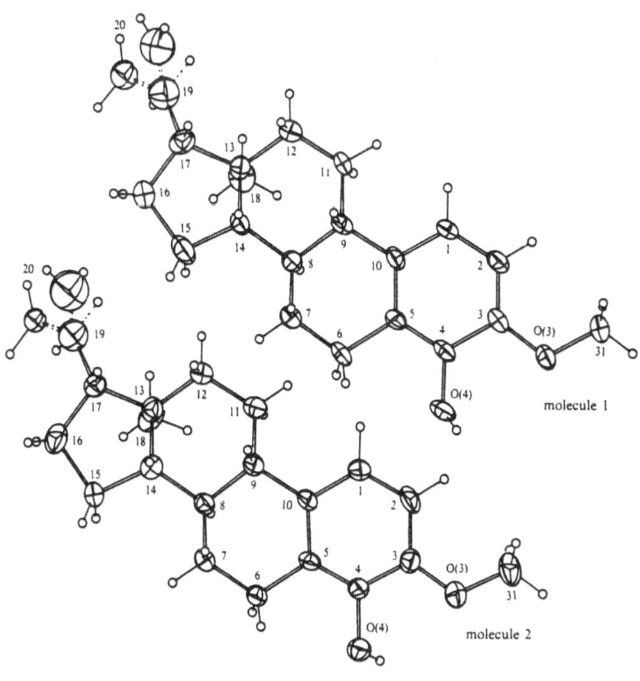

Fig. 1. $C_{21}H_{28}O_2$: the two independent molecules.

N-(3-HYDROXYESTRA-1,3,5(10)-TRIEN-17β-YL)-3-HYDROXYPROPYLAMINE (PROLAME)
$C_{21}H_{31}NO_2$

J.M. FERNANDEZ-G, M.F. RUBIO-ARROYO, M. SORIANO-GARCIA, R.A. TOSCANO, M. DEL CARMEN-PEREZ-CESAR, C. RUBIO-POO, J.J. MANDOKI, A. DE LA PENA, C. LEMINI, N. MENDOZA-PATINO and F. CRUZ, 1985. *Steroids*, **45**, 151-157.

Orthorhombic, $P2_12_12_1$, a = 8.356, b = 13.343, c = 16.119 Å, Z = 4. Cu radiation, R = 0.041 for 1249 reflections.

In the molecule (Fig. 1) ring A is planar; ring B is a half-chair with a pseudo diad axis through the C(5)-C(10) and C(7)-C(8) bonds. Ring C has a normal chair conformation, and ring D is close to a perfect envelope. Molecules are linked by O-H...O and O-H...N hydrogen bonds (O...O 2.767(4), O...N 2.861(4) Å).

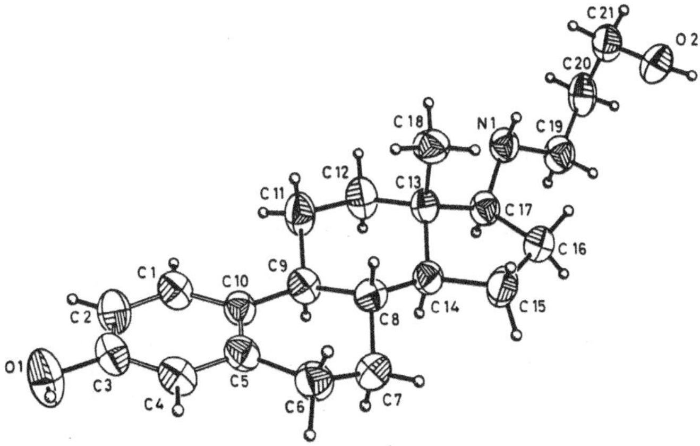

Fig. 1. A view of $C_{21}H_{31}NO_2$, prolame.

3-OXO-A-NORANDROSTANE-5β,17β-DIOL 17-ACETATE 5-((R)-METHANESULPHINATE)
$C_{21}H_{32}O_5S$

M. PRZYBYLSKA and F.R. AHMED, 1985. *Acta Cryst.*, **C41**, 1467-1470.

Monoclinic, $P2_1$, a = 12.475, b = 7.203, c = 11.510 Å, β = 90.80°, Z = 2. Cu radiation, R = 0.029 for 2239 reflections.

In the steroid moiety of the molecule (Fig. 1) the ring junction A/B is cis, while B/C and C/D are trans. Ring A adopts an envelope conformation, ring B a chair conformation, ring C a chair conformation and ring D is intermediate between a 13β-envelope form and a half-chair form. The longest $C(sp^3)$-$C(sp^3)$ bonds of 1.551(3) and 1.558(3) Å are to the fully saturated C(10) atom. Other bond lengths and angles are normal. The S=O bond is 1.474(2) Å. The geometry around the S atom, with the lone pair of electrons, is tetrahedral with chirality R. The molecules are linked by C-H...O=S hydrogen bonds into chains along *b*.

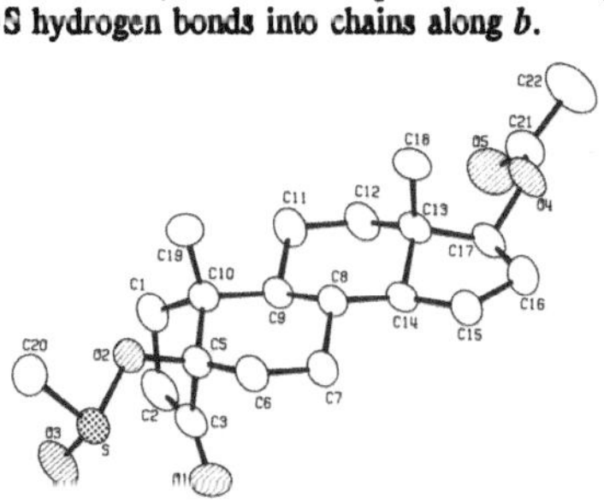

Fig. 1. $C_{21}H_{32}O_5S$: the molecular structure.

9α-BROMO-16α-METHYL-11β,17α,21-TRIHYDROXYPREGNA-1,4-DIENE-3,20-DIONE (9α-BROMO-16α-METHYLPREDNISOLONE)
$C_{22}H_{29}BrO_5$ (I)

9α-BROMO-11β,17α,21-TRIHYDROXYPREGNA-1,4-DIENE-3,20-DIONE (9α-BROMO-PREDNISOLONE)
$C_{21}H_{27}BrO_5$ (II)

J. IBALL, J.N. LOW, R.J. NASH and B.P. THOMPSON, 1985. *Acta Cryst.*, C41, 439-441.

I. Monoclinic, $P2_1$, a = 6.550, b = 14.46, c = 10.75 Å, β = 100.9°, Z = 2. Cu radiation, R = 0.072 for 1749 reflections.

II. Monoclinic, $P2_1$, a = 7.737, b = 14.92, c = 8.94 Å, β = 103.6°, Z = 2. Cu radiation, R = 0.075 for 1650 reflections.

The molecules of both compounds have very similar stereochemistries. Ring A is planar, as is the enol side chain. Rings B and C have chair conformations and ring D is a C(13) envelope. The ring junctions A/B, B/C and C/D are all trans. Bond lengths and angles in the molecule are normal. The crystal structures (Fig. 1) are stabilized by hydrogen bonds: O(3)...O(17) 2.77(2) Å in (I) and O(3)...O(17), O(3)...O(11) of lengths 2.75(2) and 2.82(2) Å respectively in (II).

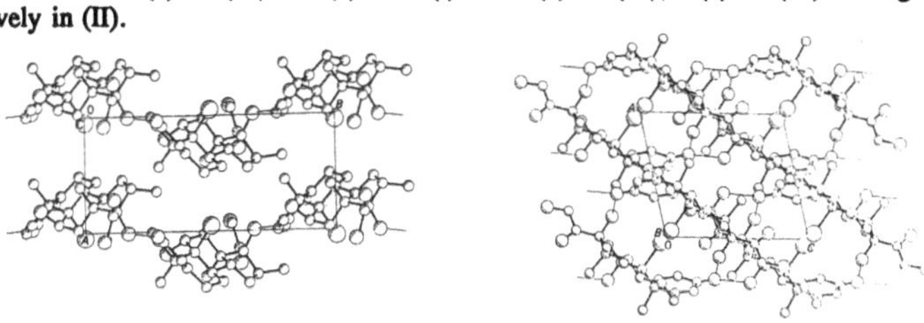

Fig. 1. The molecular packing and hydrogen bonding (left) in $C_{22}H_{29}BrO_5$ viewed along the c-axis and (right) $C_{21}H_{27}BrO_5$ viewed along the b-axis.

21-ACETOXY-17α-HYDROXY-PREGN-4-ENE-3,11,20-TRIONE (CORTISONE ACETATE)
$C_{23}H_{30}O_6$

J.A. KANTERS, A. DE KOSTER, V.J. VAN GEERESTEIN and L.A. VAN DIJCK, 1985. *Acta Cryst.*, C41, 760-763.

Orthorhombic, $P2_12_12_1$, a = 8.018, b = 12.579, c = 20.963 Å, Z = 4. Mo radiation, R = 0.049 for 1302 reflections.

In the steroid backbone of this form of cortisone acetate (Fig. 1) ring A has a 1α-sofa conformation, rings B and C are slightly distorted chairs and ring D is a 13β,14α-half-chair. The ring junctions B/C and C/D are both trans. The side chain consists of two planar four atom fragments- the acetate group and atoms C(17,20,21),O(20)- with an interplanar dihedral angle of 77.2(3)°. The carbonyl O(22) is synperiplanar with respect to C(21). Bond lengths and angles in the molecule agree with expected values. The molecules are connected into infinite chains along the b-axis through head-to-tail O(17)-H...O(2) hydrogen bonds of length 2.761(4) Å. There are several short van der Waals contacts involving carbonyl O atoms. The structure is compared with that of another form (1).

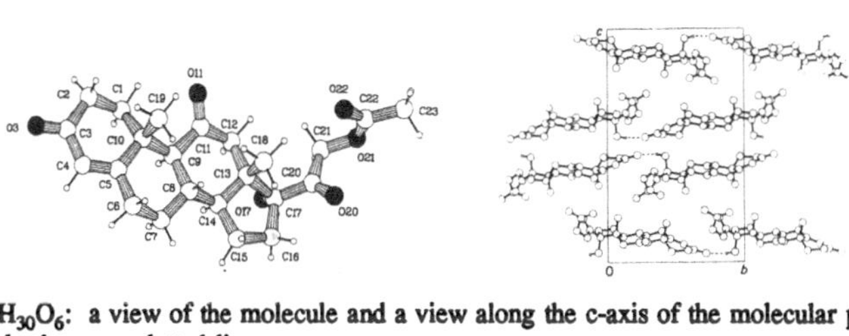

Fig. 1. $C_{23}H_{30}O_6$: a view of the molecule and a view along the c-axis of the molecular packing with hydrogen bonds shown as dotted lines.

1. Structure Reports, 38B, 547.

17α-METHYL-12a-METHYLENE-C(12a)HOMO-18-NOR-(13S)-ANDROST-5-EN-3β-YL ACETATE $C_{23}H_{34}O_2$

R. FREIRE, J.A. SALAZAR, E. SUAREZ, C. BETANCOR, D. MELIAN, T. PRANGE and C. PASCARD, 1985. *J. Chem. Res.*, S152-153, M1824-1847.

Monoclinic, $P2_1$, a = 12.282, b = 9.466, c = 9.189 Å, β = 106.9°, Z = 2. Cu radiation, R = 0.095 for 1714 reflections.

The seven-membered ring (Fig. 1) is in a boat conformation with C11 out of the mean plane of the molecule towards the α-face. The methylene C18 atom has an axial position quite similar to that of the C19 angular methyl group. The C-D ring junction is cis with the 13-H and 14-H both α and axial.

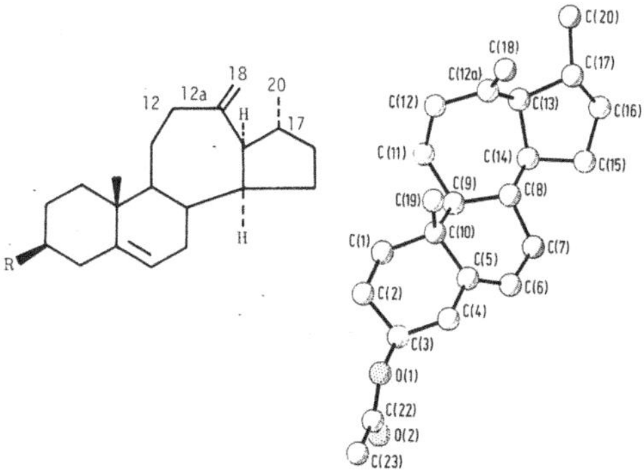

Fig. 1. The structure of $C_{23}H_{34}O_2$ (R = OAc).

11β-HYDROXY-16α,17α,21-TRIMETHYL-5α-PREGNA-1,4-DIENE-3,20-DIONE $C_{24}H_{34}O_3$ (I)

11β,17α-DIHYDROXY-21-METHYL-5α-PREGN-4-ENE-3,20-DIONE-17-BUTYRATE $C_{26}H_{38}O_5$ (II)

J.C. BARNES, J.D. PATON, R.T. LOGAN and J. REDPATH, 1985. *Acta Cryst.*, C41, 763-766.

I. Monoclinic, $P2_1$, a = 11.76, b = 11.75, c = 7.596 Å, β = 98.85°, Z = 2. Cu radiation, R = 0.070 for 1235 reflections.

II. Orthorhombic, $P2_12_12_1$, a = 15.107, b = 10.6656, c = 14.741 Å, Z = 4. Mo radiation, R = 0.044 for 1746 reflections.

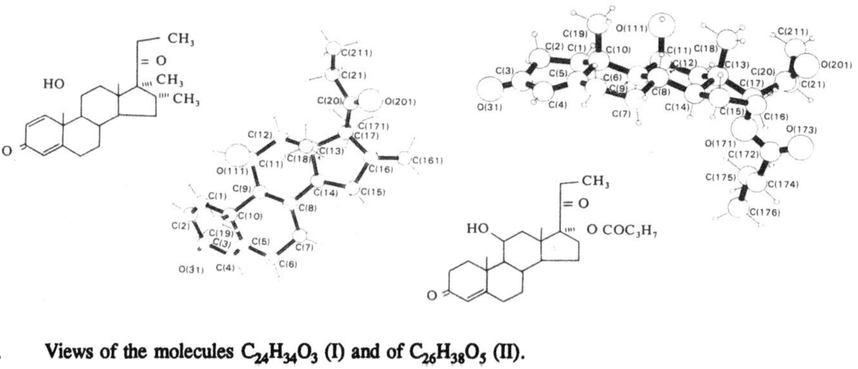

Fig. 1. Views of the molecules $C_{24}H_{34}O_3$ (I) and of $C_{26}H_{38}O_5$ (II).

Bond lengths and angles in the molecules (Fig. 1) have values that are typical for 17-substituted steroids. The principal difference in conformation between these compounds is that while in (I) the methyl and carbonyl groups in the EtCO side chain have the normal almost eclipsed conformation, In (II) they adopt an almost staggered conformation. The O(carbonyl)=C-C-C(H_3 or H_2) torsion angles are 11.8(3) and 45.9(4)° for (I) and (II) respectively. In both structures the molecules are linked by hydrogen bonds between the hydroxyl oxygen atom and O(31) and all other intermolecular distances are essentially van der Waals contacts.

METHYL (4'S,16S)-3β-ACETOXY-17-OXOSPIRO(ANDROST-5-ENE-16,3'-(4,5-DIHYDROPYRAZOLE))-4'-CARBOXYLATE
$C_{25}H_{34}N_2O_5$

P. GLUZINSKI, J.W. KRAJEWSKI, Z. URBANCZYK-LIPKOWSKA, G. ARGAY and L. PARKANYI, 1985. *Acta Cryst.*, C41, 1630-1632.

Monoclinic, $P2_1$, a = 6.414, b = 9.729, c = 19.192 Å, β = 91.09°, Z = 2. Mo radiation, R = 0.033 for 1150 reflections.

Ring A in the steroid molecule (Fig. 1) has a $^{10}C_3$ chair conformation with the O-acetyl substituent at C(3) in the equatorial position. The unsaturated ring B has the half-chair conformation $^8H_9$ with C(8) and C(9) deviating by 0.371(5) and -0.360(6) Å from the C(5,6,7,10) mean plane. Ring C has a slightly deformed $^{13}C_9$ conformation and the five-membered ring D a slightly distorted $_{14}E$ envelope conformation. The dihydropyrazole ring E is nearly planar and the D/E dihedral angle is 91.8(2)°. The bowing of ring A relative to the remainder of the steroid skeleton is given by the value, 17.03(2)°, of the dihedral angle between the mean plane of A and that of rings B, C and D. Bond lengths and angles in the molecule are in agreement with normal values and no unusually short intermolecular contacts are reported.

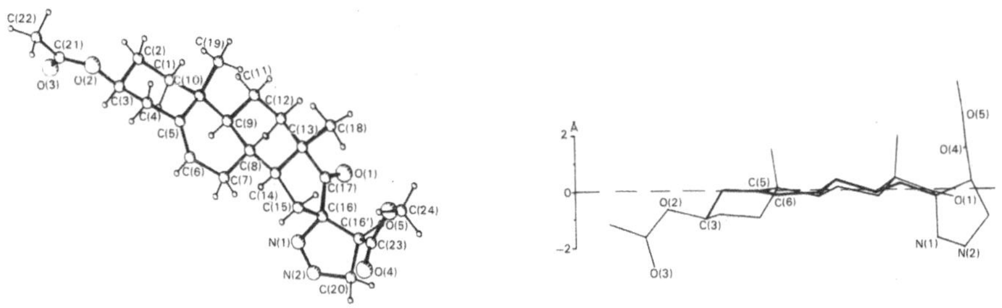

Fig. 1. Views of the $C_{25}H_{34}N_2O_5$ molecule showing the atomic labelling scheme and along the least-squares B, C, D plane the trace of which is shown by the dashed line.

16α,17α-CYCLOHEXANORPREGN-4-EN-20β-OL-3-ONE
$C_{25}H_{38}O_2$

A.E. KAPUL'SKII, V.I. SIMONOV, I.S. LEVINA, L.E. KULIKOVA and A.V. KAMERNITSKII, 1985. *Kristallografiya*, 30, 1096-1101.

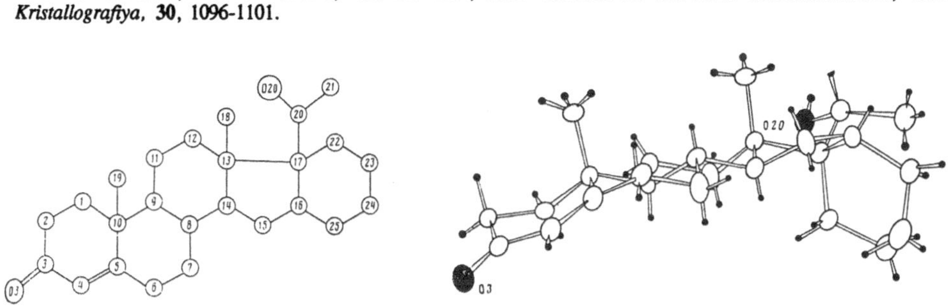

Fig. 1. The structure and conformation of $C_{25}H_{38}O_2$.

Orthorhombic, $P2_12_12_1$, a = 6.422, b = 11.280, c = 29.682 Å, Z = 4. Mo radiation, R = 0.087 for 1961 reflections.

In the molecule (Fig. 1), ring A has the 1α,2β-half-chair conformation, rings B and C have chair conformations, ring D is a 13β-envelope and the D' ring is in chair conformation. Molecular dimensions are normal. Molecules are linked by O-H...O hydrogen bonds, O...O 2.865 Å.

3β-HYDROXY-21,27-BISNOR-5α-SOLANIDANE HYDROBROMIDE MONOHYDRATE (21,27-BISNORDEMISSIDINE HYDROBROMIDE MONOHYDRATE)
$C_{25}H_{42}BrNO,H_2O$ $C_{25}H_{42}NO^+,Br^-,H_2O$

D. MILJKOVIC, K. GASI, M. KINDJER, S. STANKOVIC, B. RIBAR and G. ARGAY, 1985. *Croat. Chem. Acta*, **58**, 721-736.

Orthorhombic, $P2_12_12_1$, a = 7.463, b = 10.227, c = 31.497 Å, Z = 4. Mo radiation, R = 0.085 for 1349 reflections.

The analysis establishes the structure and conformation as shown in Fig. 1. The six-membered rings all have chair conformations and the five-membered rings have envelope conformations (ring D has C(14) as the flap and nitrogen-containing ring E has C(21) as the flap). The two methyl groups are β-axial.

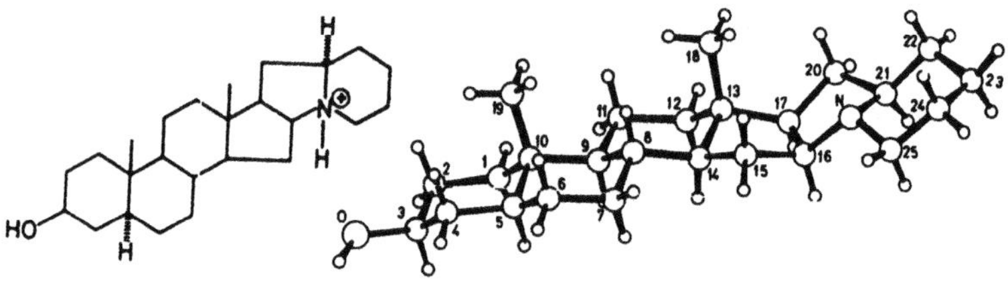

Fig. 1. The $C_{25}H_{42}NO^+$ cation and a view showing the conformation; the hydrogen atom on the N atom has been omitted from this drawing.

3-ACETOXYDAIGREDORIGENIN
$C_{26}H_{36}O_7$ (I)

DAIGREMONTIANIN
$C_{26}H_{30}O_9$ (II)

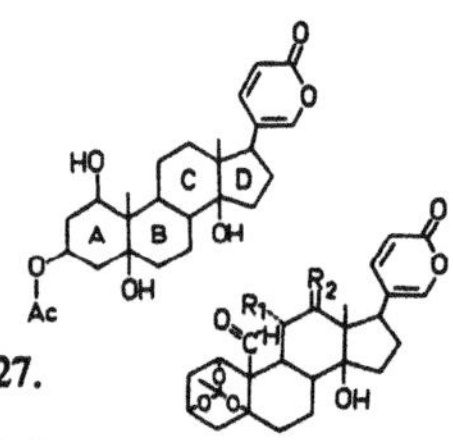

H. WAGNER, M. FISCHER and H. LOTTER, 1985. *Z. Naturforsch.*, **B40**, 1226-1227.

I. Orthorhombic, $P2_12_12_1$, a = 11.254, b = 14.199, c = 14.361 Å, Z = 4. Cu radiation, R = 0.079 for 1484 reflections.

II. Monoclinic, $P2_1$, a = 6.691, b = 18.456, c = 9.257 Å, β = 96.11°, Z = 2. Mo radiation, R = 0.0726 for 2085 reflections.

In (I) the rings A, B, C and D are connected cis/trans/cis and the lactone ring at C(17) is β-oriented. Molecule (II) is shown to have an 11α-hydroxy-12-oxo substitution pattern.

9α,10α-EPOXY-5-METHYL-19-NOR-5β-CHOLEST-1-ENE-3,6-DIONE
$C_{27}H_{40}O_3$

J.M. COXON, P.D. QUAID and P.J. STEEL, 1985. *Acta Cryst.*, **C41**, 739-741.

Orthorhombic, $P2_12_12_1$, a = 15.539, b = 19.840, c = 7.931 Å, Z = 4. Mo radiation, R = 0.046 for 1430 reflections.

In the molecule (Fig. 1) the configuration of the epoxy oxygen atom is 9α,10α and C(1)=C(2) is a double bond. The A ring has a 4α,5β-half-chair, the B ring a 5β,8β-half-boat and the C ring a chair conformation. Ring D has a conformation intermediate between a 13β-envelope and a 13β,14α-half-chair. The ring junctions are trans-syn-trans-anti-trans. The 17β-side chain exists in an extended conformation. Bond lengths and angles are normal and intermolecular distances are normal van der Waals contacts.

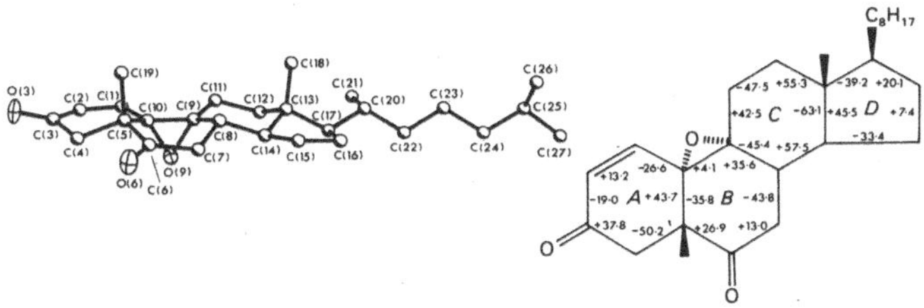

Fig. 1. A view of the $C_{27}H_{40}O_3$ molecule and the intra-annular torsion angles.

SPIROSTA-5,25(27)-DIENE-1β,3β,11α-TRIOL MONOHYDRATE (SAPOGENIN MONOHYDRATE) $C_{27}H_{40}O_5,H_2O$

A. KALMAN, G. ARGAY, B. RIBAR, D. ZIVANOV-STAKIC and S. VLADIMIROV, 1985. *Acta Cryst.*, C41, 1645-1647.

Monoclinic, $P2_1$, a = 10.085, b = 7.854, c = 31.956 Å, β = 93.11°, Z = 4. Cu radiation, R = 0.038 for 4262 reflections.

The structure contains two crystallographically independent molecules (Fig. 1) which have slight differences in their conformations. In both conformers (I) and (II), rings A and C have normal chair forms, while ring D has a half-chair form with a pseudo-diad axis bisecting the C(13)-C(14) bond. Ring B exhibits a sofa conformation in conformer (I) while in (II) this ring has an 8α-9β half-chair form. Ring E, cis-fused to ring D, in molecule (I) has a conformation intermediate between half-chair and envelope while in (II) it has a half-chair form. The independent sapogenin molecules together with the water molecules are linked by eight hydrogen bonds forming two-dimensional networks parallel to (001) and separated by c/2. The O...O separations in these bonds range from 2.600(5) to 2.879(5) Å.

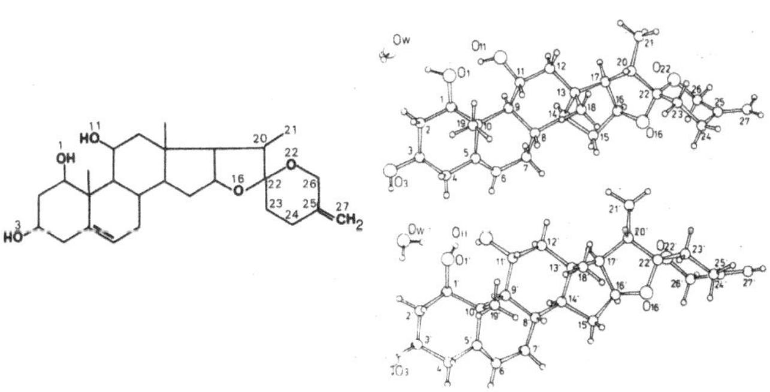

Fig. 1. $C_{27}H_{40}O_5,H_2O$: structural formula of sapogenin (left) and views of conformer (I) (top right) and of conformer (II) (bottom right).

3α,10-EPOXY-5-METHYL-19-NOR-5β,10α-CHOLESTANE $C_{27}H_{46}O$

K. BISCHOFBERGER, J.R. BULL, J.L.M. DILLEN and P.H. VAN ROOYEN, 1985. *S. Afr. J. Chem.*, 38, 129-133.

Orthorhombic, $P2_12_12_1$, a = 40.288, b = 19.131, c = 12.601 Å, Z = 16. Cu radiation, R = 0.060 for 7598 reflections.

The molecular structure of one of the four crystallographically-independent molecules is shown in Fig. 1 together with a stereoscopic view of the unit cell. Ring A adopts a slightly distorted boat conformation having a "mirror plane" through C(3), C(10), and the bridging O atom, whereas ring B adopts a twist-boat conformation having a twofold "axis of symmetry" along C(7)-C(10). Ring A is considerably more puckered than ring B as a consequence of the additional constraint imposed by the 3α,10α-epoxy bridge.

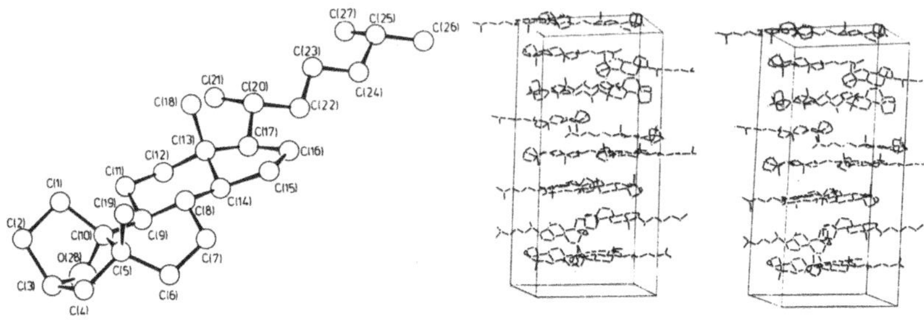

Fig. 1. Molecular structure of $C_{27}H_{46}O$, and stereoscopic view of the unit cell.

22α,23α-DIBROMO-10-METHYL-19-NORANTHRAERGOSTA-5,7,9(10),14-TETRAENE
$C_{28}H_{40}Br_2$ (I)

2β,3α,22,23-TETRABROMO-18-NOR-17-ISOERGOSTA-8(9),11,13(14)-TRIENE
$C_{27}H_{38}Br_4$ (II)

G. FERGUSON, B. KAITNER, A.F. SOMOGYVARI, V.I. BENDALL, W.B. WHALLEY, C.L. YEATES, R. EDMUNDS and J.M. MIDGLEY, 1985. *J. Chem. Soc. Perkin Trans. 1,* 1337-1342.

I. Monoclinic, *F*2, a = 31.38, b = 5.918, c = 28.469 Å, β = 102.20°, Z = 8. Mo radiation, R = 0.051 for 853 reflections.

II. Monoclinic, *P*2$_1$, a = 10.512, b = 7.682, c = 17.418 Å, β = 91.59°, Z = 2. Mo radiation, R = 0.052 for 1136 reflections.

There is some disorder in the crystals of (I). In each case the absolute stereochemistry was determined from anomalous dispersion. In (I) rings A and C have symmetrical half-chair conformations; the aromatic ring B is essentially planar and the cyclopentene ring D has a C(17) envelope conformation (Fig. 1). In (II) ring A has a distorted chair conformation with trans di-axial bromine atoms; ring B is in a sofa conformation; aromatic ring C is planar; ring D is a C(16) envelope. The bromine-containing side chain is maximally extended in both (I) and (II).

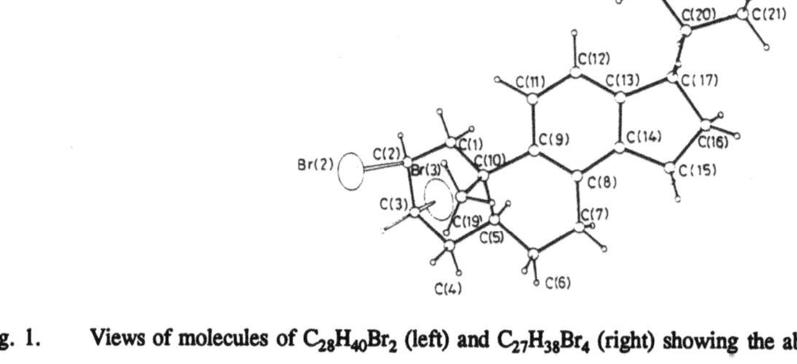

Fig. 1. Views of molecules of $C_{28}H_{40}Br_2$ (left) and $C_{27}H_{38}Br_4$ (right) showing the absolute stereochemistries.

3-ACETYL-29-NORCYASTERONE MONOHYDRATE
$C_{30}H_{44}O_9,H_2O$

F. CAMPS, J. COLL, A. CORTEL, E. MOLINS and C. MIRAVITLLES, 1985. *J. Chem. Res.*, S14-15, M241-266.

Orthorhombic, $P2_12_12_1$, a = 11.687, b = 7.453, c = 33.244 Å, Z = 4. Mo radiation, R = 0.041 for 2142 reflections.

Ring A has a chair conformation defined by the planes (Fig. 1) C(5), C(4), C(10); C(4), C(10), C(1), C(3) and C(1), C(3), C(2), all the interplanar angles being in the range 129.1(9) to 132.7(9)°. Rings B and C exhibit distorted half chair and distorted chair conformations respectively, whilst rings D and E have envelope conformations, even though ring E is almost planar.

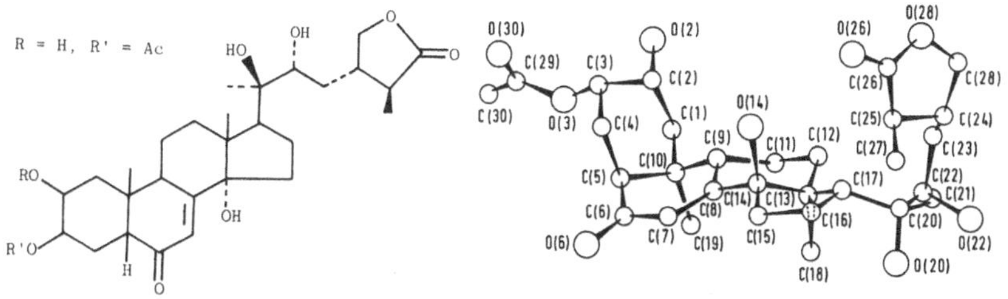

Fig. 1. The $C_{30}H_{44}O_9,H_2O$ molecule.

Δ^5-9,11-seco-GORGOSTENE-3β,11,24β-TRIOL-9-ONE
$C_{30}H_{50}O_4$

J.D. KORP, I. BERNAL, C.F. CAMPANA and G.M. SHELDRICK, 1985. *Z. Kristallogr.*, **170**, 205-212.

Monoclinic, $P2_1$, a = 11.260, b = 14.917, c = 17.292 Å, β = 103.84°, Z = 4. Mo radiation, R = 0.041 for 3930 reflections.

There are two independent molecules in the asymmetric unit; the only major conformational difference between them being with the orientation of the O(11) hydroxyl group (Fig. 1). The mean Csp_3-Csp_3 bond length is 1.533 Å. Ring A has a chair conformation, ring D is an envelope (C(13) flap) and ring B is described as a six-membered envelope with a C(8) flap. The molecules are linked by an extensive hydrogen-bond network.

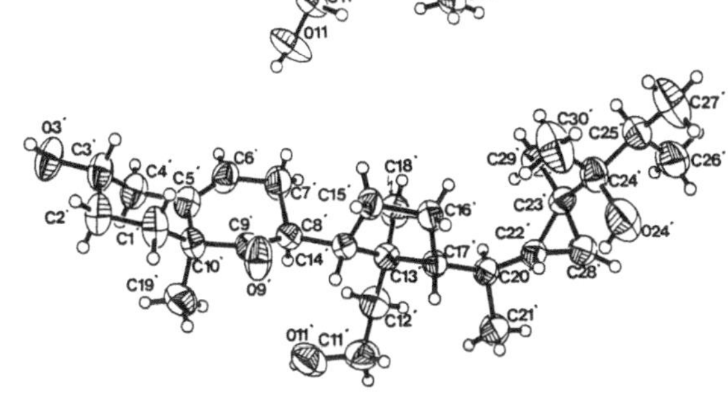

Fig. 1. Views of the two $C_{30}H_{50}O_4$ molecules.

D:C-friedo-LUP-8-EN-3β-OL ACETATE D:C-friedo-LUP-9(11)-EN-3β-OL ACETATE COMPLEX (TYLOLUPENOL A ACETATE - TYLOLUPENOL B ACETATE COMPLEX)
$C_{32}H_{52}O_2,C_{32}H_{52}O_2$

H. NAKAI, M. SHIRO, T. TOZYO, H. ISHII and ZHEN-WEN XU, 1985. *Acta Cryst.*, C41, 252-255.

Triclinic, $P1$, a = 13.562, b = 14.492, c = 7.691 Å, α = 93.61, β = 90.77, γ = 71.31°, Z = 1. Cu radiation, R = 0.044 for 4883 reflections.

The absolute configurations were determined on the basis of the β-configuration of the methyl group attached to C(10) in lupane. The ring junctions (Fig. 1) A/B, C/D and D/E are respectively trans, trans and cis for molecules A and B. Rings A-E respectively adopt the chair, envelope, envelope, chair and envelope conformations in A and the chair, skew boat, half-chair, chair and envelope in B. Molecular dimensions are not unusual and packing is governed by van der Waals interactions.

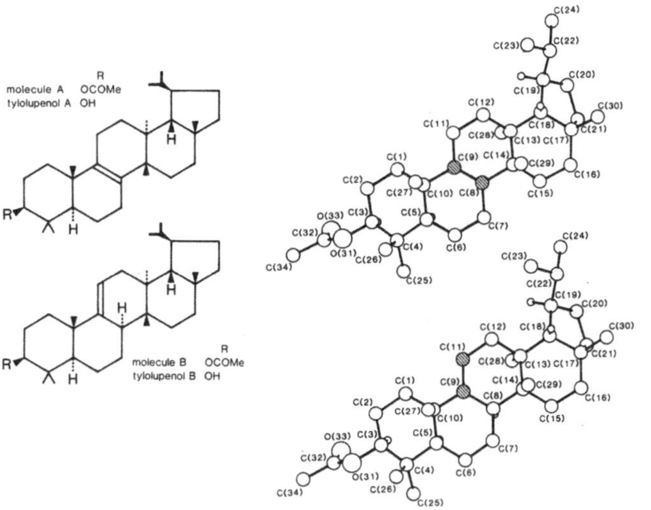

Fig. 1. $C_{32}H_{52}O_2$: the molecular skeletons and views of molecules A and B.

18-OXO-D:C-friedo-BACCHAR-9(11)-EN-3β-YL ACETATE
$C_{32}H_{52}O_3$ (I)

D-friedo-13,18-seco-BACCHAR-13-ENE-3β,18-DIYL DIBENZOATE
$C_{44}H_{62}O_4$ (II)

S. OHTA, M. TORI, K. SUEMURA, T. IROKAWA, Y. MORIYAMA, T. TSUYUKI, T. TAKAHASHI, A. ITAI and Y. IITAKA, 1985. *Bull. Chem. Soc. Jpn.*, 58, 2687-2696.

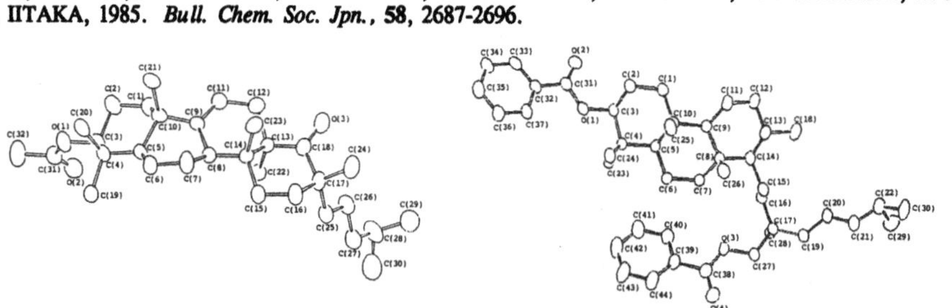

Fig. 1. Views of $C_{32}H_{52}O_3$ (I) (left) and $C_{44}H_{62}O_4$ (II) (right).

I. Monoclinic, $P2_1$, a = 17.273, b = 11.981, c = 7.323 Å, β = 99.13°, Z = 2. Cu radiation, R = 0.056 for

2349 reflections.

II. Orthorhombic, $P2_12_12_1$, a = 17.071, b = 21.191, c = 10.870 Å, Z = 4. Cu radiation, R = 0.066 for 3173 reflections.

The structures and stereochemistries of these derivatives were established as shown in Fig. 1.

3,17β-DIBENZYLOXY-9β-METHOXY-11-ESTRA-1,3,5(10)-TRIEN-11-ONE
$C_{33}H_{36}O_4$

J.R. HANSON, P.B. HITCHCOCK, LING YANG-ZHI and TAN JIA-YI, 1985. *J. Chem. Res.*, S46-47, M651-670.

Monoclinic, $P2_1$, a = 7.002, b = 12.853, c = 15.262 Å, β = 98.33°, Z = 2. Mo radiation, R = 0.106 for 902 reflections.

The crystal structure analysis established the stereochemistry of the molecule (Fig. 1).

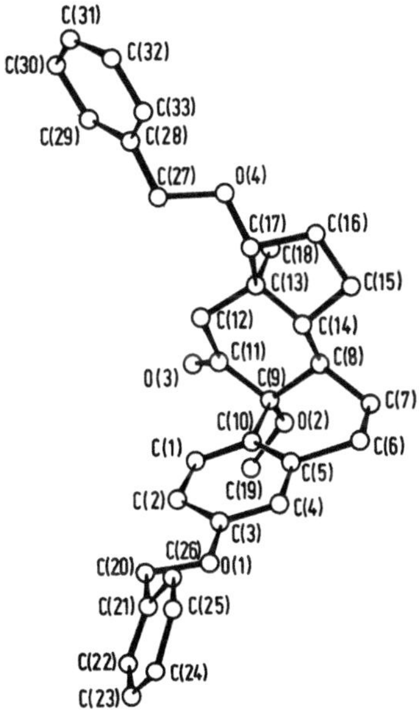

Fig. 1. Perspective drawing of a molecule of $C_{33}H_{36}O_4$ viewed from the α-face.

α-SANTONIN
$C_{15}H_{18}O_3$

C. KITABAYASHI, Y. MATSUURA, N. TANAKA, Y. KATSUBE and T. MATSUURA, 1985. *Acta Cryst.*, C41, 1779-1781.

Orthorhombic, $P2_12_12_1$, a = 34.575, b = 10.705, c = 6.991 Å, Z = 8. Mo radiation, R = 0.051 for 3129 reflections.

The two crystallographically independent molecules in the asymmetric unit (Fig. 1) have similar conformations and molecular dimensions. The dienone rings are only approximately planar and the puckered six-membered ring has a chair conformation. In the crystal, the dienone rings are nearly stacked along the *b*-axis with an interplanar angle of 28°. The carbonyl O atoms are surrounded by short intermolecular contacts.

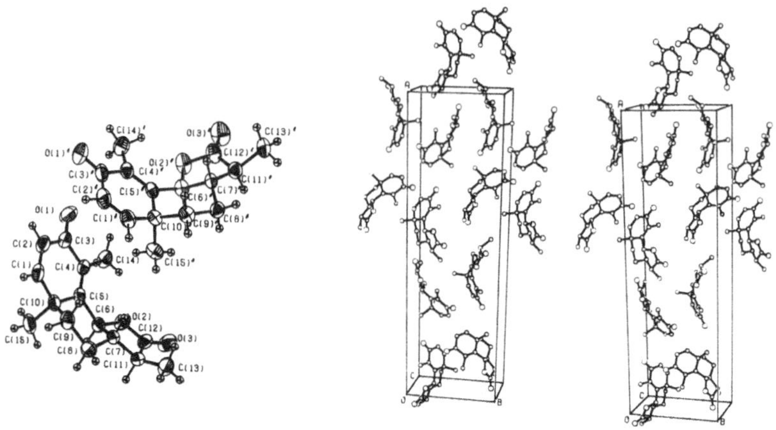

Fig. 1. $C_{15}H_{18}O_3$: the two independent molecules and a stereoview of the packing.

ASTERISCUNOLIDE A
$C_{15}H_{18}O_3$ (I)

ASTERISCUNOLIDE B
$C_{15}H_{18}O_3$ (II)

ASTERISCUNOLIDE C
$C_{15}H_{18}O_3$ (III)

ASTERISCUNOLIDE D
$C_{15}H_{18}O_3$ (IV)

A. SAN FELICIANO, A.F. BARRERO, M. MEDARDE, J.M.M. DEL CORRAL, A. ARAMBURU, A. PERALES, J. FAYOS and F. SANCHEZ-FERRANDO, 1985. *Tetrahedron*, **41**, 5711-5717.

I. Monoclinic, $P2_1$, a = 11.179, b = 5.939, c = 9.929 Å, β = 90.50°, Z = 2. Cu radiation, R = 0.047 for 1880 reflections.

II. Monoclinic, $C2$, a = 15.572, b = 6.418, c = 13.871 Å, β = 103.78°, Z = 4. Cu radiation, R = 0.055 for 1883 reflections.

III. Orthorhombic, $P2_12_12_1$, a = 13.749, b = 10.938, c = 9.244 Å, Z = 4. Cu radiation, R = 0.049 for 2394 reflections.

IV. Orthorhombic, $P2_12_12_1$, a = 14.774, b = 14.356, c = 8.985 Å, Z = 4. Cu radiation, R = 0.039 for 1367 reflections.

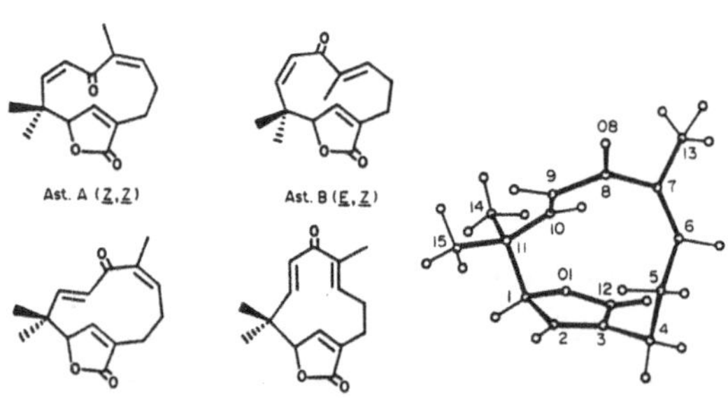

Fig. 1. The asteriscunolide A-D molecules and a view of molecule C.

The four asteriscunolides A-D are shown to have (Z,Z), (E,Z), (Z,E) and (E,E) stereochemistry (Fig. 1).

(1S,7S,8R,10S)-3-OXOGUAIA-4,11(13)-DIEN-8,12-OLIDE 8-EPIXERANTHOLIDE
$C_{15}H_{18}O_3$

W.H. WATSON, R.P. KASHYAP, M. MISKI and T.J. MABRY, 1985. *Acta Cryst.*, **C41**, 1126-1128.

Orthorhombic, $P2_12_12_1$, a = 7.558, b = 22.365, c = 7.435 Å, Z = 4. Mo radiation, R = 0.046 for 1138 reflections.

The seven-membered ring in the molecule (Fig. 1) exhibits a twist boat conformation and the cis-fused lactone ring has a flattened envelope form. The pentenone ring is virtually planar. The molecular structure is compared with that of the isomeric xerantholide molecule (1) which has a lactone trans-fused to the macrocycle which is in an almost ideal chair conformation. Bond lengths and angles in the molecule are in general agreement with normally acceptable values. Intermolecular contacts are consistent with essentially van der Waals interactions.

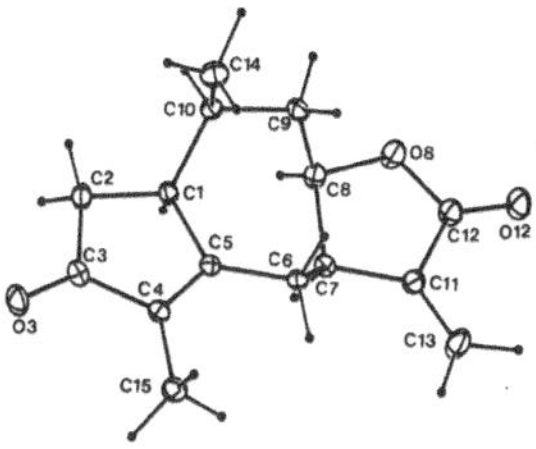

Fig. 1. A view of the $C_{15}H_{18}O_3$ molecule.

1. U. RYCHLEWSKA, 1981. Pol. J. Chem., 55, 1895.

1β,10α,4α,5β-DIEPOXY-7,8-FURANOGERMACRANE (GLECHOMAFURAN)
$C_{15}H_{20}O_3$

J.D. KORP and I. BERNAL, 1985. *J. Cryst. Spect. Res.*, **15**, 191-199.

Monoclinic, $P2_1$, a = 6.785, b = 12.325, c = 16.390 Å, β = 97.89°, Z = 4. Mo radiation, R = 0.031 for 1354 reflections.

There are two independent molecules and in each the epoxides are shown to be α,β-attached (Fig. 1), and the two associated methyl groups are positioned syn to each other. Epoxide C-O and C-C distances average 1.459(5) and 1.473(6) Å respectively. The β-methylfuran ring fused to the main ring is essentially flat and the furan ring has two localized double bonds.

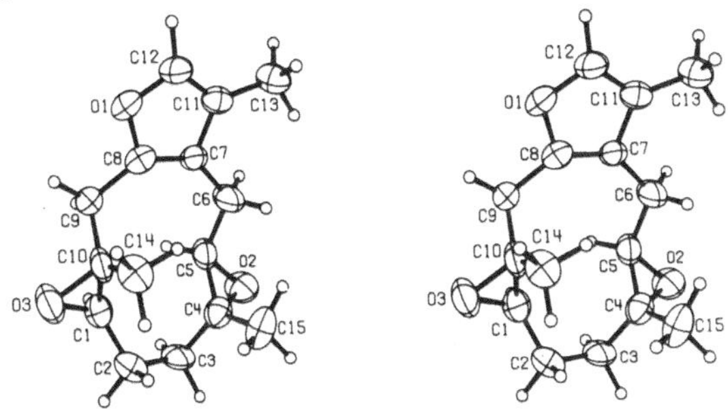

Fig. 1. A stereoview of $C_{15}H_{20}O_3$.

(4S,5S,7R,8R)-10,14-EPOXYRATIBIDA-11(13)-EN-8,12-OLIDE

$C_{15}H_{20}O_3$ (I)

(1S,5S,6R,7S,10S,11R)-1-HYDROXY-2-OXO-5-ANGELOXYXANTH-3-EN-6,12-OLIDE
$C_{20}H_{28}O_6$ (II)

W. HERZ, P. KULANTHAIVEL and V.L. GOEDKEN, 1985. *J. Org. Chem.*, **50**, 610-617.

I. Orthorhombic, $P2_12_12_1$, a = 7.102, b = 9.975, c = 18.166 Å, Z = 4. Mo radiation, R = 0.050 for 985 reflections.

II. Orthorhombic, $P2_12_12_1$, a = 10.778, b = 10.869, c = 16.649 Å, Z = 4. Mo radiation, R = 0.059 for 1286 reflections.

The crystal structure analyses established the stereochemistries of the molecules. In (I) the cyclohexane ring is in a distorted boat conformation and the cyclopentane is in a half-chair conformation (Fig. 1). In (II) the cycloheptane ring is almost a perfect boat and the lactone ring is close to a perfect envelope.

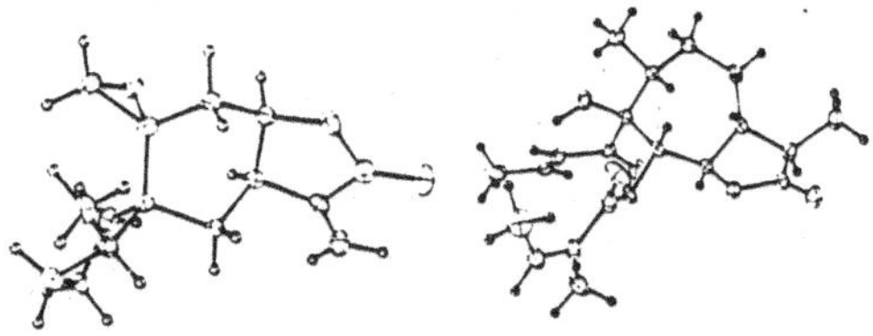

Fig. 1. Drawings of molecules of $C_{15}H_{20}O_3$ (I) (left) and $C_{20}H_{28}O_6$ (II) (right).

8-METHYLENE-4,13-DIMETHYL-11-OXABICYCLO[8.3.0]TRIDEC-4-ENE-9,12-DIONE
$C_{15}H_{20}O_3$ (I)

ISOCHAPLIATRIN
$C_{15}H_{22}O_3$ (II)

A.G. GONZALES, A. GALINDO, H. MANSILLA, A. GUTIERREZ, J.A. PALENZUELA, M. DE LOS A. GOMEZ-RODRIGUEZ, M. MARTINEZ-RIPOLL and S. GARCIA-BLANCO, 1985. *J. Org. Chem.*, **50**, 5856-5858.

I. Monoclinic, $P2_1$, a = 10.446, b = 7.8994, c = 8.8555 Å, β = 105.854°, Z = 2. Cu radiation, R = 0.037 for 1069 reflections.

II. Monoclinic, $P2_1$, a = 12.241, b = 7.5293, c = 7.619 Å, β = 92.808°, Z = 2. Cu radiation, R = 0.037 for 924 reflections.

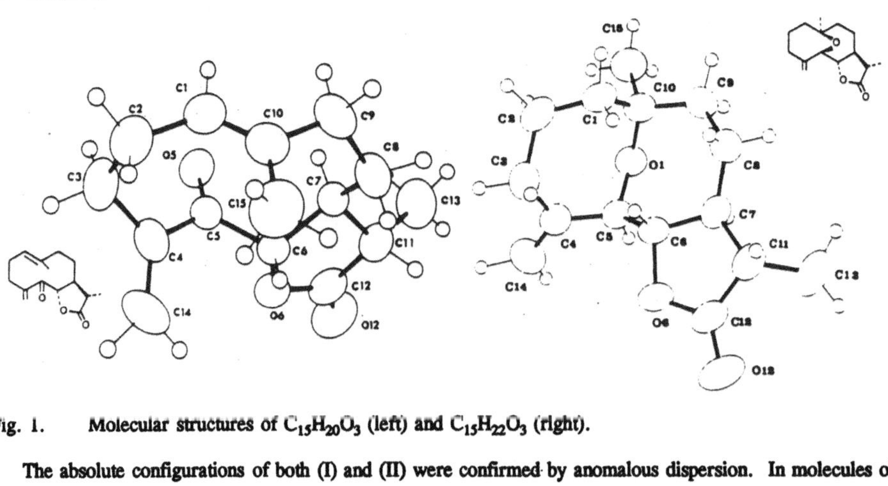

Fig. 1. Molecular structures of $C_{15}H_{20}O_3$ (left) and $C_{15}H_{22}O_3$ (right).

The absolute configurations of both (I) and (II) were confirmed by anomalous dispersion. In molecules of (I)

(Fig. 1) the substituents C(14) and C(15) are β-oriented and O(5) is α-oriented. The double bond $\Delta^{1(10)}$ is trans and the ten-membered ring approximates to a CCC conformation. The lactone ring attached to C(6) and C(7) shows an envelope conformation with C(7) as the flap. In (II) each seven-membered ring has a twist-chair conformation whilst the lactone ring attached to C(6) and C(7) has an envelope conformation with C(7) as the flap.

3-OXOGRANDOLIDE
$C_{15}H_{20}O_4$

U. RYCHLEWSKA, 1985. *Acta Cryst.*, C41, 540-542.

Orthorhombic, $P2_12_12_1$, a = 6.2941, b = 7.6749, c = 28.278 Å, Z = 4. Cu radiation, R = 0.047 for 1051 reflections.

The three-ring guaianolide system (Fig. 1) of this sesquiterpene lactone has a twist-chair exo-methylenecycloheptane moiety with a C_2 symmetry cis-fused with the cyclopentanone at C(1)-C(5) and trans-annelated with the γ-lactone at C(6)-C(7). The secondary hydroxyl is at C(9) and is β-oriented. It participates in an intermolecular hydrogen bond with the cyclopentane keto group, O...O 2.976(3) Å. The cyclopentanone ring adopts a half-chair conformation with an average endocyclic torsion angle of 22.8°. The γ-lactone ring is slightly less puckered with an average endocyclic torsion angle of 15.4°.

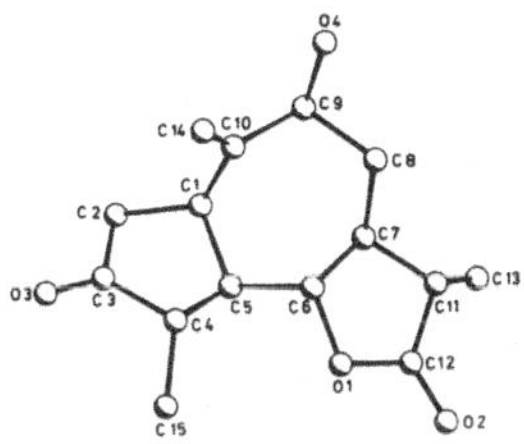

Fig. 1. $C_{15}H_{20}O_4$: the molecule.

VEROCEPHOL
$C_{15}H_{20}O_4$

M. SALMON, M. SORIANO-GARCIA, R.A. TOSCANO, J. CARDENAS, R. MIRANDA, F. VARGAS and E. ANGELES, 1985. *J. Org. Chem.*, 50, 4171-4172.

Tetragonal, $P4_32_12$, a = 9.763, c = 28.893 Å, Z = 8. Cu radiation, R = 0.046 for 2164 reflections.

The two six-membered rings (Fig. 1) have a cis junction. The cyclohexene ring A has a half-chair conformation whilst the more substituted cyclohexene ring B has a chair conformation. The five-membered lactone ring is essentially planar.

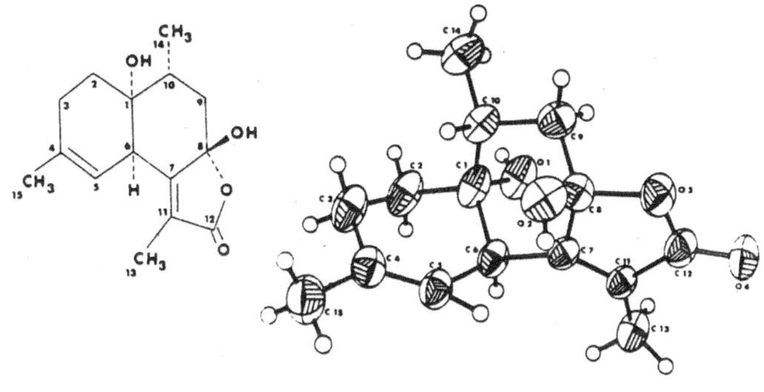

Fig. 1. The $C_{15}H_{20}O_4$ molecule.

2-METHYL-6-(4-METHYL-2-OXO-3-CYCLOHEX-1-YL)HEPT-2-ENOIC ACID
$C_{15}H_{22}O_3$

H. PREUT, W. KREISER, T. MULLER and P.G. JONES, 1985. *Acta Cryst.*, **C41**, 1480-1482.

Orthorhombic, $P2_12_12_1$, a = 8.109, b = 11.806, c = 15.132 Å, Z = 4. Mo radiation, R = 0.055 for 899 reflections.

The relative configuration of this natural 1-bisabolone is shown to be 6R,7S. The bond distances C(2)-C(3) (1.344(8)) and C(10)-C(11) (1.338(7) Å) show double bond character and the neighbouring bonds are shorter than single bonds (Fig. 1). The fragments A(C(9-13),O(2,3)), B(C(6-10)) and C(C(1-4),C(6),C(15),O(1)) are essentially planar with dihedral angles between these planes of 88.9(5) (A/B), 106.8(5) (A/C) and 23.6(5)° (B/C). In the crystal, molecules are linked by hydrogen bonds to form chains parallel to *b* (O...O 2.709(4) Å).

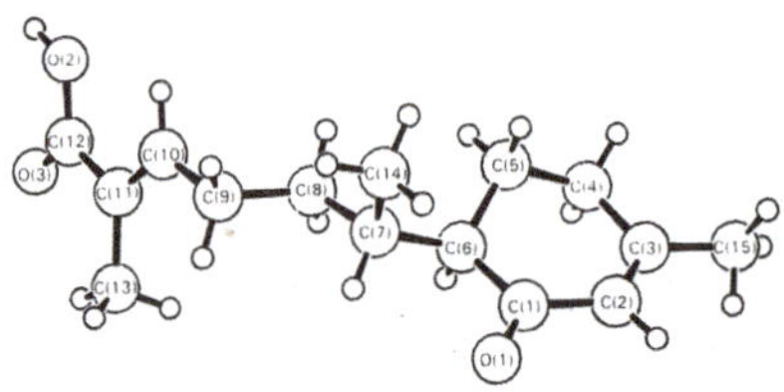

Fig. 1. A general view of the $C_{15}H_{22}O_3$ molecule.

(6R,7S,10R,11S)-1-OXO-6β,7α,10β,11βH-GERMACR-4-EN-6,12-OLIDE
$C_{15}H_{22}O_3$

M.A. GOMEZ-RODRIGUEZ, M. MARTINEZ-RIPOLL and S. GARCÍA-BLANCO, 1985. *Acta Cryst.*, **C41**, 1275-1277.

Orthorhombic, $P2_12_12_1$, a = 14.012, b = 12.986, c = 7.7028 Å, Z = 4. Cu radiation, R = 0.045 for 1250 reflections.

The ten-membered ring has a CCC conformation which is slightly distorted due to the influence of the substituents (Fig. 1). The double bond C(4)=C(5) makes the C(3)-C(4) and C(5)-C(6) bonds trans, C(14) is β-axial, C(15) is equatorial and O(1) α-axial. The lactone ring at C(6) and C(7) has an envelope conformation with C(7) as the flap.

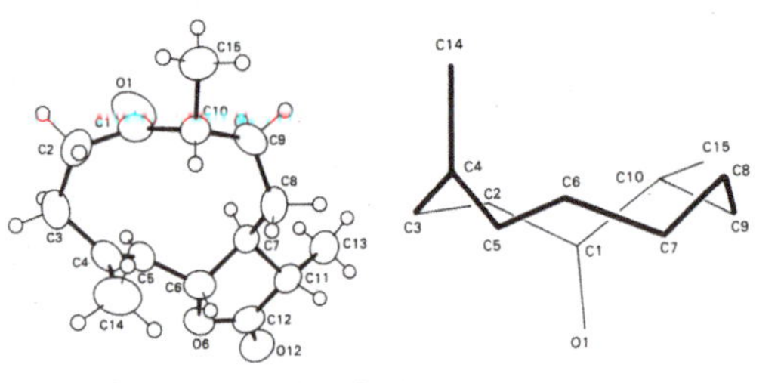

Fig. 1. $C_{15}H_{22}O_3$: the molecular structure and projection of the ten-membered ring onto the plane normal to the best ring plane.

1α-HYDROXYALLIACOLIDE
$C_{15}H_{22}O_5$

A.G. AVENT, J.R. HANSON, P.B. HITCHCOCK and BOONLENG YEOH, 1985. *J. Chem. Soc. Perkin Trans. 1*, 2749-2751.

Orthorhombic, $P2_12_12_1$, a = 8.881, b = 12.045, c = 13.279 Å, Z = 4. Mo radiation, R = 0.036 for 1282 reflections.

The crystal structure analysis unambiguously established the stereochemistry of the molecule (Fig. 1).

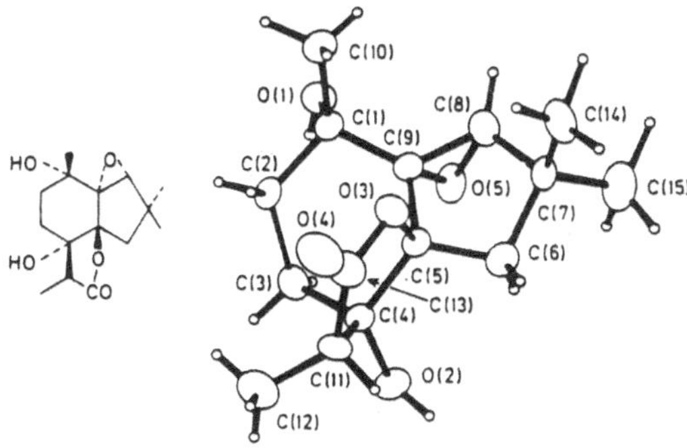

Fig. 1. Perspective drawing of a molecule of $C_{15}H_{22}O_5$.

DIMETHYL 9α-BROMO-10-OXO-8,13-EPOXYCEDRANE-12β,15-DIOATE
$C_{17}H_{21}BrO_6$ (I)

DIMETHYL 9β-BROMO-10-OXO-8,13-EPOXYCEDRANE-12β,15-DIOATE
$C_{17}H_{21}BrO_6$ (II)

W.J.J. SMEETS, J.A. KANTERS and K. VENKATASUBRAMANIAN, 1985. *Acta Cryst.*, **C41**, 102-105.

I. Tetragonal, $P4_12_12$, a = 11.904, c = 24.599 Å, Z = 8. Mo radiation, R = 0.0557 for 1899 reflections.

II. Orthorhombic, $P2_12_12_1$, a = 11.286, b = 11.558, c = 12.974 Å, Z = 4. Cu radiation, R = 0.0541 for 2350 reflections.

The molecular framework of the molecules (Fig. 1) consists of epoxycedrane skeletons. These are each composed of a six-membered ring A, two cis-fused five-membered rings B and D and a five-membered ring C containing the O bridge. Ring A has a distorted boat conformation, rings C and D have half-chair conformations and ring B is a distorted envelope in (I) and intermediate between half-chair and envelope in (II). In (I) the Br has an α configuration while in (II) it has a β configuration. All the methoxycarbonyl side chains are planar in (I) and (II) with the exception of that attached at C(2) in (I). The side chains at C(8) clearly have different orientations relative to the skeletons in (I) and (II) due to the different Br configurations. There are several slightly short intermolecular contacts in I.

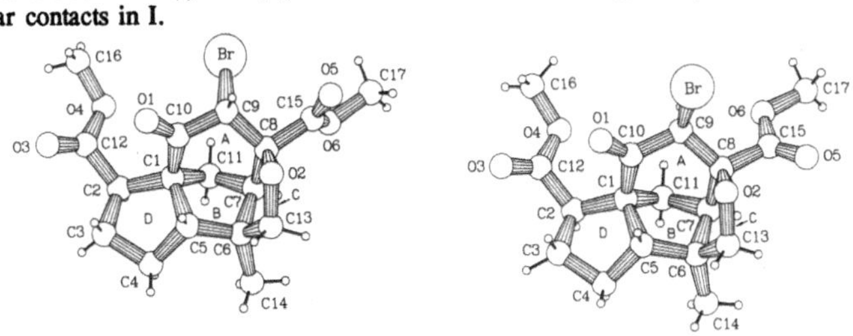

Fig. 1. $C_{17}H_{21}BrO_6$: views of (I) (left) and of (II) (right).

(E,E)-8-ACETOXYGERMACRA-1(10),4-DIEN-6,12-OLIDE
$C_{17}H_{22}O_4$

I. VICKOVIC, W.H. WATSON, M. SILVA, M.E. GARVEZ and M. HOENEISEN, 1985. *Acta Cryst.*, **C41**, 436-439.

Monoclinic, $P2_1$, a = 7.151, b = 16.421, c = 6.772 Å, β = 98.18°, Z = 2. Cu radiation, R = 0.047 for 986 reflections.

In this compound the lactone ring is cis-fused to the ten-membered germacradiene ring at the C(6)-C(7) bond (Fig. 1). This cis-fusion results in the ten-membered ring adopting an unusual chair-boat conformation. The methyl groups are syn and lie below the plane of the ten-membered ring. The molecular structure compared well with that calculated using molecular mechanics.

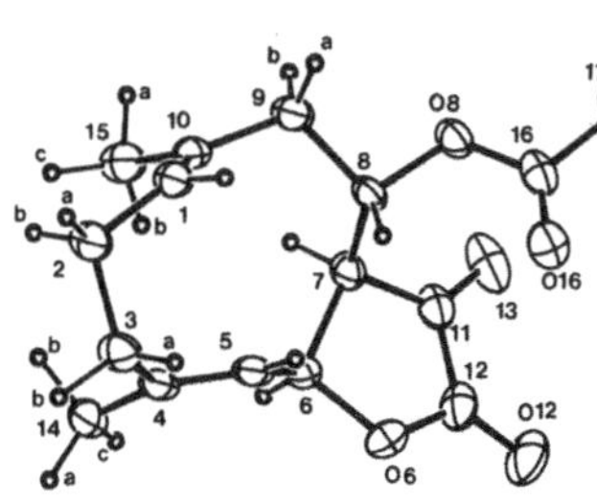

Fig. 1. $C_{17}H_{22}O_4$: a view of the molecule.

4-EPIMATRICIN
$C_{17}H_{22}O_5$ (I)

4-EPIMATRICIN 3β-HYDROXYDIHYDROCOSTUNOLIDE MONOHYDRATE
$C_{32}H_{42}O_8,H_2O$ (II) $C_{17}H_{22}O_5,C_{15}H_{22}O_3,H_2O$

G. APPENDINO, M. CALLERI, G. CHIARI and D. VITERBO, 1985. *J. Chem. Soc. Perkin Trans. 2*, 203-207.

I. Monoclinic, $P2_1$, a = 10.689, b = 7.123, c = 11.858 Å, β = 116.08°, Z = 2. Mo radiation, R = 0.039 for 1475 reflections.

II. Monoclinic, $P2_1$, a = 11.645, b = 10.981, c = 13.419 Å, β = 109.4°, Z = 2. Mo radiation, R = 0.049 for 3237 reflections.

In (I) the cycloheptene ring has a C_s chair conformation with a mirror plane through C(7) and the middle point of the double bond C(1)=C(10). The γ-lactone ring has an envelope conformation. In the adduct, (II), the cycloheptene ring has a less marked C_s symmetry and the γ-lactone ring is twisted in the epimatricin molecule as compared with (I) alone. Diagrams of epimatricin (I) and 3β-hydroxydihydrocostunolide in (II) are in Fig. 1 and Fig. 2 shows the structure of the adduct.

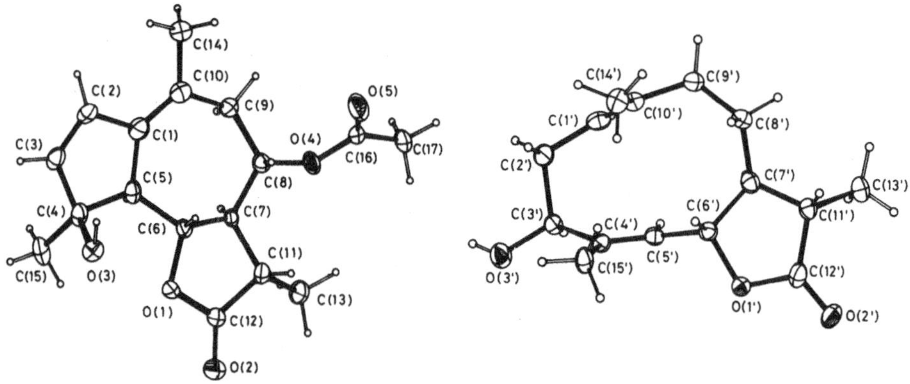

Fig. 1. ORTEP drawings of 4-epimatricin (left) and 3β-hydroxydihydrocostunolide (right).

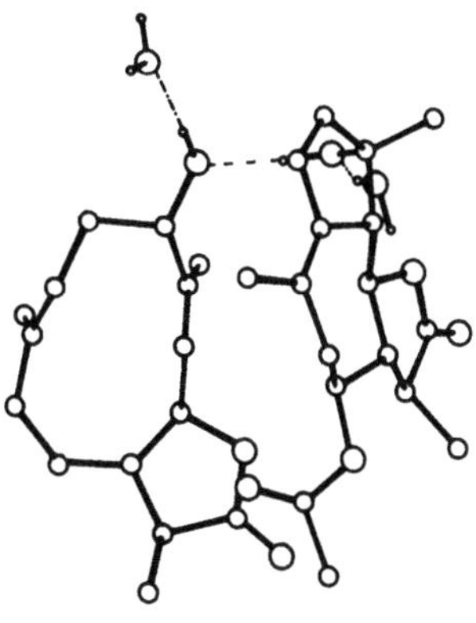

Fig. 2. Diagram of hydrogen bonding in the 4-epimatricin 3β-hydroxydihydrocostunolide adduct $C_{32}H_{44}O_9$.

(6E,10Z)-(3S,3aR,4S,11aR)-3a,5,8,11a-TETRAHYDRO-4-ACETOXY-3,6,10-TRIMETHYL-2,9(3H,4H)-CYCLODECA[b]FURANDIONE (HELIANGOLIDIN)
$C_{17}H_{22}O_5$

J.L. BRETON, F. CAMPS, J. COLL, L. EGUREN, J.A. GAVIN, A.G. GONZALEZ, X. MARTORELL, C. MIRAVITLLES, E. MOLINS and J. TORRAMILANS, 1985. *Tetrahedron*, **41**, 3141-3146.

Orthorhombic, $P2_12_12_1$, a = 9.727, b = 10.229, c = 16.772 Å, Z = 4. Mo radiation, R = 0.052 for 1163 reflections.

The cyclodecadiene ring is shown to adopt a chair-boat conformation (Fig. 1) and the γ-lactone ring has a planar conformation.

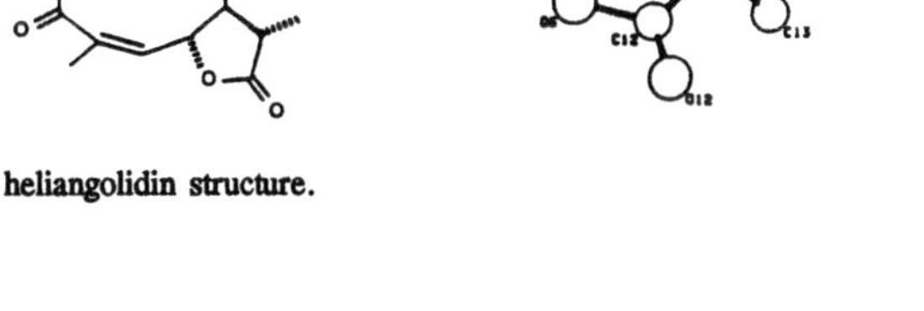

Fig. 1. The heliangolidin structure.

5-ACETOXY-3,4-DIMETHOXY-9,10-DIHYDROPHENANTHRENE (JAKINEOL)
$C_{18}H_{18}O_4$

A. MATSUO, H. NOZAKI, M. SUZUKI and M. NAKAYAMA, 1985. *J. Chem. Res.*, S174-175, M1916-1975.

Triclinic, $P\bar{1}$, a = 8.528, b = 9.873, c = 10.368 Å, α = 107.88, β = 110.97, γ = 95.22°, Z = 2. Mo radiation, R = 0.039 for 2065 reflections.

The analysis established the structure and stereochemistry of this antifungal material from the liverwort *R. jakii* to be as shown in Fig. 1.

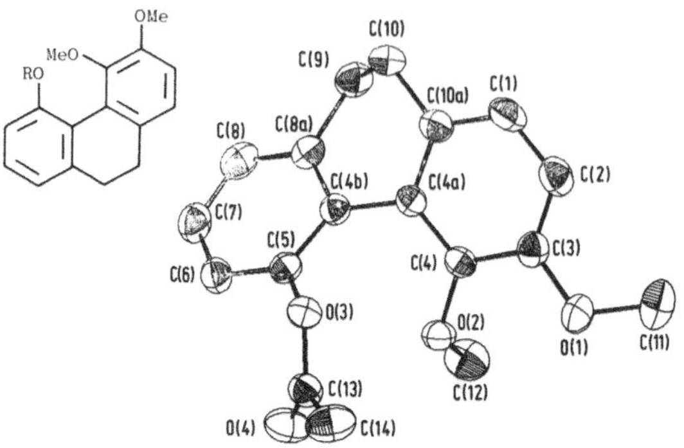

Fig. 1. The $C_{18}H_{18}O_4$ molecule (R = Ac).

2-ACETOXY-2α,3α-DIHYDRO-4α-HYDROXY-3-(2-METHYLISOCROTONOYLOXY)-(1αH,5αH,7αH,8βH)-GUAIA-10(14),11(13)-DIEN-8,12-OLIDE (OTOIMBRICATIN)
$C_{22}H_{28}O_7$

M. SORIANO-GARCIA, R.A. TOSCANO and C. GUERRERO, 1985. *Acta Cryst.*, C41, 538-540.

Orthorhombic, $P2_12_12_1$, a = 11.783, b = 13.227, c = 14.115 Å, Z = 4. Cu radiation, R = 0.072 for 1395 reflections.

Otoimbricatin A is a guaianolide-type sesquiterpene lactone containing a five-membered ring cis-fused to a seven-membered ring and a trans-fused α,β-unsaturated γ-lactone (Fig. 1). The five-membered rings both adopt envelope conformations and the seven-membered ring has a deformed chair conformation. The acetate and methylisocrotonate groups are oriented to minimise transannular repulsions between their O atoms and the cyclopentene ring. The molecules in the crystal are held together by hydrogen bonds (O(3)...O(6) 2.963(9) Å) and van der Waals interactions.

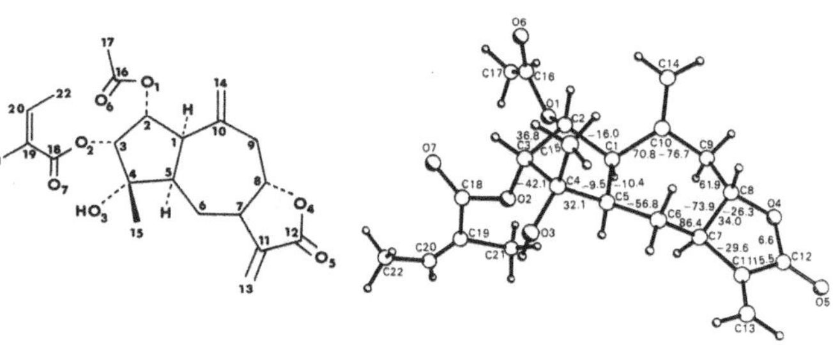

Fig. 1. $C_{22}H_{28}O_7$: the molecular skeleton and the structure showing torsion angles.

LEPTOCARPIN ACETATE
$C_{22}H_{28}O_7$

A.G. OBER, F.R. FRONCZEK and N.H. FISCHER, 1985. *J. Nat. Prod.*, 48, 302-306.

Orthorhombic, $P2_12_12_1$, a = 11.441, b = 12.915, c = 15.091 Å, Z = 4. Mo radiation, R = 0.043 for 1538 reflections.

The conformation of the α-methylene-γ-lactone ring (Fig. 1) is an intermediate form between a C(7) envelope and a half-chair with C(12) on the pseudo twofold axis. The O(2)-C(12)-C(11)-C(13) torsion angle is +7.8° and C(11)-C(7)-C(6)-O(1) is +16.7°. The absolute configuration derived from chiroptical data is as shown.

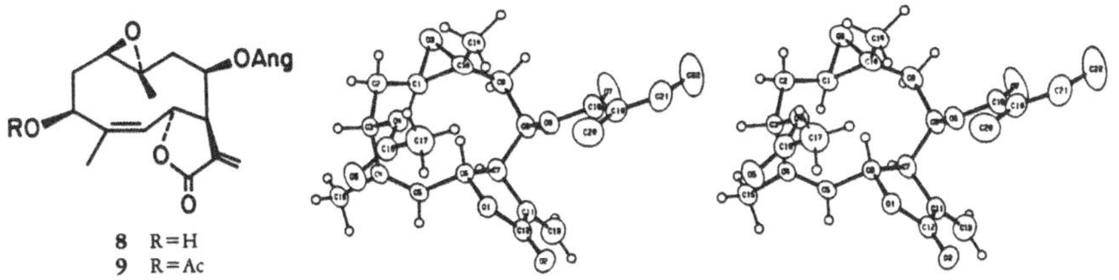

Fig. 1. Leptocarpin acetate and a stereoview.

trans,trans-GERMACRA-1(10),4-DIEN-cis-6,12-OLIDE (LASEROLIDE)
$C_{22}H_{30}O_6$

U. RYCHLEWSKA, 1985. *Acta Cryst.*, C41, 1118-1120.

Monoclinic, $P2_1$, a = 6.566, b = 12.551, c = 12.836 Å, β = 94.83°, Z = 2. Cu radiation, R = 0.044 for 1432 reflections.

The conformation of the ten-membered ring in the molecule (Fig. 1) is chair boat and the methyl groups at C(4) and C(10) have syn α orientations. The γ-lactone ring is cis-fused to the macrocycle at the C(6)-C(7) bond and introduces strain into the cyclodecadiene skeleton. This is evident from the value of the C(5)-C(6)-C(7)-C(8) torsion angle, -27.0(6)°. The γ-lactone ring has a torsion angle at the junction, O(1)-C(6)-C(7)-C(11), of -21.0° and as a consequence the C(7)-H(7) and C(6)-O(1) bonds are synclinal. This ring adopts the unusual 7-α-envelope conformation.

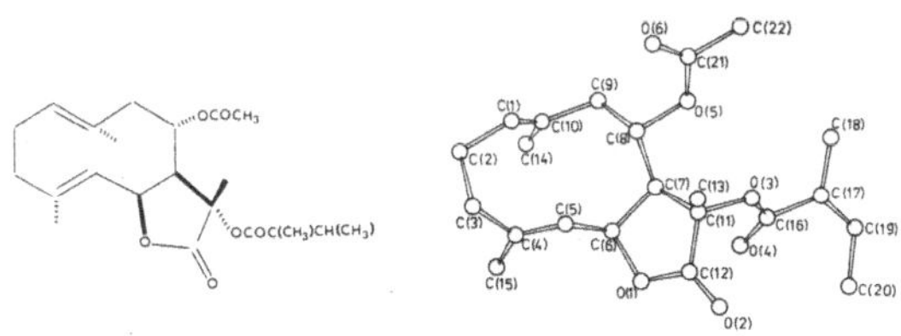

Fig. 1. Structural formula and a view of the $C_{22}H_{30}O_6$ molecule.

rel-(3R,3aR,8S,8aR)-3a-HYDROXY-1-OXO-6-DAUCEN-8-YL p-ANISATE
$C_{23}H_{30}O_5$ (I)

rel-(1S,3R,3aS,4S,8S,8aR)-1,8-DIACETOXY-3-HYDROXY-6-DAUCEN-4-YL p-ANISATE
$C_{27}H_{36}O_8$ (II)

W.H. WATSON, R.P. KASHYAP and I. TAVANAIEPOUR, 1985. *Acta Cryst.*, C41, 1650-1653.

I. Orthorhombic, $P2_12_12_1$, a = 11.723, b = 17.261, c = 10.515 Å, Z = 4. Mo radiation, R = 0.044 for 1448 reflections.

II. Orthorhombic, $P2_12_12_1$, a = 11.733, b = 26.654, c = 8.644 Å, Z = 4. Mo radiation, R = 0.052 for 1927 reflections.

Both compounds belong to the carotane class of sesquiterpenes. As a consequence of the five- and seven-membered rings being cis-fused in (I) and trans-fused in (II) (Fig. 1) the molecules have significantly different conformations. In both molecules the five-membered ring has an envelope conformation with C(4) and C(2) as flaps in (I) and (II), respectively. The seven-membered ring in (II) has an almost ideal chair conformation while in (I) it has a highly distorted twist-chair conformation. The bond common to both rings, C(1)-C(5), differs significantly being of length 1.578(5) Å in (I) and 1.552(6) Å in (II). A number of valence angles also show significant differences between the two molecules. No unusually short intermolecular contacts are reported.

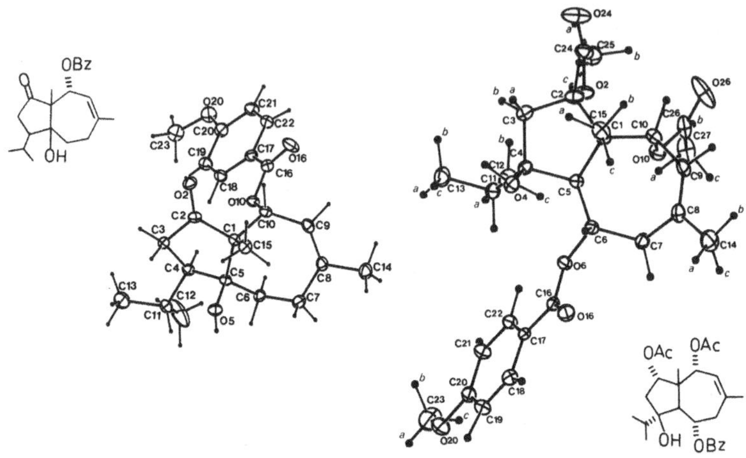

Fig. 1. Structural formula and views of $C_{23}H_{30}O_5$ (left) and of $C_{27}H_{36}O_8$ (right).

GUAYULIN A
$C_{24}H_{30}O_2$

S.F. WATKINS, F.R. FRONCZEK, G. CHIARI, G.W. REYNOLDS and E. RODRIGUEZ, 1985. *J. Nat. Prod.*, **48**, 631-633.

Tetragonal, $P4_32_12$, a = 10.426, c = 38.749 Å, Z = 8. Cu radiation, R = 0.049 for 2221 reflections.

The structure and relative stereochemistry are shown in Fig. 1. The cyclopropane ring is cis-fused to the main ring and β-oriented, while the cinnamate group is α-oriented. The endocyclic double bonds are trans-trans and are oriented in a crossed fashion. Both double bonds are twisted from planarity due to steric crowding.

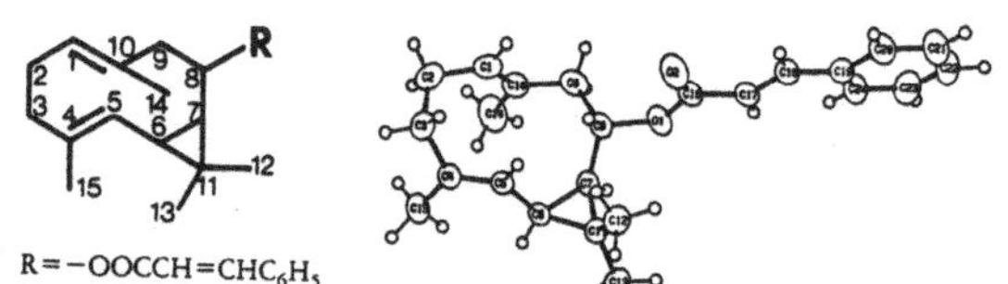

Fig. 1. The Guayulin A structure.

4-O-METHYLMELLEOLIDE METHANOL SOLVATE
$C_{24}H_{30}O_6$,$3(CH_4O)$

D.M.X. DONNELLY, F. ABE, D. COVENEY, N. FUKUDA, J. O'REILLY, J. POLONSKY and T. PRANGE, 1985. *J. Nat. Prod.*, **48**, 10-16.

Triclinic, $P1$, a = 16.252, b = 9.588, c = 9.315 Å, α = 117.1, β = 85.16, γ = 102.0°, Z = 2. Cu radiation, R = 0.081 for 2151 reflections.

The two molecules in the asymmetric unit have different conformations; one is shown in Fig. 1. In both molecules the cyclobutane rings have similar folded conformations (puckering angle around C(6)-C(5) is 20.5° in one molecule and 17.6° in the other). The cyclopentane ring in one molecule has a twist conformation and an envelope conformation in the other.

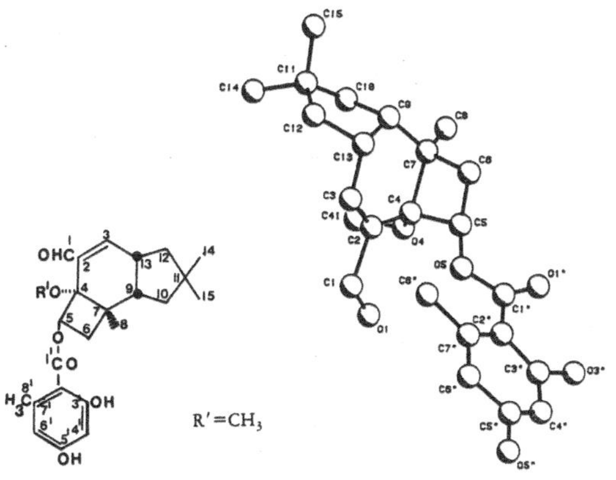

Fig. 1. The 4-O-methylmelleolide skeleton and a view of one of the molecules.

ABSINTHINE BENZENE CYCLOPENTANE SOLVATE
$C_{30}H_{40}O_6, C_6H_6, C_5H_{10}$

Z. KARIMOV, 1985. *Kristallografiya*, **30**, 682-687.

Orthorhombic, $P2_12_12_1$, a = 28.623, b = 15.277, c = 8.570 Å, Z = 4. Cu radiation, R = 0.078 for 2716 reflections.

The molecule has two identical parts linked by C(2)-C(17) and C(3)-C(20) bonds (Fig. 1). Lactone ring A is a C(22) envelope, cyclohexane ring B is a distorted chair, cyclopentane rings C and D have C(16) envelope conformations, five-membered ring E is planar, cycloheptane ring F has a chair conformation and ring G is a C(7) envelope. The rings A/B and F/G have trans configuration, the C/E rings have cis configuration.

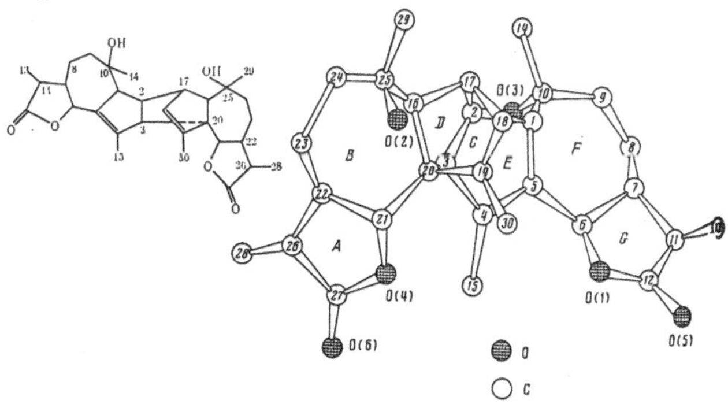

Fig. 1. The absinthine molecule.

METHYL 4-HYDROXY-3-OXO-9βH-15,16-BISNORPODOCARPA-1,4,7-TRIEN-17-OATE
$C_{16}H_{18}O_4$ (I)

METHYL 1,2-DIOXO-9βH-15,16-BISNORPODOCARPA-3,7-DIEN-17-OATE
$C_{16}H_{18}O_4$ (II)

D. PITEA, M. GASTALDI, F. ORSINI, F. PELIZZONI, A. MUGNOLI and E. ABBONDANTI, 1985. *J. Org. Chem.*, **50**, 1853-1859.

I. Monoclinic, $C2/c$, a = 14.284, b = 8.624, c = 22.639 Å, β = 93.07°, Z = 8. Mo radiation, R = 0.039 for

1373 reflections.

II. Monoclinic, $P2_1/n$, a = 9.344, b = 14.835, c = 10.179 Å, β = 100.67°, Z = 4. Mo radiation, R = 0.038 for 1841 reflections.

The crystal structure analyses established the stereospecificities of the reactions which produced these molecules (Fig. 1).

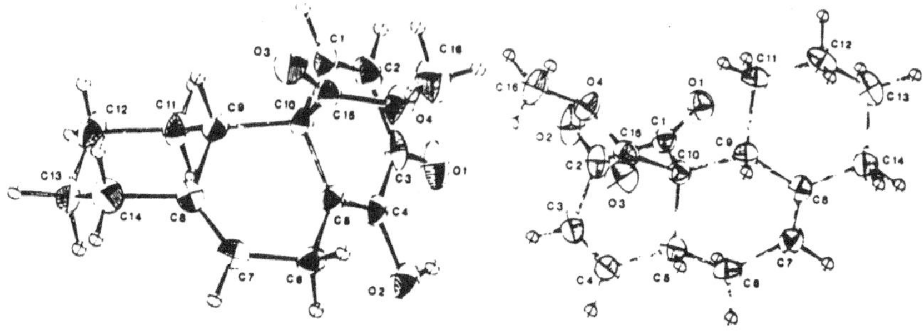

Fig. 1. Drawings of molecules of $C_{16}H_{18}O_4$: (I) (left) and (II) (right).

1-METHYL-9-(1-METHYLVINYL)-4,17-DIOXATRICYCLO[11.2.1.1^{3,6}]HEPTADEC-6(16)-ENE-5,11,14-TRIONE

$C_{19}H_{24}O_5$ (I)

6,16-EPOXY-1-METHYL-9-(1-METHYLVINYL)-4,17-DIOXATRICYCLO[11.2.1.1^{3,6}]HEPTADECANE-5,11,14-TRIONE

$C_{19}H_{24}O_6$ (II)

A. SATO, W. FENICAL, Z. QITAI and J. CLARDY, 1985. *Tetrahedron*, **41**, 4303-4308.

I. Orthorhombic, $P2_12_12_1$, a = 6.049, b = 11.889, c = 24.212 Å, Z = 4. Cu radiation, R = 0.063 for 1150 reflections.

II. Monoclinic, $P2_1$, a = 6.496, b = 13.873, c = 10.085 Å, β = 98.48°, Z = 2. Cu radiation, R = 0.068 for 1205 reflections.

The relative stereochemistries for these norcembrene diterpenoids were determined as shown in Fig. 1.

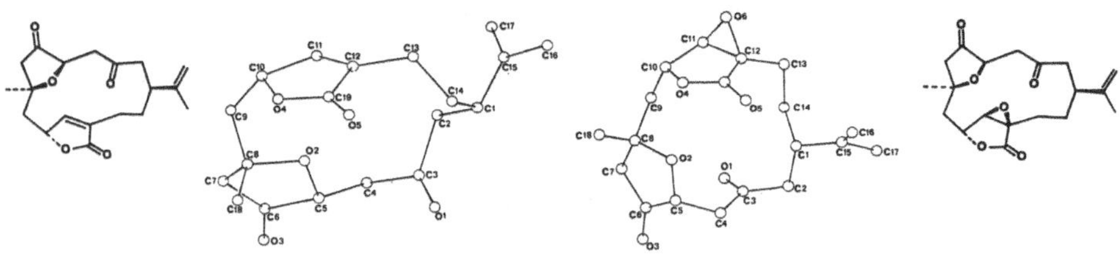

Fig. 1. Views of $C_{19}H_{24}O_5$ (I) (left) and $C_{19}H_{24}O_6$ (II).

7,11-DIHYDROXY-3,4,9,11b-TETRAMETHYL-1,2,8,9,11b-PENTAHYDROPHENANTHRO[3,2-b]FURAN-6(2H)-ONE (UNCINATONE)

$C_{20}H_{22}O_4$

A.-C. DORSAZ, A. MARSTON, H. STOECKLI-EVANS, J.D. MSONTHI and K. HOSTETTMANN, 1985. *Helv. Chim. Acta*, **68**, 1605-1609.

Orthorhombic, $P2_12_12_1$, a = 16.74, b = 12.69, c = 8.28 Å, Z = 4. Mo radiation, R = 0.057 for 952 reflections.

The crystal structure analysis established the stereochemistry of the molecule (Fig. 1). The cyclohexene ring has a half-chair conformation and the cyclohexadiene-1-one ring has a flat sofa conformation. The benzene ring is planar and the dihydrofuran ring has a flat envelope conformation with C(9) at the flap. The methyl substituent at C(12) is axial.

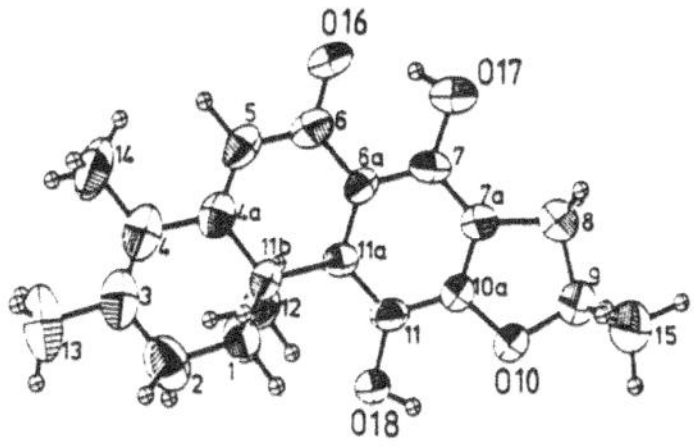

Fig. 1. Perspective drawing of a molecule of $C_{20}H_{22}O_4$.

(-)-(1R,4R,5S,7R,9S,13S,16R)-7-(3'-FURANYL)-4,13,16-TRIMETHYL-2,8,11,17-TETRAOXAPENTACYCLO-[7.6.1.$1^{1,7}$.$0^{5,16}$.$0^{9,13}$]HEPTADECA-3,12-DIONE (SAUDIN)
$C_{20}H_{22}O_7$

J.S. MOSSA, J.M. CASSADY, M.D. ANTOUN, S.R. BYRN, A.T. McKENZIE, J.F. KOZLOWSKI and P. MAIN, 1985. *J. Org. Chem.*, **50**, 916-918.

Monoclinic, $P2_1$, a = 15.891, b = 12.200, c = 9.337 Å, β = 100.01°, Z = 4. Cu radiation, R = 0.060 for 2393 reflections.

The crystal structure analysis established the stereochemistry of the molecule (Fig. 1). The lactone groups are arranged in a polyether lattice.

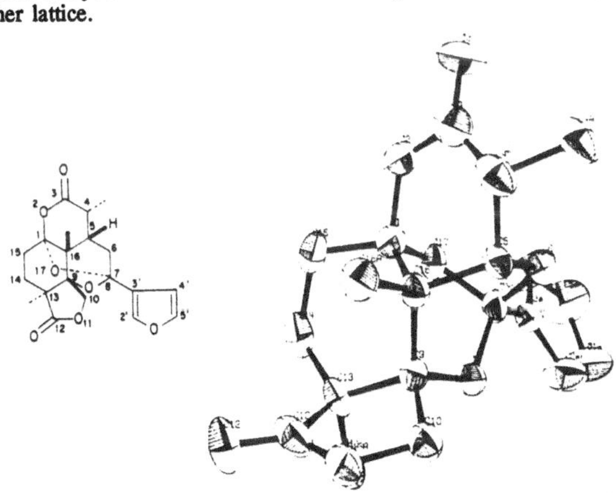

Fig. 1. The $C_{20}H_{22}O_7$ molecule.

FUJENAL
$C_{20}H_{26}O_4$

A.G. AVENT, C. CHAMBERLAIN, J.R. HANSON and P.B. HITCHCOCK, 1985. *J. Chem. Soc. Perkin Trans. 1*, 2493-2496.

Tetragonal, $P4_12_12$, a = 11.340, c = 27.037 Å, Z = 8. Mo radiation, R = 0.037 for 691 reflections.

Fujenal was shown to be a seco-ring B ent-kaurenoid diterpene (Fig. 1) in which rotation about the C(9)-C(10) bond has moved the overall conformation away from that of kaurene. Ring A has a chair conformation flattened by the 6,19-anhydride ring; ring C has a chair conformation and ring D has an envelope conformation.

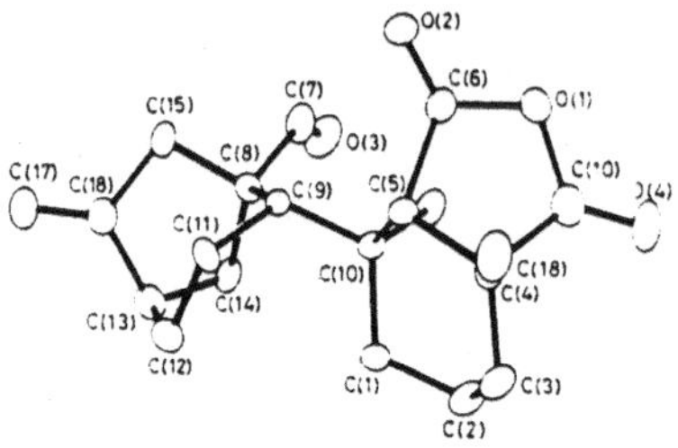

Fig. 1. Molecular structure of $C_{20}H_{26}O_4$.

ICACINOL
$C_{20}H_{26}O_7$

P. ON'OKOKO, M. VANHAELEN, R. VANHAELEN-FASTRE, J.P. DECLERCQ and M. VAN MEERSSCHE, 1985. *Tetrahedron,* **41**, 745-748.

Monoclinic, $P2_1$, a = 6.409, b = 11.931, c = 11.522 Å, β = 98.10°, Z = 2. Cu radiation, R = 0.040 for 1190 reflections.

The analysis establishes the structure and relative stereochemistry as shown in Fig. 1. There are no unusual bond lengths or angles.

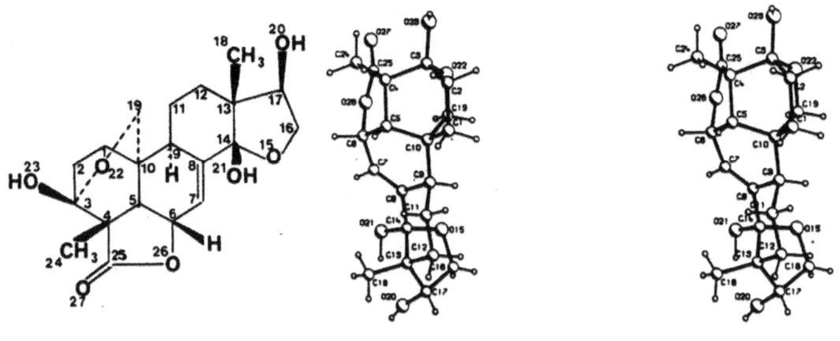

Fig. 1. The icacinol molecule and a stereoview.

ent-5β-H,6α-H-TRACHYLOBAGIBBERELLANE-7,18-DIOIC ACID
$C_{20}H_{28}O_4$

J.D. ARRAEZ, B.M. FRAGA, A.G. GONZALEZ, J.G. LUIS, J. FAYOS and A. PERALES, 1985. *J. Chem. Soc. Perkin Trans. 1,* 207-211.

Monoclinic, $P2_1$, a = 27.575, b = 7.308, c = 8.956 Å, β = 91.14°, Z = 4. Cu radiation, R = 0.067 for 2718 reflections.

There are two molecules of the acid in the crystallographic asymmetric unit. The structure was confirmed by the crystal structure analysis (Fig. 1), and its absolute configuration determined. The two independent molecules in the unit cell have identical conformations and no significant geometrical differences; they are linked by hydrogen bonds across a pseudo centre of symmetry. The hydrogen bond lengths range from 2.62 to 2.65 Å. The two molecules have a twist-envelope-twist conformation.

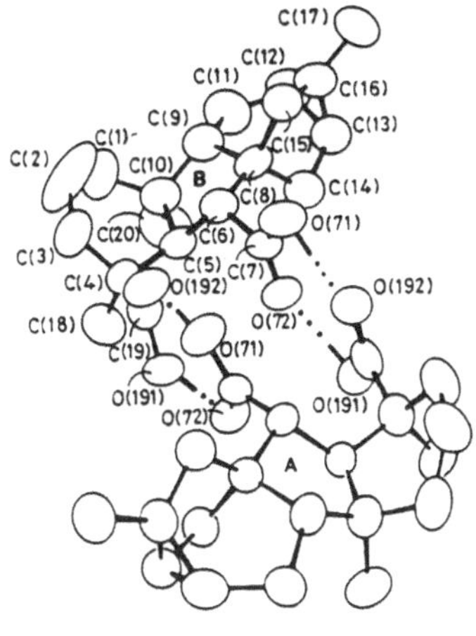

Fig. 1. The conformations of the two molecules in the hydrogen bonded dimer of $C_{20}H_{28}O_4$.

3-(13,14,14-TRIMETHYL-5,8-METHANO-2-OXATRICYCLO[4.4.3.0^{1,6}]TRIDECANE)-2-METHYL-PROP-2-ENOIC ACID
$C_{20}H_{30}O_3$

P.J. CARROLL, L.M. ENGELHARDT, E.L. GHISALBERTI, P.R. JEFFERIES, E.J. MIDDLETON, T.A. MORI and A.H. WHITE, 1985. *Aust. J. Chem.*, **38**, 1351-1363.

Monoclinic, $P2_1$, a = 11.195, b = 21.632, c = 7.859 Å, β = 108.55°, Z = 4. Mo radiation, R = 0.040 for 3860 reflections.

The asymmetric unit of this eremane diterpene comprises two molecules with similar conformations, linked in pairs by $CO_2H...HO_2C$ hydrogen bonds. The carbon skeleton (Fig. 1) incorporates a tricyclo[6.2.1.0^{1,5}]undecane system. The two six-membered rings adopt chair conformations, while the two five-membered rings are envelopes with C(10) and C(13) as flaps. The molecular dimensions are generally as expected.

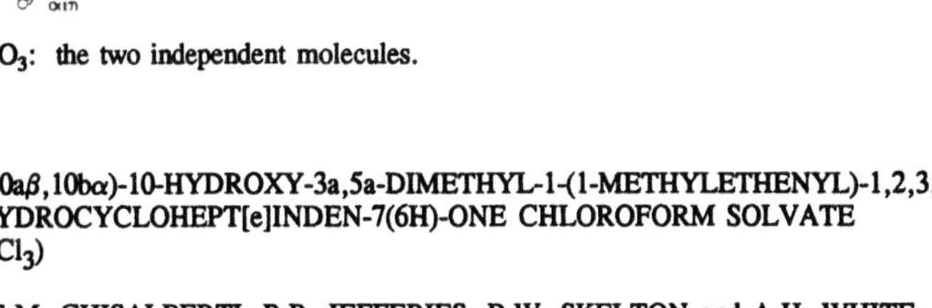

Fig. 1. $C_{20}H_{30}O_3$: the two independent molecules.

(1α,3aβ,5aα,10α,10aβ,10bα)-10-HYDROXY-3a,5a-DIMETHYL-1-(1-METHYLETHENYL)-1,2,3,3a,4,5,5a,10,-10a,10b-DECAHYDROCYCLOHEPT[e]INDEN-7(6H)-ONE CHLOROFORM SOLVATE
$C_{20}H_{30}O_3$,0.5($CHCl_3$)

M.P. CASSIDY, E.M. GHISALBERTI, P.R. JEFFERIES, B.W. SKELTON and A.H. WHITE, 1985. *Aust. J. Chem.*, **38**, 1187-1195.

Triclinic, $P\bar{1}$, a = 12.13, b = 9.34, c = 8.92 Å, α = 89.8, β = 87.6, γ = 76.4°, Z = 2. Mo radiation, R = 0.060 for 5559 reflections (at 120 K).

The asymmetric unit comprises two molecules of compound and one of solvent. In the molecule (Fig. 1), the five-membered ring adopts an envelope conformation and the six-membered ring is a fairly regular chair, with the substituent methyl groups at C(6,9) both axial. There is a substantial twist about the double bond C(11)-C(12) of the seven-membered ring, the dihedral angle C(10)-C(11)-C(12)-C(13) being -168.1, -167.3° for the two molecules.

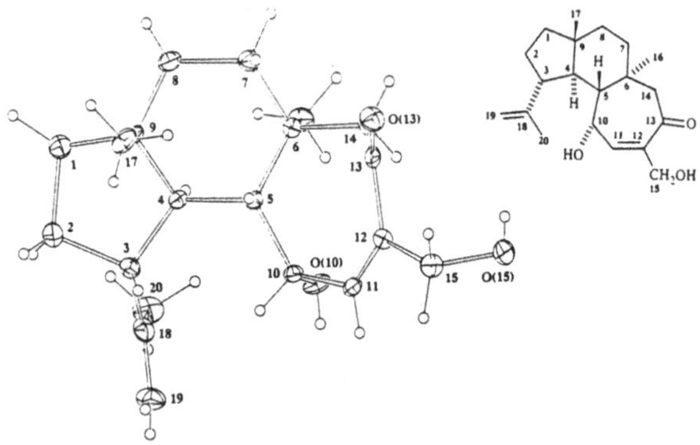

Fig. 1. $C_{20}H_{30}O_3$,0.5($CHCl_3$): the molecular structure and skeleton.

(1R*,3E,7Z,12R*)-20-HYDROXYCEMBRA-3,7,15-TRIEN-19-OIC ACID
$C_{20}H_{32}O_3$

S. KOSELA, E.L. GHISALBERTI, P.R. JEFFERIES, B.W. SKELTON and A.H. WHITE, 1985. *Aust. J. Chem.*, **38**, 1365-1370.

Monoclinic, $P2_1$, a = 16.52, b = 9.579, c = 11.84 Å, β = 91.55°, Z = 4. Mo radiation, R = 0.049 for 2649 reflections (at 120 K).

The asymmetric unit contains two crystallographically independent molecules (Fig. 1), with the same absolute configuration, but different conformations. Molecular dimensions show good agreement between the two molecules and are generally as expected.

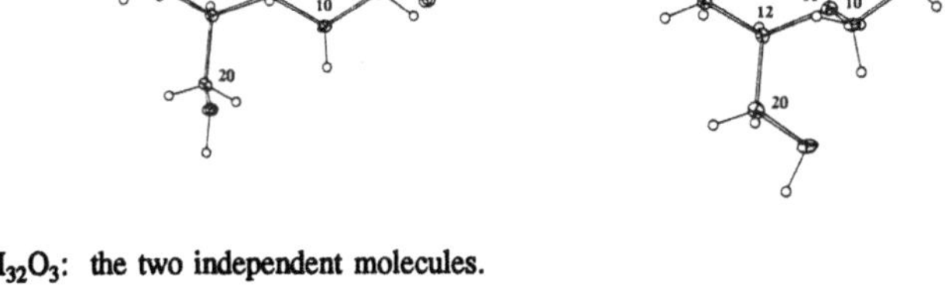

Fig. 1. $C_{20}H_{32}O_3$: the two independent molecules.

DICTYOXETANE MONOHYDRATE
$C_{20}H_{32}O_3,H_2O$

K.C. PULLAIAH, R.K. SURAPANENI, C.B. RAO, K.F. ALBIZATI, B.W. SULLIVAN, D.J. FAULKNER, HE CUN-HENG and J. CLARDY, 1985. *J. Org. Chem.*, **50**, 3665-3666.

Monoclinic, $P2_1$, a = 7.842, b = 16.320, c = 8.242 Å, β = 115.86°, Z = 2. Cu radiation, R = 0.074 for 981 reflections.

The five-membered ring (Fig. 1) has an envelope conformation with C(7) as the flap. The six-membered ring is in a chair conformation as, also, is the seven-membered ring. The oxetane ring is only slightly folded with all of the torsional angles roughly ±6°.

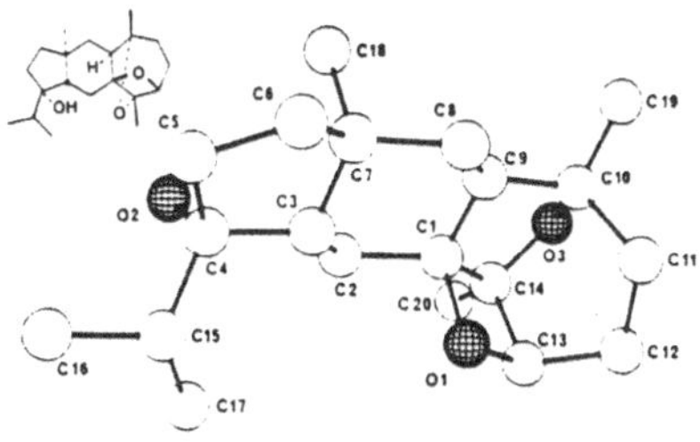

Fig. 1. Perspective drawing of a molecule of $C_{20}H_{32}O_3,H_2O$ (hydrogen atoms omitted, oxygen atoms hatched).

11-(1-(METHOXYCARBONYL)ETHYL)-4,8α,13,13-TETRAMETHYLBICYCLO[10.1.0]TRIDEC-4-ENE-7,10-DIONE (20-DEMETHYL-20-METHOXYCARBONYLBERDIMER-4-ENE-7,10-DIONE)
$C_{21}H_{32}O_4$ (I)

11-(1-(METHOXYCARBONYL)ETHYL)-4,8β,13,13-TETRAMETHYLBICYCLO[10.1.0]TRIDEC-4-ENE-7,10-DIONE (20-DEMETHYL-20-METHOXYCARBONYLBERDIMER-4-ENE-7,10-DIONE)
$C_{21}H_{32}O_4$ (II)

E.L. GHISALBERTI, P.R. JEFFERIES, T.A. MORI, B.W. SKELTON and A.H. WHITE, 1985. *Tetrahedron*, **41**, 2517-2526.

I. Monoclinic, $P2_1$, a = 10.280, b = 10.752, c = 10.196 Å, β = 114.78°, Z = 2. Mo radiation, R = 0.043 for 1920 reflections.

II. Orthorhombic, $P2_12_12_1$, a = 32.09, b = 10.158, c = 6.212 Å, Z = 4. Mo radiation, R = 0.052 for 1850 reflections.

The analyses determined the structures and conformations of these diterpenes to be as shown in Fig. 1.

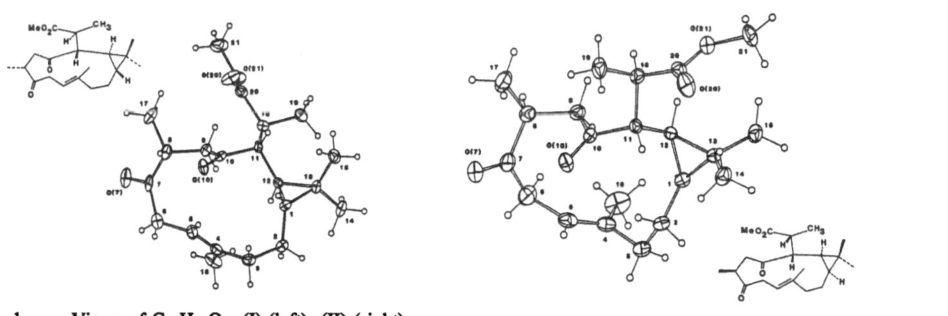

Fig. 1. Views of $C_{21}H_{32}O_4$, (I) (left), (II) (right).

7β-ACETOXY-8β,13-EPOXY-10(9-8)-abeo-17-NORLABDAN-9-ONE
$C_{21}H_{34}O_4$

P.K. GRANT, L.R. HANTON and SIEWFAH TSAI, 1985. *Aust. J. Chem.*, **38**, 1461-1470.

Orthorhombic, $P2_12_12_1$, a = 15.843, b = 16.245, c = 7.775 Å, Z = 4. Mo radiation, R = 0.056 for 1639 reflections.

The six-membered A ring of the molecule (Fig. 1) adopts a chair conformation with the C(19) and C(20) methyl groups 1,3-diaxial. These methyl groups are splayed outwards from the ring by 11° because of steric interactions. The five-membered ring B adopts an envelope conformation, with C(5) as flap. This atom lies on the side of the ring opposite the C(19) and C(20) methyl groups. The plane of the acetate side chain makes an angle of 108° with the plane of the ring. Ring C has a flattened chair conformation because of the presence of the sp^2 C atom C(9) and O(2). Molecular dimensions are normal.

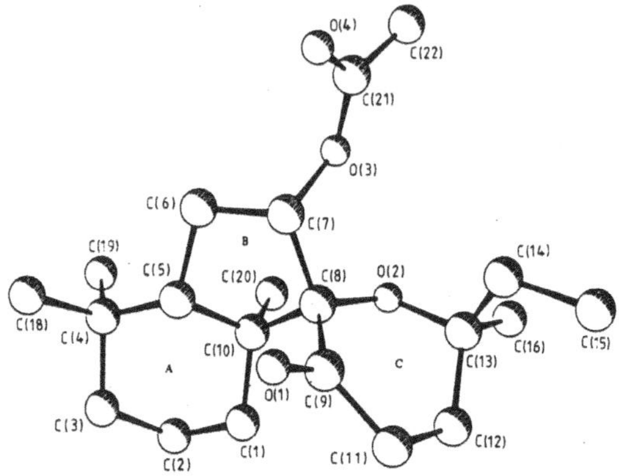

Fig. 1. $C_{21}H_{34}O_4$: the molecular structure.

SALVIGENOLIDE
$C_{22}H_{22}O_7$

B. ESQUIVEL, J. CARDENAS, A. TOSCANO, M. SORIANO-GARCIA and L. RODRIGUEZ-HAHN, 1985. *Tetrahedron,* **41**, 3213-3217.

Orthorhombic, $P2_12_12_1$, a = 8.625, b = 12.453, c = 18.052 Å, Z = 4. Mo radiation, R = 0.050 for 1225 reflections.

The molecule (Fig. 1) has an unusual system of six-seven-five trans-fused rings. The cyclohexene ring has a distorted sofa conformation. The cycloheptene ring adopts a chair-type conformation (with an approximate mirror plane through C(6)). Both γ-lactone rings have envelope conformations with flaps at C(5) and C(17). The furan ring is planar.

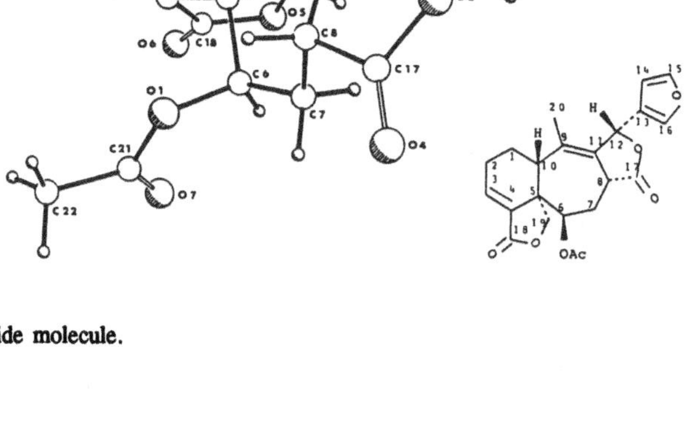

Fig. 1. The salvigenolide molecule.

KALLOLIDE A ACETATE
$C_{22}H_{26}O_5$

S.A. LOOK, M.T. BURCH, W. FENICAL, ZHENG QITAI and J. CLARDY, 1985. *J. Org. Chem.*, **50**, 5741-5746.

Orthorhombic, $P2_12_12_1$, a = 6.603, b = 15.517, c = 20.355 Å, Z = 4. Cu radiation, R = 0.052 for 1498 reflections.

The α,β-unsaturated γ-lactone ring is distorted with the C(8)-C(9)-C(10)-C(11) torsion angle equal to 167° (Fig. 1).

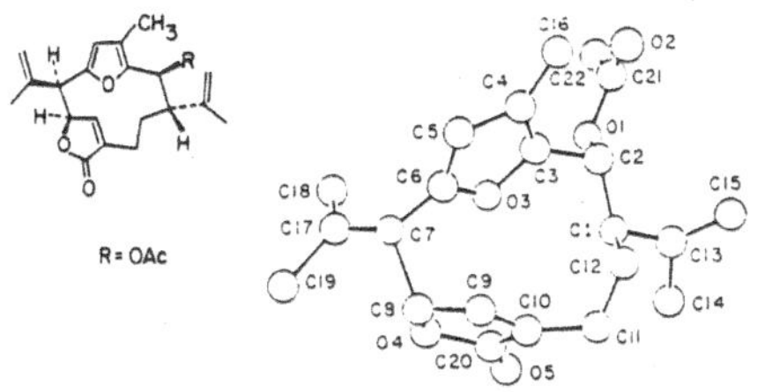

Fig. 1. Kallolide A acetate.

2,18-DIACETOXY-15,16-EPOXY-3-HYDROXY-12-OXO-19-NORLABDANE-8(17),13(16),14-TRIENE MONOHYDRATE
$C_{23}H_{30}O_7,H_2O$

A. MOSQUERA, A.V. RIVERA, E.R. DE GIL and A. BAHSAS, 1985. *Acta Cryst.*, C41, 433-436.

Orthorhombic, $P2_12_12_1$, a = 14.076, b = 21.088, c = 7.822 Å, Z = 4. Mo radiation, R = 0.067 for 1235 reflections.

The structure consists of discrete molecules (Fig. 1) forming layers parallel to (010) at y = 0 and y = 0.5, with water molecules located in channels running parallel to the c-axis. The rings of the bicyclic ring system in the molecule are in chair conformations and the furan ring is planar. All the substituents of the bicyclic ring system are in cis positions and this gives rise to steric hindrance which causes the angles of the fused ring system to widen. The angles which deviate most from the tetrahedral value are C(2)-C(1)-C(10) (117.7°) and C(1)-C(8)-C(7) (114.8°). Each water molecule forms three different hydrogen bonds with thee diterpene molecules.

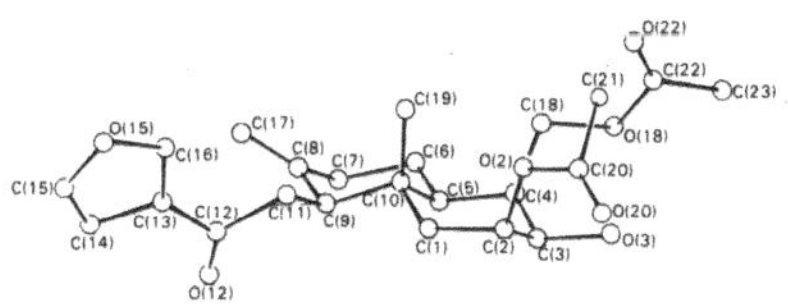

Fig. 1. A view of the $C_{23}H_{30}O_7$ molecule.

19-(p-BROMOPHENYL)-12-METHOXYPODOCARPA-8,11,13-TRIEN-19-ONE (p-BROMOPHENYL-1,2,3,4,4a,-9,10,10a-OCTAHYDRO-6-METHOXY-1,4a-DIMETHYL-1-PHENANTHRYL KETONE)
$C_{24}H_{27}BrO_2$

M.C. COULDWELL, R.A.J. SMITH and J. SIMPSON, 1985. *Acta Cryst.*, C41, 983-985.

Monoclinic, $P2_1$, a = 7.830, b = 19.049, c = 7.361 Å, β = 104.34°, Z = 2. Mo radiation, R = 0.0451 for 1660 reflections.

The structure of the diterpene (Fig. 1) shows ring A adopts a chair conformation and ring B a half-chair conformation. The aryl ketone function is 4β-axial with respect to ring A. The aromatic ring, C, is planar with the methoxy substituent at C(12) displaced from the β-face of the molecule. Carbonyl oxygen, O(1), is twisted out of the plane of the C(21)-C(26) ring, dihedral angle 41.0° (O(1),C(19),C(21)/C(19),C(21),C(22)).

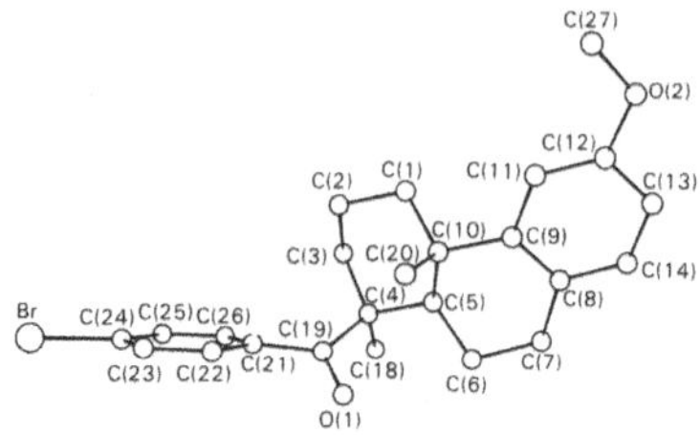

Fig. 1. The $C_{24}H_{27}BrO_2$ molecule.

7α,17β-DIHYDROXY-15,17-OXIDOSPONGIAN-16-ONE-7-BUTYRATE
$C_{24}H_{36}O_6$

F.J. SCHMITZ, J.S. CHANG, M.B. HOSSAIN and D. VAN DER HELM, 1985. *J. Org. Chem.*, **50**, 2862-2865.

Tetragonal, $P4_1$, a = 10.782, c = 18.531 Å, Z = 4. Mo radiation, R = 0.032 for 2152 reflections (at 138 K).

The crystal structure analysis established the stereochemistry of this natural product (Fig. 1). The three cyclohexane rings are all in the chair conformation; the tetrahydrofuran ring has an envelope conformation with C(17) as the flap; the γ-lactone ring is in a folded envelope conformation with C(16) as the flap.

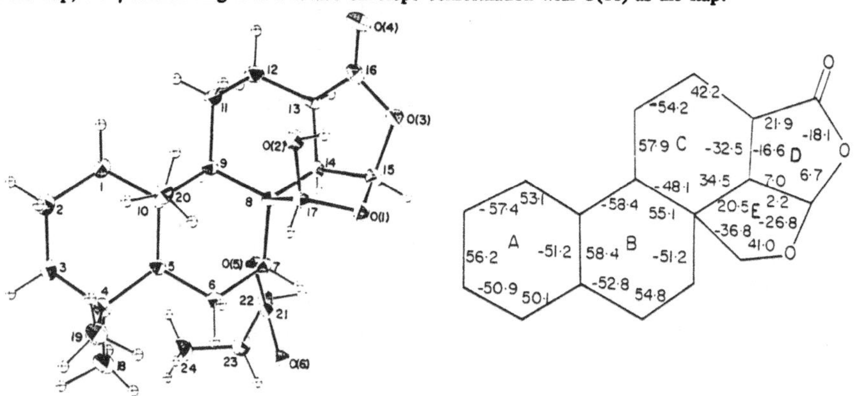

Fig. 1. The $C_{24}H_{36}O_6$ molecule and endocyclic torsion angles.

(+)-8α-ACETOXY-12-(4-BROMOBENZOYLOXY)-13,14,15,16-TETRANORLABDANE ((+)-12-(4-BROMOBENZOYLOXY)-13,14,15,16-TETRANORLABDANE-8α-YL ACETATE)
$C_{25}H_{35}BrO_4$ (I)

(-)-8α,12-DIHYDROXY-13,14,15,16-TETRANOR-9-EPILABDANE ((-)-13,14,15,16-TETRANOR-9β-LABDANE-8α,12-DIOL)
$C_{16}H_{30}O_2$ (II)

G. BERNARDINELLI and W. GIERSCH, 1985. *Acta Cryst.*, C41, 746-749.

I. Orthorhombic, $P2_12_12_1$, a = 6.1383, b = 17.093, c = 22.912 Å, Z = 4. Mo radiation, R = 0.056 for 1828 reflections.

II. Orthorhombic, $P2_12_12_1$, a = 7.235, b = 11.931, c = 17.064 Å, Z = 4. Mo radiation, R = 0.037 for 624 reflections.

In both structures (Fig. 1) rings A and B are in the chair conformation and are trans-fused. The substituents at C(8) and the side-chain at C(9) are attached in equatorial positions for (I) and axial positions for (II). In (I) the torsion angles C(9)-C(18)-C(19)-O(20) and C(18)-C(19)-O(20)-C(21) are 69 and 173° respectively. Bond lengths and angles in both compounds agree with accepted values. In (II) there are two intermolecular hydrogen bonds, O(1)...O(2), of lengths 2.754(6) and 2.879(6) Å while in (I) intermolecular distances are essentially van der Waals contacts.

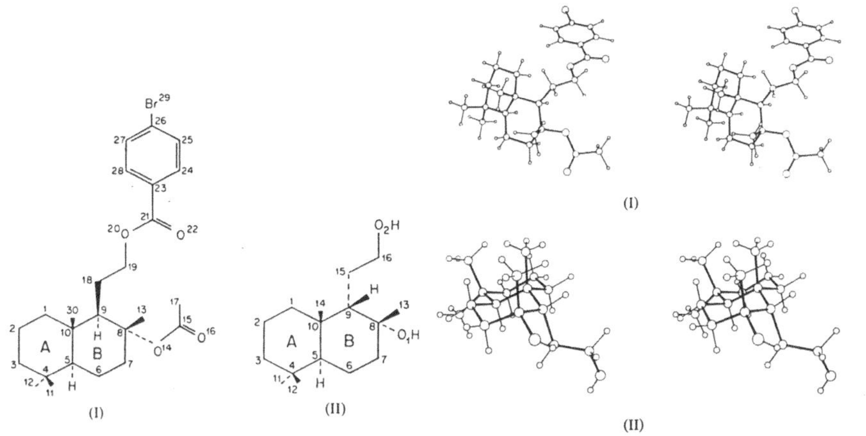

Fig. 1. Structural formulae of $C_{25}H_{35}BrO_4$ (I) and $C_{16}H_{30}O_2$ (II) and stereoviews showing the molecular configurations.

(13S,18R)-5,18:13,18-DIEPOXYSERRULAT-14-EN-8-OL p-BROMOBENZOATE
$C_{27}H_{29}BrO_4$

A.D. ABELL, E. HORN, G.P. JONES, M.R. SNOW, R.A. MASSY-WESTROPP and R. RICCIO, 1985. *Aust. J. Chem.*, **38**, 1837-1845.

Monoclinic, $P2_1$, a = 9.112, b = 9.743, c = 13.423 Å, β = 102.27°, Z = 2. Mo radiation, R = 0.038 for 1908 reflections.

The structure of this p-bromobenzoate derivative of a diterpenoid acetal isolated from 'Eremophilia rotundifolia' is shown in Fig. 1. The molecular dimensions are normal.

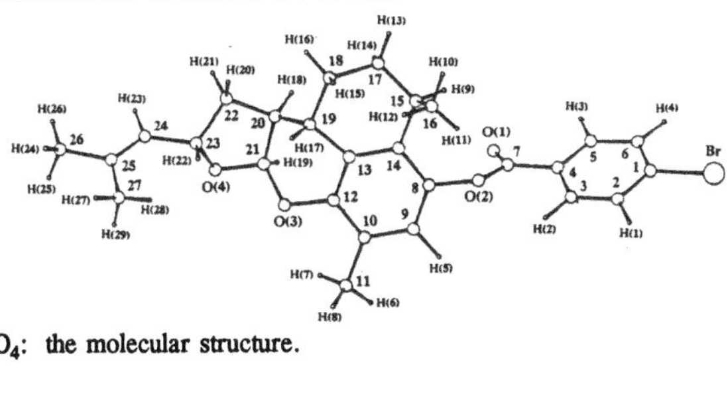

Fig. 1. $C_{27}H_{29}BrO_4$: the molecular structure.

ent-2-BENZYLIDENE-16α-HYDROXYATIS-13-EN-3-ONE HEMIHYDRATE
$C_{27}H_{34}O_2$,0.5(H_2O)

L.P.L. PIACENZA, K.H. PEGEL, M. LAING, E.S. WAIGHT, C.M. WEEKS and C.P. GORST-ALLMAN, 1985. *J. Chem. Soc. Perkin Trans. 1*, 703-709.

Trigonal, $P3_221$, a = 10.844, c = 66.653 Å, Z = 12. Cu radiation, R = 0.083 for 3996 reflections.

The water molecules are on twofold axes in the crystal and seem to be disordered. The crystal structure analysis established the structure of the molecule (Fig. 1); the asymmetric unit contains two independent molecules of the diterpene and one water molecule. There are three independent intermolecular hydrogen bonds in the crystal: O(3)...O(16) 2.95, O(3)...O(16') 2.89, O(16')...O(W) 2.91 Å.

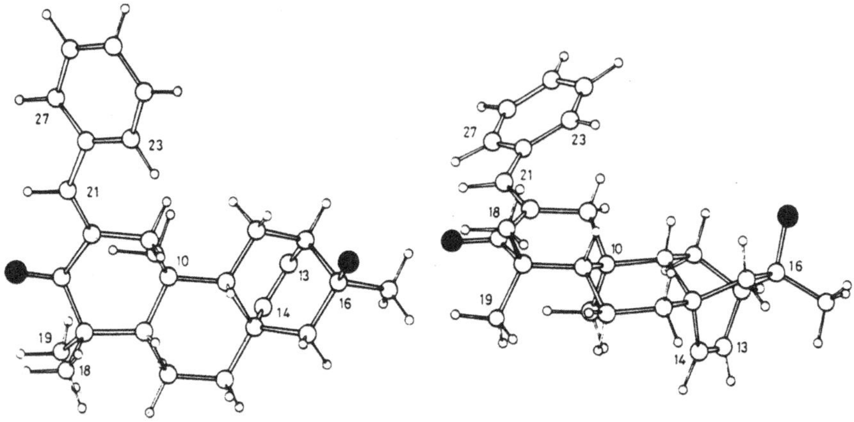

Fig. 1. ORTEP drawings of a molecule of $C_{27}H_{34}O_2$.

(E)-(-)-(2R,3R,4S,5R,7S,8R,13S,15R)-3,15-DIACETOXY-5,7-DIBENZOYLOXY-2,8-DIHYDROXYJATROPHA-6-(17),11-DIENE-9,14-DIONE (ESULONE A)
$C_{38}H_{42}O_{12}$

G.D. MANNERS and R.Y. WONG, 1985. *J. Chem. Soc. Perkin Trans. 1*, 2075-2081.

Orthorhombic, $P2_12_12_1$, a = 8.847, b = 19.496, c = 21.445 Å, Z = 4. Cu radiation, R = 0.090 for 2203 reflections.

There is evidence of disorder in the crystal of atoms C(26-30) of the phenyl group (Fig. 1). There is a short intermolecular distance between O(16)...O(20) (2.85 Å). The analysis confirmed the structure of esulone A and established that the C(4)-C(15) ring junction is trans.

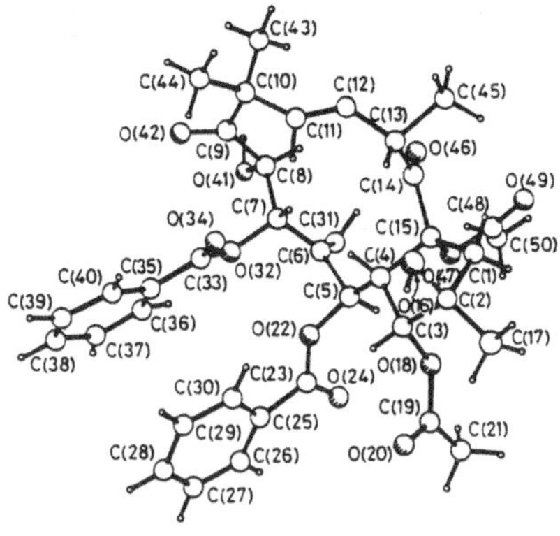

Fig. 1. Perspective view of a molecule of $C_{38}H_{42}O_{12}$.

EUPHOSCOPIN A p-BROMOBENZOATE
$C_{38}H_{43}BrO_9$ (I)

EUPHORNIN ACETAL p-BROMOBENZOATE

$C_{37}H_{45}BrO_7$ (II)

S. OHBA, M. ITO, Y. SAITO, Y. SHIZURI, S. KOSEMURA, J. OHTSUKA and S. YAMAMURA, 1985. *Acta Cryst.*, C41, 487-490.

I. Orthorhombic, $P2_12_12_1$, a = 16.271, b = 23.577, c = 9.786 Å, Z = 4. Cu radiation, R = 0.046 for 2948 reflections.

II. Monoclinic, $P2_1$, a = 17.104, b = 13.837, c = 7.611 Å, β = 99.14°, Z = 2. Mo radiation, R = 0.033 for 1793 reflections.

Molecules (I) (Fig. 1) and (II) (Fig. 2), which consist of five- and twelve-membered rings, have the jatrophone skeleton in which the configuration at the ring junction C(4)-C(15) is trans. The double bonds in the macroring adopt a trans conformation. There are seven asymmetric carbon atoms of which three, C(2), C(13) and C(14), in (I) have different absolute configurations from the corresponding ones in (II). The torsion angle C(9)-C(10)-C(11)-C(12) is -47.3(7)° in (I) and 132.4(5)° in (II), due to this difference in absolute configuration at C(13). Bond lengths and angles in the molecules have normal values. There are no unusually short intermolecular distances and the packing is governed by van der Waals forces.

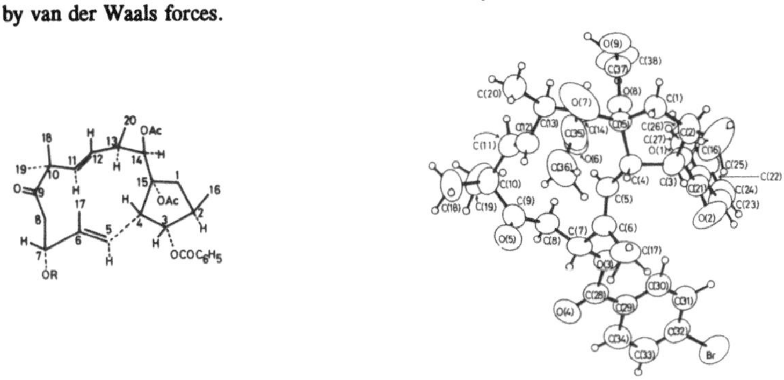

Fig. 1. A view with atomic numbering in $C_{38}H_{43}BrO_9$.

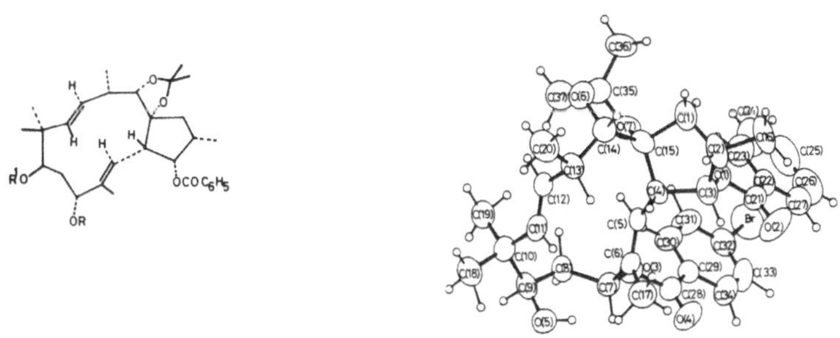

Fig. 2. A view with the numbering scheme of $C_{37}H_{45}BrO_7$.

DUKUNOLIDE A
$C_{26}H_{26}O_9$ (I)

DUKUNOLIDE C p-BROMOBENZOATE MONOHYDRATE
$C_{33}H_{29}BrO_{11},H_2O$ (II)

M. NISHIZAWA, Y. NADEMOTO, S. SASTRAPRADJA, M. SHIRO and Y. HAYASHI, 1985. *J. Org. Chem.*, 50, 5487-5490.

I. Orthorhombic, $P2_12_12_1$, a = 12.404, b = 21.483, c = 8.747 Å, Z = 4. Cu radiation, R = 0.056 for 1949

reflections.

II. Orthorhombic, $P2_12_12$, a = 8.963, b = 38.092, c = 8.817 Å, Z = 4. Cu radiation, R = 0.035 for 2631 reflections.

The absolute configuration of (II) was determined by anomalous dispersion. In (I) the ring junctions at C(1), C(2) and C(5), C(10) are cis. The γ-lactone ring and the mean plane of C(1,2,22,8,14,15) face each other with a dihedral angle of 43°. The molecules (I) and (II) are shown in Fig. 1.

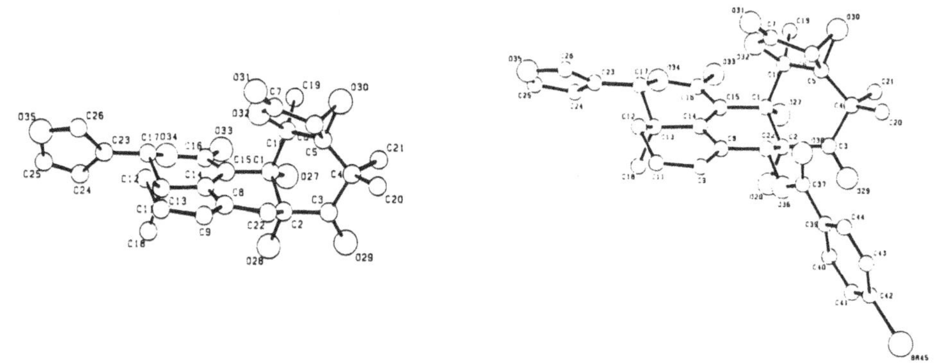

Fig. 1. ORTEP drawings of $C_{26}H_{26}O_9$ (I) (left) and $C_{33}H_{29}BrO_{11},H_2O$ (II) (right).

METHYL 24-NOR-3α,11α-DIHYDROXYLUP-20(29)-EN-28-OATE
$C_{30}H_{48}O_4$

L. KUTSCHABSKY, D. PFEIFFER, M. LISCHEWSKI, P.D. TY and G. ADAM, 1985. *Croat. Chem. Acta*, **58**, 427-434.

Orthorhombic, $P2_12_12_1$, a = 8.437, b = 22.340, c = 14.096 Å, Z = 4. Mo radiation, R = 0.072 for 1943 reflections.

The analysis establishes the structure and stereochemistry as shown in Fig. 1. The six-membered rings are in chair conformation and the five-membered ring has a 17β-envelope conformation. Molecules are linked by O-H...O hydrogen bonds O(3)...O(281') 2.863(7) Å.

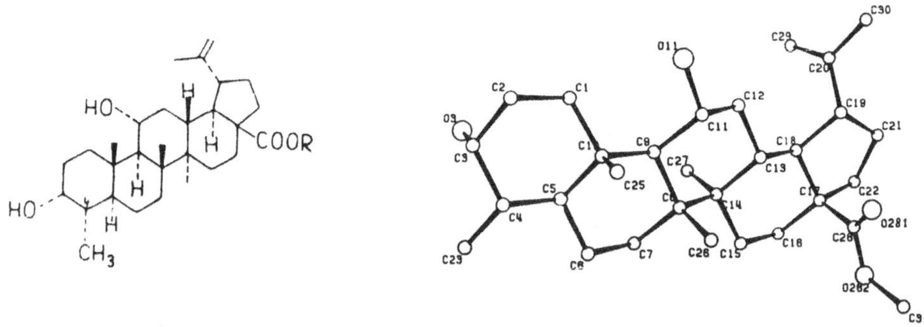

Fig. 1. The $C_{30}H_{48}O_4$ molecule (R=CH_3) and the molecular conformation.

TARAXASTEROL
$C_{30}H_{50}O,C_2H_6O$

W.F. REYNOLDS, J.F. SAWYER, R.G. ENRIQUEZ, L.I. ESCOBAR, M.A. CHAVEZ and J.N. SHOOLERY, 1985. *Canad. J. Chem.*, **63**, 1048-1054.

Orthorhombic, $P2_12_12_1$, a = 7.447, b = 17.637, c = 22.269 Å, Z = 4. Cu radiation, R = 0.0773 for 1941 reflections.

The analysis shows that the six-membered rings A-D are in flattened chair conformations and ring E is in a distorted boat conformation.

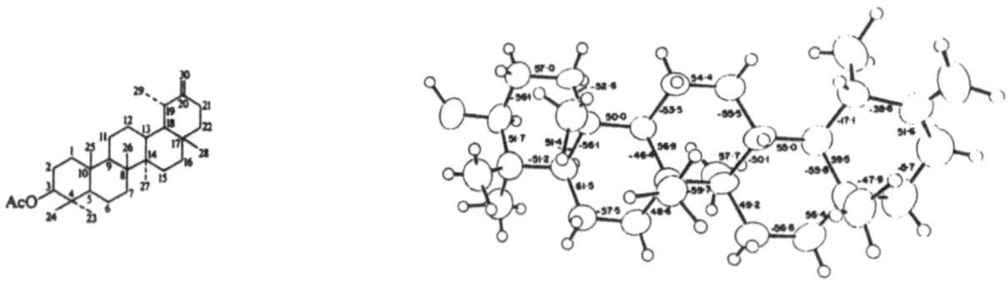

Fig. 1. The taraxasterol molecule and ring torsion angles.

(24S)-11α-ACETOXY-24,25-EPOXYDAMMAR-20(22)-EN-23-ONE-3,19-HEMIACETAL (CORDIALIN A ACETATE)
$C_{32}H_{48}O_6$

K.D. ONAN, D. LAVIE, V.V. VELDE and R. ZELNIK, 1985. *J. Chem. Soc. Perkin Trans. 1*, 323-326.

Orthorhombic, $P2_12_12_1$, a = 6.431, b = 20.530, c = 21.777 Å, Z = 4. Mo radiation, R = 0.074 for 1284 reflections.

Ring A has a distorted boat conformation; ring B is a flattened chair; ring C is a puckered chair; ring D is a distorted envelope (Fig. 1). The configuration at C(24) is S; much of the C(17) β-oriented side chain lies in a plane. There are hydrogen bonds in the crystal: O(31)...O(37)' 2.80° Å and C(3)-O(31)...O(37)' 110.9°.

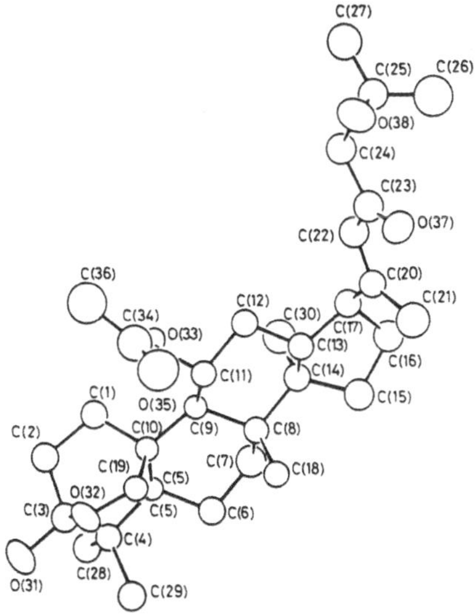

Fig. 1. Perspective view of a molecule of $C_{32}H_{48}O_6$.

5α,10α-EPOXYALNUSAN-3β-YL ACETATE
$C_{32}H_{52}O_3$

M. TAKAI, M. TORI, T. TSUYUKI, T. TAKAHASHI, A. ITAI and Y. IITAKA, 1985. *Bull. Chem. Soc. Jpn.*,

58, 185-191.

Monoclinic, *C*2, a = 13.446, b = 6.739, c = 32.257 Å, β = 103.71°, Z = 4. Cu radiation, R = 0.056 for 2951 reflections.

The structure and stereochemistry were determined as shown in Fig. 1.

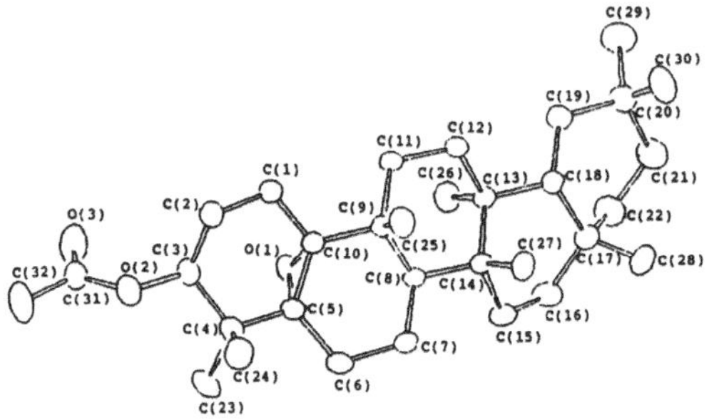

Fig. 1. A view of $C_{32}H_{52}O_3$.

(20R)-3β-ACETOXYLUPAN-29-OL
$C_{32}H_{54}O_3$

R.E. CORBETT, A.N.T. CONG, A.L. WILKINS and R.A. THOMSON, 1985. *J. Chem. Soc. Perkin Trans. 1*, - 2051-2056.

Orthorhombic, $P2_12_12_1$, a = 15.895, b = 22.14, c = 8.298 Å, Z = 4. Mo radiation, R = 0.064 for 1261 reflections.

The crystal structure analysis established the (20R) configuration (Fig. 1) with C(20) hydroxymethylene and methyl groups oriented axially and equatorially respectively with respect to the pentacyclic skeleton.

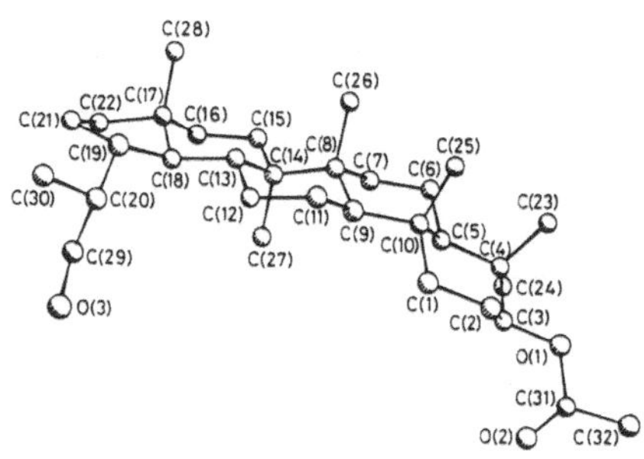

Fig. 1. Perspective drawing of a molecule of $C_{32}H_{54}O_3$.

(21R,24R)-21,24:21,25-DIEPOXY-2α,3β-ISOPROPYLIDENEDIOXYLANOST-8-EN-12α-OL (2α,3β-O-ISOPROPYLIDENEANHYDROCRUSTINOL)
$C_{33}H_{52}O_5$

G. GATTI, M.P. GIANOTTI and G. VIDARI, 1985. *Acta Cryst.*, C41, 1340-1342.

Triclinic, *P*1, a = 17.237, b = 7.287, c = 6.275 Å, α = 109.52, β = 89.92, γ = 99.15°, Z = 1. Mo radiation, R = 0.081 for 1609 reflections.

The five-, six- and seven-membered rings which constitute the bicyclo[3.2.1]octane system in the molecule (Fig. 1) adopt chair, half-chair and boat conformations, respectively. Ring A has a chair conformation flattened to reduce the severe 1,3-diaxial methyl-methyl interaction. In ring B, which has a half-chair conformation, C(5) and C(6) are displaced by -0.61(2) and 0.15(2) Å out of the C(7,8,9,10) plane. Ring C adopts a 1,2-diplanar conformation with C(12) and C(13) displaced by 0.33(2) and 0.85(2) Å above the C(8,9,11,14) plane. The cyclopentane ring D has a conformation closer to a half-chair than to the envelope form. Bond lengths and angles in the strained molecule are in general agreement with expected values.

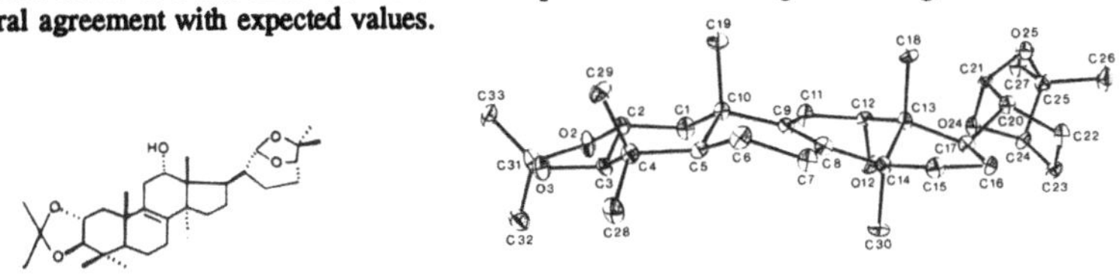

Fig. 1. $C_{33}H_{52}O_5$: structural formula and a view of the molecule.

6-METHYL-4H-(29'-OXO-12',29'-CYCLOOLEAN-2',12'-DIENO)[3,2-b]PYRAN-4-ONE
$C_{34}H_{46}O_3$

K.L. GAUL, K.G. LEWIS, B.W. SKELTON and A.H. WHITE, 1985. *Aust. J. Chem.*, **38**, 1371-1380.

Monoclinic, $P2_1$, a = 14.857, b = 11.263, c = 8.378 Å, β = 93.98°, Z = 2. Mo radiation, R = 0.055 for 1444 reflections.

The molecule (Fig. 1) is a strained enone in which dihedral angles about the C(12)-C(13) double bond are 23.2 (C(11,12,13,14)) and 25.7° (C(29,12,13,18)). The C(12)-C(13) distance is 1.33(1) Å. Other molecular dimensions are as expected.

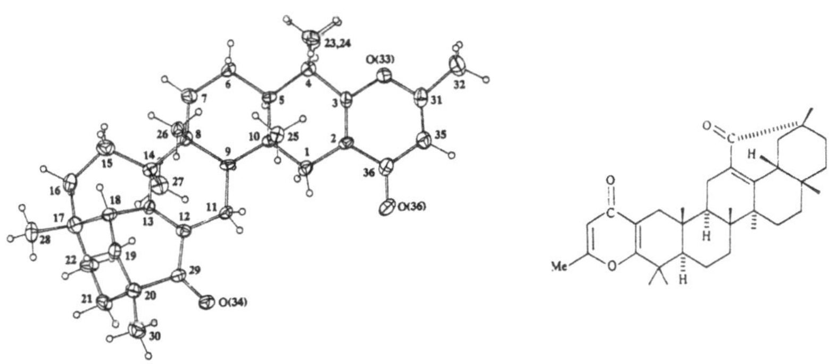

Fig. 1. $C_{34}H_{46}O_3$: the molecular structure and skeleton.

1β-HYDROXY-2,3-DIHYDRO-1H-PYRROLIZINE-7-METHANOL (DEHYDRORETRONECINE)
$C_8H_{11}NO_2$

M.F. MACKAY, P. MITRPRACHACHON and C.C.J. CULVENOR, 1985. *Acta Cryst.*, **C41**, 1104-1106.

Orthorhombic, $P2_12_12_1$, a = 7.370, b = 8.148, c = 12.847 Å, Z = 4. Cu radiation, R = 0.045 for 685 reflections.

In the molecule (Fig. 1) the pyrrolizidine nucleus is planar except for C(6) which is displaced from the plane of the other atoms to the side opposite from the C(7) hydroxyl O atom. The torsion angle C(2)-C(1)-C(9)-O(10) is 47.4(4)°. In the pyrrole system the bond lengths in the ring N(4), C(5,6,7,8) indicate extensive electron delocalization. The exocyclic angles C(3)-N(4)-C(5) and C(1)-C(8)-C(7) of values 137.2(3) and 142.1(3)° respectively, are considerably distorted from the standard trigonal value. In the crystal, hydrogen bonds involve only the hydroxyl substituents and are of length 2.739(4) Å and link the molecules into sheets perpendicular to the *c*-axis.

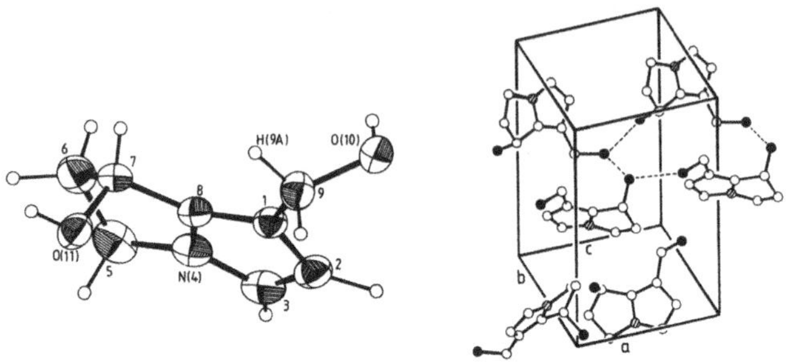

Fig. 1. $C_8H_{11}NO_2$: views of the molecule and crystal packing.

CROTANECINE
$C_8H_{13}NO_3$

J.F. RICHARDSON and C.C.J. CULVENOR, 1985. *Acta Cryst.*, C41, 1475-1477.

Orthorhombic, $P2_12_12_1$, a = 10.9526, b = 12.4628, c = 6.0816 Å, Z = 4. Mo radiation, R = 0.027 for 861 reflections.

The A ring of the molecule (Fig. 1) is exo-buckled with a pucker angle of 40.8(2)° and there is an angle of 125.2(2)° between the mean planes defined by ring B atoms and atoms C(5), N(4), C(8) and C(7) of ring A. Each molecule is linked to three others by intermolecular hydrogen bonds (O(10)...N(4) 2.72, O(11)...O(10) 2.74 and O(12)...O(11) 2.72 Å).

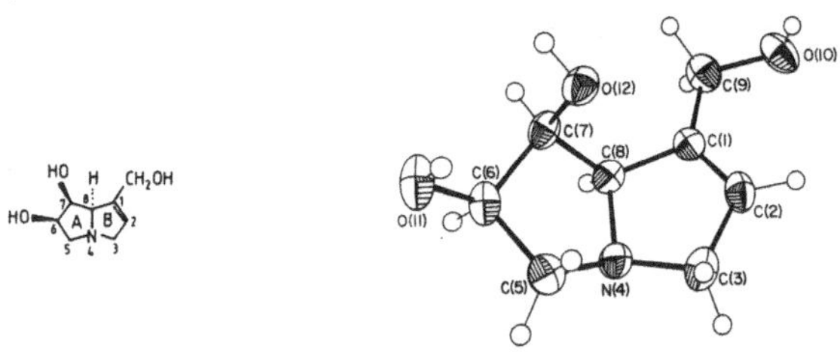

Fig. 1. $C_8H_{13}NO_3$: the molecular skeleton and structure.

β-EPOXYHELIOTRIDINE
$C_8H_{13}NO_3$ (I)

α-EPOXYHELIOTRIDINE
$C_8H_{13}NO_3$ (II)

α-EPOXYRETRONECINE
$C_8H_{13}NO_3$ (III)

J.A. GLINSKI, D. VANDERVEER and L.H. ZALKOW, 1985. *Acta Cryst.*, C41, 1345-1348.

I. Orthorhombic, $P2_12_12_1$, a = 6.2951, b = 7.7019, c = 17.722 Å, Z = 4. Mo radiation, R = 0.037 for 1389 reflections.

II. Monoclinic, $P2_1$, a = 6.251, b = 8.578, c = 8.069 Å, β = 110.68°, Z = 2. Mo radiation, R = 0.038 for 1299 reflections.

III. Monoclinic, $P2_1$, a = 6.509, b = 8.340, c = 7.799 Å, β = 105.85°, Z = 2. Mo radiation, R = 0.032 for 1395 reflections.

In all three molecules (Fig. 1), ring A consisting of atoms N(4),C(8,1,2,3) and fused to the epoxide ring has an envelope conformation. Atom N(4) is the flap and is displaced by 0.231, 0.373 and 0.032 Å from the planes of the other four atoms in (I), (II) and (III) respectively. The dihedral angle between this plane and that of the oxirane ring is 105.2, 106.1 and 105.0° in (I), (II) and (III). Ring B, consisting of atoms N(4),C(8,7,6,5), has slightly different conformations in the different isomers. They are all envelopes with C(5), C(7) and C(5) as their respective flaps. Bond lengths and angles in the molecules agree well with expected values. In the crystals there are stabilizing intermolecular hydrogen bonds N(4)...H-O(2) of lengths 2.70, 2.73 and 2.77 Å in (I), (II) and (III) respectively.

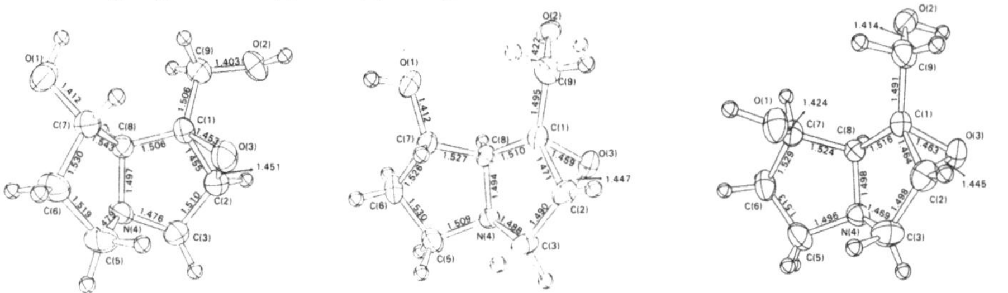

Fig. 1. $C_8H_{13}NO_3$: views of the molecule and bond lengths in (I) (left), in (II) (centre) and in (III) (right).

RETRONECINE
$C_8H_{13}NO_2$ (I)

HELIOTRIDINE
$C_8H_{13}NO_2$ (II)

L.T. GELBAUM, J.A. GLINSKI, D. VANDERVEER and L.H. ZALKOW, 1985. *Acta Cryst.*, C41, 1342-1345.

I. Orthorhombic, $P2_12_12_1$, a = 7.944, b = 8.536, c = 12.062 Å, Z = 4. Mo radiation, R = 0.040 for 825 reflections.

II. Orthorhombic, $P2_12_12_1$, a = 11.904, b = 7.620, c = 8.800 Å, Z = 4. Mo radiation, R = 0.038 for 1328 reflections.

In the fused ring systems of the diastereoisomeric molecules (Fig. 1) the dihedral angles between the planes C(5,6,7) and C(5),N(4),C(8) in (I) and in (II) are 40.7 and 40.2° respectively. In (I) the ring system is in an exo-puckered form while in (II) it has an endo-puckered conformation. The dihedral angles between the least-squares planes defined by atoms C(1,2,3),N(5),C(8) and C(5),N(4),C(8,7) are 124.4 and 121.6° for (I) and (II) respectively. In both (I) and (II) atoms O(1) and O(2) are in antiparallel conformations. Bond lengths and angles in the molecules are normal. Both crystal structures are stabilized by intermolecular hydrogen bonds N(4)...H-O(2) of lengths 2.69 and 2.72 Å in (I) and (II) respectively.

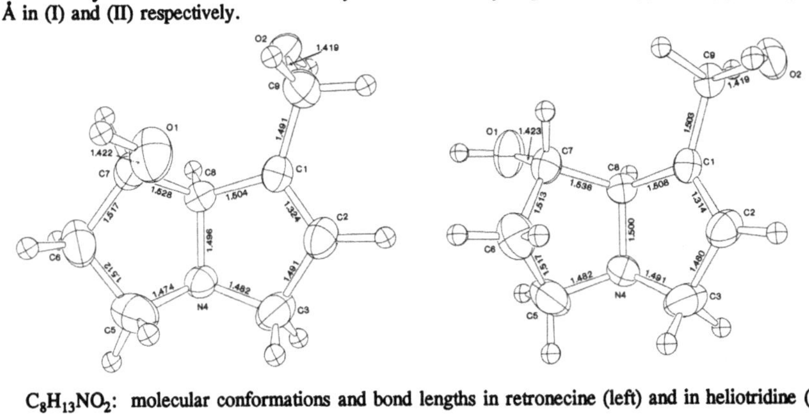

Fig. 1. $C_8H_{13}NO_2$: molecular conformations and bond lengths in retronecine (left) and in heliotridine (right).

(+)-7-METHYL-2,3,5,7a-TETRAHYDRO-1H-PYRROLIZINE-1,8-DIYL SUCCINATE

$C_{12}H_{15}NO_4$

M. BURTON, A.A. FREER and D.J. ROBINS, 1985. *Acta Cryst.*, C41, 944-946.

Orthorhombic, $P2_12_12_1$, a = 7.934, b = 10.786, c = 13.246 Å, Z = 4. Mo radiation, R = 0.043 for 1429 reflections.

The carbonyls of the ester groups of this ten-membered macrocyclic pyrrolizidine alkaloid analogue (Fig. 1) are nearly antiparallel on opposite sides of the ring system. The dihedral angle between the ester groups defined by the planes C(9),O(10),C(11),O(11) and O(14),C(14),O(15),C(7) is 159(5)°. The torsion angle C(2)-C(1)-C(9)-C(10) is -116.2(3)°. The shortest transannular distances are O(10)...O(15) 2.847(2) and O(10)...O(14) 2.867(3) Å. The degree of exo-puckering of the pyrrolizidine system as defined by the planes of atoms C(5),C(6),C(7) and C(5),N(4),C(8) is 43.7(4)°. The angle between the least-squares planes defined by atoms C(1),C(2),C(3),N(4),C(8) and C(5),N(4),C(8),C(7) of the pyrrolizidine nucleus is 126.8(3)°. The N(4)-C(8) ring fusion distance is 1.492(3) Å.

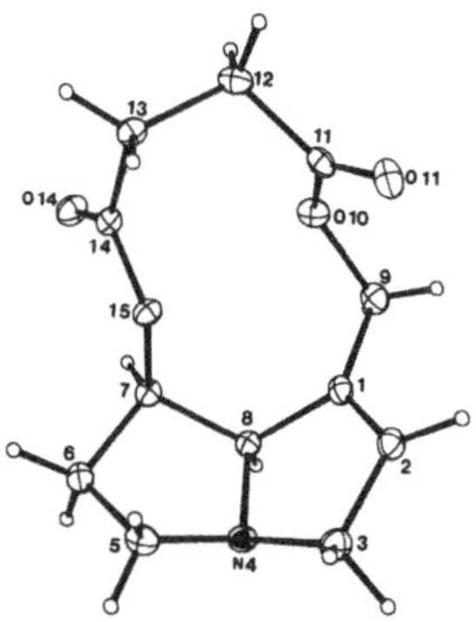

Fig. 1. $C_{12}H_{15}NO_4$: the macrocyclic alkaloid analogue.

(+)-(7R,8R)-1-HYDROXYMETHYL-5,6,7,8-TETRAHYDRO-3H-PYRROLIZINYL(Z)-2-METHYLBUT-2-ENOATE (O-7-ANGELYL RETRONECINE)
$C_{13}H_{19}NO_3$

H. WIEDENFELD, A. KIRFEL, E. RODER and G. WILL, 1985. *Arch. Pharm.*, 318, 294-299.

Orthorhombic, $P2_12_12_1$, a = 6.436, b = 8.653, c = 23.528 Å, Z = 4. Mo radiation, R = 0.057 for 814 reflections.

The angle between the two four-membered units C(1),C(3),N,C(8) and N,C(5),C(7),C(8) is 126.6°. The angles of 176° and 172° between the planes C(1),C(2),C(8),N and C(1),C(2),C(3), and C(3),C(8),N and C(1),C(2),C(8) respectively show that the five-membered ring, with the double bond between C(1) and C(2), is not planar (Fig. 1). Bond distances of interest are: C(1)=C(2) 1.301(7), C(3)-N 1.452(8), C(5)-N 1.480(8), C(8)-N 1.488(7), C(7)-O(2) 1.447(7), O(2)-C(10) 1.342(8), C(10)=O(3) 1.204(9), C(11)=C(12) 1.03(2) Å.

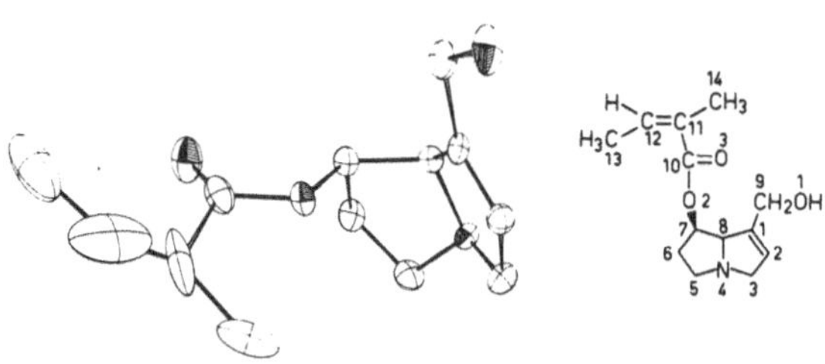

Fig. 1. Molecular structure of $C_{13}H_{19}NO_3$ with numbering scheme.

IPOHARDINE CHLORIDE
$C_{15}H_{16}ClNO$ $C_{15}H_{16}NO^{+},Cl^{-}$

YU ZHAOLOU, ZHU NAIJUE, LU RENRONG, CAI SHAOPEI, LAO XIAFEI and LIU ZHUJIN, 1985. *Jiegou Huaxue (J. Struct. Chem.)*, **4**, 156-159.

Orthorhombic, $Pc2_1b$, a = 7.308, b = 18.407, c = 19.465 Å, Z = 8. Mo radiation, R = 0.0483 for 993 reflections.

The two independent formula units are shown in Fig. 1. The dihedral angles between the six-membered rings are 38.6 and 40.3°. Molecular dimensions are normal.

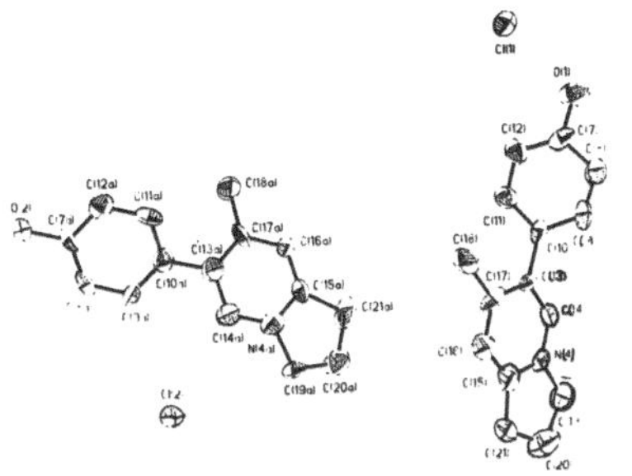

Fig. 1. The asymmetric unit for $C_{15}H_{16}ClNO$.

3,8-DIMETHYL-2-AZONIABICYCLO[2.2.2]OCT-2-EN-5-ONE PICRATE (MEARSINIUM PICRATE)
$C_{15}H_{16}N_4O_8$ $C_9H_{14}NO^{+},C_6H_2N_3O_7^{-}$

G.B. ROBERTSON and U. TOOPTAKONG, 1985. *Acta Cryst.*, **C41**, 1332-1335.

Orthorhombic, $P2_12_12_1$, a = 23.461, b = 10.8541, c = 6.6458 Å, Z = 4. Cu radiation, R = 0.037 for 2412 reflections.

The crystal structure is composed of discrete hydrogen-bonded ion pairs (Fig. 1). The mearsine cation, $[C_9H_{14}NO^{+}]$, is bicyclic with an isoquinuclidine skeleton. The half-rings C(7,8,9,10) and C(7),N(7),C(11,10) are planar and the half-ring C(10,12,13,7) is slightly nonplanar. The N(7)-H...O(1) inter-ion hydrogen bond is strong and of length 2.686(2) Å. The aromatic nucleus of the picrate anion is slightly nonplanar and the ketonic oxygen atom O(1) is displaced from the least-squares plane of the ring by 0.126(2) Å. The p-nitro group and the o-nitro groups at N(2) and N(6) are twisted out of this plane by 8.2(1), 22.5(1) and 41.3(1)° respectively. The picrate anions stack face-to-face along the *c*-axis with an interplane spacing of approximately c/2.

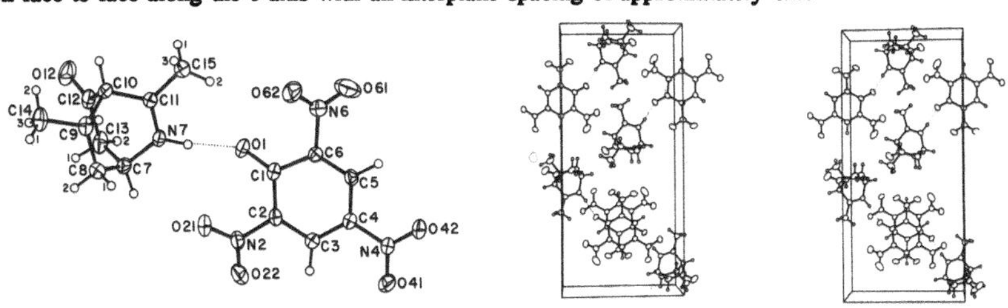

Fig. 1. $C_{15}H_{16}N_4O_8$: atomic numbering in the hydrogen-bonded ion pair and a stereoview down the *c*-axis of the crystal packing arrangement.

SUPININE
$C_{15}H_{25}NO_4$ (I)

HELEURINE

$C_{16}H_{27}NO_4$ (II)

M.F. MACKAY, P. MITRPRACHACHON, P.J. OLIVER and C.C.J. CULVENOR, 1985. *Acta Cryst.*, **C41**, 722-725.

I. Orthorhombic, $P2_12_12_1$, a = 10.068, b = 11.644, c = 13.086 Å, Z = 4. Cu radiation, R = 0.041 for 1411 reflections.

II. Orthorhombic, $P2_12_12_1$, a = 9.269, b = 11.175, c = 16.345 Å, Z = 4. Cu radiation, R = 0.048 for 1367 reflections.

Both alkaloid molecules adopt similar extended conformations (Fig. 1) apart from the pyrrolizidine nucleus which is endo-buckled in (I) and exo-buckled in (II). The pucker angles for (I) and (II) are 30.1(3) and 44.7(5)° respectively. The angles between the least-squares planes of the two five-membered rings are 121.9(4) and 117.8(5)° for (I) and (II) respectively. In (I), atoms C(1,9,11,12),O(10,18,19) are nearly coplanar while in (II) C(9,11,12),O(10,18) are coplanar and O(19) is displaced by 0.12(1) Å from this plane. Bond lengths and angles in the molecules are normal. In (I) each molecule is linked to four others by a network of O-H...N hydrogen bonds of lengths 2.758(3) and 2.801(3) Å. In (II) there is only one such interaction: O(19)-N...N(4) of length 2.812(5) Å.

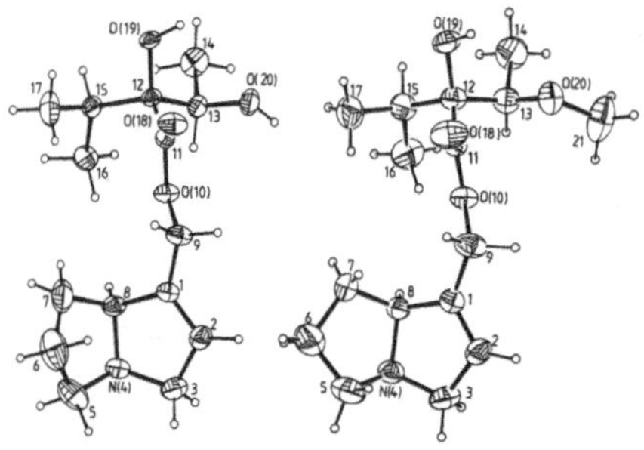

Fig. 1. Views of the molecules of (a) supinine and (b) heleurine.

N-PYRUVOYL-cyclo-(L-PHENYLALANYL-D-PROLYL) (PYROERGOTAMINE)
$C_{17}H_{18}N_2O_4$

R.O. DAY, V.W. DAY, D.M.S. WHEELER, P.A. STADLER and H.-R. LOOSLI, 1985. *Helv. Chim. Acta*, **68**, 724-733.

Monoclinic, $P2_1$, a = 6.197, b = 14.458, c = 9.025 Å, β = 106.24°, Z = 2. Mo radiation, R = 0.046 for 2193 reflections.

Each of the two peptide linkages in the diketopiperazine (Fig. 1) which exists in a folded or boat conformation (dihedral angle 42°) is nearly planar. The R-group of the phenylalanine residue occupies a quasi-axial position of the diketopiperazine ring, whilst that of the proline residue is in a quasi-equatorial position. The pyrrolidine ring has an envelope conformation.

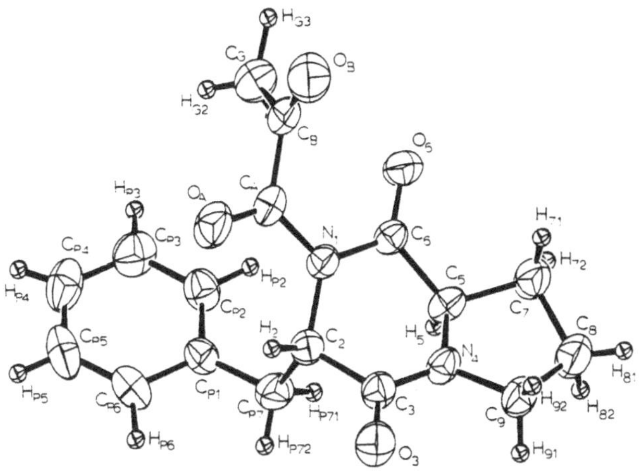

Fig. 1. ORTEP diagram of a molecule of $C_{17}H_{18}N_2O_4$.

3,4-BENZO-2,13-ETHANO-5,13-DIAZATRICYCLO[7.4.0.0^{2,6}]TRIDECAN-8-ONE (ASPIDOSPERMA U-43)

$C_{17}H_{20}N_2O$

M. NATSUME, I. UTSUNOMIYA, K. YAMAGUCHI and S.-I. SAKAI, 1985. *Tetrahedron,* **41**, 2115-2123.

Triclinic, $P\bar{1}$, a = 11.129, b = 14.624, c = 10.465 Å, α = 101.67, β = 84.67, γ = 59.73°, Z = 4. Mo radiation, R = 0.058 for 4442 reflections.

The analysis establishes the structure and stereochemistry to be as shown in Fig. 1.

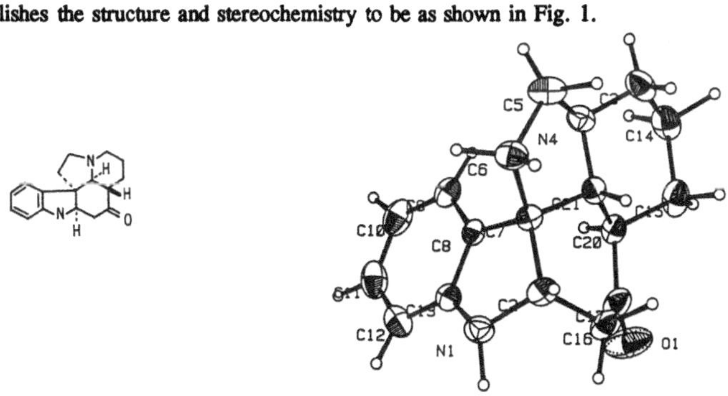

Fig. 1. The structure and stereochemistry of $C_{17}H_{20}N_2O$.

17-METHYL-4,5a-EPOXY-3,14-DIHYDROXY-6-AMINOMORPHINAN DIHYDROCHLORIDE MONOHYDRATE

$C_{17}H_{24}Cl_2N_2O_3,H_2O$ $C_{17}H_{24}N_2O_3^{2+},2(Cl^-),H_2O$

O.W. LEVER Jr., A.V. BHATIA and KWEN-JEN CHANG, 1985. *J. Med. Chem.*, **28**, 1652-1656.

Monoclinic, $P2_1$, a = 7.895, b = 11.008, c = 10.501 Å, β = 95.50°, Z = 2. Cu radiation, R = [not quoted] for 1961 reflections.

The cation is shown to have a twist-boat conformation for ring C which corresponds to the conformation observed in solution. The molecule has the characteristic T-shape of the classical opiates (Fig. 1). The piperidine ring D has a well defined chair form; the twist-boat conformation of ring C places the 7α-hydrogen in a position directly over the aromatic ring.

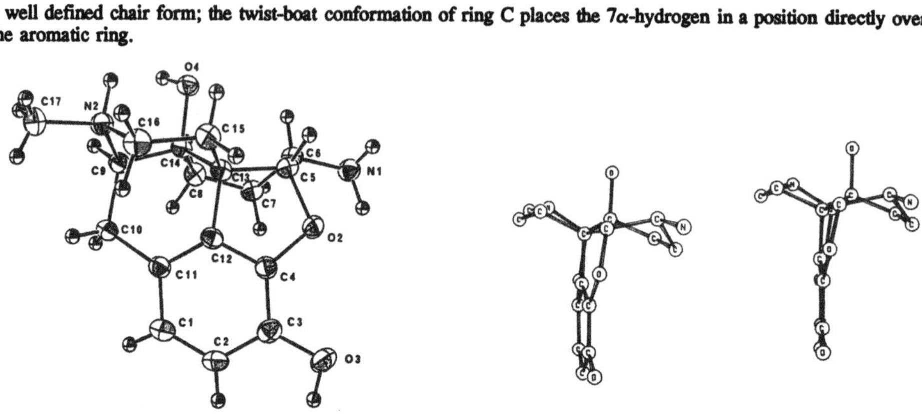

Fig. 1. The $C_{17}H_{24}N_2O_3^{2+}$ ion, and a stereoview (omitting hydrogens).

endo-2-PHENYLPROPIONIC ACID 8-METHYL-8-AZABICYCLO[3.2.1]OCT-3-YL ESTER HYDROCHLORIDE DIHYDRATE (YDRATROPYLTROPINE HYDROCHLORIDE DIHYDRATE)

$C_{17}H_{24}ClNO_2,2(H_2O)$ $C_{17}H_{24}NO_2^+,Cl^-,2(H_2O)$

R.H. JONES and T.A. HAMOR, 1985. *Acta Cryst.*, C41, 620-622.

Monoclinic, *P*2$_1$/*c*, a = 8.104, b = 19.647, c = 12.173 Å, β = 105.82°, Z = 4. Mo radiation, R = 0.0637 for 1202 reflections.

The phenyl ring of the molecule (Fig. 1) is oriented at an angle of 83(1)° to the plane of the ester group and forms, with the tropane ring system, the extremities of a claw-like arrangement, typical of anticholinergic agents. In the tropane ring system the piperidinium ring adopts a flattened chair shape. The five-membered ring C(1),-C(5-7),N has an envelope shape with the N atom as the flap (displacement 0.71(1) Å). The ester group C(3),O(1),-C(9),O(2),C10) adopts the antiperiplanar conformation typical of esters. Bond lengths and angles are normal except for one short bond (C(10)-C(11)) of 1.454(10) Å. In the crystal, cations, Cl^- ions and water molecules are linked by a network of N-H...Cl, OH...Cl and O-H...O hydrogen bonds.

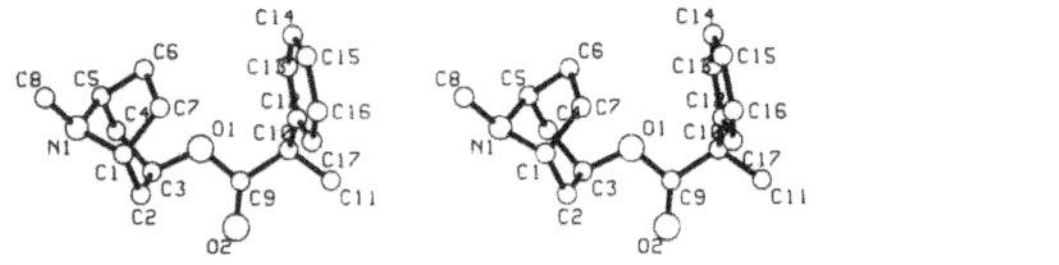

Fig. 1. $C_{17}H_{24}ClNO_2,2(H_2O)$: stereoview of the cation.

1-(3-HYDROXY-4-METHOXYBENZYL)-6,7-METHYLENEDIOXYISOQUINOLINE (ISOSEVANINE)

$C_{18}H_{15}NO_4$

I.R.C. BICK, Y.A.G.P. GUNAWARDANA, V.A. PATRICK and A.H. WHITE, 1985. *Aust. J. Chem.*, 38, 1571-1574.

Monoclinic, *P*2$_1$/*n*, a = 22.45, b = 5.746, c = 11.569 Å, β = 93.65°, Z = 4. Mo radiation, R = 0.052 for 1016 reflections.

The title alkaloid (Fig. 1) was isolated from 'Hedycarya angustifolia'. The ten atoms of the larger aromatic system are coplanar and make an angle of 88.7° with the plane of the phenyl ring and its coplanar methoxy group. The molecular dimensions are normal.

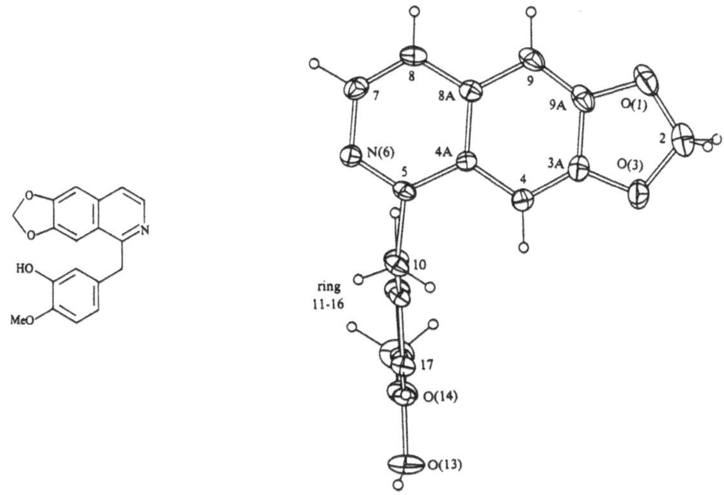

Fig. 1. $C_{18}H_{15}NO_4$: the molecular skeleton and structure.

HYGROPHYLLINE
$C_{18}H_{27}NO_6$

M.F. MACKAY, P. MITRPRACHACHON and C.C.J. CULVENOR, 1985. *Acta Cryst.*, C41, 395-397.

Orthorhombic, $P2_12_12_1$, a = 8.455, b = 9.637, c = 22.645 Å, Z = 4. Cu radiation, R = 0.056 for 1489 reflections.

In the alkaloid molecule (Fig. 1) the pyrrolizidine ring nucleus has ring A in an exo and ring B in an endo conformation. The dihedral angle between the least-squares planes of atoms C(1,3,8),N(4) and C(5,7,8),N(4) is 127.7(6)°. Atoms in the primary and secondary ester groups- respectively C(9,11,12),O(10,22) and C(7,15,16),O(17,25)- are each coplanar. The unsaturated side chain C(15)=C(20)-C(21) is in a nearly cis arrangement with atoms C(14,15,16,20,21) coplanar. The carbonyl bonds in the ester functions are anti-parallel with an angle between the bonds of 26.5(6)°. The trans-annular O(10)...O(17) distance is 2.966(5) Å and the O(23)...O(24) and O(22)...O(24) intramolecular separations of 2.818(6) and 2.701(5) Å respectively, indicate probable hydrogen bonding. The molecules are linked into helices along the a-axis through O(24)...O(22) hydrogen bonds of length 2.738(6) Å.

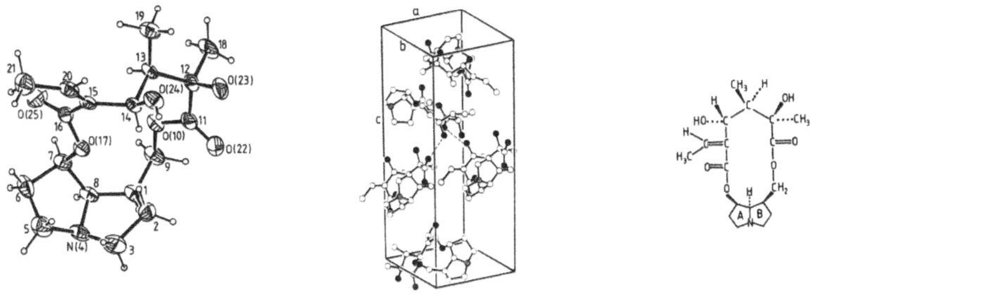

Fig. 1. $C_{18}H_{27}NO_6$: views of the molecule, the molecular packing and the structural formula.

(±)-DEMETHOXYCARBONYL-3α-NAUCLECHINE ETHYL ACETATE SOLVATE
$C_{19}H_{19}N_3O,0.5(C_4H_8O_2)$ (I)

(±)-DEMETHOXYCARBONYL-3β-NAUCLECHINE MONOHYDRATE

$C_{19}H_{19}N_3O,H_2O$ (II)

J.F. SAWYER, A. SHARIFF and S. McLEAN, 1985. *Acta Cryst.*, C41, 1810-1814.

I. Monoclinic, $P2_1/c$, a = 11.255, b = 11.515, c = 13.657 Å, β = 91.29°, Z = 4. Mo radiation, R = 0.072 for 1369 reflections.

II. Monoclinic, $P2_1/c$, a = 8.253, b = 18.449, c = 11.053 Å, β = 102.32°, Z = 4. Mo radiation, R = 0.049 for 1055 reflections.

In (I) (Fig. 1) the hydroxyl substituent is pseudoaxial while in (II) it is pseudoequatorial. In both isomers the seven-membered rings (D) have slightly distorted C_s chair conformations. Ring C adopts a half-chair conformation in both (I) and (II). Equivalent bond lengths in the two isomers are virtually the same and are normal. There are some small but significant differences in bond angles, the maximum difference involving angles C(18)-C(19)-C(20) and O(24)-C(19)-C(20) which have both opened out by 4° in (I) to accommodate an intramolecular hydrogen bond from H(241) to N(4) (1.93 Å). In the crystal of (I) there is also an N(1)-H(11)...N(23) hydrogen bond (2.26 Å). In (II) the molecules are packed in a layered arrangement which alternates with layers of water molecules in the *a* direction. Each water molecule forms a hydrogen bond to N(4) and is involved in a hydrogen bond from H(11) to give 'dimers' about centres of symmetry. These are further linked by a hydrogen bond to the pseudoequatorial O of an adjacent 'dimer' and by the hydrogen bond from the hydroxyl proton to N(23) of another 'dimer'.

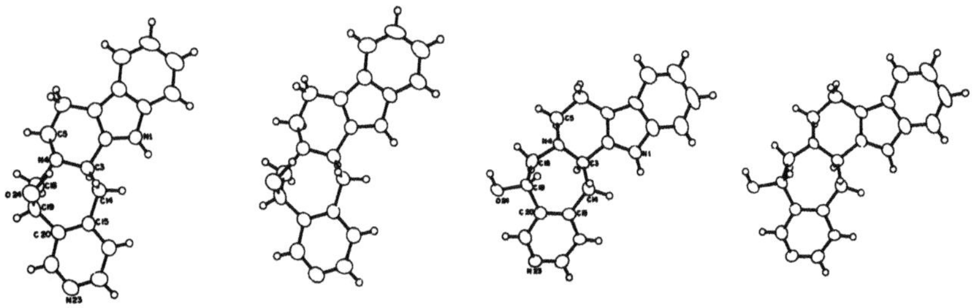

Fig. 1. Stereoviews of $C_{19}H_{19}N_3O,0.5(C_4H_8O_2)$ (left) and $C_{19}H_{19}N_3OH_2O$ (right).

2,6-LUTIDINE-ISATINECATE
$C_{19}H_{25}NO_6$

S.E. DREWES, J.S. FIELD, A.T. PITCHFORD, P.H. VAN ROOYEN and J.L.M. DILLEN, 1985. *S. Afr. J. Chem.*, 38, 90-94.

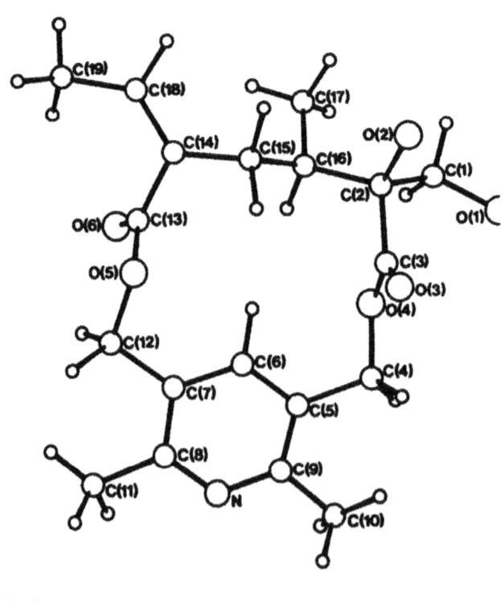

Fig. 1. Molecular structure of 2,6-lutidine-isatinecate showing atomic numbering.

Tetragonal, $P4_2$, a = 18.998, c = 5.548 Å, Z = 4. Cu radiation, R = 0.081 for 1330 reflections.

The crystal structure analysis reveals that in the 13-membered ring of the alkaloid the carbonyl groups are anti-parallel, with the α,β-unsaturated carbonyl group at C(13) directed below the plane of the macrocycle (Fig. 1). The two lactone carbonyl groups are close to collinearity, with a dihedral angle of 29.5° between them, and pointing in opposite directions. All bond lengths and angles are as expected. There is no evidence of hydrogen bonding between the hydrogen atom of the primary hydroxyl group and O(2).

(-)-3-ACETYL-N-METHYLMORPHINANIUM PERCHLORATE
$C_{19}H_{26}ClNO_5$ $C_{19}H_{26}NO^+,ClO_4^-$

E. MOHACSI, J. O'BRIEN, J. BLOUNT and J. SEPINWALL, 1985. *J. Med. Chem.*, **28**, 1177-1180.

Orthorhombic, $P2_12_12_1$, a = 7.458, b = 10.413, c = 24.42 Å, Z = 4. Cu radiation, R = 0.086 for 1345 reflections.

The structure and stereochemistry were determined to be as shown in Fig. 1.

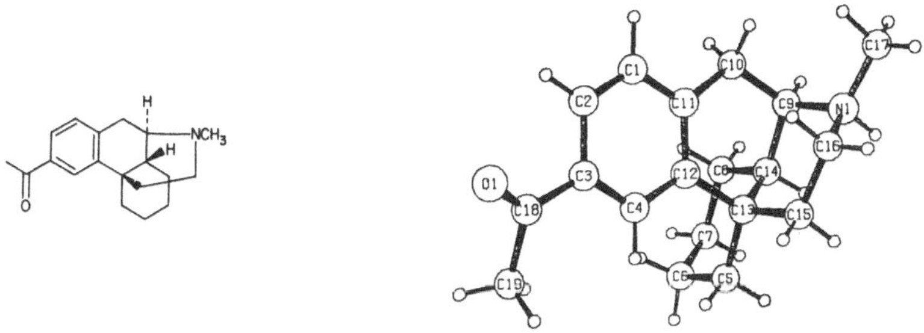

Fig. 1. The $C_{19}H_{26}NO^+$ ion and the parent molecule.

NEODIHYDROKOUMINE
$C_{20}H_{24}N_2O$

FAN ZHAOCHANG, LIU ZHUJIN and WANG CHIHAO, 1985. *Jiegou Huaxue (J. Struct. Chem.)*, **4**, 123-127.

Orthorhombic, $P2_12_12_1$, a = 12.671, b = 13.634, c = 9.653 Å, Z = 4. Mo radiation, R = 0.071 for 1053 reflections.

The piperidine and tetrahydropyran rings are cis-fused (Fig. 1), with the former in a twist conformation and the latter in a chair form. Molecules are linked by N-H...O hydrogen bonds.

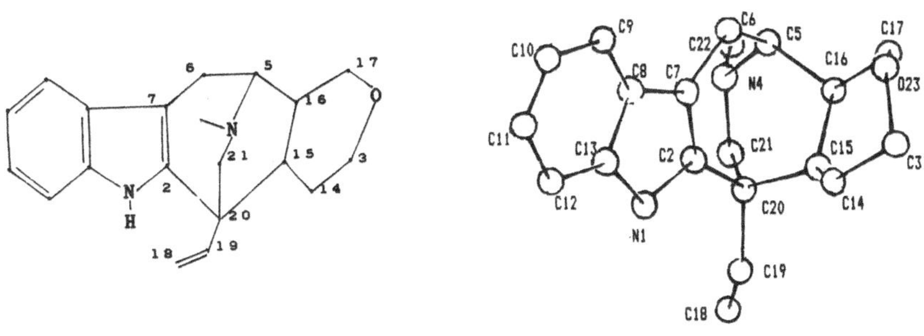

Fig. 1. The neodihydrokoumine molecule.

IPOHARDINE PICRATE
$C_{21}H_{18}N_4O_8$ $C_{15}H_{16}NO^+,C_6H_2N_3O_7^-$

YU ZHAOLOU, ZHU NAIJUE, LU RENRONG, CAI SHAOPEI, LAO XIAFEI and LIU ZHUJIN, 1985. *Jiegou Huaxue (J. Struct. Chem.)*, **4**, 152-155.

Monoclinic, $P2_1/n$, a = 10.265, b = 25.329, c = 7.995 Å, β = 90.60°, Z = 4. Mo radiation, R = 0.058 for 1958 reflections.

The cation and anion are shown in Fig. 1. The dihedral angle between the two six-membered rings of the cation is 52.1° and ions are linked by O(8)-H...O(7) hydrogen bonds (2.707 Å).

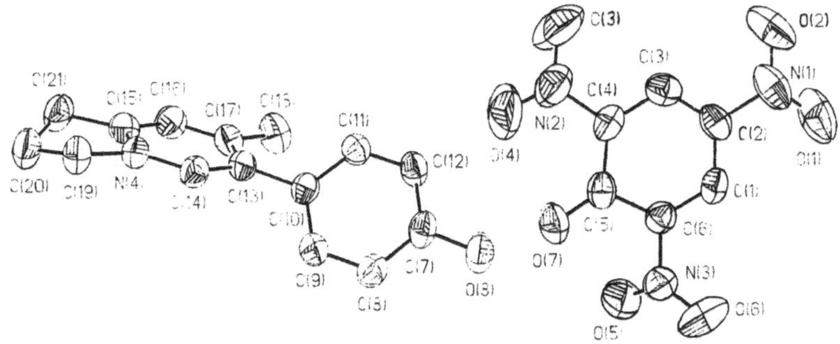

Fig. 1. The cation and anion in $C_{21}H_{18}N_4O_8$.

6,7-DIDEHYDROCOLCHICEINE
$C_{21}H_{21}NO_6$

E. MOLINS, J. RIUS, C. MIRAVITLLES and A. BLADE-FONT, 1985. *Acta Cryst.*, **C41**, 1232-1235.

Monoclinic, $P2_1/a$, a = 9.535, b = 16.660, c = 12.450 Å, β = 107.89°, Z = 4. Mo radiation, R = 0.040 for 2016 reflections.

Ring C of the molecule (Fig. 1) has a skew-boat conformation with C(10) at the prow of the boat. Ring B has a flattened boat conformation with an angle of 56.3° between the mean planes of atoms C(4a),C(5),C(6) and C(4a),C(6),C(7),C(12b). The angle between the latter mean plane and that of atoms C(7),C(7a),C(12a),C(12b) is 39.6(4)°. Ring A is planar, coplanar with the methoxy group C(3)-O(3)-C(3m) and perpendicular to the C(2)-O(2)-C(2m) group. The C(1)-O(1)-C(1m) methoxy group forms a dihedral angle of 84.3(3)° with ring A.

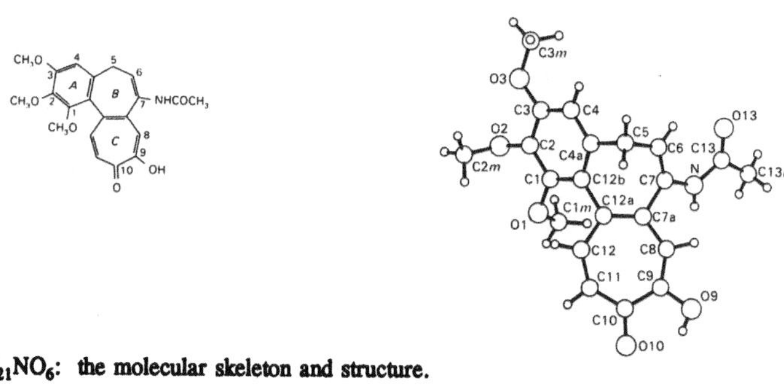

Fig. 1. $C_{21}H_{21}NO_6$: the molecular skeleton and structure.

STRYCHNINE
$C_{21}H_{22}N_2O_2$ (I)

BRUCINE ETHANOL SOLVATE DIHYDRATE
$C_{23}H_{26}N_2O_4,C_2H_6O,2(H_2O)$ (II)

S.S.B. GLOVER, R.O. GOULD and M.D. WALKINSHAW, 1985. *Acta Cryst.*, **C41**, 990-994.

I. Orthorhombic, $P2_12_12_1$, a = 11.267, b = 11.892, c = 12.105 Å, Z = 4. Mo radiation, R = 0.037 for 1195 reflections [see also the following Report].

II. Orthorhombic, $P2_12_12_1$, a = 7.723, b = 12.337, c = 25.212 Å, Z = 4. Mo radiation, R = 0.039 for 1694 reflections.

Strychnine and brucine (Fig. 1) are related indole alkaloids, the only difference being the presence of two methoxy groups (O(3),C(22) and O(4),C(23)) in (II). Most of the differences in the structures of the two compounds are insignificant and caused mainly by hydrogen bonding at O(1) in (II). However, there is a significant difference in the indole ring system between the two compounds. The benzene ring is planar in both compounds and the azole ring less so, the interplanar angle being 4.6° in (I) and 6.6° in (II). The average bond length in the indole system in (I) is 1.386(7) while in (II) it is 1.391(21) Å. A major difference occurs at C(3)-C(4) which is 1.375(8) in (I) and 1.427(5) Å in (II). Packing in (I) is governed by van der Waals interactions while in (II) the packing is dominated by hydrogen bonding involving the solvent molecules. The arrangement in (II) is best described as corrugated monolayers of brucine separated by hydrogen-bonded solvent molecules.

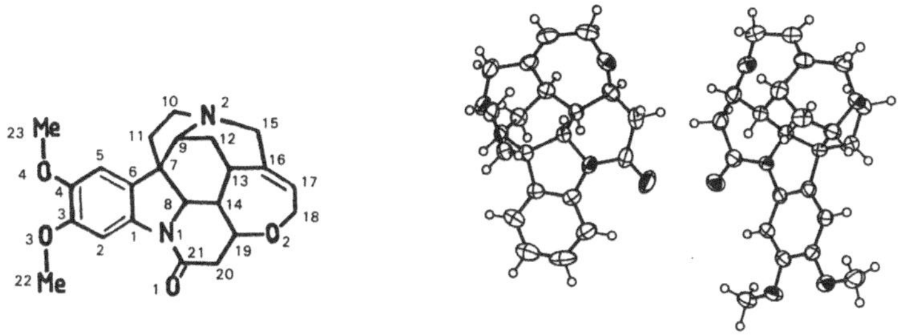

Fig. 1. The molecular skeleton (left), the molecular structure of strychnine (centre) and of brucine (right).

STRYCHNINE
$C_{21}H_{22}N_2O_2$ (I)

STRYCHNINE NITRATE
$C_{21}H_{23}N_3O_5$ (II) $C_{21}H_{23}N_2O_2^+NO_3^-$

A. MOSTAD, 1985. *Acta Chem. Scand.*, B39, 705-716.

I. Orthorhombic, $P2_12_12_1$, a = 11.321, b = 11.810, c = 12.016 Å, Z = 4. Mo radiation, R = 0.033 for 2427 reflections (at 133 K) [see also the preceding Report].

II. Orthorhombic, $P2_12_12_1$, a = 7.411, b = 8.058, c = 30.743 Å, Z = 4. Mo radiation, R = 0.045 for 2339 reflections (at 133 K).

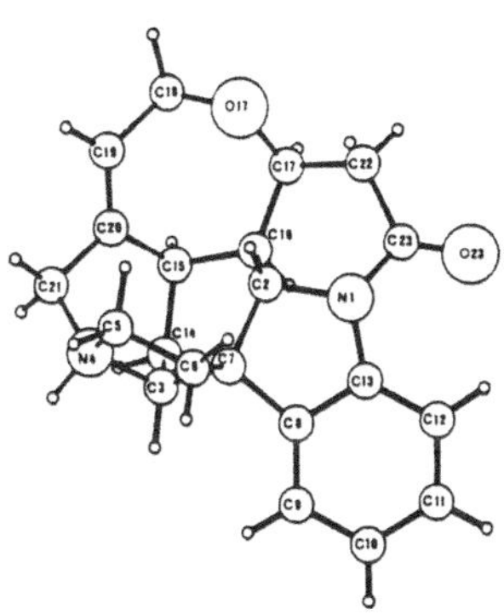

Fig. 1. The strychnine cation as it appears in the crystals of the nitrate $C_{21}H_{23}N_3O_5$ (II).

The geometry of the strychnine molecule (Fig. 1) is very similar in the two crystals. In (II) there is a strong hydrogen bond between N(4) in the strychnine molecule and O(1) in the nitrate ion: O(1)-H-N(4) 175.4°; O(1)...H 1.68 and O(1)...N(4) 2.680 Å. In (I) the C(3)-N(4) distance is 1.491(2), C(5)-N(4) 1.480(2) and C(21)-N(4) 1.474(2) Å. In (II) C(3)-N(4) 1.538(3), C(5)-N(4) 1.515(3) and C(21)-N(4) 1.510(3) Å respectively.

3-O-DEMETHYLCOLCHICINE ACETONE SOLVATE
$C_{21}H_{23}NO_6,C_3H_6O$

P.N. SHARMA, A. BROSSI and J.V. SILVERTON, 1985. *Acta Cryst.*, C41, 1069-1072.

Orthorhombic, $P2_12_12_1$, a = 8.030, b = 9.700, c = 30.710 Å, Z = 4. Cu radiation, R = 0.052 for 2788 reflections.

The bond lengths and angles in this biologically active compound (Fig. 1) are in good agreement with those observed in other comparable molecules. The molecular conformation is very similar to that observed for all colchinoids. The six-membered ring and a seven-membered ring are both planar. The flexible non-planar seven-membered ring has a boat conformation and its N-carbonyl group has an orientation very similar to that in other colchinoids. The orientations of the methoxy groups are all different and appear to be determined by intermolecular forces. The molecules are linked by hydrogen bonds N-H...O(9) and O(3)-H...O(13) of lengths 2.845(4) and 2.674(4) Å respectively. The acetone of solvation is located in vacancies between the molecules and does not appear to be involved in the hydrogen bonding within the crystal.

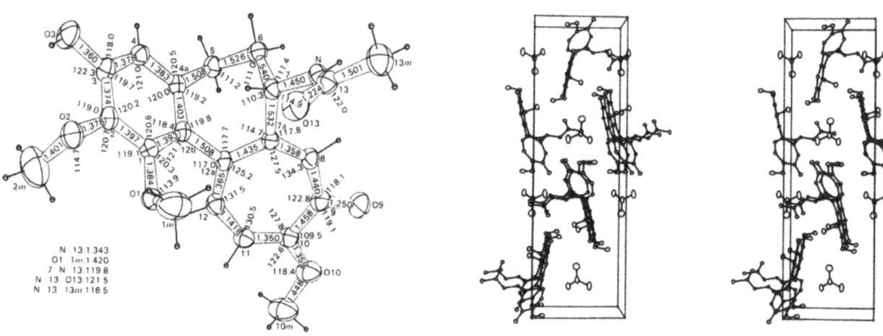

Fig. 1. $C_{21}H_{23}NO_6,C_3H_6O$: molecular conformation, bond lengths and angles and a stereoview of the molecular packing.

N-METHYL-1,2,10-TRIMETHOXY-6aα-APORPHIN-11-OL IODIDE (ISOCORYDINE IODOMETHYLATE)
$C_{21}H_{26}INO_4$ $C_{21}H_{26}NO_4^+,I^-$

S. TOURE, G. GERMAIN and L.A. DJAKOURE, 1985. *Acta Cryst.*, C41, 1827-1828.

Orthorhombic, $P2_12_12_1$, a = 7.161, b = 11.701, c = 25.632 Å, Z = 4. Mo radiation, R = 0.027 for 1770 reflections.

Isocorydine iodomethylate (Fig. 1) is a natural product with an aporphine type of structure which was isolated from 'Rhigiocarya racemifera'. In the tetracyclic moiety rings C(8-13) and C(1-3,14-16) are both aromatic. The N(6) atom carries a positive charge and the N(6)-C(5), N(6)-C(22), N(6)-C(23) and N(6)-C(6a) bond distances are 1.525(8), 1.501(10), 1.523(9) and 1.545(9) Å respectively. Crystalline cohesion is achieved through intermolecular hydrogen bonding (H(191)...O(24) 2.418, H(182)...O(17) 2.579 Å). The iodide ion is involved in short contacts to C(22) and H(51).

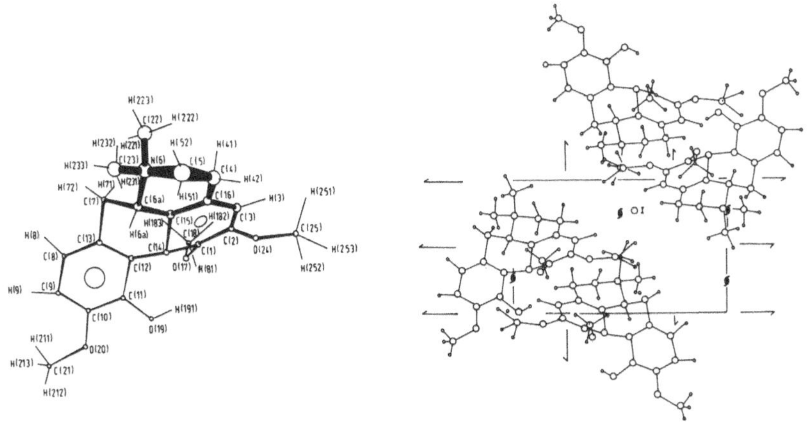

Fig. 1. $C_{21}H_{26}INO_4$: a view of the cation and the structure projected along [001].

10-HYDROXY-10-METHYL-10,11-DIHYDROQUININE
$C_{21}H_{28}N_2O_3$

A. SUSZKO-PURZYCKA, T. LIPINSKA, E. PIOTROWSKA and B.J. OLEKSYN, 1985. *Acta Cryst.*, **C41**, 977-980.

Orthorhombic, $P2_12_12_1$, a = 11.199, b = 11.237, c = 15.318 Å, Z = 4. Mo radiation, R = 0.032 for 1790 reflections.

The quinoline part of the molecule (Fig. 1) is not planar, the largest deviations from the least-squares plane being for C(14) (-0.024(2) Å) and C(16) (-0.026(2) Å). The methoxy group is coplanar with the quinoline moiety. The quinuclidine moiety has a skew conformation. The three fused rings of this group all have boat conformations with the C-N(1) bonds nearly eclipsed. The substituents at C(8) and C(3) are exo and endo, respectively. Packing is determined by intermolecular hydrogen bonds between hydroxyl groups and N atoms of the quinuclidine moiety (O(26)...N 2.775(5), O(12)...N(13) 2.869(5) Å). This leads to the formation of layers parallel to the ab plane.

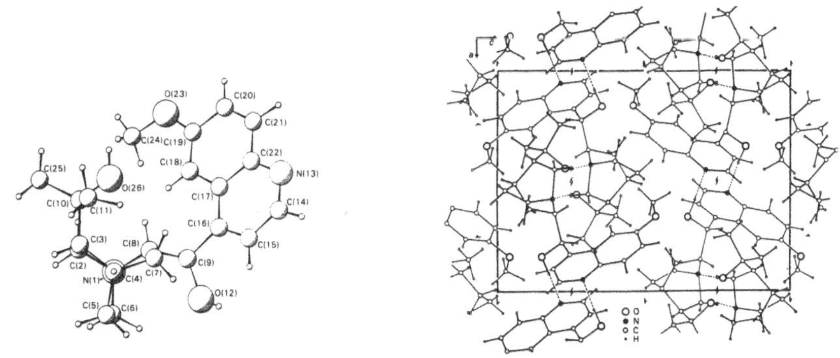

Fig. 1. $C_{21}H_{28}N_2O_3$: the molecular structure and packing.

19,20-EPOXY-15-HYDROXYICAJINE
$C_{22}H_{24}N_2O_5$

A. MOSTAD, H. MOSTAD and C. ROMMING, 1985. *Acta Chem. Scand.*, **B39**, 141-148.

Orthorhombic, $P2_12_12_1$, a = 10.314, b = 10.843, c = 16.039 Å, Z = 4. Mo radiation, R = 0.041 for 2123 reflections (at 143 K).

In the crystal there is an intermolecular hydrogen bond (2.859 Å) between O(29) of one molecule and O(27) of another. The molecule is shown in Fig. 1.

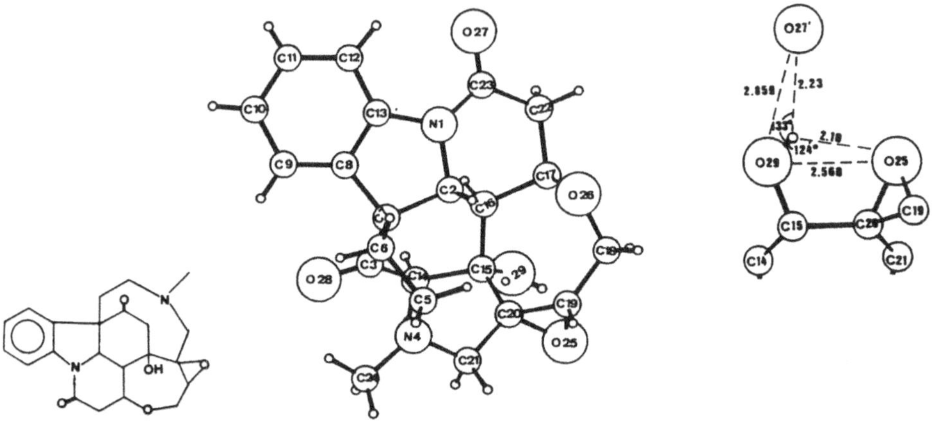

Fig. 1. Schematic drawing of a molecule of $C_{22}H_{24}N_2O_5$ (left) and the hydrogen bonding (right).

1-O-ACETYL-1-O-DEMETHYLCOLCHICINE
$C_{23}H_{25}NO_7$ (I)

2-O,3-O-DIACETYL-10-DEMETHOXY-2-O,3-O-DIDEMETHYL-10-METHYLTHIOCOLCHICINE METHANOL SOLVATE
$C_{24}H_{25}NO_7S,CH_4O$ (II)

1-O,2-O-DIACETYL-10-DEMETHOXY-1-O,2-O-DIDEMETHYL-10-METHYLTHIOCOLCHICINE MONOHYDRATE
$C_{24}H_{25}NO_7S,H_2O$ (III)

P. KEREKES, A. BROSSI, J.L. FLIPPEN-ANDERSON and C.F. CHIGNELL, 1985. *Helv. Chim. Acta,* **68**, 571-580.

I. Monoclinic, $C2$, a = 20.244, b = 7.074, c = 17.132 Å, β = 94.40°, Z = 4. Cu radiation, R = 0.051 for 1584 reflections.

II. Orthorhombic, $P2_12_12_1$, a = 9.471, b = 11.341, c = 23.702 Å, Z = 4. Cu radiation, R = 0.056 for 1755 reflections.

III. Orthorhombic, $P2_12_12_1$, a = 7.117, b = 10.764, c = 31.612 Å, Z = 4. Cu radiation, R = 0.061 for 2270 reflections.

In (III) the AcO moiety on C(6) was disordered (Fig. 1). The crystal structure analyses established the stereochemistries of the molecules. In (I) there are molecules of solvent of H_2O and acetone in a disordered arrangement; the exact proportions of each were not determined.

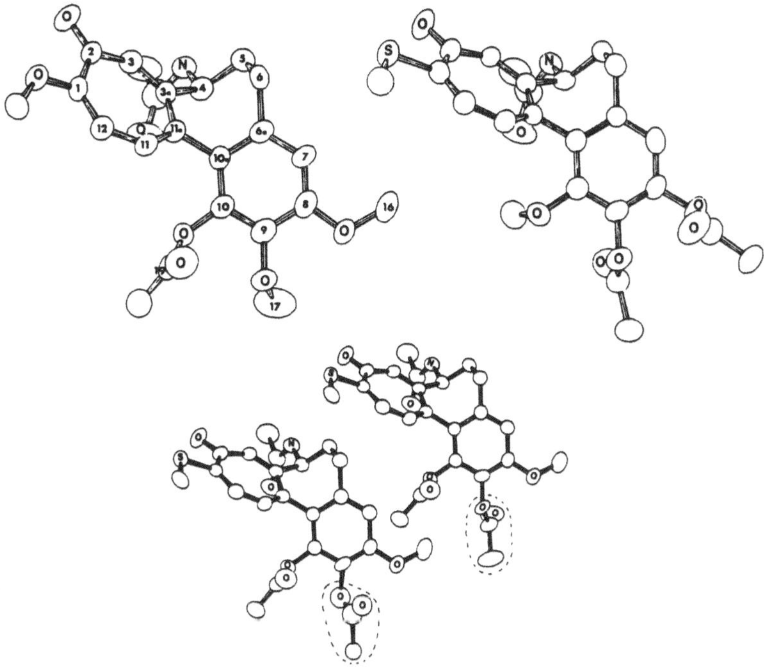

Fig. 1. ORTEP drawings of molecules of $C_{23}H_{25}NO_7$ (top left), $C_{24}H_{25}NO_7S$ (top right) and $C_{24}H_{25}NO_7S$ (III) (below) showing disordered positions for the AcO moiety.

(7S)-N-ACETYLDEMECOLCINE
$C_{23}H_{27}NO_6$

J.V. SILVERTON, P.N. SHARMA and A. BROSSI, 1985. *Acta Cryst.*, C41, 755-758.

Monoclinic, $P2_1$, a = 11.629, b = 11.748, c = 15.782 Å, β = 97.23°, Z = 4. Cu radiation, R = 0.040 for 2998 reflections.

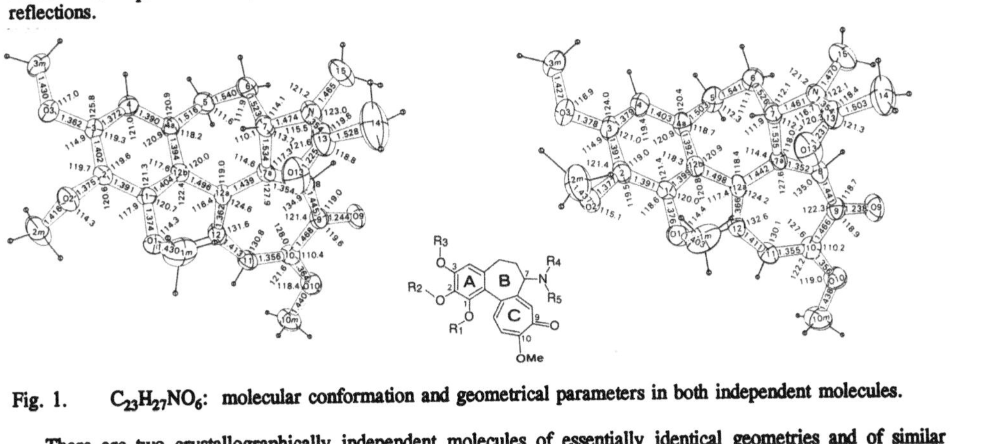

Fig. 1. $C_{23}H_{27}NO_6$: molecular conformation and geometrical parameters in both independent molecules.

There are two crystallographically independent molecules of essentially identical geometries and of similar conformations in the asymmetric unit of this active colchinoid compound (Fig. 1). The acetyl group lies on the same side of both molecules. The only distinct conformational difference between the independent molecules lies in the position of the C(2) methoxy group. These and other minor conformational differences are attributed to different packing forces acting on the different molecules. The bond lengths, angles and molecular conformations are also similar to those of reported compounds of this type (1).

1. Structure Reports, 48B, 686; 44B, 550; 45B, 641.

9-ACETYLQUITENINE ETHYL ESTER
$C_{23}H_{28}N_2O_5$

L. DUPONT, A. KONSUR, K. LEWINSKI and B. OLEKSYN, 1985. *Acta Cryst.*, C41, 616-619.

Orthorhombic, $P2_12_12_1$, a = 11.469, b = 18.575, c = 10.142 Å, Z = 4. Mo radiation, R = 0.0376 for 2094 reflections.

The quinoline ring (Fig. 1) is planar with the bonds C(21)-O(20) and C(15)-C(16) mutually trans. The conformation of the quinuclidine moiety is skewed and the ring has no exact mirror symmetry. The non-bonded C(4)...N(1) distance is 2.571(4) Å. Both acetyl and ethyl ester groups have normal geometry. The acetyl group is planar and forms an angle of 94.0° with the plane of the quinoline ring. The ester group is tilted by 15.5° from the plane of the quinoline ring. The overall geometry of the molecule is described as trans with respect to the C(9)-C(8) bond.

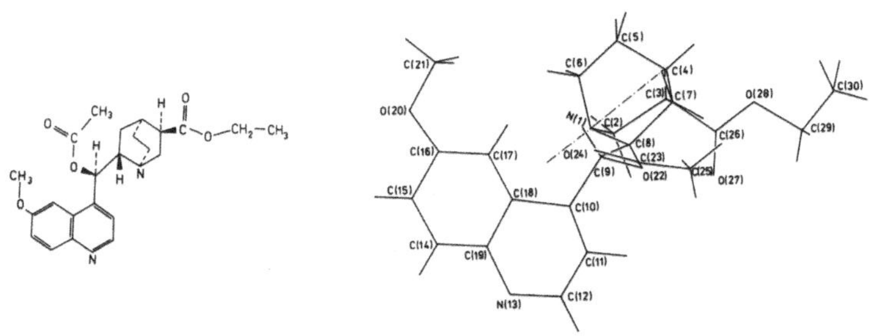

Fig. 1. $C_{23}H_{28}N_2O_5$: the molecular skeleton and structure.

1-HYDROXY-3-GERANOYLOXY-4-METHOXY-10-METHYLACRIDAN-9-ONE
$C_{25}H_{29}NO_4$

G. BAUDOUIN, F. TILLEQUIN, M. KOCH, M.-E. TRAN HUU DAU, J. GUILHEM, J. PUSSET and G. CHAUVIERE, 1985. *J. Nat. Prod.*, 48, 260-265.

Triclinic, $P\bar{1}$, a = 12.518, b = 9.914, c = 8.854 Å, α = 98.17, β = 89.53, γ = 98.95°, Z = 2. Cu radiation, R = 0.048 for 2941 reflections.

The analysis established the structure and conformation as shown in Fig. 1.

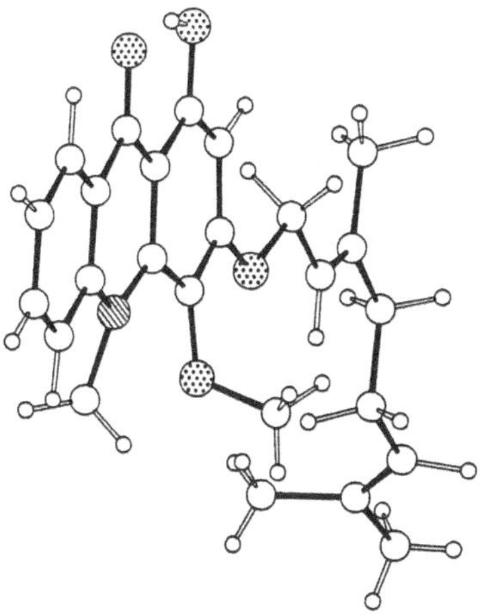

Fig. 1. The $C_{25}H_{29}NO_4$ molecule.

ANCISTROTECTORINE
$C_{26}H_{31}NO_4$

N. RUANGRUNGSI, V. WONGPANICH, P. TANTIVATANA, H.J. COWE, P.J. COX, S. FUNAYAMA and G.A. CORDELL, 1985. *J. Nat. Prod.*, **48**, 529-535.

Monoclinic, $P2_1$, a = 11.858, b = 7.043, c = 14.693 Å, β = 114.43°, Z = 2. Mo radiation, R = 0.080 for 1087 reflections.

The analysis establishes that the molecule is a 7-3' linked naphthalene-isoquinoline alkaloid as shown in Fig. 1. The C(1) and C(3) methyl groups are cis and the N-containing ring of the isoquinoline moiety adopts a sofa conformation with C(3) the out of plane atom. Steric hindrance results in a dihedral angle of 56.4° between the two aromatic rings.

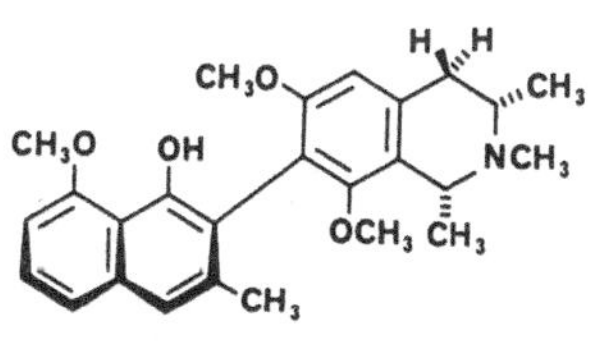

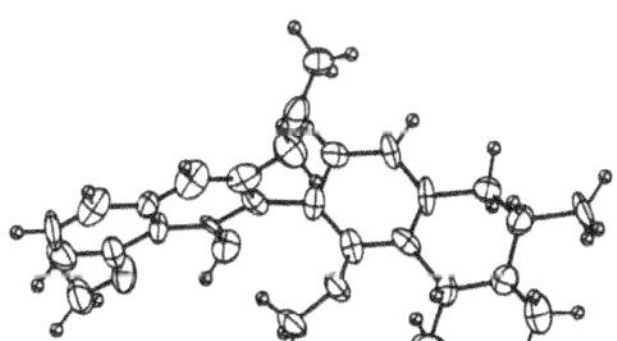

Fig. 1. The ancistrotectorine structure.

ISOBAIMONIDINE (3α,6α,20β-TRIHYDROXY-5α-CEVANINE)
$C_{27}H_{45}NO_3$

V. KETTMANN, F. PAVELCIK, I. MASTEROVA and J. TOMKO, 1985. *Acta Cryst.*, C**41**, 392-394.

Monoclinic, $P2_1$, a = 12.048, b = 7.292, c = 14.408 Å, β = 101.54°, Z = 2. Cu radiation, R = 0.042 for 1195 reflections.

The ring junctions of the C-nor-D-homo skeleton of this alkaloid molecule (Fig. 1) are A/B trans, B/C trans, C/D cis, D/E trans and E/F trans. All the six-membered rings except the D-ring are in the $^4C_1$ conformation. Ring D has a flexible boat conformation with torsion angles C(12)-C(13)-C(14)-C(15) and C(15)-C(16)-C(17)-C(13) of 10.3(7) and 20.3(7)° respectively. The five-membered ring has a distorted envelope conformation with C(9) displaced by 0.68(1) Å from the least-squares plane of the other four ring atoms. All bond lengths, except C(5)-C(10) and C(12)-C(14) at the A/B and C/D ring junctions (both 1.59(1) Å) are normal. Molecules are linked by O

H...O hydrogen bonds with O...O separations lying in the range 2.75(1) to 2.81(1) Å.

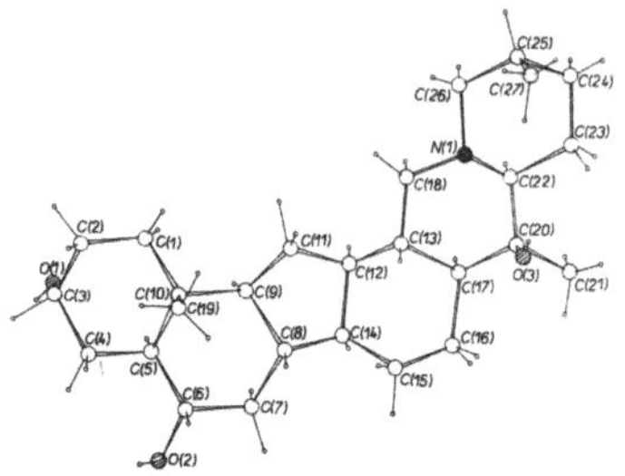

Fig. 1. $C_{27}H_{45}NO_3$: a perspective view of the molecule.

COLCHICEINE BENZOATE MONOHYDRATE
$C_{28}H_{27}NO_7,H_2O$

E. MOLINS, J. RIUS, X. SOLANS, C. MIRAVITLLES, A. BLADE-FONT and G. GERMAIN, 1985. *Acta Cryst.*, **C41**, 556-558.

Orthorhombic, $P2_12_12_1$, a = 10.324, b = 11.348, c = 22.944 Å, Z = 4. Mo radiation, R = 0.052 for 1399 reflections.

The troponoid ring C of the molecule (Fig, 1) exhibits a definite bond length alternation with an arrangement corresponding to an isocolchicine-like structure. The three longer C-C bonds are in the range 1.329(8)-1.386(8) Å and the longer bonds are in the range 1.407(8)-1.448(10) Å. The normals to the three rings A, B, C are A-B 43.7(4), B-C 36.3(4) and A-C 52.1(4)°. The molecules are held together by hydrogen bonds (O(1)...O(13a) 2.741(7), O(3)...O(1) 2.909(6) Å) involving the water molecule.

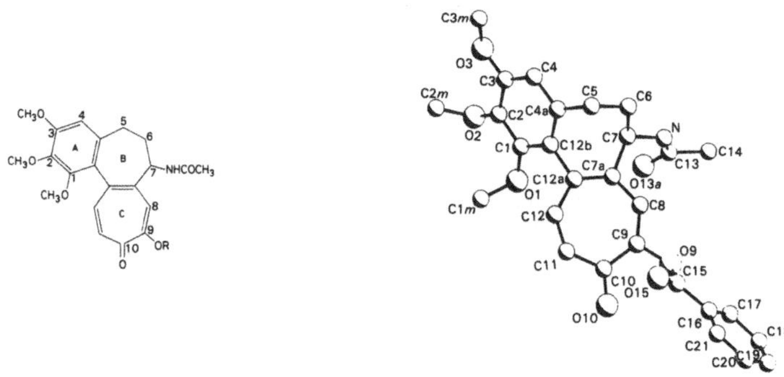

Fig. 1. $C_{28}H_{27}NO_7,H_2O$: the molecular skeleton and structure.

YUEHCHUKENE N-ACETATE
$C_{28}H_{28}N_2O$

YUN-CHEUNG KONG, CHOI-NANG LAM, KIN-FAI CHENG and T.C.W. MAK, 1985. *Jiegou Huaxue (J. Struct. Chem.)*, **4**, 30-33.

Monoclinic, $P2_1/n$, a = 8.032, b = 19.712, c = 14.460 Å, β = 101.22°, Z = 4. Mo radiation, R = 0.079 for 1523 reflections.

The alkaloid is shown to have a novel bis-indole skeleton (Fig. 1) with cis-fusion of the two rings sharing the C(10)-C(11) bond. The acetyl group is nearly coplanar with its attached indole ring and the resulting conjugation leads to the endocyclic bonds at N(2) being longer than the corresponding bonds at N(1) (N(2)-C(19) 1.397(6), N(2)-C(26) 1.419(6); N(1)-C(1) 1.388(6), N(1)-C(8) 1.372(6) Å). The alkaloid has anti-implantation activity.

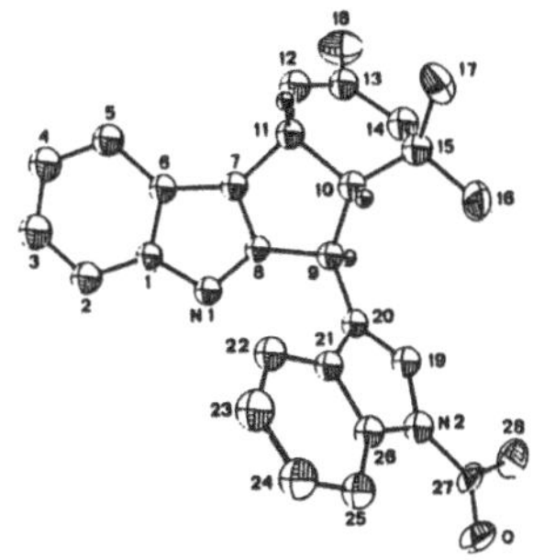

Fig. 1. A view of $C_{28}H_{28}N_2O$.

(22R,23S,25R)-22,26-EPIMINO-3β,23-DIHYDROXY-5α-CHOLESTAN-4-ONE AMMONIUM ACETATE

$C_{29}H_{52}N_2O_5$ $C_{27}H_{45}NO_3,C_2H_3O_2^-,H_4N^+$

D. GOMEZ, A.V. RIVERA, E.R. DE GIL and A. USUBILLAGA, 1985. *Acta Cryst.*, C41, 1197-1199.

Orthorhombic, $P2_12_12_1$, a = 31.862, b = 12.434, c = 7.192 Å, Z = 4. Mo radiation, R = 0.070 for 1538 reflections.

In the molecule (Fig. 1), rings A, B and C of the substituted androstane moiety and the substituted piperidine ring are all in the chair conformation. The androstane ring D has an envelope conformation with C(13) as flap. The piperidine moiety is oriented almost perpendicular to the steroid fragment of the molecule. Bond lengths N(1)-C(26) (1.507(10)) and N(1)-C(22) (1.516(11) Å) are longer than the expected 1.479(5) Å. All other bond lengths and angles lie within normal ranges. Hydrogen bonds are formed involving the substituted piperidine rings, ammonium ions and acetate ions.

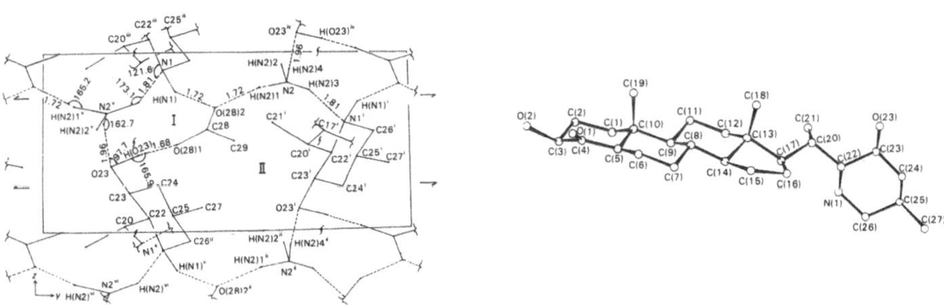

Fig. 1. $C_{29}H_{52}N_2O_5$: the molecular structure and (100) projection of the layer content of one unit cell at x = 0.

N-BENZOYL-L-ALANYL-STRYCHNINIUM DIHYDRATE

$C_{31}H_{33}N_3O_5,2(H_2O)$ $C_{21}H_{23}N_2O_2,C_{10}H_{10}NO_3,2H_2O$

R.O. GOULD, R. KELLY and M.D. WALKINSHAW, 1985. *J. Chem. Soc. Perkin Trans. 2*, 847-852.

Orthorhombic, $P2_12_12_1$, a = 10.751, b = 30.366, c = 8.608 A, Z = 4. Mo radiation, R = 0.051 for 2135 reflections.

The 1:1 complex of strychnine and N-benzoyl-L-alanine has the strychnine molecules packed in bilayers separated by hydrogen-bonded sheets of water molecules and benzoylalanine molecules. In the strychnine molecule (Fig. 1) the C(9)-N(2) bond length is 1.538(7) Å whereas the C(10)-N(2) and C(15)-N(2) distances are 1.491(8) and 1.494(8) Å respectively. In the benzoylalanine molecule the geometry is similar to that found in related compounds. The hydrogen-bonding of this structure is discussed in detail.

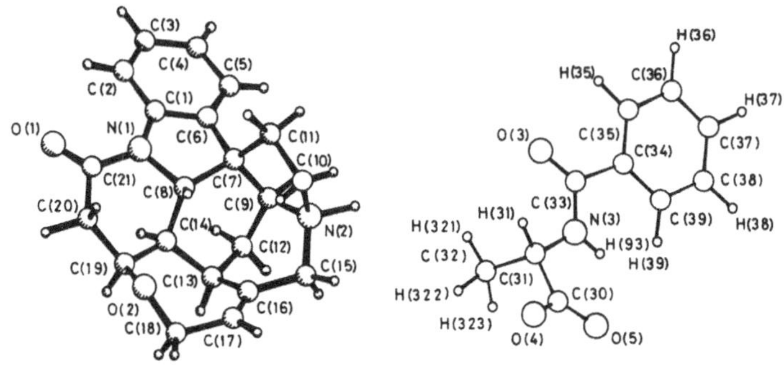

Fig. 1. Structures of $C_{21}H_{23}N_2O_2$ (left) and $C_{10}H_{10}NO_3$ (right).

N'-METHYLTHIOBINUPHARIDINE IODIDE DIHYDRATE

$C_{31}H_{45}IN_2O_2S,2(H_2O)$ $C_{31}H_{45}N_2O_2S^+,I^-,2(H_2O)$

A.E. KOZIOL, 1985. *J. Mol. Struct.*, **130**, 327-334.

Orthorhombic, $P2_12_12_1$, a = 9.499, b = 15.522, c = 23.694 Å, Z = 4. Mo radiation, R = 0.073 for 2060 reflections.

The alkaloid cation has two trans-quinolizidine moieties with equatorial methyl and furan groups (Fig. 1); the central tetrahydrothiophene ring has a conformation between an envelope and a half-chair, with the S atom 0.59(1) Å from the plane of the other four atoms. The six-membered rings all have chair conformations.

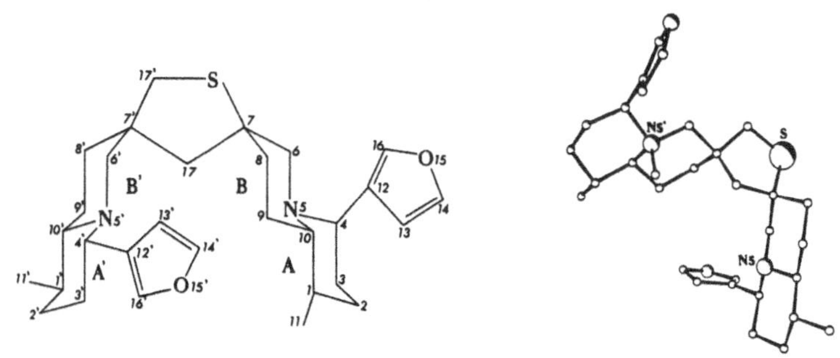

Fig. 1. The thiobinupharidine molecule and the N'-methyl cation.

EXSEROHILONE 4,11-BIS(p-BROMOBENZOATE)

$C_{34}H_{28}Br_2N_2O_8S_2$

K. SUGAWARA, F. SUGAWARA, G.A. STROBEL, YALI FU, HE CUN-HENG and J. CLARDY, 1985. *J. Org. Chem.*, **50**, 5631-5633.

Monoclinic, $P2_1$, a = 11.289, b = 11.224, c = 14.904 Å, β = 63.91°, Z = 2. Cu radiation, R = 0.047 for 2059 reflections.

The X-ray crystal structure analysis established the stereochemistry of the molecule (Fig. 1).

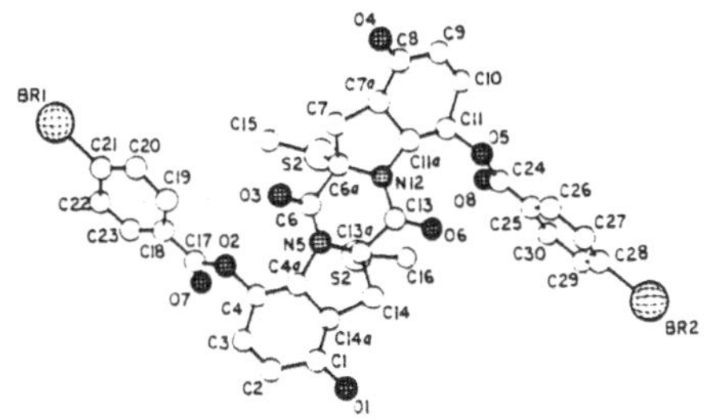

Fig. 1. The $C_{34}H_{28}Br_2N_2O_8S_2$ structure.

BRUCINE (-)-1-(o-BROMOPHENYL)-1-PHENYL-2-PROPYNOL
$C_{38}H_{37}BrN_2O_5$ (I) $C_{23}H_{26}N_2O_4,C_{15}H_{11}BrO$

(-)-SPARTEINE (-)-1-(o-BROMOPHENYL)-1-PHENYL-2-PROPYNOL
$C_{30}H_{37}BrN_2O$ (II) $C_{15}H_{26}N_2,C_{15}H_{11}BrO$

F. TODA, K. TANAKA, H. UEDA and T. OSHIMA, 1985. *Israel J. Chem.*, **25**, 338-345.

I. Orthorhombic, $P2_12_12_1$, a = 12.446, b = 33.485, c = 7.728 Å, Z = 4. Mo and Cu radiation, R = 0.054 for 2604 reflections.

II. Orthorhombic, $P2_12_12_1$, a = 13.498, b = 24.283, c = 8.156 Å, Z = 4. Mo and Cu radiation, R = 0.067 for 2100 reflections.

The analyses determined the mode of complexation and absolute configuration as shown in Fig. 1. In both cases two hydrogen bonds (O-H...N and C≡C-H...O) are important in the intermolecular packing.

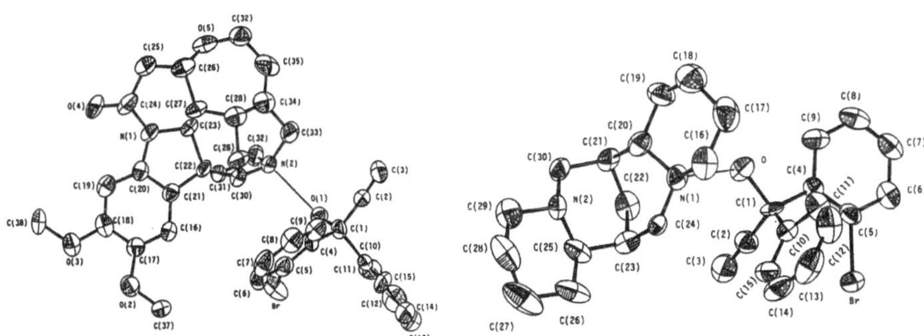

Fig. 1. Views of the brucine (left) and sparteine (right) complexes with $C_{15}H_{11}BrO$.

N,N-DIMETHYLCARACURINIUM-II DICHLORIDE OCTAHYDRATE
$C_{40}H_{44}Cl_2N_4O_2,8(H_2O)$ $C_{40}H_{44}N_4O_2^{2+},2(Cl^-),8(H_2O)$

P.E. BOURNE, S.L. GINELL, B.W. LOW and L. LESSINGER, 1985. *J. Cryst. Spect. Res.*, **15**, 453-471.

Monoclinic, $P2_1$, a = 12.695, b = 7.424, c = 21.762 Å, β = 98.03°, Z = 2. Cu radiation, R = 0.100 for 2931 reflections.

The cation (Fig. 1) has non-crystallographic twofold symmetry, a highly fused ring system and is structurally rigid. The analysis provides more accurate cation dimensions than were reported for the corresponding iodide derivative (1) but the overall features are in general agreement. In the crystal lattice, layers of alkaloid cations parallel to the bc-plane alternate with layers containing the two Cl^- and eight water molecules distributed almost randomly over ten sites. There are no hydrogen bonds between layers; the average O...Cl hydrogen-bond distance within the H_2O/Cl^- layer is 2.94 Å.

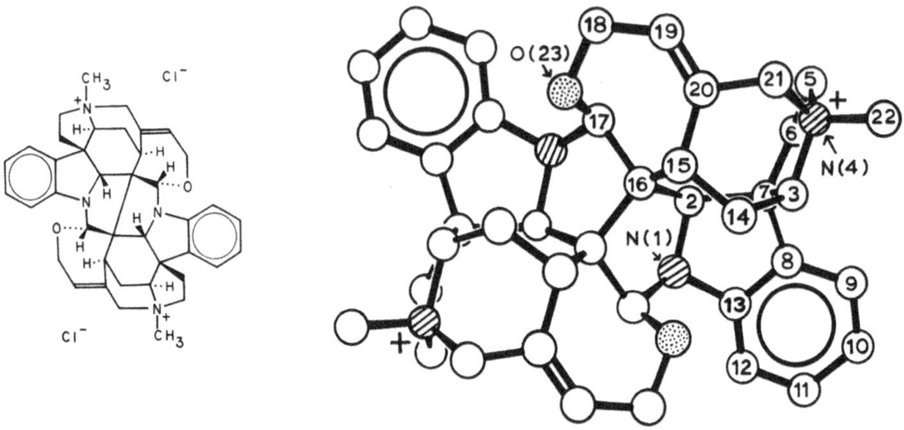

Fig. 1. The N,N-dimethylcaracurinium-II cation.

1. Structure Reports, 30B, 230.

(R)-CARNITINE
$C_7H_{15}NO_3$ (I)

(±)-ACETYLCARNITINE MONOHYDRATE
$C_9H_{17}NO_4,H_2O$ (II)

R.D. GANDOUR, W.J. COLUCCI and F.R. FRONCZEK, 1985. *Bioorg. Chem.*, **13**, 197-208.

I. Monoclinic, $P2_1$, a = 7.342, b = 6.089, c = 9.530 Å, β = 97.58°, Z = 2. Cu radiation, R = 0.034 for 889 reflections.

II. Monoclinic, $P2_1/c$, a = 12.315, b = 8.851, c = 11.109 Å, β = 95.02°, Z = 4. Mo radiation, R = 0.063 for 1136 reflections.

Views of the zwitterion structures for (I) and (II) are in Fig. 1. Carnitine (I) is in a fully extended conformation whereas the acetyl derivative (II) is folded; acetylation of the hydroxy group in (I) causes a rotation of approximately -22° about the C(3)-C(4) bond. There are significant differences in bond angles between (I) and (II), with C(2)-C(3)-C(4) 6.1° smaller in (II) than in (I) and O(3)-C(3)-C(4) 5.3° larger in (II) compared with (I). The orientation of the carboxylate groups also differs in the two structures (Fig. 1).

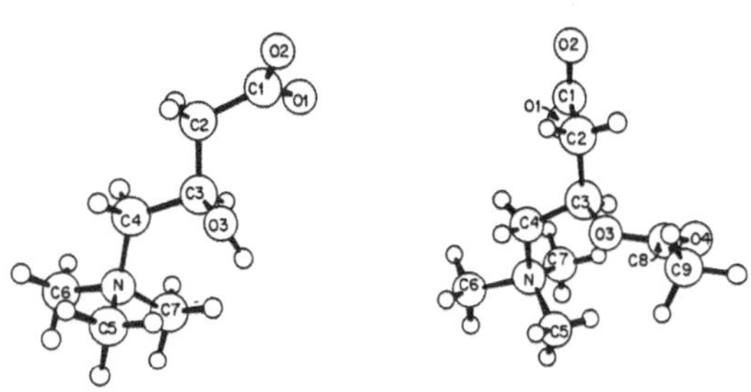

Fig. 1. The solid state conformations of carnitine (left) and acetylcarnitine (right).

7-BROMO-1,5-DIHYDROXY-2-AZABICYCLO[3.3.1]NON-6-ENE-3,8-DIONE (7-BROMOCAVERNICOLENONE)
$C_8H_8BrNO_4$

M. D'AMBROSIO, C. MEALLI, A. GUERRIERO and F. PIETRA, 1985. *Helv. Chim. Acta*, **68**, 1453-1460.

Monoclinic, $P2_1/c$, a = 10.827, b = 11.300, c = 7.328 Å, β = 94.7°, Z = 4. Mo radiation, R = 0.062 for 1227 reflections.

Although the crystals were affected by twinning it was possible to determine the structure of the molecule (Fig. 1).

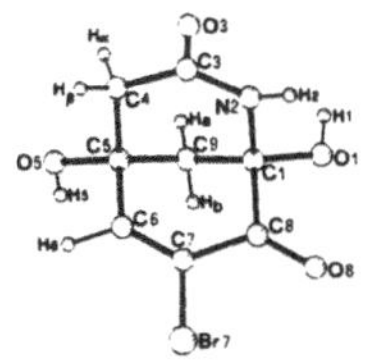

Fig. 1. Perspective drawing of a molecule of $C_8H_8BrNO_4$.

(1R)-2-ACETYL-9-AZABICYCLO[4.2.1]NON-2-ENE HYDROCHLORIDE (ANATOXIN-A HYDROCHLORIDE)

$C_{10}H_{16}ClNO$ $C_{10}H_{16}NO^+,Cl^-$

A.M.P. KOSKINEN and H. RAPOPORT, 1985. *J. Med. Chem.*, **28**, 1301-1309.

Orthorhombic, $P2_12_12_1$, a = 6.9846, b = 10.7355, c = 14.3140 Å, Z = 4. Mo radiation, R = 0.0241 for 1404 reflections.

A view of anatoxin-A hydrochloride is in Fig. 1; the seven-membered ring adopts a twist-chair conformation with C(4) pointing down and the pyrrolidine ring is folded so that the nitrogen is pushed back towards the C(7)-C(8) ethano bridge. Both the C=O and C=C double bonds are well localized. The structure is compared with that of the N-acetyl derivative (1).

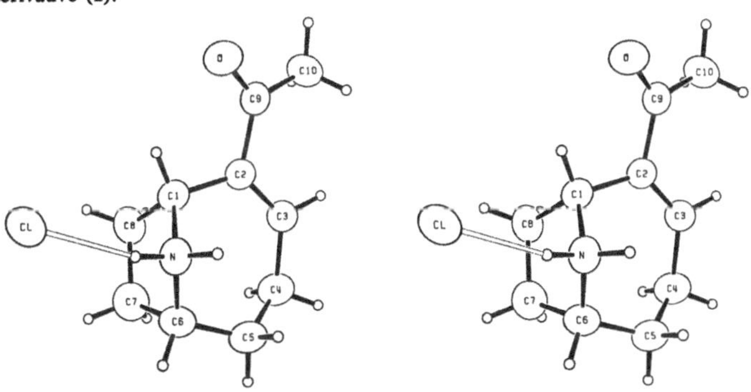

Fig. 1. A stereoview of anatoxin-A hydrochloride.

1. Structure Reports, **38B**, 322.

(-)-(1S,2'R,3R,5S)-5,2'-DIMETHYLSPIRO(2,4,6-TRIOXABICYCLO[3.3.0]OCTANE-3,3'-TETRAHYDROFURAN)-1,2'-DIOL (SPIRO-BIS-PINNAKETAL)

$C_{10}H_{16}O_6$

H. WIEDENFELD, F. KNOCH and M. KOCH, 1985. *Arch. Pharm.*, **318**, 289-293.

Triclinic, $P\bar{1}$, a = 7.391, b = 7.997, c = 10.335 Å, α = 108.10, β = 94.53, γ = 107.68°, Z = 2. Mo radiation, R = 0.034 for 1512 reflections.

Ring A (C(5),C(1),O(2),C(3),O(4)) is connected to ring B (C(3),C(2'),O(1'),C(5'),C(4')) through the spiro carbon atom C(4), (Fig. 1). The angle between the two ring planes is 93.2°. Ring B has a half-chair conformation with a folding angle of 139.8° between planes C(2'),C(3),C(4') and C(2'),O(1'),C(5'). The other five-membered ring (ring C, C(5),C(1),C(8),C(7),O(6)), makes an interplanar angle of 106.4° with ring A. The conformation of ring C is a half chair, like that of ring B, with an angle of 159.5° between C(5),O(6),C(7) and C(5),C(1),C(8) planes. Bond distances are: O(1')-C(2') 1.429(2), O(1')-C(5') 1.445(3), O(4)-C(5) 1.421(3), O(4)-C(3) 1.420(3), O(6)-C(5) 1.416(3). O(6)-C(7) 1.439(3), O(7')-C(2') 1.421(2), O(2)-C(1) 1.429(3), O(2)-C(3) 1.410(2), O(10)-C(1) 1.381(3), C(5)-C(1) 1.557(2), C(5)-C(9) 1.494(4) Å.

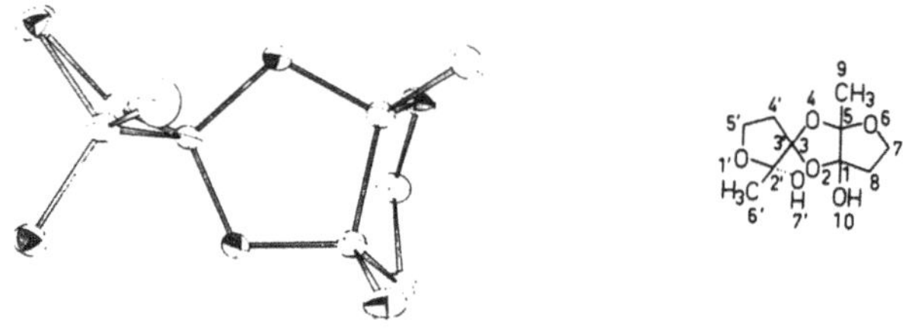

Fig. 1. Molecular structure of $C_{10}H_{16}O_6$ with numbering scheme.

CEPHALOSPOROLIDE B 3-O-METHANESULPHONATE
$C_{11}H_{16}O_6S$ (I)

CEPHALOSPOROLIDE C
$C_{10}H_{16}O_5$ (II)

CEPHALOSPOROLIDE E
$C_{10}H_{14}O_4$ (III)

M.J. ACKLAND, J.R. HANSON, P.B. HITCHCOCK and A.H. RATCLIFFE, 1985. *J. Chem. Soc. Perkin Trans. 1*, 843-847.

I. Orthorhombic, $P2_12_12_1$, a = 5.545, b = 11.272, c = 20.739 Å, Z = 4. Mo radiation, R = 0.055 for 1125 reflections.

II. Monoclinic, $P2_1$, a = 5.566, b = 7.759, c = 12.633 Å, β = 92.84°, Z = 2. Mo radiation, R = 0.051 for 789 reflections.

III. Orthorhombic, $P2_12_12_1$, a = 6.431, b = 9.967, c = 15.518 Å, Z = 4. Mo radiation, R = 0.047 for 686 reflections.

The X-ray structure analyses of (I), (II) and (III) elucidated their structures. In (I) there is a cis configuration at the double bond (Fig. 1); the crystal structure analysis showed cephalosporolide B to be 3-hydroxy-6-oxodec-4-en-9-olide. Cephalosporolide C was shown to be 3,4-dihydroxy-6-oxodecan-9-olide.

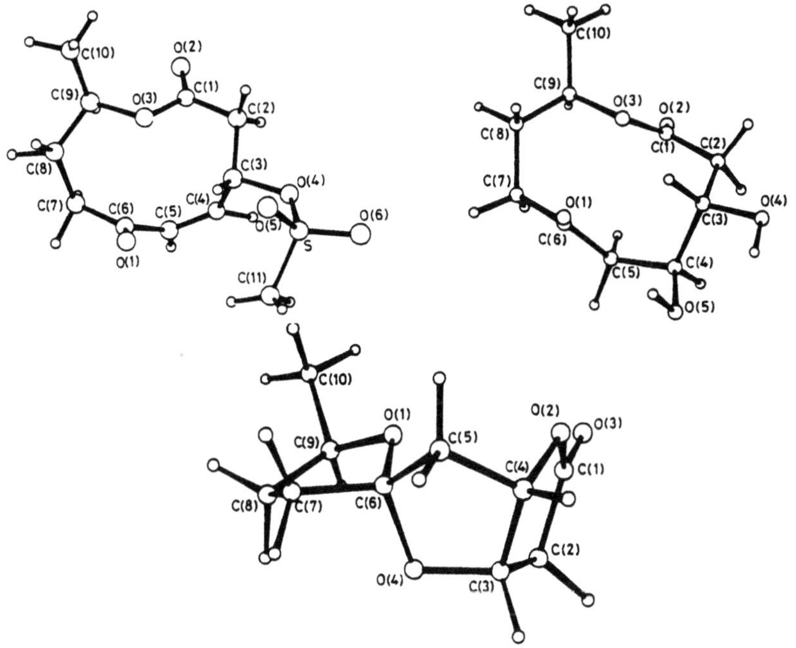

Fig. 1. Perspective drawings of molecules of: $C_{11}H_{16}O_6S$ (top left), $C_{10}H_{16}O_5$ (top right) and $C_{10}H_{14}O_4$ (bottom).

(3S,4S,5R,6R)-5-ACETOXY-3-(1-BROMOACETOXYETHYL)-3,4-DIMETHYLTETRAHYDROFURAN-2-ONE
$C_{12}H_{17}BrO_6$

H. UHR, A. ZEECK, W. CLEGG, E. EGERT, H. FUHRER and H.H. PETER, 1985. *J. Antibiot.*, **38**, 1684-1690.

Orthorhombic, $P2_12_12_1$, a = 9.846, b = 10.962, c = 13.416 Å, Z = 4. Mo radiation, R = 0.042 for 2850 reflections.

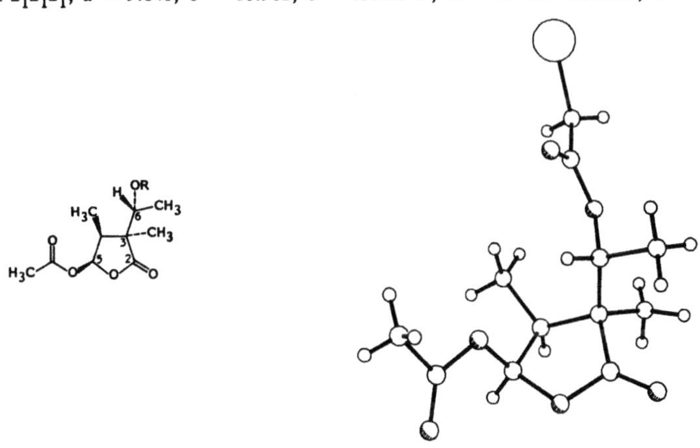

Fig. 1. The $C_{12}H_{17}BrO_6$ molecule (R = $COCH_2Br$).

The structure and absolute stereochemistry of this derivative of acetomycin were established as shown in Fig.

1.

3-ACETYL-4-HYDROXY-6-PHENYL-2-PYRONE
$C_{13}H_{10}O_4$

V.G. THAILAMBAL and V. PATTABHI, 1985. *Acta Cryst.*, C41, 802-804.

Monoclinic, $P2_1/c$, a = 5.558, b = 24.301, c = 7.982 Å, β = 92.22°, Z = 4. Mo radiation, R = 0.051 for 1008 reflections.

In the molecule (Fig. 1) the pyrone ring is planar to within 0.013 Å and the dihedral angles between this plane and (a) the phenyl ring plane and (b) the plane of the acetyl group are 3.5(2) and 2.6(3)° respectively. There is a very strong asymmetrical hydrogen bond between the hydroxyl oxygen atom, O(14), and the carbonyl oxygen atom of the acetyl group, O(17), of length 2.464(7) Å, stabilizing the near coplanarity of the pyrone ring and the acetyl moiety. Bond lengths and angles in the molecule are normal and there are several short C-H...O intermolecular contacts which are possible hydrogen bonds.

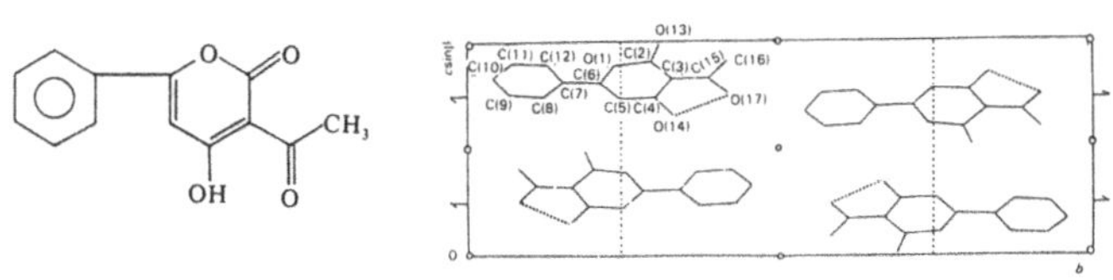

Fig. 1. $C_{13}H_{10}O_4$: the molecular formula and the molecular packing projected down the a-axis.

(3aR,5aS,7R,9aR,9bR)-3a,5a,6,7,8,9,9a,9b-OCTAHYDRO-7,9b-DIMETHYLNAPHTHO[1,2-c]FURAN-1(3H)-ONE (OBLONGOLIDE)
$C_{14}H_{20}O_2$

M.J. BEGLEY and J.F. GROVE, 1985. *J. Chem. Soc. Perkin Trans. 1*, 861-863.

Orthorhombic, $P2_12_12_1$, a = 7.854, b = 7.920, c = 20.058 Å, Z = 4. Cu radiation, R = 0.0593 for 673 reflections.

The cyclohexane ring has a chair conformation (Fig. 1) and the cyclohexene ring has a half chair conformation. The lactone ring has an envelope conformation with C(3a) 0.49 Å out of the plane of the other four atoms.

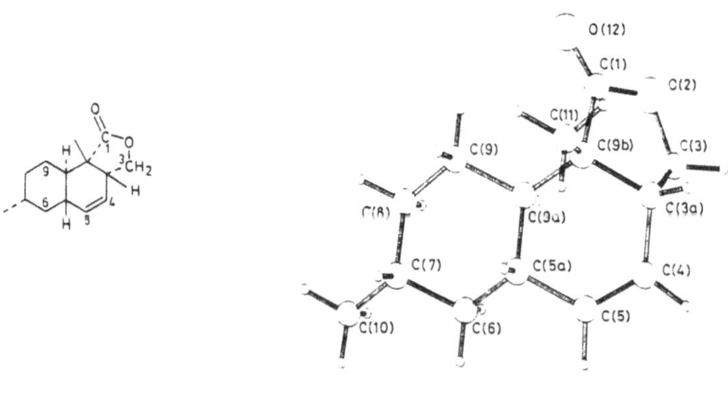

Fig. 1. Perspective drawing of a molecule of $C_{14}H_{20}O_2$.

3,4',5,7-TETRAHYDROXYFLAVYLIUM BROMIDE MONOHYDRATE (PELARGONIDIN BROMIDE MONOHYDRATE)
$C_{15}H_{11}BrO_5,H_2O$ $C_{15}H_{11}O_5^+,Br^-,H_2O$

N. SAITO and K. UENO, 1985. *Heterocycles*, **23**, 2709-2714.

Triclinic, $P\bar{1}$, a = 8.929, b = 11.360, c = 7.299 Å, α = 95.0, β = 93.61, γ = 93.35°, Z = 2. Cu radiation, R = 0.067 for 2351 reflections.

Details of dimensions and possible resonance structures are in Fig. 1. The cation is nearly planar; the phenyl ring is only rotated 3.8° out of the benzopyrylium plane. The ions and water molecules are linked by O-H...Br and O-H...O hydrogen bonds.

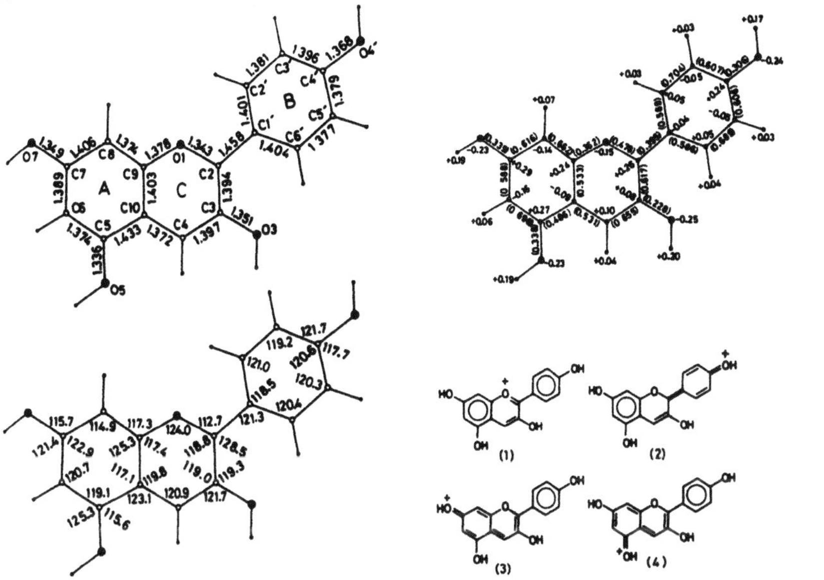

Fig. 1. Dimensions, charge densities and π-bond orders for the $C_{15}H_{11}O_5^+$ cation along with possible resonance forms.

1,3,4,9-TETRAHYDRO-8-HYDROXY-5-METHOXY-6,7-METHYLENEDIOXY-1-METHYLNAPHTHO[2,3-c]-FURAN-4,9-DIONE (VENTILONE-C)

$C_{15}H_{12}O_7$

T. HANUMAIAH, G.S.R. RAO, C.P. RAO, K.V.J. RAO, H.J. COWE, P.J. COX, R.A. HOWIE, D.S. MARSHALL and R.H. THOMSON, 1985. *Tetrahedron*, **41**, 635-642.

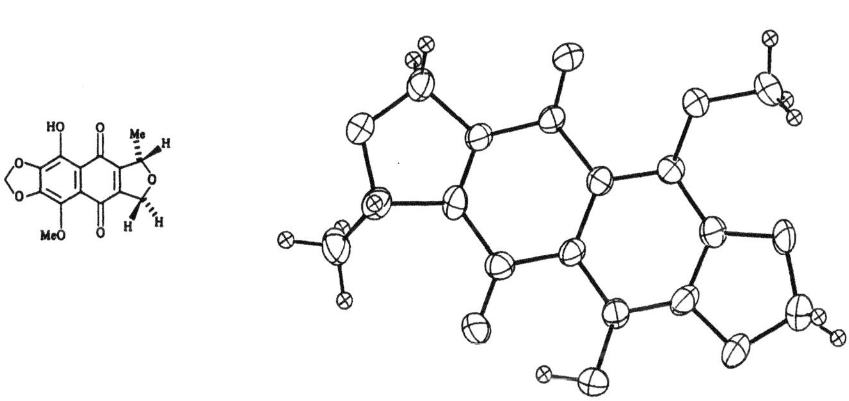

Fig. 1. The ventilone-C molecule.

Orthorhombic, $P2_12_12$, a = 6.715, b = 19.365, c = 9.658 Å, Z = 4. Mo radiation, R = 0.0533 for 820 reflections.

Ventilone-C is determined to have structure and stereochemistry as shown in Fig. 1.

(R)-ISOLAURALLENE
$C_{15}H_{20}Br_2O_2$ (I)

(S)-NEOLAURALLENE
$C_{15}H_{20}Br_2O_2$ (II)

A. FURUSAKI, S. KATSURAGI, K. SUEHIRO and T. MATSUMOTO, 1985. *Bull. Chem. Soc. Jpn.*, **58**, 803-809.

I. Orthorhombic, $P2_12_12_1$, a = 10.393, b = 27.824, c = 5.397 Å, Z = 4. Mo radiation, R = 0.076 for 1091 reflections.

II. Monoclinic, $P2_1$, a = 13.338, b = 4.929, c = 12.435 Å, β = 98.22°, Z = 2. Mo radiation, R = 0.069 for 1002 reflections.

The conformations of the nine-membered rings are similar for both compounds; ring torsion angles are in Fig. 1. The diastereomeric compounds have the same absolute configuration at each of the C atoms to which the ethyl group or Br atom is attached while they have different configurations at each of the bridgehead C atoms.

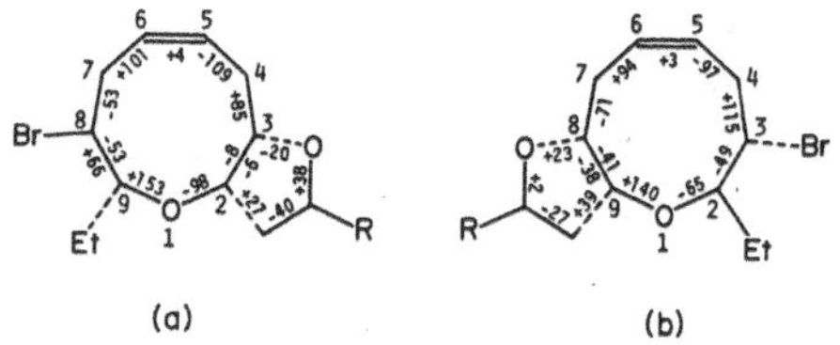

Fig. 1. Ring torsion angles in (a) R-isolaurallene and (b) S-neolaurallene.

(2Z,7E)-3,7-DIMETHYL-10-(1-METHYLETHYLIDENE)-2,7-CYCLODECADIEN-1-ONE (ISOGERMACRONE)
$C_{15}H_{22}O$

U. JACOBSSON, T. NORIN, M. WEBER and E. TSANKOVA, 1985. *Tetrahedron*, **41**, 2033-2036.

Monoclinic, $P2_1/n$, a = 6.499, b = 7.908, c = 25.750 Å, β = 89.73°, Z = 4. Mo radiation, R = 0.046 for 839 reflections.

The structure and stereochemistry of iso-germacrone are shown in Fig. 1. The torsion angle C(3)-C(4)-C(5)-C(6) is 163° indicating a strained 4E double bond. The molecule has an anti orientation of the methyl groups at C(4) and C(10).

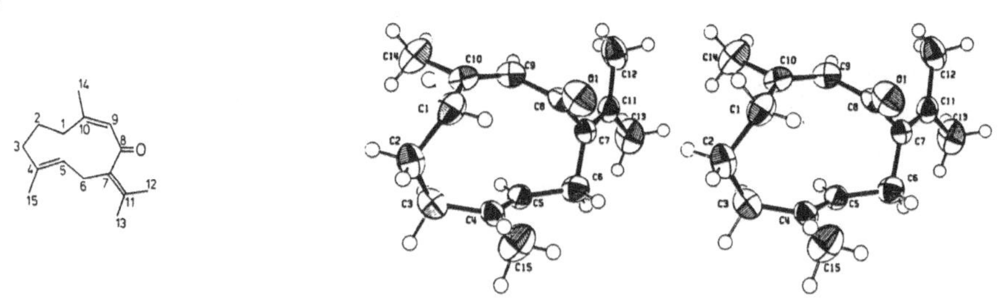

Fig. 1. The iso-germacrone molecule and a stereoview.

3,3',5,5',7-PENTAHYDROXY-4'-METHOXY-2,3-cis-FLAVANE TETRAHYDRATE
$C_{16}H_{16}O_7,4(H_2O)$

G. WEERATUNGA, L. BOHLIN and S. SUNDELL, 1985. *Acta Chem. Scand.*, **B39**, 597-601.

Monoclinic, $P2_1$, a = 8.760, b = 19.892, c = 10.251 Å, β = 97.53°, Z = 4. Cu radiation, R = 0.044 for 1931 reflections.

The crystallographic asymmetric unit consists of two independent molecules (Fig. 1) and eight water molecules of hydration. The molecules are stacked in rows extending in the a-direction with the water molecules placed between these rows and these, together with the flavane hydroxyl groups form an extensive hydrogen bond network. The conformations of the two independent molecules in the cell differ slightly in their torsion angles.

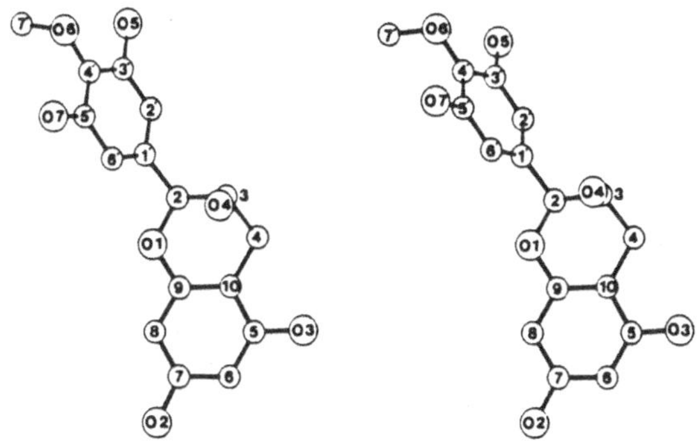

Fig. 1. Schematic drawings of the two independent molecules of $C_{16}H_{16}O_7,4(H_2O)$ in the crystal.

(3R,6S)-(+)-1,7-DIOXASPIRO[5.5]UNDEC-3-YL 3,5-DINITROBENZOATE
$C_{16}H_{18}N_2O_8$

K. MORI, H. WATANABE, K. YANAGI and M. MINOBE, 1985. *Tetrahedron*, **41**, 3663-3672.

Orthorhombic, $P2_12_12_1$, a = 9.778, b = 31.075, c = 5.768 Å, Z = 4. Cu radiation, R = 0.039 for 1489 reflections.

The structure, stereochemistry and packing of this derivative of the sex pheromone of the olive fruit fly are shown in Fig. 1. The bond C(3)-O(12) is clearly axial.

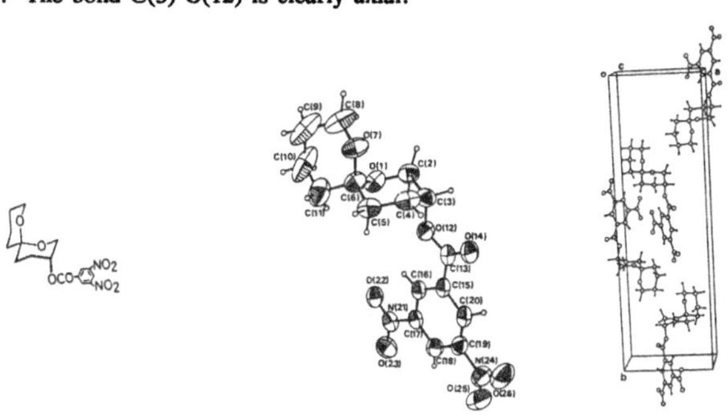

Fig. 1. The structure and packing of $C_{16}H_{18}N_2O_8$.

4-DEACETOXY-10-EPIOLGUINE
$C_{16}H_{20}O_7$

G. DELGADO, R. PEREDA-MIRANDA and A.R. DE VIVAR, 1985. *Heterocycles*, **23**, 1869-1872.

Orthorhombic, $P2_12_12_1$, a = 7.3663, b = 14.7419, c = 15.7355 Å, Z = 4. Cu radiation, R = 0.0576 for 902 reflections.

The structure and stereochemistry are as shown in Fig. 1 with H(5) axial, the C(6)-C(7) epoxide cis, and the C(8)=C(9) double bond trans. The α,β-unsaturated γ-lactone ring has a pseudo-chair conformation with C(5) at the flap and the side chain pseudo-equatorial. The configuration at the C(10)-C(11) chiral centres is shown to be *pref.*

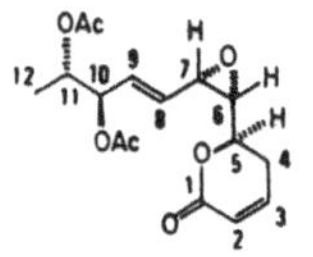

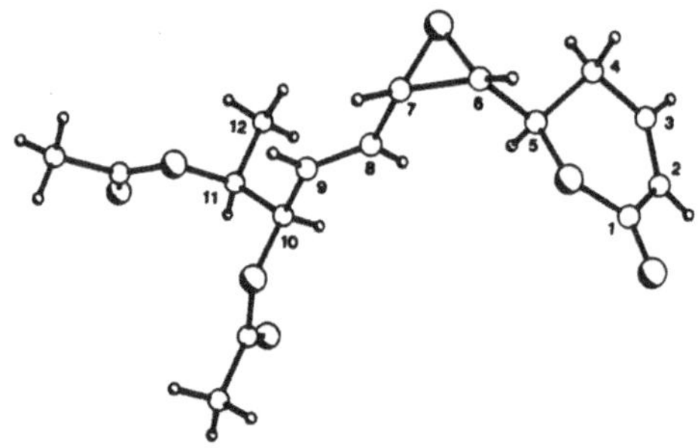

Fig. 1. 4-Deacetoxy-10-epiolguine.

1-METHOXYDECOMPOSITIN
$C_{18}H_{24}O_5$

L. RODRIGUEZ-HAHN, B. ESQUIVEL, A. ORTEGA, J. GARCIA, E. DIAZ, J. CARDENAS, M. SORIANO-GARCIA and A. TOSCANO, 1985. *J. Org. Chem.*, **50**, 2865-2867.

Orthorhombic, $P2_12_12_1$, a = 8.408, b = 12.108, c = 16.447 Å, Z = 4. Mo radiation, R = 0.052 for 1077 reflections.

The crystal structure analysis established the stereochemistry of the molecule (Fig. 1).

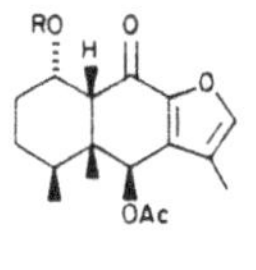

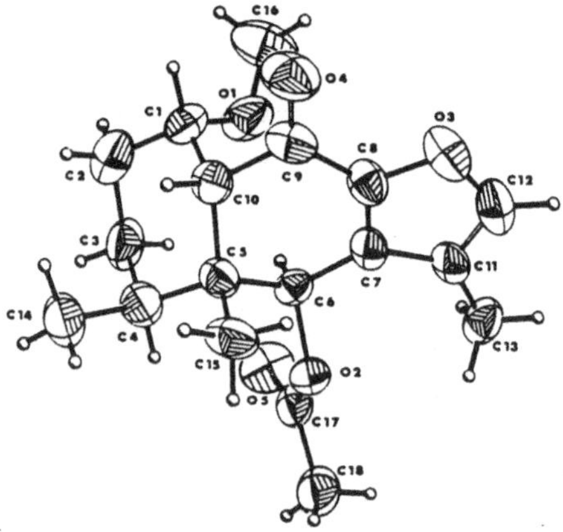

Fig. 1. The $C_{18}H_{24}O_5$ molecule (R = Me).

6R-(1R,2R,3S)-5,6-DIHYDRO-6-[1,2,3-TRIS(ACETYLOXY)HEPTYL]-2H-PYRAN-2-ONE (BORONOLIDE)
$C_{18}H_{26}O_8$

A. KJAER, R. NORRESTAM and J. POLONSKY, 1985. *Acta Chem. Scand.*, **B39**, 745-749.

Orthorhombic, $P2_12_12_1$, a = 6.111, b = 16.659, c = 18.560 Å, Z = 4. Mo radiation, R = 0.043 for 862 reflections (at 120 K). (a = 6.145, b = 17.165, c = 18.712 Å at 294 K).

Attempts to determine the absolute configuration of the molecule (Fig. 1) were unsuccessful. The three acetoxy groups of the boronolide molecule are all planar. In the heptyl chain C(6-12), four of the atoms, C(7-10), and O(5) form a planar zig-zag chain. The pyran ring (O(1),C(1-5)) has a half boat conformation with the C(5) atom deviating by 0.576 Å from the best plane through O(1), C(1-4).

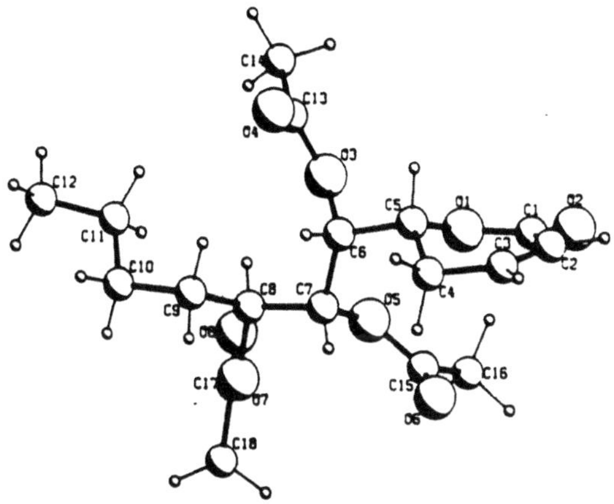

Fig. 1. A perspective drawing of a molecule of $C_{18}H_{26}O_8$.

METHYL α-(3-HYDROXY-5-OXO-4-PHENYL-2H-2-FURYLIDENE)PHENYLACETATE (VULPINIC ACID) $C_{19}H_{14}O_5$

C. BRASSY, B. BACHET, L. MOLHO and D. MOLHO, 1985. *Acta Cryst.*, C41, 781-783.

Monoclinic, $P2_1/n$, a = 26.037, b = 7.842, c = 7.837 Å, β = 94.28°, Z = 4. Mo radiation, R = 0.051 for 2377 reflections.

In the molecule (Fig. 1) the acetyl group, the furylidene ring and the attached benzene ring C(19,20...24) are almost coplanar. The dihedral angle between the ring planes is 13(1)°. The phenyl ring C(6,7...11) is also planar and the angle between its plane and that of the furylidene ring is 69(1)°. Bond lengths and angles in the molecule are normal and its conformation is stabilized by an intramolecular hydrogen bond O(18)-H...O(4), of length 2.513(5) Å, between the acetate oxygen atom and the adjacent hydroxy group. There are also two short intramolecular C-H...O contacts between the oxygen atoms attached to the furan ring and the coplanar benzene ring. Intermolecular distances are essentially van der Waals contacts.

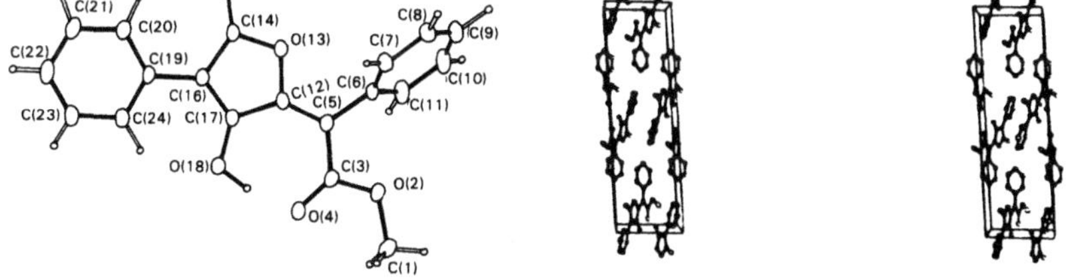

Fig. 1. $C_{19}H_{14}O_5$: a projection of the molecule showing the atomic numbering and a stereoview of the unit cell contents viewed along the b-axis.

5-HYDROXY-2-(4'-HYDROXY-3',5'-DIMETHOXYPHENYL)-3,7-DIMETHOXY-4H-1-BENZOPYRAN-4-ONE (α-form)

$C_{19}H_{18}O_8$ (I)

5-HYDROXY-2-(4'-HYDROXY-3',5'-DIMETHOXYPHENYL)-3,7-DIMETHOXY-4H-1-BENZOPYRAN-4-ONE (β-form)
$C_{19}H_{18}O_8$ (II)

I.R. CASTLEDEN, S.R. HALL, S. NIMGIRAWATH, S. THADANITI and A.H. WHITE, 1985. *Aust. J. Chem.*, **38**, 1177-1185.

I. Triclinic, $P\bar{1}$, a = 12.663, b = 9.592, c = 7.444 Å, α = 102.48, β = 101.39, γ = 91.72°, Z = 2. Mo radiation, R = 0.043 for 1181 reflections.

II. Monoclinic, $P2_1/n$, a = 17.139, b = 12.728, c = 7.845 Å, β = 95.07°, Z = 4. Mo radiation, R = 0.044 for 1421 reflections.

For each of the two phases, the asymmetric unit of the structure is a single molecule (Fig. 1). The skeletal atoms of the two fused six-membered rings are essentially coplanar and form dihedral angles of 7.8 (α) and 0.4° (β) with the associated phenyl ring. The C(2)-C(21) bond distance (1.452(5) (α), 1.470(5) Å (β)) indicates considerable inter-ring conjugation. Of the four methoxy groups, the methyl substituent is coplanar with the associated ring system except for the substituent at C(3). Packing is dominated by a parallel disposition of molecular planes which exhibit appreciable overlap. There are short hydrogen bonds involving phenolic, carbonyl and methoxy groups (O(7)...O(24) 2.884 (α), O(4)...O(24) 2.76(1) Å (β)).

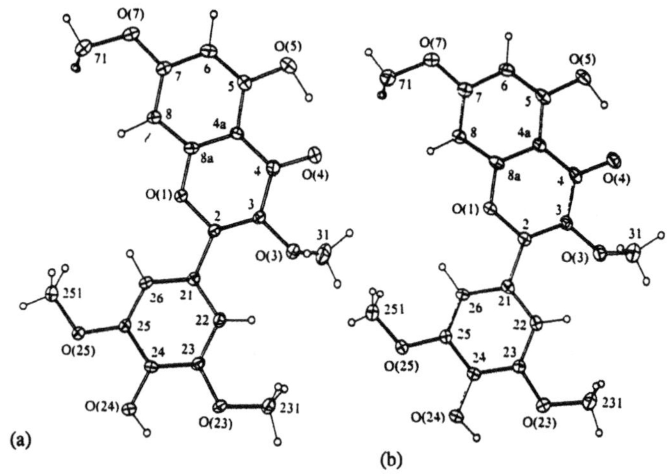

Fig. 1. $C_{19}H_{18}O_8$: (a) α-phase, (b) β-phase.

HYDROXYJATROPHONE C
$C_{20}H_{24}O_4$

V.A. SANTOPIETRO, J. DONOHUE and A.B. SMITH III, 1985. *J. Cryst. Spect. Res.*, **15**, 247-255.

Orthorhombic, $P2_12_12_1$, a = 10.222, b = 10.622, c = 16.434 Å, Z = 4. Mo radiation, R = 0.049 for 1627 reflections.

The analysis confirms the proposed structure and in particular the relative stereochemistry of the two O atoms bonded to the cyclopentene ring (Fig. 1). The geometry about C(5)-C(6) and C(8)-C(9) is *E* and the hydroxy group at C(2) is β. Molecules are linked by O-H...O hydrogen bonds (O...O 2.818 Å).

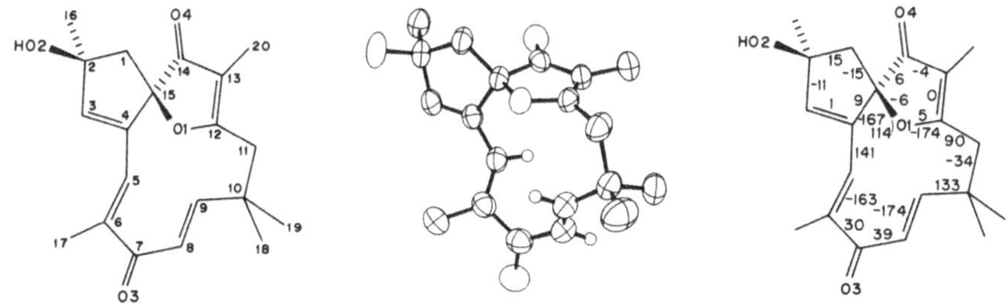

Fig. 1. Hydroxyjatrophone C: structure, conformation and endocyclic torsion angles.

RABDOSERRIN-D
$C_{20}H_{30}O_5$

WU ZIWU and CHEN YUANZHU, 1985. *Jiegou Huaxue (J. Struct. Chem.)*, **4**, 9-11.

Orthorhombic, $P2_12_12_1$, a = 7.298, b = 9.872, c = 25.143 Å, Z = 4. Mo radiation, R = 0.064 for 1424 reflections.

The three six-membered rings (Fig. 1) have chair conformation. Atom C(14) is 0.678 Å from the C(8),C(13),C(16),C(17) plane. The molecule has antitumour activity.

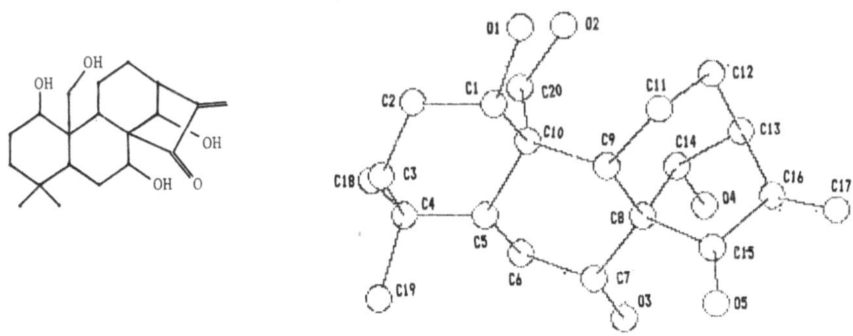

Fig. 1. The structure of rabdoserrin-D.

(2R,3R,4S(R*))-3-((4'-HYDROXY-3'-METHOXYPHENYL)METHYL)-4-((4''-HYDROXY-3''-METHOXYPHENYL)(METHOXY)METHYL)-2-METHOXYTETRAHYDROFURAN (α-INTERMEDIANOL)
$C_{22}H_{28}O_7$

R.C. CAMBIE, G.R. CLARK, P.A. CRAW, T.C. JONES, P.S. RUTLEDGE and P.D. WOODGATE, 1985. *Aust. J. Chem.*, **38**, 1631-1645.

Orthorhombic, $P2_12_12_1$, a = 5.948, b = 8.828, c = 39.517 Å, Z = 4. Mo radiation, R = 0.043 for 1478 reflections.

In the molecule (Fig. 1) the average C-C (aromatic) distance is 1.383(8) Å. All other bond lengths and angles are normal except for C(3)-C(4) which is 1.577(5) Å. The dihedral angle between H(2) and H(3) is 74.3° and the tetrahydrofuran ring is slightly distorted. There are two intermolecular hydrogen bonds in the crystal structure, O(3')...O(4') 2.89 and O(7a)...O(4'') 2.83 Å.

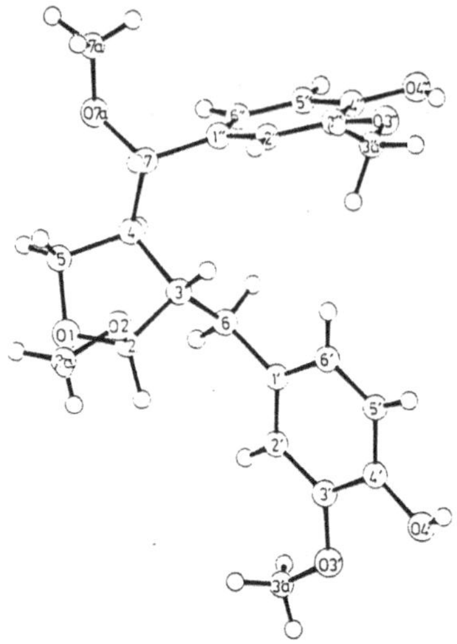

Fig. 1. $C_{22}H_{28}O_7$: the molecular structure.

COLEONOL
$C_{22}H_{34}O_7$

S. RAMAKUMAR, K. VENKATESAN, J.S. TANDON and M.M. DHAR, 1985. *Z. Kristallogr.*, **173**, 81-86.

Orthorhombic, $P2_12_12_1$, a = 20.24, b = 16.26, c = 6.69 Å, Z = 4. Cu radiation, R = 0.104 for 1440 reflections.

An investigation of this material Coleonol isolated from Coleus Forskohlii establishes that it is identical to Forskolin reported previously (1).

1. Structure Reports, **46B**, 564.

1',4-DIHYDROXY-6,7'-DIMETHOXY-2,3'-DIMETHYL(1,2'-BINAPHTHALENE)-5,5',8,8'-TETRONE 3,3'-DIMETHOXYISODIOSPYRIN
$C_{24}H_{18}O_8$

Y. WANG, M.C. CHENG, H.C. CHEN and F.C. CHEN, 1985. *Acta Cryst.*, **C41**, 1270-1271.

Orthorhombic, $P2_12_12_1$, a = 9.1422, b = 10.370, c = 21.494 Å, Z = 4. Mo radiation, R = 0.0275 for 1587 reflections.

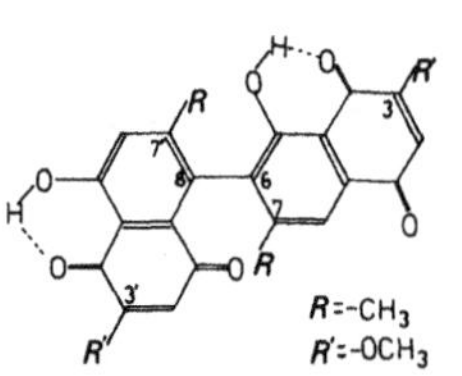

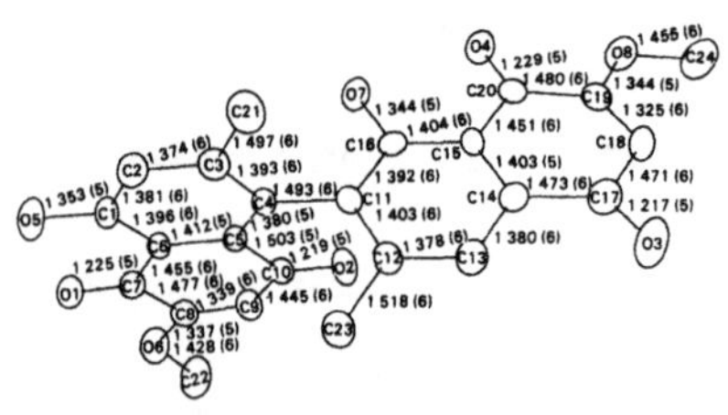

Fig. 1. $C_{24}H_{18}O_8$: the molecular skeleton and bond lengths.

The molecule (Fig. 1) can be considered as two juglone moieties connected by a C-C single bond (C(6)-C(8')). The dihedral angle between the two planes is 79.7(5)°. The hydroxyl group at the 5-position and the quinone group at the 4-position form a pseudo-six-membered ring through intramolecular hydrogen bonding (O(1)...O(5) 2.561,

O(4)...O(7) 2.583 Å). No intermolecular hydrogen bonds were observed.

(R,S)-5,7-DIHYDROXY-8-(1-(4-HYDROXYPHENYL)ETHYL)-3-METHOXY-2-(4-METHOXYPHENYL)-4H-BENZOPYRAN-4-ONE (HAPLOPAPPIN)

$C_{25}H_{22}O_7$

R. TSCHESCHE, M.S. DELHVI, S. SEPULVEDABOZA, F. ZILLIKEN, A. KIRFEL and G. WILL, 1985. *Liebigs Ann. Chem.*, 2465-2471.

Monoclinic, $P2_1/n$, a = 9.622, b = 18.114, c = 12.724 Å, β = 100.73°, Z = 4. Mo radiation, R = 0.070 for 1553 reflections.

The analysis established the structure and stereochemistry of this flavenoid haplopappin to be as shown in Fig. 1.

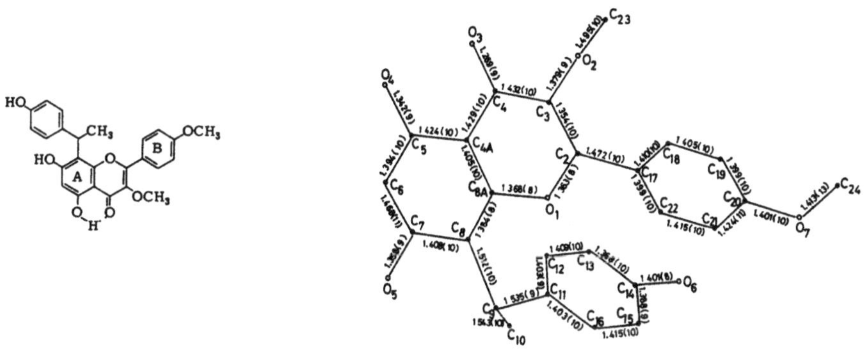

Fig. 1. The haplopappin molecule and conformation.

PENTA-O-ACETYL-(+)-CATECHIN

$C_{25}H_{24}O_{11}$

F.R. FRONCZEK, G. GANNUCH, W.L. MATTICE, R.W. HEMINGWAY, G. CHIARI, F.L. TOBIASON, K. HOUGLUM and A. SHANAFELT, 1985. *J. Chem. Soc. Perkin Trans.* 2, 1383-1386.

Monoclinic, $C2$, a = 23.200, b = 9.801, c = 11.080 Å, β = 100.64°, Z = 4. Mo radiation, R = 0.058 for 1121 reflections.

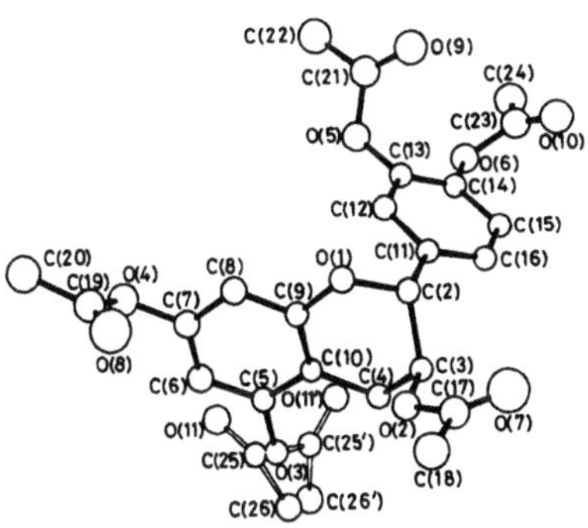

Fig. 1. Conformation of molecules of $C_{25}H_{24}O_{11}$ in the crystal: two positions are shown for the disordered acetyl group.

One of the acetyl groups is disordered; axial positions are occupied by the substituents bonded to heterocyclic ring atoms C(2) and C(3). In Fig. 1, two positions are shown for the disordered acetyl group; each position is equally populated.

PSAMMAPLYSIN-A ACETAMIDE ACETATE
$C_{25}H_{27}Br_4N_3O_8$

D.M. ROLL, C.W.J. CHANG, P.J. SCHEUER, G.A. GRAY, J.M. SHOOLERY, G.K. MATSUMOTO, G.D. VAN DUYNE and J. CLARDY, 1985. *J. Am. Chem. Soc.*, **107**, 2916-2920.

Monoclinic, $I2_1$, a = 16.340, b = 4.9619, c = 38.126 Å, β = 101.88°, Z = 4. Cu radiation, R = 0.116 for 1327 reflections.

The chemical aspects of the structure (Fig. 1) are unexceptional. However, the structure to date has been refined in the body-centred space group $I2_1$. The body-centring breaks down in the regions of the O-acetate and N-acetate fragments but $P2_1$ refinements are hampered by strong correlations between nearly body-centred atoms and few, weak non-body-centred data.

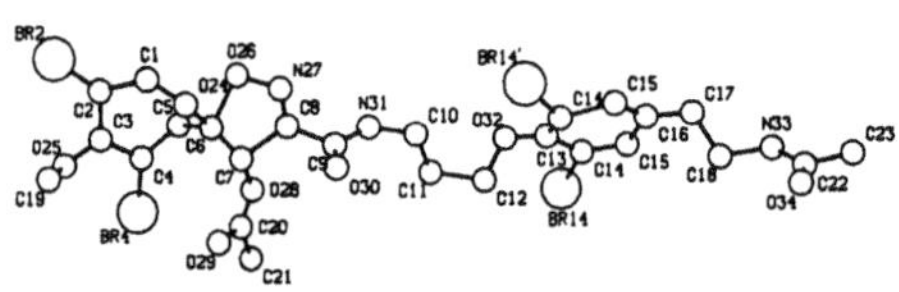

Fig. 1. $C_{25}H_{27}Br_4N_3O_8$: view of the structure.

CAESALPININE C (CELALLOCINNINE)
$C_{25}H_{31}N_3O_2$

S.B. MAHATO, N.P. SAHU, E. MULLER and P. LUGER, 1985. *J. Chem. Soc. Perkin Trans. 2*, 193-196.

Orthorhombic, $P2_12_12_1$, a = 9.963, b = 8.026, c = 28.627 Å, Z = 4. Cu radiation, R = 0.05 for 1369 and 2234 reflections.

The crystal structure analysis established the geometry of the molecule (Fig. 1). The cinnamoyl double bond has a cis configuration. The conformation of the thirteen-membered cycloalkane ring is close to [13333], C(2)-C(3) being the isolated gauche bond.

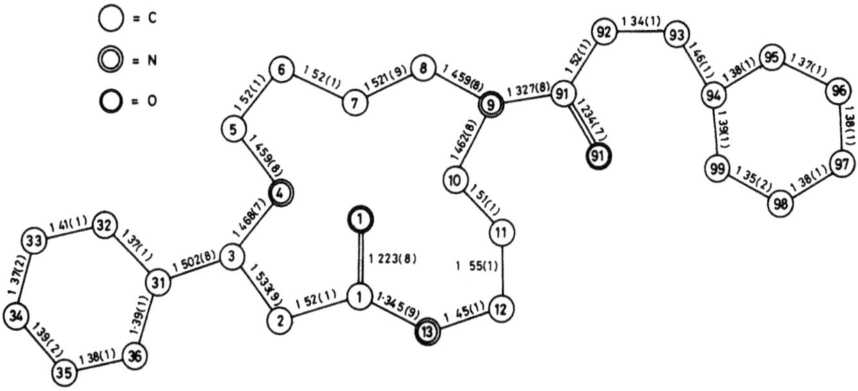

Fig. 1. Structure of $C_{25}H_{31}N_3O_2$ showing bond lengths (Å). Double circles represent nitrogen atoms.

GARVEATIN A TRIACETATE
$C_{26}H_{26}O_8$

E. FAHY, R.J. ANDERSEN, HE CUN-HENG and J. CLARDY, 1985. *J. Org. Chem.*, **50**, 1149-1150.

Monoclinic, $P2_1/n$, a = 8.1755, b = 18.0988, c = 16.0627 Å, β = 86.079°, Z = 4. Cu radiation, R = 0.0511 for 2057 reflections.

The crystal structure analysis established the stereochemistry of this antimicrobial metabolite to be a 1(4H)anthracenone as shown in Fig. 1.

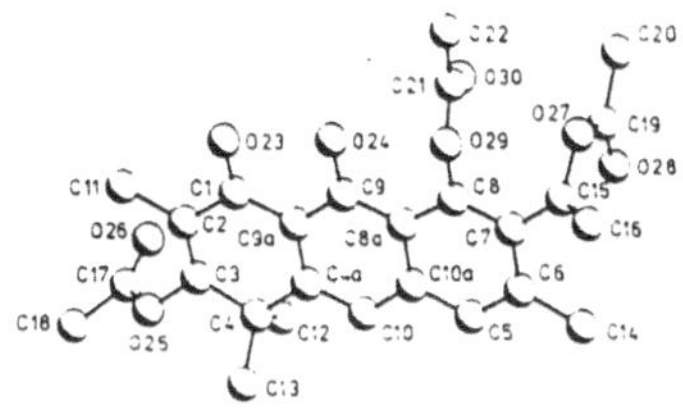

Fig. 1. A molecule of $C_{26}H_{26}O_8$.

LAMELLARIN A DIHYDRATE
$C_{30}H_{27}NO_{10},2(H_2O)$

R.J. ANDERSEN, D.J. FAULKNER, HE CUN-HENG, G.D. VAN DUYNE and J. CLARDY, 1985. *J. Am. Chem. Soc.*, **107**, 5492-5495.

Triclinic, $P\bar{1}$, a = 10.992, b = 11.587, c = 12.059 Å, α = 94.60, β = 81.97, γ = 67.74°, Z = 2. Mo radiation, R = 0.058 for 2521 reflections.

The main pentacyclic array in the molecule is essentially planar (Fig. 1) with the hydroxyl at C(5) anti to the C(13) methoxyl group.

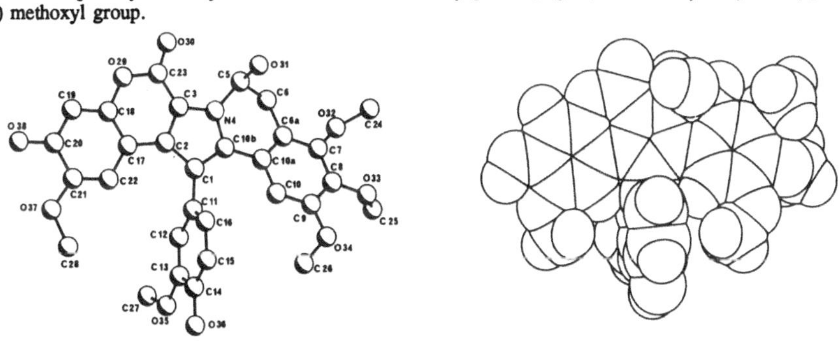

Fig. 1. Two views of Lamellarin A.

METHYL 12α-ACETOXY-20,24-DIMETHYL-16,24-DIOXOSCALARA-14,17-DIEN-25-OATE
$C_{30}H_{42}O_6$

J.P. DECLERCQ, M. VAN MEERSSCHE, J.C. BRAEKMAN and D. DALOZE, 1985. *Acta Cryst.*, **C41**, 1222-1224.

Orthorhombic, $P2_12_12_1$, a = 7.943, b = 14.198, c = 25.232 Å, Z = 4. Mo radiation, R = 0.053 for 1676 reflections.

This molecule,which was isolated from the sponge 'Carteriospongia foliasceus', exhibits a high degree of unsaturation in ring D (C(14)-C(15) 1.327(8), C(17)-C(18) 1.339(8) Å) (Fig. 1). There is an axial ethyl group at the C(4) position. O(33) is disordered.

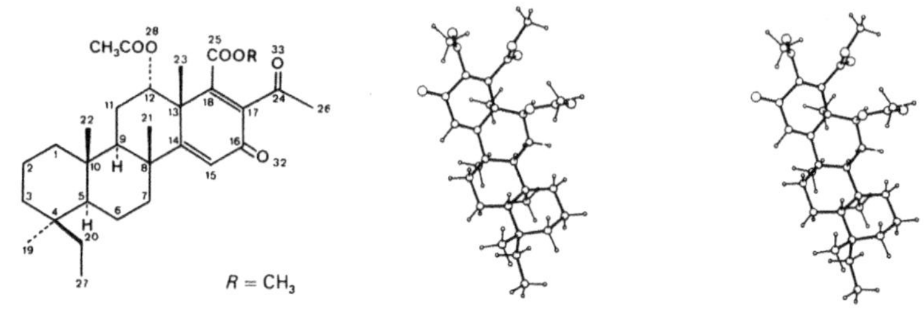

Fig. 1. $C_{30}H_{42}O_6$: the molecular skeleton and a stereoview of the structure.

ASPERLICIN ACETONITRILE SOLVATE
$C_{31}H_{29}N_5O_4,C_2H_3N$

J.M. LIESCH, O.D. HENSENS, J.P. SPRINGER, R.S.L. CHANG and V.J. LOTTI, 1985. *J. Antibiot.*, **38**, 1638-1641.

Tetragonal, $P4_12_12$, a = 20.644, c = 14.089 Å Z = 8. Cu radiation, R = 0.058 for 2055 reflections.

The analysis established the relative and absolute stereochemistry of this cholecystokinin antagonist to be as shown in Fig. 1.

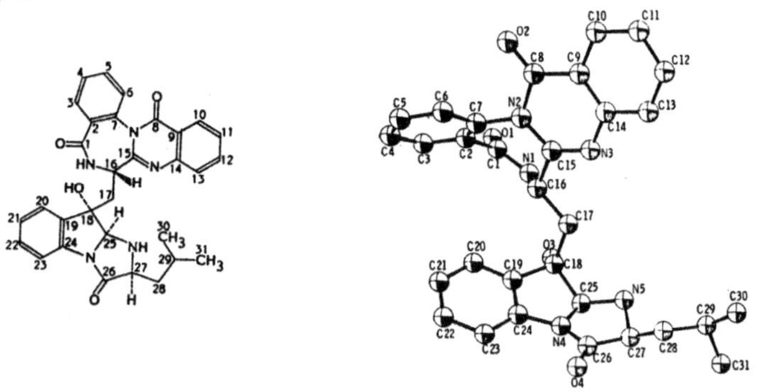

Fig. 1. Asperlicin.

7,7'-DIACETYL-6,6'-DIMETHYL-2,2'-BINAPHTHALENE-1,1',8,8'-TETRAACETATE (STYPANDROL TETRA-ACETATE)
$C_{34}H_{30}O_{10}$

S.M. COLEGATE, P.R. DORLING, C.R. HUXTABLE, B.W. SKELTON and A.H. WHITE, 1985. *Aust. J. Chem.*, **38**, 1233-1241.

Triclinic, $P\bar{1}$, a = 14.352, b = 11.633, c = 10.153 Å, α = 65.84, β = 86.35, γ = 80.99°, Z = 2. Mo radiation, R = 0.052 for 2052 reflections.

The title compound is the acetate of the tetrol responsible for the toxicity of the plant 'Stypandra imbricata'. The molecule (Fig. 1) has the expected bond lengths and angles and has a dihedral angle of 44.3(1)° between the planes of the two naphthalene rings.

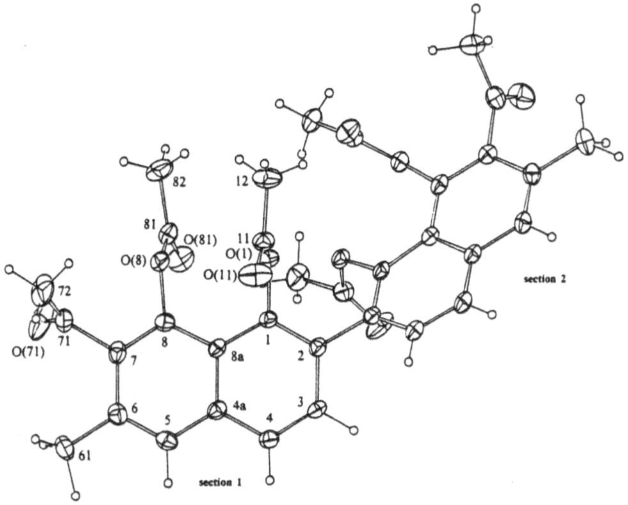

Fig. 1. $C_{34}H_{30}O_{10}$: the molecular structure.

PECTENOTOXIN-1 TRIHYDRATE
$C_{47}H_{70}O_{15}$,$3(H_2O)$

T. YASUMOTO, M. MURATA, Y. OSHIMA, M. SANO, G.K. MATSUMOTO and J. CLARDY, 1985. *Tetrahedron,* **41**, 1019-1025.

Monoclinic, *C*2, a = 31.551, b = 6.219, c = 25.350 Å, β = 97.78°, Z = 4. Cu radiation, R = 0.113 for 1905 reflections.

The structure of this shellfish toxin is as shown in Fig. 1

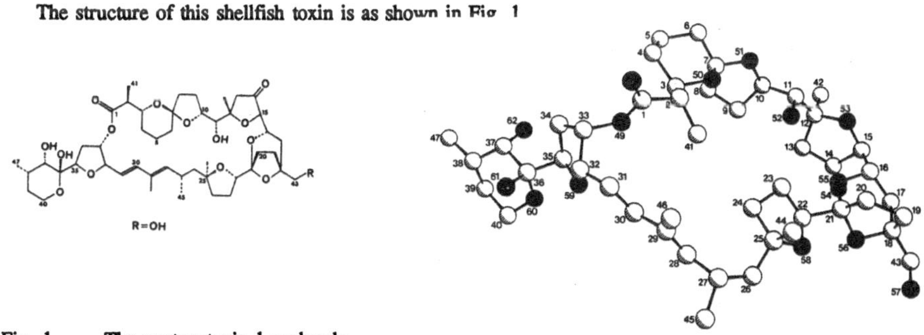

Fig. 1. The pectenotoxin-1 molecule.

NORHALICHONDRIN A p-BROMOBENZOYL METHYL ESTER
$C_{67}H_{87}BrO_{22}$

D. UEMURA, K. TAKAHASHI, T. YAMAMOTO, C. KATAYAMA, J. TANAKA, Y. OKUMURA and Y. HIRATA, 1985. *J. Am. Chem. Soc.*, **107**, 4796-4798.

Monoclinic, $P2_1$, a = 44.654, b = 9.264, c = 9.494 Å, β = 93.72°, Z = 2. Mo radiation, R = 0.1092 for 3080 reflections.

The X-ray analysis established the structure and absolute configuration as shown in Fig. 1.

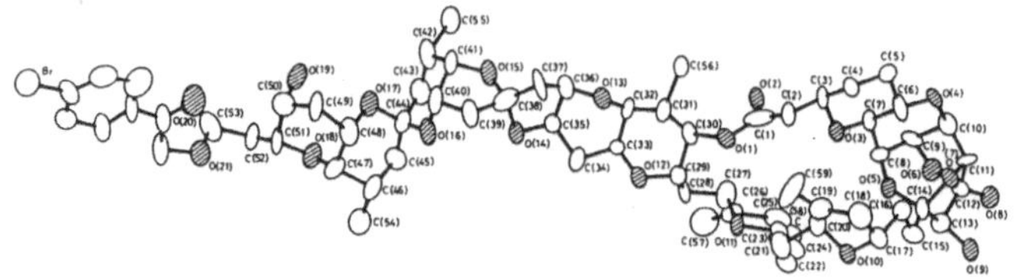

Fig. 1. A view of norhalichondrin A p-bromobenzoylmethyl ester.

2',4,4',6-TETRANITRO-2-CARBOXYBIPHENYL DIMETHYL SULPHOXIDE
$C_{13}H_6N_4O_{10},C_2H_6OS$

E.G. POPOVA, L.A. CHETKINA, V.K. BEL'SKII, A.M. ANDRIEVSKII, A.N. POPLAVSKII and K.M. DYUMAEV, 1985. *Zh. Strukt. Khim.*, **26**, 2-139-144 [J. Struct. Chem., **26**, 262-266].

Monoclinic, *P*112$_1$/*n*, a = 11.282, b = 21.866, c = 7.836 Å, γ = 99.32°, Z = 4. Mo radiation, R = 0.033 for 1893 reflections.

The 2',4,4',6-tetranitro-2-carboxybiphenyl and dimethyl sulphoxide molecules are connected in pairs by strong O-H...O hydrogen bonding (O...O 2.494 Å, see Fig. 1). The position of the hydrogen atom and the geometry of the carbonyl group and the Me_2SO molecule indicate partial ionic character of the molecules in the complex. The dihedral angle between the phenyl ring planes in the biphenyl molecule is 84.8° and the ortho NO_2 and CO_2H groups are twisted out of the respective phenyl mean planes by 6.8 and 40.3°, respectively.

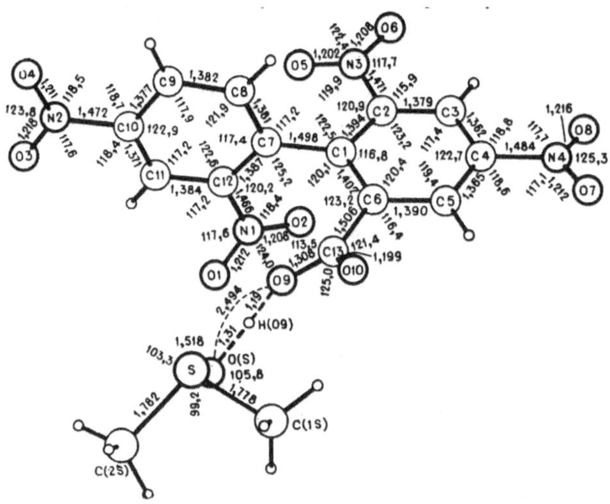

Fig. 1. Geometry of the $C_{13}H_6N_4O_{10},C_2H_6OS$ complex.

N,N-DIMETHYLANILINE HEXAFLUOROBENZENE
$C_{14}H_{11}F_6N$ $C_8H_{11}N,C_6F_6$

T. DAHL, 1985. *Acta Cryst.*, **C41**, 931-933.

Triclinic, *P*$\bar{1}$, a = 6.780, b = 8.256, c = 12.327 Å, α = 99.47, β = 92.01, γ = 106.04°, Z = 2. Mo radiation, R = 0.060 for 1581 reflections (at 120 K).

The molecules (Fig. 1) are stacked (Fig. 2) alternately in infinite columns. Although the N atom has approximately tetrahedral geometry, the overlap between adjacent molecules is approximately equal on both sides of the dimethylaniline indicating that it is a π donor. The corresponding mean interplanar distances are 3.382(2) and

3.392(2) Å.

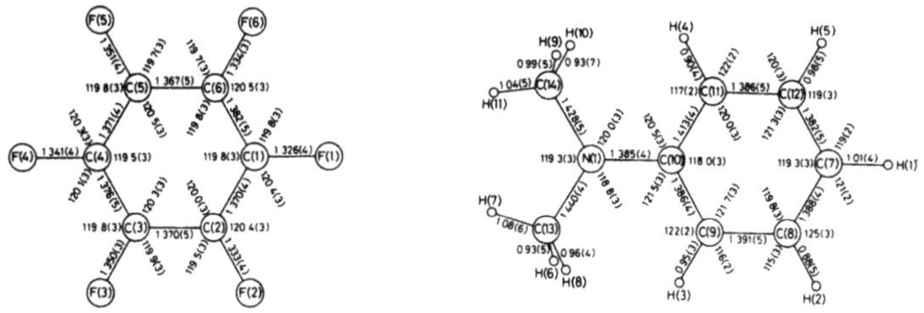

Fig. 1. $C_{14}H_{11}F_6N$: bond distances and angles.

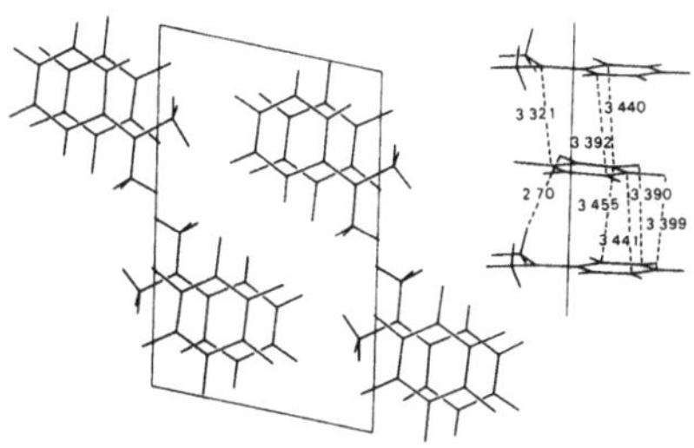

Fig. 2. The molecular packing of $C_{14}H_{11}F_6N$ viewed along [100] and perpendicular to [100].

BIS(ETHYLENEDITHIO)TETRATHIAFULVALENE PENTACYANOTETRASILVER
$C_{15}H_8Ag_4N_5S_8$ $C_{10}H_8S_8^+,C_5Ag_4N_5^-$

U. GEISER, H.H. WANG, L.E. GERDOM, M.A. FIRESTONE, L.M. SOWA, J.M. WILLIAMS and M.-H. WHANGBO, 1985. *J. Am. Chem. Soc.*, **107**, 8305-8307.

Orthorhombic, *Fddd*, a = 13.24, b = 19.48, c = 19.62 Å, Z = 8. Mo radiation, R = 0.043 for 1690 reflections.

In this complex (Fig. 1) every ET^+ cation of the 3D donor network in $(ET)Ag_4(CN)_5$ is surrounded by four ET^+ cations in a quasi-tetrahedral arrangement such that every nearest neighbour ET^+ pair interacts via two equivalent S...S contacts of 3.902(4) Å. Each ET^+ cation is located at the intersection of three twofold axes so that only one-quarter of an ET^+ is crystallographically unique. The 3D network of ET^+ is interwoven with honeycomb-like polymeric networks of Ag^+ and CN^- ions. Each ET^+ is surrounded by a silver cyanide "necklace" consisting of 10 Ag^+ and 10 CN^- ions. There are two kinds of Ag^+ ion; one is bonded to the C atoms of two CN^- ions in linear coordination while the other has trigonal planar geometry.

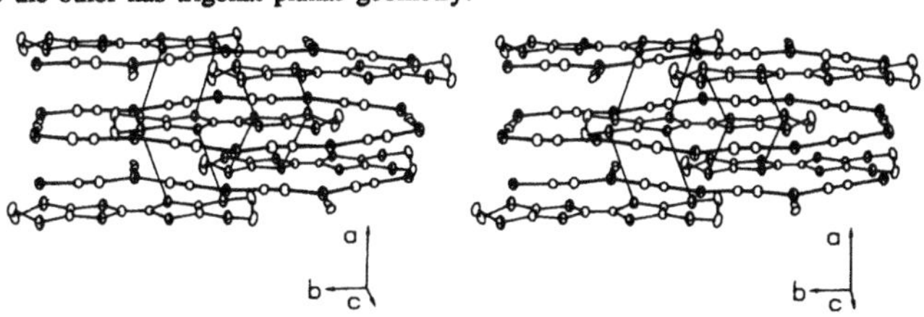

Fig. 1. A stereoview of the crystal structure of $(ET)Ag_4(CN)_5$.

BIS(PENTAFLUOROPHENOL) DIOXANE COMPLEX
$C_{16}H_{10}F_{10}O_4$ $2(C_6F_5OH),C_4H_8O_2$

T. GRAMSTAD, S. HUSEBYE and K. MAARTMANN-MOE, 1985. *Acta Chem. Scand.*, **B39**, 767-771.

Monoclinic, $P2_1/c$, a = 7.951, b = 10.279, c = 10.195 Å, β = 94.76°, Z = 2. Mo radiation, R = 0.033 for 1991 reflections (at -135°C).

The complex is centrosymmetric and sits about a crystallographic centre of symmetry. The complex is hydrogen bonded (Fig. 1) with O(1)-H(1) 0.83(2), H(1)...O(2) 1.82(2), O(1)...O(2) 2.647(1) Å, and C(1)-O(1)-H(1) 111(1) and O(1)-H(1)...O(2) 170(2)°. The plane through the aromatic ring and the plane of the carbon atoms of the dioxane molecule are almost perpendicular to one another. The oxygen and carbon atoms of the pentafluorophenol molecule are coplanar, but the fluorine atoms are not, although they all lie on the same side of this plane.

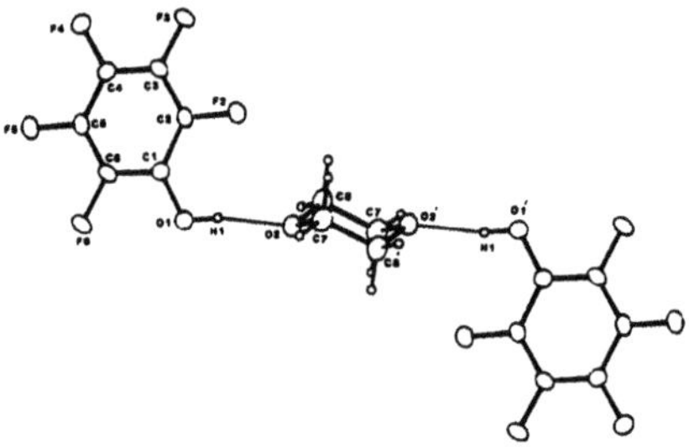

Fig. 1. The centrosymmetric 2:1 complex, $C_{16}H_{10}F_{10}O_4$, showing the benzene rings almost at right angles to the carbon plane of the dioxane molecule.

PICRIC ACID NAPHTHALENE
$C_{16}H_{11}N_3O_7$

A. BANERJEE and C.J. BROWN, 1985. *Acta Cryst.*, C41, 82-84.

Monoclinic, $P2_1/a$, a = 16.248, b = 6.871, c = 14.306 Å, β = 96.62°, Z = 4. Cu radiation, R = 0.066 for 918 reflections.

In this disordered structure both naphthalene and picric acid molecules lie approximately parallel to (010) and stack alternately with hydrogen bonds linking centrosymmetrically-related picric acid molecules (Fig. 1). The disorder arises from the uncertainty in the site of the benzene ring O attachment in the picric acid molecules. In 80% of these the sites are O(1) and in 20% the sites are O(8). The O(1)...O(1') and O(8)...O(8') possible hydrogen bonds are of length 2.84(1) and 3.08(4) Å respectively. In addition, neighbouring picric acid molecules in the layers are hydrogen bonded with a O(4)-H...O(6) distance of 2.901(9) Å. There are also π-bonding interactions within the stacks between the picric acid and naphthalene moieties with interplanar spacings of 3.4 Å (i.e. b/2).

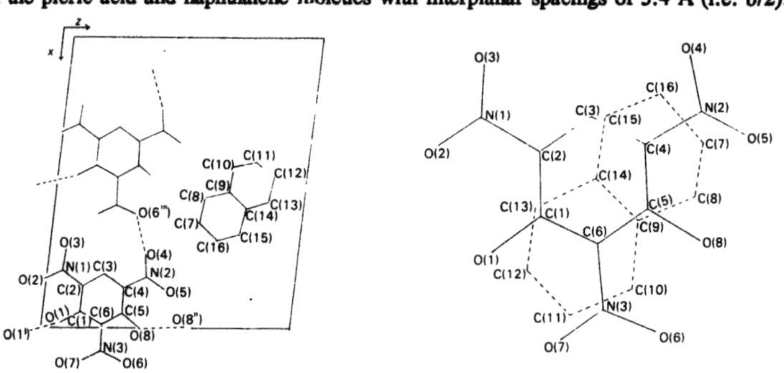

Fig. 1. $C_{16}H_{11}N_3O_7$: a projection onto (010) of the contents of part of the unit cell between y = 0 and y = 0.50 with hydrogen bonds shown as broken lines (left) and (right) the overlap of the naphthalene and picric acid molecules as viewed down the b-axis.

DURENE FLUORANIL
$C_{16}H_{14}F_4O_2$ $C_{10}H_{14}.C_6F_4O_2$

T. DAHL and B. SØRENSEN, 1985. *Acta Chem. Scand.*, B39, 423-428.

Triclinic, $P\bar{1}$, a = 6.820, b = 7.161, c = 8.908 Å, α = 109.57, β = 106.60, γ = 99.05°, Z = 1. Cu radiation, R = 0.087 for 1127 reflections.

Complexes of 1,2,3,5-tetramethylbenzene with hexafluorobenzene and fluoranil did not give sufficiently good data to allow for complete crystal structure analyses. In this complex the partner molecules are stacked alternately in infinite columns with a mean separation of 3.383(3) Å between the molecular planes (Fig. 1).

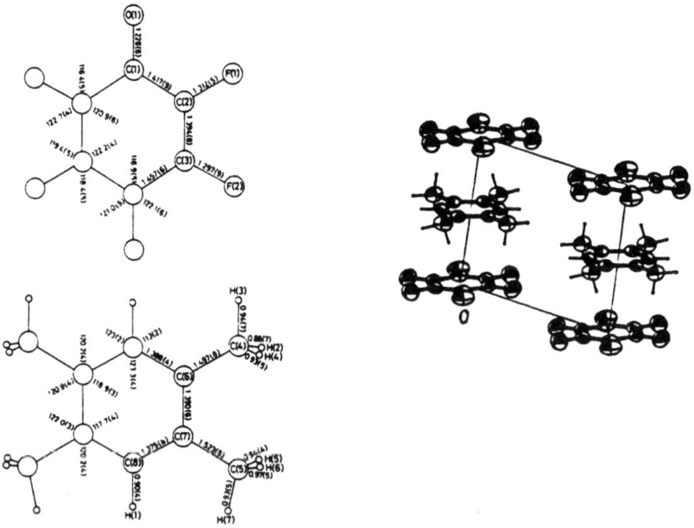

Fig. 1. Bond distances (Å) and angles (°) in the molecule of the complex $C_{16}H_{14}F_4O_2$ (left) with the packing of the molecules viewed perpendicular to (001) (right).

(OXAMIDE OXIMATO)(OXAMIDE OXIME)NICKEL(II) 7,7,8,8-TETRACYANOQUINODIMETHANIDE

$C_{16}H_{15}N_{12}NiO_4$ $C_4H_{11}N_8NiO_4^+, C_{12}H_4N_4^-$

H. ENDRES, A. BONGART, D. NOTHE, I. HENNIG, D. SCHWEITZER, H. SCHAFER, H.W. HELBERG and S. FLANDROIS, 1985. *Z. Naturforsch.*, B40, 489-495.

Triclinic, $P\bar{1}$, a = 3.7718, b = 7.436, c = 17.511 Å, α = 88.67, β = 86.93, γ = 85.05°, Z = 1. Mo radiation, R = 0.035 for 1454 reflections.

The structure has segregated regular parallel stacks of centrosymmetric planar metal complex ions (Fig. 1) and centrosymmetric planar tcnq⁻ anions, with intermolecular N-H...N hydrogen bonds (2.694(5) Å) linking cations and anions. The interplanar spacing in the cation stacks is 3.34(1) Å (3.20 Å in the anion stacks).

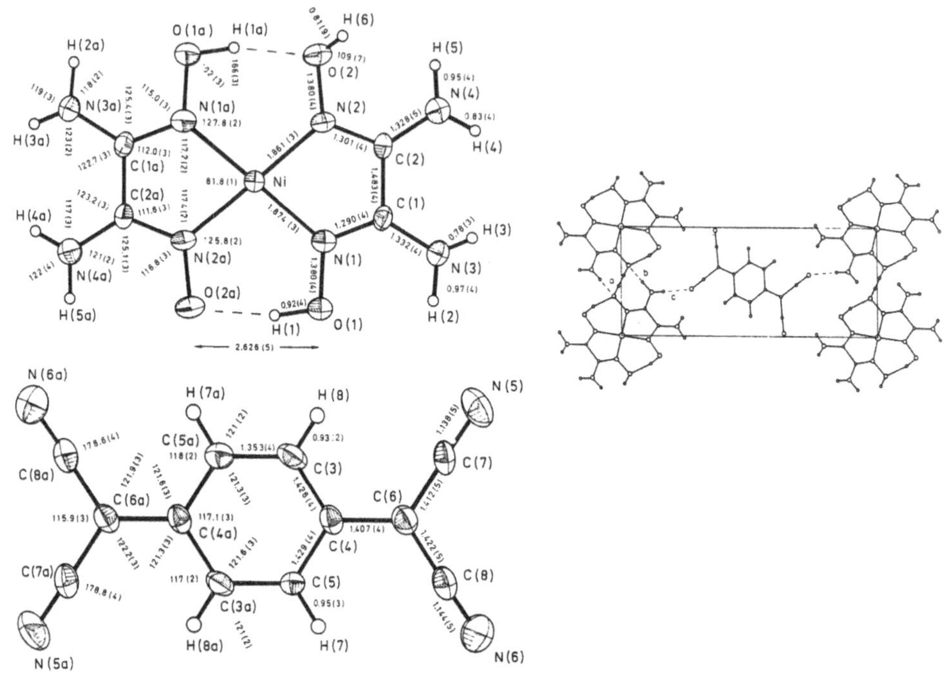

Fig. 1. Views of the cation and anion and a projection of the structure along the short *a*-axis.

(1,1'-DIMETHYL-4,4'-BIPYRIDINIUM) BIS(cis-1,2-DICYANO-1,2-ETHYLENEDITHIOLATO-S,S')NICKEL(II)
$C_{20}H_{14}N_6NiS_4$ (I) $C_8N_4NiS_4^{2-},C_{12}H_{14}N_2^{2+}$

(1,1'-DIMETHYL-4,4'-BIPYRIDINIUM) BIS(BIS(cis-1,2-DIPHENYL-1,2-ETHYLENEDITHIOLATO-S,S')-NICKEL(III))
$C_{68}H_{54}N_2Ni_2S_8$ (II) $2(C_{28}H_{20}NiS_4^-),C_{12}H_{14}N_2^{2+}$

H. KISCH, A. FERNANDEZ, Y. WAKATSUKI and H. YAMAZAKI, 1985. *Z. Naturforsch.*, **B40**, 292-297.

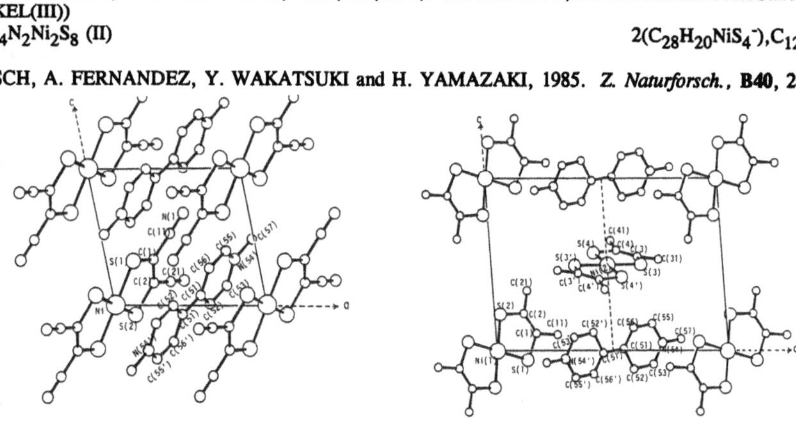

Fig. 1. Projections of $C_8N_4NiS_4^{2-},C_{12}H_{14}N_2^{2+}$ (I) (left) and $2(C_{28}H_{20}NiS_4^-),C_{12}H_{14}N_2^{2+}$ (II) (right).

I. Triclinic, $P\bar{1}$, a = 7.691, b = 11.193, c = 6.432 Å, α = 97.14, β = 96.98, γ = 90.28°, Z = 1. Mo radiation, R = 0.047 for 1257 reflections.

II. Triclinic, $P\bar{1}$, a = 13.670, b = 13.736, c = 9.760 Å, α = 105.79, β = 94.42, γ = 118.56°, Z = 1. Mo radiation, R = 0.041 for 3700 reflections.

In both complexes (Fig. 1) the donors and acceptors are all situated on inversion centres. In (I) the planar acceptor (methyl viologen) interacts with the planar dithiolene via the expected parallel arrangement, but a novel perpendicular arrangement is also found in (II). The Ni-S distances are 2.167 and 2.181(2) Å in (I) and 2.128-2.146(2) Å in (II). The shortest distance between molecular planes is 3.4 Å in (I).

BIS(BIS(ETHYLENEDITHIO)TETRATHIAFULVALENE) DIBROMOIODIDE
$C_{20}H_{16}Br_2IS_{16}$ (I) $C_{10}H_8S_8{}^+,C_{10}H_8S_8,Br_2I^-$

BIS(BIS(ETHYLENEDITHIO)TETRATHIAFULVALENE) DIIODOBROMIDE
$C_{20}H_{16}BrIS_{16}$ (II) $C_{10}H_8S_8{}^+,C_{10}H_8S_8,BrI_2^-$

H. ENDRES, M. HILLER, H.J. KELLER, K. BENDER, E. GOGU, I. HEINEN and D. SCHWEITZER, 1985. *Z. Naturforsch.*, **B40**, 1664-1671.

I. Triclinic, $P\bar{1}$, a = 6.5979, b = 8.998, c = 15.138 Å, α = 94.03, β = 95.13, γ = 110.28°, Z = 1. Mo radiation, R = 0.072 for 2794 reflections.

II. Triclinic, $P\bar{1}$, a = 6.606, b = 9.047, c = 15.229 Å, α = 94.28, β = 95.34, γ = 109.92°, Z = 1. Mo radiation, R = 0.045 for 3353 reflections.

Both isomorphous structures contain BEDT-TTF molecules in general positions and linear anions on inversion centres (Fig. 1). The different sizes of the anions lead to slightly different intra- and inter-stack S...S contacts in the two complexes. The I-Br distances are 2.764(1) in (I) and 2.852(1) in (II). For previous work on (I) see **1**.

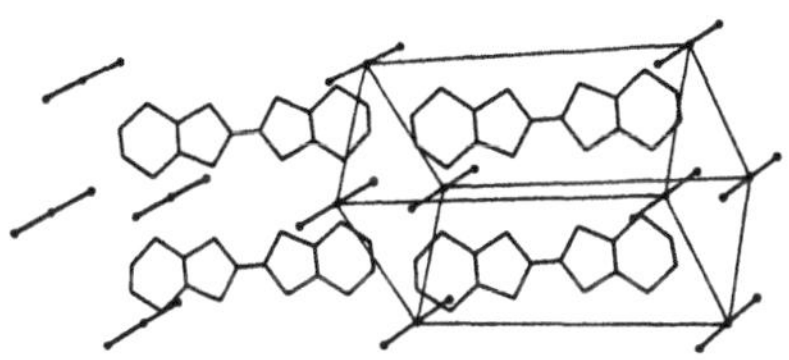

Fig. 1. A projection of the $(BEDT\text{-}TTF)_2Br_2I$ structure perpendicular to the plane through the BEDT-TTF S atoms.

1. Structure Reports, 51B, 691.

BIS(BIS(ETHYLENEDITHIO)TETRATHIAFULVALENE) TRIIODIDE
$C_{20}H_{16}I_3S_{16}$ $C_{10}H_8S_8{}^+,C_{10}H_8S_8,I_3^-$

P.C.W. LEUNG, T.J. EMGE, M.A. BENO, H.H. WANG, J.M. WILLIAMS, V. PETRICEK and P. COPPENS, 1985. *J. Am. Chem. Soc.*, **107**, 6184-6191.

Triclinic, $P\bar{1}$, a = 6.560, b = 9.011, c = 15.171 Å, α = 95.08, β = 95.94, γ = 110.28°, Z = 1. Mo radiation, R = 0.035 for 7782 reflections (at 120 K).

This organic charge-transfer complex is the first ambient-pressure sulphur-based superconductor (T_c = 1.4(1) K). A feature of the structure is the two-dimensional network of BEDT-TTF molecules which is characterized by very short interstack S...S contacts (< 3.60 Å) between adjacent molecules (Fig. 1). This corrugated sheet network of S...S interactions is largely responsible for the anisotropic electronic properties of this organic metal.

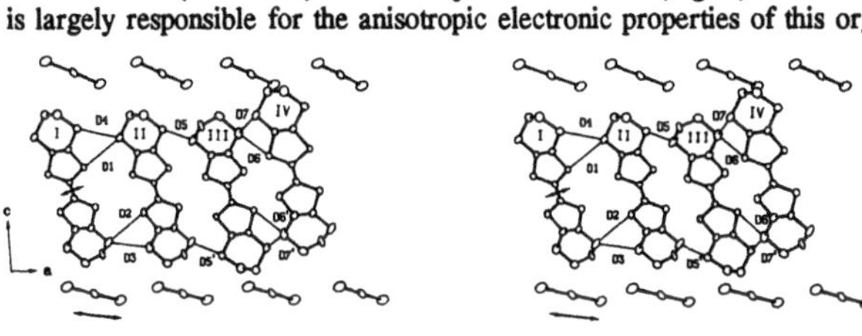

Fig. 1. A stereoview of the intermolecular contacts between adjacent BEDT-TTF molecules.

3,4-DIMETHOXYCINNAMIC ACID 2,4-DINITROCINNAMIC ACID

$C_{20}H_{18}N_2O_{10}$ $C_{11}H_{12}O_4,C_9H_6N_2O_6$

J.A.R.P. SARMA and G.R. DESIRAJU, 1985. *J. Chem. Soc. Perkin Trans. 2*, 1905-1912.

Triclinic, $P\bar{1}$, a = 8.857, b = 13.872, c = 8.578 Å, α = 101.11, β = 100.98, γ = 95.98°, Z = 2. Mo radiation, R = 0.054 for 1313 reflections.

The crystal structure analysis of this 1:1 Donor-Acceptor complex revealed that the crystal consists of interleaving donor and acceptor molecules in the stack (Fig. 1) with two unequal separations of 3.80 and 4.97 Å.

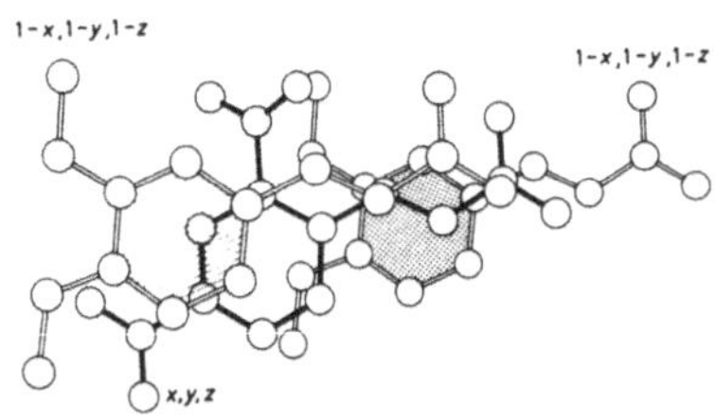

Fig. 1. Projection drawing of the $C_{20}H_{18}N_2O_{10}$ crystal structure down the stack direction.

BIS(4,5-DIMETHYL-4',5'-ETHYLENEDITHIOLOTETRATHIAFULVALENE) PERCHLORATE TETRAHYDROFURAN SOLVATE

$C_{20}H_{20}ClO_4S_{12},C_4H_8O$ $C_{10}H_{10}S_6^+,C_{10}H_{10}S_6,ClO_4^-,C_4H_8O$

R. HEID, H. ENDRES, H.J. KELLER, E. GOGU, I. HEINEN, K. BENDER and D. SCHWEITZER, 1985. *Z. Naturforsch.*, **B40**, 1703-1710.

Triclinic, $P\bar{1}$, a = 6.731, b = 7.632, c = 33.32 Å, α = 96.53, β = 92.99, γ = 103.13°, Z = 2. Mo radiation, R = 0.077 for 1844 reflections.

Sheets of cations with an average charge of +(0.5) are separated by sheets of anions and THF solvent molecules (Fig. 1). Within each cation stack the molecules are parallel (by inversion symmetry) but the cations of crystallographically different stacks are inclined at 36(1)° with respect to each other. Interplanar distances are 3.52-3.74(4) Å.

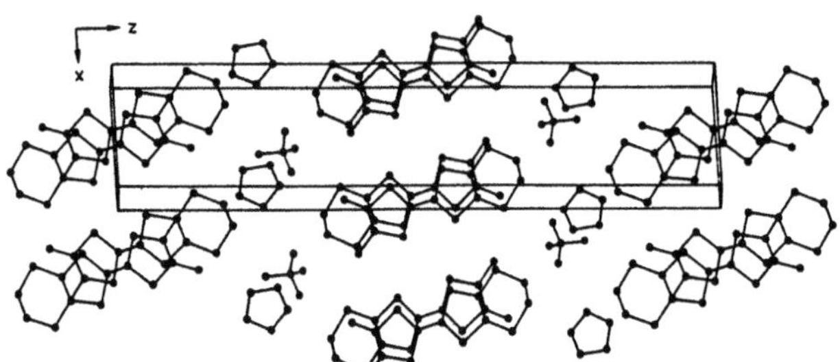

Fig. 1. A projection of the $C_{10}H_{10}S_6^+,C_{10}H_{10}S_6,ClO_4^-,C_4H_8O$ structure.

BIS(3,3',4,4'-TETRAMETHYL-2,2',5,5'-TETRASELENAFULVALENIUM) DIFLUOROPHOSPHATE

$C_{20}H_{24}F_2O_2PSe_8$ (I) (II) $C_{10}H_{12}Se_4^+,C_{10}H_{12}Se_4,F_2O_2P^-$

K. ERIKS, H.H. WANG, P.E. REED, M.A. BENO, E.H. APPELMAN and J.M. WILLIAMS, 1985. *Acta Cryst.*, C41, 257-260.

I. Triclinic, $P\bar{1}$, a = 7.271, b = 7.704, c = 13.361 Å, α = 84.28, β = 86.70, γ = 70.50°, Z = 1. Mo radiation, R = 0.046 for 1962 reflections.

II. Triclinic, $P\bar{1}$, a = 7.128, b = 7.694, c = 13.201 Å, α = 85.35, β = 87.23, γ = 69.76°, Z = 1. Mo radiation, R = 0.038 for 2005 reflections (at 125 K).

In both (I) and (II) the structure of the tetramethyltetraselenafulvalenium cation is as expected, the starting point for the refinement being the parameters reported in 1 for the corresponding SiF_5 salt. The anion, however, exhibits a different structure at the two temperatures. In (I), the $PO_2F_2^-$ ion is positioned with its phosphorus atom at the origin with P-O(1). P-O(2), P-F(1) and P-F(2) distances of 1.58(2), 1.49(1), 1.10(2) and 1.49(1) Å respectively. In (II), the anion moves its position from the origin by 0.375 Å yielding a well-resolved site separation of 0.75 Å (P-P) (Fig. 1). Here the P-O(1), P-O(2), P-F(1) and P-F(2) distances are 1.478(12), 1.412(15), 1.532(11) and 1.547(10) Å respectively.

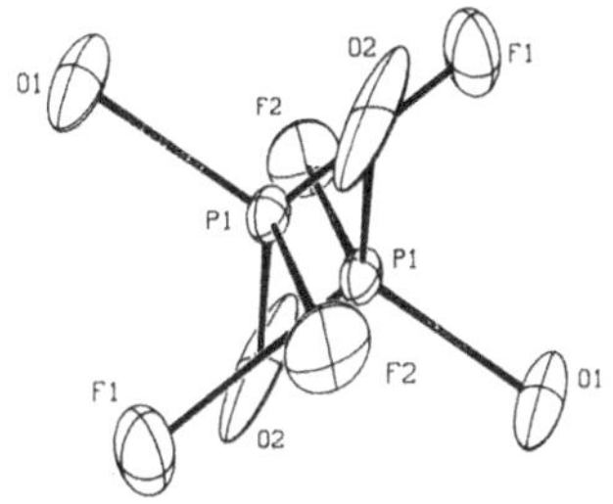

Fig. 1. $C_{20}H_{24}F_2O_2PSe_8$: the $PO_2F_2^-$ anion showing the P-P' site separation.

1. Structure Reports, 51B, 694.

BIS(ADENOSINE 5'-TRIPHOSPHATO)MANGANESE HEXAAQUAMANGANESE BIS(2,2' DIPYRIDYLAMINE) DODECAHYDRATE

$C_{20}H_{26}MnN_{10}O_{26}P_6,2(C_{10}H_9N_3),H_{12}MnO_6,12(H_2O)$

M. SABAT, R. CINI, T. HAROMY and M. SUNDARALINGAM, 1985. *Biochemistry*, **24**, 7827-7833.

Orthorhombic, $C222_1$, a = 10.234, b = 22.699, c = 31.351 Å, Z = 4. Cu radiation, R = 0.072 for 3516 reflections.

The Mn(1) atom lies on a twofold axis and is hexacoordinated by O atoms from six phosphate groups (Fig. 1); there are no metal-base interactions. The $Mn(H_2O)_6$ moiety is also on a twofold axis. The ATP molecule (Fig. 1) is protonated at N(1) and has the anti conformation (χ 66.0°). The ribofuranose ring is in the ^2_3T conformation. The adenine bases form hydrogen-bonded self-pairs across a crystallographic twofold axes. The dipyridylamine molecules are not coordinated to metal ions and there is some disorder associated with them.

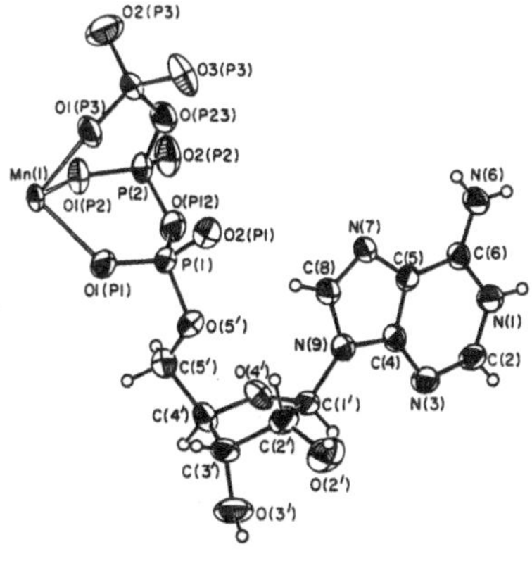

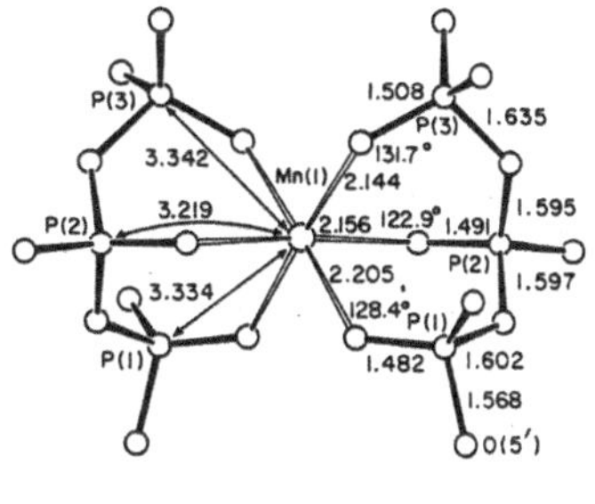

Fig. 1. The MnATP complex and details of the Mn(1) coordination.

4,7,10,13-TETRAOXA-1,16-DITHIA[16](1,1')RUTHENOCENOPHANE DICHLOROMERCURATE(II)

$C_{20}H_{28}Cl_2HgO_4RuS_2$ $C_{20}H_{28}O_4RuS_2{}^+,Cl_2Hg^-$

S. AKABORI, H. MUNEGUMI, Y. HABATA, S. SATO, K. KAWAZOE, C. TAMURA and M. SATO, 1985. *Bull. Chem. Soc. Jpn.*, **58**, 2185-2191.

Triclinic, $P\bar{1}$, a = 13.722, b = 11.678, c = 8.179 Å, α = 102.69, β = 72.82, γ = 100.73°, Z = 2. Cu radiation, R = 0.089 for 3352 reflections.

The $HgCl_2$ anion is situated at the centre of the macrocycle cavity as shown in Fig. 1. The intramolecular distances from Hg to the ring atoms are S(6) 3.034(6), O(9) 3.120(18), O(12) 2.888(17), O(15) 2.838(17), O(18) 3.048(17), S(21) 3.330(6) Å with Hg-Cl 2.327 and 2.304(7) Å. The Hg atom thus has a distorted hexagonal bipyramidal geometry. In the $Ru(Cp)_2$ portion of the ligand the Cp rings are eclipsed, with Ru-C 2.18-2.23(3) Å, mean 2.21 Å. The Ru...Hg distance is 4.846(2) Å, too long for any interaction.

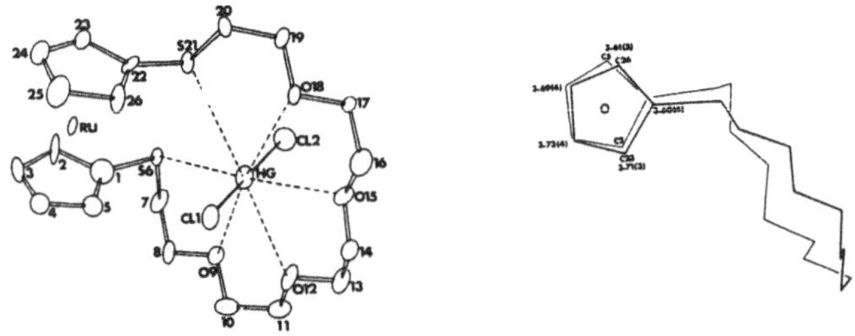

Fig. 1. A view of the $C_{20}H_{28}Cl_2HgO_4RuS_2$ structure and a side on view with the $HgCl_2$ moiety omitted.

N-METHYLPHENAZINIUM BIS(MALEONITRILEDITHIOLATO)GOLD(III)

$C_{21}H_{11}AuN_6S_4$ $C_8AuN_4S_4{}^-,C_{13}H_{11}N_2{}^+$

P. KUPPUSAMY, N. VENKATALAKSHMI and P.T. MANOHARAN, 1985. *J. Cryst. Spect. Res.*, **15**, 629-641.

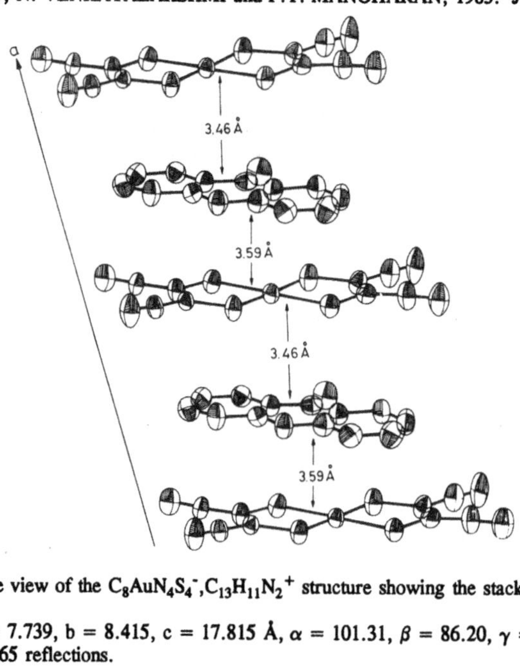

Fig. 1. A side view of the $C_8AuN_4S_4{}^-,C_{13}H_{11}N_2{}^+$ structure showing the stacking.

Triclinic, $P\bar{1}$, a = 7.739, b = 8.415, c = 17.815 Å, α = 101.31, β = 86.20, γ = 98.17°, Z = 2. Mo radiation, R = 0.046 for 3065 reflections.

Both cation and anion are individually planar and form a mixed stack along a, with slightly alternating plane-to-plane distances of 3.46(8) and 3.59(5) Å (Fig. 1). The Au-S distances in the anion are 2.310(5)-2.327(4) Å with intra-ring S-Au-S 90.6 and 90.7(8)°.

N-METHYLPHENAZINIUM BIS(MALEODINITRILEDITHIOLATO-S,S')NICKEL(III)
$C_{21}H_{11}N_2NiS_4$ $C_8N_4NiS_4^-,C_{13}H_{11}N_2^+$

P. KUPPUSAMY, P.T. MANOHARAN, C. MAHADEVAN and M. SESHASAYEE, 1985. *J. Cryst. Spect. Res.*, **15**, 359-376.

Triclinic, $P\bar{1}$, a = 8.768, b = 11.817, c = 22.131 Å, α = 97.44, β = 91.90, γ = 94.04°, Z = 4. Mo radiation, R = 0.039 for 3213 reflections.

There are two independent formula units in the crystal asymmetric unit; the Ni atoms have approximate square-planar geometry with mean Ni-S 2.138(6) Å. Anions and cations form segregated stacks along the a-direction (Fig. 1). The shortest Ni...Ni contacts are 4.111(2) and 4.208(1) Å.

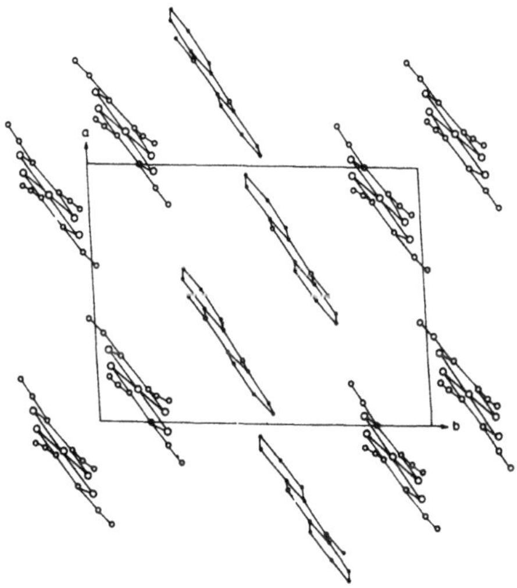

Fig. 1. A projection of the $C_8N_4NiS_4^-,C_{13}H_{11}N_2^+$ structure onto the ab-plane.

TRIS(TETRATHIAFULVALENIUM) TETRACHLORODIETHYLSTANNATE
$C_{22}H_{22}Cl_4S_{12}Sn$ $C_4H_{10}Cl_4Sn^{2-},2(C_6H_4S_4^+),C_6H_4S_4$

K. UEYAMA, G.-E. MATSUBAYASHI, R. SHIMIZU and T. TANAKA, 1985. *Polyhedron*, **4**, 1783-1789.

Tetragonal, *I4cm*, a = 11.710, c = 25.242 Å, Z = 4. Mo radiation, R = 0.082 for 479 reflections.

The crystal contains alternate layers along c of $SnEt_2Cl_4$ anions and ttf units (Fig. 1). Two Sn-Cl bonds of each anion lie along a crystal fourfold axis and the other ligands are disordered in the equatorial plane. The long axes of the ttf units are parallel to c. The ttf are grouped in threes, the planes of molecules in each group being normal to those of molecules in adjacent groups. Bond lengths include: Sn-Cl 2.58-2.67(2), Sn-C 2.21(7), C-C (central) 1.26, 1.34, S-C 1.62-1.80(7) Å.

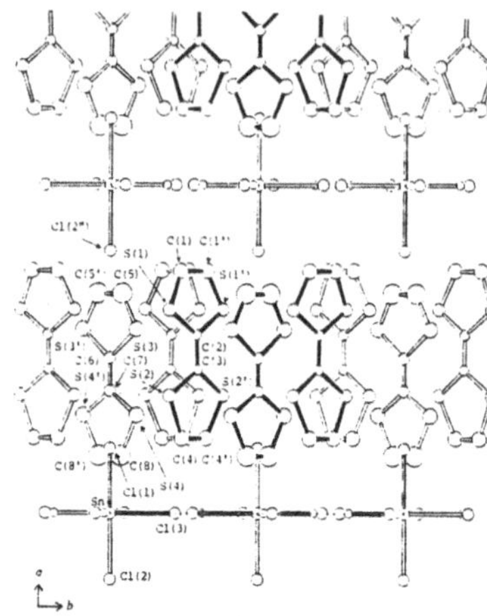

Fig. 1. $C_{22}H_{22}Cl_4S_{12}Sn$: part of the a-axis projection.

N-METHYLPHENAZINIUM BIS(BENZENE-1,2-DITHIOLATO-S,S')NICKEL(III)
$C_{25}H_{19}N_2NiS_4$ $C_{12}H_8NiS_4^-,C_{13}H_{11}N_2^+$

C. MAHADEVAN, M. SESHASAYEE, P. KUPPUSAMY and P.T. MANOHARAN, 1985. *J. Cryst. Spect. Res.*, 15, 305-316.

Triclinic, $P\bar{1}$, a = 9.363, b = 10.148, c = 13.623 Å, α = 85.64, β = 107.33, γ = 113.10°, Z = 2. Mo radiation, R = 0.036 for 2874 reflections.

There are two independent anions, each with its Ni atom on an inversion centre, and a cation in a general position. The Ni atoms have planar geometry which is almost square (Ni-S 2.145-2.161(1) Å, ring S-Ni-S 91.7 and 92.0°). One of the anions and the cations form a "mixed-stack" along the short diagonal of the ab-plane (CAC-CAC-) with C-A = A-C = 3.43 Å and C-C = 3.91 Å (Fig. 1).

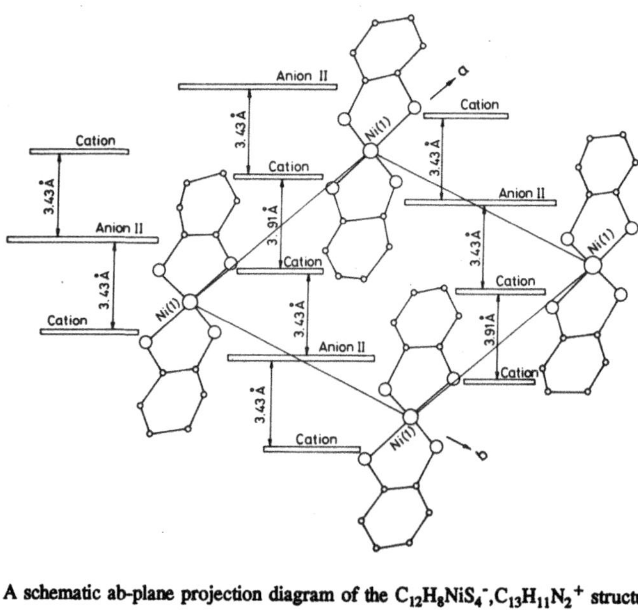

Fig. 1. A schematic ab-plane projection diagram of the $C_{12}H_8NiS_4^-,C_{13}H_{11}N_2^+$ structure.

ACRIDINIUM ACRIDINETRICHLOROCUPRATE(II) MONOHYDRATE

$C_{26}H_{19}Cl_3CuN_2,H_2O$ $C_{13}H_9Cl_3CuN^-,C_{13}H_{10}N^+,H_2O$

P.C. HEALY, C. PAKAWATCHAI and A.H. WHITE, 1985. *Aust. J. Chem.*, **38**, 669-675.

Triclinic, $P\bar{1}$, a = 16.27, b = 10.080, c = 7.236 Å, α = 88.85, β = 82.68, γ = 81.12°, Z = 2. Mo radiation, R = 0.037 for 1942 reflections.

The structure comprises columns of anions and cations stacked along *c*, with parallel arrays of acridine species in each column. Both acridine species are normal. The stereochemistry about the Cu atom is unusual (Fig. 1) being of pseudo-m symmetry; one of the Cl atoms is coplanar with the Cu atom and the acridine (Cu-Cl 2.237(2) Å, N-Cu-Cl 143.5(1)°), with the other two disposed to either side of that plane (Cu-Cl 2.252(3), 2.243(2) Å, Cl-Cu-Cl 145.62(7)°). Cu-N is 2.018(5) Å.

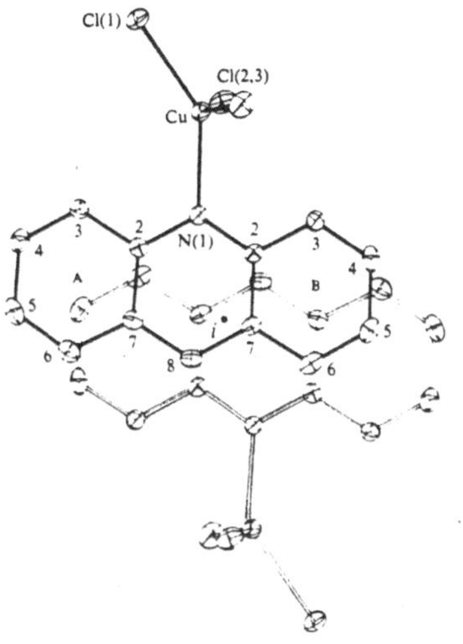

Fig. 1. The anion in $C_{26}H_{19}Cl_3CuN_2,H_2O$ projected normal to the acridine plane, together with its inversion-related counterpart.

PERYLENE BIS(DICYANOETHYLENE-1,2-DITHIOLATO-S,S')PLATINUM(II)

$C_{28}H_{12}N_4PtS_4$ $C_8N_4PtS_4^-,C_{20}H_{12}^+$

R.P. SHIBAEVA, V.F. KAMINSKII, M.A. SIMONOV, E.B. YAGUBSKII and E.E. KOSTYUCHENKO, 1985. *Kristallografiya*, **30**, 488-491.

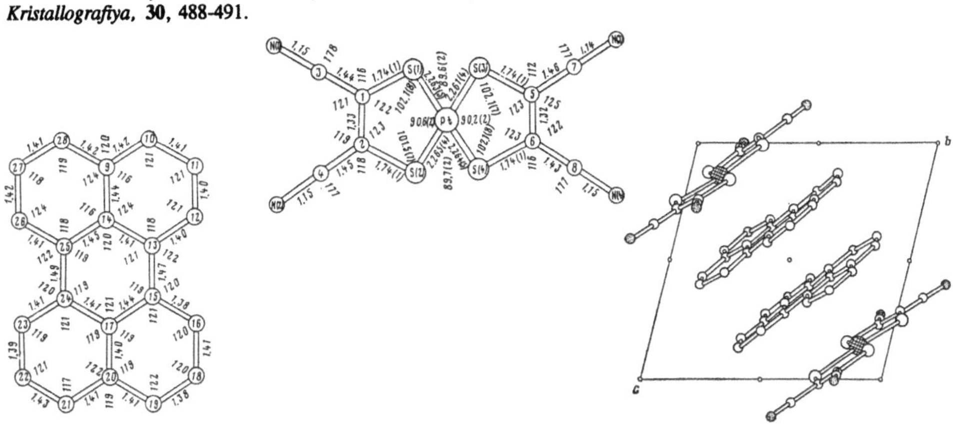

Fig. 1. The $C_{20}H_{12}^+$ and $C_8N_4PtS_4^-$ ions and the crystal packing along the [100] direction.

Triclinic, $P\bar{1}$, a = 11.296, b = 11.502, c = 11.362 Å, α = 104.01, β = 120.11, γ = 80.83°, Z = 2. Mo radiation, R = 0.078 for 2909 reflections.

The structure is characterized by the presence of mixed cation-anion stacks extending approximately along the [011] direction. The Pt coordination is square-planar. Interplanar anion-anion, cation-cation and anion-cation separations are 3.49, 3.34 and 3.34 Å respectively.

BIS(DIBENZOTETRATHIAFULVALENIUM) BIS(μ_2-BROMO)TETRABROMODICUPRATE(II)

$C_{28}H_{16}Br_6Cu_2S_8$ $2(C_{14}H_8S_4^+),Br_6Cu_2^{2-}$

M. HONDA, C. KATAYAMA, J. TANAKA and M. TANAKA, 1985. *Acta Cryst.*, **C41**, 688-690.

Triclinic, $P\bar{1}$, a = 9.982, b = 12.122, c = 7.893 Å, α = 83.70, β = 112.80, γ = 104.97°, Z = 1. Cu radiation, R = 0.039 for 2678 reflections.

The structure (Fig. 1) consists of planar cation radicals stacked to form columns along the c-axis with an interplanar spacing of about 3.5(5) Å and of centrosymmetric $Cu_2Br_6^{2-}$ dianions. In the dianions the Cu atoms, which have nearly square-planar coordination, are bridged by two Br atoms with Cu-Br bonds of length 2.433(1) and 2.416(1) Å. The terminal Br atoms have Cu-Br lengths of 2.375(1) and 2.369(1) Å. In the stacks the shortest contacts are S...S of length 3.391(2) Å. In the cation radical the central C=C bond length (1.397(6) Å) is significantly longer than the equivalent bond in the neutral molecule (1.336(2) Å).

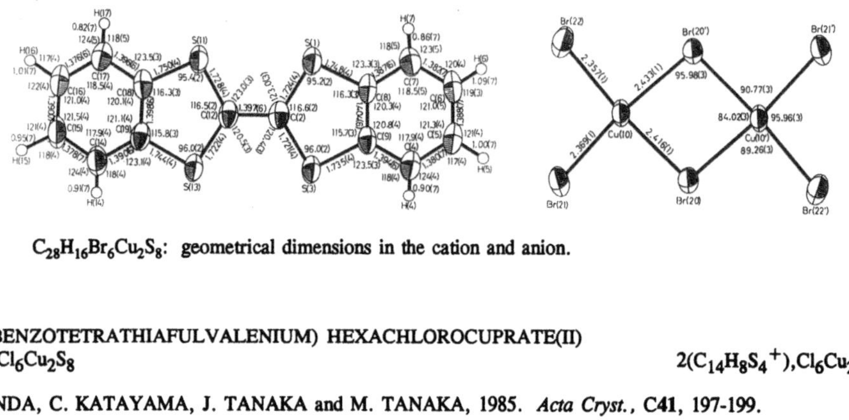

Fig. 1. $C_{28}H_{16}Br_6Cu_2S_8$: geometrical dimensions in the cation and anion.

BIS(DIBENZOTETRATHIAFULVALENIUM) HEXACHLOROCUPRATE(II)

$C_{28}H_{16}Cl_6Cu_2S_8$ $2(C_{14}H_8S_4^+),Cl_6Cu_2^{2-}$

M. HONDA, C. KATAYAMA, J. TANAKA and M. TANAKA, 1985. *Acta Cryst.*, **C41**, 197-199.

Monoclinic, *C2/m*, a = 9.224, b = 24.976, c = 7.317 Å, β = 108.28°, Z = 2. Cu radiation, R = 0.053 for 1337 reflections.

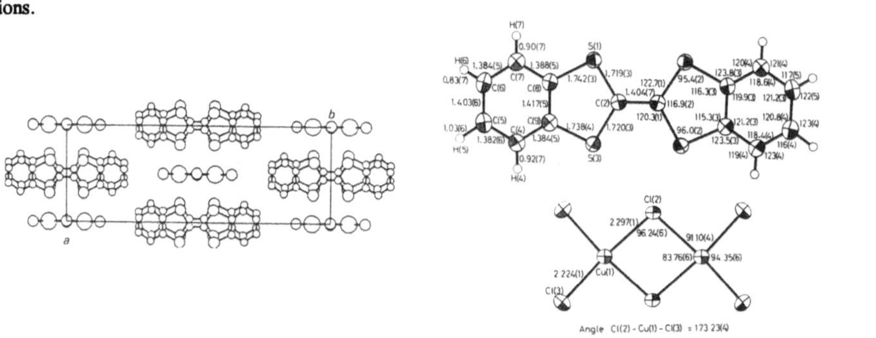

Fig. 1. $C_{28}H_{16}Cl_6Cu_2S_8$: projection of the structure along the c-axis (left) and the molecular dimensions (right).

The structure contains stacking columns of the dibenzotetrathiafulvalenium cation along the c-axis (Fig. 1). The shortest contact between adjacent cations in a column is between S(1) and S(3) (3.443(1) Å). In the cation the central

C=C bond is longer than expected (1.404(7) Å). The anion has normal geometry.

DIQUAT BIS(HEXAFLUOROPHOSPHATE) DIBENZO-30-CROWN-10 METHANOL SOLVATE
$C_{28}H_{40}O_{10},C_{12}H_{12}F_{12}N_2P_2,CH_4O$

H.M. COLQUHOUN, E.P. GOODINGS, J.M. MAUD, J.F. STODDART, J.B. WOLSTENHOLME and D.J. WILLIAMS, 1985. *J. Chem. Soc. Perkin Trans.* 2, 607-624.

Monoclinic, $P2_1/c$, a = 11.369, b = 22.218, c = 37.663 Å, β = 92.87°, Z = 8. Cu radiation, R = 0.13 for 5549 reflections.

There were two independent complex molecules ((I) and (II)) in the crystallographic asymmetric unit; the gross structural features are very similar for the independent molecules. (I) and (II) differ slightly in the conformations of their polyether chains. The hexafluorophosphate ions show severe disorder. Two views of the two complex molecules are shown in Fig. 1.

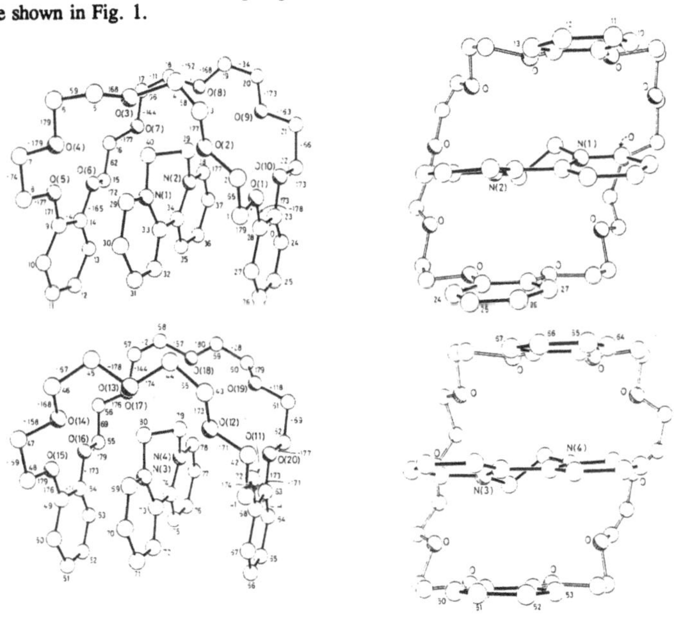

Fig. 1. Skeletal representations of the two independent complex molecules of $C_{28}H_{40}O_{10},C_{12}H_{12}F_{12}N_2P_2$ (left) with corresponding underside skeletal representations (right). (Hexafluorophosphate ions are omitted.)

N,N-DIMETHYLMORPHOLINIUM BIS(7,7,8,8-TETRACYANOQUINODIMETHANIDE)
$C_{30}H_{22}N_9O$ $C_6H_{14}NO^+,C_{12}H_4N_4^-,C_{12}H_4N_4$

J.A.M. MIDDELDORP, R.J.J. VISSER and J.L. DE BOER, 1985. *Acta Cryst.*, **B41**, 369-374.

Monoclinic, $P2_1/c$, a = 13.280, b = 26.497, c = 14.782 Å, β = 92.35°, Z = 8. Mo radiation, R = 0.051 for 12232 reflections (at 95 K).

For details of the room temperature structure see (1). At room temperature the DMM molecules are randomly disordered; on cooling to 95 K the space group changes and the disorder disappears. The DMM ordering is accompanied by slight shifts of TCNQ molecules, which changes the TCNQ stacks from dimeric to octameric.

1. Structure Reports, 48B, 753.

BIS(TETRAMETHYLTETRATHIAFULVALENE) HEXACYANOBUTADIENE
$C_{30}H_{24}N_6S_8$ (I) $C_{10}H_{12}S_4{}^+,C_{10}H_{12}S_4,C_{10}N_6{}^-$

TETRAMETHYLTETRATHIAFULVALENE HEXACYANOBUTADIENE
$C_{20}H_{12}N_6S_4$ (II) $C_{10}H_{12}S_4{}^+,C_{10}N_6{}^-$

TETRAMETHYLTHIOTETRATHIAFULVALENE HEXACYANOBUTADIENE
$C_{20}H_{12}N_6S_8$ (III) $C_{10}H_{12}S_8{}^+,C_{10}N_6{}^-$

TETRAMETHYLTHIOTETRATHIAFULVALENE
$C_{10}H_{12}S_8$ (IV)

C. KATAYAMA, M. HONDA, H. KUMAGAI, J. TANAKA, G. SAITO and H. INOKUCHI, 1985. *Bull. Chem. Soc. Jpn.*, **58**, 2272-2278.

I. Monoclinic, *C2/m*, a = 18.266, b = 9.813, c = 9.501 Å, β = 102.741°, Z = 2. Cu radiation, R = 0.042 for 1398 reflections.

II. Monoclinic, *C2/m*, a = 12.8016, b = 21.6196, c = 4.0223 Å, β = 108.062°, Z = 2. Cu radiation, R = 0.049 for 867 reflections.

III. Monoclinic, *C2/c*, a = 30.161, b = 4.046, c = 23.414 Å, β = 117.48°, Z = 4. Cu radiation, R = 0.051 for 1954 reflections.

IV. Monoclinic, *P2₁/n*, a = 15.668, b = 7.804, c = 14.010 Å, β = 106.16°, Z = 4. Cu radiation, R = 0.053 for 2593 reflections.

In the 2:1 complex (I) (Fig. 1) the TMTTF exists as a dimer cation stacked along the *c*-axis; the HCBD anions are included in the cavity of the dimer stack with their molecular planes parallel to the stacking axis. The complex (II) (Fig. 2) is comprised of 1:1 donor and acceptor with segregated cation and anion stacks with interplanar spacings of 3.61 Å in TMTTF and 3.43 Å in HCBD columns. Complex (III) has regular segregated columns with partial charge transfer (Fig. 3). The neutral molecule (IV) has a non-planar conformation (Fig. 4).

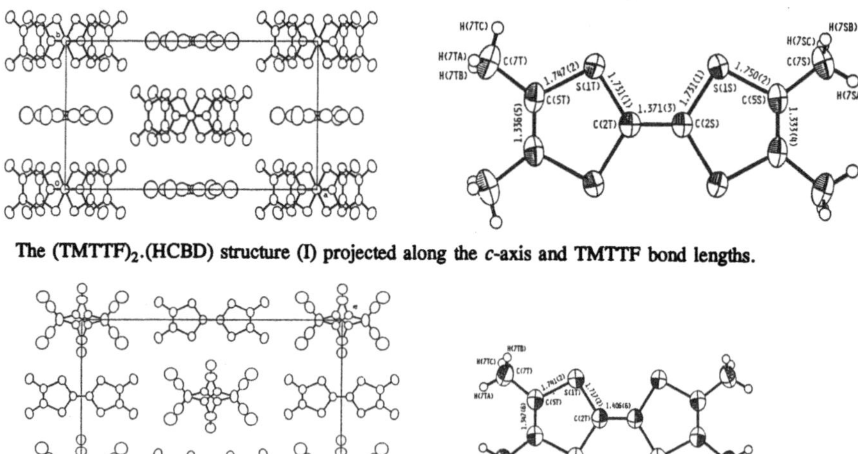

Fig. 1. The $(TMTTF)_2$.(HCBD) structure (I) projected along the *c*-axis and TMTTF bond lengths.

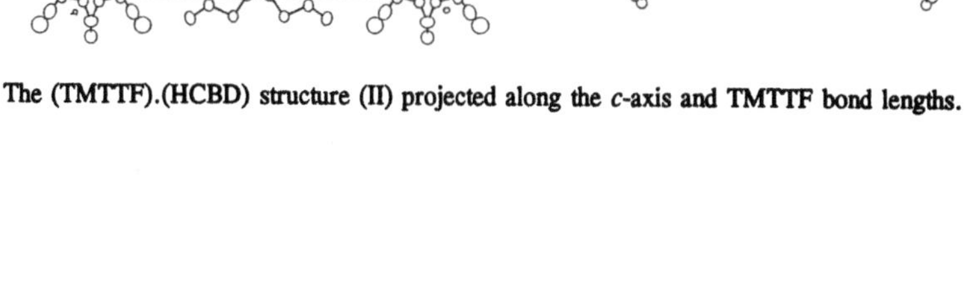

Fig. 2. The (TMTTF).(HCBD) structure (II) projected along the *c*-axis and TMTTF bond lengths.

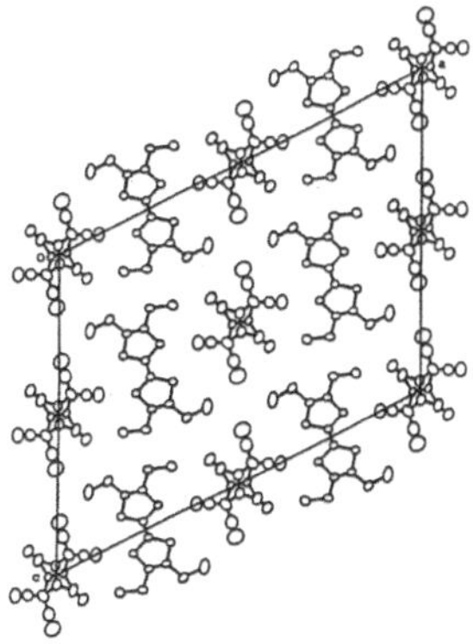

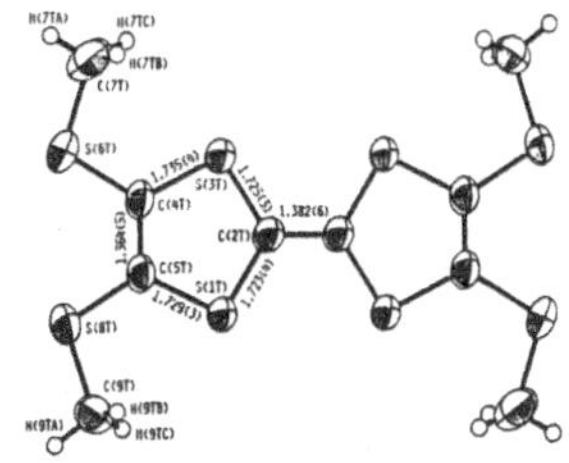

Fig. 3. The (TTMTTF).(HCBD) structure (III) projected onto the (010) plane and TTMTTF bond lengths.

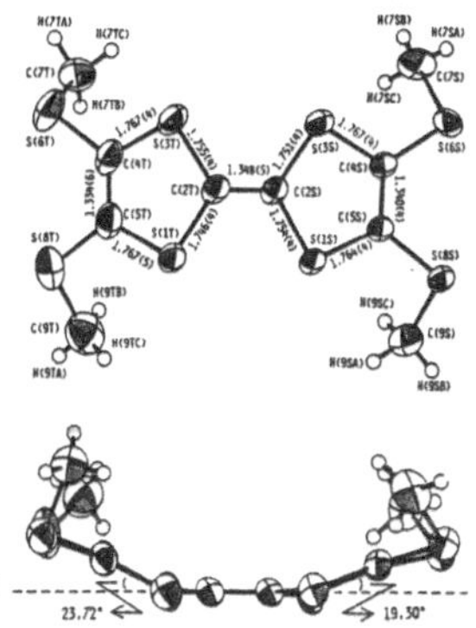

Fig. 4. The conformation and bond len

PHENOTHIAZINE BIS(PYROMELLITIC DIANHYDRIDE)

$C_{32}H_{13}NO_{12}S$ $C_{12}H_9NS,2(C_{10}H_2O_6)$

C. BRIERLEY, R. BARTON, B. ROBERTSON, R. ANTHONJ and N. KARL, 1985. *J. Chem. Phys.*, **82**, 1522-1528.

Monoclinic, $P2_1/c$, a = 7.689, b = 12.953, c = 28.14 Å, β = 104.30°, Z = 4. Mo radiation, R = 0.064 for 2863 reflections.

The structure of this (PTZ):$(PMDA)_2$ complex contains infinite donor-acceptor-donor-acceptor-stacks with the *a*-axis as the stack axis. The extra acceptor PMDA molecule sits outside the stack with its length and the normal to the plane of the molecule approximately perpendicular to the stack axis (Fig. 1). The interplanar separation in the stack is 3.425 Å. The PTZ molecules are folded slightly; the angle between the planes of the two phenyl rings are 175.7 and 176.4°.

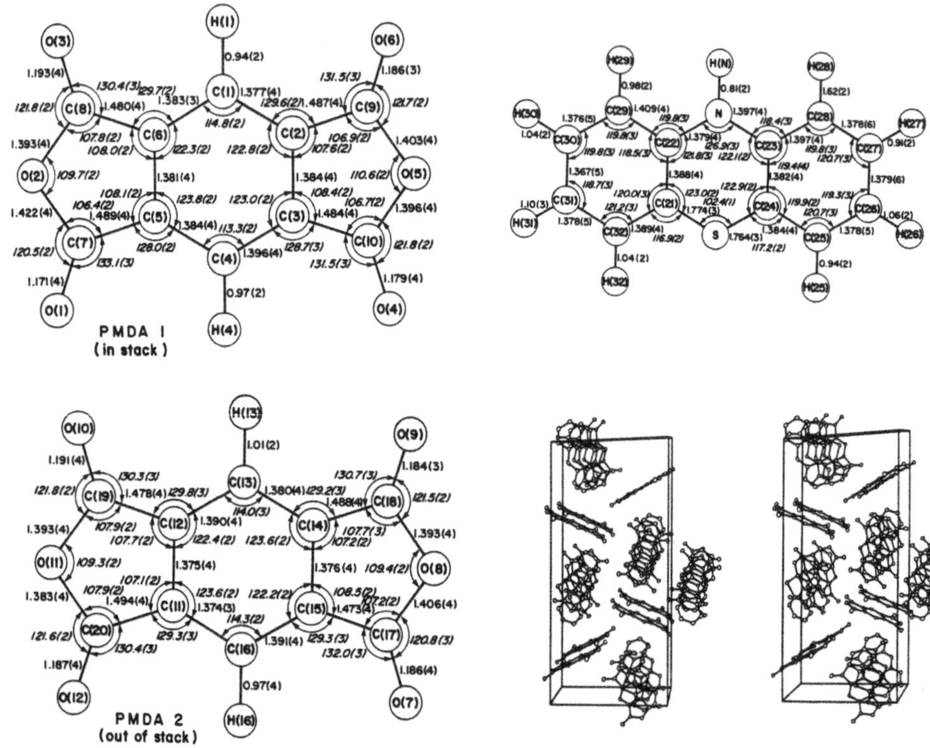

Fig. 1. The dimensions for the (PTZ):(PMDA)$_2$ complex and a stereoview of the unit cell.

[2.2](9,10)ANTHRACENOPHANE TETRACYANOETHYLENE
$C_{36}H_{24}N_4$ $C_{32}H_{24}^{+},C_6N_4^{-}$

J.M. MASNOVI, J.K. KOCHI, E.F. HILINSKI and P.M. RENTZEPIS, 1985. *J. Phys. Chem.*, **89**, 5387-5395.

Monoclinic, $P2_1/n$, a = 8.083, b = 17.823, c = 10.159 Å, β = 112.52°, Z = 2. Mo radiation, R = 0.042 for 1082 reflections.

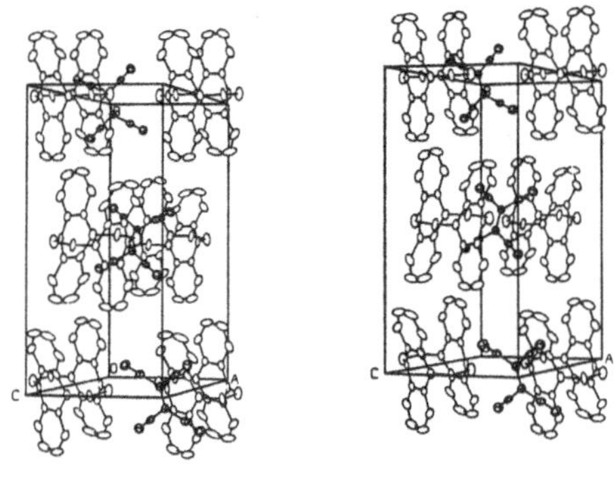

Fig. 1. A stereoview of the $C_{32}H_{24}^{+},C_6N_4^{-}$ structure.

Both ions in this complex lie about crystallographic inversion centres (Fig. 1). Infinite stacks of alternating cations and anions extend along *c*. The anthracene rings of the anthracenophane are bent by ~19° from planarity because of restrictions imposed by the ethano bridges.

TRIS(ACETYLACETONATO)SILICONIUM(IV) BIS(7,7,8,8-TETRACYANOQUINODIMETHANIDE)

$C_{39}H_{29}N_8O_6Si$ $C_{15}H_{21}O_6Si^+,C_{12}H_4N_4^-,C_{12}H_4N_4$

K. UEYAMA, G.-E. MATSUBAYASHI, I. SHIMOHARA, T. TANAKA and K. NAKATSU, 1985. *J. Chem. Res.*, S48-49, M801-854.

I. Monoclinic, $P2_1/a$, a = 34.981, b = 13.543, c = 7.837 Å, β = 95.15°, Z = 4. Mo radiation, R = 0.064 for 4610 reflections.

II. Triclinic, $P\bar{1}$, a = 7.941, b = 33.04, c = 7.775 Å, α = 95.92, β = 98.55, γ = 76.86°, Z = 2. Mo radiation, R = 0.240 for 2724 reflections.

The triclinic form has a disordered structure. In the monoclinic form the TCNQ(7,7,8,8-tetracyano-p-quinodimethanide) molecules stack in columns along the *b* axis, consisting of tetramer units with mean interplanar spacings in the range 3.14 to 3.34 Å. Adjacent columns form sheets parallel to the bc plane and successive sheets are separated along the *a* axis by the cationic species (Fig. 1).

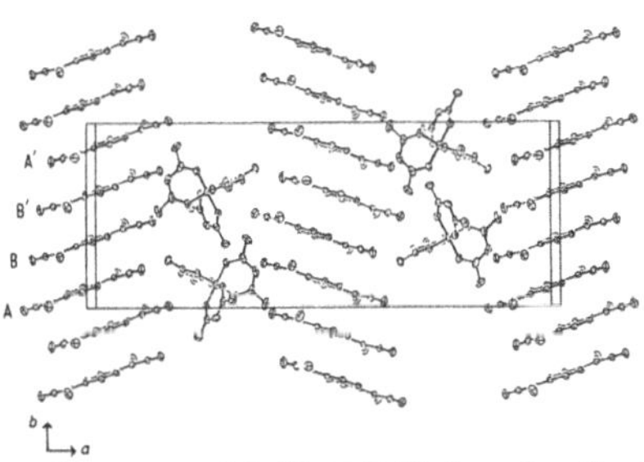

Fig. 1. Projection of the crystal structure of $C_{39}H_{29}N_8O_6Si$ along the *c** axis.

DIIODOTETRAKIS(4-METHYLPHENYL ISOCYANIDE)RHODIUM(III) 7,7,8,8-TETRACYANOQUINODIMETHANIDE

$C_{44}H_{32}I_2N_8Rh$ $C_{32}H_{28}I_2N_4Rh^+,C_{12}H_4N_4^-$

G.-E. MATSUBAYASHI, T. IINUMA, T. TANAKA, K. OKA and K. NAKATSU, 1985. *Inorg. Chim. Acta*, **102**, 145-150.

Monoclinic, $C2/c$, a = 7.469, b = 24.759, c = 22.875 Å, β = 92.95°, Z = 4. Mo radiation, R = 0.052 for 3299 reflections.

The cations, on inversion centres, are trans-isomers. They interact in the *a* direction by I...ring C contacts and ring overlap (Fig. 1). Bond lengths and angles include: Rh-C 1.965, 1.993(7), Rh-I 2.677(1) Å, C-Rh-C 87.4(3), C-Rh-I 87.8, 92.1(2)°. The radical anions, on twofold axes, partly overlap with spacing 3.18 Å so as to give stacks along a, but are well separated from the cations. Bond lengths include: N-C 1.144, 1.147(10), NC-C 1.417, 1.423(10), C(CN) C 1.403(10), ring C-C 1.429, 1.435, 1.349, 1.369(10) Å.

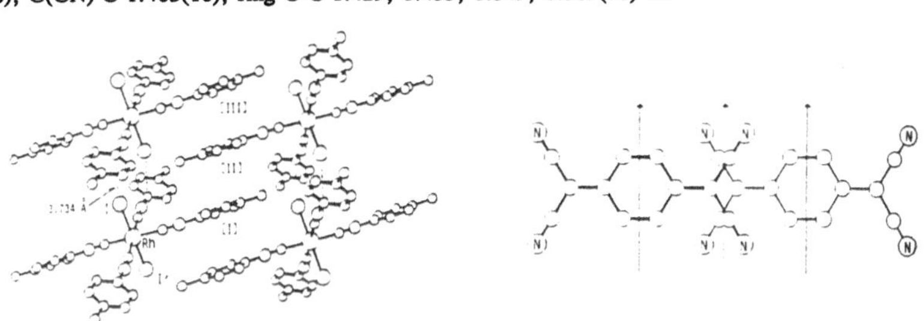

Fig. 1. $C_{44}H_{32}I_2N_8Rh$: (a) the cations, (b) the anions.

BIS(2-(α-HYDROXYIMINO-2,6-DIMETHYLBENZYL)-5-p-TOLYL-2H-TETRAZOLE) TRIETHYLAMMONIUM 5-p-TOLYLTETRAZOLIDE

$C_{48}H_{57}N_{15}O_2$ $2(C_{17}H_{17}N_5O),C_8H_7N_4^-,C_6H_{16}N^+$

M. CANNAS, G. GELLI, G. MARONGIU and G. NAVARRA, 1985. *Acta Cryst.*, **C41**, 954-956.

Triclinic, $P\bar{1}$, a = 20.10, b = 17.22, c = 7.84 Å, α = 111.2, β = 105.5, γ = 93.5°, Z = 2. Cu radiation, R = 0.089 for 4912 reflections.

In the asymmetric unit of this molecular complex (Fig. 1), the 5-p-tolyltetrazolide anion is hydrogen bonded to the triethylammonium cation and to the oximic hydroxyl groups of two molecules of the 2,5-disubstituted tetrazole. The 5-p-tolyltetrazolide molecule lies in the middle of the complex. Hydrogen bond lengths are O(18)...N(31) 2.690(7), O(28)...N(34) 2.706(8) and N(33)...N(4) 2.827(8) Å. Bond lengths and angles lie within normal ranges. The tetrazole and benzene rings are not coplanar, the angles between the mean planes ranging from 10.2(5)° to 15.1(7)°, suggesting little π-delocalisation.

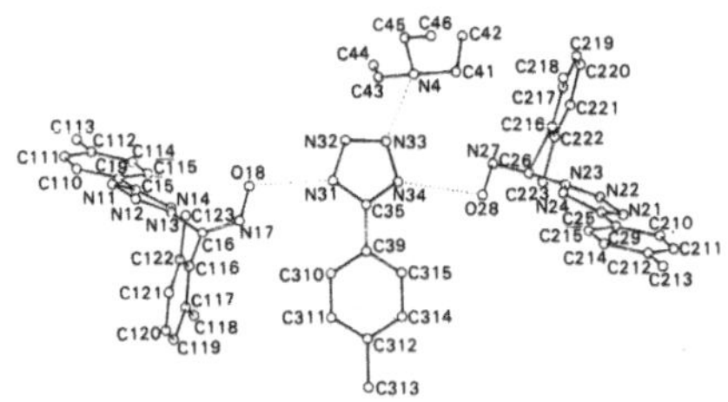

Fig. 1. A view of the $C_{48}H_{57}N_{15}O_2$ molecular complex.

BIS(TRIPHENYLPHOSPHINE)IMINIUM 7,7,8,8-TETRACYANOQUINODIMETHANE ACETONITRILE

$C_{54}H_{36}N_7P_2,C_2H_3N$ $C_{36}H_{30}NP_2^+,C_{12}H_4N_4^-,0.5(C_{12}H_4N_4),C_2H_3N$

J. HALFPENNY, 1985. *Acta Cryst.*, **C41**, 119-121.

Triclinic, $P\bar{1}$, a = 8.84, b = 16.70, c = 17.16 Å, α = 69.85, β = 84.11, γ = 85.26°, Z = 2. Mo radiation, R = 0.056 for 2833 reflections.

In the structure (Fig. 1) the cation has a non-linear structure in which the P(1)-N(1)-P(2) angle is 137.6(4)°. The average P-N bond length is 1.590(5) Å and the P-C bond lengths lie in the range 1.763 to 1.798(5) Å. There are 1.5 TCNQ molecules in the asymmetric unit-TCNQ(A) in a general position and TCNQ(B) of essentially identical geometry sited on a centre of symmetry. The TCNQ molecules are stacked in groups of three, ABA. Within the well-separated triads in the crystal the molecules are displaced so that there is overlap between the quinonoid double bond and the ring of the adjacent molecule. The spacing between the mean molecular planes in the triads is 3.30(2) Å. Contacts between the TCNQ moieties and the cation and the solvate molecule have normal van der Waals distances.

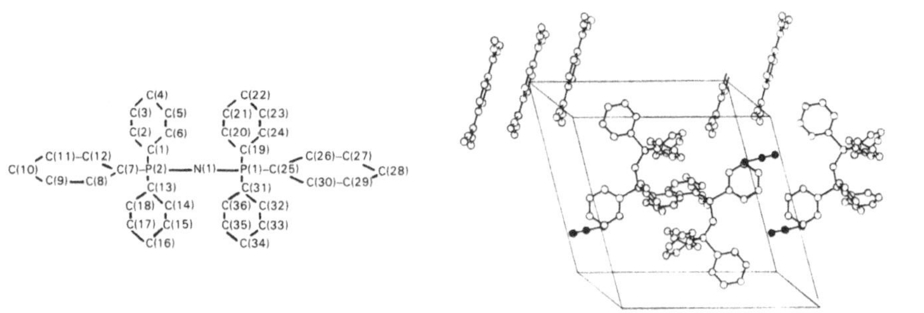

Fig. 1. $C_{54}H_{36}N_7P_2,C_2H_3N$: atomic numbering scheme in the cation and a view of the unit cell contents in which the solvate molecules are shown in black for clarity.

BIS(DIBENZOTETRATHIAFULVALENIUM) BIS(DIBENZOTETRATHIAFULVALENE) BIS(TETRAFLUOROBORATE) ETHANOL SOLVATE

$C_{56}H_{32}B_2F_8S_{16},C_2H_6O$ $2(C_{14}H_8S_4{}^+),2(C_{14}H_8S_4),2(BF_4{}^-),C_2H_6O$

M. TANAKA, M. HONDA, C. KATAYAMA and J. TANAKA, 1985. *Acta Cryst.*, **C41**, 1635-1638.

Triclinic, $P\bar{1}$, a = 10.772, b = 27.309, c = 10.623 Å, α = 85.13, β = 103.87, γ = 100.49°, Z = 2. Cu radiation, R = 0.132 for 7482 reflections.

The structure consists of two crystallographically independent stacks. Planar dibenzotetrathiafulvalenium radicals (I), (II) and (III) stacked parallel to each other form columns along the *a*-axis (Fig. 1) and have a mean interplanar spacing of 3.6 Å. The neutral dibenzotetrathiafulvalene molecules (IV), the ethanol solvate and $BF_4{}^-$ anions form a second column along the *a*-axis. In this mixed stack one ethanol and two $BF_4{}^-$ anions alternate with the dibenzotetrathiafulvalene molecules, (IV). Bond lengths and angles in the $BF_4{}^-$ anions and in the solvate molecule, though not accurately determined, appear to be normal.

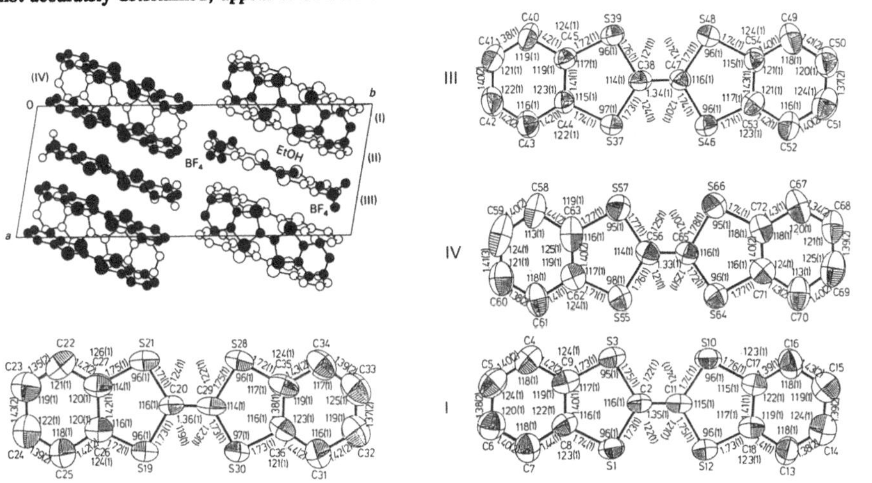

Fig. 1. $C_{56}H_{32}B_2F_8S_{16},C_2H_6O$: a projection of the crystal structure onto (001) and molecular dimensions for radicals (I), (II) and (III) for molecule (IV).

N,N'-DI-n-PROPYL-4,4'-BIPYRIDINIUM TETRAKIS(7,7,8,8-TETRACYANO-p-QUINODIMETHANIDE)

$C_{64}H_{38}N_{18}$ $C_{16}H_{22}N_2{}^{2+},2(C_{12}H_4N_4{}^-),2(C_{12}H_4N_4)$

G.J. ASHWELL and S.C. WALLWORK, 1985. *Z. Naturforsch.*, **A40**, 726-730.

Monoclinic, $P2_1/c$, a = 13.334, b = 25.934, c = 7.877 Å, β = 93.66°, Z = 2. Mo radiation, R = 0.070 for 3277 reflections.

The cations lie on inversion centres and there are two TCNQ moieties in general positions (Fig. 1). The TCNQ units stack plane-to-plane in groups of four with a favourable exocyclic bond to quinonoid ring overlap of adjacent molecules and short interplanar spacings of 3.20 and 3.23 Å.

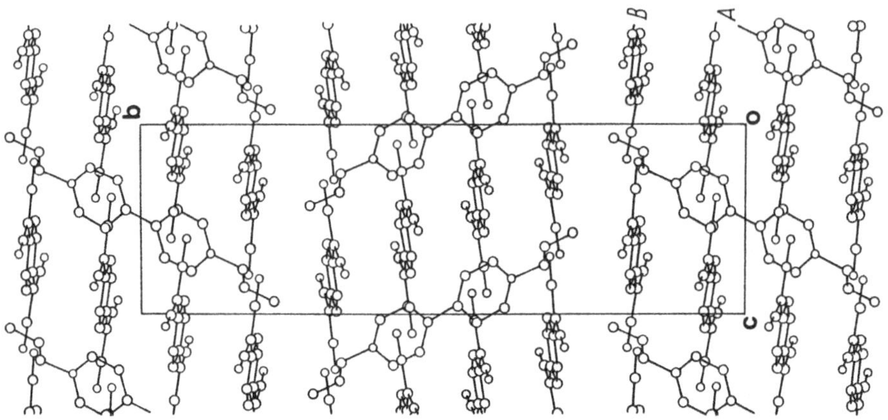

Fig. 1. A projection of the $C_{16}H_{22}N_2^{2+}$,$2(C_{12}H_4N_4^-)$,$2(C_{12}H_4N_4)$ along *a* showing the zigzag arrangement of cations and the isolated TCNQ tetrads.

BIS(9,9'-BIFLUORENYLIDENE) PYRENE
$C_{68}H_{42}$ (I) $C_{16}H_{10}$,$2(C_{26}H_{16})$

BIS(9,9'-BIFLUORENYLIDENE) PERYLENE
$C_{72}H_{44}$ (II) $C_{20}H_{12}$,$2(C_{26}H_{16})$

9,9'-BIFLUORENYLIDENE (α-form)
$C_{26}H_{16}$ (III)

9,9'-BIFLUORENYLIDENE (β-form)
$C_{26}H_{16}$ (IV)

JUNG-SI LEE and S.C. NYBURG, 1985. *Acta Cryst.*, C41, 560-567.

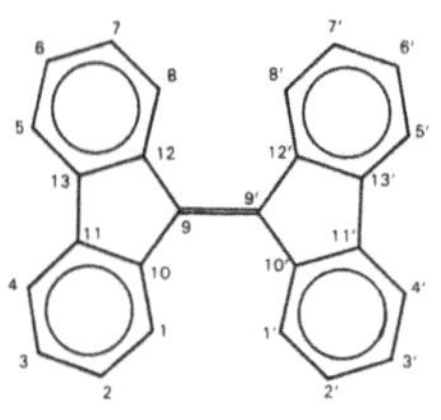

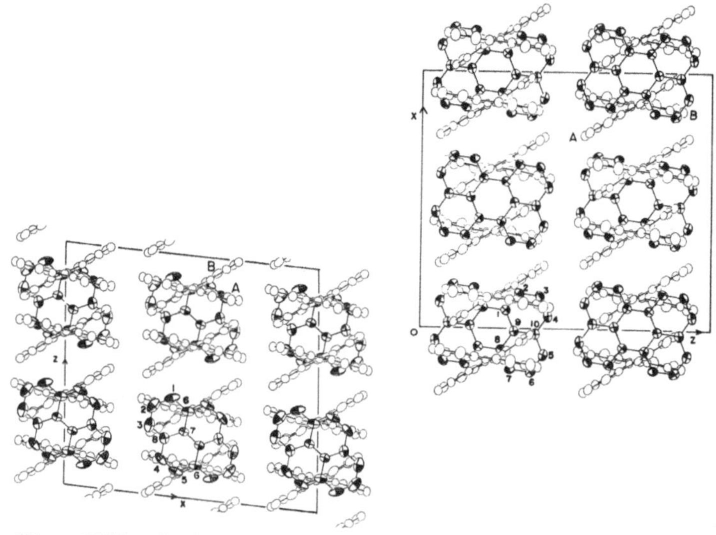

Fig. 1. The 9,9'-bifluorenylidene (BFL) molecule and views of the 2:1 BFL:pyrene and 2:1 BFL:perylene complexes.

I. Monoclinic, *C2/c*, a = 17.846, b = 15.449, c = 16.270 Å, β = 95.59°, Z = 4. Cu radiation, R = 0.054 for 2267 reflections.

II. Monoclinic, *I2/c*, a = 16.102, b = 15.342, c = 18.798 Å, β = 90.39°, Z = 4. Cu radiation, R = 0.050 for 2100 reflections.

III. Orthorhombic, *Pbcn*, a = 36.68, b = 17.080, c = 8.145 Å, Z = 12. Cu radiation, R = 0.074 for 2560 reflections.

IV. Orthorhombic, $P2_12_12_1$, a = 8.991, b = 19.350, c = 9.715 Å, Z = 4. Cu radiation, R = 0.050 for 1393 reflections.

In all four structures the 9,9'-bifluorenylidene (BFL) molecules (Fig. 1) are twisted due to intramolecular overcrowding. The average dihedral angle between the molecular half fragments is 39.3(1)°, that between a five- and a six-membered ring is 5.6(1)° and that between six-membered rings attached to the same five-membered ring is 5.6(1.4)°. Complexes (I) and (II) are made up of similar columns parallel to the y-axis with a crystallographic diad axis along their cores (Fig. 1). The BFL molecules are in general positions and the aromatic macrocycles have diad axes through their centres, normal to their mean molecular planes.

HEXAPERYLENE PERCHLORATE

$C_{120}H_{72}ClO_4$ (I) $C_{20}H_{12}^+,5(C_{20}H_{12}),ClO_4^-$

TRIPERYLENE PERCHLORATE

$C_{60}H_{36}ClO_4$ (II) $C_{20}H_{12}^+,2(C_{20}H_{12}),ClO_4^-$

DIPERYLENE HEXAFLUOROPHOSPHATE TETRAHYDROFURAN SOLVATE

$C_{40}H_{24}F_6P,0.66(C_4H_8O)$ (III) $C_{20}H_{12}^+,C_{20}H_{12},F_6P^-,0.66(C_4H_8O)$

H. ENDRES, H.J. KELLER, B. MULLER and D. SCHWEITZER, 1985. *Acta Cryst.*, C41, 607-613.

I. Triclinic, $P\bar{1}$, a = 12.571, b = 13.699, c = 13.835 Å, α = 110.43, β = 107.13, γ = 107.09°, Z = 1. Mo radiation, R = 0.063 for 3171 reflections.

II. Triclinic, $P\bar{1}$, a = 13.009, b = 13.821, c = 13.850 Å, α = 66.05, β = 83.36, γ = 63.51°, Z = 2. Mo radiation, R = 0.116 for 3523 reflections.

III. Monoclinic, *P2/m*, a = 13.04, b = 14.12, c = 13.75 Å, β = 110.80°, Z = 3. Mo radiation, R = 0.093 for 1737 reflections.

The structure of (I) contains three independent perylene dimers arranged approximately perpendicularly to each other (Fig. 1). The anions are severely disordered. In (II) tetrameric perylene stacks are surrounded on four sides by single perylene species, having their planes nearly perpendicular to the molecular planes in the stacks. In (III) nearly regular perylene stacks are surrounded by single perylene species in a way similar to (II). The solvent molecule and one perylene are disordered over two orientations.

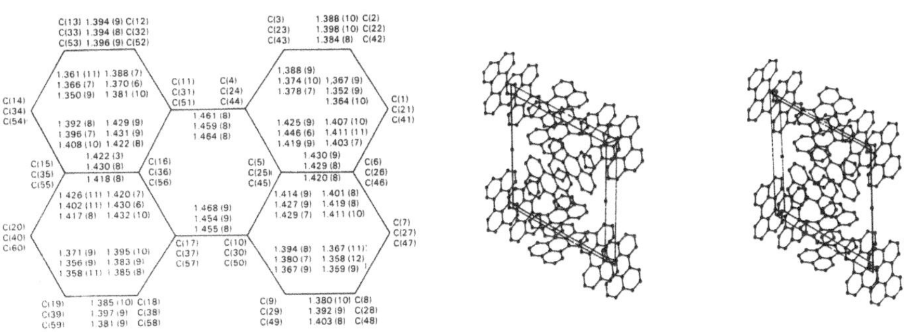

Fig. 1. $C_{120}H_{72}ClO_4$: bond distances and a stereoview along **a**.

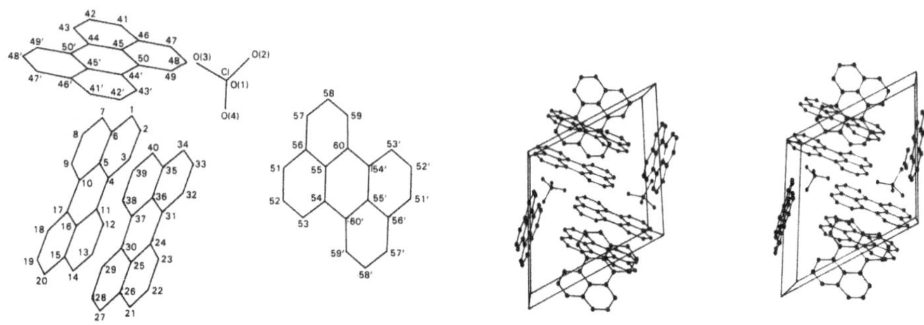

Fig. 2. $C_{60}H_{36}ClO_4$: relative orientation of the molecules and a stereoview along **c**.

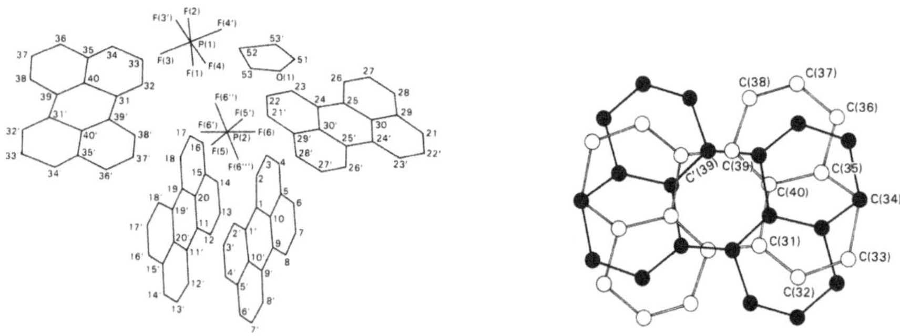

Fig. 3. $C_{40}H_{24}F_6P,0.66(C_4H_8O)$: relative orientation of the molecules and the disordered perylene molecule.

(μ_2-ETHYLENEDIAMINE-N,N')CADMIUM(II)TETRAKIS(μ_2-CYANO)CADMIUM(II) BENZENE CLATHRATE $(C_6H_8Cd_2N_6)_n,2n(C_6H_6)$ (I)

(ETHYLENEDIAMINE-N,N')CADMIUM(II)TETRAKIS(μ_2-CYANO)CADMIUM(II) $(C_6H_8Cd_2N_6)_n$ (II)

S.-I. NISHIKIORI and T. IWAMOTO, 1985. *J. Inclusion Phenomena,* 3, 283-295.

I. Tetragonal, *P*$4_2$22, a = 8.265, c = 15.512 Å, Z = 2. Mo radiation, R = 0.038 for 1632 reflections.

II. Tetragonal, *I*4_1/*acd*, a = 14.366, c = 23.771 Å, Z = 16. Mo radiation, R = 0.035 for 1181 reflections.

The three-dimensional host structure of (I) is built of infinite -Cd-en-Cd-en- chains extending along the *a*- and *b*- axes alternately, and of tetrahedral $Cd(CN)_4$ moieties arranged between the crossing -Cd-en-Cd-en- chains (Fig. 1). There are two kinds of cavities containing the benzene host molecules. In (II) (Fig. 1) the en ligand is no longer bridging but becomes bidentate, chelating to a Cd atom in its three-dimensional network linked only by CN ligands.

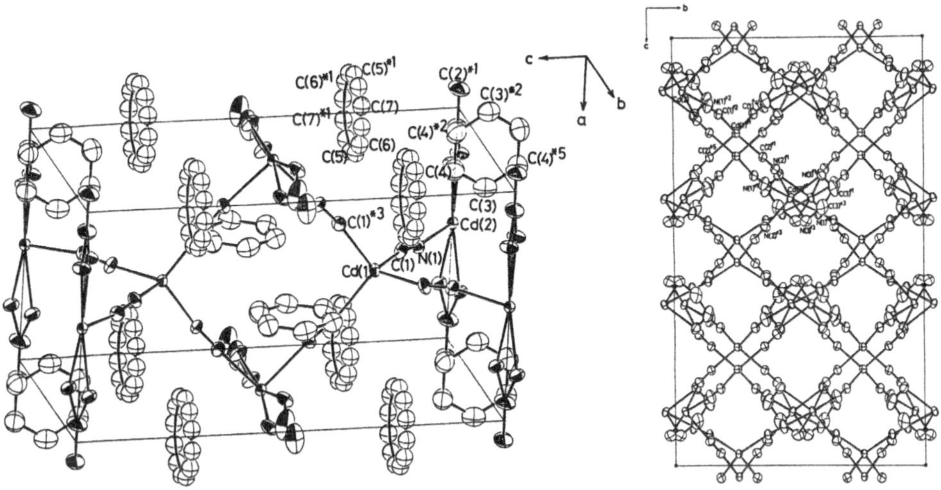

Fig. 1. Views of the crystal structure of $(C_6H_8Cd_2N_6)_n$,$2n(C_6H_6)$ (I) (left) and of $(C_6H_8Cd_2N_6)_n$ (II) (right).

CESIUM (μ_2-AZIDO-N,N)BIS(TRIMETHYLALUMINIUM) p-XYLENE SOLVATE
$C_6H_{18}CsAl_2N_3$,$2(C_8H_{10})$ (I) $C_6H_{18}Al_2N_3^-$,Cs^+,$2(C_8H_{10})$

POTASSIUM-DIBENZO-18-CROWN-6 (μ_2-AZIDO)BIS(TRIMETHYLALUMINIUM) 1-METHYLNAPHTHALENE SOLVATE
$C_{26}H_{42}Al_2KN_3O_6$,$1.5(C_{11}H_{10})$ (II) $C_6H_{18}Al_2N_3^-$,$C_{20}H_{24}KO_6^+$,$1.5(C_{11}H_{10})$

J.L. ATWOOD, W.E. HUNTER, R.D. ROGERS and J.A. WEEKS, 1985. *J. Inclusion Phenomena*, **3**, 113-123.

I. Monoclinic, *C2/m*, a = 19.143, b = 16.227, c = 10.392 Å, β = 114.06°, Z = 4. Mo radiation, R = 0.037 for 2179 reflections.

II. Monoclinic, $P2_1/c$, a = 14.176, b = 13.021, c = 25.324 Å, β = 98.23°, Z = 4. Mo radiation, R = 0.148 for 1502 reflections.

In (I) (Fig. 1) the Cs atom lies on a mirror plane and is sandwiched between two p-xylene molecules; the anion is disordered about a twofold axis. In (II) one of the methylnaphthalene molecules is disordered about an inversion centre (Fig. 2) and interacts weakly with the K^+ ion. In both (I) and (II) the anions have staggered conformations with mean Al-N

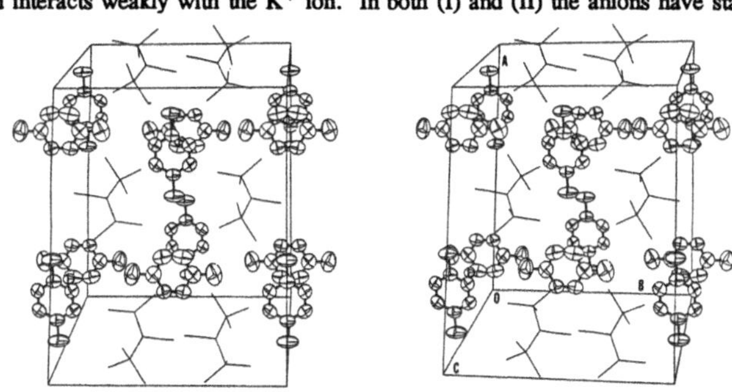

Fig. 1. A stereoview of $C_6H_{18}Al_2N_3^-$,Cs^+,$2(C_8H_{10})$ (I).

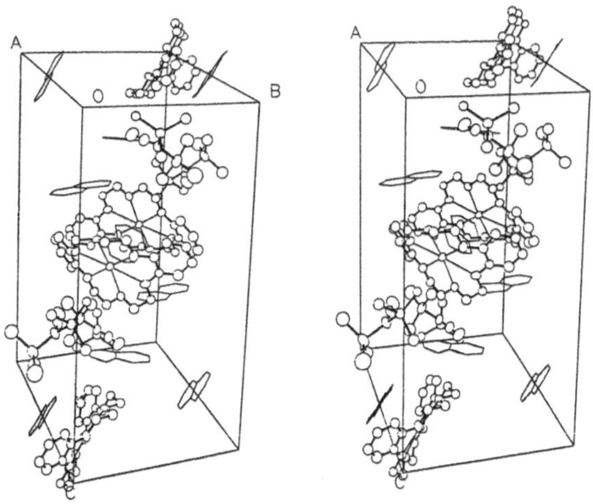

Fig. 2. A stereoview of $C_6H_{18}Al_2N_3^-$,$C_{20}H_{24}KO_6^+$,1.5($C_{11}H_{10}$) (II).

α-1,3,5-BENZENETRICARBOXYLIC ACID BROMINE CLATHRATE (α-TRIMESIC ACID BROMINE CLATHRATE)
$C_9H_6O_6$,0.16(Br_2) (I)

γ-1,3,5-BENZENETRICARBOXYLIC ACID BENZENEDICARBOXYLIC ACID (γ-TRIMESIC ACID BENZENEDICARBOXYLIC ACID)
1.05($C_9H_6O_6$),0.04($C_8H_6O_4$) (II)

F.H. HERBSTEIN, M. KAPON and G.M. REISNER, 1985. *Acta Cryst.*, **B41**, 348-354.

I. Monoclinic, *C2/c*, a = 26.510, b = 16.449, c = 26.580 Å, β = 91.80°, Z = 48. Mo radiation, R = 0.111 for 1925 reflections.

II. Orthorhombic, *I222*, a = 24.225, b = 15.364, c = 16.562 Å, Z = 24. Mo radiation, R = 0.084 for 4897 reflections.

For previous work on α-trimesic acid see (1). In (I), Fig. 1, the Br_2 molecules are accommodated in a considerably distorted manner in cavities around 0, 0.5, 0.25. Compound (II) is isostructural with trimesic acid I_5 (2); the cavity around (0.5,0,0.5) accommodates disordered trimesic acid molecules and various benzene dicarboxylic acids which are presumed to stabilize the γ-phase.

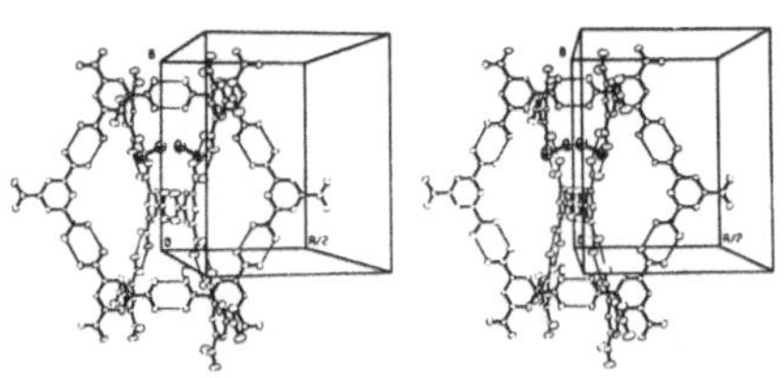

Fig. 1. Stereoview of part of the α-TMA.1/6Br_2 structure.

1. Structure Reports, **34B**, 62.
2. F.H. HERBSTEIN, M. KAPON and G.M. REISENER, 1981. *Proc. Roy. Soc. London Ser. A.*, **376**, 301-318.

1,4,7,10,13,16-HEXAOXACYCLOOCTADECANE BIS(THIOUREA) CLATHRATE (18-CROWN-6 BIS(THIOUREA) CLATHRATE)

$C_{12}H_{24}O_6,2(CH_4N_2S)$

M.G.B. DREW and D.G. NICHOLSON, 1985. *Acta Cryst.*, C41, 1358-1360.

Triclinic, $P\bar{1}$, a = 8.368, b = 8.691, c = 8.680 Å, α = 114.9, β = 106.8, γ = 71.5°, Z = 1. Mo radiation, R = 0.061 for 1829 reflections.

The 18-crown-6 molecule (Fig. 1) is centrosymmetric and has pseudo D_{3d} symmetry. All the O-C-C-C torsion angles are close to ±70° and all C-C-O-C torsion angles are close to 180°. Bond lengths and angles in the molecule and in the unique thiourea molecule in the asymmetric unit are in good agreement with normal values. Two thiourea molecules in the crystal are hydrogen-bonded on either side of the 18-crown-6 molecule through three N-H...O interactions. These N...O separations lie in the range 2.878(4) to 3.249(4) Å. The 18-crown-6- thiourea (1:2) adducts are linked by almost linear N-H...S hydrogen bonds, of average length 3.42 Å, to form a polymeric chain along the *b*-axis.

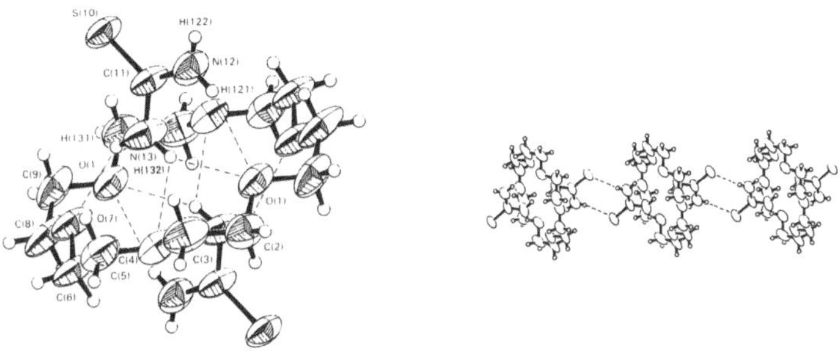

Fig. 1. $C_{12}H_{24}O_6,2(CH_4N_2S)$: a view of the adduct (left) and of the polymeric chain lying along the *b*-axis (right) with hydrogen bonds shown as broken lines.

BIS(2,6-DIMETHYLPIPERIDINE)BIS(THIOCYANATO-N)NICKEL(II) BENZENE CLATHRATE
$C_{16}H_{30}N_4NiS_2,C_6H_6$

M. KOMAN, E. DURCANSKA, M. HANDLOVIC and J. GAZO, 1985. *Acta Cryst.*, C41, 1418-1420.

Triclinic, $P\bar{1}$, a = 11.859, b = 8.494, c = 7.365 Å, α = 103.98, β = 112.00, γ = 102.59°, Z = 1. Mo radiation, R = 0.055 for 1611 reflections.

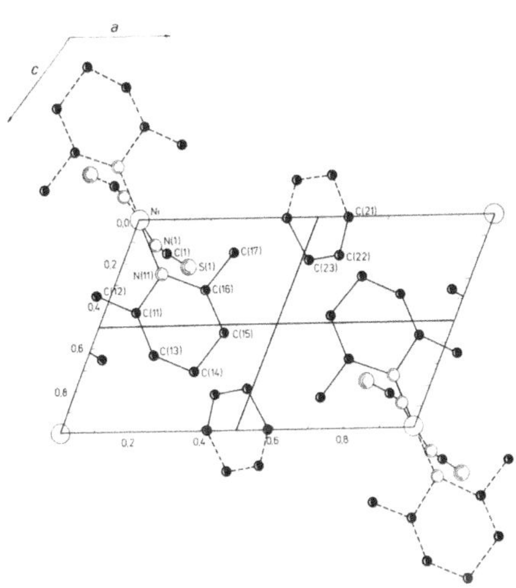

Fig. 1. $C_{16}H_{30}N_4NiS_2$: packing of the molecules projected along *b*.

The complex molecule (Fig. 1) lies on a crystallographic centre of symmetry. The coordination about the Ni atom is square-planar with Ni-N(thiocyanato) and Ni-N(piperidine) distances of 1.849(7) and 1.984(7) Å respectively. The N(NCS)-Ni-N(pip) angle is 88.59(3)°. The benzene molecule also lies on a centre of symmetry and it forms a dihedral angle of 51(3)° with the coordination plane of the Ni atom. The plane through atoms C(11),C(13),C(15),C(16) is almost perpendicular to the Ni coordination plane and forms an angle of 49(3)° with the benzene ring plane.

2,5-BIS(4-CHLOROPHENYL)HYDROQUINONE 1,4-BUTANEDIOL CLATHRATE
$C_{18}H_{12}Cl_2O_2,C_4H_{10}O_2$ (I)

trans-9,10-DIHYDROXY-9,10-DIPHENYL-9,10-DIHYDROANTHRACENE 1,4-BUTANEDIOL CLATHRATE
$C_{26}H_{20}O_2,C_4H_{10}O_2$ (II)

F. TODA, K. TANAKA and T.C.W. MAK, 1985. *J. Inclusion Phenomena*, 3, 225-233.

I. Monoclinic, *C2/c*, a = 16.186, b = 7.626, c = 16.939 Å, β = 91.32°, Z = 4. Mo radiation, R = 0.048 for 1743 reflections.

II. Monoclinic, *C2/c*, a = 11.881, b = 13.009, c = 16.689 Å, β = 110.67°, Z = 4. Mo radiation, R = 0.066 for 1783 reflections.

Both structures (Fig. 1, Fig. 2) feature centrosymmetric hydrogen bonded $(OH)_4$ rings formed by molecular components on special positions. In (I) the hydroquinone has twofold crystallographic symmetry and in (II) the dihydroanthracene lies about an inversion centre. In both structures the 1,4-butanediols lie about inversion centres. The different packing modes account for the different conformations (g^+ag^- and aaa for (I) and (II) respectively) of the 1,4-butanediol.

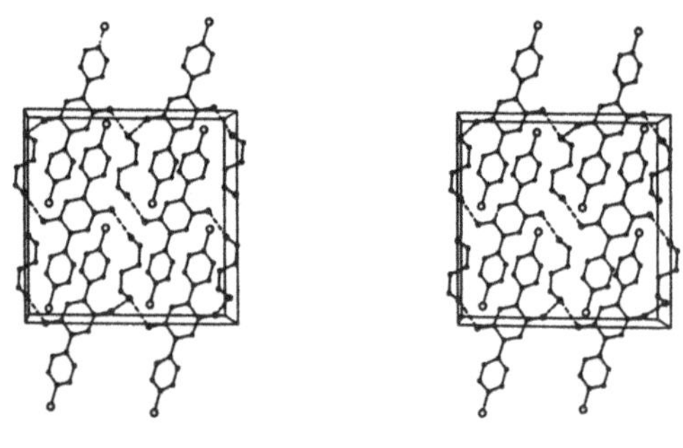

Fig. 1. A stereoview of $C_{18}H_{12}Cl_2O_2,C_4H_{10}O_2$ (I).

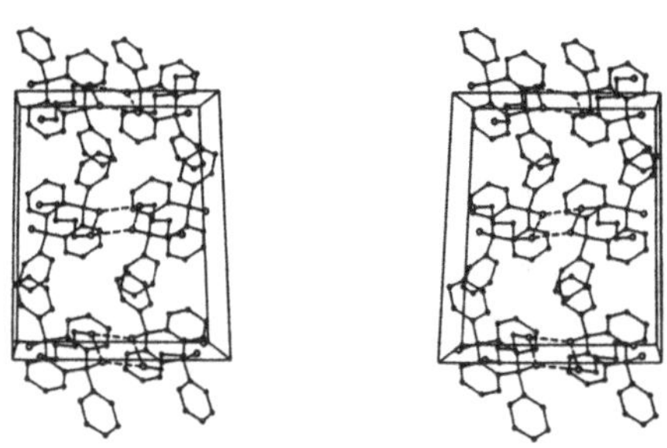

Fig. 2. A stereoview of $C_{26}H_{20}O_2,C_4H_{10}O_2$ (II).

CESIUM BENZO-15-CROWN-5 PICRATE
$C_{20}H_{22}CsN_3O_{12}$ $C_{14}H_{20}O_5,C_6H_2N_3O_7^-,Cs^+$

K. VENKATASUBRAMANIAN, K. JOSHI, N.S. POONIA, W.R. MONTFORT, S.R. ERNST and M.L. HACKERT, 1985. *J. Inclusion Phenomena*, 3, 453-459.

Triclinic, $P\bar{1}$, a = 7.377, b = 11.372, c = 14.806 Å, α = 90.31, β = 91.06, γ = 108.32°, Z = 2. Mo radiation, R = 0.055 for 3575 reflections.

The Cs^+ ion is 9-coordinate involving the five crown O atoms (Cs...O 3.00(1)-3.24(1) Å), the phenoxide (Cs...O 3.03(1) Å) and an ortho nitro group oxygen (Cs...O 3.01(1) Å) of the picrate ion and with two additional O atoms (Cs...O 3.17(1) and 3.40(1) Å) from a para nitro group of an adjacent picrate ion (Fig. 1). The Cs^+ ion lies 2.07 Å out of the mean plane of the crown O atoms.

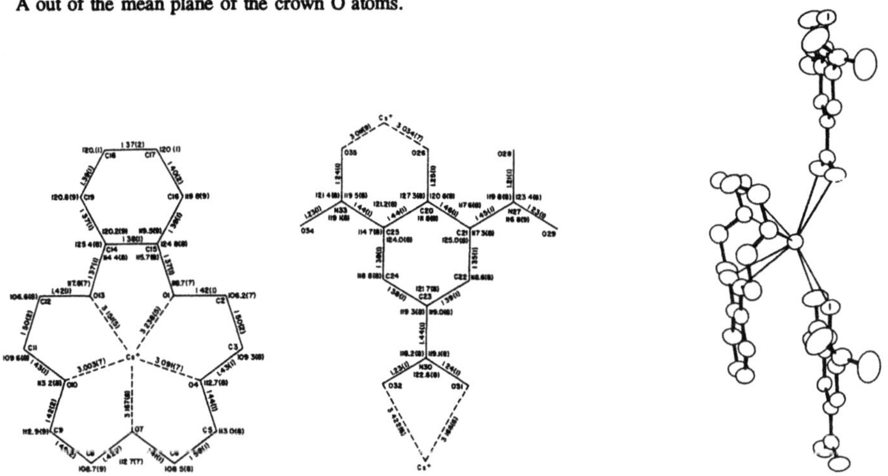

Fig. 1. Molecular dimensions and Cs^+ coordination in $C_{14}H_{20}O_5,C_6H_2N_3O_7^-,Cs^+$.

BIS(2'-HYDROXY-2,4,4,4',7-PENTAMETHYLFLAVAN) 1,4-BIS(HYDROXYMETHYL)CYCLOHEXANE CLATHRATE
$2(C_{20}H_{24}O_2),C_8H_{16}O_2$ (I)

BIS(2'-HYDROXY-2,4,4,4',7-PENTAMETHYLFLAVAN) PIPERAZINE CLATHRATE ETHANOL SOLVATE
$2(C_{20}H_{24}O_2),C_4H_{10}N_2,2(C_2H_6O)$ (II)

2'-HYDROXY-2,4,4,4',7-PENTAMETHYLFLAVAN DIETHYL ETHER CLATHRATE
$C_{20}H_{24}O_2,C_4H_{10}O$ (III)

BIS(2'-HYDROXY-2,4,4,5',6-PENTAMETHYLFLAVAN) N,N'-DIMETHYLPIPERAZINE CLATHRATE
$2(C_{20}H_{24}O_2),C_6H_{14}N_2$ (IV)

J.H. GALL, M. McCARTNEY, D.D. MacNICOL and P.R. MALLINSON, 1985. *J. Inclusion Phenomena*, 3, 421-435.

I. Triclinic, $P\bar{1}$, a = 8.137, b = 9.106, c = 14.552 Å, α = 93.56, β = 94.08, γ = 98.62°, Z = 1. Mo radiation, R = 0.052 for 2630 reflections.

II. Triclinic, $P\bar{1}$, a = 8.438, b = 10.252, c = 14.052 Å, α = 71.28, β = 89.72, γ = 83.88°, Z = 1. Cu radiation, R = 0.063 for 2749 reflections.

III. Triclinic, $P\bar{1}$, a = 8.309, b = 10.546, c = 26.664 Å, α = 102.10, β = 100.48, γ = 81.04°, Z = 4. Mo radiation, R = 0.078 for 1815 reflections.

IV. Triclinic, $P\bar{1}$, a = 7.411, b = 10.143, c = 15.109 Å, α = 98.73, β = 88.30, γ = 109.72°, Z = 1. Mo radiation, R = 0.041 for 1823 reflections.

The structure of (I) (Fig. 1) is similar to that of the 1,4-dioxan water complex (1). The complex lies about an inversion centre; the hydrogen bond distances linking the components are 2.749(2) and 2.967(2) Å. Complex (II) also lies about an inversion centre at the centre of the chair-shaped piperazine ring. Molecules are linked (Fig. 1) by O-H...N (2.83(1) Å) and ϕOH...O (2.66(1) Å) hydrogen bonds; there is also a weak (3.27(1) Å) N-H...O interaction. There are two independent formula units in (III) both with ϕOH...OEt$_2$ hydrogen bonds (2.77(1) and 2.71(1) Å) as shown in Fig. 2. Complex (IV) has the dimethylpiperazine lying about an inversion centre and takes part in O-H...N hydrogen bonds (2.759(3) Å) with the host molecule (Fig. 3).

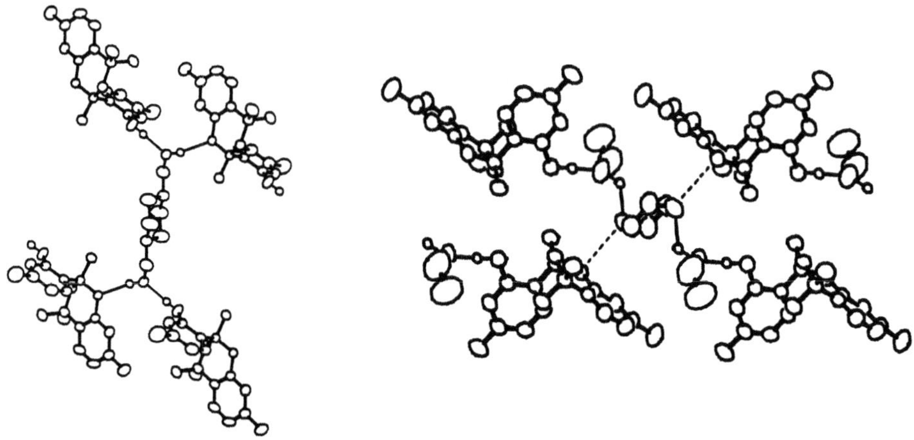

Fig. 1. Views of the $2(C_{20}H_{24}O_2),C_8H_{16}O_2$ (I) (left) and $2(C_{20}H_{24}O_2),C_4H_{10}N_2,2(C_2H_6O)$ (II) (right) clathrates.

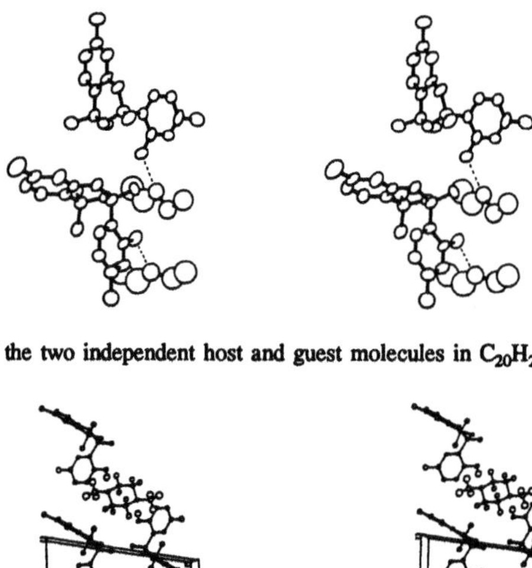

Fig. 2. A stereoview of the two independent host and guest molecules in $C_{20}H_{24}O_2,C_4H_{10}O$ (III).

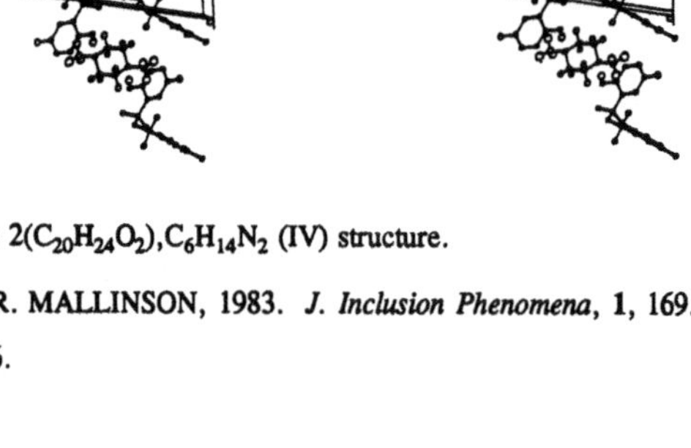

Fig. 3. A stereoview of the $2(C_{20}H_{24}O_2),C_6H_{14}N_2$ (IV) structure.

1. D.D. MacNICOL AND P.R. MALLINSON, 1983. *J. Inclusion Phenomena*, **1**, 169.
Structure Reports, **50B**, 606.

1,4,7,10,13,16-HEXAOXACYCLOOCTADECANE PHENYLACYLAMMONIUM HEXAFLUOROPHOSPHATE (18-CROWN-6 PHENYLACYLAMMONIUM HEXAFLUOROPHOSPHATE)

$C_{20}H_{34}F_6NO_7P$ $C_{12}H_{24}O_6,C_8H_{10}NO^+,F_6P^-$

J.M. MAUD, J.F. STODDART and D.J. WILLIAMS, 1985. *Acta Cryst.*, **C41**, 137-140.

Monoclinic, $P2_1/n$, a = 11.189, b = 17.461, c = 13.449 Å, β = 99.72°, Z = 4. Cu radiation, R = 0.095 for 2162 reflections.

The asymmetric unit of this structure consists of the host 18-crown-6 molecule, the guest phenacylammonium ion (Fig. 1) and a disordered PF_6^- anion. The host molecule has a conformation that is slightly distorted from the more usual all-gauche conformation with approximate D_{3d} symmetry. The sequences of torsion angles (O-C, C-C, C-O) starting from O(1) are ag$^+$a, ag$^-$a, ag$^-$a, ag$^+$a and ag$^-$a with the greatest deviations from normal values in the region C(8)-C(9) of the macrocycle. The guest cation is complexed through three N-H...O hydrogen bonds of lengths 2.96(1), 2.86(1) and 2.89(1) Å to O(1), O(13) and O(7) respectively. In the guest ion the carbonyl group, which is almost coplanar with the phenyl ring, is located above and close to the C(8)-C(9) region of the macrocycle. There are also several short C-H...F contacts between the anion and the host molecule.

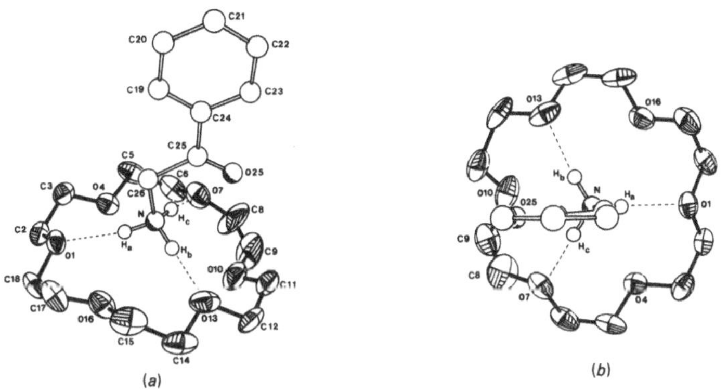

Fig. 1. $C_{20}H_{34}F_6NO_7P$: views of the host-guest complex (a) side-on and (b) from above the macrocycle with the hydrogen bonds indicated as broken lines.

2,2,3,3,11,11,12,12-OCTAMETHYL-1,4,7,10,13,16-HEXAOXACYCLOOCTADECANE BIS(BORANEAMMONIA) CLATHRATE

$C_{20}H_{52}B_2N_2O_6$ $C_{20}H_{40}O_6,2(BH_3NH_3)$

D.R. ALSTON, J.F. STODDART, J.B. WOLSTENHOLME, B.L. ALLWOOD and D.J. WILLIAMS, 1985. *Tetrahedron*, **41**, 2923-2926.

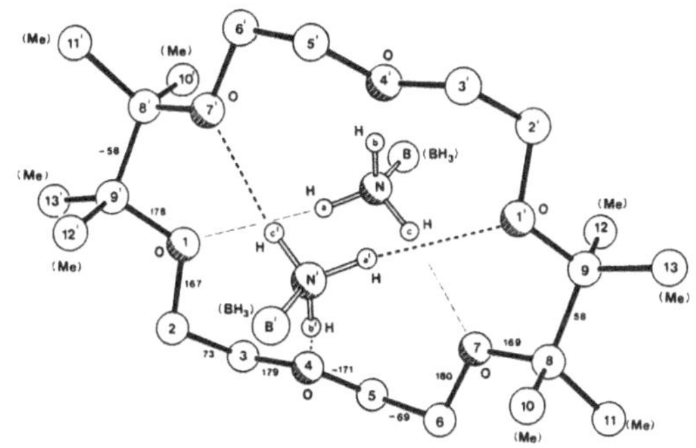

Fig. 1. The structure of $C_{20}H_{40}O_6,2(BH_3NH_3)$.

Monoclinic, $P2_1/n$, a = 8.253, b = 12.721, c = 12.973 Å, β = 91.53°, Z = 2. Cu radiation, R = 0.042 for 1302 reflections.

The complex lies about an inversion centre (Fig. 1) and has the two BH_3NH_3 moieties hydrogen bonded to the ether oxygens (N...O 3.38, 3.16, 3.04 Å). Two of the methyl groups on C(8) and C(9) are oriented pseudo-trans-diaxial (antiperiplanar) and the other two are pseudo-equatorial (synclinal). The B-N bonds (1.595(3) Å) are inclined at 7° to the plane through the O atoms of the host.

1,3-XYLYL-18-CROWN-5 S-t-BUTYLISOTHIOURONIUM PERCHLORATE CLATHRATE
$C_{21}H_{37}ClN_2O_9S$ $C_{16}H_{24}O_5,C_5H_{13}N_2S^+,ClO_4^-$

D.N. REINHOUDT, J.A.A. DE BOER, J.W.H.M. UITERWIJK and S. HARKEMA, 1985. *J. Org. Chem.*, **50**, 4809-4812.

Monoclinic, $P2_1/n$, a = 14.082, b = 17.049, c = 11.058 Å, β = 90.99°, Z = 4. Cu radiation, R = 0.078 for 4153 reflections (at 147 K).

The guest molecule is hydrogen bonded to one of the NH_2 groups to the crown ether (Fig. 1). N(25)...O(4) 2.817(4) Å, N(25)-H...O(4) 171(5)°, N(25)-O(10) 2.929(4) Å, N(25)-H...O(10) 153(5)°. The second NH_2 group does not form linear hydrogen bonds.

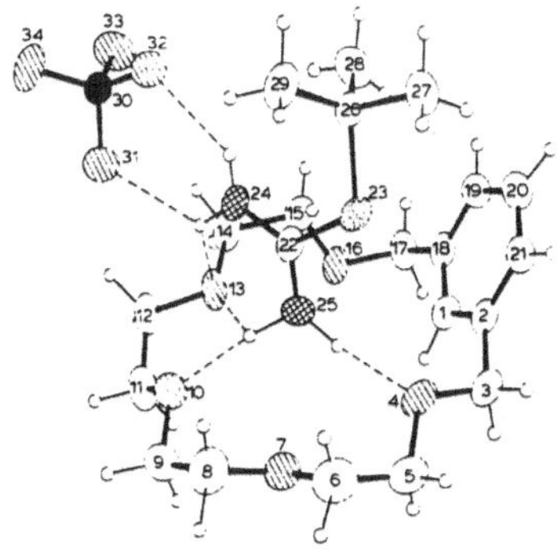

Fig. 1. Solid-state structure of $C_{21}H_{37}ClN_2O_9S$ complex.

METHYL 4,6-O-BENZYLIDENE-2,3-DIDEOXY-α-D-GALACTOPYRANOSIDO[2,3-b]-1,4,7,10,13,16-HEXAOXACYCLOOCTADECANE BORANEAMMONIA CLATHRATE
$C_{24}H_{36}O_{10},H_6BN$ (I)

METHYL 4,6-O-BENZYLIDENE-2,3-DIDEOXY-α-D-MANNOPYRANOSIDO[2,3-b]-1,4,7,10,13,16-HEXAOXACYCLOOCTADECANE BIS(BORANEAMMONIA) CLATHRATE
$C_{24}H_{36}O_{10},2(H_6BN)$ (II)

METHYL 4,6-O-BENZYLIDENE-2,3-DIDEOXY-α-D-MANNOPYRANOSIDO[2,3-b](METHYL 4',6'-o-BENZYLIDENE-2',3'-DIDEOXY-α-D-MANNOPYRANOSIDO)[2',3'-k]-1,4,7,10,13,16-HEXAOXACYCLOOCTADECANE BORANEAMMONIA CLATHRATE
$C_{36}H_{48}O_{14},H_6BN$ (III)

(1R,2R,7R,24R)-3,5,8,11,14,17,20,23,26,28-DECAOXATRICYCLO[21.4.0.0^{2,7}]OCTACOSANE BORANE-AMMONIA CLATHRATE
$C_{18}H_{32}O_{10},H_6BN$ (IV)

1,4:1',4':3,6:3',6'-TETRAANHYDRO-2,2':5,5'-BIS(O-OXYDIETHYLENE)DI-D-MANNITOL BIS(BORANE-AMMONIA) CLATHRATE
$C_{20}H_{32}O_{10},2(H_6BN)$ (V)

H. SHAHRIARIZAVAREH, J.F. STODDART, M.K. WILLIAMS, B.L. ALLWOOD and D.J. WILLIAMS, 1985. *J. Inclusion Phenomena*, **3**, 355-377.

I. Monoclinic, $P2_1$, a = 12.834, b = 8.628, c = 13.017 Å, β = 103.60°, Z = 2. Cu radiation, R = 0.035 for 1947 reflections.

II. Monoclinic, $P2_1$, a = 10.947, b = 7.696, c = 18.234 Å, β = 97.10°, Z = 2. Cu radiation, R = 0.034 for 2108 reflections.

III. Orthorhombic, $P2_12_12_1$, a = 14.071, b = 15.411, c = 18.446 Å, Z = 4. Cu radiation, R = 0.068 for 2166 reflections.

IV. Tetragonal, $I4$, a = 20.574, c = 11.849 Å, Z = 8. Cu radiation, R = 0.069 for 1628 reflections.

V. Monoclinic, $P2_1$, a = 8.414, b = 9.353, c = 17.332 Å, β = 100.27°, Z = 2. Cu radiation, R = 0.040 for 1943 reflections.

All the host molecules in these five structures are chiral. In (I) there is disorder of the macrocyclic component of the host (Fig. 1). Complexing occurs at the α-face with three-point binding of BH_3NH_3 in the perching mode. In (II) (Fig. 2) the BH_3NH_3 complexation is to the two (α and β) diastereotopic faces of the macrocycle. The BH_3NH_3 in (III) binds to the sterically congested β-face through three-point (N-H...O) hydrogen bonding (Fig. 3) and there is disorder of the macrocyclic component of the host. Complex (IV) has the BH_3NH_3 in a quasi-perching position in its three-point binding mode to the 20-membered ring (Fig. 4). The two BH_3NH_3 molecules in (V) bind to opposite faces of the macrocycle via a total of one bifurcated and four single N-H...O hydrogen bonds (Fig. 4).

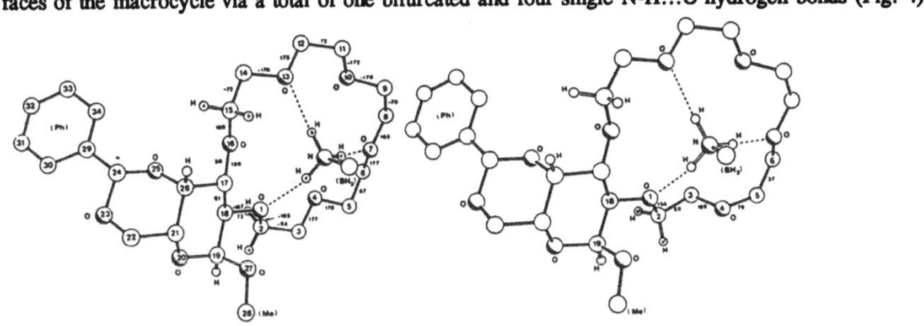

Fig. 1. The two conformers of $C_{24}H_{36}O_{10},H_6BN$ (I).

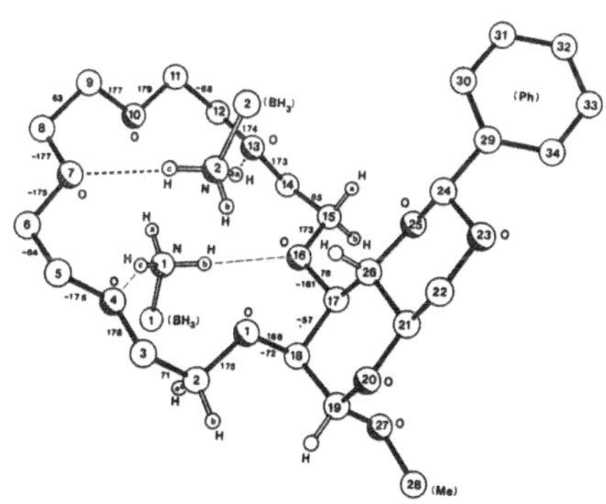

Fig. 2. The $C_{24}H_{36}O_{10},2(H_6BN)$ (II) complex.

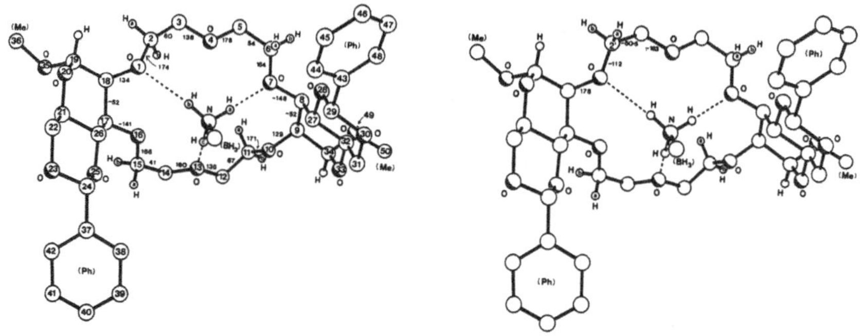

Fig. 3. The two equal occupancy conformers of $C_{36}H_{48}O_{14}$,H_6BN (III).

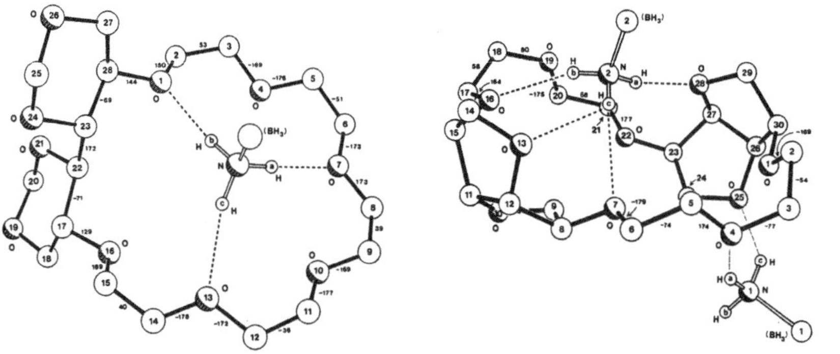

Fig. 4. Views of $C_{18}H_{32}O_{10}$,H_6BN (IV) (left) and of $C_{20}H_{32}O_{10}$,2(H_6BN) (V) (right).

BIS(3α,12α-DIHYDROXY-5β-CHOLAN-24-OIC ACID) PHENYLACETYLENE CLATHRATE (BIS(DEOXYCHOLIC ACID) PHENYLACETYLENE CLATHRATE)

2($C_{24}H_{40}O_4$),C_8H_6

E. GIGLIO, F. MAZZA and L. SCARAMUZZA, 1985. *J. Inclusion Phenomena,* **3**, 437-445.

Orthorhombic, $P2_12_12_1$, a = 25.542, b = 13.662, c = 7.227 Å, Z = 2. Cu radiation, R = 0.071 for 2168 reflections.

The crystal packing (Fig. 1) is characterized by an assembly of antiparallel bilayers of deoxycholic acid molecules which leave empty channels running along the *c*-direction. Phenylacetylene molecules are accommodated within the cavities in an ordered fashion.

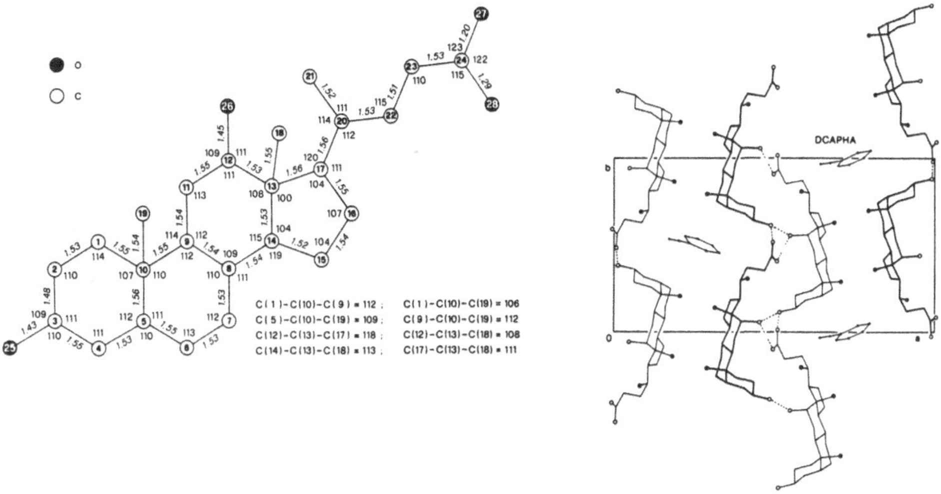

Fig. 1. The deoxycholic acid molecule and a view of the crystal structure viewed along the *c*-direction.

DEOXYCHOLIC ACID TRIDEUTEROACETOPHENONE
$C_{24}H_{40}O_4$,0.40($C_8H_5D_3O$) (I)

DEOXYCHOLIC ACID ACETOPHENONE
$C_{24}H_{40}O_4$,0.40(C_8H_8O) (II), (III)

DEOXYCHOLIC ACID m-CHLOROACETOPHENONE
$C_{24}H_{40}O_4$,0.33(C_8H_7ClO) (IV)

5β-(S)-(1-PHENYLETHANOL)DEOXYCHOLIC ACID
$C_{32}H_{48}O_5$ (V)

C.P. TANG, H.C. CHANG, R. POPOVITZ-BIRO, F. FROLOW, M. LAHAV, L. LEISEROWITZ and R.K. McMULLAN, 1985. *J. Am. Chem. Soc.*, **107**, 4058-4070.

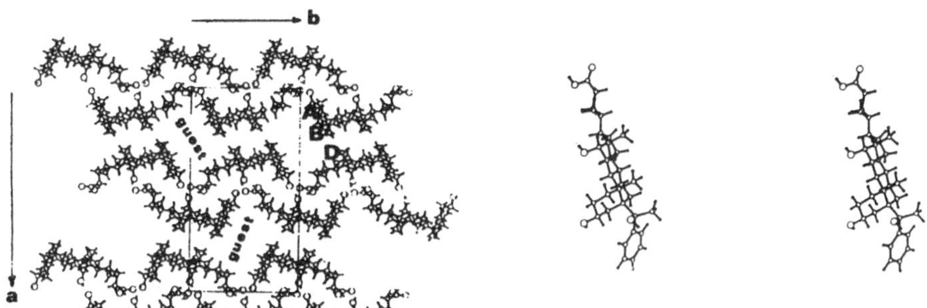

Fig. 1. The packing arrangement of deoxycholic acid viewed along the channel *c*-axis (guest molecules are not shown), and a stereoview of the photoproduct (V).

I. Orthorhombic, $P2_12_12_1$, a = 25.199, b = 13.602, c = 7.192 Å, Z = 4. Neutron radiation, R = 0.10 for 1313 reflections (at 16 K).

II. Orthorhombic, $P2_12_12_1$, a = 25.243, b = 13.606, c = 7.198 Å, Z = 4. Mo radiation, R = 0.059 for 6488 reflections (at 103 K).

III. Orthorhombic, $P2_12_12_1$, a = 25.589, b = 13.709, c = 7.247 Å, Z = 4. Cu radiation, R = 0.058 for 2770 reflections.

IV. Orthorhombic, $P2_12_12_1$, a = 25.283, b = 13.639, c = 7.198 Å, Z = 4. Mo radiation, R = 0.075 for 5812 reflections (at 103 K).

V. Monoclinic, $P2_1$, a = 12.107, b = 15.115, c = 7.522 Å, β = 94.08°, Z = 2. Cu radiation, R = 0.041 for 2712 reflections.

The host-guest channel inclusion compounds (Fig. 1) (I)-(IV) each yield on UV irradiation a photoproduct via addition of guest to the steroid tertiary carbon atom C(5) with formation of a new chiral carbon centre of S-configuration; molecule (V) is such a photoproduct (Fig. 1). The inclusion complexes each contain two independent guest molecules arranged along the channel axes such that both should form the same diasteriomeric product at C(5).

BIS(DEOXYCHOLIC ACID) PINACOLONE CLATHRATE
$C_{24}H_{40}O_4$,0.5($C_6H_{12}O$)

V.M. COIRO, F. MAZZA, G. POCHETTI, E. GIGLIO and N.V. PAVEL, 1985. *Acta Cryst.*, C41, 229-232.

Orthorhombic, $P2_12_12_1$, a = 27.132, b = 13.543, c = 14.228 Å, Z = 8. Cu radiation, R = 0.078 for 3067 reflections.

The crystal packing is characterised by an assembly of antiparallel pleated bilayers of the steroid molecules, which pack together leaving empty spaces occupied by guest molecules. The cross-section of the channels is almost square. The pinacolone assumes two different orientations in the channels. These are related by a pseudo twofold axis (Fig. 1). The geometry and the conformation of both side chain and D ring of the two deoxycholic acid molecules are comparable. The side chain is gauche and the D ring has half-chair symmetry. Bond lengths and angles are shown in the figure.

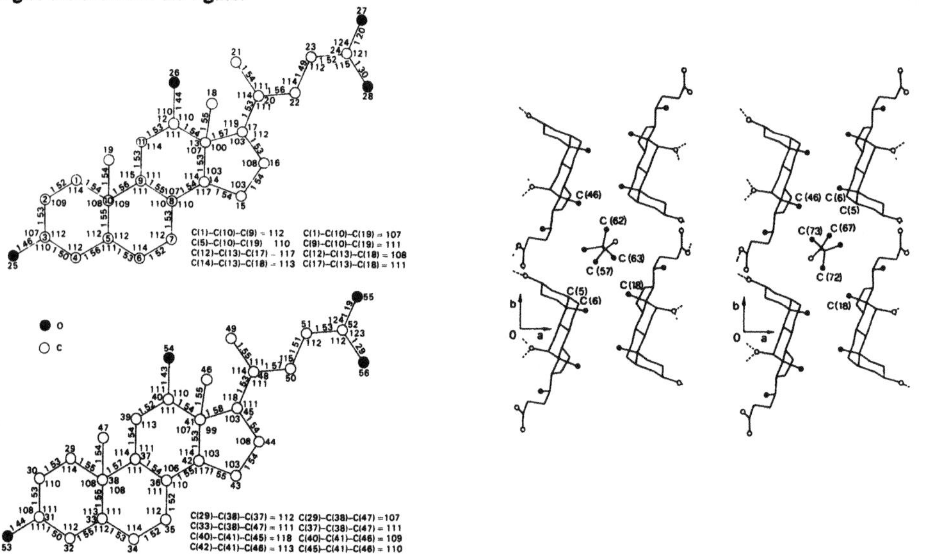

Fig. 1. $2(C_{24}H_{40}O_4),C_6H_{12}O$: bond distances and angles and the channel viewed along c.

DEOXYCHOLIC ACID ACETONE
$C_{24}H_{40}O_4$,0.6(C_3H_6O) (I), (II)

APOCHOLIC ACID ACETONE

$C_{24}H_{38}O_4,C_3H_6O$ (III)

DEOXYCHOLIC ACID DIETHYL KETONE
$C_{24}H_{40}O_4,0.5(C_5H_{10}O)$ (IV)

DEOXYCHOLIC ACID CYCLOHEXANONE
$C_{24}H_{40}O_4,0.5(C_6H_{10}O)$ (V)

DEOXYCHOLIC ACID ETHYL METHYL KETONE
$C_{24}H_{40}O_4,0.5(C_4H_8O)$ (VI), (VII)

DEOXYCHOLIC ACID METHYL PENTYL KETONE
$C_{24}H_{40}O_4,0.33(C_7H_{14}O)$ (VIII)

R. POPOVITZ-BIRO, C.P. TANG, H.C. CHANG, M. LAHAV and L. LEISEROWITZ, 1985. *J. Am. Chem. Soc.*, **107**, 4043-4058.

I. Orthorhombic, $P2_12_12_1$, a = 25.416, b = 13.514, c = 7.194 Å, Z = 4. Mo radiation, R = 0.067 for 4109 reflections (at 103 K).

II. Orthorhombic, $P2_12_12_1$, a = 25.809, b = 13.610, c = 7.233 Å, Z = 4. Cu radiation, R = 0.067 for 2505 reflections.

III. Orthorhombic, $P2_12_12_1$, a = 24.570, b = 14.264, c = 7.530 Å, Z = 4. Cu radiation, R = 0.083 for 2579 reflections.

IV. Orthorhombic, $P2_12_12_1$, a = 25.828, b = 13.560, c = 7.240 Å, Z = 4. Cu radiation, R = 0.108 for 2771 reflections.

V. Orthorhombic, $P2_12_12_1$, a = 26.990, b = 13.354, c = 14.141 Å, Z = 8. Cu radiation, R = 0.086 for 5442 reflections.

VI. Orthorhombic, $P2_12_12_1$, a = 25.462, b = 13.448, c = 7.176 Å, Z = 4. Mo radiation, R = 0.096 for 6011 reflections (at 103 K).

VII. Orthorhombic, $P2_12_12_1$, a = 25.805, b = 13.593, c = 7.228 Å, Z = 4. Cu radiation, R = 0.097 for 2770 reflections.

VIII. Orthorhombic, $P2_12_12_1$, a = 25.529, b = 13.440, c = 7.214 Å, Z = 4. Mo radiation, R = 0.056 for 4232 reflections (at 103 K).

Deoxycholic acid (DCA) forms channel inclusion complexes with acetone ((I), (II)) (host-guest ratio 5:3), diethyl ketone (IV) (2:1) and ethyl methyl ketone ((VI, (VII)) (2:1). UV irradiation of these complexes led to stereospecific photoaddition of the guests. The host structures are isomorphous; they form hydrogen-bonded bilayers which are juxtaposed by hydrophobic contacts to form inclusion channels delineated by four walls. The occluded ketones are sandwiched between the wide walls of the channel comprising the steroid A and B rings. The 3:1 complex (VIII) DCA-methyl pentyl ketone (Fig. 1) is very similar to (VII). A different motif was found in the cyclohexanone complex (V) (Fig. 1) where the guest is sandwiched between D-rings and steroid side chains. The 1:1 apocholic acid acetone complex (III) (Fig. 2) is light stable. There is a similar host bilayer structure as found for DCA but the acetone molecules are now stacked up the channel axis with the ketone C=O bond tending to be parallel to the nearest C-H of the steroid wall with O...H distances 3.5 Å.

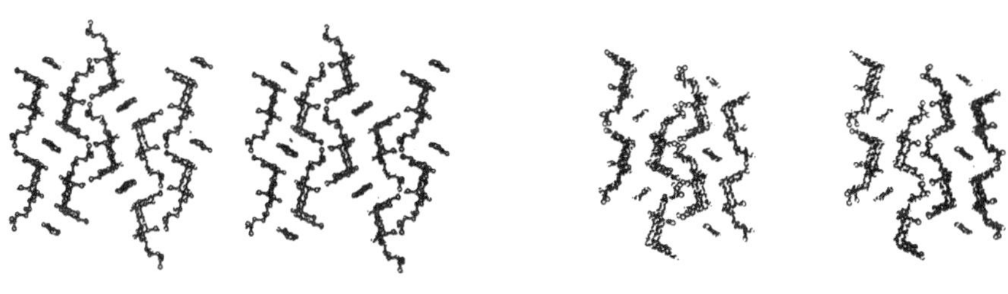

Fig. 1. Stereoviews of DCA-cyclohexanone (V) (left) and DCA-methyl pentyl ketone (VIII) (right).

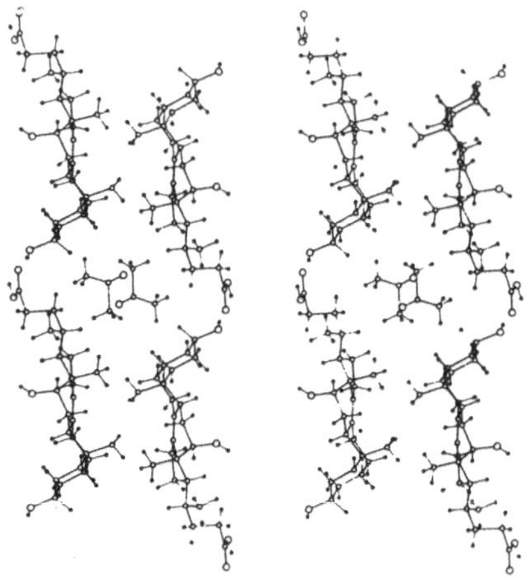

Fig. 2. A stereoview of APA-acetone (III).

trans-BIS(ISOTHIOCYANATO)TETRAKIS(4-METHYLPYRIDINE-N)CADMIUM(II) 4-METHYLPYRIDINE CLATHRATE HYDRATE
$C_{26}H_{28}CdN_6S_2,0.67(C_6H_7N),0.33(H_2O)$

N.V. PERVUKHINA, N.V. PODBEREZSKAYA, V.V. BAKAKIN, N.V. KISLIKH, G.N. CHEKHOVA and Yu.A. DYADIN, 1985. *Zh. Strukt. Khim.*, **26**, 6-120-127 [J. Struct. Chem., **26**, 934-941].

Triclinic, *P*1, a = 17.726, b = 16.942, c = 11.262 Å, α = 76.83, β = 115.06, γ = 122.50°, Z = 3. Mo radiation, R = 0.077 for 3612 reflections.

The structure (Fig. 1) is molecular-insular in nature, formed by $Cd(4\text{-}MePy)_4(NCS)_2$ host molecules and two types of guest molecules, 4-MePy and H_2O. The three independent Cd atoms have distorted octahedral coordination geometry consisting of four 4-MePy nitrogen atoms and two NCS^- nitrogen atoms with mean Cd-N 2.39 Å. The metal atoms are located at the apices of the unit-cell, in the middle of side a, and in the middle of the bc face to form a channel of varying cross-section in the centre of the cell extending along *c*. The maximum cross-section of the channel (z ≈ 0) is occupied by 4-MePy molecules and the minimum cross-section (z ≈ 1/2) is occupied by a water molecule.

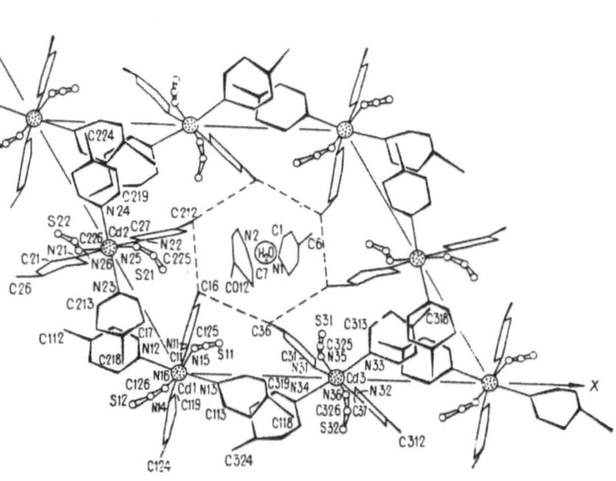

Fig. 1. Projection of the $Cd(4\text{-}MePy)_4(NCS)_2{\cdot}0.67(4\text{-}MePy){\cdot}0.33(H_2O)$ structure on the (001) plane.

18-CROWN-6 BIS(CHLOROPHENYL UREA) CLATHRATE
$C_{26}H_{38}Cl_2N_4O_8$ $C_{12}H_{24}O_6,2(C_7H_7ClN_2O)$

J. WEILER, 1985. *Bull. Soc. Chim. Belg.*, **94**, 1101-1102.

Monoclinic, *C2/c*, a = 25.590, b = 7.965, c = 15.237 Å, β = 101.64°, Z = 4. Mo radiation, R = 0.0457 for 1093 reflections.

The 18-crown-6 molecule lies about an inversion centre and the urea molecules are linked to it by three N-H...O hydrogen bonds (Fig. 1), with N(17)...O(5) 2.942, N(20)...O(2) 3.265, N(20)...O(8*) 3.145 Å.

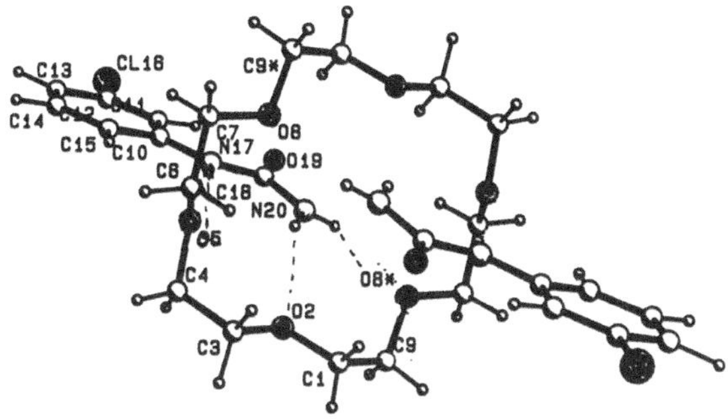

Fig. 1. A view of the $C_{12}H_{24}O_6,2(C_7H_7ClN_2O)$ complex.

N,N'-(5-t-BUTYL-1,3-PHENYLENEBIS(METHYLENE))BIS(QUINUCLIDINIUM) DIBROMIDE 1-PROPANOL MONOHYDRATE CLATHRATE
$C_{26}H_{42}Br_2N_2,C_3H_8O,H_2O$ $C_{26}H_{42}N_2^{2+},C_3H_8O,2(Br^-),H_2O$

H.-P. JOSEL, H.-G. LOHR, A. ENGEL, J. RAPP, F. VOGTLE, S. FRANKEN and H. PUFF, 1985. *J. Inclusion Phenomena*, **3**, 43-50.

Orthorhombic, *Pnma*, a = 14.696, b = 16.276, c = 12.659 Å, Z = 4. Mo radiation, R = 0.045 for 1702 reflections (at 188 K).

The cation, propanol and water molecules lie on crystallographic mirror planes (Fig. 1) and the propanol is disordered. The two ammonium side branches of the quinuclidinium salt are arranged cisoid and four host cations define a cage type cavity containing the propanol guest.

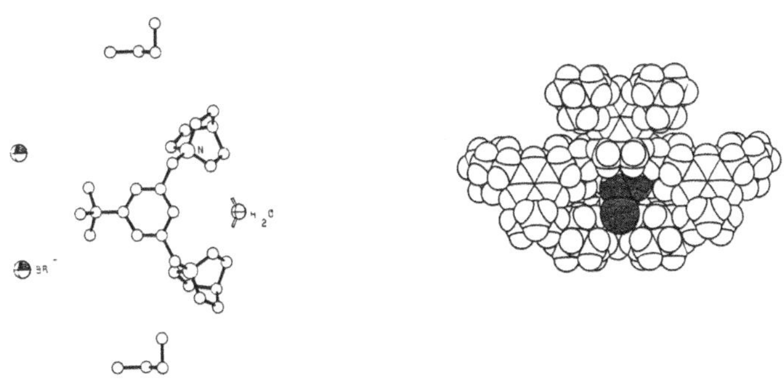

Fig. 1. The conformation of host and guest molecules in $C_{26}H_{42}N_2^{2+},C_3H_8O,2(Br^-),H_2O$, and a space filling model with the propanol marked by a dotted pattern.

2,3,7,8,12,13-HEXAMETHOXY-5,10-DIHYDRO-15H-TRIBENZO[a,d,g]CYCLONONANE HYDRATE (CYCLO

TRIVERATRYLENE HYDRATE)
$C_{27}H_{30}O_6,0.16(H_2O)$

G.I. BIRNBAUM, D.D. KLUG, J.A. RIPMEESTER and J.S. TSE, 1985. *Canad. J. Chem.*, **63**, 3258-3263.

Monoclinic, *C2/c*, a = 24.216, b = 9.636, c = 23.117 Å, β = 118.08°, Z = 8. Cu radiation, R = 0.046 for 3986 reflections.

The cyclononatriene ring has a crown conformation as shown in Fig. 1. The water molecules reside in between CTV molecules stacked along the b-direction with their O atoms 3.086 Å from methoxy oxygen O(4).

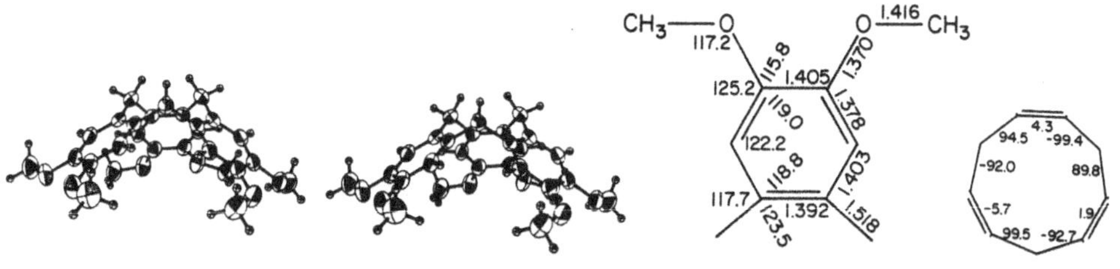

Fig. 1. A perspective view of $C_{27}H_{30}O_6$; also shown are mean dimensions and torsion angles in the nine-membered ring.

8,15,18,21,28-PENTAOXA-34,35-DIAZAPENTACYCLO[28.3.1.1^{2,6}.0^{9,14}.0^{22,27}]PENTATRIACONTA-1(35),2,-4,6(34),9,11,13,22,24,26,30,32-DODECAENE PROPANOL CLATHATE DIBENZO-(2,2'-BIPYRIDINO)-21-CROWN-5 PROPANOL CLATHRATE
$C_{28}H_{26}N_2O_5,C_3H_8O$

E. WEBER, H.-P. JOSEL, H. PUFF and S. FRANKEN, 1985. *J. Org. Chem.*, **50**, 3125-3132.

Monoclinic, *C2/c*, a = 26.652, b = 8.138, c = 26.674 Å, β = 105.61°, Z = 8. Mo radiation, R = 0.088 for 2884 reflections.

The propanol molecules are coordinated via hydrogen-bonds to one of the bipyridine nitrogen atoms (O...N 3.00 Å) and sit in the hollow of the host (Fig. 1). The packing diagram shows the host-guest topology to be a channel-like clathrate (tubulatocoordinatoclathrate).

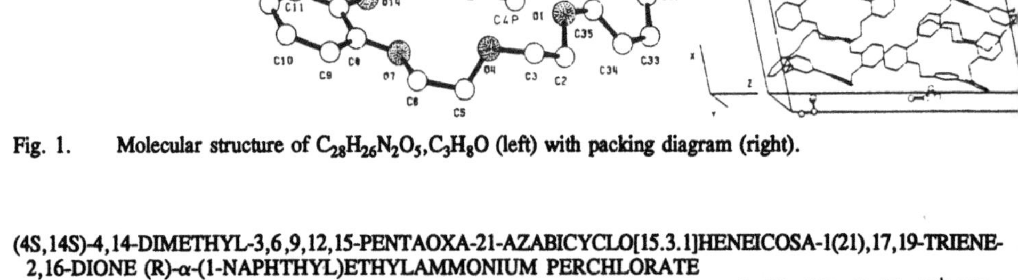

Fig. 1. Molecular structure of $C_{28}H_{26}N_2O_5,C_3H_8O$ (left) with packing diagram (right).

(4S,14S)-4,14-DIMETHYL-3,6,9,12,15-PENTAOXA-21-AZABICYCLO[15.3.1]HENEICOSA-1(21),17,19-TRIENE-2,16-DIONE (R)-α-(1-NAPHTHYL)ETHYLAMMONIUM PERCHLORATE
$C_{29}H_{37}ClN_2O_{11}$ (I) $C_{17}H_{23}NO_7,C_{12}H_{14}N^+,ClO_4^-$

(4S,14S)-4,14-DIMETHYL-3,6,9,12,15-PENTAOXA-21-AZABICYCLO[15.3.1]HENEICOSA-1(21),17,19-TRIENE-2,16-DIONE (S)-α-(1-NAPHTHYL)ETHYLAMMONIUM PERCHLORATE
$C_{29}H_{37}ClN_2O_{11}$ (II) $C_{17}H_{23}NO_7,C_{12}H_{14}N^+,ClO_4^-$

R.B. DAVIDSON, N.K. DALLEY, R.M. IZATT, J.S. BRADSHAW and C.F. CAMPANA, 1985. *Israel J. Chem.*, **25**, 33-38.

I. Orthorhombic, $P2_12_12_1$, a = 8.47, b = 11.78, c = 31.19 Å, Z = 4. Mo radiation, R = 0.170 for 1680 reflections.

II. Monoclinic, $P2_1$, a = 8.654, b = 11.954, c = 15.130 Å, β = 97.39°, Z = 2. Mo radiation, R = 0.081 for 2751 reflections.

In each case the ammonium cation is hydrogen bonded to the macrocycle via two N-H...O and one N-H...N hydrogen bond. The hydrogen bonding pattern is similar in both complexes with N...N 2.96 and 2.92 Å, N...O 2.86-3.06 Å.

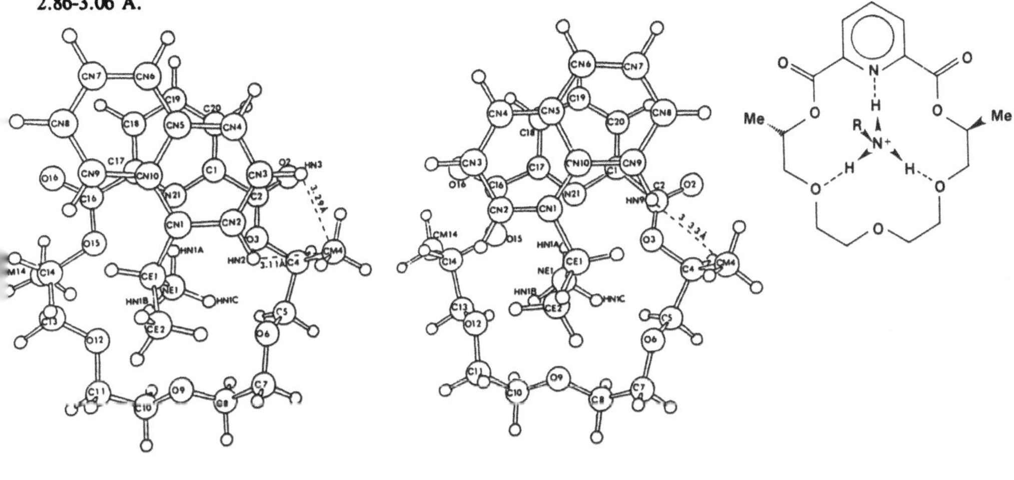

S *R*

Fig. 1. Diagrams of the S and R complexes and an indication of the hydrogen bonding scheme.

1,1,6,6-TETRAPHENYLHEXA-2,4-DIYNE-1,6-DIOL BIS(BENZYLIDENEACETOPHENONE) CLATHRATE $C_{30}H_{22}O_2,2(C_{15}H_{12}O)$ (I)

2,5-DIPHENYLHYDROQUINONE BIS(DIBENZYLIDENEACETONE) CLATHRATE $C_{18}H_{14}O_2,2(C_{17}H_{14}O)$ (II)

M. KAFTORY, K. TANAKA and F. TODA, 1985. *J. Org. Chem.*, **50**, 2154-2158.

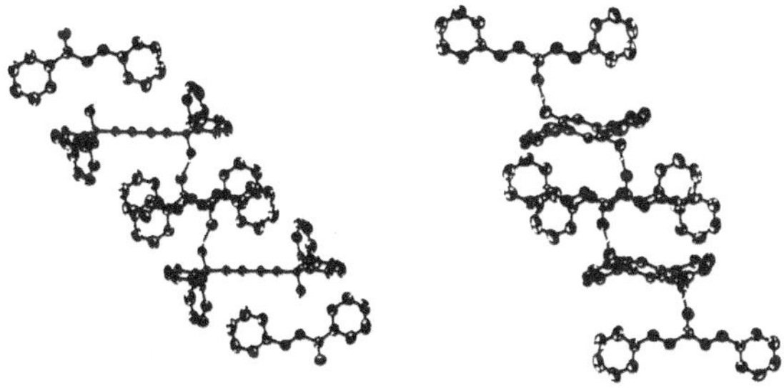

Fig. 1. Drawings of the molecules in the host-guest complexes $C_{30}H_{22}O_2,2(C_{15}H_{12}O)$ (left) and $C_{18}H_{14}O_2,2$-$(C_{17}H_{14}O)$ (right).

I. Triclinic, $P\bar{1}$, a = 12.568, b = 12.161, c = 8.532 Å, α = 109.60, β = 109.60, γ = 82.69°, Z = 1. Mo radiation, R = 0.074 for 3558 reflections.

II. Triclinic, $P\bar{1}$, a = 13.627, b = 10.099, c = 8.056 Å, α = 107.86, β = 97.59, γ = 103.15°, Z = 1. Mo

radiation, R = 0.079 for 2418 reflections.

The crystal structure analyses of these two inclusion complexes show that in both complexes (Fig. 1) two molecules of the guest are linked by hydrogen bonds between the carbonyl oxygen atom and the hydroxyl groups of the host. The molecules in both structures are packed with parallel planar guest molecules related by an inversion centre with double-bond centre-to-centre distances ranging from 3.787 to 3.497 Å.

(18-CROWN-6)POTASSIUM PERCHLOROTRIPHENYLMETHANIDE CLATHRATE

$C_{31}H_{24}Cl_{15}KO_6$ $C_{12}H_{24}KO_6^+,C_{19}Cl_{15}^-$

C. MIRAVITLLES, E. MOLINS, X. SOLANS, G. GERMAIN and J.P. DECLERCQ, 1985. *J. Inclusion Phenomena*, **3**, 27-34.

Monoclinic, $P2_1$, a = 10.491, b = 18.016, c = 11.670 Å, β = 103.11°, Z = 2. Mo radiation, R = 0.039 for 2339 reflections.

The triphenylmethanide ion has a propeller conformation, with the phenyl rings (Fig. 1) making dihedral angles of 45, 45 and 40(1)° with the C(1),C(2),C(8),C(14) plane. The potassium atom is at the centre of the 18-crown-6 ring which has approximate D_{3d} symmetry and K...O 2.74-2.85 Å.

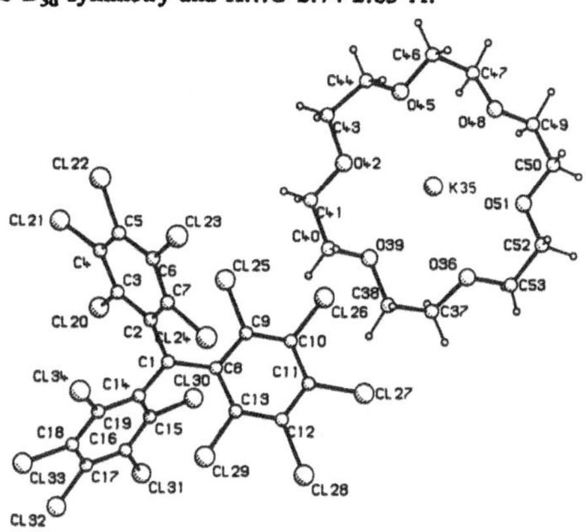

Fig. 1. The $C_{12}H_{24}KO_6^+,C_{19}Cl_{15}^-$ structure.

TETRAKIS(TETRAETHYLAMMONIUM) TETRAFLUORIDE UNDECAHYDRATE CLATHRATE

$C_{32}H_{80}F_4N_4,11(H_2O)$ $4(C_8H_{20}N^+),4(F^-),11(H_2O)$

T.C.W. MAK, 1985. *J. Inclusion Phenomena*, **3**, 347-354.

Orthorhombic, $Pna2_1$, a = 16.130, b = 16.949, c = 17.493 Å, Z = 4. Mo radiation, R = 0.091 for 2278 reflections.

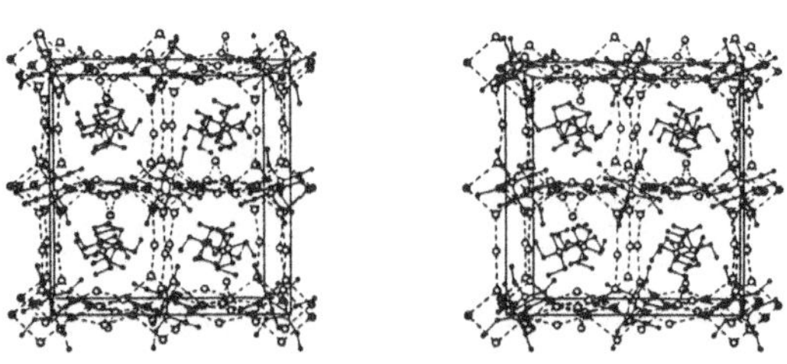

Fig. 1. A stereoview of the crystal structure of $4(Et_4NF),11(H_2O)$.

The structure (Fig. 1) is a clathrate hydrate containing infinite chains of edge-sharing $(H_2O)_4F^-$ tetrahedra extending parallel to the *a*-axis. The chains are laterally linked by bridging water molecules to form a three dimensional hydrogen-bonded anion/water framework. Ordered Et_4N^+ cations occupy the voids in two open channel systems running in the *b*- and *c*-directions. F...O distances are in the range 2.586-2.851 Å.

TRI-o-THYMOTIDE CHLOROCYCLOHEXANE CLATHRATE
$C_{33}H_{36}O_6,0.5(C_6H_{11}Cl)$

R. GERDIL and A. FREW, 1985. *J. Inclusion Phenomena,* **3**, 335-344.

Trigonal, $P3_121$, a = 13.604, c = 30.605 Å, Z = 6. Mo radiation, R = 0.059 for 2014 reflections (at 158 K).

This cage type inclusion complex contains an axial-Cl chair and an axial-Cl boat conformation of the guest distributed statistically in a 2:1 ratio over the available sites (Fig. 1).

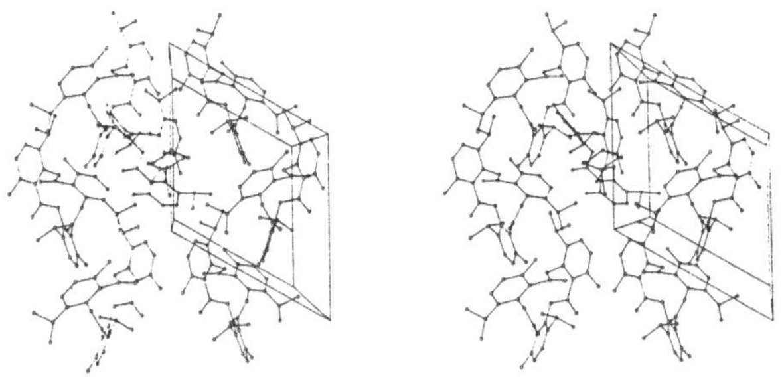

Fig. 1. A stereoview of $C_{33}H_{36}O_6,0.5(C_6H_{11}Cl)$. For clarity the minor guest conformer is not shown. The major guest conformer is shown single positioned on a twofold axis.

(2,2'-BIPYRIDYL)DIAMMINEPLATINUM(II) DIBENZO-24-CROWN-8 BIS(HEXAFLUOROPHOSPHATE) CLATHRATE
$C_{34}H_{46}F_{12}N_4O_8P_2Pt$ (I) $C_{10}H_{14}N_4Pt^{2+},C_{24}H_{32}O_8,2(F_6P^-)$

(2,2'-BIPYRIDYL)DIAMMINEPLATINUM(II) DIBENZO-30-CROWN-10 BIS(HEXAFLUOROPHOSPHATE) CLATHRATE HYDRATE
$C_{38}H_{54}F_{12}N_4O_{10}P_2Pt,0.6(H_2O)$ (II) $C_{10}H_{14}N_4Pt^{2+},C_{28}H_{40}O_{10},2(F_6P^-),0.6(H_2O)$

H.M. COLQUHOUN, S.M. DOUGHTY, J.M. MAUD, J.F. STODDART, D.J. WILLIAMS and J.B. WOLSTENHOLME, 1985. *Israel J. Chem.*, **25**, 15-26.

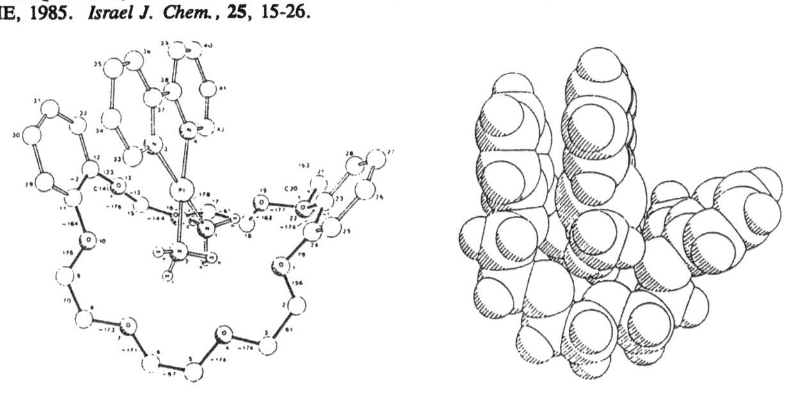

Fig. 1. The $[Pt(bipy)(NH_3)_2.DB24C8]^{2+}$ complex and a space-filling drawing.

I. Triclinic, $P\bar{1}$, a = 11.775, b = 14.578, c = 14.766 Å, α = 88.06, β = 72.80, γ = 67.92°, Z = 2. Cu radiation, R = 0.038 for 4534 reflections.

II. Monoclinic, $P2_1/n$, a = 16.081, b = 15.912, c = 18.718 Å, β = 101.34°, Z = 4. Cu radiation, R = 0.053 for 4797 reflections.

The analyses reveal that in the crown-8 complex (I) only one of the crown ether benzo rings interacts with the bipyridyl ligand in a charge-transfer sense (Fig. 1), but when the crown is enlarged as in (II), the macrocycle is sufficiently large and flexible to allow both benzo rings to encapsulate the Pt-bipyridyl complex in its U-shaped cavity (Fig. 2).

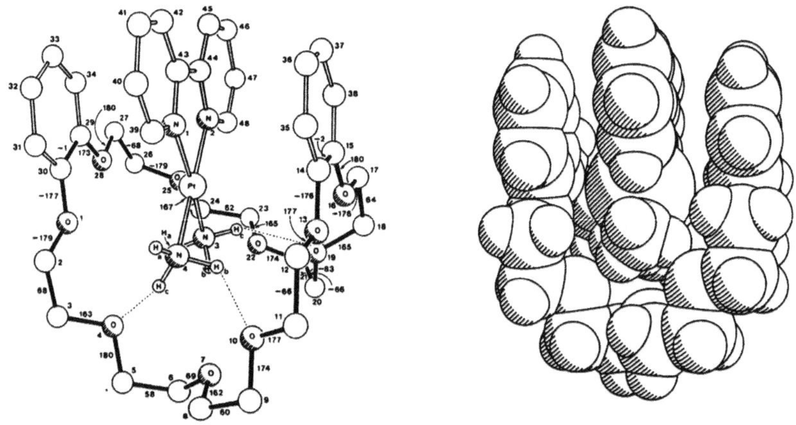

Fig. 2. The $[Pt(bipy)(NH_3)_2.DB30C8]^{2+}$ complex and a space-filling drawing.

1,1,6,6-TETRAKIS(2,4-DIMETHYLPHENYL)-2,4-HEXADIYNE-1,6-DIOL ETHANOL SOLVATE
$C_{38}H_{38}O_2,2(C_2H_6O)$ (I)

1,1-BIS(2,4-DIMETHYLPHENYL)-2-BUTYN-1-OL ETHANOL SOLVATE
$C_{20}H_{22}O,C_2H_6O$ (II)

9-HYDROXY-9-(1-PROPYNYL)FLUORENE METHANOL SOLVATE
$C_{16}H_{12}O,CH_4O$ (III)

F. TODA, K. TANAKA and T.C.W. MAK, 1985. *Bull. Chem. Soc. Jpn.*, **58**, 2221-2227.

I. Triclinic, $P\bar{1}$, a = 8.249, b = 9.522, c = 14.803 Å, α = 100.95, β = 110.82, γ = 113.08°, Z = 1. Mo radiation, R = 0.105 for 1834 reflections.

II. Triclinic, $P\bar{1}$, a = 8.142, b = 11.281, c = 12.262 Å, α = 102.76, β = 97.08, γ = 111.19°, Z = 2. Mo radiation, R = 0.073 for 2104 reflections.

III. Monoclinic, $P2_1/a$, a = 8.641, b = 10.883, c = 30.63 Å, β = 92.88°, Z = 8. Mo radiation, R = 0.154 for 1885 reflections.

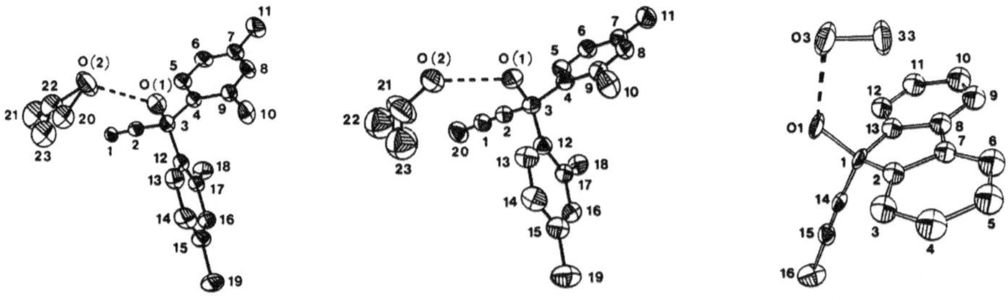

Fig. 1. The asymmetric units of (I), (II) and (III) (one pair) (left to right).

In (I) the molecule lies about an inversion centre (Fig. 1) and its hydroxyl groups are linked by hydrogen bonds to neighbouring (disordered) ethanol molecules, generating a layer structure with $(OH)_4$ rings centred at a second inversion centre (O...O 2.723 and 2.669(7) Å). In (II) two butynol and two (disordered) ethanol molecules are likewise alternately linked into a planar hydrogen-bonded $(OH)_4$ ring (O...O 2.700 and 2.686(4) Å). In (III) there are two independent sets of fluorene and methanol molecules; in each set the molecules are linked by hydrogen bonds to form infinite zig-zag chains parallel to the *a*-axis.

β-CYCLODEXTRIN 1-HYDROXYMETHYLADAMANTANE UNDECAHYDRATE
$C_{42}H_{70}O_{35},C_{11}H_{18}O,11(H_2O)$

J.A. HAMILTON, 1985. *Carbohydr. Res.*, **142**, 21-28.

Orthorhombic, $C222_1$, a = 19.162, b = 23.965, c = 32.597 Å, Z = 8. Mo radiation, R = 0.115 for 4824 reflections (at -160°C).

The β-cyclodextrin occurs as a dimer in the crystal by means of extensive hydrogen-bonding across the secondary hydroxyl ends of two molecules related by a twofold axis. The primary hydroxyl ends of two adjacent molecules are also related by twofold symmetry and form strong hydrogen bonds with the hydroxyl group of the guest molecules (Fig. 1). Other stabilizing hydrogen bonds are formed via water molecules. A unique feature is the presence of water molecules trapped in the cavity at the secondary hydroxyl interface.

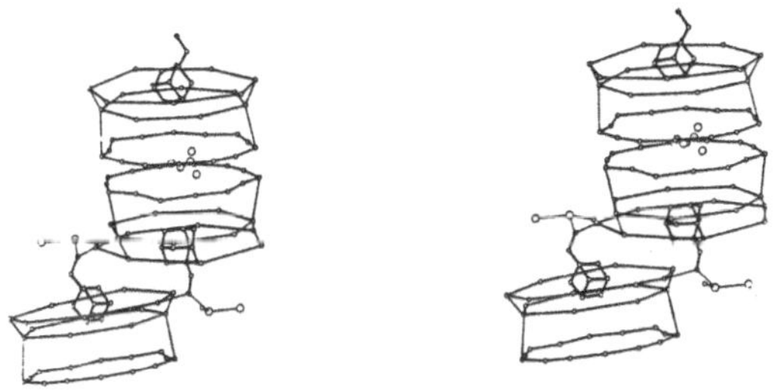

Fig. 1. A stereoview of the dimer plus one additional complex unit. The β-cyclodextrin molecules are represented by their primary and secondary hydroxyl groups and glycosidic oxygen atoms. The larger open circles represent water molecules.

β-CYCLODEXTRIN BENZYL ALCOHOL CLATHRATE PENTAHYDRATE
$C_{42}H_{70}O_{35},C_7H_8O,5(H_2O)$

K. HARATA, K. UEKAMA, M. OTAGIRI, F. HIRAYAMA and Y. OHTANI, 1985. *Bull. Chem. Soc. Jpn.*, **58**, 1234-1238.

Monoclinic, $P2_1$, a = 15.365, b = 10.101, c = 20.869 Å, β = 109.90°, Z = 2. Mo radiation, R = 0.066 for 5290 reflections.

The β-cyclodextrin molecule is somewhat elliptically distorted from a regular heptagonal symmetry because of the inclusion of a planar molecule (Fig. 1). One primary hydroxyl group has a gauche-trans conformation while the others are in gauche-gauche conformation. The secondary hydroxyl groups form intramolecular O(2)-H...O(3') or O(3')-H...O(2) hydrogen bonds which maintain the β-cyclodextrin structure. The guest molecule is fully enclosed within the host cavity.

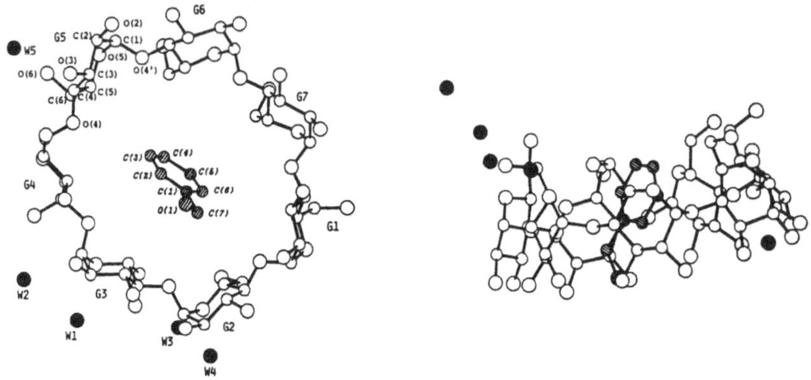

Fig. 1. Two views of the β-cyclodextrin benzyl alcohol clathrate pentahydrate structure.

N-TRITYLALANINE ANHYDRIDE DIETHYL ETHER CLATHRATE
$C_{44}H_{40}N_2O_3,C_4H_{10}O$

I. GOLDBERG, L.-T.W. LIN and H. HART, 1985. *Acta Cryst.*, **C41**, 1535-1538.

Monoclinic, *P*2_1, a = 9.092, b = 24.809, c = 9.135 Å, β = 102.60°, Z = 2. Mo radiation, R = 0.090 for 1663 reflections.

The inclusion-type structure consists of layers of host molecules (Fig. 1) separated by pseudo-channels occupied by the guest molecules. The channel axis is parallel to c. The host lattice has higher symmetry than the entire crystal. The diethyl ether molecule is disordered. Bond lengths and angles are of low precision.

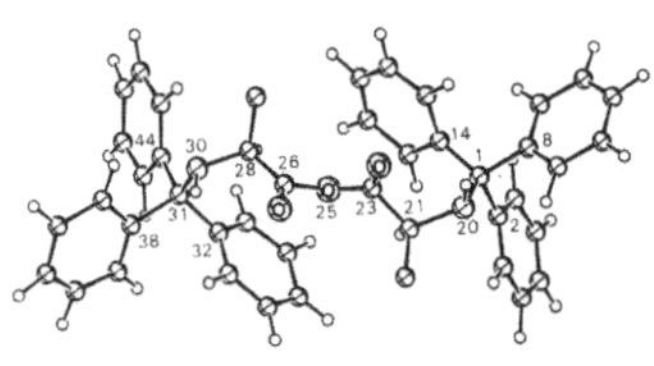

Fig. 1. $C_{44}H_{40}N_2O_3,C_4H_{10}O$: the molecular structure of the host species.

1',1''-DIMETHYLDISPIRO(1,6,20,25-TETRAOXA[6.1.6.1]PARACYCLOPHANE-13,4':32,4''-BIS(PIPERIDINE)) BENZENE CLATHRATE MONOHYDRATE
$C_{44}H_{54}N_2O_4,2(C_6H_6),H_2O$ (I)

1',1''-DIMETHYLDISPIRO(1,6,20,25-TETRAOXA[6.1.6.1]PARACYCLOPHANE-13,4':32,4''-BIS(PIPERIDINE)) p-XYLENE CLATHRATE
$C_{44}H_{54}N_2O_4,C_8H_{10}$ (II)

C. KRIEGER and F. DIEDERICH, 1985. *Chem. Ber.*, **118**, 3620-3631.

I. Triclinic, *P*$\bar{1}$, a = 7.992, b = 10.028, c = 14.736 Å, α = 90.85, β = 90.21, γ = 93.96°, Z = 1. Mo radiation, R = 0.042 for 4121 reflections.

II. Triclinic, *P*$\bar{1}$, a = 6.262, b = 12.106, c = 15.340 Å, α = 107.74, β = 90.66, γ = 86.76°, Z = 1. Mo radiation, R = 0.042 for 3515 reflections.

Both complexes lie about inversion centres and have "face-to-face" conformations (Fig. 1). In (I) one C_6H_6 ring is enclosed within the cavity of the host and the other is located in channel type intermolecular cavities in the lattice. The water molecule forms hydrogen bonds to the piperidine N atoms. In (II) (Fig. 2) the host molecules are stacked along the *a*-axis and p-xylene molecules are sandwiched by two adjacent hosts; the methyl groups of the p-xylene molecules are inserted into the cavities of the sandwiching hosts.

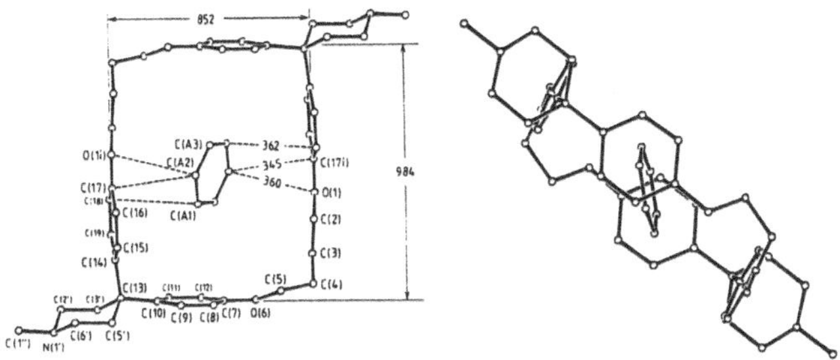

Fig. 1. The molecular structure of $C_{44}H_{54}N_2O_4,2(C_6H_6),H_2O$ (I) in views onto the best plane of the host and parallel to the best plane.

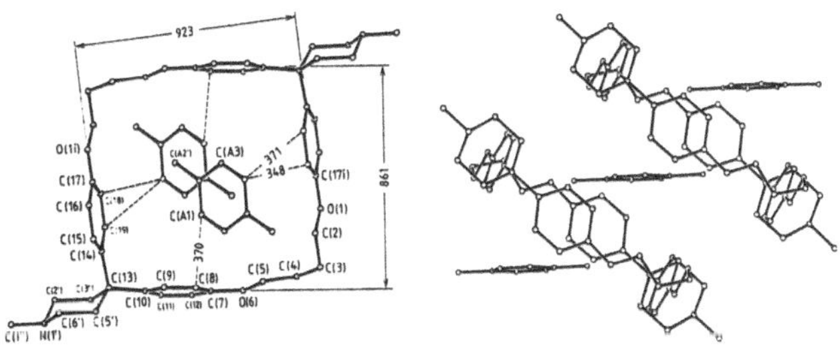

Fig. 2. The molecular structure of $C_{44}H_{54}N_2O_4,C_8H_{10}$ (II) in views onto the best plane of the host and parallel to the best plane.

p-t-BUTYLCALIX[4]ARENE ANISOLE
$C_{44}H_{56}O_4,0.5(C_7H_8O)$

R. UNGARO, A. POCHINI, G.D. ANDREETTI and P. DOMIANO, 1985. *J. Chem. Soc. Perkin Trans.* 2, 197-201.

Tetragonal, *P4/n*, a = 12.823, c = 25.618 Å, Z = 2. Cu radiation, R = 0.073 for 2088 reflections.

Two molecules of the macrocycle, in the cone conformation, face the methyl groups of the para-substituents creating a closed cavity which encapsulates one molecule of anisole (Fig. 1). The guest is disordered in the cavity in at least eight equivalent orientations.

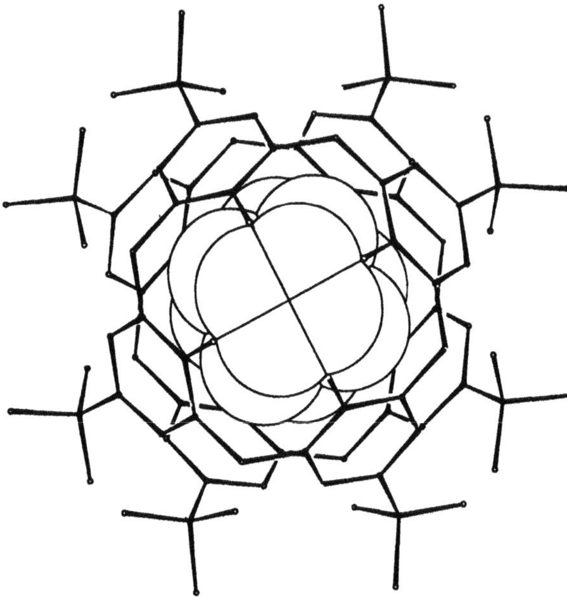

Fig. 1. Perspective view of the molecular cavity, in the crystals of $C_{44}H_{56}O_4.0.5(C_7H_8O)$, along the fourfold axis. The anisole atoms are represented by their van der Waals spheres.

6A,6D-DIDEOXY-6A-(t-BUTYLTHIO)-β-CYCLODEXTRIN TETRADECAHYDRATE
$C_{46}H_{78}O_{34}S,14(H_2O)$

K. FUJITA, A. MATSUMAGA, T. IMOTO, K. HIROTSU, S. KAMITORI and T. HIGUCHI, 1985. *J. Am. Chem. Soc.*, **107**, 1790-1791.

Orthorhombic, $P2_12_12_1$, a = 32.489, b = 15.460, c = 15.172 Å, Z = 4. Mo radiation, R = 0.120 for 2780 reflections.

The structure is isomorphous with 6-deoxy-6-(t-butylthio)-β-cyclodextrin (1) and shows no unexceptional features. The t-butyl group is included in the hydrophobic cavity of a neighbouring cyclodextrin moiety as shown in Fig. 1.

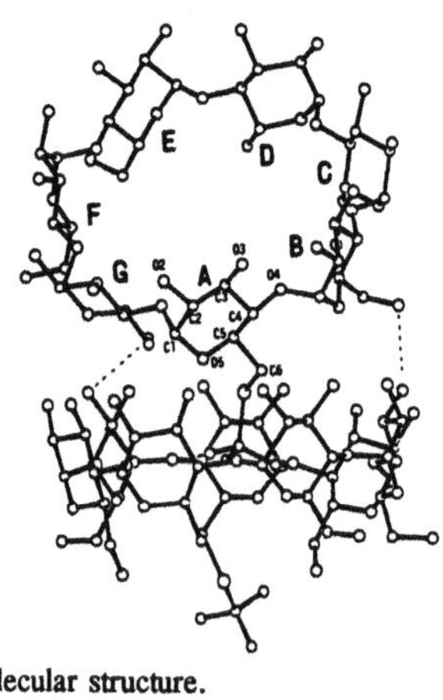

Fig. 1. $C_{46}H_{78}O_{34}S,14(H_2O)$: the molecular structure.

1. Structure Reports, 49B, 746.

GPSR Compliance

The European Union's (EU) General Product Safety Regulation (GPSR) is a set of rules that requires consumer products to be safe and our obligations to ensure this.

If you have any concerns about our products, you can contact us on ProductSafety@springernature.com

In case Publisher is established outside the EU, the EU authorized representative is:

Springer Nature Customer Service Center GmbH
Europaplatz 3
69115 Heidelberg, Germany

Batch number: 10583996

Printed by Printforce, the Netherlands